Fundamental Concepts of
EARTHQUAKE ENGINEERING

Fundamental Concepts of
EARTHQUAKE ENGINEERING

Roberto Villaverde

CRC Press
Taylor & Francis Group
Boca Raton London New York

CRC Press is an imprint of the
Taylor & Francis Group, an **informa** business

CRC Press
Taylor & Francis Group
6000 Broken Sound Parkway NW, Suite 300
Boca Raton, FL 33487-2742

© 2009 by Taylor & Francis Group, LLC
CRC Press is an imprint of Taylor & Francis Group, an Informa business

No claim to original U.S. Government works
Printed in the United States of America on acid-free paper
10 9 8 7 6 5 4 3 2 1

International Standard Book Number-13: 978-1-4200-6495-7 (Hardcover)

This book contains information obtained from authentic and highly regarded sources. Reasonable efforts have been made to publish reliable data and information, but the author and publisher cannot assume responsibility for the validity of all materials or the consequences of their use. The authors and publishers have attempted to trace the copyright holders of all material reproduced in this publication and apologize to copyright holders if permission to publish in this form has not been obtained. If any copyright material has not been acknowledged please write and let us know so we may rectify in any future reprint.

Except as permitted under U.S. Copyright Law, no part of this book may be reprinted, reproduced, transmitted, or utilized in any form by any electronic, mechanical, or other means, now known or hereafter invented, including photocopying, microfilming, and recording, or in any information storage or retrieval system, without written permission from the publishers.

For permission to photocopy or use material electronically from this work, please access www.copyright.com (http://www.copyright.com/) or contact the Copyright Clearance Center, Inc. (CCC), 222 Rosewood Drive, Danvers, MA 01923, 978-750-8400. CCC is a not-for-profit organization that provides licenses and registration for a variety of users. For organizations that have been granted a photocopy license by the CCC, a separate system of payment has been arranged.

Trademark Notice: Product or corporate names may be trademarks or registered trademarks, and are used only for identification and explanation without intent to infringe.

Library of Congress Cataloging-in-Publication Data

Villaverde, Roberto.
    Fundamental concepts of earthquake engineering / Roberto Villaverde.
       p. cm.
    Includes bibliographical references and index.
    ISBN 978-1-4200-6495-7 (alk. paper)
    1. Earthquake engineering--Textbooks. I. Title.

TA654.6.V55 2009
624.1'762--dc22
                                                                                         2008040939

Visit the Taylor & Francis Web site at
http://www.taylorandfrancis.com

and the CRC Press Web site at
http://www.crcpress.com

*Dedicated to the memory of
Jesús Villaverde Herrera
Carolina Lazo de Villaverde
Parents extraordinaire*

Contents

Preface xix
Author xxiii

Chapter 1 Introduction 1

1.1 Importance of Earthquake Engineering.. 1
1.2 Damaging Effects of Earthquakes... 1
    1.2.1 Ground Failures... 2
    1.2.2 Indirect Effects 10
    1.2.3 Ground Shaking 16
1.3 Earthquake Forces... 22
1.4 Design for Earthquake Forces... 23
1.5 Definition and Scope 24
1.6 Historical Background 27
Further Readings.. 27

Chapter 2 Seismic Regions of the World 29

2.1 Introduction... 29
2.2 World Seismicity 30
2.3 Seismicity of United States 32
2.4 Earthquake Statistics... 32
    2.4.1 Worldwide Earthquakes 33
    2.4.2 U.S. Earthquakes 34
2.5 On-Line Seismicity Information 34
Further Readings.. 35

Chapter 3 Earthquake Genesis.. 37

3.1 Introduction... 37
3.2 Types of Earthquakes.. 38
3.3 Earth Structure 39
3.4 Plate Tectonics Theory.. 40
3.5 Plate Interaction, Formation of Earth's Surface Features,
    and Earthquake Generation... 42
3.6 Cause of Plate Movement... 45
3.7 Intraplate Earthquakes 46
3.8 Earthquake Faults.. 46
3.9 Earthquake-Generation Mechanism: Elastic Rebound Theory 49
3.10 Focus, Epicenter, Rupture Surface, and Fault Slip... 53
3.11 Foreshocks and Aftershocks 54
3.12 Earthquake Prediction... 55
    3.12.1 Introduction... 55
    3.12.2 Historical Background 55
    3.12.3 Seismic Gaps and Seismic Quiescence 59
    3.12.4 Earthquake Precursors 60
    3.12.5 Current Status of Earthquake Predictions 61

Further Readings..61
Review Questions...61

Chapter 4 Earthquake Propagation...63
4.1 Introduction...63
4.2 Longitudinal Waves in Long Rod...63
    4.2.1 Differential Equation of Motion 63
    4.2.2 Solution of Differential Equation...65
    4.2.3 Solution in Terms of Initial Conditions 65
    4.2.4 Solution Interpretation 66
    4.2.5 Relationship between Particle Velocity and Wave Velocity 67
    4.2.6 Transmitted and Reflected Waves at Discontinuity...68
4.3 Shear Waves in Long Shear Beam 74
4.4 Harmonic Waves 78
    4.4.1 Definition 78
    4.4.2 Characteristics 80
4.5 Waves in Long Flexural Beam: Wave Dispersion...82
    4.5.1 Differential Equation of Motion 82
    4.5.2 Solution to Differential Equation of Motion...83
    4.5.3 Apparent Waves and Group Velocity..84
4.6 Elastic Waves in Unbounded Three-Dimensional Medium: P and S Waves...........................86
    4.6.1 Introduction..86
    4.6.2 Stresses and Strains in Three-Dimensional Solid 86
    4.6.3 Differential Equations of Motion...89
    4.6.4 First Particular Solution of Differential Equations of Motion......................................91
    4.6.5 Interpretation of First Particular Solution of Equations of Motion 94
    4.6.6 Second Particular Solution of Differential Equations of Motion 95
    4.6.7 Interpretation of Second Particular Solution of Equations of Motion..........................97
    4.6.8 General Solution of Differential Equations of Motion 98
4.7 Elastic Waves in Elastic Half-Space: Rayleigh Waves..99
    4.7.1 Introduction..99
    4.7.2 Equations of Motion for Plane Waves... 100
    4.7.3 Solution of Decoupled Equations of Motion 102
    4.7.4 Particle Motion Generated by Rayleigh Waves..105
4.8 Elastic Waves in Layered Half-Space: Love Waves..108
    4.8.1 Introduction..108
    4.8.2 Equations of Motion for Plane Waves...109
    4.8.3 Solution of Equations of Motion 110
    4.8.4 Particle Motion Generated by Love Waves...113
    4.8.5 Dispersion of Love Waves 113
4.9 Reflection and Refraction of Body Waves 115
    4.9.1 Introduction..115
    4.9.2 Elements of Ray Theory..116
    4.9.3 Relative Amplitudes of Incident, Refracted, and Reflected Waves 118
4.10 Attenuation of Wave Amplitude with Distance 122
    4.10.1 Radiation Damping..122
    4.10.2 Material Damping..123
Further Readings..125
Problems 125

Contents

Chapter 5 Measurement of Earthquakes 129

5.1 Introduction 129
5.2 Intensity Scales 129
5.3 Seismographs and Seismograms 134
  5.3.1 Historical Background 134
  5.3.2 Components and Design Features 135
  5.3.3 Seismograms 139
5.4 Location of Earthquake Hypocenter 140
5.5 Magnitude Scales 146
  5.5.1 Richter or Local Magnitude 146
  5.5.2 Surface- and Body-Wave Magnitudes 148
5.6 Seismic Moment and Moment Magnitude 149
  5.6.1 Scale Saturation 149
  5.6.2 Seismic Moment 151
  5.6.3 Moment Magnitude 153
5.7 Accelerographs and Accelerograms 154
  5.7.1 Need for Strong Motion Recording Instruments 154
  5.7.2 Accelerograph Components and Design Features 155
  5.7.3 Accelerograms 159
  5.7.4 Accelerogram Processing 162
  5.7.5 Strong Motion Instrument Arrays 165
  5.7.6 Accelerograms from Near-Fault Sites 166
Further Readings 177
Problems 177

Chapter 6 Characterization of Strong Ground Motions 181

6.1 Introduction 181
6.2 Peak Ground Acceleration 181
6.3 Strong-Motion Duration 182
6.4 Response Spectrum 182
  6.4.1 Background 182
  6.4.2 Definition 183
  6.4.3 Importance of Response Spectrum 186
  6.4.4 Tripartite Representation 189
  6.4.5 Computation of Response Spectra 193
6.5 Nonlinear Response Spectrum 196
  6.5.1 Background 196
  6.5.2 Load-Deformation Curves under Severe Cyclic Loading 197
  6.5.3 Definition 199
  6.5.4 Graphical Representation 199
  6.5.5 Computation of Nonlinear Response Spectrum 202
6.6 Fourier Spectrum 207
  6.6.1 Theoretical Background 207
    6.6.1.1 Fourier Series 207
    6.6.1.2 Exponential Fourier Series 208
    6.6.1.3 Fourier Transform 210
  6.6.2 Definition 210
  6.6.3 Total Energy and Fourier Spectrum 212
  6.6.4 Relationship between Fourier and Undamped Velocity Response Spectra 214

		6.6.5 Computation of Fourier Spectrum	216
		6.6.6 Fast Fourier Transform	218
6.7	Power Spectral Density		222
6.8	Single-Parameter Measures of Ground Motion Intensity		225
	6.8.1	Introductory Remarks	225
	6.8.2	Root Mean Square Acceleration	225
	6.8.3	Housner Spectral Intensity	225
	6.8.4	Arias Intensity	226
	6.8.5	Araya–Saragoni Intensity	228
6.9	Design-Spectrum-Compatible Time Histories		228
Further Readings			230
Problems			230

Chapter 7 Seismic Hazard Assessment 237

7.1	Introduction		237
7.2	Identification of Seismic Sources		239
7.3	Factors Affecting Ground Motion Characteristics at Site		244
7.4	Ground Motion Intensity at Given Site: Attenuation Laws		246
	7.4.1	Background	246
	7.4.2	Boore, Joyner, and Fumal Equations for Shallow Crustal Earthquakes in Active Regions	247
	7.4.3	Campbell Equations for Shallow Crustal Earthquakes in Active Regions	253
	7.4.4	Youngs, Chiou, Silva, and Humphrey Equations for Subduction-Zone Earthquakes	257
7.5	Magnitude-Recurrence Relationships		261
	7.5.1	Introduction	261
	7.5.2	Gutenberg–Richter Recurrence Law	261
	7.5.3	Bounded Gutenberg–Richter Recurrence Law	262
	7.5.4	Characteristic Earthquake Recurrence Law	263
7.6	Ground Motion Intensity in Given Time Interval		264
	7.6.1	Introduction	264
	7.6.2	Poisson Probabilistic Model	265
	7.6.3	Probability of Exceeding Given Ground Motion Intensity in Given Time Interval	267
	7.6.4	Exceedance Probability Curves	267
	7.6.5	Frequency-Intensity Curves	268
7.7	Semiprobabilistic Seismic Hazard Evaluation		270
7.8	Probabilistic Seismic Hazard Evaluation		275
	7.8.1	Introduction	275
	7.8.2	Probability Distribution of Source-to-Site Distance	275
	7.8.3	Probability Distribution of Earthquake Magnitude	277
	7.8.4	Probability of Exceeding Specified Value of Ground Motion Parameter	277
	7.8.5	Annual Probability of Exceeding Specified Ground Motion Intensity Level	278
7.9	Seismic Zonation Maps		282
7.10	Microzonation Maps		287
Further Readings			289
Problems			289

Contents

Chapter 8 Influence of Local Site Conditions 295

8.1 Problem Identification 295
8.2 Effect of Site Conditions on Ground Motion Characteristics 299
8.3 Evaluation of Site Effects Using Statistical Correlations 304
8.4 Evaluation of Site Effects Using Analytical Techniques 306
    8.4.1 Introductory Remarks 306
    8.4.2 One-Dimensional Continuous Model 306
    8.4.3 One-Dimensional Lumped-Mass Model 315
    8.4.4 One-Dimensional Continuous Model: Solution in Frequency Domain 319
        8.4.4.1 Case I: Homogeneous, Undamped Soil on Rigid Rock 319
        8.4.4.2 Case II: Homogeneous, Damped Soil on Rigid Rock 321
        8.4.4.3 Case III: Homogeneous, Damped Soil on Flexible Rock 325
        8.4.4.4 Case IV: Layered, Damped Soil on Flexible Rock 327
    8.4.5 Nonlinear Site Response 332
8.5 Determination of Shear Modulus and Damping Ratio 334
    8.5.1 Shear Modulus and Damping Ratio of Sands 336
    8.5.2 Shear Modulus and Damping Ratio of Gravels 338
    8.5.3 Shear Modulus and Damping Ratio of Saturated Clays 340
Further Readings 347
Problems 347

Chapter 9 Design Response Spectrum 353

9.1 Introduction 353
9.2 Peak Ground Acceleration and Response Spectrum Shape Method 354
    9.2.1 Elastic Design Spectrum 354
    9.2.2 Inelastic Design Spectrum 355
9.3 Newmark–Hall Approach 360
    9.3.1 Background 360
    9.3.2 Elastic Design Spectra 360
    9.3.3 Inelastic Design Spectra 364
9.4 Direct Statistical Correlations 375
Further Readings 375
Problems 376

Chapter 10 Structural Response by Response Spectrum Method 379

10.1 Introduction 379
10.2 Modal Superposition Method 379
    10.2.1 Discrete Model 379
    10.2.2 Equation of Motion 380
    10.2.3 Transformation of Equation of Motion 382
    10.2.4 Solution of rth Independent Equation 386
    10.2.5 Structural Response in Terms of Modal Responses 386
    10.2.6 Maximum Structural Response 389
10.3 Maximum Structural Response Using Response Spectra 389
    10.3.1 Maximum Modal Responses in Terms of Response Spectrum Ordinates .. 389
    10.3.2 Approximate Rules to Estimate Maximum System Response 391
        10.3.2.1 Double-Sum Combination Rule 391
        10.3.2.2 Square Root of the Sum of the Squares 393
        10.3.2.3 Rosenblueth's Rule 394

		10.3.2.4	Complete Quadratic Combination	395

- 10.3.2.4 Complete Quadratic Combination 395
- 10.3.2.5 Singh and Maldonado's Rule 395
- 10.3.2.6 Comparative Studies 397
- 10.3.2.7 Final Remarks 403
- 10.4 Properties of Modal Participation Factors 406
- 10.5 Determination of Modal Damping Ratios 407
- 10.6 Summary of Response Spectrum Method 416
- 10.7 Response Spectrum Method Based on Modal Acceleration Method 423
- 10.8 Response Spectrum Method for Multicomponent Ground Motions 431
  - 10.8.1 Introduction 431
  - 10.8.2 Equation of Motion 432
  - 10.8.3 Transformation of Equation of Motion 434
  - 10.8.4 Solution of rth Independent Equation 434
  - 10.8.5 Structural Response in Terms of Modal Responses 435
  - 10.8.6 Maximum Modal Responses in Terms of Response Spectrum Ordinates .. 436
  - 10.8.7 Approximate Rule to Estimate Maximum System Response 437
- 10.9 Response Spectrum Method for Spatially Varying Ground Motions 449
- Further Readings 462
- Problems 462

Chapter 11 Structural Response by Step-by-Step Integration Methods 469

- 11.1 Introduction 469
- 11.2 General Classification 470
- 11.3 Central Difference Method 470
- 11.4 Houbolt Method 476
- 11.5 Constant Average Acceleration Method 482
- 11.6 Linear Acceleration Method 489
- 11.7 Wilson-θ Method 495
- 11.8 Newmark-β Method 500
- 11.9 Stability and Accuracy Issues 503
  - 11.9.1 Introductory Remarks 503
  - 11.9.2 Numerical Stability 504
  - 11.9.3 Accuracy 510
- 11.10 Analysis of Nonlinear Systems 513
  - 11.10.1 Introduction 513
  - 11.10.2 Incremental Equations of Motion 513
  - 11.10.3 Solution of Incremental Equation of Motion 516
    - 11.10.3.1 Solution by Central Differences Method 516
    - 11.10.3.2 Solution by Newmark-β Method 518
  - 11.10.4 Updating of Stiffness Matrix: Elastoplastic Case 521
  - 11.10.5 Sources of Error and Corrective Techniques 530
- 11.11 Final Remarks 537
- Further Readings 539
- Problems 539

Chapter 12 Structural Response by Equivalent Lateral Force Procedure 541

- 12.1 Introduction 541
- 12.2 Derivation of Procedure 541
  - 12.2.1 Overview 541
  - 12.2.2 Maximum Modal Lateral Forces 542

	12.2.3	Modal Base Shear	543
	12.2.4	Modal Lateral Forces in Terms of Base Shear	544
	12.2.5	Basic Assumptions	545
	12.2.6	Approximate Maximum Lateral Forces	545
	12.2.7	Approximate Fundamental Natural Period	546
12.3		Application of Procedure	547
12.4		Limitations of Procedure	564
12.5		Nonlinear Static Procedure	565
Further Readings			568
Problems			568

Chapter 13 Structural Response Considering Soil–Structure Interaction ... 571

13.1	Problem Definition	571
13.2	Kinematic and Inertial Interaction	572
13.3	Vibration of Foundations on Elastic Halfspace	574
	13.3.1 Introductory Remarks	574
	13.3.2 Displacements under Static Point Loads	574
	13.3.3 Displacements and Rotations under Static Distributed Loads	575
	13.3.4 Soil Spring Constants	582
	13.3.5 Displacements and Rotations under Dynamic Loads	583
	13.3.6 Vibrations of Rectangular Foundations	590
13.4	Lumped-Parameter Models	591
	13.4.1 Hsieh's Equations	591
	13.4.2 Lysmer and Hall Analogs	593
13.5	Coupled Rocking and Horizontal Vibrations of Rigid Circular Foundations	598
13.6	Vibration of Rigid Circular Foundations Supported by Elastic Layer	603
13.7	Vibration of Embedded Foundations	605
13.8	Foundation Vibrations by Method of Impedances	610
13.9	Material or Internal Damping in Soils	621
13.10	Vibrations of Foundations on Viscoelastic Halfspace	624
13.11	Simplified Equivalent Single-Degree-of-Freedom System Method	627
13.12	Advanced Methods of Analysis	637
	13.12.1 Introduction	637
	13.12.2 Direct Method	638
	13.12.3 Substructure Method	642
	13.12.4 Computer Programs	645
13.13	Experimental Verification	645
Further Readings		646
Problems		647

Chapter 14 Seismic Response of Nonstructural Elements ... 649

14.1	Introduction	649
14.2	Importance of Nonstructural Elements	649
14.3	General Physical Characteristics	653
14.4	General Response Characteristics	654
14.5	Modeling of Nonstructural Elements	654
14.6	Methods of Analysis	655
14.7	Floor Response Spectrum Method	656
14.8	Modal Synthesis Method	663

	14.8.1	Introduction 663
	14.8.2	Mode Shapes and Natural Frequencies of Combined System 663
		14.8.2.1 Primary System 664
		14.8.2.2 Secondary System 665
		14.8.2.3 Compatibility Requirements 668
		14.8.2.4 Eigenvalue Problem 668
		14.8.2.5 Mode Shapes of Combined System in Terms of Modal Coordinates 670
	14.8.3	Approximate Natural Frequencies and Mode Shapes of Combined System 674
	14.8.4	Damping Ratios of Combined System 680
	14.8.5	Maximum Elastic Response of Nonstructural Element 681
	14.8.6	Effect of Structural and Nonstructural Nonlinear Behavior 683
Further Readings 685
Problems 686

Chapter 15 Seismic Protection with Base Isolation 689

15.1 Introduction 689
15.2 Basic Concept 689
15.3 Historical Perspective 691
15.4 Isolation Bearings 695
    15.4.1 Introduction 695
    15.4.2 Laminated Rubber Bearings 695
        15.4.2.1 Low-Damping Rubber Bearings 696
        15.4.2.2 Lead–Rubber Bearings 697
        15.4.2.3 High-Damping Rubber Bearings 698
    15.4.3 Sliding Bearings 700
        15.4.3.1 Electricité-de-France Bearing 700
        15.4.3.2 TASS Bearing 701
        15.4.3.3 Friction Pendulum Bearing 702
    15.4.4 Helical Steel Springs 705
15.5 Methods of Analysis 706
    15.5.1 Linear Two-Degree-of-Freedom Model 706
        15.5.1.1 Equations of Motion 706
        15.5.1.2 Natural Frequencies 709
        15.5.1.3 Mode Shapes 712
        15.5.1.4 Modal Participation Factors 713
        15.5.1.5 Uncoupled Equations of Motion 715
        15.5.1.6 Damping Ratios 716
        15.5.1.7 Maximum Response 717
    15.5.2 Rigid Structure Single-Degree-of-Freedom Model 721
    15.5.3 Linear Multi-Degree-of-Freedom Model 721
        15.5.3.1 Equations of Motion 721
        15.5.3.2 Natural Frequencies and Mode Shapes 724
        15.5.3.3 Approximate Maximum Response 727
    15.5.4 Linear Multi-Degree-of-Freedom Model with Nonclassical Damping 732
    15.5.5 Equivalent Linear Model 746
        15.5.5.1 Effective Stiffness 746
        15.5.5.2 Effective Damping Ratio 746

Contents xv

15.6	Implementation Issues	747
Further Readings		748
Problems		749

Chapter 16 Seismic Protection with Energy Dissipating Devices ... 753

- 16.1 Introduction ... 753
- 16.2 Basic Concepts ... 754
- 16.3 Energy Dissipating Devices ... 760
  - 16.3.1 Introduction ... 760
  - 16.3.2 Friction Dampers ... 760
  - 16.3.3 Viscoelastic Dampers ... 764
  - 16.3.4 Fluid Dampers ... 771
  - 16.3.5 Yielding Metallic Dampers ... 776
- 16.4 Analysis of Structures with Added Dampers ... 785
  - 16.4.1 Overview ... 785
  - 16.4.2 Seismic Response of Simple Structure with Added Dampers ... 786
    - 16.4.2.1 Viscoelastic Damper ... 787
    - 16.4.2.2 Fluid Damper ... 787
    - 16.4.2.3 Friction or Yielding Steel Damper ... 788
  - 16.4.3 Seismic Response of Multistory Structures with Added Dampers ... 800
    - 16.4.3.1 Viscoelastic Dampers ... 801
    - 16.4.3.2 Fluid Dampers ... 802
    - 16.4.3.3 Friction or Yielding Steel Dampers ... 804
- 16.5 Implementation Issues ... 810
- Further Readings ... 812
- Problems ... 812

Chapter 17 Seismic Code Provisions ... 815

- 17.1 Introduction ... 815
- 17.2 General Requirements ... 816
  - 17.2.1 Allowable Seismic-Force-Resisting Systems ... 816
  - 17.2.2 Load Combinations ... 819
  - 17.2.3 Story Drift Limits ... 821
  - 17.2.4 Direction of Earthquake Loads ... 821
- 17.3 Design Ground Motion ... 821
  - 17.3.1 Overview ... 821
  - 17.3.2 Ground Motion Intensity Maps ... 822
  - 17.3.3 Site-Specific Response Spectra ... 822
  - 17.3.4 Site Classes and Site Coefficients ... 822
  - 17.3.5 Design Spectral Accelerations ... 831
  - 17.3.6 Design Response Spectrum ... 831
- 17.4 Analysis Procedures ... 835
  - 17.4.1 Overview ... 835
  - 17.4.2 Occupancy Categories and Importance Factors ... 835
  - 17.4.3 Seismic Design Category ... 835
  - 17.4.4 Structural Irregularities ... 836
  - 17.4.5 Allowable Analysis Procedures ... 837
  - 17.4.6 Mathematical Model ... 840
- 17.5 Simplified Method ... 840

17.6	Equivalent Lateral Force Procedure		841
	17.6.1	Overview	841
	17.6.2	Base Shear	841
	17.6.3	Natural Period	842
	17.6.4	Equivalent Lateral Forces	844
	17.6.5	Horizontal Shear Distribution	844
	17.6.6	Overturning Moment	845
	17.6.7	Story Drift Determination	846
	17.6.8	P-Delta Effects	846
17.7	Modal Response Spectrum Analysis		850
	17.7.1	Modal Properties	850
	17.7.2	Modal Base Shears	851
	17.7.3	Modal Lateral Forces and Drifts	851
	17.7.4	Design Values	852
17.8	Linear Response Time-History Analysis		852
17.9	Nonlinear Response Time-History Analysis		854
17.10	Soil–Structure Interaction Effects		862
	17.10.1	Introduction	862
	17.10.2	Modifications to Equivalent Lateral Force Procedure	863
		17.10.2.1 Base Shear	863
		17.10.2.2 Effective Building Period	864
		17.10.2.3 Effective Damping Ratio	866
		17.10.2.4 Vertical Distribution of Base Shear	867
		17.10.2.5 Story Shears, Overturning Moments, and Torsional Moments	867
		17.10.2.6 Lateral Displacements	868
	17.10.3	Modifications to Modal Analysis Procedure	871
		17.10.3.1 Modal Base Shears	871
		17.10.3.2 Modal Lateral Forces, Story Shears, and Overturning Moments	872
		17.10.3.3 Modal Lateral Displacements	872
		17.10.3.4 Design Values	872
17.11	Nonstructural Components		872
17.12	Base-Isolated Structures		880
	17.12.1	Overview	880
	17.12.2	General Requirements	880
	17.12.3	Effective Natural Periods and Minimum Displacements	881
	17.12.4	Equivalent Lateral Force Procedure	883
		17.12.4.1 Conditions for Use	883
		17.12.4.2 Minimum Displacements	883
		17.12.4.3 Minimum Lateral Forces	884
		17.12.4.4 Drift Limits	884
	17.12.5	Dynamic Analysis	885
		17.12.5.1 Choice of Procedure	885
		17.12.5.2 Input Earthquake	885
		17.12.5.3 Response Spectrum Analysis	885
		17.12.5.4 Time-History Analysis	885
		17.12.5.5 Minimum Displacements	886
		17.12.5.6 Minimum Lateral Forces	886
		17.12.5.7 Drift Limits	887
	17.12.6	Required Testing of Isolation System Components	887
	17.12.7	Effective Stiffness and Effective Damping of Isolating System	887
	17.12.8	Peer Review	888

17.13	Structures with Added Damping Devices		891
	17.13.1	Overview	891
	17.13.2	General Requirements	891
	17.13.3	Methods of Analysis	892
	17.13.4	Equivalent Lateral Force Procedure	894
		17.13.4.1 Conditions for Use	894
		17.13.4.2 Modeling	894
		17.13.4.3 Base Shear	894
		17.13.4.4 Design Lateral Forces	896
		17.13.4.5 Floor Deflections, Story Drifts, and Story Velocities	897
		17.13.4.6 Story Drift Limits	898
		17.13.4.7 Design Forces in Damping System	899
	17.13.5	Response Spectrum Procedure	899
		17.13.5.1 Conditions for Use	899
		17.13.5.2 Modeling	899
		17.13.5.3 Base Shear	899
		17.13.5.4 Design Lateral Forces	900
		17.13.5.5 Floor Deflections, Story Drifts, and Story Velocities	900
		17.13.5.6 Story Drift Limits	901
		17.13.5.7 Design Forces in Damping System	901
	17.13.6	Nonlinear Procedures	902
	17.13.7	Response Modification Due to Damping Increase	903
	17.13.8	Effective Damping	903
		17.13.8.1 Inherent Damping	905
		17.13.8.2 Hysteretic Damping	905
		17.13.8.3 Viscous Damping	905
	17.13.9	Effective Ductility Demand	906
	17.13.10	Combination of Load Effects	907
	17.13.11	Required Testing of Damping Devices	912
	17.13.12	Peer Review	912
Further Readings			920
Problems			921

Appendix 929
Index 939

Preface

Earthquake engineering is a relatively new discipline. Not long ago, earthquakes were believed to be acts of God, and mitigation measures consisted mainly of prayers rather than the application of scientific and engineering principles. Notwithstanding its comparatively young existence, today we can say that earthquake engineering has come of age. It is a discipline that embraces a series of concepts and procedures that are well established and proven successful in mitigating the effect of earthquakes in our built environment. It is an important component of civil engineering practice and research and is taught in virtually all civil engineering programs in the United States and many other parts of the world. Many states within the United States now include seismic provisions in their building codes and require civil engineers to exhibit knowledge of earthquake engineering principles to obtain their professional license. Surprisingly however, up to now there has not been a suitable textbook to introduce senior or first-year graduate students to the subject. There have been several excellent books during the past four decades devoted to earthquake engineering, but, for the most part, they have been written as reference books directed to practicing professionals or advanced graduate students. They emphasize a general coverage with the purpose of disseminating the current knowledge and current practice, leaving the fundamentals and the details to a comprehensive list of references. As such, they usually overwhelm the uninitiated, who are left unclear as to what are the fundamental issues, and what are the more controversial and less durable ones.

Fundamental Concepts of Earthquake Engineering has been conceived as a textbook to introduce beginners to the fundamental aspects of earthquake engineering. It is written primarily for students who have had little or no exposure to the subject, although it is also written to serve as a reference document for practicing engineers and those who need to prepare for the seismic portion of a civil engineering professional exam. It presupposes no previous knowledge of any of the aspects of earthquake engineering, and every attempt has been made to present all pertinent background information. However, the level at which the text has been written presumes familiarity with basic mathematics such as calculus, linear algebra, probability theory, and differential equations. It also presumes knowledge of the material covered in basic undergraduate engineering courses such as statics, dynamics, structural analysis, and matrix methods of analysis. Furthermore, it presumes a formal training in and thorough understanding of the basic principles of structural dynamics.

Fundamental Concepts of Earthquake Engineering has been modeled after textbooks that cover other, more established disciplines, whose main purpose is to present basic concepts, not all that is known about the discipline or recent advances and state-of-the-art procedures. In other words, concepts whose mastery would allow students to understand the essence of the subject, awaken their interest for topics not discussed in this book, and facilitate the learning of these topics by self-study. In this spirit, a great effort has been made to identify the most relevant aspects of the subject, present them clearly and thoroughly, and integrate them in a logical sequence. Similarly, the coverage is at the deepest level that is possible and practical. This means the presentation of theoretical derivations and disclosure of all applicable assumptions and limitations. Additionally, several pedagogical features are introduced. First, many numerical examples are included to illustrate, clarify, and reinforce the covered material, and many end-of-chapter problems are given to provide the students with the hands-on experience that is needed to master them. Second, many photographs are inserted throughout the text as a means to inspire and motivate students, annotating each photograph with a detailed description of the subject or concept it is supposed to illustrate. Third, some of the discussed concepts and methods are summarized in a series of highly visible boxes to draw attention to them and facilitate their review. Fourth, relevant terms and definitions are written in italics the first time they are introduced to highlight their relevance and also indicate they are part of the established earthquake engineering language. Finally, ample anecdotal and historical facts are disseminated throughout this book to make it more interesting and pleasant to read. Overall, the intention is to provide students with the background that is needed to understand the strengths and weaknesses of current methods of analysis and seismic code provisions, interpret and properly implement these provisions, and follow with ease specialized technical literature, advances in the field, and the unavoidable code changes.

This book emphasizes the concepts and procedures that are used in current practice to assess the earthquake effects that control the seismic design of a structure. As, in general, these concepts and procedures involve (a) investigation of regional seismicity, (b) definition of seismic hazard at the construction site, (c) quantification of local site effects, (d) definition of expected ground motion characteristics at the construction site, (e) estimation of structural response under expected ground motions, and (f) compliance with seismic code provisions, *Fundamental Concepts of Earthquake Engineering* is organized around these topics. Accordingly, Chapter 1 describes the purpose of earthquake engineering, summarizes the type and extent of damage an earthquake can produce, gives an overview of the main aspects of earthquake engineering, and presents some historical information. The following four chapters, Chapters 2 through 5, are devoted to some basic concepts of seismology. Knowledge of these concepts is deemed necessary to understand where earthquakes are likely to occur, how the mechanism of earthquake generation and propagation may affect the ground motion characteristics at a site, how earthquakes are measured, and what the difference is between the various scales used to quantify the size of an earthquake and its potential to cause damage. Chapter 6 describes the different ways earthquake ground motions may be characterized for engineering purposes. Chapter 7 deals with the procedures used to determine the characteristics of the expected earthquake ground motions at a given site, whereas Chapter 8 explains how local soil conditions may affect these characteristics and describes the traditional methods used to quantify such effects. Chapter 9 introduces the design spectrum, a fundamental concept that constitutes the basis of many of the methods used in current practice for the design of earthquake-resistant structures. The following four chapters, Chapters 10 through 13, describe the conventional methods used in earthquake engineering to compute the response of structures to earthquake ground motions. Approximate and "exact" and linear and nonlinear methods, as well as some of the methods employed to account for soil–structure interaction effects, are considered in these chapters. Concepts dealing with the seismic design of nonstructural elements (equipment and architectural components) and some of the methods that can be used to estimate their response to earthquake ground motions are presented in Chapter 14. Because of the heavy damage experienced by nonstructural elements in past earthquakes, the seismic design of nonstructural components has become a subject of increasing interest. The following two chapters, Chapters 15 and 16, are also devoted to topics of increasing interest. These chapters provide an introductory description of two modern techniques for the protection of structures against earthquake effects: base isolation and damping enhancement with added energy-dissipating devices. Finally, Chapter 17 describes current seismic provisions and correlates these provisions with the concepts presented in previous chapters. The appendix contains a comprehensive worldwide list of historical earthquakes, which dramatically illustrates the frequency with which large earthquakes occur around the world and the catastrophic effect these earthquakes have had throughout the ages.

Clearly, the material goes beyond what may be covered in a quarter or semester of study. It offers, thus, flexibility in the selection of topics for an introductory course, a graduate course, supplementary reading, or self-study. For a first 15-week graduate course, it is suggested to include topics from Chapters 1 through 10, Chapter 12, and Chapter 17.

Many individuals and institutions have contributed to the completion of this book. The author is grateful to the many authors and publishers who generously granted permission to reproduce some of their tables, figures, and photographs. The contribution of the students, who throughout the years served as a testbed for the effectiveness of the lecture notes that eventually led to this book, is hereby acknowledged with a sincere word of appreciation. In particular, heartfelt thanks are extended to Samit Ray Chaudhury, who through his lively discussions and expert opinions provided much needed feedback. Finally, the author wishes to express his deepest gratitude to the institutions that introduced him to the thrills and challenges of earthquake engineering and furnished him with the rounded knowledge that made this book possible: Universidad Nacional Autónoma de México, Japan's International Institute of Seismology and Earthquake Engineering, and University of Illinois at Urbana-Champaign.

A textbook, as is well known, is a compilation of knowledge generated by a large number of individuals other than the author. Acknowledgment, however, of all the sources from which this knowledge is taken becomes, given the required long lists of citations, an unnecessary distraction to the reader. Traditionally, therefore, textbooks omit such an acknowledgment. *Fundamental Concepts of Earthquake Engineering* has followed this tradition and also omits, for the most part, referring to the publications on which it is based. For this, the author extends his sincere apologies to all those affected, and hopes that the inclusion of their work will be considered by itself a testimony of the value of their contributions.

Being the first edition of a fairly large volume that contains materials from a wide range of disciplines, it is inevitable that it will contain some errors and inconsistencies. The author will be grateful to any reader who brings to his attention any of these errors and inconsistencies. Suggestions for improvement are also most welcomed. Earthquake engineering is a field that is being actively researched and advances are taking place at a rapid pace. Hence, it is likely that the book will be revised from time to time.

Roberto Villaverde
Irvine, California

Author

Dr. Roberto Villaverde is professor emeritus of civil engineering at the University of California, Irvine, where he was an active faculty member from 1982 to 2004. He received his PhD degree in civil engineering from the University of Illinois at Urbana-Champaign and is an alumnus of Japan's International Institute of Seismology and Earthquake Engineering. The late Nathan M. Newmark, a prominent earthquake engineer and coauthor of the classical book *Fundamentals of Earthquake Engineering*, published by Prentice-Hall in 1971, was his dissertation advisor at the University of Illinois.

Before joining the University of California, Dr. Villaverde was a researcher at the Engineering Institute of the National University of Mexico and design engineer at various engineering firms and governmental organizations. He has taught courses in structural analysis, geotechnical engineering, steel design, structural dynamics, earthquake engineering, soil dynamics, random vibrations, and wind engineering. He is a registered civil engineer in the State of California and has participated in reconnaissance surveys of the devastating earthquakes in Mexico City in 1985; Loma Prieta, California, in 1989; Northridge, California, in 1994; and Kobe, Japan, in 1995. He has also participated in numerous research projects dealing with many and diverse earthquake engineering topics such as shear wall–frame interaction, seismic risk, torsional response of structures, dynamic response of soils, seismic response of nonstructural components, nonlinear structural analysis, fragility of equipment in electric power substations, structural collapse under earthquake loads, and passive control of structures. He has published extensively in the areas of structural dynamics and earthquake engineering and is a coauthor of the *International Handbook of Earthquake Engineering*, published by Chapman & Hall in 1994, and *Earthquake Engineering: From Engineering Seismology to Performance-Based Engineering*, published by CRC Press in 2004.

1 Introduction

The suddenness of an Earthquake that comes at an instant, unthought of, without warning, that seems to bring unavoidable death along with it, is able to touch an adamantine heart. To see death stalking o'er a great city, ready to sweep us all away, in an instantaneous ruin without a single moment to recollect our thoughts; this is Fear without remedy; this is Fear beyond battle and pestilence. The lighting and thunderbolt, the arrow that flieth by day, may suddenly take off an object or two, and leave no space for repentance; but what horror can equal that when above a million people are liable to be buried, in one common grave.

The Reverend William Stukeley

1.1 IMPORTANCE OF EARTHQUAKE ENGINEERING

Earthquakes are one of nature's greatest hazards to life and property. Throughout historic times, they have caused the destruction of countless cities and villages around the world and inflicted the death of thousands of people (see Appendix). In the last 30 years alone, thousands of people were injured or lost their lives, and many more were left homeless, by earthquakes. The totally unexpected and nearly instantaneous devastation they may cause produces a unique psychological impact and a fear on modern societies that it is unsurpassed by any other natural hazard. This devastation, however, is owed almost entirely to the effect of earthquakes on civil engineering structures and the ground that supports them. It is the collapse of bridges, buildings, dams, and other structures that, together with the indirect effects of these collapses, causes extensive damage and loss of life during earthquakes. In principle, therefore, with the effective application of scientific and engineering principles and techniques, societies can minimize, if not completely eliminate, earthquake catastrophes.

The threat that earthquakes pose to life and property is not the only concern to modern societies. The economic impact is the other one. Earthquakes constitute one of the most powerful forces to which most civil engineering structures will ever be subjected, and thus designing structures to resist these forces represents an economic burden. Conversely, not designing them to resist such forces can bring serious economic consequences in the event of a strong earthquake. However, strong earthquakes at any given location occur, on average, only once every 50–100 years. The probability that any given structure will be affected by a major earthquake during its lifetime is thus extremely low. Therefore, in dealing with earthquakes, modern societies are also faced with the great challenge of deciding what is the optimum investment to protect themselves against a hazard that, on the one hand, is capable of producing great economic losses but, on the other, may never materialize.

1.2 DAMAGING EFFECTS OF EARTHQUAKES

In general, an earthquake can damage a structure in three different ways: (a) by causing a ground failure, (b) by producing other effects that may indirectly affect the structure, and (c) by shaking the ground on which the structure rests.

1.2.1 Ground Failures

Possible ground failures are (a) surface faulting, (b) ground cracking, (c) ground subsidence, (d) landslides, and (e) soil liquefaction. An earthquake fault is a geological feature (ground fissure) associated with the generation of earthquakes. During an earthquake, the two sides of a fault may slip relative to one another. If a structure lies across a surface fault, then the structure may be damaged when the fault slips during an earthquake. Although this type of damage rarely occurs, it has happened in previous earthquakes (see Figure 1.1). Long bridges, dams, and pipelines are particularly susceptible to this type of damage.

Ground cracking is possible when the soil at the surface loses its support and sinks, or when it is transported to a different location. It occurs because when displaced, a soil layer breaks, causing fissures, scarps, horsts, and grabens on the ground surface. As described in the following paragraphs, ground cracking occurs mainly as the result of ground subsidence, a slope failure, or liquefaction of an underlying soil layer. Ground cracking may induce differential settlements, and thus it may severely disrupt bridges, pipelines, embankments, and other structures with long foundations.

Ground subsidence is a phenomenon in which the ground surface of a site settles or depresses as a result of the compaction induced by the vibratory effect of earthquakes. Sites with loose or compressible soils are the most likely to experience ground subsidence. Uncompacted fills and reclaimed lands are in this category. As it involves a relatively uniform soil deformation, ground subsidence usually causes only minor damage to building structures. This damage may be in the form of cracks and perhaps some tilting of the building. In contrast, it may substantially damage elongated structures such as bridges, pipelines, channels, and road embankments as the area affected by the subsidence may not cover the entire structure. Spectacular cases of ground subsidence have been observed in past earthquakes. One of such cases is the ground subsidence and the corresponding settlement that some buildings experienced during the 1985 Mexico City earthquake. Because of the severity of the subsidence, the first stories of these buildings permanently became basements after the earthquake (see Figure 1.2). Another case is the permanent sea flooding of a coastal village caused by a ground subsidence of 3 m following the 1976 Tangshan earthquake in China.

Landslides are often triggered by strong earthquakes (see Figure 1.3). These landslides represent the failure of slopes that are marginally stable before the earthquake and become unstable as a result of the violent shaking generated by the earthquake. A structure may be damaged by a

FIGURE 1.1 House damaged by fault scarp (foreground, right) after scarp goes through the house during the 1959 Hebgen Lake, Montana, earthquake. (Photograph courtesy of Earthquake Engineering Research Center, University of California, Berkeley, CA.)

FIGURE 1.2 Settlement of building in Mexico City induced by ground subsidence during the 1985 Michoacán earthquake.

FIGURE 1.3 Landslide at Turnagain Heights in Anchorage following the 1964 Alaska earthquake. A layer of sand below the surface liquefied as a result of the strong shaking. With its supporting soil weakened, the hill slumped downward and slid toward the sea. The ground broke into large chunks that turned and twisted as they moved along. Large cracks and fissures appeared between the earth blocks, and scarps as high as 15 m were formed. Houses and their occupants traveled along, turning and breaking up in the way. Over 70 buildings were dislodged and destroyed by the slide. The sliding area was ~2 km long and 300 m in width. (Photograph reproduced with permission of Earthquake Engineering Research Institute from Seed, H.B. and Idriss, I.M., *Ground Motions and Soil Liquefaction during Earthquakes*, Earthquake Engineering Research Institute, Oakland, CA, 1983.)

landslide if the structure, or part of it, happens to be on top of the soil mass that comes down during the landslide (see Figure 1.4), or when it lies below the landslide area and the sliding soil lands on top of it (see Figure 1.5). For the most part, earthquake-induced landslides are small, so the damage they induce is usually localized. However, there have been some instances in which landslides have buried entire towns and villages. A most unfortunate case in point is the landslide originated by the

FIGURE 1.4 Affluent home in Pacific Palisades in Los Angeles, California, area partially destroyed by a landslide during the 1994 Northridge earthquake. (Photograph courtesy of Earthquake Engineering Research Center, University of California, Berkeley, CA.)

FIGURE 1.5 Rockslide near the Okushiri Port in Hokkaido, Japan, that buried the Yo Yo Hotel during the 1993 Hokkaido-Nansei-Oki earthquake. (Photograph by E. L. Harp and T. L. Youd. Reproduced with permission of Earthquake Engineering Research Institute from Hokkaido-Nansei-Oki Earthquake and Tsunami of July 12, 1993, Reconnaissance Report, *Earthquake Spectra*, Supplement A to Volume 11, April 1995.)

May 31, 1970, earthquake near the city of Chimbote, Peru. High on the slopes of Mount Huascaran, ~130 km from the epicenter, the earthquake loosened rocks and ice and gave rise to a massive landslide. Increasing its speed and mass in its way down, the slide quickly acquired enormous proportions, filling a long valley with rock, ice, and mud. In the process, it partially destroyed the town of Ranrahirca, located 12 km away from the mountain and, after branching off to one side, totally buried the village of Yungay (see Figure 1.6). Totally, 18,000 people were killed by this single landslide.

Soil liquefaction is a phenomenon by which fine saturated granular soils temporarily change from a solid to a liquid state and as a result lose their ability to carry loads or remain stable (see Box 1.1 for a detailed explanation of this phenomenon). It occurs when a deposit of loose soil is vigorously shaken or vibrated, and thus it is commonly observed during earthquakes. It is caused

(a)

(b)

FIGURE 1.6 Overview of the village of Yungay, Peru, before (a) and after (b) being buried by a massive landslide that originated on the slopes of Mount Huascaran. The landslide was triggered by the 1970 Chimbote earthquake. (Photographs courtesy of Earthquake Engineering Research Center, University of California, Berkeley, CA.)

by a water pressure buildup that is generated when a saturated soil is compacted by the effect of the earthquake vibrations. As a result of this water pressure buildup, an upwardly water flow is generated, and the soil particles are separated from one another. The soil is then turned into a mixture of water and soil, with the soil particles floating in the water. Miniature sand or silt craters are left at the locations where the water reaches the surface (see Figures 1.7 and 1.8). Soil liquefaction is one of the most dramatic causes of earthquake damage and has been responsible for a substantial amount of damage in previous earthquakes (see Figures 1.9 through 1.11).

FIGURE 1.7 Sand boils on the surface of liquefied soil following the 1964 Niigata, Japan, earthquake. (Photograph courtesy of Earthquake Engineering Research Center, University of California, Berkeley, CA.)

FIGURE 1.8 Sand boils on the surface of the soil in the Marina District in San Francisco, California, after the 1989 Loma Prieta earthquake. (Photograph courtesy of Earthquake Engineering Research Center, University of California, Berkeley, CA.)

Introduction 7

FIGURE 1.9 Tilting and settlement of apartment buildings caused by soil liquefaction following the 1964 Niigata, Japan, earthquake. The movement took place slowly so no one was hurt, although reportedly the experience was horrifying for the building residents. The buildings suffered no structural damage. (Photograph courtesy of Earthquake Engineering Research Center, University of California, Berkeley, CA.)

FIGURE 1.10 Buried oil-storage tank floated out of the ground after the backfill around the tank liquefied during the 1993 Hokkaido-Nansei-Oki earthquake in Japan. The tank was ~1/3 full at the time of the earthquake. Lightweight, buried objects may float and be pulled to the surface when the soil in which they are buried liquefies. (Photograph by T. L. Youd. Reproduced with permission of Earthquake Engineering Research Institute from Hokkaido-Nansei-Oki Earthquake and Tsunami of July 12, 1993, Reconnaissance Report, *Earthquake Spectra*, Supplement A to Volume 11, April 1995.)

FIGURE 1.11 Severe liquefaction destroyed much of the city's port during the 1995 Kobe, Japan, earthquake. The gantry cranes suffered derailing and structural damage. Note the lateral movement of the quay walls and the ejected silt over the paved surface. (Reproduced with permission from Mainichi Newspapers.)

Box 1.1 Soil Liquefaction

Soil liquefaction is a phenomenon by which some soils temporarily change from a solid to a liquid state and lose, as a result, their ability to carry loads or remain stable. Liquefaction occurs in water-saturated soils and takes place when a soil deposit is shaken or vibrated. Therefore, soil liquefaction commonly occurs during earthquakes in low-lying areas near rivers, lakes, bays, marshlands, and coastlines. Liquefaction can affect the integrity of buildings, bridges, buried tanks and pipelines, and many other constructed facilities. For example, liquefaction can produce the sinking or tilting of heavy structures, the floating of buried lightweight structures, and the failure of retaining structures. It can also cause the slumping of slopes, landslides, ground subsidence, and the settlement of buildings. Liquefaction has caused extensive damage and some spectacular failures during the earthquakes of Alaska in 1964; Niigata, Japan, in 1964; Tangshan, China, in 1976; Guatemala in 1976; San Juan, Argentina, in 1977; Miyagiken-Oki, Japan, in 1978; Valparaiso, Chile, in 1985; Loma Prieta, California, in 1989; Kobe, Japan, in 1995; and Chi-Chi, Taiwan, in 1999.

Strong shaking can induce liquefaction of a saturated soil because a steady vibration of the soil progressively induces the collapse of its loosely packed granular structure (the way the soil particles are originally arranged) and compacts it (see Figure B1.1a). As a result of this compaction, the pressure in the water contained between the soil particles (pore water) rises. If the pressure in the pore water eventually rises to a level approaching that exerted

Introduction 9

by the weight of the overlying soil, and if the soil has no drainage outlets, then this water pressure generates an upwardly water flow, separates the soil particles from one another, and turns the soil into a mixture of water and floating soil particles (see Figure B1.1b). A common manifestation of liquefaction is fountains of water with sand spouted up from the ground, which leave holes surrounded by low craters (called *sand boils*). The presence of sand boils following an earthquake (see Figures 1.7 and 1.8) is therefore a common indication that a soil has liquefied.

The ease with which a soil may liquefy depends on the characteristics of the soil and the intensity and duration of ground shaking. In accordance with the preceding explanation of

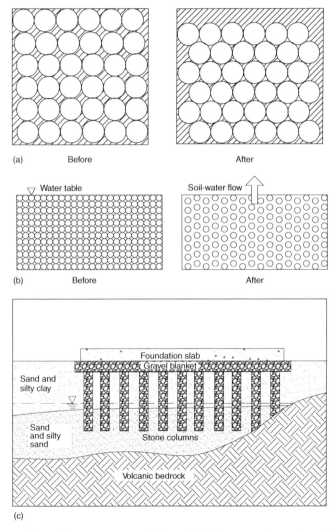

FIGURE B1.1 (a) Schematic arrangement of soil particles in saturated granular soil before and after strong shaking. (b) Saturated granular soil before and after liquefaction. (c) Vibroreplacement is a soil improvement technique in which a vibrator is used to penetrate a soil deposit to a desired depth and according to a predetermined grid pattern. The resulting cavities are then filled with a coarse-grained material, such as gravel or crushed rock, and compacted to form stone columns. Because they are compacted and formed with coarse-grained materials, these stone columns simultaneously reduce the soil compressibility and provide a drainage system to avoid the water pressure buildup that gives rise to liquefaction.

the phenomenon, liquefaction mainly occurs in saturated soils that have a relatively loose structure, are uncemented, have a low permeability, and are subjected to an intense ground motion with a sufficiently long duration. The soils most susceptible to liquefaction are thus loose sands and silts under the water table, with no clay content, a small particle size (fine-grained soils), a uniform grain size distribution, and restricted drainage in areas of high seismicity. Dense soils, included well-compacted fills, have low susceptibility to liquefaction. Liquefaction is most commonly observed at sites where groundwater is within a few meters from the ground surface.

Flow failures, lateral spreading, and ground subsidence are ground failures associated with liquefaction. Flow failures are the most catastrophic among these ground failures. In this type of failure, large masses of soil are laterally displaced tens of meters (in some instances tens of kilometers) down steep long slopes. Flows may be comprised of completely liquefied soil or blocks of solid material riding on a layer of liquefied soil. Flow failures have been responsible for the collapse of dams, embankments, and foundations. Flow failures caused the damage to the Sheffield Dam during the 1925 Santa Barbara earthquake and the Lower San Fernando Dam during the 1971 San Fernando earthquake. Lateral spreading generally develops in gentle slopes with a free face and involves the lateral displacement of large surficial blocks of soil as a result of liquefaction of an underlying soil layer. Lateral spreading occurs in response to a combination of gravity forces and the inertial forces generated by an earthquake. The lateral displacement is oriented toward the free face and may amount up to several meters. The displaced ground usually breaks up internally, causing fissures, scarps, horsts, and grabens on the ground surface. Lateral spreading may severely disrupt buildings, bridges, and pipelines if the ground failures occur across their foundations. A case in point is more than 200 bridges that were destroyed during the 1964 Alaska earthquake by the lateral spreading of flood plains toward their river beds. The major pipeline breaks during the 1906 San Francisco, California, earthquake are other examples. Lateral spreading also occurred along quay walls in part of the extensive port facilities in Kobe, Japan, during the 1995 earthquake, rendering some of them inoperative and causing the disruption or collapse of cranes. Ground subsidence associated with liquefaction occurs when the excess water pressure generated by the earthquake dissipates and the soil consolidates. Since it takes some time to dissipate this excess water pressure, the subsidence usually occurs well after the ground shaking ceases. Liquefaction caused the ejection of large volumes of silt and a ground subsidence of 30–50 cm in the islands of Rokko and Portopia, which are reclaimed land in Osaka Bay, during the 1995 Kobe earthquake.

Several ground modification techniques are available to minimize the potential for liquefaction in loose saturated soils. Some of these techniques include (a) replacement of liquefiable soils; (b) densification of the *in situ* material by vibroflotation (a technique in which a vibrating device is inserted into the ground to make a loose soil settle), dynamic compaction (impact densification or heavy tamping), or the use of compaction piles; (c) improvement of soil by grouting or chemical stabilization; and (d) the construction of vertical gravel drains (relief wells) to prevent the buildup of the upward pore water pressure that makes a soil liquefy (see Figure B1.1c).

1.2.2 Indirect Effects

Common indirect effects that may damage a structure are (a) tsunamis, (b) seiches, and (c) fires. Tsunamis are large sea waves generated by a sudden depression of the ocean floor (see Box 1.2 for a detailed explanation of this phenomenon). The dislocation of the ocean floor produced by the slippage of undersea earthquake faults is a common source of tsunamis. When a tsunami reaches a coastal area, its height may increase to catastrophic levels and strike the area with a tremendous force. To make matters worse, tsunamis can travel great distances across the oceans at a high speed and strike far away lands with undiminished fury. Tsunami run-ups as high as 30 m have been reported. Tsunamis are one of the most terrifying phenomena associated with earthquakes (see Figures 1.12 and 1.13 and Box 1.3). Reference to the devastating effects of tsunamis can be found throughout recorded history all over the world.

FIGURE 1.12 Tsunami generated by the 1946 Aleutian Islands, Alaska, earthquake as it approaches the beach front of the island of Hawaii. The tsunami reached a height of 6.1 m. (Photograph from Geological Hazard Collection, National Geophysical Data Center, National Oceanic and Atmospheric Administration.)

FIGURE 1.13 Aftermath of tsunami that struck Aonae, Japan, after the 1993 Hokkaido-Nansei-Oki earthquake. (Photograph by J. Preuss. Reproduced with permission of Earthquake Engineering Research Institute from Hokkaido-Nansei-Oki Earthquake and Tsunami of July 12, 1993, Reconnaissance Report, *Earthquake Spectra*, Supplement A to Volume 11, April 1995.)

Box 1.2 Tsunamis

A tsunami is a large sea wave generated, mostly, by an undersea earthquake. More specifically, a tsunami is a sea wave that is formed by a sudden vertical dislocation of the ocean floor induced by the slippage of an earthquake fault under the ocean (see Figure B1.2). Tsunamis can also be generated by underwater landslides and volcanic activity. Tsunami is a Japanese word that literally means *harbor wave*. Although in this literal sense, this word refers to the destination of the wave, not to its origin, the word is now universally accepted to denote a seismic sea wave. It is likely that the name originated from the fact that large seismic waves are observed only inside a bay or a harbor. Often, these seismic sea waves are also called *tidal waves*, but seismologists dislike this term because they are unrelated to the tides, which are created by the moon and the sun. Only a small fraction of undersea earthquakes generate significant tsunamis. A tsunami can reach a nearby coast soon after the occurrence of the earthquake. In this case, it is considered to be a local tsunami. A tsunami can also travel great distances across the oceans at high speed—normally at a speed between 600 and 800 km/h—and reach far away coastal areas. In such a case, the tsunami is called a *distant tsunami*. Depending on where it originates, a tsunami can have periods (time between two consecutive waves) that range from a few minutes to nearly 2 h. Because of its long periods, a tsunami can produce successive waves at intervals of more than an hour. Both local and distant tsunamis are capable of producing a great loss of life and devastating property damage.

A tsunami usually starts with a sudden vertical displacement of the ocean floor. The greater this vertical displacement, the higher the wave, and the greater the tsunami's destructive power. Such a vertical displacement gives rise to a wave that, in the middle of the ocean, can have a crest-to-trough height of no more than 1 m and a distance between crests (wavelength) of the order of 100 km. Indeed, a tsunami can be imperceptible to people on a boat or a ship. However, as the wave approaches the shoreline and the water depth starts decreasing, the wave height increases significantly and topples with tremendous force on the shore. Tsunamis with run-up heights of up to 30 m have been reported.

Although not all tsunamis show the same behavior, according to some accounts the first manifestation of a tsunami on the coast is the slow and silent rising of the sea well beyond the normal high tide mark. Then the water retreats making a sucking sound and leaving rocks, reefs, and sunken wrecks completely exposed and anchored ships stranded. Soon after that, the water becomes choppy and a fierce turbulence rolls up the sand from the ocean floor. Finally, a huge wave traveling at a high speed and making a terrifying noise sweeps in toward

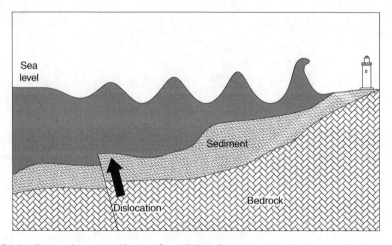

FIGURE B1.2 Tsunami generated by sea floor dislocation.

the land engulfing and swallowing everything in its path. The water remains high for some time, 10–15 min or even longer, and then with the same sucking noise it withdraws taking with it anything in its course. Some time later, more waves, moving at progressively slower speeds, strike the coastline again.

Ever since historical times, tsunamis have inflicted great losses of life and property in many coastal areas around the world. In 1896, a tsunami struck the eastern coast of Japan with 15 m high waves destroying more than 10,000 homes and drowning 26,000 people. In 1946 at Unimak Island, in the eastern Aleutian Islands, near Alaska, a magnitude 7.2 earthquake set in motion a seismic sea wave powerful enough to wipe out a lighthouse 32 ft above sea level and its radio antenna at an elevation of 103 ft The wave then traveled to the Hawaiian Islands, where, 5 h later, it killed 173 people (96 in Hilo alone) and caused millions of dollars in damage. In 1960, a tsunami originated by a large earthquake off the Arauco Province in Chile struck the coasts of Chile, Hawaii, and Japan. It inflicted 61 deaths in Hilo, Hawaii, and 120 deaths in the Sanriku coast of Japan. In 1964, the great Alaska earthquake generated a tsunami that caused several deaths (23 in Chenega Island) and considerable damage to seaports and coastlines of Alaska. Then, it traveled to Crescent City, on the far north cost of California, where, sweeping 30 m up the hillside, it killed 11 people, destroyed hundreds of houses, and sunk or capsized 23 boats that were anchored on the harbor. More recently, a tsunami generated by an earthquake in 1983, off the west coast of the main Japanese island of Honshu, killed 105 people near the city of Akita, and another, triggered by an earthquake in 1993 in the Japan Sea, off southwest the Japanese island of Hokkaido, killed 196 people in the small island of Okushiri, located 30 km west of Hokkaido. Both tsunamis also caused substantial property damage.

The coastal harbors of Japan, especially those along the northeast coast of Honshu Island, are frequently the target of damaging tsunamis. Destructive tsunamis have also occurred along the western coast of North America, the Aleutian Islands, the Philippines, New Guinea, Indonesia, the islands of the South Pacific, Hawaii, Peru, Chile, and Central America. Tsunamis have also been observed, although they are not as common, in the Caribbean Sea, eastern Atlantic Ocean, Mediterranean Sea, and Indian Ocean.

A tsunami warning system was developed for the Pacific Ocean after the tsunami that devastated Hilo, Hawaii, in 1946. This warning system, known as the Pacific Tsunami Warning System, is based at the Honolulu Observatory and is operated by the National Oceanographic and Atmospheric Administration (NOAA). This warning system is composed of an international network of stations in countries around the edges of the Pacific Ocean that monitor the ocean waves to detect tsunamis. All the stations are equipped with seismographs, tide gauges, and advanced communication systems. When an earthquake occurs anywhere within the Pacific Ocean, all the stations in the network are alerted. If the earthquake is a large one and is located where it could generate a tsunami, a *tsunami watch* is issued. When information is received that a tsunami has indeed been formed, a *tsunami warning* is sent out to all the areas that could be affected by the tsunami, together with estimated arrival times. After a tsunami warning is received, inhabitants of coastal regions are warned by the local officials of the impending tsunami and, if merited, evacuated to high ground.

Box 1.3 Indian Ocean Tsunami of December 26, 2004

On December 26, 2004, a massive tsunami struck the costs of the Indian Ocean, devastating many communities in Southeast Asia and East Africa. The countries that suffered major casualties and extensive damage were Indonesia, Sri Lanka, India, Thailand, the Maldives, Somalia, Myanmar, Malaysia, and Seychelles (see Figure B1.3a). It reached and caused serious damage as far as Port Elizabeth in South Africa, some 8000 km away from the epicenter of the causative earthquake, where the tsunami reached a height of 1.5 m about 16 h after the occurrence of the earthquake. The tsunami generated a succession of several waves with heights of up to 30 m, occurring in retreat and rise cycles with a period of over 30 min between each cycle (see Figures B1.3c and B1.3d). The third wave was the most powerful and with the greatest height, occurring about an hour and a half after the arrival of the first wave.

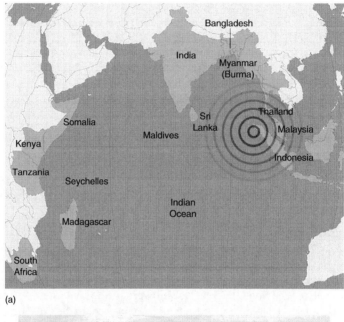

(a)

(b)

FIGURE B1.3 (a) Countries affected by the 2004 Indian Ocean tsunami. (b) Inundated village near the coast of Sumatra, a few days after being struck by the tsunami. (Map from Wikipedia; Photograph courtesy of U.S. Navy.) (c)–(f) Sequence of photographs showing the arrival of the tsunami at Khao Lak beach in Thailand: (c) water recedes from the beach, (d) a wave approaches the beach, (e) the wave gets closer to the beach and increases in height, and (f) the wave crashes ashore. (Photographs courtesy of the family of John and Jackie Knill.)

FIGURE B1.3 (Continued)

In many places, the waves reached as far as 2 km inland. Because of the distances involved, the time the tsunami took to reach the various coastlines varied between 15 min and several hours after the occurrence of the earthquake. The northern regions of the Indonesian island of Sumatra were hit quickly, whereas Sri Lanka and the east coast of India were hit 90 min to 2 h later. The tsunami killed at least 30,900 people in Sri Lanka, 10,700 in India, 5,300 in Thailand, 150 in Somalia, 90 in Myanmar, 82 in the Maldives, 68 in Malaysia, 10 in Tanzania, 3 in Seychelles, 2 in Bangladesh, and 1 in Kenya. All together, it left 186,983 people dead and 42,883 missing, making it the worst tsunami and one of the deadliest natural catastrophes in modern history.

The earthquake that triggered the tsunami originated in the Indian Ocean just north of Simeulue Island, off the western coast of northern Sumatra, Indonesia. It occurred at 07:58:53 local time and had a moment magnitude of 9.1 (see Section 5.6.3 for a definition of moment magnitude). It was the fourth largest earthquake ever recorded worldwide. The hypocenter was ~160 km west of Sumatra at a depth of 30 km below mean sea level. The earthquake itself caused severe damage and casualties in northern Sumatra and the Nicobar Islands in India.

Seiches are long-period oscillating waves generated by distant earthquakes in enclosed bodies of water such as bays, lakes, reservoirs, and even swimming pools. Seiches occur when the natural frequency of a water body matches the frequency of the incoming earthquake waves, that is, when the water body resonates with the earthquake waves. Seiches may last for

FIGURE 1.14 Over 100 fires broke out within minutes in the densely built-up areas of the city after the 1995 Kobe, Japan, earthquake. Overall, more than 1 million m² of the city were burned. (Reproduced with permission from Mainichi Newspapers.)

hours. When large, seiches may damage structures located near the water. Damaging waves of up to 1.5 m in height were generated by the 1964 Alaska earthquake in lakes in Louisiana and Arkansas.

Fires, by far, have been the most devastating indirect effect of earthquakes. Fires are started when, for example, an earthquake breaks gas pipes or destroys oil-storage tanks, and overturned stoves, furnaces, and heaters, or downed power lines, ignite the gas or spilled oil. Earthquake-induced fires have devastated cities and killed thousands of people in the past. Two dramatic examples are the fires following the 1906 San Francisco and the 1923 Kanto earthquakes. After the 1906 San Francisco earthquake, fires broke out in several places and spread for 3 days, burning 508 blocks of the city. Similarly, after the 1923 Kanto earthquake, the cities of Tokyo and Yokohama burned to ashes by fires started by the earthquake and fanned by high winds. In Tokyo, nearly two-thirds of the entire city was wiped out. In Yokohama, 70,000 homes were consumed by the fire. Overall, 140,000 people were killed by the earthquake and the ensuing fire. More recently, the 1995 Kobe earthquake has been a reminder that modern cities are still vulnerable to earthquake-induced fires. Several wards throughout the city were ravaged by fires started by this earthquake. It is estimated that the fire burned an area of over 1 million m² (see Figure 1.14).

1.2.3 Ground Shaking

Ground shaking may be considered the earthquake effect that is the most damaging to structures. During an earthquake, as is commonly known, the ground moves vertically and horizontally, at times strongly and violently. This motion, in turn, makes a structure lying on the shaking ground oscillate back and forth and up and down and makes the structure experience large stresses and deformations in this process. If strong enough, ground shaking may cause the partial or total collapse of the structure. Furthermore, since an earthquake makes the ground shake over extensive areas of the ground surface, ground shaking may simultaneously affect a large number of structures. Thus, if an earthquake occurs near an urban center, ground shaking may bring widespread destruction in a matter of seconds. Similarly, since virtually all structures rest on the earth's surface,

Introduction

they may be seriously affected by ground shaking. In fact, ground shaking has been responsible for the collapse of, or serious damage to, buildings, bridges, dams, retaining structures, chimneys, pipelines, subway structures, tanks, equipment on buildings, and equipment in industrial plants during past earthquakes (see Figures 1.15 through 1.25). Needless to say, ground shaking is the primary concern of structural engineers in the seismic regions of the world.

FIGURE 1.15 Total collapse of a 22-story steel frame building in Pino Suárez Complex during the earthquake that affected Mexico City in 1985. The collapsed building was practically identical to the building standing on background. (Photograph from Geological Hazard Collection, National Geophysical Data Center, National Oceanic and Atmospheric Administration.)

FIGURE 1.16 Collapse of mid-rise building during the 1999 Chi-Chi earthquake in Taiwan. (Photograph courtesy of NSF-sponsored Geo-Engineering Earthquake Reconnaissance (GEER) Association.)

FIGURE 1.17 Collapse of the three spans of Agua Caliente Bridge as a result of relative motion between piers during the 1976 Guatemala earthquake. (Photograph courtesy of Earthquake Engineering Research Center, University of California, Berkeley, CA.)

FIGURE 1.18 Dislocation of bridge deck caused by failure of bridge pier during the 1999 Chi-Chi earthquake in Taiwan. (Photograph courtesy of NSF-sponsored Geo-Engineering Earthquake Reconnaissance (GEER) Association.)

FIGURE 1.19 Collapse of the elevated Hanshin Expressway during the 1995 Kobe, Japan, earthquake. (Reproduced with permission from Mainichi Newspapers.)

FIGURE 1.20 Street depression caused by underground collapse of the Daikai subway station during the 1995 Kobe, Japan, earthquake. (Photograph courtesy of Thomas D. O'Rourke, Cornell University.)

FIGURE 1.21 Upstream slope of 42 m high Lower San Fernando Dam after a portion of it failed and slid down under the water during the 1971 San Fernando, California, earthquake. (Photograph courtesy of Earthquake Engineering Research Center, University of California, Berkeley, CA.)

FIGURE 1.22 Failure of quay wall in Rokko Island during the 1995 Kobe, Japan, earthquake. (Photograph courtesy of José Pires.)

Introduction

FIGURE 1.23 (a) Collapse of elevated water tank during the 1979 El Centro, California, earthquake and (b) tank similar to the collapsed tank. (Photographs courtesy of Earthquake Engineering Research Center, University of California, Berkeley, CA.)

FIGURE 1.24 Failure of circuit breaker at electric power substation during the 1996 North Palm Springs, California, earthquake. (Photograph by L. E. Escalante. Reproduced with permission of Earthquake Engineering Research Institute, *Reducing Earthquake Hazards: Lessons Learned from Earthquakes*, Publication No. 86-02, Earthquake Engineering Research Institute, November 1986.)

FIGURE 1.25 Collapse of 17 t scoreboard at Anaheim Stadium (black object below stadium's A logo and surrounded by billboards) during the 1994 Northridge, California, earthquake.

1.3 EARTHQUAKE FORCES

For structural engineers and from a conceptual point of view, earthquakes represent just another force for which structures need be designed. Earthquake forces, however, possess several characteristics that make them unique in comparison with any other forces, such as gravity, wind, or thermal forces. Earthquake forces, the result of a back and forth, and up and down, motion of the ground that supports a structure, can be exceptionally large in magnitude, can change rapidly and erratically during the duration of the earthquake, and may be radically different from earthquake to earthquake, from one site to another, from one type of foundation soil to another, and from one structure to another. Furthermore, earthquake forces depend on the properties of the structure. This means that if one modifies such properties, one also modifies the earthquake forces that will affect the structure. It also means that they can—and usually do—change if the earthquake damages the structure. Most importantly, earthquake forces are unpredictable. The reason is that little is known today about the mechanism that generates earthquakes, and not much more is known about the factors that shape the earthquake–generated ground shaking that gives rise to earthquake forces. As a result, the magnitude and characteristics of earthquake forces can only be, at best, roughly estimated.

Another unique feature of earthquake forces is the likelihood associated with their occurrence. At any given seismic region, small earthquakes are likely to occur often, moderate earthquakes once every few years, large earthquakes every 50–100 years, and extremely large earthquakes perhaps every 1000 or 2000 years. In a given time interval, there is thus a high probability that a structure will be affected by small earthquake forces and a negligibly small probability that it will be affected by extremely large earthquake forces. Therefore, since it is economically unfeasible to design every structure to resist the largest earthquake forces ever possible, the design earthquake forces necessarily have to be those for which there is an appreciable probability of occurrence during their lifetime. What is an appreciable probability is usually decided by the designer in consultation with the owner of the structure and depends, among other factors, on the importance of the structure. In conjunction with such probability of occurrence, there is also a probability that the magnitude of the earthquake forces for which a structure is designed be exceeded in a given time interval. Consequently, implicitly or explicitly, the specification of earthquake forces for the design of a structure is always accompanied with an associated probability of exceedance.

Earthquake forces are also distinct from other forces in the sense that they affect the strength and behavior of structural materials. That is, the properties of structural materials under earthquake loads are different from the properties that are considered when designing, for example, for gravity loads. This is owed to the fact that earthquake forces are applied suddenly, are relatively short, and change in direction many times during the earthquake. Thus, the magnitude of the earthquake forces is only part of the information a structural engineer needs to know to properly design a structure against these forces.

1.4 DESIGN FOR EARTHQUAKE FORCES

Because of the unpredictability of earthquake forces, the uncertainty of their occurrence, and the devastating effects they may produce, the design of an earthquake-resistant structure is an elaborate process that requires the participation of architects, seismologists, geologists, soil engineers, foundation engineers, and structural engineers. In general, it involves many of the following steps:

1. Identification of the sources where future earthquakes are likely to occur with the aid of historical information, seismological data, and geological studies
2. Determination of the probable size of future earthquakes on the basis of the attributes of the identified seismic sources
3. Definition of the distance and orientation of each seismic source with respect to the structure's location
4. Establishment of semiempirical equations that correlate ground motion characteristics with earthquake size, seismic source orientation and distance, and site soil conditions with the help of instrumental and observational records from previous earthquakes
5. Dynamic analysis of the soil deposits at the structure's site to quantify the ground motion amplification that may be induced as a result of their flexibility
6. Selection or modification of structural configuration, structural system, and structural materials to minimize undesirable structural responses and best resist the expected earthquake forces
7. Dynamic analysis of the structure and its components to estimate the maximum values of the internal forces and deformations that may be generated by a ground motion with the established characteristics
8. Analysis of foundation soil to assess its susceptibility to earthquake effects
9. Verification of analytical results using laboratory tests of scaled models using shaking tables, or field tests of full-scale models using artificial means to generate ground vibrations
10. Configuration, proportioning, and detailing of the members and connections of the structure in accordance with the estimated maximum internal forces and deformations
11. Improvement of foundation soil properties to reduce soil's susceptibility to earthquake effects

A design that involves all these steps is certainly expensive and time consuming. Therefore, such a design is usually limited to structures of critical importance, such as high-rise buildings, major dams, nuclear power plants, and long-span bridges. For ordinary structures, the design is usually carried out in accordance with the requirements of a seismic code or a combination of seismic code requirements and some of the described steps. In most areas where earthquakes are likely, cities issue official seismic codes that specify the earthquake forces for which structures and their components should be designed according to the location, soil conditions, type of structure, and geometric characteristics. Most seismic codes also specify a set of minimum requirements to ensure safe designs and offer a series of guidelines to facilitate the design process.

1.5 DEFINITION AND SCOPE

Earthquake engineering may thus be considered the branch of civil engineering that provides the principles and procedures for the planning, analysis, and design of structures and facilities that are capable of resisting, to a preselected extent, the effects of earthquakes. More specifically, earthquake engineering provides the principles and procedures for (a) the selection of the proper location of structures to minimize their exposure to earthquake hazards; (b) the estimation of the earthquake forces that may affect structures and their surrounding environment in a given time interval; (c) the analysis of structures and surrounding environment under the effect of such forces to determine the maximum stresses and deformations that may be imposed upon them; (d) the configuration, proportioning, and detailing of structures to make them resist such stresses and deformations without the collapse or failure of any of their components; and (e) the improvement of soils and the stabilization of natural slopes to guarantee the stability of structures supported on weak soils or slopes. It is based on concepts from seismology, geology, probability theory, geotechnical engineering, structural engineering, and structural dynamics. Seismology plays a role in defining the source of possible earthquakes in a given region, their likely magnitude and likely frequency of occurrence, the location and orientation of these sources, and the characteristics of past earthquakes. Geology complements seismological information by providing maps of known geological faults, geomorphological studies that permit the identification of earthquake activity long before the available historical seismological record, and the use of trenching, boreholes, age-dating, and geophysical surveys to quantify past earthquake activity. Concepts from probability theory are used in the derivation of realistic models to assess the magnitude and frequency of the earthquakes expected at a given seismic region and to account for the large uncertainties involved in the determination of the aforementioned earthquake forces. Concepts from geotechnical engineering are applied to quantify the influence of local soil conditions on the characteristics of the earthquake-generated motion at the ground surface of a specified site. They are also applied to identify soil deposits that are susceptible to liquefaction or ground subsidence and potentially unstable slopes. In addition, geotechnical engineering provides the techniques to design embankments and retaining structures against earthquake forces and strengthen soil deposits susceptible to liquefaction or ground subsidence and marginally stable slopes. Structural engineering furnishes the rules and methods to select the structural system that is best suited to resist the expected earthquake forces and perform the structural analysis and component proportioning and detailing referred to earlier. Finally, as earthquake forces are intrinsically dynamic forces, that is, forces whose magnitude varies with time, the principles of structural dynamics are necessarily applied in many of the aspects of earthquake engineering. For example, structural dynamic principles are used in the analysis of structures, embankments, and slopes, and in the investigation of the influence of soil conditions on the earthquake ground motion at a site.

Because of the complexity, unpredictability, and large magnitude of earthquake forces and the catastrophic consequences of earthquakes, earthquake engineering is a discipline that has evolved beyond the traditional establishment of principles and procedures for the planning, analysis, and design of earthquake-resistant structures and facilities. Thus, earthquake engineering also deals nowadays with, for example, (a) postearthquake field damage investigations, (b) the development and deployment of a network of instruments to record earthquake ground motions and the response of structures to earthquakes, (c) the elaboration of seismic zonation and microzonation maps, (d) the formulation of reliability and engineering decision models, (e) the techniques for laboratory testing of scale models on shaking tables and field testing of full-size structures, (f) system identification techniques, (g) the development of seismic-protective systems, and (h) the methods to repair structures that have suffered earthquake damage or upgrade structures in need of seismic strengthening.

Postearthquake damage field investigations are an important component of earthquake engineering. They are essential to document the consequences of different earthquakes on different soils, structural systems, and structural materials; to verify or correct assumptions previously made about these soils, structural systems, and structural materials; and to uncover unusual damaging earthquake effects. It is through field investigations that earthquake engineering has progressed and succeeded in protecting life and property from earthquakes. It is also through field investigations that earthquake engineers learn new lessons every time a major earthquake strikes a large urban center and realize that the earthquake problem is far from being solved.

The deployment of a network of instruments to record ground and structural motions during earthquakes is fundamental to learn how earthquake ground motions vary from earthquake to earthquake, from one geographical region to another, and from one type of soil to another, and how earthquake ground motions attenuate with distance. It is also fundamental to learn how soils and structures behave under the effect of these ground motions. It is by means of the records from these instruments that earthquake engineers identify the factors that influence earthquake ground motions and develop ground motion attenuation relationships and design recommendations in regard to the magnitude and characteristics of the earthquake forces expected at different locations. These records are also used to calibrate or verify results from analytical studies. Several of these networks presently exist in different parts of the world, and the number of instruments in them continues to grow. Valuable records have been obtained during the last three decades from worldwide earthquakes.

Seismic zonation maps are charts that delineate the variation of the seismicity or earthquake hazard in a large geographical region. These maps are developed by using a predefined measure of earthquake intensity as a delineation index and by dividing the geographical region under consideration into zones or provinces of equal or similar earthquake intensity. A zonation map for a small region, such as a city, with consideration of local differences in topographical and geological conditions is called a microzonation map. Zonation and microzonation maps are important from the point of view that they give an overview of the seismicity in a given region and facilitate the assessment of the seismic hazard in a given area. As such, they are useful for land-use planning and site selection. They are also useful to specify the earthquake intensity for which structures should be designed in the different zones of a geographical region. Seismic codes ordinarily make use of these maps for such a purpose.

As noted earlier, the design of structures and facilities to resist strong earthquakes significantly increases the cost of construction. However, a strong earthquake at any given location is a rare phenomenon whose time of occurrence and characteristics cannot be predicted with any degree of certainty. Therefore, the substantial resources that are invested to protect a whole community against earthquake effects may be totally wasted if such a strong earthquake never occurs. In contrast, if such an investment is not made, the community will be unnecessarily exposed to a risk of disastrous consequences. It is clear, thus, that the design of earthquake-resistant structures and facilities requires a suitable balance between safety and economy. To find this balance in terms of the uncertainties involved in the definition of the earthquake hazard and the economic priorities of a society, earthquake engineers use models based on reliability theory and decision analysis. The formulation of reliability and engineering decision models has thus become an essential component of earthquake engineering. Some of the recommendations in seismic codes are implicitly or explicitly based on these models.

As it is difficult to understand and predict the behavior of soils and structures under earthquake ground motions by theoretical means alone, laboratory and field tests are frequently performed to establish the capacity of soils, structures, or structural components to resist earthquake loads; to determine their dynamic properties; or to verify the results of analytical studies. However, because of the large scale of civil engineering structures and the difficulties in reproducing field conditions, laboratory and field tests are a problematic and expensive undertaking. Therefore, the special methods used for the testing of soils and structures under simulated earthquake loads have become by themselves an important branch of earthquake engineering. In general, laboratory tests of structural systems are carried out using computer-controlled shaking tables to simulate earthquake ground motions. At times, hydraulic actuators, with the capability to rapidly reverse their direction of action, are used to directly simulate the forces and deformations induced by an earthquake on a structure or structural component instead of the simulation of the earthquake ground motion itself. Common techniques to simulate earthquake vibrations in the field are the use of mechanical shakers, explosives, or simply the pulling and sudden release of a structure by means of a rope or a cable. Laboratory tests of soils and soil structures are usually carried out using dynamic triaxial machines and occasionally small shaking tables enclosed in a centrifuge. Field tests using ambient vibrations (vibrations induced by wind, vehicular traffic, etc.) and sensitive electronic equipment to measure these vibrations are nowadays a popular method to determine the dynamic properties of actual soils and structures.

Structural or system identification is a technique whereby the physical properties of a structure are determined using the recorded motions of the structure during earthquakes or experimental investigations. As recordings at numerous points are needed to have good estimates of such properties, and as such numerous recordings are usually not available, an important problem in structural identification is that of obtaining reliable estimates of the physical properties in question using only a minimum number of such recordings. Structural identification techniques are useful to determine the actual properties of existing structures that are in need of upgrading, or those that have been seriously damaged during an earthquake and need to be repaired. Structural identification techniques are also useful in the interpretation of results from experimental tests.

Seismic-protective systems are devices that are added to a structure to help the structure resist earthquake effects. In broad terms, a protective system limits the amount of the earthquake energy that is transmitted to the structure by modifying the properties of the structure, absorbs the earthquake energy by means of energy-dissipating devices, or counteracts the earthquake energy by means of an external source of energy. Seismic-protective systems that are currently being investigated and in some cases implemented in real structures include (a) base isolation systems, (b) supplemental dampers, (c) tuned mass dampers, and (d) active control systems. In a base isolation system, a structure is separated from the ground by means of flexible elements, usually rubber pads, in a way that makes earthquakes distort these flexible elements instead of the structure. Supplemental dampers are energy-dissipating devices that are installed at strategic points of a structure with the purpose of absorbing earthquake energy. The most commonly used types are (a) friction dampers (sliding steel plates covered with friction pads), (b) viscous dampers (similar to shock absorbers in automobiles), (c) viscoelastic dampers (stacked steel plates separated by an inert polymer), and (d) hysteretic dampers (shaped steel plates designed to yield and deform beyond their elastic range under small loads). Tuned mass dampers are systems composed of a mass, a flexible element, and a damper and are usually installed on the top of a building. A tuned mass damper moves opposite to the building where it is installed and thus counteracts the motion of the building. In an active control system, actuators are used to apply forces that balance the earthquake forces in a structure or forces that change the properties of the structure. As earthquake forces change randomly and rapidly, active control systems use a computer to control the magnitude and direction of the applied forces and sensors to determine the magnitude and direction of the earthquake forces.

In spite of all the advances in earthquake engineering, a large number of the existing structures and facilities in the world's seismic regions were built using techniques and seismic code provisions that are below current standards. Undoubtedly, many of these structures and facilities need to be upgraded if catastrophic failures are to be avoided in future earthquakes. Undoubtedly, too, many of those structures and facilities will be damaged in the event of a strong earthquake as time and financial resources will always be an obstacle to upgrade them all in a timely fashion. Procedures are thus needed to upgrade such structures in an effective and economically viable way. Effective and economical methods to repair structures damaged by an earthquake are also needed. Because of the growing awareness of the vulnerability of cities, the extensive damage produced by earthquakes in the last few decades, the high cost involved in the upgrading or repairing of structures, and the unmatched challenge that these two activities represent to structural engineers, seismic strengthening and the repair of structures damaged by earthquakes have recently become an important subdiscipline of earthquake engineering.

1.6 HISTORICAL BACKGROUND

Robert Mallet, an Irish civil engineer, is often cited as the first earthquake engineer and his report on the 1857 Naples earthquake is considered to be the first scientific investigation that included observations of the seismological, geological, and engineering aspects of an earthquake. Modern research on earthquake-resistant structures, however, began in Japan in 1891, the year of the Nobi earthquake (7000 deaths; also known as the Mino-Owari earthquake), with the formation of an earthquake investigation committee set up by the Japanese government. It was this committee who first proposed the use of a lateral force equal to a fraction of the total weight of a building to account for the forces exerted on buildings by earthquakes. Similar developments in Italy after the devastating Messina earthquake in 1908 (58,000 deaths) led to the appointment of a committee composed of practicing and academic engineers to study the earthquake and the formulation of practical recommendations for the seismic design of buildings. In its report, this committee recommended that the first story of a building be designed for a horizontal force equal to 1/12 of the building weight above and that its second and third stories be designed for a horizontal force equal to 1/8 of the building weight above. These Japanese and Italian disasters thus gave birth to practical considerations for the earthquake design of structures and to earthquake engineering as a new branch of engineering.

In the United States, interest in earthquakes and earthquake engineering began after the 1906 earthquake in San Francisco, California (1000 deaths), which caused great damage and loss of lives. At that time, however, California was still sparsely populated and, therefore, the interest generated by this earthquake was not enough to motivate public officials to develop earthquake design regulations. It was only after the 1933 earthquake in Long Beach, California, that American engineers became fully aware of the dangers of earthquakes, and a great impetus was given to the study of seismology and earthquake-resistant designs. As they became fully interested, the first inquiry was to find out the nature of the motion of the ground during an earthquake. Special instruments were designed and deployed at various areas of high seismicity to record such a motion permanently. Congress charged the U.S. Coast and Geodetic Survey with the responsibility to study and report strong earthquake motions. About the same time, new building codes were drawn up and enforced. The California Legislature passed the Field Act, which made it mandatory for all school buildings to be designed and built to resist earthquakes. Shortly after, the State of California adopted the Riley Act, which made it mandatory to design most buildings in the state for a lateral load equal to 2% of the sum of their dead and live loads. The Pacific Coast Building Officials (to become later the International Conference of Building Officials) published the nation's first seismic design provisions in 1927 in its Uniform Building Code.

Ever since, earthquake engineering has unfolded at a steady pace and its principles spread all over the world. It has rapidly evolved into a science-based discipline, with a large body of knowledge and institutionalized research and educational programs. Although learning takes place at a very slow pace due to the infrequency of large earthquakes, advances in methods of dynamic analysis and experimental research have provided engineers with valuable data to gain, year after year, a further understanding of earthquakes and the effects of earthquakes in civil engineering structures and facilities, and to develop new devices and techniques to protect these structures and facilities from such effects. As a result, cities around the world and the people living in them are little by little becoming less vulnerable to the devastating effect of earthquakes.

FURTHER READINGS

1. Bolt, B. A., *Earthquakes*, 5th Edition, W. H. Freeman and Co., New York, 2004, 378 p.
2. Eiby, G. A., *Earthquakes*, Van Nostrand Reinhold Co., New York, 1980, 209 p.
3. Gere, J. M. and Shah, H. C., *Terra Non Firma*, W. H. Freeman and Co., New York, 1984, 203 p.

4. Halacy, D. S., Jr., *Earthquakes*, Bobbs-Merrill Co., Indianapolis, IN, 1974, 162 p.
5. Hodgson, J. H., *Earthquakes and Earth Structure*, Prentice-Hall, Englewood Cliffs, NJ, 1964, 166 p.
6. Hudson, D. E., Nine milestones on the road to earthquake safety, *Proceedings of Ninth World Conference on Earthquake Engineering*, Vol. 2, Muruzen Co., Ltd, Tokyo, Japan, 1989, pp. 3–11.
7. Housner, G. W., Historical view of earthquake engineering, *Proceedings of the Eighth World Conference on Earthquake Engineering*, Post-Conference Volume, Prentice-Hall, Englewood Cliffs, NJ, 1984, pp. 25–39.
8. Kramer, S. L., Liquefaction (Chapter 9) in *Geotechnical Earthquake Engineering*, Prentice-Hall, Upper Saddle River, NJ, 1996, pp. 348–422.
9. Verney, P., *The Earthquake Handbook*, Paddington Press, Ltd, New York and London, 1979, 224 p.

2 Seismic Regions of the World

What is an earthquake? It is not a bunch of wiggles on a sheet of paper, at least not to the person involved in it. To him it is the most terrifying of experiences, in which the earth ceases to be the solid foundation that instinct, bred of generations of experience, has led him to expect, so that he becomes unable to trust his own senses. It is an experience which may leave him homeless, take away his loved ones, destroy his economy. It is a terror which has led the Roman Catholic Church to include in its Litany: From the scourge of the earthquake, O Lord, deliver us.

John H. Hodgson, 1965

2.1 INTRODUCTION

Earthquakes do not occur in all geographical locations around the world. Rather, they take place within certain limited areas. Fortunately, earthquakes have been recorded instrumentally since the beginning of the twentieth century at a large number of seismographic stations distributed throughout the world (see Box 2.1). Therefore, it is now possible to know where and when earthquakes have occurred in the past, how large they have been, and where in the world earthquakes are likely to occur again. The description of the time, location, size, and frequency of the earthquakes that have occurred in a region is referred to as the description of the *seismicity* of the region. Conventionally, this seismicity is portrayed in the form of *seismicity maps*. Seismicity maps show the geographical location where earthquakes have occurred during a specified time interval and describe the intensity of these earthquakes by means of circles or dots of different sizes. A seismicity map may describe the geographic distribution of earthquakes around the world, a country, or a specific region. Seismicity maps, the cooperative work of hundreds of seismologists throughout many years, have contributed in a fundamental way to define the seismicity of the earth and are nowadays an essential tool for the planners, geologists, engineers, and government officials involved in earthquake mitigation activities. Seismicity maps have also played a major role in the evolution of the plate tectonic theory described in Chapter 3.

Box 2.1   The Global Seismographic Network

The U.S. Geological Survey operates in cooperation with a consortium of more than 90 universities, an international network of seismographic stations named the Global Seismographic Network (GSN). This network, designed to obtain high quality data in digital form and be able to access these data by computer via the World Wide Web, monitors and records earthquakes, volcanic eruptions, and nuclear explosions throughout the world. The network is composed of 128 stations located in more than 80 countries in all continents. The map in Figure B2.1 shows the location of these stations.

The growth of the international network of seismographic stations has been somewhat haphazard. It started in 1898 when the British Association for the Advancement of Science recommended the deployment of a uniform network of stations distributed throughout the

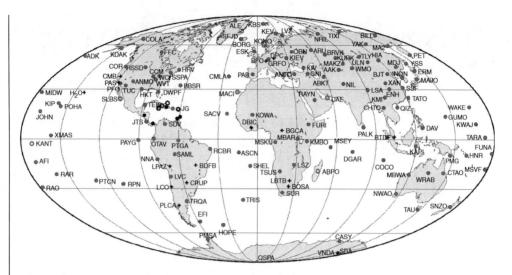

FIGURE B2.1 Global Seismographic Network. (After National Earthquake Information Center, U.S. Geological Survey.)

world. The instruments used were those designed by John Milne, a British seismologist who had worked for many years in Japan, and were operated by observatories throughout the British Empire. Soon after, many other national and regional groups of stations began to be set up. However, the setup of these stations was without coordination and with no uniformity in the type of instruments used. The resulting network was nearly chaotic. The records coming from this network were often difficult and sometimes impossible to compare.

By 1964, the World Wide Standardized Seismographic Station Network (WWSSN) was set up. This network was initiated as a result of a ban on the testing of nuclear weapons and the need to discriminate between explosions and earthquakes at great distances as a means to enforce this ban. A standard package of six seismographs controlled by precise chronometers that could be calibrated by radio signals was distributed to 120 stations in six nations around the globe. This network made possible a rapid assessment of the global characteristics of significant earthquakes for the first time. It greatly improved the quality, coverage, and quantity of the earthquake data recorded at seismographic stations.

The WWSSN has now been replaced by the state-of-the-art GSN. As noted earlier, this upgrade significantly improves the earthquake reporting and research capabilities of seismologists and earthquake specialists all over the world.

2.2 WORLD SEISMICITY

Figure 2.1 shows the distribution of earthquakes around the globe in the period between 1963 and 1988. It may be seen from this map that the world's areas of high seismicity are

1. A zone that extends from the Aleutian Islands through Alaska; the Pacific side of Canada, the United States, and Mexico; Central America; the Pacific side of Colombia, Ecuador, and Peru all the way down to Chile
2. A zone that goes from the Kamchatka Peninsula in Russia; through the Kuril Islands, Japan, Taiwan, the Philippines; to New Guinea, Indonesia, and New Zealand
3. An east–west trans-Asiatic zone running from Burma through the Himalayan Mountains and the Middle East to the Caucasus Mountains and the Mediterranean Sea

Seismic Regions of the World

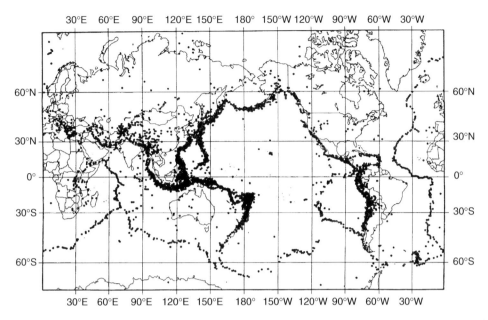

FIGURE 2.1 Seismicity of the earth between 1963 and 1988. (After National Earthquake Information Center, U.S. Geological Survey.)

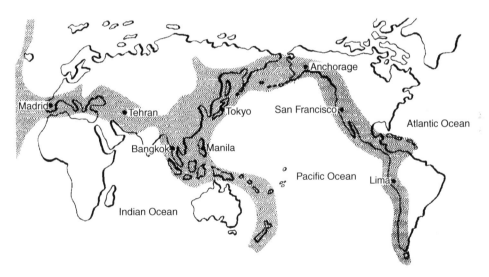

FIGURE 2.2 Seismic belts of the world. (After Hansen, W.R. and Eckel, E.B. in *A Summary Description of the Alaska Earthquake—Its Settings and Effects*, Professional Paper No. 541, U.S. Geological Survey, 1966.)

4. A zone that follows the submarine Mid-Atlantic Ridge, which extends along the full length of the Atlantic Ocean and appearing above the surface in Iceland, the Azores Island, and Tristan da Cunhan

The first two zones constitute what is known as the circum-Pacific seismic belt or the *Ring of Fire* (see Figure 2.2). Approximately 80% of the world's largest earthquakes take place in this belt. The third one is called the Eurasian or Alpine–Himalayan Belt. Although it only accounts for ~17% of the world's largest earthquakes, some of the most destructive earthquakes have occurred in that belt. Earthquakes

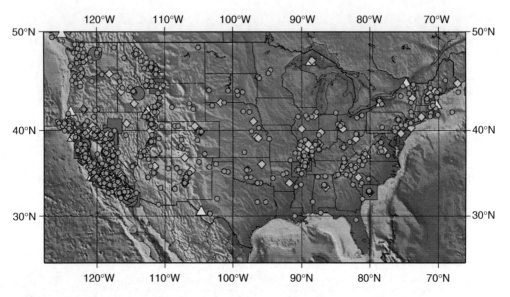

FIGURE 2.3 Location of damaging earthquakes in the United States between 1750 and 1996. (After National Earthquake Information Center, U.S. Geological Survey.)

along the Mid-Atlantic Ridge are obviously not destructive, but they are important because they suggest a correlation between the occurrence of earthquakes there and the existence of the ridge.

It may also be seen from Figure 2.1 that many areas of the world are almost free from earthquakes. Notable examples are the eastern part of South America, the central and northern regions of Canada, much of Siberia, West Africa, and large parts of Australia. Northern Europe is, for the most part, not seismically active, although destructive earthquakes have occurred in Germany, Austria, Switzerland, the North Sea region, and Scandinavia.

2.3 SEISMICITY OF UNITED STATES

The seismicity of the United States is depicted in Figure 2.3. As shown in this figure, earthquakes have occurred virtually in all 50 states. The areas of high seismicity include

1. The states of California, Nevada, and Washington
2. The states of Utah, Idaho, and Montana
3. The northern part of New York State
4. The states of Missouri (New Madrid area) and South Carolina
5. Alaska, Puerto Rico, and Hawaii

California and Alaska are the states that have had the largest number of earthquakes. Florida and North Dakota are the states with the smallest number. Historically, California is the state that has suffered the most damaging earthquakes, although damaging shocks have also occurred in Alaska; New Madrid, Missouri; and Charleston, South Carolina.

2.4 EARTHQUAKE STATISTICS

A few statistics can give the uninitiated a good perspective of the overall frequency of earthquake occurrence, the frequency of large magnitude earthquakes, and where such large earthquakes occur. For this purpose, some statistical data concerning the worldwide and the U.S. earthquakes are presented in this section.

2.4.1 Worldwide Earthquakes

Table 2.1 summarizes, based on the observations since 1900, the average frequency with which earthquakes occur every year around the world. As it may be seen from this table, the earth is in a constant state of seismic activity. As expected, large earthquakes are not as frequent as small ones, but still great and major earthquakes occur with a frequency that is worrisome. Fortunately, great and major earthquakes do not always take place in populated areas, so they are not always catastrophic.

The location, dates of occurrence, and magnitudes of the largest earthquakes that have occurred in the world since 1900 are shown in Table 2.2. A comprehensive list of the most significant earthquakes in the world throughout historical times is presented in Appendix A.

TABLE 2.1
Frequency of Occurrence of Worldwide Earthquakes Based on Observations Since 1900

Category	Magnitude[a]	Number of Earthquakes/Year
Great	8 and higher	1
Major	7–7.9	18
Strong	6–6.9	120
Moderate	5–5.9	800
Light	4–4.9	6,200
Minor	3–3.9	49,000
Very minor	2–3	365,000
Barely perceptible	1–2	17,885,000

[a] Magnitude, defined formally in Chapter 5, is a measure of an earthquake's strength. Earthquakes are considered small if they have a magnitude of <4, moderate if they have a magnitude between 4 and 6, strong if they have a magnitude between 6 and 7.5, and very strong if they have a magnitude >7.5.

Source: National Earthquake Information Center, U.S. Geological Survey.

TABLE 2.2
Largest Earthquakes in the World Since 1900

Number	Location	Date	Magnitude
1	Chile	May 22, 1960	9.5
2	Prince William Sound, Alaska	March 28, 1964	9.2
3	Northern Sumatra, Indonesia	December 26, 2004	9.1
4	Kamchatka, Russia	November 4, 1952	9.0
5	Off Ecuador Coast	January 31, 1906	8.8
6	Rat Islands, Alaska	February 4, 1965	8.7
7	Northern Sumatra, Indonesia	March 28, 2005	8.6
8	Andreanof Islands, Alaska	March 9, 1957	8.6
9	Assam–Tibet	August 15, 1950	8.6
10	Kuril Islands, Russia	October 13, 1963	8.5
11	Banda Sea, Indonesia	February 1, 1938	8.5
12	Kamchatka, Russia	February 3, 1923	8.5

Source: National Earthquake Information Center, U.S. Geological Survey.

TABLE 2.3
Frequency of Occurrence of Earthquakes in the United States Based on Observations Since 1900

		Number of Earthquakes			
Category	Magnitude	Western States	Eastern States	Alaska	Hawaii
Great	8 and higher	1	0	7	0
Major	7–7.9	18	0	84	1
Strong	6–6.9	129	1	411	15
Moderate	5–5.9	611	41	1886	36
Light	4–4.9	3171	335	8362	315

Source: National Earthquake Information Center, U.S. Geological Survey.

TABLE 2.4
Largest Earthquakes in the United States

Number	Magnitude	Date (UTC)	Location
1	9.2	March 28, 1964	Prince William Sound, Alaska
2	9.0	January 26, 1700	Cascadia Subduction Zone
3	8.7	February 4, 1965	Rat Islands, Alaska
4	8.6	March 9, 1957	Andreanof Islands, Alaska
5	8.2	November 10, 1938	East of Shumagin Islands, Alaska
6	8.1	April 14, 1946	Unimak Islands, Alaska
7	8.1	December 16, 1811	New Madrid, Missouri
8	8.0	September 10, 1899	Yakuta Bay, Alaska
9	8.0	February 7, 1812	New Madrid, Missouri
10	7.9	November 3, 2002	Denali Fault, Alaska
11	7.9	November 30, 1987	Gulf of Alaska
12	7.9	May 7, 1986	Andreanof Islands, Alaska
13	7.9	September 4, 1899	Near Cape Yakataga, Alaska
14	7.9	April 3, 1868	Ka'u District, Island of Hawaii
15	7.9	January 9, 1857	Fort Tejon, California
16	7.9	November 17, 2003	Rat Islands, Alaska
17	7.8	June 10, 1996	Andreanof Islands, Alaska
18	7.8	April 18, 1906	San Francisco, California
19	7.8	February 24, 1892	Imperial Valley, California
20	7.8	January 23, 1812	New Madrid, Missouri

Source: National Earthquake Information Center, U.S. Geological Survey.

2.4.2 U.S. Earthquakes

In the United States, the frequency of earthquake occurrence, based on the observations since 1900, is shown in Table 2.3. The geographical region, dates of occurrence, and magnitudes of the largest earthquakes in record are listed in Tables 2.4 and 2.5.

2.5 ON-LINE SEISMICITY INFORMATION

The National Earthquake Information Center of the U.S. Geological Survey maintains a Web site with current and historical information about the occurrence of earthquakes and their location. This information includes a near-real time list of the last 21 earthquakes anywhere in the world, useful

TABLE 2.5
Largest Earthquakes in the Contiguous United States

Number	Magnitude	Date	Location
1	9.0	January 26, 1700	Cascadia Subduction Zone
2	8.1	December 16, 1811	New Madrid, Missouri
3	8.0	February 7, 1812	New Madrid, Missouri
4	7.9	January 9, 1857	Fort Tejon, California
5	7.8	April 18, 1906	San Francisco, California
6	7.8	February 24, 1892	Imperial Valley, California
7	7.8	January 23, 1812	New Madrid, Missouri
8	7.6	March 26, 1872	Owens Valley, California
9	7.3	June 28, 1992	Landers, California
10	7.3	August 18, 1959	Hebgen Lake, Montana
11	7.3	June 21, 1952	Kern County, California
12	7.3	January 31, 1922	Eureka, California
13	7.3	September 1, 1886	Charleston, South Carolina
14	7.3	November 23, 1873	California–Oregon Coast
15	7.3	December 15, 1872	North Cascades, Washington

Source: National Earthquake Information Center, U.S. Geological Survey.

seismological and statistical data, and seismicity maps of the world, the United States, and several other regions. For example, regional seismicity maps are available for Africa, Alaska, Australia and Indonesia, Canada, Central America, Central Asia, Central Pacific Ocean, Europe, Indian Ocean, Japan and Kuril Islands, Middle East, North Atlantic Ocean, North Pole, Pacific Rim, South America, South Atlantic Ocean, and South Pole. Interested readers may access this site at the address: http://earthquake.usgs.gov/eqcenter/

FURTHER READINGS

1. Bolt, B. A., *Earthquakes*, 5th Edition, W. H. Freeman and Co., New York, 2004, 378 p.
2. Hu, Y. X., Liu, S. C., and Dong, W., *Earthquake Engineering*, E & FN Spon, London, 1996, 410 p.
3. Lay, T. and Wallace, T. C., *Modern Global Seismology*, Academic Press, San Diego, 1995, 521 p.
4. Lomnitz, C., *Global Tectonics and Earthquake Risk*, Elsevier Scientific Publishing Co., New York, 1974, 320 p.
5. Richter, C. F., *Elementary Seismology*, W. H. Freeman and Co., San Francisco, CA, 1958, 768 p.

3 Earthquake Genesis

Theological preoccupations during the Middle Ages put a temporary end to speculation about the origin of earthquakes. The Wrath-of-God theory of earthquakes had its origin in Old Testament interpretations of natural disasters, such as the destruction of Sodom and Gomorrah, and the fall of Jericho, which may have been early instances of seismic activity in the Jordan Trough. Naturalistic explanations of earthquakes were formally banned as heresy by a father of the Church in the 5th Century.

Cinna Lomnitz, 1974

3.1 INTRODUCTION

Ever since human beings first felt the earth shake, they have had the desire to know why that happens and developed, as a result, numerous explanations for it. For the most part, however, these explanations have been based on superstitious beliefs. For example, the ancient Japanese believed that the islands of Japan rested on the back of a giant catfish whose movements made the earth shake. The Algonquin Indians of North America believed that a giant tortoise supported the earth, which shook whenever the tortoise shifted from one foot to another. For the ancient Mexicans, the earth was a divine being with monster features, generally with those of a reptile and a fish, which caused earthquakes when it moved. In a similar fashion, a frog has been the culprit in parts of Asia, a giant mole in India, and an ox in China. Earthquakes were also often interpreted as a form of punishment from angry gods. In Greek mythology, Poseidon, ruler of the sea, caused earthquakes when he was angry. His counterpart in ancient Rome was Neptune, who not only could instill fear into people with earthquakes, but also could punish them with floods over the land and waves onto the shore. Even eighteenth-century European clergymen tended to view earthquakes from a moralistic standpoint. In 1752, a London journalist wrote: "Earthquakes generally happen to great cities and towns. The chastening rod is directed where there are inhabitants, the objects of its monition, not to bare cliffs and uninhabited beach." After the famous Lisbon earthquake of 1755, which caused a great loss of life from a sequence of several shocks and a giant tsunami, a clergyman in England chastised the people of Lisbon for their lewdness and debauchery, whereas others blamed the dreadful inquisition and noted that the Palace of the Inquisition was one of the first buildings destroyed. An early attempt for a scientific explanation was made by Aristotle, who found an explanation for the cause of earthquakes in the interior of the earth. Aristotle theorized that the winds of the atmosphere were drawn into the caverns and passageways in the interior of the earth and that earthquakes and the eruption of volcanoes were caused by these winds as they were agitated by fire and moved about trying to escape.

Today, after the numerous scientific developments of the twentieth century and the many years of geological and seismological studies, there seems to be a clear understanding of what causes earthquakes, and where and how often they may occur. This chapter, thus, introduces the modern theories that explain the mechanisms that give birth to earthquakes and the phenomena that are deemed responsible for these mechanisms. It will also describe the observed correlation between such earthquake-generating mechanisms and some prominent features on the earth's surface, the locations where earthquakes occur, and the frequency of earthquake occurrence. Additionally, a brief account will be given of the efforts being made to use the current understanding of the earthquake-generating mechanism to develop techniques for the near-term prediction of the size, time, and location of future earthquakes.

3.2 TYPES OF EARTHQUAKES

In the most general sense of the word, an earthquake is nothing else but a phenomenon that involves the motion or shaking of the earth's crust. In this general sense, an earthquake may be, therefore, caused by (a) tectonic forces (the forces involved in the formation of the earth's features), (b) volcanic activity, (c) conventional and nuclear explosions, (d) the sliding or fall of a large soil or rock mass (such as in the case of landslides and the collapse of mines and caverns), (e) a meteorite impact, and (f) the filling of reservoirs and wells (see Box 3.1). However, the earthquakes produced by volcanic activity, a soil or rock mass collapse, a conventional explosion, a meteorite impact, or the filling of reservoirs and wells are, for the most part, of a relatively small size and affect only an area of limited extent. Earthquakes generated by a nuclear explosion may be strong, but the factors that generate these earthquakes are known and may be controlled. In contrast, the earthquakes generated by tectonic forces may be exceptionally large and may affect a large geographical region at once. In fact, most of the catastrophic earthquakes that have occurred through historical times have been of the tectonic type. The discussion in this chapter will be, therefore, limited to the generation of tectonic earthquakes.

Box 3.1 Earthquakes Triggered by Filling of Reservoirs and Wells

There exists nowadays plenty of evidence which demonstrates that earthquakes can be generated by the filling of great reservoirs. One of the earliest instances in record took place when the Hoover Dam on the Colorado River was completed and Lake Mead began to fill. Before the construction of the dam, no seismic activity had been recorded in the region around Lake Mead. After the lake began to fill in 1936, earthquakes began to occur. The largest (magnitude 5) one occurred in 1939 when the lake was ~80% full. The maximum lake level was reached in 1941, and low-level seismic activity has continued ever since. The largest reservoir-induced earthquake took place at Koyna Dam, near Bombay, India. Koyna Dam was also constructed in a nonseismic zone. However, soon after the reservoir began to fill in 1962, small earthquakes shook the region around it. The reservoir was full in 1965. In 1967, several significant earthquakes occurred. The largest one, with a magnitude of 6.4, caused 177 deaths and extensive damage in a nearby village. After that earthquake, the seismic activity decreased and seems to have stopped. Similar events have occurred in China (Hsinfengkiang Dam), France (Monteynard Dam), Zimbabwe-Rhodesia (Kariba Dam), Greece (Kremasta Dam), California (Oroville Dam), and Egypt (Aswan Dam). Reservoir-induced earthquakes are usually small, although several have had magnitudes between 5 and 6, and three have had magnitudes >6.

The filling of wells can also generate earthquakes. This phenomenon was first noted at the Rocky Mountain Arsenal, located northeast of Denver, Colorado, on the outskirts of the city. A boring ~3 km deep was drilled for the purpose of disposing of contaminated water. The wastewater was pumped in the boring beginning in 1962, and almost immediately earthquakes began to occur. Pumping continued at various rates during the next 4 years but was discontinued when it became evident that the number of earthquakes was closely related to the amount of fluid injected into the earth. The earthquakes were felt by the residents of Denver. Most of the earthquakes were small, although the largest, which occurred a year after the filling was stopped, had a magnitude of 5.2. An opportunity to verify this phenomenon came a few years later near Rangely, a small city in northwestern Colorado. Oil production at the Rangely oil field began in 1945; the oil was taken from a sandstone formation ~1.5 km below the surface. Fluid pressures in the rocks decreased as a result of the oil removal until 1957, at which time water was pumped back into the wells to facilitate secondary oil recovery. Beginning in 1969, the U.S. Geological Survey pumped water in and out of these wells. It was found that small earthquakes (magnitudes <3.5) were occurring frequently in the injection zone and that the amount of seismic activity was correlated with the pumping and the increase in pore water pressure. When injection took place and pore water pressure increased, the number of earthquakes increased. In contrast, when the water was pumped out the number decreased.

The occurrence of earthquakes induced by the filling of reservoirs and wells has been explained on the basis of the additional stresses that may be generated by the weight of water on nearby geological faults and the increase in water pressure, which may reduce the shear strength between the sides of these faults.

3.3 EARTH STRUCTURE

The earth is roughly spherical, with an equatorial diameter of 12,740 km and a polar diameter of 12,700 km, the higher equatorial diameter caused by the higher velocities at the equator due to the earth's rotation. Its mass is $\sim 4.9 \times 10^{21}$ kg, which implies an average specific gravity of 5.5. As the specific gravity of the rocks at the surface of the earth is between 2.7 and 3, it may be, thus, inferred that the materials in the interior have higher specific gravities.

In a macroscopic scale, the interior of the earth is divided into four concentric layers: (a) inner core, (b) outer core, (c) mantle, and (d) crust (see Figure 3.1). The inner core lies at the center of the earth with a radius of ~1216 km. It is thought to be solid, is composed of nickel and iron, and has a specific gravity of ~15. Further toward the surface is the outer core with a thickness of ~2270 km. It appears to be in a liquid state and is composed mainly of iron, oxygen, and silicon. Its specific gravity is estimated to be between 9 and 12. The mantle extends from the base of the crust to a depth

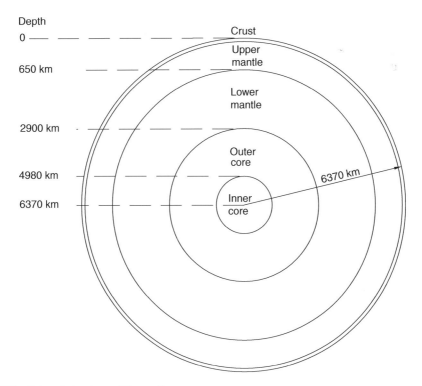

FIGURE 3.1 Internal structure of the earth.

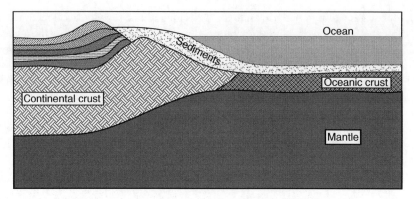

FIGURE 3.2 Cross section of the earth's crust under the oceans and continents.

of 2885 km and is composed of dense granitic and basaltic rocks in a viscous, semimolten state, the result of temperatures as high as 2000°C (for comparison steel melts at 1500°C). It has a specific gravity between 4 and 5 and is divided into the upper mantle and the lower mantle to reflect the existence of a discontinuity in the properties of its composing materials. The crust thickness ranges between 25 and 60 km under the continents and between 4 and 6 km under the oceans. In addition to being thinner, the oceanic crust is also, in general, more uniform and denser than the continental crust. The crust is composed of granitic and basaltic rocks (see Figure 3.2).

Independently from the actual structure of the earth, the layer that embodies its outermost 100 km (which includes the crust) is considered to be a layer of rigid rock and is identified as the *lithosphere* (from the Greek lithos: stone). Similarly, it is considered that there is a layer of ~400 km in thickness below the lithosphere with relatively softer rocks. This layer is called the *asthenosphere* (from the Greek asthenes: without strength). The rock materials making up the lithosphere and asthenosphere are of a similar type, with the demarcation between the two established on the basis of their different rigidity.

3.4 PLATE TECTONICS THEORY

The theory of plate tectonics postulates that the earth's crust is fractured and thus divided into a small number of large and rigid pieces, referred to as plates. The size of these plates varies from a few hundred to many thousands of kilometers. Their location, as well as their given names, is shown in Figure 3.3. The theory of plate tectonics also postulates that these plates *float* on the semimolten asthenosphere and that they move relative to one another. At some locations, these plates are moving apart (diverge) and at others the plates are moving toward each other (converge) or sliding past each other, as shown in Figure 3.3. Plate tectonics theory also asserts that plate motion is responsible for the long-term, large-scale formation and changes occurring on the earth's surface and for most of the seismic and volcanic activity around the world. It claims, further, that earthquakes take place near these plate boundaries as a result of the stresses that build up in the earth's crust as the plates tend to move and interact with one another.

The theory of plate tectonics has evolved from the theory of continental drift originally proposed by the German scientist Alfred Wegener in 1915. Wegener's theory of continental drift proclaimed that the earth's surface was not static, but dynamic, and that the oceans and continents are in constant motion. He based his assertion on the similarity between the coastlines, geology, and life forms of eastern South America and western Africa, and the southern part of India and northern part of Australia. He believed that 200 million years ago the earth had only one large continent that he called Pangaea (see Figure 3.4) and that this large continent broke into pieces that slowly drifted toward the current position of the current continents.

Earthquake Genesis

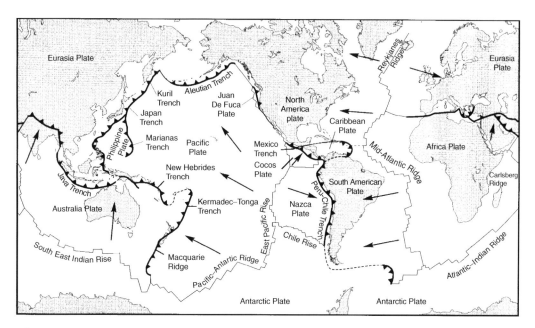

FIGURE 3.3 Major tectonic plates of the earth and their general direction of movement. (After Fowler, C.M.R., *The Solid Earth: An Introduction to Global Geophysics*, Cambridge University Press, Cambridge, U.K., 1990.)

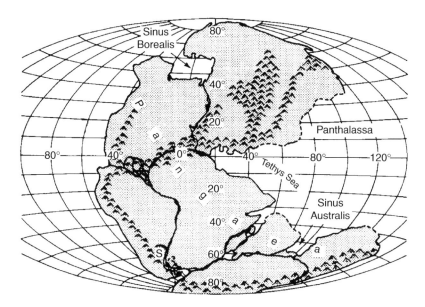

FIGURE 3.4 Position of continents 200 million years ago. (Reproduced with permission of N.H. Prentiss, from Dietz, R.S. and Holden, J.C., *Scientific American*, October 1970.)

Although the theory of continental drift was received with contempt when it was originally proposed, today this theory and that of plate tectonics have become widely accepted and acknowledged as one of the greatest advances in the earth sciences. Their acceptance has come as a result of studies conducted after the 1960s, which, with the help of a modern worldwide network of earthquake-recording instruments, the developing of new techniques such as deep-water echo sounding, and a detailed exploration of the ocean floor, have provided a strong supporting evidence of the historical movement of the continents postulated by the theory of continental drift. Moreover, precise geodetic measurements have shown that the plates identified by the plate tectonics theory are indeed moving relative to one another and that this motion is between 1 and 13 cm/year.

3.5 PLATE INTERACTION, FORMATION OF EARTH'S SURFACE FEATURES, AND EARTHQUAKE GENERATION

As a result of the differential motion between them, tectonic plates interact with one another in three distinct ways, giving rise to an assortment of effects on the earth's surface. When two plates diverge, a *rift* is developed, creating a condition where molten basaltic magma from the asthenosphere rises to lift up the plate edges and create new crust (see Figure 3.5). Most of the known plate divergence occurs in what is now ocean area. This process of plate divergence, known as *sea-floor spreading*, has been responsible for the formation of the mid-oceanic *ridges* and *rises* (submarine mountain chains) as well as the deep valleys in between. Examples are the Mid-Atlantic Ridge, East Pacific Rise, and South Indian Rise (see Figure 3.3). Earthquake and volcano activity is involved along the boundaries of these diverging plates, although the earthquakes that occur there are usually of low magnitude. Earthquakes are triggered when the tensional forces that produce the plate divergence fracture the newly formed crust. Spreading ridges may protrude above the ocean as in the case of the island of Iceland, where, with its 150 volcanoes, volcanic activity is nearly continuous.

When two plates converge, the plates either collide head on or one dives beneath the other. Where an oceanic plate and a continental plate converge, the oceanic plate, being thinner and heavier, tends to be pushed below the continental plate (dips) to form what is known as a *subduction zone* (see Figure 3.6). In this process, the edge of the oceanic plate below the continental one melts and becomes part of the asthenosphere. The creation of new crust where plates diverge is thus balanced by an equivalent loss at a subduction zone, which together complete a continuous cycle that replaces the ocean floor every 200 million years or so. Subduction zones exist off the coasts of Mexico and Chile, south of the Aleutian Islands, off the eastern coast of Japan, and off the coasts

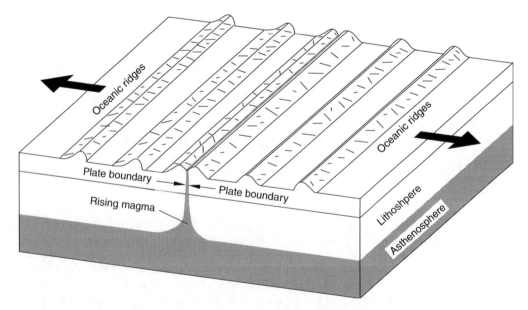

FIGURE 3.5 Schematic view of mid-oceanic ridges and spreading oceanic plates causing the rising of magma from the asthenosphere to form new ocean floor.

Earthquake Genesis

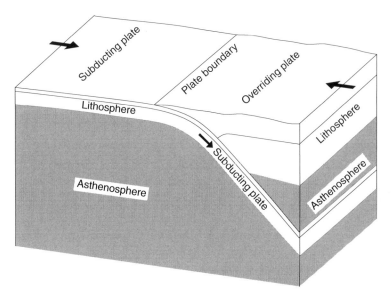

FIGURE 3.6 Subduction of oceanic plate under continental plate.

of Washington and Oregon (see Figure 3.3). Plate subduction has given birth to the deep trenches that are found under the oceans. The Marianas Trench with a depth of 10,915 m (deepest ocean spot) and the Peru–Chile Trench with a depth of 8063 m are both located at the boundaries where an oceanic plate subducts under a continental one. Plate subduction also gives rise to some of the world's most powerful earthquakes. In fact, almost 90% of the seismic energy released by tectonic plate movement comes from earthquakes generated in subduction zones. Earthquakes occur above the subducting plate in regions of the continental plate and at the interface between the subducting and overriding plates. The latter occurs along a well-defined area, known as the *Benioff zone*, which outlines the subducting plate (see Figure 3.7). Earthquakes in these zones are known to take place at depths as great as 700 km. It is believed that earthquakes cannot occur at depths >700 km because beyond this depth a subducting plate melts. Subduction zones are also known to be regions of active volcanism (for instance, the Andes Mountains of South America). Apparently, the volcanic activity is related to the surface breakthrough of a rising, low-density magma, created when the ocean crust, carried into the asthenosphere by the subducting plate, melts (see Figure 3.8).

When two plates carrying continental crust converge, there is no subduction as it occurs at the ocean trenches. Being lightweight and thick, continental crust is too buoyant to sink into the asthenosphere. In this case, the plates simply collide head on and over time fold up like an accordion (see Figure 3.9). This process has created some of the world's mightiest mountain ranges, such as the Alps and the Himalayas. Regions near the boundary where two continent-supporting plates converge are usually seismically but not volcanically active. Earthquakes are generated at these boundaries when rock blocks near the boundaries fracture under the high compressional forces between the colliding plates and slip.

Where plate edges slide past each other, crust is neither created nor destroyed, nor do changes occur on the surface of the earth (see Figure 3.10). The boundaries where this type of interaction occurs are often called *transform faults*. A transform fault develops where the axis of a spreading ridge or a subduction zone has been offset. As it may be observed from the jagged lines and erratic curves that identify the spreading ridges and subduction zones in Figure 3.3, transform faults are a rather common occurrence, a condition brought about by the irregular fracturing of the lithosphere. The sliding of one plate against another generates earthquake activity but no volcanism. Earthquakes in these boundaries typically occur at shallow depths, that is, between 5 and 40 km below the surface. The San Andreas Fault near the west coast of the United States has been identified

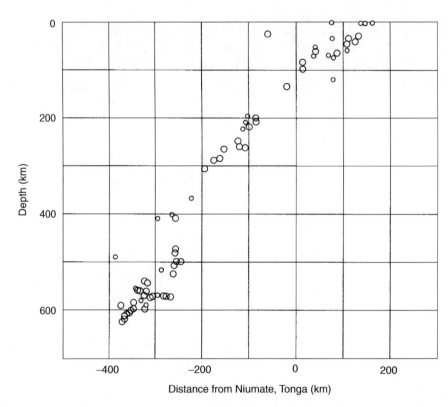

FIGURE 3.7 Location of earthquakes that occurred in 1965 under the Tonga Trench in the Southwest Pacific, delineating a subducting plate that reaches a depth of more than 600 km at an angle of ~45°. (After Isacks, B., Oliver, J., and Sykes, R.S., *Journal of Geophysical Research*, 73, 18, 5855–5899, 1968. © American Geophysical Union. Reproduced with permission from American Geophysical Union.)

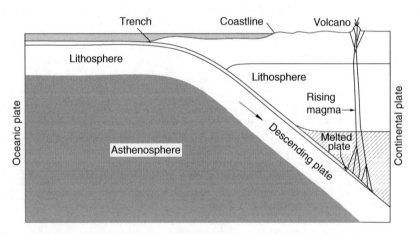

FIGURE 3.8 Formation of volcanoes in subduction zones.

as a transform fault. Other important transform faults are the Motagua Fault that runs from the Caribbean Sea to the mountains west of Guatemala City, the Alpine Fault in New Zealand, and the Dead Sea Fault that connects the Read Sea to the Bitlis Mountains in Turkey.

The overall interrelationship between the relative motions of the tectonic plates, together with some of the most important features on the earth's surface, is illustrated in Figure 3.11.

Earthquake Genesis

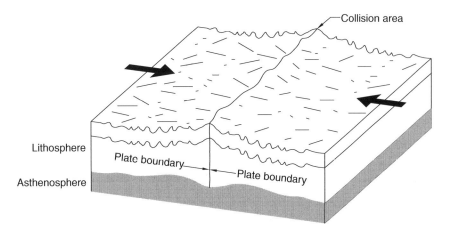

FIGURE 3.9 Convergence of two continental plates and consequent formation of mountain range.

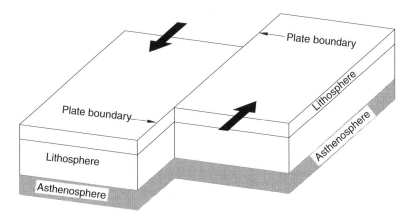

FIGURE 3.10 Plates sliding past each other and formation of transform fault.

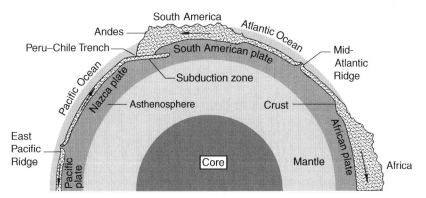

FIGURE 3.11 Earth's cross section showing the direction of plate motion and the relationship of this motion with some of the most important earth features.

3.6 CAUSE OF PLATE MOVEMENT

Many theories have been offered to explain the plate movement. The prevalent one so far is the convection current theory. This theory, or group of theories, as there are many variants, postulates that heat is being generated within the earth by radioactive disintegration and that this heat is being

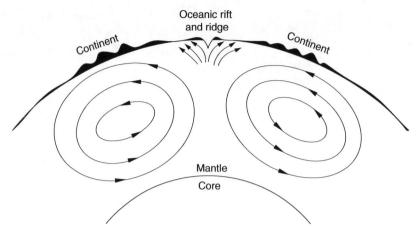

FIGURE 3.12 Schematic representation of convection currents in the interior of the earth.

carried to the surface by gigantic convection currents within the mantle, similarly to the way heated air moves by convection through a cool room. As it is known that the earth's mantle is solid, it is hypothesized that these convection currents must move very slowly if they exist at all. A rate of 0.5 m a century has been suggested. This, however, does not present much of a problem as geological processes are incredibly slow.

Different proponents draw these convection cells in different ways. One example is provided in Figure 3.12. According to this model, the currents are rising and separating under the mid-ocean ridges to provide the observed tension and driving toward the boundary of the continent with the thrust necessary to push up mountains. A problem with this theory arises when one attempts to go from stylized drawings such as that in Figure 3.12 to placing the convection cells in the actual earth in a way that would explain the formation of all the existing mountains, island chains, and mid-oceanic ridges. The pattern of these convection cells is very complex, and a convection model that explains even the majority of these features has never been produced.

3.7 INTRAPLATE EARTHQUAKES

Although the vast majority of earthquakes occur at plate boundaries, some seismic regions of the earth are far away from plate boundaries. This seismicity is referred to as intraplate seismicity and constitutes a manifestation of the internal deformation of a plate. The nature of intraplate seismicity is often quite complicated, and the tectonic driving mechanisms are presently poorly understood. It is speculated that the origin of the stress field that causes intraplate earthquakes stems mainly from the driving forces of plate tectonics and that this stress field is probably the result of localized weaknesses in the structure of the plate. The lack of a satisfactory explanation of intraplate earthquakes is one of the weaknesses of the plate tectonic theory. Well-known examples of intraplate earthquakes are those that have occurred in the vicinity of New Madrid, Missouri, in 1811–1812; Charleston, South Carolina, in 1886; Tangshan, China, in 1976; and Marathawada, India, in 1993.

3.8 EARTHQUAKE FAULTS

When the boundaries between the earth's tectonic plates manifest themselves on the surface of the earth, they are seen as long uneven fractures or fissures on a rock formation whose sides have moved relative to each other. Geologists call these fractures or fissures as *faults* and identify them by the abrupt discontinuities on the structure of the adjacent rock and the irregularities on the earth's surface features along the fault line. Faults may range in length from several meters to hundreds of kilometers, extend to considerable depths, and exhibit displacements of several meters. In many instances, faults are not characterized by a single fissure, but by a major fissure and an intricate series of fractures that branch out beyond the edges of the tectonic plates. They resemble the failure plane of a concrete cylinder tested in the laboratory and the multiple cracks that surround this failure plane. A well-known example is the San Andreas Fault and its complex network of subsidiary faults (see map in Figure 3.13).

The existence of a fault at some location on the earth's surface is indicative that a relative motion took place between its two sides at some time in the past. In some faults, this motion takes place gradually and in others intermittently during several intervals. The motion takes place intermittently when the rock along the fault locks and undergoes a displacement only at those times when the rock breaks apart. Earthquakes occur at these faults and thus they are identified as earthquake faults. In some of these faults, however, the last displacement has occurred tens of thousands of years ago and are thus considered to be *inactive faults*.

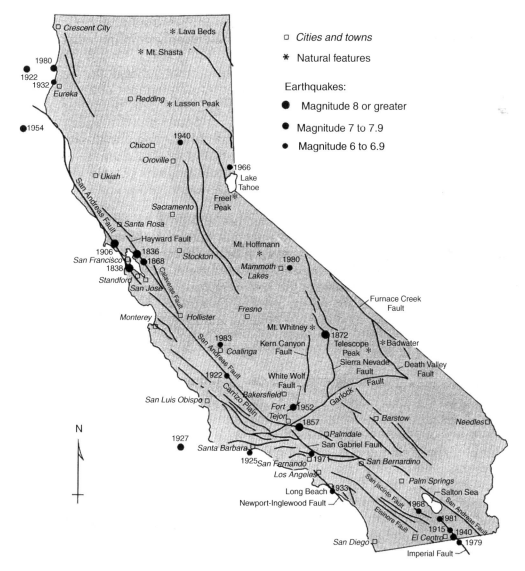

FIGURE 3.13 Main faults in the State of California and the location of major past earthquakes. (Reproduced with permission from Gere, J.M. and Shah, H.C., *Terra Non Firma*, W.H. Freeman and Co., New York, 1980.)

When the two sides of a fault move past each other, the relative motion is horizontal and the fault is called a *transcurrent* or *strike-slip fault*. Strike is the angle measured from to the north made by the horizontal line defined by the intersection of a fault plane with the earth's surface. Thus, the relative motion (slip) in strike-slip faults occurs along the strike of the fault. A strike-slip fault is considered to be of the *left-lateral* type if, as seen from either side, the other side of the fault slips toward the left. Conversely, the fault is considered to be of the *right-lateral* type if it slips toward the right (see Figure 3.14). Examples of strike-slip faults are the San Andreas Fault in California and the Motagua Fault in Guatemala. However, the San Andreas Fault is of the right-lateral type, whereas the Motagua Fault is of the left-lateral type. Slippage of a strike-slip fault leaves offset streams, trees, and fences (see Figure 3.15).

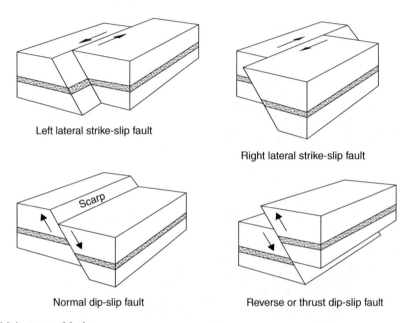

FIGURE 3.14 Main types of faults.

FIGURE 3.15 Offset of rows in lettuce field resulting from the slippage of the San Andreas Fault during the October 15, 1979, El Centro, California, earthquake. (Photograph from Geological Hazard Collection, National Geophysical Data Center, National Oceanic and Atmospheric Administration.)

FIGURE 3.16 Scarp left by fault rupture during the December 7, 1988, Armenian earthquake. (Photograph from Geological Hazard Collection, National Geophysical Data Center, National Oceanic and Atmospheric Administration.)

If, on the other hand, the two sides of a fault press against each other or pull away from each other, then the relative motion is primarily vertical. In this case, the fault is called a *dip-slip* fault because the slip occurs along the direction of the fault's dip. Dip is the angle formed by the plane of the fault with respect to the surface of the earth. If in a dip fault the upper rock block moves downward, the fault is called a *normal fault*. If, on the other hand, the upper rock block moves upward, then the fault is considered to be a *reverse or thrust fault* (see Figure 3.14). The slip in normal faults occurs in response to extensional strains and results in a horizontal lengthening of the crust. In contrast, the slip in a reverse fault occurs in response to compressional strains and produces a horizontal shortening of the crust. Examples of normal faults are the faults where the 1954 Dixie Valley, Nevada, and the 1959 Hebgen Lake, Montana, earthquakes occurred. Examples of reverse faults are those responsible for the 1952 Kern County and the 1971 San Fernando earthquakes in California. Slippage in a dip-slip fault produces an exposed steep slope called the fault's *scarp* (see Figure 3.16).

3.9 EARTHQUAKE-GENERATION MECHANISM: ELASTIC REBOUND THEORY

It was seen in the foregoing sections that most earthquakes occur near the boundaries between the earth's tectonic plates and that earthquake occurrence is a consequence of the relative motion between these plates. It should be recognized, however, that this relative motion cannot provide, by itself, a satisfactory explanation for the origin of earthquakes as a steady relative motion between plates cannot produce the violent vibrations of the earth's crust that characterize an earthquake. The complete explanation is given, instead, by the elastic rebound theory formulated by Harry F. Reid of Johns Hopkins University shortly after the 1906 San Francisco earthquake in California.

For many years, scientists had observed the rupture of the ground during earthquakes and the permanent vertical or horizontal displacements left along extensive lengths at the surface of the earth near where the earthquakes occurred. But for the most part, scientists believed that these displacements were part of the damage caused by an earthquake. It was not until 1891, after the devastating Mino-Owari earthquake in Japan that year left a ground rupture that extended almost completely across the main Japanese island of Honshu, when Bunjiro Koto, a Japanese geologist, suggested for the first time that the rupture of the ground during an earthquake was not caused by the earthquake but, instead, the rupture of the ground was responsible for the generation of the earthquake. Even so, it was not until after the 1906 San Francisco earthquake that it was amply recognized that earthquakes were caused by the slippage of a fault in the earth's crust. This recognition came about as a result of the detailed investigation conducted by Harry F. Reid of that earthquake. In this investigation, Reid discovered that for hundreds of kilometers along the San Andreas Fault, fences and roads crossing the fault had displaced as much as 6 m. From the analysis of the precise geodetic surveys conducted before and after the earthquake, he also discovered that the rocks parallel to the fault had been strained and sheared. It was on the basis of these findings that Reid postulated his now famous theory of elastic rebound.

Reid's theory of elastic rebound asserts that earthquakes are the result of the sudden release of stresses that gradually accumulate in the rocks on the opposite sides of a fault when these opposite sides tend to move relative to one another, but the motion is initially prevented by frictional forces. When the accumulated stresses become larger than the frictional resistance between the two sides, then the fault starts to slip at its weakest point and quickly ruptures along its entire surface. A few seconds afterward, the rock on both sides of the fault rebounds—or springs back—to an unstrained position and the stored elastic energy is released in the form of heat and radiating seismic waves.

Thus, according to the elastic rebound theory, earthquakes are generated according to the following process (see Figure 3.17):

1. Stresses are generated and gradually accumulated along the sides of a fault as a result of the relative motion between such two sides and the friction forces that resist this motion.
2. Stresses along the sides of the fault overcome the frictional resistance of the fault.
3. Fault suddenly slips.
4. The two sides of the fault rebound to an unstressed state causing a disturbance.
5. Disturbance is propagated in the form of radial waves.

Nowadays, there is plenty of evidence that confirms the elastic rebound theory. This evidence includes geodetic observations of the gradual accumulation of strain at known faults and the horizontal displacement of the ground that is left permanently after the occurrence of a major earthquake (see Figures 3.15 and 3.16). Faults and fault offsets have been found even at the site of intraplate earthquakes for which it was previously believed that no fault mechanism was involved (see Figure 3.18). Furthermore, surveys in stable areas of the continents such as in the Canadian and Australian shields have shown that little change has taken place in recent times.

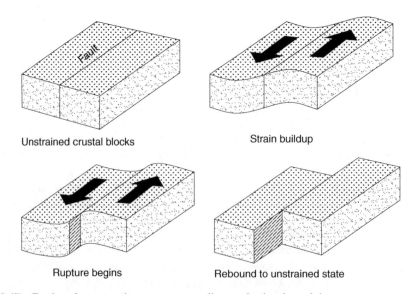

FIGURE 3.17 Earthquake-generation process according to elastic rebound theory.

Earthquake Genesis

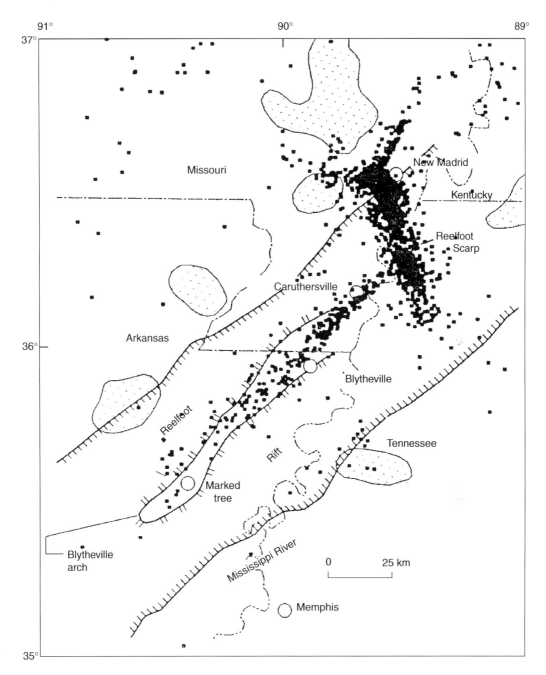

FIGURE 3.18 Buried rift under the central Mississippi Valley at the site of a series of major earthquakes that occurred in the New Madrid area in 1811–1812. The rift and the fault-bounded, down-dropped block of crust in the rift lie between the single-hachured lines. The location of recent earthquakes is indicated by the solid squares (after U.S. Geological Survey).

The concepts behind the elastic rebound theory serve not only to explain the mechanism that generates earthquakes, but also to develop a technique to determine the orientation of the causative fault. With reference to Figure 3.19, it may be noted that after the rupture of a fault and the rebound of its two sides, the rock adjacent to the fault will be subjected to the distribution of tensile and compressive stresses shown in this figure. That is, compressive stresses in two diametrically opposed

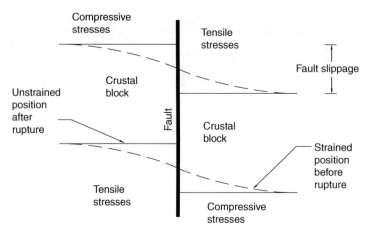

FIGURE 3.19 Distribution of stresses around a vertical earthquake fault right after the fault's rupture.

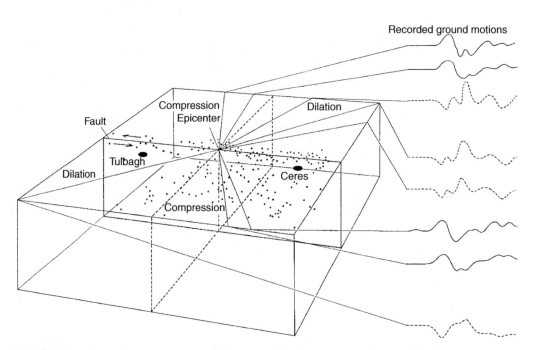

FIGURE 3.20 Records of ground motion detected at different stations during an earthquake near Ceres in South Africa showing how the direction of the first motion changes with the azimuth between the earthquake center and the recording station. Upward motions (away from the source) are compressional, and downward motions (toward the source) are dilatational. Orientation of the fault can be detected from the pattern of compressional and dilatational motions observed around the center of the earthquake and the location of the aftershocks (dots), which tend to lie along a single plane. (Reproduced with permission of Iselin, A.D. from Boore, D.M., *Scientific American*, December 1977.)

quadrants and tension stresses in the other two. It may be noted too that two orthogonal lines separate these tension and compression quadrants; one is the fault line and the other is a line perpendicular to it. Thus, if observations are made about the direction of the first wave that arrives at a series of recording stations (see Figure 3.20), the earthquake fault orientation can be established by plotting the distribution of tensile and compressive stresses and by identifying the two orthogonal lines that separate the four quadrants. Notice, however, that since any of these two lines may represent the fault line, additional information is needed to resolve the ambiguity. This ambiguity is usually eliminated by using information from the known local geology, or, alternatively, from the pattern of aftershocks (see Section 3.11) that usually occur, scattered, along the fault plane.

3.10 FOCUS, EPICENTER, RUPTURE SURFACE, AND FAULT SLIP

According to the elastic rebound theory, earthquakes originate right after a fault suddenly ruptures. This rupture, however, is neither instantaneous nor concentrated at a single point. Rather, it begins at a point below the earth's surface, spreads across the fault plane with a velocity V, and finally stops after the rupture extends to cover an area with an average length L and average width W. In the process, the two fault surfaces are offset by a finite amount referred to as the *fault slip*. Thus, earthquakes are characterized by the location of their focus or epicenter, their focal depth, the size and orientation of the associated fault's rupture surface, and the fault slip. With reference to Figure 3.21 and for the purpose of this characterization, the *focus* or *hypocenter* of an earthquake is defined as the point where the rupture of the associated fault originates. Correspondingly, *epicenter* is defined as the projection of the hypocenter on the earth's surface. In a similar fashion, rupture surface is considered to be the area within the fault plane that is displaced during an earthquake, and fault slip is the corresponding relative displacement between the two sides of the fault.

The ruptured area during an earthquake may range from a few to thousands of square kilometers. As an example, the ruptured area during the 1989 Loma Prieta earthquake in California was ~40 km in length and 20 km in depth. The velocity with which a fault ruptures has been estimated to be between 2 and 3 km/s. The focal depth of past earthquakes has been between a few kilometers and all the way up to 680 km. According to their focal depth, earthquakes are thus classified as (a) shallow, if the depth is <60 km; (b) intermediate, if the depth is between 60 and 300 km; and (c) deep, if the depth is between 300 and 700 km. The amount of displacement during a fault rupture greatly differs among faults and earthquake size. It can amount to a few millimeters or several meters. For example, the fault displacement in the 1906 San Francisco earthquake was nearly 6 m and in the great Chilean earthquake of 1960 was close to 20 m.

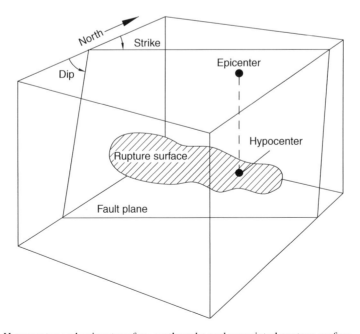

FIGURE 3.21 Hypocenter and epicenter of an earthquake and associated rupture surface.

TABLE 3.1
Average Slip Rate in Some Known Faults

Fault	Slip Rate (mm/year)
Fairweather, Alaska	38–74
San Andreas, California	20–53
Bocono, Venezuela	8–10
Motagua, Guatemala	5–7
Wasatch, Utah	0.9–1.8
Newport-Inglewood, California	0.1–1.2
Upper Rhine, Germany	0.15
Atlantic Coast, United States	0.0002

Source: Idriss, I.M., *Proceedings of the 11th International Conference on Soil Mechanics and Foundation Engineering*, San Francisco, CA, August 12–16, 1985.

A parameter related to fault displacement that is often used to characterize the activity of earthquake faults is the *slip rate*. For a given fault, slip rate is calculated by dividing the cumulative displacement of the fault, determined from offset geological or geomorphic features, by the estimated age at which the earliest displacement took place. Thus, slip rate is an average value over a geological time interval. The slip rates for a few selected faults are listed in Table 3.1. Note that, as expected, highly active faults such as the San Andreas Fault have a much higher slip rate than minor faults.

3.11 FORESHOCKS AND AFTERSHOCKS

Large earthquakes are usually preceded and followed by a sequence of tremors of smaller size that originate at approximately the same location. The tremors preceding the so-called main shock are known as *foreshocks*. Those following the main shock are called *aftershocks*. Foreshocks are normally small in size and only a few occur before an earthquake. Moreover, not all earthquakes are preceded by a foreshock. Aftershocks may be relatively large in number and size, but their frequency and size gradually decreases with time. Foreshocks are believed to be the precursors of the fault rupture that generates the main earthquake, whereas aftershocks are considered to be the result of adjustments to the stress imbalances produced by such rupture. Aftershocks almost as large as the main event have been observed in the past. For example, the 1985, magnitude[*] 8.1, earthquake in Mexico City was followed by a magnitude 7.5 aftershock the following day. The occurrence of aftershocks may last for weeks and sometimes for months after the occurrence of the main event. For instance, aftershocks with magnitudes of 6.1, 5.5, 4.8, and 5 occurred 16, 22, 25, and 32 days, respectively, after the 1985, magnitude 7.9, earthquake in central Chile. Another example is provided by the aftershocks recorded after the 1989, magnitude 7.1, Loma Prieta earthquake in the Santa Cruz, California, mountains. Fifty-one aftershocks of magnitude 3 or greater occurred during the first day following the earthquake and 16 during the second day. A total of 87 aftershocks were recorded after 21 days (see Figure 3.22). The magnitude of the largest aftershock during this earthquake was 5 and occurred 33 h after the main shock.

Sometimes it is difficult to know which shock is the foreshock and which is the main one. In some cases, the first shock is a major earthquake, yet the shock that takes place a few days later is equally large. In these cases, there is no way to determine whether the first shock is a foreshock of the second or the second an aftershock of the first.

---

[*] Earthquake magnitude is formally defined in Section 5.5.

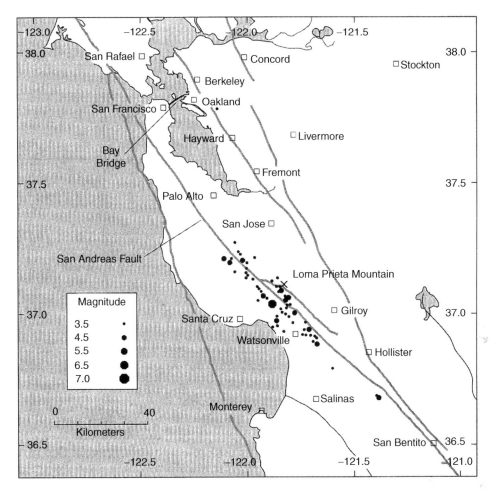

FIGURE 3.22 Sequence and magnitude of recorded aftershocks after the 1989 Loma Prieta earthquake in the Santa Cruz, California, mountains. (From Bolt, B.A., *Earthquakes*, W.H. Freeman and Co., New York, 2004. With permission.)

3.12 EARTHQUAKE PREDICTION

3.12.1 Introduction

Many efforts have been made throughout the years to be able to predict the size, location, and time of occurrence of future earthquakes. Although unable to prevent the occurrence of the event or damage to the built environment, these efforts have been made in the belief that earthquake predictions could save many lives and millions of dollars in economic losses. So far, however, an accurate prediction of seismic events remains elusive. There has been, however, some progress in the discovery and understanding of effects that are likely to take place before the occurrence of an earthquake, giving hope that in the near future earthquake prediction will become a reality.

3.12.2 Historical Background

Early efforts in earthquake prediction were conducted in the former Soviet Union after a catastrophic earthquake killed 12,000 people in the mountainous region of Tadzhikistan on July 10, 1949. The Soviets had been affected by another catastrophic earthquake the previous year, and thus Soviet officials were determined to search for clues that could allow them to predict future earthquakes and avoid additional calamities. They organized a geological study of the afflicted area on a scale never before attempted. They conducted detailed surveys of the region and made measurements of the gravity, magnetism, and electric resistance of the ground. They also devoted their efforts to obtain accurate records from the relatively mild but continuing earthquake activity.

After 15 years of gathering data, the Russians found evidence of what appeared to be a consistent early-warning signal of earthquakes. They discovered that for some time before the occurrence of a sizable earthquake, seismic records would show a fluctuation in the traveling speed of seismic waves. The anomaly would last for a few days, a few months, or a few years, and then it would return to normal. Shortly afterward, there would be a relatively large tremor. Equally significant, they also found that the longer the duration of the anomaly, the greater the magnitude of the ensuing tremor.

Upon learning of the Russian discovery, seismologists at the Lamont-Doherty Geological Observatory began an intensive review of the seismic records from their seismographic station near Blue Mountain Lake in upper New York State. Soon, they reported that, in fact, there had been changes in the seismic wave velocities before the occurrence of earthquakes. They further noted that the duration and scale of the changes seemed to have a direct relationship to the magnitude of the earthquakes. Also motivated by the Russian discovery, geophysicists at the California Institute of Technology reviewed their seismic records from the 1971 San Fernando earthquake, a magnitude 6.6 event, and found that the velocity of the seismic waves had changed for almost 3 years before the earthquake.

The evidence of the change in the traveling speed of seismic waves was accumulating rapidly, but there was no understanding as to why it was happening. The first hint came in the mid-1960s from William Brace, a geologist at the Massachusetts Institute of Technology, who made experimental studies of the properties of rock under stress. He found that just before a rock was fractured, the rock tended to dilate and, along with this dilation, several changes in the rock properties took place. The rock became more porous, conducted electricity more easily, and slowed down the passage of high frequency waves.

The connection between the dilatancy of rocks and the change in the velocity of seismic waves was first made by Amos Nur, a geophysicist at Stanford University. He theorized that as the stress in the bedrock approaches the point of rupture, it causes multiple fissures and dilatancy. He also reasoned that as a result of the fissures, the speed of the seismic waves dropped. However, just before fracture, the fissures would be filled with groundwater, permitting thus a return to the normal velocity. The greater the dilated region, the longer it would take for the groundwater to seep in and return to the normal wave velocity. He suggested that the change in wave velocity could last for a few months before a magnitude 6 earthquake, a year before a magnitude 7, and as much as 10 years before a magnitude 8.

Using this hypothesis, Christopher Scholz, a geophysicist at Lamont-Doherty Observatory, suggested that dilatancy might also explain several of the phenomena observed by the Russians. That is, the increase in volume due to dilatancy accounted for the crustal uplift and tilting observed before many earthquakes. Likewise, the fissures in the rock increased the amount of water in it and hence its electrical conductivity. In the same way, the fissures in the rock increased the area of contact between rock and water, and thus the amount of radon* observed in well water before earthquakes.

Equipped with a new set of early-warning signals and an understanding of the mechanism that could trigger these signals, many seismologists devoted their research efforts to earthquake prediction and began to make a series of predictions afterward. Unfortunately, some of these predictions were successful (see Box 3.2) and some others were not. The most successful and well-known prediction was that of the magnitude 7.3, February 4, 1975, Haicheng earthquake in

---

* Radon is a colorless, radioactive gas with a half-life of 3.82 days formed by the radioactive decay of radium. It is found in small quantities in rock and soil and produces most of the normal background radioactivity.

northeast China (see detailed account in Box 3.2). Chinese seismologists had been monitoring a series of precursors for a period of over five years and continued to do so until only a few hours before the occurrence of the main shock. The monitored precursors included patterns of large and small tremors and anomalous changes in tilt, the earth's magnetic field, water levels in wells, and animal behavior. A swarm of apparent foreshocks prompted officials to issue a few hours earlier a warning that a strong earthquake was imminent. The earthquake indeed came and devastated the city. The warning, however, helped to evacuate people from buildings and communes and save thus thousands of lives. This success caused great excitement in the scientific community and injected a renewed vigor to earthquake-prediction research.

Box 3.2   Successful Earthquake Predictions

One of the first successful earthquake predictions came from Yash Aggarwal, a doctoral student at the Lamont-Doherty Observatory. Pursuing the Soviet line of research, he acted on evidence that he had found in the seismic history of the Blue Mountain Lake area in upper New York State. Early in July of 1973, he added to the existing network of the seismological station at Blue Mountain Lake seven portable recording instruments capable of detecting microearthquakes and determining the location of any changes in seismic wave velocities. Toward the end of the month, he detected signs of reduced wave velocity in the area. On August 1, he found that this velocity had returned to normal. It seemed to him a classical example of the pattern found in Tadzhikistan, so he called the Lamont-Doherty Observatory with a flat prediction, a magnitude 2.5 earthquake could be expected in a couple of days. He based his estimate of the earthquake magnitude on the size of the area in which the wave velocity had changed, and his prediction of the day of occurrence on the length of time before the velocity returned to normal. Two days later, precisely on schedule, a tremor of the predicted magnitude shook the Blue Mountain Lake area.

Four months later, another prediction was attempted by James H. Whitcomb of the California Institute of Technology, who observed that in 1972 seismic wave velocities had dropped in the area around Riverside, California, and stayed below normal trough for most of 1973. When the wave velocity began to rise again in November, he predicted that a magnitude 5.5 earthquake would occur within 3 months. When the tremor hit, on January 30, its magnitude was only 4.1, but the forecast was accurate enough to consider another successful prediction.

Another prediction took place in 1974 near Hollister, California, in the region where the Calaveras Fault merges with the San Andreas Fault. Data from magnetometers revealed that changes in the earth's magnetic field were taking place, and tiltmeters showed that the ground was moving. Seismologists from the U.S. Geological Survey began studying the area and on November 27, 1974, they met to assess the accumulated data. They concluded that the area was about to have an earthquake, but they were not sufficiently confident to issue a public warning. On the day after the meeting, a magnitude 5.2 earthquake occurred. Later, they learned that seismic wave velocities had been changing at the same time that the magnetic and tilting changes were occurring. If the seismologists had had the wave velocity data at the time of their meeting, they would undoubtedly have made a public prediction.

The world's most dramatic and famous prediction took place in 1975 in the city of Haicheng in northeastern China. The region around Haicheng had been under careful study by seismologists for several years before 1975 because there were indications that a large earthquake might occur there in the near future. Instruments were installed to record tilting of the land surface, fluctuations in the magnetic field, and changes in the electric resistance of the ground. The populace was asked to record the water levels in wells and to report any strange animal behavior. By 1974, a large force of specialists were in the area, running surveys, mapping fault lines, and installing geomagnetic monitors and tiltmeters. The scientists discovered that the region had been uplifted and tilted toward the northwest. In addition, they found that the strength of the magnetic field was increasing. During the first 5 months of 1974, there was a significant increase in the number of minor tremors up to five times the normal frequency. The State Seismological Bureau in Beijing issued a tentative forecast: A moderate to strong earthquake would occur within 2 years. On December 22, another swarm of tremors began, the strongest registering a magnitude of 4.8. Soon after, the Seismological Bureau refined its prediction: There would be an earthquake of magnitude 5.5–6 somewhere in the vicinity of the major industrial port of Yingkow during the first 6 months of 1975.

At the beginning of February 1975, the tremors became more frequent and more worrisome by the day. Wells started bubbling, rats and mice were observed leaving their holes, and snakes crept out of winter hibernation and lay frozen on the icy roads. A swarm of minor tremors began, with more than 500 recorded in 72 h, culminating in a 4.8 jolt in the morning of February 4. Then there was an eerie quiet. The Seismological Bureau told the local party committees to prepare for a cataclysm. At 2 p.m., a military commander in a commune near Yingkow went on the radio with an urgent broadcast to the area's residents: "There probably will be a strong earthquake tonight. We require all people to leave their homes." A similar announcement was made at Haicheng, a city of 90,000 some 30 miles inland, and was telephoned to other towns and communes in the threatened area. Altogether, the warning went to three million people.

The earthquake hit the area at 7:36 p.m. that day. Light flashed across the sky and the ground shook violently. Roads buckled, bridges twisted and crashed, 4 m jets of water and sand shot into air, and rural communes were flattened. Almost 90% of Haicheng's buildings were severely damaged or destroyed. Yet, in an area of 3 million people where a quake of this violence could kill tens of thousands, the estimated toll was only ~30. The advanced warning had averted a catastrophe.

After many years of observations, Soviet seismologists made another successful prediction when on November 1, 1978, they announced that a large earthquake would occur within the next 24 h in the area around Garm, Tadzhikistan. The prediction was based on several kinds of evidences, including increases in seismicity, deformations shown by tiltmeters and strain meters, changes in seismic wave velocities, and changes in water wells. An earthquake of magnitude 7 came ~6 h after the announcement. It was located between the Pamir and the Tien Shan Mountains, ~150 km east of Garm.

Based on the concept of seismic gap (see Section 3.12.3), scientists at the Geophysics Laboratory of the University of Texas at Galveston concluded that a 185 mile stretch of plate boundary along the coast of Oaxaca, Mexico, exhibited all the signs of a seismic gap in the alpha phase. After reviewing seismic records, they found that moderate earthquake activity along that section of the subduction zone had stopped altogether in the middle of 1973. After further study, the Texan geophysicists published a paper identifying the phenomenon as a probable precursor to a large earthquake. The length of the fault's dormant section suggested a main shock that could reach a magnitude between 7.25 and 7.75. As it happened, Karen McNally, a senior research fellow from the California Institute of Technology, was at Mexico City's Institute of Geophysics in August 1978 when a moderate tremor took place in Oaxaca, the second quake to hit the region in 5 weeks. The Oaxaca seismic gap appeared to be coming alive. McNally sensed an opportunity to study the beta phase of the seismic gap theory and proposed to establish a network of portable recording instruments arrayed near the possible fracture. By November 8, a seven-station network was in place ready to record Oaxaca tremors. For a week the instruments were quiet. Then came a series of small tremors, which dwindled after November 15. On November 28, the calm was broken with a second series of tremors. There was another quiet period, lasting 18 h. And then, just as forecasted by the Texan seismologists, a magnitude 7.8 earthquake rocked a large area of Mexico.

Notwithstanding the successful prediction of the Haicheng earthquake, just a few months later Chinese seismologists failed to alert the inhabitants of Tangshan, an industrial and coal-mining city of 1 million people, 150 km east of Beijing, of the incoming magnitude 8 earthquake that leveled the city and reportedly killed as many as 250,000 people on July 28, 1976. Chinese seismologists had observed a series of possible precursors. For example, some medium tremors had been recorded in the area in 1975. They had also found some long-term changes in the gravity field and the electrical resistance of the ground and in January 1976 recorded a sharp change in the magnetism of the area. As a result, an open-ended long-term alert was issued. Late in July, they also observed sudden variations in groundwater levels and obtained reports of strange animal behavior. However, they found these early signals to be too scattered and ambiguous to support an amended and more urgent forecast.

The final blow to the optimism expressed by a number of researchers in the recent past came as a result of the August 1979, magnitude 5.7 earthquake that struck an area 10 km southeast of San Francisco, California, near Coyote Lake on the Calaveras Fault, a major branch of the San Andreas Fault. The area had been heavily instrumented with a dense network of monitoring equipment centered in Hollister, 25 km away, so researchers had an opportunity to learn more about earthquake precursors. Instead, they learned that the earthquake did not show any of the known precursors that could have been used for a short-term prediction. No swarms of small tremors, no tilting of the earth's surface, and no changes in the geophysical properties of the crust were observed during this earthquake. Disappointedly, they concluded that the Coyote Lake earthquake was a living proof that some earthquakes were just impossible to predict.

The Tangshan and the Coyote Lake earthquakes brought thus an abrupt end to the growing belief that reliable earthquake forecasting was near. It became evident that many events identified by the Chinese and the Russians as sure signs of an impending tremor were not so reliable after all. They appeared before some earthquakes but not before others. Even dilatancy, which had been so rapidly gaining credibility as a key precursor, failed the test of universality. It was clear that the cause–effect relationship of precursors was far from being fully understood.

3.12.3 SEISMIC GAPS AND SEISMIC QUIESCENCE

Another approach to earthquake prediction explored in the recent past has been the identification of seismic gaps and periods of seismic quiescence. The theory of plate tectonics suggests that major fractures can be expected to occur with approximate regularity along the great earthquake belts. Consequently, if a fault has slipped recently, it will probably remain quiet for a number of years. Furthermore, the longer the time interval since the last rupture, the greater the likelihood that a fracture will occur again soon. Thus, by studying the amount of earthquake activity that has been occurring in the past years, seismologists can determine which regions have been slipping and which regions have been relatively quiet. A region along a plate boundary where the seismicity has been significantly less than that along adjacent segments of the boundary is called a *seismic gap*. In this way, seismic gaps have been identified along the San Andreas Fault in California, off the coasts of Alaska (see Figure 3.23), Mexico, and Japan, as well as in other places around the world.

The concept of a seismic gap is useful in long-range prediction of earthquakes but is of no help in predicting the time or magnitude of a particular earthquake at a particular location. However, a refinement in the concept of the seismic gap has advanced the hopes for a close range prediction. In this refinement, it is considered that there are two phases that prelude a major event: an alpha phase during which there is no seismic activity whatsoever and a beta phase during which there are frequent small foreshocks just before the main fracture. Evidence of the second phase in a seismic gap may be thus used for the short-term production of an imminent earthquake (see Box 3.2).

Related to the concept of a seismic gap, it has also been found that once the aftershock sequence of a major earthquake is concluded, the seismicity level drops to a low background level. Prior to the next event, the background activity sometimes drops almost to zero for an interval of months or years. This phenomenon is known as *seismic quiescence* and is fairly common. The quiescence

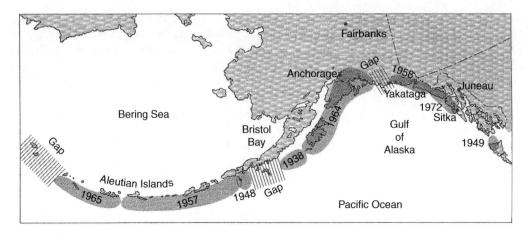

FIGURE 3.23 Rupture areas of large earthquakes and seismic gaps along the Alaska–Aleutian arc plate boundary. (Reproduced with permission from Gere, J.M. and Shah, H.C., *Terra Non Firma*, W.H. Freeman and Co., New York, 1980.)

is often broken by a buildup of activity, with swarms of activity before the main shock, which is called a preshock sequence. Seismic quiescence is usually terminated by foreshock activity. Of earthquakes with a magnitude >7, almost 70% are preceded by foreshocks, even if no quiescence occurs. The nature of foreshock activity varies greatly, but it typically begins between 5 and 10 days before the main shock.

3.12.4 Earthquake Precursors

As noted in the foregoing sections, many phenomena appear to be precursory to earthquakes; that is, many phenomena seem to give clues, in one way or another, that the occurrence of an earthquake is imminent. Earthquake precursors that have been reported in the recent past range from odd animal behavior, muddy ponds, bubbling wells, and an unnatural glow in the sky to those that involve precise scientific measurements. However, the majority of the most promising precursors are related to the changes caused by the strain accumulation in the earth's crust that leads up to earthquakes. These are (a) strains in the rocks in the region surrounding an earthquake fault, (b) large-scale ground deformations, (c) land subsidence and uplifts, (d) land surface tilting, (e) rise or fall of water levels in wells, (f) changes in the wave velocity of the crustal rock, (g) localized changes in the earth's magnetic field, (h) changes in the electrical resistance of the rocks near the earthquake fault, (i) changes in the radon content in ground water, and (j) a sudden swarm of small earthquakes a few days before the main shock.

A plausible explanation that has been given for the occurrence of these precursors is based on the events that lead to and are known to occur before an earthquake. First, there is a slow buildup of the elastic strain in the crustal rock near the fault zone produced by ongoing tectonic forces. After this strain reaches a certain level, cracks develop, causing, as a result, swelling or dilatation of the crustal rock. As cracks open, the wave velocity of the rock drops, the ground surface rises and rotates, radon gas escapes, the electric resistivity decreases, and the earth's magnetic field undergoes localized changes. Thereafter, water flows from the surrounding rocks into the cracks, leading to conditions favorable for rock slippages and an increase in the number of small earthquakes in the fault's vicinity. After water fills the cracks, the wave propagation velocity begins to increase, the uplift of the ground ceases, emission of radon from the fresh cracks dwindles, and the electrical resistivity decreases further. Immediately thereafter, an earthquake occurs.

All of these listed precursors have been observed before earthquakes, but not in a consistent manner. That is, sometimes they have been observed but no earthquake followed, and sometimes they were not observed yet an earthquake occurred. Thus, these precursors seem to offer a promise but presently are not completely reliable to base earthquake predictions on them.

3.12.5 Current Status of Earthquake Predictions

In regard to the viability of earthquake predictions, it seems that at the present time the unified opinion of seismologists is that many more years of study and data acquisition are needed before earthquake predictions will be reliable enough to be of any practical value. Yet, there is a group of researchers who quite bluntly believe that earthquakes are inherently unpredictable.[*] Their sentiment is that earthquakes are so complex that there are simply no means to be able to issue a reliable forecast of any earthquake anywhere. Another group of researchers with a more moderate view claims that earthquake prediction is difficult and not possible yet, but not impossible. They believe that the difficulty of earthquake prediction is partly due to the inherent complexity of the earthquake phenomenon and an incomplete understanding of the earthquake-generation process. They also believe that reports of physical precursors prior to large earthquakes are by far too many to be ignored. They contend that, while it is certain that presently earthquakes cannot be predicted accurately and consistently, statements indicating that earthquakes are unpredictable presume that the mechanics of the earthquake process has been completely understood.[†]

In spite of this apparent pessimism, many countries continue nowadays their research efforts toward earthquake prediction. A proof is the many new precursors and innovative observations that have been reported in the *Journal of Earthquake Prediction Research*, a journal jointly sponsored by the State Seismological Bureau of China and the Academy of Science of Russia.

FURTHER READINGS

1. Bolt, B. A., *Earthquakes*, 5th edition, W. H. Freeman and Co., New York, 2004, 378 p.
2. Eiby, G. A., *Earthquakes*, Van Nostrand Reinhold Co., New York, 1980, 209 p.
3. Gere, J. M. and Shah, H. C., *Terra Non Firma*, W. H. Freeman and Co., New York, 1984, 203 p.
4. Gubbins, D., *Seismology and Plate Tectonics*, Cambridge University Press, Cambridge, U.K., 1990, 339 p.
5. Hodgson, J. H., *Earthquakes and Earth Structure*, Prentice-Hall, Englewood Cliffs, NJ, 1964, 166 p.
6. Ikeya, M., *Earthquakes and Animals: From Folk Legends to Science*, World Scientific Publishing Co., Singapore, 2004, 295 p.
7. Lay, T. and Wallace, T. C., *Modern Global Seismology*, Academic Press, San Diego, 1995, 521 p.
8. Lomnitz, C., *Global Tectonics and Earthquake Risk*, Elsevier Scientific Publishing Co., New York, 1974, 320 p.
9. Sieh, K. and LeVay, S., *The Earth in Turmoil*, W. H. Freeman and Co., New York, 1998, 324 p.

REVIEW QUESTIONS

1. In the most general sense of the word, what is an earthquake?
2. What are the different factors that may cause an earthquake?
3. What is the mechanism that causes tectonic earthquakes?
4. What is the explanation given for the occurrence of earthquakes induced by the filling of reservoirs and wells?
5. What are the basic concepts postulated by the plate tectonics theory?
6. In what directions are the earth's tectonic plates moving in relation to one another?
7. What is the relationship between plate tectonics and the formation of ocean trenches, mid-oceanic ridges, and mountain ranges?
8. What is a subduction zone?

---

[*] See, for instance, Geller, R. J. et al., Earthquakes cannot be predicted, *Science*, 275, March 14, 1997.
[†] See Presidential Address by Wesnousky, S. G., *Seismological Research Letters*, 68, 5, September/October 1997.

9. Why are subduction zones regions of active volcanism?
10. What is the deepest depth at which earthquakes have occurred in subduction zones?
11. What is a transform fault?
12. Where do transform faults commonly develop?
13. Is volcanism associated with transform faults?
14. What is the prevalent theory that explains the movement of tectonic plates?
15. What is an intraplate earthquake?
16. What is the explanation given for the occurrence of intraplate earthquakes?
17. What is an earthquake fault?
18. When it is said that an earthquake fault is inactive?
19. What are the different types of earthquake faults?
20. What is the direction of motion in a strike-slip fault?
21. If during an earthquake a fault ruptures through your property, would you gain or lose land if the fault type is (a) strike-slip, (b) normal, or (c) thrust?
22. What kind of fault is associated with an extensional strain environment?
23. What is a fault's scarp?
24. What are the main concepts behind the elastic rebound theory?
25. How can one determine the orientation of an earthquake-generating fault if one knows the direction of the first waves that arrive at a series of recording stations in the epicentral area?
26. What is the difference between the focus and the epicenter of an earthquake?
27. Where is the hypocenter of an earthquake in relation to a fault's rupture surface?
28. What is a fault's rupture surface?
29. What is a fault's slip?
30. What is the estimated velocity with which a fault ruptures?
31. What are some of the known earthquake precursors that have been used in earthquake prediction?
32. How can the concept of seismic gap be used for earthquake prediction?
33. What is the difference among a foreshock, an aftershock, and the main shock?
34. What is a seismic gap?

4 Earthquake Propagation

> At half-past two o'clock of a moonlit morning in March, I was awakened by a tremendous earthquake, and though I had never before enjoyed a storm of this sort, the strange thrilling motion could not be mistaken, and I ran out of my cabin, both glad and frightened, shouting, "A noble earthquake! A noble earthquake!" feeling sure I was going to learn something.
>
> **John Muir, 1872**

4.1 INTRODUCTION

When an earthquake fault ruptures, the rupture does not affect a large area simultaneously. Instead, the rupture introduces a localized disturbance in the area surrounding the fault, and then this disturbance propagates in the form of spherical waves to other points throughout the earth. In this way, an earthquake disturbance is always carried to great distances, often thousands of kilometers away from the fault where it originates. An analogy that is often used to explain this phenomenon is that of a small pebble dropped into a large deposit of water. After the pebble is dropped, one can see that waves are generated and propagated in all directions, whereas if a small floating object is placed on the water surface, the object will oscillate about its original position.

An understanding of the motions generated by an earthquake at distant points from the source, the techniques used to determine the geographical location where the earthquake originates, and the classification of the earthquake size requires an understanding of wave propagation. This chapter, therefore, is devoted to describe some of the basic concepts of wave propagation. It begins with a detailed description of wave propagation theory for the case of one-dimensional waves. Then, this description is expanded to the case of three-dimensional waves. Thereafter, waves that propagate in a bounded media are discussed.

The theory of wave propagation is developed on the basis of equations that describe the motion of continuous bodies. Although waves can propagate in inelastic media, the deformations of concern in seismology and earthquake engineering are small enough to make the assumption of an elastic medium reasonable. The discussion in this chapter is thus limited to the propagation of waves in an elastic medium.

4.2 LONGITUDINAL WAVES IN LONG ROD

4.2.1 Differential Equation of Motion

Consider the bar shown in Figure 4.1 and let the bar be subjected to an axial disturbance. Assume that the length of the bar is much longer than the dimensions of its cross section, that the resultant stress is uniform over this cross section, and that the bar's cross sections remain plane at all times. In addition, let A denote its cross-section area, E Young's modulus of elasticity, ρ the density of the bar (mass per unit volume), and $u(x, t)$ the longitudinal displacement of a point on the bar at time t and distance x from the origin of the system of coordinates. At any time after the application of the disturbance, the stresses on the longitudinal faces of a differential element of the bar are as

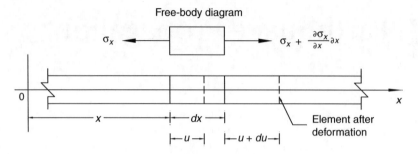

FIGURE 4.1 Infinitely long bar and free-body diagram of differential element.

indicated in the free-body diagram shown in Figure 4.1. By virtue of Newton's second law, the motion of such a differential element is described by

$$\sum F_x = m a_x \qquad (4.1)$$

where m is the mass of the element, a_x its acceleration along the x-axis, and F_x a force acting on the element along the x-axis. In terms of stresses and displacements, the equation of motion for the differential element under consideration may be thus expressed as

$$-\sigma_x A + \left(\sigma_x + \frac{\partial \sigma}{\partial x} dx\right) A = (\rho A\, dx) \frac{\partial^2 u(x, t)}{\partial t^2} \qquad (4.2)$$

But by Hooke's law and the definition of strain, it is possible to express the stress σ_x as

$$\sigma_x = E \varepsilon_x = E \frac{\partial u(x, t)}{\partial x} \qquad (4.3)$$

Upon substitution of this equation, Equation 4.2 may therefore be put into the form

$$\frac{\partial^2 u(x, t)}{\partial t^2} = v_c^2 \frac{\partial^2 u(x, t)}{\partial x^2} \qquad (4.4)$$

where v_c is a constant defined as

$$v_c = \sqrt{\frac{E}{\rho}} \qquad (4.5)$$

It is of interest to note that when the bar is constrained to deform laterally, the relationship between σ_x and ε_x is given by the constrained modulus of elasticity M, defined as

$$M = \frac{E(1-\mu)}{(1+\mu)(1-2\mu)} \qquad (4.6)$$

where μ denotes Poisson ratio. In such a case, therefore, the constant v_c in Equation 4.4 is given by

$$v_c = \sqrt{\frac{E(1-\mu)}{\rho(1+\mu)(1-2\mu)}} \qquad (4.7)$$

Earthquake Propagation

Equation 4.4 is the equation that describes the motion of the bar at any time and is a partial differential equation known as the *one-dimensional wave equation*.

4.2.2 Solution of Differential Equation

D'Alambert found that the wave equation given by Equation 4.4 is satisfied by any function of $x - v_c t$, that is, by

$$u(x, t) = F(x - v_c t) \tag{4.8}$$

where F is any arbitrary function. To prove that this solution indeed satisfies Equation 4.4, it is convenient to introduce first a new variable $\xi = x - v_c t$. Then, one can take the second partial derivative of u with respect to x and t to obtain

$$\frac{\partial^2 u(x, t)}{\partial x^2} = \frac{\partial}{\partial x}\left[\frac{\partial F(\xi)}{\partial \xi}\frac{\partial \xi}{\partial x}\right] = \frac{\partial}{\partial x}[F'(\xi)] = F''(\xi) \tag{4.9}$$

$$\frac{\partial^2 u(x, t)}{\partial t^2} = \frac{\partial}{\partial t}\left[\frac{\partial F(\xi)}{\partial \xi}\frac{\partial \xi}{\partial t}\right] = \frac{\partial}{\partial t}[-v_c F'(\xi)] = v_c^2 F''(\xi) \tag{4.10}$$

where F' and F'' represent the first and second derivatives of F with respect to the argument of the function. Thus, these two equations show that $\partial^2 u(x, t)/\partial t^2 = v_c^2 \partial^2 u(x, t)/\partial x^2$ and, hence, that Equation 4.8 satisfies Equation 4.4.

Proceeding similarly, it may be shown that the function

$$u(x, t) = G(x + v_c t) \tag{4.11}$$

also satisfies Equation 4.4. Consequently, the general solution to the differential equation in question is of the form

$$u(x, t) = F(x - v_c t) + G(x + v_c t) \tag{4.12}$$

4.2.3 Solution in Terms of Initial Conditions

The functions F and G in the general solution to the wave equation represent arbitrary functions. The precise form of these functions may be defined only with the help of the initial conditions that cause the motion of the bar. For example, if at $t = 0$ the bar is subjected to displacements and velocities described by given functions $\varphi(x)$ and $\psi(x)$, respectively, then one can write

$$u(x, 0) = F(x) + G(x) = \varphi(x) \tag{4.13}$$

$$\left.\frac{\partial u(x, t)}{\partial t}\right|_{t=0} = \left.\frac{\partial u(x, t)}{\partial \xi}\frac{\partial \xi}{\partial t}\right|_{t=0} = -v_c F'(x) + v_c G'(x) = \psi(x) \tag{4.14}$$

where, as before, F' and G' denote the first derivatives of F and G with respect to the argument of the functions. Furthermore, if this last equation is integrated from x_0 to x, where x_0 is any arbitrary value of x, the equation may also be expressed as

$$-v_c F(x) + v_c G(x) = \int_{x_0}^{x} \psi(\zeta) d\zeta \tag{4.15}$$

where ζ is a dummy variable. As a result, because Equations 4.13 and 4.15 are two simultaneous equations in the two unknowns $F(x)$ and $G(x)$, one can use these equations to solve for $F(x)$ and $G(x)$ to obtain

$$F(x) = \frac{1}{2}\left[\varphi(x) - \frac{1}{v_c}\int_{x_0}^{x}\psi(\zeta)d\zeta\right] \quad (4.16)$$

$$G(x) = \frac{1}{2}\left[\varphi(x) + \frac{1}{v_c}\int_{x_0}^{x}\psi(\zeta)d\zeta\right] \quad (4.17)$$

After a variable change from x to $x - v_c t$ in Equation 4.16 and from x to $x + v_c t$ in Equation 4.17, the substitution of these two equations into Equation 4.12 thus leads to

$$u(x,t) = \frac{1}{2}\left[\varphi(x - v_c t) - \frac{1}{v_c}\int_{x_0}^{x-v_c t}\psi(\zeta)d\zeta\right] + \frac{1}{2}\left[\varphi(x + v_c t) + \frac{1}{v_c}\int_{x_0}^{x+v_c t}\psi(\zeta)d\zeta\right] \quad (4.18)$$

which, by combining the two integrals into a single one, may also be expressed as

$$u(x,t) = \frac{1}{2}\left[\varphi(x - v_c t) + \varphi(x + v_c t)\right] + \frac{1}{2v_c}\int_{x-v_c t}^{x+v_c t}\psi(\zeta)d\zeta \quad (4.19)$$

It may be seen, thus, that Equation 4.19 fully defines the longitudinal displacements in the bar if the initial displacement and velocity functions are known.

4.2.4 Solution Interpretation

As the principle of superposition is valid for linear systems, it is possible to consider one term at a time in Equation 4.19 and in the interpretation of the solution of the differential equation in question. Accordingly, it will be assumed first that

$$u(x,t) = \frac{1}{2}\varphi(x - v_c t) \quad (4.20)$$

Consider now the solution function at two specific times, say $t = 0$ and t_0:

$$u(x,0) = \frac{1}{2}\varphi(x) \quad (4.21)$$

$$u(x,t_0) = \frac{1}{2}\varphi(x - v_c t_0) = \frac{1}{2}\varphi(\bar{x}) \quad (4.22)$$

If $u(x,0)$ and $u(x,t_0)$ are then plotted as shown in Figure 4.2, it may be seen that the shape of the displacement function does not change with time. Instead, it just moves to the right a distance equal to $v_c t_0$. Furthermore, the speed with which this movement takes place is equal to $v_c t_0/t_0 = v_c$. Consequently, $\varphi(x - v_c t_0)/2$ represents a wave with the shape of the initial displacement $\varphi(x)$ traveling in the positive direction of the x-axis with a velocity equal to the constant v_c defined by Equation 4.5.

Using similar arguments, it is easy to show that the second term in Equation 4.19 also represents a wave with the shape of the initial displacement $\varphi(x)$ and traveling speed v_c. The exception is that now the direction of travel is in the negative direction of the x-axis (see Figure 4.3). Furthermore, it can be shown that the third term in Equation 4.19 portrays similar waves, except that the shape of these waves is defined by the functions that result from integrating the initial velocity function from 0 to $x - v_c t_0$ and from 0 to $x + v_c t_0$.

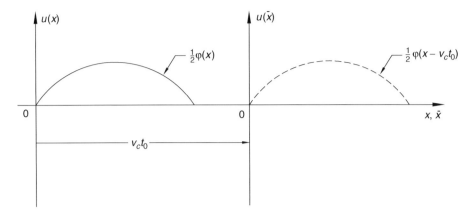

FIGURE 4.2 Shape and position of disturbance $\varphi(x - v_c t)/2$ at two different times.

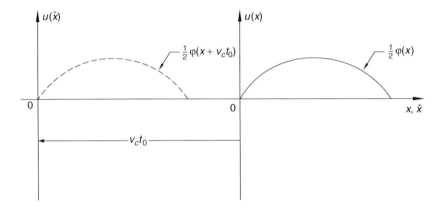

FIGURE 4.3 Shape and position of disturbance $\varphi(x + v_c t)/2$ at two different times.

Note that according to these findings, waves propagate with a velocity that depends only on the properties of the bar material and not on the amplitude of the initial disturbances.

4.2.5 Relationship between Particle Velocity and Wave Velocity

In dealing with the propagation of waves, it is important to distinguish between particle velocity and wave propagation velocity. Particle velocity is the velocity with which a single point within the rod will move as the wave passes through it. On the other hand, wave velocity is the velocity with which a disturbance travels along the length of the rod. If the rod displacements are defined, for example, by Equation 4.20, it may be noted then that the relationship between particle velocity and wave velocity is given by

$$v_x(x,\, t) = \frac{\partial u(x,\, t)}{\partial t} = -\frac{1}{2} v_c \varphi'(x - v_c t) \tag{4.23}$$

However, by virtue of Hooke's law and the relationship between axial strains and axial displacements, it is possible to write

$$\sigma_x = E\varepsilon_x = E\frac{\partial u(x,\, t)}{\partial x} = \frac{1}{2} E\varphi'(x - v_c t) \tag{4.24}$$

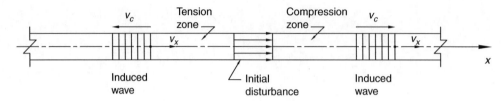

FIGURE 4.4 Relation between wave velocity and particle velocity in bar subjected to initial axial stress.

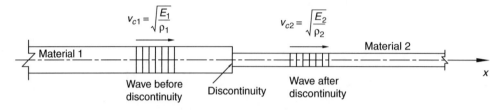

FIGURE 4.5 Wave traveling along two different rod sections.

Hence, the relationship between particle velocity and wave velocity may also be expressed as

$$v_x(x,\ t) = -\frac{\sigma_x}{E} v_c \qquad (4.25)$$

Note that according to this relationship, the particle velocity depends on the magnitude of the stress applied to the rod. Note, too, that the direction of the particle velocity is in the direction of propagation of the wave when a compressive stress is applied and opposite to this direction when a tension stress is applied. The relative direction between wave velocity and particle velocity can be seen in the example shown in Figure 4.4.

4.2.6 Transmitted and Reflected Waves at Discontinuity

It has been seen that the velocity of wave propagation depends on the properties of the medium within which a wave travels. Hence, if a rod is composed of two different materials, it is expected that the velocity of a wave traveling within the rod will change, as shown in Figure 4.5, after it goes from one material to the other. This change in the wave velocity, however, produces an imbalance in the magnitude of the displacements and forces at the interface that separates one material from the other. This imbalance is owed to the fact that, because both displacements and forces depend on the wave velocity (recall that $\sigma_x = E\ \partial u/\partial x$ and u is a function of v_c), the displacements and forces immediately before the interface will be different than those immediately after the interface. Thus, an additional wave is formed in the rod to compensate for this imbalance. This wave has the same shape as the original wave but travels in the opposite direction (see Figure 4.6). For this reason, it is called the *reflected wave*. To distinguish this wave from the original one, the latter is called the *incident wave* before it crosses the discontinuity and the *transmitted wave* right after.

The relationship among the incident, transmitted, and reflected waves can be found by establishing the continuity requirement that the forces and velocities at the two sides of the interface where the rod changes from one material to the other be exactly the same. For this purpose, let u_i, u_t, and u_r, respectively, denote the displacements, and P_i, P_t, and P_r, respectively, represent the forces generated by the incident, transmitted, and reflected waves. Assume, in addition, that the incident, transmitted, and reflected waves are of the form

$$u_i = F_i(x - v_{c1}t) \qquad (4.26)$$

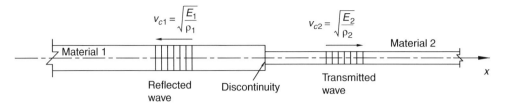

FIGURE 4.6 Transmitted and reflected waves after incident wave crosses rod discontinuity.

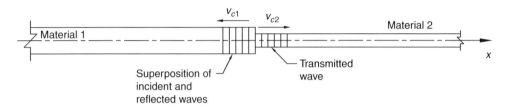

FIGURE 4.7 Incident, reflected, and transmitted waves at discontinuity.

$$u_t = F_t(x - v_{c2}t) \tag{4.27}$$

$$u_r = F_r(x + v_{c1}t) \tag{4.28}$$

where v_{c1} and v_{c2}, respectively, denote the wave velocity in the left- and right-hand sides of the rod. Right after the incident wave crosses the interface, the incident and the reflected waves affect the left-hand side of the rod, while only the transmitted wave affects the right-hand side (see Figure 4.7). To preserve the continuity between the two parts of the rod, one has thus that at that precise time the velocity and force generated by the incident and the reflected waves on the left-hand side of the rod should be the same as those on the right-hand side. Mathematically, these two conditions may be expressed as

$$\frac{\partial u_i}{\partial t} + \frac{\partial u_r}{\partial t} = \frac{\partial u_t}{\partial t} \tag{4.29}$$

$$P_i + P_r = P_t \tag{4.30}$$

Note, however, that the first derivative in Equation 4.29 may be written as

$$\frac{\partial u_i}{\partial t} = -v_{c1}F_i'(x - v_{c1}) = -v_{c1}\frac{\partial u_i}{\partial x} = -v_{c1}\varepsilon_{xi} = -v_{c1}\frac{\sigma_{xi}}{E_1} = -\frac{v_{c1}}{A_1 E_1}P_i \tag{4.31}$$

and similarly the other two derivatives may be expressed as

$$\frac{\partial u_t}{\partial t} = -\frac{v_{c2}}{A_2 E_2}P_t \tag{4.32}$$

$$\frac{\partial u_r}{\partial t} = \frac{v_{c1}}{A_1 E_1}P_r \tag{4.33}$$

Consequently, Equation 4.29 can also be put into the form

$$-\frac{v_{c1}}{A_1 E_1} P_i + \frac{v_{c1}}{A_1 E_1} P_r = \frac{v_{c2}}{A_2 E_2} P_t \tag{4.34}$$

from which one finds that

$$P_t = \alpha(P_i - P_r) \tag{4.35}$$

where

$$\alpha = \frac{v_{c1}}{v_{c2}} \frac{A_2 E_2}{A_1 E_1} = \sqrt{\frac{m_2 E_2 A_2}{m_1 E_1 A_1}} \tag{4.36}$$

in which m_1 and m_2, respectively, denote the mass per unit length of the left-hand and right-hand side sections of the rod.

Thus, by substitution of Equation 4.35 into Equation 4.30 and solving afterward for P_r, one obtains

$$P_r = \frac{\alpha - 1}{\alpha + 1} P_i \tag{4.37}$$

which, upon substitution into Equation 4.35, in turn leads to

$$P_t = \frac{2\alpha}{\alpha + 1} P_i \tag{4.38}$$

If in the light of Equations 4.31 through 4.33, Equations 4.37 and 4.38 are now written in terms of displacements as

$$\frac{A_1 E_1}{v_{c1}} \frac{\partial u_r}{\partial t} = \frac{\alpha - 1}{\alpha + 1} \frac{A_1 E_1}{v_{c1}} \frac{\partial u_i}{\partial t} \tag{4.39}$$

$$-\frac{A_2 E_2}{v_{c2}} \frac{\partial u_t}{\partial t} = \frac{2\alpha}{\alpha + 1} \frac{A_1 E_1}{v_{c1}} \frac{\partial u_i}{\partial t} \tag{4.40}$$

then, after solving for u_r and u_t from these two equations, one similarly gets

$$u_r = -\frac{\alpha - 1}{\alpha + 1} u_i \tag{4.41}$$

$$u_t = \frac{2}{\alpha + 1} u_i \tag{4.42}$$

Equations 4.37 and 4.38 in combination with Equations 4.41 and 4.42 provide the desired relationships between the forces and displacements induced by incident, transmitted, and reflected waves. Note that by means of these relationships one can determine the amplitudes of the forces and displacements generated by the transmitted and reflected waves if one knows the amplitudes of the displacements and forces generated by the incident wave.

Earthquake Propagation

EXAMPLE 4.1 AMPLITUDES OF TRANSMITTED AND REFLECTED WAVES

Determine the amplitudes of the transmitted and reflected waves that are generated in a rod after an incident longitudinal compressive wave (1) crosses a boundary between two different sections, each characterized by the same *mEA* product; (2) reaches a fixed end; and (3) reaches a free end. Express the answers in terms of u_i, the amplitude of the incident wave.

Solution

1. As the product *mEA* is the same for both sections, then according to Equation 4.36 the parameter α is equal to unity. In view of Equations 4.41 and 4.42, in this case the amplitudes of the transmitted and reflected waves are

$$u_t = u_i \qquad u_r = 0$$

   That is, the incident wave crosses the boundary without changing its amplitude and without generating a reflected wave.

2. A fixed end can be simulated with a rod section whose modulus of elasticity is infinitely large. In the case of a fixed end, therefore, one has that $m_2 E_2 A_2$ is equal to infinity and the parameter α defined by Equation 4.36 is also equal to infinity. As a result, Equations 4.41 and 4.42 give in this case

$$u_t = 0 \qquad u_r = -u_i$$

   This means that when an incident wave reaches a fixed end, the wave is not transmitted (as expected) and is reflected in the form of a compressive wave of the same amplitude.

3. If a free end is simulated by a rod section with a null modulus of elasticity, then in the case of a free end the factor α becomes zero and, correspondingly, Equations 4.41 and 4.42 yield

$$u_t = 2u_i \qquad u_r = -(-u_i)$$

   In this case, therefore, the incident wave doubles its amplitude as it reaches the free end and is reflected in the form of a tensile wave of the same amplitude.

EXAMPLE 4.2 ANALYSIS OF STRESS WAVES GENERATED BY HAMMER IMPACT ON PILE

During the process of driving a 150 ft long reinforced concrete pile into a deposit of soft clay, a pile-driving hammer hits the top of the pile with a force $P(t)$ whose magnitude and variation with time is as shown in Figure E4.2a. Determine the stress distribution along the length of the pile (1) right after the end of the loading pulse and (2) 0.019 s after the hammer first comes in contact with the pile. Consider a pile with modulus of elasticity of 3×10^6 psi, a cross-section area of 400 in.², and a unit weight of 150 pcf.

Solution

1. Being long in comparison with the dimensions of its cross section, the pile can be considered to be a one-dimensional long rod subjected to an initial disturbance that propagates along the length of the rod in the form of a wave. The axial stress in the pile at any distance x from the pile top and at any time t may therefore be expressed as

$$\sigma(x, t) = E \frac{\partial u(x, t)}{\partial x} = E \frac{\partial [F(x - v_c t)]}{\partial t} = f(x - v_c t)$$

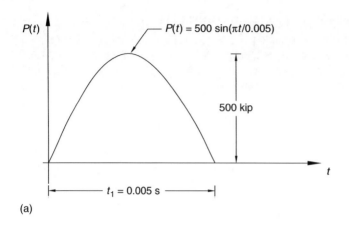

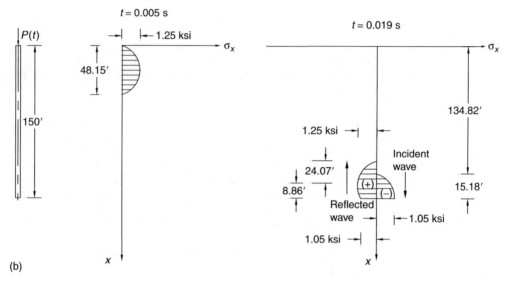

FIGURE E4.2 (a) Loading pulse in Example 4.2 and (b) distribution of stresses in pile at $t = 0.005$ and 0.019 s.

where E represents the modulus of elasticity of the pile, f denotes an unknown function, and v_c is the velocity of wave propagation, which according to Equation 4.5 and the given values of the unit weight and modulus of elasticity of the pile is given by

$$v_c = \sqrt{\frac{3 \times 10^6 (12^3)(386.4)}{150}} = 115{,}559 \text{ in./s} = 9630 \text{ ft/s}$$

The form of the function f, however, may be determined from the known initial conditions as follows. Consider, first, that, according to the given initial loading pulse, the stress generated by the hammer at the top of the pile (i.e., at $x = 0$) may be expressed as

$$\sigma(0,t) = -\frac{P(t)}{A} = -\frac{500}{400}\sin\left(\frac{\pi t}{0.005}\right) = -1.25\sin\left(\frac{\pi t}{0.005}\right)$$

where the negative sign indicates that the generated stress is a compressive one. Note, then, that if the argument of the sine function in this equation is multiplied and divided by $-v_c$ or its numerical equivalent, the equation can also be written as

$$\sigma(0,t) = f(-v_c t) = -1.25\sin\left[\frac{-\pi}{(9630)(0.005)}(-v_c t)\right] = -1.25\sin\left[\frac{-\pi}{48.15}(-v_c t)\right]$$

Earthquake Propagation

Finally, introduce a change of variable from $-v_c t$ to $x - v_c t$ to obtain

$$\sigma(x,t) = f(x - v_c t) = -1.25 \sin\left[\frac{-\pi}{48.15}(x - v_c t)\right]$$

This expression constitutes the equation that yields the value of the axial stress in the pile at any distance x from the pile top and any time t in the interval between the application of the loading pulse and the time the wave reaches the pile tip. To determine thus the stresses in the pile at the end of the pulse, one can simply substitute $t = 0.005$ s into this equation. In this way, one finds that such stresses are given by

$$\sigma(x, 0.005) = -1.25 \sin \pi \left(1 - \frac{x}{48.15}\right) \text{ ksi}$$

A plot of the corresponding stress distribution is shown in Figure E4.2b.

2. The time it takes for the wave front to reach the pile tip from the moment the hammer hits the pile top is given by

$$t_{x=150} = \frac{150}{9630} = 0.01558 \text{ s}$$

Similarly, the time it takes for the wave front to bounce back and reach the pile top again is another 0.01558 s. Hence, at $t = 0.019$ s, the wave is traveling upward and its front has moved away from the pile tip a distance x' given by

$$x' = 9630(0.019 - 0.01558) = 32.93 \text{ ft}$$

Consequently, the stresses in the pile at $t = 0.019$ s are given by the superposition of the incident and the reflected waves.

Now, according to the equation developed earlier for $\sigma(x, t)$, the stresses generated by the incident wave at $t = 0.019$ s are given by

$$\sigma(x, 0.019) = -1.25 \sin\left\{\frac{-\pi}{48.15}[x - 9630(0.019)]\right\} = -1.25 \sin\left[\frac{\pi}{48.15}(182.97 - x)\right]$$

From this equation, one finds thus that the stress induced by the incident wave at the pile tip is given by

$$\sigma(150, 0.019) = -1.25 \sin\left[\frac{\pi}{48.15}(182.97 - 150)\right] = -1.05 \text{ ksi}$$

Similarly, one finds that the distance from the pile top to the point where such a stress is zero, that is, where $(182.97 - x)/48.15 = 1$, is given by

$$x = 182.97 - 48.15 = 134.82 \text{ ft}$$

The reflected wave has the same shape and travels with the same velocity as the incident wave. Its type, however, depends on the boundary conditions, as discussed in Example 4.1. As the pile is being driven into soft clay, then for all practical purposes it may be considered that the pile displacements are unrestricted and assumed thus that the pile tip is a free end. Under this assumption, the reflected wave may therefore be considered to be a tensile wave. Consequently, if written in terms of the distance $x' = x - 150$, which corresponds to a distance measured from the pile tip, and the time $t' = t - 0.01558$, which is the time measured from the time at which the reflected wave first appears, the equation that gives the stresses generated by the reflected wave may be written as

$$\sigma(x', t') = 1.25 \sin\left[\frac{\pi}{48.15}(9630t' - x')\right] \text{ ksi}$$

Hence, the stresses at $t = 0.019$ s, which corresponds to $t' = 0.00342$ s, are given by

$$\sigma(x', 0.00342) = 1.25 \sin\left\{\frac{\pi}{48.15}[9630(0.00342) - x']\right\} = 1.25 \sin\left[\frac{\pi}{48.15}(32.93 - x')\right]$$

From this equation, one has thus that at $t' = 0.00342$ s the reflected wave stress at the pile tip is given by

$$\sigma(0, 0.0034) = 1.25 \sin\left[\frac{\pi}{48.15}(32.93)\right] = 1.05 \text{ ksi}$$

Similarly, it may be seen that the value of this stress is equal to zero when $x' = 32.93$ ft and that the maximum value of 1.25 occurs when $(32.93 - x')/48.15 = 1/2$, or when $x' = 8.86$ ft.

The distribution of the stresses generated by the incident and the reflected waves at $t = 0.019$ s is thus as depicted in Figure E4.2b. Note that the resultant stress at $x = 141.14$ ft (i.e., $x' = 8.86$ ft) is given by

$$\sigma(141.14, 0.019) = 1.25 - 1.25 \sin\left\{\frac{-\pi}{48.15}[141.14 - 9630(0.019)]\right\} = 1.25 - 0.50 = 0.75 \text{ ksi}$$

It is worthwhile to observe that tensile stresses may be developed during the pile-driving process.

4.3 SHEAR WAVES IN LONG SHEAR BEAM

A shear beam is a structural member in which the magnitude of its shear deformations is much greater than the magnitude of its flexural deformations. A shear beam, therefore, is mainly distorted as opposed to being bent. A typical example of a shear beam are short *shear walls*, which as a result of their large moment of inertia are difficult to bend and thus their main mode of deformation is in shear. Uniform soil deposits are another example of a system that is easier to distort than to bend. As a result, soil deposits are in some instances idealized as one-dimensional shear beams.

To derive the differential equation that describes the motion of a shear beam, consider the bar shown in Figure 4.8, and let the bar be subjected to a lateral disturbance, that is, a disturbance applied in the direction perpendicular to the axis of the beam. As in the case of the longitudinal rod discussed in the previous section, assume that the length of the bar is much longer than the dimension of its cross section, that the resultant stress is uniform over the cross section, and that the beam's cross sections remain plane at all times. Similarly, let A denote the cross-section area, G the shear modulus of elasticity, ρ the density of the bar, and $v(x, t)$ the transverse displacement of the beam at time t and at distance x from the origin of the system of coordinates. After the application of the disturbance, the shear stresses acting on the faces of a differential element of the beam are as indicated in the free-body diagram shown in Figure 4.8. By application of Newton's second law along the direction of the y-axis, the motion of the differential element may be expressed as

$$\sum F_y = ma_y \tag{4.43}$$

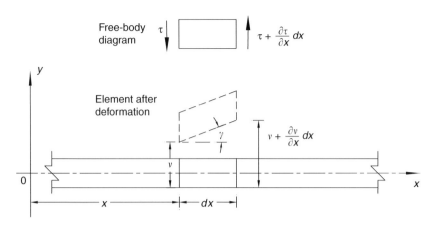

FIGURE 4.8 Long shear beam and free-body diagram of differential element.

where m is the mass of the element, a_y its acceleration along the y direction, and ΣF_y the forces acting on the element and also along the y direction. However, in terms of stresses and displacements, this equation may also be written as

$$-\tau_x A + \left(\tau_x + \frac{\partial \tau}{\partial x} dx\right) A = (\rho A\, dx) \frac{\partial^2 v(x,t)}{\partial t^2} \tag{4.44}$$

Furthermore, it may be seen from Figure 4.8 that the shear strain in the element is given by

$$\gamma = \frac{\partial v(x,t)}{\partial x} \tag{4.45}$$

and that by virtue of Hooke's law for shearing stresses one has that

$$\tau_x = G\gamma = G\frac{\partial v(x,t)}{\partial x} \tag{4.46}$$

Consequently, Equation 4.44 may also be put into the form

$$\frac{\partial^2 v(x,t)}{\partial t^2} = v_s^2 \frac{\partial^2 v(x,t)}{\partial x^2} \tag{4.47}$$

where v_s is a constant defined as

$$v_s = \sqrt{\frac{G}{\rho}} \tag{4.48}$$

A comparison of Equation 4.47 with Equation 4.4 reveals that these two equations are of the same form. This means that the motion introduced by a lateral disturbance in a shear beam is also governed by the one-dimensional wave equation and has the same type of solution. That is, a solution of the form

$$v(x,t) = F(x - v_s t) + G(x + v_s t) \tag{4.49}$$

As a result, the initial lateral disturbance is propagated along the length of the beam in the form of waves propagating with a velocity v_s. Note, however, there are some differences between these

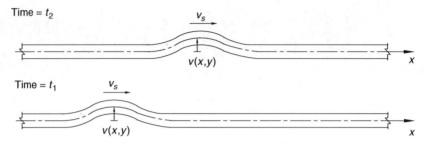

FIGURE 4.9 Position of shear wave and corresponding particle velocities at two different times.

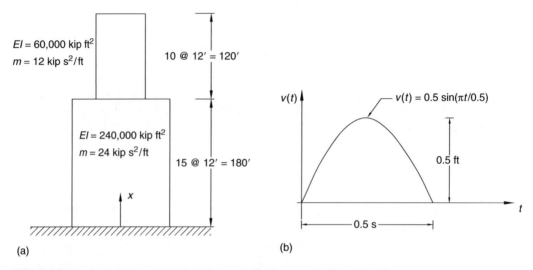

FIGURE E4.3 (a) Building and (b) initial ground displacement in Example 4.3.

waves and the waves discussed in the previous section. First, the direction of the disturbance in the shear beam is perpendicular to the direction along which the waves propagate (see Figure 4.9). In other words, the particle velocity is perpendicular to the direction of wave propagation. Second, the velocity of propagation is different. In the case of the shear beam, this velocity depends on the shear modulus of elasticity as opposed to Young's modulus in the case of the longitudinal waves. Finally, because $G = E/[2(1 + \mu)]$, where μ denotes Poisson ratio, G is always less than E, and v_s is always less than v_c. That is, waves in a shear beam travel at a lower speed than the waves in a longitudinal rod.

Example 4.3 Shear force in a building subjected to ground displacement pulse

A 25 story building, with a setback and the properties indicated in Figure E4.3a, is subjected to the lateral ground displacement pulse shown in Figure E4.3b. Modeling the building as a shear beam, obtain the shear force at the level just below the setback 2 s after the application of the pulse. The values of m and EI given in the figure, respectively, denote story mass and the rigidity of all the columns within each story.

Solution

The ground displacement pulse will generate a wave that will travel upward along the height of the building. Because of the existing discontinuity, this wave will be transmitted and reflected when it

Earthquake Propagation

reaches such a discontinuity. In addition, the transmitted wave will also be reflected when it reaches the top of the building. To solve this problem, one needs, thus, to know the wave velocities in the two sections of the building, the relative positions of the incident, transmitted, and reflected waves at the time of 2 s, and the relative amplitude of the displacements generated by these three waves.

To determine the wave velocities, one can consider that (a) the lateral rigidity of a prismatic member with flexural rigidity EI, length L, and fixed at both ends is equal to $12EI/L^3$; (b) this lateral rigidity is equal to V/Δ, where V is the applied shear force and Δ the induced deformation; and (c) an equivalent shear modulus of elasticity for the member is accordingly given by

$$G = \frac{\tau}{\gamma} = \frac{V/A}{\Delta/L} = \frac{12EI}{AL^2}$$

where τ denotes shear stress, γ shear strain, A the member's cross-section area, and L its length. Using this formula and considering that mass in terms of density ρ is equal to ρAL, the shear wave velocity in the lower and upper sections of the building are thus, respectively, equal to

$$v_{s1} = \sqrt{\frac{12(240,000)}{12(24)}} = 100 \text{ ft/s}$$

$$v_{s2} = \sqrt{\frac{12(60,000)}{12(12)}} = 70.71 \text{ ft/s}$$

On the basis of these wave velocities, it may be seen that it takes for the initial incident wave 1.8 s to reach the level of the building's setback and that it takes for the reflected wave 1.8 s to reach the ground. Consequently, at a time of 2 s, the wave has been reflected, and the reflected wave is somewhere between the setback and the ground. It may also be seen that the incident and the reflected waves have each a length of $100(0.5) = 50$ ft, which is the distance traveled by the displacement pulse during the duration of the pulse. On the basis of this information, it may be concluded that at a time of 2 s the shear force at the setback level is given by the superposition of the incident and the reflected waves.

The relative amplitude between the incident and the reflected waves may be determined by making use of equations that enforce the compatibility of forces and displacements at the level of the discontinuity as it was done for longitudinal waves in a long rod. It may be shown, however, that for shear beams these compatibility equations also lead to equations of the form 4.36, 4.41, and 4.42, except that in the case of shear beam the parameter α is a function of the shear modulus of elasticity G and shear wave velocities v_{s1} and v_{s2} as opposed to the modulus of elasticity E and shear wave velocities v_{c1} and v_{c2}. For this problem, one has thus

$$\alpha = \frac{100}{70.71} \frac{A_2}{A_1} \frac{12(6\times 10^4)/(12^2 A_2)}{12(24\times 10^4)/(12^2 A_1)} = 0.354$$

$$v_r = -\frac{0.354 - 1}{0.354 + 1} = 0.477 v_i$$

where v_i and v_r, respectively, represent the amplitudes of the incident and the reflected waves.

Now, because the shear force at any level is given by

$$V(x,t) = A\tau = AG\frac{\partial v(x,t)}{\partial x} = \frac{12EI}{L^2}\frac{\partial v(x,t)}{\partial x}$$

the computation of the desired shear force requires the derivation of equations for the lateral displacements generated by the incident and the reflected waves and the calculation of the corresponding first derivatives with respect to x, where x is the coordinate along the height of the building with origin at ground level. Proceeding as in Example 4.2, one can then derive these equations as follows.

Consider first the incident wave. At $x = 0$, the equation for the incident wave is given by

$$v_i(0, t) = 0.5\sin\left(\frac{\pi t}{0.5}\right) = -0.5\sin\left[\frac{-\pi v_{s1} t}{(0.5)(100)}\right] = -0.5\sin\left(\frac{-\pi v_{s1} t}{50}\right)$$

Hence

$$v_i(x, t) = -0.5\sin\left[\frac{\pi}{50}(x - v_{s1} t)\right]$$

and

$$\frac{\partial v_i(x, t)}{\partial x} = -\frac{\pi}{100}\cos\left[\frac{\pi}{50}(x - v_{s1} t)\right]$$

At $x = 180$ ft and $t = 2$ s, one has thus that the shear force due to the incident wave is given by

$$V_i = -\frac{12(240{,}000)}{12^2}\frac{\pi}{100}\cos\left\{\frac{\pi}{50}[180 - 100(2.0)]\right\} = -194.2 \text{ k}$$

Consider now the reflected wave. As the amplitude of the reflected wave is 0.477 times that of the incident one, then for the reflected wave one can write

$$v_r(x', t') = -0.239\sin\left[\frac{\pi}{50}(x' - v_{s1} t')\right]$$

and

$$\frac{\partial v_r(x, t)}{\partial x} = -\frac{0.239\pi}{50}\cos\left[\frac{\pi}{50}(x' - v_{s1} t')\right]$$

where $x' = x - 180$ is a new coordinate with origin at the level of the setback and $t' = t - 1.8$ a new time variable that becomes zero when the incident wave reaches the setback. Thus, at $x' = 0$ and $t' = 2 - 1.8 = 0.2$ s, the shear force due to the reflected wave results as

$$V_r = -\frac{12(240{,}000)}{12^2}\frac{0.239\pi}{50}\cos\left\{\frac{\pi}{50}[0 - 100(0.2)]\right\} = -92.8 \text{ k}$$

The resultant shear force just below the level of the setback is therefore given by

$$V = V_i + V_r = -194.2 - 92.8 = -287 \text{ k}$$

4.4 HARMONIC WAVES

4.4.1 Definition

A wave of special interest in seismology and earthquake engineering is the harmonic wave. As implied by its name, a harmonic wave is a wave whose amplitude varies with time and distance according to a sine or a cosine function. For example, a harmonic displacement wave is described by

$$u(x, t) = \sin(kx - \omega t) \qquad (4.50)$$

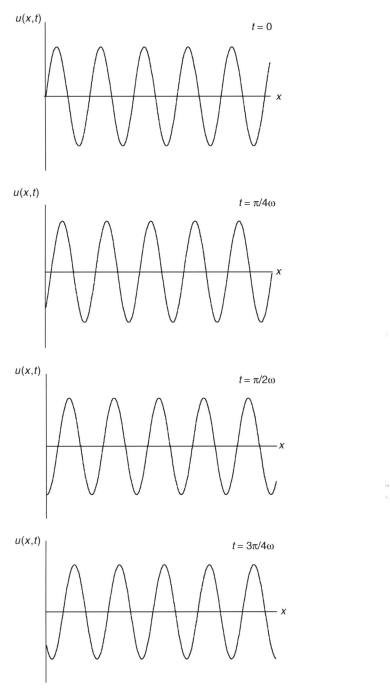

FIGURE 4.10 Variation of harmonic wave amplitude with time and distance.

where $u(x, t)$ denotes displacement at time t and distance x from the origin of the system of coordinates, and k and ω are known constants. The amplitude of a harmonic wave varies with time and distance as shown in Figure 4.10.

The importance of harmonic waves in seismology and earthquake engineering lies on the fact that through a Fourier series decomposition (see Section 6.6) any elastic wave can be represented by a superposition of harmonic waves. Therefore, many of the concepts in wave propagation theory are developed specifically for harmonic waves, with the implicit understanding that the general case may be obtained by the superposition of an infinite number of harmonic waves.

4.4.2 Characteristics

Harmonic waves possess several special characteristics. To describe these characteristics, consider the displacement wave described by Equation 4.50. Note that for $t = 0$, this equation is reduced to

$$u(x, 0) = \sin(kx) \tag{4.51}$$

which when plotted leads to the curve shown in Figure 4.11. It may be seen thus that a harmonic wave reaches its maximum amplitude repeatedly each time it travels a distance λ. This distance is a characteristic of each particular wave and is called *wavelength*. It may be seen too that the argument of the sine function varies from 0 to 2π every time the wave travels a wavelength. That is, the argument of the sine function is equal to 2π whenever $x = \lambda$. Consequently, for a harmonic wave one has that

$$k\lambda = 2\pi \tag{4.52}$$

from which it follows that

$$k = 2\pi/\lambda \tag{4.53}$$

Thus, if it is considered that 2π radians are equal to one cycle, it may be seen that the constant k in Equation 4.50 represents the number of cycles completed every time the wave travels a unit length. This constant is called *wave number*. Note that if expressed in units of cycles per unit length as opposed to radians per unit length, k becomes equal to the inverse of the wavelength.

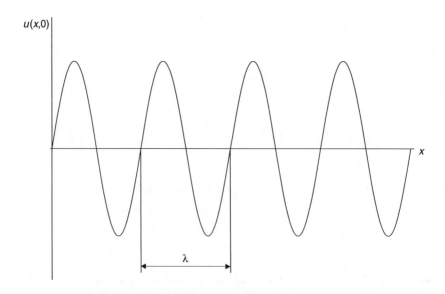

FIGURE 4.11 Variation of harmonic wave amplitude with distance at time $t = 0$.

Earthquake Propagation

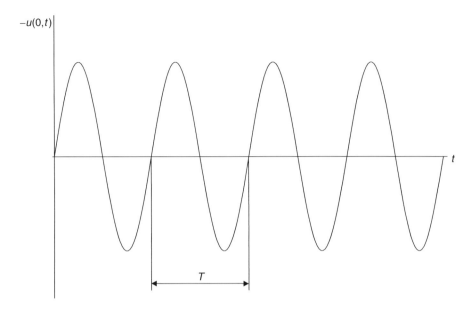

FIGURE 4.12 Variation of harmonic wave amplitude with time at distance $x = 0$.

Consider now the displacement wave defined by Equation 4.50 at $x = 0$, in which case the equation is reduced to

$$u(0,t) = \sin(-\omega t) = -\sin(\omega t) \tag{4.54}$$

which when plotted yields the curve shown in Figure 4.12. It may be seen from this figure that the wave repeatedly reaches its maximum amplitude every T seconds. T is thus called the *period* of the wave and represents the time it takes for the wave to complete a full cycle. As with the wavelength, T is a characteristic of each particular wave. As every time the wave completes a full cycle, the argument of the sine function varies from 0 to 2π; it may be seen from Equation 4.54 that

$$\omega T = 2\pi \tag{4.55}$$

from which one may conclude that the constant ω in Equation 4.50 is given by

$$\omega = \frac{2\pi}{T} \tag{4.56}$$

and represents the number of cycles completed by the wave per unit time. This constant is called the *circular frequency* of the wave and is expressed in radians per second. If a frequency f is defined as the number of cycles completed by the wave every second, then one has that

$$f = \frac{1}{T} = \frac{\omega}{2\pi} \tag{4.57}$$

where f is expressed in cycles per second or hertz.

Finally, note that by substitution of Equation 4.50 into Equation 4.4, the wave equation for longitudinal displacements, one finds that $\omega = v_c k$. Therefore, the velocity of propagation of harmonic waves is given by

$$v_c = \frac{\lambda}{T} = \frac{\lambda \omega}{2\pi} = \lambda f = \frac{\omega}{k} \tag{4.58}$$

Also note that by virtue of Equations 4.53 and 4.58 a harmonic wave may be alternatively expressed as

$$u(x, t) = \sin(kx - \omega t) = \sin k(x - v_c t) = \sin \frac{2\pi}{\lambda}(x - v_c t) \tag{4.59}$$

4.5 WAVES IN LONG FLEXURAL BEAM: WAVE DISPERSION

4.5.1 Differential Equation of Motion

Consider the beam shown in Figure 4.13 after it has been subjected to a sudden lateral disturbance. As in the case of the long rod and shear beam discussed in the foregoing sections, assume that the length of the beam is much longer than the dimensions of its cross section. In addition, assume that its flexural deformations are much more important than its shear deformations; that is, assume that its shear deformations are negligibly small. Let A denote its cross-section area, I its moment of inertia with respect to an axis perpendicular to the x-axis, E Young's modulus of elasticity, and ρ the beam's density. Similarly, let $v(x, t)$ and $\theta(x, t)$, respectively, represent the lateral displacement of the beam and the rotation of its cross section at time t and distance x from the origin of the system of coordinates. If Newton's second law for forces and moments is applied to the differential element shown in Figure 4.13, then the motion of the element may be described by

$$\sum F_y = m a_y \tag{4.60}$$

$$\sum M_0 = I_m \alpha \tag{4.61}$$

where m is the mass of the element, I_m its mass moment of inertia, a_y its translational acceleration along the direction of the y-axis, α its rotational acceleration, and ΣF_y and ΣM_0, respectively, the resultants of all the forces and all the moments about point O acting on the element. Note, however, that if written explicitly in terms of the translational and angular displacements and the forces and moments that act on the element, and if it is considering that $I_m = \rho J \, dx$, where J denotes the polar moment of inertia of the element's cross section, these equations may also be expressed as

$$\frac{\partial V}{\partial x} dx = \rho A \frac{\partial^2 v(x, t)}{\partial t^2} dx \tag{4.62}$$

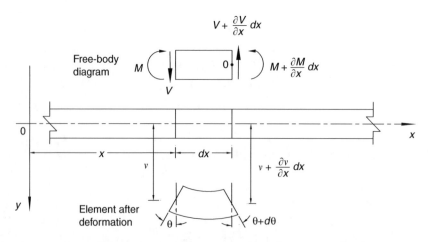

FIGURE 4.13 Long flexural beam and free-body diagram of infinitesimal element.

$$\frac{\partial M}{\partial x} dx - V\,dx = \rho J \frac{\partial^2 \theta}{\partial t^2} dx \qquad (4.63)$$

Furthermore, by assuming that the beam's rotational accelerations are negligibly small (i.e., neglecting the beam's rotational inertia), and because from the elementary theory of strength of materials one has that

$$M = -EI \frac{\partial^2 v(x,\,t)}{\partial x^2} \qquad (4.64)$$

it is possible to write Equation 4.63 approximately as

$$V = \frac{\partial M}{\partial x} = -EI \frac{\partial^3 v(x,\,t)}{\partial x^3} \qquad (4.65)$$

which in turn allows one to express the first derivative of V with respect to x as

$$\frac{\partial V}{\partial x} = -EI \frac{\partial^4 v(x,\,t)}{\partial x^4} \qquad (4.66)$$

Substitution of this last equation into Equation 4.62 thus leads to the following differential equation in terms of the lateral displacement $v(x,\,t)$ alone:

$$\frac{\partial^4 v}{\partial x^4} + \frac{1}{a^2}\frac{\partial^2 v(x,\,t)}{\partial t^2} = 0 \qquad (4.67)$$

where

$$a = \sqrt{\frac{EI}{\rho A}} \qquad (4.68)$$

4.5.2 Solution to Differential Equation of Motion

It may be seen from the inspection of Equation 4.67 that the vibrations in a long flexural beam are not governed by the wave equation. This means that the solution of the differential equation that describes the vibrations in a flexural beam cannot be expressed in terms of a disturbance that propagates with a constant velocity and without changing shape. In other words, this differential equation is not satisfied by a solution of the form

$$v(x,\,t) = F(x - v_f t) \qquad (4.69)$$

Although, in general, the solution to the equation of motion for a flexural beam cannot be represented by a traveling wave, there is, nevertheless, a particular type of wave that satisfies such an equation. This particular type of wave is the harmonic wave. However, the harmonic wave that satisfies the differential equation in question is not an ordinary one. It is, rather, one that possesses a peculiar characteristic.

To demonstrate that a harmonic wave indeed satisfies Equation 4.67 and find what such a characteristic is, consider a harmonic wave of the form

$$v(x,\,t) = \sin\left[\frac{2\pi}{\lambda}(x - v_f t)\right] \qquad (4.70)$$

When the lateral displacement $v(x, t)$ is given by this equation, its fourth derivative with respect to x and second derivative with respect to t result as

$$\frac{\partial^4 v}{\partial x^4} = \left(\frac{2\pi}{\lambda}\right)^4 \sin\left[\frac{2\pi}{\lambda}(x - v_f t)\right] \tag{4.71}$$

$$\frac{\partial^2 v}{\partial t^2} = -\left(\frac{2\pi}{\lambda}\right)^2 v_f^2 \sin\left[\frac{2\pi}{\lambda}(x - v_f t)\right] \tag{4.72}$$

Hence, upon substitution of these two equations into Equation 4.67, one gets

$$\left(\frac{2\pi}{\lambda}\right)^4 \left\{\sin\left[\frac{2\pi}{\lambda}(x - v_f t)\right] - \frac{v_f^2}{a^2}\left(\frac{\lambda}{2\pi}\right)^2 \sin\left[\frac{2\pi}{\lambda}(x - v_f t)\right]\right\} = 0 \tag{4.73}$$

from which it may be concluded that a harmonic wave satisfies the differential equation of motion for a flexural beam only if

$$\frac{v_f^2}{a^2}\left(\frac{\lambda}{2\pi}\right)^2 = 1 \tag{4.74}$$

or, what is tantamount, if the wave velocity is given by

$$v_f = \frac{2\pi}{\lambda} a = \frac{2\pi}{\lambda}\sqrt{\frac{EI}{\rho A}} \tag{4.75}$$

It may be seen, thus, that in a flexural beam a harmonic initial disturbance will propagate in the form of a wave. However, the velocity of propagation of this wave will depend not only on the properties of the propagation medium, but also on its wavelength. This means that waves with different wavelengths will travel with different velocities. This also implies that if a disturbance with an arbitrary shape is represented by a superposition of harmonic waves with different wavelengths, each of the component waves will travel with a different velocity, each will arrive at a different time at a particular site, and the shape of the original disturbance, defined by the superposition of its harmonic components, will change as the disturbance propagates away from the location where it originated. Whenever the velocity of a wave depends on its wavelength, it is said that the wave undergoes *dispersion* and that the wave is a dispersive one.

It is worthwhile to note that according to Equation 4.75, the velocity of propagation of waves with a very short wavelength, which in view of the relationship between wavelength and frequency also means waves with a very high frequency (see Equation 4.58), will tend to infinity. However, from the physical point of view, this is not possible. This anomaly arises because for wavelengths approaching the depth of a beam's cross section, rotatory inertia and shear deformation effects cannot be neglected and the elementary beam theory used in the derivation of Equation 4.75 is no longer valid.

4.5.3 Apparent Waves and Group Velocity

Wave dispersion gives rise to a reinforcement and an interference of waves in a way that may lead to the formation of an apparent wave that travels with a velocity that is different from the velocities of the component waves. To gain an idea of the effect of dispersion in this context, consider as an example two harmonic waves defined by

$$v_1(x, t) = \sin\left[\frac{2\pi}{\lambda_1}(x - v_{f1} t)\right] = \sin(k_1 x - \omega_1 t) \tag{4.76}$$

Earthquake Propagation

$$v_2(x, t) = \sin\left[\frac{2\pi}{\lambda_2}(x - v_{f2}t)\right] = \sin(k_2 x - \omega_2 t) \tag{4.77}$$

where λ_1, λ_2; ω_1, ω_2; k_1, k_2; v_{f1}, v_{f2}; respectively, denote the wavelengths; frequencies; wave numbers; and wave velocities of such two waves. Consider, in addition, the combined effect of these two waves, which mathematically may be expressed as

$$v(x, t) = v_1(x, t) + v_2(x, t) = \sin(k_1 x - \omega_1 t) + \sin(k_2 x - \omega_2 t) \tag{4.78}$$

which, in turn, by virtue of the trigonometric identity

$$\sin A + \sin B = 2 \sin\frac{A+B}{2} \cos\frac{A-B}{2} \tag{4.79}$$

may also be written as

$$v(x, t) = 2 \sin\left(\frac{k_1 + k_2}{2} x - \frac{\omega_1 + \omega_2}{2} t\right) \cos\left(\frac{k_1 - k_2}{2} x - \frac{\omega_1 - \omega_2}{2} t\right) \tag{4.80}$$

If the relationship between the wave numbers and the relationship between the frequencies of the two individual waves are such that

$$k_2 = k_1 + \Delta k \tag{4.81}$$

$$\omega_2 = \omega_1 + \Delta\omega \tag{4.82}$$

where Δk and $\Delta\omega$, respectively, represent small wave number and frequency increments, then one has that

$$\frac{k_1 + k_2}{2} = \frac{2k_1 + \Delta k}{2} \cong k_1 \tag{4.83}$$

$$\frac{\omega_1 + \omega_2}{2} = \frac{2\omega_1 + \Delta\omega}{2} \cong \omega_1 \tag{4.84}$$

As a result, Equation 4.80 may be written approximately as

$$v(x, t) = 2 \sin(k_1 x - \omega_1 t) \cos\left(\frac{\Delta k}{2} x - \frac{\Delta\omega}{2} t\right) \tag{4.85}$$

Furthermore, if this equation is rewritten as

$$v(x, t) = A(x, t) \sin(k_1 x - \omega_1 t) \tag{4.86}$$

where

$$A(x, t) = 2 \cos\left(\frac{\Delta k}{2} x - \frac{\Delta\omega}{2} t\right) \tag{4.87}$$

then it may be seen that the superposition of such two waves may be interpreted as a single wave with wave number k_1, frequency ω_1, and an amplitude that varies with time and distance according to the function that defines $A(x, t)$. This interpretation is shown graphically in Figure 4.14 for a time

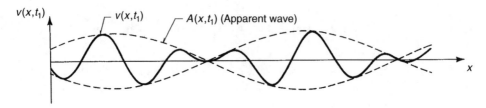

FIGURE 4.14 Apparent wave formed by superposition of two similar harmonic waves.

$t = t_1$. Observe from this figure that although a nonharmonic disturbance* does not propagate in the form of a wave, the cyclic variation in the amplitude of the propagated disturbance may give the appearance of a traveling wave with an apparent wavelength. It may also be observed that the velocity of propagation of the apparent wave—in this case equal to $\Delta\omega/\Delta k$—is different from the propagation velocities of the superimposed harmonic components. To distinguish these velocities from one another, the velocity of the apparent wave is referred to as the *group velocity* and the velocity of the superimposed waves as the *phase velocity*. Phase is the name commonly used to designate the harmonic components of a nonharmonic disturbance.

4.6 ELASTIC WAVES IN UNBOUNDED THREE-DIMENSIONAL MEDIUM: P AND S WAVES

4.6.1 INTRODUCTION

Waves propagate in a three-dimensional medium in much the same way as they do in the one-dimensional bars and beams treated in the preceding sections. Consequently, the equations that describe the propagation of waves in a three-dimensional medium may be formulated in a manner similar to that used in such one-dimensional cases, that is, by making use of Newton's second law and the relationships between stresses and strains and strains and displacements. The major difference is that in the three-dimensional case three equations of motion need to be considered (one along each of the medium's three orthogonal directions) as opposed to only one in the one-dimensional cases and that the relationships between stresses and strains and strains and displacements involve more variables than those considered in the one-dimensional problems. In this section, therefore, an extension of the concepts presented in the foregoing sections is made to describe how a localized disturbance propagates in the particular case of an unbounded elastic three-dimensional medium. Accordingly, this section presents the derivation of the equations of motion for such a case, the solution to these equations, and the conclusions that may be drawn from the interpretation of this solution. For completeness, however, a brief review is made first of the notation used in the description of stresses in a three-dimensional medium, as well as of the equations that describe the aforementioned relationships between stresses and strains and strains and displacements.

4.6.2 STRESSES AND STRAINS IN THREE-DIMENSIONAL SOLID

In the most general case, a point within a stressed solid is subjected to nine components of stress. As shown in Figure 4.15 for an infinitesimally small element with dimensions dx, dy, and dz along the direction of the x, y, and z axes in a Cartesian system of coordinates, three of these components are normal stresses whereas the other six are shearing stresses. The normal stresses acting along the direction of the x, y, and z axes are denoted by the symbols σ_x, σ_y, and σ_z, respectively. The shearing

---

* The superposition of two harmonic waves with different wavelengths and periods yields a nonharmonic wave.

Earthquake Propagation

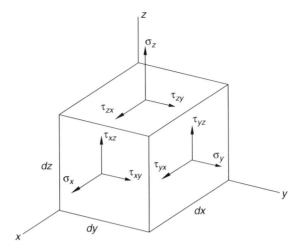

FIGURE 4.15 Normal and shear stresses in differential element.

stresses are denoted by the symbols τ_{xy}, τ_{yx}, τ_{yz}, τ_{zy}, τ_{zx}, and τ_{xz}, where the first subscript indicates the coordinate axis normal to the plane on which the stress acts, whereas the second indicates the coordinate axis to which the stress is parallel. Thus, for example, τ_{zy} is a shear stress acting on a plane perpendicular to the z-axis along the direction of the y-axis. By establishing the moment equilibrium of the element in Figure 4.15, it can be easily shown that

$$\tau_{xy} = \tau_{yx} \qquad \tau_{xz} = \tau_{zx} \qquad \tau_{yz} = \tau_{zy} \tag{4.88}$$

from which it may be seen that only six independent components of stress are required to define the element's state of stress.

A point in a stressed solid is also subjected to six independent components of strain. The relationship among the normal strains ε_x, ε_y, ε_z, the shearing strains γ_{xy}, γ_{yz}, γ_{zx}, and the three components of displacement u, v, w, all, respectively, along the direction of the x, y, z axes, are given by

$$\varepsilon_x = \frac{\partial u}{\partial x} \qquad \varepsilon_y = \frac{\partial v}{\partial y} \qquad \varepsilon_z = \frac{\partial w}{\partial z} \tag{4.89}$$

$$\gamma_{xy} = \frac{\partial v}{\partial x} + \frac{\partial u}{\partial y} \qquad \gamma_{yz} = \frac{\partial w}{\partial y} + \frac{\partial v}{\partial z} \qquad \gamma_{zx} = \frac{\partial u}{\partial z} + \frac{\partial w}{\partial x} \tag{4.90}$$

Similarly, the relationship between the three components of displacement u, v, w and the element's rigid-body rotations about the x, y, and z axes are defined by

$$\bar{\omega}_x = \frac{1}{2}\left(\frac{\partial w}{\partial y} - \frac{\partial v}{\partial z}\right) \qquad \bar{\omega}_y = \frac{1}{2}\left(\frac{\partial u}{\partial z} - \frac{\partial w}{\partial x}\right) \qquad \bar{\omega}_z = \frac{1}{2}\left(\frac{\partial v}{\partial x} - \frac{\partial u}{\partial y}\right) \tag{4.91}$$

These relationships may be easily visualized by considering the two-dimensional differential element $ABCD$ shown in Figure 4.16. This element has a square shape before any deformation takes place. After its deformation, the element is displaced, distorted, and rotated into the shape $A'B'C'D'$. From the figure and the definition of strain, it may be seen that the normal strains along the direction of the x and y axes are indeed as indicated in Equation 4.89. Similarly, if shear strain is defined as the total distortion in the element as measured by the sum of the angles α and β, then it may be seen that the shear strain in the x–y plane is as given by Equation 4.90. Finally, it may be seen that

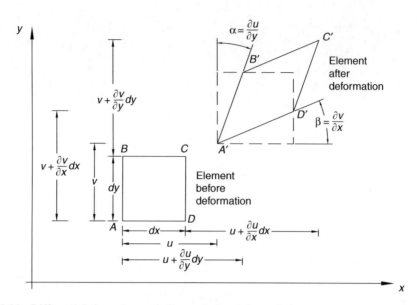

FIGURE 4.16 Differential plane element before and after deformation.

the rotation of the element about the z-axis, defined as the average of the rotation angles α and β and considering counterclockwise rotations positive and clockwise rotations negative, is indeed as given by Equation 4.91.

The relationship between stresses and strains in a three-dimensional solid is given by the generalized version of Hooke's law for elastic, homogeneous, and isotropic materials. The relationship between normal strains and normal stresses is given by

$$\varepsilon_x = \frac{1}{E}[\sigma_x - \mu(\sigma_y + \sigma_z)] \quad \varepsilon_y = \frac{1}{E}[\sigma_y - \mu(\sigma_x + \sigma_z)] \quad \varepsilon_z = \frac{1}{E}[\sigma_z - \mu(\sigma_x + \sigma_y)] \quad (4.92)$$

where E is Young's modulus and μ Poisson ratio. Similarly, the relationship between shear strains and shear stresses is given by

$$\gamma_{xy} = \frac{\tau_{xy}}{G} \quad \gamma_{yz} = \frac{\tau_{yz}}{G} \quad \gamma_{xz} = \frac{\tau_{xz}}{G} \quad (4.93)$$

where G is the shear modulus of elasticity.

Equations 4.92 may be solved to express explicitly the normal stresses in terms of the normal strains. The resulting equations are

$$\sigma_x = \lambda \bar{\varepsilon} + 2G\varepsilon_x \quad \sigma_y = \lambda \bar{\varepsilon} + 2G\varepsilon_y \quad \sigma_z = \lambda \bar{\varepsilon} + 2G\varepsilon_z \quad (4.94)$$

where

$$\lambda = \frac{\mu E}{(1+\mu)(1-2\mu)} \quad (4.95)$$

$$G = \frac{E}{2(1+\mu)} \quad (4.96)$$

$$\bar{\varepsilon} = \varepsilon_x + \varepsilon_y + \varepsilon_z \quad (4.97)$$

Earthquake Propagation

The parameters λ and G defined by Equations 4.95 and 4.96 are known as *Lame's constants*. When small deformations are involved, $\bar{\varepsilon}$ represents a *volumetric strain*, because in such a case second-order differentials may be neglected and thus

$$\frac{\Delta V}{V_0} = \frac{(dx + \varepsilon_x dx)(dx + \varepsilon_y dy)(dx + \varepsilon_z dz) - dx\,dy\,dz}{dx\,dy\,dz} \cong \varepsilon_x + \varepsilon_y + \varepsilon_z \qquad (4.98)$$

where V_0 represents the initial volume of the element and ΔV the change in volume it experiences as a result of the deformation. $\bar{\varepsilon}$ is sometimes also called *dilatation*.

4.6.3 Differential Equations of Motion

To derive the equations that describe the motion in an unbounded elastic three-dimensional medium after a disturbance is introduced at a localized point of the medium, consider a differential element with dimensions dx, dy, and dz at any arbitrary point of the medium (see Figure 4.17). The application of Newton's second law along each of the three coordinate axes gives

$$\Sigma F_x = ma_x \qquad (4.99)$$

$$\Sigma F_y = ma_y \qquad (4.100)$$

$$\Sigma F_z = ma_z \qquad (4.101)$$

where m denotes the mass of the element, ΣF_x, ΣF_y, and ΣF_z, respectively, represent the resultants of all the forces exerted on the element along its x, y, and z directions, and a_x, a_y, and a_z are the corresponding accelerations. The resultants ΣF_x, ΣF_y, and ΣF_z may be expressed, however, in terms of the normal and shearing stresses acting on the element, which for a three-dimensional element are as shown in Figure 4.18. Similarly, the accelerations a_x, a_y, and a_z may be expressed as the second-time derivatives of the displacements u, v, and w, and the mass of the element in terms of its density ρ and its volume $dV = dx\,dy\,dz$. In terms of stresses, the displacement u, and the density ρ, Equation 4.99 may be thus written as

$$\left[\left(\sigma_x + \frac{\partial \sigma_x}{\partial x}dx\right) - \sigma_x\right]dy\,dz + \left[\left(\tau_{yx} + \frac{\partial \tau_{yx}}{\partial y}dy\right) - \tau_{yx}\right]dx\,dz \\ + \left[\left(\tau_{zx} + \frac{\partial \tau_{zx}}{\partial z}dz\right) - \tau_{zx}\right]dx\,dy = \rho\,dx\,dy\,dz\,\frac{\partial^2 u}{\partial t^2} \qquad (4.102)$$

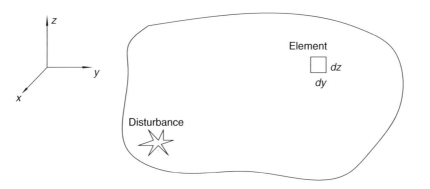

FIGURE 4.17 Differential element in unbounded medium affected by initial disturbance.

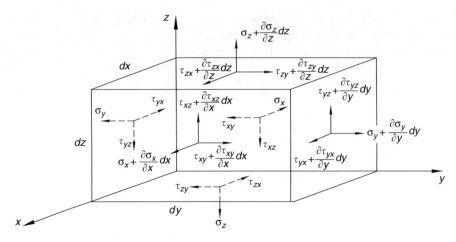

FIGURE 4.18 Free-body diagram of three-dimensional differential element.

which after its simplification may be reduced to

$$\frac{\partial \sigma_x}{\partial x} + \frac{\partial \tau_{yx}}{\partial y} + \frac{\partial \tau_{zx}}{\partial z} = \rho \frac{\partial^2 u}{\partial t^2} \quad (4.103)$$

In the same way, Equations 4.100 and 4.101 may be expressed as

$$\frac{\partial \sigma_y}{\partial y} + \frac{\partial \tau_{xy}}{\partial x} + \frac{\partial \tau_{zy}}{\partial z} = \rho \frac{\partial^2 v}{\partial t^2} \quad (4.104)$$

$$\frac{\partial \sigma_z}{\partial z} + \frac{\partial \tau_{xz}}{\partial x} + \frac{\partial \tau_{yx}}{\partial y} = \rho \frac{\partial^2 w}{\partial t^2} \quad (4.105)$$

For their solution, however, each of these equations needs to be written in terms of a single dependent variable. To this end, then, one can make use of the stress–strain and strain–displacement relationships introduced in Section 4.6.2. Accordingly, by substitution of Equations 4.93 and 4.94, and by recalling that $\tau_{yx} = \tau_{xy}$ and $\tau_{zx} = \tau_{xz}$, Equation 4.103 may also be put into the form

$$\frac{\partial}{\partial x}(\lambda \bar{\varepsilon} + 2G\varepsilon_x) + \frac{\partial}{\partial y}(G\gamma_{xy}) + \frac{\partial}{\partial z}(G\gamma_{xz}) = \rho \frac{\partial^2 u}{\partial t^2} \quad (4.106)$$

Upon substitution of the corresponding strain–displacement relationships given by Equations 4.89 and 4.90, one is then led to

$$\lambda \frac{\partial}{\partial x} \bar{\varepsilon} + 2G \frac{\partial}{\partial x}\left(\frac{\partial u}{\partial x}\right) + G \frac{\partial}{\partial y}\left(\frac{\partial v}{\partial x} + \frac{\partial u}{\partial y}\right) + G \frac{\partial}{\partial z}\left(\frac{\partial u}{\partial z} + \frac{\partial w}{\partial x}\right) = \rho \frac{\partial^2 u}{\partial t^2} \quad (4.107)$$

or to

$$\lambda \frac{\partial \bar{\varepsilon}}{\partial x} + G\left(\frac{\partial^2 u}{\partial x^2} + \frac{\partial^2 v}{\partial x \partial y} + \frac{\partial^2 w}{\partial x \partial z}\right) + G\left(\frac{\partial^2 u}{\partial x^2} + \frac{\partial^2 u}{\partial y^2} + \frac{\partial^2 u}{\partial z^2}\right) = \rho \frac{\partial^2 u}{\partial t^2} \quad (4.108)$$

But

$$\frac{\partial^2 u}{\partial x^2} + \frac{\partial^2 v}{\partial x \partial y} + \frac{\partial^2 w}{\partial x \partial z} = \frac{\partial}{\partial x}\left(\frac{\partial u}{\partial x} + \frac{\partial v}{\partial y} + \frac{\partial w}{\partial z}\right) = \frac{\partial \bar{\varepsilon}}{\partial x} \quad (4.109)$$

Earthquake Propagation

Therefore, Equation 4.108 may also be written as

$$(\lambda + G)\frac{\partial \bar{\varepsilon}}{\partial x} + G\left(\frac{\partial^2 u}{\partial x^2} + \frac{\partial^2 u}{\partial y^2} + \frac{\partial^2 u}{\partial z^2}\right) = \rho \frac{\partial^2 u}{\partial t^2} \qquad (4.110)$$

or, in compact form, as

$$(\lambda + G)\frac{\partial \bar{\varepsilon}}{\partial x} + G\nabla^2 u = \rho \frac{\partial^2 u}{\partial t^2} \qquad (4.111)$$

where ∇^2 is an operator defined as

$$\nabla^2 = \frac{\partial^2}{\partial x^2} + \frac{\partial^2}{\partial y^2} + \frac{\partial}{\partial z^2} \qquad (4.112)$$

and known as the Laplacian operator.*

Proceeding similarly with Equations 4.104 and 4.105, one arrives to the following corresponding equations for the y and z directions:

$$(\lambda + G)\frac{\partial \bar{\varepsilon}}{\partial y} + G\nabla^2 v = \rho \frac{\partial^2 v}{\partial t^2} \qquad (4.113)$$

$$(\lambda + G)\frac{\partial \bar{\varepsilon}}{\partial z} + G\nabla^2 w = \rho \frac{\partial^2 w}{\partial t^2} \qquad (4.114)$$

4.6.4 First Particular Solution of Differential Equations of Motion

Equations 4.111, 4.113, and 4.114 represent in their classical form the differential equations that govern the motion of an unbounded elastic medium. It should be noted, however, that in this form the equations are still not in terms of a single dependent variable as the volumetric strain $\bar{\varepsilon}$ is a function of the displacements u, v, and w. Thus, their solution requires further mathematical manipulation.

There are two ways by means of which one can convert the aforementioned differential equations into equations in terms of a single dependent variable, each allowing the determination of a particular solution. The first involves differentiating Equations 4.111, 4.113, and 4.114, respectively, with respect to x, y, and z, and adding afterward the three resulting equations, as these operations lead to

$$(\lambda + G)\left(\frac{\partial^2 \bar{\varepsilon}}{\partial x^2} + \frac{\partial^2 \bar{\varepsilon}}{\partial y^2} + \frac{\partial^2 \bar{\varepsilon}}{\partial z^2}\right) + G\nabla^2\left(\frac{\partial u}{\partial x} + \frac{\partial v}{\partial y} + \frac{\partial w}{\partial z}\right) = \rho \frac{\partial^2}{\partial t^2}\left(\frac{\partial u}{\partial x} + \frac{\partial v}{\partial y} + \frac{\partial w}{\partial z}\right) \qquad (4.115)$$

which, in view of the definition of volumetric strain (see Equation 4.97), may also be written as

$$(\lambda + 2G)\nabla^2 \bar{\varepsilon} = \rho \frac{\partial^2 \bar{\varepsilon}}{\partial t^2} \qquad (4.116)$$

or as

$$\frac{\partial^2 \bar{\varepsilon}}{\partial t^2} = v_p^2 \nabla^2 \bar{\varepsilon} \qquad (4.117)$$

---

* In some texts, the symbol ∇ is identified as *nabla*, and thus the operator ∇^2 is sometimes also called *nabla square*.

which is an equation where $\bar{\varepsilon}$, the volumetric strain, is the only dependent variable. Note that in Equation 4.117 v_p is a constant defined as

$$v_p = \sqrt{\frac{\lambda + 2G}{\rho}} \tag{4.118}$$

An alternative way to derive Equation 4.117 is by setting the three components of rotation equal to zero. That is, by making

$$\bar{\omega}_x = \bar{\omega}_y = \bar{\omega}_z = 0 \tag{4.119}$$

where these components of rotations are defined by Equation 4.91. To demonstrate that this condition indeed leads to Equation 4.117, consider first that when the three components of rotation are zero, the displacements u, v, and w can be expressed in terms of a potential function ϕ as

$$u = \frac{\partial \phi}{\partial x} \qquad v = \frac{\partial \phi}{\partial y} \qquad w = \frac{\partial \phi}{\partial z} \tag{4.120}$$

because in such a case one has that

$$\bar{\omega}_x = \frac{1}{2}\left(\frac{\partial w}{\partial y} - \frac{\partial v}{\partial z}\right) = \frac{1}{2}\left(\frac{\partial^2 \phi}{\partial y \partial z} - \frac{\partial^2 \phi}{\partial y \partial z}\right) = 0 \tag{4.121}$$

and similarly for $\bar{\omega}_y$ and $\bar{\omega}_z$. Consider, then, that if the displacements u, v, and w are expressed in terms of the potential function ϕ, the volumetric strain results as

$$\bar{\varepsilon} = \frac{\partial u}{\partial x} + \frac{\partial v}{\partial y} + \frac{\partial w}{\partial z} = \frac{\partial^2 \phi}{\partial x^2} + \frac{\partial^2 \phi}{\partial y^2} + \frac{\partial^2 \phi}{\partial z^2} = \nabla^2 \phi \tag{4.122}$$

and its first derivatives with respect to x, y, and z as

$$\frac{\partial \bar{\varepsilon}}{\partial x} = \frac{\partial^2 (\partial \phi / \partial x)}{\partial x^2} + \frac{\partial^2 (\partial \phi / \partial x)}{\partial y^2} + \frac{\partial^2 (\partial \phi / \partial x)}{\partial z^2} = \nabla^2 u \tag{4.123}$$

$$\frac{\partial \bar{\varepsilon}}{\partial y} = \frac{\partial^2 (\partial \phi / \partial y)}{\partial x^2} + \frac{\partial^2 (\partial \phi / \partial y)}{\partial y^2} + \frac{\partial^2 (\partial \phi / \partial y)}{\partial z^2} = \nabla^2 v \tag{4.124}$$

$$\frac{\partial \bar{\varepsilon}}{\partial z} = \frac{\partial^2 (\partial \phi / \partial z)}{\partial x^2} + \frac{\partial^2 (\partial \phi / \partial z)}{\partial y^2} + \frac{\partial^2 (\partial \phi / \partial z)}{\partial z^2} = \nabla^2 w \tag{4.125}$$

Hence, by substitution of Equations 4.123 through 4.125 into Equations 4.111, 4.113, and 4.114 and consideration of Equation 4.118, one obtains

$$\frac{\partial^2 u}{\partial t^2} = v_p^2 \nabla^2 u \tag{4.126}$$

$$\frac{\partial^2 v}{\partial t^2} = v_p^2 \nabla^2 v \tag{4.127}$$

$$\frac{\partial^2 w}{\partial t^2} = v_p^2 \nabla^2 w \tag{4.128}$$

Earthquake Propagation

Finally, take the derivative of Equations 4.126 through 4.128, respectively, with respect to x, y, and z, and add the three resulting equations. The result is Equation 4.117, which is precisely what one wanted to demonstrate.

To facilitate the interpretation and the solution of Equations 4.126 through 4.128, it is convenient to transform them to spherical coordinates. For this purpose, consider the transformation

$$r^2 = x^2 + y^2 + z^2 \tag{4.129}$$

where r is the radial distance measured from the origin of the system of coordinates. Consider, then, that after differentiating both sides of this equation with respect to x, one arrives to

$$\frac{\partial r}{\partial x} = \frac{x}{r} \tag{4.130}$$

which, by making use of the chain rule, in turn allows one to write $\partial^2 u/\partial x^2$ in terms of the spherical coordinate r as

$$\frac{\partial^2 u}{\partial x^2} = \frac{\partial}{\partial x}\left(\frac{\partial u}{\partial x}\right) = \frac{\partial}{\partial x}\left(\frac{\partial u}{\partial r}\frac{\partial r}{\partial x}\right) = \frac{\partial}{\partial x}\left(\frac{\partial u}{\partial r}\frac{x}{r}\right) = \frac{x}{r}\frac{\partial^2 u}{\partial x \partial r} + \frac{\partial u}{\partial r}\frac{\partial}{\partial x}\left(\frac{x}{r}\right) = \frac{x}{r}\frac{\partial}{\partial r}\left(\frac{\partial u}{\partial x}\right) + \frac{\partial u}{\partial r}\frac{r - x(\partial r/\partial x)}{r^2}$$

$$= \frac{x}{r}\frac{\partial}{\partial r}\left(\frac{\partial u}{\partial r}\frac{\partial r}{\partial x}\right) + \frac{\partial u}{\partial r}\frac{1}{r}\frac{r^2 - x^2}{r^2} = \frac{x}{r}\left(\frac{\partial r}{\partial x}\frac{\partial^2 u}{\partial r^2} + \frac{\partial u}{\partial r}\frac{\partial^2 r}{\partial r \partial x}\right) + \frac{\partial u}{\partial r}\frac{1}{r}\frac{r^2 - x^2}{r^2} \tag{4.131}$$

or, if it is noted that

$$\frac{\partial^2 r}{\partial r \partial x} = \frac{\partial}{\partial x}\left(\frac{\partial r}{\partial r}\right) = 0 \tag{4.132}$$

as

$$\frac{\partial^2 u}{\partial x^2} = \frac{x^2}{r^2}\frac{\partial^2 u}{\partial r^2} + \frac{1}{r}\frac{r^2 - x^2}{r^2}\frac{\partial u}{\partial r} \tag{4.133}$$

Similarly, $\partial^2 u/\partial y^2$ and $\partial^2 u/\partial z^2$ may be expressed as

$$\frac{\partial^2 u}{\partial y^2} = \frac{y^2}{r^2}\frac{\partial^2 u}{\partial r^2} + \frac{1}{r}\frac{r^2 - y^2}{r^2}\frac{\partial u}{\partial r} \tag{4.134}$$

$$\frac{\partial^2 u}{\partial z^2} = \frac{z^2}{r^2}\frac{\partial^2 u}{\partial r^2} + \frac{1}{r}\frac{r^2 - z^2}{r^2}\frac{\partial u}{\partial r} \tag{4.135}$$

Thus, upon substitution of these last three equations, Equation 4.126 may be transformed into

$$\frac{\partial^2 u}{\partial t^2} = v_p^2\left[\frac{\partial^2 u}{\partial r^2}\left(\frac{x^2 + y^2 + z^2}{r^2}\right) + \frac{1}{r}\frac{\partial u}{\partial r}\frac{r^2 - x^2 + r^2 - y^2 + r^2 - z^2}{r^2}\right] = v_p^2\left(\frac{\partial^2 u}{\partial r^2} + \frac{2}{r}\frac{\partial u}{\partial r}\right) \tag{4.136}$$

which may also be written as

$$r\frac{\partial^2 u}{\partial t^2} = v_p^2\left(r\frac{\partial^2 u}{\partial r^2} + 2\frac{\partial u}{\partial r}\right) \tag{4.137}$$

or, if it is considered that

$$r\frac{\partial^2 u}{\partial r^2} + 2\frac{\partial u}{\partial r} = \frac{\partial^2(ru)}{\partial r^2} \tag{4.138}$$

as

$$\frac{\partial^2(ru)}{\partial t^2} = v_p^2 \frac{\partial^2(ru)}{\partial r^2} \qquad (4.139)$$

Likewise, it may be demonstrated that under the same transformation Equations 4.127 and 4.128 lead to

$$\frac{\partial^2(rv)}{\partial t^2} = v_p^2 \frac{\partial^2(rv)}{\partial r^2} \qquad (4.140)$$

$$\frac{\partial^2(rw)}{\partial t^2} = v_p^2 \frac{\partial^2(rw)}{\partial r^2} \qquad (4.141)$$

A comparison of Equations 4.139 through 4.141 with Equation 4.4 reveals that these differential equations have the form of the classical one-dimensional wave equation discussed in Section 4.2. Therefore, their solution is of the form

$$ru = F_{pu}(r - v_p t) + G_{pu}(r + v_p t) \qquad (4.142)$$

$$rv = F_{pv}(r - v_p t) + G_{pv}(r + v_p t) \qquad (4.143)$$

$$rw = F_{pw}(r - v_p t) + G_{pw}(r + v_p t) \qquad (4.144)$$

where, as in the one-dimensional case, F_{pu}, F_{pv}, F_{pw} and G_{pu}, G_{pv}, G_{pw}, respectively, denote arbitrary functions of $r - v_p t$ and $r + v_p t$ whose specific forms depend on the initial conditions of the problem at hand. From these equations, the displacements u, v, and w at time t and distance r from the origin of the system of coordinates are thus given by

$$u = \frac{1}{r}[F_{pu}(r - v_p t) + G_{pu}(r + v_p t)] \qquad (4.145)$$

$$v = \frac{1}{r}[F_{pv}(r - v_p t) + G_{pv}(r + v_p t)] \qquad (4.146)$$

$$w = \frac{1}{r}[F_{pw}(r - v_p t) + G_{pw}(r + v_p t)] \qquad (4.147)$$

where v_p is given by Equation 4.118.

4.6.5 Interpretation of First Particular Solution of Equations of Motion

It may be seen from Equations 4.145 through 4.147 that an initial disturbance in an unbounded elastic medium propagates equally in all directions in the form of a wave that travels with a velocity equal to the constant v_p given by Equation 4.118. It may also be inferred from these equations that the displacements generated by the wave decay in proportion to the inverse of the distance traveled by the wave. Additionally, because substitution of Equations 4.95 and 4.96 into Equation 4.118 allows one to express v_p as

$$v_p = \sqrt{\frac{E(1 - \mu)}{\rho(1 + \mu)(1 - 2\mu)}} \qquad (4.148)$$

Earthquake Propagation

it may be seen that the velocity of propagation of the wave depends on the modulus of elasticity, Poisson ratio, and the density of the medium and is equal to the velocity of propagation of longitudinal waves in a laterally constrained bar (see Equation 4.7).

Because Equations 4.145 through 4.147 are valid only for the case when the components of rotation $\bar{\omega}_x$, $\bar{\omega}_y$, and $\bar{\omega}_z$ are all equal to zero and the volumetric strain $\bar{\varepsilon}$ is different from zero, this wave is characterized by a particle motion that generates a volumetric expansion or contraction but no rotations. That is, the wave induces particle motion in the direction of propagation but not in any of the two perpendicular directions. To reflect its character, this wave is sometimes referred to as a longitudinal, dilatational, compressional, or irrotational wave, although strictly speaking it should only be called an irrotational wave. To avoid a name that may misrepresent its true character, it is now customary to call it a *primary wave* or simply a *P wave*. Figure 4.19a depicts the type of motion generated by a P wave.

4.6.6 Second Particular Solution of Differential Equations of Motion

The second way one can transform Equations 4.111, 4.113, and 4.114 into an equation in terms of a single dependent variable entails subtracting Equation 4.113 from Equation 4.114 after Equation 4.113 is differentiated with respect to z and Equation 4.114 with respect to y, a manipulation that allows one to obtain

$$(\lambda + G)\left[\frac{\partial \bar{\varepsilon}}{\partial z \partial y} - \frac{\partial \bar{\varepsilon}}{\partial y \partial z}\right] + G\nabla^2\left(\frac{\partial w}{\partial y} - \frac{\partial v}{\partial z}\right) = \rho \frac{\partial^2}{\partial t^2}\left(\frac{\partial w}{\partial y} - \frac{\partial v}{\partial z}\right) \qquad (4.149)$$

which, after simplifying and taking Equation 4.91 into account, may also be written as

$$\frac{\partial^2 \bar{\omega}_x}{\partial t^2} = v_s^2 \nabla^2 \bar{\omega}_x \qquad (4.150)$$

where $\bar{\omega}_x$ is the component of rotation about the x-axis and v_s a constant defined as

$$v_s = \sqrt{\frac{G}{\rho}} \qquad (4.151)$$

By a similar type of manipulation, it is also possible to arrive to the following two additional equations:

$$\frac{\partial^2 \bar{\omega}_y}{\partial t^2} = v_s^2 \nabla^2 \bar{\omega}_y \qquad (4.152)$$

$$\frac{\partial^2 \bar{\omega}_z}{\partial t^2} = v_s^2 \nabla^2 \bar{\omega}_z \qquad (4.153)$$

Here, $\bar{\omega}_y$ and $\bar{\omega}_z$ are the components of rotation about the y and z axes, respectively.

Equations 4.150, 4.152, and 4.153 are expressed in terms of a single variable and may be thus used to obtain explicit solutions for the components of rotation $\bar{\omega}_x$, $\bar{\omega}_y$, and $\bar{\omega}_z$. To facilitate, however, the visualization of the generated motion, it is convenient to obtain equations that are expressed explicitly in terms of the components of displacement u, v, and w. To do so, one can differentiate Equations 4.150, 4.152, and 4.153, respectively, with respect to x, y, and z, express $\bar{\omega}_x$, $\bar{\omega}_y$, and $\bar{\omega}_z$ in terms of the displacements u, v, and w, and add the first two equations and subtract the third one. In this way, one gets

$$\frac{\partial^2}{\partial t^2}\left(\frac{\partial^2 u}{\partial y \partial z}\right) = v_s^2 \nabla^2\left(\frac{\partial^2 u}{\partial y \partial z}\right) \qquad (4.154)$$

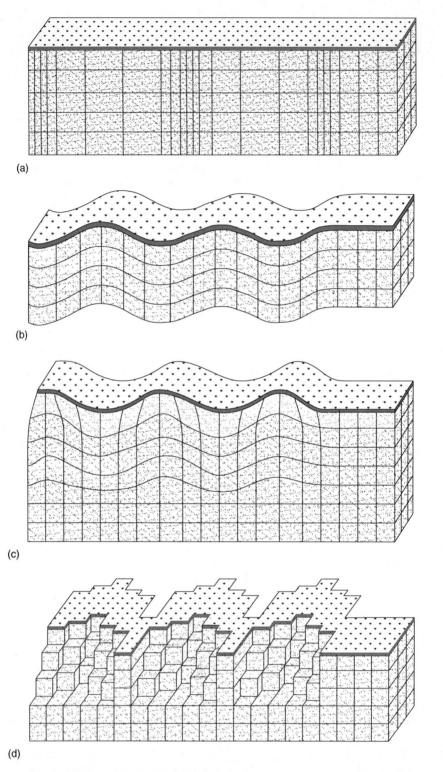

FIGURE 4.19 Motion generated by P, S, Love, and Rayleigh waves as waves propagate from left to right: (a) primary waves compress and stretch the medium, (b) secondary waves generate up-and-down and side-to-side oscillations, (c) Rayleigh waves rotate the top of the medium in an elliptic pattern like an ocean breaker, and (d) Love waves distort the top of the medium side to side.

Earthquake Propagation

which after both sides are integrated with respect to y and z may also be written as

$$\frac{\partial^2 u}{\partial t^2} = v_s^2 \nabla^2 u \qquad (4.155)$$

In the same fashion, the following two equations may be obtained:

$$\frac{\partial^2 v}{\partial t^2} = v_s^2 \nabla^2 v \qquad (4.156)$$

$$\frac{\partial^2 w}{\partial t^2} = v_s^2 \nabla^2 w \qquad (4.157)$$

Note that Equations 4.155 through 4.157 may be obtained alternatively by simply substituting $\bar{\varepsilon} = 0$ into Equations 4.111, 4.113, and 4.114, respectively, and considering that, according to Equation 4.151, $G/\rho = v_s^2$. Note too that Equations 4.155 through 4.157 are of the same form as Equations 4.126 through 4.128. Therefore, by using the procedure established in Section 4.6.4, they may also be expressed in spherical coordinates as

$$\frac{\partial^2 (ru)}{\partial t^2} = v_s^2 \frac{\partial^2 (ru)}{\partial r^2} \qquad (4.158)$$

$$\frac{\partial^2 (rv)}{\partial t^2} = v_s^2 \frac{\partial^2 (rv)}{\partial r^2} \qquad (4.159)$$

$$\frac{\partial^2 (rw)}{\partial t^2} = v_s^2 \frac{\partial^2 (rw)}{\partial r^2} \qquad (4.160)$$

which have the form of the classical wave equation. Correspondingly, their solution is given by

$$u = \frac{1}{r}[F_{su}(r - v_s t) + G_{su}(r + v_s t)] \qquad (4.161)$$

$$v = \frac{1}{r}[F_{sv}(r - v_s t) + G_{sv}(r + v_s t)] \qquad (4.162)$$

$$w = \frac{1}{r}[F_{sw}(r - v_s t) + G_{sw}(r + v_s t)] \qquad (4.163)$$

where, as before, r signifies radial distance, v_s denotes the constant defined by Equation 4.151, and F_{su}, F_{sv}, F_{sw} and G_{su}, G_{sv}, G_{sw} are, respectively, functions of $r - v_s t$ and $r + v_s t$ whose form may be determined from the initial conditions of the problem under consideration.

4.6.7 Interpretation of Second Particular Solution of Equations of Motion

Equations 4.161 through 4.163 reveal that an initial disturbance in an unbounded elastic medium also gives rise to a second type of wave and that the characteristics of this wave are different from the characteristics of the primary wave discussed in Section 4.6.5. This second type of wave also propagates in all directions and decays in proportion to the inverse of the traveled distance. It propagates, however, with a different velocity. This velocity is equal to the constant v_s defined by Equation 4.151, depends on the shear modulus of elasticity and the density of the medium, and is equal to the velocity of propagation of shear waves in a one-dimensional shear beam (see Equation 4.48). Furthermore, such a velocity is always lesser than the velocity of propagation of the primary wave as, according to Equations 4.96, 4.148, and 4.151, the relationship between the two velocities is given by

$$\frac{v_p}{v_s} = \sqrt{\frac{2(1-\mu)}{1-2\mu}} \qquad (4.164)$$

and, thus, for any possible value of μ this ratio is always greater than unity. For example, the velocity v_s for a Poisson ratio of 0.3 is equal to 0.53 times the velocity of the primary wave.

Given that Equations 4.161 through 4.163 are valid only for the case when the volumetric strain $\bar{\varepsilon}$ is equal to zero and the rotations $\bar{\omega}_x$, $\bar{\omega}_y$, and $\bar{\omega}_z$ are different from zero, it may also be concluded that this second wave generates a particle motion that involves distortions but no expansions or contractions. In other words, it generates a particle motion that introduces no volume changes. As such, it produces particle motion only in the two directions that are perpendicular to the direction of propagation. To indicate its nature and distinguish it from the primary wave, this type of wave is at times called a transverse, shear, or distortional wave. The most appropriate name, however, is equivoluminal wave because this name accurately reflects its true character. For simplicity and to point out that it is always the second one to arrive at a given site, this wave is now conventionally called *secondary wave* or just *S wave*. Figure 4.19b illustrates the motion involved with this type of wave.

For convenience, an S wave is often resolved into two perpendicular components, one along each of two directions that are perpendicular to the direction of propagation. In such a case, one of the components is called the SH wave and the other the SV wave. In the case of a plane wave* propagating along the direction of a horizontal axis, an SH wave induces particle motion that is contained in a horizontal plane. In contrast, the particle motion generated by an SV wave is contained in a vertical plane.

4.6.8 General Solution of Differential Equations of Motion

Only the two particular solutions discussed in the preceding sections satisfy the differential equations that describe the motion of an unbounded elastic medium. According to the theory of differential equations, the general solution to these differential equations is therefore given by a linear combination of such two particular solutions. In practical terms, this means that the motion induced by an initial, localized disturbance in an elastic three-dimensional medium is, in general, given by the superposition of P and S waves. It should be noted, however, that because P and S waves propagate with different velocities, the two waves quickly separate from each other as they travel away from the location where the initial disturbance originated. For this reason, the two waves are seldom superimposed and it is customary to consider them separately.

In summary, it can be stated that an initial disturbance in an unbounded elastic medium gives rise to two waves. The first type travels with a velocity $v_p = \sqrt{(\lambda + 2G)/\rho}$ and generates particle motion that involves an expansion or contraction of the medium but no rotations as the wave passes by. The second type travels with a velocity $v_s = \sqrt{G/\rho}$ and produces particle motion characterized by up-and-down and side-to-side distortions but no expansions or contractions, that is, no volume changes. Both v_p and v_s depend on the properties of the medium but are different from each other, with v_p being substantially faster than v_s. To distinguish them from the surface waves discussed in the following sections, collectively these two waves are referred to as *body waves*. Tables 4.1 and 4.2 give typical values of the P and S wave velocities in some rocks and soils. Because of the close correlation between shear wave velocity and shear modulus of elasticity, shear wave velocities are often used to characterize the stiffness of soils and rocks.

---

* A plane wave is a wave whose front is a plane as opposed to a spherical or a cylindrical surface. Waves whose front is sufficiently away from the source of the disturbance may be approximated as plane waves.

TABLE 4.1
Typical Velocities of P Waves (v_p) and S Waves (v_s) in Rock

Rock	v_p (m/s)	v_s (m/s)
Limestone	6030	3030
Granite	5640	2870
Diorite	5780	3060
Basalt	6400	3200
Gabbro	6450	3420
Dunite	8000	4370

TABLE 4.2
Typical Velocities of P Waves (v_p) and S Waves (v_s) in Soil

Soil	v_p (m/s)	v_s (m/s)
Moist clay	1500	150
Loess at natural moist	800	260
Dense sand and gravel	480	250
Fine-grained sand	300	110
Medium-grained sand	550	160
Medium size gravel	750	180

Note: $v_p = 1463$ m/s in water.

EXAMPLE 4.4 TIME LAG BETWEEN ARRIVAL OF P AND S WAVES AND OCCURRENCE OF EARTHQUAKE

The early warning system in Mexico City relies on sensors located 320 km away to warn the city that an earthquake has occurred along the west coast of Mexico. Assuming that a radio signal can be transmitted 10 s after the initiation of an earthquake and that it takes 2 s for the radio signal to reach Mexico City, determine (1) the time gap between the time an earthquake warning is received and the time the earthquake reaches Mexico City and (2) the time gap between the time the earthquake warning is received and the time the more damaging S waves arrive. Consider average P and S wave velocities of 5 and 3 km/s, respectively.

Solution

1. The radio signal reaches the city 12 s after an earthquake is detected in the west coast. The P waves, which are the first to arrive, will reach the city in 320/5 = 64 s. The time gap between the arrival of the radio signal and the arrival of the first earthquake wave is thus equal to 64 − 12 = 52 s.
2. It takes 320/3 = 106.7 s for the first S wave to reach the city. Hence, the time gap between the arrival of the first S wave and the reception of the radio signal is equal to 106.7 − 12 = 94.7 s.

4.7 ELASTIC WAVES IN ELASTIC HALF-SPACE: RAYLEIGH WAVES

4.7.1 INTRODUCTION

If people are asked to describe the sensations experienced during a moderately strong earthquake, it is likely that their description will be similar to that made once by a casual observer:

> First there was a sudden jolt that made me lose my balance for a second. Then, I could feel the ground moving, and a second, stronger jolt came. After a few seconds of shaking, a rolling and swaying motion started, like being in a boat. The swaying lasted until the earthquake ended. There was noise all the time.*

This description clearly identifies the arrival of P and S waves. The P wave is the first to arrive, and the S wave is the one that produces the second, stronger jolt. The description, however, includes an additional third wave that produces a motion—rolling and swaying—that is different from the motion generated by a P or an S wave. The arrival of such a third wave may also be identified in a record from an earthquake ground motion, such as in the one shown in Figure 4.20. This different type of wave intrigued some observers for many years, but an explanation for it could not be given until 1885, when Lord Rayleigh (John William Strutt), a British mathematical physicist, provided

---

* Quoted in Gere, J. M. and Shah, H. C., *Terra Non Firma*, W. H. Freeman and Co., New York, 1984.

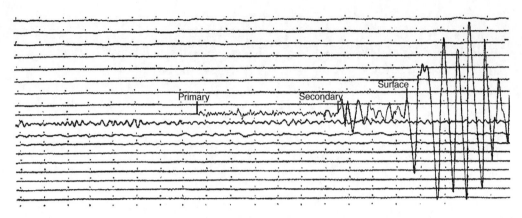

FIGURE 4.20 Ground motion recorded in Palisades, New York, generated by an earthquake in South America on May 28, 1976, showing the arrival of primary, secondary, and surface waves.

a mathematical proof of its existence. This wave can exist only in a homogeneous medium with a boundary (as opposed to an unbounded medium) and is distinguished by the fact that the particle motion it generates is confined to locations near the ground surface. It is one of the waves that are nowadays referred to as *surface waves* and is known as *Rayleigh wave*.

Rayleigh waves are of primary importance in earthquake engineering. Their importance is owed to the fact that, as it will be shown later on, Rayleigh waves attenuate with distance at a slower rate than body waves. As a result, the earthquake motions observed at the ground surface of sites away from the location where the earthquake originates are mainly due to Rayleigh waves. This section, therefore, is devoted to a thorough discussion of Rayleigh waves. It is shown first how the boundary conditions associated with a bounded medium lead to an additional solution to the equations of motion established in the previous section and that this additional solution represents a wave with characteristics that are different from the characteristics of P and S waves. Then, equations are derived to determine its velocity of propagation and the particle motion it generates. Thereafter, these equations are interpreted to characterize such velocity and particle motion. For simplicity, however, the discussion is limited to the case of plane waves.

4.7.2 Equations of Motion for Plane Waves

Consider a semi-infinite elastic medium with a plane free surface (often called *half-space*) and a plane wave traveling within this medium along the direction of the x-axis. That is, assume that the displacement component v is equal to zero and that the components u and w do not vary along the direction of the y-axis (see Figure 4.21). Consider, in addition, that the positive direction of the z-axis is downward and that all particle motion takes place in planes parallel to the x–z plane.

Assume now that the displacements u and w are of the form

$$u = \frac{\partial \Phi}{\partial x} + \frac{\partial \Psi}{\partial z} \tag{4.165}$$

$$w = \frac{\partial \Phi}{\partial z} - \frac{\partial \Psi}{\partial x} \tag{4.166}$$

where Φ and Ψ are potential functions that allow the decoupling of the equations of motion. In terms of these potential functions, the volumetric strain $\bar{\varepsilon}$ may be expressed as (see Equation 4.97)

$$\bar{\varepsilon} = \frac{\partial u}{\partial x} + \frac{\partial w}{\partial z} = \frac{\partial}{\partial x}\left(\frac{\partial \Phi}{\partial x} + \frac{\partial \Psi}{\partial z}\right) + \frac{\partial}{\partial z}\left(\frac{\partial \Phi}{\partial z} - \frac{\partial \Psi}{\partial x}\right) = \frac{\partial^2 \Phi}{\partial x^2} + \frac{\partial^2 \Phi}{\partial z^2} = \nabla^2 \Phi \tag{4.167}$$

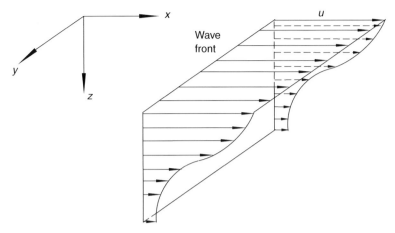

FIGURE 4.21 Front of plane wave propagating in the direction of x-axis.

where, as before, ∇^2 is the Laplacian operator defined by Equation 4.112 without the derivative with respect to y.

After substitution of Equations 4.165 through 4.167, the differential equations that describe the motion of an elastic medium along its x and z directions (i.e., Equations 4.111 and 4.114 in Section 4.6.3) may be therefore written as

$$(\lambda + G)\frac{\partial}{\partial x}(\nabla^2\Phi) + G\nabla^2\left(\frac{\partial \Phi}{\partial x} + \frac{\partial \Psi}{\partial z}\right) = \rho\frac{\partial}{\partial t^2}\left(\frac{\partial \Phi}{\partial x} + \frac{\partial \Psi}{\partial z}\right) \quad (4.168)$$

$$(\lambda + G)\frac{\partial}{\partial z}(\nabla^2\Phi) + G\nabla^2\left(\frac{\partial \Phi}{\partial z} - \frac{\partial \Psi}{\partial x}\right) = \rho\frac{\partial}{\partial t^2}\left(\frac{\partial \Phi}{\partial z} - \frac{\partial \Psi}{\partial x}\right) \quad (4.169)$$

which after rearrangement may also be put into the form

$$(\lambda + 2G)\frac{\partial}{\partial x}(\nabla^2\Phi) + G\frac{\partial}{\partial z}(\nabla^2\Psi) = \rho\frac{\partial}{\partial x}\left(\frac{\partial^2\Phi}{\partial t^2}\right) + \rho\frac{\partial}{\partial z}\left(\frac{\partial^2\Psi}{\partial t^2}\right) \quad (4.170)$$

$$(\lambda + 2G)\frac{\partial}{\partial z}(\nabla^2\Phi) - G\frac{\partial}{\partial x}(\nabla^2\Psi) = \rho\frac{\partial}{\partial z}\left(\frac{\partial^2\Phi}{\partial t^2}\right) - \rho\frac{\partial}{\partial x}\left(\frac{\partial^2\Psi}{\partial t^2}\right) \quad (4.171)$$

Differentiating now Equations 4.170 and 4.171, respectively, with respect to z and x and adding first and then subtracting the resulting equations reduce the aforementioned equations of motion to

$$(\lambda + 2G)\frac{\partial^2}{\partial x \partial z}(\nabla^2\Phi) = \rho\frac{\partial^2}{\partial x \partial z}\left(\frac{\partial^2\Phi}{\partial t^2}\right) \quad (4.172)$$

$$G\frac{\partial^2}{\partial x \partial z}(\nabla^2\Psi) = \rho\frac{\partial^2}{\partial x \partial z}\left(\frac{\partial^2\Psi}{\partial t^2}\right) \quad (4.173)$$

which after both sides are integrated with respect to x and z may also be expressed as

$$\frac{\partial^2\Phi}{\partial t^2} = v_p^2\left(\frac{\partial^2\Phi}{\partial x^2} + \frac{\partial^2\Phi}{\partial z^2}\right) \quad (4.174)$$

$$\frac{\partial^2 \Psi}{\partial t^2} = v_s^2 \left(\frac{\partial^2 \Psi}{\partial x^2} + \frac{\partial^2 \Psi}{\partial z^2} \right) \tag{4.175}$$

where, as before,

$$v_p = \sqrt{\frac{\lambda + 2G}{\rho}} \tag{4.176}$$

and

$$v_s = \sqrt{\frac{G}{\rho}} \tag{4.177}$$

Notice, thus, that through the transformations given by Equations 4.165 and 4.166, the original equations of motions are reduced to a set of decoupled equations in terms of the potential functions Φ and Ψ.

4.7.3 Solution of Decoupled Equations of Motion

The solution of the decoupled equations of motion obtained earlier may be expressed in the form of an infinite sum of harmonic waves of the form

$$\Phi = F(z)e^{i(kx - \omega t)} \tag{4.178}$$

$$\Psi = G(z)e^{i(kx - \omega t)} \tag{4.179}$$

where $F(z)$ and $G(z)$ are functions of z that describe the variation of the wave amplitudes with depth, and k, ω, and λ, respectively, denote wave number, circular frequency, and wavelength. Note that for mathematical convenience these harmonic waves are expressed in terms of the complex-valued exponential function, but they could have been expressed as well in terms of a sine or cosine function. Substitution of Equations 4.178 and 4.179 into Equations 4.174 and 4.175 thus yields

$$-\omega^2 F(z) = v_p^2 [-k^2 F(z) + F''(z)] \tag{4.180}$$

$$-\omega^2 G(z) = v_s^2 [-k^2 G(z) + G''(z)] \tag{4.181}$$

where $F''(z)$ and $G''(z)$, respectively, signify the second derivatives of $F(z)$ and $G(z)$ with respect to z. After rearrangement, these equations may also be put into the form

$$F''(z) - q^2 F(z) = 0 \tag{4.182}$$

$$G''(z) - s^2 G(z) = 0 \tag{4.183}$$

where q^2 and s^2 are defined as

$$q^2 = k^2 - \omega^2 / v_p^2 \tag{4.184}$$

$$s^2 = k^2 - \omega^2 / v_s^2 \tag{4.185}$$

which in view of the fact that (see Equation 4.58)

$$k = \frac{\omega}{v_R} \quad (4.186)$$

where v_R denotes the propagation velocity of the waves under investigation, may also be expressed as

$$q^2 = \frac{\omega^2}{v_R^2}\left(1 - \frac{v_R^2}{v_p^2}\right) = \frac{\omega^2}{v_R^2}(1 - \alpha^2 V^2) \quad (4.187)$$

$$s^2 = \frac{\omega^2}{v_R^2}\left(1 - \frac{v_R^2}{v_s^2}\right) = \frac{\omega^2}{v_R^2}(1 - V^2) \quad (4.188)$$

in which α and V are given by

$$\alpha^2 = \frac{v_s^2}{v_p^2} = \frac{G}{\lambda + 2G} = \frac{1 - 2\mu}{2(1 - \mu)} \quad (4.189)$$

$$V = \frac{v_R}{v_s} \quad (4.190)$$

Equations 4.182 and 4.183 represent two independent ordinary differential equations with constant coefficients. As such, the solutions to these equations are of the form

$$F(z) = A_1 e^{-qz} + A_2 e^{qz} \quad (4.191)$$

$$G(z) = B_1 e^{-sz} + B_2 e^{sz} \quad (4.192)$$

It may be noted, however, that the coefficients A_2 and B_2 in these equations necessarily have to be equal to zero. The reason is that the displacements u and w (both are functions of Φ and ψ, as indicated by Equations 4.165 and 4.166) cannot grow up to infinity with depth as suggested by the terms $A_2 e^{qz}$ and $B_2 e^{sz}$ in Equations 4.191 and 4.192. In other words, if a disturbance is produced near the surface of the bounded medium, it is expected that the amplitude of the disturbance will decay, as discussed in Sections 4.6.5 and 4.6.7, in proportion to the distance traveled by the wave. On the basis of this argument, the solutions of Equations 4.182 and 4.183 may therefore be written as

$$F(z) = A_1 e^{-qz} \quad (4.193)$$

$$G(z) = B_1 e^{-sz} \quad (4.194)$$

which, when combined with Equations 4.178 and 4.179, lead to the following solutions for the potential functions Φ and Ψ:

$$\Phi = A_1 e^{-qz} e^{i(kx - \omega t)} \quad (4.195)$$

$$\Psi = B_1 e^{-sz} e^{i(kx - \omega t)} \quad (4.196)$$

The consideration of a bounded medium implies the existence of a boundary where neither normal nor shear stresses can exist. Therefore, the displacement solution in the bounded medium needs to satisfy these boundary conditions in addition to the corresponding equations of motion. For the case of plane waves, the two stresses that need to be equal to zero at the boundary are σ_z and τ_{zx}; that is, the normal stress along the direction of the z-axis and the shear stress acting along the direction of the x-axis on a plane perpendicular to the z-axis. For the case under investigation, the boundary conditions may therefore be expressed as (see Equations 4.93 and 4.94)

$$\sigma_z = \left|\lambda\bar{\varepsilon} + 2G\varepsilon_z\right|_{z=0} = \left|\lambda\left(\frac{\partial u}{\partial x} + \frac{\partial w}{\partial z}\right) + 2G\frac{\partial w}{\partial z}\right|_{z=0} = 0 \quad (4.197)$$

$$\tau_{zx} = \left|G\gamma_{zx}\right|_{z=0} = \left|G\left(\frac{\partial u}{\partial z} + \frac{\partial w}{\partial x}\right)\right|_{z=0} = 0 \quad (4.198)$$

or, after substitution of Equations 4.165 and 4.166, as

$$\left|\lambda\frac{\partial^2\Phi}{\partial x^2} + (\lambda + 2G)\frac{\partial^2\Phi}{\partial z^2} - 2G\frac{\partial^2\Psi}{\partial x\partial z}\right|_{z=0} = 0 \quad (4.199)$$

$$\left|2\frac{\partial^2\Phi}{\partial x\partial z} + \frac{\partial^2\Psi}{\partial z^2} - \frac{\partial^2\Psi}{\partial x^2}\right|_{z=0} = 0 \quad (4.200)$$

which, after making use of Equations 4.195 and 4.196 and setting $z = 0$, in turn lead to

$$A_1[(\lambda + 2G)q^2 - \lambda k^2] - 2iB_1Gks = 0 \quad (4.201)$$

$$2iA_1kq + B_1(s^2 + k^2) = 0 \quad (4.202)$$

From these two equations, one finds thus that

$$\frac{A_1}{B_1} = \frac{2iGks}{(\lambda + 2G)q^2 - \lambda k^2} \quad (4.203)$$

$$\frac{A_1}{B_1} = -\frac{s^2 + k^2}{2ikq} \quad (4.204)$$

and, by equating the right-hand sides of these two equations, that

$$4Gk^2sq = (s^2 + k^2)[(\lambda + 2G)q^2 - \lambda k^2] \quad (4.205)$$

which, after both sides are squared and Equations 4.184 through 4.190 are considered, may also be written as

$$16G^2k^8(1 - \alpha^2V^2)(1 - V^2) = k^4(2 - V^2)^2 k^4[(\lambda + 2G)(1 - \alpha^2V^2) - \lambda]^2 \quad (4.206)$$

or as

$$V^6 - 8V^4 - (16\alpha^2 - 24)V^2 - 16(1 - \alpha^2) = 0 \quad (4.207)$$

which represents a cubic equation in V^2.

Equation 4.207 is a characteristic equation that can be solved to find real solutions for V. Note that because α is a function of Poisson ratio only (see Equation 4.189), V also depends only on this ratio. Table 4.3 gives the values of V for various values of Poisson ratio. Observe from the values of

TABLE 4.3
Values of V for Different Poisson Ratios

μ	$V = v_R/v_s$
0.25	0.919
0.29	0.926
0.33	0.933
0.40	0.943
0.50	0.955

V in this table that the propagation velocity of Rayleigh waves is slightly slower than the propagation velocity of S waves.

Using the solutions for the potential functions obtained in the previous section (i.e., Equations 4.195 and 4.196) and the equations that define the displacements u and w in terms of these potential functions (i.e., Equations 4.165 and 4.166), it is possible to express such displacements as

$$u = -A_1 ik e^{-qz} e^{i(kx-\omega t)} - B_1 s e^{-sz} e^{i(kx-\omega t)} = -(ikA_1 e^{-qz} + B_1 s e^{-sz}) e^{i(kx-\omega t)} \quad (4.208)$$

$$w = -A_1 q e^{-qz} e^{i(kx-\omega t)} + B_1 ik e^{-sz} e^{i(kx-\omega t)} = -(A_1 q e^{-qz} - iB_1 k e^{-sz}) e^{i(kx-\omega t)} \quad (4.209)$$

which, after substitution of Equation 4.204, may also be written as

$$u = A_1 ki \left(-e^{-qz} + \frac{2qs}{s^2 + k^2} e^{-sz} \right) e^{i(kx-\omega t)} \quad (4.210)$$

$$w = A_1 q \left(-e^{-qz} + \frac{2k^2}{s^2 + k^2} e^{-sz} \right) e^{i(kx-\omega t)} \quad (4.211)$$

The displacements u and w given by these equations are complex valued. To obtain, then, displacements that are real valued, one needs to consider only the real parts on each side of the equations. Proceeding accordingly and noticing that according to Euler's relationship

$$e^{i(kx-\omega t)} = \cos(kx - \omega t) + i \sin(kx - \omega t) \quad (4.212)$$

Equations 4.210 and 4.211 lead to

$$u = A_1 k[e^{-qz} - 2qs(s^2 + k^2)^{-1} e^{-sz}] \sin(kx - \omega t) \quad (4.213)$$

$$w = A_1 q[-e^{-qz} + 2k^2(s^2 + k^2)^{-1} e^{-sz}] \cos(kx - \omega t) \quad (4.214)$$

where A_1 is a constant of integration that can be determined from initial conditions and s and q are, respectively, given by Equations 4.187 and 4.188.

4.7.4 Particle Motion Generated by Rayleigh Waves

It may be observed that the amplitude of the displacements induced by Rayleigh waves is a function of the depth z, the velocities of propagation v_p, v_s, and v_R, and the wave number k (or, what is tantamount, wavelength λ). It may also be noted that both u and w decay exponentially with depth. This means that the only significant displacements are those that occur near the free surface; that is why these types of waves are called surface waves. Furthermore, note that Equations 4.213 and 4.214

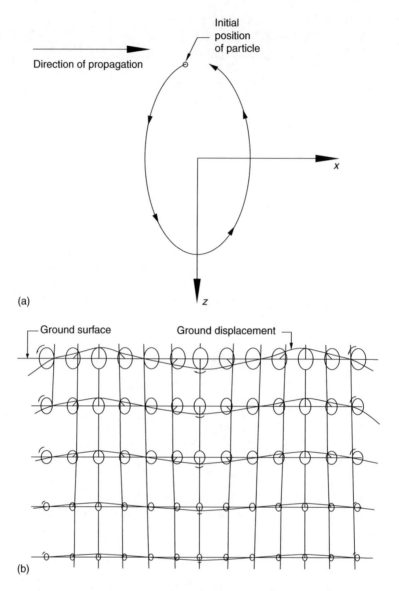

FIGURE 4.22 Particle motion induced by Rayleigh waves: (a) motion of a single particle and (b) motion of segment below the earth's surface showing how displacements decrease with depth.

are the parametric equations of an ellipse whose major axis is normal to the free surface, that is, parallel to the z-axis. Hence, the particle motion induced by Rayleigh waves is in the form of a retrograde ellipse as shown in Figure 4.22. A graphical description of the motion generated by Rayleigh waves is given in Figure 4.19c.

Figure 4.23 shows the variation with depth of the amplitude of the vertical and horizontal displacements induced by Rayleigh waves for various values of Poisson ratio. Several interesting aspects of Rayleigh waves may be observed from this figure:

1. The amplitude of the horizontal and vertical displacements decays rapidly with depth, and the depth below which this amplitude becomes small is roughly between one and one-and-a-half wavelengths.

Earthquake Propagation

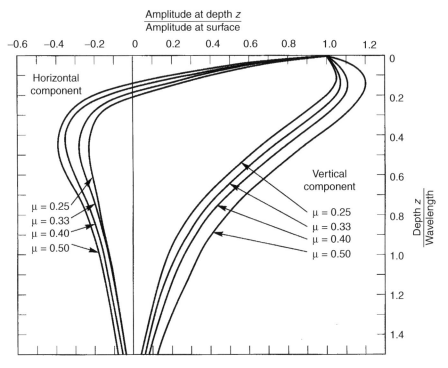

FIGURE 4.23 Variation with depth of amplitude of horizontal and vertical displacements induced by Rayleigh waves in media with different Poisson ratios. (After Richart, F.E. et al., *Vibrations of Soils and Foundations*, Prentice-Hall, Inc., Englewood Cliffs, NJ, 1970. Reprinted with permission from Pearson Education, Inc., Upper Saddle River, NJ.)

2. The maximum amplitudes do not occur at the surface, but at depths slightly below it.
3. Below a certain depth, the horizontal displacements are in the opposite direction from those observed at the surface.

EXAMPLE 4.5   VELOCITY OF PROPAGATION AND DISPLACEMENTS INDUCED BY RAYLEIGH WAVE

Determine the velocity of propagation of a Rayleigh wave with a frequency of 50 Hz in a medium characterized by a modulus of elasticity of 350,000 kPa, a Poisson ratio of 0.35, and a unit weight of 20 kN/m³. Determine also the amplitude of the displacements induced by the wave at the level of the ground surface.

Solution

According to Equation 4.189, α^2 in Equation 4.207 for a Poisson ratio of 0.35 is given by

$$\alpha^2 = \frac{1 - 2(0.35)}{2(1 - 0.35)} = 0.231$$

In this case, therefore, the characteristic equation that gives the velocity of propagation of Rayleigh waves is of the form

$$V^6 - 8V^4 + 20.31V^2 - 12.31 = 0$$

from which one finds that

$$V = \frac{v_R}{v_s} = 0.935$$

Thus, because for $E = 350{,}000$ kPa, $\mu = 0.35$, and $\gamma = 20$ kN/m^3, Equations 4.177 and 4.164 yield

$$v_s = \sqrt{\frac{350{,}000(9.81)}{2(1+0.35)(20)}} = 252 \text{ m/s}$$

$$v_p = \sqrt{\frac{2(1-0.35)}{1-2(0.35)}}(252) = 525 \text{ m/s}$$

one has that

$$v_R = 0.935(252) = 236 \text{ m/s}$$

To determine the displacements, it is necessary to compute first the wave number k and the parameters q and s defined by Equations 4.184 and 4.185. Accordingly, using Equation 4.186, the given wave frequency, and the determined value of v_R, one obtains

$$k = \frac{2\pi(50)}{236} = 1.331 \text{ rad/m}$$

which corresponds to a wavelength of 4.72 m. Similarly, Equations 4.184 and 4.185 yield

$$q = \sqrt{1.331^2 - (2\pi)^2(50)^2/525^2} = 1.189$$

$$s = \sqrt{1.331^2 - (2\pi)^2(50)^2/252^2} = 0.466$$

Thus, after substitution of the determined values of k, q, and s and making $z = 0$, Equations 4.213 and 4.214 lead to the following expressions to determine the displacement components u and v at the ground surface in terms of time and distance:

$$u = A_1(1.331)\left[1 - \frac{2(1.189)(0.466)}{0.466^2 + 1.331^2}\right]\sin(1.331x - 314.16t) = 0.589 A_1 \sin(1.331x - 314.16t)$$

$$v = A_1(1.189)\left[-1 + \frac{2(1.331)^2}{0.466^2 + 1.189^2}\right]\cos(1.331x - 314.16t) = 1.394 A_1 \cos(1.331x - 314.16t)$$

4.8 ELASTIC WAVES IN LAYERED HALF-SPACE: LOVE WAVES

4.8.1 Introduction

Rayleigh's theory helped to explain some of the motions observed on the surface of the earth during earthquakes but not all of them. In studying records from sites far away from the earthquake source, seismologists consistently noted at the beginning of the twentieth century that the vertical motion was not as large as the horizontal one, contrary from what one would expect from Rayleigh waves. They concluded, therefore, that there was an additional surface wave appearing in earthquake records, although they could not give an explanation for its existence.

Based on such observations, A. E. H. Love, a British mathematician, worked out in 1911 a mathematical model to solve the enigma. He found that a different type of wave could be accounted for by considering a two-layer medium, such as the medium that characterizes the interior of the earth near its surface, and the additional boundary conditions at the interface between such two layers. He also found that this wave propagated through the surface of the medium as in the case of Rayleigh waves, but it produced transverse motion like an SH wave. Furthermore, the wave propagated with a velocity that was different from the propagation velocity of P, S, and Rayleigh waves. This section presents the theory that explains the existence of Love waves and allows one to determine the velocity with which they propagate and the type of particle motion they generate. As in the case of Rayleigh waves, the discussion is limited to the case of plane waves.

4.8.2 Equations of Motion for Plane Waves

Consider a homogeneous surficial layer of thickness H overlying a homogeneous half-space as shown in Figure 4.24. Consider, in addition, that the surficial layer is softer than the half-space, that is, that $v_{s1} < v_{s2}$, where v_{s1} and v_{s2}, respectively, denote the shear wave velocities of these two layers.

Assume now the existence of a plane wave that travels in the direction of the x-axis and that this wave produces motion only in the direction of the y-axis. Furthermore, assume that this motion decays rapidly with depth. It is the intention of the derivation that follows to demonstrate that the existence of this wave is indeed possible and obtain the conditions that are necessary for its existence.

By setting ε and $\partial^2 v/\partial y^2$ equal to zero in Equation 4.113, the equations that describe the motion in the two layers under consideration when such motion occurs only along the direction of the y-axis may be written as

$$\frac{\partial^2 v_1}{\partial t^2} = \frac{G_1}{\rho_1}\left(\frac{\partial^2 v_1}{\partial x^2} + \frac{\partial^2 v_1}{\partial z^2}\right) \quad \text{for } 0 \leq z \leq H \tag{4.215}$$

$$\frac{\partial^2 v_2}{\partial t^2} = \frac{G_2}{\rho_2}\left(\frac{\partial^2 v_2}{\partial x^2} + \frac{\partial^2 v_2}{\partial z^2}\right) \quad \text{for } z \geq H \tag{4.216}$$

where subscripts 1 and 2 are used to denote the displacements and properties of the upper and lower layers, respectively. Assume that the solution of these equations is given by an infinite sum of harmonic waves of the form

$$v_1 = F_1(z)e^{i(kx-\omega t)} \tag{4.217}$$

$$v_2 = F_2(z)e^{i(kx-\omega t)} \tag{4.218}$$

where $F_1(z)$ and $F_2(z)$ represent displacement amplitudes that vary with depth z, and k and ω, respectively, denote the wave number and circular frequency of one of the harmonic components of the wave. Note that, as with Rayleigh waves, for mathematical convenience these harmonic components are expressed in terms of a complex-valued exponential function and that the real

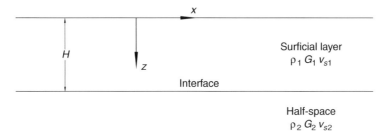

FIGURE 4.24 Layered medium consisting of surficial layer over a half-space.

displacements will be obtained at the end of the derivation by considering only the real parts of these equations.

By substitution of Equations 4.217 and 4.218 and considering that $G_1/\rho_1 = v_{s1}^2$ and $G_2/\rho_2 = v_{s2}^2$, Equations 4.215 and 4.216 lead to

$$-\omega^2 F_1(z) = v_{s1}^2[-k^2 F_1(z) + F_1''(z)] \quad \text{for } 0 \leq z \leq H \tag{4.219}$$

$$-\omega^2 F_2(z) = v_{s2}^2[-k^2 F_2(z) + F_2''(z)] \quad \text{for } z \geq H \tag{4.220}$$

which after rearrangement may also be expressed as

$$F_1''(z) - s_1^2 F_1(z) = 0 \quad \text{for } 0 \leq z \leq H \tag{4.221}$$

$$F_2''(z) - s_2^2 F_2(z) = 0 \quad \text{for } 0 \leq z \leq H \tag{4.222}$$

where s_1^2 and s_2^2 are defined as

$$s_1^2 = k^2 - \frac{\omega^2}{v_{s1}^2} \tag{4.223}$$

$$s_2^2 = k^2 - \frac{\omega^2}{v_{s2}^2} \tag{4.224}$$

or, in view of the relationship between frequency and wave velocity (see Equation 4.58), as

$$s_1^2 = k^2\left(1 - \frac{v_L^2}{v_{s1}^2}\right) = \omega^2\left(\frac{1}{v_L^2} - \frac{1}{v_{s1}^2}\right) \tag{4.225}$$

$$s_2^2 = k^2\left(1 - \frac{v_L^2}{v_{s2}^2}\right) = \omega^2\left(\frac{1}{v_L^2} - \frac{1}{v_{s2}^2}\right) \tag{4.226}$$

where v_L represents the velocity of the postulated wave, henceforth referred to as Love wave.

4.8.3 Solution of Equations of Motion

Equations 4.221 and 4.222 represent second-order differential equations with constant coefficients. Therefore, their solution is of the form

$$F_1(z) = A_1 e^{-s_1 z} + B_1 e^{s_1 z} \quad \text{for } 0 \leq z \leq H \tag{4.227}$$

$$F_2(z) = A_2 e^{-s_2 z} + B_2 e^{s_2 z} \quad \text{for } z \geq H \tag{4.228}$$

Notice, however, that the second term on the right-hand side of Equation 4.228 makes the displacement amplitude $F_2(z)$ tend to infinity as z becomes infinitely large. This contradicts the postulated condition that the wave under investigation rapidly decays with depth. Thus, the existence of the postulated wave requires that $B_2 = 0$. In view of Equations 4.217, 4.218, 4.227, and 4.228, the displacements v_1 and v_2 may therefore be expressed as

$$v_1 = (A_1 e^{-s_1 z} + B_1 e^{s_1 z}) e^{i(kx - \omega t)} \quad \text{for } 0 \leq z \leq H \tag{4.229}$$

$$v_2 = A_2 e^{-s_2 z} e^{i(kx - \omega t)} \quad \text{for } z \geq H \tag{4.230}$$

Earthquake Propagation

To satisfy the boundary conditions, it is necessary that the induced displacements generate zero stresses at the free surface. In other words, it is required that at $z = 0$,

$$\sigma_z = 0 \qquad \tau_{zy} = 0 \qquad \tau_{zx} = 0 \tag{4.231}$$

In the case under consideration, the first and third equations in Equation 4.231 are satisfied automatically because there are no displacements along the x and z directions. The second one requires that at $z = 0$,

$$\tau_{zy} = G_1 \gamma_{zy} = G_1 \frac{\partial v_1}{dz} = 0 \tag{4.232}$$

which upon substitution of Equation 4.229 leads to

$$\left.\frac{\partial v_1}{\partial z}\right|_{z=0} = (-A_1 s_1 e^{-s_1 z} + B_1 s_1 e^{s_1 z}) e^{i(kx-\omega t)}\bigg|_{z=0} = 0 \tag{4.233}$$

This equation, in turn, allows one to conclude that

$$A_1 = B_1 \tag{4.234}$$

In consequence, Equations 4.229 and 4.230 may be rewritten as

$$v_1 = A_1(e^{-s_1 z} + e^{s_1 z}) e^{i(kx-\omega t)} \quad \text{for } 0 \leq z \leq H \tag{4.235}$$

$$v_2 = A_2 e^{-s_2 z} e^{i(kx-\omega t)} \quad \text{for } z \geq H \tag{4.236}$$

Because in the case of a two-layer medium there exists a discontinuity at the interface between the two layers, the displacements induced by the postulated wave also have to satisfy the requirement of equal displacements and stresses at this interface. That is, it is required that

$$v_1\big|_{z=H} = v_2\big|_{z=H} \tag{4.237}$$

$$G_1 \frac{\partial v_1}{dz}\bigg|_{z=H} = G_2 \frac{\partial v_2}{dz}\bigg|_{z=H} \tag{4.238}$$

which, after substitution of Equations 4.235 and 4.236, may also be written as

$$A_1(e^{-s_1 H} + e^{s_1 H}) = A_2 e^{-s_2 H} \tag{4.239}$$

$$G_1 A_1(-s_1 e^{-s_1 H} + s_1 e^{s_1 H}) = -G_2 A_2 s_2 e^{-s_2 H} \tag{4.240}$$

Furthermore, as in view of Euler's relationship one has that

$$e^{-s_1 H} = e^{i(is_1 H)} = \cos(is_1 H) + i \sin(is_1 H) \tag{4.241}$$

$$e^{s_1 H} = e^{-i(is_1 H)} = \cos(is_1 H) - i \sin(is_1 H) \tag{4.242}$$

Equations 4.239 and 4.240 may also be put into the form

$$2A_1 \cos(is_1 H) = A_2 e^{-s_2 H} \tag{4.243}$$

$$2iG_1 A_1 \sin(is_1 H) = G_2 A_2 s_2 e^{-s_2 H} \tag{4.244}$$

Thus, from Equation 4.243 one can write

$$A_2 = \frac{2\cos(is_1 H)}{e^{-s_2 H}} A_1 \tag{4.245}$$

Similarly, upon the substitution of this last equation, Equation 4.244 yields

$$2iG_1 A_1 \sin(is_1 H) = \frac{2A_1 G_2 s_2 \cos(is_1 H) e^{-s_2 H}}{e^{-s_2 H}} \tag{4.246}$$

which after rearrangement leads to

$$\tan(is_1 H) = \frac{G_2 s_2}{iG_1 s_1} \tag{4.247}$$

from which one obtains that

$$\tan\left(H\sqrt{-s_1^2}\right) = \frac{G_2}{G_1} \sqrt{\frac{s_2^2}{-s_1^2}} \tag{4.248}$$

which in the light of Equations 4.225 and 4.226 may also be written as

$$\tan\left(kH\sqrt{v_L^2/v_{s1}^2 - 1}\right) = \frac{G_2}{G_1} \sqrt{\frac{1 - v_L^2/v_{s2}^2}{v_L^2/v_{s1}^2 - 1}} \tag{4.249}$$

Equation 4.249 represents the characteristic equation that defines the propagation velocity of the postulated wave. It leads to real solutions, so the postulated wave exists. Observe from this equation that v_L depends on the wave number k or, what is tantamount, the wavelength. Observe also that for an infinitely long wavelength, k tends to zero and thus v_L tends to v_{s2}. Similarly, for extremely short wavelengths, k tends to infinity and hence v_L approaches v_{s1}. It may be concluded, therefore, that, in general, the velocity of propagation of Love waves lies in between the values of the shear wave velocities of the two strata that compose the propagation medium. That is,

$$v_{s1} \leq v_L \leq v_{s2} \tag{4.250}$$

It may also be concluded that Love waves propagate with a velocity that is faster than the velocity of Rayleigh waves because the latter travels with a velocity that is slower than the velocity of shear waves.

Equation 4.245 may also be used to express the displacement v_2 in terms of the constant A_1. Accordingly, by substitution of this equation into Equation 4.236, one obtains

$$v_2 = A_1 \frac{2\cos(is_1 H)}{e^{-s_2 H}} e^{-s_2 z} e^{i(kx - \omega t)} = 2A_1 \cos\left(H\sqrt{-s_1^2}\right) e^{-s_2(z-H)} e^{-i(kx - \omega t)} \tag{4.251}$$

Hence, the displacements v_1 and v_2 may be expressed as

$$v_1 = A_1 [e^{i(is_1 z)} + e^{-i(is_1 z)}] e^{i(kx - \omega t)} \quad \text{for } 0 \leq z \leq H \tag{4.252}$$

$$v_2 = 2A_1 \cos\left(H\sqrt{-s_1^2}\right) e^{-s_2(z-H)} e^{i(kx - \omega t)} \quad \text{for } z \geq H \tag{4.253}$$

or, if only the real parts are considered, as

$$v_1 = 2A_1 \cos\left(z\sqrt{-s_1^2}\right)\cos(kx - \omega t) \quad \text{for } 0 \leq z \leq H \tag{4.254}$$

$$v_2 = 2A_1 \cos\left(H\sqrt{-s_1^2}\right) e^{-s_2(z-H)} \cos(kx - \omega t) \quad \text{for } z \geq H \tag{4.255}$$

Finally, after substitution of Equations 4.225 and 4.226, one arrives to

$$v_1 = 2A_1 \cos\left(\omega z \sqrt{\frac{1}{v_{s1}^2} - \frac{1}{v_L^2}}\right)\cos(kx - \omega t) \quad \text{for } 0 \leq z \leq H \tag{4.256}$$

$$v_2 = 2A_1 \cos\left(\omega H \sqrt{\frac{1}{v_{s1}^2} - \frac{1}{v_L^2}}\right) e^{-\omega(z-H)\sqrt{\frac{1}{v_L^2} - \frac{1}{v_{s2}^2}}} \cos(kx - \omega t) \quad \text{for } z \geq H \tag{4.257}$$

4.8.4 Particle Motion Generated by Love Waves

It may be observed from Equations 4.256 and 4.257 that the particle motion generated by Love waves varies sinusoidally with depth in the surficial layer and decays exponentially with depth in the underlying half-space as shown in Figure 4.25. Note, therefore, that the motion generated by Love waves occurs mostly near the surface and thus they are of the surface type. That is, Love waves are surface waves. Note, too, that the most significant part of the motion is concentrated in the surficial layer. For this reason, Love waves are often thought of as SH waves trapped by multiple reflections within this surficial layer. Note, further, that the particle motion generated by a Love wave occurs only along the direction of the y-axis when one assumes that the wave travels along the direction of the x-axis. Hence, it may be concluded that, in general, Love waves induce a horizontal motion that is perpendicular to the direction along which the wave propagates. This type of motion may indeed be seen in actual earthquake ground motion records such as that shown in Figure 4.26, where one can observe the arrival of the Love wave in the north–south record but not in the other two. The general nature of the motion generated by Love waves is depicted in Figure 4.19d.

4.8.5 Dispersion of Love Waves

It was mentioned earlier and may be seen from the inspection of Equation 4.249 that the propagation velocity of Love waves depends on their wavelength. This means that Love waves with different

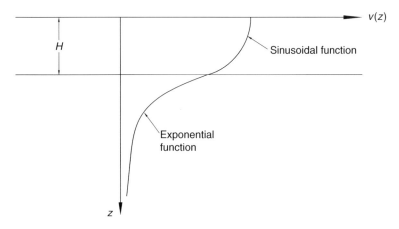

FIGURE 4.25 Variation with depth of particle displacements generated by Love waves.

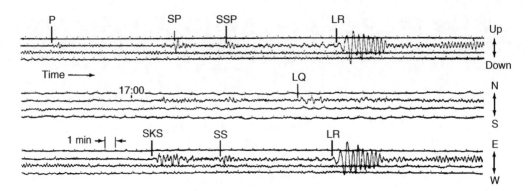

FIGURE 4.26 Ground motion records obtained at Berkeley, California, from an earthquake in the South Pacific where the arrival of Rayleigh (LR) and Love (LQ) waves can be seen. Observe that the arrival of Love wave is detected in only the N–S direction. (From Bolt, B.A., *Earthquakes*, W.H. Freeman and Co., New York, 2004. With permission.)

wavelengths (or different frequencies) will travel with different velocities and arrive at a particular site at different times. It also means that, as in the case of waves in a flexural beam (see Section 4.5.2), Love waves undergo dispersion and are therefore considered to be dispersive. Because of dispersion, Love waves with different wavelengths may interfere with one another, form an apparent wave, and give the impression that they travel with a different velocity, the *group velocity* discussed in Section 4.5.3.

Although for the ideal case of a homogenous medium Love waves show dispersion and Rayleigh waves do not, it is important to keep in mind that near the earth's surface Rayleigh waves also show dispersion. The reason is that the stiffness of the soils and rocks near the earth's surface normally increases with depth and thus the depth to which a Rayleigh wave generates significant displacements increases in proportion to its wavelength (see Figure 4.23). As a result, the average stiffness of the propagation medium is higher for waves with long wavelengths than for waves with short wavelengths. This means that Rayleigh waves with short wavelengths will travel faster than those with long wavelengths. In the case of earthquakes, therefore, the low-frequency components of both Love and Rayleigh waves will usually arrive at a site before their high-frequency counterparts.

EXAMPLE 4.6 VELOCITY OF PROPAGATION AND DISPLACEMENTS INDUCED BY LOVE WAVE

A 30 m layer of soil with a unit weight of 20 kN/m³, a shear modulus of elasticity of 0.8×10^6 kN/m², and a Poisson ratio of 0.25 rests on what may be idealized as a homogeneous semi-infinite rock formation with a unit weight of 28 kN/m³, a shear modulus of elasticity of 12×10^6 kN/m², and a Poisson ratio of 0.25. Determine (1) the velocity with which a Love wave with a period of 0.3 s would propagate in this medium and (2) the amplitude of the displacement generated by this wave at a depth of 0, 10, 20, 30, 40, and 50 m.

Solution

1. The propagation velocity of a Love wave depends on the shear wave velocities of the medium within which the wave propagates and its wave number. Consider then that according to Equation 4.151 and the given data, the shear wave velocities of the upper and lower layers are, respectively, given by

$$v_{s1} = \sqrt{\frac{0.8 \times 10^6 \times 9.81}{20}} = 626.4 \text{ m/s}$$

$$v_{s2} = \sqrt{\frac{12 \times 10^6 \times 9.81}{28}} = 2050 \text{ m/s}$$

TABLE E4.6
Displacements Generated by Love Wave with a Period of 0.3 s

Depth (m)	Displacement/A_1 (m)
0	2
10	1.90
20	1.61
30	1.16
40	1.12
50	1.09

Similarly, according to Equation 4.58 the wave number of a wave with a period of 0.3 s is given by

$$k = \frac{2\pi}{v_L(0.3)} = \frac{20.944}{v_L}$$

Upon substitution of these expressions into Equation 4.249, it may be seen, therefore, that the propagation velocity of a 0.3 s Love wave is given by

$$\tan\left(\frac{20.944 \times 30}{v_L}\sqrt{\frac{v_L^2}{626.4^2} - 1}\right) = \frac{12 \times 10^6}{0.8 \times 10^6}\sqrt{\frac{1 - v_L^2/2050^2}{v_L^2/626.4^2 - 1}}$$

from which, by trial and error, it is found that

$$v_L = 1969 \text{ m/s}$$

2. Observing that for the determined values of v_L, v_{s1}, and v_{s2}, one has that

$$\sqrt{\frac{1}{v_{s1}^2} - \frac{1}{v_L^2}} = \sqrt{\frac{1}{626.4^2} - \frac{1}{1969^2}} = 0.00151$$

$$\sqrt{\frac{1}{v_L^2} - \frac{1}{v_{s2}^2}} = \sqrt{\frac{1}{1969^2} - \frac{1}{2050^2}} = 0.00014$$

then, according to Equations 4.256 and 4.257 and the fact that $\omega = 2\pi/0.3 = 20.944$ rad/s, the displacements generated by the wave under consideration are given by

$$v_1 = 2A_1\cos(20.944 \times 0.00151z) = 2A_1\cos(0.0317z) \quad \text{for } 0 \leq z \leq 30 \text{ m}$$

$$v_2 = 2A_1\cos(20.944 \times 0.00151 \times 30)e^{-20.944(0.00014)(z-30)} = 1.155\,A_1 e^{-0.00293(z-30)} \quad \text{for } z \geq 30 \text{ m}$$

Consequently, the displacements at $z = 0$, 10, 20, 30, 40, and 50 m are as indicated in Table E4.6.

4.9 REFLECTION AND REFRACTION OF BODY WAVES

4.9.1 INTRODUCTION

As in the one-dimensional case, when a three-dimensional wave crosses a boundary or discontinuity, the wave is transmitted and reflected. There are, however, two major differences between the reflection and refraction of one- and three-dimensional waves. The first is that the transmitted three-dimensional wave does not travel along the same direction as the incident wave. Rather, it bends or refracts with respect to the direction of the incident wave. The second is that after crossing a boundary or discontinuity, a single three-dimensional P or S wave may generate two transmitted and reflected waves, one as a P wave and the other as an S wave.

This section is thus devoted to describe the relationship among incident, transmitted, and reflected waves in a three-dimensional medium. As some of the methods that are used to determine this relationship are based on ray theory, some elements of ray theory are presented first.

4.9.2 Elements of Ray Theory

The concept of a ray used in wave propagation theory is based on *Fermat's principle* of least time. This principle states that the time it takes for a disturbance to propagate between two arbitrary points *A* and *B* is equal to the time it takes for the disturbance to travel along the shortest continuous path that connects *A* and *B*. This concept is illustrated in Figure 4.27.

On the basis of Fermat's principle, a *ray path* is thus defined as the path that produces such a minimum travel time. Similarly, a *ray* is defined as a vector in the direction of a ray path. Furthermore, a *wave front* is defined as a surface of equal travel time, such that a ray and a ray path are always perpendicular to the wave front (see Figure 4.28). Note that according to these definitions, rays are a family of parallel lines in the case of plane waves and a family of spokes radiating from a source in the case of spherical waves.

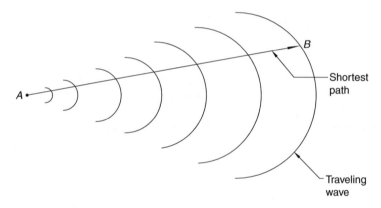

FIGURE 4.27 Shortest continuous path between two points through which a wave travels.

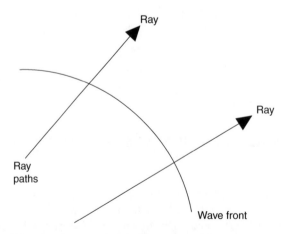

FIGURE 4.28 Relationship among ray, ray path, and wave front.

Earthquake Propagation

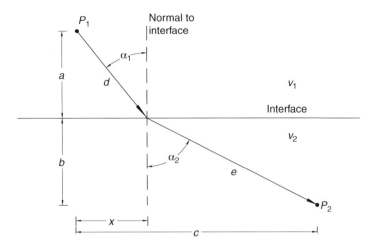

FIGURE 4.29 Refraction of incident ray after crossing discontinuity.

The angle of inclination of a wave (or a ray path) after it crosses a discontinuity may be obtained using Fermat's principle as follows. Consider a point P_1 in a medium with a wave propagation velocity v_1, and a point P_2 in a medium with a wave propagation velocity v_2 across an interface (see Figure 4.29). Let α_1 be the angle that the incident ray makes with the normal to the interface, and α_2 be the angle that defines the inclination of the refracted ray. In terms of the geometry of Figure 4.29, the time it takes for a wave to travel from point P_1 to point P_2 may be expressed as

$$t_{1-2} = \frac{d}{v_1} + \frac{e}{v_2} = \frac{\sqrt{a^2 + x^2}}{v_1} + \frac{\sqrt{b^2 + (c-x)^2}}{v_2} \tag{4.258}$$

But according to Fermat's principle, this time should correspond to the minimum path. Therefore, the refraction angle α_2 may be found by making the first derivative of this expression with respect to x equal to zero. Proceeding accordingly, one obtains

$$\frac{dt_{1-2}}{dx} = \frac{x}{v_1\sqrt{a^2 + x^2}} - \frac{c - x}{v_2\sqrt{b^2 + (c-x)^2}} = 0 \tag{4.259}$$

However, it may be seen from Figure 4.29 that

$$\frac{x}{\sqrt{a^2 + x^2}} = \sin \alpha_1 \quad \text{and} \quad \frac{c - x}{\sqrt{b^2 + (c-x)^2}} = \sin \alpha_2 \tag{4.260}$$

Hence, upon substitution of Equation 4.260 into Equation 4.259, one gets the following relationship between the angles α_1 and α_2:

$$\frac{\sin \alpha_1}{v_1} = \frac{\sin \alpha_2}{v_2} \tag{4.261}$$

This relationship is known as *Snell's law*. It implies that the ratio of the sine of the angle of inclination of a wave ray to its velocity of propagation remains constant regardless of the medium through which it travels. Consequently, it is sometimes enunciated as

$$\frac{\sin i}{v} = p \tag{4.262}$$

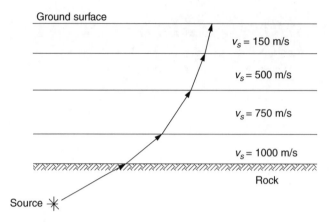

FIGURE 4.30 Refraction of shear wave as it goes through a series of successively softer soil layers.

where i is a wave ray's angle of incidence measured with respect to the normal to an interface, v the velocity of wave propagation in the medium under consideration, and p a constant. It also implies that waves are refracted after they cross a discontinuity.

It is of interest to note that according to Snell's law, waves propagating upward through horizontal layers of successively lower velocity will be refracted closer and closer to a vertical path as shown in Figure 4.30. This is in fact the justification in some of the methods of analysis employed in earthquake engineering for the approximate consideration of earthquake excitations as a train of vertical waves.

4.9.3 Relative Amplitudes of Incident, Refracted, and Reflected Waves

As in the case of one-dimensional waves, one can find the characteristics of the reflected and transmitted waves that are generated when an incident wave crosses the interface between two media by satisfying the continuity requirement that the stresses and displacements be the same at points right before and right after such an interface. In the case of three-dimensional waves, however, the stresses and displacements that need to satisfy this requirement are those normal to the interface as well as those parallel to it. In addition, it is necessary to consider, in general, the combined effect of reflected and refracted P and S waves.

To obtain the equations that give the relative amplitudes of incident, reflected, and refracted waves, consider two media, one characterized by a P-wave velocity v_{p1}, S-wave velocity v_{s1}, and density ρ_1, and the other characterized by a P-wave velocity v_{p2}, S-wave velocity v_{s2}, and density ρ_2. Consider, in addition, that to satisfy the aforementioned continuity requirement, it is necessary to transform the incident wave into a combination of P and S waves after it crosses the interface between the two media. The exception is for incident SH waves because an SH wave does not generate motion perpendicular to the plane of the interface and thus a P-wave is not needed across the interface to satisfy the continuity requirement. Specifically, consider that after crossing the aforementioned interface incident P, SH, and SV waves generate the refracted and reflected waves shown in Figure 4.31. Consider, further, that the directions of the transmitted and reflected waves in relation to the direction of the incident wave are given by Snell's law. That is, consider the when the incident wave is a P wave, those directions of the transmitted and reflected waves are given by (see Figure 4.31a)

$$\alpha_2 = \alpha_1 \tag{4.263}$$

$$\frac{\sin\beta_2}{v_{s1}} = \frac{\sin\alpha_3}{v_{p2}} = \frac{\sin\beta_3}{v_{s2}} = \frac{\sin\alpha_1}{v_{p1}} \tag{4.264}$$

Earthquake Propagation

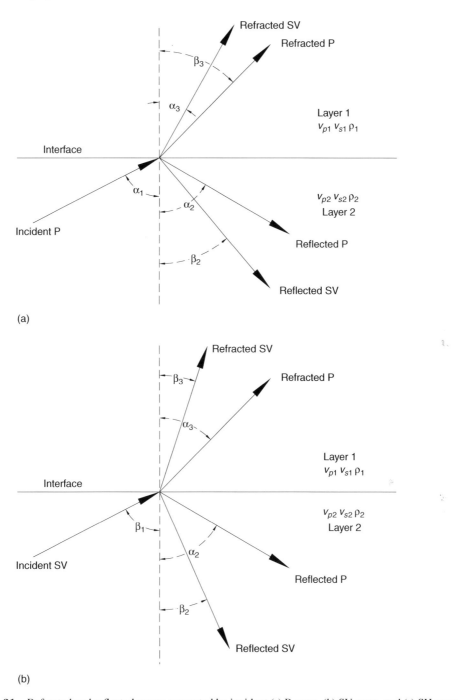

FIGURE 4.31 Refracted and reflected waves generated by incident (a) P wave, (b) SV wave, and (c) SH wave.

Similarly, conside that when the incident wave is an SH wave, the directions in question are defined by (see Figure 4.31b)

$$\beta_2 = \beta_1 \tag{4.265}$$

$$\frac{\sin \beta_3}{v_{s2}} = \frac{\sin \beta_1}{v_{s1}} \tag{4.266}$$

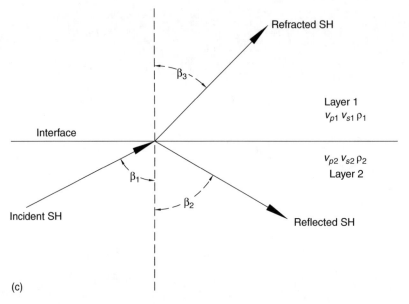

FIGURE 4.31 (Continued)

and when the incident wave is an SV wave, by (see Figure 4.31c)

$$\beta_2 = \beta_1 \tag{4.267}$$

$$\frac{\sin\alpha_2}{v_{p1}} = \frac{\sin\beta_3}{v_{s2}} = \frac{\sin\alpha_3}{v_{p2}} = \frac{\sin\beta_1}{v_{s1}} \tag{4.268}$$

Determine, then, the displacements and stresses produced by the waves shown in Figure 4.31. In general, two components of displacement and two components of stress need be considered. One of such components is the displacement or stress normal to the plane of the interface and the other parallel to it. There are, therefore, four continuity equations in each case. In general, too, all the stresses can be expressed in terms of displacements using the relationships between stresses and strains and strains and displacements introduced in Section 4.6.2, making it possible to write all four continuity equations in terms of displacements. Thus, if the incident wave is a harmonic wave, and if A, B, C, D, E, and F, respectively, denote the displacement amplitudes of an incident P, incident S, reflected P, reflected S, refracted P, and refracted S wave, such continuity equations for the case of an incident P wave result as

$$(A - C)\sin\alpha_1 + D\cos\beta_1 - E\sin\alpha_3 + F\cos\beta_3 = 0 \tag{4.269}$$

$$(A + C)\cos\alpha_1 + D\sin\beta_1 - E\cos\alpha_3 - F\sin\beta_3 = 0 \tag{4.270}$$

$$-(A+C)\sin 2\alpha_3 + D\frac{v_{p1}}{v_{s1}}\cos 2\beta_3 + E\frac{\rho_1}{\rho_2}\left(\frac{v_{s2}}{v_{s1}}\right)^2 \frac{v_{p1}}{v_{p2}}\sin 2\alpha_3 - F\frac{\rho_1}{\rho_2}\left(\frac{v_{s2}}{v_{s1}}\right)^2 \frac{v_{p1}}{v_{s2}}\cos 2\beta_3 = 0 \tag{4.271}$$

$$-(A-C)\cos 2\beta_1 + D\frac{v_{s1}}{v_{p1}}\sin 2\beta_1 + E\frac{\rho_1}{\rho_2}\frac{v_{p2}}{v_{p1}}\cos 2\beta_3 - F\frac{\rho_1}{\rho_2}\frac{v_{s2}}{v_{p1}}\sin 2\beta_3 = 0 \tag{4.272}$$

Earthquake Propagation

from which one can find the amplitudes of the refracted and reflected waves in terms of the known direction and amplitude of the incident wave. Similarly, when the incident wave is an SV wave, the corresponding equations are

$$(B + D)\sin\beta_1 + C\cos\alpha_1 - E\cos\alpha_3 - F\sin\beta_3 = 0 \quad (4.273)$$

$$(B - D)\cos\beta_1 + C\sin\alpha_1 + E\sin\alpha_3 - F\cos\beta_3 = 0 \quad (4.274)$$

$$(B + D)\cos 2\beta_1 - C\frac{v_{E1}}{v_{p1}}\sin 2\alpha_1 + E\frac{\rho_1}{\rho_2}\frac{v_{s2}^2}{v_{s1}v_{p2}}\sin 2\alpha_3 - F\frac{\rho_1}{\rho_2}\frac{v_{s2}}{v_{s1}}\cos 2\beta_3 = 0 \quad (4.275)$$

$$-(B - D)\sin 2\beta_1 + C\frac{v_{p1}}{v_{s1}}\cos 2\beta_1 + E\frac{\rho_1}{\rho_2}\frac{v_{p2}}{v_{s1}}\cos 2\beta_3 + F\frac{\rho_1}{\rho_2}\frac{v_{s2}}{v_{s1}}\sin 2\beta_3 = 0 \quad (4.276)$$

A similar set of equations can be found for the case of an incident SH wave. However, because in such a case only SH waves are reflected and refracted, the resulting equations are much simpler than those in the previous two cases. It is, therefore, possible to obtain explicit expressions to determine the relative amplitudes of the refracted and reflected waves in terms of the amplitudes of the incident one. These expressions are

$$D = \left(\frac{1 - (\rho_1/\rho_2)(v_{s2}/v_{s1})(\cos\beta_3/\cos\beta_1)}{1 + (\rho_1/\rho_2)(v_{s2}/v_{s1})(\cos\beta_3/\cos\beta_1)}\right)B \quad (4.277)$$

$$F = \left(1 + \frac{1 - (\rho_1/\rho_2)(v_{s2}/v_{s1})(\cos\beta_3/\cos\beta_1)}{1 + (\rho_1/\rho_2)(v_{s2}/v_{s1})(\cos\beta_3/\cos\beta_1)}\right)B \quad (4.278)$$

It may be inferred from the foregoing discussion that the reflection and the refraction of waves make the propagation of waves a complex process when they cross multiple discontinuities, as is the case when earthquake waves travel through the multiple layers and discontinuities of the earth's crust. Because of this phenomenon, a site far away from an earthquake source may be affected by waves that travel through different paths and reach the site at different times. The end result is a scattering effect that gives rise to a ground motion that is longer and more irregular than the ground motion that would be observed at a site near the source (see Figure 4.32).

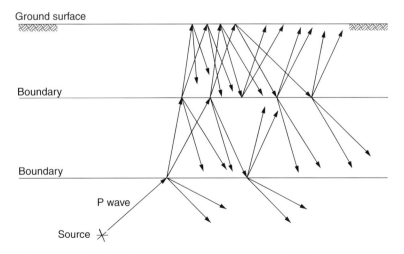

FIGURE 4.32 Scattering of incident P wave as it goes through a stratified soil deposit.

4.10 ATTENUATION OF WAVE AMPLITUDE WITH DISTANCE

4.10.1 Radiation Damping

It has been seen in the previous sections that the energy introduced by a disturbance in an elastic medium is propagated away from the location where the disturbance originates in the form of a wave and that this energy is absorbed by the material through which the wave passes in the form of elastic deformations. It has also been seen that a wave propagates equally in all directions, which means that the volume of the material affected simultaneously by a wave increases with the distance traveled by the wave. As a disturbance releases a fixed amount of energy (or fixed amount per unit time), it may then be concluded that the elastic energy per unit volume absorbed by the medium will decrease with such a distance. As a result, it may also be concluded that the amplitude of the displacements and stresses induced by a wave will also decrease as the wave moves away from its source. This conclusion is in fact corroborated by Equations 4.145 through 4.147 as these equations indicate that the displacements induced by a wave indeed decay with distance. Another conclusion that can be drawn from this analysis is that the decay with distance of the displacements and stresses generated by a wave takes place even in the absence of any damping mechanism in the propagation medium. That is, the aforementioned reduction in energy per unit volume occurs despite the fact that no damping forces were introduced in the derivation of Equations 4.145 through 4.147 and, thus, solely as a result of a change in the volume of the material deformed by the wave. The reduction in the amplitude of displacements and stresses generated by a wave due to the spreading of a constant source of energy over an increasingly larger volume of material is different from, and is in addition to, the reduction that may be attained as a result of the energy dissipation mechanisms that inherently exist in all materials. For this reason, the former type of reduction is considered to be the effect of what is often referred to as *radiation damping* or *geometric damping*.

Another aspect of the reduction with distance of the amplitude of the displacements and stresses generated by a wave that is important to keep in mind is that body and surface waves have a different rate of geometric decay. This is owed to the fact that body waves propagate in all directions, whereas surface waves propagate only horizontally. For example, if a disturbance is idealized as a point source, then body waves propagate outwardly along a spherical front (see Figure 4.33a), whereas surface waves propagate outwardly along a cylindrical front (see Figure 4.33b). If it is considered that the deformation energy in an elastic medium is proportional to the square of the induced deformation and that the surface area of a sphere is equal to 4π times the square of its radius, then it may be concluded that the displacements induced by a body wave are proportional to the inverse of the square root of the surface area of the wave front, or that they are proportional to $1/r$, where r denotes distance from the disturbance source. Similarly, it may be concluded that

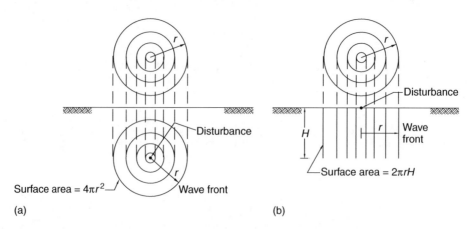

FIGURE 4.33 Schematic representation of propagation of (a) body and (b) surface waves.

Earthquake Propagation

the displacements induced by a surface wave are proportional to $1/\sqrt{r}$. It may be seen, thus, that surface waves attenuate geometrically more slowly than body waves. It may also be seen that at long distances from the location where the disturbance that generates the waves occurs, the ground motion observed will be mainly due to the effect of surface waves as opposed to body waves. In fact, this difference in the rate of amplitude decay helps to explain why in earthquake records obtained at long distances from the center of the earthquake, the surface wave is the wave that exhibits, as is the case for the record shown in Figure 4.20, the largest ground motion amplitude.

4.10.2 Material Damping

As mentioned earlier, the discussion of wave propagation in the previous sections has been limited to elastic solids with no energy dissipation mechanism involved. However, this type of behavior cannot occur in real materials because, as is well known, all materials possess a form of internal damping that makes them dissipate energy when deformed. In real materials, therefore, part of the energy transmitted by a wave to the propagation medium is dissipated in the form of heat or permanent deformations. This in turn means that as a wave spreads out away from its source, the transmitted energy and the displacements and stresses induced at points far away from where the wave originates will be dramatically reduced. In other words, waves will attenuate with distance as a result of internal energy losses in the propagation medium. The dissipation of energy associated with the deformation of a material is called *internal damping* or *material damping*.

The process of wave attenuation due to the internal damping in a propagation medium is a complex one that defies mathematical modeling. Consequently, this process is usually described using a simplified phenomenological model. The simplest of such phenomenological models is the so-called *Kelvin–Voight* model (see Section 13.9). In this model, the material that constitutes a medium is represented by a series of linear springs and viscous dampers in parallel, so as to have the stresses in the medium represented (for the one-dimensional case) by a relationship of the form

$$\sigma_x(x, t) = E\varepsilon_x + C\dot{\varepsilon}_x(x, t) \tag{4.279}$$

where all symbols are as defined before and C is a damping constant.

On the basis of this model and in similarity with the solution for a damped single-degree-of-freedom dynamic system, the displacement amplitude generated by the passing of a wave may be expressed as

$$A(r, t) = A_0 e^{-\xi\omega t} \tag{4.280}$$

where r signifies distance from the wave-generating source, ξ a damping factor that defines the internal damping in the propagation medium, ω the frequency of the wave, and A_0 the displacement amplitude generated by the wave in the absence of any damping.

At times, the damping in a propagation medium is characterized in terms of a *quality factor* defined as

$$Q = \frac{1}{2\xi} \tag{4.281}$$

In terms of this quality factor, Equation 4.280 may be thus written as

$$A(r, t) = A_0 e^{-\omega t/2Q} \tag{4.282}$$

which, by considering that $t = r/v_\omega$, where v_ω is the wave velocity, and $\omega = 2\pi f$, f being the wave frequency in hertz, may also be expressed as

$$A(r, t) = A_0 e^{-(\pi f/Q v_\omega)r} \tag{4.283}$$

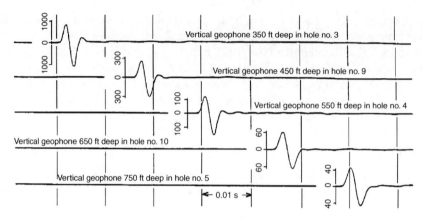

FIGURE 4.34 Widening of propagated pulse resulting from the attenuation with distance of pulse's high-frequency components. (After McDonal, F.J. et al., *Geophysics*, 23, 421–439, 1958.)

TABLE 4.4
Quality Factor for Various Rocks

Rock	Q for P waves	Q for S waves
Shale	30	10
Sandstone	58	31
Granite	250	70–150

Source: Lay, T. and Wallace, T.C., *Modern Global Seismology*, Academic Press, Inc., San Diego, 1995.

It is apparent from this equation that, for a constant value of the quality factor Q, a high-frequency wave will attenuate more rapidly than a low-frequency one. This is something that should be expected because, for a given distance, a high-frequency wave would go through more oscillations than a low-frequency one. Indeed, this phenomenon has been observed in field measurements. Consider, for example, the records shown in Figure 4.34, which depict the displacements induced at different depths by a pulse generated at the ground surface, and compare the width of the pulse at depths of 350 and 750 ft. It may be seen from this figure that as a result of the faster attenuation of high-frequency waves, the width of the pulse (period) increases as the wave travels from a depth of 350 to 750 ft.

As expected, the magnitude of the quality factor Q depends on the material of the propagation medium. What is not so evident, though, is that the quality factor is systematically larger for P waves than for S waves. As with any other damping parameter, quality factors are determined empirically from field recordings. Typical values are shown in Table 4.4. As Q is inversely proportional to the damping factor ξ, note that the greater the quality factor is, the smaller the attenuation rate will be.

EXAMPLE 4.7   ATTENUATION OF SHEAR WAVE WITH DISTANCE

A harmonic, one-dimensional shear wave with a frequency of 1 Hz travels through a viscoelastic material with a shear modulus of elasticity of 2×10^6 psf, a unit weight of 120 pcf, and a quality factor of 10. Determine the distance after which the wave amplitude is reduced to 10% of the initial amplitude.

Solution

Let r_1 and r_2, respectively, denote the distances from the source where the wave originates to two arbitrary points. According to Equation 4.283, the amplitudes generated by the wave at these two points are given by

$$v(r_1, t_1) = A_0 e^{-(\pi f/Qv_s)r_1}$$

$$v(r_2, t_2) = A_0 e^{-(\pi f/Qv_s)r_2}$$

Thus, if the amplitude $v(r_2, t_2)$ is equal to 10% of the amplitude $v(r_1, t_1)$, one has that

$$\frac{v(r_2, t_2)}{v(r_1, t_2)} = e^{-(\pi f/Qv_s)(r_2 - r_1)} = 0.10$$

from which one finds that

$$r_2 - r_1 = -\frac{\ln(0.10)}{\pi f} Qv_s$$

Consequently, since in this case the wave velocity is given by

$$v_s = \sqrt{\frac{2 \times 10^6 (32.2)}{120}} = 732.6 \text{ ft/s}$$

the wave amplitude is reduced to 10% when it travels a distance equal to

$$r_2 - r_1 = -\frac{\ln(10)}{\pi(1.0)} (10)(732.6) = 5369 \text{ ft}$$

FURTHER READINGS

1. Achenbach, J. D., *Wave Propagation in Elastic Solids*, American Elsevier Pub. Co., New York, 1973.
2. Bullen, K. E. and Bolt, B. A., *An Introduction to the Theory of Seismology*, 4th Edition, Cambridge University Press, Cambridge, U.K., 1985.
3. Ewing, W. M., Jardestsky, W. S., and Press, F., *Elastic Waves in Layered Media*, McGraw-Hill, New York, 1957.
4. Kolsky, H., *Stress Waves in Solids*, Dover Publications, Inc., New York, 1963.
5. Lay, T. and Wallace, T. C., *Modern Global Seismology*, Academic Press, Inc., San Diego, 1995.
6. Richart, F. E., Hall, J. R., Jr., and Woods, R. D., *Vibrations of Soils and Foundations*, Prentice-Hall, Inc., Englewood Cliffs, NJ, 1970.

PROBLEMS

4.1 Determine the velocity of propagation of longitudinal waves traveling along a laterally constrained rod when the rod is made of (a) steel, (b) cast iron, and (c) concrete with $f'_c = 4000$ psi.

4.2 A rod of infinite length is subjected to an initial longitudinal displacement given by

$$u_0 = 2(1 - x) \quad 0 \leq x \leq 1$$

$$u_0 = 2 + x \quad -2 \leq x \leq 0$$

Draw plots of the rod's longitudinal displacement u against the position variable x at times $t = 1, 2, 3,$ and 4 s. Consider that the velocity of propagation of longitudinal waves in the rod is equal to 0.5 m/s.

4.3 Repeat Problem 2 considering an initial longitudinal velocity instead of an initial displacement and that this initial velocity is given by

$$v_0 = A \quad -2 \leq x \leq 2$$
$$v_0 = 0 \quad \text{elsewhere}$$

where A is a constant.

4.4 A long bar with a density of 7850 kg/m³ and a modulus of elasticity of 7.85 kPa is subjected to an initial axial disturbance defined by

$$u_0(x) = 0$$

$$v_0(x) = \begin{cases} \cos x & \text{if } |x| \leq \pi/2 \\ 0 & \text{elsewhere} \end{cases}$$

where u_0 and v_0, respectively, denote initial displacement and initial velocity. Determine the displacement induced by the disturbance at a distance $x = -\pi/3$ and time $t = 2\pi/3$.

4.5 During the process of driving a 40 m long reinforced concrete pile into a foundation soil, a pile-driving hammer imparts a force impulse that can be assumed to vary as a half sine wave with an amplitude of 2500 kN and a duration of 0.012 s. To monitor the stresses in the pile during the driving process, strain gauges are attached at the middle of the pile and at 5 m from its tip. Determine the maximum tensile and compressive stresses recorded by the gauges (a) when the pile is being driven through a soft soil that offers no resistance to the penetration of the pile and (b) when the pile tip encounters rigid bedrock. Consider a modulus of elasticity of 20,000 MPa, a diameter of 500 mm, and a density of 2300 kg/m³.

4.6 The 25 story building shown in Figure E4.3a and considered in Example 4.3 is subjected to the lateral ground displacement pulse shown in Figure P4.6. Modeling the building as a shear beam, obtain the lateral displacements at the top of the building and at the level of its setback 2 s after the application of the pulse.

4.7 Assume that during an earthquake the horizontal motion at the ground surface of a deep deposit of soil is generated by a shear wave traveling parallel to the ground surface. Assume further that records of the associated ground velocity have been obtained at a point on the surface along two perpendicular directions. Derive an equation to determine the rotational motion experienced by the ground about an axis perpendicular to the ground surface passing through the point where the records were obtained. Express the desired equation in terms of the two horizontal components of the ground velocity and the shear wave velocity of the soil deposit.

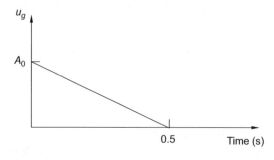

FIGURE P4.6 Loading pulse in Problem 4.6.

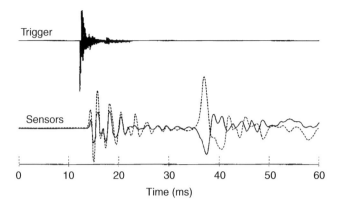

FIGURE P4.9 Ground motion records referred to in Problem 4.9. (After Kramer, S.L., *Geotechnical Earthquake Engineering*, Prentice-Hall, Inc., Upper Saddle River, NJ, 1996. Reprinted with permission from Pearson Education, Inc., Upper Saddle River, NJ.)

4.8 Consider two harmonic displacement waves, both with a unit amplitude. The first wave has a wavelength of 10 m and a period of 1 s. The corresponding parameters of the second wave are 11 m and 1.1 s, respectively. Plot the variation with distance for the first 300 m of the displacement induced by the superposition of the two waves 2 s after the initiation of the waves, and determine the wavelength of the apparent wave from this plot.

4.9 The ground motion records shown in Figure P4.9 are obtained from sensors located at the same depth in two vertical boreholes spaced 5 m apart. The recorded motions resulted from downward (solid line) and upward (dashed line) mechanical blows imparted at the level of the sensors in a third borehole that was colinear with the other boreholes. The trigger that started the recording of the motions was installed in the borehole that was the nearest to the borehole where the disturbance was applied. Determine the average shear wave velocity of the soil between the sensors.

4.10 Find the amplitude of the vertical and horizontal displacements at depths of 10 and 150 m below the ground surface generated by the passing of a Rayleigh wave with a wavelength of 100 m. Consider a medium with a Poisson ratio of 0.30.

4.11 A 40 m layer of soil has a unit weight of 20 kN/m^3, a modulus of elasticity of 2×10^6 kPa, and a Poisson ratio of 0.20. Below this layer of soil there is a homogeneous semi-infinite rock formation with a unit weight of 25 kN/m^3, a modulus of elasticity of 25×10^6 kPa, and a Poisson ratio of 0.25. Determine the wavelength of the Love wave that would travel through this two-layer medium with a velocity equal to the average of the shear wave velocities of the two layers.

4.12 A harmonic one-dimensional shear wave with a frequency of 2 Hz travels through sandstone with a shear modulus of elasticity of 2×10^6 psf, a unit weight of 118 pcf, and a quality factor of 30. Determine the percentage by which the amplitude of the wave is reduced after it travels a distance of 10,000 ft.

5 Measurement of Earthquakes

It is our thesis that even the worst earthquake has some virtues—that by studying it we can learn something about the structure of the earth, about the causes of earthquakes, and about how we may protect ourselves against their damage.

John H. Hodgson, 1964

5.1 INTRODUCTION

As is well known to those who have experienced them, not all earthquakes have the same intensity. Some are barely felt, some are felt strongly but cause only moderate damage, and yet some others are so strong that are capable of producing widespread and catastrophic damage. From the engineering point of view, it is thus important to have a scale with which one can measure or quantify the intensity of earthquakes. This chapter will describe the different scales that throughout the years have been devised to measure the size of earthquakes and are still of relevance today. It will also give a description of the instruments that are employed nowadays to record the ground motions generated by earthquakes and collect the information that is needed to determine the earthquake size and the location where earthquakes originate. It will present, in addition, some techniques to determine this location.

5.2 INTENSITY SCALES

Intensity scales are among the first measurement systems devised to characterize the strength of earthquakes. These scales are based on a qualitative description of the damage caused by an earthquake to the natural and built environment at a particular location and the associated human reaction. The use of an intensity scale to measure the strength of an earthquake dates back to 1564 with the introduction of the Gastaldi scale. Of a more recent vintage are the intensity scales developed in the 1880s by M. S. de Rossi in Italy and Francois Forel in Switzerland for European conditions, and a refined version of these scales devised by the Italian seismologist Giuseppe Mercalli in 1902. In recent times, the most widely used intensity scale in North America and other parts of the world is a modified version of the Mercalli scale introduced by Harry O. Wood and Frank Newman in 1931 for U.S. conditions. This scale, known as the *Modified Mercalli Intensity scale* or *MMI scale* and described in Box 5.1, is also based on an assessment of the local destructiveness induced by an earthquake and the way people react to it. It is composed of 12 grades, ranging from Grade I for an earthquake that is not felt by the people to Grade XII for an earthquake that causes total destruction. Other modern intensity scales are the 8-grade scale of the Japanese Meteorological Agency (JMA), developed in 1949 for Japanese conditions (see Box 5.2), and the 12-grade Medvedev–Sponheuer–Karnik (MSK) scale, introduced in 1964 and intended for international use (see Box 5.3).

Box 5.1 Modified Mercalli Intensity (MMI) Scale

  I. Not felt except by a very few under especially favorable circumstances.
 II. Felt only by a few persons at rest, especially on upper floors of buildings. Delicately suspended objects may swing.
III. Felt quite noticeably indoors, especially on upper floors of buildings, but many people do not recognize it as an earthquake. Standing motor cars may rock slightly. Vibration like passing of truck. Duration estimated.
 IV. During the day felt indoors by many, outdoors by few. At night some awakened. Dishes, windows, doors disturbed; walls make cracking sound. Sensation like heavy truck striking building. Standing motor cars rocked noticeably.
  V. Felt by nearly everyone, many awakened. Some dishes, windows, etc., broken; a few instances of cracked plaster; unstable objects overturned. Disturbances of trees, poles, and other tall objects sometimes noticed. Pendulum clocks may stop.
 VI. Felt by all, many frightened and run outdoors. Some heavy furniture moved; a few instances of fallen plaster or damaged chimneys. Damage slight.
VII. Everybody runs outdoors. Damage negligible in buildings of good design and construction; slight to moderate in well-built ordinary structures; considerable in poorly built or badly designed structures; some chimneys broken. Noticed by persons driving motor cars.
VIII. Damage slight in specially designed structures; considerable in ordinary substantial buildings, with partial collapse; great in poorly built structures. Panel walls thrown out of frame structures. Fall of chimneys, factory stacks, columns, monuments, walls. Heavy furniture overturned. Sand and mud ejected in small amounts. Changes in well water. Persons driving motor cars disturbed.
 IX. Damage considerable in specially designed structures; well-designed frame structures thrown out of plumb; great in substantial buildings, with partial collapse. Buildings shifted off foundations. Ground cracked conspicuously. Underground pipes broken.
  X. Some well-built wooden structures destroyed; most masonry and frame structures destroyed with foundations; ground badly cracked. Rails bent. Landslides considerable from river banks and steep slopes. Shifted sand and mud. Water splashed over banks.
 XI. Few, if any, (masonry) structures remain standing. Bridges destroyed. Broad fissures in ground. Underground pipelines completely out of service. Earth slumps and land slips in soft ground. Rails bent greatly.
XII. Damage total. Waves seen on ground surface. Lines of sight and level are distorted. Objects thrown into the air.

Box 5.2 Japanese Seismic Intensity Scale

  0. No sensation: registered by seismographs but no perception by the human body
  I. Slight: felt by persons at rest or persons especially sensitive to earthquakes
 II. Weak: felt by most persons; slight rattling of doors and Japanese latticed paper sliding doors (shoji)
III. Rather strong: shaking of houses and buildings; heavy rattling of doors and shoji; swinging of chandeliers and other hanging objects; movement of liquids in vessels
 IV. Strong: strong shaking of houses and buildings; overturning of unstable objects; spilling of liquids out of vessels four-fifths full
  V. Very strong: cracking of plaster of walls; overturning of tombstones and stone lanterns; damage to masonry chimneys and mud-plastered warehouses

VI. Disastrous: demolition of up to 30% of Japanese wooden houses; numerous landslides and embankment failures; fissures on flat ground
VII. Ruinous: demolition of more than 30% of Japanese wooden houses

---

Box 5.3   Medvedev–Sponheuer–Karnik Intensity (MKS) Scale

I. Not noticeable. Intensity of vibration is below limit of sensibility. Tremor is detected and recorded by seismographs only.

II. Scarcely noticeable. Vibration is felt only by individual people at rest in houses, especially on upper floors of buildings.

III. Weak, partially observed only. Earthquake is felt indoors by a few people, outdoors only in favorable circumstances. Vibration is like that due to the passing of a light truck. Attentive observers notice a slight swinging of hanging object, somewhat more heavily on upper floors.

IV. Largely observed. Earthquake is felt indoors by many people, outdoors by few. Here and there people awake, but no one is frightened. Vibration is like that due to the passing of a heavily loaded truck. Windows, doors, and dishes rattle. Floors and walls creak. Furniture begins to shake. Hanging objects swing slightly. Liquids in open vessels are slightly disturbed. In standing motor cars the shock is noticeable.

V. Awakening. Earthquake is felt indoors by all, outdoors by many. Many sleeping people awake. A few run outdoors. Animals become uneasy. Buildings tremble throughout. Hanging objects swing considerably. Pictures knock against walls or swing out of place. Occasionally pendulum clocks stop. Few unstable objects may be overturned or shifted. Open doors and windows are thrust open and slam back again. Liquids spill in small amounts from well-filled open containers. The sensation of vibration is like that due to a heavy object falling inside the building. Slight damage of Grade[1] 1 in buildings of Type[2] A is possible. Sometimes change in flow of springs.

VI. Frightening. Felt by most indoors and outdoors. Many people in buildings are frightened and run outdoors. A few persons lose their balance. Domestic animals run out of their stalls. In few instances dishes and glassware may break, books fall down. Heavy furniture may possibly move and small steeple bells may ring. Damage of Grade 1 is sustained in single buildings of Type B and in many of Type A. Damage in few buildings of Type A is of Grade 2. In few cases cracks up to widths of 1 cm possible in wet ground. In mountains occasional land slips; changes in flow of springs and in level of well water are observed.

VII. Damage to buildings. Most people are frightened and run outdoors. Many find it difficult to stand. The vibration is noticed by persons driving motor cars. Large bells ring. In many buildings of Type C damage of Grade 1 is caused; in many buildings of Type B damage is of Grade 2. Many buildings of Type A suffer damage of Grade 3, few of Grade 4. In single instances land slips of roadway on steep slopes; cracks in roads; seams of pipelines damaged; cracks in stone walls. Waves are formed on water, and water is made turbid by mud stirred up. Water levels in wells change, and the flow of springs changes. In few cases dry springs have their flow restored and existing springs stop flowing. In isolated instances parts of sandy or gravelly banks slip off.

VIII. Destruction of buildings. Fright and panic; also persons driving motor cars are disturbed. Here and there branches of trees break off. Even heavy furniture moves and partly overturns. Hanging lamps are in part damaged. Many buildings of Type C suffer damage of Grade 2, few of Grade 3. Many buildings of Type B suffer damage of Grade 3 and few of Grade 4, and many buildings of Type A suffer damage of Grade 4 and few of Grade 5.

Occasional breakage of pipe seams. Memorials and monuments move and twist. Tombstones overturn. Stone walls collapse. Small land slips in hollows and on banked roads on steep slopes; cracks in ground up to widths of several centimeters. Water in lakes becomes turbid. Dry wells refill and existing wells become dry. In many cases change in flow and level of water.

IX. General damage to buildings. General panic; considerable damage to furniture. Animals run to and fro in confusion and cry. Many buildings of Type C suffer damage of Grade 3, a few of Grade 4. Many buildings of Type B show damage of Grade 4, a few of Grade 5. Monuments and columns fall. Considerable damage to reservoirs; underground pipes partly broken. In individual cases railway lines are bent and roadways damaged. On flat land overflow of water, sand, and mud is often observed. Ground cracks to widths of up to 10 cm, on slopes and river banks more than 10 cm; furthermore a large number of slight cracks in ground; fall of rocks, many landslides, and earth flows; large waves on water. Dry wells renew their flow and existing wells dry up.

X. General destruction of buildings. Many buildings of Type C suffer damage of Grade 4, a few of Grade 5. Many buildings of Type B show damage of Grade 5; most of Type A have destruction Type 5; critical damage to dams and dikes and severe damage to bridges. Railway lines are bent slightly. Underground pipes are broken or bent. Road paving and asphalt show waves. In ground, cracks up to widths of several decimeters, sometimes up to 1 m. Parallel to water courses broad fissures occur. Loose ground slides from steep slopes. From river banks and steep coasts considerable landslides are possible. In coastal areas displacement of sand and mud; change of water level in wells; water from canals, lakes, rivers, etc., thrown on land. New lakes occur.

XI. Catastrophe. Severe damage even to well-built buildings, bridges, water dams, and railway lines; highways become useless; underground pipes destroyed. Ground considerably distorted by broad cracks and fissures, as well as by movement in horizontal and vertical directions; numerous land slips and falls of rock.

XII. Landscape changes. Practically all structures above and below ground are greatly damaged or destroyed. The surface of the ground is radically changed. Considerable ground cracks with extensive vertical and horizontal movements are observed. Falls of rock and slumping of river banks over wide areas; lakes are dammed; waterfalls appear, and rivers are deflected.

[1]Damage to buildings is classified into the following five grades:
  Grade 1: Slight damage: Fine cracks in plaster; fall of small pieces of plaster
  Grade 2: Moderate damage: Small cracks in walls; fall of fairly large pieces of plaster; particles slip off; cracks in chimneys; parts of chimneys fall down
  Grade 3: Heavy damage: Large and deep cracks in walls; fall of chimneys
  Grade 4: Destruction: Gaps in walls; parts of buildings may collapse; separate parts of building lose their cohesion; inner walls and filled-in walls of the frame collapse
  Grade 5: Total damage: Total collapse of buildings

[2]Buildings are classified into the following three types:
  Structure A: Buildings in field stone, rural structures, adobe houses, clay houses
  Structure B: Ordinary brick buildings, buildings of the large block and prefabricated type, half-timbered structures, buildings in natural hewn stone
  Structure C: Reinforced buildings, well-built wooden structures

In dealing with intensity scales, it is important to keep in mind that they do not involve a precise scientific measurement of the severity of earthquakes and are therefore of limited value. The problem with these scales is that they depend on subjective factors such as (a) previous experience of people with earthquakes, (b) local design and construction practices, (c) whether or not the earthquake occurs in an inhabited region, and (d) the population density. For example, the description *many frightened* in the MMI scale will depend on the location of the earthquake. A tremor that would alarm the residents of Cleveland, Ohio, would most likely be ignored by people in Tokyo or Los Angeles. Likewise, the collapse of buildings, a key factor in determining an intensity rating, may not only reflect the power of an earthquake but also whether or not the collapsed buildings were designed to resist seismic loads. Intensity scales cannot therefore by their own nature be accurate.

Despite their limitations, intensity scales may be useful to estimate the size and location of earthquakes that occurred prior to the development of modern seismic instruments. Because qualitative descriptions of the effects of earthquakes are often available through historical records, intensity scales may be used to characterize the rate of earthquake recurrence at the locations wherever these historical records are available. Intensity scales may also be useful to describe the distribution of damage in a region, to identify areas of poor soils, and to approximately locate the earthquake epicenter. For this reason, even in modern times, contours of equal intensities, or *isoseisms*, are routinely plotted over a map of the geographical regions where strong earthquakes occur. Examples of these plots, called *isoseismal maps*, are shown in Figures 5.1 and 5.2 for the 1989 Loma Prieta and the 1994 Northridge earthquakes in California.

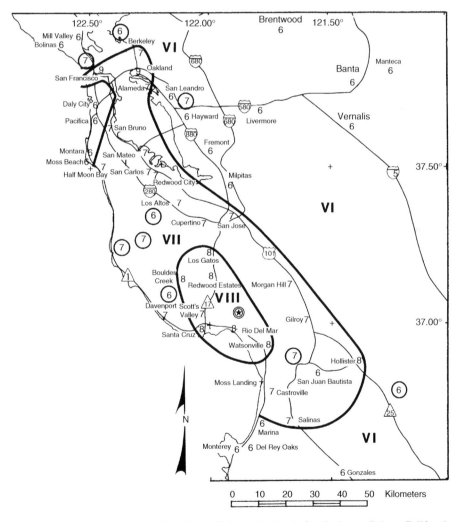

FIGURE 5.1 Isoseismal map of Modified Mercalli Intensity levels for the Loma Prieta, California, earthquake of October 17, 1989. Location of epicenter is shown by circled star. (Reproduced from Performance of Structures During the Loma Prieta Earthquake of October 17, 1989, NIST Special Publication 778, National Institute of Standards and Technology, 1990.)

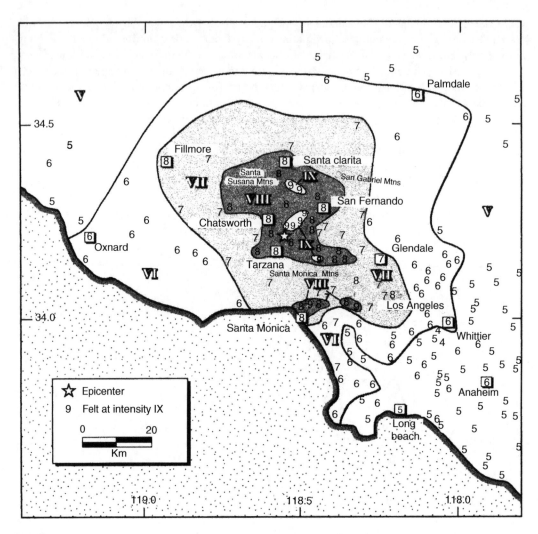

FIGURE 5.2 Isoseismal map of Modified Mercalli Intensity levels for the Northridge, California, earthquake of January 17, 1994. (After Dewey, J.W., Reagor, B.G., and Dengler, L., Isoseismal map of the Northridge, California, earthquake of January 17, 1994, 89th Annual Meeting of Seismological Society of America, Pasadena, CA, 1994. © Seismological Society of America.)

5.3 SEISMOGRAPHS AND SEISMOGRAMS

5.3.1 HISTORICAL BACKGROUND

Aware that intensity scales were based on subjective appraisals of damage submitted by a wide variety of observers, scientists studying earthquakes at the end of the nineteenth century realized that an understanding of the earthquake phenomenon required accurate and consistent physical measurements. They felt, thus, the need for advanced devices that would record and preserve the ground motion generated by earthquakes.

One of the first of such devices was developed by Filippo Cecchi in Italy in 1875. Cecchi's instrument was designed to start a clock and a recording device at the first sign of shaking. It then recorded the relative motion between a pendulum and the shaking ground as a function of time. The oldest known record produced by this instrument is dated February 23, 1887. Afterward, John Milne, an English geologist working in Japan in the late 1800s, developed with the help of James Ewing, a professor of mechanical engineering and physics, and Thomas Gray, a professor of telegraphic engineering, the first instrument to record the motion of the ground in all of its three directions: up and down, back and forth, and side to side. This was accomplished with three independent pendulums, each with an attached stylus that inscribed the motion of the pendulum on a roll of smoked paper. In addition, the instrument was implemented with a clock mechanism that enabled each device to operate for 24 h at a time and stamp the paper with the precise time of the first wave arrival. This instrument became known as the *Milne seismograph* and was for many years the standard equipment for seismologists around the world. With it, a new era in the study of earthquakes began.

As remarkable as it was at that time, the seismograph that Milne, Ewing, and Gray developed at the end of the nineteenth century fell short of meeting the demands of a science that had more questions than answers. One of its shortcomings was that it responded only to motions with periods within a narrow range. Another was the tendency of the seismograph's pendulum to keep swinging indefinitely once in motion. Without a way to control this free-vibration motion, the seismograph was unable to record accurately the motion generated by late-arrival waves. A substantial improvement came in 1898, when Emil Wiechert in Germany introduced a viscous damping mechanism that restrained the seismograph pendulum and greatly improved its accuracy.

The next advance in the development of the modern seismograph was made in 1914 by Boris Galitzin, a Russian seismologist. Galitzin introduced a design that did away with the need for a mechanical linkage between pendulum and recorder. He mounted a wire coil on the seismograph's pendulum and suspended the coil and the pendulum between the poles of a magnet fixed to the ground. This way, the motion of the pendulum generated an electric current that was proportional to the pendulum's velocity, which Galitzin used to rotate a galvanometer coil. Light reflected from a mirror on the galvanometer coil was then recorded on photographic paper. This was a development that dominated the seismograph design throughout the twentieth century and had profound implications for seismology. With it, designers were able to use smaller pendulum masses and eliminate the unwanted friction between a stylus and a recording medium. Equally important, it meant the ability to transmit the recordings from remote seismographs to a central location, eliminating thus the need for traveling periodically from site to site to monitor the instruments.

5.3.2 COMPONENTS AND DESIGN FEATURES

As it may be inferred from the previous discussion, seismographs are devices that record the variation with time of the ground displacements generated by earthquakes. As shown schematically in Figure 5.3, they are ordinarily composed of a pendulum, a damping element, a stylus attached to

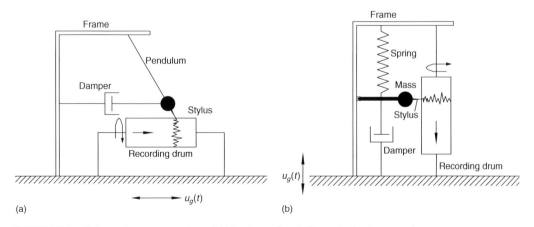

FIGURE 5.3 Schematic representation of (a) horizontal and (b) vertical seismographs.

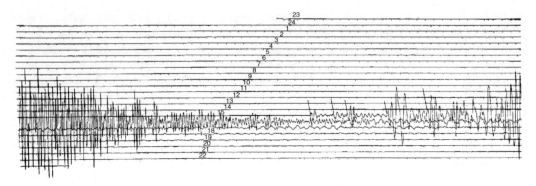

FIGURE 5.4 Record from seismograph obtained May 18–19, 1962, in Alert, northern Canada, of an earthquake that originated off the coast of Mexico. (After Hodgson, J.H., *Earthquakes and Earth Structure*, Prentice-Hall, Englewood Cliffs, NJ, ©1964. Reprinted with permission from Pearson Education, Inc., Upper Saddle River, NJ.)

the pendulum mass, and a recording drum. To be able to detect distant earthquakes, seismographs are also implemented with a mechanical, optical, or electromagnetic system that magnifies the movement of the pendulum up to hundred-, thousand-, or million-fold, respectively. The damping element is normally an air, oil, or electromagnetic damper and is introduced into the system to damp out the motion of the pendulum as soon as the ground motion stops. Dampers with a damping ratio of the order of 80% of critical are used for this purpose. The recording drum holds wrapped around a waxed or smoked paper on which the stylus inscribes the movement of the pendulum. As mentioned in the previous section, in some designs this motion is recorded on photographic paper or magnetic tape. The recording drum continuously rotates at a fixed rate and moves from left to right on its shaft in a period of 24 h, producing thus a continuous record of the pendulum motion. However, when the paper is removed from the drum and is laid flat, the record appears as a number of parallel lines, as it can be seen in the typical record shown in Figure 5.4. Another design feature of seismographs is the time marks they introduce on the record at regular intervals. These time marks are produced by deflecting the trace for a second or two, usually at the end of every minute (see Figure 5.4). By convention, seismographs are marked in terms of Universal Time Coordinated (UTC) [also called Greenwich Mean Time (GMT)], not local time. Seismographs operate 24 h a day, 365 days a year. As earthquake-generated ground motions are neither purely vertical nor purely horizontal, in practice a vertical seismograph and two horizontal seismographs are used to record three components of ground motion along three orthogonal directions. This way the motion of the ground at any instant may be completely characterized by the vector sum of the motion recorded along each of these three directions. Figure 5.5 shows the interior of a modern electromagnetic seismograph.

As it may be observed from Figure 5.3, a seismograph can only measure the relative motion between its pendulum mass and the recording drum, which is fixed to the moving ground. To gain, then, an insight as to how a seismograph measures the motion of the ground, consider the following simplified formulation. Assume first that the earthquake ground motion is predominantly sinusoidal with amplitude u_0 and dominant period T_g, and that the displacement, velocity, and acceleration of the ground may be expressed approximately as

$$u = u_0 \sin\left(\frac{2\pi}{T_g}t\right) \quad v = v_0 \cos\left(\frac{2\pi}{T_g}t\right) \quad a = a_0 \sin\left(\frac{2\pi}{T_g}t\right) \tag{5.1}$$

Measurement of Earthquakes

FIGURE 5.5 Modern electromagnetic three-component, surface-mount broadband seismometer. This seismometer weighs 7.7 kg. (17 lb), is 190.5 mm (7 1/2 in.) in diameter, and is 304.8 mm. (12 in.) in height including leveling feet and handle. The instrument is designed for use in a well-insulated surface vault or any other environment with a stable temperature. (Photograph courtesy of Geotech Instruments, LLC.)

where $v_0 = (2\pi/T_g)u_0$ and $a_0 = (2\pi/T_g)^2 u_0$. Consider then that in terms of such assumptions the relative motion between the pendulum and the ground may be expressed by the differential equation of motion

$$m\ddot{u} + c\dot{u} + ku = -m\frac{4\pi^2}{T_g^2}u_0 \sin\left(\frac{2\pi}{T_g}t\right) \tag{5.2}$$

where m denotes the mass of the pendulum, c is the damping coefficient of the seismograph's damper, k the corresponding spring constant, and u the displacement of the pendulum relative to the ground surface. Consider, in addition, that the solution of Equation 5.2 is of the form

$$u(t) = -u_0 \frac{\sin\left(\frac{2\pi}{T_g}t - \theta\right)}{\sqrt{\left[1 - \left(\frac{T_g}{T_n}\right)^2\right]^2 + \left[2\xi\frac{T_g}{T_n}\right]^2}} \tag{5.3}$$

where T_n and ξ, respectively, denote the seismograph's natural period and damping ratio, and θ is a phase angle. Hence, the displacement amplitude recorded by a seismograph divided over that of the ground may be expressed as

$$\frac{u}{u_0} = \frac{1}{\sqrt{\left[1-\left(\frac{T_g}{T_n}\right)^2\right]^2 + \left[2\xi\frac{T_g}{T_n}\right]^2}} \tag{5.4}$$

Similarly, by dividing both sides of Equation 5.4 by $2\pi/T_g$ and $(2\pi/T_g)^2$, respectively, the recorded displacement amplitude over the amplitude of the ground velocity v_0 or ground acceleration a_0 may be written as

$$\frac{u}{v_0} = \frac{1}{2\pi} \frac{T_g}{\sqrt{\left[1-\left(\frac{T_g}{T_n}\right)^2\right]^2 + \left[2\xi\frac{T_g}{T_n}\right]^2}} \tag{5.5}$$

$$\frac{u}{a_0} = \frac{1}{4\pi^2} \frac{T_g^2}{\sqrt{\left[1-\left(\frac{T_g}{T_n}\right)^2\right]^2 + \left[2\xi\frac{T_g}{T_n}\right]^2}} \tag{5.6}$$

A plot of Equations 5.4 through 5.6 in a tripartite graph for the particular case of $T_n = 1.0$ s and damping ratios of 0.25, 0.50, and 1.0 are shown in Figure 5.6. The magnification factor u/v_0 is measured along the vertical axis, whereas u/u_0 and u/a_0 are measured along the axes inclined at 45°. Observe, thus, that for values of the damping ratio ξ around 50%, the amplification factor u/u_0 is approximately constant over the range of periods for which T_g is smaller than T_n. Similarly, observe that for large damping ratios, the amplification factor u/v_0 is approximately constant in the range of periods where T_g is roughly equal to T_n. Finally, observe that for intermediate values of the damping ratio ξ, the amplification factor u/a_0 is approximately constant for the range of periods where T_g is longer than T_n. This means that the motion recorded by a seismograph is proportional to the ground displacement if T_g is smaller than T_n, proportional to the ground velocity if T_g is roughly equal to T_n, and proportional to the ground acceleration if T_g is longer than T_n.

Consequently, if the predominant periods of earthquake ground motions are more or less known, the natural frequency and damping ratio of an instrument may be selected so as to measure a motion that is proportional to the displacement, velocity, or acceleration of the ground. In such a case, the only difference between the recorded motion and the actual ground motion would be a constant of proportionality, which can be easily determined from calibration tests. Seismographs are normally designed to measure displacements and are therefore constructed with a high damping ratio and a natural period that is short in comparison with the dominant period of the motions they are supposed to measure. For example, the *Wood–Anderson seismograph*, a seismograph used by Richter to develop his magnitude scale (introduced in Section 5.5), has a natural period of 0.8 s and a damping ratio of 80%.

If carefully designed, seismographs can respond equally to excitations with a considerable range of periods. But earthquake motions can have periods ranging from a few tenths of a second to many minutes, and no dynamic system can respond equally to excitations with such a period range. For this reason, some seismographs are designed to respond well to short-period (SP) waves

Measurement of Earthquakes

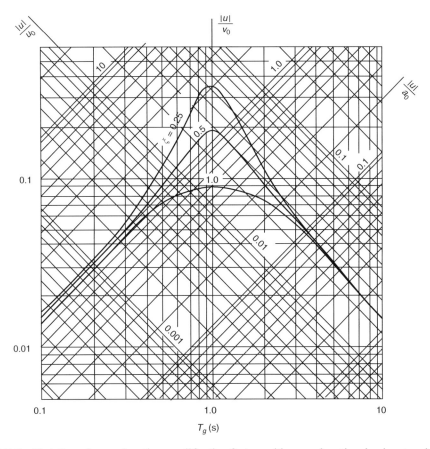

FIGURE 5.6 Variation of ground motion amplification factors with ground motion dominant period for the case of seismograph with natural period of 1.0 s.

(i.e., periods from one-fifth of a second to two seconds), while some others are designed for long-period waves (i.e., periods between 15 and 100 s). Typical dynamic response curves for such instruments are shown in Figure 5.7, where the magnification factor shown represents the ratio of recorded amplitude to actual ground displacement. A well-equipped seismographic station will have thus more than one set of instruments (see Figure 5.8).

5.3.3 Seismograms

The records obtained from a seismograph are called *seismograms*. A seismogram is thus a record of the variation with time of the displacement of the ground, magnified by the magnification factor of the seismograph, at the location where the seismograph is installed. A typical seismogram is shown in Figure 5.4. The numbers in the middle of the record indicate the hours referred to the GMT. The small deflections at regular intervals along the trace are time marks at 1 min intervals. There are 60 such marks in each line, so each line represents the motion recorded during 1 h.

As explained in Chapter 4, the arrival of P, S, and surface waves may be detected in a seismogram by considering that the P, S, and surface waves arrive at different times, have different amplitudes, and have different dominant periods (i.e., interval between peaks). This way, the arrival of the P and S waves can be clearly identified in the seismogram of Figure 5.4. In Figure 5.9, which shows an enlargement of the initial portion of the seismogram in Figure 5.4, these arrivals are identified and marked with the labels P and S.

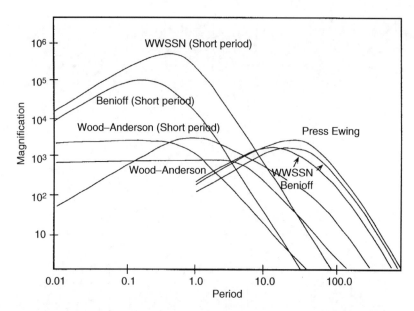

FIGURE 5.7 Magnification curves for typical short-period and long-period seismographs. (Courtesy of Kanamori, H., California Institute of Technology.)

FIGURE 5.8 Typical seismographic station.

It should be noted that the traces in a seismogram are never without some little ripples. These little ripples show up in a seismogram because seismographs are so sensitive that they are able to detect the ever-present background noise of the earth. They are called *microseisms* and arise from local disturbances such as traffic on the streets, effect of winds on trees, breaking of the surf on the beach, and other natural and human-made disturbances.

5.4 LOCATION OF EARTHQUAKE HYPOCENTER

As currently determined by seismologists, the location of an earthquake hypocenter involves a complex three-dimensional seismic velocity model of the earth and error minimization techniques. However, a rough estimate can be obtained using the arrival times of the P and S waves recorded in

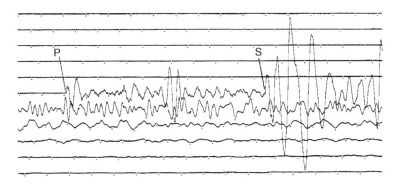

FIGURE 5.9 Enlarged initial portion of the seismogram in Figure 5.4 showing the arrival of P and S waves. (After Hodgson, J.H., *Earthquakes and Earth Structure*, Prentice-Hall, Englewood Cliffs, NJ, ©1964. Reprinted with permission from Pearson Education, Inc., Upper Saddle River, NJ.)

the seismograms of the earthquake. This technique is based on the principle that it takes longer for S waves than for P waves to arrive to an observation point. That is, it considers the fact that there is always a time lag between the arrival of P and S waves at any observation point, and that this time lag is directly proportional to the distance from the observation point to the earthquake hypocenter. As in terms of the velocities of propagation of P and S waves this time lag may be expressed as

$$t_{sp} = \frac{\Delta}{V_s} - \frac{\Delta}{V_p} = \left(\frac{1}{V_s} - \frac{1}{V_p}\right)\Delta \tag{5.7}$$

where V_p is average P-wave velocity; V_s average S-wave velocity; and Δ distance from earthquake focus to observation point (hypocentral distance), then the distance from earthquake focus to an observation point can be determined from

$$\Delta = \frac{t_{sp}}{\frac{1}{V_s} - \frac{1}{V_p}} \tag{5.8}$$

or, approximately, from

$$\Delta = (8 \text{ km/s}) t_{sp} \tag{5.9}$$

if it is taken into account that the average P-wave and S-wave velocities in rock are approximately equal to 6.0 and 3.5 km/s, respectively.

On the basis of such a time-lag concept, the epicenter of an earthquake may be thus located by (a) identifying the time lag between the arrival of the P and S waves in the earthquake seismograms recorded at several seismographic stations, (b) computing the respective hypocentral distances using the identified time lags and either Equation 5.8 or 5.9, and (c) drawing on a map of the region circles with center in each station and radii equal to the calculated hypocentral distances. The epicenter of the earthquake will be located at the point where the drawn circles intersect (see Figure 5.10).

Note that this procedure requires the seismograms from at least three stations as less than three circular arcs cannot define a unique intersection point. However, as ordinarily many stations contribute to an epicentral determination, this requirement does not constitute a problem. Notice also that seldom the arcs will intersect at a single point. Most likely, they will define a small

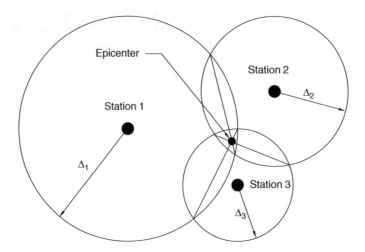

FIGURE 5.10 Earthquake epicenter defined by intersection of circles with radii equal to hypocentral distances.

triangle instead. In such a case, the location of the earthquake epicenter will be defined by the intersection of the cords of the intersecting arcs (see Figure 5.10).

In practice, epicentral distances are not determined using an equation of the form of Equation 5.8 or 5.9 but, rather, from average *travel-time curves* or tables that directly give the time it takes for a P or an S wave to travel a given distance through the surface or the interior of the earth. These travel-time curves are developed by correlating the wave arrival times recorded at a large number of seismographic stations around the world, and estimates of the distances from each of these stations to the epicenters of the recorded earthquake, and by averaging these correlations over thousands of earthquakes. Travel-time curves are remarkably accurate and, for teleseismic distances, they can predict travel times with an accuracy of a few seconds. Figure 5.11 shows typical average travel-time curves for P, S, Love, and Rayleigh waves for surface events.

An alternative procedure that can give accurate estimates of an earthquake epicenter when seismograms from a large number of stations are available involves the determination of the earthquake initiation time using a simple graphical technique known as *Wadati diagram*. In this diagram, the time lag between the arrival of P and S waves observed at several seismographic stations are plotted against the corresponding arrival times of the P waves. As this time lag goes to zero at the hypocenter, a straight-line fit of the data points and a reading of the intercept of this line with the P-wave arrival time axis will give the approximate time at which the earthquake began. Figure 5.12 shows an example of a Wadati diagram. Once the earthquake initiation time has been estimated, the epicentral distances from each of the stations are determined by multiplying the corresponding travel times of the P wave by an estimate of the average P-wave velocity. That is, if t_{pi} represents the arrival time of the P wave to the ith station, t_0 is the time at which the earthquake begins, and V_p is the average P-wave velocity, the epicentral distance from the ith station is given by

$$\Delta_i = (t_{pi} - t_0)V_p \tag{5.10}$$

Once the epicentral distances from various stations are determined this way, the epicenter is determined as with the previous method.

The methods outlined earlier for the location of an earthquake's epicenter and the determination of the earthquake's origin time presume a flat-layered earth. For the case of a spherical earth, the following trial and error procedure is used. First, the seismograms from several of the globally distributed stations that recorded the earthquake are collected. Second, noting that for each station

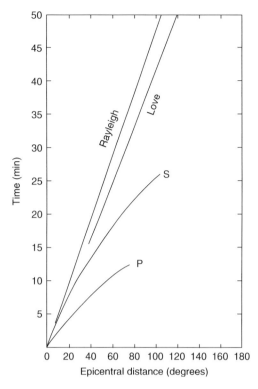

FIGURE 5.11 Average travel-time curves for P, S, Rayleigh, and Love waves for surface seismic events. (*Note*: 1° = 111 km.)

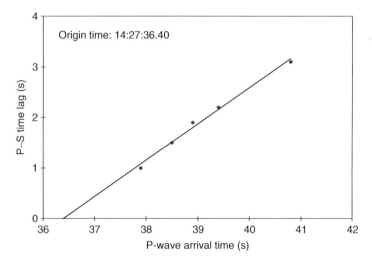

FIGURE 5.12 Example of Wadati diagram to determine initiation time of local earthquake.

the three coordinates of the hypocenter and the origin time are unknown, a value for each of these four unknowns is guessed and, on the basis of these guessed values, the expected arrival times of the P wave are calculated. These predictions are then compared to the observed times, and a calculation is made of how much in error the guessed values were. The initial guess is thus corrected and the process repeated until the difference between calculated and observed arrival times is acceptably small.

The focal depth of an earthquake is determined using the trial and error procedure just described, or a refined procedure that involves the identification of seismic waves that travel directly up from the earthquake hypocenter and are reflected back down before traveling to other parts of the earth. However, an estimate of it may be obtained by using data from a nearby station as follows. First, the location of the earthquake epicenter is established using one of the methods described earlier. Second, the hypocentral distance from the nearby station is determined in terms of the recorded time lag between the arrival of the P and S waves at that station or the travel time of the P wave. In addition, the distance between the earthquake epicenter and the station is computed considering the known location of the earthquake epicenter and the known location of the station. As it may be seen from Figure 5.13, the earthquake depth can then be determined from the simple trigonometric relationship between this depth and the epicentral and hypocentral distances from the nearby station. Note that this procedure necessarily requires the data from a nearby station as for a faraway station the hypocentral and epicentral distances are approximately the same.

EXAMPLE 5.1 LOCATION OF EARTHQUAKE EPICENTER

The arrival times of the P and S waves of an earthquake at three different seismological stations are listed in Table E5.1a. On the basis of this information, determine the location of the earthquake epicenter.

Solution

In terms of the arrival times listed in Table E5.1a and Equation 5.9, the time lag between the arrival of the P and S waves at each of the stations and the corresponding hypocentral distances are as indicated in Table E5.1b. Drawing arcs on a regional map from each station with radii equal to the found hypocentral distances, the earthquake epicenter is located, as shown in Figure E5.1, at the point where these three arcs intersect.

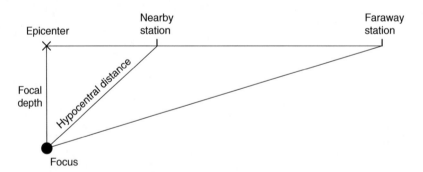

FIGURE 5.13 Focal depth in terms of epicentral and hypocentral distances from a nearby station.

TABLE E5.1A
Arrival Times of P and S Waves at Three Different Stations

Station	P Wave			S Wave		
	h	min	s	h	min	s
A	14	29	56.1	14	45	33.6
B	14	19	06.3	14	26	36.3
C	14	16	36.7	14	22	14.2

TABLE E5.1B
Time Lag between Arrival of P and S Waves at Three Different Stations and Corresponding Epicentral Distances

Station	Time Lag (s)	Δ (km)
A	937.5	7500
B	450.0	3600
C	337.5	2700

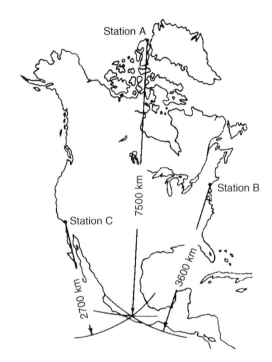

FIGURE E5.1 Location of epicenter of earthquake in Example 5.1 using hypocentral distances from three seismographic stations.

EXAMPLE 5.2 ESTIMATION OF FOCAL DEPTH

Based on the information from several seismographic stations, it is established that the epicenter of an earthquake is located at 39° N and 119.8° W. A station nearby the earthquake epicenter is located at 38.5° N and 120° W. The calculated hypocentral distance from this nearby station is 86.4 km. Find the earthquake's focal depth. Consider that one degree of latitude corresponds to a distance of 111 km and that one degree of longitude at latitude L corresponds to a distance of $111 \cos L$ km.

Solution

According to the provided coordinates, the distance between the nearby station and the earthquake epicenter is approximately equal to

$$\sqrt{(39.0 - 38.5)^2(111)^2 + (120.0 - 119.8)^2(111 \cos 38.75°)^2} = 58.1 \text{ km}$$

Hence, as the hypocentral distance from the station is equal to 86.4 km, the focal depth results as

$$\sqrt{86.4^2 - 58.1^2} = 64 \text{ km}$$

5.5 MAGNITUDE SCALES

5.5.1 Richter or Local Magnitude

Besides providing information for the location of earthquakes, seismograms also provide the information that is needed to estimate the size or strength of an earthquake in terms of what is called *earthquake magnitude*. This instrumentally quantified measure of earthquake strength is widely used nowadays by seismologists, engineers, and even the general public. Although in some cases it fails to give an accurate representation of the true strength of an earthquake, it is still routinely used to characterize the intensity of earthquakes and remains a key parameter in earthquake hazard analysis.

The concept of earthquake magnitude was introduced by Charles Richter in 1935 in an effort to overcome the limitations of the intensity scales, the only method used back then to describe and compare earthquakes. Following a fundamental idea first used by K. Wadati in Japan, Richter based his magnitude scale on a measurement of the wave motion recorded by a seismograph. He borrowed the term *magnitude* from astronomy as the relative brightness of stars (stellar magnitude) is referred to as magnitude. However, the analogy stops there because in astronomy a smaller magnitude means an increased brightness.

Richter defined his scale in terms of the peak amplitude of the trace recorded by the then standard Wood–Anderson seismograph, which, as observed earlier, has a magnification factor of 2800, a natural period of 0.8 s, and a damping ratio of 80%. However, because such an amplitude can vary significantly from earthquake to earthquake, he used the logarithm of it, as opposed to the amplitude itself, to compress the range of the scale. Similarly, as the amplitude of seismic waves decreases with distance from the earthquake epicenter, he set the measurement of this amplitude at a standard distance of 100 km. Furthermore, he described such a peak trace amplitude in relation to the peak trace amplitude that would be generated by a zero-magnitude earthquake; that is, a barely perceptible earthquake. For this purpose, he defined a zero-magnitude earthquake as that which theoretically would produce a seismogram with a peak trace of 1 μm (10^{-6} m) at a distance of 100 km. As introduced by Richter, earthquake magnitude is thus defined as the logarithm to base ten of the peak wave amplitude measured in micrometers recorded by a Wood–Anderson seismograph at a distance of 100 km from the earthquake epicenter. That is,

$$M = \log \frac{A}{A_0} \tag{5.11}$$

where A is the peak amplitude in micrometers in a seismogram (magnified ground displacement) recorded at 100 km from the earthquake epicenter, and A_0 the peak amplitude of a zero-magnitude earthquake, defined as 1 μm at a distance of 100 km.

This magnitude scale is known nowadays as *Richter magnitude* or *local magnitude* scale and is identified by the symbol M_L. It is used by seismologists to characterize shallow earthquakes recorded at distances of <600 km. Because it is a logarithmic scale, a one unit increase in magnitude means a 10-fold increase in ground motion amplitude. For example, an earthquake of magnitude 6.0 exhibits ten times the ground motion amplitude of a magnitude 5.0 earthquake, a hundred times the ground motion amplitude of a magnitude 4.0 earthquake, and thousand times the ground motion amplitude of a 3.0 earthquake. Note also that as it is based on calculated values, the scale has no upper limit and may even give negative magnitudes. The largest magnitude in record has been 8.9, corresponding to earthquakes recorded in the Columbia–Ecuador border in 1906 and off the coast of Japan in 1933. The great Chile earthquake in 1960 had a Richter magnitude of 8.25, while the Alaska earthquake in 1964 registered one of 8.6.

Example 5.3 Calculation of Earthquake Magnitude

The seismogram from an earthquake recorded by a seismograph located exactly at 100 km from the earthquake epicenter exhibits a peak amplitude of 1 mm. Determine the magnitude of the earthquake.

Solution

According to the definition of earthquake magnitude,

$$M = \log \frac{A}{A_0} = \log \frac{1000\,\mu m}{1\,\mu m} = \log 10^3 = 3$$

Therefore, the magnitude for the recorded earthquake is 3.0. Notice that an earthquake of magnitude 3.0 induces a ground displacement of 1 mm/2800 at a distance of 100 km from its epicenter.

Example 5.4 Ground Displacement Induced by Magnitude 9.0 Earthquake

Determine the peak ground displacement induced by an earthquake with a magnitude of 9.0 at a distance of 100 km from the earthquake epicenter.

Solution

According to the definition of earthquake magnitude, for an earthquake with a magnitude of 9.0,

$$M = \log \frac{A}{A_0} = \log \frac{A}{1\,\mu m} = 9$$

from which one obtains

$$A = 10^9\,\mu m = 1000\text{ m}$$

Hence, as the magnification factor of a Wood–Anderson seismograph is 2800, an earthquake of magnitude 9.0 induces a ground displacement of 1000 m/2800 = 0.38 m at a distance of 100 km from its epicenter.

As no seismograph is likely to be located exactly at 100 km from an earthquake's epicenter, an extrapolation or correction is needed to be able to determine earthquake magnitudes from seismograms obtained at epicentral distances other than 100 km. A common procedure is that of constructing a curve that defines the variation of ground motion amplitude with distance for the zero-magnitude earthquake. This curve, known as the A_0 curve, is determined empirically using data from a large number of earthquakes and is unique for each seismological station. They reflect the geological and geophysical conditions that surround each station. By means of this empirical curve and adopting the assumption that the ratio of the peak ground motion amplitudes at two given distances is the same independently of earthquakes magnitude, it is then possible to compute the magnitude of an earthquake using the original definition and data from any seismogram. To do so, one simply considers that A represents the peak amplitude in the selected seismogram and A_0 the zero-magnitude amplitude that corresponds to the epicentral distance to the station from which the seismogram is obtained. Richter, for example, developed the following empirical equation to define the variation with distance of the amplitude A_0 for earthquakes in Southern California:

$$\log A_0 = 5.12 - 2.56 \log \Delta \qquad (5.12)$$

In this equation, A_0 is in micrometers, Δ epicentral distance in kilometers, and $10 < \Delta < 600$ km.

Substitution of Equation 5.12 into Equation 5.11 and allowing for the nominal magnification factor of the Wood–Anderson seismograph, the magnitude of Southern California earthquakes may be thus computed by means of

$$M = \log a + 2.56 \log \Delta - 1.67 \qquad (5.13)$$

where a is the actual ground motion amplitude in micrometers.

EXAMPLE 5.5   EARTHQUAKE MAGNITUDE IN TERMS OF A_0 EMPIRICAL EQUATION

Determine the magnitude of an earthquake recorded in Southern California at a seismological station located 147 km from the earthquake epicenter. The recorded peak amplitude was 66 mm.

Solution

According to Richter's empirical equation, the zero-magnitude amplitude at a distance of 147 km is given by

$$\log A_0 = 5.12 - 2.56 \log(147) = -0.43$$

from which one obtains that $A_0 = 0.37$ μm. Hence, the magnitude of the earthquake is equal to

$$M = \log \frac{A}{A_0} = \log \frac{66000}{0.37} = 5.2$$

5.5.2   Surface- and Body-Wave Magnitudes

Richter's original definition of earthquake magnitude cannot be used for distant earthquakes, that is, earthquakes with epicentral distances over 600 km. The reason is that body waves decay more rapidly than surface waves and, therefore, the peak amplitude in a seismogram from a station far from the earthquake epicenter would be measured on a surface wave as opposed to a body wave, which would be the case for a nearby station. As a result, a large amplitude at a distant station may be so not because the earthquake is a large one but because the measurement is made on a slowly decaying surface wave. A related complication is that the attenuation of body and surface waves is essentially different. Richter's magnitude, in addition, is based on readings from seismograms recorded by a Wood–Anderson seismograph and cannot be calculated directly using the tracings made by the other seismographs operating around the world.

Because Richter's original scale is limited to local earthquakes at epicentral distances of no more than 600 km and recorded in only one kind of instrument, the desire for the global characterization of earthquake size made it necessary for the introduction of a new definition of earthquake magnitude. This new definition was the *surface-wave magnitude* introduced by B. Gutenberg and Richter himself in 1936. The definition of this magnitude is the same as that originally introduced by Richter, except that the measured amplitude corresponds, by convention, to that of a surface wave with a period of ~20 s. Beyond ~600 km, the seismograms of shallow earthquakes recorded by long-period seismographs are dominated by surface waves with a period of ~20 s. In general, therefore, surface-wave magnitude is determined using seismograms from long-period seismographs. In general, too, the record from the vertical component of motion is used in this definition. The symbol used to identify surface-wave magnitude is M_s. Using the format of Equation 5.13, surface-wave magnitude is at times alternatively defined by

$$M_s = \log a + 1.66 \log \Delta + 2.0 \qquad (5.14)$$

Measurement of Earthquakes

where a is the peak ground displacement measured in micrometers and Δ epicentral distance measured in degrees (360° corresponding to the circumference of the earth). Note that the surface-wave magnitude is based on the measurement of ground displacement and not the trace amplitude in a particular seismograph. It can be therefore quantified from seismograms obtained from any type of seismograph. Surface-wave magnitude is most commonly used to describe the size of moderate to large shallow earthquakes for which surface waves provide the largest motion in a seismogram, that is, earthquakes with epicentral distances of >1000 km and focal depths of <70 km.

Although widely used to characterize large worldwide earthquakes, surface-wave magnitude also has a limitation because deep focus earthquakes do not generate significant surface waves. As a result, another magnitude scale was developed to scale distant earthquakes with deep foci. This scale, introduced by Gutenberg in 1945, is also based on the original definition of magnitude, but considers the peak amplitude of the first few cycles of the P wave. It is called *body-wave magnitude* and is denoted by the symbol m_b. As many short-period instruments have a peak response near 1 Hz, in practice the period at which the body-wave magnitude is determined is 1 s. In the format of Equations 5.13 and 5.14, this magnitude is expressed by the equation

$$m_b = \log(a/T) + Q(h, \Delta) \tag{5.15}$$

where a is the actual ground motion amplitude in micrometers, T the measured wave period, and Q an empirical function of focal depth h and epicentral distance Δ.

Because the P wave is a distinctive phase in the seismograms of earthquakes with deep foci, body-wave magnitude is the best scale to characterize deep earthquakes. It is nevertheless limited because it relies on direct P waves and these direct waves are best observed at distances >1000 km. In addition, it cannot fully describe the strength of earthquakes that are the result of many kilometers of fault rupture as it is determined from short period waves. As a result of these limitations, body-wave magnitude is not used to characterize local earthquakes or extremely large events.

In dealing with the magnitude scales introduced earlier, it should be kept in mind that none of the three scales coincides. Furthermore, it should be realized that none of the scales is very effective to describe the strength of very large earthquakes, although it seems that the surface-wave magnitude is the best among the three of them.

EXAMPLE 5.6 CALCULATION OF SURFACE-WAVE MAGNITUDE

The peak amplitude of the Rayleigh wave (reduced to ground displacement) in a seismogram from a seismographic station located at a distance of 28° from an earthquake's epicenter is 4.3 μm. Determine using this information the earthquake's surface-wave magnitude.

Solution

According to the given data and the definition of surface-wave magnitude, as given by Equation 5.14, the earthquake's surface-wave magnitude is equal to

$$M_s = \log(4.3) + 1.66\log(28) + 2.0 = 5.0$$

5.6 SEISMIC MOMENT AND MOMENT MAGNITUDE

5.6.1 SCALE SATURATION

It has been observed that the magnitude of large earthquakes is always between 8.0 and 8.9, independently of the size of the ruptured fault area and independently of how strong they are felt. That is, it has been observed that the magnitude scales saturate for large earthquakes.

The saturation point for Richter magnitude is $\sim M_L = 7$, while that for the body-wave magnitude (m_b) is about the same value. In contrast, the surface-wave magnitude, which is defined in terms of the amplitude of 20 s surface waves (wavelength ~60 km), saturates at $\sim M_s = 8$. As an example of the saturation problem of the magnitude scales, Figure 5.14 shows a graphical comparison between two large earthquakes, the 1906 San Francisco earthquake and the 1960 Chilean earthquake. Both earthquakes had a magnitude M_s of 8.3. However, the area that ruptured in the San Francisco earthquake was ~15 km deep and 400 km long, whereas the area that ruptured in the Chilean earthquake was almost equal to the area of the state of California. Evidently, the Chilean earthquake was a much larger event, but according to the magnitude scale, both earthquakes were about the same size.

The reason for the saturation has been explained on the grounds that extremely large earthquakes are the result of the slippage of faults hundreds of kilometers long and the inability of the magnitude scales to measure the power in waves with a wavelength of the order of hundreds of kilometers. That is, Richter magnitude measures seismic waves in a period range between 0.1 and 2 s (see graph corresponding to Wood–Anderson seismograph in Figure 5.7). This range corresponds approximately to wavelengths of 300 m to 6 km (recall that wavelength equals wave velocity times wave period). Similarly, surface-wave magnitude measures seismic waves with a period of 20 s, which corresponds to a wavelength of ~60 km. Hence, as the wavelength of the waves considered by the magnitude scales is much shorter than the length of the faults that generate extremely large earthquakes (equal to the wavelength of the seismic waves that correspond to the quasi-static deformation of the ground around the fault), the wave train that the magnitude scale measures corresponds to that emitted from only a fraction of the fault's area rather than from the entire fault. The result is an underestimation of the true strength of such earthquakes.

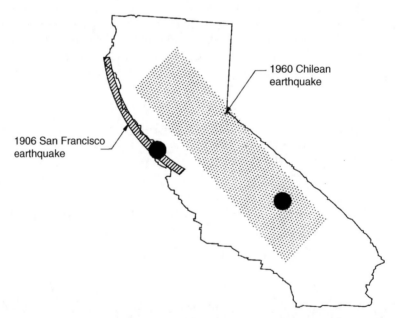

FIGURE 5.14 Size of fault-ruptured areas during the 1906 San Francisco and 1960 Chilean earthquakes relative to the size of the state of California. Black dots indicate locations of earthquake epicenters relative to corresponding ruptured areas. (Reproduced with permission of Iselin, A.D. from Boore, D.M., The motion of the ground in earthquakes, *Scientific American*, December 1977.)

5.6.2 Seismic Moment

To overcome the aforementioned weakness of the magnitude scales, a new measure of earthquake size has been introduced. This new measure is based on the size of the earthquake source and is known as *seismic moment*. It is defined as

$$M_0 = GAD \tag{5.16}$$

where M_0 is the seismic moment in dyne-centimeter or Newton-meter, G the shear modulus or rigidity of rock adjacent to the earthquake fault (approximately equal to 3×10^{11} dyne/cm² or 3×10^{10} N/m²), A the area ruptured along fault plane, and D the average amount of fault slip.

Seismic moment is not as easy to determine as Richter's magnitude, but it offers the advantage that it provides a measure of the physical mechanism that generates an earthquake and the energy released from the entire fault, not just a portion of it. As such, it does not saturate at the upper end of the scale as Richter's magnitude does. Seismic moment provides thus a clear distinction of earthquake size. Table 5.1 provides a comparison between the seismic moments and magnitudes of some large earthquakes.

The concept of seismic moment comes from two considerations. First, it is considered that the deformation pattern in the rock adjacent to an earthquake fault in an imminent state of rupture is mathematically equivalent to the deformation field generated by two equal but opposite couples

TABLE 5.1
Seismic Moment and Magnitude of Some Great Earthquakes

Date	Region	$M_0 / 10^{27}$ (dyne-cm)	M_s	M_w
July 9, 1905	Mongolia	45	8.2	8.4
January 31, 1906	Ecuador	204	8.6	8.8
April 18, 1906	San Francisco	10	8.2	7.9
January 3, 1911	Turkestan	4	8.4	7.7
December 16, 1920	Kansu, China	6	8.5	7.8
September 1, 1923	Kanto, Japan	8	8.2	7.9
March 2, 1933	Sanrika	45	8.5	8.4
May 24, 1940	Peru	22	8.0	8.2
April 6, 1943	Chile	22	7.9	8.2
August 15, 1950	Assam	100	8.6	8.6
November 4, 1952	Kamchatka	350	8.0	9.0
March 9, 1957	Aleutian Islands	585	8.0	9.1
November 6, 1958	Kurile Islands	32	8.7	8.3
May 22, 1960	Chile	2000	8.3	9.5
March 28, 1964	Alaska	820	8.4	9.2
October 17, 1966	Peru	16	7.5	8.1
August 11, 1969	Kurile Islands	22	7.8	8.2
October 3, 1974	Peru	16	7.6	8.1
July 27, 1976	China	2	8.0	7.5
August 16, 1976	Mindanao	16	8.2	8.1
March 3, 1985	Chile	2	7.8	7.5
September 19, 1985	Mexico	11	8.1	8.0

Source: Bolt, B.A. in *The Seismic Design Handbook*, Naeim, F., ed., Kluwer Academic Publishers, Boston, MA, 2001.

embedded in an unruptured elastic medium (known as *double-couple source model*). Two equal but opposite couples deform the medium but keep the system in equilibrium. Second, it is considered that the magnitude of these couples may be a good measure of the strength of the earthquake generated when such a deformation field is released. Seismic moment is thus defined as the magnitude of these couples.

To prove that the magnitude of such equivalent couples is indeed given by Equation 5.16, consider the vertical earthquake fault shown in Figure 5.15a and assume that the fault is in an imminent state of rupture. Assume also that the fault zone is filled with the same material that surrounds the fault and that there is continuity of displacement across the fault. Consider, further, that as a result of an acting couple, a differential element of the fault's surface area, dA, is subjected to shearing stresses τ as indicated in Figure 5.15b, and that it has deformed in shear an amount D. As such, the moment exerted by these shearing stresses in each element of fault area is given by

$$dM_0 = (\tau \, dA)\Delta x \tag{5.17}$$

as $\Delta x \to 0$, where Δx is the distance that separates the two faces of the fault. Note, however, that in view of Hooke's law for shearing stresses, Equation 5.17 may also be written as

$$dM_0 = \gamma G \, dA \Delta x \tag{5.18}$$

where γ denotes shear strain and G is the shear modulus of elasticity of the rock adjacent to the fault. Furthermore, note from Figure 5.15b that the shear strain in the differential element is equal to

$$\gamma = \frac{D}{\Delta x} \tag{5.19}$$

Therefore, the moment generated by the shearing forces on the faces of the differential element dA may be expressed as

$$dM_0 = GD \, dA \tag{5.20}$$

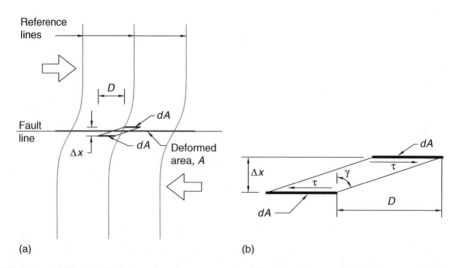

FIGURE 5.15 (a) Vertical fault in an imminent state of rupture. (b) Close-up of differential element.

Measurement of Earthquakes

and, by integration, the moment over the entire deformed fault area as

$$M_0 = G \int_A D \, dA = GA\bar{D} \tag{5.21}$$

where $\bar{D}$ is the average amount of shear deformation along the fault's surface area, which will be equal to the average amount of fault slip after the fault rupture takes place.

Three methods are commonly used by seismologists to estimate seismic moment. The first involves the direct use of Equation 5.16 and estimates of the fault dimensions and fault slip. These estimates are obtained from field measurements or inferences from the analysis of the wavelength of the waves generated by the earthquake. The second entails the use of formulas derived from mathematical models of a seismic source. The third uses the long-period asymptote in the amplitude Fourier spectrum (see Section 6.6) of the displacements recorded in the far field, as seismic moment is proportional to this amplitude.

EXAMPLE 5.7   DETERMINATION OF SEISMIC MOMENT USING FAULT DIMENSIONS AND SLIP

Estimate the seismic moment of the 1906 San Francisco, California, earthquake considering that the fault's ruptured length was ~400 km, the ruptured depth 15 km, and the average fault offset 5 m.

Solution

Considering Equation 5.16 and that the shear modulus of rock is approximately equal to 3×10^{11} dyne/cm², the desired seismic moment is equal to

$$M_0 = 3 \times 10^{11}(400 \times 10^5)(15 \times 10^5)(500) = 9 \times 10^{27} \text{ dyne-cm}$$

5.6.3 MOMENT MAGNITUDE

To take advantage of the superiority of the seismic moment as a measure of earthquake size, but to count with a scale that gives values that are close to the magnitudes obtained with Richter's scale, T. C. Hanks and H. Kanamori proposed in 1979 a measure of earthquake size based on seismic moment. This measure is known as *moment magnitude* and is denoted by the symbol M_w. It is defined by the empirical equation

$$M_w = \frac{2}{3} \log M_0 - 10.7 \tag{5.22}$$

where M_0 is seismic moment in dyne-centimeter, or

$$M_w = \frac{2}{3} \log M_0 - 6.0 \tag{5.23}$$

when M_0 is expressed in Newton-meter.

The moment magnitude scale does not saturate for great earthquakes the way the other magnitude scales do. The reason is that it is tied directly to seismic moment and, as seen earlier, seismic moment does not saturate for large earthquakes. It has been observed that M_w is equal to M_L in the range $3 \leq M_w \leq 5$; equal to M_s in the range $5 \leq M_w \leq 7.5$; and always $>M_L$ and M_s whenever $M_w > 7.5$. Moment magnitude is slowly replacing Richter magnitude and is nowadays routinely reported by seismological stations to describe the size of earthquakes.

Table 5.1 lists the moment magnitudes for several great worldwide earthquakes and compares these magnitudes with the corresponding surface-wave magnitudes. Note, for example, that the 1960 Chilean earthquake, the largest earthquake of the twentieth century, had a moment magnitude of 9.5, but a surface-wave magnitude of only 8.3.

EXAMPLE 5.8   COMPUTATION OF MOMENT MAGNITUDE

The south-central segment of the San Andreas Fault in California has been creeping at an average rate of 37 mm/year. This fault segment has a length and width of ~325 and 12 km, respectively, and ruptures on an average every 50 years. What is the moment magnitude of the largest earthquake that can be generated at this fault segment?

Solution

According to the given data, the fault's ruptured area and the fault slip in 50 years are equal to

$$A = (325 \times 10^5)(12 \times 10^5) = 3.9 \times 10^{13} \text{ cm}^2$$

$$\bar{D} = 3.7 \text{ cm/year} \times 50 \text{ year} = 185 \text{ cm}$$

Hence, considering a rock rigidity of 3×10^{11} dyne/cm², the seismic moment results as

$$M_0 = (3 \times 10^{11})(3.9 \times 10^{13})(185) = 2.165 \times 10^{27} \text{ dyne-cm}$$

Using Equation 5.22, the moment magnitude is thus equal to

$$M_w = \frac{2}{3}\log(2.165 \times 10^{27}) - 10.7 = 7.5$$

5.7 ACCELEROGRAPHS AND ACCELEROGRAMS

5.7.1 NEED FOR STRONG MOTION RECORDING INSTRUMENTS

Although seismographs and the records obtained from seismographs are quite useful to study some of the features of earthquakes and determine their hypocenters and magnitudes, seismographs are of limited use for engineering applications. There are several reasons for this limitation. First, seismographs cannot record the motion generated by strong earthquakes near epicentral regions, which are precisely the motions that can cause widespread damage and are therefore of engineering interest. Because of the large amplification factor with which a seismograph magnifies the ground motion that is intended to record, strong earthquakes displace the stylus off the recording film or paper in the seismographs located near the earthquake epicenter. Second, seismographs are invariably installed on solid rock as seismologists have a primordial interest in measuring earth crust movements. Hence, seismographs cannot give information about the effect of local soil conditions and the extent to which a soil deposit can amplify the ground motion observed in rock. Finally, the design of seismographs emphasizes accuracy in the timing of wave arrivals, rather than in the amplitude of the waves. However, in engineering the amplitude of motion is of greater significance than the time at which waves arrive.

To fill the gap left by seismographs, a different type of instrument was designed specifically to record strong earthquake ground motions and installed in urban areas and other locations where they could furnish useful information for engineering applications. This different type of instrument was introduced in the 1930s and is called *strong motion accelerograph*. Throughout the years, thousands have been deployed worldwide.

Accelerographs are designed to measure directly the variation of ground acceleration with time. There are two reasons why they are designed to measure ground acceleration and not ground displacement like in the case of seismographs. The first is that acceleration meters are easier to build than displacement meters for measurements in the frequency range that is typical of strong ground motions. The second has to do with the fact that, in general, numerical integration is more accurate than numerical differentiation. Hence, it is better to record acceleration and integrate to obtain velocity and displacement rather than record displacement and differentiate to obtain velocity and acceleration.

5.7.2 Accelerograph Components and Design Features

As seismographs, accelerographs are composed of a pendulum, a damper, a recording device, and an internal clock. Modern accelerographs are also equipped with Global Positioning System (GPS) timing and a cellular modem communication system for remote interrogation. As seismographs, accelerographs record three orthogonal components of motion, two horizontal and one vertical. Different from seismographs, however, accelerographs do not record the ground motion continuously. As accelerographs are used to obtain records of strong earthquakes, and as strong earthquakes occur only after so many years at a given site, a continuous recording would involve an inefficient use of the recording medium and technical personnel to monitor the instrument. Thus, accelerographs are implemented with a device that triggers the recording system when the ground acceleration reaches a specified level. Most accelerographs are designed to be triggered by the vertical component of motion as the first motion to arrive to a station is predominantly vertical. As an earthquake ordinarily subsides within a few minutes, the device is also designed to stop the recording after the ground shaking falls again to imperceptible levels. The ground acceleration that triggers the operation of an accelerograph is usually small enough to detect all important events but, at the same time, is large enough to avoid having the instrument triggered by local events or ambient vibrations. As they are triggered after the ground motion begins, some accelerographs fail to record the portion of the motion that takes place before the recording starts. Current generation accelerographs, however, are outfitted with a solid-state short memory module (known as *pre-event memory*) that preserves the first few seconds of motion before the triggering device starts the permanent recording. These instruments acquire data continuously, always saving the last few seconds of data in a buffer so that the motion preceding the instrument triggering can be captured with the rest of the record. Also differently from seismographs, accelerographs operate with batteries as strong earthquakes often disrupt electric power. Batteries are charged up from either a solar panel or an AC supply.

There are two types of accelerographs in common use today: the torsion pendulum accelerograph and the force balance or servo accelerograph. The torsion pendulum accelerograph uses a coil of wire that swings like a door about a vertical suspension axis (see Figure 5.16). As the coil rotates, it moves in the magnetic field setup by a magnet attached to the ground and the induced eddy currents in the coil setup viscous damping forces. Amplification of the coil's motion is carried out optically with a light source and a system of cylindrical lenses and prisms. This amplified motion is then recorded on photographic film. The force balance or servo accelerograph is composed of a pendulum, a small plate capacitor, a pendulum actuator, and a signal amplifier (see Figure 5.17). When the pendulum is displaced from its equilibrium position under the action of a ground motion, the capacitor detects the displacement and generates an electric current that is amplified and fed back to the pendulum actuator. Then, the pendulum actuator applies a force that brings the pendulum back to its equilibrium position. The current that is fed from the signal amplifier to the pendulum actuator is proportional to the acceleration of the pendulum. The ground acceleration is thus obtained by converting this current into an analog signal and recording this signal on photographic film or magnetic tape. The merit of force balance accelerographs is that in them the displacement of the pendulum is so small that all undesirable characteristics such as the nonlinearity of the spring become insignificant.

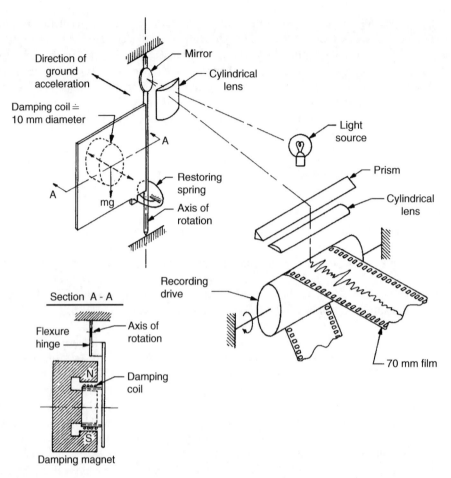

FIGURE 5.16 Schematic representation of torsion pendulum strong motion accelerograph. (Reproduced with permission of Earthquake Engineering Research Institute from Hudson, D.E., *Reading and Interpreting Strong-Motion Accelerograms*, Earthquake Engineering Research Institute, Oakland, CA, 1979.)

As is in the case of seismographs, the design of accelerographs is based on the response analysis of a single-degree-of-freedom system subjected to a sinusoidal ground acceleration. However, the natural periods and damping ratios of accelerographs are selected to measure a response that is proportional to ground acceleration instead of ground displacement. As discussed in Section 5.3.2, these natural periods and damping ratios can be selected on the basis of the following analysis:

From structural dynamics, it is known that the displacement response of a single-degree-of-freedom system subjected to a sinusoidal ground acceleration $\ddot{u}_g(t) = \ddot{u}_0 \sin \omega_g t$ is given by

$$u(t) = -\frac{1}{\omega_n^2} \ddot{u}_0 D \sin(\omega_g t - \theta) \tag{5.24}$$

where $\ddot{u}_0$ denotes ground acceleration amplitude, ω_n is the natural frequency of the system, ω_g the frequency of the excitation, θ a phase angle, and D a dynamic magnification factor defined as

$$D = \frac{1}{\sqrt{\left[1 - \left(\frac{\omega_g}{\omega_n}\right)^2\right]^2 + \left[2\xi \frac{\omega_g}{\omega_n}\right]^2}} \tag{5.25}$$

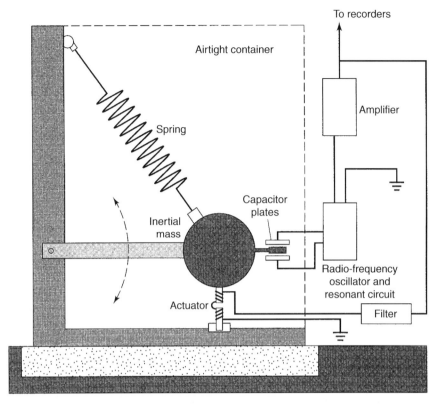

FIGURE 5.17 Schematic representation of vertical force-balance strong motion accelerograph. (Reproduced with permission of The State of Irving Geis from Press, F. Resonant vibrations of the earth, *Scientific American*, 28–37, November 1965.)

However, as

$$\ddot{u}_0 \sin(\omega_g t - \theta) = \ddot{u}_0 \sin\left[\omega_g\left(t - \frac{\theta}{\omega_g}\right)\right] = \ddot{u}_g\left(t - \frac{\theta}{\omega_g}\right) \quad (5.26)$$

then the displacement response $u(t)$ may also be expressed as

$$u(t) = -\frac{D}{\omega_n^2} \ddot{u}_g\left(t - \frac{\theta}{\omega_g}\right) \quad (5.27)$$

from which it may be concluded that

$$\ddot{u}_g\left(t - \frac{\theta}{\omega_g}\right) = -\frac{\omega_n^2}{D} u(t) \quad (5.28)$$

Observe thus that, as ω_n is a constant, the ground acceleration is directly proportional to the instrument's displacement response whenever the dynamic magnification factor D is also a constant.

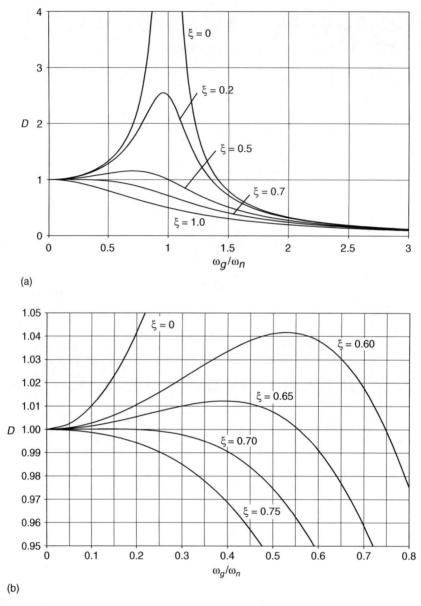

FIGURE 5.18 (a) Variation of dynamic magnification factor with damping ratio and frequency ratio, and (b) close-up of it in range $0 \leq \omega_g/\omega_n \leq 0.8$.

Observe also from the plots of the magnification factor D shown in Figure 5.18 that for a damping ratio of 60%, D is approximately constant within an error of ±5% in the range $0 < \omega_g/\omega_n < 1.0$ or, what is tantamount, in the range $0 < T_n/T_g < 1.0$, where T_n and T_g, respectively, denote the instrument's natural period and the excitation's dominant period. Thus, if $T_n = 0.1$ s, D is approximately constant in the range $T_g > 0.10$ s, which covers most cases of practical interest in earthquake engineering. Most accelerographs are thus constructed with a natural period of <0.1 s and a damping ratio of ~60%. The characteristics of commercial accelerographs vary among the different types available, but typical characteristics are natural frequency = 25 Hz; damping ratio = 60–70%; range = ±2g; starting acceleration = 0.05g; start time delay = 0.05–0.10 s; size = 40 × 40 × 20 cm; and weight = 10 kg.

The recording system in an accelerograph can be of the analog or digital type. A recording system is said to be of the analog type if the data obtained in the field are recorded in analog form, such as on photographic film or smoked paper. In contrast, it is considered to be of the digital type if the field data are recorded in digital form. In a digital recorder, the ground motion signal is converted into digital form by means of an analog-to-digital (A/D) converter and recorded on a magnetic tape. Analog recorders are simpler and less expensive than digital recorders, which result in a reduced field maintenance. They require, however, a laborious laboratory data processing. Digital recorders, on the other hand, extend the frequency and dynamic range of the instrument, make it relatively easy to provide the accelerograph with the aforementioned data storage module to record the onset of ground motion, make it possible to use the data directly for computer processing, and reduce considerably the time and effort required for data processing. The disadvantages are an increased complexity and cost, a larger standby power requirement, which reduces unattended time intervals and increases maintenance problems, and the need for personnel of a higher level of training and experience. Another disadvantage is the possibility of the malfunctioning of the tape cassette, a major source of maintenance problems for instruments under adverse environmental conditions. Current generation accelerographs, however, have replaced the mechanically driven digital magnetic tape by a solid-state memory.

5.7.3 Accelerograms

The records obtained from accelerographs are called *accelerograms*. The first accelerogram ever recorded was obtained during the March 10, 1933, Long Beach earthquake in Southern California. Ever since, thousands of accelerograms have been obtained worldwide, providing useful information of engineering relevance. Typical accelerograms from representative earthquakes are shown in Figure 5.19. This figure illustrates the wide variability in intensity, duration, and dominant period that accelerograms may have. In this regard, it is of interest to note that the largest horizontal ground acceleration in record has been 1.82g, where g represents the acceleration of gravity. This acceleration was recorded at the Cedar Hill Nursery station in the city of Tarzana, California (6.5 km from epicenter) during the 1994 (M_L = 6.4) Northridge earthquake (see Figure 5.20). The largest vertical acceleration ever recorded has been over 2g. This acceleration was recorded by an accelerograph located directly above the hypocenter some 8–10 km below during the 1985 (M_s = 6.9) Nahanni, Canada, earthquake* (see record in Figure 5.19). It is also of interest to note that the horizontal peak ground acceleration during some notable earthquakes has been 0.17g during the 1985 (M_L = 8.1) Mexico City earthquake; 0.629g during the 1989 (M_L = 7.1) Loma Prieta, California, earthquake; 0.849g during the 1995 (M_L = 7.2) Kobe earthquake; 0.41g during the 1999 (M_w = 7.4) Kocaeli, Turkey, earthquake; and 1.0g during the 1999 (M_w = 7.6) Chi-Chi, Taiwan, earthquake.

An accelerogram from a modern accelerograph typically includes seven traces, as shown in Figure 5.21. The meaning of each of these traces is as follows:

Trace No.1: Code indicating absolute time
Trace No.2: Horizontal acceleration in the direction of the long dimension of accelerograph case
Trace No.3: Fixed trace
Trace No.4: Vertical acceleration
Trace No.5: Fixed trace
Trace No.6: Horizontal acceleration in the direction of the short direction of accelerograph case
Trace No.7: Time marks (with a rate of two pulses per second) generated by internal time circuit

---

* The peak vertical trace in this record was actually off-scale, but experts agreed that the peak vertical ground acceleration was indeed at least equal to 2g.

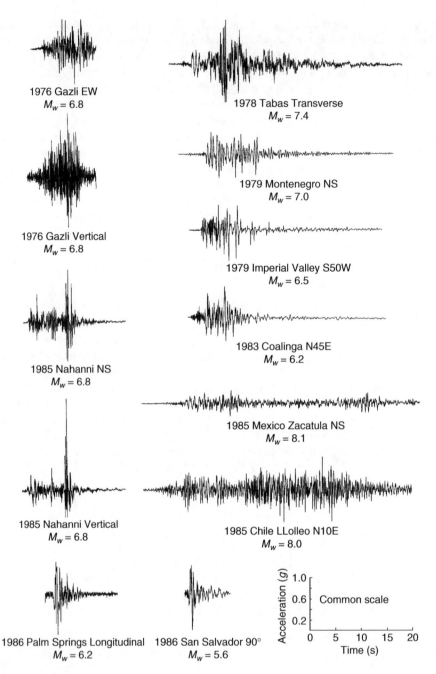

FIGURE 5.19 Accelerograms from different earthquakes recorded in epicentral regions, plotted using same horizontal and vertical scales.

The absolute time code in Trace 1 is received from a signal transmitted by the radio station WWVB operated by the National Institute of Standards and Technology from Fort Collins, Colorado. The signal is broadcasted interference-free on a frequency of 60 kHz over a radius of ~2000 miles. The time code in Trace 1 serves two purposes. First, it makes possible by comparison with data from seismological stations a positive identification of the measured earthquake. Second, it provides an accurate way of synchronizing various accelerographs hundreds of kilometers apart.

Measurement of Earthquakes

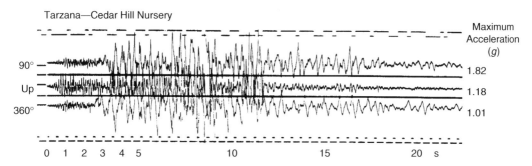

FIGURE 5.20 Accelerogram recorded at Cedar Hill Nursery station in the city of Tarzana, California, during the 1994 Northridge earthquake. (Reproduced with permission of Earthquake Engineering Research Institute from Northridge Earthquake January 17, 1994, Preliminary Reconnaissance Report, Earthquake Engineering Research Institute, 1994.)

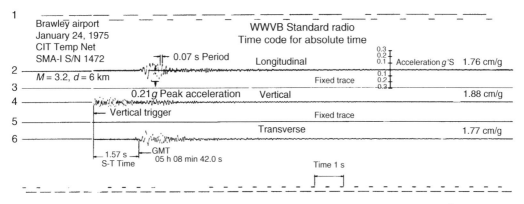

FIGURE 5.21 Typical seven traces in a typical accelerogram. (Reproduced with permission of Earthquake Engineering Research Institute from Hudson, D.E., *Reading and Interpreting Strong-Motion Accelerograms*, Earthquake Engineering Research Institute, Oakland, CA, 1979.)

This way of synchronizing accelerographs in a network is usually less expensive than hardwire or telephone links. The two fixed traces (No.3 and No.5) provide a means for correcting certain spurious motions that may affect an instrument, such as film distortion or transverse slip of the film produced by oversize guiding holes. As fixed and variable traces would be affected the same by such spurious motions, a correction can be made to the variable trace by subtracting the motion observed in the fixed trace.

Accelerograms are routinely digitized (i.e., converted into a finite sequence of time-acceleration points) to make them computer readable. In the beginning, digitization was performed by hand using only an engineering scale. In the late 1970s, semiautomatic digitizers were employed. When these semiautomatic digitizers are used, an operator moves a lens with crosshairs across an accelerogram mounted on an illuminated table. By pressing a foot-operated switch, the coordinates of the crosshairs position are then recorded on paper or magnetic tape. This form of digitization involves tiring and inaccurate work in which operator accuracy and fatigue significantly affects the process. Nowadays, accelerograms are digitized either in the field through A/D converters or in the laboratory using fully automatic equipment. Typically, digitization is performed at rates of 200 samples/s. Many of the earlier accelerograms, however, were digitized at a rate of 50 samples/s, that is, at time intervals of 0.02 s.

Accelerograph records from past earthquakes can be obtained from a number of sources. The National Geophysical Data Center of the National Oceanic and Atmospheric Administration (Web site: http://www.ngdc.noaa.gov) offers for sale a three-CD collection of over 15,000 accelerograms from earthquakes in the United States and other parts of the world. A number of strong motion databases can also be downloaded from the Internet. These databases can be accessed through the links to various sources of ground motion data provided in the World Wide Web sites maintained by the University of California at Berkeley (Web site: http://peer.berkeley.edu/smcat), the State University of New York at Buffalo (Web site: http://mceer.buffalo.edu/links/agrams.asp), and the National Geophysical Data Center itself.

5.7.4 Accelerogram Processing

There are several sources of error in the recording and digitization of accelerograms that make the digitized record deviate from the actual acceleration of the ground. Background noise is one of such sources. Background noise is always present in an accelerogram. Ocean waves, traffic, construction activities, wind, and even atmospheric pressure changes are a constant source of low- and high-frequency noise that may be detected by an accelerograph. Some other error sources are: (a) the uncertain location of the record baseline, (b) the nonuniform frequency response of the instrument, (c) the elongation and meandering of the recording paper or film, and (d) the reading and digitization of the record.

The error introduced by the uncertain location of the record baseline is always present in records from analog accelerographs. This error arises because analog accelerographs are triggered after the ground acceleration reaches a specified level and stops recording the motion after the acceleration falls again below a certain acceleration level. As a consequence, the accelerograph misses the pretriggering portion of the ground motion and the end of it, making the location of the zero-acceleration baseline unknown. The baseline of accelerograms has thus to be located arbitrarily at some point. However, small errors in the location of the accelerogram baseline lead to large errors in the calculated values of the ground displacements. Figure 5.22 illustrates this problem. This figure shows how an error in the location of the baseline as small as $0.001g$ at the beginning of a 20 s long accelerogram would introduce an error in the displacement of 196 cm at the end of the motion. This exaggerated error occurs because integration of a constant acceleration error produces a velocity error that increases linearly with time and a displacement error that increases quadratically with time.

In view of this problem, incomplete accelerograms always require a cautious adjustment of their baseline. This adjustment is known as *baseline correction*. Early methods for baseline correction considered the baseline to have the form of a parabola and adjusted this parabola to minimize in the least squares sense the mean square of the velocities. This method of baseline correction had no physical justification, but served to avoid having a disproportionally large value at the end of the integrated displacement record. Modern methods are based on the use of high-pass filters as explained later.

To correct for all the errors described earlier, improve their resolution, and increase their frequency range, the raw accelerograms collected from the field are carefully processed before using them in any application. This processing normally consists of (a) the application of a correction scheme to correct for the nonuniform response of the instrument (see Figure 5.18); (b) the use of high-pass and low-pass filters (see Figure 5.23) to remove unwanted noise in the signal, especially that produced by the digitization process; and (c) integration to obtain the velocity and displacement time histories. The instrument correction is usually performed by modeling the accelerograph as a single-degree-of-freedom system and using this model to obtain the true ground acceleration from the known instrument response, that is, the recorded relative displacement. Ordinarily, the instrument correction is only important for frequencies above the usual range of engineering interest. However, this correction may be of significance for accelerographs located in buildings

Measurement of Earthquakes

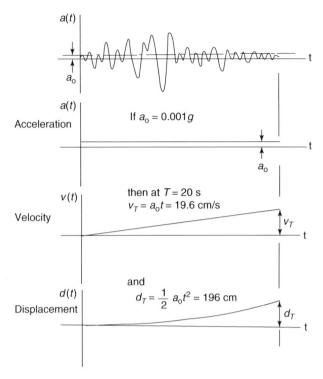

FIGURE 5.22 Error at the end of displacement time history introduced by double integration of an error of 0.001g in location of accelerogram baseline. (Reproduced with permission of Earthquake Engineering Research Institute from Hudson, D.E., *Reading and Interpreting Strong-Motion Accelerograms*, Earthquake Engineering Research Institute, Oakland, CA, 1979.)

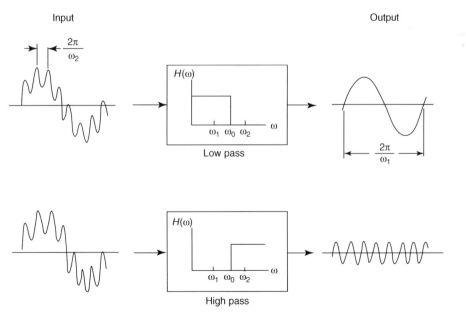

FIGURE 5.23 Schematic representation of low- and high-pass filters where $H(\omega)$ represents transfer function and ω frequency. (Reproduced with permission of Earthquake Engineering Research Institute from Hudson, D.E., *Reading and Interpreting Strong-Motion Accelerograms*, Earthquake Engineering Research Institute, Oakland, CA, 1979.)

or near the abutments of dams and bridges as the motion in these structures may be dominated by high-frequency components. The low-pass filter is necessary if instrument correction is performed because the instrument correction strongly amplifies any high-frequency noise present in the acceleration record. The high-pass filter is necessary because any low-frequency noise present in the acceleration record is strongly amplified when the record is double integrated to obtain the displacement time history.

The details of current procedures for accelerogram processing are contained in the documentation of a computer program developed by the U.S. Geological Survey (USGS) for the processing of uncorrected ground motion records. This program, named Basic Accelerogram Processing (BAP), makes a linear baseline correction, applies instrument correction, and filters high or low frequencies from an input acceleration time series. It also generates the corresponding velocity and displacement time histories and plots the results after each processing step. The program and its documentation can be downloaded from the USGS' National Strong-Motion Program Web site (http://nsmp.wr.usgs.gov).

Typical results from the processing of an accelerogram are shown in Figure 5.24. The top curve shows the corrected acceleration time history plotted at 0.02 s intervals for one of the horizontal components of an earthquake recorded in the city of El Centro, California, in 1940. The second curve is the corrected ground velocity. Note that this curve does not start exactly at zero, which is a consequence of the data processing technique. The lowest trace in the figure corresponds to the displacement time history obtained by double integration of the corrected acceleration record. A slight distortion at the beginning of the record is involved here too. Note, however, that even after 50 s the displacement curve stays close to the zero axis, thus showing

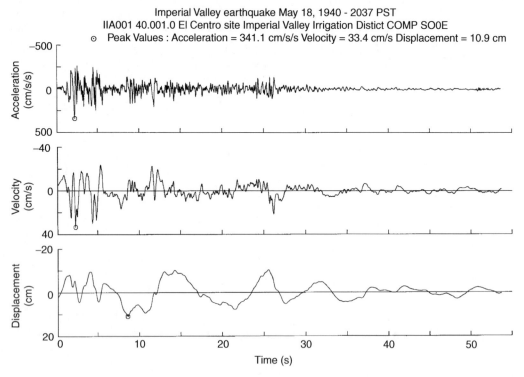

FIGURE 5.24 Corrected ground acceleration and integrated velocity and displacement time histories for typical earthquake ground motion. (Reproduced with permission of Earthquake Engineering Research Institute from Hudson, D.E., *Reading and Interpreting Strong-Motion Accelerograms*, Earthquake Engineering Research Institute, Oakland, CA, 1979.)

5.7.5 STRONG MOTION INSTRUMENT ARRAYS

Throughout the years, the recording of strong earthquake ground motions has provided fundamental data for the study of earthquakes as well as the design of earthquake-resistant structures, contributing extensively to the progress of earth science and earthquake engineering. However, single isolated instruments cannot provide by themselves sufficient information to attain a clear understanding of the factors that may influence the characteristics of strong ground motions. For this reason, one-, two-, and three- dimensional strong motion accelerograph arrays linked together by a common time base have been installed and tailored to the acquisition of some specific information in some seismic regions of the world. In general, two kinds of arrays have been set. The first kind has been installed within the near-field regions of strong earthquakes to obtain a greater understanding of the physical processes involved in the generation and transmission of seismic energy. The second kind has been established at sites with particular characteristics to study the way ground motions are affected by not only source characteristics and wave propagation path between source and site, but also local factors such as topographic and soil features, soil-structure interaction effects, and soil liquefaction. Observation of strong motion earthquakes by instrument arrays began in the late 1970s and has provided to date valuable data.

One such array, called SMART-1, has been operating since 1980 near Lotung in the northeast corner of Taiwan, an area of high seismicity. As shown in Figure 5.25, this is a two-dimensional

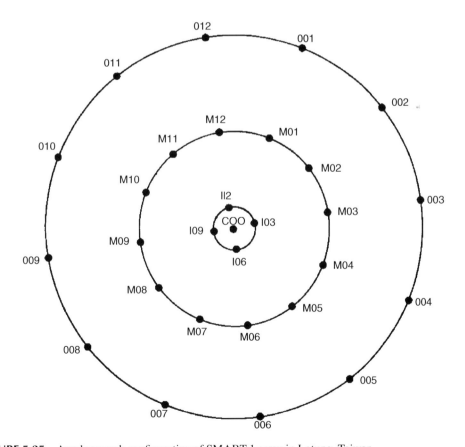

FIGURE 5.25 Accelerograph configuration of SMART-1 array in Lotung, Taiwan.

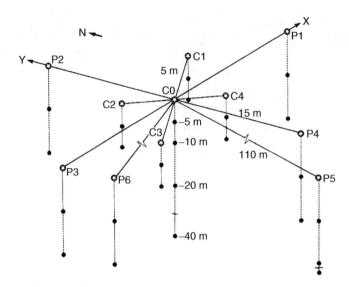

FIGURE 5.26 General configuration of three-dimensional accelerometer array in Chiba, Japan.

array consisting of a central recorder and 37 accelerographs arranged in three concentric rings with radii of 100 m, 1 km, and 2 km. The accelerographs are all of the three-component analog type and have a common time base. They have a range of ±2g, are connected to a digital tape recorder, and may be triggered by vertical or horizontal motion. During the first 6 months of operation, the array recorded nine earthquakes with local magnitudes ranging from 3.8 to 6.9.

Japan has also installed several of these arrays; one of the most significant being the three-dimensional array at the Chiba Experimental Station of the University of Tokyo. This array is composed of 44 three-component accelerographs, 15 of which are at the ground surface and the rest at depths of 5, 10, 20, and 40 m (see Figure 5.26). The stations C0-P5-P6 form a large triangular array with each side of the triangle measuring ~110 m. Eight accelerometers are densely placed around the central station C0. Four of them are only 5 m from this central station, and the other four are 15 m apart. The array has been operating successfully since 1982 and has recorded over 160 earthquakes.

In the United States, accelerograph arrays have been established along the San Andreas, San Jacinto, and Imperial faults. One of the most important is the one-dimensional El Centro array, which crosses the Imperial and Brawley faults in the Imperial Valley in California (see Figure 5.27). This array is 45 km long and includes 13 stations. It also includes the El Centro differential array, a dense array of six three-component accelerographs configured along a 305 m line. Shortly after installation, El Centro array recorded the 1979 Imperial Valley earthquake ($M_s = 6.9$), which occurred only 5.6 km away, producing valuable information about near-field ground motions.

5.7.6 Accelerograms from Near-Fault Sites

The availability of recordings from accelerograph networks has permitted the identification of a characteristic that distinguishes many of the accelerograms recorded in the proximity (within ~15 km) of an earthquake fault. This characteristic is a single or double high-amplitude acceleration pulse—called *fling* by seismologists—that appears at the beginning of the record. The period of the pulse may vary from 0.5 to 5 s or more. Such a pulse is observed, for example, in the accelerogram recorded in the N65E direction at the Cholame # 2 station during the 1966 Parkfield earthquake ($M_L = 5.6$) and the accelerogram recorded at Pacoima Dam in the S16E direction during the 1971 San Fernando earthquake ($M_L = 6.2$). These accelerograms, which are some of the

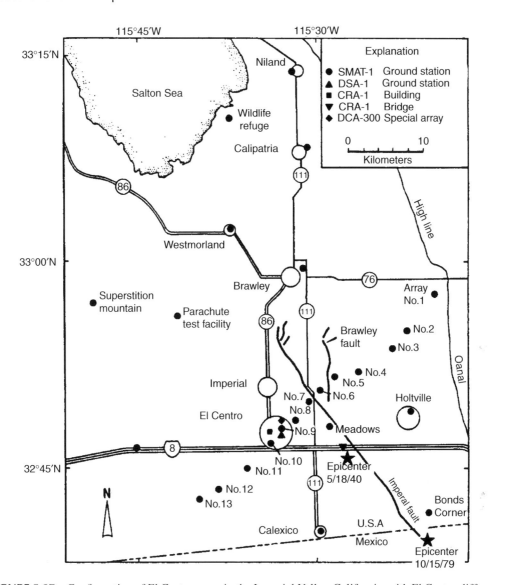

FIGURE 5.27 Configuration of El Centro array in the Imperial Valley, California, with El Centro differential array located near Station 9. (After Porcella, R.L. and Matthiesen, R.B. Reproduced with permission of Earthquake Engineering Research Institute from Seed H.B. and Idriss, I.M., *Ground Motions and Soil Liquefaction during Earthquakes*, Earthquake Engineering Research Institute, Oakland, CA, 1983.)

first in which this characteristic was observed, are shown in Figure 5.28, together with the velocity and displacement time histories derived from them. The Cholame # 2 record was obtained at a distance of ~60 m from the fault trace and is characterized by three large acceleration pulses. The Pacoima Dam record was obtained ~3.3 km from the causative fault and is characterized by a large velocity pulse ~3 s after the instrument was triggered. Large long-duration pulses have also been observed in some of the near-source recordings from the 1979 Imperial Valley, 1992 Landers, and 1994 Northridge earthquakes in California, as well as in the 1995 Kobe earthquake in Japan, the 1999 Izmit earthquake in Turkey, and the 1999 Chi-Chi earthquake in Taiwan. As the area under an acceleration pulse represents an incremental velocity, and the area under a velocity pulse represents an incremental displacement, an alternative means of distinguishing accelerograms with

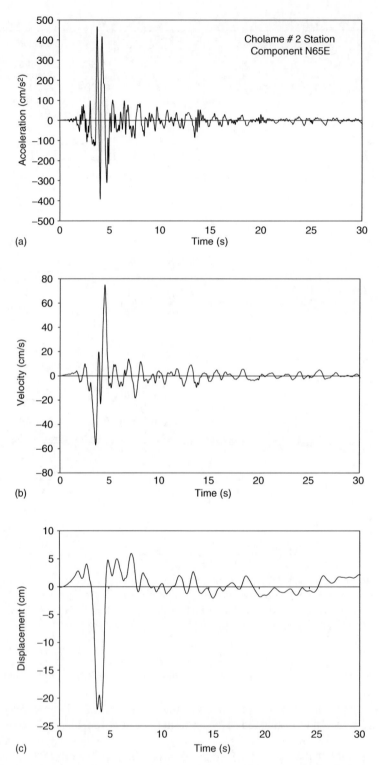

FIGURE 5.28 Acceleration, velocity, and displacement time histories corresponding to the near-source accelerograms obtained during the (a–c) 1966 Parkfield and the (d–f) 1971 San Fernando earthquakes in California.

Measurement of Earthquakes

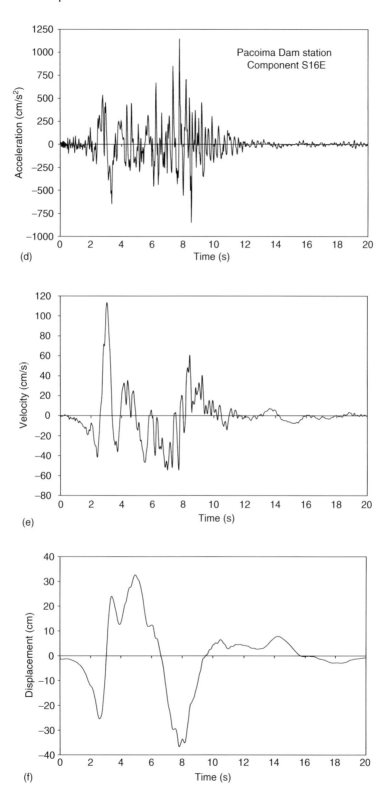

FIGURE 5.28 (Continued)

such a characteristic is the existence of a large velocity or a large displacement pulse in the velocity and displacement time histories derived from them. It should be noted that not all accelerograms from stations near seismic sources exhibit the aforementioned pulse. All depends on earthquake magnitude, type and length of the fault rupture, and azimuth of the recording station. Near-source pulses are important from the engineering point of view because they may affect adversely long-period structures.

Apparently, what causes the long-period pulses in near-source ground motions are (a) the constructive interference of the earthquake waves due to the directivity of the fault rupture, and (b) the static displacement associated with the permanent offset of the ground during an earthquake. Rupture directivity effects result from the fact that large earthquakes are produced by a rupture that begins at a point on a fault (hypocenter) and spreads with a velocity that is ~85% of the corresponding shear wave velocity. The large pulse occurs when the fault ruptures toward the recording site, and the slip on the fault plane is aligned with the rupture direction. An explanation of this phenomenon was first given by H. Benioff in 1955 during his investigation of the intensity patterns observed in the recordings from the 1952 Kern County, California, earthquake. He considered that the propagation of a dislocation along a fault represents a moving source and that, as in the case of sound waves emitted from a moving acoustic source, this moving source generates different signals at the opposite ends of the fault. The signal at a site toward which the rupture propagates would exhibit a large-amplitude pulse. In contrast, the signal at a site for which the rupture propagates away would exhibit small amplitude and long-duration. He demonstrated these directivity effects for the case of a radiating source moving along a straight line using circular wave fronts emitted at equal time intervals (see Figure 5.29). This explanation for the constructive interference of earthquake waves is based on the assumption that the fault rupture propagates only in one direction. It should be realized, however, that for some earthquakes the rupture starts in the center

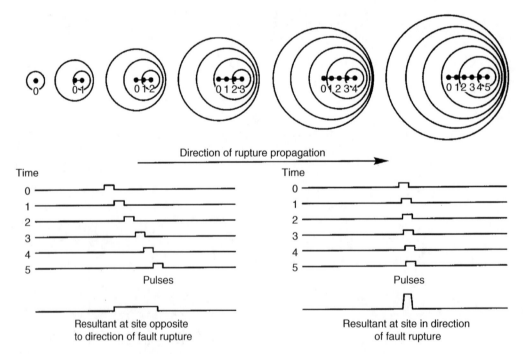

FIGURE 5.29 Schematic illustration of effect of moving radiating source on amplitude and duration of ground motions at sites in and opposite to the direction of fault rupture. (After Benioff, H. Reproduced with permission of Earthquake Engineering Research Institute from Singh, J.P., *Earthquake Spectra*, 1, 2, 1985.)

of a fault segment and spreads in both directions and that for such earthquakes directivity effects may be less obvious.

Directivity effects have been observed, for example, in some of the recordings from the earthquake that occurred in the Imperial Valley, California, on October 15, 1979, with an epicenter located ~1 km south of the United States–Mexico border (see Figure 5.27). This earthquake was generated on the strike-slip Imperial fault and had a moment magnitude of 7.2. The fault rupture propagated in a northwesterly direction into the United States. About 25 accelerograph stations located within 15 km from the fault recorded the earthquake. The location of these stations, some of them part of the El Centro array that crosses the Imperial and Brawley faults described in Section 5.7.5, is indicated in Figure 5.27. The directivity effects during this earthquake can be seen by comparing the accelerograms shown in Figure 5.30, which were recorded in the direction transverse to the fault (S50W) at the El Centro Array # 7 (located ~1 km from the fault trace) and Bonds Corner (located ~3 km from the fault trace) stations. It may be observed that in the accelerogram from El Centro Array # 7 station, located in the direction of the rupture propagation, there are three long-duration, large-amplitude pulses. In contrast, in the record from the Bonds Corner station, located opposite to the direction of propagation, there are no long-duration acceleration pulses. Furthermore, large long-duration pulses may also be seen in the velocity and displacement time histories from El Centro Array # 7 station but not in those from the Bonds Corner station.

The schematic illustration in Figure 5.29 shows that rupture directivity may lead to large acceleration, velocity, or displacement pulses at a site when two conditions are met: (a) the rupture propagates toward the site and (b) the fault-slip direction is aligned with the site. These conditions are met in most strike-slip faults as in a strike-slip fault the slip is oriented horizontally in the direction along the fault's strike, and the rupture also propagates horizontally along the fault's strike. They are also met in reverse and normal dip-slip faults when the rupture propagates up the fault plane, and the slip is also directed up the fault plane. This means that sites located near the surface exposure of a dip-slip fault are likely to experience large acceleration, velocity, and displacement pulses when an earthquake occurs in that fault. It may also be inferred that not all near-fault sites will experience large pulses during an earthquake. In other words, rupture directivity may cause spatial variations in ground motion amplitude and duration around a fault. Furthermore, it may produce systematic differences between the fault-normal and fault-parallel components of the horizontal ground motion as the radiation pattern of the shear waves in the near field caused by the shear dislocation on a fault causes that the large long-duration pulses occur in the component of motion perpendicular to the fault. This may be seen, for example, by comparing the two horizontal components of ground acceleration shown in Figure 5.31, which were recorded at the Rinaldi station ~8 km from the fault during the 1994 Northridge, California, earthquake ($M_w = 6.7$). This earthquake was produced on a blind thrust fault, with the surface projection of the fault rupture having a strike of ~122°. It may be seen that the component of motion perpendicular to the strike of the fault (fault normal) exhibits a distinctive long-duration pulse and that this type of pulse is not present in the parallel component (fault parallel).

Large displacement pulses due to the permanent offset of the ground may occur at a site when the site is located close to an earthquake fault with a large surface rupture and may show up on the ground displacement component parallel to the slip direction. Thus, for strike-slip earthquakes, the pulse due to the rupture directivity and the pulse due to the permanent ground offset will appear in two different components of the horizontal ground motion. This is observed, for example, in the horizontal ground velocity time histories shown in Figure 5.32, which correspond to the accelerograms recorded at the Lucerne Valley station (located ~1.1 km from the surface rupture of the fault) during the 1992 Landers, California, earthquake ($M_w = 7.3$). In this case, the fault-normal component exhibits a rupture directivity pulse, while the fault-parallel component exhibits a long-period

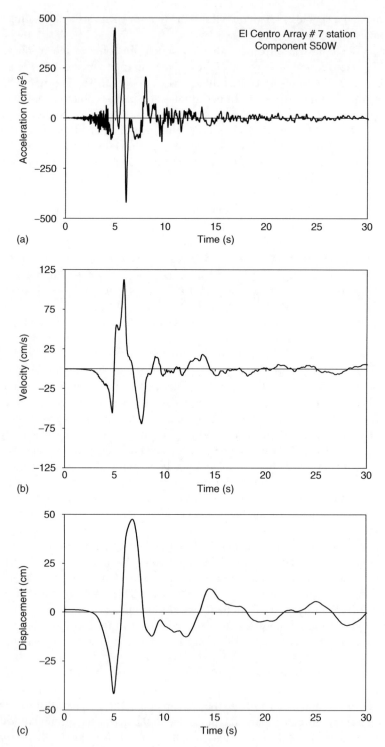

FIGURE 5.30 Acceleration, velocity, and displacement time histories from (a–c) El Centro Array # 7 and (d–f) Bonds Corner stations corresponding to the accelerograms obtained in the S50W direction during the 1979 Imperial Valley, California, earthquake.

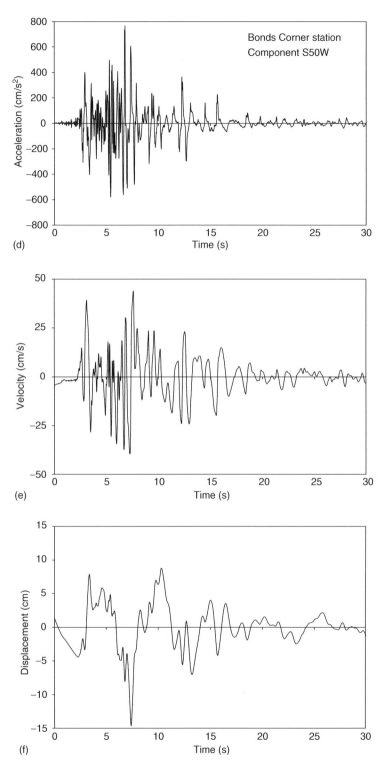

FIGURE 5.30 (Continued)

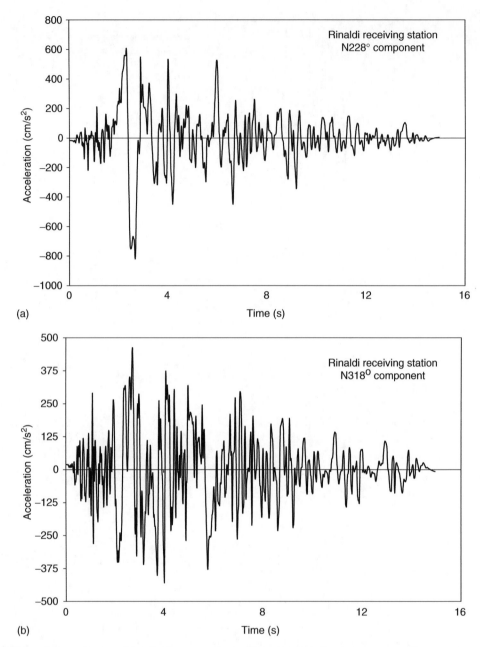

FIGURE 5.31 (a) Fault-normal (N228°) and (b) fault-parallel (N318°) components of horizontal ground acceleration recorded at Rinaldi receiving station during the 1994 Northridge, California, earthquake.

permanent offset pulse. This resulted from the fact that this earthquake occurred on a strike-slip fault oriented roughly in the north–south direction. For dip-slip earthquakes, those two pulses may appear on the same horizontal ground motion component because a permanent ground offset may occur on the horizontal component perpendicular to the strike of the fault at locations directly up dip from the hypocenter.

An example of the large pulses produced by the permanent offset of the ground during an earthquake are those manifested in the acceleration, velocity, and displacement time histories obtained from the accelerogram recorded in the north–south direction at the Taichung–Shigang School

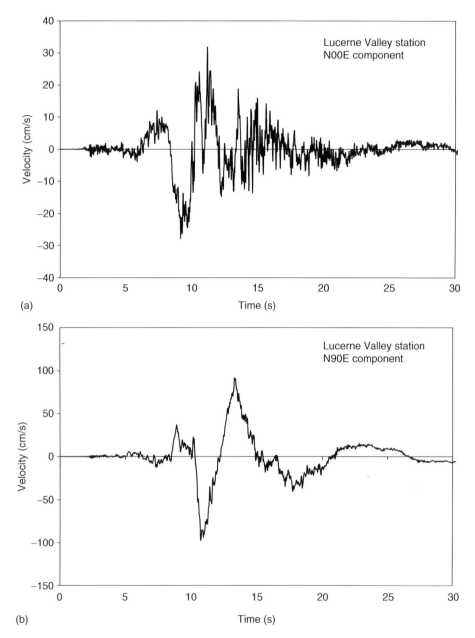

FIGURE 5.32 (a) Fault-parallel (N00E) and (b) fault-normal (N90E) components of ground horizontal velocity corresponding to the ground acceleration records obtained at the Lucerne Valley station during the 1992 Landers, California, earthquake.

station (TCU068) during the 1999 Chi-Chi earthquake ($M_w = 7.6$) in Taiwan (see Figure 5.33). This earthquake occurred in a thrust fault with a strike of ~3° and a dip of ~29°. The rupture of the fault, which propagated from south to north produced an average slip of ~1 m but as large as 8 m at some locations. The TCU068 station was located near the northern end of the fault trace. It may be seen that the velocity and displacement pulses in these time histories are distinctively large and long, indicating thus that they are closely related to the fault displacement. It should be mentioned that the peak displacements obtained by integration from the strong motion data recorded during this earthquake are in good agreement with field fault scarp measurements.

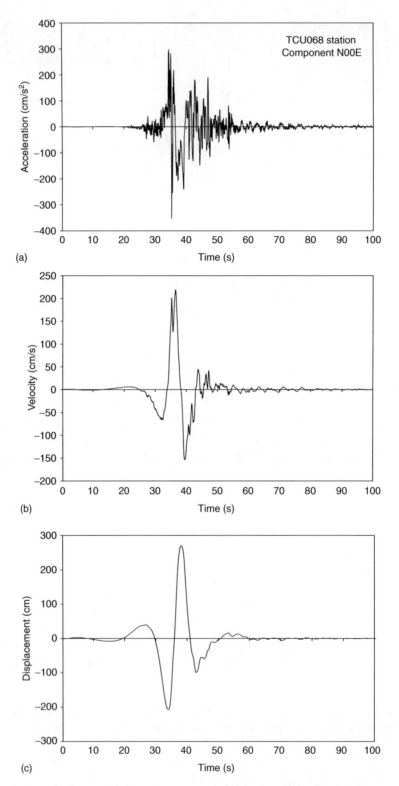

FIGURE 5.33 (a) Acceleration, (b) velocity, and (c) displacement time histories corresponding to the accelerogram recorded in the north–south direction at the TCU068 station during the 1999 Chi-Chi, Taiwan, earthquake.

Measurement of Earthquakes

FURTHER READINGS

1. Bolt, B. A., Tsai, Y. B., Yeh, K., and Hsu, M. K., Earthquake strong motions recorded by a large near-source array of digital seismographs, *Earthquake Engineering and Structural Dynamics*, 10, 1982, 561–573.
2. Bolt, B. A., *Earthquakes*, 5th Edition, W. H. Freeman and Co., New York, 2004, 378 p.
3. Bolt, B. A., Engineering Seismology, Chapter 2 in *Earthquake Engineering: From Engineering Seismology to Performance-Based Design*, Bozorgnia, Y. and Vertero, V. V., eds., CRC Press LLC, Boca Raton, FL, 2004.
4. Boore, D. M., The motion of the ground in earthquakes, *Scientific American*, December 1977.
5. Converse, A. M., *BAP: Basic Strong Motion Accelerogram Processing Software, Version 1.0*, Open-File Report 92-296A, U.S. Geological Survey, U.S. Department of Interior, Denver, CO, 1992.
6. Hodgson, J. H., *Earthquakes and Earth Structure*, Prentice-Hall, Englewood Cliffs, NJ, 1964, 166 p.
7. Hudson, D. E., *Reading and Interpreting Strong Motion Accelerograms*, Earthquake Engineering Research Institute, Oakland, CA, 1979, 112 p.
8. Lay, T. and Wallace, T. C., *Modern Global Seismology*, Academic Press, San Diego, 1995, 517 p.
9. Medvedev, S. V. and Sponheuer, W., Scale of seismic intensity, *Proceedings of the 4th World Conference on Earthquake Engineering*, 1, A2 143–153, Asociación Chilena de Sismología e Ingeniería Antisísmica, Santiago, Chile, 1969.
10. Nagata, S., Katayama, T., Yamazaki, F., Lu, L., and Turker, T., A dense seismograph array in Chiba, Japan and its strong motion database, *Proceedings of 4th U.S. National Conference on Earthquake Engineering*, 1, 357–366, Earthquake Engineering Research Institute, Oakland, CA, 1990.
11. Okamoto, S., *Introduction to Earthquake Engineering*, 2nd Edition, University of Tokyo Press, Tokyo, 1984, 629 p.
12. Porcella, R. and Matthiesen, R., *Strong Motion Records, Oct. 15, 1979 Imperial Valley Earthquake*, USGS Open-File Report 79-1654, 1979, 42 p.

PROBLEMS

5.1 The P and S waves from an earthquake in central California arrived at three different stations at the times indicated in Table P5.1. Using the map of central California given in Figure P5.1, determine and plot in the map the location of the earthquake epicenter.

5.2 Figure P5.2 shows the initial portion of a seismogram recorded in Australia from an earthquake in Oaxaca, Mexico. Identify in the figure the arrival times of the P, S, and surface waves and explain the reasons for the selection.

5.3 The seismogram shown in Figure P5.3 was obtained from a Wood–Anderson seismograph located 100 km from the hypocenter of the recorded earthquake. Determine (a) the local magnitude of the earthquake and (b) the average shear wave velocity for the region surrounding the recording station considering that the earthquake started at 07:19:32 and the P wave arrived at the recording site at 07:19:45.2.

TABLE P5.1
Arrival Times of P and S Waves at Three Different Stations in Central California

Station	P Wave			S Wave		
	h	min	s	h	min	s
A	15	45	54.2	15	46	07.1
B	15	46	07.6	15	46	28.0
C	15	46	04.5	15	46	25.5

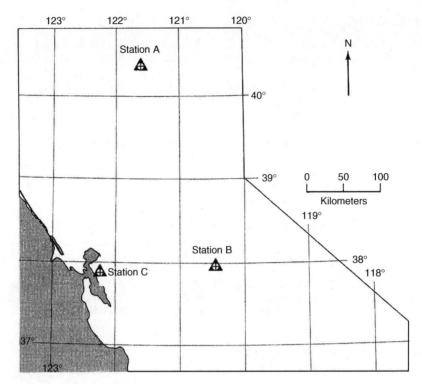

FIGURE P5.1 Map of central California showing location of seismographic stations A, B, and C.

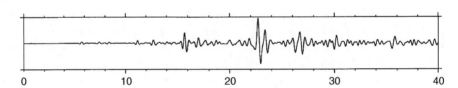

FIGURE P5.2 Seismogram considered in Problem 5.2.

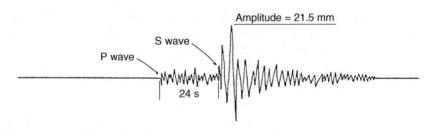

FIGURE P5.3 Seismogram considered in Problem 5.3

5.4 A seismogram from an earthquake exhibits a peak trace amplitude of 75 mm. The seismogram is obtained from a Wood–Anderson seismograph at a seismological station 500 km away from the epicenter of the earthquake. If the distance correction equation for this station is given by

$$-\log A_0 = 1.110 \log\left(\frac{r}{100}\right) + 0.00189(r - 100) + 3$$

Measurement of Earthquakes

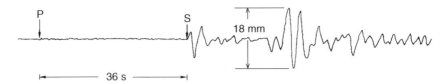

FIGURE P5.5 Seismogram considered in Problem 5.5.

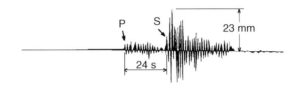

FIGURE P5.7 Seismogram considered in Problem 5.7.

where A_0 denotes amplitude in millimeters and r hypocentral distance in kilometers, what was the local magnitude of the recorded earthquake?

5.5 The seismogram shown in Figure P5.5 was obtained in Southern California from a Wood–Anderson seismograph located 75 km from the hypocenter of the recorded earthquake. Determine (a) the local magnitude of the earthquake using Richter's empirical attenuation curves for Southern California and (b) the velocity of propagation of the P waves assuming that the Poisson ratio for the medium along which the earthquake waves traveled is equal to 0.30.

5.6 A seismogram from an earthquake shows a peak trace amplitude of 125 mm. The seismogram was obtained from a seismograph with a magnification factor of 1000 at a seismological station 600 km away from the earthquake epicenter. Determine the local magnitude of the recorded earthquake using a distance correction defined by

$$-\log A_0 = 1.110 \log\left(\frac{r}{100}\right) + 0.00189(r - 100) + 3$$

where A_0 denotes amplitude in millimeters and r hypocentral distance in kilometers.

5.7 The seismogram shown in Figure P5.7 was obtained in Southern California with a Wood–Anderson seismograph at a seismographic station located 165 km from the earthquake epicenter. (a) Determine the local magnitude of the earthquake using Richter's attenuation equation for Southern California. (b) If the earthquake started at 07:19:32, and if the first P wave arrived at the recording site at 07:19:45.2, what is the average Poisson ratio for the region surrounding the recording station?

5.8 The peak ground displacement recorded by a long-period seismograph located at a distance of 3000 km from an earthquake's epicenter was 7.6 mm. Determine the earthquake's surface-wave magnitude.

5.9 Estimate the moment magnitude of an earthquake generated at a fault that slips, on average, 9.2 m and ruptures an area 44 km wide and 75 km long.

5.10 Estimate the moment magnitude of earthquakes from faults whose ruptured area is >30 km long and 25 km deep and slip >1.0 m.

5.11 A building is to be constructed in Southern California at a site whose location with respect to nearby earthquake faults is indicated in Table P5.11. Also given in this table is the local magnitude of the largest earthquake that is likely to occur at these faults. Find the maximum ground displacement for which the building should be designed.

TABLE P5.11
Magnitudes and Distances to Sites Considered in Problem 5.11

Fault	M_L	Closest Distance to Site (km)
San Andreas	8.0	78
San Jacinto	7.5	81
Santa Monica	7.25	26
Whittier	7.0	32
Newport-Inglewood	7.0	7

5.12 The undamped natural frequency and damping ratio of a commercial accelerograph are 25 Hz and 60%, respectively. What would be the error introduced by the accelerograph in the recording of a sinusoidal ground acceleration with a frequency of 12.5 Hz due to the nonuniform frequency response of the instrument?

5.13 Download an uncorrected acceleration record from the Internet and integrate it to obtain the velocity and displacement time histories. Determine and comment on the value of the displacement obtained at the end of the displacement time history.

6 Characterization of Strong Ground Motions

The inability of a single parameter, whether the peak ground acceleration, the effective acceleration, or any other variable, to characterize properly an earthquake and its potential effects on structures has been recognized for a very long time. Yet the use of a single parameter has withstood the attacks of time and the research community and continues to be accepted in seismic design.

José M. Roesset, 1994

6.1 INTRODUCTION

As it may be seen from a brief review of the accelerograms presented in the previous chapter, different earthquakes produce ground motions with different characteristics, that is, ground motions with different intensity, duration, and dominant periods. Thus, the question arises as to what is the best way to characterize ground motions for the purpose of seismic design. In other words, what is the best way to describe or specify the characteristics of the expected ground motions at a given seismic region? A satisfactory answer to this question has not been found yet, as manifested by the continuing research efforts made toward that goal. The problem is that a future ground motion cannot be represented by an accelerogram from a previous earthquake as in all likelihood a future earthquake will not reproduce the ground motion from a previous earthquake, even if the earthquake occurs at the same location and the ground motion is recorded at the same site. Another problem is that not a single ground motion characteristic, or set of characteristics, is sufficient to describe the effect that a ground motion may have on the response of a structure. Hence, a design ground motion cannot be specified, at least accurately, in terms of such ground motion characteristic or characteristics.

In current practice, several parameters are used to characterize ground motions for design purposes. These parameters include peak acceleration, peak velocity, ground motion duration, response spectrum, Fourier spectrum, and some other parameters that are defined in terms of these spectra. This chapter describes, thus, these ground motion parameters, points out their advantages and disadvantages, and illustrates their use in design procedures.

6.2 PEAK GROUND ACCELERATION

As indicated in Section 5.7, information on the characteristics of strong earthquake ground motions is obtained from acceleration records from past earthquakes. In addition, acceleration records can be integrated with respect to time to obtain, first, time histories of ground velocity and, then, time histories of ground displacement. Thus, from ground acceleration records it is possible to extract some of the main characteristics of recorded ground motions, such as peak ground acceleration, peak ground velocity, peak ground displacement, and ground motion duration. As structural response is directly proportional to ground acceleration, traditionally the peak value observed in a ground acceleration record (in absolute value terms) is used to describe the intensity of the ground motion at the recording site. Consequently, for many years peak ground acceleration has been used as a parameter to characterize ground motion intensity and as a measure of ground motion severity. In accordance with this tradition, many building codes utilize, explicitly or implicitly, peak ground acceleration values to characterize the ground motion intensity expected at given seismic regions. It should be noted, however, that peak ground acceleration alone does not reflect the intensity of structural response because duration and dominant period may also influence the response of a structure significantly. For example, a high acceleration may appear to be potentially dangerous, but if it occurs only for a short time, it will cause little damage to some structures. On the other hand, a relatively small acceleration that continues with uniform frequency for a number of seconds can cause considerable damage in some other structures. Peak ground acceleration, therefore, needs to be supplemented with additional information to accurately characterize ground motion severity.

In recent years, peak ground velocity has been preferred over peak ground acceleration to characterize ground motion severity. This is so because it has been found that peak ground velocity correlates better than peak ground acceleration with observed structural damage, particularly for structures within the intermediate range of the typical natural frequencies.

6.3 STRONG-MOTION DURATION

Although peak ground acceleration and peak ground velocity characterize the intensity of ground motion in a useful way, many researchers believe that damage to structures is more closely related to the duration of strong ground shaking than to any particular ground motion peak value. Indeed, as structural damage strongly depends on the number of load reversals that may occur during an earthquake, a motion of short duration may not generate enough load reversals to damage a structure, even if the amplitude of the motion is high. Conversely, a motion with a moderate amplitude but a long duration can produce enough load reversals to cause significant damage.

Accelerograms contain all the recorded ground motion accelerations from the time the recording starts until the time the accelerations return back to the level of background noise. For engineering purposes, however, only the strong part is of interest, that is, the portion during which significant structural response develops. Thus, several procedures have been proposed to define the strong-motion part in an acceleration record. In 1969, B. A. Bolt proposed the *bracketed duration*, which is the elapsed time between the first and the last peaks with an acceleration greater than a specified value (usually 0.05g). In 1975, M. D. Trifunac and A. G. Brady defined the strong-motion duration as the time interval between the points at which the integral of the acceleration square (i.e., $\int a^2 \, dt$) is equal to 5 and 95% of the total value. A third procedure was suggested by M. W. McCann and H. C. Shah in 1979 based on the rate of change of an accelerogram's cumulative root mean square (see Section 6.8.2 for the definition of root mean square acceleration). In this procedure, the end of the strong-motion part is considered to be the point at which the rate of change of the cumulative root mean square becomes negative and remains so for the remainder of the record. The initial time is obtained in the same manner except that the search is performed starting from the end of the record.

Table 6.1 shows the duration of the strong-motion part for several records computed according to the procedures introduced by Bolt, Trifunac and Brady, and McCann and Shah. It may be observed from these results that, as expected, these procedures lead to different strong-motion durations. Therefore, as no definition of strong-motion duration has been agreed upon as being the most realistic, this parameter cannot be used objectively to characterize strong ground motions despite its recognized importance.

6.4 RESPONSE SPECTRUM

6.4.1 BACKGROUND

A response spectrum is another, more meaningful way to characterize earthquake ground motions. The concept of response spectrum was born out of an instrument made by K. Suyehiro at the Earthquake Research Institute at the University of Tokyo in 1926. This instrument consisted of

TABLE 6.1
Strong-Motion Duration for Eight Earthquake Records According to Several Definitions

			Effective Duration (s)		
Record	Component	Total Duration (s)	Bolt	Trifunac and Brady	McCann and Shah
El Centro, 1940	S00E	53.74	25.86	24.42	25.44
	S90W	53.46	25.40	24.54	25.82
Taft, 1952	N21E	54.34	19.50	30.54	34.32
	S69E	54.38	15.12	28.86	32.96
El Centro, 1934	S00W	90.28	12.86	21.10	21.96
	S90W	90.22	18.12	20.28	18.48
Olympia, 1949	N04W	89.06	22.30	25.80	22.94
	N86E	89.02	21.04	18.08	21.52

Source: Mohraz, B. and Sadek, F., in *The Seismic Design Handbook*, 2nd edition, Naeim, F. ed., Kluwer Academic Publishers, Boston, MA, 2001.

13 unidirectional pendula with periods of vibration ranging from 0.22 to 1.81 s. It was devised to observe how different pendula respond differently to different earthquake ground motions. This concept was further developed by H. Benioff, a seismologist at the California Institute of Technology, and extended to earthquake engineering by M. A. Biot, an engineering graduate student at this same institution. In his 1932 doctoral thesis, Biot proposed the calculation of what it is now called response spectrum. In 1941, well before the advent of electronic digital computers, Biot also built a mechanical analog to generate the first complete response spectrum.

The response spectrum concept has proved so useful in earthquake engineering that a response spectrum is now routinely computed and published virtually for all the strong earthquake ground motions that are recorded anywhere in the world. Enough of them have been obtained to get a reasonable idea of the average characteristics and the way factors such as fault type, regional geology, and local soil conditions influence these characteristic.

6.4.2 Definition

The concept of a response spectrum may be introduced with reference to the single-degree-of-freedom structure with mass m, stiffness k, and damping constant c shown in Figure 6.1. Let t denote time, $u_g(t)$ the displacement of the ground with respect to a fixed reference frame, $u(t)$ the relative displacement of the mass of the structure with respect to its base, and $y(t) = u_g(t) + u(t)$ the displacement of the mass with respect to the fixed reference frame. Similarly, let $\omega_n = (k/m)^{1/2}$ be the circular natural frequency of the structure, $\xi = c/2m\omega_n$ its damping ratio or percentage of critical damping, and $\omega_d = \omega_n(1 - \xi^2)^{1/2}$ its damped natural frequency.

The differential equation of motion for such a system may be expressed as

$$m\ddot{u}(t) + c\dot{u}(t) + ku(t) = -m\ddot{u}_g(t) \tag{6.1}$$

which in terms of the natural frequency ω_n and damping ratio ξ may also be put into the form

$$\ddot{u}(t) + 2\omega_n\xi\dot{u}(t) + \omega_n^2 u(t) = -\ddot{u}_g(t) \tag{6.2}$$

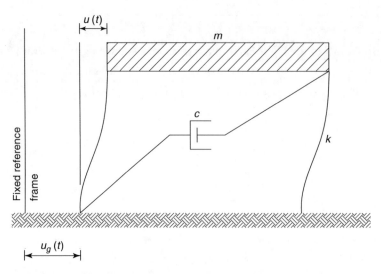

FIGURE 6.1 Single-degree-of-freedom system under base excitation.

Using Duhamel's integral and considering zero initial conditions, the relative displacement $u(t)$ may then be written as

$$u(t) = \frac{-1}{\omega_d} \int_0^t \ddot{u}_g(\tau) e^{-\xi \omega_n (t-\tau)} \sin \omega_d (t-\tau) d\tau \tag{6.3}$$

where τ is a dummy variable. As a result, the absolute value of the system's maximum relative displacement, may be expressed as

$$\max |u(t)| = SD(\omega_n, \xi) = \left| -\frac{1}{\omega_d} \int_0^t \ddot{u}_g(\tau) e^{-\xi \omega_n (t-\tau)} \sin \omega_d (t-\tau) d\tau \right|_{max} \tag{6.4}$$

and denoted as $SD(\omega_n, \xi)$. Similarly, after taking the first and the second derivatives of this expression, the absolute value of the system's maximum relative velocity and maximum absolute acceleration (denoted, respectively, as $SV(\omega_n, \xi)$ and $SA(\omega_n, \xi)$) may be written as

$$\max |\dot{u}(t)| = SV(\omega_n, \xi) = \left| -\int_0^t \ddot{u}_g(\tau) e^{-\xi \omega_n (t-\tau)} \left[\cos \omega_d (t-\tau) - \frac{\xi}{\sqrt{1-\xi^2}} \sin \omega_d (t-\tau) \right] d\tau \right|_{max} \tag{6.5}$$

$$\max |\ddot{u}(t) + \ddot{u}_g(t)| = SA(\omega_n, \xi) = \left| \omega_d \int_0^t \ddot{u}_g(\tau) e^{-\xi \omega_n (t-\tau)} \left[\left(1 - \frac{\xi^2}{1-\xi^2}\right) \sin \omega_d (t-\tau) + \frac{2\xi}{\sqrt{1-\xi^2}} \cos \omega_d (t-\tau) \right] d\tau \right|_{max} \tag{6.6}$$

where in the derivation of the last two equations it has been considered that for any function $f(\tau, t)$

$$\frac{d}{dt} \left[\int_0^t f(\tau, t) d\tau \right] = \int_0^t \frac{\partial f(\tau, t)}{dt} d\tau + f(\tau, t) \big|_{\tau=t} \tag{6.7}$$

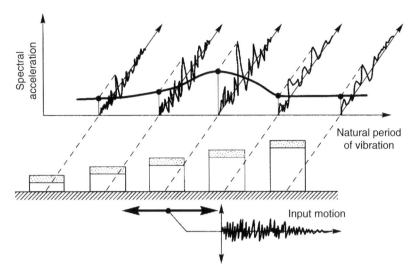

FIGURE 6.2 Schematic interpretation of response spectrum. (After Kramer, S.L., *Geotechnical Earthquake Engineering*, Prentice-Hall, Inc., Upper Saddle River, NJ, 1996. Reprinted with permission from Pearson Education, Inc., Upper Saddle River, NJ.)

It may be seen, thus, that the response of a single-degree-of-freedom structure to an earthquake ground motion, be it in terms of relative displacement, relative velocity, or absolute acceleration, depends on only two of the structure's parameters: its undamped natural frequency and its damping ratio. Hence, a good indicator of the level of response that may be induced by a particular ground motion in a set of different structures is curves that describe the variation of such a response with the structures' natural frequencies and damping ratios. These curves constitute what is known as *response spectrum*, and the ordinates in these curves are known as *spectral ordinates*.

A schematic interpretation of the concept of response spectrum is shown in Figure 6.2. Note that for a particular ground acceleration time history (input motion) and a structure with a particular natural frequency ω_n and damping ratio ξ, one can determine the response time history of that particular structure to that particular ground motion and obtain the maximum value of such a response. Then, one can repeat the procedure for a series of different structures with different natural periods but the same damping ratio and determine the corresponding maximum values of their response. The plot of such maximum values yields the response spectrum for the considered ground acceleration time history and the chosen damping ratio.

A response spectrum may be thus formally defined as a graphical representation of the variation with natural frequency (or natural period) and damping ratio of the absolute value of the maximum response of a single-degree-of-freedom system to a given ground acceleration time history. If the response being considered is the absolute acceleration, then the response spectrum is denoted as *acceleration response spectrum*. Similarly, if the is response under consideration is the relative velocity or relative displacement, then the response spectrum is referred to as *velocity response spectrum* or *displacement response spectrum*, respectively. Traditionally, a response spectrum is displayed as a family of curves, one for each of several damping ratios. The horizontal axis may display either natural frequencies in hertz or natural periods in seconds. Figure 6.3 shows some examples of a response spectrum. Acceleration, velocity, and displacement response spectra of one of the horizontal components of the ground motions recorded during the 1985 Mexico City and 1989 Loma Prieta, California, earthquakes are shown, respectively, in Figures 6.3a through c and d through f. It is important to note that according to the aforementioned definition, Equations 6.4 through 6.6 constitute the mathematical representation of a spectral displacement, a spectral velocity, and a spectral acceleration, respectively.

6.4.3 Importance of Response Spectrum

As it may be inferred from its definition, a response spectrum is a convenient tool to assess the level of response a given ground motion will induce in structures that can be modeled as a single-degree-of-freedom system. It also provides a means to identify which structures will be affected the most by the given ground motion. Furthermore, it is shown in Chapter 10 that a response spectrum is also useful to estimate the maximum response of multi-degree-of-freedom systems in a simple way, despite the fact that a response spectrum is defined only for single-degree-of-freedom systems. However, the usefulness of a response spectrum lies beyond the estimation of maximum response. As different ground motion time histories yield markedly different response spectra, a response spectrum may also be used to characterize earthquake ground motions. That is, the differences in the ground motions recorded at different sites and on different soils may be evaluated by a comparison of

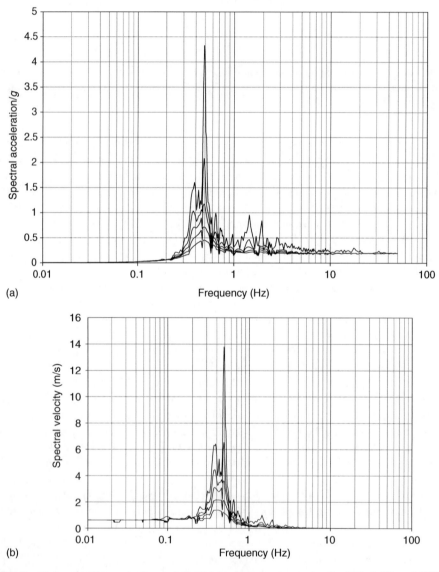

FIGURE 6.3 (a) Acceleration, velocity, and displacement response spectra for 0, 2, 5, 10, and 20% damping corresponding to (a–c) September 19, 1985, Mexico City earthquake, SCT station, S60E component and (d–f) October 17, 1989, Loma Prieta, California, earthquake, Foster City station, E–W component.

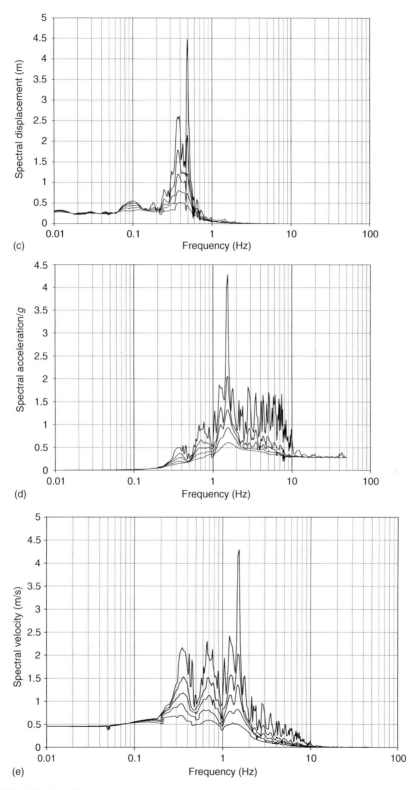

FIGURE 6.3 (Continued)

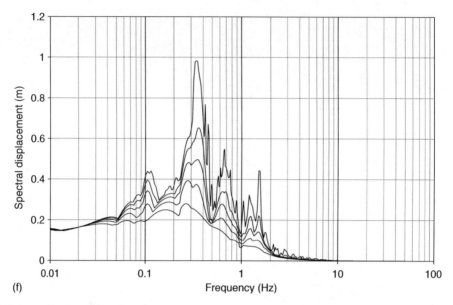

(f)

FIGURE 6.3 (Continued)

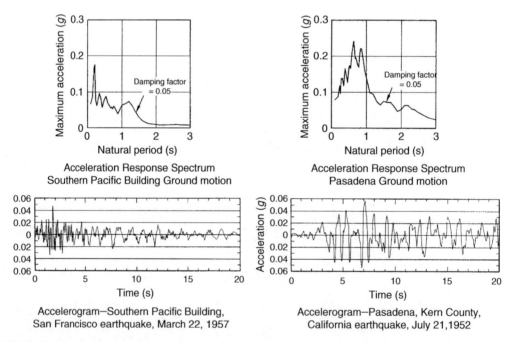

FIGURE 6.4 Accelerograms and corresponding acceleration response spectra of two ground motions with nearly the same peak ground acceleration. (Reproduced with permission of Earthquake Engineering Research Institute from Seed, H.B. and Idriss, I.M., *Ground Motion and Soil Liquefaction During Earthquakes*, Earthquake Engineering Research Institute, Oakland, CA, 1982.)

their response spectra. Consider, for example, the two ground motion records and the corresponding response spectra shown in Figure 6.4. Observe that these two records exhibit about the same peak ground acceleration but their response spectra are totally different. From the comparison of the two response spectra, it may be then concluded that the two records in question are radically different and will affect structures differently, something that would be difficult to do by a direct comparison

Characterization of Strong Ground Motions

of the records themselves. Because of its usefulness, the concept of response spectrum constitutes, directly or indirectly, the basis of many of the methods used in the analysis of earthquake-resistant structures and the formulation of the design recommendations in building codes.

6.4.4 TRIPARTITE REPRESENTATION

An alternative and convenient way to plot a response spectrum is by means of the so-called tripartite or tetralogarithmic representation. If a new spectral response referred to as *spectral pseudovelocity* is defined as

$$PSV(\omega_n, \xi) = \left| \int_0^t \ddot{u}_g(\tau) e^{-\xi \omega_n (t-\tau)} \sin \omega_n (t-\tau) d\tau \right|_{max} \quad (6.8)$$

then, for small damping ratios (see Equations 6.4 and 6.6), one has that

$$SA(\omega_n, \xi) \approx \omega_n PSV(\omega_n, \xi) \quad (6.9)$$

$$SD(\omega_n, \xi) \approx \frac{PSV(\omega_n, \xi)}{\omega_n} \quad (6.10)$$

where these approximate values of the spectral acceleration and spectral displacement are sometimes referred to as *pseudoacceleration* and *pseudodisplacement*, respectively, to reflect the fact that they are not equal to the true values. Thus, a simple approximate relationship among the spectral displacement, spectral acceleration, and pseudovelocity is

$$SA = \omega_n \quad PSV = \omega_n^2 SD \quad (6.11)$$

Using the last two terms in this relationship and taking logarithms on both sides of the equation, one can then write the relationship between *PSV* and *SD*, alternatively, as

$$\log PSV = \log \omega_n + \log SD \quad (6.12)$$

Consequently, if *SD* is considered constant, it is easy to see that this expression represents the equation of straight line with a slope of +1 and that lines of constant *SD* are straight lines with an inclination of 45°. Similarly, it can be shown that lines of constant *SA* are straight lines with an inclination of −45°. Thus, it is possible to read spectral displacements, pseudovelocities, and spectral accelerations versus natural frequency or period in a single plot. All one has to do is plot pseudovelocity versus natural frequency on vertical and horizontal logarithmic scales and draw additional logarithmic scales making an angle of 45° and −45° with respect to the vertical axis. In this way, pseudovelocities are read from the vertical scale, spectral displacements from the scale at −45°, and spectral accelerations from the scale at 45°. Figure 6.5 gives an example of a response spectrum using such a tripartite representation. Note that when the plot is drawn in terms of natural periods as opposed to natural frequencies, the spectral accelerations are read from the scale at −45° and the spectral displacements from the scale at 45°.

In using response spectra in a tripartite plot, it is important to observe the following:

1. For zero damping, the relationships $SD = PSV/\omega_n$ and $SA = \omega_n PSV$ are exact (see Equations 6.4 and 6.6).
2. As pseudovelocity and spectral velocity are, respectively, defined in terms of a sine and a cosine function (see Equations 6.5 and 6.8), these two parameters are, in general, not equal to each other (see Figure 6.6). However, it has been argued that pseudovelocity is related to the maximum strain energy stored in a single-degree-of-freedom system and is, therefore, a useful parameter by itself.

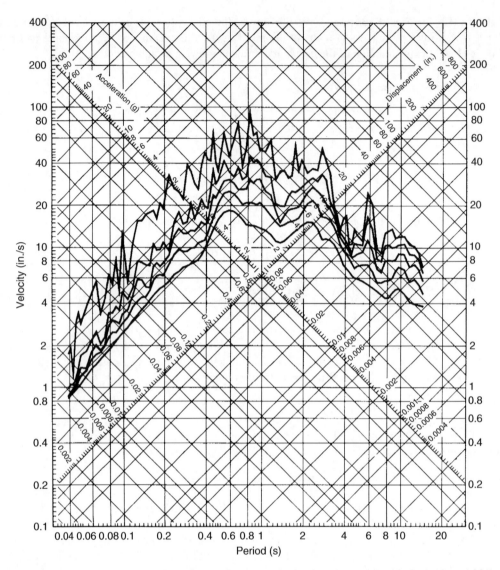

FIGURE 6.5 Response spectrum in tripartite representation for damping ratios of 0, 2, 5, 10, and 20% of the N–S component of the ground motion recorded at El Centro station during the May 18, 1940, Imperial Valley earthquake. (Reproduced with permission of Earthquake Engineering Research Institute from Hudson, D.E., *Reading and Interpreting Strong Motion Accelerograms*, Earthquake Engineering Research Institute, Oakland, CA, 1979.)

3. When the natural period of a single-degree-of-freedom structure is very long, the mass of the structure is very large compared with its stiffness (see Figure 6.7a). Hence, it will hardly move under any ground excitation, and the relative displacement of the mass with respect to the ground will be nearly equal to the displacement of the ground. The spectral displacement for a structure with a long natural period is, thus, approximately equal to the peak ground displacement.
4. When the natural period of a single-degree-of-freedom structure is very small, the stiffness of the structure is very large compared with its mass (see Figure 6.7b). In such a case, then, the structure will move almost as a rigid body under any ground excitation, and the absolute acceleration of its mass will be nearly equal to the ground acceleration.

Characterization of Strong Ground Motions

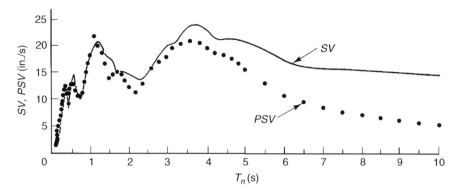

FIGURE 6.6 Five-percent-damping spectral velocity and spectral pseudovelocity for N11W component of ground motion recorded during the Eureka, California, earthquake of December 21, 1954. (Reproduced from Berg, G.V., *Elements of Structural Dynamics*, Prentice-Hall, Inc., Englewood Cliffs, NJ, 1989. Reprinted with permission from Pearson Education, Inc., Upper Saddle River, NJ.)

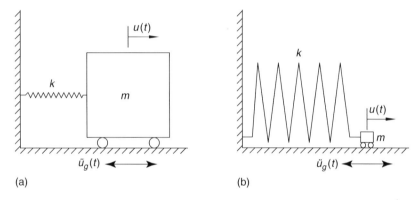

FIGURE 6.7 Single-degree-of-freedom systems with (a) long and (b) short natural periods.

Consequently, the spectral acceleration for a structure with a small natural period is approximately equal to the peak ground acceleration.

5. In view of items (3) and (4), it is possible to read the ground motion's peak acceleration and peak displacement from a ground motion's response spectrum. The peak ground acceleration is obtained from the value of the spectral acceleration for very short periods, whereas the peak ground displacement is determined from the value of the spectral displacement for very long periods (see Figure 6.8).

The relationship between pseudovelocity and maximum strain energy may be demonstrated as follows:

Consider first that the maximum strain energy stored in a single-degree-of-freedom system with mass m and stiffness k is given by

$$E_s = \frac{1}{2} k [u(t)]_{max}^2 = \frac{1}{2} k (SD)^2 \tag{6.13}$$

Then, consider that per unit mass this energy is given by

$$\frac{E_s}{m} = \frac{1}{2} \frac{k}{m} (SD)^2 = \frac{1}{2} (\omega_n SD)^2 \tag{6.14}$$

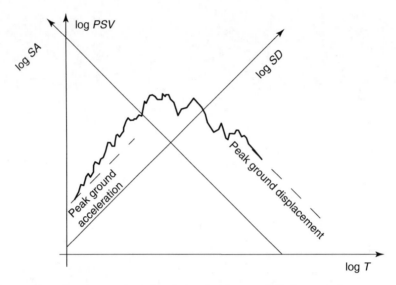

FIGURE 6.8 Reading of peak ground acceleration and peak ground displacement from response spectrum.

which in view of Equation 6.10 may also be expressed as

$$\frac{E_s}{m} = \frac{1}{2}(PSV)^2 \qquad (6.15)$$

EXAMPLE 6.1 EARTHQUAKE RESPONSE OF A SINGLE-DEGREE-OF-FREEDOM SYSTEM

A bin is mounted on a rigid platform supported by four columns 6 m long. The weight of the platform, bin, and contents, which may be assumed concentrated at a point at a distance of 1.5 from the bottom of the platform, is 680 kN (see Figure E6.1). The columns are braced in the longitudinal direction (normal to the plane of the drawing) but are unbraced in the transverse direction. The modulus of elasticity and moment of inertia of the columns are, respectively, equal to 201 GPa and 4×10^{-4} m⁴. Assuming the damping ratio of the structure is 5%, find the maximum lateral displacement of the system and the maximum bending moment in the columns due to an earthquake ground motion characterized by the response spectrum depicted in Figure 6.5. Assume the columns are rigidly connected to the base and the rigid platform.

Solution

Assuming the rotations at the end of the columns are zero, the lateral stiffness of each of the structure's four columns is given by

$$k = \frac{12EI}{L^3} = \frac{12(201 \times 10^6)(4 \times 10^{-4})}{6^3} = 4466.7 \text{ kN/m}$$

Therefore, the natural period of the system results as

$$T = 2\pi\sqrt{\frac{W}{kg}} = 2\pi\sqrt{\frac{680}{4(4466.7)(9.81)}} = 0.391 \text{ s}$$

From the response spectrum in Figure 6.5 for a natural period of 0.391 s and a damping ratio of 5%, one obtains

$$SD(0.391, 0.05) = 0.95 \text{ in.} = 0.024 \text{ m}$$

Characterization of Strong Ground Motions 193

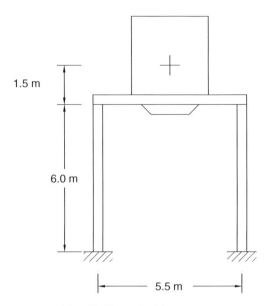

FIGURE E6.1 Bin and structure considered in Example 6.1.

Hence, under the given earthquake ground motion, the center of mass of the system displaces 0.95 in. or 0.024 m. In a similar manner, the bending moment induced at the ends of the columns results as

$$M = \frac{6EI}{L^2}\Delta = \frac{6(201 \times 10^6)(4 \times 10^{-4})}{6^2}(0.024) = 322 \text{ kN m}$$

6.4.5 Computation of Response Spectra

Several methods have been used in the past for the calculation of response spectra. One of the first was based on the numerical integration of the differential equation of motion that describes the motion of single-degree-of-freedom systems. Without the availability of modern digital computers, this method, however, turned out to be a formidable task. As a result, other more practical techniques were devised. One of these techniques has a semigraphical approach. Another was the mechanical analog used by Biot to compute the first complete response spectrum mentioned earlier. Still another one was the passive electrical analog system developed by Biot himself, which speeded up the calculations and increased the accuracy. Nowadays, response spectra are invariably computed with digital computers, integrating numerically the differential equation of motion for single-degree-of-freedom systems.

For the numerical integration of such an equation of motion, the third-order Runge–Kutta integration scheme has been used and is still preferred by many owing to its accuracy, stability, and self-starting features. This scheme is considered to be one of the best methods of integration when the accelerogram is not digitized at equal time intervals. For standardized accelerograms, which are normally digitized at equally spaced time intervals of 0.02 s and are defined by a series of successive linear segments, an approach proposed by N. C. Nigam and P. C. Jennings in 1969 appears to be the most efficient. This section describes Nigam–Jennings approach.

Let a single-degree-of-freedom system be characterized by its natural frequency ω and damping ratio ξ. In addition, let the system be subjected to a ground acceleration $a(t)$ represented by the piecewise linear function shown in Figure 6.9. If, in view of the linear variation of the ground acceleration within each segment, the ground acceleration at any time t within the time interval $t_i \leq t \leq t_{i+1}$ is expressed as

$$a(t) = a_i + \frac{\Delta a_i}{\Delta t}(t - t_i) \tag{6.16}$$

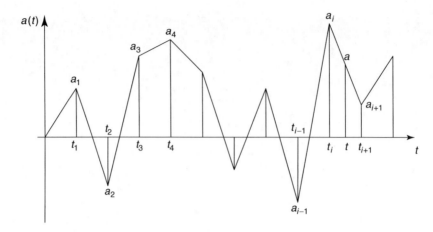

FIGURE 6.9 Acceleration time history defined by a series of successive linear segments at equal time intervals.

where

$$\Delta t = t_{i+1} - t_i \qquad (6.17)$$

and

$$\Delta a_i = a_{i+1} - a_i \qquad (6.18)$$

then the equation of motion for such a single-degree-of-freedom system in the aforementioned time interval may be written as

$$\ddot{u} + 2\omega\xi\dot{u} + \omega^2 u = -a_i - \frac{\Delta a_i}{\Delta t}(t - t_i) \qquad (6.19)$$

Equation 6.19 is a linear differential equation with constant coefficients, and as such its general solution is given by the sum of its homogenous solution and a particular solution. Consequently, the solution of Equation 6.19 is of the form

$$u(t) = e^{-\xi\omega(t-t_i)}[C_1 \sin\omega_d(t - t_i) + C_2 \cos\omega_d(t - t_i)] - \frac{a_i}{\omega^2} + \frac{2\xi}{\omega^3}\frac{\Delta a_i}{\Delta t} - \frac{1}{\omega^2}\frac{\Delta a_i}{\Delta t}(t - t_i) \qquad (6.20)$$

where $\omega_d = \omega(1 - \xi^2)^{1/2}$ and C_1 and C_2 are constants of integration. Upon taking the first derivative of Equation 6.20, the corresponding velocity is similarly given by

$$\dot{u}(t) = e^{-\xi\omega(t-t_i)}\{\omega_d[C_1 \cos\omega_d(t - t_i) - C_2 \sin\omega_d(t - t_i)] - \xi\omega[C_1 \sin\omega_d(t - t_i) + C_2 \cos\omega_d(t - t_i)]\} - \frac{1}{\omega^2}\frac{\Delta a_i}{\Delta t} \qquad (6.21)$$

After setting the initial conditions $u = u_i$ and $\dot{u} = \dot{u}_i$ at $t = t_i$, the constants of integration C_1 and C_2 become

$$C_1 = \frac{1}{\omega_d}\left(\xi\omega u_i + \dot{u}_i - \frac{2\xi^2 - 1}{\omega^2}\frac{\Delta a_i}{\Delta t} + \frac{\xi}{\omega}a_i\right) \qquad (6.22)$$

$$C_2 = u_i - \frac{2\xi}{\omega^3}\frac{\Delta a_i}{\Delta t} + \frac{a_i}{\omega^2} \qquad (6.23)$$

Setting t equal to t_{i+1} and substitution of C_1 and C_2 into Equations 6.20 and 6.21 lead, thus, to the recurrence relationship for the displacement and velocity of the system at time t_{i+1}

$$\begin{Bmatrix} u_{i+1} \\ \dot{u}_{i+1} \end{Bmatrix} = [A(\xi,\omega,\Delta t)] \begin{Bmatrix} u_i \\ \dot{u}_i \end{Bmatrix} + [B(\xi,\omega,\Delta t)] \begin{Bmatrix} a_i \\ a_{i+1} \end{Bmatrix} \quad (6.24)$$

where the elements of the matrices A and B are given by

$$a_{11} = e^{-\xi\omega\Delta t}\left(\frac{\xi}{\sqrt{1-\xi^2}}\sin\omega_d\Delta t + \cos\omega_d\Delta t\right) \quad (6.25)$$

$$a_{12} = \frac{e^{-\xi\omega\Delta t}}{\omega_d}\sin\omega_d\Delta t \quad (6.26)$$

$$a_{21} = -\frac{\omega}{\sqrt{1-\xi^2}}e^{-\xi\omega\Delta t}\sin\omega_d\Delta t \quad (6.27)$$

$$a_{22} = e^{-\xi\omega\Delta t}\left(\cos\omega_d\Delta t - \frac{\xi}{\sqrt{1-\xi^2}}\sin\omega_d\Delta t\right) \quad (6.28)$$

$$b_{11} = e^{-\xi\omega\Delta t}\left[\left(\frac{2\xi^2-1}{\omega^2\Delta t}+\frac{\xi}{\omega}\right)\frac{\sin\omega_d\Delta t}{\omega_d}+\left(\frac{2\xi}{\omega^3\Delta t}+\frac{1}{\omega^2}\right)\cos\omega_d\Delta t\right]-\frac{2\xi}{\omega^3\Delta t} \quad (6.29)$$

$$b_{12} = -e^{-\xi\omega\Delta t}\left[\left(\frac{2\xi^2-1}{\omega^2\Delta t}+\frac{\xi}{\omega}\right)\frac{\sin\omega_d\Delta t}{\omega_d}+\frac{2\xi}{\omega^3\Delta t}\cos\omega_d\Delta t\right]-\frac{1}{\omega^2}+\frac{2\xi}{\omega^3\Delta t} \quad (6.30)$$

$$b_{21} = e^{-\xi\omega\Delta t}\left[\left(\frac{2\xi^2-1}{\omega^2\Delta t}+\frac{\xi}{\omega}\right)\left(\cos\omega_d\Delta t - \frac{\xi}{\sqrt{1-\xi^2}}\sin\omega_d\Delta t\right)\right.$$
$$\left.-\left(\frac{2\xi}{\omega^3\Delta t}+\frac{1}{\omega^2}\right)(\omega_d\sin\omega_d\Delta t + \xi\omega\cos\omega_d\Delta t)\right]+\frac{1}{\omega^2\Delta t} \quad (6.31)$$

$$b_{22} = -e^{-\xi\omega\Delta t}\left[\frac{2\xi^2-1}{\omega^2\Delta t}\left(\cos\omega_d\Delta t - \frac{\xi}{\sqrt{1-\xi^2}}\sin\omega_d\Delta t\right)\right.$$
$$\left.-\frac{2\xi}{\omega^3\Delta t}(\omega_d\sin\omega_d\Delta t + \xi\omega\cos\omega_d\Delta t)-\frac{1}{\omega^2\Delta t}\right] \quad (6.32)$$

It may be seen, thus, that if the displacement and the velocity of the system are known at time t_i, the displacement and the velocity at time t_{i+1} may be determined through the application of Equation 6.24. Similarly, if the system's displacement and velocity are known at time t_{i+1}, then the absolute acceleration at time t_{i+1} may be obtained directly from the equation of motion (Equation 6.19) after setting $t = t_{i+1}$. This way, as shown in the algorithm presented in Box 6.1, the time histories of the system's displacement, velocity, and acceleration response may be computed through a step-by-step application of this procedure. A complete response spectrum may then be obtained by repeating the procedure for different natural frequencies (or natural periods) and damping ratios and identifying the absolute maximum values in such time histories.

> **Box 6.1 Algorithm to Determine Response Spectrum by Nigam–Jennings Approach**
>
> 1.0 Initialize and read input data:
>   1.1 Set value of time step Δt
>   1.2 Read ground acceleration values a_i, $i = 1, 2, ..., N$, where N = number of points in accelerogram
>   1.3 Set value of ξ
>   1.4 Choose value of ω and compute value of ω_d
>   1.5 Set $u_0 = 0$ and $\dot{u}_0 = 0$
>   1.6 Compute the elements of matrices $[A]$ and $[B]$ using Equations 6.25 through 6.32.
> 2.0 For the ith time step:
>   2.1 Compute the values of the displacement and velocity response at time t_{i+1} according to
>
> $$\begin{Bmatrix} u_{i+1} \\ \dot{u}_{i+1} \end{Bmatrix} = [A(\omega, \xi, \Delta t)] \begin{Bmatrix} u_i \\ \dot{u}_i \end{Bmatrix} + [B(\omega, \xi, \Delta t)] \begin{Bmatrix} a_i \\ a_{i+1} \end{Bmatrix}$$
>
>   2.2 Compute the value of the absolute acceleration response at time t_{i+1} according to
>
> $$\ddot{y}_{i+1} = \ddot{u}_{i+1} + a_{i+1} = -2\omega\xi\dot{u}_{i+1} - \omega^2 u_{i+1}$$
>
>   2.3 Set $i = i + 1$ and go back to Step 2.1.
> 3.0 After performing the calculation for the targeted number of time steps, identify and store the largest displacement, largest velocity, and largest acceleration, all in absolute value terms.
> 4.0 Repeat Steps 1.0 through 3.0 for different values of ω.
> 5.0 Repeat Steps 1.0 through 4.0 for different values of ξ.

Note that an advantage of this method is the fact that for a constant time interval Δt, the matrices $[A]$ and $[B]$ are calculated only once, as they depend only on ξ, ω, and Δt. It should also be noted that if indeed a ground acceleration time history varies linearly within each time interval, the calculation of a response spectrum using this approach involves no approximations other than those introduced by round-off errors. In this sense, it may be considered to be an exact method. It represents, thus, the best approach to obtain the response spectra of accelerograms that have been digitized at equally spaced time intervals and defined by a series of successive linear segments.

6.5 NONLINEAR RESPONSE SPECTRUM

6.5.1 BACKGROUND

As the definition of response spectrum given in Section 6.3 was based on the equations for the response of a linear single-degree-of-freedom system, it was implicitly considered that such a definition was only valid for linear systems. That is, it was implicitly assumed that the spring or resisting element in the system had a linear force-deformation behavior. In the case of earthquakes, however, structures are not designed to resist loads in their linear range of behavior. According to the current earthquake-resistant design philosophy, structures are designed to resist

- Minor earthquakes elastically without damage
- Moderate earthquakes elastically without structural damage, but possibly with some nonstructural damage
- Strong earthquakes inelastically with some structural and nonstructural damage, but without a collapse

Thus, implicitly or explicitly, the design of structures against earthquake ground motions requires a nonlinear analysis.

An approximate procedure used by many building codes to perform a nonlinear analysis is based on the concept of a nonlinear response spectrum. As it will be seen later on, a nonlinear response spectrum, like its linear counterpart, depends on the characteristics of the ground motion being considered and directly displays the effect of a ground motion in structures with specified properties. Consequently, a nonlinear response spectrum also represents a convenient alternative for ground motion characterization.

6.5.2 Load-Deformation Curves under Severe Cyclic Loading

When a structure is subjected to a low-level excitation, it behaves as a linear elastic structure for all practical purposes. That is, the deformations in the structure are proportional to the applied forces. In contrast, when the same structure is subjected to severe cyclic loading, the structure typically behaves as shown in Figure 6.10 for steel structures and Figure 6.11 for reinforced concrete ones. In such a case, the following may be observed:

1. The deformations in the structure may surpass the yield deformation of the material used for the construction of the structure without the collapse or failure of the structure.
2. The slope of the load-deformation curve (i.e., the structure's stiffness) is reduced when the deformations exceed the yield deformation of the structure, and in the case of reinforced concrete structures this slope is further reduced with additional loading cycles.
3. The removal of the applied load does not take the structure to the original condition of zero deformation existing before the application of the load; that is, the structure is left with a permanent deformation or permanent set.
4. Because of the load reversals and the permanent deformations, the load-deformation behavior of a structure beyond its linear range is defined by a series of loops known as *hysteretic loops*.

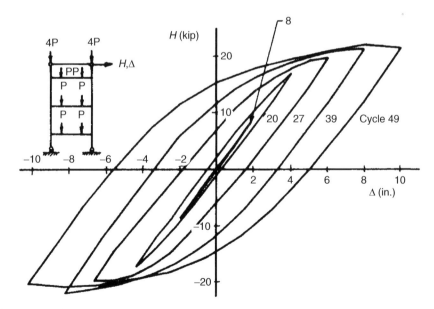

FIGURE 6.10 Typical nonlinear load-deformation curve for steel structures. (After Carpenter, L.D. and Lu, L.W., *Proceedings of 4th World Conference on Earthquake Engineering*, Santiago, Chile, Vol. I, B2-125–136, 1969. Reproduced with permission from Asociación Chilena de Sismología e Ingeniería Sísmica.)

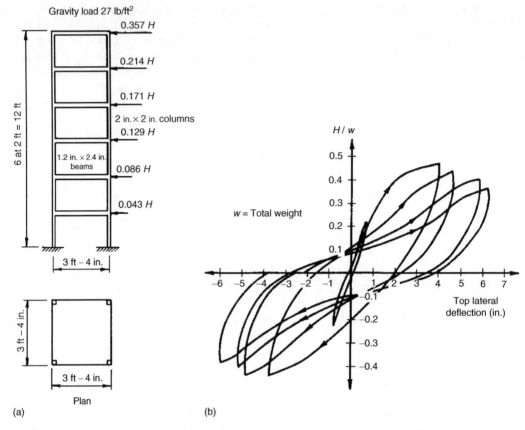

FIGURE 6.11 Typical nonlinear load-deformation curve for reinforced concrete structures (a) reinforced concrete building model and (b) lateral load-deflection curves. (After Park, R. and Pauley, T., *Reinforced Concrete Structures*, John Wiley & Sons, 1975. (c) John Wiley & Sons, Limited, New York. Reproduced with permission.)

5. The stiffness and strength of steel structures remains practically unaltered after the application of several load cycles whenever local and global instabilities are avoided and P-delta effects are not significant. In contrast, the stiffness and strength of reinforced concrete structures is reduced after each deformation cycle. This loss of stiffness and strength is, respectively, known as *stiffness degradation* and *strength degradation.*
6. Earthquake energy is dissipated when a permanent deformation is left in the structure; this dissipated energy is equal to the area under each loop in the load-deformation curve and is often referred to as *hysteretic damping.*

On the basis of these observations, it may be seen that by incurring into a structure's nonlinear range of behavior

1. Energy dissipation takes place in the form of hysteretic damping, reducing thus the response of the structure.
2. Energy dissipation takes place at the cost of damaging the structure and leaving permanent deformations.
3. Special care is necessary in the design and detailing of the structure to insure that it can withstand large deformations and several cycles of loading without a significant stiffness and strength deterioration.

Characterization of Strong Ground Motions

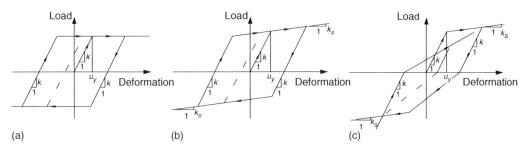

FIGURE 6.12 Idealized nonlinear load-deformation curves for structures under severe cyclic loading (a) elastoplastic, (b) bilinear, and (c) stiffness degrading.

In view of the complex behavior observed when structures are subjected to severe cyclic loadings, their nonlinear analysis is customarily carried out using simplified load-deformation models instead of the actual curves. Some of the simplified models that are frequently used in current practice are shown in Figure 6.12.

6.5.3 Definition

Since a response spectrum is defined as a set of curves that display the maximum response of single-degree-of-freedom systems to a given ground excitation, in principle the concept of response spectrum does not have to be limited to linear systems. It is possible, therefore, to extend this concept to nonlinear systems. It should be noted, however, that such an extension requires the specification of a load-deformation behavior, as the response of a structure depends on its load-deformation behavior. It also requires the consideration of a natural frequency or period defined in terms of the initial, elastic proprieties of the structure. Thus, a nonlinear response spectrum may be defined formally as the graphical representation of the variation with initial natural frequency or period of the absolute value of the maximum response of a single-degree-of-freedom system with a given load-deformation behavior and damping ratio to a specified ground acceleration time history.

6.5.4 Graphical Representation

Nonlinear response spectra may be plotted in much the same way as linear response spectra, that is, plotting maximum response versus natural period or frequency for various values of damping. Conventionally, however, they are plotted in a tripartite graph for a single damping ratio and various values of the *ductility factor* μ, defining this factor as

$$\mu = \frac{u_{max}}{u_y} \qquad (6.33)$$

where u_{max} denotes the maximum deformation induced in the spring or resisting element of the system by the earthquake ground motion under consideration, and u_y the corresponding yield deformation (see Figure 6.13). A ductility factor describes how many times the maximum deformation in a structure exceeds the structure's yield deformation, providing thus a more meaningful measure of the severity of the deformation than the value of the deformation itself. Similarly, nonlinear response spectra are plotted in terms of either total deformation (see Figure 6.14) or yield deformation (see Figure 6.15). In the total deformation spectrum, the product of the maximum deformation u_{max} and the initial natural frequency in radians per second ω is plotted in the vertical axis for several values of the ductility factor μ. In this way, for a given value of μ, one can read the maximum

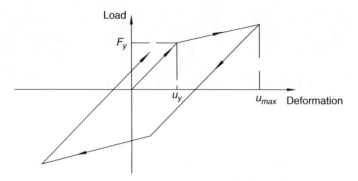

FIGURE 6.13 Maximum and yield deformations in a bilinear single-degree-of-freedom system.

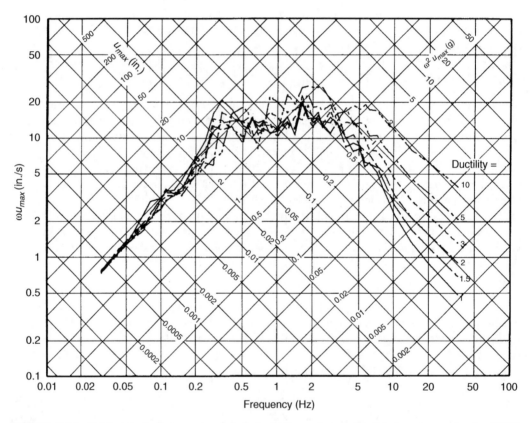

FIGURE 6.14 Total deformation spectra for elastoplastic systems with 5% damping under the N86E accelerogram from the 1949 Olympia, Washington, earthquake. (After Riddell, R. and Newmark, N.M., *Statistical Analysis of the Response of Nonlinear Systems Subjected to Earthquakes*, Structural Research Series No. 468, Department of Civil Engineering, University of Illinois, Urbana, IL, August 1979.)

deformation u_{max} from the displacement scale and the product $\omega^2 u_{max}$ from the acceleration scale. Note, however, that in a nonlinear response spectrum ωu_{max} and $\omega^2 u_{max}$ have no physical meaning. In the yield deformation spectrum, the quantity that is plotted in the vertical axis for several values of the ductility factor μ is the product of the yield deformation u_y and the initial natural frequency ω. In a yield deformation spectrum, therefore, the displacement and acceleration scales, respectively, give the yield deformation u_y and the product $\omega^2 u_y$. As in the case of a total deformation spectrum,

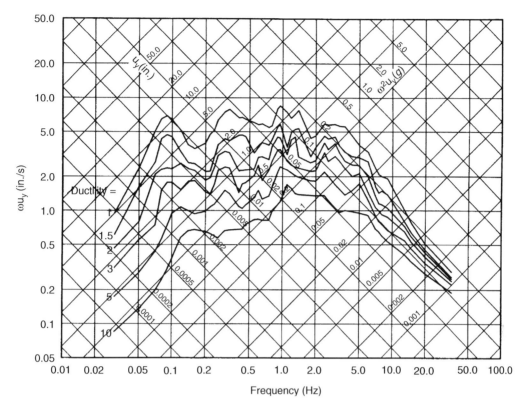

FIGURE 6.15 Yield deformation spectra for elastoplastic systems with 10% damping under the N10W Santiago accelerogram from the 1971 Central Chile earthquake. (After Riddell, R. and Newmark, N.M., *Statistical Analysis of the Response of Nonlinear Systems Subjected to Earthquakes*, Structural Research Series No. 468, Department of Civil Engineering, University of Illinois, Urbana, IL, August 1979.)

the product ωu_y has no physical meaning. However, $\omega^2 u_y$ represents the mass acceleration at which the system yields and is equal to the system's yield strength over its mass. To find the value of the maximum or total deformation from a yield deformation spectrum, one enters with the value of the initial natural frequency to read a value of u_y from the curve corresponding to the selected ductility factor μ. Then, one multiplies this value of u_y by μ, the used value of the ductility factor, if the system's load-deformation behavior is assumed elastoplastic (see Figure 6.12a), or by a similar relationship if any other load-deformation behavior is considered.

It should be noted that from the design point of view, the yield deformation response spectrum is more practical than the total deformation one. This is so because in current design practice, structures are designed to resist maximum forces instead of maximum deformations. Accordingly, the parameter of interest in the design of a structure is the maximum force induced in the structure by the design ground motion. However, as in the case of elastoplastic systems such maximum force is equal to the product of the yield deformation and the structural stiffness (see Figure 6.12a), the design parameter of interest in the case of elastoplastic systems is the yield deformation that is needed to resist the design ground motion without excessive inelastic deformations. Thus, in an ordinary design, one decides first what is the level of inelastic deformation a structure is capable to resist, that is, an acceptable ductility factor. Then, with this ductility factor and the initial natural frequency of the structure, one finds the corresponding yield deformation u_y from a yield deformation spectrum. Thereafter, the value of the design maximum force is obtained by multiplying this yield deformation by the structural stiffness k.

EXAMPLE 6.2 DESIGN YIELD FORCE FOR A SINGLE-DEGREE-OF-FREEDOM SYSTEM

A single-degree-of-freedom steel structure has a stiffness of 472 kip/in., a mass of 2.95 kip s²/in., and a damping ratio of 10%. Determine the design force that will make the structure resist the design earthquake without failure. Consider that (a) the design ground motion is the accelerogram recorded along the S16E direction at Pacoima Dam during the San Fernando earthquake of February 9, 1971 (yield deformation spectrum given in Figure 6.15), (b) the load-deformation behavior of the structure is elastoplastic, and (c) steel structures can withstand inelastic deformations as large as five times their yield deformation.

Solution

According to the given properties, the initial natural frequency of the structure is given by

$$f_n = \frac{1}{2\pi}\sqrt{\frac{472}{2.95}} = 2 \text{ Hz}$$

and its ductility factor is equal to 5. Hence, from the yield-deformation response spectrum in Figure 6.15 for a natural frequency of 2 Hz, a damping ratio of 10%, and a ductility factor of 5, the corresponding yield deformation is

$$u_y = 0.14 \text{ in.}$$

As a result, the yield force for which the structure should be designed for is given by

$$F_y = 472(0.14) = 66.1 \text{ kips}$$

6.5.5 COMPUTATION OF NONLINEAR RESPONSE SPECTRUM

Consider a single-degree-of-freedom system with the bilinear load-deformation behavior shown in Figure 6.16, where R denotes load and u deformation. Let u_y be the system's initial yield deformation, u_{yp} and u_{yn} its current positive and negative yield deformations, s the permanent set remaining in it after an excursion of yielding and subsequent unloading, k its stiffness when the deformation in the structure is less than its yield deformation, and α the ratio of the postyield to the elastic stiffness. Observe that initially $s = 0$, $u_{yp} = u_y$, and $u_{yn} = -u_y$. Note also that a region of a linearly elastic deformation of magnitude $2u_y$ separates the current positive and negative yield deformations.

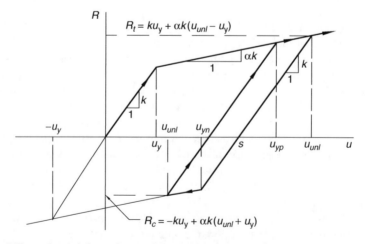

FIGURE 6.16 Bilinear force-deformation model and related parameters.

Under a bilinear force-deformation behavior, the restoring force in the structure is defined by two distinct curves: (a) the elastic loading or unloading curve with stiffness k and (b) the postyielding curve with stiffness αk. Consider first the linear elastic curve that follows unloading. The linear elastic response that follows unloading is given by the equation of motion

$$\ddot{u} + 2\xi\omega\dot{u} + \omega^2(u - s) = -a(t) \tag{6.34}$$

where u, $\dot{u}$, and $\ddot{u}$, respectively, denote relative displacement, velocity, and acceleration; ω and ξ, respectively, represent the system's undamped natural frequency and damping ratio; and $a(t)$ depicts ground acceleration. Note, however, that under the assumption of a ground acceleration that varies linearly between successive discrete values, $a(t)$ may be expressed as

$$a(t) = a_i + \frac{\Delta a_i}{\Delta t}(t - t_i) \tag{6.35}$$

where

$$\Delta t = t_{i+1} - t_i \tag{6.36}$$

$$\Delta a_i = a_{i+1} - a_i \tag{6.37}$$

in which a_i and a_{i+1} signify the ground accelerations at times t_i and t_{i+1}. As a result, during the time interval $t_i \leq t \leq t_{i+1}$, the system's equation of motion may be expressed as

$$\ddot{u} + 2\xi\omega\dot{u} + \omega^2 u = -a_i + \omega^2 s - \frac{\Delta a_i}{\Delta t}(t - t_i) \tag{6.38}$$

or

$$\ddot{u} + 2\xi\omega\dot{u} + \omega^2 u = -\bar{a}_i - \frac{\Delta \bar{a}_i}{\Delta t}(t - t_i) \tag{6.39}$$

where

$$\bar{a}_i = a_i - \omega^2 s \tag{6.40}$$

$$\bar{a}_{i+1} = a_{i+1} - \omega^2 s \tag{6.41}$$

and $\Delta \bar{a}_i = \bar{a}_{i+1} - \bar{a}_i = \Delta a_i$.

Equation 6.39 is of the same form as Equation 6.19, the equation of motion of a linear single-degree-of-freedom system under a piecewise linear ground acceleration. As such, its solution may also be expressed as

$$\begin{Bmatrix} u_{i+1} \\ \dot{u}_{i+1} \end{Bmatrix} = [A(\omega, \xi, \Delta t)] \begin{Bmatrix} u_i \\ \dot{u}_i \end{Bmatrix} + [B(\omega, \xi, \Delta t)] \begin{Bmatrix} \bar{a}_i \\ \bar{a}_{i+1} \end{Bmatrix} \tag{6.42}$$

where the elements of the matrices $[A]$ and $[B]$ are given by Equations 6.25 through 6.32. The permanent set s appearing in this equation is computed at the instant of unloading. Following an excursion of positive yielding, this set is given by

$$s = (1 - \alpha)(u_{unl} - u_y) \tag{6.43}$$

whereas after an excursion of negative yielding it is equal to

$$s = (1-\alpha)(u_{unl} + u_y) \tag{6.44}$$

In the foregoing equations, u_{unl} is the relative displacement computed at the instant of unloading (see Figure 6.16). If unloading is detected at this point, then the yield levels u_{yp} and u_{yn} are updated. Following a positive yield excursion and unloading, the current yield levels become $u_{yp} = u_{unl}$ and $u_{yn} = u_{unl} - 2u_y$. Similarly, after a negative yield excursion and unloading, these yield levels are $u_{yp} = u_{unl} + 2u_y$ and $u_{yn} = u_{unl}$.

Consider now the portion of the curve that follows a loading beyond the current yield levels. The equation of motion when the system's relative displacement is greater than the current positive yield level may be expressed as

$$\ddot{u} + 2\xi\omega\dot{u} + \omega^2(u_{yp} - s) + \alpha\omega^2(u - u_{yp}) = -a(t) \tag{6.45}$$

which is valid for $u > u_{yp}$ until unloading is detected, or when the product $\dot{u}_{i+1} \times \dot{u}_i$ is negative. Upon substitution of Equation 6.35, Equation 6.45 may also be written as

$$\ddot{u} + 2\bar{\xi}\bar{\omega}\dot{u} + \bar{\omega}^2 u = -\bar{a}_i - \frac{\Delta\bar{a}_i}{\Delta t}(t - t_i) \tag{6.46}$$

where

$$\bar{\omega} = \omega\sqrt{\alpha} \tag{6.47}$$

$$\bar{\xi} = \frac{\xi}{\sqrt{\alpha}} \tag{6.48}$$

$$\bar{a}_i = a_i - \omega^2 s + \omega^2 u_{yp}(1-\alpha) \tag{6.49}$$

$$\bar{a}_{i+1} = a_{i+1} - \omega^2 s + \omega^2 u_{yp}(1-\alpha) \tag{6.50}$$

and, as before, $\Delta\bar{a}_i = \bar{a}_{i+1} - \bar{a}_i = \Delta a_i$. Note that $\bar{\omega}$ and $\bar{\xi}$, the equivalent properties associated with the strain-hardening branch of the force-deformation curve, are defined only for $\alpha > 0$. After an excursion of negative yielding, that is, when $u < u_{yn}$, Equation 6.46 applies with the modification

$$\bar{a}_i = a_i + \omega^2 s - \omega^2 u_{yn}(1-\alpha) \tag{6.51}$$

$$\bar{a}_{i+1} = a_{i+1} + \omega^2 s - \omega^2 u_{yn}(1-\alpha) \tag{6.52}$$

The solution of Equation 6.46 is also given by Equation 6.42, provided $\bar{\omega}$ and $\bar{\xi}$ are substituted by ξ and ω in the elements of the matrices $[A]$ and $[B]$.

To maintain satisfactory accuracy in the response computations, the points at which the character of the solution changes from linear elastic to yielding, or from yielding to unloading, must be detected with a reasonable precision. This may be accomplished in different ways, but a convenient technique is to repeat the calculations with a series of reduced time steps after a yielding or unloading event is detected. These reduced time steps may be conveniently selected to be a fraction of the original time step, as for example, $\Delta t/10$, $\Delta t/100$, or $\Delta t/1000$. This way the matrices $[A]$ and $[B]$ have to be calculated only a few times more.

Characterization of Strong Ground Motions

With this technique, when yielding or unloading is detected within a time step Δt, the largest fractional time step is used first to locate the time subinterval during which yielding or unloading occurs. Once this subinterval is obtained, the second fractional time step is used to further refine the location of the event. This scheme is repeated until the smallest time step is used, or until the desired times of yielding or unloading are determined with some preestablished accuracy. This method of determining with precision the times at which yielding and unloading take place requires the calculation of additional sets of the elements of the matrices $[A]$ and $[B]$. However, since this calculation is performed only for a few extra time steps and involves only arithmetic operations, it will not increase the computational time significantly.

An algorithm to obtain a nonlinear response spectrum using the method just described is presented in Box 6.2. It should be noted that, as presented, this algorithm shows the way to determine nonlinear response spectra for given values of the yield deformation u_y. If spectra for a preassigned ductility factor are desired, then it is necessary to perform the calculations for several values of the yield deformation and interpolate the results to find the value of the yield deformation that leads to the desired ductility factor. It should also be noted that the algorithm does not incorporate the technique described earlier to determine with precision the times at which yielding and unloading events occur.

Box 6.2 Algorithm to Determine Nonlinear Response Spectrum

1.0 Initialize and read input data:

1.1 Read ground acceleration values a_i, $i = 1, 2, \ldots, N$, where N = number of points in accelerogram

1.2 Select values of α and time step Δt

1.3 Choose values of ω, ξ, and u_y

1.4 Set $u_0 = 0$, $\dot{u}_0 = 0$, $s = 0$, $u_{yp} = u_y$, $u_{yn} = -u_y$, and $\bar{a}_0 = 0$

1.5 Compute two sets of the elements in matrices $[A]$ and $[B]$ using Equations 6.25 through 6.32. For the first set consider the actual values of ω and ξ. For the second set, substitute ω and ξ by $\bar{\omega} = \omega\sqrt{\alpha}$ and $\bar{\xi} = \xi\sqrt{\alpha}$, respectively.

2.0 For the ith time step:

2.1 Compute the value of $\bar{a}_{i+1}$ as

$$\bar{a}_{i+1} = a_{i+1} - \omega^2 s \quad \text{if } u_{yn} \leq u_i \leq u_{yp}$$

$$\bar{a}_{i+1} = a_{i+1} - \omega^2 s + \omega^2 u_{yp}(1-\alpha) \quad \text{if } u_i < u_{yp}$$

$$\bar{a}_{i+1} = a_{i+1} - \omega^2 s + \omega^2 u_{yn}(1-\alpha) \quad \text{if } u_i > u_{yn}$$

2.2 Compute the displacement and velocity response at time t_{i+1} according to

$$\begin{Bmatrix} u_{i+1} \\ \dot{u}_{i+1} \end{Bmatrix} = [A(\omega, \xi, \Delta t)] \begin{Bmatrix} u_i \\ \dot{u}_i \end{Bmatrix} + [B(\omega, \xi, \Delta t)] \begin{Bmatrix} \bar{a}_i \\ \bar{a}_{i+1} \end{Bmatrix}$$

if $u_{yn} \leq u_i \leq u_{yp}$, or

$$\begin{Bmatrix} u_{i+1} \\ \dot{u}_{i+1} \end{Bmatrix} = [A(\bar{\omega}, \bar{\xi}, \Delta t)] \begin{Bmatrix} u_i \\ \dot{u}_i \end{Bmatrix} + [B(\bar{\omega}, \bar{\xi}, \Delta t)] \begin{Bmatrix} \bar{a}_i \\ \bar{a}_{i+1} \end{Bmatrix}$$

if $u_{yn} > u_i > u_{yp}$.

2.3 If the system's displacement u_i lies in one of the elastic branches of its load-deformation curve at the beginning of the time step, check according to the following criteria whether the system remains in the same branch at the end of the time step or yielding has taken place:
   if $u_{yn} < u_{i+1} < u_{yp}$, consider the system remains in the elastic range
   if $u_{i+1} > u_{yp}$, consider the system has yielded in tension
   if $u_{i+1} < u_{yn}$, consider the system has yielded in compression
2.4 If the system's displacement u_i lies in one of the postelastic branches of its load-deformation curve at the beginning of the time step, check according to the following criteria whether the system remains in the same branch at the end of the time step or was unloaded to one of the elastic branches:
   if $\dot{u}_{i+1} \times \dot{u}_i \geq 0$, consider the system remains in the same postelastic branch
   if $\dot{u}_{i+1} \times \dot{u}_i < 0$, consider the system was unloaded to one of the elastic branches
2.5 If the system remains in an elastic branch or yields in tension or compression, set $i = i + 1$ and go back to Step 2.1.
2.6 If the system was unloaded following an excursion of yielding in tension, set

$$u_{unl} = u_{i+1}$$

compute the permanent set s as

$$s = (1 - \alpha)(u_{unl} - u_y)$$

and update the yield levels to

$$u_{yp} = u_{unl} \quad \text{and} \quad u_{yn} = u_{unl} - 2u_y$$

Afterward, set $i = i + 1$ and go back to Step 2.1.
2.7 If the system was unloaded following an excursion of yielding in compression, set

$$u_{unl} = u_{i+1},$$

compute the permanent set s as

$$s = (1 - \alpha)(u_{unl} + u_y)$$

and update the yield levels to

$$u_{yp} = u_{unl} + 2u_y \quad \text{and} \quad u_{yn} = u_{unl}$$

Afterward, set $i = i + 1$ and go back to Step 2.1.
3.0 After performing the calculation for the targeted number of time steps, select the displacement with the largest value in absolute terms.
4.0 Repeat Steps 1.0 through 3.0 for different values of ω, ξ, and u_y.

6.6 FOURIER SPECTRUM

6.6.1 Theoretical Background

6.6.1.1 Fourier Series

If a function of time $f(t)$ is periodic with period T, that is, if the function repeats itself every T seconds, or it repeats itself with a frequency $\omega_0 = 2\pi/T$, then $f(t)$ can be expanded into an infinite series of cosine and sine functions of the form

$$f(t) = \frac{1}{2}a_0 + \sum_{m=1}^{\infty}(a_m \cos\omega_m t + b_m \sin\omega_m t) \qquad (6.53)$$

where

$$\omega_m = m\omega_0 \qquad (6.54)$$

$$a_0 = \frac{2}{T}\int_{-T/2}^{T/2} f(t)\,dt \qquad (6.55)$$

$$a_m = \frac{2}{T}\int_{-T/2}^{T/2} f(t)\cos\omega_m t\,dt \qquad (6.56)$$

$$b_m = \frac{2}{T}\int_{-T/2}^{T/2} f(t)\sin\omega_m t\,dt \qquad (6.57)$$

Such an infinite series is known as *Fourier series* and is said to be the Fourier series representation or expansion of the function $f(t)$. If, for example, a function of time is defined as

$$f(t) = \begin{cases} 1 + \dfrac{4t}{T} & -\dfrac{T}{2} < t \le 0 \\ 1 - \dfrac{4t}{T} & 0 \le t \le \dfrac{T}{2} \end{cases} \qquad (6.58)$$

then

$$a_0 = \frac{2}{T}\int_{-T/2}^{0}\left(1 + \frac{4t}{T}\right)dt + \frac{2}{T}\int_{0}^{T/2}\left(1 - \frac{4}{T}\right)dt = 0 \qquad (6.59)$$

$$a_m = \frac{2}{T}\int_{-T/2}^{0}\left(1 + \frac{4t}{T}\right)\cos\omega_m t\,dt + \frac{2}{T}\int_{0}^{T/2}\left(1 - \frac{4t}{T}\right)\cos\omega_m t\,dt = \frac{4}{m^2\pi^2}(1 - \cos m\pi)$$

$$= \begin{cases} 0 & \text{if } m \text{ even} \\ \dfrac{8}{m^2\pi^2} & \text{if } m \text{ odd} \end{cases} \qquad (6.60)$$

$$b_m = \frac{2}{T}\int_{-T/2}^{0}\left(1 + \frac{4t}{T}\right)\sin\omega_m t\,dt + \frac{2}{T}\int_{0}^{T/2}\left(1 - \frac{4t}{T}\right)\sin\omega_m t\,dt = 0 \qquad (6.61)$$

Hence, its Fourier series expansion is given by

$$f(t) = \frac{8}{\pi^2}\left(\cos\omega_0 t + \frac{1}{3^2}\cos 3\omega_0 t + \frac{1}{5^2}\cos 5\omega_0 t + \cdots\right) \quad (6.62)$$

Note thus that based on the concept of Fourier series, $f(t)$ may be considered composed of an infinite number of harmonic functions, each with a different amplitude and frequency, and whose superposition gives the function itself. These harmonic functions are referred to as the harmonic components of $f(t)$, and those with the largest amplitudes are referred to as the main or dominant components of the function. The periods of these dominant components are in turn referred to as the dominant periods of the function.

6.6.1.2 Exponential Fourier Series

As from Euler's relationship one has that

$$\cos\omega_m t = \frac{1}{2}(e^{i\omega_m t} + e^{-i\omega_m t}) \quad \text{and} \quad \sin\omega_m t = \frac{1}{2i}(e^{i\omega_m t} - e^{-i\omega_m t}) \quad (6.63)$$

where i is the unit imaginary number and e the exponential function, then substitution of these two expressions into Equation 6.53 leads to

$$f(t) = \frac{1}{2}a_0 + \sum_{m=1}^{\infty}\left[a_m \frac{1}{2}(e^{i\omega_m t} + e^{-i\omega_m t}) + b_m \frac{1}{2i}(e^{i\omega_m t} - e^{-i\omega_m t})\right] \quad (6.64)$$

which, after rearranging terms and recalling that $1/i = -i$, may be rewritten as

$$f(t) = \frac{1}{2}a_0 + \sum_{m=1}^{\infty}\left[\frac{1}{2}(a_m - ib_m)e^{i\omega_m t} + \frac{1}{2}(a_m + ib_m)e^{-i\omega_m t}\right] \quad (6.65)$$

Letting

$$c_0 = \frac{1}{2}a_0 \quad c_m = \frac{1}{2}(a_m - ib_m) \quad c_{-m} = \frac{1}{2}(a_m + ib_m) \quad (6.66)$$

Equation 6.65 then becomes

$$f(t) = c_0 + \sum_{m=1}^{\infty}[c_m e^{i\omega_m t} + c_{-m} e^{-i\omega_m t}] \quad (6.67)$$

which may also be expressed as

$$f(t) = c_0 + \sum_{m=1}^{\infty} c_m e^{i\omega_m t} + \sum_{m=-1}^{-\infty} c_m e^{i\omega_m t} = \sum_{m=-\infty}^{\infty} c_m e^{i\omega_m t} \quad (6.68)$$

Expressing now the coefficients c_m in terms of Equations 6.55 through 6.57, one gets

$$c_0 = \frac{1}{2}a_0 = \frac{1}{T}\int_{-T/2}^{T/2} f(t)dt \quad (6.69)$$

Characterization of Strong Ground Motions

$$c_m = \frac{1}{2}(a_m - ib_m) = \frac{1}{T}\int_{-T/2}^{T/2} f(t)\cos\omega_m t\, dt - i\frac{1}{T}\int_{-T/2}^{T/2} f(t)\sin\omega_m t\, dt$$

$$= \frac{1}{T}\left[\int_{-T/2}^{T/2} f(t)(\cos\omega_m t - i\sin\omega_m t)dt\right] \quad (6.70)$$

$$= \frac{1}{T}\int_{-T/2}^{T/2} f(t)e^{-i\omega_m t}dt$$

and similarly

$$c_{-m} = \frac{1}{T}\int_{-T/2}^{T/2} f(t)e^{i\omega_m t}dt \quad (6.71)$$

Note, however, that if $f(t)$ is real, then

$$c_{-m} = c_m^* \quad (6.72)$$

where * indicates complex conjugate. Consequently, Equations 6.69 through 6.71 may be combined to obtain the following single formula:

$$c_m = \frac{1}{T}\int_{-T/2}^{T/2} f(t)e^{-i\omega_m t}dt \quad m = 0, \pm 1, \pm 2, \ldots \quad (6.73)$$

Thus, the expansion of a periodic function $f(t)$ into a Fourier series may be expressed alternatively as

$$f(t) = \sum_{m=-\infty}^{\infty} c_m e^{i\omega_m t} \quad (6.74)$$

where

$$c_m = \frac{1}{T}\int_{-T/2}^{T/2} f(t)e^{-i\omega_m t}dt \quad (6.75)$$

Note that $|c_m|$ represents the amplitude of the component with frequency ω_m in the series in Equation 6.74 and that this amplitude is a function of ω_m alone. A plot of $|c_m|$ versus ω_m, such as that shown in Figure 6.17, may therefore be useful to identify the dominant components in the Fourier series that defines a periodic function.

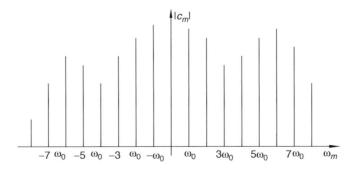

FIGURE 6.17 Variation of $|c_m|$ with ω_m.

6.6.1.3 Fourier Transform

If a function $f(t)$ is not periodic, but it is defined for $-\infty < t < \infty$, and if this function is such that

$$\int_{-\infty}^{\infty} |f(t)|\, dt < \infty \tag{6.76}$$

then it is possible to express the function in terms of a Fourier expansion by interpreting it as a periodic function with an infinitely long period. In this case, however, the expansion is in terms of an integral as opposed to a series as shown later.

With reference to the exponential form of Fourier series given by Equations 6.74 and 6.75, it may be noted that if one substitutes Equation 6.75 into Equation 6.74 and considers that $1/T = \omega_0/2\pi$, one obtains

$$f(t) = \sum_{m=-\infty}^{\infty} \left[\frac{1}{2\pi} \int_{-T/2}^{T/2} f(x) e^{-i\omega_m x}\, dx \right] \omega_0 e^{i\omega_m t} \tag{6.77}$$

where the dummy variable x has been introduced to avoid confusion with the variable t. It may be noted then that if $T \to \infty$, ω_0 becomes infinitesimally small (i.e., $\omega_0 \to d\omega$), the discrete variable ω_m turns into the continuous variable ω, and the summation becomes an integral over the continuous variable ω. As a result, when $T \to \infty$, Equation 6.77 becomes

$$f(t) = \frac{1}{2\pi} \int_{-\infty}^{\infty} \left[\int_{-\infty}^{\infty} f(x) e^{-i\omega x}\, dx \right] e^{i\omega t}\, d\omega \tag{6.78}$$

Accordingly, $f(t)$ can also be written as

$$f(t) = \frac{1}{2\pi} \int_{-\infty}^{\infty} F(\omega) e^{i\omega t}\, d\omega \tag{6.79}$$

where

$$F(\omega) = \int_{-\infty}^{\infty} f(t) e^{-i\omega t}\, dt \tag{6.80}$$

The integral in Equation 6.80 is called the *Fourier transform* of the function $f(t)$, and the integral in Equation 6.79 its *inverse Fourier transform*. $F(\omega)$ is a function of frequency only and thus it is said that $F(\omega)$ represents the function $f(t)$ in the frequency domain. The two integrals form a transform pair, which allows the transformation of $f(t)$ back and forth between the time and the frequency domains.

6.6.2 Definition

A Fourier spectrum constitutes the representation of a time history into the frequency domain. Hence, the Fourier spectrum of a ground motion time history is simply defined as the Fourier transform of the ground motion time history. That is, if $\ddot{u}_g(t)$ denotes ground acceleration in the time domain, the Fourier spectrum of $\ddot{u}_g(t)$ is defined as

$$F(\omega) = \int_{-\infty}^{\infty} \ddot{u}_g(t) e^{-i\omega t}\, dt \tag{6.81}$$

Note, however, that since $e^{-i\omega t} = \cos\omega t - i\sin\omega t$ and $\ddot{u}_g(t) = 0$ when $t \geq t_d$ or $t \leq 0$, where t_d denotes the duration of the ground motion, the Fourier spectrum of $\ddot{u}_g(t)$ may also be defined as

$$F(\omega) = \int_0^{t_d} \ddot{u}_g(t)\cos\omega t\, dt - i\int_0^{t_d} \ddot{u}_g(t)\sin\omega t\, dt \tag{6.82}$$

Consequently, the amplitude and the phase of the Fourier spectrum of $\ddot{u}_g(t)$ may be expressed, respectively, as

$$AFS = |F(\omega)| = \sqrt{\left[\int_0^{t_d} \ddot{u}_g(t)\cos\omega t\, dt\right]^2 + \left[\int_0^{t_d} \ddot{u}_g(t)\sin\omega t\, dt\right]^2} \tag{6.83}$$

$$PFS = \text{Arg}[F(\omega)] = \arctan\left(-\frac{\int_0^{t_d} \ddot{u}_g(t)\sin\omega t\, dt}{\int_0^{t_d} \ddot{u}_g(t)\cos\omega t\, dt}\right) \tag{6.84}$$

The Fourier spectrum of a ground motion thus portrays the ground motion in the frequency domain and serves to analyze the composition of the ground motion in terms of harmonic components. In particular, the amplitude Fourier spectrum is used to identify the harmonic components of the ground motion that possess the largest amplitudes. As these harmonic components are in general identified in terms of their frequencies, this type of analysis is referred to as a *frequency analysis*. By the same token, the description of the frequency composition of a ground motion is known as the analysis of its *frequency content*. Amplitude Fourier spectra is the quantity most frequently studied by seismologists in their investigations of earthquake mechanisms, as the frequency content of a ground motion is a function of source mechanism, focal depth, epicentral distance, travel path, site-soil conditions, and earthquake magnitude.

Examples of amplitude and phase Fourier spectra are shown in Figures 6.18 and 6.19. Note that in engineering applications, it is customary to plot Fourier spectra as a function of period instead of frequency. Note too that the units of the ordinates in an acceleration Fourier spectrum are those of velocity.

EXAMPLE 6.3   AMPLITUDE FOURIER SPECTRUM OF RECTANGULAR PULSE

Find the amplitude Fourier spectrum of the rectangular pulse defined by

$$\ddot{u}_g(t) = \begin{cases} 1 & |t| < \dfrac{t_d}{2} \\ 0 & |t| > \dfrac{t_d}{2} \end{cases}$$

Solution

From the application of Equation 6.81, one has that

$$F(\omega) = \int_{-t_d/2}^{t_d/2} e^{-i\omega t}\, dt = \frac{1}{-i\omega}e^{-i\omega t}\bigg|_{-t_d/2}^{t_d/2} = -\frac{1}{i\omega}(e^{i\omega t_d/2} - e^{-i\omega t_d/2}) = -\frac{2}{\omega}\sin\frac{\omega t_d}{2}$$

Hence

$$|F(\omega)| = t_d \frac{\sin(\omega t_d/2)}{\omega t_d/2}$$

The graphical representation of this expression is shown in Figure E6.3.

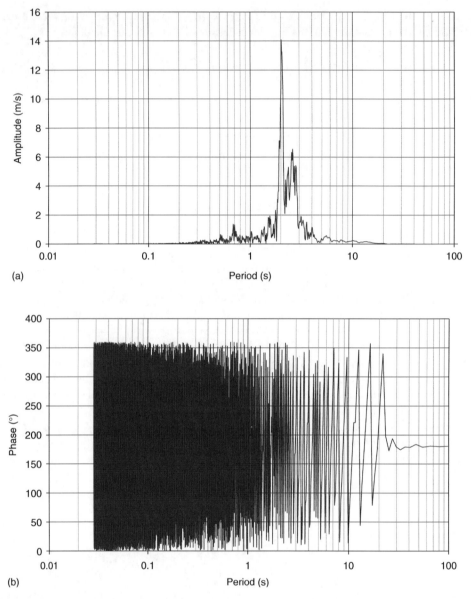

FIGURE 6.18 (a) Amplitude and (b) phase Fourier spectra of S60E component of ground acceleration recorded at SCT station during the September 19, 1985, Mexico City earthquake.

6.6.3 Total Energy and Fourier Spectrum

The amplitude Fourier spectrum of an acceleration ground motion may be interpreted as a measure of the total energy contained at the end of the ground motion in an undamped single-degree-of-freedom system subjected to that ground motion. To demonstrate this assertion, consider that at time t the total energy (kinetic plus deformation energy) in a single-degree-of-freedom system with mass m and stiffness constant k is given by

$$E_T(t) = \frac{1}{2} m\dot{u}^2(t) + \frac{1}{2} k u^2(t) \tag{6.85}$$

Characterization of Strong Ground Motions 213

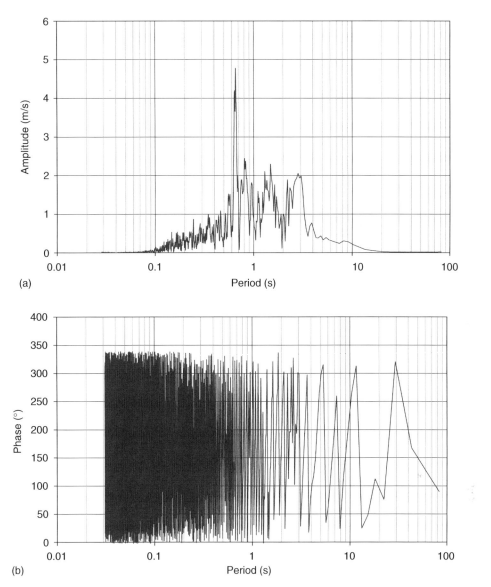

FIGURE 6.19 (a) Amplitude and (b) phase Fourier spectra of E–W component of ground acceleration recorded at Foster City station during the October 17, 1989, Loma Prieta, California, earthquake.

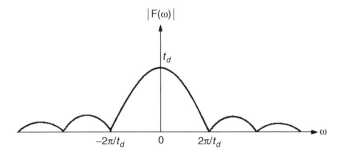

FIGURE E6.3 Amplitude Fourier spectrum of rectangular pulse in Example 6.3.

where, as before, $u(t)$ and $\dot{u}(t)$, respectively, denote the system's displacement and velocity responses at time t. Consequently, the square root of twice this total energy per unit mass may be written as

$$\sqrt{\frac{2E_T(t)}{m}} = \sqrt{\dot{u}^2(t) + \omega_n^2 u^2(t)} \qquad (6.86)$$

where ω_n denotes the system's natural frequency. Note, however, that from the equations that describe the response of a single-degree-of-freedom system, one has that for zero damping (see Equations 6.4 and 6.5)

$$\omega_n u(t) = \int_0^t \ddot{u}_g(t) \sin\omega_n(t-\tau) d\tau = \sin\omega_n t \int_0^t \ddot{u}_g(t)\cos\omega_n \tau\, d\tau - \cos\omega_n t \int_0^t \ddot{u}_g(t)\sin\omega_n \tau\, d\tau \qquad (6.87)$$

$$\dot{u}(t) = \int_0^t \ddot{u}_g(t) \cos\omega_n(t-\tau) d\tau = \cos\omega_n t \int_0^t \ddot{u}_g(t)\cos\omega_n \tau\, d\tau + \sin\omega_n t \int_0^t \ddot{u}_g(t)\sin\omega_n \tau\, d\tau \qquad (6.88)$$

and thus

$$\dot{u}^2(t) + \omega_n^2 u^2(t) = \left[\cos\omega_n t \int_0^t \ddot{u}_g(t)\cos\omega_n \tau\, d\tau + \sin\omega_n t \int_0^t \ddot{u}_g(t)\sin\omega_n \tau\, d\tau\right]^2$$

$$+ \left[\sin\omega_n t \int_0^t \ddot{u}_g(t)\cos\omega_n \tau\, d\tau - \cos\omega_n t \int_0^t \ddot{u}_g(t)\sin\omega_n \tau\, d\tau\right]^2 = \left[\int_0^t \ddot{u}_g(t)\cos\omega_n \tau\, d\tau\right]^2$$

$$+ \left[\int_0^t \ddot{u}_g(t)\sin\omega_n \tau\, d\tau\right]^2 \qquad (6.89)$$

Therefore, after setting $\omega_n = \omega$, Equation 6.86 may also be written as

$$\sqrt{\frac{2E_T(t)}{m}} = \sqrt{\left[\int_0^t \ddot{u}_g(t)\cos\omega\tau\, d\tau\right]^2 + \left[\int_0^t \ddot{u}_g(t)\sin\omega\tau\, d\tau\right]^2} \qquad (6.90)$$

which for the particular case when t is equal to the ground motion duration (i.e., when $t = t_d$) becomes

$$\sqrt{\frac{2E_T(t_d)}{m}} = \sqrt{\left[\int_0^{t_d} \ddot{u}_g(t)\cos\omega\tau\, d\tau\right]^2 + \left[\int_0^{t_d} \ddot{u}_g(t)\sin\omega\tau\, d\tau\right]^2} = |F(\omega)| \qquad (6.91)$$

It may be seen, thus, that, indeed, the ordinates in an amplitude Fourier spectrum describe the relative energy carried by each of the harmonic components that make up the motion.

6.6.4 Relationship between Fourier and Undamped Velocity Response Spectra

There exists a close relationship between amplitude Fourier spectrum and velocity response spectrum for zero damping. This relationship can be disclosed as follows:

As indicated earlier, for zero damping the velocity response of a single-degree-of-freedom system is given by

$$\dot{u}(t) = \int_0^t \ddot{u}_g(t) \cos\omega_n(t-\tau) d\tau = \cos\omega_n t \int_0^t \ddot{u}_g(t)\cos\omega_n \tau\, d\tau + \sin\omega_n t \int_0^t \ddot{u}_g(t)\sin\omega_n \tau\, d\tau \qquad (6.92)$$

Characterization of Strong Ground Motions

However, since $A \sin \omega t + B \cos \omega t = \sqrt{A^2 + B^2} \cos(\omega t - \theta)$, where $\theta = \tan^{-1} B/A$, this velocity response can also be expressed as

$$\dot{u}(t) = \sqrt{\left[\int_0^t \ddot{u}_g(t) \cos \omega_n \tau \, d\tau\right]^2 + \left[\int_0^t \ddot{u}_g(t) \sin \omega_n \tau \, d\tau\right]^2} \cos(\omega t - \theta) \quad (6.93)$$

and, thus, the undamped spectral velocity, which by definition corresponds to the maximum value of the velocity response, is given by

$$SV(\omega_n, 0) = |\dot{u}(t)|_{max} = \sqrt{\left[\int_0^{t_{max}} \ddot{u}_g(t) \cos \omega_n \tau \, d\tau\right]^2 + \left[\int_0^{t_{max}} \ddot{u}_g(t) \sin \omega_n \tau \, d\tau\right]^2} \quad (6.94)$$

where t_{max} denotes the time at which the maximum value of $\dot{u}(t)$ occurs.

A comparison between Equation 6.94 and the equation that defines amplitude Fourier spectrum (Equation 6.83) shows that, if $\omega_n = \omega$, the only difference between Fourier and undamped velocity spectrum is the upper limits in the two integrals. In the expression for the Fourier spectrum, the upper limit is equal to t_d, the ground motion duration, whereas in the expression for the response spectrum it is equal to t_{max}, the time at which the maximum response occurs. As, in general, t_d is not equal to t_{max}, the velocity response $\dot{u}(t)$ at time t_d will, in general, be less than the maximum value. Consequently, the ordinates in a Fourier amplitude spectrum will always be smaller than those in the corresponding undamped velocity spectrum. The exception is only for those rare occasions in which t_{max} coincides with t_d in which case the two spectra would be equal to each other. Notwithstanding this limitation, the existing link between the two spectra is useful as it allows an approximate assessment of the frequency content in a ground motion from its undamped velocity response spectrum.

The spectra shown in Figure 6.20 show the similarity between Fourier and velocity response spectra for zero damping and the fact that the latter provides an upper bound to the former.

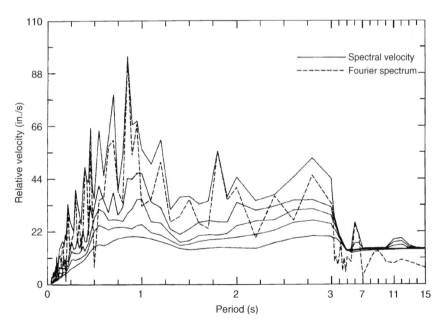

FIGURE 6.20 Comparison of Fourier and undamped velocity response spectra (top curve) for the S00E accelerogram recorded during the 1940 El Centro earthquake. (Reproduced with permission of Earthquake Engineering Research Institute from Hudson, D.E., *Reading and Interpreting Strong-Motion Accelerograms*, Earthquake Engineering Research Institute, Oakland, CA, 1979.)

6.6.5 Computation of Fourier Spectrum

In principle, the original definition can be used directly to obtain the Fourier spectrum of a ground motion $\ddot{u}_g(t)$. In practice, however, the integration cannot be carried out analytically as in most cases $\ddot{u}_g(t)$ cannot be expressed in analytical form. In these cases, therefore, a discrete form of the Fourier transform is necessary. One form of such a discrete Fourier transform may be obtained as follows.

According to the original definition, the Fourier spectrum of an acceleration function $\ddot{u}_g(t)$ whose duration is t_d may be expressed as

$$F(\omega) = \int_0^{t_d} \ddot{u}_g(t) e^{-i\omega t}\, dt \tag{6.95}$$

where $\ddot{u}_g(t)$ represents ground acceleration at time t. However, if $\ddot{u}_g(t)$ is defined only at N discrete points and these discrete points correspond to $t_j = j\,\Delta t$, $j = 0, 1, 2, \ldots, (N-1)$, where $\Delta t = t_d/N$ is the time interval between points, then the integral in Equation 6.95 may be replaced by the sum of N areas, each equal to $\ddot{u}_g(t_j)e^{-i\omega t_j}\Delta t$. As such, in the discrete case Equation 6.95 becomes

$$F(\omega) = \sum_{j=0}^{N-1} \ddot{u}_g(t_j) e^{-i\omega t_j}\Delta t = t_d \frac{1}{N}\sum_{j=0}^{N-1} \ddot{u}_g(t_j) e^{-i\omega j(t_d/N)} \tag{6.96}$$

If, in addition, $F(\omega)$ is obtained only at the discrete values $\omega_m = m\,\Delta\omega$, $m = 0, 1, 2, \ldots, (N-1)$, where $\Delta\omega$ is any arbitrary frequency increment, then the discrete values of $F(\omega)$ are given by

$$F(\omega_m) = t_d \frac{1}{N}\sum_{j=0}^{N-1} \ddot{u}_g(t_j) e^{-im\Delta\omega j(t_d/N)} \quad m = 0, 1, 2, \ldots, (N-1) \tag{6.97}$$

where it should be noted that no information is lost by restricting the allowable range of the integer m, because, as it will be shown shortly, the values of $F(\omega_m)$ for frequencies beyond a certain limiting value are meaningless. It may be shown, however, that to exactly regain the values of $\ddot{u}_g(t_j)$ through the application of the corresponding inverse Fourier transform to $F(\omega_m)$, it is necessary to make $\Delta\omega$ equal to

$$\Delta\omega = \frac{2\pi}{t_d} \tag{6.98}$$

To satisfy such a requirement, Equation 6.97 needs, thus, to be written as

$$F(\omega_m) = t_d \frac{1}{N}\sum_{j=0}^{N-1} \ddot{u}_g(t_j) e^{-i2\pi(jm/N)} \quad m = 0, 1, 2, \ldots, (N-1) \tag{6.99}$$

which may also be expressed as

$$F(\omega_m) = t_d C_m \quad m = 0, 1, 2, \ldots, (N-1) \tag{6.100}$$

where C_m is defined as

$$C_m = \frac{1}{N}\sum_{j=0}^{N-1} \ddot{u}_g(t_j) e^{-i2\pi(jm/N)} \quad m = 0, 1, 2, \ldots, (N-1) \tag{6.101}$$

and represents the so-called *discrete Fourier transform*, that is, the discretized version of the formula to compute the coefficients c_m in the exponential form of Fourier series (see Equation 6.75).

Equation 6.99 may be considered alternatively the discretized form of the Fourier transform and the expression by means of which the Fourier spectrum of a discrete acceleration function $\ddot{u}_g(t_j)$ may be determined. It should be noted, however, that, in practice, Fourier spectra are obtained through the use of a highly efficient algorithm for the calculation of the discrete Fourier transform. This algorithm, known as the *fast Fourier transform*, will be presented in the following section. It is important to realize too that in the generation of a Fourier spectrum using Equation 6.99 or 6.101, the only frequencies for which an ordinate can be computed are those corresponding to $0 \leq m \leq N/2$. The reason for this is that the coefficients C_m obtained for $m > N/2$ are just repetitions of those determined for $m \leq N/2$. The explanation for this occurrence is that C_m is equal to $C_{-(N-m)}$ and thus, as $C_{-m} = C_m^*$ and hence $|C_{-m}| = |C_m|$, it follows that $|C_m| = |C_{(N-m)}|$. The fact that C_m is equal to $C_{-(N-m)}$ may be shown by simply substituting $-(N-m)$ for m in Equation 6.101 as for $C_m = C_{-(N-m)}$ one obtains

$$C_{-(N-m)} = \frac{1}{N}\sum_{j=0}^{N-1}\ddot{u}_g(t_j)e^{i2\pi j((N-m)/N)} = \frac{1}{N}\sum_{j=0}^{N-1}\ddot{u}_g(t_j)e^{i2\pi j}e^{-i2\pi(jm/N)} = C_m \quad (6.102)$$

since $e^{2\pi i j} = \cos 2\pi j - i \sin 2\pi j = 1$ for all integer values of j.

The fact that the coefficients C_m for $m > N/2$ are equal in absolute value to the coefficients C_m for $m \leq N/2$ restricts the number of the harmonic components that may be represented in a discretized ground acceleration $\ddot{u}_g(t_j)$. There is thus a maximum frequency that is represented in this discretized function. This maximum frequency is known as the *Nyquist frequency* or *folding frequency*. In radians per second, this frequency is equal to (see Equation 6.98)

$$\omega_f = \frac{2\pi}{t_d}\left(\frac{N}{2}\right) = \frac{2\pi N/2}{N\Delta t} = \frac{\pi}{\Delta t} \quad (6.103)$$

whereas in hertz is equal to

$$f_f = \frac{\omega_f}{2\pi} = \frac{1}{2\Delta t} \quad (6.104)$$

where, as before, Δt is the time interval between discretized points.

Finally, it is important to realize that the existence of harmonic components beyond the Nyquist frequency in the original function $\ddot{u}_g(t)$ will introduce distortions in the lower harmonic components of $\ddot{u}_g(t_j)$ and hence in the generated Fourier spectrum. This phenomenon is called *aliasing*. In the light of this fact, it is recommended that the number of sample points used to define the original function be at least twice the number that is necessary to capture the highest harmonic component present in the original function and avoid thus spurious results due to aliasing.

The amplitude Fourier spectrum of a ground motion may be determined alternatively in terms of the corresponding displacement and velocity responses of an undamped single-degree-of-freedom system according to

$$|F(\omega)| = \sqrt{\dot{u}^2(t_d) + \omega^2 u^2(t_d)} \quad (6.105)$$

where, as before, t_d denotes the duration of the ground motion and $u(t_d)$ and $\dot{u}(t_d)$, respectively, represent the displacement and velocity responses of the system at the end of the ground motion. This formula is obtained from the consideration of the relationship between Fourier spectrum and total energy discussed in Section 6.6.3. Specifically, it is derived by combining Equations 6.86 and 6.91. Observe, thus, that as $u(t_d)$ and $\dot{u}(t_d)$ are calculated during the computation of a response spectrum, a response spectrum algorithm may also be used to generate amplitude Fourier spectra.

6.6.6 Fast Fourier Transform

The fast Fourier transform is a numerical algorithm that accelerates the calculation of the discrete Fourier transform. Introduced by J. W. Cooley and J. W. Tukey in 1965 (allegedly discovered independently by others, starting with Gauss back in 1805), this algorithm has made frequency-domain analyses a practical reality.

The concepts on which the fast Fourier transform are based may be described as follows. As seen in the foregoing section, the discrete Fourier transform of a discretized function of time or finite sequence $\{x_j\}$, $j = 1, 2, ..., (N - 1)$, where N is the number of data points, is another finite sequence $\{X_m\}$, where

$$X_m = \frac{1}{N} \sum_{j=0}^{N-1} x_j e^{-i(2\pi mj/N)} \quad m = 0, 1, 2, \ldots, (N-1) \tag{6.106}$$

Suppose that $\{x_j\}$ is the sequence shown in Figure 6.21a, where N is an even number. Upon partitioning of this sequence into two shorter sequences $\{y_j\}$ and $\{z_j\}$, one gets

$$y_j = x_{2j} \quad j = 0, 1, 2, \ldots, \left(\frac{N}{2} - 1\right) \tag{6.107}$$

$$z_j = x_{2j+1} \quad j = 0, 1, 2, \ldots, \left(\frac{N}{2} - 1\right) \tag{6.108}$$

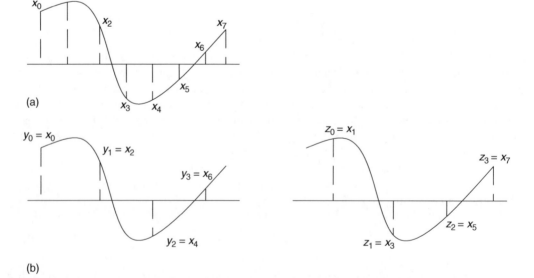

FIGURE 6.21 Schematic representation of (a) sequence $\{x_r\}$ and (b) sequence $\{x_r\}$ partitioned into half sequences $\{y_r\}$ and $\{z_r\}$.

Characterization of Strong Ground Motions

Notice, thus, that the discrete Fourier transform of these two shorter sequences are the sequences $\{Y_m\}$ and $\{Z_m\}$, where

$$Y_m = \frac{1}{N/2} \sum_{j=0}^{(N/2)-1} y_j e^{-i(2\pi mj/(N/2))} \quad m = 0, 1, 2, \ldots, \left(\frac{N}{2} - 1\right) \tag{6.109}$$

$$Z_m = \frac{1}{N/2} \sum_{j=0}^{(N/2)-1} z_j e^{-i(2\pi mj/(N/2))} \quad m = 0, 1, 2, \ldots, \left(\frac{N}{2} - 1\right) \tag{6.110}$$

However, if the elements of the original discrete Fourier transform of the original sequence $\{x_j\}$ are separated into odd and even terms, this discrete Fourier transform may also be written as

$$X_m = \frac{1}{N} \left\{ \sum_{j=0}^{N/2-1} x_{2j} e^{-i(2\pi m(2j)/N)} + \sum_{j=0}^{N/2-1} x_{2j+1} e^{-i(2\pi m(2j+1)/N)} \right\} \tag{6.111}$$

which in view of Equations 6.107 and 6.108 may also be put into the form

$$X_m = \frac{1}{N} \left\{ \sum_{j=0}^{(N/2)-1} y_j e^{i(2\pi mj/(N/2))} + e^{-i(2\pi m/N)} \sum_{j=0}^{(N/2)-1} z_j e^{-i(2\pi mj/(N/2))} \right\} \tag{6.112}$$

Thus, by comparison with Equations 6.109 and 6.110, X_m may also be expressed as

$$X_m = \frac{1}{2}\left(Y_m + e^{-i(2\pi m/N)} Z_m\right) \quad m = 1, 2, \ldots, \left(\frac{N}{2} - 1\right) \tag{6.113}$$

Equation 6.113 applies for the values of m only between 0 and $(N/2) - 1$, that is, for half the elements in the $\{X_m\}$ sequence. The additional half may be obtained by taking advantage of the fact that, as explained in Section 6.6.5, Y_m and Z_m are periodic in m and repeat themselves with a period $N/2$. That is

$$Y_{m-(N/2)} = Y_m \tag{6.114}$$

$$Z_{m-(N/2)} = Z_m \tag{6.115}$$

Consequently, the elements of the entire sequence may be determined according to

$$X_m = \frac{1}{2}\left(Y_m + e^{-i(2\pi m/N)} Z_m\right) \quad m = 1, 2, \ldots, \left(\frac{N}{2} - 1\right) \tag{6.116}$$

and

$$X_m = \frac{1}{2}\left(Y_{m-(N/2)} + e^{-i(2\pi m/N)} Z_{m-(N/2)}\right) \quad m = \frac{N}{2}\left(\frac{N}{2} + 1\right), \ldots, (N-1) \tag{6.117}$$

or, if m runs only from 0 to $(N/2) - 1$,

$$X_m = \frac{1}{2}(Y_m + e^{-i(2\pi m/N)}Z_m) \quad m = 1, 2, \ldots, \left(\frac{N}{2} - 1\right) \tag{6.118}$$

and

$$X_{(N/2)+m} = \frac{1}{2}(Y_m + e^{-i(2\pi(m+N/2)/N)}Z_m) \quad m = 0, 1, \ldots, \left(\frac{N}{2} - 1\right) \tag{6.119}$$

which, by noticing that $e^{-i\pi} = -1$, may be simplified to

$$X_{(N/2)+m} = \frac{1}{2}(Y_m - e^{-i(2\pi m/N)}Z_m) \quad m = 0, 1, \ldots, \left(\frac{N}{2} - 1\right) \tag{6.120}$$

Finally, introducing a new variable defined as

$$W = e^{-i(2\pi/N)} \tag{6.121}$$

one finds that the elements in the sequence $\{X_m\}$ may be expressed as

$$X_m = \frac{1}{2}(Y_m + W^m Z_m) \quad m = 1, 2, \ldots, \left(\frac{N}{2} - 1\right) \tag{6.122}$$

$$X_{(N/2)+m} = \frac{1}{2}(Y_m - W^m Z_m) \quad m = 1, 2, \ldots, \left(\frac{N}{2} - 1\right) \tag{6.123}$$

Notice, thus, that the discrete Fourier transform of the original sequence $\{x_j\}$ may be obtained from the discrete Fourier transforms of the two half-sequences $\{y_j\}$ and $\{z_j\}$. Equations 6.122 and 6.123 constitute the essence of the fast Fourier transform. For, if the original number of elements N in the sequence $\{x_j\}$ is a power of 2, then the sequences $\{y_j\}$ and $\{z_j\}$ may themselves be partitioned into half sequences and continue the partitioning until the last subsequences contain only one element each. Furthermore, observe that if a sequence $\{s_j\}$ has only one term, then for this sequence $N = 1$, $j = 0$, and $m = 0$, and according to the definition of a discrete Fourier transform (see Equation 6.106)

$$X_0 = \frac{1}{1}\sum_{j=0}^{0} x_0 e^{-i(2\pi(0)(0))/1} = x_0 \tag{6.124}$$

That is, the discrete Fourier transform of a single-element sequence is the element itself. Therefore, the advantage of reducing a given sequence to a series of single-element sequences is that single-element sequences represent their own discrete Fourier transforms. Once these single-element sequences are found, the discrete Fourier transform of the original sequence may be obtained by simply applying Equations 6.122 and 6.123 repeatedly.

The process to find the discrete Fourier transform of a given sequence $\{x_j\}$ may be thus summarized as follows. First, the original sequence is partitioned recursively into half sequences until a series of one-element sequences are obtained. Second, the discrete Fourier transforms of these one-term sequences are made equal to the obtained one-term sequences. Third, Equations 6.122 and 6.123 are applied successively to this series of one-term sequences until the discrete Fourier transform of the original sequence $\{x_j\}$ is found.

Characterization of Strong Ground Motions

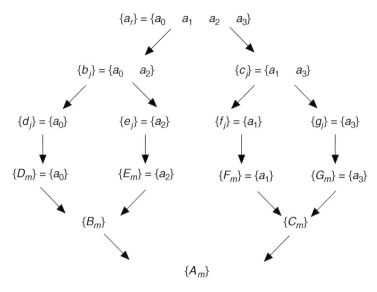

FIGURE 6.22 Steps in application of fast Fourier transform to a four-term sequence.

As an illustration of the application of the fast Fourier technique, consider a sequence $\{a_j\}$ with only four terms and partition this sequence to one-term subsequences as shown in Figure 6.22. Using the notation in this figure, the elements of the discrete Fourier transforms of the single-term sequences are

$$D_0 = a_0 \qquad E_0 = a_2 \qquad F_0 = a_1 \qquad G_0 = a_3 \qquad (6.125)$$

Thus, the combination of these discrete Fourier transforms according to Equations 6.122 and 6.123 yields

$$B_0 = \frac{1}{2}(a_0 + a_2) \qquad B_1 = \frac{1}{2}(a_0 - a_2) \qquad (6.126)$$

$$C_0 = \frac{1}{2}(a_1 + a_3) \qquad C_1 = \frac{1}{2}(a_1 - a_3) \qquad (6.127)$$

where it has been considered that for the case of $N/2 = 1$, $w = e^{-i\pi} = -i$. Similarly, a second application of Equations 6.122 and 6.123 with $N/2 = 2$ and $w = e^{-i\pi/2} = -i$ leads to

$$A_0 = \frac{1}{4}(a_0 + a_2 + a_1 + a_3) \qquad (6.128)$$

$$A_1 = \frac{1}{4}\left[a_0 - a_2 - i(a_1 - a_3)\right] \qquad (6.129)$$

$$A_2 = \frac{1}{4}\left[a_0 + a_2 - (a_1 + a_3)\right] \qquad (6.130)$$

$$A_3 = \frac{1}{4}\left[a_0 - a_2 + i(a_1 - a_3)\right] \qquad (6.131)$$

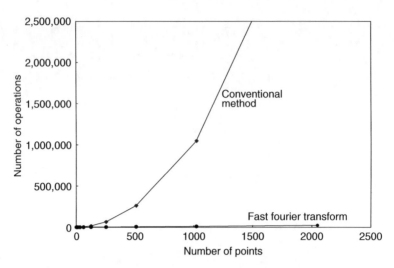

FIGURE 6.23 Variation of required number of operations with number of points using conventional method and fast Fourier transform to compute the Fourier transform of discretized functions.

By referring to Equation 6.106, it may be easily verified that this is the result one would have obtained had one used the conventional definition of the discrete Fourier transform.

The advantage of the fast Fourier transform over the direct approach may now be appreciated. The direct computation of the discrete Fourier transform requires approximately N^2 arithmetic operations. In contrast, the fast Fourier transform reduces this number to approximately $N \log_2 N$. Thus, for example, for $N = 2^{15}$ the direct approach requires $\sim 1.1 \times 10^9$ operations, whereas the fast Fourier transform involves 4.9×10^5 operations, that is, $\sim 1/2000$ the number of operations required by the direct approach. The fast Fourier transform offers thus an enormous reduction in computational time, particularly for large values of N. Indeed, as it may be seen from Figure 6.23, the direct approach consumes so much computer time that without the fast Fourier transform frequency-domain analyses involving large data sets would be just unfeasible. As an added benefit, the fast Fourier transform offers an increase in accuracy as well. As fewer operations are required, it reduces the round-off error due to the limited word size in computers.

6.7 POWER SPECTRAL DENSITY

If a nonperiodic real function $a(t)$ with a finite duration t_d represents, say, the voltage across a 1 Ohm resistor, then the average power delivered by $a(t)$ is given by

$$\bar{P} = \frac{1}{t_d} \int_0^{t_d} a^2(t) dt \qquad (6.132)$$

which, incidentally, also represents the mean square of $a(t)$. From Parseval's theorem, however, one has that

$$\int_{-\infty}^{\infty} a^2(t) dt = \frac{1}{2\pi} \int_{-\infty}^{\infty} |F(\omega)|^2 d\omega = \int_{-\infty}^{\infty} |F(f)|^2 df \qquad (6.133)$$

where $F(\omega)$ is the Fourier transform of $a(t)$, ω depicts frequency in radians per second, and f denotes frequency in hertz. Therefore, Equation 6.132 may also be expressed as

$$\bar{P} = \frac{1}{t_d} \int_{-\infty}^{\infty} |F(f)|^2 df \qquad (6.134)$$

Characterization of Strong Ground Motions

or, as

$$\bar{P} = \int_{-\infty}^{\infty} G(f) df \qquad (6.135)$$

where

$$G(f) = \frac{|F(f)|^2}{t_d} \qquad (6.136)$$

The average power delivered by $a(t)$ in the time interval t_d is thus given by the area under the curve of the function of frequency defined by Equation 6.136. As such, this function is referred to as the temporal *power spectrum* or *power spectral density function* of the time function $a(t)$ and is sometimes employed to identify the power contained in earthquake ground motions. As is closely related to the amplitude Fourier spectrum, the power spectral density function is also utilized to characterize the frequency content of ground motions. Furthermore, it is used to extract important characteristics of ground motions. Some of these characteristics are, for example, their root mean square acceleration a_{rms}, central frequency Ω (a measure that indicates where the power spectral density concentrates), and shape factor δ (a measure of the spectrum frequency bandwidth) defined, respectively, as

$$a_{rms} = \sqrt{\lambda_0} \qquad (6.137)$$

$$\Omega = \sqrt{\frac{\lambda_2}{\lambda_0}} \qquad (6.138)$$

$$\delta = \sqrt{1 - \frac{\lambda_1^2}{\lambda_0 \lambda_2}} \qquad (6.139)$$

where λ_0, λ_1, and λ_2 denote spectral moments of order 0, 1, and 2, respectively, given by

$$\lambda_n = \int_0^{\infty} f^n G(f) df \quad n = 0, 1, 2 \qquad (6.140)$$

A typical temporal power spectrum, corresponding to the N–S accelerogram recorded during the 1940 El Centro, California, earthquake, is shown in Figure 6.24.

In current practice, the concept of a spectral density function is used primarily in conjunction with probabilistic methods for computing the response of structures to random excitations and in some related applications, such as in the generation of artificial accelerograms. In these probabilistic methods, though, use is not made of the spectral densities of individual records. Instead, use is made of an average power spectral density determined by averaging the temporal power spectral densities of an ensemble of a large number of representative accelerograms. That is, a spectral density determined as

$$G(f) = \frac{1}{N} \sum_{i=1}^{N} G_i(f) \qquad (6.141)$$

where $G_i(f)$ is the power spectral density of the ith sample or record and N the number of records considered in the ensemble.

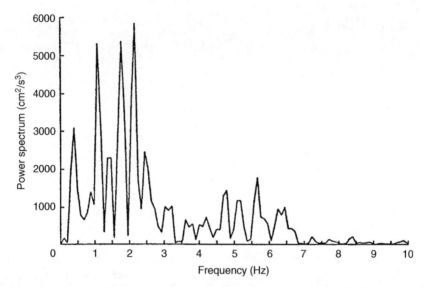

FIGURE 6.24 Power spectral density function of the N–S component of the 1940 El Centro earthquake. (After Watabe, M., *Theory of Spectral Analysis and its Application to Design Seismic Waves*, Seminar Notes, 1990.)

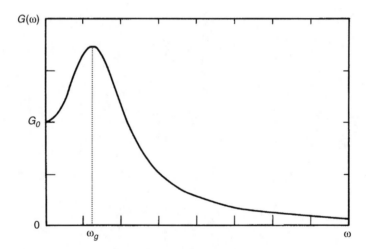

FIGURE 6.25 Kanai–Tajimi power spectral density function.

In many cases, the characteristics of the expected earthquake ground motions at a given site are represented by a smooth analytical model of their spectral density function. A well-known such model is that proposed by H. Tajimi in 1960 based on a semiempirical formula suggested by K. Kanai to determine the vibrational characteristics of soil sites. This model, known as the *Kanai–Tajimi spectral density function*, is of the form (see Figure 6.25)

$$G(\omega) = \frac{1 + 4\xi_g^2(\omega/\omega_g)^2}{[1 - (\omega/\omega_g)^2]^2 + (2\xi_g\omega/\omega_g)^2} G_0 \tag{6.142}$$

where G_0, ω_g, and ξ_g are the parameters of the function, which may be obtained for a given site by equating the root mean square acceleration, the central frequency, and the shape factor of the

Characterization of Strong Ground Motions

smooth and raw spectral densities. It represents the spectral density of a ground excitation with a constant spectral density equal to G_0 over the entire frequency range, after it has been filtered through a soil deposit characterized as a single-degree-of-freedom system with natural frequency ω_g and damping ratio ξ_g. Another well-known model is the Clough–Penzien spectral density function. Clough–Penzien spectral density function modifies the Kanai–Tajimi one to avoid the numerical difficulties that arise with this model in the neighborhood of $\omega = 0$ when the spectral density is divided by ω^2 and ω^4 to obtain the spectral densities of ground velocity and ground displacement. It is obtained by passing the Kanai–Tajimi spectral density through an additional filter with parameters ω_1 and ξ_1. Clough–Penzien spectral density function is given by

$$G(\omega) = \frac{1 + 4\xi_g^2(\omega/\omega_g)^2}{[1-(\omega/\omega_g)^2]^2 + (2\xi_g\omega/\omega_g)^2} \frac{(\omega/\omega_1)^4}{[1-(\omega/\omega_1)^2]^2 + 4\xi_1^2(\omega/\omega_1)^2} G_0 \tag{6.143}$$

where all symbols are as defined earlier.

6.8 SINGLE-PARAMETER MEASURES OF GROUND MOTION INTENSITY

6.8.1 Introductory Remarks

Quantification of the severity or intensity of earthquake ground motions in terms of a single parameter is desirable for several reasons. It facilitates a comparison between different ground motions, characterizes their damage potential in a simple way, and simplifies the establishment of design limit states. Thus, several single-parameter intensity measures have been proposed throughout the years. Some of these measures are presented in the following subsections, together with an assessment of their ability to reflect the three major characteristics of earthquake ground motions: peak amplitude, frequency content, and strong-motion duration. It should be noted, however, that presently there is no consensus as to which of these intensity measures best describes the intensity or damage potential of earthquakes.

6.8.2 Root Mean Square Acceleration

A single parameter that reflects the amplitude and duration characteristics of a strong ground motion is the *root mean square acceleration*. This parameter is defined as

$$a_{rms} = \sqrt{\frac{1}{t_d}\int_0^{t_d}[a(t)]^2 dt} \tag{6.144}$$

where $a(t)$ denotes acceleration at time t and t_d the ground motion duration. Because it depends on amplitude as well as duration, the root mean square acceleration is useful to characterize strong ground motions for engineering purposes.

6.8.3 Housner Spectral Intensity

In 1952, G. W. Housner introduced a measure of earthquake intensity based on the pseudovelocity spectrum of a linear oscillator with viscous damping. This measure of earthquake intensity, known nowadays as Housner's spectral intensity, is defined as the area enclosed by the pseudovelocity spectrum between the periods of 0.1 and 2.5 s. That is, it is defined by the integral

$$I_H = \int_{0.1}^{2.5} PSV(T,\xi) dT \tag{6.145}$$

where $PSV(T, \xi)$ is the pseudovelocity corresponding to a natural period T and damping ratio ξ. Housner's spectral intensity has the units of length. As the maximum force produced by an earthquake in a single-degree-of-freedom system with low damping is given by

$$F_{max} = kSD(T, \xi) = \frac{k}{\omega} PSV(T, \xi) = \sqrt{km}\ PSV(T, \xi) \tag{6.146}$$

where k, m, and ω are, respectively, the mass, stiffness, and natural frequency of the oscillator, and $SD(T, \xi)$ the spectral displacement corresponding to a natural period T and damping ratio ξ. Housner's spectral intensity represents a measure of the average maximum stress experienced by the structures in the $0.1 \leq T \leq 2.5$ period range, a range that covers most structures of practical interest. Consequently, it also represents a measure of earthquake destructiveness.

Housner's intensity captures the amplitude and frequency content of a ground motion in a single parameter as response spectrum ordinates directly depend on the intensity and the frequency content of a ground motion. However, it does not account for the effective strong-motion duration. The reason is that a response spectrum is insensitive to the ground motion duration. That is, the value of a response spectrum ordinate is the same independently of the length of the ground motion after the time at which the maximum value of the response is reached.

6.8.4 Arias Intensity

A parameter closely related to the root mean square acceleration is the Arias intensity. Arias intensity, introduced in 1970 by A. Arias, is defined as

$$I_A = \frac{\pi}{2g} \int_0^{t_d} a^2(t)dt \tag{6.147}$$

where $a(t)$ denotes ground acceleration, g the acceleration of gravity, and t_d the total ground motion duration. The units of Arias intensity are those of velocity.

Arias intensity is based on the premise that the damage experienced by a structure during an earthquake is proportional to the energy per unit weight dissipated by the structure during the total duration of the earthquake. It is defined as the energy per unit weight jointly dissipated by a collection of single-degree-of-freedom structures whose natural frequencies are uniformly distributed in the interval $(0, \infty)$. Thus, Arias intensity is given by

$$I_A = \int_0^{\infty} E_1\, d\omega \tag{6.148}$$

where E_1 is the energy dissipated per unit weight of a structure with natural frequency ω and damping ratio ξ when the base of the structure is excited by a horizontal ground acceleration $a(t)$. However, if it is considered that the response of a structure excited by a ground acceleration is equivalent to the response of an equivalent fixed-base structure with a force equal to ma applied to its mass, such an energy per unit weight may be expressed as

$$E_1 = \frac{1}{g} \int_0^{t_d} a(t)\dot{u}(t, \omega)dt \tag{6.149}$$

where $\dot{u}(t, \omega)$ denotes the structure's velocity response at time t. As a result, Equation 6.148 may also be written as

$$I_A = \frac{1}{g} \int_0^{\infty} \int_0^{t_d} a(t)\dot{u}(t, \omega)dt\, d\omega \tag{6.150}$$

Characterization of Strong Ground Motions

or, upon interchanging the order of integration, as

$$I_A = \frac{1}{g}\int_0^{t_d} a(t)\left(\int_0^\infty \dot{u}(t,\omega)d\omega\right)dt \qquad (6.151)$$

The inner integral may be evaluated using the expression to compute the displacement response of a single-degree-of-freedom system in terms of Duhamel's integral as follows. Consider first that

$$u(t,\omega) = \int_0^t a(\tau)e^{-\xi\omega(t-\tau)}\sin\omega_d(t-\tau)d\tau \qquad (6.152)$$

where $\omega_d = \omega(1-\xi^2)^{1/2}$. Consider then the integral

$$\int_0^\infty u(t,\omega)d\omega = \int_0^\infty \int_0^t a(\tau)e^{-\xi\omega(t-\tau)}\sin\omega_d(t-\tau)d\tau\, d\omega \qquad (6.153)$$

which, after changing the order of integration and introducing the change of variable $y = \omega_d(t-\tau)$, leads to

$$\int_0^\infty u(t,\omega)d\omega = \int_0^t \frac{a(\tau)}{\sqrt{1-\xi^2}}\left\{\int_0^\infty e^{-y\sqrt{\xi^2/(1-\xi^2)}}\frac{\sin y}{y}dy\right\}d\tau$$

$$= \int_0^t \frac{a(\tau)}{\sqrt{1-\xi^2}}\cos^{-1}\xi\, d\tau = \frac{\cos^{-1}\xi}{\sqrt{1-\xi^2}}\int_0^t a(\tau)d\tau = \frac{\cos^{-1}\xi}{\sqrt{1-\xi^2}}v(t) \qquad (6.154)$$

where $v(t)$ is the ground velocity at time t. Differentiate now both sides of Equation 6.154 to obtain

$$\int_0^\infty \dot{u}(t,\omega)d\omega = \frac{\cos^{-1}\xi}{\sqrt{1-\xi^2}}a(t) \qquad (6.155)$$

Thus, after substitution of this result into Equation 6.151 and considering that $a(t) = 0$ for $t > t_d$, one arrives to

$$I_A = \frac{\cos^{-1}\xi}{g\sqrt{1-\xi^2}}\int_0^{t_d} a^2(t)dt \qquad (6.156)$$

which may be approximated by the expression for the particular case of zero damping as

$$I_A = \frac{\pi}{2g}\int_0^{t_d} a^2(t)dt \qquad (6.157)$$

given that for small damping ratios the function $\cos^{-1}\xi/\sqrt{1-\xi^2}$ varies slowly with ξ.

It is of interest to observe that there is a close relationship between Arias intensity and Fourier spectrum. From Parseval's theorem it is known that

$$\int_0^\infty a^2(t)dt = \frac{1}{\pi}\int_0^\infty |F(\omega)|^2\, d\omega \qquad (6.158)$$

where $F(\omega)$ is the Fourier transform of $a(t)$. Therefore, Arias intensity may also be expressed as

$$I_A = \frac{1}{2g} \int_0^{t_d} |F(\omega)|^2 \, d\omega \qquad (6.159)$$

from which it may be seen that Arias intensity is proportional to the area under the square of the Fourier amplitude spectrum of the ground acceleration $a(t)$.

Arias intensity is dependent on ground acceleration amplitude, ground motion duration, and, indirectly, frequency content. It has been found, however, that it is not very sensitive to the frequency content of the excitation and long acceleration pulses.

6.8.5 Araya–Saragoni Intensity

R. Araya and G. R. Saragoni found from the studies of ductility demands in single-degree-of-freedom systems that earthquake destructiveness depends not only on total dissipated energy as proposed by Arias, but also on the frequency of zero crossings. In 1984, they therefore proposed a definition of earthquake intensity that modifies Arias intensity to incorporate such zero-crossing frequency. Araya–Saragoni intensity is defined as

$$I_{AS} = \frac{I_A}{v_0^2} \qquad (6.160)$$

where v_0 is the number of zero acceleration crossings per second. It has been found that Araya–Saragoni intensity correlates well with the earthquake damage described by the modified Mercalli intensity scale.

6.9 DESIGN-SPECTRUM-COMPATIBLE TIME HISTORIES

In the solution of some problems such as the response of inelastic structures (discussed in Chapter 12) or the generation of floor response spectra (discussed in Chapter 14), it is necessary to use an acceleration time history as input. In these cases, one has the option of using an accelerogram recorded on a site whose geological and tectonic conditions match those of the site of interest, or a modified acceleration time history whose response spectrum envelops the design spectrum generated for the site. As accelerograms that match a given set of geological and tectonic conditions are not always available, the use of modified ground motions that envelop a design spectrum has become the preferred alternative in design practice. Figure 6.26 illustrates this concept. This figure shows an original acceleration record that has been modified to match a specific design spectrum, and the response spectra that are obtained before and after it is modified. The original record is shown in part (a) of the figure, whereas part (b) shows the modified record. The target design spectrum and the response spectra of the original and modified records are displayed in part (c). It may be observed that there is a close match between the response spectrum of the modified record and the target design spectrum.

Several methods have been used in the past to generate design-spectrum-compatible time histories. Among the most widely used methods are (a) the modification of actual ground motion records and (b) the manipulation of artificial time histories generated in the frequency domain. In the first method, an accelerogram recorded at a site that matches as much as possible the geological and tectonic conditions of the site under investigation is first selected. Then this accelerogram is modified by multiplying the abscissas and ordinates of the accelerogram by scale factors that, through an iterative procedure, make the response spectrum of the modified accelerogram match as closely as possible the target design spectrum. In the second method, an artificial accelerogram

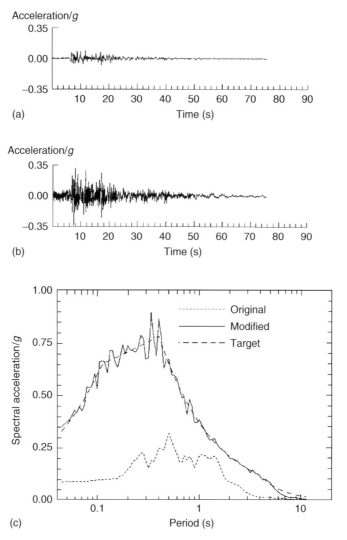

FIGURE 6.26 (a) Original and (b) modified acceleration time histories, and (c) comparison of their acceleration response spectra with target design spectrum. (Reprinted from Mukherjee, S. and Gupta, V.K., *Soil Dynamics and Earthquake Engineering*, 22, 799–804, 2002. With permission from Elsevier.)

is generated by combining in the frequency domain a Fourier amplitude spectrum with a Fourier phase spectrum. Then the ordinates of the Fourier amplitude spectrum are iteratively adjusted until a time history consistent with the target design spectrum is produced. The Fourier spectra used in this procedure may be selected from actual ground motions or may be obtained through some theoretical means. A disadvantage of this frequency-domain procedure, however, is that, because harmonics are added or subtracted along the entire length of the original accelerogram, the resulting time history seldom resembles an actual earthquake ground motion. More recently, a wavelet-based procedure has been proposed* in which the modification of the original accelerogram is carried out without altering the temporal variation of its frequency composition. In this procedure, a recorded accelerogram is decomposed into a desired number of time histories with no overlapping

---

* See Mukherjee, S. and Gupta, V. K., Wavelet based generation of spectrum compatible time-histories, *Soil Dynamics and Earthquake Engineering*, 22, 9–12, 2002, 799–804.

frequencies, and then each of these time histories is appropriately scaled to match the response spectrum of the modified accelerogram with a specified design spectrum.

It is important to keep in mind when using design-spectrum-compatible time histories that these time histories are not unique. That is, different time histories can match the same target design spectrum but may provide different answers to the same problem. It is also important to ensure that the developed time histories have characteristics that are consistent with the characteristics of actual earthquake ground motions. Many time histories may appear reasonable in the time domain but not when examined in the frequency domain, and vice versa. For example, a modified accelerogram may produce after integration unreasonable velocity and displacement histories.* Design-spectrum-compatible time histories should therefore be used with caution.

FURTHER READINGS

1. Kramer, S. L., *Geotechnical Earthquake Engineering*, Prentice-Hall, Inc., Upper Saddle River, NJ, 1996, 653 p.
2. Humar, J. L., *Dynamics of Structures*, Prentice-Hall, Inc., Englewood Cliffs, NJ, 1990, 780 p.
3. Mohraz, B. and Sadek, F., Earthquake ground motion and response spectra, Chapter 2 in *The Seismic Design Handbook*, 2nd edition, Naeim, F. ed., Kluwer Academic Publishers, Boston, MA, 2001, pp. 47–124.
4. Nau, J. M., Computation of inelastic response spectra, *Journal of Engineering Mechanics*, ASCE, 109, 1, February 1983, 279–288.
5. Newland, D. E., *An Introduction to Random Vibrations and Spectral Analysis*, Longman Group Ltd., London, 1975, 285 p.
6. Newmark, N. M. and Hall, W. J., *Earthquake Spectra and Design*, Earthquake Engineering Research Institute, Oakland, CA, 1989, 103 p.
7. Nigam, N. C. and Jennings, P. C., Calculation of response spectra from strong-motion earthquake records, *Bulletin Seismological Society of America*, 59, 1969, 909–922.
8. Paz, M. and Leigh, W., *Structural Dynamics: Theory and Computation*, 5th edition, Kluwer Academic Publishers, Boston, MA, 2004, 812 p.
9. Uang, C. M. and Bertero, V. V., Implications of Recorded Earthquake Ground Motions on Seismic Design of Building Structures, Report No. UCB/EERC-88/13, University of California, Berkeley, CA, November 1988.

PROBLEMS

6.1 Suppose the predominant horizontal ground shaking during an earthquake can be assumed sinusoidal with a frequency of 5 Hz. A rigid body with a mass of 1 kg rests freely on the ground and just starts slipping during an earthquake. If the coefficient of friction between the ground and the body is around 0.5, estimate the peak displacement of the ground during the earthquake.

6.2 Figure P6.2 shows the elastic response spectrum for the S38W accelerogram recorded at the Union Bank Building in Los Angeles, California, during the 1971 San Fernando earthquake. Estimate the maximum ground displacement at the recording site during that earthquake from this response spectrum.

6.3 The total-deformation response spectrum for elastoplastic systems with 5% damping of the N21E accelerogram recorded at the Castaic station during the 1971 San Fernando earthquake is shown in Figure P6.3. The recording station was located at a distance of 48 km from the earthquake epicenter. On the basis of this spectrum and Richter's attenuation equation for the magnitude of earthquakes in Southern California, estimate the local magnitude of the earthquake.

---

* For a discussion in this regard see Naeim, F. and Lew, M., On the use of design-spectrum-compatible time histories, *Earthquake Spectra*, 11, 1, 1995, 111–127.

Characterization of Strong Ground Motions

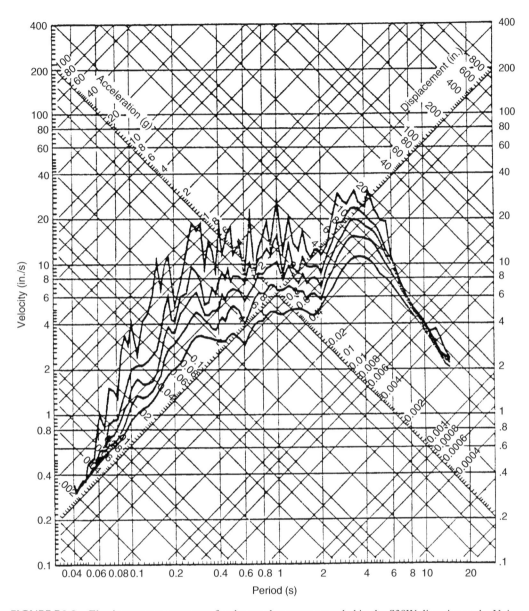

FIGURE P6.2 Elastic response spectrum for the accelerogram recorded in the S38W direction at the Union Bank Building in Los Angeles, California, during the 1971 San Fernando earthquake. (Reproduced from Analyses of Strong-Motion Earthquake Accelerograms, Vol. III, Part C, Report No. EERL 73-81, California Institute of Technology, Pasadena, CA, 1973.)

6.4 Using the response spectrum in Figure P6.4a, construct a design spectrum by drawing an envelope to the 5% damping response spectrum, that is, by drawing lines perpendicular to the displacement, velocity, and acceleration axes through the points of maximum spectral displacement, maximum spectral velocity, and maximum spectral acceleration for 5% damping. On the basis of this design spectrum, determine the design lateral earthquake force for the power transformer shown in Figure P6.4b. The transformer is supported by a steel pedestal and can be idealized as a single-degree-of-freedom system.

6.5 Two identical 30 story towers are to be erected adjacent to each other. Their fundamental periods are estimated to be ~3 s, and their lateral force–resisting structures are designed

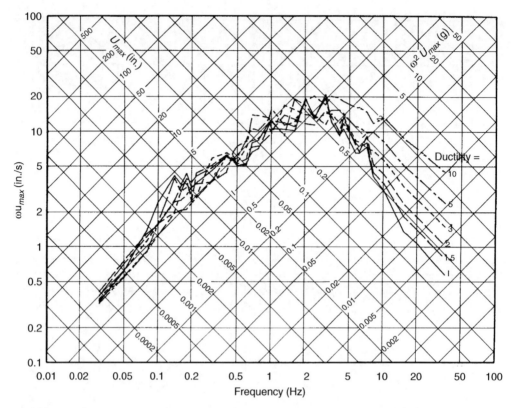

FIGURE P6.3 Total-deformation nonlinear response spectrum for elastoplastic systems with 5% damping corresponding to the N21E accelerogram recorded at the Castaic station during the 1971 San Fernando earthquake. (After Riddell, R. and Newmark, N.M., *Statistical Analysis of the Response of Nonlinear Systems Subjected to Earthquakes*, Structural Research Series No. 468, Department of Civil Engineering, University of Illinois, Urbana, IL, August 1979.)

to withstand inelastic deformations of up to six times their yield values. Using the design spectrum in Figure P6.5, which represents an envelope to yield-deformation spectra in a tetralogarithmic representation for a ductility factor of six, estimate the minimum clear distance that must be left between the two towers to avoid collision.

6.6 An elastoplastic single-degree-of-freedom system with a mass of 2.95 kip s²/in., a stiffness constant of 472 kip/in., a damping ratio of 5%, and a yield strength of 330 kip is subjected to an earthquake ground motion whose yield-deformation inelastic response spectrum is shown in Figure P6.6. If this spectrum is for elastoplastic systems with 5% damping, what is the maximum relative displacement that the system's mass will experience under this ground motion?

6.7 The yield-deformation inelastic response spectrum shown in Figure P6.7 is representative of the earthquake ground motions expected in a given area. On the basis of this response spectrum, find the yield strength that will limit the displacements of an elastoplastic single-degree-of-freedom system with a mass of 2.95 kip s²/in., a stiffness constant of 472 kip/in., and a damping ratio of 5% to values no greater than five times the displacement at which it yields.

Characterization of Strong Ground Motions

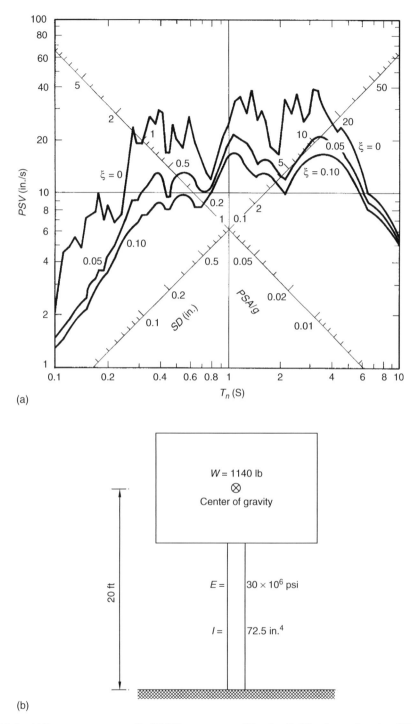

FIGURE P6.4 (a) Response spectrum for N11W component of Eureka, California, earthquake of December 21, 1954 and (b) simplified model of power transformer. (Response spectrum from Berg, G.V., *Elements of Structural Dynamics*, Prentice-Hall, Inc., Englewood Cliffs, NJ, 1989. Reproduced with permission from Pearson Education, Inc., Upper Saddle River, NJ.)

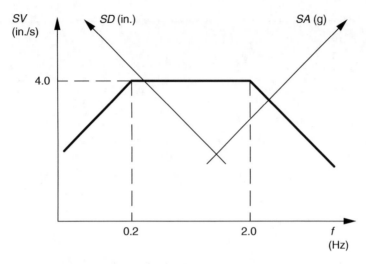

FIGURE P6.5 Design spectrum considered in Problem 6.5.

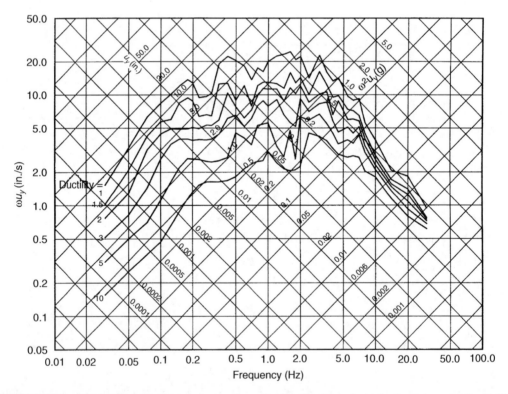

FIGURE P6.6 Yield-deformation response spectrum for 5% damping elastoplastic systems of the E–W accelerogram recorded during the 1972 Managua earthquake. (After Riddell, R. and Newmark, N.M., *Statistical Analysis of the Response of Nonlinear Systems Subjected to Earthquakes*, Structural Research Series No. 468, Department of Civil Engineering, University of Illinois, Urbana, IL, August 1979.)

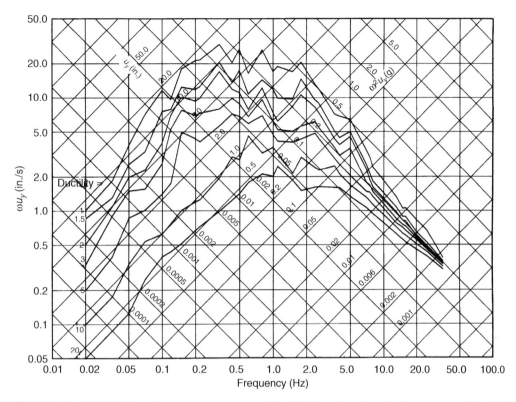

FIGURE P6.7 Yield-deformation response spectrum for 5% damping elastoplastic systems of the E-W accelerogram recorded during the 1940 El Centro earthquake. (After Riddell, R. and Newmark, N.M., *Statistical Analysis of the Response of Nonlinear Systems Subjected to Earthquakes*, Structural Research Series No. 468, Department of Civil Engineering, University of Illinois, Urbana, IL, August 1979.)

6.8 Determine the ordinate corresponding to a frequency of 1 Hz in the amplitude Fourier spectrum of a ground motion whose accelerations in meters per second squared are defined by

$$\ddot{u}_g(t) = \begin{cases} e^{-t} & \text{for } t > 0 \\ 0 & \text{otherwise} \end{cases}$$

6.9 During the calculation of the response spectrum ordinates for a given accelerogram, it is found that at the end of the accelerogram the displacement and velocity responses of an undamped single-degree-of-freedom system with a natural frequency of 2 Hz are 30 cm and 18 cm/s, respectively. What is the ordinate corresponding to a period of 0.5 s in the amplitude Fourier spectrum of the same accelerogram?

7 Seismic Hazard Assessment

SEISMORISKOLOGY

The seismologist is a gentle soul,
Who knows how much the Earth will roll
And shake and jump and twist and bend;
Some day we will even know when.
We first defined magnitude: M_s, m_b,
M_L, m_c, and even m_{Lg};
Then we figured seismic moment should be
A product of parameters: $M_0 = \mu A d$.
When an earthquake occurs, and others fret,
The seismologist comes and sets up a net;
To see what fault made the Earth creak,
Was it strike-slip, thrust, normal, or oblique?
The geologist thinks in terms of eons,
And converses in words like "post-Ordovician,"
The seismologist thinks in microtime,
And talks of Q, and V_s, and PP'.
The engineer has the very simple notion
That earthquakes consists of a design ground motion.
He doesn't know propagation or the seismic source,
And really needs to take a seismology course.
San Andreas, Santa Rosa, San Jacinto, San Gregorio,
San Joaquin, Santa Susana, San Fernando, San Gorgonio.

Using logic that would make the Pope in Rome faint,
The seismologist likes to name a fault after a saint.
When "the big one" comes, that others fear,
The real seismologist wants to be near,
To see the beast, to know its length,
To touch his belly, to measure his strength.
To look for some precursor signs,
So next time we can read the beast's mind.
And some day with luck, broadcast a warning,
That the big one is coming the following morning.
For seismologists, I submit, it's really quite easy
To compute seismic risk using probability.
There are no units, and before we've begun,
We know the answer lies between zero and one.
As a risk expert, don't say anything absurd,
And learn to use each new buzzword,
Like probability, uncertainty, randomness, and likelihood;
Then seismoriskology will be perfectly understood

Robin McGuire, 1992

7.1 INTRODUCTION

The analysis of a structure under a given ground motion is a relatively simple task that may be accomplished using well-established principles of structural dynamics. However, as discussed in previous chapters, earthquake ground motions may be radically different from one earthquake to another and from one site to another. Furthermore, it is impossible to forecast how strong a future earthquake will be and when and where it will occur. Therefore, the characteristics of future earthquake ground motions are, for the most part, unpredictable and, hence, the selection of the ground motion characteristics for which structures should be designed constitutes a difficult and elaborate undertaking that involves the use of historical information, statistical data, geological inferences, probabilistic models, empirical correlations, and, more often than not, engineering judgment. This process of evaluating for the purpose of seismic design the likely characteristics of future earthquake ground motions in a given seismic area is a critical step in the seismic design of structures called *seismic hazard assessment*.*

As seen in the preceding chapter, several parameters may be used to characterize earthquake ground motions, but by far the most widely used have been peak ground acceleration, peak ground velocity, and response spectrum ordinates. As a result, the objective of a seismic hazard assessment in the past has been, and still is, an estimation of the peak ground acceleration, peak ground velocity, or response spectrum ordinates of the ground motions expected at a given site as a result of the earthquakes that can be generated, within a given time interval, in the vicinity of the site. Thus, in general terms, an assessment of the seismic hazard at a specific site involves the following steps:

1. Identification and characterization of all earthquake sources capable of producing significant ground motions at the site
2. Estimation of the magnitude and the frequency of the earthquakes that can be generated at these sources
3. Evaluation of the distance and orientation of each source with respect to the site
4. Establishment of statistical correlations between earthquake magnitude, earthquake source characteristics, distance from source to site, and ground motion intensity (e.g., peak ground acceleration, peak ground velocity, or response spectrum ordinates)
5. Estimation of the expected ground motion intensity at the site using these correlations in terms of (a) the expected earthquake magnitudes at the identified sources, (b) the known characteristics of these sources, and (c) the estimated distances from the sources to the site

There are obviously several ways by which such an assessment can be made. In the early days of earthquake engineering, it was made deterministically without due consideration of all the uncertainties involved in the evaluation process and the fact that earthquakes are, for the most part, random events. That is, an assessment of the seismic hazard at a particular site was made by assuming the occurrence of an earthquake of a presumed possible largest magnitude at a nearby, previously known, earthquake fault. Today, these assessments are invariably made, explicitly or implicitly, within a probabilistic framework. In one approach, called in this text the semiprobabilistic approach, some variables, such as distance to seismic source, are assumed deterministic while some others, such as frequency of occurrence, are assumed random with a given probability distribution. In another approach, referred herein as the probabilistic approach, all variables are considered to be random and are defined in terms of given probability distributions.

This chapter describes some of the techniques in current use to identify seismic sources, the main factors that affect the form of the ground motions expected at a site, and some of the empirical correlations that have been proposed to estimate ground motion parameters in terms of these factors. A description will also be made of some recurrence relationships that may be used to specify the average rate at which an earthquake of a specific size may occur at a given source. Then, based on these empirical correlations and recurrence relationship, the aforementioned semiprobabilistic and probabilistic methodologies of seismic hazard assessment will be introduced. Finally, a description will be made of how these seismic hazard-assessment methods have been used to generate the zonation and microzonation maps incorporated into some building codes and seismic provisions.

---

* Although the words *risk* and *hazard* are virtually synonyms, by convention the terms *seismic hazard* and *seismic risk* have different meanings in the earthquake engineering literature; seismic hazard is used to describe the earthquake phenomenon while seismic risk is used to characterize the possible loss from earthquake effects.

7.2 IDENTIFICATION OF SEISMIC SOURCES

The identification and characterization of all possible sources of seismic activity (earthquake faults, for the most part) and the evaluation of their potential for generating significant ground motions at a particular site or region is usually accomplished using information obtained from the records from seismographic stations, historical accounts, and geological evidence. This section will briefly describe how each of these techniques may be used for such a purpose.

Records from the hundreds of seismographic stations that detect and record the occurrence of large earthquakes around the world provide quantitative data about the size, location, depth, and time of occurrence. They also provide some of the characteristics of the sources that produce these earthquakes, such as fault type, fault orientation, ruptured area, and amount of fault slip. Thus, with the help of these records, one can learn where and when earthquakes have occurred in the past, identify what sources of seismic activity are near a site, and establish the likely characteristics of the earthquakes that may be generated in those sources. However, compared to the time intervals at which large earthquakes occur, seismographs and seismographic networks have been available only in recent times (since around 1900). Besides, the fact that no earthquake has been recorded in a particular region is no guaranty that earthquakes have not occurred in the past or that they will not occur in the future. Therefore, the information obtained from instrumental records is necessarily a small part of the information needed to establish the seismic hazard at a site, and has to be complemented with the information obtained from investigations based on historical accounts and geological evidence of earthquake activity in the distant past.

Historical accounts of earthquake effects are one useful way to detect the occurrence of past earthquakes, to estimate their magnitude, and to identify their geographical location. Printed historical records may extend back at least a few hundred years in some regions of the world and even thousands of years in some others. For example, printed historical records extend back ~300 years in the United States, 2000 years in Japan and the Middle East, and 3000 years in China. Therefore, whenever historical records provide sufficient data about the intensity of the earthquakes and damage distribution, they can be used to estimate the magnitude of past earthquakes and the location of their epicenters. Although the accuracy of these magnitudes and epicenters strongly depend on the population density at the time of the earthquakes, historical data provide nonetheless strong evidence for the existence of seismic sources and help to establish the location of past earthquake activity. Moreover, as most historical accounts are dated, historical data are also useful to evaluate the rate of earthquake recurrence.

Geological evidence is also helpful to identify seismic sources and the frequency of past earthquakes. According to the plate tectonics and elastic rebound theories, earthquakes occur as a result of years of strain energy accumulation and the sudden relative motion between two tectonic plates. Therefore, it is expected that throughout the years past earthquakes have left a written geological record of their occurrence in the form of an offset (relative displacement) of soil and rock strata (see Figure 7.1), folding of these strata (see Figure 7.2), or a visible fault scarp on the earth's surface (see Figure 7.3). The branch of science that studies the geological record of past earthquakes is called *paleoseismology*.

A variety of tools and techniques are available to identify earthquake faults using geological evidence. These tools and techniques include the review of published literature and geological maps, air photographs and remote sense imagery (infrared photography), geophysical methods, and field reconnaissance studies, which may include the logging of trenches and test pits and borings. The following are examples of geological features that hint the existence of an earthquake fault (see Figure 7.4):

1. Disruption of the ground surface and evidence of movement and grinding along the two sides of the disruption
2. Juxtaposition of strata with dissimilar materials, missing or repeated strata, or truncation of strata

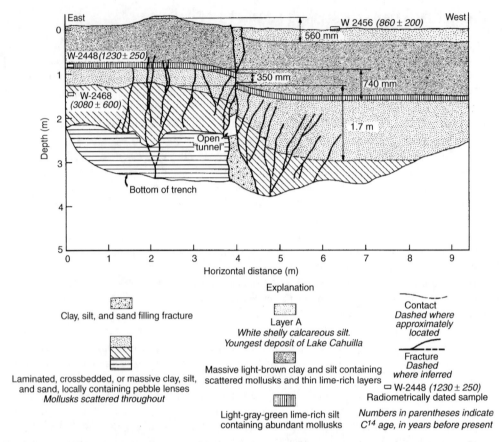

FIGURE 7.1 Displaced strata in trench across fault that produced the 1968 Borrego Mountain, California, earthquake, showing offsets that took place in this and previous earthquakes. (After Clark, M.M., Grantz, H., and Robin, M., Professional Paper 787, U.S. Geological Survey, 1972.)

FIGURE 7.2 Buckling of rock formations caused by tectonic compression, exposed by a road cut on State Route 14 across the San Andreas Fault 1.6 km southwest of the city of Palmdale and 100 km north of Los Angeles, CA.

FIGURE 7.3 Surface scarp 37 km long developed during the 1983 Borah Peak earthquake in Idaho. The vertical displacement is ~1.2 m and the left-lateral slip is ~0.5 m. (Photograph courtesy of U.S. Geological Survey.)

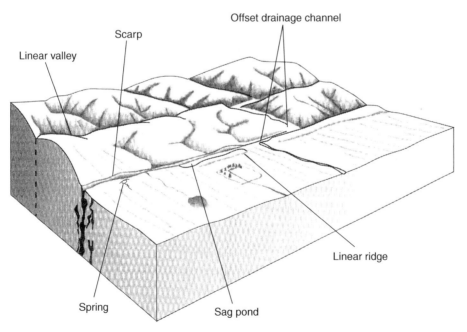

FIGURE 7.4 Topographical and geomorphic indicators of earthquake activity in the vicinity of a fault. (After Wesson, R.L. et al., Professional Paper 941-A, U.S. Geological Survey, 1975.)

3. Scarps or triangular facets on ridges, offset streams or drainage, tilting or changes in elevation of terraces or shorelines, sag ponds (see Figure 7.5), and anomalous stream gradients
4. Abrupt changes in levels, gradients, and chemical composition of groundwater; abrupt changes in the alignment of springs or volcanic vents; and presence of hot springs
5. Lineaments on remote sensing imagery caused by topographic, vegetation, or tonal contrasts

FIGURE 7.5 Sag pond formed when upthrust fault movement created a steep slope and cut off drainage. (Photograph courtesy of U.S. Geological Survey.)

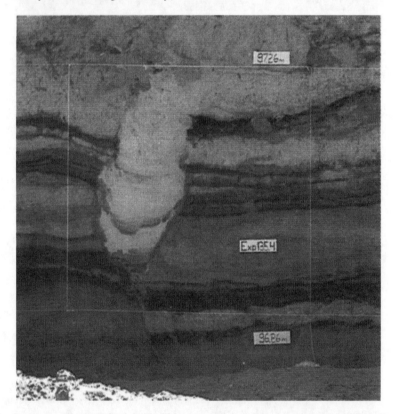

FIGURE 7.6 Remnants of a liquefaction sand blow caused by an earthquake around 1700 discovered in an old California stream bed. (Photograph courtesy of Kerry Sieh California Institute of Technology.)

6. Steep linear gravity or magnetic gradients, differences in seismic wave velocities, and offset of seismic reflection horizons in a geophysical survey
7. Changes in geodetic surveys such as in tilting and the distance between fixed points
8. Traces of sand liquefaction along the depth of excavated trenches (see Figure 7.6)

FIGURE 7.7 Using a trenching technique, seismologists look for evidence of the occurrence of strong earthquakes in the distant past. (a) A 5 m deep excavation that uncovered evidence of 12 major earthquakes, which took place along a segment of the San Andreas Fault at intervals averaging 131 years between the sixth century and 1857. (b) Geologist Kerry Sieh pointing to one of several layers of peat that were displaced on or about the indicated dates. (Photographs courtesy of Kerry Sieh, California Institute of Technology.)

Trenching across a known fault is becoming a widely used method to estimate the occurrence of earthquakes through geological times. Trenching permits to uncover past fault displacements, which may be measured to estimate the magnitude of the associated earthquakes and dated using radiocarbon techniques to determine when they occurred. For example, seismologists in the United States have discovered using this technique that strong earthquakes have occurred along a segment of the San Andreas Fault in the years AD 260, 350, 590, 735, 845, 935, 1015, 1083, 1350, 1550, 1720, and 1857 (see Figure 7.7).

The way geological evidence is used to estimate the magnitude of past earthquakes is by means of established empirical correlations between the fault characteristics and magnitude of recorded earthquakes, such as the following, which have been proposed by Wells and Coppersmith[*] to predict the moment magnitude of earthquakes as a function of surface rupture length, rupture area, or maximum surface fault displacement:

$$M_w = 5.08 + 1.16 \log L \quad (SD = 0.28) \tag{7.1}$$

$$M_w = 4.07 + 0.98 \log A \quad (SD = 0.24) \tag{7.2}$$

$$M_w = 6.69 + 0.74 \log D \quad (SD = 0.40) \tag{7.3}$$

In these equations, M_w is the expected value of moment magnitude, L the surface rupture length in kilometers, A the rupture area in kilometer square, D the maximum surface displacement in meters, and SD signifies standard deviation.

Note that as a result of the unavoidable scarcity of instrumental data, the uncertainty in correlations of this type is always considerably high. Hence, it is important to account for this uncertainty in the application of these correlations.

7.3 FACTORS AFFECTING GROUND MOTION CHARACTERISTICS AT SITE

Mark Twain, with no scientific training but an acute observer nonetheless, made the following description of the varied nature of earthquake ground motions after experiencing several of them while in California in the 1860s:

> I have tried a good many of them here, and of several varieties, some that came in the form of a universal shiver; others that gave us two or three sudden upward heaves from below; others that sway grandly and deliberately from side to side; and still others that came rolling and undulating beneath our feet like a great wave of the sea.

As observed by Mark Twain then, it is known nowadays that the characteristics of earthquake ground motions may vary substantially from earthquake to earthquake, from region to region, and even from site to site during the same earthquake. Earthquake ground motions may vary in duration, maximum intensity, the way this intensity changes with time, and the relationship between their vertical and horizontal components. It is also known that the reason ground motions may be radically different from one site to another is simply because a large number of factors may affect their characteristics. According to the concepts discussed in the foregoing chapters, an earthquake ground motion felt at a site is produced as a result of a complicated process that may be explained more or less as follows: First, a rupture takes place at a small area within the crustal rocks of a fault (what seismologist identify as the earthquake focus). Then, this rupture spreads outwardly in all directions over an extensive area (called the rupture zone) until it reaches a rock that is not sufficiently strained to permit a further extension of the rupture (see Figure 7.8). After the rupture stops, the adjacent sides of the fault spring back to an unstrained position giving rise to stress waves. These stress waves then propagate in all directions, traveling through a complex stratification of the earth's crust, changing their propagation velocity, sustaining multiple reflections and refractions along the way, and attenuating as they get farther away from their source (see Figure 7.9). Finally, as they reach the earth's surface, the waves undergo further modifications as a result of the soft soil deposits they may have to cross and the additional reflections and refractions caused by any structures embedded or laying on top of these soil deposits.

Thus, the characteristics of an earthquake ground motion at a given site may be affected, among others, by the following factors:

1. Type of fault where the earthquake is generated (e.g., dip slip or strike slip)
2. Fault orientation with respect to site (angle between source-to-site vector and fault line)

---

[*] *Bulletin Seismological Society of America*, Vol. 84, No. 4, 1994, pp. 974–1002.

Seismic Hazard Assessment

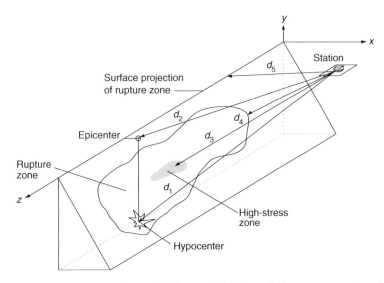

FIGURE 7.8 Fault-rupture propagation and different definitions of *distance to earthquake source* used in various attenuation equations: d_1 = hypocentral distance or focal distance; d_2 = epicentral distance; d_3 = distance to high-stress zone; d_4 = closest distance to rupture zone; d_5 = closest distance to surface projection of rupture zone.

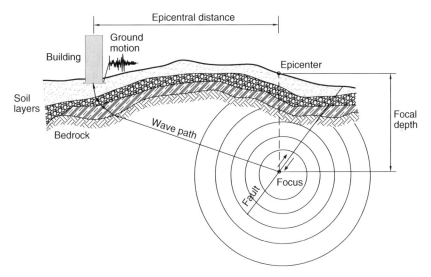

FIGURE 7.9 Schematic representation of factors affecting characteristics of ground motion at building site.

3. Direction of fault rupture with respect to site (i.e., toward or away from site)
4. Dimensions of ruptured area
5. Depth of ruptured area
6. Earthquake magnitude
7. Distance from fault to site
8. Geological characteristics of propagation medium
9. Local soil properties and topography
10. Size and type of structures on site

Obviously, it is difficult, if not virtually impossible, to quantify by analytical means the effect of all these factors. First of all, the mechanism involved in the generation of an earthquake is still not well understood, and second, the medium through which the earthquake waves propagate is intricate and difficult to survey. Therefore, such a quantification is presently done by means of empirical relationships derived on the basis of statistical data and regression analyses. It should be borne in mind, however, that by their own nature the goodness of these empirical relationships is limited by the scarcity of the available data, and that they constantly change as additional data is collected and incorporated into their derivation.

7.4 GROUND MOTION INTENSITY AT GIVEN SITE: ATTENUATION LAWS

7.4.1 BACKGROUND

As discussed in the previous section, several empirical equations, sometimes referred to as *attenuation laws* or *attenuation relationships*, have been developed in the past to correlate ground motion intensity at a given site with some of the factors that affect this intensity. In their conventional form, these relationships correlate a measure of ground motion intensity with earthquake magnitude, hypocentral distance, and local soil conditions. In the past few years, however, the availability of additional strong-motion data has allowed the incorporation of additional parameters such as type of fault, fault orientation, and faulting style (e.g., strike-slip versus reverse fault) into these equations. Traditionally, peak ground acceleration or peak ground velocity has been used as a measure of ground motion intensity, although, recently, acceleration response spectrum ordinates are becoming a common ground motion parameter. Correlations with some other parameters such as duration or Fourier spectrum ordinates have also been proposed. Because of the problems associated with the determination of peak ground displacements, correlations with this parameter have been avoided altogether.

Most attenuation relationships are developed by a regression analysis using data from recorded strong ground motions. However, the functional form of the predictive equations is usually selected to reflect as closely as possible the mechanism that gives rise to ground motions. This minimizes the number of empirical coefficients and provides some confidence in their application to conditions that are poorly represented in the database. The functional form of many of the attenuation equations developed in the past are based on the following observations:

1. Peak values of ground motion parameters are approximately lognormally distributed. Consequently, if Y represents one of such peak values, the regression is usually performed on the logarithm of Y instead of Y itself.
2. Earthquake magnitude is typically defined as the logarithm of some peak ground motion parameter (ground displacement in most cases). Therefore, it is common to assume that $\log Y$ is proportional to M, where M denotes the earthquake magnitude.
3. The spreading of stress waves as they travel away from the source of an earthquake makes the amplitude of body waves decrease according to $1/R$, where R is the distance to the source, and the amplitude of surface waves according to $1/R^{1/2}$. Hence, most predictive equations assume ground motions attenuate with distance according to either $1/R$ or $1/R^{1/2}$.
4. Some of the energy carried by stress waves is absorbed by the materials through which they travel. This material damping causes the amplitude of ground motions to decrease exponentially with distance R and may become important at distances >100 km.
5. Ground motion parameters are influenced by site soil characteristics (e.g., hard rock versus alluvial soil) and source characteristics (e.g., strike-slip versus reverse faulting).

Putting together all these observations, many of the proposed attenuation relationships have the form

$$\ln Y = C_1 + C_2 M + C_3 \ln(R + C_4) + C_5 R + C_6 f(source) + C_7 f(soil) \tag{7.4}$$

where C_i, $i = 1, \ldots, 7$, are constants obtained through a regression analysis with the available ground motion data, $f(source)$ is a function of source parameters, and $f(soil)$ a function of soil characteristics.

Although attenuation equations of this type attempt to account for as many factors as possible, the scatter in the regression analysis is still invariably large. It is therefore important to calculate the uncertainty involved in the use of these equations and consider it when the equations are applied. Traditionally, the standard deviation of ln Y is used to measure this uncertainty.

When using a predictive relationship, it is important to know how parameters such as magnitude and site-to-source distance are defined and use them in a consistent manner. In the past, these relationships were developed in terms of local magnitude, but modern relationships invariably use moment magnitude. Similarly, different investigators use different definitions of source-to-site distance. As illustrated in Figure 7.8, definitions of source-to-site distance used in the past include (a) hypocentral distance, (b) epicentral distance, (c) distance to high-stress zone, (d) closest distance to rupture zone, and (e) closest distance to surface projection of rupture zone. By the same vein, as different tectonic environments give rise to different ground motion attenuation relationships, it is important to recognize before using it for which tectonic environment an attenuation relationship has been derived. Currently, three different categories are considered: (a) shallow crustal earthquakes in active tectonic regions (e.g., western North America), (b) shallow crustal earthquakes in stable continental regions (e.g., central and eastern North America), and (c) subduction zones (e.g., northwest North America, western South America, and Japan).

More than 30 attenuation relationships have been proposed since 1965. Many of these relationships are updates obtained after new data become available following the occurrence of a significant earthquake. Many others, however, differ from one another in the functional form adopted, the number of ground motion records used, the geographical region for which the relationships are derived, and the variables considered in their derivation. For example, some are developed for a specific region such as western North America, Italy, or Japan, while others, using worldwide data, are applicable to many regions around the world. Similarly, some account for fault characteristics while others consider only the basic variables of magnitude, distance to the source, and soil characteristics. Furthermore, some differentiate between the tectonic environments in which earthquakes may be generated while others do not. So far, the largest amount of strong ground motion records comes from shallow earthquakes in active tectonic regions and, hence, most of the available attenuation relationships have been developed for this tectonic environment. Nevertheless, as a result of an increasingly larger data set, recent relationships have increasingly become more refined although, at the same time, more complicated. As expected, different results are obtained with each of these relationships, and the difference may at times amount to as much as one order of magnitude. As expected, too, each is best suited for seismic sources whose characteristics are similar to those considered in its derivation.

In view of the impracticality of a review of all these relationships in an elementary textbook such as this, and, more importantly, in view of their short useful life, only three of such relationships will be presented here. Two of these are for shallow crustal earthquakes in an active tectonic environment and the other for earthquakes generated in subduction zones. Among those for shallow crustal earthquakes, one is derived using data from western North America and the other using worldwide data. These three relationships will suffice to illustrate the form of current attenuation equations and the significance of differentiating between regional and worldwide data and the aforementioned tectonic environments. Readers interested in attenuation relationships other than those presented here may find the appropriate references in some of the publications listed at the end of the chapter.

7.4.2 Boore, Joyner, and Fumal Equations for Shallow Crustal Earthquakes in Active Regions[*]

Back in 1981, Joyner and Boore[†] proposed attenuation relationships to predict peak ground acceleration and peak ground velocity. These relationships became well known and have been used widely by the profession. Since then, however, they have been updated and expanded to predict response

---

[*] *Seismological Research Letters*, Vol. 68, No.1, 1997, pp. 128–153.
[†] *Bulletin Seismological Society of America*, Vol. 71, No. 6, 1981, pp. 2011–2038.

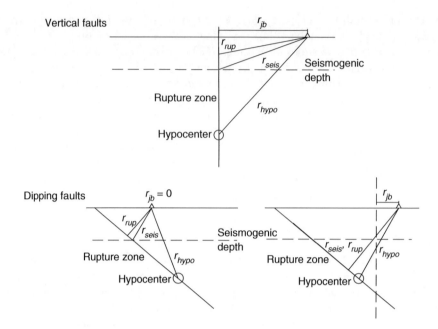

FIGURE 7.10 Source-to-site distances considered in different attenuation relationships: r_{hypo} = hypocentral distance; r_{jb} = closest distance to surface projection of rupture zone; r_{rup} = closest distance to rupture zone; r_{seis} = closest distance to seismogenic part of rupture zone (After Abrahamson, N.A. and Shedlock, K.M., *Seismological Research Letters*, 68, 1, 9–23, 1997. © Seismological Society of America.)

spectrum ordinates. In this section, therefore, the updated relationships developed by Boore, Joyner, and Fumal in 1997 will be presented.

The relationships proposed by Boore, Joyner, and Fumal predict horizontal peak ground acceleration and pseudoacceleration response spectrum ordinates for a damping ratio of 5%. For the derivation of these equations, they use strong ground motion data from shallow earthquakes with a moment magnitude >5.0, a depth of <20 km, and recorded at free-field stations in western North America. They are expressed in terms of moment magnitude and a source-to-site distance defined as the closest horizontal distance from the recording station to a point on the earth's surface that lies directly above the fault rupture (see Figure 7.10). The equation for horizontal peak ground acceleration is derived in terms of the geometric mean (the mean of the logarithm) of the two horizontal components obtained at each recording station. All the earthquakes in the data set, with the exception of one, were generated at either strike-slip or reverse faults. The soil characteristics at the recording sites were defined in terms of the average shear wave velocity over the top 30 m of the soil deposits at the recording site, computed by dividing 30 m by the time it takes an S wave to travel from the surface to a depth of 30 m.

The proposed equation and the regression coefficients to estimate 5% damping pseudoacceleration response spectrum ordinates for the horizontal component of earthquake ground motions are summarized in Table 7.1. These regression coefficients are given for 46 natural periods ranging from 0.01 to 2 s. Note that the corresponding coefficients to estimate peak ground acceleration are listed in this same Table as the entries for a zero natural period. Both the peak ground acceleration and the response spectrum ordinates are expressed as a fraction of g, the acceleration of gravity. The standard deviation of the error in using these equations, $\sigma_{\ln Y}$, is also given in this table. The validity of the equation is limited to moment magnitudes between 5.5 and 7.5 and distances not greater than 80 km. Figures 7.11 and 7.12 illustrate the variation of peak ground acceleration and response spectrum ordinates with magnitude and distance according to this equation.

TABLE 7.1
Boore, Joyner, and Fumal's Attenuation Equation and Regression Coefficients for Shallow Earthquakes in Western North America to Estimate Horizontal Peak Ground Acceleration and Pseudoacceleration Response Spectrum Ordinates for 5% Damping (Valid for $5.5 \leq M_w \leq 7.5$ and $r \leq 80$ km)

$$\ln Y = b_1 + b_2 (M_w - 6) + b_3 (M_w - 6)^2 + b_5 \ln r + b_V \ln(V_s/V_A)$$

Period	b_{1SS}	b_{1RV}	b_{1ALL}	b_2	b_3	b_5	b_V	V_A	h	$\sigma_{\ln Y}$
0.000	−0.313	−0.117	−0.242	−0.527	−0.000	−0.778	−0.371	1396.	5.57	0.520
0.100	1.006	1.087	1.059	0.753	−0.226	−0.934	−0.212	1112.	6.27	0.479
0.110	1.072	1.164	1.130	0.732	−0.230	−0.937	−0.211	1291.	6.65	0.481
0.120	1.109	1.215	1.174	0.721	−0.233	−0.939	−0.215	1452.	6.91	0.485
0.130	1.128	1.246	1.200	0.711	−0.233	−0.939	−0.221	1596.	7.08	0.486
0.140	1.135	1.261	1.208	0.707	−0.230	−0.938	−0.228	1718.	7.18	0.489
0.150	1.128	1.264	1.204	0.702	−0.228	−0.937	−0.238	1820.	7.23	0.492
0.160	1.112	1.257	1.192	0.702	−0.226	−0.935	−0.248	1910.	7.24	0.495
0.170	1.090	1.242	1.173	0.702	−0.221	−0.933	−0.258	1977.	7.21	0.497
0.180	1.063	1.222	1.151	0.705	−0.216	−0.930	−0.270	2037.	7.16	0.499
0.190	1.032	1.198	1.122	0.709	−0.212	−0.927	−0.281	2080.	7.10	0.501
0.200	0.999	1.170	1.089	0.711	−0.207	−0.924	−0.292	2118.	7.02	0.502
0.220	0.925	1.104	1.019	0.721	−0.198	−0.918	−0.315	2158.	6.83	0.508
0.240	0.847	1.033	0.941	0.732	−0.189	−0.912	−0.338	2178.	6.62	0.511
0.260	0.764	0.958	0.861	0.744	−0.180	−0.906	−0.360	2173.	6.39	0.514
0.280	0.681	0.881	0.780	0.758	−0.168	−0.899	−0.381	2158.	6.17	0.518
0.300	0.598	0.803	0.700	0.769	−0.161	−0.893	−0.401	2133.	5.94	0.522
0.320	0.518	0.725	0.619	0.783	−0.152	−0.888	−0.420	2104.	5.72	0.525
0.340	0.439	0.648	0.540	0.794	−0.143	−0.882	−0.438	2070.	5.50	0.530
0.360	0.361	0.570	0.462	0.806	−0.136	−0.877	−0.456	2032.	5.30	0.532
0.380	0.286	0.495	0.385	0.820	−0.127	−0.872	−0.472	1995.	5.10	0.536
0.400	0.212	0.423	0.311	0.831	−0.120	−0.867	−0.487	1954.	4.91	0.538
0.420	0.140	0.352	0.239	0.840	−0.113	−0.862	−0.502	1919.	4.74	0.542
0.440	0.073	0.282	0.169	0.852	−0.108	−0.858	−0.516	1884.	4.57	0.545
0.460	0.005	0.217	0.102	0.863	−0.101	−0.854	−0.529	1849.	4.41	0.549
0.480	−0.058	0.151	0.036	0.873	−0.097	−0.850	−0.541	1816.	4.26	0.551
0.500	−0.122	0.087	−0.025	0.884	−0.090	−0.846	−0.553	1782.	4.13	0.556
0.550	−0.268	−0.063	−0.176	0.907	−0.078	−0.837	−0.579	1710.	3.82	0.562
0.600	−0.401	−0.203	−0.314	0.928	−0.069	−0.830	−0.602	1644.	3.57	0.569
0.650	−0.523	−0.331	−0.440	0.946	−0.060	−0.823	−0.622	1592.	3.36	0.575
0.700	−0.634	−0.452	−0.555	0.962	−0.053	−0.818	−0.639	1545.	3.20	0.582
0.750	−0.737	−0.562	−0.661	0.979	−0.046	−0.813	−0.653	1507.	3.07	0.587
0.800	−0.829	−0.666	−0.760	0.992	−0.041	−0.809	−0.666	1476.	2.98	0.593
0.850	−0.915	−0.761	−0.851	1.006	−0.037	−0.805	−0.676	1452.	2.92	0.598
0.900	−0.993	−0.848	−0.933	1.018	−0.035	−0.802	−0.685	1432.	2.89	0.604
0.950	−1.066	−0.932	−1.010	1.027	−0.032	−0.800	−0.692	1416.	2.88	0.609
1.000	−1.133	−1.009	−1.080	1.036	−0.032	−0.798	−0.698	1406.	2.90	0.613
1.100	−1.249	−1.145	−1.208	1.052	−0.030	−0.795	−0.706	1396.	2.99	0.622
1.200	−1.345	−1.265	−1.315	1.064	−0.032	−0.794	−0.710	1400.	3.14	0.629
1.300	−1.428	−1.370	−1.407	1.073	−0.035	−0.793	−0.711	1416.	3.36	0.637
1.400	−1.495	−1.460	−1.483	1.080	−0.039	−0.794	−0.709	1442.	3.62	0.643
1.500	−1.552	−1.538	−1.550	1.085	−0.044	−0.796	−0.704	1479.	3.92	0.649
1.600	−1.598	−1.608	−1.605	1.087	−0.051	−0.798	−0.697	1524.	4.26	0.654

(Continued)

TABLE 7.1 (Continued)

$$\ln Y = b_1 + b_2 (M_w - 6) + b_3(M_w - 6)^2 + b_5 \ln r + b_V \ln(V_s/V_A)$$

Period	b_{1SS}	b_{1RV}	b_{1ALL}	b_2	b_3	b_5	b_V	V_A	h	$\sigma_{\ln Y}$
1.700	−1.634	−1.668	−1.652	1.089	−0.058	−0.801	−0.689	1581.	4.62	0.660
1.800	−1.663	−1.718	−1.689	1.087	−0.067	−0.804	−0.679	1644.	5.01	0.664
1.900	−1.685	−1.763	−1.720	1.087	−0.074	−0.808	−0.667	1714.	5.42	0.669
2.000	−1.699	−1.801	−1.743	1.085	−0.085	−0.812	−0.655	1795.	5.85	0.672

Y = peak ground acceleration or 5% damping pseudoacceleration response spectrum ordinate divided by g.
M_w = moment magnitude.
$r = \sqrt{R^2 + h^2}$.
R = closest horizontal distance to surface projection of fault-rupture zone (see Figure 7.10).
h = fictitious depth to rupture zone determined from regression analysis.
V_S = average over the upper 30 m of shear wave velocities in m/s in site soil.
$b_1 = b_{1SS}$ for strike-slip faults; $= b_{1RS}$ for reverse-slip faults; $= b_{1ALL}$ if fault type not specified.

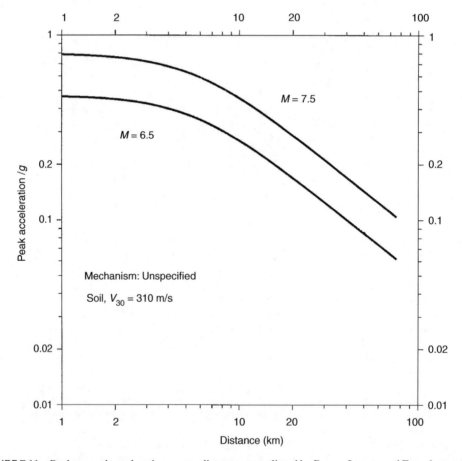

FIGURE 7.11 Peak ground acceleration versus distance as predicted by Boore, Joyner, and Fumal attenuation equations for earthquakes with magnitudes of 6.5 and 7.5, at soil sites, and an unspecified fault type. (After Boore, D.M., Joyner, W.B., and Fumal, T.E., *Seismological Research Letters*, 68, 1, 128–153, 1997. © Seismological Society of America.)

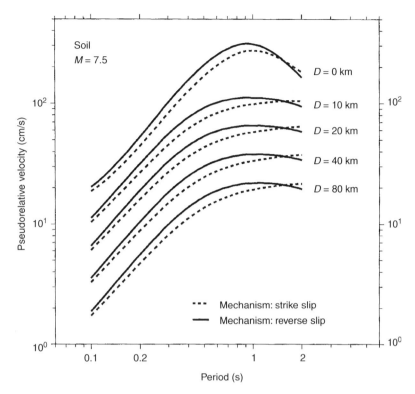

FIGURE 7.12 A 5% damping pseudovelocity response spectra as predicted by Boore, Joyner, and Fumal attenuation equations for strike- and reverse-slip earthquakes with a magnitude of 7.5, at soil sites, and distances of 0, 10, 20, 40, and 80 km. (After Boore, D.M., Joyner, W.B., and Fumal, T.E., *Seismological Research Letters*, 68, 1, 128–153, 1997. © Seismological Society of America.)

TABLE 7.2
Joyner and Boore Attenuation Equation for Shallow Earthquakes in Western North America to Estimate Horizontal Peak Ground Velocity (Valid for $5.0 \leq M_w \leq 7.7$ and $r \leq 370$ km)

$$\log V_H = 2.09 + 0.49(M_w - 6) - \log r - 0.0026r + 0.17\,S$$
$$\sigma_{\log V_H} = 0.33$$

V_H = peak horizontal ground velocity in cm/s.
$\sigma_{\log V_H}$ = standard deviation of $\log V_H$.
M_w = moment magnitude.
$r = (R^2 + 4^2)^{1/2}$.
R = closest horizontal distance in kilometers to vertical projection of fault rupture on earth's surface.
S = 0 for rock; = 1 for soil.

An update to Joyner and Boore's original attenuation relationship to predict the peak value of horizontal ground velocity was made back in 1988.[*] Although this relationship does not include the recent strong-motion data incorporated into their 1997 equation to predict peak ground acceleration and response spectrum ordinates, it is also presented in this section for completeness. The relationship is shown in Table 7.2. A graphical representation of the equation is presented in Figure 7.13.

---

[*] *Earthquake Engineering and Soil Dynamics II: Recent Advances in Ground Motion Evaluation*, Geotechnical Special Publication 20, ASCE, New York, 1988, pp. 43–102.

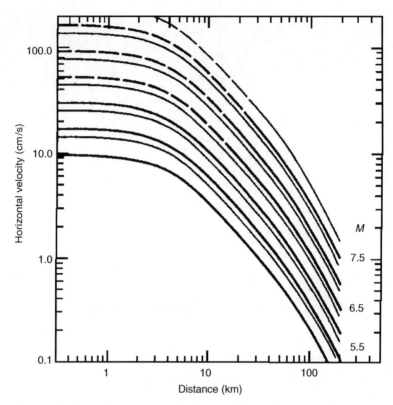

FIGURE 7.13 Variation of mean peak ground horizontal velocity with distance and magnitude according to Joyner and Boore's 1988 attenuation relationship: heavy lines are for deep rock sites and light lines for soil sites. (Reproduced with permission of ASCE from Joyner, W.B. and Boore, D.M., *Earthquake Engineering and Soil Dynamics II—Recent Advances in Ground-Motion Evaluation*, Geotechnical Special Publication No. 20, ASCE, 43–102, 1988.)

The equation proposed by Joyner and Boore to predict peak ground velocity is derived using strong ground motion data from earthquakes in western North America with a moment magnitude between 5.0 and 7.7, focal depths of <20 km, and recorded at distances of <370 km. This equation is therefore valid for earthquakes that satisfy these characteristics.

EXAMPLE 7.1   PEAK GROUND ACCELERATION AND VELOCITY AT A SITE USING BOORE, JOYNER, AND FUMAL'S EQUATION

Find the mean and the mean plus one standard deviation values of the peak horizontal ground acceleration and peak horizontal ground velocity at a site that overlies a soil deposit and is located 38.5 km from a known earthquake fault. It is estimated that earthquakes with a moment magnitude of up to 7.5 may occur at this fault. Similarly, it is estimated that the focal depths of these earthquakes will not be >20 km. The average shear wave velocity for the soil deposit at the site may be assumed equal to 310 m/s. Use Boore, Joyner, and Fumal's equation to estimate the peak ground acceleration and Joyner and Boore's to estimate the peak ground velocity.

Solution:

Both Boore, Joyner, and Fumal and Joyner and Boore's equations are expressed in terms of the closest horizontal distance to the surface projection of the fault-rupture zone. As in this case the location and the dimensions of the fault-rupture zone are not known, it will be assumed that the required distance is equal to the given distance to the fault. Accordingly, r, as defined in Table 7.1, will be considered equal to

$$r = (38.5^2 + 5.57^2)^{1/2} = 38.9 \text{ km}$$

Thus, according to the attenuation equation in Table 7.1 and the corresponding regression coefficients for peak ground acceleration (zero period), the mean value of $\ln A_H$ and its standard deviation are:

$$\ln A_H = b_1 + b_2(M_W - 6) + b_3(M_W - 6)^2 + b_5 \ln r + b_V \ln(V_s/V_A)$$

$$= -0.242 + 0.527(7.5 - 6) - 0.778 \ln(38.9) - 0.371 \ln(310/1396)$$

$$= -1.742$$

$$\sigma_{\ln A_H} = 0.520$$

Hence, the mean and the mean plus one standard deviation value of the peak ground acceleration are

$$(A_H)_{0.50} = e^{-1.742} = 0.175g$$

$$(A_H)_{0.84} = e^{-1.742+0.520} = 0.295g$$

Similarly, using the equations in Table 7.2 for peak ground velocity, one obtains

$$r = (38.5^2 + 4.0^2)^{1/2} = 38.7$$

$$\log V_H = 2.09 + 0.49(M_W - 6) - \log r - 0.0026r + 0.17 \, s$$

$$= 2.09 + 0.49(7.5 - 6) - \log 38.7 - 0.0026(38.7) + 0.17(1)$$

$$= 1.307$$

$$\sigma_{\log V_H} = 0.33$$

Therefore, the mean and mean plus one standard deviation values of the peak ground velocity are

$$(V_H)_{0.50} = 10^{1.307} = 20.3 \text{ cm/s}$$

$$(V_H)_{0.84} = 10^{1.307+0.33} = 43.3 \text{ cm/s}$$

7.4.3 Campbell Equations for Shallow Crustal Earthquakes in Active Regions[*]

Campbell has also developed empirical attenuation relationships for shallow crustal earthquakes in active seismic regions. Campbell equations predict free-field horizontal and vertical components of peak ground acceleration, peak ground velocity, and 5% damping pseudoacceleration spectral ordinates. For the derivation of these equations, he uses strong ground motion data from worldwide earthquakes generated at strike-slip, reverse, and thrust faults with moment magnitudes between 4.7 and 8.1, source-to-site distances between 3 and 60 km, depths to fault-rupture zone of <25 km, and recorded at stations on sites overlying deposits of more than 10 m of firm soil, soft rock, or hard rock. Source-to-site distance is defined as the shortest distance between the recording site and the presumed seismogenic rupture on the causative fault, assuming that the part of the fault rupture

---

[*] *Seismological Research Letters*, Vol. 68, No.1, 1997, pp. 154–179.

TABLE 7.3
Campbell's Attenuation Equations and Regression Coefficients for Worldwide Shallow Earthquakes to Estimate Horizontal Peak Ground Acceleration, Horizontal Peak Ground Velocity, and Pseudoacceleration Response Spectrum Ordinates for 5% Damping (Valid for $4.7 \leq M_w \leq 8.1$ and $3.0 \leq r \leq 60$ km)

$$\ln(A_H) = -3.512 + 0.904 M_w - 1.328 \ln \sqrt{r^2 + [0.149 \exp(0.647 M_w)]^2}$$
$$+ (1.125 - 0.112 \ln r - 0.0957 M_w) F + (0.440 - 0.171 \ln r) S_{SR} + (0.405 - 0.222 \ln r) S_{HR}$$

$$\sigma_{\ln A_H} = 0.55 \quad \text{if } A_H < 0.068g$$
$$= 0.173 - 0.140 \ln(A_H) \quad \text{if } 0.068g \leq A_H \leq 0.21g$$
$$= 0.39 \quad \text{if } A_H$$

$$\ln(V_H) = \ln(A_H) + 0.26 + 0.29 M_w - 1.44 \ln[r + 0.0203 \exp(0.958 M_w)] + 1.89 \ln[r + 0.361 \exp(0.576 M_w)]$$
$$+ (0.0001 - 0.000565 M_w) r - 0.12 F - 0.15 S_{SR} - 0.30 S_{HR} + 0.75 \tanh(0.51 D)(1 - S_{HR}) + f_V(D)$$

$$\sigma_{\ln(V_H)} = \sqrt{\sigma_{\ln A_H}^2 + 0.06^2}$$

$$\ln(SA_H) = \ln(A_H) + c_1 + c_2 \tanh[c_3(M_w - 4.7)] + (c_4 + c_5 M_w) r$$
$$+ 0.5 c_6 S_{SR} + c_6 S_{HR} + c_7 \tanh(c_8 D)(1 - S_{HR}) + f_{SA}(D)$$

$$\sigma_{\ln(SA_H)} = \sqrt{\sigma_{\ln A_H}^2 + 0.27^2}$$

Period (s)	c_1	c_2	c_3	c_4	c_5	c_6	c_7	c_8
0.05	0.05	0	0	−0.0011	0.000055	0.20	0	0
0.075	0.27	0	0	−0.0024	0.000095	0.22	0	0
0.1	0.48	0	0	−0.0024	0.000007	0.14	0	0
0.15	0.72	0	0	−0.0010	−0.00027	−0.02	0	0
0.2	0.79	0	0	0.0011	−0.00053	−0.18	0	0
0.3	0.77	0	0	0.0035	−0.00072	−0.40	0	0
0.5	−0.28	0.74	0.66	0.0068	−0.00100	−0.42	0.25	0.62
0.75	−1.08	1.23	0.66	0.0077	−0.00100	−0.44	0.37	0.62
1.0	−1.79	1.59	0.66	0.0085	−0.00100	−0.38	0.57	0.62
1.5	−2.65	1.98	0.66	0.0094	−0.00100	−0.32	0.72	0.62
2.0	−3.28	2.23	0.66	0.0100	−0.00100	−0.36	0.83	0.62
3.0	−4.07	2.39	0.66	0.0108	−0.00100	−0.22	0.86	0.62
4.0	−4.26	2.03	0.66	0.0112	−0.00100	−0.30	1.05	0.62

V_H = mean value of horizontal peak ground velocity in cm/s.
SA_H = mean value of ordinate in pseudoacceleration-response spectrum divided by g, corresponding to the horizontal component of ground motion, a damping ratio of 5%, and a selected natural period.
$\sigma_{\ln A_H}$ = standard deviation in estimation of the logarithm of peak ground acceleration.
σ_{V_H} = standard deviation in estimation of peak ground velocity.
σ_{SA_H} = standard deviation in estimation of pseudoacceleration response spectrum values.
M_w = moment magnitude.
A_H = mean value of horizontal peak ground acceleration divided by g.
r = shortest distance in kilometers to the zone of seismogenic rupture (see Figure 7.10).
D = depth to basement rock (crystalline rock with P wave velocity of at least 5 km/s).
F = 0 for strike-slip and normal faults; = 1 for reverse, reverse-oblique, and thrust-oblique faults.
S_{SR} = 1 for soft rock sites; = 0 for alluvium or firm soil sites.
S_{HR} = 1 for hard rock sites; = 0 for alluvium or firm soil sites.
$f_V(D) = 0$ if $D \geq 1$ km; $= -0.30(1 - S_{HR})(1 - D) - 0.15(1 - D) S_{SR}$ if $D < 1$ km.
$f_{SA}(D) = 0$ if $D \geq 1$ km; $= c_6(1 - S_{HR})(1 - D) + 0.5 c_6(1 - D) S_{SR}$ if $D < 1$ km.

Seismic Hazard Assessment

within the soft sediments in the upper part of the earth's crust is not seismogenic (see Figure 7.10). According to this definition, the source-to-site distance cannot be less than the depth to the top of the seismogenic part of the earth's crust, which in turn cannot be shallower than approximately 2–4 km. To derive the equations for the horizontal ground motion parameters, Campbell uses the geometric mean (i.e., the mean of the logarithm) of the peak values of the two horizontal components recorded at each station. The equations proposed by Campbell and the corresponding regression coefficients are summarized in Table 7.3. Expressions that give the dispersion (standard deviation) obtained in the regression analysis of each of the considered ground motion parameters are also presented in this table.

Figure 7.14 illustrates the variation of peak ground acceleration with magnitude, distance to the source, and type of fault according to Campbell's equations. Similarly, Figure 7.15 shows the pseudoacceleration-response spectra that result from using Campbell's equations for sites on firm soil, soft rock, and hard rock, a moment magnitude of 6.5, a distance to the source of 25 km, and a depth to the basement rock of 5 km.

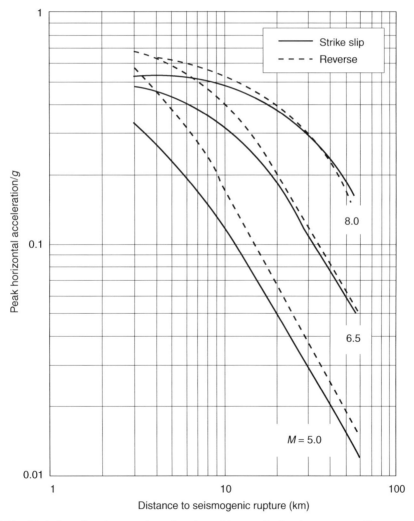

FIGURE 7.14 Variation of peak ground acceleration with magnitude, site-to-source distance, and type of fault according to Campbell's attenuation relationship. (After Campbell, K.W., *Seismological Research Letters*, 68, 1, 154–179, 1997. © Seismological Society of America.)

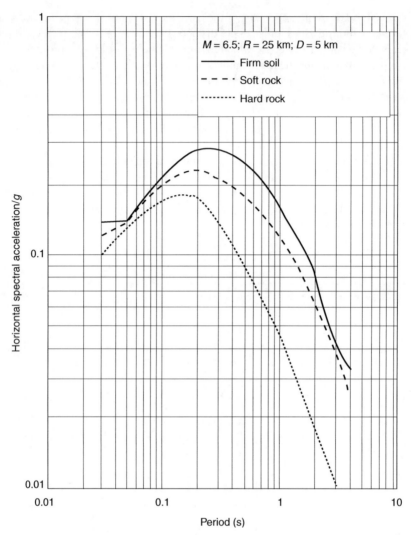

FIGURE 7.15 Pseudoacceleration-response spectra obtained with Campbell's attenuation relationships for earthquakes with moment magnitude of 6.5 and sites at a distance of 25 km from seismic source and basement rock at a depth of 5 km. (After Campbell, K.W., *Seismological Research Letters*, 68, 1, 154–179, 1997. © Seismological Society of America.)

Example 7.2 Peak ground acceleration and velocity at a site using Campbell's equations

Using Campbell's equations find the mean and the mean plus one standard deviation of the peak horizontal ground acceleration and peak horizontal ground velocity at a site that overlies an alluvium deposit and is located 38.5 km away from a vertical strike-slip earthquake fault. The thickness of the nonseismogenic layer of the earth's crust in the area near the fault is 2.5 km. Consider an earthquake with a moment magnitude of 7.5.

Solution

Campbell's equations are expressed in terms of the shortest distance to the seismogenic part of a fault's rupture zone. Hence, as the given distance of 38.5 km corresponds to the distance to the surface projection of the fault, not the distance to the seismogenic part of the fault, the use of Campbell's equations requires the determination of this distance first. Accordingly, considering that in this example the distance from the surface to the top of the seismogenic part of the earth's crust is 2.5 km, the shortest distance from the site to the seismogenic part of the fault is

$$r = \sqrt{38.5^2 + 2.5^2} = 38.6 \text{ km}$$

Note also that for the case under consideration, $M_w = 7.5$ and $F = S_{HR} = S_{SR} = 0$. Thus, upon substitution of these values into the formulas in Table 7.3 for peak ground acceleration, one obtains

$$\ln(A_H) = -3.512 + 0.904 M_w - 1.328 \ln\sqrt{r^2 + [0.149\exp(0.647 M_w)]^2}$$
$$+ (1.125 - 0.112 \ln r - 0.0957 M_w) F + (0.440 - 0.171 \ln r) S_{SR} + (0.405 - 0.222 \ln r) S_{HR}$$
$$= -3.512 + 0.904(7.5) - 1.328 \ln\sqrt{(38.6)^2 + \{0.149\exp[0.647(7.5)]\}^2} = -1.729$$

$$\sigma_{\ln A_H} = 0.173 - 0.140 \ln(A_H)$$
$$= 0.173 - 0.140(-1.729)$$
$$= 0.415$$

In consequence, the mean and the mean plus one standard deviation of the peak ground acceleration at the site are

$$(A_H)_{0.50} = e^{-1.729} = 0.178 g$$

$$(A_H)_{0.84} = e^{-1.729 + 0.415} = 0.269 g$$

Similarly, observing that for a depth of 2.5 km to the top of the seismogenic part of the earth's crust $D \geq 1$ and thus $f_V(D) = 0$, the formulas for peak ground velocity in Table 7.3 give

$$\ln(V_H) = \ln(A_H) + 0.26 + 0.29 M_w - 1.44 \ln[r + 0.0203\exp(0.958 M_w)] + 1.89 \ln[r + 0.361\exp(0.576 M_w)]$$
$$+ (0.0001 - 0.000565 M_w) r - 0.12 F - 0.15 S_{SR} - 0.30 S_{HR} + 0.75 \tanh(0.51 D)(1 - S_{HR}) + f_V(D)$$
$$= -1.729 + 0.26 + 0.29(7.5) - 1.44 \ln\{38.6 + 0.0203\exp[0.958(7.5)]\}$$
$$+ 1.89 \ln\{38.6 + 0.361\exp[0.576(7.5)]\} + [0.0001 - 0.000565(7.5)]38.6 + 0.75 \tanh[0.51(2.5)] = 3.079$$

$$\sigma_{\ln(V_H)} = \sqrt{\sigma_{\ln A_H}^2 + 0.06^2}$$
$$= \sqrt{0.415^2 + 0.06^2}$$
$$= 0.419$$

As a result, the mean and mean plus one standard deviation of the peak ground velocity at the site are

$$(V_H)_{0.50} = e^{3.079} = 21.8 \text{ cm/s}$$

$$(V_H)_{0.84} = e^{3.079 + 0.419} = 33.0 \text{ cm/s}$$

7.4.4 YOUNGS, CHIOU, SILVA, AND HUMPHREY EQUATIONS FOR SUBDUCTION-ZONE EARTHQUAKES*

There have been much fewer recordings of earthquakes at subduction zones than from shallow earthquakes in active tectonic regions. However, the sparse data collected so far have shown that strong ground motions from earthquakes generated at subduction zones attenuate at a different rate than those from earthquakes generated at other tectonic environments. Furthermore, it has been

---

* *Seismological Research Letters*, Vol. 68, No.1, 1997, pp. 58–73.

found that even among the earthquakes that generate at subduction zones, there is a difference in the attenuation of ground motions between intraslab and interface earthquakes (interface earthquakes are shallow events that occur at the interface between a subducting and an overriding plate, while intraslab earthquakes are deep events that occur within a subducting oceanic plate; see Figure 3.8).

TABLE 7.4
Youngs, Chiou, Silva, and Humphrey's Attenuation Equations and Regression Coefficients for Subduction Zone Earthquakes to Estimate Horizontal Peak Ground Acceleration and Acceleration Response Spectrum Ordinates for 5% Damping (Valid for $M_w \geq 5.0$ and $10 \leq r \leq 500$ km)

For ground motions on rock:

$$\ln y = 0.2418 + 1.414 M_w + C_1 + C_2 (10 - M_w)^3 + C_3 \ln[r + 1.7818 \exp(0.554 M_w)]$$
$$+ 0.00607 H + 0.3846 Z_T \quad \sigma_{\ln y} = C_4 + C_5 M_w$$

Period (s)	C_1	C_2	C_3	C_4*	C_5*
0	0.00	0.00	−2.552	1.45	−0.1
0.075	1.275	0.00	−2.707	1.45	−0.1
0.1	1.188	−0.0011	−2.655	1.45	−0.1
0.2	0.722	−0.0027	−2.528	1.45	−0.1
0.3	0.246	−0.0036	−2.454	1.45	−0.1
0.4	−0.115	−0.0043	−2.401	1.45	−0.1
0.5	−0.40	−0.0048	−2.360	1.45	−0.1
0.75	−1.149	−0.0057	−2.286	1.45	−0.1
1.0	−1.736	−0.0064	−2.234	1.45	−0.1
1.5	−2.634	−0.0073	−2.160	1.50	−0.1
2.0	−3.328	−0.0080	−2.107	1.55	−0.1
3.0	−4.511	−0.0089	−2.033	1.65	−0.1

For ground motions on soil:

$$\ln y = -0.6687 + 1.438 M_w + C_1 + C_2 (10 - M_w)^3 + C_3 \ln[r + 1.097 \exp(0.617 M_w)]$$
$$+ 0.00648 H + 0.3643 Z_T \quad \sigma_{\ln y} = C_4 + C_5 M_w$$

Period (s)	C_1	C_2	C_3	C_4*	C_5*
0	0.00	0.00	−2.329	1.45	−0.1
0.075	2.400	−0.0019	−2.697	1.45	−0.1
0.1	2.516	−0.0019	−2.697	1.45	−0.1
0.2	1.549	−0.0019	−2.464	1.45	−0.1
0.3	0.793	−0.0020	−2.327	1.45	−0.1
0.4	0.144	−0.0020	−2.230	1.45	−0.1
0.5	−0.438	−0.0035	−2.140	1.45	−0.1
0.75	−1.704	−0.0048	−1.952	1.45	−0.1
1.0	−2.870	−0.0066	−1.785	1.45	−0.1
1.5	−5.101	−0.0114	−1.470	1.50	−0.1
2.0	−6.433	−0.0164	−1.290	1.55	−0.1
3.0	−6.672	−0.0221	−1.347	1.65	−0.1
4.0	−7.618	−0.0235	−1.272	1.65	−0.1

y = spectral acceleration or peak ground acceleration divided by g.
$\sigma_{\ln y}$ = standard deviation of logarithm of ground motion parameter y.
M_w = moment magnitude.
r = closest distance to rupture zone in km (see Figure 7.10).
H = focal depth in kilometers.
Z_T = source-type indicator = 0 for interface earthquake; = 1 for intraslab earthquakes.
*Standard deviation for magnitude greater than 8 is equal to the value for magnitude equal to 8.

Seismic Hazard Assessment

Intraslab earthquakes tend to produce larger peak ground motions than interface earthquakes for the same magnitude and same distance. Several attenuation relationships have thus been derived specifically for ground motions generated by earthquakes at subduction zones, and some of these relationships make the distinction between interface and intraslab earthquakes. For the purpose of illustrating the difference between attenuation relationships for shallow crustal and subduction-zone earthquakes, only those proposed by Youngs, Chiou, Silva, and Humphrey in 1997 will be presented here.

The attenuation relationships proposed by Youngs et al. were developed specifically to predict peak ground acceleration and 5% damping acceleration response spectrum ordinates of ground motions generated by subduction-zone earthquakes. The strong ground motion data utilized in the derivation of these relationships are from earthquakes recorded at the subduction zones of Alaska, Chile, the Cascadia region of the United States, Japan, Mexico, Peru, and the Solomon Islands. Most of the data comes from events recorded at large distances because subduction-zone events tend to occur offshore or at large depths. The exception is the recordings from the 1985 Michoacán, Mexico, earthquake, which were obtained at distances as short as 13 km. Youngs et al.'s relationships differentiate between interface and intraslab earthquakes. This differentiation is made on the basis of the focal depth, considering that earthquakes with a focal depth of >50 km are intraslab earthquakes. Source-to-site distance was measured in terms of the closest distance to the rupture surface (see Figure 7.10) when a rupture surface was identified and hypocentral distance in all other cases. The data were restricted to free-field recordings from earthquakes with a magnitude of 5.0 or greater. Site conditions were classified in three groups: (a) rock, (b) shallow stiff soil, and (c) deep soil. The geometrical mean of the two ground acceleration components is used in the regression analysis for peak ground acceleration.

The equations proposed by Youngs et al. are summarized in Table 7.4. Figure 7.16 illustrates the variation of peak ground acceleration with magnitude, distance, and soil conditions as predicted by these equations. Similarly, Figure 7.17 depicts the corresponding mean acceleration response spectrum shapes. The relationships of Youngs et al. are considered appropriate for earthquakes with a moment magnitude of 5.0 or greater at distances between 10 and 500 km.

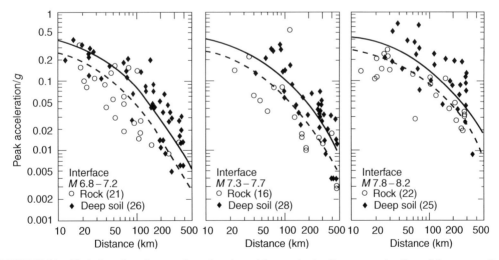

FIGURE 7.16 Variation of peak ground acceleration with magnitude, distance, and soil conditions according to the attenuation relationships of Youngs et al.: solid lines are for deep soil sites and dashed lines for rock sites. (After Youngs, R.R., Chiou, S.J., Silva, W.J., and Humphrey, J.R., *Seismological Research Letters*, 68, 1, 58–73, 1997. © Seismological Society of America.)

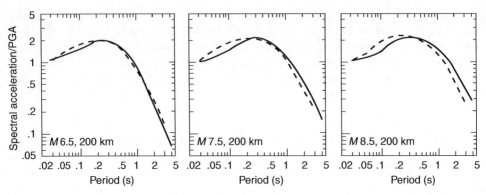

FIGURE 7.17 Acceleration response spectrum shapes (spectral acceleration/peak ground acceleration) for 5% damping obtained using the attenuation relationships of Youngs et al.: solid lines are for deep soil sites and dashed lines for rock sites. (After Youngs, R.R., Chiou, S.J., Silva, W.J., and Humphrey, J.R., *Seismological Research Letters*, 68, 1, 58–73, 1997. © Seismological Society of America.)

EXAMPLE 7.3 PEAK GROUND ACCELERATION AT A SITE USING THE EQUATIONS OF YOUNGS ET AL.

Using the attenuation equations of Youngs et al. find the mean and the mean plus one standard deviation of the peak horizontal ground acceleration at a site overlying deposits of soft soil and located 375 km from an active subduction zone. Consider an earthquake with a moment magnitude of 8.1 and a focal depth of 13 km.

Solution

As the earthquake specified in this problem occurs at a shallow depth, it may be considered to be an interface earthquake. Consequently, one may assume the source-type indicator Z_T in the attenuation equations of Youngs et al. equal to zero. Likewise, it may be considered that $r = 375$ km, $H = 13$ km, and $M_w = 8.1$. As a result, the equation for soil sites in Table 7.4 yields

$$\ln A_H = -0.6687 + 1.438 M_w + C_1 + C_2(10 - M_w)^3 + C_3 \ln[r + 1.097 \exp(0.617 M_w)] + 0.00648 H + 0.3643 Z_T$$
$$= -0.6687 + 1.438(8.1) - 2.329 \ln\{375 + 1.097 \exp[0.617(8.1)]\} + 0.00648(13)$$
$$= -3.579$$

$$\sigma_{\ln A_H} = C_4 + C_5 M_w$$
$$= 1.45 - 0.1(8.0)$$
$$= 0.650$$

where it is noted that a magnitude of 8.0 is used in the computation of the standard deviation of $\ln A_H$ as Youngs et al. specify that the standard deviation for magnitudes >8.0 be considered equal to the value for a magnitude of 8.0.

The mean and the mean plus one standard deviation of the peak ground acceleration at the site result, thus, equal to

$$(A_H)_{0.50} = e^{-3.579} = 0.028 g$$

$$(A_H)_{0.84} = e^{-3.579 + 0.650} = 0.053 g$$

7.5 MAGNITUDE-RECURRENCE RELATIONSHIPS

7.5.1 Introduction

A key element in the assessment of the seismic hazard in a region is the estimation of the recurrence intervals of earthquakes of different magnitudes. Many models have been proposed for this purpose, but so far no one appears to be the preferred one. For the purpose of introducing the concept of magnitude-recurrence relationships and illustrating the role of these models in a seismic hazard assessment, only three of such models will be discussed in this chapter. It should be borne in mind, however, that these models may not necessarily be the best for all seismic regions.

7.5.2 Gutenberg–Richter Recurrence Law

In 1944, B. Gutenberg and C. F. Richter collected data from Southern California earthquakes over a period of many years and organized the data according to the number of earthquakes that exceeded a given magnitude over that period. Based on these data, they made a plot against earthquake magnitude m of the logarithm of the mean annual rate of exceedance, λ_m, of earthquakes of magnitude m. This mean annual rate of exceedance λ_m was defined as the number of earthquakes that exceed a given magnitude m in the considered time interval divided by this time interval. Gutenberg and Richter found that the relationship between such two parameters was approximately linear and thus they proposed to describe the recurrence of earthquakes in a given region as

$$\log \lambda_m = a - bm \tag{7.5}$$

where a and b are constants obtained by a regression of the seismicity data from the region of interest and, as expected, vary from region to region. Equation 7.5 is known as the *Gutenberg–Richter earthquake recurrence law*. The number 10^a is interpreted as the mean yearly number of earthquakes of magnitude equal or greater than zero. Similarly, b (the slope of the line, known in the literature as the *b value*) is interpreted as the relative likelihood between large and small earthquakes, as when b increases, the number of large magnitude earthquakes decreases and the number of small magnitude earthquakes increases. As an example, Figure 7.18 shows Gutenberg–Richter law for a south central segment of the San Andreas Fault. Values of a and b for different regions of the world are listed in Table 7.5.

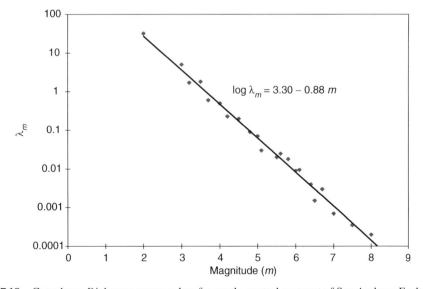

FIGURE 7.18 Gutenberg–Richter recurrence law for south–central segment of San Andreas Fault.

TABLE 7.5
Values of Regression Coefficients *a* and *b* in Gutenberg–Richter Law Determined for Shallow Earthquakes during a Time Interval of 14 Years for Various Seismic Regions

Region	a	b
Japan	6.86	1.22
New Guinea	7.83	1.35
New Zealand	—	1.04
W. Canada	5.05	1.09
W. United States	5.94	1.14
E. United States	5.79	1.38
Central America	7.36	1.45
Colombia–Peru	5.60	1.11
N. Chile	4.78	0.88
S. Chile	4.46	0.92
Mediterranean	5.45	1.10
Iran–Turkmenia	6.02	1.18
Java	5.37	0.94
E. Africa	3.80	0.87

Source: Kaila, K.L. and Narain, H., *Bulletin of the Seismological Society of America*, 61, 1275–1291, 1971.

Note that by conversion from decimal to natural logarithms Gutenberg–Richter law may also be expressed as

$$\lambda_m = 10^{a-bm} = e^{\alpha-\beta m} \quad (7.6)$$

where $\alpha = 2.303a$ and $\beta = 2.303b$. Gutenberg–Richter law implies, therefore, that the probability distribution of earthquake magnitude is an exponential one.

7.5.3 Bounded Gutenberg–Richter Recurrence Law

The standard Gutenberg–Richter law covers earthquake magnitudes from zero to infinity. For engineering purposes, however, small earthquakes are of little interest. Hence, it is common to disregard those that are not capable of causing significant damage. If earthquakes with a magnitude smaller than a lower-threshold magnitude m_0 are neglected, then a bounded Gutenberg–Richter law may be written as

$$\lambda_m = \lambda_{m_0} \exp[-\beta(m - m_0)] \quad m > m_0 \quad (7.7)$$

where $\lambda_{m_0} = \exp(\alpha - \beta m_0)$. Normally, the lower-threshold magnitude is set to values between 4.0 and 5.0 as earthquakes with magnitudes <4.0 or 5.0 seldom cause significant damage.

Using Equation 7.7 and considering that probability may be interpreted as a relative frequency, the cumulative probability distribution of earthquake magnitude in terms of the bounded Gutenberg–Richter law may be then expressed as

$$F_M(m) = P[M < m \mid M > m_0] = \frac{\lambda_{m_0} - \lambda_m}{\lambda_{m_0}} = 1 - e^{-\beta(m-m_0)} \quad (7.8)$$

Seismic Hazard Assessment

from which it is found that the corresponding probability density function is given by

$$f_M(m) = \frac{d}{dm} F_M(m) = \beta e^{-\beta(m-m_0)} \tag{7.9}$$

The standard Gutenberg–Richter law predicts nonzero mean rates of exceedance for earthquakes with a magnitude up to infinity, in spite of the fact that earthquakes with magnitudes greater than ~9.0 have never been observed. If Gutenberg–Richter law is limited to earthquakes with a magnitude less than a maximum value m_{max}, then the mean annual rate of exceedance results as

$$\lambda_m = \lambda_{m_0} \frac{\exp[-\beta(m-m_0)] - \exp[-\beta(m_{max}-m_0)]}{1 - \exp[-\beta(m_{max}-m_0)]} \quad m_0 \leq m \leq m_{max} \tag{7.10}$$

and the corresponding cumulative distribution function and probability density function become

$$F_M(m) = P[M < m \mid m_0 \leq m \leq m_{max}] = \frac{1 - \exp[-\beta(m-m_0)]}{1 - \exp[-\beta(m_{max}-m_0)]} \tag{7.11}$$

$$f_M(m) = \frac{\beta \exp[-\beta(m-m_0)]}{1 - \exp[-\beta(m_{max}-m_0)]} \tag{7.12}$$

The bounded recurrence law given by Equation 7.10 is shown in Figure 7.19 for the case when it is assumed that the rate of recurrence of earthquakes with a magnitude of m_0 is constant. In the application of this recurrence law, it is important to note that the selection of the upper bound m_{max} is a difficult problem and that the results of a hazard assessment are affected significantly by the value of this upper bound.

7.5.4 Characteristic Earthquake Recurrence Law

Gutenberg–Richter law was derived using seismicity data that included many different seismic sources. Therefore, when the law is applied to individual seismic sources, discrepancies exist between earthquake recurrence intervals obtained with this law and those obtained using geological data. Several other models have been thus proposed over the past several years to accommodate such discrepancies. One of such models is the one developed in 1985 by R. R. Youngs and K. J. Coppersmith based on the concept of *characteristic earthquake*. This concept of a characteristic earthquake may be explained as follows:

Recent paleoseismicity studies seem to indicate that individual faults and fault segments tend to slip approximately the same amount during each earthquake. As such, it has been suggested that individual faults periodically generate earthquakes of the same or similar magnitude (hence the name characteristic earthquake) and that this magnitude is the maximum that earthquakes can attain at a given fault. In general, the recurrence rates attained by dating these characteristic earthquakes have given different occurrence rates than those obtained by an extrapolation of Gutenberg–Richter law.

In an effort to match the recurrence rates based on paleoseismicity studies, Youngs and Coppersmith developed a recurrence relationship considering an exponential distribution for earthquakes with low magnitude and a uniform distribution for earthquakes with magnitudes near the magnitude of the characteristic earthquake. This relationship is shown in Figure 7.20, where it is compared with the bounded Gutenberg–Richter recurrence relationship for the same m_{max}, b value, and slip rate. The use of the characteristic earthquake model rather than the Gutenberg–Richter law leads thus to a significant reduction in the rate of occurrence of moderate-magnitude earthquakes and a modest increase in the rate of events with magnitudes near the magnitude of the characteristic earthquake.

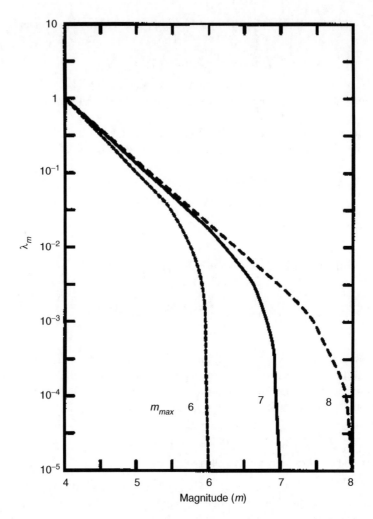

FIGURE 7.19 Bounded Gutenberg–Richter recurrence law for $m_0 = 4$ and $m_{max} = 6, 7,$ or 8. (After Youngs, R.R. and Coppersmith, K.J., *Bulletin of the Seismological Society of America*, 75, 4, 939–964, 1985. © Seismological Society of America.)

7.6 GROUND MOTION INTENSITY IN GIVEN TIME INTERVAL

7.6.1 Introduction

For a given distance to a fault and a given earthquake magnitude, attenuation relationships can be used to estimate the peak ground acceleration, peak ground velocity, or response spectrum ordinates at a site. However, in selecting the design ground motion for the design of a building or any other structure, the question arises as to what is the magnitude that should be used in the design. Should one design for the maximum magnitude in record and accept the risk that an earthquake of a greater magnitude may occur during the lifetime of the structure? Or should one design for the greatest magnitude tectonically possible and design the structure for an earthquake that most likely will occur every one or two thousand years? Obviously, there is no easy answer to this question, but it is clear that the answer should involve a time interval and a measure of chance, that is, a probability of occurrence.

Seismic Hazard Assessment

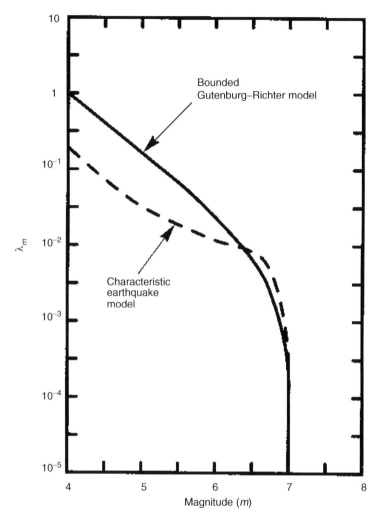

FIGURE 7.20 Recurrence relationships according to bounded Gutenberg–Richter law and Youngs–Coppersmith characteristic earthquake model. (After Youngs, R.R. and Coppersmith, K.J., *Bulletin of the Seismological Society of America*, 75, 4, 939–964, 1985. © Seismological Society of America.)

7.6.2 POISSON PROBABILISTIC MODEL

The occurrence of an earthquake of a given magnitude in a given time interval is usually modeled with a Poisson process. If Δt denote a short time subinterval; $t = n\Delta t$ a given time interval of finite length; and X_t the number of occurrences of an event in time interval t, in a Poisson process it is assumed that

1. An event may occur at any time within the interval t.
2. The occurrence of an event in a subinterval Δt is independent of that in any other subinterval.
3. The probability of occurrence, p, of an event in Δt is proportional to Δt; that is,

$$p = \lambda \, \Delta t \tag{7.13}$$

where λ is a proportionality constant with the same numerical value for all subintervals Δt. In a Poisson process, therefore, as $n \to \infty$ and $\Delta t \to 0$, the probability that the number of occurrences of an event in a time interval t be equal to x is given by

$$P(X_t = x) = \frac{(\lambda t)^x}{x!} e^{-\lambda t} \quad x = 0, 1, 2, \ldots \tag{7.14}$$

and the mean value and the variance of the number of occurrences X_t are equal to

$$E(X_t) = \lambda t \tag{7.15}$$

$$Var(X_t) = \lambda t \tag{7.16}$$

From Equation 7.15, it may be noted that in a Poisson process

1. λt = average number of events in time interval t
2. λ = average number of occurrences of the event per unit time, or mean occurrence rate
3. $1/\lambda = T_R$ = mean return period, that is, the average time (in days, months, or years) at which the event is repeated

Modern seismic hazard evaluations are often described in terms of a mean annual rate of occurrence or a return period. Consequently, the understanding of these two concepts is fundamental in the understanding of seismic hazard assessment methods. The following example will further illustrate their meaning.

EXAMPLE 7.4 MEAN OCCURRENCE RATE PER UNIT TIME AND RETURN PERIOD

If every 20 years in a time interval of 100 years the number of earthquakes of a magnitude >7.0 is as indicated in the Table E7.4, find (a) the mean or average occurrence rate per unit time of earthquakes with a magnitude >7.0 and (b) the corresponding mean or average return period.

Solution

In this case, the occurrence rate of earthquakes with a magnitude >7.0 for the first, second, third, fourth, and fifth 20 year time intervals is 1/20, 0/20, 2/20, 1/20, and 0/20, respectively. Therefore, the average value, or mean recurrence rate, is

$$\lambda = (1/20 + 0/20 + 2/20 + 1/20 + 0/20)/5 = 0.04 \text{ events/year}$$

Alternatively, this same average may be obtained by simply summing up the total number of earthquakes of magnitude >7.0 in the time interval of 100 years and dividing by the number of years in this interval. That is,

$$\lambda = (1 + 0 + 2 + 1 + 0)/100 = 0.04 \text{ events/year}$$

The average return period is the inverse of the mean occurrence rate. Hence,

$$T_R = 1/\lambda = 1/0.04 = 25 \text{ years}$$

TABLE E7.4
Earthquake Data Considered in Example 7.4

Period	Number of Occurrences
1880–1899	1
1900–1919	0
1920–1939	2
1940–1959	1
1960–1979	0

7.6.3 PROBABILITY OF EXCEEDING GIVEN GROUND MOTION INTENSITY IN GIVEN TIME INTERVAL

If the event in a Poisson process is made to signify the occurrence of an earthquake that generates at a site a ground motion intensity greater than y, and if X_t is the number of occurrences of such an event in a time interval t, then according to Equation 7.14 the probability that the ground motion intensity at the site, Y, be greater than a given value y (i.e., the probability that the event occurs at least once in a time interval t) may be expressed as

$$P_t(Y > y) = P(X_t \geq 1) = 1 - P(X_t = 0) = 1 - e^{-\lambda t} \qquad (7.17)$$

where all symbols are as previously defined. The following example will illustrate the use of this equation to calculate such a probability.

EXAMPLE 7.5 PROBABILITY OF EXCEEDING A GIVEN PEAK GROUND ACCELERATION

From the analysis of statistical data on the occurrence of earthquakes near a construction site over the past 100 years, it is found that a peak ground acceleration of 0.01g is exceeded five times in those 100 years. What is the probability that a peak ground acceleration of 0.01g be exceeded in a time interval of 45 years?

Solution

From the given statistical data, the average annual recurrence of peak ground accelerations of 0.01g or greater is equal to

$$\lambda = \frac{5}{100} = 0.05 \text{ events/year}$$

which is equivalent to a return period of 1/0.05=20 years. Then, in view of Equation 7.17, the probability of exceeding a peak ground acceleration of 0.01g in 45 years is

$$P_{45}(Y > 0.01g) = P(X_{45} \geq 1) = 1 - e^{-0.05(45)} = 0.895$$

It should be noted that the assumption made in a Poisson process that the occurrence of an event at a given time is independent of the occurrence of events at any other time contradicts the physical process by which earthquakes are generated. That is, according to the elastic rebound theory, earthquakes occur as a result of the release of the strain energy that is built up over an extended time interval. In reality, therefore, the occurrence of a large earthquake substantially reduces the probability occur at another large earthquake that the same source shortly thereafter. Hence, this probability should be a function of the time of occurrence, size, and location of preceding events.

In spite of this inadequacy, it is generally accepted that a Poisson process is still useful to characterize the occurrence of earthquakes in regions where the seismic hazard is not dominated by a single seismic source. Besides, there is presently insufficient data to justify the use of models that account for prior seismicity. Because of these reasons and its simplicity, the Poisson process is still used widely in modern seismic hazard assessments.

7.6.4 EXCEEDANCE PROBABILITY CURVES

From the foregoing discussion, it may be seen that the probability of exceeding a given ground motion intensity at a given region depends on the statistical data that reflect the seismicity of the region and the time interval of interest. In practice, thus, the ground motion intensity for the design of a structure is often selected on the basis of (a) a predetermined time interval—sometimes called the *exposure time* and usually considered equal to the life expectancy of the structure, (b) statistical data from the area where the structure is to be built, and (c) an acceptable probability of exceeding

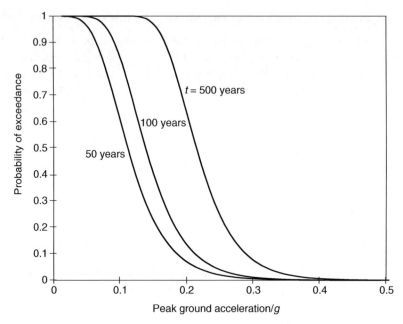

FIGURE 7.21 Typical exceedance probability curves.

the selected design ground motion intensity. This selection process is facilitated by the use of exceedance probability curves such as those shown in Figure 7.21. For, once these curves are available, the design ground motion intensity may be easily obtained for any value of an acceptable probability of exceedance and life expectancy of the structure. Curves of this type are sometimes referred to as *seismic hazard curves*.

It should be considered when using these curves that the acceptable probability of exceedance is always selected on the basis of the importance of the structure under consideration. For example, the intensity levels recommended in many building codes for ordinary structures are based on a 10% probability of exceedance in 50 years. In contrast, critical structures such as dams and nuclear power plants are designed for a probability of exceedance of the order of 10^{-4}.

EXAMPLE 7.6   USE OF EXCEEDANCE PROBABILITY CURVES

Using the curves in Figure 7.21, determine the design peak ground acceleration for the site for which these curves were constructed if the acceptable probability of exceeding the value of the selected peak ground acceleration in 50 years is 30%.

Solution

From the curve for $t = 50$ years and an exceedance probability of 0.3, one finds
 Design peak ground acceleration = $0.15g$.

7.6.5 FREQUENCY-INTENSITY CURVES

As the determination of the probability of exceeding a given ground motion intensity at a specific site requires the calculation of the corresponding mean annual frequency of occurrence, it is also possible to select a design ground motion intensity on the basis of curves that directly relate ground motion intensity with annual frequency of occurrence or return period. These curves,

Seismic Hazard Assessment

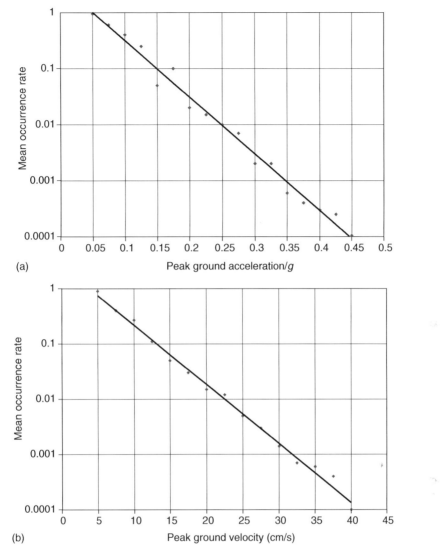

FIGURE 7.22 Typical frequency-intensity curves for (a) peak ground acceleration and (b) peak ground velocity.

known as frequency-intensity curves, may be generated directly using the available statistical data for the site or region of interest, as it will be shown in Section 7.7, or indirectly on the basis of a probabilistic formulation, as it will be shown in Section 7.8. Examples of frequency-intensity curves for peak ground acceleration and peak ground velocity are depicted in Figure 7.22.

When frequency-intensity curves are used to select the design ground motion intensity, the selection is made on the basis of an acceptable annual occurrence rate or a return period. If desired, the probability of exceeding the selected ground motion level in a time interval t is computed by substituting into Equation 7.17 the value of the mean annual occurrence rate λ corresponding to the selected ground motion level and the value of the time interval t. Common values of the return period T_R in design are 2500 years for ordinary structures and 10,000 years for critical or very important ones.

7.7 SEMIPROBABILISTIC SEISMIC HAZARD EVALUATION

Earthquakes, as indicated several times before, are random events. As such, there exist large uncertainties associated with the prediction of parameters such as the size (magnitude), location, and time of occurrence of future earthquakes. These uncertainties, and the need to account for them properly, make mandatory the use of probabilistic methods in any seismic hazard assessment. In some cases, however, the uncertainty associated with the prediction of some of the variables that influence the ground motion intensity level at a given site is small in comparison with that of some others. Hence, for simplification purposes, the uncertainty in the prediction of such variables may be neglected, and the variables can be treated as deterministic. In some seismic regions, for example, the location of the seismic sources is relatively well known, so that the source-to-site distance may be considered as a deterministic variable. Procedures in which some variables are considered deterministic while some others are considered random are herein designated as semiprobabilistic. Naturally, there are several ways a semiprobabilistic seismic hazard evaluation can be carried out. Therefore, this section will describe only a version of it just to illustrate the concept of such an approach.

As with any other type of seismic hazard assessment, the objective of a semiprobabilistic seismic hazard evaluation is the estimation of a ground motion parameter such as peak ground acceleration, peak ground velocity, or a response spectrum ordinate to characterize the ground motions expected at a given site resulting from the earthquakes that can occur within a given time interval, in the vicinity of the site. Similarly, it involves (a) the identification and characterization of all earthquake sources capable of producing significant ground motions at the site, (b) the establishment using seismological data of the magnitude and the frequency with which earthquakes may be generated at these sources (i.e., establishment of magnitude-recurrence relationships for each of the sources), (c) the determination of the distance from each source to the site, (d) the use of attenuation relationships to establish a correlation between ground motion intensity and earthquake magnitude, source characteristics, distance from source to site, and soil conditions, (e) the construction of frequency-intensity curves using the magnitude-recurrence and attenuation relationships, and (f) the selection of the ground motion intensity at the site on the basis of a specified return period or probability of exceedance in a given time interval. However, in the semiprobabilistic approach, it is assumed that the source-to-site distances are deterministic variables and that ground motion intensity, as given by the attenuation relationships, is uniquely defined by its mean value or its mean and one standard deviation; that is, the uncertainty in the determination of ground motion intensity as a function of magnitude, distance, etc., is either neglected or considered equal to one standard deviation. Thus, the only sources of randomness accounted for in this approach are the earthquake magnitude and the time of earthquake occurrence.

In another variation from the formal probabilistic approach that will be discussed in Section 7.8, the frequency-intensity curves are obtained directly from the available seismological data instead of using any of the magnitude-recurrence relationships reviewed in Section 7.5. That is, the frequency-intensity curves are generated by (a) calculating using a valid attenuation law the values of the selected ground motion parameter that corresponds to each of the earthquakes comprised in the seismological data, (b) determining using this information the maximum value attained by the ground motion parameter each year during the period for which there exists seismological data, (c) computing the average annual frequency with which such maximum is exceeded (i.e., the number of years the ground motion parameter exceeds a given value over the total number of years in the time interval under consideration), (d) plotting the yearly maxima against the corresponding frequencies of occurrence, and (e) fitting a straight line through the points so obtained. This approach is somewhat equivalent to obtaining first magnitude-frequency relationships on the basis of the available statistical data and then using attenuation equations to find frequency-ground motion intensity relationships. However, it offers the advantage that it is not necessary to obtain a magnitude-frequency fitting for each of the seismic sources considered in the analysis.

Seismic Hazard Assessment

In many cases, the available historical seismological data set is not large enough or does not include earthquakes of large magnitudes (say, >7.0). In those cases, therefore, the procedure described in the foregoing paragraph may not lead to realistic frequency-intensity curves; it may significantly underestimate or overestimate the frequency of occurrence of large earthquakes. In such cases, it is appropriate to fit a prescribed probability distribution to the frequency-ground motion intensity points obtained with the seismological data instead of a simple straight line as previously suggested. As a probability distribution characterizes a phenomenon in physical terms, the frequency-intensity curves obtained by fitting a probability distribution will be more realistic than those obtained on the basis of seismological data alone. A Type II extreme-value probability distribution may be used for such a purpose as in the development of some probabilistic methods it has been found that the probability of exceeding a given ground motion intensity is described well by a Type II extreme-value probability distribution. A Type II extreme-value probability distribution is given by

$$F_Y(y) = P[Y \leq y] = \exp\left\{-\left(\frac{y}{\sigma}\right)^{-\gamma}\right\} \tag{7.18}$$

where σ and γ are the parameters of the distribution.

A procedure to fit given statistical data to a Type II extreme-value distribution and obtain the parameters σ and γ of the distribution may be established as follows: First, define a transformation variable s (called standard variable for Type II extreme-value probability distribution) as

$$s = -\ln\left(\frac{y}{\sigma}\right)^{-\gamma} \tag{7.19}$$

Then, by means of this variable, transform the original equation for a Type II probability distribution into the equation

$$F_Y(y) = \exp\{-e^{-s}\} \tag{7.20}$$

from which, after solving for s, one obtains

$$s = -\ln(-\ln F_Y) \tag{7.21}$$

Finally, rewrite Equation 7.19 as

$$\ln y = \ln \sigma + \frac{1}{\gamma} s \tag{7.22}$$

Observe from Equation 7.22 that $\ln y$ and s are linearly related. Hence, the parameters σ and γ of the distribution may be found by fitting by least squares a straight line to a series of $(s, \ln y)$ points defined using available statistical data and Equation 7.21. Observe also that according to Equation 7.22, γ is equal to the inverse of the slope of the fitted straight line and $\ln \sigma$ is equal to the intercept with the $\ln y$-axis; that is, σ is equal to the value of y when $s = 0$, or what is tantamount, to the value of y when $F_y = e^{-1} = 0.368$ (see Equation 7.20).

The procedure to fit peak ground acceleration data to a Type II extreme-value probability distribution according to the approach just described is summarized in Box 7.1, where it is assumed that $F_Y(y)$ is approximately equal to the relative frequency with which a value y of the ground motion parameter Y occurs in the given statistical data. Note that although the procedure in Box 7.1 is described for the particular case of peak ground acceleration, the procedure is equally valid to fit a Type II extreme-value distribution to any other ground motion parameter. Note also that once the parameters of the probability distribution are determined, the corresponding frequency-intensity curves may be generated directly using this distribution as the mean annual frequency of exceeding a particular value y of the ground motion parameter Y is given by

$$\lambda_y = 1 - F_Y(y) = 1 - P[Y \leq y] = 1 - \exp\left\{-\left(\frac{y}{\sigma}\right)^{-\gamma}\right\} \qquad (7.23)$$

where, as defined before, $F_Y(y)$ denotes the cumulative probability distribution of Y, and γ and σ are the parameters of a Type II extreme-value distribution.

Box 7.1   Procedure to Fit a Type II Extreme-Value Probability Distribution to Given Peak Ground Acceleration Data

1. Tabulate given peak ground acceleration data in ascending order
2. Estimate cumulative probabilities according to

$$F_{A_{max}}(\hat{a}) = P(A_{max} \leq \hat{a}) = \frac{j}{N+1}$$

   where $\hat{a}$ is data point in acceleration table, j numerical order of $\hat{a}$ in table, N total number of data points in table

3. Calculate the values of the standard variable s of the distribution using the following equation:

$$s = -\ln\left[-\ln\left(\frac{j}{N+1}\right)\right]$$

4. Using the least squares technique, fit a straight line between the values of $\ln a$ and s, where $\ln a$ is the dependent variable and s is the independent one
5. With the equation of the obtained straight line expressed as

$$\ln a = b + ms$$

   determine the parameters of the distribution as $\gamma = 1/m$ (i.e., the inverse of the straight line slope) and σ = value of a corresponding to $s = 0$ (i.e., $\sigma = e^b$)
6. Substituting the found values of γ and σ, obtain the equation of the distribution as

$$F_{A_{max}} = \exp\left\{-\left(\frac{a}{\sigma}\right)^{-\gamma}\right\}$$

The procedure to generate frequency-intensity curves for a given site on the basis of the semi-probabilistic approach presented earlier is summarized in Box 7.2.

Box 7.2   Procedure to Generate Frequency-Intensity Curves for a Given Site Using Semiprobabilistic Approach

1. Estimate influence area where the occurrence of earthquakes may significantly affect site (normally defined by a radius of 300–400 km).

Seismic Hazard Assessment

2. Collect seismological, geological, and statistical data about historical earthquakes in influence area. For each earthquake in the data set, define
   a. Magnitude
   b. Source characteristics
   c. Source-to-site distance.
3. Use an attenuation law valid for the area under consideration to calculate for each earthquake in the data set the corresponding value of the ground motion intensity (e.g., peak ground acceleration, peak ground velocity, or response spectrum ordinate) that would be observed at the given site.
4. Identify for each year covered by the data set the attained maximum value, y_j, of the ground motion intensity at the site.
5. Determine the cumulative frequencies of occurrence of the maximum values y_j, obtained in Step 4. To calculate these frequencies, tabulate first these values in an ascending order. Then, use the equation

$$p_j = \frac{j}{N+1}$$

where p_j represents the aforementioned cumulative frequency corresponding to the jth value in the list, j is the number that defines the order of this value in the list, and N the total number of years covered by the data set.
6. Use the procedure described in Box 7.1 to find the parameters of the Type II extreme-value probability distribution that best fits the data points (p_j, y_j).
7. Generate the desired frequency-intensity curves using the equation

$$\lambda_y = 1 - \exp\left\{-\left(\frac{y}{\sigma}\right)^{-\gamma}\right\}$$

where λ_y denotes the mean annual frequency of exceeding a particular value y of the ground motion intensity at the given site (mean annual rate of occurrence) and γ and σ are the parameters of the Type II extreme-value distribution obtained in Step 6.

EXAMPLE 7.7 ANNUAL PROBABILITY OF EXCEEDANCE USING SEMIPROBABILISTIC APPROACH

Using the seismological data recorded in a region around a given site between 1966 and 1989 and an attenuation relationship valid for this region, it is found that the peak ground accelerations that would be observed at the site during that time interval are those listed in Table E7.7a. Using these data and the semiprobabilistic approach described in Section 7.7, find (1) an expression to compute the mean annual frequency of exceeding a given peak ground acceleration value at the site and (2) the peak ground acceleration corresponding to a return period of 1000 years.

Solution

1. According to the procedure established in Box 7.2, the first step in the solution to this problem is the tabulation in an ascending order of the yearly maximum values of the peak ground acceleration in each of the 24 years covered by the statistical data and the determination of the cumulative frequencies of occurrence of such maximum values. After this, it is necessary to define for each of the 24 years a point $(s, \ln a_j)$, where a_j is the maximum value of the peak ground acceleration during the jth year, and s_j is the corresponding value of the standard variable for a Type II extreme-value probability distribution, calculated using Equation 7.21. Proceeding accordingly, one obtains the yearly maximum value of the peak ground acceleration, the cumulative frequency of occurrence,

TABLE E7.7A
Peak Ground Acceleration Data in Example 7.7

Year	Peak Ground Acceleration/g	Year	Peak Ground Acceleration/g	Year	Peak Ground Acceleration/g
1966	0.504	1974	0.405	1982	0.000
1967	0.423	1975	0.441	1983	0.585
1968	0.000	1976	0.504	1984	0.423
1969	0.387	1977	0.468	1985	0.486
1970	0.531	1977	0.217	1986	0.423
1971	0.432	1978	0.387	1987	0.000
1972	0.423	1979	0.441	1988	0.540
1972	0.213	1980	0.486	1989	0.118
1973	0.585	1981	0.531	1989	0.468

TABLE E7.7B
Peak Ground Accelerations Arranged in Ascending Order (Expressed as a Fraction of g) and Associated Occurrence Frequencies

j	a_j	$\ln a_j$	$j/25$	s_j	j	a_j	$\ln a_j$	$j/25$	s_j
1	0.	—	0.04	−1.169	13	0.441	−0.819	0.52	0.425
2	0.	—	0.08	−0.927	14	0.468	−0.759	0.56	0.545
3	0.	—	0.12	−0.752	15	0.468	−0.759	0.60	0.672
4	0.387	−0.949	0.16	−0.606	16	0.486	−0.722	0.64	0.807
5	0.387	−0.947	0.20	−0.476	17	0.486	−0.722	0.68	0.953
6	0.405	−0.904	0.24	−0.356	18	0.504	−0.685	0.72	1.113
7	0.423	−0.860	0.28	−0.241	19	0.504	−0.685	0.76	1.293
8	0.423	−0.860	0.32	−0.131	20	0.531	−0.633	0.80	1.500
9	0.423	−0.860	0.36	−0.021	21	0.531	−0.633	0.84	1.747
10	0.423	−0.860	0.40	0.087	22	0.540	−0.616	0.88	2.057
11	0.432	−0.839	0.44	0.197	23	0.585	−0.536	0.92	2.484
12	0.441	−0.819	0.48	0.309	24	0.585	−0.536	0.96	3.199

and the (s, $\ln a_j$) points for each of the 24 years under consideration listed in Table E7.7b. Next, it is necessary to find by least squares the equation of the straight line that best fits these 24 (s, $\ln a_j$) points. With the help of a spreadsheet program, it is found that the slope of this line is equal to 0.120 and that its intercept with the ln y-axis is equal to −0.851. In this case, therefore, γ and σ, the parameters of the Type II extreme-value distribution that best fits the given data, result as

$$\gamma = 1/0.120 = 8.33$$

$$\sigma = e^{-0.851} = 0.427$$

According to Equation 7.23, the mean annual frequency of exceeding a given value of the peak ground acceleration at the site may be, thus, written as

$$\lambda_a = 1 - P[A_H \leq a] = 1 - \exp\left\{-\left(\frac{a}{0.427}\right)^{-8.33}\right\}$$

where a represents a peak ground acceleration value expressed as a fraction of g, and λ_a the probability of exceeding this peak ground acceleration in 1 year.

2. As return period is the inverse of the annual frequency of exceedance, then for a return period of 1000 years one has that $\lambda_a = 1/1000 = 0.001$. By substitution of this value into the foregoing equation, the peak ground acceleration at the site corresponding to a return period of 1000 years is thus equal to

$$a = 0.427[-\ln(1-0.001)]^{-\frac{1}{8.33}} = 0.978g$$

7.8 PROBABILISTIC SEISMIC HAZARD EVALUATION

7.8.1 INTRODUCTION

The determination of the seismic hazard at a given site using the procedure described earlier requires a relatively precise characterization of the relevant seismic sources and large amounts of local statistical data which in many cases may not be available. In such cases, therefore, the seismic hazard evaluation at a site may be more appropriately carried out using whatever scarce seismological data are available together with probabilistic models that conform to physical reasoning. These probabilistic models provide a rational framework to identify, quantify, and combine the uncertainties involved in the definition of the size, location, source characteristics, and rate of occurrence of the earthquakes expected in a given seismic region as well as those involved in the description of the variation of ground motion characteristics with these parameters (e.g., attenuation relationships).

A probabilistic seismic hazard assessment involves obtaining through a formal mathematical calculation the level of a ground motion parameter that has a selected probability of being exceeded during a specified time interval. It normally includes the following steps:

1. The adoption of probability distributions to describe
   a. The number of earthquakes in a given time interval
   b. The distance to nearby faults
   c. Ground motion intensity given earthquake magnitude and distance to source
2. The selection of an attenuation relationship to correlate magnitude, source-to-site distance, and source characteristics with ground motion intensity at site
3. The integration of the adopted probability distributions and the selected attenuation equation to produce an explicit expression for the probability of exceeding a specified ground motion intensity at a given site or region during a selected time interval

7.8.2 PROBABILITY DISTRIBUTION OF SOURCE-TO-SITE DISTANCE

For the purpose of a seismic hazard assessment, a seismic source may be idealized as a point source if the zone where earthquakes originate is small compared with the source-to-site distance; as a linear source if earthquakes occur at well-defined fault planes that are long but shallow (so that variation in fault depth has little influence in source-to-site distance); or as a volumetric source if earthquakes occur at unidentified fault planes zones, or where faulting is so extensive as to make it difficult to procure a distinction between individual faults (see Figure 7.23). For the purpose of

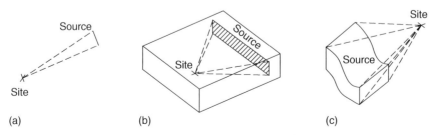

FIGURE 7.23 Examples of (a) point, (b) linear, and (c) volumetric seismic sources.

illustrating the procedures used in a probabilistic hazard assessment, only a line source will be considered here.

As previously seen, the attenuation relationships used to predict ground motion parameters are function of some measure of a source-to-site distance. Earthquakes, however, may occur anywhere within a fault zone. For a fault zone with finite dimensions, such a source-to-site distance has thus an inherent variability and this variability is for the most part random. Hence, the variability in a source-to-site distance needs to be described in terms of a probability density function.

Earthquakes may be considered equally likely to occur at any location within a fault zone. Therefore, it may be assumed that the occurrence of an earthquake within a fault zone has a uniform probability distribution. A uniform distribution of the occurrence of earthquakes within a fault zone does not mean, however, that the distance from a site to a seismic source is also uniformly distributed. The probability distribution for this distance may be derived instead as follows:

Consider the linear source shown in Figure 7.24. The probability that an earthquake occurs on a small segment of the fault between $L = l$ and $L = l + dl$ is the same as the probability that the distance from site to source be between $R = r$ and $R = r + dr$, that is,

$$f_L(l)dl = f_R(r)dr \qquad (7.24)$$

where $f_L(l)$ and $f_R(r)$, respectively, denote the probability density functions for the variables L and R. Hence,

$$f_R(r) = f_L(l)\frac{dl}{dr} \qquad (7.25)$$

Thus, as under the assumption of a uniform probability distribution for L one has that $f_L(l) = 1/L_f$, and as $l^2 = r^2 - r_{min}^2$ (see Figure 7.24), the probability density function of the source-to-site distance R results as

$$f_R(r) = \frac{r}{L_f\sqrt{r^2 - r_{min}^2}} \qquad (7.26)$$

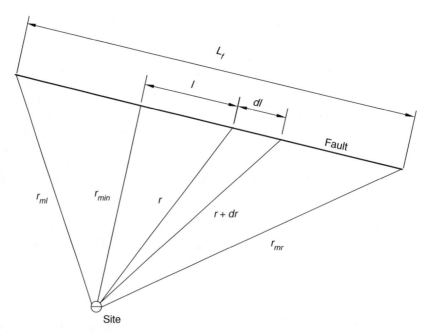

FIGURE 7.24 Variation of distance from site to linear seismic source.

where L_f is the fault length and r_{min} represents the shortest distance from the site to the fault.

It may be observed from Figure 7.24 that it is possible to have in some cases the same value of r for two different points along the fault. This is something that it is important to consider when Equation 7.26 is used to determine the corresponding probability distribution. For example, for the particular case when $r_{ml} \leq r_{mr}$ (refer to Figure 7.24), the cumulative probability distribution of r should be written as

$$P[R \leq r] = \frac{2}{L_f} \int_{r_{min}}^{r} \frac{r}{\sqrt{r^2 - r_{min}^2}} dr \quad \text{if } r \leq r_{ml} \qquad (7.27)$$

or as

$$P[R \leq r] = \frac{2}{L_f} \int_{r_{min}}^{r_{ml}} \frac{r}{\sqrt{r^2 - r_{min}^2}} dr + \frac{1}{L_f} \int_{r_{ml}}^{r} \frac{r}{\sqrt{r^2 - r_{min}^2}} dr \quad \text{if } r > r_{ml} \qquad (7.28)$$

7.8.3 Probability Distribution of Earthquake Magnitude

In general, a seismic source may randomly produce earthquakes of different magnitudes up to a maximum, with small earthquakes occurring more frequently than large earthquakes. Thus, in a probabilistic hazard evaluation, the magnitude of the earthquakes that may be generated at a source in a given time interval is also described in terms of a probability distribution. As seen in Section 7.5, this probability distribution may be derived directly from the magnitude-recurrence relationship that may be developed for a given seismic source. As several sources may affect the seismic hazard at a site, and as the magnitude-recurrence relationships that describe these sources are generally different from one another, it is necessary to consider in the derivation of these probability distributions that each of the possible sources may have a different magnitude-recurrence relationship. For the purpose of illustrating the formulation of a probabilistic hazard assessment, it will be assumed in the discussion that follows that the probability distribution of earthquake magnitude at a source is given by that derived in Section 7.5.3 from the bounded Gutenberg–Richter magnitude-recurrence relationship, that is, Equation 7.9.

7.8.4 Probability of Exceeding Specified Value of Ground Motion Parameter

As seen in Section 7.4, the ground motion intensity at a given site is usually determined by means of an attenuation relationship in terms of earthquake magnitude, source-to-site distance, source characteristics, and soil conditions. In most cases, these relationships are obtained by performing a regression analysis to data from a limited set of strong ground motion records. As such, these relationships are far from being accurate and invariably show a considerable amount of scatter. In a probabilistic hazard assessment, the uncertainty associated with the use of these relationships is also accounted for in the form of a probability distribution.

Most attenuation relationships give the mean of the ground motion parameter they intend to predict as well as a measure of the scatter in the data, usually in terms of the standard deviation of the logarithm of the predicted parameter. As most ground motion parameters may be assumed to be lognormally distributed (the logarithm of the parameter is normally distributed), then the probability distribution of the ground motion parameter may be easily obtained once its mean and the standard deviation of its logarithm are known. For example, the probability that a particular ground motion parameter Y exceeds a certain value y at a site at a distance r from a seismic source when an earthquake of magnitude M occurs at this source may be expressed as

$$P[Y > y \mid m, r] = 1 - F_Y(y) = 1 - \frac{1}{\sqrt{2\pi}} \int_{-\infty}^{z} e^{-s^2/2} ds = \frac{1}{\sqrt{2\pi}} \int_{-\infty}^{-z} e^{-s^2/2} ds \qquad (7.29)$$

where $P[Y > y|m,r]$ = probability that Y exceeds y given m and r; $F_y(y)$ = cumulative probability function of Y, assumed here to be lognormally distributed; s = dummy variable, and

$$z = \frac{\ln y - E[\ln y]}{\sigma_{\ln y}} \qquad (7.30)$$

in which $E[\ln y]$ and $\sigma_{\ln y}$, respectively, denote the mean and standard deviation of the ground motion parameter Y as determined with an attenuation relationship for a given distance r and earthquake magnitude m.

It should be noted that because of the unbounded characteristics of the lognormal distribution, the adoption of this distribution to describe the variability of a ground motion parameter may signify nonzero probabilities to unrealistic values of the parameter. For this reason, some investigators prefer the use of probability distributions that impose an upper bound to the value that the ground motion parameter may take. Nevertheless, for the sake of simplicity it will be assumed in the discussion that follows that ground motion parameters may be described adequately with a lognormal distribution.

7.8.5 Annual Probability of Exceeding Specified Ground Motion Intensity Level

The results of a probabilistic hazard assessment may be expressed in different ways. However, a convenient and common way to do it is by means of the frequency-intensity curves introduced in Section 7.6.5. This convenience is owed to the fact that by means of these curves one can readily select the design level of ground motion intensity on the basis of a return period, which is a good indicator of the risk involved in the use of the selected intensity level. It is also owed to the fact that, as previously seen, there is a direct association between these curves and the probability of exceeding a given ground motion intensity in a given time interval.

The generation of frequency-intensity curves for a given site on the basis of a probabilistic hazard assessment requires the estimation of the annual probability of exceeding a specified level of ground motion intensity as a result of the earthquakes that may be generated at all the possible seismic sources near the site. In turn, this annual probability of exceedance may be obtained by simply adding the individual annual probabilities of exceedance corresponding to each of the considered seismic sources. That is, if the site of interest is in a region where there are N_S potential seismic sources, and if λ_{iy} represents the annual probability of exceeding a particular value y of a ground motion parameter Y as a result of the earthquakes that may be generated at the ith seismic source, then the annual probability of exceeding that particular value y of the ground motion parameter Y as a result of the earthquakes that may be generated at the N_S seismic sources is given by

$$\lambda_y = \sum_{i=1}^{N_S} \lambda_{iy} \qquad (7.31)$$

It may be considered, however, that the probability of exceeding a particular value y of the ground motion parameter Y due to a single seismic source, herein denoted as $P[Y > y]$, is the same as the relative occurrence frequency of earthquakes whose magnitude is larger than the magnitude m that is necessary to produce a level y of the ground motion parameter Y at the site of interest. That is,

$$P[Y > y] = \frac{\lambda_m}{\lambda_{m_0}} \qquad (7.32)$$

where λ_m is given by Equation 7.7 and defined in Section 7.5.3, $\lambda_{m_0} = \exp(\alpha - \beta_{m_0})$, and the magnitude m_0 associated with λ_{m_0} is the lower-threshold magnitude of the earthquakes that do not significantly

Seismic Hazard Assessment

contribute to the seismic hazard at the site. As a result, the annual probability of exceedance due to an individual seismic source may be expressed as

$$\lambda_{iy} = \lambda_{m_0} P[Y > y] \tag{7.33}$$

It may also be considered that for a given earthquake occurrence at the ith seismic source, the probability that the ground motion parameter Y exceed the particular value y may be written in terms of conditional probabilities as

$$P[Y > y] = P[Y > y \mid X_1, X_2, \ldots, X_n] P[X_1 = x_1; X_2 = x_2; \ldots; X_n = x_n] \tag{7.34}$$

where $X_1, X_2, \ldots, X_n$, denote the random variables, such as earthquake magnitude and source-to-site distance, that may have an influence on the ground motion intensity at the site. Furthermore, if it is assumed that the random variables $X_1, X_2, \ldots, X_n$ are statistically independent, such a probability may be written alternatively as

$$P[Y > y] = P[Y > y \mid X_1, X_2, \ldots, X_n] P[X_1 = x_1] P[X_2 = x_2] \ldots P[X_n = x_n] \tag{7.35}$$

or as

$$P[Y > y] = \iint \ldots \int P[Y > y \mid X_1, X_2, \ldots, X_n] f_{X_1}(x_1) f_{X_2}(x_2) \ldots f_{X_n}(x_n) dx_1 dx_2 \ldots dx_n \tag{7.36}$$

if the probabilities $P[X_i = x_i]$ in Equation 7.35 are expressed in terms of their respective probability density functions.

Thus, if it is assumed that the earthquake magnitude m and the source-to-site distance r are the only random variables (i.e., the uncertainty in these two variables is significantly larger than the uncertainty in the others), the probability that the ground motion parameter Y exceed the particular value y when an earthquake occurs at the ith seismic source is given by

$$P[Y > y] = \iint P[Y > y \mid m, r] f_M(m) f_R(r) dm dr \tag{7.37}$$

which after substitution of Equations 7.9, 7.26, and 7.29 may also be put into the form

$$P[Y > y] = \frac{2\beta}{\sqrt{2\pi} L_f} \int_{r=r_{min}}^{r_{max}} \int_{m=m_0}^{m_{max}} \int_{s=-\infty}^{-z} \frac{r e^{-\left[\frac{s^2}{2} + \beta(m-m_0)\right]}}{\sqrt{r^2 - r_{min}^2}} ds\, dm\, dr \tag{7.38}$$

where

$$z = \frac{\ln y - E[\ln y]}{\sigma_{\ln y}} \tag{7.39}$$

where all other symbols are as defined before, and where for simplicity it has been assumed that r_{min} corresponds to the geometric center of the fault, that is, $r_{ml} = r_{mr} = r_{max}$ (see Figure 7.24). The expected value and the standard deviation of $\ln y$ that appear in Equation 7.39 are determined directly from the attenuation relationships selected for the analysis for given values of the magnitude m and the source-to-site distance r. The triple integral in Equation 7.38 is normally evaluated by numerical integration given its complexity. Moreover, this numerical integration is usually carried out using one of the available mathematical software packages to facilitate its evaluation.

Finally, the annual probability of exceeding at a given site a specified level of ground motion intensity as a result of the earthquakes that may be generated at multiple seismic sources near the site may be calculated by putting together Equations 7.31, 7.33, and 7.38. The corresponding frequency-intensity curves may be generated by simply computing the annual probability of exceedance for a range of ground motion intensities. The procedure to obtain these frequency-intensity curves is summarized in Box 7.3 and the following example will illustrate its application.

Box 7.3 Procedure to Generate Frequency-Intensity Curves for a Given Site Using Probabilistic Approach

1. Estimate influence area where the occurrence of earthquakes may significantly affect site (normally defined by a radius of 300–400 km).
2. Collect seismological, geological, and statistical data about historical earthquakes in the influence area and identify potential sources of seismic activity.
3. Establish the magnitude-recurrence relationship of each potential seismic source according to Gutenberg–Richter law (i.e., Equation 7.7).
4. Determine for each potential seismic source the values of the constants α and β that define their magnitude-recurrence relationships, their lower-threshold magnitudes m_0, and their likely maximum magnitudes M_{max}. Compute, in addition, the corresponding mean annual number of earthquakes with magnitude m_0 or greater according to

$$\lambda_{m_0} = \exp(\alpha - \beta m_0)$$

5. For each source, determine the fault length, L_f, and the minimum and maximum distances to the site, that is, r_{min} and r_{max} (see Figure 7.24).
6. Select for each of the seismic sources a valid attenuation law to calculate for a given magnitude m and source-to-site distance r the mean value and the standard deviation of Y, the ground motion parameter of interest (e.g., peak ground acceleration, peak ground velocity, or response spectrum ordinate).
7. Select a value y of the ground motion parameter Y and, for each of the potential seismic sources, calculate using numerical integration and according to Equation 7.38, the probability that the selected value y of the ground motion parameter Y be exceeded.
8. Determine for each source and according to Equation 7.33 the mean annual probability that the ground motion parameter Y exceed the selected y value.
9. Considering the combined effect of all seismic sources, determine using the values obtained in Step 8 and according to Equation 7.31, the mean annual probability λ_y that the ground motion parameter under consideration exceeds the selected y value.
10. Repeat Steps 7–9 for different values of y.
11. Generate the desired frequency-intensity curve by plotting the obtained values of λ_y against the corresponding values of y.

EXAMPLE 7.8 PROBABILITY OF EXCEEDING GIVEN PEAK GROUND ACCELERATION BY PROBABILISTIC APPROACH

The site shown in Figure E7.8 is located in the western United States and, as shown in this figure, two active faults are near the site. The closest distance from the site to the surface projection of Fault 1 is 10 km and to the surface projection of Fault 2 is 20 km. The maximum distances are 18 and 31 km, respectively. The estimated fault lengths are 30 km for Fault 1 and 65 km for Fault 2. From previous investigations of the area seismicity it has been determined that the maximum magnitude of the earthquakes expected at Fault 1 is 7.5 and that of the earthquakes expected at Fault 2 is 8.5. Similarly, it has been determined that the a and b constants that define Gutenberg–Richter law are 2.25 and 0.75 for Fault 1 and

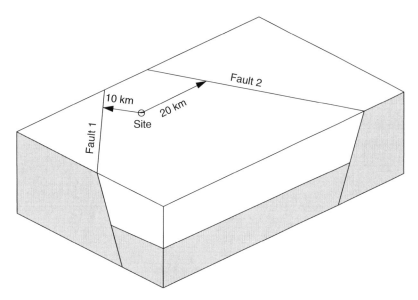

FIGURE E7.8 Location of seismic sources with respect to site in Example 7.8.

3.30 and 0.88 for Fault 2. Find the probability of exceeding a peak ground acceleration of 0.30g at the site in 100 years using the probabilistic approach described in Section 7.8 and considering a lower-threshold magnitude of 5.0 for both faults. Assume an average shear wave velocity of 1396 m/s for the soil deposits at the site. Assume also that the closest distances to the faults correspond to their geometric centers.

Solution

The solution to this problem may be obtained by closely following the procedure outlined in Box 7.3. Accordingly, one may start by calculating for each of the two faults near the site of the mean annual frequency of earthquakes of magnitude equal or greater than m_0, the lower-threshold magnitude, which in this case is equal to 5.0. As seen in Section 7.5.3 this annual frequency is given by

$$\lambda_{m_0} = \exp(\alpha - \beta m_0)$$

where $\alpha = 3.303a$ and $\beta = 3.303b$, in which a and b are the constants in the Gutenberg–Richter law. In consequence, one has that

$$\lambda_{m_0} = \exp[2.303(2.25) - 2.303(0.75)(5.0)] = 0.032 \text{ events/year} \quad \text{for Fault 1}$$

$$\lambda_{m_0} = \exp[2.303(3.30) - 2.303(0.88)(5.0)] = 0.079 \text{ events/year} \quad \text{for Fault 2}$$

One then may proceed with the development of an equation to express the standard variable z that appears in Equation 7.38 explicitly in terms of earthquake magnitude and fault-to-site distance. For this purpose, one can use the attenuation equations proposed by Boore, Joyner, and Fumal, which, for peak ground acceleration and considering that in this case $V_s = 1396$ m/s, may be written as

$$E[\ln A_H] = -0.242 + 0.527(M_w - 6) - 0.778 \ln\sqrt{R^2 + 5.57^2}$$

$$\sigma_{\ln A_H} = 0.520$$

After the substitution of these two expressions into Equation 7.39, the standard variable z may therefore be expressed in terms of magnitude and source-to-site distance as

$$z = \frac{\ln A_H - [-0.242 + 0.527(M_w - 6) - 0.778 \ln\sqrt{R^2 + 5.57^2}]}{0.520}$$

or, after considering that for the problem under consideration $A_H = 0.30g$, as

$$z = -1.850 - 1.013(m-6) + 1.496\ln\sqrt{r^2 + 31.025}$$

where, for consistency with the form of Equation 7.38, M_w has been substituted by m and R by r.

Now, the probability that the peak ground acceleration at the site exceed $0.30g$ will be computed individually for each fault. To this end, Equation 7.38 will be used in conjunction with the above expression for z, establishing the integration limits in this equation according to characteristic specified for each fault. The resulting equations and probability values are

$$P[Y > 0.30g] = \frac{2(1.727)}{\sqrt{2\pi}(30)} \int_{r=10}^{18} \int_{m=5}^{7.5} \int_{s=-\infty}^{-z} \frac{re^{-\left[\frac{s^2}{2} + 1.727(m-5)\right]}}{\sqrt{r^2 - 10^2}} ds\, dm\, dr = 0.01539$$

for Fault 1, and

$$P[Y > 0.30g] = \frac{2(2.027)}{\sqrt{2\pi}(65)} \int_{r=20}^{31} \int_{m=5}^{8.5} \int_{s=-\infty}^{-z} \frac{re^{-\left[\frac{s^2}{2} + 2.027(m-5)\right]}}{\sqrt{r^2 - 20^2}} ds\, dm\, dr = 0.00182$$

for Fault 2.

With the use of Equation 7.33 and the values of λ_{m_0} obtained earlier, one finds thus that

$$\lambda_{1y} = (0.032)(0.01539) = 4.925 \times 10^{-4} \text{ events/year} \quad \text{for Fault 1}$$

$$\lambda_{2y} = (0.079)(0.00182) = 1.438 \times 10^{-4} \text{ events/year} \quad \text{for Fault 2}$$

which, upon substitution into Equation 7.31, lead to a mean annual occurrence rate for the site under investigation given by

$$\lambda_y = 4.925 \times 10^{-4} + 1.438 \times 10^{-4} = 6.363 \times 10^{-4}$$

As a result and according to Equation 7.17, the probability of exceeding in 100 years a peak ground acceleration of $0.30g$ at the site is

$$P_{100}[A_H > 0.30\,g] = P[X_{100} \geq 1] = 1 - e^{-(6.363 \times 10^{-4})(100)} = 0.062$$

7.9 SEISMIC ZONATION MAPS

An alternative and common way to select the design level of ground motion intensity for a given site is by means of the seismic zonation maps contained in building codes or published by governmental agencies. These maps divide a region or country into areas of different seismicity and provide for each of these areas the design value of the ground motion parameter or parameters used by the code to characterize ground motion intensity. For example, the 1997 version of the Uniform Building Code divides the United States into six zones (see Figure 7.25) and gives the values of the peak ground accelerations expected at these zones.

Seismic zonation maps are usually constructed on the basis of historical seismological data, following one of the procedures described in the foregoing sections to establish the seismic hazard at a site. Specifically, seismic zonation maps are obtained as follows:

1. A grid is drawn over the region under consideration.
2. A value of the ground motion parameter of interest is calculated for each intersection point in the grid by identifying all significant seismic sources around each point, developing

Seismic Hazard Assessment

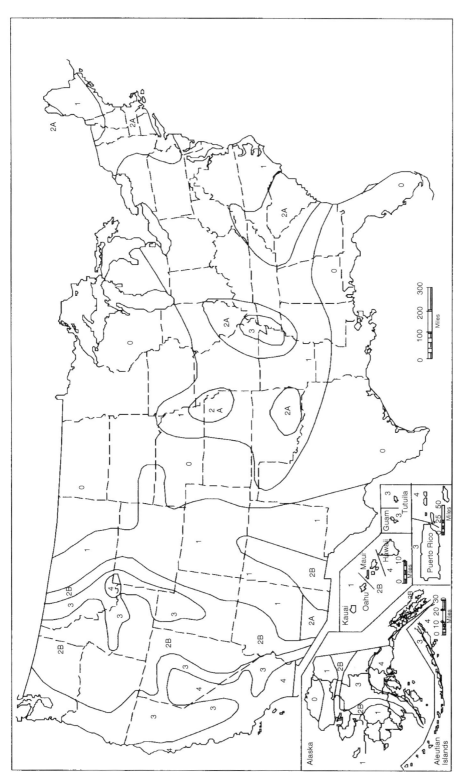

FIGURE 7.25 Seismic zonation map of the United States in the 1997 edition of the Uniform Building Code. (Reproduced with permission from the 1997 Uniform Building Code. © International Code Council, Inc. (ICC), Washington, D.C. All rights reserved.)

FIGURE 7.26 Contours of mean peak ground acceleration on rock in the United States (expressed as a percentage of g) with 10% probability of exceedance in 50 years (After Algermissen, S.T., et al., Miscellaneous Field Studies Map MF-212, U.S. Geological Survey, 1990.)

Seismic Hazard Assessment

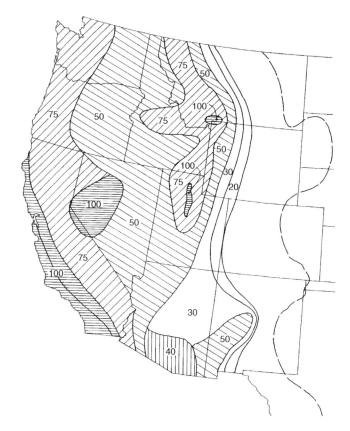

FIGURE 7.27 Zonation map of the western United States to obtain pseudoacceleration response spectrum ordinates corresponding to a natural period of 0.3 s, a damping ratio of 5%, and a 10% probability of being exceeded in 50 years. (Reproduced from *1994 NEHRP Recommended Provisions for Seismic Regulations for New buildings*, Federal Emergency Management Agency Report No. 222, Washington, D.C., 1995.)

exceedance probability curves in terms of past seismicity data and point-to-source distances, and by reading from these curves the value of the ground motion parameter that corresponds to a selected time interval and an acceptable probability of exceedance.
3. Contours of equal values of the ground motion parameter being considered are drawn, as shown in Figure 7.26 for peak ground acceleration in the United States, on the basis of the values obtained for each intersection point.
4. Zones in which the ground motion parameter varies within preselected ranges are identified.

The ground motion parameter that has been used the most in the seismic zonation maps of building codes is peak ground acceleration. Some codes also include zonation maps for peak ground velocity. However, with the availability of attenuation equations to generate site-specific response spectra, the trend now is toward the use of zonation maps in terms of response spectrum ordinates. Thus, modern building codes substitute zonation maps based on peak ground acceleration or peak ground velocity for zonation maps that are based on response spectrum ordinates.

Examples of zonation maps based on response spectrum ordinates are shown in Figures 7.27 and 7.28. These maps, developed by the Building Seismic Safety Commission in 1993 and included in the 1994 edition of the National Earthquake Hazard Reduction Program (NEHRP) provisions,[*]

---

[*] NEHRP Recommended Provisions for Seismic Regulations for New Buildings, Federal Emergency Management Agency Report No. 222, Washington, D.C., 1995.

FIGURE 7.28 Zonation map of the western United States to obtain pseudoacceleration response spectrum ordinates corresponding to a natural period of 1 s, a damping ratio of 5%, and a 10% probability of being exceeded in 50 years. (Reproduced from *1994 NEHRP Recommended Provisions for Seismic Regulations for New buildings*, Federal Emergency Management Agency Report No. 222, Washington, D.C., 1995.)

show a zonation based on response spectrum ordinates for the western United States and give for the different areas in the region the pseudoacceleration response spectrum ordinates (expressed as a percentage of the acceleration of gravity) that correspond to a damping ratio of 5% and a probability of exceedance of 10% in 50 years. The spectral ordinates in Figure 7.27 are for a natural period of 0.3 s and those in Figure 7.28 for a natural period of 1 s. Zonation maps that give the value of two response spectrum ordinates such as these are useful to develop, in combination with some empirical equations and for design purposes, an envelope to the entire response spectrum.

Zonation maps that establish the distribution of the seismic hazard in the United States and other parts of the world are developed and distributed by the National Seismic Hazard Mapping Project of the U.S. Geological Survey. Current versions of these maps may be obtained by accessing a Web site that the U.S. Geological Survey maintains for this purpose. The Web site address is http://earthquake.usgs.gov/research/hazmaps/.

Although it is always possible to establish the seismic hazard at a given site or region by means of seismic zonation maps in regions or countries where they are available, it is important to keep in mind that the ground motion intensities given in these maps correspond to an established acceptable risk. Typically, these ground motion intensities normally correspond to levels intended to prevent structural collapses but not necessarily structural and nonstructural damage. It is also important to keep in mind that the zonation of a large region can only give a rough estimate of the seismic hazard at a site as zonations are based on crude averages and some important factors such as site conditions may not be taken into account. Seismic zonation maps are thus useful for the design of ordinary structures but not for the design of critical structures such as large dams and nuclear power plants.

7.10 MICROZONATION MAPS

When a city or urban center is large and shows areas with markedly different soil, geological, and topographic conditions, the characteristics of earthquake ground motions within each of these areas may vary significantly (see Chapter 8). In these cases, it is convenient to develop a seismic zonation of the city to define in more detail than what can be obtained from a regional seismic map the ground motion characteristics expected at different areas of the city. Maps that divide a city or urban center into zones of different seismicity and give the design values of a ground motion parameter for each of these zones are known as *microzonation maps.*

Microzonation maps are produced in the same way as seismic zonation maps, except that the effect of soil, geological, and topographic conditions is accounted for more accurately than what is normally considered in the development of regional zonation maps. Microzonation maps may also include the identification of areas susceptible to ground failures such as liquefaction, landslides, and surface faulting, or geological data such as the location of old rivers and lakes and the depth and stiffness of soil deposits.

Examples of microzonation maps are shown in Figures 7.29 and 7.30. Figure 7.29 shows a microzonation map for Mexico City, while Figure 7.30 shows a microzonation map for the city of

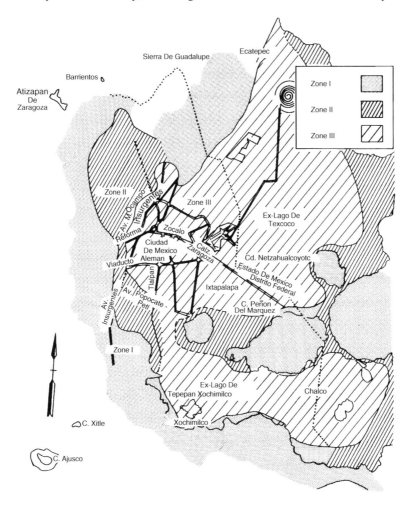

FIGURE 7.29 Microzonation map of Mexico City: Zone I = Hill Zone; Zone II = Transition Zone; Zone III = Lake Zone. (Reproduced from *Normas Tecnicas Complementarias para Diseño y Construccion de Cimentaciones*, Gaceta Oficial del Departamento del Distrito Federal, Mexico, 1987.)

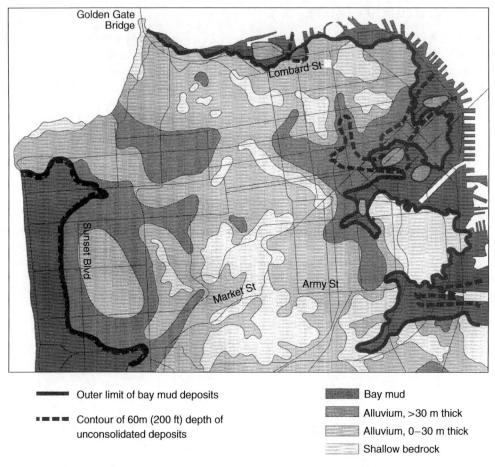

FIGURE 7.30 Microzonation map of the city of San Francisco, California. (Reproduced from *Performance of Structures During the Loma Prieta Earthquake of October 17, 1989*, NIST Special Publication 778, National Institute of Standards and Technology, 1990.)

San Francisco, California. Both of these areas have shown significant site effects during earthquakes. In the microzonation of Mexico City, the city is divided into three zones with different subsurface conditions. These zones are identified as the Foothill Zone, the Lake Zone, and the Transition Zone. The Foothill Zone is characterized by shallow, compact deposits of mostly granular soil, basalt, or volcanic tuff. In contrast, the Lake Zone is characterized by deposits of soft soils formed from the pluviation of airborne silt, clay, and ash from nearby volcanoes that extend to a considerable depth. The Transition Zone lies between the Foothill and Lake Zones. In the Transition Zone the soil deposits are thin and interspersed with alluvial deposits. In the microzonation of the San Francisco Bay area, the area is divided into four zones, also according to soil characteristics. These four zones are (a) a zone with deep deposits of normally consolidated and highly compressible alluvial clays and silty to sandy clays (known as San Francisco Bay mud), (b) a zone with deposits of alluvium with depths of >30 ft, (c) a zone with shallow deposits of alluvium with depths of <30 ft, and (d) a zone with rock and shallow residual soils. The San Francisco Bay mud is generally found near the coastal areas, where the soil thickness varies from zero up to several tens of feet.

FURTHER READINGS

1. Abrahamson, N. A. and Shedlock, K. M., Overview, *Seismological Research Letters*, 68(1), 1997, pp. 9–23.
2. Idriss, I. M., Evaluating seismic risk in engineering practice, *Proceedings of 11th International Conference on Soil Mechanics and Foundation Engineering*, San Francisco, CA, August 12–16, 1985, Vol. 1, pp. 255–320.
3. McGuire, R. K., *Seismic Hazard and Risk Analysis*, Monograph MNO-10, Earthquake Engineering Research Institute, Oakland, CA, 2004, 221p.
4. McGuire, R. K. and Arabasz, W. J., An introduction to probabilistic seismic hazard analysis, in *Geotechnical and Environmental Geophysics*, Ward, S. H. ed., Society of Exploration Geophysicists, 1990, 1, pp. 333–353.
5. Reiter, L., *Earthquake Hazard Analysis—Issues and Insights*, Columbia University Press, New York, 1990, 254 p.
6. Schwartz, D. P. and Coppersmith, K. J., Seismic hazards: New trends in analysis using geological data, in *Active Tectonics*, Wallace, R. E. ed., Academic Press, Orlando, FL, 1986, pp. 215–230.
7. Schwartz, D. P., Geologic characterization of seismic sources: Moving into the 1990s, *Earthquake Engineering and Soil Dynamics II: Recent Advances in Ground Motion Evaluation*, Geotechnical Special Publication 20, ASCE, New York, 1988, pp. 1–42.
8. Wells, D. L. and Coppersmith, K. J., New empirical relationships among magnitude, rupture length, rupture width, rupture area, and surface displacement, *Bulletin Seismological Society of America*, 84, 4, 1994, pp. 974–1002.

PROBLEMS

7.1 It is determined from paleoseismicity studies that a nearby fault has slipped up to 5 m in the past. On the basis of this information, estimate the maximum moment magnitude of the earthquakes that can occur at this fault.

7.2 A site in Southern California is overlain by soil deposits with an average shear wave velocity of 225 m/s over the top 30 m. The site is located 25 km away from a known earthquake fault. Find the mean and the mean plus one standard deviation of the peak horizontal ground acceleration and peak horizontal ground velocity that would be observed at the site in the event of a shallow earthquake with a moment magnitude of 7.0 at that fault.

7.3 Generate the expected 5% damping pseudoacceleration response spectrum for the site described in Problem 7.2 and earthquakes with a moment magnitude of 7.0.

7.4 An earthquake generated at a strike-slip fault in Manjil, Iran, had a moment magnitude of 7.4. Find the mean and the mean plus one standard deviation of the peak horizontal ground acceleration and peak horizontal ground velocity produced by this earthquake at a site overlain by soft rock and located 4 km from the fault. The basement rock at the area near the fault is estimated to be at a depth of 1400 m.

7.5 Obtain the expected 5% damping pseudoacceleration-response spectrum for the site described in Problem 7.4 and earthquakes with a moment magnitude of 7.4.

7.6 Find the mean and the mean plus one standard deviation of the peak horizontal ground acceleration at a site overlying deposits of firm soil and located at a distance of 100 km from the subduction zone in central Chile if an earthquake with a moment magnitude of 7.9 and a focal depth of 31 km occurs at this subduction zone.

7.7 Determine the expected 5% damping acceleration response spectrum for the site described in Problem 7.6 and earthquakes with a moment magnitude of 7.9.

7.8 The number of earthquakes recorded in a seismically active region over a period of 120 years is shown in Table P7.8. Based on the central values of the given magnitude ranges,
   a. Estimate the parameters of the Gutenberg–Richter law for the region and
   b. Calculate neglecting earthquakes with a magnitude of <4.0 the probability that an earthquake in the region will have a magnitude between 6.5 and 7.5

TABLE P7.8
Earthquake Data for Problem 7.8

Magnitude Range	Number of Earthquakes in 120 years
3.5–4.5	420
4.5–5.5	27
5.5–6.5	4
6.5–7.5	1

TABLE P7.12
Earthquake Data for Problem 7.12

Magnitude Range	Number of Earthquakes in 100 years
4.75–5.75	250
5.75–6.75	78
6.75–8.25	2

7.9 The average number of earthquakes exceeding a magnitude M_w in a year in a seismic region is given by

$$\log N = 4.97 - 0.87 M_w$$

What is the probability that an earthquake of magnitude 5.0 or greater occur at least once in a 50 year interval in that region?

7.10 It is found from the analysis of the occurrences of earthquakes near a construction site over a period of 200 years and the use of an attenuation equation to estimate peak ground accelerations that eight times over the considered period the peak ground acceleration at the site is $>0.25g$. What is the probability that the site be subjected to a peak ground acceleration $>0.25g$ in a time interval of 100 years?

7.11 A regression analysis between the peak ground acceleration (expressed as a fraction of the acceleration of gravity) at a site and the frequency with which this acceleration has been exceeded in the past yields the empirical equation

$$a = 0.33 \log T_R - 0.323$$

where a represents peak ground acceleration and T_R denotes return period. Using this equation and assuming that the number of occurrences of an earthquake with given intensity is given by a Poisson process, determine
  a. The probability that the ground acceleration at the given site exceed $1.0g$ in a time interval of 100 years and
  b. The peak ground acceleration corresponding to a probability of exceedance of 1% in a time interval of 50 years

7.12 The frequency with which earthquakes occur at a fault 40 km from a construction site is given in Table P7.12 in terms of their moment magnitude. Using this information and the approach described in Section 7.6, construct a frequency-peak ground acceleration curve, and estimate on the basis of this curve the peak ground acceleration at the construction site that has a 30% probability of being exceeded in a time interval of 50 years. Assume an average shear wave velocity for the site soil of 30 m/s. Base your calculations on the central values of the given magnitude ranges.

7.13 The data shown in Table P7.13 represent the estimated values of the peak ground accelerations induced by the earthquakes that have occurred in the vicinity of a construction site over a period of 20 years. Using these data and the approach described in Section 7.6, find
  a. An expression to determine the mean annual frequency with which a given peak ground acceleration would be exceeded at the site

TABLE P7.13
Peak Ground Acceleration Data for Problem 7.13

Year	Peak Ground Acceleration/g	Year	Peak Ground Acceleration/g	Year	Peak Ground Acceleration/g
1978	0.392	1984	0.444	1991	0.432
1979	0.343	1985	0.000	1992	0.000
1980	0.523	1986	0.423	1993	0.586
1981	0.623	1986	0.333	1993	0.000
1982	0.387	1987	0.354	1994	0.427
1982	0.098	1988	0.000	1995	0.419
1983	0.413	1989	0.345	1995	0.118
1983	0.312	1989	0.498	1996	0.405
1984	0.367	1990	0.600	1997	0.478

b. The peak ground acceleration at the site corresponding to a return period of 500 years and
c. The peak ground acceleration at the site corresponding to an exceedance probability of 1% in 50 years

7.14 A multistory building is to be built in the city of Anaheim, California, which lies over deposits of alluvium with shear wave velocities of 790, 1000, and 1670 ft/s at depths between 0 and 38 ft, 38 and 80 ft, and 80 and 95 ft, respectively. The historical seismic activity within a 100 mile radius from the building site is summarized in Table P7.14, where the magnitude of all the earthquakes in record and the corresponding epicentral distances from the site to the causative seismic sources are listed. Using these data, the attenuation laws proposed by Boore, Joyner, and Fumal, and the semiprobabilistic approach described in Section 7.7, construct frequency-intensity curves for the expected peak ground accelerations at the site. Then, on the basis of these curves and considering that the life expectancy of the building is 45 years
   a. Draw exceedance probability curves for time intervals of 50, 100, and 500 years
   b. Obtain the peak ground acceleration at the building site for an exceedance probability of 10% and
   c. Determine the peak ground accelerations at the building site and the associated probabilities of exceedance corresponding to return periods of 100, 500, and 1000 years

7.15 The site shown in Figure P7.15 is overlain by deposits of firm soil and located near two seismic sources: Fault A and Fault B. Fault A is a strike-slip fault while Fault B is a reverse one. The location of these two faults with respect to the site is as shown in the figure, in which the given coordinates are in kilometers. The moment magnitudes of the largest earthquakes that can be produced at these two faults are 8.0 for Fault A and 9.0 for Fault B. Their seismicity is described, respectively, by the following Gutenberg–Richter laws:

$$\log \lambda_m = 4.0 - 0.7 M_w$$

$$\log \lambda_m = 3.0 - 0.75 M_w$$

TABLE P7.14
Seismicity Data for Problem 7.14

Latitude North	Longitude West	Date	Depth (km)	Earthquake Magnitude	Site Intensity (MM)	Approximate Distance, mi (km)
33.000	117.300	11/22/1800	0.000	6.500	IV	69 (111)
33.700	117.900	12/08/1812	0.000	6.900	IX	11 (017)
34.000	119.000	09/24/1827	0.000	5.500	III	64 (102)
32.700	117.200	05/27/1862	0.000	5.900	III	90 (144)
34.100	116.700	02/07/1889	0.000	5.300	III	71 (144)
34.100	117.900	08/28/1889	0.000	5.200	VI	17 (027)
33.400	116.300	02/09/1890	0.000	6.300	III	98 (157)
34.300	118.600	04/04/1893	0.000	5.400	IV	50 (081)
34.100	119.400	05/19/1893	0.000	5.500	II	87 (140)
34.300	117.600	07/30/1894	0.000	5.900	V	35 (057)
32.800	116.800	10/23/1894	0.000	5.700	II	97 (156)
34.200	117.400	7/22/1899	0.000	5.500	V	37 (060)
34.300	117.500	7/22/1899	0.000	6.500	VI	38 (062)
33.800	117.000	12/25/1899	0.000	6.600	V	52 (084)
34.200	117.100	9/2/1907	0.000	5.300	IV	52 (083)
33.700	117.400	4/11/1910	0.000	5.000	V	31 (050)
33.700	117.400	5/13/1910	0.000	5.000	V	31 (050)
33.700	117.400	5/15/1910	0.000	6.000	VI	31 (050)
33.500	116.500	9/30/1916	0.000	5.000	II	84 (136)
34.900	118.900	10/23/1916	0.000	5.200	II	92 (148)
34.700	119.000	10/23/1916	0.000	5.200	II	85 (137)
33.750	117.000	4/21/1918	0.000	6.600	V	52 (084)
33.750	117.000	6/6/1918	0.000	5.000	III	52 (084)
33.200	116.700	1/1/1920	0.000	5.000	II	83 (133)
34.000	117.250	7/23/1923	0.000	6.000	V	39 (062)
34.000	119.500	2/18/1926	0.000	5.000	II	92 (148)
34.000	118.500	8/4/1927	0.000	5.000	IV	35 (057)
34.180	116.920	1/16/1930	0.000	5.200	III	61 (098)
34.180	116.920	1/16/1930	0.000	5.100	III	61 (098)
33.950	118.632	8/31/1930	0.000	5.200	IV	42 (068)
33.617	117.967	3/11/1933	0.000	6.300	VII	17 (027)
33.750	118.083	3/11/1933	0.000	5.000	VI	13 (020)
33.750	118.083	3/11/1933	0.000	5.100	VII	13 (020)
33.750	118.083	3/11/1933	0.000	5.000	VI	13 (020)
33.700	118.067	3/11/1933	0.000	5.100	VI	14 (023)
33.575	117.983	3/11/1933	0.000	5.200	VI	20 (032)
33.683	118.050	3/11/1933	0.000	5.500	VII	15 (024)
33.700	118.067	3/11/1933	0.000	5.100	VI	14 (023)
33.750	118.083	3/11/1933	0.000	5.100	VII	13 (020)
33.850	118.267	3/11/1933	0.000	5.000	V	21 (033)
33.750	118.083	3/13/1933	0.000	5.300	VII	13 (020)
33.617	118.017	3/14/1933	0.000	5.100	VI	18 (029)
33.783	118.133	10/2/1933	0.000	5.400	VII	14 (023)
34.100	116.800	10/24/1935	0.000	5.100	III	65 (105)
33.408	116.261	3/25/1937	10.000	6.000	II	99 (160)
33.699	117.511	5/31/1938	10.000	5.500	VI	25 (040)
34.083	116.300	5/18/1940	0.000	5.400	II	93 (150)

TABLE P7.14 (Continued)

Latitude North	Longitude West	Date	Depth (km)	Earthquake Magnitude	Site Intensity (MM)	Approximate Distance, mi (km)
34.067	116.333	5/18/1940	0.000	5.200	II	91 (147)
34.067	116.333	5/18/1940	0.000	5.000	II	91 (147)
34.867	118.933	9/21/1941	0.000	5.200	II	91 (147)
33.783	118.250	11/14/1941	0.000	5.400	VI	20 (33)
34.267	116.967	8/29/1943	0.000	5.500	III	61 (98)
33.976	116.721	6/12/1944	10.000	5.100	III	68 (110)
33.994	116.712	6/12/1944	10.000	5.300	III	69 (111)
33.950	116.850	9/28/1946	0.000	5.000	III	61 (98)
34.017	116.500	7/24/1947	0.000	5.500	III	81 (131)
34.017	116.500	7/25/1947	0.000	5.000	II	81 (131)
34.017	116.500	7/25/1947	0.000	5.200	II	81 (131)
34.017	116.500	7/26/1947	0.000	5.100	II	81 (131)
33.267	119.450	11/18/1947	0.000	5.000	I	98 (157)
33.933	116.383	12/4/1948	0.000	6.500	III	87 (141)
32.817	118.350	12/26/1951	0.000	5.900	III	76 (123)
34.950	118.867	7/21/1952	0.000	5.300	II	93 (150)
35.000	118.833	7/23/1952	0.000	5.400	II	95 (153)
35.000	119.833	7/23/1952	0.000	5.200	II	95 (153)
34.900	118.950	8/1/1952	0.000	5.100	II	93 (150)
34.519	118.198	8/23/1952	13.100	5.000	III	49 (078)
35.150	118.633	1/27/1954	0.000	5.000	I	98 (158)
34.983	118.983	5/23/1954	0.000	5.100	I	99 (159)
34.941	118.987	11/15/1961	10.700	5.000	I	97 (156)
34.932	118.976	3/1/1963	13.900	5.000	I	96 (155)
33.710	116.925	9/23/1963	16.500	5.000	III	57 (092)
34.712	116.503	9/25/1965	10.600	5.200	II	99 (160)
33.343	116.346	4/28/1969	20.000	5.800	II	96 (155)
33.291	119.193	10/24/1969	10.000	5.100	II	84 (135)
34.270	117.540	9/12/1970	8.000	5.400	V	35 (057)
34.411	118.401	2/9/1971	8.400	6.400	V	48 (077)
34.411	118.401	2/9/1971	8.000	5.800	IV	48 (077)
34.411	118.401	2/9/1971	8.000	5.800	IV	48 (077)
34.411	118.401	2/9/1971	8.000	5.300	IV	48 (077)
34.308	118.454	2/9/1971	6.200	5.200	IV	44 (071)
34.065	119.035	2/21/1973	8.000	5.900	IV	66 (107)
33.986	119.475	8/6/1973	16.900	5.000	II	90 (145)
34.516	116.495	6/1/1975	4.500	5.200	II	92 (149)
33.944	118.681	1/1/1979	11.300	5.000	IV	45 (072)
34.327	116.445	3/15/1979	2.500	5.200	II	90 (144)
33.501	116.513	2/25/1980	13.600	5.500	II	84 (135)
33.671	119.111	9/4/1981	0.000	5.400	III	70 (113)

Number of records: 88.
Maximum magnitude during time interval: 6.9.
Maximum intensity during time interval: IX.
Distance from site to nearest historical earthquake: 11 mi (17 km).
Number of years in considered time interval: 186.

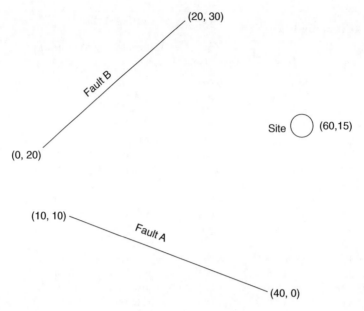

FIGURE P7.15 Site and nearby faults in Problem 7.15 (coordinates in kilometers).

Using the probabilistic approach described in Section 7.8, Campbell's attenuation equations, and a lower-threshold magnitude of 5.0 for both faults
a. Determine the mean annual probabilities that the peak ground acceleration at the site exceed 0.10g and 0.40g
b. Construct an approximate frequency-peak ground acceleration curve for the site using these two probabilities and
c. Using this frequency-intensity curve, estimate the probability that the peak ground acceleration at the site exceed 0.25g in 50 years

8 Influence of Local Site Conditions

All analyses are based on some assumptions which are not quite in accordance with the facts. From this, however, it does not follow that the conclusions of the analysis are not close to the facts.

Hardy Cross, 1926

8.1 PROBLEM IDENTIFICATION

It is nowadays a well-accepted fact that topography and soil properties may significantly affect the characteristics of the earthquake ground motions that might take place at a given site. As far back as 1819, J. MacMurdo observed, in describing the effects of an earthquake in India, that "buildings situated on rock were not by any means so much affected by the earthquake as those whose foundations did not reach to the bottom of the soil." Similarly, R. Mallet in his report on the 1857 Neapolitan earthquake noted a correlation between local geology and damage, and H. O. Wood and H. F. Reid showed that the distribution of the ground shaking intensity during the 1906 San Francisco earthquake was related to local soil and geological conditions. Even more, as earlier as 1927, B. Gutenberg had already developed amplification factors for sites with different subsurface conditions.

The reason for such an influence is presently well understood. The flexibility of a soil deposit, ridge, or hill makes it vibrate like a flexible structure and not as a rigid body, as it would be the case for a flat formation of stiff rock. The extent of this influence depends, of course, on the inherent stiffness, density, and damping characteristics of the soil deposit, hill, or ridge, in much the same way the dynamic characteristics of a conventional structure influence its dynamic response. One should expect, thus, that the ground motions induced by an earthquake at the base of the building shown in Figure 8.1 would be different if the building stands on rock, the surface of a soft soil deposit, the top of a hill, or the surface of a stratified soil with a mixture of soft and firm soils. By the same token, if two soil deposits behave elastically, have the same geometry, and are subjected to the same input motion, but one contains soft soils (Site A) and the other stiff soils (Site B), then it would be expected that the softer deposit would amplify the low-frequency motions more whereas the stiff deposit would amplify the high-frequency motions more. That is, their frequency response functions would be as shown schematically in Figure 8.2.

A dramatic example of the effect of local soil conditions on ground motion characteristics was observed in Mexico City during the 1985 Michoacán earthquake. The downtown area of Mexico City, lying above a deep deposit of soft, lacustrine, volcanic clays (see Figure 8.3), experienced a ground motion that was significantly more intense than the ground motion recorded at the surrounding hills during this earthquake. The extent of the difference can be appreciated by comparing the records of the ground motion obtained at a site on the downtown area and at a site on the hills (see Figures 8.4 and 8.5) or by comparing the corresponding acceleration response spectra (see Figures 8.6 and 8.7). The peak ground acceleration observed at the hill site was $0.034g$, whereas that on the downtown area was $0.158g$. Likewise, the peak spectral accelerations for 0% damping at the hill and downtown sites were $0.432g$ and $3.54g$, respectively.

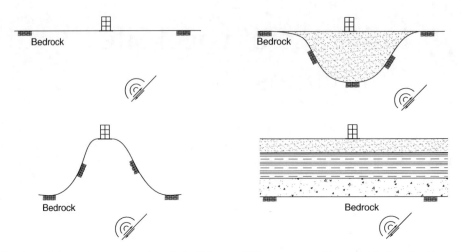

FIGURE 8.1 Schematic representation of a building on different topographic, geological, and soil conditions.

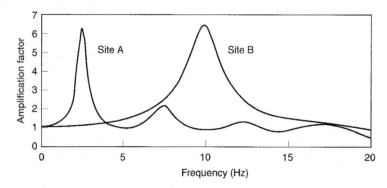

FIGURE 8.2 Frequency response functions for hypothetical soft (Site A) and stiff (Site B) soil deposits.

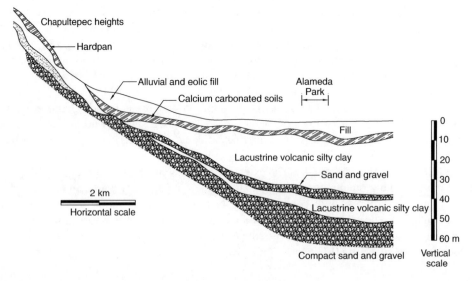

FIGURE 8.3 East–west geological profile of Mexico City Valley.

Influence of Local Site Conditions

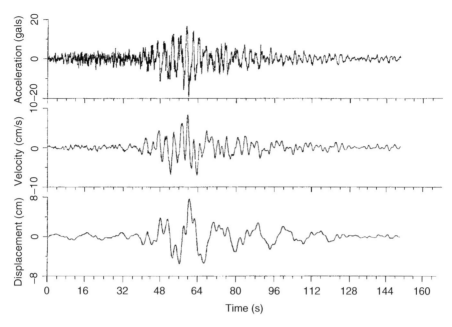

FIGURE 8.4 Ground motion recorded at hill site in Mexico City during the 1985 Michoacán earthquake.

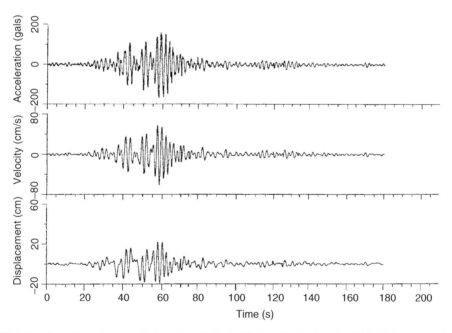

FIGURE 8.5 Ground motion recorded at soft-soil site in Mexico City during the 1985 Michoacán earthquake.

Another example of ground motion amplification owing to soil conditions was that observed during the 1989 Loma Prieta earthquake at the surface of the soft soils existing around the San Francisco Bay in northern California. Figures 8.8 and 8.9 show the acceleration records obtained on rock and soft-soil sites during that earthquake. Both sites are situated ~80 km from the earthquake epicenter. However, the soft-soil site is underlain by a sandy fill, a layer of young bay mud, and layers of dense to very dense sands and stiff to hard clays. The fill has a loose to medium

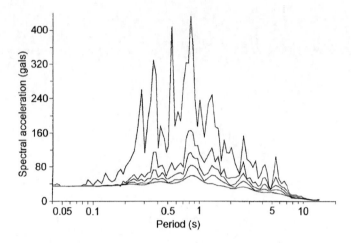

FIGURE 8.6 Acceleration response spectra for 0, 2, 5, 10, and 20% damping corresponding to ground motion recorded at hill site in Mexico City during the 1985 Michoacán earthquake.

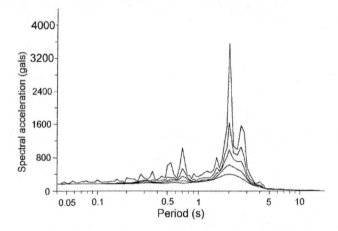

FIGURE 8.7 Acceleration response spectra for 0, 2, 5, 10, and 20% damping corresponding to ground motion recorded at soft-soil site in Mexico City during the 1985 Michoacán earthquake.

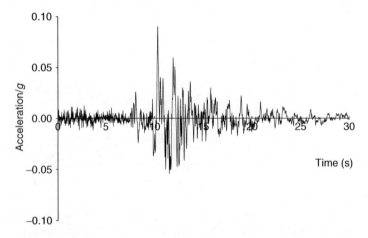

FIGURE 8.8 East–west component of ground acceleration recorded at Rincon Hill station (rock site) during the 1989 Loma Prieta earthquake.

Influence of Local Site Conditions

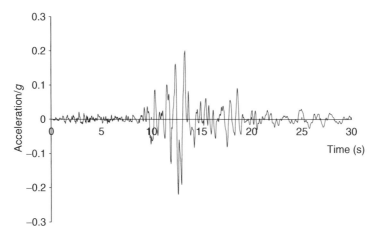

FIGURE 8.9 East–west component of ground acceleration recorded at Outer Harbor Wharf, Oakland, station (soft-soil site) during the 1989 Loma Prieta earthquake.

density. The shear wave velocities of young bay mud range from ~150 to 500 ft/s. Rock is encountered at a depth of between 300 and 500 ft As a result, the soft-soil site experienced a much higher acceleration than the rock site. The peak ground acceleration recorded at the soft site was 0.220g, whereas that at the rock site was 0.090g.

Topographic irregularities have also been identified as the source of ground motion amplifications. Perhaps the best known example is the high accelerations recorded on top of the abutment of Pacoima Dam during the 1971 San Fernando (M_L = 6.4) earthquake in Southern California. The recorded peak horizontal acceleration was ~1.25g, a value that was considerably higher than the values expected for an earthquake of this magnitude. The accelerograph, however, was located at the crest of a narrow rock ridge adjacent to the dam. Subsequent investigations have attributed this unusually high peak acceleration to the dynamic response of the ridge itself. Another apparently unmistakable example of topographic ground motion amplification is the extent of the damage observed in several groups of similar houses located at topographically different sites during the 1985 Chile (M_s = 7.8) earthquake. The houses on hill tops or ridges were heavily damaged whereas those on nearly flat sites were only slightly damaged.

It may be seen, thus, that local site conditions play an important role in the characterization of ground motions for design purposes. In general, the effect of local geological and soil conditions has been accounted for in the past by means of either statistical correlations or some simple analytical techniques. This chapter describes these two approaches for the evaluation of site effects. A general description of the way site conditions may affect ground motion characteristics is also given.

8.2 EFFECT OF SITE CONDITIONS ON GROUND MOTION CHARACTERISTICS

The effect of subsurface conditions on peak ground acceleration and peak ground velocity has been investigated by conducting statistical studies of the relative values observed at sites with different subsurface conditions. In one such study, acceleration records from western U.S. earthquakes with a magnitude of ~6.5 were considered. The records were obtained at the surface of four different site types: (a) rock sites, (b) sites underlain by cohesionless soils or stiff soils extended up to depths of ~200 ft, (c) sites underlain by deep cohesionless soils with depths >250 ft, and (d) sites underlain by soft to medium-stiff clays and sands. On the basis of this study, it was found that, on

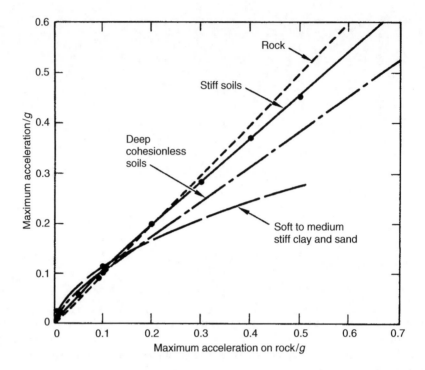

FIGURE 8.10 Empirical relationships between peak accelerations recorded on rock and soil sites. (Reproduced with permission of Earthquake Engineering Research Institute from Seed, H.B. and Idriss, I.M., *Ground Motions and Soil Liquefaction during Earthquakes*, Earthquake Engineering Research Institute, Oakland, CA, 1983.)

average, the peak ground accelerations were slightly greater at the surface of soil deposits than on rock for low acceleration levels and somewhat smaller for high acceleration levels (see Figure 8.10). In a more recent study that included data from the 1985 Mexico City and 1989 Loma Prieta earthquakes, a different trend was revealed. In this study, it was found that at low and moderate levels the peak ground accelerations at soft-soil sites were significantly higher than those at rock sites (see Figure 8.11). That is, soft soils exhibited peak ground accelerations as large as four times those observed on rock in the acceleration range between $0.05g$ and $0.1g$. For rock accelerations $>0.1g$, the ground motion amplification factor tended to decrease to ~3.0 at low accelerations and ~1.0 for rock accelerations of ~$0.4g$. It was concluded, thus, that low rock accelerations might be amplified several times at soil sites, especially those containing soft clays. By contrast, large rock accelerations are amplified to a lesser degree and may even be slightly decreased at very high levels. Undoubtedly, the latter observation is an indication that at high acceleration levels the nonlinear behavior of soft soils prevents any ground motion amplification.

In regard to the influence of site conditions on peak ground velocity, it seems to be a general agreement that site conditions have a pronounced influence on this parameter. In general, the peak ground velocities recorded at the surface of soil deposits are greater than those recorded on rock, with the values for soil deposits being typically about twice than those obtained on rock sites (see Figure 8.12).

Statistical studies have also been conducted to measure the effect of local site conditions on the frequency content of the motions recorded at the surface of soil deposits and, hence, the shape of their response spectra. In the first study reported earlier for the investigation of the effect of soil conditions on peak ground acceleration, normalized response spectra were also determined for the acceleration records examined in the study and averaged for the four site conditions considered.

Influence of Local Site Conditions

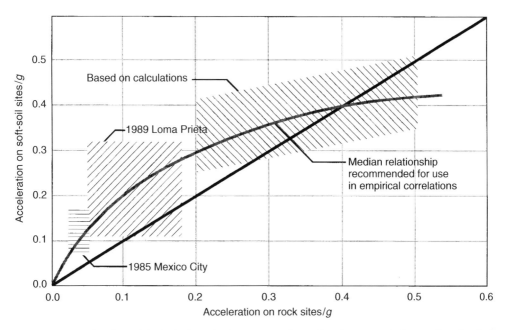

FIGURE 8.11 Updated empirical relationships between peak accelerations recorded on rock and soil sites. (From Idriss, I.M., *H. Bolton Seed Memorial Symposium Proceedings*, BiTech Publishers Ltd., Vancouver, BC, 1990. With permission.)

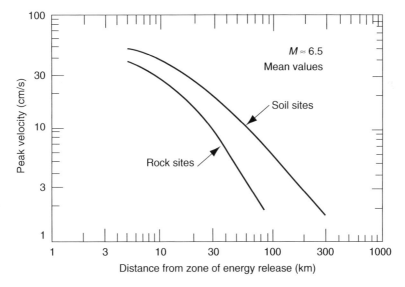

FIGURE 8.12 Variation of peak ground velocity with distance for soil and rock sites. (Reproduced with permission of Earthquake Engineering Research Institute from Seed, H.B. and Idriss, I.M., *Ground Motions and Soil Liquefaction during Earthquakes*, Earthquake Engineering Research Institute, Oakland, CA, 1983.)

The normalization was accomplished by dividing the actual response spectra by their corresponding peak ground accelerations or zero-period ordinates. The purpose of the normalization was to investigate the effect of local soil conditions on response spectrum shape alone. The results of this study are presented in Figure 8.13. It may be seen from this figure that local soil conditions have a significant influence on the shape of response spectra. At periods >0.5 s, the spectral amplifications are

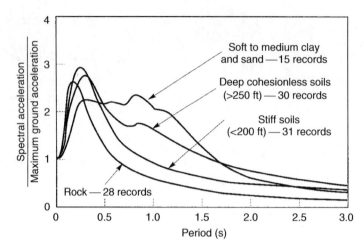

FIGURE 8.13 Average normalized 5% damping response spectra for different subsurface conditions. (After Seed, H.B. et al., *Bulletin of the Seismological Society of America*, 66, 221–243, 1976. © Seismological Society of America.)

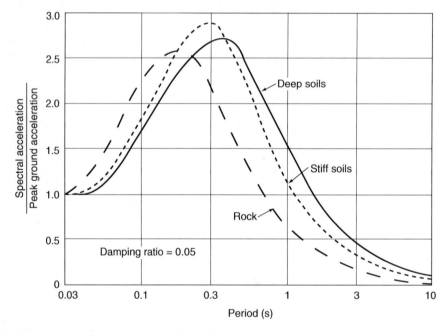

FIGURE 8.14 Average spectral shapes for three different soil conditions. (After Idriss, I.M., *Proceedings of the 11th International Conference on Soil Mechanics and Foundation Engineering*, San Francisco, CA, 1985, 255–320. © 1985 Taylor & Francis Group. With permission.)

much higher for soil sites than for rock sites. At longer periods, the spectral amplification increases with decreasing subsurface soil stiffness. Deep and soft soil deposits thus generate significant long-period motions.

In a subsequent study, data from the 1979 Imperial Valley ($M_s = 6.8$) earthquake were incorporated into the data considered in the aforementioned study, and a new set of average spectral shapes was obtained. The resulting curves, smoothed and extended to periods of up to 10 s, are shown in Figure 8.14. It may be noted from these curves that the ordinates for deep soils are of the order of 3–4.5 times those for rock in the period ranging between 3.0 and 10 s.

Influence of Local Site Conditions

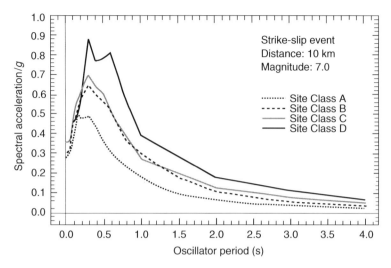

FIGURE 8.15 Average 5% damping response spectra for different subsurface conditions. (Reproduced with permission of Earthquake Engineering Research Institute from Crouse, C.B. and McGuire, J.W., *Earthquake Spectra*, 12, 407–439, 1996.)

The aforementioned results have been confirmed by another more recent study. In this study, a regression analysis was performed with the response spectra of 238 horizontal ground acceleration records from 16 earthquakes with surface-wave magnitudes >6.0. The purpose was to formulate an empirical relationship in terms of various earthquake parameters to determine the mean spectral pseudovelocity at a given site. The subsurface conditions of the recording sites were classified into four groups: (a) rock (Site Class A), (b) soft rock or stiff soil (Site Class B), (c) medium-stiff soil (Site Class C), and (d) soft clay (Site Class D). The average 5% damping acceleration response spectra obtained with such a relationship for the four subsurface conditions considered and the particular case of a strike-slip earthquake with a 7.0 magnitude at a distance of 10 km from the source are shown in Figure 8.15. As in the previous study, it may be seen that the spectral ordinates for records obtained at soft-soil sites are higher than those for records obtained at rock sites.

The effect of topographic conditions on peak ground acceleration has also been studied by several investigators. One of the earliest investigations was performed by D. M. Boore in 1972. Prompted by the amplifications observed on the Pacoima Dam abutment during the 1971 San Fernando earthquake, he estimated the localized ground motion variations in a ridge under SH waves using an analytical model. He found that ground motions may be amplified up to 100% at the ridge top and that both amplification and abatement may occur along the slope of the ridge. In another study conducted in 1974, A. M. Rogers and coworkers performed scale model experiments to study the amplification of P waves in a ridge. Amplifications of the order of 50% were found for broadband input motions and 200% for band-limited input motions. However, when compared with the amplifications from field measurements reported by L. L. Davis and L. R. West in 1973, qualitative, but not quantitative, agreement was found. The field studies showed amplifications of 400% for peak velocities and amplifications as high as 20 times from base to crest in terms of spectral velocity. In a further study, the peak ground accelerations recorded on a mountain ridge in Matsuzaki, Japan, during five earthquakes were analyzed to investigate the ground motion variation along the ridge slope. Figure 8.16 shows the mean values and error bars of such peak ground accelerations, after normalizing them with respect to the peak ground acceleration recorded at the crest. It was found that the average peak crest acceleration was ~2.5 times the average base acceleration.

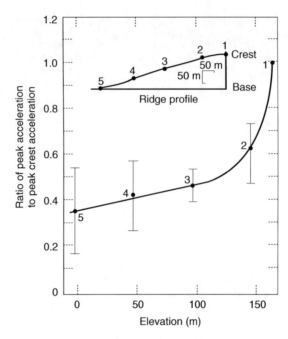

FIGURE 8.16 Mean and error bars of normalized peak ground accelerations recorded on a ridge during five earthquakes in Matsuzaki, Japan. (After Jibson, R.W., Open-File Report 87-268, U.S. Geological Survey, 1987.)

The effect of local site conditions on the characteristics of ground motions may be thus summarized as follows:

1. Accelerations at the surface of soft soil deposits are likely to be greater than those on rock sites when the accelerations are low and somewhat less when the accelerations are high.
2. Subsurface conditions have a large influence on peak ground velocity, and this influence may be much larger than the influence on peak ground acceleration.
3. Subsurface conditions have a significant influence on response spectrum shape.
4. Topographical irregularities may have an important effect on ground motion characteristics.

8.3 EVALUATION OF SITE EFFECTS USING STATISTICAL CORRELATIONS

The simplest way to take into account the influence of local site conditions on the expected ground motion characteristics at a site is by using statistical correlations similar to those presented in the foregoing section. As explained before, these correlations may be obtained by averaging the characteristics of ground motions recorded on sites with different geological characteristics. By means of such correlations, one may then establish relationships to determine the expected peak ground acceleration, peak ground velocity, or response spectrum shape at a site in terms of the site's soil properties or characteristics. Examples of these correlations are the attenuation equations presented in Chapter 7. Alternatively, one can establish average response spectra for different site conditions, such as those shown in Figures 8.13 and 8.14. In fact, this is the approach that building codes have used for many years to account for soil conditions in the selection of the design ground motion (see Figure 8.17). Furthermore, one can establish modification factors to affect the ground motion values or response spectrum shapes estimated for rock sites, the way modern building codes introduce soil condition effects into their recommendations for the selection of the design ground

Influence of Local Site Conditions

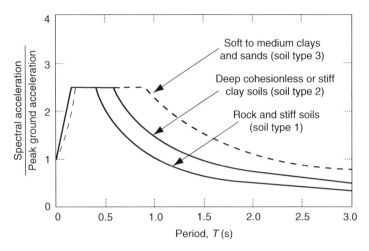

FIGURE 8.17 Normalized response spectra specified by the 1994 edition of Uniform Building Code for different soil types. (Reproduced with permission of Earthquake Engineering Research Institute from Seed, H.B. and Idriss, I.M., *Ground Motions and Soil Liquefaction during Earthquakes*, Earthquake Engineering Research Institute, Oakland, CA, 1983.)

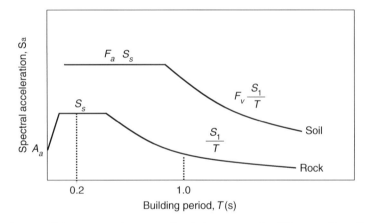

FIGURE 8.18 Two-coefficient approach to prescribe design spectra for different soil conditions.

motion. For example, the International Building Code (discussed in Chapter 17) accounts for the effect of local site conditions by affecting the spectral ordinates determined or specified for rock sites on the basis of site coefficients that depend on soil type and shaking intensity. This concept is illustrated in Figure 8.18 in which F_a is the site coefficient for short periods and F_v is the site coefficient for long periods. S_s and S_1, the spectral accelerations for periods of 0.2 and 1.0 s on rock sites, respectively, account for the level of shaking. F_a and F_v are both functions of soil type and shaking intensity. It may be noted, thus, that this two-coefficient approach allows the consideration of the fact that, as suggested by recent evidence (see Figure 8.15), short-period spectral accelerations may be amplified several times at soft-soil sites under low levels of shaking.

Although simple to use, it should be kept in mind that statistical correlations are only as good as the data used to derive them. In many cases, these data do not include a sufficiently large number of records from sites with a particular soil type or desired ground shaking intensity and may thus lead to dubious correlations. In fact, this is the reason why building codes make changes to their recommendations to assess local site conditions every time a significant new earthquake generates a new set of data.

Statistical correlations to account for the effect of topographic irregularities on ground motion characteristics have not been developed yet. Correspondingly, with the exception of the French building code, building codes have not incorporated explicit provisions to consider possible topographic amplifications.

8.4 EVALUATION OF SITE EFFECTS USING ANALYTICAL TECHNIQUES

8.4.1 INTRODUCTORY REMARKS

Site effects may also be evaluated using analytical techniques given that such an evaluation may be reduced to the problem of determining the response of a soil deposit or rock formation to the motion of the bedrock immediately around or beneath it. In principle, therefore, the finite element method can be used to estimate the ground motion characteristics at the surface of a given site if the ground motion at the bedrock level is known. As illustrated in Figure 8.19, if a soil deposit is discretized into finite elements and the bedrock motion is known, then it is a simple task to compute the acceleration response at the surface of the soil deposit and, hence, the characteristics of the corresponding motion at the level of the ground surface. In practice, however, such a solution is seldom utilized, as the large number of elements needed to formulate the problem makes it too time consuming and too expensive for practical purposes. As a consequence, over the years, a number of techniques have been developed to simplify the problem. This section describes some of the better-known simplified models.

8.4.2 ONE-DIMENSIONAL CONTINUOUS MODEL

If the ground surface and the surface of the underlying bedrock are basically horizontal, then the lateral extent of a soil deposit has no influence on the dynamic response of the soil and the deposit may be considered to have an infinite lateral extent. If it is assumed, in addition, that such a response is caused by SH waves propagating vertically from the underlying rock, then it may be considered that the motion of the soil deposit is the result of the shear deformations induced in the soil by a horizontal motion acting at its base. In such a case, the motion generated by a known bedrock motion along the depth of the soil deposit and at the ground surface may be determined on the basis of a simple one-dimensional shear-beam model with distributed mass, rigidity, and damping. Moreover, if it is assumed that the soil possesses linear elastic properties, then a closed-form solution may be formulated as follows.

Consider the homogeneous soil deposit shown in Figure 8.20 and let this soil deposit be characterized by its shear modulus of elasticity G, density ρ, and damping constant c (force/unit velocity/

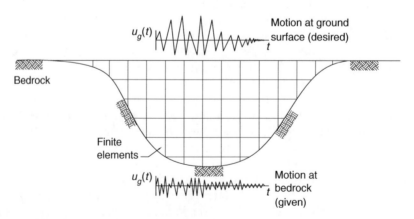

FIGURE 8.19 Finite element model of soil deposit.

Influence of Local Site Conditions

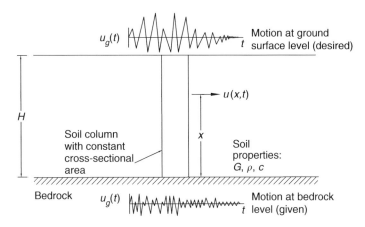

FIGURE 8.20 Profile of horizontal deposit of homogeneous soil.

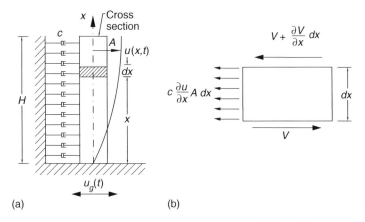

FIGURE 8.21 (a) Shear-beam model and (b) free-body diagram of a differential element.

unit volume). Consider, in addition, a column of such a soil deposit and denote the cross-sectional area of this column by A. Furthermore, let such a column be represented, as shown in Figure 8.21a, by a one-dimensional shear beam fixed at its base, free at its other end, restrained by uniformly distributed external viscous dampers, and subjected to a base displacement $u_g(t)$. Let $u(x, t)$ denote the displacement at a distance x from the base of the beam at time t, and let V depict the corresponding shear force.

At any time after the application of the base excitation, the forces acting on a differential element of the beam are as indicated in the element's free-body diagram shown in Figure 8.21b. Therefore, by application of Newton's second law, the motion of such a differential element may be described by

$$-V + V + \frac{\partial V}{\partial x}dx - cA\,dx\,\frac{\partial u(x,t)}{\partial t} = \rho A\,dx\,\frac{\partial^2 [u_g(t) + u(x,t)]}{\partial t^2} \tag{8.1}$$

which, after rearranging terms and simplifying, may also be written as

$$\rho A \frac{\partial^2 u(x,t)}{\partial^2 t} - \frac{\partial V}{\partial x} + cA\frac{\partial u(x,t)}{\partial t} = -\rho A \frac{\partial^2 u_g(t)}{\partial t^2} \tag{8.2}$$

However, by Hooke's law for shear stresses, the shear force V may be expressed as

$$V = A\tau = AG\gamma = AG\frac{\partial u}{\partial x} \tag{8.3}$$

where τ and γ denote shear stress and shear strain, respectively. Consequently, $\partial V/\partial x$ may be written as

$$\frac{\partial V}{\partial x} = AG\frac{\partial^2 u}{\partial x^2} \tag{8.4}$$

and Equation 8.2 as

$$\rho\frac{\partial^2 u(x,t)}{\partial^2 t} + c\frac{\partial u(x,t)}{\partial t} - G\frac{\partial^2 u(x,t)}{\partial^2 x} = -\rho\frac{\partial^2 u_g(t)}{\partial t^2} \tag{8.5}$$

For the free-vibration case, this equation is reduced to

$$\rho\frac{\partial^2 u(x,t)}{\partial^2 t} + c\frac{\partial u(x,t)}{\partial t} - G\frac{\partial^2 u(x,t)}{\partial^2 x} = 0 \tag{8.6}$$

which is a partial differential equation that may be solved by the method of separation of variables. For this purpose, let $u(x,t)$ be of the form

$$u(x,t) = U(x)T(t) \tag{8.7}$$

where $U(x)$ represents a function of x alone and $T(t)$ a function of t alone. Accordingly, substitution of this equation into Equation 8.6 leads to

$$\rho U(x)\frac{d^2 T(t)}{dt^2} + cU(x)\frac{dT(t)}{dt} = GT(t)\frac{d^2 U(x)}{dx^2} \tag{8.8}$$

which may be rewritten as

$$\frac{1}{T(t)}\frac{d^2 T(t)}{dt^2} + \frac{c}{\rho}\frac{1}{T(t)}\frac{dT(t)}{dt} = \frac{G}{\rho}\frac{1}{U(x)}\frac{d^2 U(x)}{dx^2} \tag{8.9}$$

Note, however, that the left-hand side of the equation is a function of time alone and the right-hand side a function of x alone. Therefore, it may be concluded that each side should be equal to a constant. By denoting this constant as $-\omega_n^2$, one can then obtain two ordinary differential equations as follows. From the left-hand side of Equation 8.9, one gets

$$\ddot{T} + 2\xi_n\omega_n\dot{T} + \omega_n^2 T = 0 \tag{8.10}$$

where ξ_n is defined as

$$\xi_n = \frac{c}{2\omega_n\rho} \tag{8.11}$$

and a dot over a variable signifies differentiation with respect to t. Similarly, from the right-hand side of Equation 8.9, one obtains

$$U'' + \frac{\omega_n^2}{v_s^2}U = 0 \tag{8.12}$$

where

$$v_s = \sqrt{\frac{G}{\rho}} \tag{8.13}$$

and a prime denotes differentiation with respect to x.

From elementary structural dynamics, the solution to Equation 8.10 is given by

$$T(t) = e^{-\xi_n \omega_n t}(A_1 \sin \omega_{dn} t + A_2 \cos \omega_{dn} t) = A_n e^{-\xi_n \omega_n t} \cos(\omega_n t - \theta_n) \tag{8.14}$$

where A_n is a constant that represents vibration amplitude and equal to $(A_1^2 + A_2^2)^{1/2}$, and θ_n, is a phase angle equal to $\tan^{-1}(A_2/A_1)$. In addition, $\omega_{dn} = \omega_n(1 - \xi_n^2)^{1/2}$ and ω_n is the constant introduced earlier, which may now be interpreted as an oscillation frequency and hence the beam's natural frequency. Likewise, ξ_n may now be interpreted as a damping ratio. The constants A_1 and A_2 may be determined from the initial conditions. The solution of Equation 8.12 is of the form

$$U(x) = B_1 \sin \frac{\omega_n}{v_s} x + B_2 \cos \frac{\omega_n}{v_s} x \tag{8.15}$$

where B_1 and B_2 are constants that can be determined from the boundary conditions. For the case under consideration, these boundary conditions are as follows. At $x = 0$

$$u(x,t) = T(t)U(x) = 0 \tag{8.16}$$

from which it may be concluded that

$$U(0) = 0 \tag{8.17}$$

At $x = H$

$$\tau = G\frac{\partial u}{\partial x} = GT(t)U'(x) = 0 \tag{8.18}$$

from which it may be determined that

$$U'(H) = 0 \tag{8.19}$$

Upon substitution of $x = 0$ into Equation 8.15 and consideration of Equation 8.17, one thus finds that

$$B_2 = 0 \tag{8.20}$$

and, as a result, that

$$U(x) = B_1 \sin \frac{\omega_n}{v_s} x \qquad (8.21)$$

Similarly, as

$$U'(x) = \frac{\omega_n}{v_s} B_1 \cos \frac{\omega_n}{v_s} x \qquad (8.22)$$

substitution of $x = H$ into this equation and consideration of Equation 8.19 lead to

$$\frac{\omega_n}{v_s} B_1 \cos \frac{\omega_n}{v_s} H = 0 \qquad (8.23)$$

This equation is satisfied by either $B_1 = 0$ or by making the angle of the cosine function in it equal to $(2n-1)\pi/2$, where n is any integer. However, $B_1 = 0$ leads to a trivial solution (i.e., $u(x,t) = 0$). Consequently, from Equation 8.23 it may be concluded that for a solution other than the trivial one it is necessary that

$$\frac{\omega_n}{v_s} H = (2n-1)\frac{\pi}{2} \quad n = 1, 2, \ldots, \infty \qquad (8.24)$$

from which one finds that

$$\omega_n = \frac{2n-1}{2} \pi \frac{v_s}{H} \quad n = 1, 2, \ldots, \infty \qquad (8.25)$$

Similarly, after substitution of this equation into Equation 8.21, one finds that

$$U(x) = B_1 \sin\left(\frac{2n-1}{2} \pi \frac{x}{H}\right) \quad n = 1, 2, \ldots, \infty \qquad (8.26)$$

Notice that as the constant B_1 in this equation cannot be determined, $U(x)$ may be interpreted as a mode shape. That is, $U(x)$ can be defined in terms of relative amplitudes but not absolute values. This also means that $U(x)$ can be normalized with respect to any constant and that the value of B_1 may be selected arbitrarily as 1.0. Thus, Equations 8.25 and 8.26, respectively, define the natural frequencies and mode shapes of the soil deposit under consideration. Note, too, that, according to Equation 8.25, the fundamental natural period of the soil deposit is given by

$$T_1 = \frac{2\pi}{\omega_1} = \frac{2\pi}{\pi/2} \frac{H}{v_s} = 4\frac{H}{v_s} \qquad (8.27)$$

which indicates that such a fundamental natural period may be considered as being equal to four times the time it takes a shear wave to travel along the depth of the soil deposit. Finally, note that, according to Equations 8.7, 8.14, and 8.26, in each mode there is a solution of the form

$$u_n(x,t) = A_n e^{-\xi_n \omega_n t} \sin\left(\frac{2n-1}{2} \pi \frac{x}{H}\right) \cos(\omega_n t - \theta_n) \qquad (8.28)$$

Influence of Local Site Conditions

The general solution of Equation 8.6 is therefore given by the linear combination of all such solutions.

Equation 8.12, the differential equation that defines $U(x)$, is satisfied by Equation 8.26. That is, Equation 8.12 is satisfied by any of the mode shapes of the system. Therefore, for the rth and sth modes, Equation 8.12 may be written as

$$GU_r''(x) + \rho\omega_r^2 U_r(x) = 0 \tag{8.29}$$

$$GU_s''(x) + \rho\omega_s^2 U_s(x) = 0 \tag{8.30}$$

where Equation 8.13 has been used, and $U_r(x)$ and $U_s(x)$, respectively, denote the rth and sth mode shapes of the system and ω_r^2 and ω_s^2 are the corresponding natural frequencies. If, however, Equations 8.29 and 8.30 are, respectively, multiplied through by $U_s(x)$ and $U_r(x)$ and both are integrated over H, the depth of the soil deposit, one obtains

$$\int_0^H GU_r''(x)U_s(x)dx + \omega_r^2 \int_0^H \rho U_r(x)U_s(x)dx = 0 \tag{8.31}$$

$$\int_0^H GU_s''(x)U_r(x)dx + \omega_s^2 \int_0^H \rho U_s(x)U_r(x)dx = 0 \tag{8.32}$$

One may note, furthermore, that the integrals on the left-hand sides of these equations may be integrated by parts (recalling that $\int uv'\, dx = uv - \int vu'\, dx$) and thus that these integrals may be expressed alternatively as

$$\int_0^H GU_r''(x)U_s(x)dx = GU_r'(x)U_s(x)\big|_0^H - \int_0^H GU_r'(x)U_s'(x)dx \tag{8.33}$$

$$\int_0^H GU_s''(x)U_r(x)dx = GU_s'(x)U_r(x)\big|_0^H - \int_0^H GU_s'(x)U_r'(x)dx \tag{8.34}$$

which, after recalling that $U_r(0) = U_s(0) = U_r'(H) = U_s'(H) = 0$, may be reduced to

$$\int_0^H GU_r''(x)U_s(x)dx = -\int_0^H GU_r'(x)U_s'(x)dx \tag{8.35}$$

$$\int_0^H GU_s''(x)U_r(x)dx = -\int_0^H GU_s'(x)U_r'(x)dx \tag{8.36}$$

As a result, Equations 8.31 and 8.32 may also be written as

$$-\int_0^H GU_r'(x)U_s'(x)dx + \omega_r^2 \int_0^H \rho U_r(x)U_s(x)dx = 0 \tag{8.37}$$

$$-\int_0^H GU_s'(x)U_r'(x)dx + \omega_s^2 \int_0^H \rho U_s(x)U_r(x)dx = 0 \tag{8.38}$$

Furthermore, after subtracting Equation 8.38 from Equation 8.37, one gets

$$(\omega_r^2 - \omega_s^2)\int_0^H \rho U_r(x)U_s(x)dx = 0 \tag{8.39}$$

which for $\omega_r \neq \omega_s$ implies that

$$\int_0^H \rho U_r(x)U_s(x)dx = 0 \quad r \neq s \tag{8.40}$$

Similarly, after substituting this result into Equation 8.31, it is found that

$$\int_0^H GU_r''(x)U_s(x)dx = 0 \quad r \neq s \tag{8.41}$$

Equations 8.40 and 8.41 show that the mode shapes of the system under consideration are orthogonal with either the rigidity or the density of the system as the weighting factor. As shown next, this property of the mode shapes may be used conveniently to simplify the analysis of the system under forced vibrations.

In view of the existence of classical mode shapes, it is possible to express in terms of these mode shapes the solution of the differential equation that describes the motion of the shear beam under study when the beam is subjected to a base excitation. Returning to the solution of Equation 8.5, let, then, this solution be expressed in terms of normal coordinates as

$$u(x,t) = \sum_{r=1}^{\infty} U_r(x)\eta_r(t) \tag{8.42}$$

where, as before, $U_r(x)$ denotes the rth mode shape of the system and $\eta_r(t)$ is, at this point, an unknown modal coordinate. Upon substitution of this transformation, one may therefore write Equation 8.5 as

$$\rho \sum_{r=1}^{\infty} U_r(x)\ddot{\eta}_r(t) + c\sum_{r=1}^{\infty} U_r(x)\dot{\eta}_r(t) - G\sum_{r=1}^{\infty} U_r''(x)\eta_r(t) = -\rho \ddot{u}_g(t) \tag{8.43}$$

Moreover, if each term in this last equation is multiplied by $U_s(x)$ and integrated over the depth H, Equation 8.5 may be expressed alternatively as

$$\sum_{r=1}^{\infty}\int_0^H \rho U_r(x)U_s(x)dx\ddot{\eta}_r(t) + \sum_{r=1}^{\infty}\int_0^H cU_r(x)U_s(x)dx\dot{\eta}_r(t) - \sum_{r=1}^{\infty}\int_0^H GU_r''(x)U_s(x)dx\eta_r(t)$$
$$= -\ddot{u}_g(t)\int_0^H \rho U_s(x)dx \tag{8.44}$$

Note, however, that in view of the orthogonality properties of the mode shapes (see Equations 8.40 and 8.41), all terms in the first and third infinite series in this equation are zero except those for which $r = s$. Furthermore, the same can be said about the second term if it is observed that, according to Equation 8.11, the damping constant c is proportional to ρ and, as a result, the mode shapes are also orthogonal with respect to c. Accordingly, Equation 8.44 may be reduced to

$$\ddot{\eta}_r(t)\int_0^H \rho U_r^2(x)dx + \dot{\eta}_r(t)\int_0^H cU_r^2(x)dx - \eta_r(t)\int_0^H GU_r''(x)U_r(x)dx = -\ddot{u}_g(t)\int_0^H \rho U_r(x)dx \tag{8.45}$$

which in view of Equations 8.11 and 8.29 may also be put into the form

$$\ddot{\eta}_r(t)\int_0^H \rho U_r^2(x)dx + 2\xi_r\omega_r\dot{\eta}_r(t)\int_0^H \rho U_r^2(x)dx + \omega_r^2\eta_r(t)\int_0^H \rho U_r^2(x)dx = -\ddot{u}_g(t)\int_0^H \rho U_r(x)dx \tag{8.46}$$

or

$$\ddot{\eta}_r(t) + 2\xi_r\omega_r\dot{\eta}_r(t) + \omega_r^2\eta_r(t) = -\alpha_r\ddot{u}_g(t) \tag{8.47}$$

where α_r is a participation factor defined as

$$\alpha_r = \frac{\int_0^H \rho U_r(x)dx}{\int_0^H \rho U_r^2(x)dx} \tag{8.48}$$

Equation 8.47 represents the differential equation that describes the response of a single-degree-of-freedom system to a ground excitation $\alpha_r\ddot{u}_g(t)$. As such, its solution may be obtained using any of the methods employed in elementary structural dynamics. In particular, for zero initial conditions and in terms of Duhamel integral, the solution to Equation 8.47 may be expressed as

$$\eta_r(t) = -\frac{\alpha_r}{\omega_{dr}}\int_0^t \ddot{u}_g(t)e^{-\xi_r\omega_r(t-\tau)}\sin\omega_{dr}(t-\tau)d\tau \tag{8.49}$$

where, as mentioned earlier, $\omega_{dr} = \omega_r(1-\xi_r^2)^{1/2}$. Notice also that, after making use of Equation 8.26 and carrying out the integrations, the participation factor α_r becomes

$$\alpha_r = \frac{\int_0^H \sin((2r-1)/2)\pi(x/H)dx}{\int_0^H \sin^2((2r-1)/2)\pi(x/H)dx} = \frac{4}{(2r-1)\pi} \tag{8.50}$$

Thus, after substituting Equations 8.49 and 8.50 into Equation 8.42, the displacement response of the soil deposit under consideration may be finally expressed as

$$u(x,t) = -\sum_{r=1}^{\infty} \frac{4}{(2r-1)\pi}\frac{U_r(x)}{\omega_{dr}}\int_0^t \ddot{u}_g(t)e^{-\xi_r\omega_r(t-\tau)}\sin\omega_{dr}(t-\tau)d\tau \tag{8.51}$$

where ω_r and $U_r(x)$, respectively, denote the rth natural frequency and mode shape of the system, which according to Equations 8.25 and 8.26 are given by

$$\omega_r = \frac{(2r-1)\pi}{2H}\sqrt{\frac{G}{\rho}} \quad r = 1, 2, \ldots, \infty \tag{8.52}$$

$$U_r(x) = \sin\left(\frac{2r-1}{2}\pi\frac{x}{H}\right) \quad r = 1, 2, \ldots, \infty \tag{8.53}$$

and where ω_{dr} signifies the rth damped natural frequency. The corresponding velocity and acceleration responses may be obtained by simply taking the first and second derivatives of $u(x, t)$ with respect to t. Similarly, the shear stress response may be determined on the basis of Equation 8.3 and the first derivative of $u(x, t)$ with respect to x.

Strictly speaking, the solution presented here is valid only for horizontal soil deposits with homogeneous and linear elastic properties. In general, however, soils do not behave as a linear elastic solid and their properties are likely to vary with depth, even in deposits composed of a single material. Nonetheless, useful results may be obtained with it when used in combination with average values of the varying properties and equivalent linear properties. Being a closed-form solution, perhaps its greatest value lies on its usefulness as a tool to assess the accuracy of other techniques.

EXAMPLE 8.1   SURFACE MOTION GIVEN MOTION AT THE BASE OF SOIL DEPOSIT

A soil deposit, considered to be homogeneous, linear elastic, and with a horizontal free surface and rock interface, has a depth of 50 ft, a unit weight of 120 pcf, a shear modulus of elasticity of 2.41×10^6 psf, a Poisson ratio of 0.45, and a damping ratio of 20% in all modes. During an earthquake, the base of the deposit is subjected to an acceleration that may be characterized by a harmonic function with an amplitude of 0.2g and a frequency of 25 rad/s. Determine the acceleration induced by this ground motion at the level of the deposit's free surface. Determine also the factor by which the ground accelerations at the base are amplified by the soil deposit. Consider only the response of the deposit in its first two modes.

Solution

According to Equations 8.50, 8.52 and 8.53, the natural frequencies, mode shapes, and participation factors in the first two modes of the soil deposit are

$$\omega_1 = \frac{(2-1)\pi}{2H}\sqrt{\frac{G}{\rho}} = \frac{\pi}{2(50)}\sqrt{\frac{(2.41 \times 10^6)(32.2)}{120}} = 25.3 \text{ rad/s}$$

$$\omega_2 = \frac{(4-1)\pi}{2H}\sqrt{\frac{G}{\rho}} = 3(25.3) = 75.8 \text{ rad/s}$$

$$U_1(x) = \sin\left(\frac{2-1}{2}\pi\frac{x}{H}\right) = \sin\left(\frac{\pi}{100}x\right)$$

$$U_2(x) = \sin\left(\frac{4-1}{2}\pi\frac{x}{H}\right) = \sin\left(\frac{3\pi}{100}x\right)$$

$$\alpha_1 = \frac{4}{(2-1)\pi} = \frac{4}{\pi} \qquad \alpha_2 = \frac{4}{(4-1)\pi} = \frac{4}{3\pi}$$

Similarly, as for the case under consideration the acceleration at the base of the soil deposit is given by

$$\ddot{u}_g(t) = 0.2g \sin(25t)$$

and for a harmonic displacement the normal coordinate η_r is given by (see Equation 8.47)

$$\eta_r(t) = -\alpha_r \frac{1}{(\omega_r)^2} \frac{\ddot{u}_g(t)\sin(\Omega t - \theta_r)}{\sqrt{[1-(\Omega/\omega_r)^2]^2 + [2\xi_r\Omega/\omega_r]^2}}$$

where Ω is the excitation frequency and θ_r a phase angle given by

$$\theta_r = \tan^{-1}\left[\frac{2\xi_r\Omega/\omega_r}{1-(\Omega/\omega_r)^2}\right]$$

Influence of Local Site Conditions

the modal coordinates for the first two modes of the system are

$$\eta_1(t) = -\frac{4}{\pi}\frac{1}{25.3^2}\frac{0.2g\sin(25t-\theta_1)}{\sqrt{[1-(25/25.3)^2]^2+[2(0.2)25/25.3]^2}} = -1.00\times10^{-3}g\sin(25t-1.5)$$

$$\eta_2(t) = -\frac{4}{3\pi}\frac{1}{75.8^2}\frac{0.2g\sin(25t-\theta_2)}{\sqrt{[1-(25/75.8)^2]^2+[2(0.2)25/75.8]^2}} = -1.64\times10^{-5}g\sin(25t-0.15)$$

Thus, according to Equation 8.42 and considering only the first two modes, the displacement response of the soil deposit at the level of its free surface ($H = 50$ ft) results as

$$u(50,t) = -1.00\times10^{-3}g\sin(25t-1.5) + 1.64\times10^{-5}g\sin(25t-0.15)$$

where it has been considered that $U_1(50) = 1.0$ and $U_2(50) = -1.0$. Consequently, the acceleration response at the same level is given by

$$\ddot{u}(50,t) = (25)^2 1.00\times10^{-3}g\sin(25t-1.5) - (25)^2 1.64\times10^{-5}g\sin(25t-0.15)$$

$$= 0.625g\sin(25t-1.5) - 0.010g\sin(25t-0.15)$$

from which it may be seen that the accelerations at the surface of the soil deposit are greater than the accelerations at its base. If the contribution of the second mode is neglected, the amplification factor is equal to

$$AF = \frac{\max[\ddot{u}_g(50,t)+\ddot{u}_g(t)]}{\max[\ddot{u}_g(t)]} = \frac{\sqrt{0.625^2+0.244^2}\,g}{0.2g} = 3.35$$

8.4.3 One-Dimensional Lumped-Mass Model

If, as mentioned earlier, the properties of a soil deposit vary with depth, or if it is composed of several layers with different properties, then its seismic response cannot be determined using a uniform shear-beam model. In such a case, it is necessary to resort to a lumped-mass model. In particular, if the ground surface and the layers in a stratified soil are basically horizontal as shown in Figure 8.22, then it is possible to model the soil as a one-dimensional shear building with lumped masses.

In a one-dimensional, lumped-mass model, a column of soil with a unit cross-sectional area is also assumed to behave as a vertical shear beam that is fixed at its base and free at its other end.

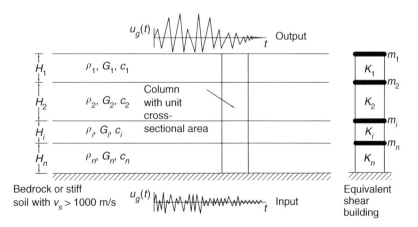

FIGURE 8.22 Horizontally stratified soil profile and equivalent shear building.

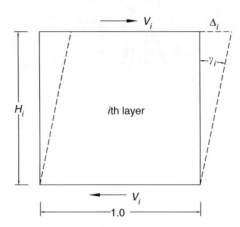

FIGURE 8.23 The ith layer of soil deposit under shear forces.

But instead of analyzing it as a continuous system, it is considered to be composed of a series of discrete masses interconnected by springs that resist lateral deformations. With this model, the mass of the soil is first discretized at a series of points. After this discretization is performed, it is then analyzed as a shear building. To determine the masses of this fictitious shear building, the mass of each floor is considered to be equal to half the mass of the layer above plus half the mass of the layer below (see Figure 8.22). That is, if ρ_i and H_i, respectively, represent the density and thickness of the ith soil layer, and if ρ_{i+1} and H_{i+1} denote those of the $(i+1)$th layer, the mass of the ith floor is calculated as

$$m_i = \frac{1}{2}\rho_i H_i + \frac{1}{2}\rho_{i+1}H_{i+1} \tag{8.54}$$

Likewise, it may be seen from Figure 8.23 and Hooke's law for shear stresses that under the action of shear forces V_i, the shear stresses at the boundaries of the ith layer of the column of soil being considered are given by

$$\tau_i = \frac{V_i}{(1.0)(1.0)} = G_i \gamma_i = G_i \frac{\Delta_i}{H_i} \tag{8.55}$$

where G_i is the shear modulus of the layer, and γ_i and Δ_i are, respectively, the shear strain and lateral displacement induced by the shear forces V_i. From the definition of stiffness coefficient or spring constant, the lateral stiffness of the layer under consideration is therefore given by

$$K_i = \frac{V_i}{\Delta_i} = \frac{G_i}{H_i} \tag{8.56}$$

As in the case of buildings, damping in a soil profile may be taken into account by specifying modal damping ratios. The values of the shear modulus and damping ratios for each layer may be determined from appropriate laboratory or field tests, or estimated as suggested in Section 8.5.

Once the masses, spring constants, and damping ratios are defined, the system's equation of motion is established following the approach described in most structural dynamics books for spring-mass systems or shear buildings. In turn, this equation of motion may be solved using any of the step-by-step integration procedures described in Chapter 11. Observe, thus, that after the discretization is made, the analysis of a soil profile can be easily carried out using any of the techniques or computer programs that are available for the analysis of structural systems. It should be

Influence of Local Site Conditions

noted, however, that the discretization procedure introduces an error whose magnitude depends on the number of lumped masses considered. To minimize this error, it is recommended to subdivide thick layers into a series of thin sublayers. It should also be noted that an additional advantage of the lumped-mass model over the continuous one is that the lumped-mass model allows for the consideration of soil profiles with a nonlinear stress–strain behavior.

EXAMPLE 8.2 DETERMINATION OF SURFACE MOTION USING LUMPED-MASS APPROACH

Using a computer program for the response analysis of shear buildings, determine the acceleration time history at the surface of the layered soil deposit shown in Figure E8.2a when the bedrock below the soil

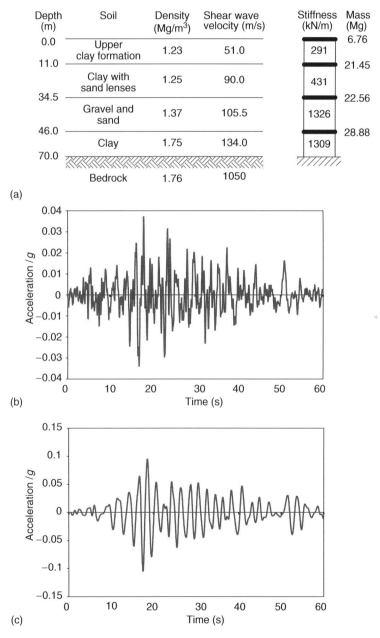

FIGURE E8.2 (a) Stratified soil profile and equivalent shear building considered in Example 8.2 and ground motions at (b) the base of soil deposit and (c) at the surface of soil deposit.

deposit is excited horizontally by the ground motion shown in Figure E8.2b. Use a lumped-mass, one-dimensional model, assuming elastic behavior and a damping ratio of 10% in the first mode of the system.

Solution

According to Equation 8.54 and the soil densities given in Figure E8.2a, the masses of an equivalent shear building are given by

$$m_1 = \frac{1}{2}(1.75)(70.0 - 46.0) + \frac{1}{2}(1.37)(46.0 - 34.5) = 28.88 \text{ Mg}$$

$$m_2 = \frac{1}{2}(1.37)(46.0 - 34.5) + \frac{1}{2}(1.25)(34.5 - 11.0) = 22.56 \text{ Mg}$$

$$m_3 = \frac{1}{2}(1.25)(34.5 - 11.0) + \frac{1}{2}(1.23)(11.0) = 21.45 \text{ Mg}$$

$$m_4 = \frac{1}{2}(1.23)(11.0) = 6.76 \text{ Mg}$$

Similarly, according to Equations 8.13 and 8.56 and the given shear wave velocities, the spring constants of the equivalent shear building are

$$K_1 = \frac{(1.75)(134.0)^2}{70.0 - 46.0} = 1309 \text{ kN/m}$$

$$K_2 = \frac{(1.37)(105.5)^2}{46.0 - 34.5} = 1326 \text{ kN/m}$$

$$K_3 = \frac{(1.25)(90)^2}{34.5 - 11.0} = 431 \text{ kN/m}$$

$$K_4 = \frac{(1.23)(51.0)^2}{11.0} = 291 \text{ kN/m}$$

Hence, the soil deposit under consideration may be substituted with the shear building shown in the right-hand side of Figure E8.2a. The analysis of this equivalent shear building when the base of the building is excited by the given bedrock motion thus leads to the top floor acceleration response shown in Figure E8.2c. This top floor response represents the accelerations that would be observed at the surface of the soil deposit when the bedrock experiences the ground motion shown in Figure E8.2b. Note that as a result of its particular soil properties and geometry, this soil profile would amplify the given ground motion by a factor of ~2.8.

It should be noted at this point that the continuous and the lumped-mass models described earlier consider the base of the equivalent shear beam to be fixed. In other words, both models assume that the bedrock below the soil deposit is infinitely rigid. In actuality, the bedrock in all cases has some flexibility and it is thus capable of transmitting waves. The consequence of this capability is that when the waves reflected from the free surface reach the interface between the bedrock and a soil deposit, part of the energy carried by these waves are transmitted back into the bedrock. As a result, energy is lost in the process and less energy is available to excite the soil deposit. Fixed-base models do not account for this energy loss and may thus lead to an overestimation of the motion generated at the free surface. A way to take into account such an energy loss is by introducing viscous dampers at the base of the fixed-base models. These viscous dampers absorb some of the energy carried by the waves that reach the soil–rock interface and, thus, simulate the effect of the energy loss that results from the transmission of waves back into the bedrock. The characteristics of these viscous dampers may be determined as discussed in Section 13.12.2.

8.4.4 One-Dimensional Continuous Model: Solution in Frequency Domain

A solution in the frequency domain offers an *exact* method for the evaluation of site effects when a soil deposit is composed of several layers with different properties. In this method, the bedrock motion is first represented in the frequency domain in the form of a Fourier series, usually through the use of the fast Fourier transform (see Section 6.6.6). Then, each term in this Fourier series is multiplied by the transfer function of the soil to obtain the Fourier series of the motion at the ground surface. Thereafter, the ground surface motion is expressed in the time domain through the use of the inverse fast Fourier transform. The key to this approach is thus the evaluation of the transfer function of the soil deposit under investigation. To illustrate, then, how this transfer function can be determined, four simple cases are studied, each with an increasing degree of complexity. An example is also presented to show how a transfer function and a frequency-domain approach can be used to determine the motion at the surface of a soil deposit given the motion observed on the bedrock underneath.

8.4.4.1 Case I: Homogeneous, Undamped Soil on Rigid Rock

To introduce in a simple way the basic concept behind the evaluation of the transfer function of a soil profile for its use in conjunction with the frequency-domain technique outlined in the foregoing paragraph, consider first the ideal case of a homogeneous linear elastic soil layer with no damping resting on a rigid bedrock (see Figure 8.24). If the bedrock is subjected to a horizontal harmonic motion, then this motion will produce incident and reflected vertically propagating waves in the soil above. Using the exponential form to represent harmonic waves (see Section 4.4), the horizontal displacement resulting from these waves may be expressed as

$$u(z, t) = Ae^{i(\omega t + kz)} + Be^{i(\omega t - kz)} \tag{8.57}$$

where ω denotes the waves' circular frequency, k represents their wave number ($= \omega/v_s$), and A and B are the amplitudes of the incident and reflected waves, respectively. At the free surface ($z = 0$), the shear stress must vanish; hence, from this condition it may be established that

$$\tau(0, t) = G\gamma = G\frac{\partial u}{\partial z}\bigg|_{z=0} = 0 \tag{8.58}$$

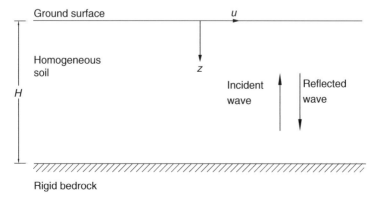

FIGURE 8.24 Homogeneous soil layer of thickness H overlying rigid bedrock.

which, upon substitution of Equation 8.57, leads to

$$ik\left[Ae^{i(\omega t+kz)} - Be^{i(\omega t-kz)}\right]_{z=0} = ik(A - B)e^{i\omega t} = 0 \tag{8.59}$$

which, for a nontrivial solution, is satisfied when $A = B$. As a result, the displacement under consideration may be written as

$$u(z,t) = 2A\frac{e^{ikz} + e^{-ikz}}{2}e^{i\omega t} = 2Ae^{i\omega t}\cos kz \tag{8.60}$$

This result and the definition of transfer function may now be used to define the transfer function that relates the amplitudes of the displacements at the top and bottom of the soil layer. Accordingly, this transfer function may be expressed as

$$H(\omega) = \frac{u(0,t)}{u(H,t)} = \frac{2Ae^{i\omega t}\cos(k0)}{2Ae^{i\omega t}\cos(kH)} = \frac{1}{\cos(kH)} = \frac{1}{\cos(\omega H/v_s)} \tag{8.61}$$

from which it may be seen that the modulus of this function (amplification function) is given by

$$|H(\omega)| = \frac{1}{|\cos(\omega H/v_s)|} \tag{8.62}$$

Notice, thus, that for this ideal case the surface displacement is always at least as large as the bedrock displacement and may, at certain frequencies, be much larger. In the extreme case, as $\omega H/v_s$ approaches $(2n-1)\pi/2$, where $n = 1, 2, \ldots, \infty$, the denominator of Equation 8.62 approaches zero and the amplification function becomes infinitely large (see Figure 8.25). That is, resonance is developed whenever $\omega H/v_s = (2n-1)\pi/2$, $n = 1, 2, \ldots, \infty$. As in an undamped system, an infinitely large amplification results whenever the excitation frequency matches one of the natural frequencies of the system, then it may be concluded that the natural frequencies of the soil layer under consideration are given by

$$\omega_n = \frac{2n-1}{2}\pi\frac{v_s}{H} \quad n = 1, 2, \ldots, \infty \tag{8.63}$$

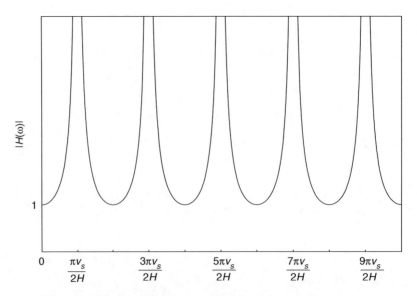

FIGURE 8.25 Amplification function of undamped linear elastic soil profile.

Influence of Local Site Conditions

which coincides with the formula given by Equation 8.25.

It may be seen, thus, that the dynamic response of a soil deposit is highly dependent on the dominant frequency of the bedrock motion and that the excitation frequency for which the amplification becomes the largest depends on the soil geometry (thickness) and rigidity (shear wave velocity).

8.4.4.2 Case II: Homogeneous, Damped Soil on Rigid Rock

Consider now that the soil layer shown in Figure 8.24 possesses internal damping and assume that it behaves as a viscoelastic solid. According to the discussion in Section 13.9, a viscoelastic solid may be characterized by the complex shear modulus

$$G^* = G(1 + 2\xi i) \tag{8.64}$$

where G is the conventional shear modulus of elasticity, ξ the damping ratio, and i the unit imaginary number. The shear wave velocity in a viscoelastic solid may be therefore expressed as

$$v_s^* = \sqrt{\frac{G^*}{\rho}} = \sqrt{\frac{G(1+2\xi i)}{\rho}} \tag{8.65}$$

which for small damping ratios may be approximated as

$$v_s^* = \sqrt{\frac{G}{\rho}}(1 + \xi i) = v_s(1 + \xi i) \tag{8.66}$$

As a result, the wave number for a viscoelastic solid is of the form

$$k^* = \frac{\omega}{v_s^*} = \frac{\omega}{v_s(1+\xi i)} \tag{8.67}$$

which for small damping ratios is approximately equal to

$$k^* = \frac{\omega}{v_s}(1 - \xi i) = k(1 - \xi i) \tag{8.68}$$

Similarly, the displacements induced by vertical incident and reflected waves may be expressed as

$$u(z, t) = A e^{i(\omega t + k^* z)} + B e^{i(\omega t - k^* z)} \tag{8.69}$$

Furthermore, after substituting k^* for k in Equation 8.61, the transfer function for the soil layer under consideration may be written as

$$H(\omega) = \frac{1}{\cos[k(1-\xi i)H]} = \frac{1}{\cos[\omega(1-\xi i)H/v_s]} \tag{8.70}$$

and, as $|\cos(x + iy)|^2 = \cos^2 x + \sinh^2 y$, the corresponding amplification function may be expressed as

$$|H(\omega)| = \frac{1}{\sqrt{\cos^2 kH + \sinh^2 k\xi H}} \tag{8.71}$$

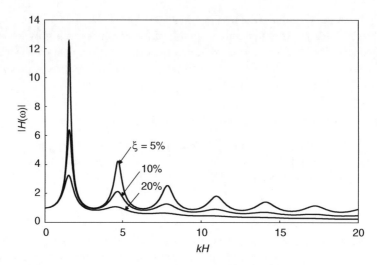

FIGURE 8.26 Amplification function of damped linear elastic soil layer.

or, as for small damping ratios $\sinh^2 k\xi H \approx k\xi H$, as

$$|H(\omega)| = \frac{1}{\sqrt{\cos^2 kH + (k\xi H)^2}} = \frac{1}{\sqrt{\cos^2(\omega H/v_s) + (\xi\omega H/v_s)^2}} \quad (8.72)$$

A plot of this amplification function is shown in Figure 8.26 for three values of the damping ratio ξ. Observe from this figure that whenever kH is nearly equal to $(2n-1)\pi/2$, $n = 1, 2, \ldots, \infty$, a local maximum is attained. As expected, this maximum never reaches an infinite value because for $\xi > 0$ the denominator on the right-hand side of Equation 8.72 never becomes zero. As expected, too, such local maxima approximately correspond to the undamped natural frequencies of the soil layer. Finally, observe that, in contrast with the undamped case, the amplification is larger at low frequencies than at high frequencies. This means that the largest amplification in a soil deposit will always occur for excitations whose dominant frequencies are approximately equal to the lowest natural frequency of the soil deposit.

EXAMPLE 8.3 DETERMINATION OF SURFACE MOTION USING FREQUENCY-DOMAIN APPROACH

Consider a homogeneous soil deposit with a depth of 100 m, a shear modulus of elasticity of 77.0 MN/m², and a density of 1924 kg/m³. In addition, assume that the soil deposit behaves as a viscoelastic solid with a damping ratio of 20% and that the bedrock below is infinitely rigid. Determine using the frequency-domain approach described in Section 8.4.4, the acceleration time history at the surface of this soil deposit when the bedrock below is excited horizontally by the acceleration time history shown in Figure E8.3a.

Solution

According to its geometric and elastic properties, the shear wave velocity of the given soil deposit is equal to

$$v_s = \sqrt{\frac{G}{\rho}} = \sqrt{\frac{77.0 \times 10^6}{1924}} = 200 \text{ m/s}$$

Influence of Local Site Conditions

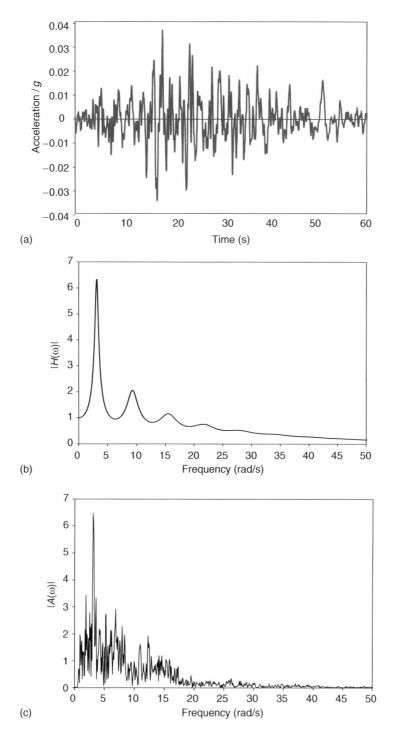

FIGURE E8.3 (a) Acceleration time history at the base of soil deposit, (b) amplification function of soil deposit in Example 8.3, (c) modulus of Fourier transform of acceleration time history in (a), (d) modulus of product of transfer function $H(\omega)$ and acceleration function $A(\omega)$, and (e) acceleration time history at the surface of soil deposit.

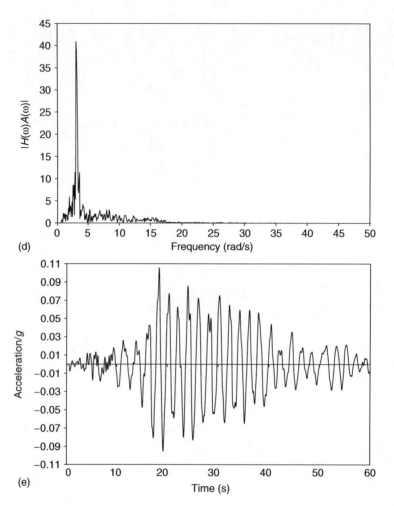

FIGURE E8.3 (Continued)

Thus, in view of Equation 8.70 and the assumption of a viscoelastic solid with a damping ratio of 10%, the transfer function of the soil deposit is given by

$$H(\omega) = \frac{1}{\cos[\omega(1-\xi i)H/v_s]} = \frac{1}{\cos[\omega(1-0.2i)(100/200)]} = \frac{1}{\cos[(0.5-0.1i)\omega]}$$

where it has been considered that for this case the transfer function for accelerations is the same as for displacements. The modulus of this transfer function varies with frequency as shown in Figure E8.3b.

Now, according to the procedure established in Section 8.4.4, the frequency-domain representation of the acceleration at the surface of the soil deposit may be obtained by multiplying this transfer function by the frequency-domain representation of the given rock acceleration. Consequently, one needs to determine first the Fourier transform of the rock acceleration shown in Figure E8.3a and then multiply this Fourier transform by the transfer function found earlier. Proceeding accordingly, one obtains functions whose moduli are shown in Figures E8.3c and E8.3d. As the aforementioned product represents the Fourier transform of the acceleration at the surface of the soil deposit, then the representation of this acceleration in the time domain is obtained by taking the inverse Fourier transform of it. The result is presented in Figure E8.3e.

A comparison of the graphs in Figures E8.3a and E8.3e shows that the soil deposit significantly amplifies the rock accelerations.

8.4.4.3 Case III: Homogeneous, Damped Soil on Flexible Rock

In the two cases considered earlier, it was assumed that the soil layer was supported by rigid bedrock. In consequence, no interaction takes place between the soil layer and the bedrock. That is, the presence of the overlying soil does not modify the bedrock motion and the bedrock does not influence the dynamic properties of the soil layer. Instead, the bedrock acts as a fixed boundary. This means, as mentioned in Section 8.4.3, that any waves coming from the soil toward the bedrock will be fully reflected back toward the ground surface, keeping all the elastic energy within the soil layer. In contrast, if the bedrock is considered flexible, waves can travel through the bedrock, and any waves traveling downward toward the soil–bedrock interface will not only be reflected but also transmitted when they reach this interface. Consequently, part of the wave energy will be transmitted away from the soil–bedrock interface through the rock. If the rock is deep enough, then the elastic energy of the transmitted waves will be, for all practical purposes, removed from the soil layer. The reason is that if the rock is deep, these waves will not return soon enough or with an appreciable amplitude to influence significantly the soil response. This effect, related to the radiation damping concept discussed in Section 4.10, will cause the motion observed at the ground surface to be smaller than in the case when rigid bedrock is assumed.

Bedrock flexibility may be incorporated into the analysis of the damped homogeneous elastic layer considered in the previous case as follows. Consider the soil layer and bedrock shown in Figure 8.27, and let subscripts r and s, respectively, identify rock and soil parameters. Consider also that the displacements induced in the soil and in the elastic bedrock by vertically propagating shear waves may be expressed as

$$u_s(z_s, t) = A_s e^{i(\omega t + k_s^* z_s)} + B_s e^{i(\omega t - k_s^* z_s)} \tag{8.73}$$

$$u_r(z_r, t) = A_r e^{i(\omega t + k_r^* z_r)} + B_r e^{i(\omega t - k_r^* z_r)} \tag{8.74}$$

where A_s and A_r, and B_s and B_r, are constants representing the amplitudes of the incident and reflected waves, respectively, and all other symbols are as defined before. As in Case II, however, the stresses should vanish at the ground surface and therefore in this case too one has that $A_s = B_s$. Similarly, from the compatibility of the displacements at the soil–rock interface, one has that

$$u_s(H, t) = u_r(0, t) \tag{8.75}$$

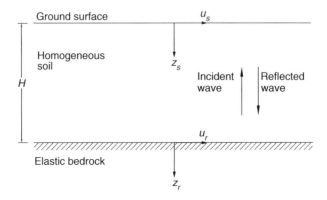

FIGURE 8.27 Homogeneous soil layer of thickness H overlying elastic bedrock.

which in terms of Equations 8.73 and 8.74 leads to

$$A_s(e^{ik_s^* H} + e^{-ik_s^* H}) = A_r + B_r \qquad (8.76)$$

In the same manner, from the compatibility of the shear stresses at the soil–rock interface, one can write

$$\tau_s(H,t) = G_s^* \left. \frac{\partial u_s}{\partial z_s} \right|_{z_s=H} = \tau_r(0,t) = G_r^* \left. \frac{\partial u_r}{\partial z_r} \right|_{z_r=0} \qquad (8.77)$$

and, thus, after substitution of Equations 8.73 and 8.74, one gets

$$A_s G_s^* k_s^* (e^{ik_s^* H} - e^{-ik_s^* H}) = G_r^* k_r^* (A_r - B_r) \qquad (8.78)$$

which may also be expressed as

$$A_s \alpha_z^* (e^{ik_s^* H} - e^{-ik_s^* H}) = A_r - B_r \qquad (8.79)$$

where α_z^* is a *complex impedance ratio* defined as

$$\alpha_z^* = \frac{G_s^* k_s^*}{G_r^* k_r^*} = \frac{\rho_s v_{ss}^*}{\rho_r v_{sr}^*} \qquad (8.80)$$

in which Equations 8.65 and 8.67 have been used and v_{ss}^* and v_{sr}^*, respectively, denote the soil and rock complex shear wave velocities. Solving Equations 8.76 and 8.79 simultaneously for A_r and B_r, one thus arrives to

$$A_r = \frac{A_s}{2}[(1+\alpha_z^*)e^{ik_s^* H} + (1-\alpha_z^*)e^{-ik_s^* H}] \qquad (8.81)$$

$$B_r = \frac{A_s}{2}[(1-\alpha_z^*)e^{ik_s^* H} + (1+\alpha_z^*)e^{-ik_s^* H}] \qquad (8.82)$$

This formulation may now be used to define the transfer function between the motion in the bedrock and the motion at the ground surface. This transfer function will be defined as the ratio of the displacement at the surface of the soil layer to the bedrock displacement when the soil layer is not present. For this purpose, note that if the soil layer is not present, the displacement at the surface of the bedrock will be equal to $2A_r e^{i\omega t}$ (recall that at a free surface $B_r = A_r$). Similarly, the displacement at the ground surface equals $2A_s e^{i\omega t}$. Consequently, the transfer function under consideration may be expressed as

$$H(\omega) = \frac{A_s}{A_r} \qquad (8.83)$$

which, in view of Equation 8.81, may also be written as

$$H(\omega) = \frac{2}{[(1+\alpha_z^*)e^{ik_s^* H} + (1-\alpha_z^*)e^{-ik_s^* H}]} \qquad (8.84)$$

Influence of Local Site Conditions

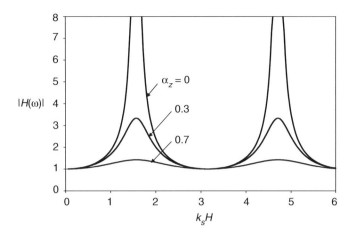

FIGURE 8.28 Amplification function of undamped soil layer overlying flexible rock.

or, if the exponential functions are expanded in terms of sine and cosine functions, as

$$H(\omega) = \frac{1}{\cos k_s^* H + i\alpha_z^* \sin k_s^* H} = \frac{1}{\cos(\omega H/v_{ss}^*) + i\alpha_z^* \sin(\omega H/v_{ss}^*)} \quad (8.85)$$

Thus, the amplification function for the case under consideration is given by

$$|H(\omega)| = \frac{1}{\left|\cos(\omega H/v_{ss}^*) + i\alpha_z^* \sin(\omega H/v_{ss}^*)\right|} \quad (8.86)$$

To understand the significance of rock flexibility on the soil response, consider, for simplicity, the case of no soil damping. In such a case, Equation 8.86 may be reduced to

$$|H(\omega)| = \frac{1}{\sqrt{\cos^2 k_s H + \alpha_z^2 \sin^2 k_s H}} \quad (8.87)$$

a plot of which is shown in Figure 8.28 for different values of the impedance ratio α_z. It may be noted, thus, that the effect of rock flexibility as reflected by the impedance ratio is similar to that of damping. That is, it reduces the amplitude of motion at the ground surface and prevents an infinitely large displacement to occur, even when the soil is undamped. Rock flexibility is thus a parameter of significant practical importance in the analysis of soil response.

8.4.4.4 Case IV: Layered, Damped Soil on Flexible Rock

The assumption of a homogeneous soil is useful to introduce the technique under consideration in a simple way and isolate the most important parameters that may influence the response of a soil deposit. Actual soils, however, are seldom homogeneous. Normally, they are layered, with each layer exhibiting different dimensions and properties, and the boundaries of these layers are the source of multiple wave reflections. Consequently, a practical soil response analysis requires the consideration of a transfer function for layered soils. Such a transfer function may be derived as follows.

Consider the layered soil deposit shown in Figure 8.29. This soil deposit is supposed to be resting on flexible bedrock and each layer is considered to be horizontal. Furthermore, each

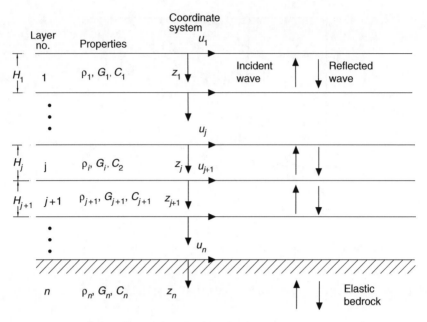

FIGURE 8.29 Layered soil deposit overlying elastic bedrock.

layer is assumed to behave as a viscoelastic solid with a complex shear modulus G_j^*, density ρ_j, and damping constant c_j. If, as in the previous case, it is considered that the displacements in the soil and in the flexible bedrock result from the effect of vertically propagating shear waves, then these displacements may be expressed as

$$u(z,t) = Ae^{i(\omega t + k^* z)} + Be^{i(\omega t - k^* z)} \tag{8.88}$$

where A and B are constants representing the amplitudes of waves traveling upward and downward, respectively, and all other symbols are as previously defined. As before, too, the corresponding shear stress is given by

$$\tau(z,t) = G^* \frac{\partial u}{\partial z} = ik^* G^* [Ae^{i(\omega t + k^* z)} - Be^{i(\omega t - k^* z)}] \tag{8.89}$$

Introducing a local system of coordinates for each layer as shown in Figure 8.29, the displacements and shear stresses at the top and bottom of the jth layer may be therefore written as

$$u_j(0,t) = (A_j + B_j)e^{i\omega t} \tag{8.90}$$

$$u_j(H_j,t) = (A_j e^{ik^* H_j} + B_j e^{-ik^* H_j})e^{i\omega t} \tag{8.91}$$

$$\tau_j(0,t) = ik_j^* G_j^* (A_j - B_j)e^{i\omega t} \tag{8.92}$$

$$\tau_j(H_j,t) = ik_j^* G_j^* (A_j e^{ik_j^* H_j} - B_j e^{-ik_j^* H_j})e^{i\omega t} \tag{8.93}$$

Influence of Local Site Conditions

To satisfy continuity, the displacements above and below any layer boundary must be the same. Hence, the displacement at the top of the $(j+1)$th layer must be equal to the displacement at the bottom of the jth layer. That is,

$$u_{j+1}(0,t) = u_j(H_j, t) \tag{8.94}$$

which in terms of Equations 8.90 and 8.91 may be written as

$$A_{j+1} + B_{j+1} = A_j e^{ik^* H_j} + B_j e^{-ik^* H_j} \tag{8.95}$$

Similarly, as the stresses must also be compatible at the layer boundaries, one has that

$$\tau_{j+1}(0,t) = \tau_j(H_j, t) \tag{8.96}$$

which, after introducing Equations 8.92 and 8.93, may also be expressed as

$$k^*_{j+1} G^*_{j+1}(A_{j+1} - B_{j+1}) = k^*_j G^*_j (A_j e^{ik^*_j H_j} - B_j e^{-ik^*_j H_j}) \tag{8.97}$$

or as

$$A_{j+1} - B_{j+1} = \alpha^*_j (A_j e^{ik^*_j H_j} - B_j e^{-ik^*_j H_j}) \tag{8.98}$$

where α^*_j is defined as

$$\alpha^*_j = \frac{G^*_j k^*_j}{G^*_{j+1} k^*_{j+1}} = \frac{\rho_j v^*_{sj}}{\rho_{j+1} v^*_{s(j+1)}} \tag{8.99}$$

and represents the complex impedance ratio between the jth and $(j+1)$th layers. In particular, note that because at a free surface the shear stress must vanish, applying Equation 8.92 to the top layer and making $\tau_1(0,t) = 0$ in the resulting equation leads to $A_1 = B_1$.

Solving Equations 8.95 and 8.98 simultaneously for A_{j+1} and B_{j+1} leads to the following recursive formulas to determine the amplitudes of the incident and reflected waves at the $(j+1)$th layer in terms of the amplitudes of the incident and reflected waves at the jth layer:

$$A_{j+1} = \frac{A_j}{2}(1 + \alpha^*_j)e^{ik^*_j H_j} + \frac{B_j}{2}(1 - \alpha^*_j)e^{-ik^*_j H_j} \tag{8.100}$$

$$B_{j+1} = \frac{A_j}{2}(1 - \alpha^*_j)e^{ik^*_j H_j} + \frac{B_j}{2}(1 + \alpha^*_j)e^{-ik^*_j H_j} \tag{8.101}$$

By applying these recursive formulas repeatedly from the first to the jth layer and considering that $A_1 = B_1$, it is then possible to express the amplitudes of the incident and reflected waves in the jth layer in terms of the amplitudes of the incident and reflected waves in the first layer. That is,

$$A_j = a_j(\omega) A_1 \tag{8.102}$$

$$B_j = b_j(\omega) B_1 \tag{8.103}$$

where $a_j(\omega)$ and $b_j(\omega)$ are known functions of the excitation frequency ω. Similarly, the transfer function that relates the displacement at the top of the ith layer to that at the top of the jth layer may be written as

$$H_{ij}(\omega) = \frac{u_i(0,t)}{u_j(0,t)} = \frac{(A_i + B_i)e^{i\omega t}}{(A_j + B_j)e^{i\omega t}} = \frac{a_i(\omega)A_1 + b_i(\omega)B_1}{a_j(\omega)A_1 + b_j(\omega)B_1} \qquad (8.104)$$

which in view of the fact that $A_1 = B_1$ may also be written as

$$H_{ij}(\omega) = \frac{a_i(\omega) + b_i(\omega)}{a_j(\omega) + b_j(\omega)} \qquad (8.105)$$

Equation 8.105 indicates that the motion in any layer can be determined from the motion at any other layer. Hence, if the motion at any level of the soil profile is known, through the use of Equation 8.105 one can determine the motion at any other level. In particular, note that the transfer function between the first and the jth layers is given by

$$H_{1j}(\omega) = \frac{u_1(0,t)}{u_j(0,t)} = \frac{2}{a_j(\omega) + b_j(\omega)} \qquad (8.106)$$

as $a_1(\omega) = b_1(\omega) = 1.0$. Also, because in a bedrock outcrop (i.e., bedrock away from the influence of the soil deposit) the amplitudes of the incident and reflected waves are the same (concluded from the fact that at a free surface the shear stress is zero), then

$$A_{n'} = a_{n'}(\omega)A_1 = B_{n'} = b_{n'}(\omega)B_1 \qquad (8.107)$$

where the subscript n' identifies the parameters corresponding to such a bedrock outcrop. Furthermore, from Equation 8.107, together with the fact that the amplitude of the incident wave is the same independent of whether or not the bedrock is overlain by a soil deposit, it is found that

$$a_{n'}(\omega) = b_{n'}(\omega) = a_n(\omega) \qquad (8.108)$$

where, once again, it has been considered that $A_1 = B_1$. Consequently, the transfer function between the top layer and the bedrock outcrop (free field) may be expressed as

$$H_{1n}(\omega) = \frac{u_1(0,t)}{u_{n'}(0,t)} = \frac{(A_1 + B_1)e^{i\omega t}}{(A_{n'} + B_{n'})e^{i\omega t}} = \frac{A_1 + B_1}{a_{n'}(\omega)A_1 + b_{n'}(\omega)B_1} = \frac{1}{a_n(\omega)} \qquad (8.109)$$

Equation 8.109 may be thus used to determine the ground motion at the surface of a soil deposit in terms of the ground motion recorded at a point not influenced by the soil deposit (free-field ground motion).

EXAMPLE 8.4 TRANSFER FUNCTION OF A LAYERED SOIL DEPOSIT

Find the transfer function of the layered soil deposit shown in Figure E8.4a. Assume that the deposit rests on an elastic bedrock and that each layer behaves as a viscoelastic solid with the unit weights (γ), shear wave velocities (v_s), and damping ratios (ξ) shown in the figure.

Influence of Local Site Conditions

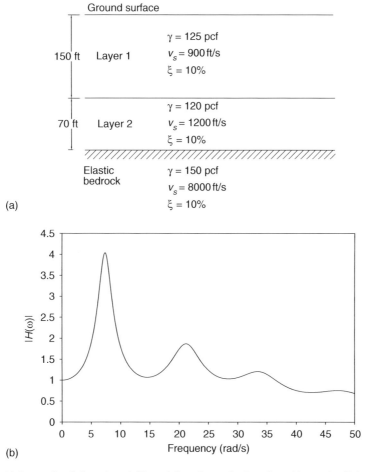

FIGURE E8.4 (a) Layered soil deposit and (b) modulus of transfer function of layered soil deposit considered in Example 8.4.

Solution

According to the given soil properties and Equation 8.65, the complex shear wave velocities for the two layers that compose the soil deposit and the underlying bedrock are equal to

$$v_{s1}^* = v_{s1}(1+\xi_1 i) = 900(1+0.1i)$$

$$v_{s2}^* = v_{s2}(1+\xi_2 i) = 1200(1+0.1i)$$

$$v_{s3}^* = v_{s3}(1+\xi_3 i) = 8000(1+0.1i)$$

Therefore, the complex impedance ratios between the layers are (see Equation 8.99)

$$\alpha_1^* = \frac{\rho_1 v_{s1}^*}{\rho_2 v_{s2}^*} = \frac{120(900)(1+0.1i)}{125(1200)(1+0.1i)} = 0.720$$

$$\alpha_2^* = \frac{\rho_2 v_{s2}^*}{\rho_3 v_{s3}^*} = \frac{125(1200)(1+0.1i)}{150(8000)(1+0.1i)} = 0.125$$

In turn, these complex impedance ratios and Equations 8.100 and 8.101 lead to the following amplitudes of the incident and reflected waves in the second layer and the bedrock in terms of the amplitude of the incident wave in the first layer:

$$A_2 = \frac{A_1}{2}(1+\alpha_1^*)e^{ik_1^*H_1} + \frac{B_1}{2}(1-\alpha_1^*)e^{-ik_1^*H_1}$$

$$= \frac{A_1}{2}(1+0.720)e^{i(150/900)(1-0.1i)\omega} + \frac{B_1}{2}(1-0.720)e^{-i(150/900)(1-0.1i)\omega}$$

$$= A_1[0.860 e^{(0.0167+0.167i)\omega} + 0.140 e^{-(0.0167+0.167i)\omega}]$$

$$B_2 = \frac{A_1}{2}(1-\alpha_1^*)e^{ik_1^*H_1} + \frac{B_1}{2}(1+\alpha_1^*)e^{-ik_1^*H_1}$$

$$= A_1[0.140 e^{(0.0167+0.167i)\omega} + 0.860 e^{-(0.0167+0.167i)\omega}]$$

$$A_3 = \frac{A_2}{2}(1+\alpha_2^*)e^{ik_2^*H_2} + \frac{B_2}{2}(1-\alpha_2^*)e^{-ik_2^*H_2}$$

$$= \frac{A_2}{2}(1+0.125)e^{i(70/1200)(1-0.1i)\omega} + \frac{B_2}{2}(1-0.125)e^{-i(70/1200)(1-0.1i)\omega}$$

$$= A_2(0.563)e^{(0.0058+0.058i)\omega} + B_2(0.438)e^{-(0.0058+0.058i)\omega}$$

where it has been considered that $B_1 = A_1$ and $k^* = \omega(1-\xi i)/v_s$.

In view of Equation 8.102 and the foregoing equations, one has thus that

$$a_3(\omega) = \frac{A_2}{A_1}(0.563)e^{(0.0058+0.058i)\omega} + \frac{B_2}{A_1}(0.438)e^{-(0.0058+0.058i)\omega}$$

and, according to Equation 8.109, that

$$H_{13'}(\omega) = \frac{1}{(A_2/A_1)(0.563)e^{(0.0058+0.058i)\omega} + (B_2/A_1)(0.438)e^{-(0.0058+0.058i)\omega}}$$

which represents the transfer function between the displacement at the surface of the soil deposit and the displacement at the bedrock surface when no soil is present. A plot of the modulus of this transfer function is shown in Figure E8.4b.

Although powerful, it is important to keep in mind that a frequency-domain solution relies on the principle of superposition and is thus limited to the analysis of linear soils. It is possible, however, to use it in combination with equivalent linear properties and an iterative procedure to account for the nonlinear behavior of soils in an approximate way.

8.4.5 Nonlinear Site Response

Under high strain levels, and hence under severe ground motions, most soils exhibit, as shown in Figure 8.30 for a typical clayey soil, a nonlinear stress–strain behavior. Thus, it is important to recognize and take into account this nonlinearity in the analysis of site response and the

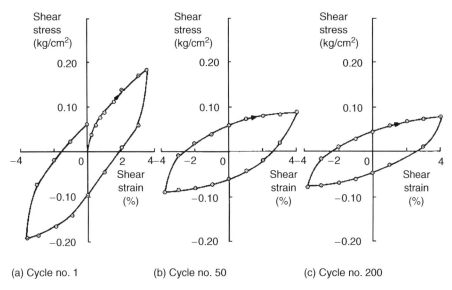

FIGURE 8.30 (a)–(c) Stress–strain curves obtained in cyclic tests of samples from San Francisco Bay mud. (Reproduced with permission of ASCE from Thiers, G.R. and Seed, H.B., *Journal of the Soil Mechanics and Foundations Division, ASCE*, 94, 555–569, 1968.)

estimation of the ground motion characteristics at the surface of a soil deposit. In current practice, soil nonlinearity is accounted for using either equivalent linearization techniques, in which case an equivalent stiffness and an equivalent damping ratio are used, or step-by-step procedures together with the adoption of a stress–strain hysteretic model. This section briefly describes these two approaches.

In the equivalent linearization technique, the actual nonlinear stress–strain behavior of a cyclically loaded soil is approximated by an equivalent shear modulus G and an equivalent damping ratio ξ. The equivalent shear modulus is taken as a secant shear modulus and the equivalent damping ratio as the damping ratio that produces the same energy loss in a single cycle as the actual hysteresis loop (see Section 8.5). As in the linear approach it is assumed that G and ξ remain constant at all times, this technique requires that the assumed values be consistent with the induced level of strain. However, as the computed level of strain depends on the assumed equivalent properties, an iterative procedure is required to ensure that the considered properties are consistent with the computed strain levels in all layers. This iterative procedure normally proceeds as follows:

1. An initial estimate of G and ξ is obtained for each layer (low-strain values are often used as the initial estimate).
2. The soil response is computed on the basis of such initial estimates and a linear elastic analysis.
3. New values of G and ξ are determined for each layer on the basis of the computed strains.
4. Steps 2 and 3 are repeated until the difference between the assumed shear moduli and damping ratios in two successive iterations becomes less than a preestablished tolerable percentage.

Although the just-described iterative procedure provides reasonable results for many practical problems, it is important to bear in mind that the method is still based on a linear analysis. That is, it assumes that the strain-compatible soil properties remain constant throughout the duration of the excitation, independent of the large variations that are likely to occur in the strain level during the event. As such, the method is incapable of representing the actual property changes that take place during an earthquake. Despite these limitations, this equivalent linearization technique constitutes the basis of SHAKE (see reference at the end of the chapter), a widely used computer program for the response analysis of layered soils.

An alternative approach to consider the nonlinear behavior of a soil deposit is a step-by-step numerical integration of the governing nonlinear differential equation of motion in the time domain. Because the integration is performed in small time steps, this approach allows the establishment of the appropriate soil properties according to the induced strain level throughout the duration of the input base motion. Ordinarily, a lumped-mass model of the type described in Section 8.4.3 and an idealized stress–strain relationship are used to perform such nonlinear analysis. The way the pertinent equations of motion are derived for the nonlinear case and the numerical techniques used to find the solutions to these nonlinear equations are discussed in Chapter 11. In general, the method involves the following steps:

1. The soil profile is divided into N layers and the material properties that describe the adopted stress–strain model are defined for each layer.
2. An equivalent shear-building model is established according to the criterion introduced in Section 8.4.3, defining its stiffness constants and damping coefficients on the basis of low-strain levels.
3. The solution time is divided into a series of small time intervals or time steps and the input base motion is defined at the beginning of each time step.
4. The system's response is determined at the end of the first time step.
5. The system's stiffness constants and damping coefficients are updated on the basis of the strains computed in the previous step and the adopted stress–strain model.
6. The system's response is computed at the end of the next step and the procedure repeated until all the time steps are covered.

It should be noted that, although apparently more accurate than the equivalent linearization technique, nonlinear methods require a reliable stress–strain model. However, the parameters that describe such models are difficult to obtain and may require, in many instances, a substantial field and laboratory testing program.

8.5 DETERMINATION OF SHEAR MODULUS AND DAMPING RATIO

As it may be inferred from the discussion in the foregoing section, the evaluation of site effects using analytical techniques requires the knowledge of the shear modulus of elasticity and damping ratio of the different layers that compose a soil profile. An accurate evaluation of these soil parameters is therefore an important step in any site response analysis. There exist a variety of laboratory and field tests that can be used to estimate the shear moduli and damping characteristics of soils, ranging from a simple force vibration test on a soil sample in the laboratory to an elaborate *in situ* measurement of the velocity of propagation of induced waves. These laboratory and field tests are well established, but given the broad field they represent, their description is beyond the scope of this book. The reader is referred to the books by Kramer and Ishihara listed at the end of this chapter for a comprehensive discussion of these tests. Also available are a series of empirical correlations that have been obtained during the last three decades from the results of laboratory and field tests. This section summarizes some of these empirical correlations with the intention of providing means to obtain quick estimates of the two parameters in question, which might prove helpful during some preliminary studies. Caution should be exercised, however, in the use of these correlations, as the scatter in their derivation is normally high and the deviation from the true values may be significantly large. In a final analysis, the determination of shear modulus and damping ratio should always be based on a direct measurement through field or laboratory tests, particularly for large or critical projects.

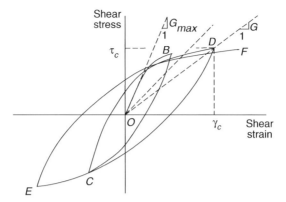

FIGURE 8.31 Typical stress–strain behavior of soils and variation of shear modulus with strain level.

As indicated in the previous section and shown in Figure 8.31 for a typical soil, most soils exhibit a nonlinear hysteretic behavior when loaded cyclically under high strains. This behavior is characterized by a backbone or initial loading curve (branch *OB* in Figure 8.31) and unloading (branches *BC* and *DE* in Figure 8.31) and reloading (branches *CD* and *EF* in Figure 8.31) curves that form hysteresis loops. It may be observed, thus, that, in general, the shear modulus G of a given soil is not a constant but it changes with strain level. It exhibits its maximum value at a very low shear strain (i.e., $G = G_{max}$ when the strain is close to zero) and decreases with an increasing strain level. For the purpose of a nonlinear step-by-step analysis, it suffices to define G_{max} and use it in combination with an adopted analytical model to represent the hysteretic behavior of the soil. An example of such an analytical model is that suggested by R. L. Kondner in 1963, which consists of the hyperbolic function

$$\tau = \frac{\gamma}{(1/G_{max}) + (\gamma/\tau_{max})} \tag{8.110}$$

that defines the backbone curve (τ, γ, and τ_{max}, respectively, denoting shear stress, shear strain, and shear strength), as well as a series of basic rules that govern the unloading and reloading behavior. For the purpose of an equivalent linear analysis, it is customary to define the shear modulus variation with strain amplitude in terms of a secant modulus defined as $G = \tau_c/\gamma_c$, where τ_c is the shear stress amplitude corresponding to a shear-strain amplitude γ_c; that is, the slope of the line that joins the two ends of the hysteretic loop formed during a complete stress reversal (see Figure 8.31). For the purpose of an equivalent linear analysis, too, the damping ratio of the soil is defined as (see Section 13.9)

$$\xi = \frac{\Delta W}{4\pi W} = \frac{1}{2\pi} \frac{A_L}{G\gamma_c^2} \tag{8.111}$$

where ΔW is the energy per unit volume dissipated during a single deformation cycle, W the corresponding deformation energy stored in a perfectly elastic material, and A_L the area enclosed by the hysteresis loop formed when the soil is subjected to the shear-strain γ_c. Because G depends on the strain amplitude γ_c, it is also customary to characterize the stress–strain relationship of a soil by its low-strain modulus G_{max} together with a G/G_{max} versus γ_c curve. This characterization is convenient because G_{max} can be obtained from *in situ* low-strain tests, as, for example, from shear wave velocity measurements. Then, the *in situ* G_{max} can be combined with the G/G_{max} versus γ_c

curve to obtain G for any strain level. It should also be noted that the shear modulus and damping ratio of a soil and their variation with strain level depends, as expected, on the type of soil. Therefore, different empirical correlations have been derived for different soil types. The description of these empirical correlations is thus given separately for sands, gravels, and saturated clays.

8.5.1 Shear Modulus and Damping Ratio of Sands

Based on measurements of shear wave velocities in laboratory samples, in 1968 B. O. Hardin and W. L. Black suggested the following relationship to determine the low-strain shear modulus of sands:

$$G_{max} = \frac{1230(2.973 - e)^2}{1 + e} \bar{\sigma}_0^{1/2} \tag{8.112}$$

In this equation, G_{max} denotes shear modulus for low amplitudes of vibration in pounds per square inch, $\bar{\sigma}_0$ the mean effective principal stress, also in pounds per square inch, and e void ratio. Equation 8.112 gives values of G_{max} that may be too low for void ratios close to 2 and increase for $e > 2.973$. Consequently, later on, in 1978, Hardin proposed to replace the void ratio function $(2.973 - e)^2/1 + e$ by $1.97/(0.3 + 0.7e^2)$, which leads to monotonically decreasing values of G_{max} with increasing values of e. For a soil element subjected to unequal major, intermediate, and minor effective stresses, $\bar{\sigma}_0$ is equal to

$$\bar{\sigma}_0 = \frac{1}{3}(\bar{\sigma}_1 + \bar{\sigma}_2 + \bar{\sigma}_3) \tag{8.113}$$

where $\bar{\sigma}_1$, $\bar{\sigma}_2$, and $\bar{\sigma}_3$, respectively, represent such major, intermediate, and minor effective stresses. As for field conditions at any depth,

$$\bar{\sigma}_1 = \bar{\sigma}_v = \text{effective vertical stress} \tag{8.114}$$

$$\bar{\sigma}_2 = \bar{\sigma}_3 = K_0 \bar{\sigma}_v \tag{8.115}$$

where K_0 denotes the at-rest earth pressure coefficient (recall $K_0 \approx 1 - \sin \phi$, where ϕ is the drained angle of internal friction), then for field conditions $\bar{\sigma}_0$ may be computed as

$$\bar{\sigma}_0 = \frac{1}{3}[\bar{\sigma}_v + 2\bar{\sigma}_v(1 - \sin \phi)] = \frac{\bar{\sigma}_v}{3}(3 - 2\sin \phi) \tag{8.116}$$

The foregoing expressions permit the estimation of the shear modulus of sands at low-strain levels. To obtain the value for any strain level, in 1972 H. B. Seed and I. M. Idriss proposed to use

$$G = 1000 K_2 \bar{\sigma}_0^{1/2} \tag{8.117}$$

where K_2 is a modulus coefficient that accounts for the influence of void ratio (or relative density) and strain amplitude, and G and $\bar{\sigma}_0$ are both expressed in pounds per square foot. Furthermore, considering that for any sand the modulus coefficient reaches its maximum, $K_{2(max)}$, at low strains, they also suggested to use a similar expression to estimate the shear modulus at low-strain amplitudes. That is,

$$G_{max} = 1000 K_{2(max)} \bar{\sigma}_0^{1/2} \tag{8.118}$$

Influence of Local Site Conditions

As, after dividing Equation 8.118 by Equation 8.117, one obtains

$$\frac{K_2}{K_{2(max)}} = \frac{G}{G_{max}} \tag{8.119}$$

then note that Equation 8.117 may be expressed alternatively as

$$G = 1000 \frac{G}{G_{max}} K_{2(max)} \bar{\sigma}_0^{1/2} \tag{8.120}$$

Based on the results from laboratory tests, Seed and Idriss recommended to use the values of $K_{2(max)}$ in Table 8.1, where they are given as a function of void ratio or relative density. Plotting the test results from several investigators, they also obtained the curves depicted in Figure 8.32, which show the average variation of G/G_{max} with shear strain and the corresponding upper and lower bounds. Considering that such upper and lower bounds fall within a narrow band, they

TABLE 8.1
Values of $K_{2(max)}$ as a Function of Void Ratio or Relative Density

Void Ratio	$K_{2(max)}$	Relative Density (%)	$K_{2(max)}$
0.4	70	30	34
0.5	60	40	40
0.6	51	45	43
0.7	44	60	52
0.8	39	75	61
0.9	34	90	70

Source: Seed, H. B. and Idriss, I. M., Soil Moduli and Damping Ratios for Dynamic Response Analyses, Report No. EERC 70-10, Earthquake Engineering Research Center, University of California, Berkeley, CA, 1970.

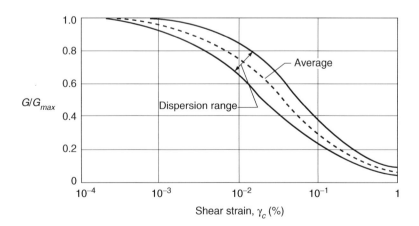

FIGURE 8.32 Variation of normalized shear modulus with shear strain for sands. (After Seed, H.B. and Idriss, I.M., Soil Moduli and Damping Ratios for Dynamic Response Analyses, Report No. EERC 70-10, Earthquake Engineering Research Center, University of California, Berkeley, CA, 1970.)

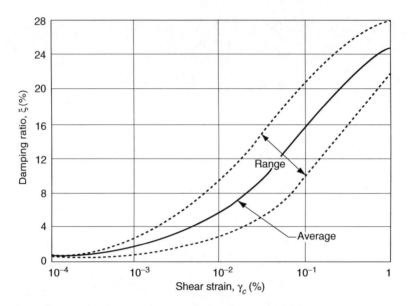

FIGURE 8.33 Variation of damping ratio with shear strain for sands. (After Seed, H.B. and Idriss, I.M., Soil Moduli and Damping Ratios for Dynamic Response Analyses, Report No. EERC 70-10, Earthquake Engineering Research Center, University of California, Berkeley, CA, 1970.)

recommended using the average curve in this figure in all problems of practical interest. Equation 8.120, Table 8.1, and Figure 8.32 provide the means to estimate the shear modulus of sands at any strain level.

Several studies have shown that the damping ratio of sands depends on factors such as (a) grain size distribution, (b) degree of saturation, (c) void ratio, (d) earth pressure coefficient at rest, (e) angle of internal friction, (f) number of stress cycles, (g) confining pressure, and (h) strain level. The last two factors, however, seem to have the largest effect. Therefore, for practical purposes, Seed and Idriss proposed the use of the curves shown in Figure 8.33 to estimate the damping ratio of sands. These curves, which show the average variation of damping ratio with strain level as well as upper and lower bounds, were obtained from the compilation of a large number of laboratory tests and after observing that the effect of confining pressure on damping ratio was small in comparison with the effect of shear-strain level.

8.5.2 Shear Modulus and Damping Ratio of Gravels

In 1986, H. B. Seed and coworkers conducted laboratory tests to measure the shear moduli and damping ratios of gravelly soils. From their results, they found that Equation 8.117 could also be used to estimate the shear modulus of gravels, although with different values of the modulus coefficient K_2. For gravels with relative densities of approximately 75 and 95%, they obtained the curves shown in Figure 8.34, which depict the variation of K_2 with shear-strain amplitude. For comparison purposes, this figure also includes the corresponding curves for typical sands. Seed and coworkers also observed that the shear modulus decay with shear strain for gravels differed from that for sands. For gravels, therefore, they proposed to use the G/G_{max} versus γ_c curves shown in Figure 8.35. Based on their test data, they concluded, additionally, that the magnitude and variation of damping ratio with strain for gravelly soils was approximately the same as for sands. They suggested, thus, also using Figure 8.33 to determine the damping ratio of gravels.

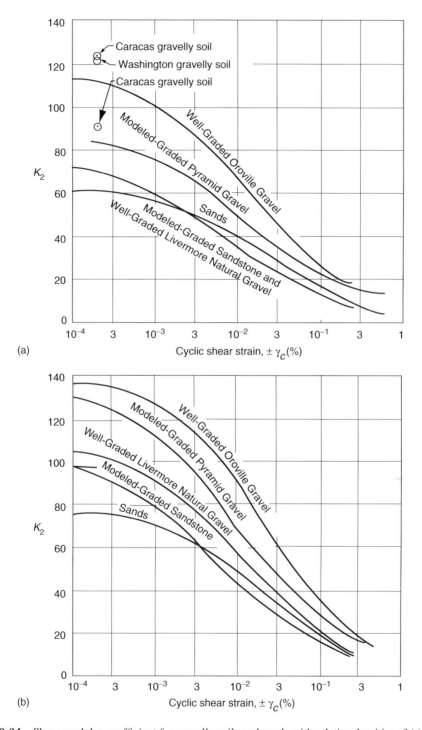

FIGURE 8.34 Shear modulus coefficient for gravelly soils and sands with relative densities of (a) 75% and (b) 95%. (Reproduced with permission of ASCE from Seed, H.B. et al., *Journal of Geotechnical Engineering, ASCE*, 112, 1016–1032, 1986.)

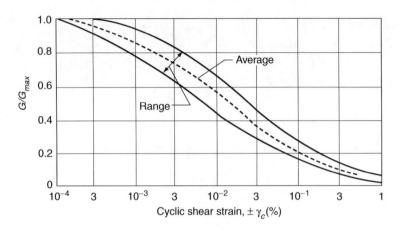

FIGURE 8.35 Variation of normalized shear modulus with shear strain for gravelly soils. (Reproduced with permission of ASCE from Seed, H.B. et al., *Journal of Geotechnical Engineering, ASCE*, 112, 1016–1032, 1986.)

TABLE 8.2
Values of *K* as a Function of Plasticity Index

Plasticity Index (%)	K
0	0
20	0.18
40	0.30
60	0.41
80	0.48
≥100	0.50

Source: Hardin, B. O. and Drnevich, V. P., *Journal of the Soil Mechanics and Foundations Divisions, ASCE*, 98, 667–692, 1972.

8.5.3 Shear Modulus and Damping Ratio of Saturated Clays

Hardin and Black also suggested an expression similar to Equation 8.112 to determine the low-strain shear modulus of normally consolidated clays. This equation is of the form

$$G_{max} = \frac{1230(2.973 - e)^2}{1+e}(OCR)^K \bar{\sigma}_0^{1/2} \qquad (8.121)$$

where *OCR* denotes overconsolidation ratio and *K* is a constant given in Table 8.2 for different plasticity indices. As in the case of sands, the void ratio function in this equation may be replaced by $1.97/(0.3 + 0.7e^2)$, which leads to improved estimates of G_{max} for large values of *e*. As in the case of sands, too, for field conditions $\bar{\sigma}_0$ is given by

$$\bar{\sigma}_0 = \tfrac{1}{3}(\bar{\sigma}_v + 2K_0\bar{\sigma}_v) \qquad (8.122)$$

except that for clays the earth pressure coefficient at rest is given by

$$K_0 = \begin{cases} 0.4 + 0.007(PI) & \text{for } 0 \leq PI \leq 40\% \\ 0.68 + 0.001(PI - 40) & \text{for } 40 \leq PI \leq 80\% \end{cases} \qquad (8.123)$$

if normally consolidated, and by

$$K_{0(overconsolidated)} = K_0 \sqrt{OCR} \qquad (8.124)$$

if overconsolidated. *PI* in Equation 8.123 denotes plasticity index expressed in percentage, and K_0 in Equation 8.124 is given by Equation 8.123.

To determine the shear modulus and damping ratio at any strain level, Seed and Idriss also collected available experimental results for the shear modulus and damping ratio of saturated cohesive soils and generated the shear modulus ratio and damping ratio versus shear-strain curves shown in Figures 8.36 and 8.37. Together with Equation 8.121 or low-strain field or laboratory

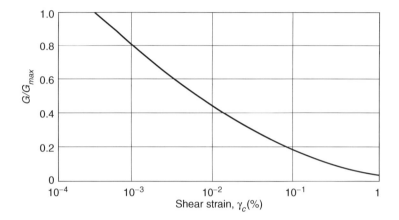

FIGURE 8.36 Average variation of normalized shear modulus with shear strain for saturated clays. (After Seed, H.B. and Idriss, I.M., Soil Moduli and Damping Ratios for Dynamic Response Analyses, Report No. EERC 70-10, Earthquake Engineering Research Center, University of California, Berkeley, CA, 1970.)

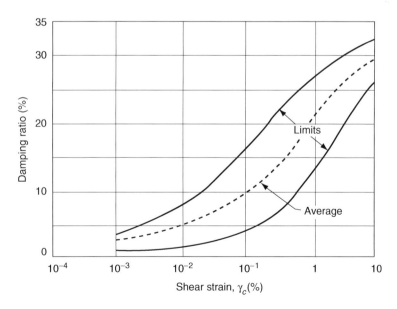

FIGURE 8.37 Variation of damping ratio of saturated clays with shear-strain level. (After Seed, H.B. and Idriss, I.M., Soil Moduli and Damping Ratios for Dynamic Response Analyses, Report No. EERC 70-10, Earthquake Engineering Research Center, University of California, Berkeley, CA, 1970.)

measurements to determine G_{max}, these curves may be used to obtain estimates of the shear modulus and damping ratio of saturated clays that might prove sufficiently accurate for many practical situations.

In 1991, M. Vucetic and R. Dobry reviewed available experimental data encompassing sands and normally and overconsolidated clays and concluded that plasticity index was the main variable controlling the form of the normalized shear modulus and damping ratio versus strain curves. On the basis of their study, they generated the curves shown in Figure 8.38, which correlate G/G_{max} and damping ratio with plasticity index. Because plasticity index is routinely determined in most geotechnical investigations, these curves can be useful to estimate the dynamic characteristics of fine-grained soils in a most practical way. It is of interest to note that the curves for $PI = 0$ in Figure 8.38 are nearly the same as the average curves for sands shown in Figures 8.32 and 8.33.

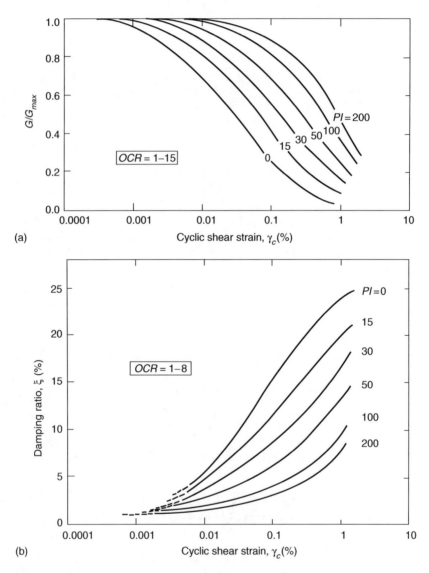

FIGURE 8.38 Variation of (a) normalized shear modulus and (b) damping ratio with shear strain and plasticity index. (Reproduced with permission of ASCE from Vucetic, M. and Dobry, R., *Journal of Geotechnical Engineering, ASCE*, 117, 89–107, 1991.)

Influence of Local Site Conditions

It should also be noted that the strain range for which a clayey soil behaves linearly (i.e., the strain range for which $G/G_{max} \approx 1.0$) is greater for high-plasticity soils than for soils with low plasticity. Furthermore, it should be noted that the damping ratios of soils with high plasticity are lower than those of low-plasticity soils at the same strain amplitude.

Finally, an expression to determine the shear modulus ratio that combines the effects of the mean effective confining pressure and plasticity index was proposed by I. Ishibashi and X. Zhang in 1993. This expression is of the form

$$\frac{G}{G_{max}} = K\bar{\sigma}_0^{m-m_0} \tag{8.125}$$

where $\bar{\sigma}_0$ denotes effective confining stress expressed in kilopascals, and

$$K = 0.5\left\{1 + \tanh\left[\ln\left(\frac{0.000102 + n(PI)}{\gamma_c}\right)^{0.492}\right]\right\} \tag{8.126}$$

$$m - m_0 = 0.272\left\{1 - \tanh\left[\ln\left(\frac{0.000556}{\gamma_c}\right)^{0.4}\right]\right\}\exp(-0.0145PI^{1.3}) \tag{8.127}$$

$$n = \begin{cases} 0 & \text{for } PI = 0 \\ 3.37 \times 10^{-6} PI^{1.404} & \text{for } 0 < PI \leq 15 \\ 7.0 \times 10^{-7} PI^{1.976} & \text{for } 15 < PI \leq 70 \\ 2.7 \times 10^{-5} PI^{1.115} & \text{for } PI > 70 \end{cases} \tag{8.128}$$

Equation 8.125 is plotted in Figure 8.39 for two values of the plasticity index, $PI = 0$ and 50. This figure clearly shows that the effect of the confining pressure $\bar{\sigma}_0$ on the modulus ratio G/G_{max} is not large for soils with high plasticity, but it may be significant for soils with low plasticity.

Ishibashi and Zhang also obtained an expression to determine the damping ratio of soils in terms of mean effective confining pressure and plasticity index. This expression is given by

$$\xi = 0.333\frac{1 + \exp(-0.0145PI^{1.3})}{2}\left[0.586\left(\frac{G}{G_{max}}\right)^2 - 1.547\frac{G}{G_{max}} + 1\right] \tag{8.129}$$

where the modulus ratio G/G_{max} is calculated according to Equation 8.125.

EXAMPLE 8.5 SHEAR MODULI AND DAMPING RATIOS OF SOIL DEPOSIT

Determine the moduli and damping ratios for the two layers that compose the soil profile shown in Figure E8.5 when the anticipated strain level in the soil profile is 0.12%. For each layer, consider that the confining pressure at midheight is the representative of the confining pressure over the entire layer.

Solution

For the sand layer, the submerged unit weight is equal to

$$\gamma_b = \gamma_{sat} - \gamma_w = \frac{G_s - 1}{1+e}\gamma_w = \frac{(2.65-1)(62.4)}{1+0.6} = 64.35 \text{ lb/ft}^3$$

344   Fundamental Concepts of Earthquake Engineering

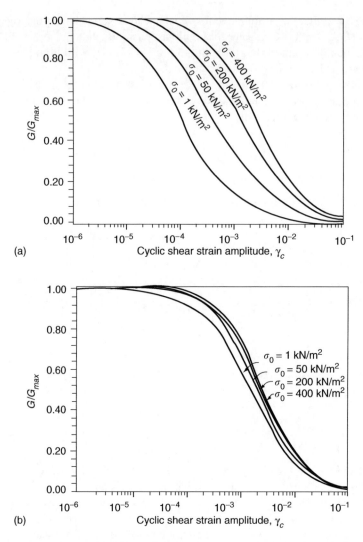

FIGURE 8.39 Variation of normalized shear modulus with shear strain and effective confining pressure for soils with (a) low plasticity ($PI = 0$) and (b) high plasticity ($PI = 50$). (Reproduced with permission of ASCE from Ishibashi, I., *Journal of Geotechnical Engineering, ASCE*, 118, 830–832, 1992.)

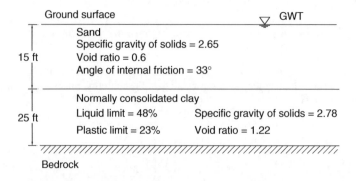

FIGURE E8.5 Soil profile considered in Example 8.5.

where γ_{sat} and γ_w, respectively, denote saturated unit weight and unit weight of water. Hence, the vertical effective stress at the middle of the layer is

$$\bar{\sigma}_v = 64.35(7.5) = 482.6 \text{ lb/ft}^2$$

and, according to Equation 8.116, the mean effective confining pressure results as

$$\bar{\sigma}_0 = \frac{\bar{\sigma}_v}{3}(3 - 2\sin\phi) = \frac{482.6}{3}(3 - 2\sin 33°) = 307.4 \text{ lb/ft}^2 = 2.13 \text{ lb/in.}^2$$

Similarly, using Equation 8.112 one obtains

$$G_{max} = \frac{1230(2.973 - e)^2}{1 + e}\bar{\sigma}_0^{1/2} = \frac{1230(2.973 - 0.6)^2}{1 + 0.6}(2.13)^{1/2} = 6318 \text{ lb/in.}^2$$

Alternatively, one can determine G_{max} using Equation 8.118 and the value of $K_{2(max)}$ in Table 8.1 for a void ratio of 0.6. That is,

$$G_{max} = 1000 K_{2(max)}\bar{\sigma}_0^{1/2} = 1000(51)(307.4)^{1/2} = 894{,}174 \text{ lb/ft}^2 = 6209 \text{ lb/in.}^2$$

The shear modulus for a strain of 0.12% may now be determined by reading from Figure 8.32 the G/G_{max} ratio corresponding to this strain level and then multiplying this ratio by the value of G_{max}. Proceeding accordingly, from the *average* curve in Figure 8.32 it is found that for $\gamma_c = 0.12\%$,

$$G/G_{max} = 0.25$$

and, hence,

$$G = 0.25(6318) = 1578 \text{ lb/in.}^2$$

Reading directly from the average curve in Figure 8.33, one finds similarly that the damping ratio for a strain of 0.12% is approximately equal to

$$\xi = 17\%$$

For the clay layer, one has that

$$\gamma_b = \gamma_{sat} - \gamma_w = \frac{G_s - 1}{1 + e}\gamma_w = \frac{(2.78 - 1)(62.4)}{1 + 1.22} = 50.03 \text{ lb/ft}^3$$

$$PI = LL - PL = (48 - 23)\% = 25\%$$

$$OCR = 1.0$$

and, according to Equation 8.123, for $PI = 25\%$,

$$K_0 = 0.4 + 0.007(PI) = 0.4 + 0.007(25) = 0.58$$

Therefore, for the middle of the clay layer,

$$\bar{\sigma}_v = 64.35(15) + 50.03(12.5) = 1591 \text{ lb/ft}^2$$

$$\bar{\sigma}_0 = \frac{1}{3}(1 + 2K_0)\bar{\sigma}_v = \frac{1}{3}[1 + 2(0.58)](1591) = 1146 \text{ lb/ft}^2 = 7.96 \text{ lb/in.}^2 = 54.9 \text{ kPa}$$

and, upon substitution into Equation 8.121,

$$G_{max} = \frac{1230(2.973 - e)^2}{1 + e}(OCR)^K \bar{\sigma}_0^{1/2} = \frac{1230(2.973 - 1.22)^2}{1 + 1.22}(1.0)(7.96)^{1/2} = 4804 \text{ lb/in.}^2$$

Now, from Figure 8.36 for a strain of 0.12%,

$$G/G_{max} = 0.18$$

and thus,

$$G = 0.18(4804) = 865 \text{ lb/in.}^2$$

Likewise, from the average curve in Figure 8.37 for a strain of 0.12%,

$$\xi = 11\%$$

It is instructive to compare the values of G/G_{max} and ξ that one obtains using alternatively Vucetic and Dobry's charts and Ishibashi and Zhang's equations. From Figure 8.38 for a strain of 0.12% and a plasticity index of 25%, one gets

$$G/G_{max} = 0.45$$

$$\xi = 11\%$$

Similarly, Equations 8.125 through 8.128 lead to

$$n = 7.0 \times 10^{-7} PI^{1.976} = 7.0 \times 10^{-7}(0.25)^{1.976} = 4.52 \times 10^{-8}$$

$$m - m_0 = 0.272\left\{1 - \tanh\left[\ln\left(\frac{0.000556}{\gamma_c}\right)^{0.4}\right]\right\}\exp(-0.0145 PI^{1.3})$$

$$= 0.272\left\{1 - \tanh\left[\ln\left(\frac{0.000556}{0.0012}\right)^{0.4}\right]\right\}\exp[-0.0145(0.25)^{1.3}] = 0.35$$

$$K = 0.5\left\{1 + \tanh\left[\ln\left(\frac{0.000102 + n(PI)}{\gamma_c}\right)^{0.42}\right]\right\}$$

$$= 0.5\left\{1 + \tanh\left[\ln\left(\frac{0.000102 + 4.52 \times 10^{-8}(0.25)}{0.0012}\right)^{0.42}\right]\right\} = 0.081$$

$$\frac{G}{G_{max}} = K\bar{\sigma}_0^{m-m_0} = 0.081(54.9)^{0.35} = 0.33$$

and Equation 8.129 to

$$\xi = 0.333 \frac{1 + \exp(-0.0145 PI^{1.3})}{2} \left[0.586 \left(\frac{G}{G_{max}} \right)^2 - 1.547 \frac{G}{G_{max}} + 1 \right]$$

$$= 0.333 \frac{1 + \exp[-0.0145(0.25)^{1.3}]}{2} [0.586(0.33)^2 - 1.547(0.33) + 1] = 18\%$$

FURTHER READINGS

1. Davis, L. L. and West, L. R., Observed effects of topography on ground motion, *Bulletin of the Seismological Society of America*, 63, 1973, 283–298.
2. Field, E. H., Kramer, S., Elgamal, A. W. et al., Nonlinear site response: Where we are at (A report from a SCEC/PEER seminar and workshop), *Seismological Research Letters*, 69, 3, 1998, 230–234.
3. Geli, L., Bard, P., and Jullien, B., The effect of topography on earthquake ground motion: A review and new results, *Bulletin of the Seismological Society of America*, 78, 1, 1988, 42–63.
4. Hardin, B. O. and Drnevich, V. P., Shear modulus and damping in soils: Design equations and curves, *Journal of the Soil Mechanics and Foundations Division, ASCE*, 98, SM7, 1972, 667–692.
5. Ishihara, K., *Soil Behaviour in Earthquake Geotechnics*, Oxford University Press, New York, 1996.
6. Jibson, R., Summary of Research on the Effects of Topographic Amplification of Earthquake Shaking on Slope Stability, Open-File Report 87-268, U.S. Geological Survey, 1987.
7. Kramer, S. L., *Geotechnical Earthquake Engineering*, Prentice-Hall, Upper Saddle River, NJ, 1996.
8. Roesset, J. M., Fundamentals of soil amplification, in *Seismic Design for Nuclear Power Plants*, Hansen, R. J. ed., The MIT Press, Cambridge, MA, 1970, pp. 183–244.
9. Schnabel, P. B., Lysmer, J., and Seed, H. B., SHAKE, A Computer Program for Earthquake Response Analysis of Horizontally Layered Sites, Report No. EERC 72-12, University of California, Berkeley, CA, 1972.
10. Seed, H. B. and Idriss, I. M., *Ground Motion and Soil Liquefaction during Earthquakes*, Earthquake Engineering Research Institute, Oakland, CA, 1982.
11. Seed, H. B. and Idriss, I. M., Soil Moduli and Damping Ratios for Dynamic Response Analyses, Report No. EERC 70-10, Earthquake Engineering Research Center, University of California, Berkeley, CA, 1970.
12. Seed, H. B., Wong, R. T., Idriss, I. M., and Tokimatsu, K., Moduli and damping factors for dynamic analyses of cohesionless soils, *Journal of Geotechnical Engineering, ASCE*, 112, 11, 1986, 1016–1032.
13. Vucetic, M. and Dobry, R., Effect of soil plasticity on cyclic response, *Journal of Geotechnical Engineering, ASCE*, 117, 1, 1991, 89–107.

PROBLEMS

8.1 A soil deposit may be considered to be homogeneous, elastic, and with a horizontal free surface and rock interface. It has a depth of 100 ft, a unit weight of 120 pcf, a shear modulus of elasticity of 4800 psi, and a damping ratio of 15% in all modes. During an earthquake, the base of the deposit is subjected to an acceleration that may be approximated by a harmonic function with an amplitude of $0.5g$ and a frequency of 36 rad/s. Determine (a) the acceleration induced by this ground motion at the level of the free surface and (b) the factor by which the peak acceleration at the base is amplified by the soil deposit. Consider that the deposit's response is significant only in its first three modes.

8.2 The response spectrum for the ground motion recorded at a rock site is shown in Figure P8.2a. Estimate the maximum acceleration that would be observed at the surface of the soil deposit shown in Figure P8.2b and the factor by which the soil deposit would

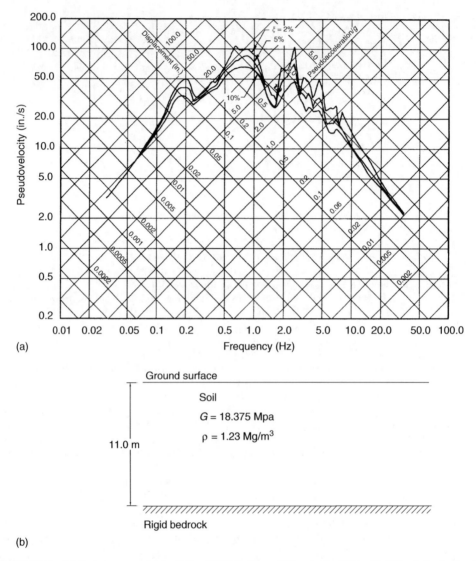

FIGURE P8.2 (a) Response spectrum (After Riddell, R. and Newmark, N.M., Statistical Analysis of the Response of Nonlinear Systems Subjected to Earthquakes, Department of Civil Engineering, University of Illinois, Urbana, IL, 1979.) and (b) soil deposit considered in Problem 8.2.

amplify the ground motion recorded on rock. Assume a linear behavior and a damping ratio of 5%.

8.3 Determine the natural frequencies and mode shapes of the layered soil profile shown in Figure P8.3. Use a one-dimensional lumped-mass model and assume each layer exhibits a linear force-deformation behavior.

8.4 Determine the 5% damping acceleration response spectrum for the ground motion at the bottom and surface of the layered soil deposit shown in Figure P8.3, when the bedrock is excited by the N–S component of the ground acceleration recorded in Foster City during the Loma Prieta, California, earthquake of October 17, 1989 (record may be downloaded from one of the World Wide Web sites suggested in Chapter 5). Use a one-dimensional

Influence of Local Site Conditions

Depth (m)	Material	Density (Mg/m³)	Shear wave velocity (m/s)
0.0			
	Upper clay formation	1.23	153
11.0			
	Clay with sand lenses	1.25	270
34.5			
	Gravel and sand	1.37	316
46.0			
	Clay	1.75	402
70.0			
	Bedrock	1.76	1050

FIGURE P8.3 Soil profile considered in Problem 8.3.

Depth (ft)	Unit weight (lb/ft³)	Shear wave velocity (ft/s)
0		
	100	714
10		
	120	897
160		
	125	1200
230		
	125	1300
730		
	135	1500
1130		
Bedrock	150	8000

FIGURE P8.5 Soil deposit considered in Problem 8.5.

lumped-mass model, assuming that each layer behaves linearly, and that the damping ratio of the system is 10% in all modes.

8.5 Determine the peak acceleration at the surface of the layered soil deposit shown in Figure P8.5 when the bedrock is excited by a ground motion represented by the elastic response spectrum shown in Figure P8.2a. Use a lumped-mass, one-dimensional model, assuming the soil deposit behaves linearly with 10% damping in all modes.

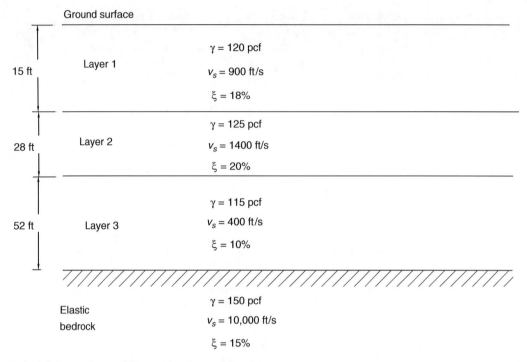

FIGURE P8.7 Soil profile considered in Problem 8.7.

8.6 Find and plot the amplification function of the soil deposit considered in Problem 8.2 when the bedrock beneath is (a) perfectly rigid and (b) elastic with a shear wave velocity of 1500 m/s and a density of 1.75 Mg/m^3. Assume that both the soil and the bedrock behave as a viscoelastic solid with a damping ratio of 20%.

8.7 Find and plot the amplification function of the layered soil deposit shown in Figure P8.7. Assume that each layer and the rock behave as a viscoelastic solid with the properties indicated in the same figure.

8.8 A site is underlain by 550 ft of soil, which in turn is underlain by rigid rock. The average shear wave velocity of the soil is 1500 ft/s^2 and its average unit weight is 125 lb/ft^3. Determine using the frequency-domain approach described in Section 8.4.4 the acceleration time history at the surface of this site when the rock underneath is excited horizontally by the N–S component of the ground acceleration recorded in the Corralitos station during the Loma Prieta, California, earthquake of October 17, 1989. Assume that the soil behaves as a viscoelastic solid with a damping ratio of 10%.

8.9 Estimate and plot the variation with depth of the maximum value of the shear modulus for the soil profile shown in Figure P8.3.

8.10 The groundwater in a 50 ft deep deposit of sand is located 10 ft below the ground surface. The unit weight of the sand above the groundwater table is 100 lb/ft^3. Below the groundwater table, the saturated unit weight is 120 lb/ft^3. Assuming that the sand's void ratio and effective angle of internal friction are 0.6 and 36°, respectively, estimate and plot the variation with depth of the maximum shear modulus for this sand deposit.

8.11 A 24 m thick layer of sand is underlain by bedrock. The groundwater table is located at a depth of 4 m below the ground surface. Estimate the maximum shear modulus and damping ratio of the sand at a depth of 12 m below the ground surface considering a

Influence of Local Site Conditions

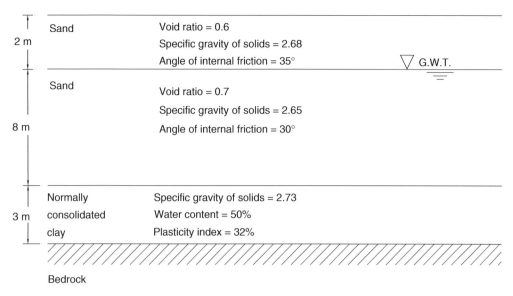

FIGURE P8.13 Soil profile considered in Problem 8.13.

shear-strain level of 0.05%. The sand has a void ratio of 0.6, a specific gravity of solids of 2.68, and an angle of internal friction of 36°.

8.12 A clay deposit extends to a depth of 60 ft below the ground surface. The clay has a void ratio of 1.0, a specific gravity of solids of 2.78, a plasticity index of 50%, and an overconsolidation ratio of 2.0. The groundwater table coincides with the ground surface. Evaluate the deposit's shear modulus and damping ratio at a depth of 30 ft at a strain level of 0.1%.

8.13 Evaluate the moduli and damping ratios for the layers of the soil profile shown in Figure P8.13 for shear strains of 10^{-5} and 10^{-2}. For each layer, consider that the confining pressure at midheight is representative of the confining pressure over the entire layer.

9 Design Response Spectrum

In dealing with earthquakes, we must contend with appreciable probabilities that failure will occur in the near future. Otherwise, all the wealth of this world would prove insufficient to fill our needs: the most modest structures would be fortresses. We must also face uncertainty on a large scale, for it is our task to design engineering systems—about whose pertinent properties we know little—to resist earthquakes and tidal waves—about whose characteristics we know even less.

N. M. Newmark and E. Rosenblueth, 1971

9.1 INTRODUCTION

The attenuation relationships described in Chapter 7 may be used to estimate the expected peak ground acceleration or peak ground velocity at a given site. For design purposes, however, the specification of peak ground acceleration or velocity is not enough to determine the response of a structure to the expected ground motions. What is needed, instead, is the specification of an acceleration time history or a response spectrum. Even more, for design purposes it is necessary to specify an ensemble of representative acceleration time histories or response spectra that averages or envelops the salient features of the response spectra of such an ensemble. The reason is that, as mentioned several times before, acceleration time histories and their response spectra vary significantly from one earthquake to another and, thus, it is highly unlikely that the characteristics of future earthquakes at a site will be similar to those recorded in the past. Additionally, it is necessary to smooth such average response spectra to eliminate peaks and valleys. Peaks and valleys are not suitable for design given the uncertainties involved in the determination of a structure's natural periods and the fact that these periods may change during and after an earthquake exposure. For engineering applications, therefore, response spectra from accelerograms that exhibit certain similarities are averaged and then smoothed before they are specified for design. The smooth response spectrum that is used for the design of structures in a given seismic area is known as *design response spectrum* or simply *design spectrum*.

The first design spectrum was proposed by G. W. Housner in 1959. This spectrum, shown in Figure 9.1, was derived by averaging and smoothing the response spectra of eight accelerograms. The accelerograms used were the two horizontal components of the ground accelerations recorded during the earthquakes of 1934 and 1940 in El Centro, California; 1949 in Olympia, Washington; and 1952 in Taft, California. In addition, the individual spectra were normalized to a peak ground acceleration of $0.2g$. Hence, the curves in Figure 9.1 correspond to a peak ground acceleration of $0.2g$. To use them for any other peak ground acceleration, one simply has to multiply the values read from them by the ratio of the desired acceleration to $0.2g$.

In general, three procedures have been used in the past to construct design response spectra. These procedures are: (a) a method based on the use of a specified peak ground acceleration and a characteristic response spectrum shape, (b) a method introduced by N. M. Newmark and W. J. Hall in 1969, and (c) a method based on statistical correlations to derive directly average response spectrum ordinates. The following sections will describe each of these methods in some detail.

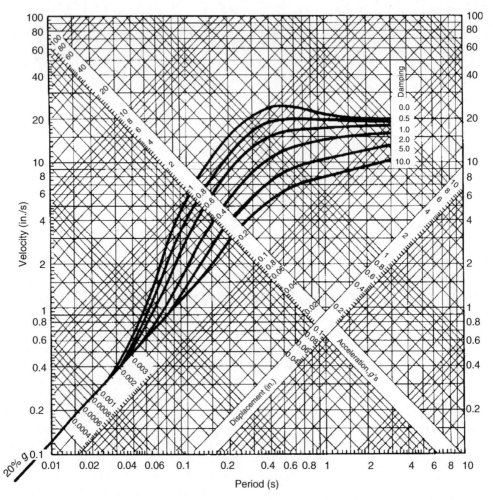

FIGURE 9.1 Housner's design spectrum. (After Housner, G.W., Chapter 5 in *Earthquake Engineering*, Wiegel, R.L., ed., Prentice-Hall, Inc., Englewood Cliffs, 1970. Reprinted by permission of Pearson Education, Inc., Upper Saddle River, NJ.)

9.2 PEAK GROUND ACCELERATION AND RESPONSE SPECTRUM SHAPE METHOD

9.2.1 ELASTIC DESIGN SPECTRUM

In this approach, an average response spectrum for a given area is obtained by multiplying an average response spectrum shape by the peak ground acceleration expected at the given area. The expected peak ground acceleration is estimated as described in Chapter 7, while the average response spectrum shape is determined by

1. Obtaining the response spectra of an ensemble of ground motions representative of those that may occur in the given area.
2. Normalizing the obtained response spectra with respect to their peak ground acceleration (i.e., zero-period acceleration), as illustrated in Figure 9.2 for a typical response spectrum.
3. Computing for each natural period (or frequency) in a preselected natural period range the mean value, or the mean plus one standard deviation, of the ordinates in the normalized response spectra.

Design Response Spectrum

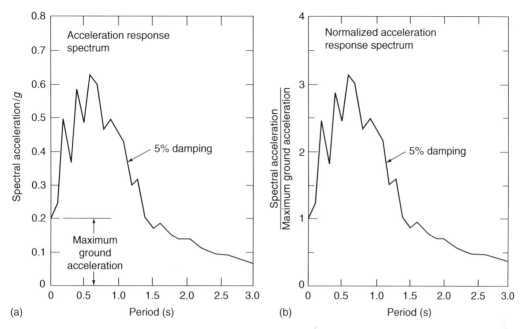

FIGURE 9.2 (a) Original and (b) normalized response spectra. (Reproduced with permission of Earthquake Engineering Research Institute from Seed, H.B. and Idriss, I.M., *Ground Motions and Soil Liquefaction during Earthquakes*, Earthquake Engineering Research Institute, Oakland, CA, 1983.)

4. Plotting the computed mean ordinates, or mean ordinates plus one standard deviation, against natural period or frequency.

Generally, different spectral shapes are defined for different soil types because, as mentioned in Section 8.2, local soil conditions have a significant influence on the shape of response spectra. These different spectral shapes are obtained by considering separately the response spectra of accelerograms that belong to specific soil groups. An example of mean spectral shapes for different soil conditions is shown in Figure 9.3, where the curves shown represent the mean spectral shapes for (a) rock sites, (b) sites with stiff soils that are <200 ft deep, (c) sites with cohesionless soils that have depths of >250 ft, and (d) sites that are underlain by soft to medium stiff clay deposits. As noted earlier, however, for engineering purposes it is convenient to use smooth spectral shapes that envelop those obtained with the procedure just described. Examples of such smooth design spectra are the curves shown in Figure 9.4, which correspond to the average spectral shapes in Figure 9.3. These smooth spectral shapes, obtained for elastic systems with 5% damping, were specified for seismic design in the 1985, 1988, 1991, and 1994 editions of the Uniform Building Code.

9.2.2 Inelastic Design Spectrum

The procedure described in the foregoing section is used to generate a design spectrum for linear elastic structures. As mentioned before, however, most structures are designed taking into consideration their capacity to resist inelastic deformations. Therefore, inelastic design spectra are needed for the design of these structures.

A simplified method by means of which a yield acceleration inelastic design spectrum may be obtained from an elastic design spectrum may be developed as follows. First, assume that (a) the force-deformation behavior of structures is approximately elastoplastic and (b) the maximum inelastic deformation in these structures under a given excitation is approximately equal to the

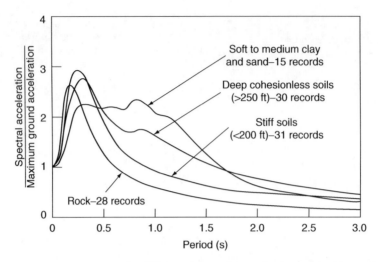

FIGURE 9.3 Average response spectra for different site conditions. (Reproduced with permission of Earthquake Engineering Research Institute from Seed, H.B. and Idriss, I.M., *Ground Motions and Soil Liquefaction during Earthquakes*, Earthquake Engineering Research Institute, Oakland, CA, 1983.)

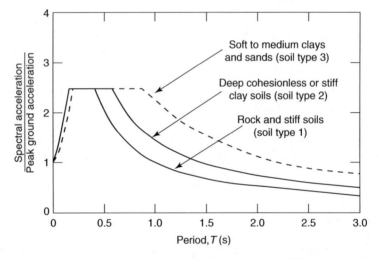

FIGURE 9.4 Smooth response spectrum shapes specified in some building codes. (Reproduced with permission of Earthquake Engineering Research Institute from Seed, H.B. and Idriss, I.M., *Ground Motions and Soil Liquefaction during Earthquakes*, Earthquake Engineering Research Institute, Oakland, CA, 1983.)

maximum deformation attained when the structures remain linear at all times (see Figure 9.5). Then, on the basis of these assumptions, the definition of ductility factor introduced in Section 6.5, and Figure 9.5, consider that for an elastoplastic system its yield strength, F_y, is given by

$$F_y = K\mu_y = K\frac{u_{max}}{\mu} \tag{9.1}$$

where K, u_y, μ, and u_{max}, respectively, denote stiffness constant, yield deformation, ductility factor, and maximum deformation. But in view of the assumption that the maximum deformation in an inelastic system is equal to the maximum deformation in its elastic counterpart, u_{max} also represents the maximum deformation in an equivalent linear system and hence the ordinate in the displacement response spectrum of the excitation corresponding to the system's natural frequency and damping

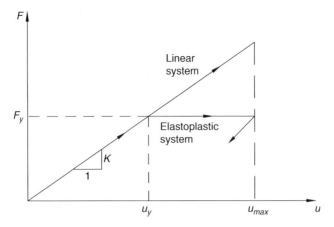

FIGURE 9.5 Linear and elastoplastic force-deformation curves.

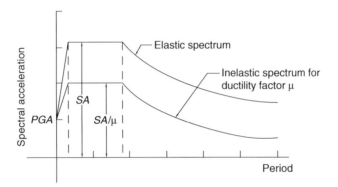

FIGURE 9.6 Inelastic acceleration design spectrum derived from elastic counterpart.

ratio. That is, $u_{max} = SD(\omega_n, \xi)$, where ω and ξ, respectively, depict such natural frequency and damping ratio, and $SD(\omega_n, \xi)$ is the aforementioned spectral ordinate. Consequently, Equation 9.1 may also be written as

$$F_y = K \frac{SD(\omega_n, \xi)}{\mu} = \omega_n^2 m \frac{SD(\omega_n, \xi)}{\mu} \cong m \frac{SA(\omega_n, \xi)}{\mu} \quad (9.2)$$

where the approximate relationship that exists between the spectral acceleration $SA(\omega_n, \xi)$ and the spectral displacement $SD(\omega_n, \xi)$ (see Section 6.4.4) has been used. From Equation 9.2, the yield acceleration may be thus expressed as

$$\ddot{u}_y = \frac{F_y}{m} \cong \frac{SA(\omega_n, \xi)}{\mu} \quad (9.3)$$

which, in words, indicates that the yield acceleration spectrum is approximately equal to the elastic spectrum divided by the ductility factor μ.

It may be seen, thus, that an inelastic response spectrum may be obtained from an elastic response spectrum by simply dividing the ordinates of the elastic spectrum by a ductility factor that is representative of the inelastic deformations a structure can withstand. This procedure is illustrated in Figure 9.6. Note that an exception to this rule is for short periods because for a zero period the spectral acceleration should be equal to the peak ground acceleration, independent of the assumed force-deformation behavior. For short periods, therefore, the inelastic response spectrum is defined by a transition line between the value of the peak ground acceleration at a zero period and the inelastic spectral acceleration at the period that defines the left corner of the elastic spectrum. The procedure to construct a design spectrum for a given site using this approach is summarized in Box 9.1.

Box 9.1 Procedure to Construct Design Spectrum Using Peak Ground Acceleration and Response Spectrum Shape

1. Estimate the peak ground acceleration expected at the site under consideration for an acceptable probability of exceedance.
2. Select a spectral shape according to the site's soil conditions.
3. Multiply the ordinates of the selected spectral shape by the estimated peak ground acceleration to obtain an elastic design spectrum.
4. Divide the ordinates of the elastic design spectrum by a preselected ductility factor to determine the corresponding inelastic design spectrum.
5. Draw a transition line between the elastic spectral acceleration at zero period and the inelastic spectral acceleration at the period that defines the left corner of the elastic spectrum.

If in Equation 9.3 the spectral acceleration is expressed as a fraction of g, the acceleration of gravity, then the lateral force that would make a single-degree-of-freedom structure yield may be expressed as the product of a coefficient and the structure's weight. That is,

$$F_y = \frac{SA(\omega_n, \xi)/g}{\mu} W = cW \tag{9.4}$$

where W denotes weight and

$$c = \frac{SA(\omega_n, \xi)/g}{\mu} \tag{9.5}$$

The coefficient c represents the *seismic coefficient* specified in many building codes for the seismic design of structures.

It should be noted at this point that the accuracy of the procedure just described to generate an inelastic design spectrum depends heavily on the accuracy of the assumption made that the maximum deformation in an elastoplastic system is approximately the same as in an equivalent linear system. It may be seen from Figure 9.7, however, that this assumption does not hold for individual acceleration records, but it is approximately true (except for small periods) for the average of a large number of records. The procedure may be thus considered to be adequately accurate on the average sense.

EXAMPLE 9.1 INELASTIC DESIGN SPECTRUM FOR A GIVEN SITE

A site is located near a seismically active fault over a deposit of stiff soil. The peak ground acceleration at the site for a probability of exceedance of 10% in 50 years has been estimated at $0.51g$. On the basis of this information and the response spectrum shapes shown in Figure 9.4, construct an inelastic design spectrum for the site. Consider a damping ratio of 5% and a ductility factor of 4.

Solution

According to the procedure established in the foregoing section, the ordinates of the elastic design spectra for the site under consideration may be obtained by multiplying the ordinates in the spectral

Design Response Spectrum

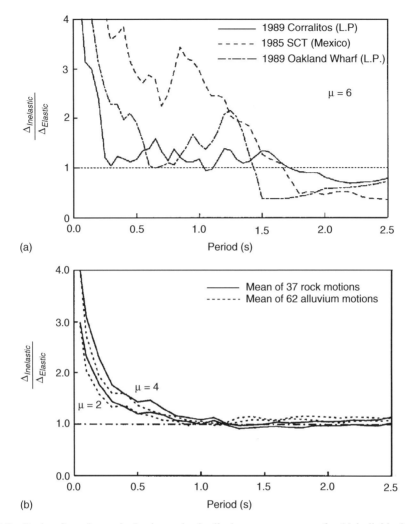

FIGURE 9.7 Ratio of maximum inelastic to elastic displacement response for (a) individual records and (b) a large number of records, on average. (Reproduced with permission of Earthquake Engineering Research Institute from Miranda, E., *Earthquake Spectra*, 9, 233–250, 1993.)

shape given in Figure 9.4 for stiff soils by the given peak ground acceleration. Thus, for example, as the largest ordinate in the spectral shape corresponding to stiff soils in Figure 9.4 is 2.5, then the largest ordinate in the elastic design spectrum for the site is equal to

$$(SA)_{elastic} = 2.5(0.51g) = 1.28g$$

Similarly, the ordinates of the inelastic design spectrum are obtained by dividing the ordinates in the elastic design spectrum by the selected or given ductility factor. Accordingly, for a ductility factor of 4, the largest ordinate in the desired inelastic design spectrum is

$$(SA)_{inelastic} = 1.28g/4 = 0.32g$$

Proceeding similarly for all other ordinates and adding a transition line as specified in Box 9.1, one obtains the inelastic design spectrum shown in Figure E9.1.

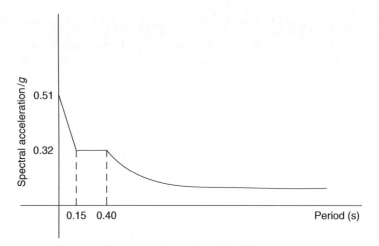

FIGURE E9.1 Inelastic design spectrum obtained in Example 9.1.

9.3 NEWMARK–HALL APPROACH

9.3.1 Background

As discussed in Section 6.2, the response of a structure to a ground motion depends, in general, on the amplitude, duration, and frequency content of the ground motion. Therefore, it cannot be expected that a design spectrum built on the basis of peak ground acceleration alone will envelop the response spectra obtained directly from ground motion records. For instance, two accelerograms may have the same peak ground acceleration, but their maximum spectral acceleration may be substantially different. Consequently, if a design spectrum were generated by multiplying the spectral shape by the peak ground acceleration in each of these accelerograms, one would obtain the same design spectrum when one should be obtaining a different one for each accelerogram, each enveloping its own response spectrum. Thus, the normalization of response spectra on the basis of peak ground acceleration is equivalent to assuming that all the ordinates in a response spectrum are affected equally by peak ground acceleration, which is far from the truth. In fact, in 1973 N. M. Newmark and W. J. Hall observed that some spectral ordinates are affected more by peak ground velocity or peak ground displacement rather than by peak ground acceleration. To take this fact into consideration, Newmark and Hall suggested a normalization procedure in which different regions of a spectrum are normalized by different measures of ground motion intensity.

9.3.2 Elastic Design Spectra

Basically, what Newmark and Hall did for the derivation of their recommendations for the construction of elastic design spectra was to draw the response spectra of a number of ground motions using a tripartite representation, but with the displacement, velocity, and acceleration scales, respectively, normalized by peak ground displacement, peak ground velocity, and peak ground acceleration. An example of such normalized spectra is shown in Figure 9.8, which corresponds to the N–S accelerogram from the 1940 El Centro, California, earthquake. From such spectra, they noted that

1. For ultralow frequencies, that is, frequencies <0.05 Hz, the spectral displacements are approximately equal to the peak ground displacement.
2. For low frequencies, that is, frequencies between 0.1 and 0.3 Hz, the spectral displacements are approximately equal to the peak ground displacement times a constant amplification factor.

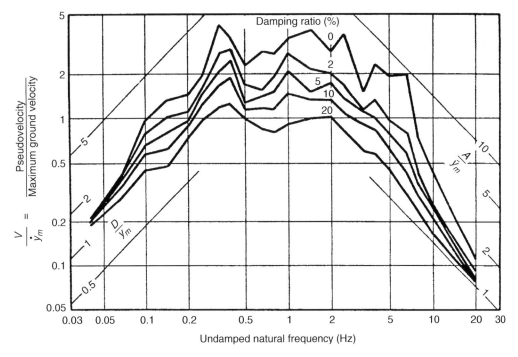

FIGURE 9.8 Normalized elastic response spectrum in tripartite representation of N–S accelerogram recorded during the 1940 El Centro, California, earthquake. (Reproduced with permission of Earthquake Engineering Research Institute from Newmark, N.M. and Hall, W.J., *Earthquake Spectra and Design*, Earthquake Engineering Research Institute, Oakland, CA, 1982.)

3. For intermediate frequencies, that is, frequencies between 0.3 and 2 Hz, the spectral velocities are approximately equal to the peak ground velocity times a constant amplification factor.
4. For high frequencies, that is, frequencies between 2 and 8 Hz, the spectral accelerations are approximately equal to the peak ground acceleration times a constant amplification factor.
5. For ultrahigh frequencies, that is, frequencies >20 Hz, the spectral accelerations are approximately equal to the peak ground acceleration.

Using the ground motions considered in their study, Newmark and Hall also calculated for different damping ratios the mean values, and mean values plus one standard deviation, of such amplification factors. These amplification factors are presented in Table 9.1, where **A**, **V**, and **D**, respectively, denote the factors by which the peak ground acceleration, peak ground velocity, and peak ground displacement are amplified in the spectral regions identified earlier.

On the basis of their study, Newmark and Hall suggested the procedure summarized in Box 9.2 for the construction of elastic design spectra. In the application of this procedure, they suggested to obtain the required values of the peak ground acceleration, peak ground velocity, and peak ground displacement for the site under consideration directly from available attenuation relationships, such as those introduced in Chapter 7. For those cases where only peak ground acceleration is available, they proposed the use of the following empirical relationships between peak ground acceleration, peak ground velocity, and peak ground displacement:

$$v/a = 36 \text{ in./s/g for rock}$$
$$v/a = 48 \text{ in./s/g for competent soils} \quad (9.6)$$
$$ad/v^2 = 6.0$$

TABLE 9.1
Newmark–Hall Amplification Factors for the Construction of Elastic Design Spectra

Damping Ratio(%)	Mean			Mean Plus One Standard Deviation		
	A	V	D	A	V	D
0.5	3.68	2.59	2.01	5.10	3.84	3.04
1	3.21	2.31	1.82	4.38	3.38	2.73
2	2.74	2.03	1.63	3.66	2.92	2.42
3	2.46	1.86	1.52	3.24	2.64	2.24
5	2.12	1.65	1.39	2.71	2.30	2.01
7	1.89	1.51	1.29	2.36	2.08	1.85
10	1.64	1.37	1.20	1.99	1.84	1.69
20	1.17	1.08	1.01	1.26	1.37	1.38

Source: Newmark, N. M. and Hall, W. J., *Earthquake Spectra and Design*, Earthquake Engineering Research Institute, Oakland, CA, 1982.

In these relationships, a, v, and d, respectively, denote peak ground acceleration expressed as a fraction of g, peak ground velocity in inches per second, and peak ground displacement in inches. In the last expression, a is expressed in inches per second square so as to make the right-hand side of the equation dimensionless.

Box 9.2 Procedure to Construct Elastic Design Spectrum Using Newmark–Hall Approach

1. Estimate for the site under consideration the expected peak values of ground acceleration, ground velocity, and ground displacement.
2. Draw in a tripartite graph lines perpendicular to the spectral displacement, spectral pseudovelocity, and spectral acceleration axes, with ordinates, respectively, equal to the
   a. Peak ground displacement times **D**
   b. Peak ground velocity times **V**
   c. Peak ground acceleration times **A**
   where **D**, **V**, and **A** are the amplification factors given in Table 9.1.
3. Draw a line perpendicular to the spectral acceleration axis with an ordinate equal to the peak ground acceleration.
4. Draw a transition line between the frequencies of 8 and 33 Hz, linking the lines drawn in Step 2 with the line drawn in Step 3.
5. Connect all five lines.

The relationships between peak ground parameters originally proposed by Newmark and Hall were derived in 1973 considering a relatively small number of ground motion records (28) and without separating them according to the geological conditions of the recording stations. With the availability of additional data, new similar expressions have been proposed, such as those proposed by B. Mohraz in 1976. Mohraz considered a large number of records—a total of 162 accelerograms from 54 stations recorded during 16 earthquakes—and four different site conditions. He arrived to the average v/a and ad/v^2 ratios listed in Table 9.2. He also obtained the average d/a ratios included

TABLE 9.2
Average v/a, ad/v², and d/a Ratios Obtained in 1976 Mohraz Study

Soil Type	v/a (in./s/g)	ad/v²	d/a (in./g)
Rock	24	5.3	8
<30 ft of alluvium underlain by rock	30	4.5	11
30–200 ft of alluvium underlain by rock	30	5.1	12
Alluvium	48	3.9	23

Source: Mohraz, B. and Sadek, F., *The Seismic Design Handbook*, 2nd Edition, Kluwer Academic Publishers, Boston, MA, 2001.

in the same table. In another study, Mohraz and coworkers computed the average v/a and ad/v^2 ratios of the acceleration records from the Loma Prieta, California, earthquake of October 17, 1989. In this study, mean v/a ratios of 51 and 49 in./s/g and mean ad/v^2 ratios of 2.8 and 2.6 for rock and alluvium, respectively, were obtained. In a similar study conducted in 1979 with 10 ground motions from worldwide earthquakes recorded on rock and alluvium sites, R. Riddell and N. M. Newmark arrived to average v/a and ad/v^2 ratios of 35 and 5.9, respectively. The differences in the v/a and ad/v^2 ratios determined in different investigations clearly show that these ratios significantly depend on magnitude, epicentral distance, and duration of the recorded earthquakes, as well as the soil properties at the recording stations.

In the application of the available relationships between ground motion parameters to estimate peak ground velocity and peak ground displacement in terms of peak ground acceleration, it should be kept in mind that some studies (e.g., the study by W. S. Dunbar and R. G. Charlwood listed at the end of the chapter) have shown that this practice can lead to significant magnitude bias in the resulting design spectrum as these relationships strongly depend on magnitude. It has also been found that the dependence of spectral shape on magnitude and site conditions can be accounted for by using independent estimates of peak ground acceleration, velocity, and displacement in the construction of design spectra. It is preferred, thus, to use Newmark–Hall approach using independent estimates of the peak ground motion parameters.

As it may be observed from the foregoing discussion, design spectra generated with Newmark–Hall approach depend not only on peak ground acceleration, but also on peak ground velocity and peak ground displacement. As such, Newmark–Hall approach represents an improvement over the procedure introduced in Section 9.2. However, it should be noted that design spectra determined by Newmark–Hall approach exhibit lower amplifications at short periods and larger amplifications at long periods when compared to spectra generated by direct statistical correlations (discussed in the following section). These discrepancies have been explained on the grounds that Newmark–Hall amplification factors are biased toward the spectra of large magnitude earthquakes and that large magnitude earthquakes generate ground motions rich in low frequencies while small magnitude earthquakes generate ground motions rich in high frequencies. The bias toward large magnitude earthquakes resulted from the fact that of the nine earthquakes considered in the derivation of the Newmark–Hall amplification factors seven had a magnitude >6.0. Bias also resulted from the use of multiple spectra from the same event, giving each of these multiple spectra the same weight. For example, eight of the 28 spectra used in the aforementioned derivation were recorded during the 1971 M6.6 San Fernando, California, earthquake. An additional bias resulted from the fact that most of the accelerograms considered were recorded on alluvial sites. This bias also contributed to produce larger amplification factors at long periods and lower amplification factors at short periods because the spectra from soil sites tend to exhibit these characteristics when compared to those obtained at rock sites.

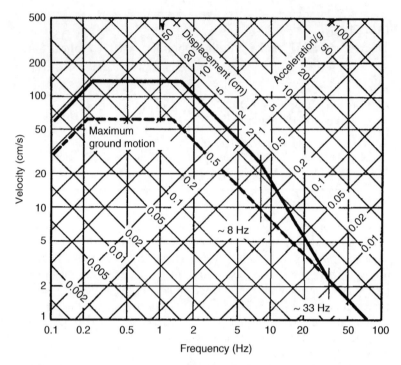

FIGURE E9.2 Elastic design spectrum obtained in Example 9.2. (Reproduced with permission of Earthquake Engineering Research Institute from Newmark, N.M. and Hall, W.J., *Earthquake Spectra and Design*, Earthquake Engineering Research Institute, Oakland, CA, 1982.)

EXAMPLE 9.2 ELASTIC DESIGN SPECTRUM USING NEWMARK–HALL APPROACH

The peak ground acceleration, velocity, and displacement expected at a given site are estimated to be 0.5g, 61 cm/s, and 45 cm, respectively. Use Newmark–Hall approach to construct an elastic design spectrum for 5% damping. Use the amplification factors corresponding to the mean values plus one standard deviation.

Solution

From Table 9.1, the mean plus one standard deviation amplification factors corresponding to 5% damping are 2.71, 2.30, and 2.01, respectively, for the constant acceleration, constant velocity, and constant displacement regions of the spectrum. Thus, the desired spectrum may be obtained by drawing lines perpendicular to the acceleration, velocity, and displacement axes through ordinates equal to 0.5g × 2.71 = 1.35g, 61 × 2.30 = 140 cm/s, and 45 × 2.01 = 90 cm, respectively, plus a transition line between the spectral acceleration defined by the previous lines at 8 Hz and the spectral acceleration at 33 Hz, which may be assumed equal to the peak ground acceleration of 0.5g. The design spectrum so generated is shown in Figure E9.2.

9.3.3 Inelastic Design Spectra

Newmark and Hall also established a procedure to construct a nonlinear design spectrum from a linear one. As with their linear design spectra, they arrived to this procedure by plotting yield-deformation and total-deformation nonlinear response spectra in tripartite graphs with their scales normalized with respect to peak ground acceleration, peak ground velocity, and peak ground displacement. Examples of such spectra are shown in Figures 9.9 and 9.10, which correspond to the

Design Response Spectrum

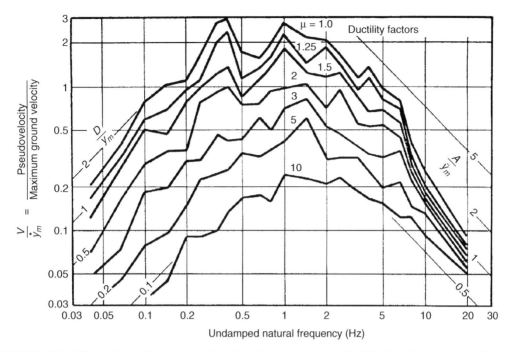

FIGURE 9.9 Yield-deformation spectra for elastoplastic systems with 2% damping corresponding to the N–S accelerogram from the 1940 El Centro earthquake. (Reproduced with permission of Earthquake Engineering Research Institute from Newmark, N.M. and Hall, W.J., *Earthquake Spectra and Design*, Earthquake Engineering Research Institute, Oakland, CA, 1982.)

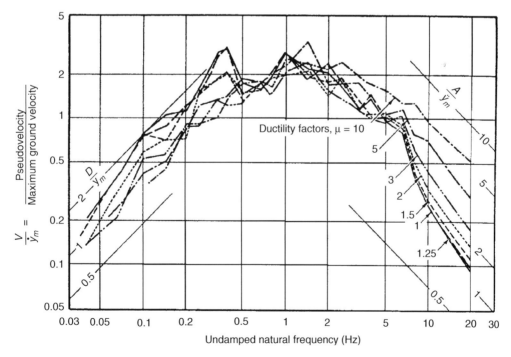

FIGURE 9.10 Total-deformation spectra for elastoplastic systems with 2% damping corresponding to the N–S accelerogram from the 1940 El Centro earthquake. (Reproduced with permission of Earthquake Engineering Research Institute from Newmark, N.M. and Hall, W.J., *Earthquake Spectra and Design*, Earthquake Engineering Research Institute, Oakland, CA, 1982.)

N–S accelerogram recorded during the 1940 El Centro, California, earthquake. From the average of such plots, Newmark and Hall noted that:

1. For frequencies <0.3 Hz (i.e., region of constant elastic spectral displacements), the total inelastic displacement response is approximately equal to the elastic displacement response (i.e., $\mu = 1$) for all μ.
2. For frequencies between 0.3 and 2 Hz (i.e., region of constant elastic spectral velocities), the total inelastic displacement response is approximately equal to the elastic displacement response for all μ.
3. For frequencies between 2 and 8 Hz (i.e., region of constant elastic spectral accelerations), the total inelastic displacement response is not equal to the elastic displacement response, but depends on μ instead.
4. For frequencies >33 Hz, the yield acceleration response is nearly the same for all ductility factors and equal to the peak ground acceleration.

Based on these observations, Newmark and Hall established that in the regions of constant elastic spectral displacements and constant elastic spectral velocities, the relationship between a yield-deformation spectrum and its elastic counterpart may be expressed, on the average, as indicated by Equation 9.2. That is,

$$u_y = \frac{SD(\omega_n, \xi)}{\mu} \tag{9.7}$$

where, as before, $SD(\omega_n, \xi)$ denotes the elastic spectral displacement corresponding to a natural frequency ω_n and damping ratio ξ, and μ represents ductility factor. Similarly, after noticing that the foregoing equation cannot be applied to the region of constant elastic spectral accelerations, they derived an alternative relationship based on energy concepts as follows.

Consider, first, that (a) the deformation energies in elastic and inelastic systems are the same, (b) the deformation energy in a system is given by the area under its force-deformation curve, and (c) the area under the force-deformation curve of an elastic system is therefore equal to the area under the force-deformation curve of its elastoplastic counterpart. Based on these considerations and with reference to Figure 9.11, the relationship between the maximum elastic and inelastic deformations may be written as

$$\frac{1}{2}Ku_0^2 = \frac{1}{2}Ku_y^2 + (u_{max} - u_y)Ku_y \tag{9.8}$$

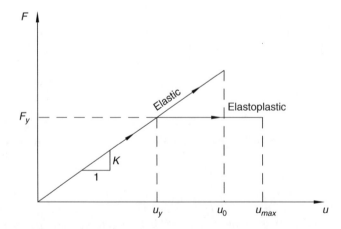

FIGURE 9.11 Force-deformation curves for linear and elastoplastic systems.

Design Response Spectrum

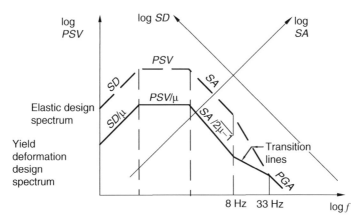

FIGURE 9.12 Construction of inelastic design spectrum using Newmark–Hall approach.

where u_y, u_0, and u_{max}, respectively, denote the yield deformation in the elastoplastic system, the maximum deformation in the elastic system, and the maximum deformation in the elastoplastic system. As before, K represents a stiffness constant or, what is tantamount, the slope of the line that defines the elastic force-deformation curve and the initial branch of the elastoplastic force-deformation curve. Note, however, that Equation 9.8 may be reduced to

$$\frac{u_0^2}{u_y^2} = 1 + 2\left(\frac{u_{max}}{u_y} - 1\right) u_y = 2\mu - 1 \qquad (9.9)$$

where the fact that u_{max}/u_y is equal to the ductility factor μ has been used. Hence, after solving for u_y from this equation, one obtains

$$u_y = \frac{u_0}{\sqrt{2\mu - 1}} \qquad (9.10)$$

which may also be expressed as

$$u_y = \frac{SD(\omega_n, \xi)}{\sqrt{2\mu - 1}} \qquad (9.11)$$

given that the maximum deformation in the elastic system is equal to spectral displacement corresponding to the system's natural frequency and damping ratio. Equation 9.11 is the relationship proposed by Newmark and Hall to obtain a yield-deformation spectrum from an elastic one in the region of constant elastic spectral accelerations.

On the basis of the preceding observations and Equations 9.7 and 9.11, Newmark and Hall proposed the procedure summarized in Box 9.3 and illustrated in Figure 9.12 to construct for a given ductility factor μ a yield-deformation design spectrum from the corresponding elastic one.

Box 9.3 Procedure to Construct Inelastic Design Spectrum Using Newmark–Hall Approach

1. Draw an elastic design spectrum in a tripartite graph following the procedure summarized in Box 9.2.
2. Select an acceptable value for the ductility factor μ.

3. Draw lines perpendicular to the spectral displacement, spectral pseudovelocity, and spectral acceleration axes, with ordinates respectively equal to the
   a. Elastic spectral displacement in the region of constant spectral displacements divided by the selected ductility factor μ
   b. Elastic spectral pseudovelocity in the region of constant spectral pseudovelocities divided by the selected ductility factor μ
   c. Elastic spectral acceleration in the region of constant spectral accelerations divided by $\sqrt{2\mu - 1}$
4. Draw a line perpendicular to the spectral acceleration axis with an ordinate equal to the peak ground acceleration.
5. Draw a transition line between the frequencies of 8 and 33 Hz, linking the lines drawn in Step 4 with the line drawn in Step 5.
6. Connect all five lines.

EXAMPLE 9.3 INELASTIC DESIGN SPECTRUM USING NEWMARK–HALL APPROACH

Construct using Newmark–Hall approach an inelastic design spectra for 5% damping and a ductility factor of 3 for a soil site where the expected peak ground acceleration is estimated to be 0.16g. Use the amplification factors corresponding to the mean plus one standard deviation.

Solution

Using Newmark and Hall relationships between peak ground acceleration and peak ground velocity and displacement (Equation 9.6), one finds that for a peak ground acceleration of 0.16g,

$$v = 48(0.16) = 7.7 \text{ in./s}$$

$$d = \frac{6.0(7.7)^2}{0.16(386.4)} = 5.8 \text{ in.}$$

Similarly, from Table 9.1, the mean amplification factors plus one standard deviation for 5% damping are 2.71, 2.30, and 2.01, respectively, for the constant acceleration, constant velocity, and constant displacement regions of the spectrum. As a result, the envelopes for the elastic spectral accelerations, velocities, and displacements are

$$SA = 2.71(0.16g) = 0.43g$$

$$SV = 2.30(7.7) = 17.7 \text{ in./s}$$

$$SD = 2.01(5.8) = 11.7 \text{ in.}$$

and, according to Equations 9.7 and 9.11, the envelopes for the inelastic spectra for a ductility factor of 3 are

$$(SA)_{in} = \frac{0.43\,g}{\sqrt{2(3)-1}} = 0.22g$$

$$(SV)_{in} = \frac{17.7}{3} = 5.9 \text{ in./s}$$

$$(SD)_{in} = \frac{11.7}{3} = 3.9 \text{ in.}$$

The resulting elastic and inelastic design spectra are shown, together with their corresponding transition lines, in Figure E9.3.

Design Response Spectrum

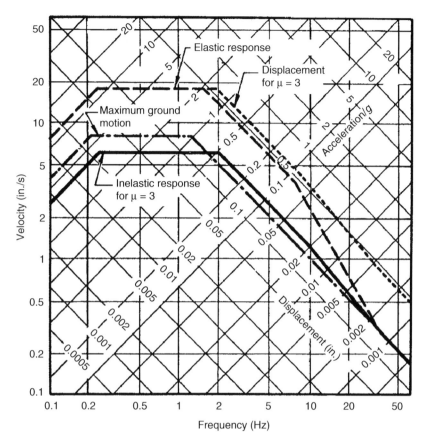

FIGURE E9.3 Inelastic response spectrum obtained in Example 9.3. (Reproduced with permission of Earthquake Engineering Research Institute from Newmark, N.M. and Hall, W.J., *Earthquake Spectra and Design*, Earthquake Engineering Research Institute, Oakland, CA, 1982.)

The procedure proposed by Newmark and Hall for the construction of inelastic design spectra was derived considering elastoplastic systems and assuming that damping has the same effect on the response of elastic and inelastic systems. Furthermore, it was derived without using a statistical analysis consistent with the analysis carried out for elastic spectra and without accounting for the dependence of spectral shapes on soil properties. To investigate the adequacy of the Newmark–Hall approach and the influence of damping and force-deformation model on inelastic response, in 1979 R. Riddell and N. M. Newmark conducted a statistical study with an ensemble of 10 acceleration records from worldwide earthquakes. In this study, they derived deamplification factors to obtain inelastic spectra from their elastic counterparts assuming elastoplastic, bilinear, and bilinear with stiffness degradation force-deformation models (Figure 6.12) and considering alternatively damping ratios of 2, 5, and 10%. These deamplification factors were defined as the number by which one can multiply an elastic spectral ordinate to obtain the corresponding ordinate in the inelastic response spectrum; that is, the ratio of an inelastic spectral ordinate to the corresponding elastic spectral ordinate. From the results of their study, they concluded, first of all, that the assumed force-deformation model does not have a significant influence on the mean ordinates of inelastic spectra and that the elastoplastic model provides, in almost every case, conservative estimates. They also concluded that the effect of damping on inelastic response becomes less important as the ductility factor increases. For example, the ordinates of the average elastic spectrum in the region of maximum spectral amplifications are reduced by 41% when the damping ratio increases from 2 to 10%, but this reduction is only 16%

for the ordinates of the average elastoplastic spectrum for a ductility factor of 10. In regard to the adequacy of the original Newmark–Hall reduction factors, they found that these factors provide a good approximation, although they are not always conservative. In particular, they provide a good match for a damping ratio of 5%. A comparison between the reduction factors obtained by Riddell and Newmark for elastoplastic systems and the original Newmark–Hall rules is shown in Figure 9.13, separately for each of the three spectral regions considered by Newmark and Hall.

In a similar manner, E. Miranda obtained in 1993 reduction factors to obtain inelastic design spectra from their elastic counterparts taking into consideration site conditions. For this purpose, he employed 124 accelerograms from 13 worldwide earthquakes and classified them according to the soil properties at the recording stations. The soil categories considered were: (a) rock, (b) alluvium, and (c) soft soils. The systems considered were assumed to have a single degree-of-freedom, a

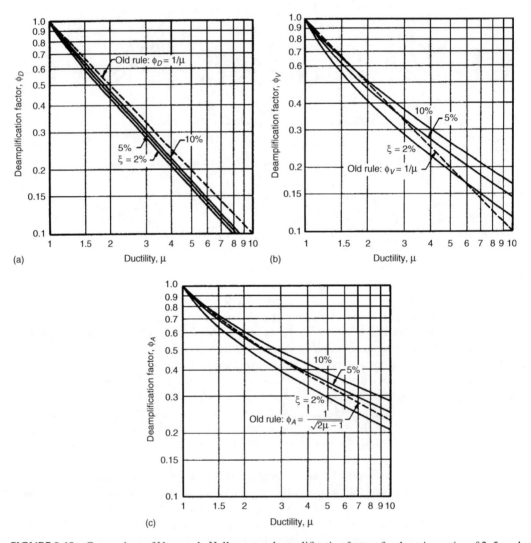

FIGURE 9.13 Comparison of Newmark–Hall average deamplification factors for damping ratios of 2, 5, and 10% against those obtained in Riddell–Newmark study: (a) displacement region, (b) velocity region, and (c) acceleration region. (After Riddell, R. and Newmark, N.M., *Statistical Analysis of the Response of Nonlinear Systems Subjected to Earthquakes*, University of Illinois, Urbana, IL, 1979.)

bilinear force-deformation behavior with a postelastic stiffness of 3% of the initial stiffness, and a damping ratio of 5%. The reduction factors were defined as

$$R_\mu = \frac{F_y(T, \mu = 1)}{F_y(T, \mu = \mu_i)} = \frac{SA(T, \mu = 1)}{SA(T, \mu = \mu_i)} \qquad (9.12)$$

where $F_y(T, \mu = 1)$ represents the lateral yield strength required to keep a system with natural period T within its elastic range of behavior, $F_y(T, \mu = \mu_i)$ denotes the lateral yield strength required to limit the system's inelastic displacements to values that are less than or equal to a preselected maximum expressed in terms of a target ductility factor μ_i, and $SA(T, \mu = 1)$ and $SA(T, \mu = \mu_i)$, respectively, depict the corresponding spectral ordinates in the elastic and inelastic response spectra of the ground motion under consideration. That is, R_μ represents the maximum reduction in strength that can be employed to limit the displacement ductility factor μ in a single-degree-of-freedom system to a preselected value μ_i. Two limits are worth noting in the consideration of this parameter. When $T \to 0$, $R_\mu \to 1$ and when $T \to \infty$, $R_\mu \to \mu$. The existence of the first limit is owed to the fact that in the case of very rigid systems, a small reduction in the yield strength results in large inelastic deformations. Hence, for these systems any strength reduction would lead to inelastic deformations in excess of the specified ductility factor, independently of the selected value. The existence of the second limit is due to the fact that for very flexible systems the maximum relative displacement is equal to the peak ground displacement regardless of the considered yield strength. In accordance with Equation 9.2, the inelastic strength demand for these systems is therefore equal to their elastic strength divided by the selected ductility factor and, hence, the corresponding reduction factor is equal to this ductility factor.

The mean reduction factors obtained by Miranda are presented in Figure 9.14. The dispersion associated with these reduction factors, measured in terms of coefficients of variation (COV), is shown in Figure 9.15 for the reduction factors for rock and alluvium sites. It should be observed that the reduction factors for the soft soil category were computed for period ratios T/T_g, where T_g denotes the ground motion predominant period, as opposed to values of the natural period T. This variation was introduced in response to (a) the observed influence of such a predominant period on the response spectra of soft soils and (b) the fact that computing the average reduction factors for soft soils in terms of T as opposed to T/T_g using records with significantly different dominant periods would have resulted in a poor description of such an influence. It is also interesting to observe from Figure 9.14 that the reduction factors for soft soils are greater than the target ductility factor for periods near the ground motion predominant period, smaller than this target ductility factor for periods of less than two-thirds of the ground motion predominant period, and approximately equal to the target ductility factor for periods longer than one-and-a-half times the ground motion predominant period.

It is worthwhile to note that Miranda's reduction factors were obtained using multiple recordings from the same event. For example, he used 18 records from the 1989 M_s 7.1 Loma Prieta earthquake and 15 records from the 1987 M_L 6.1 Whittier-Narrows earthquake. As discussed earlier, the use of multiple recordings from single event may result in a bias toward the event with the largest number of recordings, unless the recordings from the same event are weighted in proportion to the number of such recordings. It is also worthwhile to note that Miranda's reduction factors compare well with those obtained by other investigators using different data sets. Figure 9.16 shows a comparison between the mean reduction factors obtained by Miranda and those determined by R. Riddell in a 1995 study with 72 acceleration records, of which the majority were from the 1985 M_s 7.8 Chile earthquake and its aftershocks. The soil types considered by Riddell were rock (Type I), dense sands and gravels (Type II), and medium density sands and gravels (Type III), which he considered to approximately correspond to the rock, alluvium, and soft soil types considered by Miranda. The general agreement between the two sets of curves is remarkable given that the two studies employed acceleration records from earthquakes that occurred in different tectonic environments. Most of the

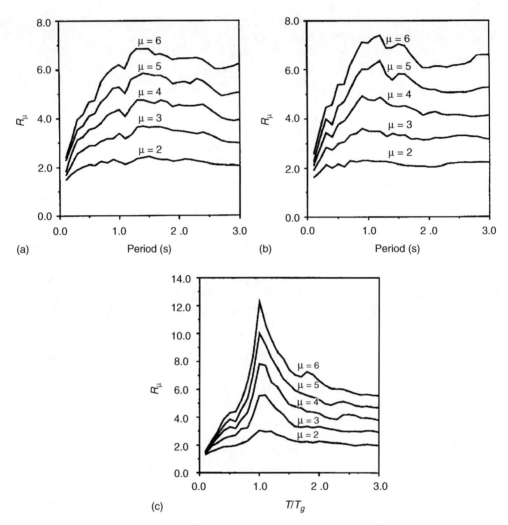

FIGURE 9.14 Miranda's mean reduction factors for (a) rock, (b) alluvium, and (c) soft soil sites. (Reproduced with permission of ASCE from Miranda, E., *Journal of Structural Engineering, ASCE*, 119, 3503–3519, 1993.)

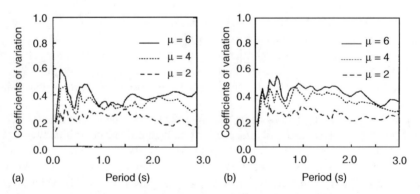

FIGURE 9.15 Coefficients of variation obtained in Miranda's statistical analysis of reduction factors for (a) rock and (b) alluvial soils. (Reproduced with permission of ASCE from Miranda, E., *Journal of Structural Engineering, ASCE*, 119, 3503–3519, 1993.)

Design Response Spectrum

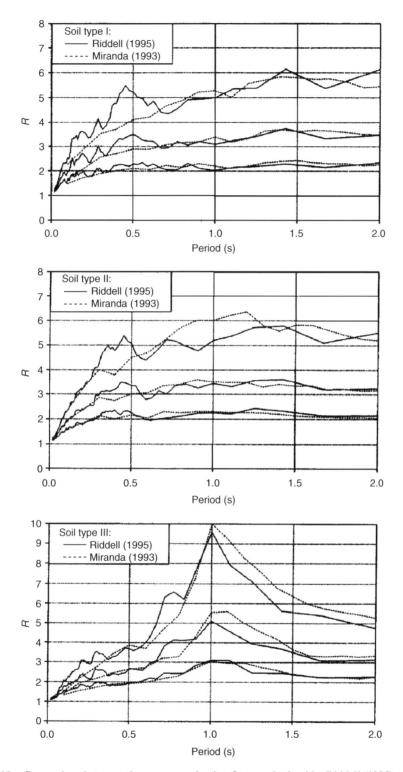

FIGURE 9.16 Comparison between the average reduction factors obtained by Riddell (1995) and Miranda (1993) for ductility factors of 2, 3, and 5 and three different site conditions. (After Riddell, R., *Earthquake Engineering and Structural Dynamics*, 24, 1491–1510, 1995. © John Wiley & Sons Limited. Reproduced with permission.)

records considered by Miranda were from strike-slip Californian earthquakes while those considered by Riddell were all from subduction Chilean earthquakes.

Through a regression analysis for each of the site-soil conditions considered in his study, Miranda derived an empirical equation to estimate analytically the reduction factors under consideration. In terms of initial fundamental period T, ductility ratio μ, and given site-soil conditions, the equation developed by Miranda is of the form

$$R_\mu = 1 + \frac{\mu - 1}{\Phi} \tag{9.13}$$

where

$$\Phi = 1 + \frac{1}{(10-\mu)T} - \frac{1}{2T}\exp\left[-\frac{3}{2}\left(\ln T - \frac{3}{5}\right)^2\right] \tag{9.14}$$

for rock sites,

$$\Phi = 1 + \frac{1}{(12-\mu)T} - \frac{2}{5T}\exp\left[-2\left(\ln T - \frac{1}{5}\right)^2\right] \tag{9.15}$$

for alluvium sites, and

$$\Phi = 1 + \frac{T_g}{3T} - \frac{3T_g}{4T}\exp\left[-3\left(\ln\frac{T}{T_g} - \frac{1}{4}\right)^2\right] \tag{9.16}$$

for soft soil sites. In Equation 9.16, T_g is the predominant period of the ground motions expected at the site under consideration, defined as the period that corresponds to the peak ordinate in the elastic velocity response spectra with 5% damping of these ground motions.

Example 9.4 Lateral strength using Newmark–Hall elastic spectrum and Miranda reduction factors

Determine the design lateral strength of a single-degree-of-freedom structure with a total mass of 10,000 kg, a natural period of 0.5 s, and a damping ratio of 5%, considering that the structure is capable of resisting inelastic deformations of up to 5 times its yield deformation. The structure will be built on a site that is characterized by alluvial soils and the 5% damping elastic design spectrum shown in Figure E9.2.

Solution

From Figure E9.2 for a natural period of 0.5 s, one obtains

$$SA = 1.35g$$

Similarly, for a natural period of 0.5 s, a ductility factor of 5, and alluvial soil conditions, Equation 9.15 leads to

$$\Phi = 1 + \frac{1}{(12-\mu)T} - \frac{2}{5T}\exp\left[-2\left(\ln T - \frac{1}{5}\right)^2\right] = 1 + \frac{1}{(12-5)(0.5)} - \frac{2}{5(0.5)}\exp\left[-2\left(\ln 0.5 - \frac{1}{5}\right)^2\right] = 1.12$$

and thus Equation 9.13 yields the following reduction factor:

$$R_\mu = 1 + \frac{\mu - 1}{\Phi} = 1 + \frac{5-1}{1.12} = 4.57$$

As a result, the desired design lateral strength is

$$F_y = \frac{mSA}{R_\mu} = \frac{10000(1.35g)}{4.57} = 29.0\,\text{kN}$$

9.4 DIRECT STATISTICAL CORRELATIONS

Design spectra may also be obtained from empirical equations that lead directly to a response spectrum. An example is the attenuation relations described in Section 7.4 to determine the mean and the standard deviation of spectral pseudovelocities for 5% damping in terms of earthquake magnitude, fault type, distance to the source, and site-soil conditions. As in the case of the attenuation laws developed to predict peak ground acceleration and peak ground velocity at a given site, these empirical equations are derived by (a) generating the response spectra of ground motion records with common characteristics and (b) correlating the ordinates in these response spectra to the characteristics of the earthquakes that generated the used ground motions records and the soils conditions at the sites where they were recorded.

The use of statistical correlations to generate a design spectrum offers several advantages over the two procedures discussed in the foregoing sections. First, the spectrum is obtained directly in one step, obviating the need to determine beforehand the values of peak ground motion parameters. Second, it does not require the normalization of the response spectra in the data set, making thus the question as to what is the best normalization parameter irrelevant. Third, the dependence of spectral shape on magnitude, hypocentral distance, and source characteristics is directly accounted for. Lastly, all uncertainties are lumped together, which eliminates the possibility of counting some of these uncertainties twice. The procedure, however, lacks the flexibility of the other two approaches as the available equations are derived for a specific period range, damping value, and load-deformation behavior. This makes the consideration of other cases cumbersome. Note also that when the empirical equations in question are used, the earthquake hazard is defined in terms of earthquake magnitude as opposed to peak ground motion intensity.

FURTHER READINGS

1. Dunbar, W. S. and Charlwood, R. G., Empirical methods for the prediction of response spectra, *Earthquake Spectra*, 7, 3, 1991, 333–353.
2. Miranda, E. and Bertero, V. V., Evaluation of strength reduction factors for earthquake-resistant design, *Earthquake Spectra*, 10, 2, 1994, 357–379.
3. Mohraz, B., A study of earthquake response spectra for different geological conditions, *Bulletin of the Seismological Society of America*, 66, 1976, 915–935.
4. Mohraz, B. and Sadek, F., Earthquake ground motion and response spectra, Chapter 2 in *The Seismic Design Handbook*, 2nd Edition, Naeim, F. ed., Kluwer Academic Publishers, Boston, MA, 2001.
5. Newmark, N. M. and Hall, W. J., *Earthquake Spectra and Design*, Earthquake Engineering Research Institute, Oakland, CA, 1982.
6. Riddell, R. and Newmark, N. M., *Statistical Analysis of the Response of Nonlinear Systems Subjected to Earthquakes*, Structural Research Series No. 468, Department of Civil Engineering, University of Illinois, Urbana, IL, 1979.
7. Seed, H. B. and Idriss, I. M., *Ground Motion and Soil Liquefaction during Earthquakes*, Earthquake Engineering Research Institute, Oakland, CA, 1982.

PROBLEMS

9.1 Construct using the peak ground acceleration and response spectrum shape approach an inelastic design spectrum for a site underlain by soft clays considering a damping ratio of 5% and a ductility factor of 6. The maximum ground acceleration that future earthquakes may generate at the site is estimated to be $0.20g$ for a probability of exceedance of 10% in 50 years.

9.2 A site is located 50 km from an active normal fault over a deposit of alluvial soils. It is estimated that the fault is capable of generating an earthquake with a magnitude of 8.0. On the basis of this information and the ground acceleration-response spectrum shape approach, construct an inelastic design spectrum for the site considering a damping ratio of 5% and a ductility factor of 2.

9.3 Construct employing the peak ground acceleration-response spectrum shape approach a design spectrum for elastoplastic structures with 5% damping, a ductility factor of 4, and located at a site that is 100 km from an active fault. It is estimated that the fault is capable of generating an earthquake with a moment magnitude of 7.5. In addition, the site is underlain by soft clays for which the average shear wave velocity over the upper 30 m is 600 m/s. Use Joyner–Boore–Fumal attenuation equation introduced in Chapter 7 to estimate the peak ground acceleration at the site. List the spectral accelerations corresponding to the natural periods of 0.1, 0.5, and 3.0 s.

9.4 The graph shown in Figure P9.4 represents a site-specific average design spectrum for elastoplastic structures with 2% damping and a ductility factor of 5. On the basis of this design spectrum, estimate the lateral strength for which a single-degree-of-freedom structure with a weight of 10 kips, a natural period of 2.5 s, and a damping ratio of 2% should be designed if (a) the structure cannot resist inelastic deformations and (b) the structure is capable of withstanding inelastic deformations as large as 5 times its yield deformation.

9.5 The peak ground acceleration at a given rock site is estimated to be $0.6g$ for a probability of exceedance of 10% in 100 years. Construct an inelastic design spectrum for 2% damping and a ductility factor of 8 using Newmark–Hall approach. Consider the average v/a and ad/v^2 ratios proposed by Mohraz to estimate the peak ground velocity and peak ground displacement at the site. Tetralogarithmic graph paper is provided in Figure P9.5.

9.6 A site in California is underlain by stiff soils and is located ~60 km from the San Andreas Fault. Considering that there are no other nearby faults and that the largest earthquake that

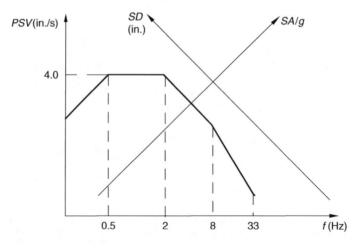

FIGURE P9.4 Inelastic design spectrum considered in Problem 9.4.

Design Response Spectrum

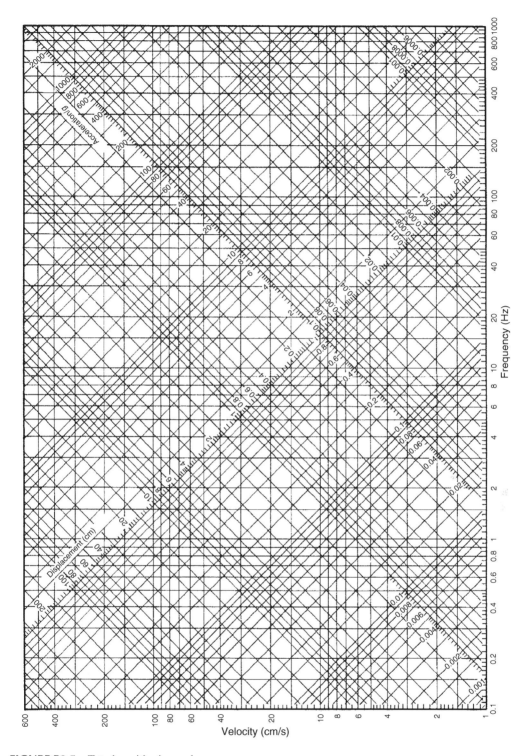

FIGURE P9.5 Tetralogarithmic graph paper.

can occur at the San Andreas Fault is one with a moment magnitude of 8.0 and a focal depth of 20 km, construct a 5% damping design spectrum for this site and the design of buildings capable of withstanding inelastic deformations as large as 5 times their yield deformations. The average shear wave velocity over the upper 30 m of the soils beneath the site is 900 m/s. Use Joyner–Boore–Fumal attenuation equation and Newmark–Hall approach. Tetralogarithmic graph paper is provided in Figure P9.5.

9.7 Construct an elastic design spectrum for the site described in Problem 9.6 using alternatively (a) Newmark–Hall approach and the average v/a and ad/v^2 ratios proposed by Mohraz and (b) Joyner–Boore–Fumal attenuation equation presented in Chapter 7. Plot the two spectra in a single spectral acceleration versus natural period graph with arithmetic scales.

9.8 The peak ground acceleration, peak ground velocity, and peak ground displacement expected at a given alluvial site are estimated to be 0.843g, 50.74 in./s, and 12.81 in., respectively. Construct an inelastic design spectrum for this site using Newmark–Hall average amplification factors to construct the elastic spectrum and Miranda's reduction factors to obtain the inelastic spectrum from the elastic spectrum. Consider a damping ratio of 3% and a ductility factor of 4. Tetralogarithmic graph paper is provided in Figure P9.5.

9.9 A single-degree-of-freedom structure has a total weight of 10 kips, a natural period of 1.5 s, and a damping ratio of 5%. The structure will be built on a site that is characterized by an expected peak ground acceleration of 0.4g and the elastic response spectrum shape shown in Figure 9.4 for rock and stiff soils. Determine the lateral strength for which this structure should be designed assuming that the structure will have the capability of resisting inelastic deformations of up to 5 times its yield strength. Use alternatively the criterion established in Section 9.2 and Miranda's reduction factors to account for the structure's ability to withstand inelastic deformations.

10 Structural Response by Response Spectrum Method

> Unfortunately, there is considerable resistance by design practitioners to adopting new methodologies and analyses. Often guided by only code rules, experienced design practitioners are hesitant to change direction from those methods established during decades of successful professional practice.
>
> **Tom Pauley, 2002**

10.1 INTRODUCTION

Once a design spectrum for a site is specified, the problem at hand now becomes determining the effect that the earthquake ground motion represented by the specified design spectrum may have on a structure built at the site. This can be accomplished by a variety of techniques. However, the response spectrum method described in this chapter is certainly the most convenient one when the excitation is given in the form of a design spectrum. This is so because this method obviates the need to convert the design spectrum into a compatible acceleration time history or a power spectral density function. This chapter presents, thus, the response spectrum method in some detail and illustrates its application to estimate the displacements, accelerations, shear forces, bending moments, and axial forces in a multi-degree-of-freedom structure when the base of the structure is excited by a specified earthquake ground motion.

In the discussion that follows, a linear elastic structural behavior is assumed. This assumption is adopted despite the fact that, as discussed in Section 6.5, earthquake-resistant structures are designed to undergo significant inelastic deformation under the effect of strong earthquakes. The justification for the use of linear elastic methods in the analysis of earthquake-resistant structures is the fact that, as explained in Chapter 9, under current practice earthquake-resistant structures are designed to just yield under the design forces and remain linear elastic before they yield. It is also assumed that no gravitational loads act simultaneously with the seismic loads. Although it is recognized that all structures are subjected to gravitational loads at the time an earthquake excites them, this assumption is based on the fact that for linear elastic structures the principle of superposition is valid. Therefore, the combined effect of gravitational and seismic loads may be determined from a separate analysis under each of these loads and the superposition of the results from these two separate analyses. Finally, it is assumed that the energy dissipation mechanism of the structure may be represented by linear viscous dampers. That is, it is assumed that the damping forces acting on the structure may be represented by viscous dampers that provide resisting forces that are proportional to the velocities acting on them.

10.2 MODAL SUPERPOSITION METHOD

10.2.1 Discrete Model

As learned in structural dynamics, inertial forces are the essential characteristic of structural systems subjected to dynamic loads. Also learned in structural dynamics is the fact that the magnitude of these inertial forces depends on the calculated structural displacements, which in turn depend on the magnitude of the inertial forces. In other words, the magnitude of the inertial forces depends on the dynamic response of the structure. This means that the full set of inertial forces acting on a structure can be determined only by calculating the translational and rotational displacements of every particle of the structure, which for any real structure obviously represents a large computational task. It also means that the dynamic analysis of a structure can be greatly simplified if the mass of the structure is concentrated at a discrete number of points given that in such a case inertial forces could be developed only at these points.

There are several ways by means of which one can discretize the mass of a structure, although by far the simplest and most common way is the lumped-mass approach. In this approach, it is assumed that the entire mass of the structure is concentrated at its nodes or joints. The values of the discrete masses at each node are thus determined by distributing the mass of each structural element to the nodes of the element using basic principles of statics. Another approach is the consistent mass method. In this method, the discrete masses at the nodes of an element are obtained in a manner that is similar to the method used to determine the element's stiffness coefficients. In any case, independent of the discretization method used, it will be assumed in the derivation that follows that the mass of the structure has been lumped at its nodes and that the displacements and rotations of the nodes completely define the deformed configuration of the structure. Similarly, it will be considered that the physical properties of the structure are completely defined by the mass matrix $[M]$, damping matrix $[C]$, and stiffness matrix $[K]$ of such a discrete model.

10.2.2 Equation of Motion

Consider a typical flexible multi-degree-of-freedom structure such as the plane frame shown in Figure 10.1. Consider, in addition, that the structure has three degrees of freedom per node: two translational and one rotational. That is, consider that the nodes of the structure are free to rotate and displace horizontally and vertically. As discussed earlier, assume also that the mass of the structure is concentrated at its nodes and that, consequently, the deformed configuration of the structure is fully defined by the displacements and rotations of its nodes. Finally, assume that the structure is subjected to a horizontal ground motion and that this ground motion is contained in the plane of the structure. With reference to Figure 10.1, let N denote the number of degrees of freedom of the structure, $u_g(t)$ the displacement of the ground at time t as measured from the structure's original position, $u_j(t)$ the

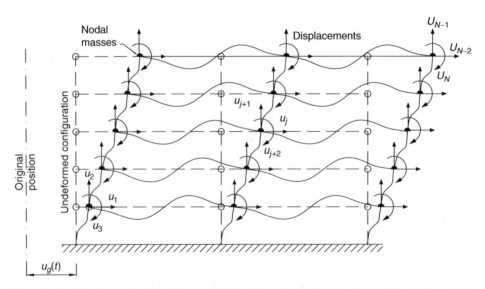

FIGURE 10.1 Displaced and deformed multi-degree-of-freedom structure subjected to earthquake ground motion.

Structural Response by Response Spectrum Method

jth nodal displacement or rotation at time t when measured relative to the base of the structure, and $y_j(t)$ the jth nodal displacement or rotation at time t when measured with respect to the structure's original position. Furthermore, let $\{u(t)\}$ be a vector that contains the relative nodal displacements and rotations $u_j(t)$, that is,

$$\{u(t)\} = \begin{Bmatrix} u_1(t) \\ u_2(t) \\ \vdots \\ u_N(t) \end{Bmatrix} \quad (10.1)$$

and $\{y(t)\}$ a vector that contains the absolute nodal displacements and rotations $y_j(t)$, that is,

$$\{y(t)\} = \begin{Bmatrix} y_1(t) \\ y_2(t) \\ \vdots \\ y_N(t) \end{Bmatrix} \quad (10.2)$$

Note that in view of the fact that the absolute displacements are equal to the relative displacements plus the displacement of the ground, $\{y(t)\}$ may also be expressed as

$$\{y(t)\} = \begin{Bmatrix} u_1(t) + u_g(t) \\ u_2(t) \\ u_3(t) \\ \vdots \\ u_{N-2}(t) + u_g(t) \\ u_{N-1}(t) \\ u_N(t) \end{Bmatrix} = \begin{Bmatrix} u_1(t) \\ u_2(t) \\ u_3(t) \\ \vdots \\ u_{N-2}(t) \\ u_{N-1}(t) \\ u_N(t) \end{Bmatrix} + \begin{Bmatrix} 1 \\ 0 \\ 0 \\ \vdots \\ 1 \\ 0 \\ 0 \end{Bmatrix} u_g = \{u(t)\} + \{J\}u_g(t) \quad (10.3)$$

where it has been assumed that the elements of $\{u(t)\}$ and $\{y(t)\}$ are arranged in a way in which for each node the horizontal displacement is listed first, the vertical displacement second, and the rotation third, and $\{J\}$ is a vector that contains ones and zeroes, with the ones assigned to the horizontal degrees of freedom and the zeroes to the vertical and rotational degrees of freedom. That is, $\{J\}$ is a vector of the form

$$\{J\} = \begin{Bmatrix} 1 \\ 0 \\ 0 \\ \vdots \\ 1 \\ 0 \\ 0 \end{Bmatrix} \quad (10.4)$$

and is called *influence vector*. It defines the displacements induced in a structure when the base of the structure is subjected to a static unit displacement in the direction of the ground motion $u_g(t)$.

When the structure is excited by an earthquake ground motion, the structure is displaced and deformed as shown schematically in Figure 10.1. Restoring forces generated by the elasticity of its members and resisting damping forces generated by the viscous dampers that represent the energy dissipation mechanism of the structure are also activated. By Newton's second law and using matrix notation, the equation of motion of the structure when its base is subjected to a horizontal ground motion $u_g(t)$ may then be expressed as

$$-\{F_s(t)\} - \{F_D(t)\} = [M]\{\ddot{y}(t)\} \tag{10.5}$$

where $\{F_s(t)\}$ and $\{F_D(t)\}$ denote vectors that contain the aforementioned elastic and damping forces, respectively, and, as before, a dot over a variable signifies its derivative with respect to t. However, these vectors may be expressed in terms of the structure's stiffness and damping matrices as

$$\{F_s(t)\} = [K]\{u(t)\} \tag{10.6}$$

$$\{F_D(t)\} = [C]\{\dot{u}(t)\} \tag{10.7}$$

and, according to Equation 10.3, the vector $\{\ddot{y}(t)\}$ may be expressed as

$$\{\ddot{y}(t)\} = \{\ddot{u}(t)\} + \{J\}\ddot{u}_g(t) \tag{10.8}$$

Therefore, Equation 10.5 may also be written as

$$-[C]\{\dot{u}(t)\} - [K]\{u(t)\} = [M][\{\ddot{u}(t)\} + \{J\}\ddot{u}_g(t)] \tag{10.9}$$

or as

$$[M]\{\ddot{u}(t)\} + [C]\{\dot{u}(t)\} + [K]\{u(t)\} = -[M]\{J\}\ddot{u}_g(t) \tag{10.10}$$

10.2.3 Transformation of Equation of Motion

Equation 10.10 represents a coupled system of second-order differential equations in which the independent variable is the time t and the dependent variables are the displacements contained in the vector $\{u(t)\}$. By coupled, it is meant that two or more of the dependent variables appear in each of the equations of the system. A variety of methods can be used to solve such a system of differential equations, but for the purpose of finding a solution in terms of a response spectrum the most convenient method is the so-called *modal superposition method*. In the modal superposition method, advantage is taken of the orthogonality properties of the mode shapes of the structure; that is, the fact that

$$\{\phi\}_r^T [M] \{\phi\}_s = 0 \quad r \neq s \tag{10.11}$$

and

$$\{\phi\}_r^T [K] \{\phi\}_s = 0 \quad r \neq s \tag{10.12}$$

where $\{\phi\}_r$ and $\{\phi\}_s$ denote two distinct mode shapes of the structure, the superscript T signifies transpose, and, as before, $[M]$ and $[K]$ are the structure's mass and stiffness matrices, respectively.

Structural Response by Response Spectrum Method

The advantage such orthogonality properties offer is that by means of them it is possible to transform the original coupled system of equations into a system of uncoupled equations, that is, a system of equations in which each equation is expressed in terms of a single dependent variable. The way these orthogonality properties can be used to uncouple the original equation of motion is similar to the way the orthogonality of the sine and cosine functions is used to expand a periodic function into a Fourier series representation. That is, the original equation of motion may be uncoupled by expressing the displacement vector $\{u(t)\}$ as a linear combination of the structural mode shapes, or, what is tantamount, by introducing the transformation of coordinates

$$\{u(t)\} = [\Phi]\{\eta(t)\} = \sum_{r=1}^{N} \{\phi\}_r \eta_r(t) \tag{10.13}$$

where

$$[\Phi] = [\{\phi\}_1 \{\phi\}_2 \cdots \{\phi\}_N] \tag{10.14}$$

is a matrix that contains the mode shapes of the structure and is called *modal matrix*. Similarly,

$$\{\eta(t)\} = \{\eta_1(t) \quad \eta_2(t) \cdots \eta_N(t)\}^T \tag{10.15}$$

is a vector that contains the new (unknown at this point) coordinates $\eta_r(t)$, $r = 1, 2, \ldots, N$, which are often referred to as *generalized coordinates* to reflect the fact that they are coordinates in the mathematical sense but not in the physical sense. Sometimes they are also called *normal coordinates* because they lead to uncoupled equations of motion.

To demonstrate that such a transformation of coordinates indeed leads to a system of uncoupled equations, note that by substituting Equation 10.13 into Equation 10.10, one obtains

$$[M][\Phi]\{\ddot{\eta}(t)\} + [C][\Phi]\{\dot{\eta}(t)\} + [K][\Phi]\{\eta(t)\} = -[M]\{J\}\ddot{u}_g(t) \tag{10.16}$$

which, if both sides of the equation are premultiplied by the transpose of the modal matrix $[\Phi]$, may also be written as

$$[\Phi]^T[M][\Phi]\{\ddot{\eta}(t)\} + [\Phi]^T[C][\Phi]\{\dot{\eta}(t)\} + [\Phi]^T[K][\Phi]\{\eta(t)\} = -[\Phi]^T[M]\{J\}\ddot{u}_g(t) \tag{10.17}$$

But the triple matrix product $[\Phi]^T[M][\Phi]$ may also be expressed as

$$\begin{aligned}
[\Phi]^T[M][\Phi] &= [\{\phi\}_1 \{\phi\}_2 \cdots \{\phi\}_N]^T [M][\{\phi\}_1 \{\phi\}_2 \cdots \{\phi\}_N] \\
&= \begin{Bmatrix} \{\phi\}_1^T[M] \\ \{\phi\}_2^T[M] \\ \vdots \\ \{\phi\}_N^T[M] \end{Bmatrix} [\{\phi\}_1 \{\phi\}_2 \cdots \{\phi\}_N] \\
&= \begin{bmatrix} \{\phi\}_1^T[M]\{\phi\}_1 & \{\phi\}_1^T[M]\{\phi\}_2 & \cdots & \{\phi\}_1^T[M]\{\phi\}_N \\ \{\phi\}_2^T[M]\{\phi\}_1 & \{\phi\}_2^T[M]\{\phi\}_2 & \cdots & \{\phi\}_2^T[M]\{\phi\}_N \\ \vdots & \vdots & \ddots & \vdots \\ \{\phi\}_N^T[M]\{\phi\}_1 & \{\phi\}_N^T[M]\{\phi\}_2 & \cdots & \{\phi\}_N^T[M]\{\phi\}_N \end{bmatrix}
\end{aligned} \tag{10.18}$$

and, thus, by virtue of the orthogonality properties of the mode shapes, is equal to

$$[\Phi]^T[M][\Phi] = \begin{bmatrix} M_1^* & 0 & \cdots & 0 \\ 0 & M_2^* & \cdots & 0 \\ \vdots & \vdots & \ddots & \vdots \\ 0 & 0 & \cdots & M_N^* \end{bmatrix} \quad (10.19)$$

where

$$M_r^* = \{\phi\}_r^T[M]\{\phi\}_r, \quad r = 1, 2, \ldots, N \quad (10.20)$$

is a scalar referred to as the *generalized mass* of the system in its rth mode of vibration. Proceeding similarly and incorporating Equation 10.12, it can also be shown that the triple matrix product $[\Phi]^T[K][\Phi]$ is equal to

$$[\Phi]^T[K][\Phi] = \begin{bmatrix} K_1^* & 0 & \cdots & 0 \\ 0 & K_2^* & \cdots & 0 \\ \vdots & \vdots & \ddots & \vdots \\ 0 & 0 & \cdots & K_N^* \end{bmatrix} \quad (10.21)$$

where

$$K_r^* = \{\phi\}_r^T[K]\{\phi\}_r, \quad r = 1, 2, \ldots, N \quad (10.22)$$

which is also a scalar and is known as the *generalized stiffness* of the system in its rth mode of vibration. Furthermore, if it is assumed that the damping matrix $[C]$ is of the Rayleigh type, that is, a matrix of the form

$$[C] = \alpha[M] + \beta[K] \quad (10.23)$$

where α and β are constants whose value may be determined as shown in Section 10.5, then it is possible to express the triple matrix product $[\Phi]^T[C][\Phi]$ as

$$[\Phi]^T[C][\Phi] = \begin{bmatrix} C_1^* & 0 & \cdots & 0 \\ 0 & C_2^* & \cdots & 0 \\ \vdots & \vdots & \ddots & \vdots \\ 0 & 0 & \cdots & C_N^* \end{bmatrix} \quad (10.24)$$

where

$$C_r^* = \{\phi\}_r^T[C]\{\phi\}_r = \alpha M_r^* + \beta K_r^*, \quad r = 1, 2, \ldots, N \quad (10.25)$$

is a scalar as well and is called the *generalized damping constant* of the system in its rth mode of vibration. Finally, the triple matrix product $[\Phi]^T[M]\{J\}$ may be written alternatively as

$$[\Phi]^T[M]\{J\} = \begin{Bmatrix} \{\phi\}_1^T \\ \{\phi\}_2^T \\ \vdots \\ \{\phi\}_N^T \end{Bmatrix} [M]\{J\} = \begin{Bmatrix} \{\phi\}_1^T[M]\{J\} \\ \{\phi\}_2^T[M]\{J\} \\ \vdots \\ \{\phi\}_N^T[M]\{J\} \end{Bmatrix} \qquad (10.26)$$

Consequently, Equation 10.17 is an equation of the form

$$\begin{bmatrix} M_1^* & 0 & \cdots & 0 \\ 0 & M_2^* & \cdots & 0 \\ \vdots & \vdots & \ddots & \vdots \\ 0 & 0 & \cdots & M_N^* \end{bmatrix} \begin{Bmatrix} \ddot{\eta}_1(t) \\ \ddot{\eta}_2(t) \\ \vdots \\ \ddot{\eta}_N(t) \end{Bmatrix} + \begin{bmatrix} C_1^* & 0 & \cdots & 0 \\ 0 & C_2^* & \cdots & 0 \\ \vdots & \vdots & \ddots & \vdots \\ 0 & 0 & \cdots & C_N^* \end{bmatrix} \begin{Bmatrix} \dot{\eta}_1(t) \\ \dot{\eta}_2(t) \\ \vdots \\ \dot{\eta}_N(t) \end{Bmatrix}$$

$$+ \begin{bmatrix} K_1^* & 0 & \cdots & 0 \\ 0 & K_2^* & \cdots & 0 \\ \vdots & \vdots & \ddots & \vdots \\ 0 & 0 & \cdots & K_N^* \end{bmatrix} \begin{Bmatrix} \eta_1(t) \\ \eta_2(t) \\ \vdots \\ \eta_N(t) \end{Bmatrix} = \begin{Bmatrix} \{\phi\}_1^T[M]\{J\} \\ \{\phi\}_2^T[M]\{J\} \\ \vdots \\ \{\phi\}_N^T[M]\{J\} \end{Bmatrix} \ddot{u}_g(t) \qquad (10.27)$$

which represents a system of N independent equations, that is, a system of equations in which each equation depends only on a single dependent variable ($\eta_r(t)$ in the case of the rth equation) and does not depend on the values of the other dependent variables. In compact form, Equation 10.27 may be written alternatively as

$$M_r^*\ddot{\eta}_r(t) + C_r^*\dot{\eta}_r(t) + K_r^*\eta_r(t) = -\{\phi\}_r^T[M]\{J\}\ddot{u}_g(t) \quad r = 1, 2, \ldots, N \qquad (10.28)$$

or, if it is considered that this equation represents the equation of motion of a single-degree-of-freedom system with mass M_r^*, stiffness K_r^*, and damping constant C_r^*, as

$$\ddot{\eta}_r(t) + 2\xi_r\omega_r\dot{\eta}_r(t) + \omega_r^2\eta_r(t) = -\Gamma_r\ddot{u}_g(t) \quad r = 1, 2, \ldots, N \qquad (10.29)$$

where it has been considered that the undamped natural frequency ω_r and damping ratio ξ_r of such a single-degree-of-freedom system are equal to

$$\omega_r = \sqrt{\frac{K_r^*}{M_r^*}} \qquad (10.30)$$

$$\xi_r = \frac{C_r^*}{2\omega_r M_r^*} \qquad (10.31)$$

and Γ_r is defined as

$$\Gamma_r = \frac{\{\phi\}_r^T[M]\{J\}}{M_r^*} = \frac{\{\phi\}_r^T[M]\{J\}}{\{\phi\}_r^T[M]\{\phi\}} \qquad (10.32)$$

This parameter Γ_r is known as the *participation factor* in the rth mode of the system and represents one of the fundamental variables in the analysis of multi-degree-of-freedom systems. Its properties are discussed at length in Section 10.4.

Notice, thus, that by applying the transformation of coordinates indicated by Equation 10.13, it is possible to reduce the original system of coupled differential equations into a system of uncoupled equations in which each equation represents the equation of motion of a single-degree-of-freedom system with natural frequency ω_r and damping ratio ξ_r and subjected to a ground motion equal to Γ_r times the original ground acceleration $\ddot{u}_g(t)$. Notice also that the problem is simplified considerably because it is easier to solve N independent equations (one for each mode of vibration) than a system of N coupled equations. Notice, further, that the solution to each of the independent equations is a relatively straightforward task given that many techniques are available to solve the equation of motion of a single-degree-of-freedom system and that these techniques are well established. Finally, notice that a solution based on this method requires the prior determination of the mode shapes and the natural frequencies of the structure.

10.2.4 Solution of rth Independent Equation

As previously mentioned, the solution to the independent equations of motion may be found using any of the methods available for the solution of the equation of motion of a single-degree-of-freedom system. Therefore, if Duhamel integral is used, and if it is considered that, as established earlier, Equation 10.29 represents the equation of motion of a single-degree-of-freedom system with natural frequency ω_r and damping ratio ξ_r and subjected to a ground acceleration $\Gamma_r \ddot{u}_g(t)$, then the solution to such an equation for the case of zero initial conditions may be written as

$$\eta_r(t) = \frac{\Gamma_r}{\omega_{dr}} \int_0^t \ddot{u}_g(\tau) e^{-\xi_r \omega_r (t-\tau)} \sin \omega_{dr}(t-\tau) d\tau \tag{10.33}$$

where $\omega_{dr} = \omega_r(1 - \xi_r^2)^{1/2}$ is the damped natural frequency of the system in its rth mode of vibration and all other variables are as defined before. Alternatively, it may be expressed as

$$\eta_r(t) = \Gamma_r z_r(t) \tag{10.34}$$

where $z_r(t)$ denotes the relative displacement response of a single-degree-of-freedom system with natural frequency ω_r and damping ratio ξ_r to a ground acceleration $\ddot{u}_g(t)$.

10.2.5 Structural Response in Terms of Modal Responses

Once the generalized coordinates $\eta_r(t)$ are determined by means of Equation 10.33 or any other similar equation, the structural displacement vector $\{u(t)\}$ may be obtained by going back to the transformation of coordinates that led to uncoupled equations of motion, that is, Equation 10.13. Accordingly, after substitution of Equation 10.34 into Equation 10.13, the displacement vector $\{u(t)\}$, which represents the solution to the original system of differential equations, is given by

$$\{u(t)\} = \sum_{r=1}^{N} \{\phi\}_r \eta_r(t) = \sum_{r=1}^{N} \Gamma_r \{\phi\}_r z_r(t) \tag{10.35}$$

which may be considered to be the final solution to the problem of finding the displacements induced in a structure when its base is subjected to an earthquake ground motion characterized by the ground acceleration $\ddot{u}_g(t)$. Note, however, that it is convenient to express Equation 10.35 alternatively as

$$\{u(t)\} = \sum_{r=1}^{N} \{u(t)\}_r \tag{10.36}$$

where $\{u(t)\}_r$ is a vector defined as

$$\{u(t)\}_r = \Gamma_r\{\phi\}_r z_r(t) \tag{10.37}$$

which depends only on the parameters that characterize the rth mode of vibration and may be thus thought of as a vector that contains the displacements induced in the structure when only the rth mode is excited. Furthermore, these displacements may be interpreted as the contribution of the rth mode of vibration to the total displacement response.

It may be considered, thus, that the displacement response of a multi-degree-of-freedom structure is given by the superposition of the displacements in each of its modes of vibration, where these modal displacements are given by the product of the mode's participation factor, the corresponding mode shape, and the displacement response of a single-degree-of-freedom system with natural frequency ω_r and damping ratio ξ_r when subjected to a ground acceleration $\ddot{u}_g(t)$. For this reason, this procedure to determine the response of a multi-degree-of-freedom structure is referred to as the *modal superposition method* or a method that is based on a *modal analysis*.

The method can be extended to also express other measures of structural response in terms of a modal superposition. For the case of relative velocities and relative accelerations, this may be accomplished by simply taking the first and second derivatives of Equation 10.35. That is,

$$\{\dot{u}(t)\} = \sum_{r=1}^{N} \Gamma_r\{\phi\}_r \dot{z}_r(t) \tag{10.38}$$

$$\{\ddot{u}(t)\} = \sum_{r=1}^{N} \Gamma_r\{\phi\}_r \ddot{z}_r(t) \tag{10.39}$$

where, in similarity with the definition of $z_r(t)$, $\dot{z}_r(t)$ and $\ddot{z}_r(t)$, respectively, denote the relative velocity response and relative acceleration response of a single-degree-of-freedom system with natural frequency ω_r and damping ratio ξ_r to a ground acceleration $\ddot{u}_g(t)$. The extension to the absolute accelerations is not so trivial. To extend the method to the absolute accelerations it is necessary to proceed as follows.

According to the relationship between relative and absolute displacements established by Equation 10.3, the absolute accelerations of a structure may be expressed as

$$\{\ddot{y}(t)\} = \{\ddot{u}(t)\} + \{J\}\ddot{u}_g(t) \tag{10.40}$$

which in view of Equation 10.39 may also be written as

$$\{\ddot{y}(t)\} = \sum_{r=1}^{N} \Gamma_r\{\phi\}_r \ddot{z}_r(t) + \{J\}\ddot{u}_g(t) \tag{10.41}$$

The vector $\{J\}$, however, may be expressed in terms of normal coordinates as

$$\{J\} = [\Phi]\{\lambda\} = \sum_{r=1}^{N} \{\phi\}_r \lambda_r \tag{10.42}$$

where λ_r, $r = 1, 2, \ldots, N$, denote such normal coordinates and may be determined by premultiplying both sides of Equation 10.42 by $\{\phi\}_r^T[M]$ and making use of the orthogonality of the mode shapes with respect to the mass matrix. Proceeding accordingly, one obtains

$$\{\phi\}_r^T[M]\{J\} = \sum_{r=1}^{N}\{\phi\}_r^T[M]\{\phi\}_r \lambda_r = \{\phi\}_r^T[M]\{\phi\}_r \lambda_r \tag{10.43}$$

from which one finds that

$$\lambda_r = \frac{\{\phi\}_r^T[M]\{J\}}{\{\phi\}_r^T[M]\{\phi\}_r} = \Gamma_r \qquad (10.44)$$

where the definition of the participation factor given by Equation 10.32 has been used. In the light of Equations 10.42 and 10.44, Equation 10.41 may be thus put into the form

$$\{\ddot{y}(t)\} = \sum_{r=1}^{N} \Gamma_r \{\phi\}_r \ddot{z}_r(t) + \sum_{r=1}^{N} \Gamma_r \{\phi\}_r \ddot{u}_g(t) \qquad (10.45)$$

which may be rewritten as

$$\{\ddot{y}(t)\} = \sum_{r=1}^{N} \Gamma_r \{\phi\}_r [\ddot{z}_r(t) + \ddot{u}_g(t)] \qquad (10.46)$$

where, in view of the definition of $\ddot{z}_r(t)$, the sum $\ddot{z}_r(t) + \ddot{u}_g(t)$ may be interpreted as the absolute acceleration response of a single-degree-of-freedom system with natural frequency ω_r and damping ratio ξ_r to a ground acceleration $\ddot{u}_g(t)$.

Finally, it should be noted that the method can also be extended to determine the internal forces in the elements of the structure because these internal forces are directly related to the structural displacements. That is, if the displacement vector $\{u(t)\}$ is known, one can readily calculate the axial forces, shear forces, and bending moments in any of the elements of the structure by simply multiplying the stiffness matrix of the element by the vector that contains the element's nodal displacements and rotations. For instance, the internal forces in the ith element of a structure at any time t are given by

$$\{F_i(t)\} = [K_i'][T_i]\{u_i(t)\} \qquad (10.47)$$

where $\{F_i(t)\}$ represents a vector whose elements are such internal forces, $[K_i']$ the element's stiffness matrix expressed in its local system of coordinates, $\{u_i(t)\}$ a vector that contains the element's nodal displacements and rotations referred to the structure or global system of coordinates, and $[T_i]$ a transformation matrix that converts these displacements and rotations to the element's local system of coordinates. Furthermore, if the displacements $\{u_i(t)\}$ are expressed in terms of the modal superposition solution indicated by Equation 10.35, $\{F_i(t)\}$ may be alternatively written as

$$\{F_i(t)\} = [K_i'][T_i]\sum_{r=1}^{N} \Gamma_r \{\phi_i\}_r z_r(t) \qquad (10.48)$$

where $\{\phi_i\}_r$ is the part of the rth mode shape $\{\phi\}_r$ that comprises the element's degrees of freedom.

In similarity with the interpretation given to the displacement response, the relative velocity, absolute acceleration, and internal force responses of a multi-degree-of-freedom structure may also be considered given by the superposition of the corresponding responses in each of the vibration modes of the structure, where, as indicated by Equations 10.38, 10.46, and 10.48, these modal responses are given by

$$\{\dot{u}(t)\}_r = \Gamma_r \{\phi\}_r \dot{z}_r(t) \qquad (10.49)$$

$$\{\ddot{y}(t)\}_r = \Gamma_r \{\phi\}_r [\ddot{z}_r(t) + \ddot{u}_g(t)] \qquad (10.50)$$

$$\{F_i(t)\}_r = \Gamma_r [K_i'][T_i]\{\phi_i\}_r z_r(t) \qquad (10.51)$$

It should be mentioned at this point that one of the advantages of the modal superposition method is that in many cases good approximate solutions can be obtained by superimposing only the responses in a few of the lower modes of the structure. This advantage results from the fact that, in general, the lower modes represent the major contribution to the total response and, thus, neglecting the contribution of the higher modes introduces only a small error. In this same regard, it is worthwhile to keep in mind that it is also advisable to limit the number of modes considered in the superposition as the mathematical idealization of complex structural systems leads in many cases to unreliable predictions of the higher modes of vibration.

10.2.6 Maximum Structural Response

Usually, the ultimate objective of a seismic analysis is the determination of the maximum response values when a structure is subjected to a given earthquake ground motion. As the modal superposition equations presented earlier give the response quantities at any time t, these maximum values may be obtained by (a) computing the values of the responses at a series of discrete times that cover a time interval slightly longer than the duration of the ground motion being considered and (b) identifying the maximum values of such a response from the individual time histories for each of the degrees of freedom. For example, if the interest is to obtain maximum displacements, then the time histories of the displacement responses are computed and examined to determine the maximum displacement experienced by each degree of freedom. These displacements may be thereafter collected in a vector of the form

$$\max\{u(t)\} = \begin{Bmatrix} \max[u_1(t)] \\ \max[u_2(t)] \\ \vdots \\ \max[u_N(t)] \end{Bmatrix} \tag{10.52}$$

It should be emphasized that such a maximum response vector does not define the response at any one particular time. Instead, its elements represent the maximum values experienced by each degree of freedom, independent of the time at which they occur. From experience, however, it is known that, in general, these maximum values do not occur at the same instant of time. In fact, even for the same degree of freedom the maximum values of different types of response do not occur at the same time (i.e., $\max[u_i(t)]$ and $\max[\ddot{u}_i(t)]$). It should also be emphasized that the maximum response values obtained by the way just described represent the true maximum responses as no approximations (other than round-off errors) are involved in their calculation.

10.3 MAXIMUM STRUCTURAL RESPONSE USING RESPONSE SPECTRA

10.3.1 Maximum Modal Responses in Terms of Response Spectrum Ordinates

It is possible to determine, conveniently and expeditiously, the maximum values of modal responses using the response spectrum of the excitation directly. To do so, one only has to consider that, according to the definition of response spectrum, the maximum value of $z_r(t)$ in Equation 10.37, for example, is equal to the ordinate corresponding to a natural frequency ω_r and damping ratio ξ_r in the displacement response spectrum of the ground acceleration $\ddot{u}_g(t)$. Therefore, the maximum values of the displacements in the rth mode of a structure may be expressed as

$$\max\{u(t)\}_r = \Gamma_r\{\phi\}_r SD(\omega_r, \xi_r) \tag{10.53}$$

where $SD(\omega_r, \xi_r)$ denotes the aforementioned response spectrum ordinate. Or, if the approximate relationship between a displacement and an acceleration response spectrum (see Section 6.4) is used, it may be expressed alternatively as

$$\max\{u(t)\}_r \approx \frac{\Gamma_r}{\omega_r^2}\{\phi\}_r SA(\omega_r, \xi_r) \tag{10.54}$$

in which $SA(\omega_r, \xi_r)$ represents the ordinate in the acceleration response spectrum of the excitation corresponding to a natural frequency ω_r and damping ratio ξ_r. Similarly, as the maximum values of $\dot{z}_r(t)$ and $\ddot{z}_r(t) + \ddot{u}_g(t)$ in Equations 10.49 and 10.50, respectively, represent the ordinates in the velocity and acceleration response spectra of the excitation, the maximum values of the modal relative velocities and absolute accelerations may be determined according to

$$\max\{\dot{u}(t)\}_r = \Gamma_r\{\phi\}_r SV(\omega_r, \xi_r) \tag{10.55}$$

$$\max\{\ddot{y}(t)\}_r = \Gamma_r\{\phi\}_r SA(\omega_r, \xi_r) \tag{10.56}$$

where $SV(\omega_r, \xi_r)$ and $SA(\omega_r, \xi_r)$ represent such ordinates. Furthermore, if the same argument is applied to Equation 10.51, then it is possible to express the maximum value of the internal forces in the ith element of the structure as

$$\max\{F_i(t)\}_r = \Gamma_r[K_i'][T_i]\{\phi_i\}_r SD(\omega_r, \xi_r) \approx \frac{\Gamma_r}{\omega_r^2}[K_i'][T_i]\{\phi_i\}_r SA(\omega_r, \xi_r) \tag{10.57}$$

Although Equations 10.53 through 10.57 are useful to determine the maximum modal responses of a structure using the response spectra of the excitation, it should be recognized that these modal responses cannot be superimposed according to Equation 10.35, 10.38, 10.46, or 10.48 to obtain the structure's maximum responses. This is because, as indicated before and illustrated in Figure 10.2, the maximum modal responses are not attained, in general, at the same time and the aforementioned

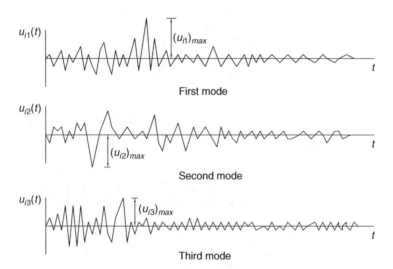

FIGURE 10.2 Typical time histories of the displacement at the ith degree of freedom of a structure in its first three modes.

equations are valid only when all the terms in them are considered at the same time t. Thus, one has that in general

$$\max\{u(t)\} \neq \sum_{r=1}^{N} \max\{u(t)\}_r \tag{10.58}$$

and similarly for the other types of responses. In words, one cannot consider that the maximum response of a structure is equal to the sum of the maximum responses in each of its modes. Furthermore, as a response spectrum does not provide information about the time at which each maximum occurs, it is not possible to derive an exact formula to compute the maximum response in terms of the modal maxima. As a result, one has to resort to the use of some approximate relationships, usually referred to as rules to combine modal responses, if one wants to determine such maximum responses in terms of the response spectra of the excitation. Some of these rules are introduced in the following section.

10.3.2 Approximate Rules to Estimate Maximum System Response

10.3.2.1 Double-Sum Combination Rule

Throughout the years, several formulas have been proposed to estimate the maximum response of structures in terms of their modal maxima. Some are simpler than others, but most importantly, all are adequate for some cases and inadequate for some others. As a general rule, the most complicated ones are also the most accurate. In what follows, some of the best-known rules to combine modal responses are described and their advantages and disadvantages are discussed. For such a purpose, these rules are expressed in terms of a generic vector $\{R(t)\}$, which is intended to signify a vector that comprises the values of any response parameter such as displacement, velocity, acceleration, or internal force. Also, $\max\{R(t)\}_r = \max\{[R(t)]_r\}$ denotes the vector that contains the maximum response values in the rth mode of the structure, whereas $\max\{R(t)\}$ represents the vector that contains the estimated maximum values of the total response. The maximum modal responses are considered together with their signs, which are the signs each attains when it reaches its maximum. As before, N represents the number of degrees of freedom of the structure.

One of the simplest approximations to the problem of how to estimate the maximum response of a system in terms of its maximum modal responses is the direct sum of the absolute values of such modal responses. This rule is known as the *absolute sum of the maxima rule* and mathematically is expressed as

$$\max\{R(t)\} = \sum_{r=1}^{N} \max\{|R(t)|\}_r \tag{10.59}$$

As this rule is equivalent to assuming that all the modal maxima occur at the same time and with the same sign, it provides an upper bound to the true maximum response. It is, however, too conservative a rule as it is highly unlikely that all the modal maxima will occur at the same time and with the same sign. Consequently, this rule is seldom used in practice, except when it is necessary to establish an upper bound of the maximum response being investigated.

Most of the other available rules have been derived using concepts of random vibrations, which facilitate the treatment of the problem. As such, they are supposed to give estimates of the maximum response in a statistical sense. That is, they are supposed to give good estimates of the average maximum response to an ensemble of earthquake ground motions. This, however, does not constitute a problem as the earthquake excitations considered in engineering designs are based on statistical averages, anyway. Nonetheless, they are also adequately accurate to estimate the maximum response under individual ground motions.

Such other rules are based on essentially two assumptions. The first is that the maximum response associated with any fixed probability of exceedance and the corresponding maximum modal responses are all proportional to their respective root mean squares. This assumption arises from the consideration of earthquake ground motions as outcomes of a stationary random process with zero mean. The second is that all the proportionality constants are equal to one another. This latter assumption is justified on the grounds that the true values of such constants exhibit only small variations from one another. On the basis of these two assumptions, it may be considered, thus, that

$$\max\{R(t)\} = \{c\sqrt{E[R^2(t)]}\} \tag{10.60}$$

and

$$\max\{R(t)\}_r = \max\{[R(t)]_r\} = \{c\sqrt{E[R^2(t)]_r}\} \tag{10.61}$$

in which $E[R^2(t)]$ denotes the mean square of the total response $R(t)$, $E[R^2(t)]_r$ the mean square of the modal response $[R(t)]_r$, and c the aforementioned proportionality constant. Note, however, that in terms of a modal superposition solution $E[R^2(t)]$ may be expressed as

$$E[R^2(t)] = E[R(t_1)R(t_2)]_{t_1=t_2=t} = E\left[\sum_{m=1}^{N}[R(t)]_m \sum_{n=1}^{N}[R(t)]_n\right] \tag{10.62}$$

which may be rewritten as

$$E[R^2(t)] = E\left[\sum_{m=1}^{N}\sum_{n=1}^{N}[R(t)]_m[R(t)]_n\right] = \sum_{m=1}^{N}\sum_{n=1}^{N}E[[R(t)]_m[R(t)]_n]$$
$$= \sum_{m=1}^{N}\sum_{n=1}^{N}\rho_{mn}\sqrt{E[R^2(t)]_m}\sqrt{E[R^2(t)]_n} \tag{10.63}$$

where ρ_{mn} is called *modal correlation coefficient* and is introduced merely for the purpose of correlating $E[[R(t)]_m[R(t)]_n]$ with the root mean squares of $[R(t)]_m$ and $[R(t)]_n$. It is defined as

$$\rho_{mn} = \frac{E[[R(t)]_m[R(t)]_n]}{\sqrt{E[R^2(t)]_m}\sqrt{E[R^2(t)]_n}} \tag{10.64}$$

As a result, Equation 10.60 may be expressed as

$$\max\{R(t)\} = \left\{\sqrt{\sum_{m=1}^{N}\sum_{n=1}^{N}\rho_{mn}c\sqrt{E[R^2(t)]_m}c\sqrt{E[R^2(t)]_n}}\right\} \tag{10.65}$$

which in view of Equation 10.61 may also be written as

$$\max\{R(t)\} = \left\{\sqrt{\sum_{m=1}^{N}\sum_{n=1}^{N}\rho_{mn}\max[R(t)]_m\max[R(t)]_n}\right\} \tag{10.66}$$

or, if the terms for which $m = n$ are separated from those for which $m \neq n$, and if it is considered from the definition of modal correlation coefficient that $\rho_{mm} = \rho_{nn} = 1$, as

$$\max\{R(t)\} = \left\{\sqrt{\sum_{r=1}^{N}(\max[R(t)]_r)^2 + \sum_{\substack{m=1\\m\neq n}}^{N}\sum_{n=1}^{N}\rho_{mn}\max[R(t)]_m\max[R(t)]_n}\right\} \tag{10.67}$$

It is of interest to observe that the absolute value of the modal correlation coefficient ρ_{mn} varies between one and zero. It is equal to unity when the mth and nth modal responses have exactly the same statistical properties (i.e., they are perfectly correlated) and zero when they are statistically independent (i.e., they are uncorrelated). It is also of interest to observe that Equation 10.67 leads to a simple algebraic sum if it is assumed that $\rho_{mn} = 1$ for all modes. This is consistent with the fact that when two modes are perfectly correlated, the vibrations in these two modes are in phase and thus their maximum values occur at the same time. It is also important to note that the cross terms in Equation 10.67 may assume positive or negative values, depending on the relative signs of the modal responses. The consideration of the signs of the modal responses is therefore an important step in the application of this equation.

The relationship between maximum response and maximum modal responses expressed by Equation 10.66 is known as the *double-sum combination rule*. It constitutes the basis for the derivation of several well-known formulas to combine modal responses. As it will be seen later on, the main difference between these formulas lies on the modal correlation coefficient ρ_{mn} considered, which in turn depends on the model selected to represent earthquake ground motions. Some of the rules that are based on this equation are presented in the following subsections.

10.3.2.2 Square Root of the Sum of the Squares

One of the first rules used for the combination of modal responses was that proposed by L. E. Goodman, E. Rosenblueth, and N. M. Newmark in 1953. This rule was derived by assuming that no significant correlation existed between the maximum responses in two different modes (which is equivalent to assume that they are statistically independent) and that the modal correlation coefficient ρ_{mn} in Equation 10.67 could be thus assumed equal to zero for $m \neq n$. As result of this assumption, the cross terms in Equation 10.67 are all equal to zero and the expression is reduced to

$$\max\{R(t)\} = \left\{ \sqrt{\sum_{r=1}^{N} (\max[R(t)]_r)^2} \right\} \tag{10.68}$$

This formula is known as the *square root of the sum of the squares rule* (SRSS rule, in short). It is a rule that is widely used in current practice, and for many years it has been the rule specified in building codes for the combination of modal responses. As it will be shown later on, this rule yields fairly accurate results for regular flexible structures such as building structures with no irregularities in plan or elevation. However, it may lead to inaccurate solutions if used in the analysis of structures that possess closely spaced natural frequencies, that is, natural frequencies that have similar values. Closely spaced natural frequencies are common in buildings with torsional modes, buildings with appendages, and soil-structure systems. An obvious example in which the rule fails is in the case of a system with two modes that have exactly the same natural frequencies and damping ratios. In such a case, the time histories of the responses in such two modes are in phase. Consequently, the maximum response values in these two modes occur at the same time and should be added directly. However, one cannot do that with the SRSS rule because this rule does not include the cross terms that correlate such two modes. The rule is also inaccurate if applied to structures with significant higher modes as the modal responses in high-frequency modes are strongly correlated with the low-frequency modal responses. Such a strong correlation stems from the fact that when the modal response in a low-frequency mode is near its peak, the modal response in the high-frequency mode oscillates several times around this peak producing thus an additive effect. The contribution from higher modes is known to be significant in the calculation of support forces, bending moments, torsional moments, and axial forces. This occurs because in each mode these responses are proportional to the square of the mode's natural frequency and thus high natural frequencies lead to large modal responses in such cases. For the same reason, the contribution from higher modes is also significant in the case of rigid structures whose dominant modes have frequencies that are outside the range of the frequencies contained in the input ground motion. Nuclear power plants are an example of this type of structure.

10.3.2.3 Rosenblueth's Rule

Observing that the correlation between modal responses could not be neglected in the case of structures with closely spaced natural frequencies, in 1968 E. Rosenblueth proposed to use the double-sum rule to combine the modal responses of such structures. For this purpose, he modeled earthquake ground motions as a white noise process (a motion with a constant power spectral density in the entire frequency range [see Section 6.7]) to obtain a simple expression for the modal correlation coefficient ρ_{mn}. He also introduced a correction to this expression to approximately account for the reduction in response due to the transitory (nonstationary) character of real earthquake ground motions. The formula suggested by Rosenblueth to compute the modal correlation coefficients in the double-sum combination rule is of the form

$$\rho_{mn} = \left[1 + \left(\frac{\omega_{dn} - \omega_{dm}}{\xi'_m \omega_m + \xi'_n \omega_n}\right)^2\right]^{-1} \tag{10.69}$$

where, as before, ω_m and ω_n represent the mth and nth natural frequencies of the structure, ξ_m and ξ_n the corresponding damping ratios, and ω_{dm} and ω_{dn} the corresponding damped natural frequencies. ξ'_m and ξ'_n denote equivalent damping ratios that introduce the aforementioned correction and are defined according to

$$\xi'_r = \xi_r + \frac{2}{\omega_r s} \tag{10.70}$$

in which s denotes the duration of the strong part of the excitation (i.e., the part of a ground motion that can be realistically modeled as a white noise process). However, if Equations 10.69 and 10.70 are combined, Rosenblueth's formula may also be written as

$$\rho_{mn} = \left[1 + \left(\frac{\omega_{dn} - \omega_{dm}}{\xi_m \omega_m + \xi_n \omega_n + 4/s}\right)^2\right]^{-1} \tag{10.71}$$

from which it may be seen that the influence of the ground motion duration on the modal correlation coefficient is small for long durations. Observe also that ρ_{mn} rapidly becomes small as the frequency ratio ω_n/ω_m deviates from 1.0, that is, when ω_n and ω_m are well separated from one another. For example, for $s = 20$ s, $\xi_m = \xi_n = 0.02$, and $\omega_n/\omega_m = 0.9$, Equation 10.71 yields $\rho_{mn} = 0.85$, but if ω_n/ω_m is reduced to 0.8 one gets $\rho_{mn} = 0.58$. It is also of interest to note that with Equation 10.71, one obtains

$$\rho_{mn} = 0 \quad \text{when} \quad \omega_m - \omega_n \gg 0$$

$$\rho_{mn} = 1.0 \quad \text{when} \quad \omega_m = \omega_n$$

$$\rho_{mn} = 0 \quad \text{when} \quad s \to \infty \quad \text{and} \quad \rho_{mn} = 1.0 \quad \text{when} \quad s \to 0 \quad \text{if} \quad \xi_m = \xi_n = 0$$

The double-sum combination rule in combination with Rosenblueth's modal correlation coefficients leads thus to the SRSS rule when the natural frequencies of the structure are well separated from one another or when the duration of the ground motion is long and the damping ratios in all modes are close to zero. In contrast, it leads to an algebraic sum when the damping ratios and the ground motion duration are close to zero.

It has been shown through comparisons with time-history results that Rosenblueth's modal combination rule provides, in general, accurate estimates of a structure's maximum responses even if the structure possesses closely spaced natural frequencies. Its main limitation, however, is that its application requires the definition of an equivalent ground motion duration and that there is no established method to determine this equivalent duration in a consistent manner. As demonstrated later on, the rule also leads to inaccurate results when the structure has significant higher modes as in the case of rigid structures.

10.3.2.4 Complete Quadratic Combination

In 1980, A. Der Kiureghian proposed an alternative expression to determine the modal correlation coefficients ρ_{mn} in the double-sum combination rule. He arrived to his expression by also modeling earthquake ground motions as a white noise process, but, differently from what Rosenblueth did, he did not introduce a correction to account for the nonstationary nature of earthquake ground motions. As such, the two formulas lead to results that differ when the ground motion duration is short but yield very similar results when the ground motion duration is long, that is, when $4/s \ll \xi_m \omega_m$ or $\xi_n \omega_n$. The expression developed by Der Kiureghian is of the form

$$\rho_{mn} = \frac{8\sqrt{\xi_m \xi_n}(\xi_n + r_{mn}\xi_m)r_{mn}^{3/2}}{(1 - r_{mn}^2)^2 + 4\xi_m \xi_n r_{mn}(1 + r_{mn}^2) + 4r_{mn}^2(\xi_m^2 + \xi_n^2)} \tag{10.72}$$

where $r_{mn} = \omega_m/\omega_n$. For the particular case of equal damping ratios, that is, $\xi_m = \xi_n = \xi$, the formula simplifies to

$$\rho_{mn} = \frac{8\xi^2(1 + r_{mn})r_{mn}^{3/2}}{(1 - r_{mn}^2)^2 + 4\xi^2 r_{mn}(1 + r_{mn})^2} \tag{10.73}$$

When Der Kiureghian's formula is used in combination with the double-sum rule, the resulting expression is known in the literature as the *complete quadratic combination (CQC) rule*. As in the case of Rosenblueth's rule, this combination rule gives accurate estimates of the maximum response of structures with and without closely spaced natural frequencies. The exception is for damping ratios close to zero, in which case the correlation coefficients given by Equation 10.72 approach zero and thus lead to the SRSS rule. It also yields inaccurate results when applied to structures with significant higher modes.

10.3.2.5 Singh and Maldonado's Rule

Recognizing the limitations of the formulations based on the representation of earthquake ground motions by an ideal white noise process, in 1991 M. P. Singh and G. O. Maldonado developed a double-sum procedure for the combination of modal responses that do not rely on such a representation. Even more, they made no assumption whatsoever about the character of the input ground motion. The only assumptions made in their formulation are those involved in the derivation of the double-sum equation. That is, the assumptions that earthquake ground motions constitute a stationary random process and that the relationship between the maximum responses and their corresponding mean squares is the same for the total response and all modal responses. Singh and Maldonado's (S–M) formulation leads to the following expression to determine the modal correlation coefficient in Equation 10.66:

$$\rho_{mn} = \frac{|A_{mn}SD_m^2 + B_{mn}(SV_m^2 - SV_n^2) + C_{mn}SD_n^2|}{SD_m SD_n} \tag{10.74}$$

In this expression, $SD_r = SD(\omega_r, \xi_r)$ and $SV_r = SV(\omega_r, \xi_r)$, $r = m, n$, respectively, denote the ordinates in the displacement and velocity response spectra of the excitation corresponding to the natural frequency and damping ratio in the rth mode of vibration. A_{mn}, B_{mn}, and C_{mn} are constants defined as

$$A_{mn} = [r_{mn}^2 + 4\xi_m\xi_n r_{mn}^3 - (1 - 4\xi_m^2)r_{mn}^4]/D_{mn} \tag{10.75}$$

$$B_{mn} = (r_{mn}^2 - 1)/\omega_n^2 D_{mn} \tag{10.76}$$

$$C_{mn} = [-1 + 4\xi_n^2 + 4\xi_m\xi_n r_{mn} + r_{mn}^2]/D_{mn} \tag{10.77}$$

in which

$$D_{mn} = 1 + 4\xi_m\xi_n r_{mn} - 2(1 - 2\xi_m^2 - 2\xi_n^2)r_{mn}^2 + 4\xi_m\xi_n r_{mn}^3 + r_{mn}^4 \tag{10.78}$$

and, as before, $r_{mn} = \omega_m/\omega_n$.

It may be noted that the S–M expression for the modal correlation coefficient ρ_{mn} depends not only on the modal natural frequencies and damping ratios, but also on the characteristics of the input ground motion, as reflected by the response spectrum ordinates SD_m, SV_m, SD_n, and SV_n. This shows that Singh and Maldonado's formulation accounts for the true nature of earthquake ground motions and is superior to those that model earthquake ground motions as an unrealizable white noise process. It may also be noted that consistently with the definition of modal correlation coefficient, Equation 10.74 leads to $\rho_{mn} = 1.0$ when $m = n$, and $\rho_{mn} = 0$ when $r_{mn} \to \infty$, that is, when $\omega_n \ll \omega_m$. More importantly, note that differently from those obtained with the expressions proposed by Rosenblueth and Der Kiureghian, the values of ρ_{mn} calculated with Equation 10.74 do not decay rapidly as r_{mn} becomes different from 1.0. This may be appreciated in the graph presented in Figure 10.3, which shows for the case of $\xi_m = \xi_n = 0.02$ the variation of the modal correlation coefficient ρ_{mn} with the frequency ratio $r_{mn} = \omega_m/\omega_n$, when the input ground motion is modeled alternatively with a broadband Kanai–Tajimi (see Section 6.7) spectral density with a central frequency ω_j and a cut-off frequency of 20 Hz (equivalent to using S–M formula) and an ideal white noise process (equivalent to using of Rosenblueth or Der Kiuerghian formula). It may be observed from this graph that in some cases the expression developed by Singh and Maldonado yields high values of the modal correlation coefficient for modes whose frequencies are well separated from one another, whereas those developed by Rosenblueth and Der Kiureghian lead to values close to zero. This difference becomes important when the higher modes contribute significantly toward the total response of a structure. That is the why the S–M rule yields, as shown later on, accurate results in the analysis of rigid structures whereas Rosenblueth's and Der Kiureghian's do not.

A major disadvantage of the S–M rule is that it depends on spectral velocities in addition to spectral displacements, or what is tantamount, spectral accelerations. This becomes a problem when the earthquake input is defined in terms of an acceleration design spectrum. In such a case, one can determine the spectral displacements but not the spectral velocities as there is no direct relationship between spectral velocities and spectral accelerations. Although in principle velocity design spectra may be derived in much the same way as acceleration design spectra, this is rarely done and thus they are not readily available. One way to overcome the problem is to use any of the empirical relationships that are available to compute spectral velocities in terms of spectral accelerations or spectral pseudovelocities. One such relationship was obtained by Sadek et al. in 2000.[*]

---

[*] Sadek, F., Mohraz, B., and Riley, M. A., Linear procedures for structures with velocity-dependant dampers, *Journal of Structural Engineering, ASCE*, 128, 8, 2000, 887–895.

Structural Response by Response Spectrum Method

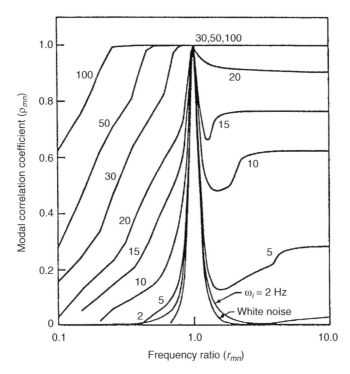

FIGURE 10.3 Modal correlation coefficient for damping ratios of 2% computed alternatively for ground motions represented by a white noise and a Kanai–Tajimi spectral density with a central frequency ω_j and a cut-off frequency of 20 Hz. (Reproduced with permission of John Wiley & Sons, Limited, from Singh, M.P. and Mehta, K.B., *Earthquake Engineering and Structural Dynam*, 11, 771–783, 1983.

This relationship was developed using the response spectra of 72 accelerograms for natural periods ranging from 0.1 to 4.0 s. It is of the form

$$SV = a_v T^{b_v} PSV \tag{10.79}$$

where SV signifies spectral velocity, PSV spectral pseudovelocity, T natural period, and a_v and b_v constants defined as

$$a_v = 1.095 + 0.647\xi - 0.382\xi^2 \tag{10.80}$$

$$b_v = 0.193 + 0.838\xi - 0.621\xi^2 \tag{10.81}$$

in which ξ denotes damping ratio.

10.3.2.6 Comparative Studies

Several investigators have performed comparative studies to assess the relative accuracy of the modal combination rules introduced in the preceding sub-sections. In one of such studies, the maximum deformations of a series of two-degree-of-freedom appendages supported by a three-story structure were computed using alternatively the absolute sum, SRSS, CQC, and Rosenblueth's rules and the results compared against the solutions obtained from a time-history analysis. The masses of the appendages as well as their attachment configurations were varied to obtain a total of 18 different structure–appendage systems. The attachment configurations considered are depicted in Figure 10.4. The damping ratios were also varied to assess the influence of this parameter on the accuracy of the combination rules. The damping ratios assumed were 0.1, 1.0, 2.0, and 10% of critical, all specified for the fundamental mode.

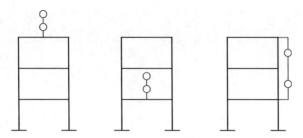

FIGURE 10.4 Structure–appendage systems considered in comparative study.

TABLE 10.1
First Three Natural Frequencies in Hertz of Structure–Appendage Systems Considered in Comparative Study

System	Mode 1	Mode 2	Mode 3	System	Mode 1	Mode 2	Mode 3	System	Mode 1	Mode 2	Mode 3
1	0.924	1.073	1.726	7	0.566	0.935	1.083	13	1.000	1.981	2.019
2	0.974	1.024	1.726	8	0.575	0.976	1.25	14	1.000	1.987	2.013
3	0.963	1.038	1.411	9	0.701	0.984	1.023	15	1.000	1.995	2.005
4	0.996	1.943	2.062	10	0.976	1.023	1.731	16	0.576	0.977	1.025
5	0.999	1.960	2.040	11	0.992	1.001	1.731	17	0.577	0.992	1.008
6	0.999	1.985	2.015	12	0.998	1.012	1.414	18	0.706	0.994	1.007

TABLE 10.2
Approximate to Exact Ratios of the Maximum Response of the Appendages in Figure 10.4 Obtained Using Different Modal Combination Rules

		Approximate Over Exact Response			
Damping Ratio (%)	Statistic	Absolute Sum	Square Root of the Sum of the Squares	Complete Quadratic Combination	Rosenblueth
0.1	Mean	3.37	2.21	2.07	1.07
	Maximum	18.20	12.63	9.66	1.27
1.0	Mean	4.72	3.11	1.53	0.99
	Maximum	20.36	14.13	4.26	1.12
2.0	Mean	6.11	4.07	1.32	0.93
	Maximum	29.76	20.72	2.62	1.07
10.0	Mean	11.16	7.44	1.09	1.02
	Maximum	58.81	41.02	1.38	1.24

In all cases, the structures and appendages were modeled as shear beams. The natural frequencies in the first three modes of the analyzed systems are given in Table 10.1, which shows that in all cases two of these natural frequencies are closely spaced. The ground motions considered were the first 10 s of the following earthquake records: (a) N–S 1940 El Centro, (b) N21E 1952 Taft, and (c) S16E 1971 Pacoima Dam. The effective durations considered in the application of Rosenblueth's rule were determined by matching the pseudovelocity response spectra of these ground motions to the pseudovelocities of a finite segment of a white noise process. The results of the study are summarized in Table 10.2. For each of the damping ratios considered, this table shows the averages and maximum values of the approximate to exact ratios of the responses obtained for all 18 cases. In the calculation of these ratios, the time-history solutions were considered to be the "exact" values. A ratio equal to unity indicates that the approximate solution is equal to the exact one, whereas a ratio of less or greater than unity indicates that the approximate response is, respectively, smaller or larger than the exact solution.

It may be seen from the results presented in Table 10.2 that for systems with closely spaced natural frequencies, the absolute sum and SRSS rules yield excessively large errors. It may be observed, in particular, that the absolute sum rule may overpredict the true maximum response with an error as large as 5800%. Similarly, the SRSS rule may lead to response predictions with an error as large as 4100%. The magnitude of these errors is obviously unacceptable and should not, thus, be used in the analysis of such systems. It may also be seen that the CQC and Rosenblueth's rules predict, on the average, the correct solutions with a tolerable error, and that these two rules give comparable results for damping ratios of the order of 10%. One can observe, however, that Rosenblueth's rule is considerably more accurate than the CQC rule for damping ratios of <2%. This confirms what was stated earlier that the CQC rule approaches the SRSS rule for damping ratios close to zero and cannot, therefore, lead to accurate results in such cases.

In another study, B. F. Maison, C. F. Neuss, and K. Kasai performed in 1983 an evaluation of the absolute sum, SRSS, Rosenblueth's and CQC rules by computing the peak responses of a high-rise building using these combination rules and comparing them against the peak responses obtained by an "exact" time-history analysis. The building considered was the 15-story east Health Sciences Building of the University of California Medical Center in San Francisco, California. It has a height of 195 ft and is square in plan with an outer dimension of 115 ft and 3 in. The structure of the building is formed with steel moment-resisting frames along its north–south and east–west directions. Two models were considered in the analysis of this building. In the first, the centers of mass and stiffness coincide and thus no torsional response was present. In the second, each floor mass is offset from the stiffness center of the building giving rise, as a result, to a torsional response as well as a response in the direction orthogonal to the direction of the excitation. Both models were assumed fixed at the ground level and with a damping ratio of 5% in all modes. The natural periods of the two models are given in Tables 10.3 and 10.4, from which it may be seen that the natural periods of the symmetric

TABLE 10.3
Natural Periods (in One Direction Only) of Symmetric Building Model in the Maison et al. Study

Mode	Period (s)
1	1.113
2	0.386
3	0.222
4	0.154

TABLE 10.4
Natural Periods of Eccentric Building Model in the Maison et al. Study

Mode	Period (s)	Mode	Period (s)
1	1.167	7	0.238
2	1.121	8	0.225
3	0.773	9	0.166
4	0.409	10	0.165
5	0.390	11	0.156
6	0.278	12	0.124

TABLE 10.5
Error in the Computation of Maximum Responses of Symmetric Model with Various Modal Combination Rules

			Error (%)			
			Double Sum			
Direction	Response	Statistic	Rosenblueth	Complete Quadratic Combination	Square Root of the Sum of the Squares	Absolute Sum
Parallel to excitation (E–W)	Deflection	Average	6	6	6	24
		Maximum	15	15	12	54
	Shear	Average	8	8	7	41
		Maximum	19	18	17	91
	Overturning moment	Average	6	6	6	33
		Maximum	19	19	18	92

Source: Maison, B. F. et al., *Earthquake Engineering and Structural Dynamics*, 11, 623–647, 1983.

model are well separated, whereas those of the eccentric model are closely spaced. The excitations considered were the S00E Pacoima Dam and N00W Orion Blvd. accelerograms recorded during the 1971 San Fernando earthquake, as well as the S00E El Centro accelerogram from the 1940 Imperial Valley earthquake. They were scaled to a peak ground acceleration of 0.2 g and applied along the east–west direction of the building. An effective ground motion duration of 10 s was considered in the application of Rosenblueth's rule. A statistical summary of the results is presented in Table 10.5 for the symmetrical model and Table 10.6 for the eccentric one.

Table 10.5 shows that the average errors in the estimation of the maximum response values of the symmetric model using the SRSS, Rosenblueth's, and CQC rules are more or less the same and that these errors are relatively small, meaning that the maximum response values predicted by these rules are relatively close to those determined by the time-history analysis. This was a consequence of the fact that the natural periods of the symmetric model were well separated and thus the modal correlation coefficients and the cross terms in the double-sum combination rules were all virtually equal to zero. As expected, the errors using the absolute sum rule are large although not excessively large. In contrast, Table 10.6 shows that the average errors in the prediction of the peak responses of the eccentric model using the SRSS rule are unacceptably large whereas those attained using Rosenblueth's and CQC rules are relatively small. Once again, this is so because the eccentric model possesses closely spaced natural periods and thus there exists a significant correlation between its modal responses, which the SRSS rule neglects, but Rosenblueth's and CQC rules do not. It may also be observed that the average errors in the use of Rosenblueth's and CQC rules are very similar, which it is not surprising as both are based on essentially the same assumptions. As in the case of the symmetric model, the errors attained using the absolute sum rule are large, except that in this case these errors are exaggeratedly large. As in the previous study, Rosenblueth's rule gives, overall, the best results.

Finally, it is instructive to examine the results of a study conducted by Singh and Maldonado to investigate the accuracy of the combination rules under discussion in the analysis of systems with significant modal responses in their higher modes. In this study, the accuracy of the absolute sum, SRSS, CQC, and the S–M rules was evaluated by comparing the maximum responses of the continuous steel beam shown in Figure 10.5 determined using these combination rules against results obtained by a time-history analysis. For the analysis, the beam was discretized into 26 finite

TABLE 10.6
Error in the Computation of Maximum Responses of Eccentric Model with Various Modal Combination Rules

			Error (%)			
			Double Sum			
Direction	Response	Statistic	Rosenblueth	Complete Quadratic Combination	Square Root of the Sum of the Squares	Absolute Sum
Parallel to excitation (E–W)	Deflection	Average	7	6	18	27
		Maximum	19	17	26	67
	Shear	Average	8	8	22	49
		Maximum	20	19	35	122
	Overturning moment	Average	6	7	25	39
		Maximum	18	18	34	120
Normal to excitation (N–S)	Deflection	Average	18	32	251	491
		Maximum	33	67	350	800
	Shear	Average	17	24	217	528
		Maximum	31	55	307	661
	Overturning moment	Average	16	25	218	520
		Maximum	25	51	299	658
Torsional	Torque	Average	9	7	13	137
		Maximum	27	26	40	288

Source: Maison, B. F. et al., *Earthquake Engineering and Structural Dynamics*, 11, 623–647, 1983.

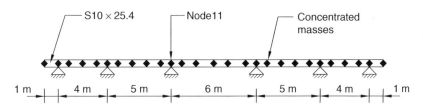

FIGURE 10.5 Continuous beam considered in the Singh–Maldonado study.

elements considering a total of 48 degrees of freedom. In addition to its weight, the beam supports masses lumped at its nodes. A damping ratio of 3% was assumed in all its modes of vibration. Two cases were considered: a stiff beam and a flexible beam. The first 10 natural frequencies of the beam in each of these two cases are listed in Table 10.7. It is worthwhile to note that in both cases the beam's natural frequencies are well separated from one another. The response parameter analyzed was the maximum bending moment at node 11 when the supports of the beam were subjected to an earthquake ground motion. The earthquake ground motion considered was represented by acceleration and velocity response spectra that are compatible with the acceleration design spectrum specified by the Nuclear Regulatory Commission for the design of nuclear power plants. This design spectrum has a cut-off frequency of 33 Hz. The results of the study are summarized in Table 10.8 in which the errors attained with each of the evaluated combination rules in comparison with the results assumed to be the exact ones are presented.

TABLE 10.7
Natural Frequencies of Beams Considered in the Singh and Maldonado Study

Mode	Frequency (Hz) Flexible Beam	Frequency (Hz) Stiff Beam	Mode	Frequency (Hz) Flexible Beam	Frequency (Hz) Stiff Beam
1	5.006	19.969	6	17.714	73.031
2	7.038	28.556	7	18.656	81.050
3	7.880	32.378	8	19.763	83.454
4	9.905	41.406	9	26.937	108.375
5	10.689	44.166	10	28.860	115.999

Source: Singh, M. P. and Maldonado, G. O., *Earthquake Engineering and Structural Dynamics*, 20, 621–635, 1991.

TABLE 10.8
Error in the Calculation of Maximum Bending Moment at Node 11 of Beam in Figure 10.5 Using Various Modal Combination Rules

Rule	Error (%) Stiff Beam	Error (%) Flexible Beam
Absolute sum	7	49
Square root of the sum of the squares	47	30
Complete quadratic combination	45	20
Singh–Maldonado	0	0

Source: Singh, M. P. and Maldonado, G. O., *Earthquake Engineering and Structural Dynamics*, 20, 621–635, 1991.

It may be seen from Table 10.8 that the errors in the results obtained with the SRSS and CQC rules are relatively high, particularly for the stiff beam. This is so despite the fact that the beam has no closely spaced natural frequencies. In contrast, the results obtained using the S–M rule are virtually the same as the exact ones. In the CQC rule, the error is comparatively large because, as discussed earlier, the expression for the modal correlation coefficient considered in this rule yields nearly zero values for the correlation between low- and high-frequency modes. This is also the reason why the error attained with the CQC and SRSS rules in the case of the stiff beam is almost the same. In contrast, for the stiff beam the absolute sum rule provides better results than the CQC rule. This is so because the stiff beam has significant high-frequency modes and because, as discussed earlier too, these modes are strongly correlated with the low-frequency modes. On the other hand, the absolute sum rule yields large errors for the flexible beam as flexible systems do not possess, in general, significant high-frequency modes. Observe, thus, that S–M rule is far superior to the CQC rule (as well as Rosenblueth's rule as both produce similar results) for the analysis of stiff structural systems.

In summary, the results of the comparative studies presented here show that

1. The absolute sum rule gives in most cases excessively large errors.
2. The SRSS rule is adequately accurate for flexible systems without closely spaced natural frequencies or significant higher-mode responses, but it leads to gross errors when these conditions are not met.
3. Rosenblueth's and CQC rules give fairly accurate results for most structures, with the exception of rigid structures with significant high-frequency modes.

4. The CQC rule leads to inaccurate results for low damping ratios, that is, <2%.
5. S–M rule is accurate for all structures, including those with closely spaced natural frequencies and rigid structures with large high-frequency modal responses.

10.3.2.7 Final Remarks

As a final point in regard to the application of the modal combination rules described earlier, it is important to keep in mind that one cannot estimate the maximum values of a response parameter in terms of the maximum values of another response parameter. For instance, one cannot estimate the maximum values of the internal forces in a structural element in terms of its maximum displacements, as Equation 10.47 would suggest. That is, one cannot assume that $\max\{F_i(t)\} = [K_i'][T_i]\max\{u_i(t)\}$. This so because the relationships that define a response parameter in terms of another often involve additions and subtractions and thus if maximum values were used in these relationships, one would be adding or subtracting maximum values. But because such maximum values do not, in general, occur at the same time, adding or subtracting maximum values would lead to an overprediction or underprediction of the desired response. For example, the elastic forces in a two-story shear building are given by

$$\begin{Bmatrix} F_1(t) \\ F_2(t) \end{Bmatrix} = \begin{bmatrix} k_1 + k_2 & -k_2 \\ -k_2 & k_2 \end{bmatrix} \begin{Bmatrix} u_1(t) \\ u_2(t) \end{Bmatrix} = \begin{Bmatrix} k_1 u_1(t) + k_2[u_1(t) - u_2(t)] \\ -k_2[u_1(t) - u_2(t)] \end{Bmatrix} \quad (10.82)$$

and thus these forces depend on the difference $u_1(t) - u_2(t)$. As the maximum values of $u_1(t)$ and $u_2(t)$ do not occur at the same time, then the maximum value of the difference $u_1(t) - u_2(t)$ is obviously greater than the difference between the maximum values of $u_1(t)$ and $u_2(t)$. Hence, the substitution of the maximum values of $u_1(t)$ and $u_2(t)$ into Equation 10.82 to estimate the maximum values of $F_1(t)$ and $F_2(t)$ would lead to an underestimation of these maximum values.

Accordingly, the maximum values of a response parameter that depends on another response parameter should be determined, instead, by calculating first the maximum modal values of the desired response and then combining these modal maximum using one of the approximate rules introduced in the preceding sections. For instance, if the desired maximum response are the maximum values of the internal forces in the ith element of a structural system, and if the SRSS rule is an adequate procedure to combine the modal responses of this system, then one would first compute the maximum modal responses according to Equation 10.57 and then one would combine these maximum modal response using the SRSS rule.

EXAMPLE 10.1   ACCELERATION RESPONSE OF A PENTHOUSE ON A THREE-STORY BUILDING BY THE RESPONSE SPECTRUM METHOD

A three-story shear building supports a two-story penthouse as shown in Figure E10.1a. Its natural frequencies, mode shapes, and participation factors are listed in Table E10.1. The building is subjected to a ground motion represented by the response spectrum shown in Figure E10.1b. Determine the maximum accelerations at the penthouse floors using the response spectrum method and the S–M formula to compute the system's modal correlation coefficients. Assume a damping ratio of 2% for all modes and consider only the first three modes of vibration.

Solution

From the response spectrum in Figure E10.1b and considering only the first three modes, the spectral accelerations, spectral displacements, and pseudovelocities corresponding to the natural frequencies and damping ratios of the system are

$SA_1 = SA(5.844, 0.02) = 0.36\ g \quad SA_2 = SA(6.784, 0.02) = 0.40\ g \quad SA_3 = SA(10.917, 0.02) = 0.82\ g$

$SD_1 = SD(5.844, 0.02) = 0.103\ m \quad SD_2 = SD(6.784, 0.02) = 0.085\ m \quad SD_3 = SD(10.917, 0.02) = 0.068\ m$

$PSV_1 = PSV(5.844, 0.02) = 0.60\ m/s \quad PSV_2 = PSV(6.784, 0.02) = 0.58\ m/s \quad PSV_3 = PSV(10.917, 0.02) = 0.74\ m/s$

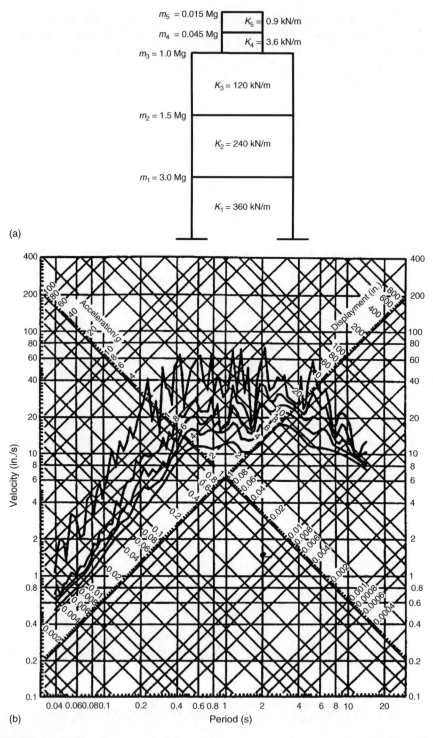

FIGURE E10.1 (a) The three-story building with two-story penthouse considered in Example 10.1 and (b) response spectra for 0, 2, 5, 10, and 20% damping of E–W component of ground acceleration recorded during the 1940 El Centro, California, earthquake. (Reproduced from Analyses of Strong Earthquake Accelerograms, Vol. III, Response Spectra, Part A, Report EERL-72-80, California Institute of Technology, Pasadena, CA, 1972.)

TABLE E10.1
Dynamic Properties of Building in Example 10.1

Mode	1	2	3	4	5
Frequency (rad/s)	5.844	6.784	10.917	12.797	18.986
Participation factor	1.583	1.423	0.347	0.893	0.314
Mode shape	0.1502	0.1824	0.1149	0.4063	0.3149
	0.3114	0.3511	0.1161	0.1840	−0.6317
	0.5009	0.4864	−0.0545	−0.6373	0.3215
	2.0637	−1.2203	−3.9383	0.9766	−0.1003
	4.7910	−5.2390	3.9933	−0.5647	0.0200

Similarly, according to the empirical formulas proposed by Sadek et al. to determine spectral velocities, (Equations 10.79 through 10.81) one has that,

$$a_v = 1.095 + 0.647\xi - 0.382\xi^2 = 1.095 + 0.647(0.02) - 0.382(0.02)^2 = 1.108$$

$$b_v = 0.193 + 0.838\xi - 0.621\xi^2 = 0.193 + 0.838(0.02) - 0.621(0.02)^2 = 0.210$$

$$SV_1 = a_v T_1^{b_v} PSV_1 = 1.108 \left(\frac{2\pi}{5.844}\right)^{0.210} (0.60) = 1.125(0.60) = 0.67 \text{ m/s}$$

$$SV_2 = a_v T_2^{b_v} PSV_2 = 1.108 \left(\frac{2\pi}{6.784}\right)^{0.210} (0.58) = 1.090(0.58) = 0.63 \text{ m/s}$$

$$SV_3 = a_v T_3^{b_v} PSV_3 = 1.108 \left(\frac{2\pi}{10.917}\right)^{0.210} (0.74) = 0.987(0.74) = 0.73 \text{ m/s}$$

Therefore, in accordance with Equation 10.56 and the mode shapes and participation factors in Table E10.1, the maximum penthouse accelerations in each of the modes of the building–penthouse system are equal to

$$\max \begin{Bmatrix} \ddot{y}_4(t) \\ \ddot{y}_5(t) \end{Bmatrix}_1 = \Gamma_1 \{\phi\}_1 SA(\omega_1,\xi_1) = 1.583 \begin{Bmatrix} 2.0637 \\ 4.7910 \end{Bmatrix} 0.36\,g = \begin{Bmatrix} 1.18 \\ 2.73 \end{Bmatrix} g$$

$$\max \begin{Bmatrix} \ddot{y}_4(t) \\ \ddot{y}_5(t) \end{Bmatrix}_2 = \Gamma_2 \{\phi\}_2 SA(\omega_2,\xi_2) = 1.423 \begin{Bmatrix} -1.2203 \\ -5.2390 \end{Bmatrix} 0.40\,g = \begin{Bmatrix} -0.69 \\ -2.98 \end{Bmatrix} g$$

$$\max \begin{Bmatrix} \ddot{y}_4(t) \\ \ddot{y}_5(t) \end{Bmatrix}_3 = \Gamma_3 \{\phi\}_3 SA(\omega_3,\xi_3) = 0.347 \begin{Bmatrix} -3.9383 \\ 3.9933 \end{Bmatrix} 0.83\,g = \begin{Bmatrix} -1.13 \\ 1.15 \end{Bmatrix} g$$

Note also that for the natural frequencies and damping ratios of the system, the constants defined by Equations 10.75 through 10.78, arranged in matrix form, result as

$$[D_{mn}] = \begin{bmatrix} 0.006 & 0.071 & 0.511 \\ 0.129 & 0.006 & 0.379 \\ 6.223 & 2.544 & 0.006 \end{bmatrix} \quad [A_{mn}] = \begin{bmatrix} 0.5 & 2.711 & 0.401 \\ -3.576 & 0.5 & 0.626 \\ -1.391 & -1.611 & 0.5 \end{bmatrix}$$

$$[B_{mn}] = \begin{bmatrix} 0 & -0.079 & -0.012 \\ 0.079 & 0 & -0.014 \\ 0.012 & 0.014 & 0 \end{bmatrix} \quad [C_{mn}] = \begin{bmatrix} 0.5 & -3.576 & -1.391 \\ 2.711 & 0.5 & -1.611 \\ 0.401 & 0.626 & 0.5 \end{bmatrix}$$

and thus, by virtue of Equation 10.74 and the determined spectral displacements and velocities, the modal correlation coefficients, also arranged in matrix form, are equal to

$$[\rho_{mn}] = \begin{bmatrix} 1.0 & 0.132 & 0.171 \\ 0.132 & 1.0 & 0.186 \\ 0.171 & 0.186 & 1.0 \end{bmatrix}$$

Hence, upon substitution of the calculated maximum modal responses and these modal correlation coefficients into Equation 10.66, the maximum accelerations, squared, at the level of the penthouse floors are given by

$$\max \begin{Bmatrix} \ddot{y}_4^2(t) \\ \ddot{y}_5^2(t) \end{Bmatrix} = \{(\max[\ddot{y}(t)]_1)^2\} + \{(\max[\ddot{y}(t)]_2)^2\} + \{(\max[\ddot{y}(t)]_3)^2\}$$

$$+ 2\rho_{12}\{\max[\ddot{y}(t)]_1 \max[\ddot{y}(t)]_2\} + 2\rho_{13}\{\max[\ddot{y}(t)]_1 \max[\ddot{y}(t)]_3\} + 2\rho_{23}\{\max[\ddot{y}(t)]_2 \max[\ddot{y}(t)]_3\}$$

$$= \left(\begin{Bmatrix} 1.18^2 \\ 2.73^2 \end{Bmatrix} + \begin{Bmatrix} 0.69^2 \\ 2.98^2 \end{Bmatrix} + \begin{Bmatrix} 1.13^2 \\ 1.15^2 \end{Bmatrix} + 2(0.132) \begin{Bmatrix} (1.18)(-0.69) \\ (2.73)(-2.98) \end{Bmatrix} + 2(0.171) \begin{Bmatrix} (1.18)(-1.13) \\ (2.73)(1.15) \end{Bmatrix} \right.$$

$$\left. + 2(0.186) \begin{Bmatrix} (-0.69)(-1.13) \\ (-2.98)(1.15) \end{Bmatrix} \right) g^2 = \begin{Bmatrix} 2.76 \\ 15.31 \end{Bmatrix} g^2$$

and thus,

$$\max \begin{Bmatrix} \ddot{y}_4(t) \\ \ddot{y}_5(t) \end{Bmatrix} = \begin{Bmatrix} 1.66 \\ 3.91 \end{Bmatrix} g$$

10.4 PROPERTIES OF MODAL PARTICIPATION FACTORS

The modal participation factors defined in Section 10.2.3 exhibit some interesting properties that are of relevance in the application of the response spectrum method. One of these properties is that for the degrees of freedom in the direction of the ground motion that excites a structure

$$\sum_{r=1}^{N} \Gamma_r \phi_{ir} = 1 \qquad (10.83)$$

where ϕ_{ir} denotes the element corresponding to the ith degree of freedom in the rth mode shape of the structure and, as before, Γ_r represents the structure's rth modal participation factor. Such a property follows directly from Equations 10.42 and 10.44, because these two equations lead to

$$\sum_{r=1}^{N} \Gamma_r \{\phi\}_r = \{J\} \qquad (10.84)$$

and the elements of $\{J\}$ are equal to unity for the degrees of freedom in the direction of the excitation. Note, thus, that for the particular case when all the mode shapes are normalized so as to make ϕ_{ir} equal to unity, the following relationship applies:

$$\sum_{r=1}^{N} \Gamma_r = 1 \qquad (10.85)$$

Structural Response by Response Spectrum Method

That is, all the modal participation factors add up to unity. It is important to note that this relationship applies only to mode shapes that have been normalized as described. If normalized in a different way the sum of all the participation factors will add up to a different number.

Equation 10.83 is of relevance because it allows an assessment of the importance of the contribution of each modal response toward the total response. With reference to Equation 10.35, if for any instant of time t the function $z_r(t)$ is of the same order of magnitude in all modes, then the modes for which the elements of the $\Gamma_r\{\phi\}_r$ vector are large compared to unity will contribute significantly toward the total response at that time whereas those modes for which the elements of the $\Gamma_r\{\phi\}_r$ vector are close to zero will contribute little. In any modal superposition, one may thus neglect the modes for which the elements of the $\Gamma_r\{\phi\}_r$ vector are small.

Another property of the participation factors that is of relevance is that a mode shape multiplied by the participation factor that corresponds to that mode shape yields a mode shape with a unit participation factor. This property follows from the definition of participation factor. That is, if $\{\phi'\}_r = \Gamma_r\{\phi\}_r$ is the mode shape that results from multiplying an original mode shape by its participation factor, then from the definition of participation factor one has that the participation factor of the normalized mode shape is equal to

$$\Gamma'_r = \frac{\{\phi'\}_r^T[M]\{J\}}{\{\phi'\}_r^T[M]\{\phi'\}_r} = \frac{\Gamma_r\{\phi\}_r^T[M]\{J\}}{\Gamma_r^2\{\phi\}_r^T[M]\{\phi\}_r} = \frac{\Gamma_r^2}{\Gamma_r^2} = 1.0 \tag{10.86}$$

The relevance of this latter property lies on the fact that for mode shapes with unit participation factors Equation 10.84 leads to

$$\sum_{r=1}^{N} \{\phi'\}_r = \{J\} \tag{10.87}$$

which shows that for mode shapes with unit participation factors the modal amplitudes corresponding to the degrees of freedom in the direction of the excitation add up to one. Thus, the elements of a unit-participation-factor mode shape give a direct indication of the importance of that mode in a modal superposition. If these elements are close to unity, then the mode will contribute significantly toward the total response, but if they are close to zero, the contribution will be negligible.

10.5 DETERMINATION OF MODAL DAMPING RATIOS

It may be seen from Equations 10.53 through 10.57 that the computation of a system's maximum modal responses requires the knowledge or specification of the damping ratios in each of its the modes. This section is devoted, thus, to review some of the methods that may be used to determine these damping ratios, the magnitude of the damping values that have been measured in actual structures during field tests, and the recommendations that have emerged from these tests in regard to the damping ratios that are reasonable to use in design applications.

It was assumed in the derivation of the uncoupled equations of motion that led to the formulation of the modal superposition method that the damping matrix of the system under consideration was of the Rayleigh type, that is, a matrix of the form

$$[C] = \alpha[M] + \beta[K] \tag{10.88}$$

where $[M]$ and $[K]$ represent the mass and stiffness matrices of the system and α and β two unknown constants. If transformed into a scalar equation in terms of damping ratios, this equation may then be used to derive a relationship whereby one can determine the damping ratios in some of the modes of the system when the damping ratios in some other modes are known. Such a transformation may be attained by premultiplying Equation 10.88 by the transpose of the mode shape $\{\phi\}_r$ and postmultiplying it by $\{\phi\}_r$ so as to obtain

$$\{\phi\}_r^T[C]\{\phi\}_r = \alpha\{\phi\}_r^T[M]\{\phi\}_r + \beta\{\phi\}_r^T[K]\{\phi\}_r \tag{10.89}$$

which by virtue of the definitions of generalized mass, generalized stiffness, and generalized damping constant introduced in Section 10.2.3 may also be written as

$$C_r^* = \alpha M_r^* + \beta K_r^* \tag{10.90}$$

If, in view of Equations 10.30 and 10.31, the generalized damping constant and the generalized stiffness are expressed as $C_r^* = 2\xi_r\omega_r M_r^*$ and $K_r^* = \omega_r^2 M_r^*$, then Equation 10.90 may be written alternatively as

$$2\xi_r\omega_r M_r^* = \alpha M_r^* + \beta \omega_r^2 M_r^* \tag{10.91}$$

from which one can solve for the damping ratio ξ_r to obtain

$$\xi_r = \frac{1}{2}\left(\frac{\alpha}{\omega_r} + \beta\omega_r\right) \tag{10.92}$$

Equation 10.92 defines the damping ratio in the rth mode of the system in terms of the natural frequency in the same mode and the constants α and β. These two constants, however, are not known. To determine the values of α and β, it is necessary to define or assume two of the modal damping ratios of the system and use Equation 10.92 to set two simultaneous equations in α and β. Then, one needs to solve for these two constants from the two simultaneous equations. Proceeding accordingly, one obtains

$$\alpha = \frac{2\omega_i\omega_j(\xi_i\omega_j - \xi_j\omega_i)}{\omega_j^2 - \omega_i^2} \qquad \beta = \frac{2(\xi_j\omega_j - \xi_i\omega_i)}{\omega_j^2 - \omega_i^2} \tag{10.93}$$

where it has been presumed that the known damping ratios are those in the ith and jth modes of the system. Observe, thus, that once the constants α and β are determined, Equation 10.92 itself may be used to determine the damping ratios in all other modes. Or, if desired, Equation 10.88 may be employed to construct a damping matrix. The application of the procedure will now be illustrated by means of the following example.

EXAMPLE 10.2   DETERMINATION OF MODAL DAMPING RATIOS WHEN THE DAMPING MATRIX IS OF THE RAYLEIGH TYPE

The natural frequencies of a five-degree-of-freedom system are 11.6, 27.5, 40.8, 51.5, and 80.4 rad/s. Determine the damping ratios for the third, fourth, and fifth modes assuming that the damping matrix of the system is of the Rayleigh type and the damping ratios in its first and second modes are 5 and 10%, respectively.

Solution

Substitution of the given natural frequencies and damping ratios for the first two modes into Equation 10.93 yields

$$\alpha = \frac{2\omega_1\omega_2(\xi_1\omega_2 - \xi_2\omega_1)}{\omega_2^2 - \omega_1^2} = \frac{2(11.6)(27.5)[(0.05)(27.5) - (0.10)(11.6)]}{(27.5)^2 - (11.6)^2} = 0.2206$$

$$\beta = \frac{2(\xi_2\omega_2 - \xi_1\omega_1)}{\omega_2^2 - \omega_1^2} = \frac{2[(0.10)(27.5) - (0.05)(11.6)]}{(27.5)^2 - (11.6)^2} = 0.0070$$

Hence, after substitution of these values of α and β into Equation 10.92, one obtains

$$\xi_r = \frac{1}{2}\left(\frac{0.2206}{\omega_r} + 0.0070\omega_r\right)$$

from which, after substitution of the natural frequencies for the third, fourth, and fifth modes, one arrives to

$$\xi_3 = \frac{1}{2}\left[\frac{0.2206}{40.8} + 0.0070(40.8)\right] = 0.146 = 14.6\%$$

$$\xi_4 = \frac{1}{2}\left[\frac{0.2206}{51.5} + 0.0070(51.5)\right] = 0.182 = 18.2\%$$

$$\xi_5 = \frac{1}{2}\left[\frac{0.2206}{80.4} + 0.0070(80.4)\right] = 0.283 = 28.3\%$$

Equation 10.88 represents the general expression to construct a damping matrix of the Rayleigh type, but it is not the only one that can be used to decouple the equation of motion and, hence, to define the modal damping ratios of a system. It may also be assumed that the damping matrix is proportional to either the mass matrix alone or the stiffness matrix alone. That is, $[C] = \alpha[M]$ or $[C] = \beta[K]$. When the damping matrix is assumed proportional to the mass matrix alone, then the corresponding expression to determine the modal damping ratios of the system is

$$\xi_r = \frac{1}{2}\frac{\alpha}{\omega_r} \qquad (10.94)$$

Similarly, when the damping matrix is assumed to be proportional to the stiffness matrix alone, the corresponding expression is

$$\xi_r = \tfrac{1}{2}\beta\omega_r \qquad (10.95)$$

It may be noted that in these two cases the modal damping ratios depend on only one constant and thus only one damping ratio needs to be known or specified to determine the value of such a constant. These alternative formulations might prove to be more convenient than Equation 10.92 when, as is usually the case, only the damping ratio for the fundamental mode of the system is known or specified.

The variation of the modal damping ratios with natural frequency is illustrated in Figure 10.6 for the cases in which the damping matrix is assumed proportional to both the mass and the stiffness

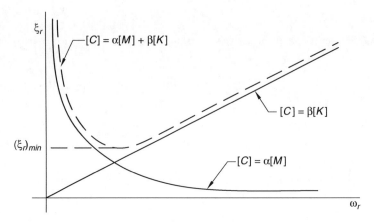

FIGURE 10.6 Variation of damping ratio with natural frequency according to various forms of Rayleigh damping.

matrices, proportional to mass matrix alone, and proportional to the stiffness matrix alone. It is of interest to observe from this figure that the modal damping ratios decrease with natural frequency for the case when the damping matrix is assumed proportional to the mass matrix alone and increase when it is assumed proportional to the stiffness matrix alone. These variations of the modal damping ratios with natural frequency are not consistent with experimental data and thus neither of these two models may be appropriate for design purposes. However, given the uncertainties associated with the calculation of the modal properties of a system in its higher modes, sometimes it may be advantageous to assume that the damping matrix is proportional to the stiffness matrix alone to deliberately damp out the system's higher modes.

An alternative approach that allows one to assume a damping matrix proportional to the mass and the stiffness matrices but requires knowing the value of only one damping ratio to determine the constants α and β and define the modal damping ratios of a structure is that based on the selection of the minimum value of ξ_r. In this approach, the equations that define α and β are obtained by taking the first derivative of Equation 10.92 with respect to ω_r and then setting this derivative equal to zero. Following this procedure, it is not too difficult to show that the constants α and β that make ξ_r a minimum are

$$\alpha = \xi_r \omega_r \qquad \beta = \xi_r / \omega_r \tag{10.96}$$

It may be noted that with this approach, one can obtain the values of α and β and the modal damping ratios of a structure by simply specifying the damping ratio in one of the modes of the structure. Note, also, that by using Equation 10.96 one makes the damping ratio employed in the evaluation of α and β to be the damping ratio with the least value (see Figure 10.6).

It should also be mentioned that there is still another combination of the mass and stiffness matrices that is different from the one that defines Rayleigh damping and also decouples the equations of motion. This combination is of the form

$$[C] = [M] \sum_{i=0}^{n-1} a_i ([M]^{-1}[K])^i \tag{10.97}$$

where n is any arbitrary number that defines the number of terms in the series, and, for each term, a_i is a constant. Damping defined by a damping matrix of this form is known as *Caughey damping*. It is considered to be an extension of Rayleigh damping because Rayleigh damping corresponds to Caughey damping when $n = 2$, the special case when only the terms for $i = 0$ and $i = 1$ are included in the series. It is shown in structural dynamics books that the damping ratios associated with Caughey damping are given by

$$\xi_r = \frac{1}{2\omega_r} \sum_{i=0}^{n-1} a_i \omega_r^{2i} \qquad (10.98)$$

Hence, Equation 10.98 may be used to define the modal damping ratios of a structure once the constants a_i are known. In turn, these constants may be determined using this same equation by setting a system of simultaneous equations in terms of the modal damping ratios that are known or specified. Because the series in Equation 10.98 may include as many terms as the number of modes in a structure, Caughey damping provides one with the flexibility of being able to specify the damping ratios in as many modes as desired. It should be noted, however, that the higher the number of modes considered, the higher will be the order of the system of equations that needs to be solved to find the constants a_i. Furthermore, the resulting algebraic equations may be numerically ill-conditioned as the coefficients ω_r^{-1}, ω_r, ω_r^3, ω_r^5, etc., may differ by orders of magnitude. Another problem with Caughey damping is that it may yield negative damping ratios in some of the higher modes of system, which would make the response in these modes increase without limit and lead thus to false results.

It is worthwhile to note that Rayleigh and Caughey damping are unnecessary approximations as it is possible to decouple a system's equations of motion even if the damping matrix of the system is not of the Rayleigh or Caughey type; that is, the damping matrix is of any general form. However, because in a system with a damping matrix that is not diagonalized after it is premultiplied and postmultiplied by its modal matrix (a nonclassical damping matrix) its various components vibrate out of phase, the transformation that decouples the equations of motion of such a system involves the use of complex-valued mode shapes. That is, mode shapes that describe the motion of a system in terms of not only relative amplitudes but also phase angles. That means having to solve an eigenvalue problem and a system of differential equations that are double the size of the eigenvalue problem and system of differential equations that one needs to solve in a classical solution with a damping matrix of the Rayleigh or Caughey type. Because of the large computational effort it entails, such a method is seldom used in the analysis of conventional structures. Another reason is that it has been observed that for small damping ratios a classical solution yields a fairly good approximation. Studies have shown that a classical solution may be justified for damping ratios up to 20% of critical. For the damping ratios found in modern structures, it seems therefore that the use of the classical solution is acceptable. It should be borne in mind, however, that there are some cases in which a classical solution may lead to significant errors. In general, those cases involve systems formed with two or more components that exhibit significantly different levels of damping. Examples are structure-equipment, soil-structure, and fluid-structure systems and systems with added energy-dissipating devices or isolators. These systems are referred to as systems with nonclassical damping or systems without classical modes of vibration and require the use of the nonclassical solution method for their analysis. A detailed description of this method is presented in Section 15.5.4. Interested readers may also consult the structural dynamics books by Cheng and Hart and Wong listed at the end of the chapter.

The procedures outlined in the preceding paragraphs are helpful to define the modal damping ratios of a structure when at least one of these modal damping ratios is known. But there are still two questions that need to be answered before one can completely define all the modal damping ratios of a structure. One is in regard to the determination of the damping ratios that are required to apply those procedures. The other is in regard to the relationship that exists between the damping ratios in the different modes of a structure. These, however, are not easy questions to answer. As the damping ratios of a structure cannot be determined in terms of the properties of the elements of the structure in much the same way as its mass and stiffness matrices are, and as no computation procedure is therefore available for evaluating these damping ratios, these questions can only be answered by extrapolation of results from field and laboratory tests. This approach, however, involves a great deal of judgment in interpreting test results and deciding whether or not the available test results may be applied to structures that are different from those tested and will be subjected to excitation levels that differ from those used in the tests. The problem is that the degree of damping in a structure depends on the materials used, the form of the structure, the characteristics of the foundation soil, the intensity of motion and stress level, and the extent of deterioration from previous events. Another problem is that most of the available data come from experimental tests that are performed with low-amplitude vibrations and that, as already mentioned, the degree of damping in a structure strongly depends on the excitation level.

To gain an idea about the magnitude of the damping ratios in actual structures, how they vary from mode to mode, and the variability involved in them, it is instructive to review some of the damping data from field tests and the analysis of earthquake records that have been reported in the literature. In 1975, G. C. Hart and R. Vasudevan determined the damping ratios in the three first modes of 12 Southern California high-rise buildings using acceleration time histories recorded during the 1971 San Fernando earthquake. The results of this study are summarized in Table 10.9. This table lists the computed damping ratios along the buildings' two horizontal directions and

TABLE 10.9
Damping Ratios in Buildings Estimated from Acceleration Records from the 1971 San Fernando, California, Earthquake

Building	Stories	Structural Material	Direction	Mode 1 T (s)	Mode 1 SV (in./s)	Mode 1 Damping Ratio (%)	Mode 2 T (s)	Mode 2 SV (in./s)	Mode 2 Damping Ratio (%)	Mode 3 T (s)	Mode 3 SV (in./s)	Mode 3 Damping Ratio (%)
1	17	RC	1	1.04	18	4.3	0.34	15	2.3	0.22	15	3.5
			2	1.55	13	4.0	0.47	30	7.4	0.30	30	8.7
2	22	RC	1	1.84	10	2.8	1.35	30	6.9	0.59	14	2.0
			2	2.17	12	4.5	1.34	20	3.1	0.58	9	1.7
3	12	RC	1	2.38	45	10.4	0.67	27	6.0	0.49	24	4.8
			2	2.94	40	9.0	0.98	32	8.0	0.59	25	5.3
4	19	Steel	1	3.34	45	11.3	1.17	28	6.3	0.80	33	7.4
			2	3.27	40	8.9	1.11	30	6.6	0.75	27	5.4
5	20	RC	1	2.13	16	4.9	0.66	25	5.0	0.25	20	4.3
			2	2.27	15	4.1	0.72	32	8.4	0.30	36	8.4
6	7	RC	1	1.49	38	9.7	0.49	30	9.2	0.25	30	9.5
			2	1.26	80	16.4	0.60	68	13.0	0.31	55	12.0
7	7	RC	1	1.03	28	9.0	*	*	*	*	*	*
			2	1.17	36	8.8	*	*	*	*	*	*
8	19	Steel	1	3.41	25	5.9	0.81	20	4.0	0.40	15	2.4
			2	3.43	33	8.4	1.39	18	4.0	0.78	6	2.0
9	34	Steel	1	2.90	24	3.2	1.07	25	3.0	0.60	25	4.0
			2	3.38	28	4.9	1.19	21	3.5	0.68	20	3.6
10	13	RC	1	1.14	10	1.9	*	*	*	*	*	*
			2	1.19	45	13.7	*	*	*	*	*	*
11	28	Steel	1	4.26	27	6.5	1.41	12	2.5	*	*	*
			2	4.27	25	5.2	1.42	13	2.0	*	*	*
12	27	Steel	1	5.40	32	7.4	*	*	*	*	*	*
			2	6.06	20	3.0	*	*	*	*	*	*

Note: T = natural period; SV = zero-damping pseudovelocity response at modal period; RC = reinforced concrete; * = not determined.

Source: Hart, G. C. and Vasudevan, R., *Journal of the Structural Division*, ASCE, 101, 11–30, 1975.

Structural Response by Response Spectrum Method

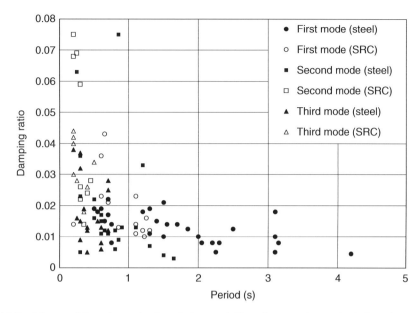

FIGURE 10.7 Measured damping ratios in existing steel (S) and composite steel-reinforced concrete (SRC) buildings. (Aoyama, H., *Proceedings U.S.A.—Japan Seminar on Composite and Mixed Construction Systems*, Gihodo Shuppan, Co., Tokyo, 1980, 171–181.)

the buildings' natural periods in the associated modes. Also listed are the zero-damping spectral velocities corresponding to the accelerations recorded at the buildings' basements and the buildings' first three natural periods, which permit a comparative assessment of the level of ground shaking in each case. These results are of significance because they provide information on the values of the damping ratios the buildings exhibit under an earthquake of moderate intensity, which by far exceed the low-amplitude levels induced in most field tests. The results also show that damping increases with the amplitude of motion and that at the level of response generated by an earthquake there is only a slight difference in the damping ratios of steel and reinforced concrete buildings (mean values for the first modes are 6.5 and 7.4%, respectively). They indicate, further, that there is no discernible difference between the damping values in different modes of the buildings, which seem to suggest that the same damping ratios should be used in all modes. In another study conducted in 1980, H. Aoyama reported the damping ratios measured during vibration tests of existing steel and composite steel-reinforced concrete buildings. The obtained damping ratios are presented in Figure 10.7. It may be observed from this figure that the measured damping ratios scatter over a wide range and that there is no an apparent relationship between the damping ratios in different modes. In a similar study published in 1986, A. P. Jeary reported the damping ratios measured from forced-vibration tests in eight tall reinforced concrete buildings. The damping ratios obtained in the fundamental modes of the buildings are listed in Table 10.10. These results confirm that the degree of damping in buildings increase with motion amplitude and that the measurements from low-level excitations yield considerably lower damping ratios than those reported by Hart and Vasudevan, for example. More recently, the Damping Evaluation Committee of the Architectural Institute of Japan collected and reported damping data for 137 steel-framed buildings, 25 reinforced concrete buildings, and 43 composite steel-reinforced concrete buildings. Most of the data were obtained from low-amplitude forced-vibration or free-vibration tests. The height of the buildings ranged from 50 to 150 m in the case of the steel buildings and from 50 to 100 m in the case of the reinforced concrete and composite buildings. Figure 10.8 shows the reported damping ratios in the fundamental modes of the buildings. It may be observed from this figure that these damping ratios are <2% in

TABLE 10.10
Measured First-Mode Damping Ratios in Forced-Vibration Tests of Buildings

Building	Direction	Frequency (Hz)	Damping Ratio (%)	Building	Direction	Frequency (Hz)	Damping Ratio (%)
1	1	1.31	2.40	5	1	0.97	1.15
	2	1.33	2.75		2	1.16	1.44
2	1	1.45	2.60	6	1	0.67	1.18
	2	2.21	3.24		2	0.85	1.43
3	1	1.46	2.77	7	1	0.84	1.39
	2	1.73	1.49		2	0.90	1.96
4	1	1.28	2.31	8	1	1.39	2.30
	2	1.49	2.74		2	1.65	3.40

Source: Jeary, A. P., *Earthquake Engineering and Structural Dynamics*, 14, 733–750, 1986.

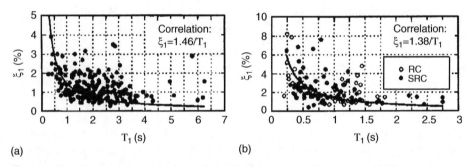

FIGURE 10.8 Measured damping ratios (ξ_1) versus natural period (T_1) in the first modes of (a) steel (S) and (b) reinforced concrete (RC) and composite steel-reinforced concrete (SRC) high-rise buildings. (Reproduced with permission of ASCE from Satake, N. et al., *Journal of Structural Engineering, ASCE*, 129, 470–477, 2003.)

TABLE 10.11
Average Relationships between Measured Damping Ratios in Adjacent Modes

Building Type	ξ_2/ξ_1	ξ_3/ξ_2
Steel	1.35	1.31
Reinforced concrete	1.37	1.42
Composite	1.84	1.69

Source: Satake, N. et al., *Journal of Structural Engineering, ASCE*, 129, 470–477, 2003.

many of the steel-framed buildings and >2% in many of the reinforced concrete and composite buildings. Furthermore, it may be seen that that they are even >5% in some of the shorter reinforced concrete and composite buildings. The study also reports the average relationships between the damping ratios in the second and first modes (ξ_2/ξ_1) and the damping ratios in the third and second modes (ξ_3/ξ_2). These average relationships are given in Table 10.11.

Average values for different types of structures obtained from the compilation of available experimental data have also been reported in the literature. In 1976, R. Haviland analyzed data from reinforced concrete, steel, and composite buildings and computed the mean values and coefficients of variation of the reported damping ratios. For this purpose, he classified the data into two groups: small- and large-amplitude excitations. He found that for reinforced concrete buildings the mean value and coefficient of variation of the measured damping ratios were, respectively, 4.26% and 0.76 for small-amplitude excitations, and 6.63% and 0.64 for large-amplitude excitations. For steel buildings, the corresponding values were 1.68% and 0.65 for low-amplitude excitations, and 5.65% and 0.45 for large-amplitude excitations. A similar study was conducted in 1965 by T. Katayama to evaluate the damping characteristics of different types of bridges. His findings are presented in Table 10.12. He also obtained the mean damping ratios listed in Table 10.13 for single-span, simply supported bridges. Table 10.12 shows that, on average, bridges exhibit considerably lower damping ratios than buildings. In 1973, N. M. Newmark and W. J. Hall also compiled information on damping ratios from experimental tests of structures and classified this information according to type of structure and stress level. Based on their study, they recommended using for design purposes the values listed in Table 10.14. Observe that according to Newmark and Hall recommendations the degree of damping in structures depends significantly on the type of structure, connection method, and stress level.

It may be seen that the results from the investigations presented earlier are in agreement, generally speaking, with the range of damping values recommended by Newmark and Hall. The damping ratios in Table 10.14 may be thus used as a guide to select the modal damping ratios in the fundamental modes of structures. The higher values may be selected for ordinary structures, and the lower values for special structures that merit conservative designs. As structures are designed to undergo high stresses when excited by a strong earthquake, normally the values recommended for

TABLE 10.12
Average Damping Ratios Measured in Bridges

Bridge Type	Damping Ratio (%)
Simple span	1.85
Arch, cantilever, and continuous	1.11
Suspension	0.90

Source: Katayama, T., A Review of Theoretical and Experimental Investigations of Damping in Structures, UNICIV Report No. 1–4, University of New South Wales, Sydney, Australia, 1965.

TABLE 10.13
Average Damping Ratios Measured in Single-Span Simply Supported Bridges

Structural Material	Bridge Type	Damping Ratio (%)
Steel	Composite girder	1.9
	Noncomposite girder	2.1
	Plate girder	1.2
	Rolled-beam girder	2.8
Concrete	Prestressed concrete	1.3
	Reinforced concrete	2.4

Source: Katayama, T., A Review of Theoretical and Experimental Investigations of Damping in Structures, UNICIV Report No. 1–4, University of New South Wales, Sydney, Australia, 1965.

TABLE 10.14
Recommended Damping Ratios for Different Types of Structures and Stress Levels

Stress Level	Type and Condition of Structure	Damping Ratio (%)
Working stress, no more than ~1/2 of yield point	Welded steel, prestressed concrete, well-reinforced concrete (only slight cracking)	2–3
	Reinforced concrete with considerable cracking	3–5
	Bolted and riveted steel, wood structures with nailed or bolted joints	5–7
At or just below yield point	Welded steel, prestressed concrete (without complete loss in prestress)	5–7
	Prestressed concrete with no prestress left	7–10
	Reinforced concrete	7–10
	Bolted and riveted steel, wood structures with bolted joints	10–15
	Wood structures with nailed joints	15–20

Source: Newmark, N. M. and Hall, W. J., *Earthquake Spectra and Design*, Earthquake Engineering Research Institute, Oakland, CA, 1982.

high stress levels should be adopted in design applications. Also, because no definitive correlation seems to exist between the damping ratios in the different modes of a structure, it is recommended to use formulations that yield approximately the same damping ratios in all modes, or slightly higher damping ratios for the higher modes.

10.6 SUMMARY OF RESPONSE SPECTRUM METHOD

The concepts presented in the foregoing sections may now be put all together to establish a step-by-step procedure to estimate the maximum response of a structure when the base of the structure is excited by an earthquake ground motion defined in terms of an acceleration response spectrum. Such a procedure is presented in Box 10.1 and its application is illustrated by means of the following example.

Box 10.1 Response Spectrum Method—Summary

1. Generate the structure's mass matrix $[M]$ and stiffness matrix $[K]$.
2. Solve the associated eigenvalue problem to determine the structure's natural frequencies ω_r and mode shapes $\{\phi\}_r$.
3. Select the damping ratios for some of the lower modes of the structure according to structural type.
4. Determine the damping ratios for the higher modes of the structure according to a selected damping model. If damping in the system is assumed of the Rayleigh type, calculate these damping ratios using

$$\xi_r = \frac{1}{2}\left(\frac{\alpha}{\omega_r} + \beta\omega_r\right)$$

where α and β are defined by Equation 10.93 or 10.96. If damping in the system is assumed of the Caughey type, then compute them according to

$$\xi_r = \frac{1}{2\omega_r}\sum_{i=0}^{n-1} a_i \omega_r^{2i}$$

Structural Response by Response Spectrum Method

where the constants a_i are determined by setting and solving a system of simultaneous equations using this equation and the known or specified damping ratios.

5. For each of the modes, read the ordinates $SA(\omega_r, \xi_r)$ that correspond to the natural frequencies and damping ratios of the structure from the acceleration response spectrum of the excitation.
6. Calculate the structure's modal participation factors according to

$$\Gamma_r = \frac{\{\phi\}_r^T [M]\{J\}}{\{\phi\}_r^T [M]\{\phi\}_r}$$

7. For each mode, determine the structure's maximum modal displacements using the following equation:

$$\max\{u_r(t)\} = \frac{\Gamma_r}{\omega_r^2} \{\phi\}_r SA(\omega_r, \xi_r)$$

8. Estimate the structure's maximum displacement response using the double-sum modal combination rule

$$\max\{u(t)\} = \left\{ \sqrt{\sum_{m=1}^{N}\sum_{n=1}^{N} \rho_{mn} \max[u(t)]_m \max[u(t)]_n} \right\}$$

where the modal correlation coefficients ρ_{mn} may be determined according to Equation 10.71, 10.72, or 10.74, or may be assumed equal to unity for $m = n$ and zero for $m \neq n$ if the structure has neither closely spaced natural frequencies nor significant higher modes.

9. Estimate similarly other desired measures of response by determining first the maximum modal responses according to Equations 10.55 through 10.57 and combining then these modal responses using the double-sum modal combination rule.

EXAMPLE 10.3 MAXIMUM RESPONSES OF A THREE-STORY SHEAR BUILDING BY THE RESPONSE SPECTRUM METHOD

Determine using the response spectrum method the maximum displacements, lateral forces, shearing forces, and overturning moments of the three-story shear building shown in Figure E10.3a when the base of the building is excited by the N–S component of the ground accelerations recorded during the 1940 El Centro earthquake. The response spectrum corresponding to this acceleration record is shown in Figure 6.5. On the basis of the calculated story shears and overturning moments, determine also the maximum values of the shearing forces, bending moments, and axial forces acting on the columns of the building. Assume a damping ratio of 2% for the first mode of the building and that the damping matrix of the structure is of the Rayleigh type.

Solution

Mass and Stiffness Matrices

Following the procedure established in Box 10.1, the first step in the solution to this problem is the formulation of the mass and stiffness matrices of the system. Accordingly, for the shear building in Figure E10.3a these matrices are

$$[M] = \begin{bmatrix} 2.0 & 0 & 0 \\ 0 & 1.5 & 0 \\ 0 & 0 & 1.0 \end{bmatrix} \quad [K] = \begin{bmatrix} 300 & -120 & 0 \\ -120 & 180 & -60 \\ 0 & -60 & 60 \end{bmatrix}$$

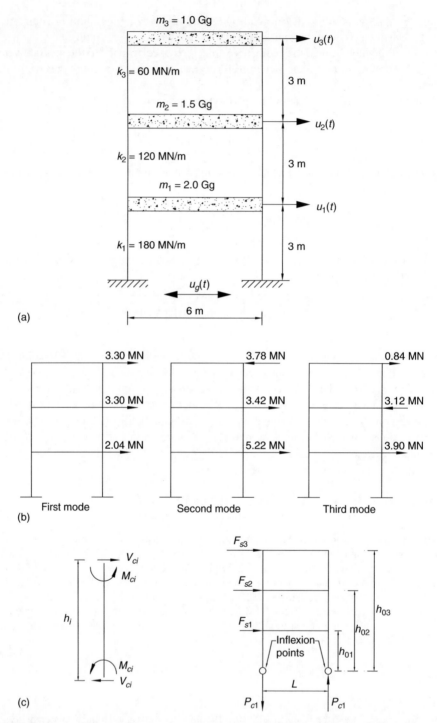

FIGURE E10.3 (a) Shear building considered in Example 10.3, (b) equivalent lateral forces in each mode of vibration and (c) free-body diagrams of ith-story column and structure above inflection points of first story.

TABLE E10.3
Dynamic Properties of the Three-Story Shear Building in Example 10.3

Mode	1	2	3
Frequency (rad/s)	4.592	9.818	14.578
Period (s)	1.37	0.64	0.43
Mode shape	1.000	1.000	1.000
	2.148	0.894	−1.042
	3.313	−1.472	0.410

Natural Frequencies and Mode Shapes

Once the mass and stiffness matrices are established, one can then proceed to solve the associated eigenvalue problem to obtain the natural frequencies and mode shapes of the system. Thus, from the solution of this eigenvalue problem one obtains the natural frequencies and mode shapes listed in Table E10.3.

Modal Damping Ratios

Next, the damping ratios for the second and third modes of the system will be determined by assuming that damping in the system is of the Rayleigh type and the damping ratio in its first mode is the lowest one. According to these assumptions and in view of Equations 10.92 and 10.96, one has that $\alpha = \xi_1 \omega_1$, $\beta = \xi_1/\omega_1$, and

$$\xi_r = \frac{1}{2}\left(\frac{\alpha}{\omega_r} + \beta\omega_r\right) = \frac{1}{2}\left(\frac{\omega_1}{\omega_r} + \frac{\omega_r}{\omega_1}\right)\xi_1$$

Consequently, the damping ratios for the second and third modes result as

$$\xi_2 = \frac{1}{2}\left(\frac{\omega_1}{\omega_2} + \frac{\omega_2}{\omega_1}\right)\xi_1 = \frac{1}{2}\left(\frac{4.592}{9.818} + \frac{9.818}{4.592}\right)0.02 = 0.026$$

$$\xi_3 = \frac{1}{2}\left(\frac{\omega_1}{\omega_3} + \frac{\omega_3}{\omega_1}\right)\xi_1 = \frac{1}{2}\left(\frac{4.592}{14.578} + \frac{14.578}{4.592}\right)0.02 = 0.035$$

Modal Participation Factors

The modal participation factors are given by Equation 10.32. However, as for shear buildings the influence vector $\{J\}$ is equal to the unit vector, the modal participation factors for the building under consideration may be calculated according to

$$\Gamma_r = \frac{\{\phi\}_r^T[M]\{J\}}{\{\phi\}_r^T[M]\{\phi\}} = \frac{\sum_{i=1}^{N} m_i \phi_{ri}}{\sum_{i=1}^{N} m_i \phi_{ri}^2}$$

where m_i is the mass lumped at the ith floor of the building and ϕ_{ri} the amplitude corresponding to the ith floor in the rth mode shape. Hence, the desired modal participation factors are given by

$$\Gamma_1 = \frac{2.0(1.0) + 1.5(2.148) + 1.0(3.313)}{2.0(1.0)^2 + 1.5(2.148)^2 + 1.0(3.313)^2} = 0.429$$

$$\Gamma_2 = \frac{2.0(1.0) + 1.5(0.894) + 1.0(-1.472)}{2.0(1.0)^2 + 1.5(0.894)^2 + 1.0(-1.472)^2} = 0.348$$

$$\Gamma_3 = \frac{2.0(1.0) + 1.5(-1.042) + 1.0(0.410)}{2.0(1.0)^2 + 1.5(-1.042)^2 + 1.0(0.410)^2} = 0.223$$

It may be noted that in this particular case the modal participation factors add up to 1.0. As discussed in Section 10.4, this is so because the three mode shapes of the system have been normalized to have a unit amplitude at the level of the first floor.

Response Spectrum Ordinates

As the natural frequencies and damping ratios are already known for each of the modes of the system, it is possible to define now the response spectrum ordinates corresponding to each mode. Reading these spectral ordinates from the response spectrum in Figure 6.5, one thus gets

$$SA(4.592, 0.02) = 0.24g = 2.35 \text{ m/s}^2$$

$$SA(9.818, 0.026) = 0.75g = 7.36 \text{ m/s}^2$$

$$SA(14.578, 0.035) = 0.90g = 8.38 \text{ m/s}^2$$

Maximum Modal Displacement Responses

The maximum modal displacement responses are given by Equation 10.54. Therefore, after substituting into this equation the natural frequencies, mode shapes, modal participation factors, and spectral ordinates determined earlier, one arrives to the following maximum modal displacement responses:

$$\max\{u(t)\}_1 = \frac{\Gamma_1}{\omega_1^2}\{\phi\}_1 SA(\omega_1, \xi_1) = \frac{0.429}{(4.592)^2}\begin{Bmatrix} 1.0 \\ 2.148 \\ 3.313 \end{Bmatrix} 2.35 = \begin{Bmatrix} 0.048 \\ 0.103 \\ 0.158 \end{Bmatrix} \text{m}$$

$$\max\{u(t)\}_2 = \frac{\Gamma_2}{\omega_2^2}\{\phi\}_2 SA(\omega_2, \xi_2) = \frac{0.348}{(9.818)^2}\begin{Bmatrix} 1.0 \\ 0.894 \\ -1.472 \end{Bmatrix} 7.36 = \begin{Bmatrix} 0.027 \\ 0.024 \\ -0.039 \end{Bmatrix} \text{m}$$

$$\max\{u(t)\}_3 = \frac{\Gamma_3}{\omega_3^2}\{\phi\}_3 SA(\omega_3, \xi_3) = \frac{0.223}{(14.578)^2}\begin{Bmatrix} 1.0 \\ -1.042 \\ 0.410 \end{Bmatrix} 8.83 = \begin{Bmatrix} 0.009 \\ -0.010 \\ 0.004 \end{Bmatrix} \text{m}$$

Maximum Displacement Responses

These modal responses may now be combined to get an estimate of the maximum displacement response. As the structure under consideration is a flexible one and has natural frequencies that are well separated from one another, the SRSS rule may be used to carry out this combination with adequate accuracy. Proceeding accordingly, one finds that the maximum displacements in the system are approximately equal to

$$\max\{u(t)\} = \begin{Bmatrix} \sqrt{0.048^2 + 0.027^2 + 0.009^2} \\ \sqrt{0.103^2 + 0.024^2 + 0.010^2} \\ \sqrt{0.158^2 + 0.039^2 + 0.004^2} \end{Bmatrix} = \begin{Bmatrix} 0.056 \\ 0.106 \\ 0.163 \end{Bmatrix} \text{m}$$

Structural Response by Response Spectrum Method

Maximum Lateral Forces

From the definition of stiffness matrix, the external forces applied to a structure are equal to the product of the stiffness matrix and the vector that contains the displacements induced by the external forces. Hence, for a shear building an equivalent set of lateral external forces that generate the same displacements and stresses as an earthquake ground motion is given by the product of its stiffness matrix and the displacements induced by the ground motion. Recall, however, that to obtain an estimate of the maximum value of any response derived from the displacement response, it is necessary to determine first the maximum values of the desired response in each mode and then combine these modal responses according to an established rule. Thus, for the shear building under consideration, the lateral forces in each of the three modes of the system are given by

$$\max\{F_s(t)\}_1 = [K]\max\{u(t)\}_1 = \begin{bmatrix} 300 & 120 & 0 \\ -120 & 180 & -60 \\ 0 & -60 & 60 \end{bmatrix} \begin{Bmatrix} 0.048 \\ 0.103 \\ 0.158 \end{Bmatrix} = \begin{Bmatrix} 2.04 \\ 3.30 \\ 3.30 \end{Bmatrix} \text{MN}$$

$$\max\{F_s(t)\}_2 = [K]\max\{u(t)\}_2 = \begin{bmatrix} 300 & 120 & 0 \\ -120 & 180 & -60 \\ 0 & -60 & 60 \end{bmatrix} \begin{Bmatrix} 0.027 \\ 0.024 \\ -0.039 \end{Bmatrix} = \begin{Bmatrix} 5.22 \\ 3.42 \\ -3.78 \end{Bmatrix} \text{MN}$$

$$\max\{F_s(t)\}_3 = [K]\max\{u(t)\}_3 = \begin{bmatrix} 300 & 120 & 0 \\ -120 & 180 & -60 \\ 0 & -60 & 60 \end{bmatrix} \begin{Bmatrix} 0.009 \\ -0.010 \\ 0.004 \end{Bmatrix} = \begin{Bmatrix} 3.90 \\ -3.12 \\ 0.84 \end{Bmatrix} \text{MN}$$

which after combining them with the SRSS rule lead to the following maximum values of the equivalent lateral forces acting on the system:

$$\max\{F_s(t)\} = \begin{Bmatrix} \sqrt{2.04^2 + 5.22^2 + 3.90^2} \\ \sqrt{3.30^2 + 3.42^2 + 3.12^2} \\ \sqrt{3.30^2 + 3.78^2 + 0.84^2} \end{Bmatrix} = \begin{Bmatrix} 6.83 \\ 5.69 \\ 5.09 \end{Bmatrix} \text{MN}$$

The magnitude and sense of the equivalent lateral forces in each of the modes of the system are depicted in Figure E10.3b.

Maximum Story Shears

On the basis of the computed modal lateral forces and considering the building as a vertical cantilever beam, it is possible to compute now the maximum values of the story shears acting on the building in each of its three modes. Accordingly, these maximum modal story shears result as

$$\max\{V(t)\}_1 = \begin{Bmatrix} 3.30 + 3.30 + 2.04 \\ 3.30 + 3.30 \\ 3.30 \end{Bmatrix} = \begin{Bmatrix} 8.64 \\ 6.60 \\ 3.30 \end{Bmatrix} \text{MN}$$

$$\max\{V(t)\}_2 = \begin{Bmatrix} -3.78 + 3.42 + 5.22 \\ -3.78 + 3.42 \\ -3.78 \end{Bmatrix} = \begin{Bmatrix} 4.86 \\ -0.36 \\ -3.78 \end{Bmatrix} \text{MN}$$

$$\max\{V(t)\}_3 = \begin{Bmatrix} 0.84 - 3.12 + 3.90 \\ 0.84 - 3.12 \\ 0.84 \end{Bmatrix} = \begin{Bmatrix} 1.62 \\ -2.28 \\ 0.84 \end{Bmatrix} \text{MN}$$

whereas, if the SRSS rule is used once again, the corresponding maximum values are equal to

$$\max\{V(t)\} = \begin{Bmatrix} \sqrt{8.64^2 + 4.86^2 + 1.62^2} \\ \sqrt{6.60^2 + 0.36^2 + 2.28^2} \\ \sqrt{3.30^2 + 3.78^2 + 0.84^2} \end{Bmatrix} = \begin{Bmatrix} 10.0 \\ 6.99 \\ 5.09 \end{Bmatrix} \text{MN}$$

Maximum Overturning Moments

The maximum modal overturning moments and the corresponding maximum values may be computed in a similar way. Hence,

$$\max\{M(t)\}_1 = \begin{Bmatrix} 3.30(9) + 3.30(6) + 2.04(3) \\ 3.30(6) + 3.30(3) \\ 3.30(3) \end{Bmatrix} = \begin{Bmatrix} 55.6 \\ 29.7 \\ 9.90 \end{Bmatrix} \text{MN m}$$

$$\max\{M(t)\}_2 = \begin{Bmatrix} -3.78(9) + 3.42(6) + 5.22(3) \\ -3.78(6) + 3.42(3) \\ -3.78(3) \end{Bmatrix} = \begin{Bmatrix} 2.16 \\ -12.4 \\ -11.3 \end{Bmatrix} \text{MN m}$$

$$\max\{M(t)\}_3 = \begin{Bmatrix} 0.84(9) - 3.12(6) + 3.90(3) \\ 0.84(6) - 3.12(3) \\ 0.84(3) \end{Bmatrix} = \begin{Bmatrix} 0.54 \\ -4.32 \\ 2.52 \end{Bmatrix} \text{MN m}$$

and correspondingly,

$$\max\{M(t)\} = \begin{Bmatrix} \sqrt{55.6^2 + 2.16^2 + 0.54^2} \\ \sqrt{29.7^2 + 12.4^2 + 4.32^2} \\ \sqrt{9.90^2 + 11.3^2 + 2.52^2} \end{Bmatrix} = \begin{Bmatrix} 55.6 \\ 32.5 \\ 15.3 \end{Bmatrix} \text{MN}$$

Maximum Shearing Forces, Bending Moments, and Axial Forces in Columns

The maximum values of the shearing forces, bending moments, and axial forces exerted on the columns of the structure when the base of the structure is excited by the earthquake ground motion being considered may now be computed using simple concepts from statics and the story shears and overturning moments determined earlier. To compute the shearing forces, it may be noted that each story of the structure has only two columns, and thus the shearing forces in the columns are simply equal to the corresponding story shear divided by two. Accordingly, these shearing forces are equal to

$$\max\{V_c(t)\} = \begin{Bmatrix} V_{c1} \\ V_{c2} \\ V_{c3} \end{Bmatrix} = \begin{Bmatrix} 10.0/2 \\ 6.99/2 \\ 5.09/2 \end{Bmatrix} = \begin{Bmatrix} 5.00 \\ 3.50 \\ 2.55 \end{Bmatrix} \text{MN}$$

Similarly, from the static equilibrium of the internal forces acting on a simple column (see free-body diagram for one of the columns of the ith story in Figure E10.3c), it is found that the bending moments in the columns are equal to $V_{ci}\, h_i/2$, where h_i represents the column height and, as before, V_{ci} denotes the shear force at the end of the column. Hence, the maximum values of the bending moments in the columns are equal to

$$\max\{M_c(t)\} = \begin{Bmatrix} M_{c1} \\ M_{c2} \\ M_{c3} \end{Bmatrix} = \begin{Bmatrix} 5.00(3.0)/2 \\ 3.50(3.0)/2 \\ 2.55(3.0)/2 \end{Bmatrix} = \begin{Bmatrix} 7.50 \\ 5.25 \\ 3.83 \end{Bmatrix} \text{MN m}$$

Finally, to determine the axial forces in the columns of the building, one can consider the static equilibrium of the moments about the inflection points of the columns of the forces acting above such inflection points (see free-body diagram in Figure E10.3c). In this way, it may be determined, for example, that the axial forces in the columns of the first story are given by

$$P_{c1}(t) = \frac{F_{s1}(t)h_{01} + F_{s2}(t)h_{02} + F_{s3}(t)h_{03}}{L} \psi$$

where all symbols are as defined in Figure E10.3c. As a conservative approximation, however, it may be assumed that the inflection points of the columns in each story are located at the bottom of the story, in which case the axial forces are equal to the overturning moment for the story divided by the horizontal distance between the two columns. On the basis of this approximation, the axial forces in the columns are thus approximately equal to

$$\max\{P_c(t)\} = \begin{Bmatrix} P_{c1} \\ P_{c2} \\ P_{c3} \end{Bmatrix} = \begin{Bmatrix} 55.6/6 \\ 32.5/6 \\ 15.3/6 \end{Bmatrix} = \begin{Bmatrix} 9.28 \\ 5.41 \\ 2.54 \end{Bmatrix} \text{MN}$$

10.7 RESPONSE SPECTRUM METHOD BASED ON MODAL ACCELERATION METHOD

The dynamic analysis of complex structures with a large number of degrees of freedom requires the solution of a large eigenvalue problem and a large number of differential equations. Therefore, any reduction in the size of the eigenvalue problem and the number of equations that need to be solved is a highly desirable goal to increase the computational efficiency of the analysis. In the traditional modal superposition method, this reduction is achieved by considering the response in only a few of the lower modes of the structure. That is, the original modal superposition series is replaced by the series

$$\{u(t)\} = \sum_{r=1}^{P} \{\phi\}_r \eta_r(t) \tag{10.99}$$

where P is a number that is smaller than N, the number of degrees of freedom of the structure. The omission of the higher modes in the modal superposition method is known as the truncation of modes and the error caused by it is often called *the missing mass effect*. The error introduced by the truncation of modes is usually acceptable, but in some cases this error may be significantly large. This can happen, for example, in the calculation of the internal forces of the structure. This is so because, as mentioned earlier, the modal internal forces are proportional to the square of the modal frequency and thus the internal forces in modes with high frequencies may be appreciably large. In these cases, therefore, it is necessary to include a large number of modes in the modal superposition if one wants to attain a solution with a high degree of accuracy.

A method that greatly increases the accuracy of the modal superposition technique when only a few modes are considered is the *modal acceleration method* introduced by D. Williams in 1949. In this method, the modal superposition series is reformulated by solving for the normal coordinate $\eta_r(t)$ from Equation 10.29 to get

$$\eta_r(t) = -\frac{\Gamma_r}{\omega_r^2}\ddot{u}_g(t) - \frac{1}{\omega_r^2}\ddot{\eta}_r(t) - \frac{2\xi_r}{\omega_r}\dot{\eta}_r(t) \tag{10.100}$$

and then substituting this expression into Equation 10.99 to obtain

$$\{u(t)\} = -\sum_{r=1}^{P}\{\phi\}_r\left[\frac{\Gamma_r}{\omega_r^2}\ddot{u}_g(t) + \frac{1}{\omega_r^2}\ddot{\eta}_r(t) + \frac{2\xi_r}{\omega_r}\dot{\eta}_r(t)\right] \tag{10.101}$$

However, when all modes are considered the first term on the right-hand side of this equation is equal to

$$\sum_{r=1}^{N}\{\phi\}_r\frac{\Gamma_r}{\omega_r^2}\ddot{u}_g(t) = \sum_{r=1}^{N}\{\phi\}_r\frac{1}{\omega_r^2}\frac{\{\phi\}_r^T[M]\{J\}}{M_r^*}\ddot{u}_g(t) = \sum_{r=1}^{N}\frac{\{\phi\}_r\{\phi\}_r^T[M]\{J\}}{K_r^*}\ddot{u}_g(t)$$

$$= \left[\frac{1}{K_1^*}\{\phi\}_1^T\{\phi\}_1 + \frac{1}{K_2^*}\{\phi\}_2^T\{\phi\}_2 + \cdots + \frac{1}{K_N^*}\{\phi\}_N^T\{\phi\}_N\right][M]\{J\}\ddot{u}_g(t)$$

$$= [\{\phi\}_1 \quad \{\phi\}_2 \quad \cdots \quad \{\phi\}_N]\begin{bmatrix}1/K_1^* & 0 & \cdots & 0 \\ 0 & 1/K_2^* & 0 & 0 \\ \vdots & 0 & \ddots & 0 \\ 0 & 0 & 0 & 1/K_N^*\end{bmatrix}\begin{Bmatrix}\{\phi\}_1^T \\ \{\phi\}_2^T \\ \vdots \\ \{\phi\}_N^T\end{Bmatrix}[M]\{J\}\ddot{u}_g(t)$$

$$\tag{10.102}$$

where the definition of modal participation factor and the relationship between generalized stiffness and generalized mass given by Equations 10.32 and 10.30, respectively, have been used. Note, though, that the square matrix on the right-hand side of Equation 10.102 represents the inverse of a diagonal matrix whose elements are the generalized stiffnesses K_r^*. Note also that according to Equation 10.21, this matrix may be expressed as the inverse of the triple matrix product $[\Phi]^T[K][\Phi]$. Consequently, Equation 10.102 may also be expressed as

$$\sum_{r=1}^{N}\{\phi\}_r\frac{\Gamma_r}{\omega_r^2}\ddot{u}_g(t) = [\Phi]([\Phi]^T[K][\Phi])^{-1}[\Phi]^T[M]\{J\}\ddot{u}_g(t)$$

$$= [\Phi][\Phi]^{-1}[K]^{-1}([\Phi]^T)^{-1}[\Phi]^T[M]\{J\}\ddot{u}_g(t) \tag{10.103}$$

$$= [K]^{-1}[M]\{J\}\ddot{u}_g(t)$$

It may be seen, thus, that when all modes are considered the first term on the right-hand side of Equation 10.101 is equal to the displacements induced in the structure when the inertial forces are

Structural Response by Response Spectrum Method

considered as static forces. For this reason, this term is considered to represent the *static* component of the solution. It may also be seen that this term is independent of the modal properties of the structure and therefore, although all modes are taken into account, its evaluation does not require the solution of the full eigenvalue problem. Accordingly, the accuracy of a truncated solution may be improved significantly without unduly increasing the computational effort if all modes are considered in the evaluation of such a first term.

Thus, if all modes are considered in the static component of Equation 10.101, this equation becomes

$$\{u(t)\} = -\sum_{r=1}^{N}\{\phi\}_r \frac{\Gamma_r}{\omega_r^2}\ddot{u}_g(t) - \sum_{r=1}^{P}\{\phi\}_r\left[\frac{1}{\omega_r^2}\ddot{\eta}_r(t) + \frac{2\xi_r}{\omega_r}\dot{\eta}_r(t)\right] \quad (10.104)$$

which, by virtue of Equation 10.103, may also be expressed as

$$\{u(t)\} = -[K]^{-1}[M]\{J\}\ddot{u}_g(t) - \sum_{r=1}^{P}\{\phi\}_r\left[\frac{1}{\omega_r^2}\ddot{\eta}_r(t) + \frac{2\xi_r}{\omega_r}\dot{\eta}_r(t)\right] \quad (10.105)$$

Equation 10.105 represents the classical modal acceleration method. The name *modal acceleration* comes from the fact that, in the undamped case, the solution is expressed in terms of the modal accelerations alone as opposed to the modal displacements. The advantage of modal acceleration method over the conventional modal superposition method is that with the modal acceleration method the summation may be carried out over only a few of the lower modes of the structure without introducing a significant error in the calculated response. In other words, the truncation of the higher modes has a much smaller effect on the accuracy of the calculated response when the modal acceleration method is used than when the conventional modal superposition method is used. Of course, when all modes are considered both approaches lead exactly to the same results.

An alternative form of the modal acceleration method may be derived as follows. From Equation 10.29, one has that

$$\frac{1}{\omega_r^2}\ddot{\eta}_r(t) + \frac{2\xi_r}{\omega_r}\dot{\eta}_r(t) = -\eta_r(t) - \frac{\Gamma_r}{\omega_r^2}\ddot{u}_g(t) \quad (10.106)$$

and thus, after substitution of this expression into Equation 10.105, one obtains

$$\{u(t)\} = -[K]^{-1}[M]\{J\}\ddot{u}_g(t) + \sum_{r=1}^{P}\{\phi\}_r\left[\eta_r(t) + \frac{\Gamma_r}{\omega_r^2}\ddot{u}_g(t)\right] \quad (10.107)$$

which, after rearranging terms, may also be written as

$$\{u(t)\} = \sum_{r=1}^{P}\{\phi\}_r\eta_r(t) - \left([K]^{-1}[M]\{J\} - \sum_{r=1}^{P}\frac{\Gamma_r}{\omega_r^2}\{\phi\}_r\right)\ddot{u}_g(t) \quad (10.108)$$

In Equation 10.108, the first term on the right-hand side represents the conventional modal superposition method when only the first P modes are considered. The additional term represents the so-called static correction that approximately accounts for the truncated modes. Note also that according to Equation 10.103, when $P = N$, the second term in the expression enclosed in parentheses is equal to the first one, and thus the equation is reduced to the conventional mode superposition method. The advantage of this alternative form of the modal acceleration method is that with it no explicit calculation of the modal velocities and accelerations is required. Besides the series in it is exactly of the same form as the series in the conventional modal superposition method. This facilitates the extension of existing computational procedures for the implementation of the modal acceleration method.

The response spectrum method may also be formulated on the basis of the modal acceleration method. To this end, it is necessary to develop a modal combination rule that is based on this method. Singh and Maldonado derived such a rule by applying the same technique that was used to obtain the combination rule defined by Equations 10.66 and 10.74, that is, by computing the root mean square of the response $\{u(t)\}$ using Equation 10.108 and introducing the assumptions indicated by Equations 10.60 and 10.61. In this way, they arrived at an expression of the form

$$\max\{u(t)\} = \left\{\left[\sum_{m=1}^{N}\sum_{n=1}^{N}\rho_{mn}\max[u(t)]_m\max[u(t)]_n + U_s^2[\max \ddot{u}_g(t)]^2 - 2U_sV_P\right]^{1/2}\right\} \quad (10.109)$$

where U_s and V_P represent, respectively, the elements of the vectors

$$\{U_s\} = [K]^{-1}[M]\{J\} - \sum_{r=1}^{P}\frac{\Gamma_r}{\omega_r^2}\{\phi\}_r \qquad \{V_P\} = \sum_{r=1}^{P}\Gamma_r\{\phi\}_r(SV_r^2 - \omega_r^2SD_r^2) \quad (10.110)$$

and max $\ddot{u}_g(t)$ denotes the maximum value of the ground acceleration. As before, SD_r and SV_r, respectively, represent the ordinates in the displacement and velocity response spectra of the excitation corresponding to the natural frequency and damping ratio of the system in its rth mode of vibration, ρ_{mn} is the modal correlation coefficient defined by Equation 10.74, and all other symbols are as previously introduced.

Equation 10.109 has been developed specifically for the maximum displacement response. Similar relationships may be obtained for other types of response by deriving first the desired response from Equation 10.108 and then applying the same technique and assumptions used to obtain Equations 10.66 and 10.109 to the resulting expression.

EXAMPLE 10.4 DISPLACEMENT RESPONSE OF A FOUR-STORY SHEAR BUILDING BY MODAL ACCELERATION METHOD

The four-story shear building shown in Figure E10.4a has the natural frequencies, mode shapes, and participation factors listed in Table E10.4. Determine the building's displacement response using the conventional response spectrum method when the building is subjected to a ground motion represented by the acceleration response specturm shown in Figure E10.4b. Determine also the building's displacement response using the modal acceleration method considering only the first mode. Assume a damping ratio of 5% for all modes.

Solution

Conventional Response Spectrum Method

From the given response spectrum, the spectral displacements corresponding to the natural frequencies and damping ratios of the system are

$$SD(13.294, 0.05) = 2.3 \text{ in.} \qquad SD(29.660, 0.05) = 0.45 \text{ in.}$$

$$SD(41.070, 0.05) = 0.18 \text{ in.} \qquad SD(55.882, 0.05) = 0.08 \text{ in.}$$

Structural Response by Response Spectrum Method

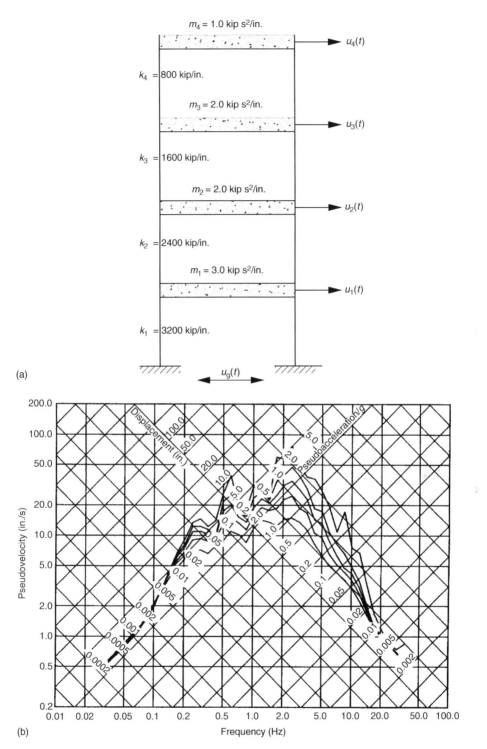

FIGURE E10.4 (a) Four-story shear building considered in Example 10.4 and (b) response spectra for 0, 2, 5, 10, and 20% damping of the N–S component of the ground acceleration recorded during the 1973 Managua, Nicaragua, earthquake. (After Riddell, R. and Newmark, N.M., Statistical Analysis of the Response of Nonlinear Systems Subjected to Earthquakes, Structural Research Series No. 468, Department of Civil Engineering, University of Illinois, Urbana, IL, 1979.)

TABLE E10.4
Dynamic Properties of Building in Example 10.4

Mode	1	2	3	4
Frequency (rad/s)	13.294	29.660	41.079	55.882
Participation factor	2.511	1.079	0.643	0.342
Mode shape	0.1387	0.2966	0.3388	0.3337
	0.2930	0.3659	0.0759	−0.5240
	0.4597	0.0675	−0.4786	0.2348
	0.5900	−0.6777	0.4314	−0.0809

Therefore, in accordance with Equation 10.53 and the given mode shapes and participation factors, the maximum displacement response in each of the modes of the system is given by

$$\max\{u(t)\}_1 = \Gamma_1\{\phi\}_1 SD(\omega_1,\xi_1) = 2.511 \begin{Bmatrix} 0.1387 \\ 0.2930 \\ 0.4597 \\ 0.5900 \end{Bmatrix} 2.3 = \begin{Bmatrix} 0.801 \\ 1.692 \\ 2.655 \\ 3.407 \end{Bmatrix} \text{in.}$$

$$\max\{u(t)\}_2 = \Gamma_2\{\phi\}_2 SD(\omega_2,\xi_2) = 1.079 \begin{Bmatrix} 0.2966 \\ 0.3659 \\ 0.0675 \\ -0.6777 \end{Bmatrix} 0.45 = \begin{Bmatrix} 0.144 \\ 0.178 \\ 0.033 \\ -0.329 \end{Bmatrix} \text{in.}$$

$$\max\{u(t)\}_3 = \Gamma_3\{\phi\}_3 SD(\omega_3,\xi_3) = 0.643 \begin{Bmatrix} 0.3388 \\ 0.0759 \\ -0.4786 \\ 0.4314 \end{Bmatrix} 0.18 = \begin{Bmatrix} 0.039 \\ 0.009 \\ -0.055 \\ 0.050 \end{Bmatrix} \text{in.}$$

$$\max\{u(t)\}_4 = \Gamma_4\{\phi\}_4 SD(\omega_4,\xi_4) = 0.342 \begin{Bmatrix} 0.3337 \\ -0.5240 \\ 0.2348 \\ -0.0809 \end{Bmatrix} 0.08 = \begin{Bmatrix} 0.009 \\ -0.014 \\ 0.006 \\ -0.002 \end{Bmatrix} \text{in.}$$

Similarly, if for simplicity the modal correlation coefficients are calculated using Der Kuireghian's formula, then according to Equation 10.73 and the given natural frequencies and damping ratios, the system's modal correlation coefficients, arranged in matrix form, are given by

$$[\rho_{mn}] = \begin{bmatrix} 1.0 & 0.13 & 0.006 & 0.003 \\ 0.13 & 1.0 & 0.084 & 0.022 \\ 0.006 & 0.084 & 1.0 & 0.094 \\ 0.003 & 0.022 & 0.094 & 1.0 \end{bmatrix}$$

Hence, if the double-sum modal combination rule is used (Equation 10.66), the building's maximum displacements, squared, are

$$\max\{u^2(t)\} = \left\{\sum_{m=1}^{4}\sum_{n=1}^{4}\rho_{mn}\max[u(t)]_m\max[u(t)]_n\right\}$$

$$= \{(\max[u(t)]_1)^2\} + \{(\max[u(t)]_2)^2\} + \{(\max[u(t)]_3)^2\} + \{(\max[u(t)]_4)^2\}$$

$$+ 2\rho_{12}\{\max[u(t)]_1\max[u(t)]_2\} + 2\rho_{13}\{\max[u(t)]_1\max[u(t)]_3\} + 2\rho_{14}\{\max[u(t)]_1\max[u(t)]_4\}$$

$$+ 2\rho_{23}\{\max[u(t)]_2\max[u(t)]_3\} + 2\rho_{24}\{\max[u(t)]_2\max[u(t)]_4\} + 2\rho_{34}\{\max[u(t)]_3\max[u(t)]_4\}$$

$$= \begin{Bmatrix} 0.801^2 \\ 1.692^2 \\ 2.655^2 \\ 3.407^2 \end{Bmatrix} + \begin{Bmatrix} 0.144^2 \\ 0.178^2 \\ 0.033^2 \\ 0.329^2 \end{Bmatrix} + \begin{Bmatrix} 0.039^2 \\ 0.009^2 \\ 0.055^2 \\ 0.050^2 \end{Bmatrix} + \begin{Bmatrix} 0.009^2 \\ 0.014^2 \\ 0.006^2 \\ 0.002^2 \end{Bmatrix} + 2(0.013)\begin{Bmatrix} (0.801)(0.144) \\ (1.692)(0.178) \\ (2.655)(0.033) \\ (3.407)(-0.329) \end{Bmatrix}$$

$$+ 2(0.006)\begin{Bmatrix} (0.801)(0.039) \\ (1.692)(0.009) \\ (2.655)(-0.055) \\ (3.407)(0.050) \end{Bmatrix} + 2(0.003)\begin{Bmatrix} (0.801)(0.009) \\ (1.692)(-0.014) \\ (2.655)(0.006) \\ (3.407)(-0.002) \end{Bmatrix} + 2(0.084)\begin{Bmatrix} (0.144)(0.039) \\ (0.178)(0.009) \\ (0.033)(-0.055) \\ (-0.329)(0.050) \end{Bmatrix}$$

$$+ 2(0.022)\begin{Bmatrix} (0.144)(0.009) \\ (0.178)(-0.014) \\ (0.033)(0.006) \\ (-0.329)(-0.002) \end{Bmatrix} + 2(0.094)\begin{Bmatrix} (0.039)(0.009) \\ (0.009)(-0.014) \\ (-0.055)(0.006) \\ (0.050)(-0.002) \end{Bmatrix} = \begin{Bmatrix} 0.668 \\ 2.903 \\ 7.053 \\ 11.688 \end{Bmatrix}$$

and thus

$$\max\{u(t)\} = \max\begin{Bmatrix} u_1(t) \\ u_2(t) \\ u_3(t) \\ u_4(t) \end{Bmatrix} = \begin{Bmatrix} 0.818 \\ 1.704 \\ 2.656 \\ 3.419 \end{Bmatrix} \text{ in.}$$

Modal Acceleration Method Considering Only the First Mode

If only the first mode is considered in the application of the modal acceleration method, then according to Equation 10.109 the maximum displacement response of the system is approximately given by

$$\max\{u(t)\} = \left\{\left[\sum_{m=1}^{1}\sum_{n=1}^{1}\rho_{mn}\max[u(t)]_m\max[u(t)]_n + U_s^2[\max\ddot{u}_g(t)]^2 - 2U_sV_P\right]^{1/2}\right\}$$

where U_s and V_P are the elements of the vectors

$$\{U_s\} = [K]^{-1}[M]\{J\} - \sum_{r=1}^{1}\frac{\Gamma_r}{\omega_r^2}\{\phi\}_r \qquad \{V_P\} = \sum_{r=1}^{1}\Gamma_r\{\phi\}_r(SV_r^2 - \omega_r^2 SD_r^2)$$

But for the system under consideration, one has that

$$\left\{\sum_{m=1}^{1}\sum_{n=1}^{1} p_{mn} \max[u(t)]_m \max[u(t)]_n\right\} = \{(\max[u(t)]_1)^2\} = \begin{Bmatrix} 0.801^2 \\ 1.692^2 \\ 2.655^2 \\ 3.407^2 \end{Bmatrix} = \begin{Bmatrix} 0.642 \\ 2.863 \\ 7.049 \\ 11.608 \end{Bmatrix}$$

$$[K] = \begin{bmatrix} 5600 & -2400 & 0 & 0 \\ -2400 & 4000 & -1600 & 0 \\ 0 & -1600 & 2400 & -800 \\ 0 & 0 & -800 & 800 \end{bmatrix} \quad [M] = \begin{bmatrix} 3.0 & 0 & 0 & 0 \\ 0 & 2.0 & 0 & 0 \\ 0 & 0 & 2.0 & 0 \\ 0 & 0 & 0 & 1.0 \end{bmatrix}$$

$$[K]^{-1} = \frac{1}{9600} \begin{bmatrix} 3 & 3 & 3 & 3 \\ 3 & 7 & 7 & 7 \\ 3 & 7 & 13 & 13 \\ 3 & 7 & 13 & 15 \end{bmatrix}$$

$$[K]^{-1}[M]\{J\} = \frac{1}{9600} \begin{bmatrix} 3 & 3 & 3 & 3 \\ 3 & 7 & 7 & 7 \\ 3 & 7 & 13 & 13 \\ 3 & 7 & 13 & 15 \end{bmatrix} \begin{bmatrix} 3.0 & 0 & 0 & 0 \\ 0 & 2.0 & 0 & 0 \\ 0 & 0 & 2.0 & 0 \\ 0 & 0 & 0 & 1.0 \end{bmatrix} \begin{Bmatrix} 1 \\ 1 \\ 1 \\ 1 \end{Bmatrix} = \begin{Bmatrix} 0.0025 \\ 0.0046 \\ 0.00646 \\ 0.0077 \end{Bmatrix}$$

$$\sum_{r=1}^{1} \frac{\Gamma_r}{\omega_r^2}\{\phi\}_r = \frac{2.511}{13.294^2} \begin{Bmatrix} 0.1387 \\ 0.2930 \\ 0.4597 \\ 0.5900 \end{Bmatrix} = \begin{Bmatrix} 0.0020 \\ 0.0042 \\ 0.00653 \\ 0.0084 \end{Bmatrix}$$

$$\{U_s\} = [K]^{-1}[M]\{J\} - \sum_{r=1}^{1} \frac{\Gamma_r}{\omega_r^2}\{\phi\}_r = \begin{Bmatrix} 0.0025 \\ 0.0046 \\ 0.00646 \\ 0.0077 \end{Bmatrix} - \begin{Bmatrix} 0.0020 \\ 0.0042 \\ 0.00653 \\ 0.0084 \end{Bmatrix} = \begin{Bmatrix} 0.0005 \\ 0.0004 \\ -0.00007 \\ -0.0007 \end{Bmatrix}$$

From the high-frequency region of the given response spectrum, the relationship between pseudovelocity and spectral displacement, and Equations 10.79 through 10.81, one also has that

$$\max \ddot{u}_g(t) = 0.32g = 0.32(386.4) = 123.6 \text{ in./s}^2$$

$$PSV_1 = \omega_1 SD_1 = 13.294(2.3) = 30.58 \text{ in./s}$$

$$a_v = 1.095 + 0.647\xi - 0.382\xi^2 = 1.095 + 0.647(0.05) - 0.382(0.05)^2 = 1.126$$

$$b_v = 0.193 + 0.838\xi - 0.621\xi^2 = 0.193 + 0.838(0.05) - 0.621(0.05)^2 = 0.233$$

$$SV_1 = a_v T_1^{b_v} PSV_1 = 1.126 \left(\frac{2\pi}{13.294}\right)^{0.233} (30.58) = 0.946(30.58) = 28.92 \text{ in./s}$$

and thus,

$$\{V_p\} = \sum_{r=1}^{1}\Gamma_r\{\phi\}_r(SV_r^2 - \omega_r^2 SD_r^2) = 2.511\begin{Bmatrix} 0.1387 \\ 0.2930 \\ 0.4597 \\ 0.5900 \end{Bmatrix}(28.92^2 - 30.58^2) = -\begin{Bmatrix} 34.40 \\ 72.67 \\ 114.01 \\ 146.33 \end{Bmatrix}$$

Consequently,

$$\max\{u^2(t)\} = \begin{Bmatrix} 0.642 \\ 2.863 \\ 7.049 \\ 11.608 \end{Bmatrix} + \begin{Bmatrix} 0.0005^2 \\ 0.0004^2 \\ 0.00007^2 \\ 0.0007^2 \end{Bmatrix}(123.6)^2 + 2\begin{Bmatrix} (0.0005)(34.40) \\ (0.0004)(72.67) \\ (-0.00007)(114.01) \\ (-0.0007)(146.33) \end{Bmatrix} = \begin{Bmatrix} 0.680 \\ 2.924 \\ 7.033 \\ 11.411 \end{Bmatrix}$$

from which one obtains

$$\max\{u(t)\} = \begin{Bmatrix} 0.825 \\ 1.710 \\ 2.652 \\ 3.378 \end{Bmatrix} \text{in.}$$

Note that despite the fact that only one mode was considered, this solution is a close approximation to the displacement response found earlier using the conventional response spectrum method and considering all four modes.

10.8 RESPONSE SPECTRUM METHOD FOR MULTICOMPONENT GROUND MOTIONS

10.8.1 INTRODUCTION

In the derivation of the modal superposition and response spectrum methods presented in Sections 10.2 and 10.3, only one component of ground motion was considered. In general, however, the ground motion at a site has six components, three translational and three rotational. In general, therefore, all structures will be affected by the simultaneous effect of these six components of ground motion. If a structure is perfectly symmetric, and if the rotational components are negligibly small, then the ground motion in one direction will not produce stresses and deformations in the other two directions. In such a case, it is adequate to perform a separate analysis along each of the three axes of symmetry of the structure under a single-component ground motion. That is, the analysis of the structure may be performed using the methods introduced in Sections 10.2 and 10.3 considering one direction at a time. For irregular structures, however, such an approach is inadequate as in these structures the lack of one or more axes of symmetry makes the ground motion in one direction induce stresses and deformations in the other two directions. Examples of such irregular structures are piping systems, curved bridges, dams, and irregular buildings with no axes of symmetry. Irregular structures require, thus, a three-dimensional analysis under the simultaneous effect of a multicomponent ground motion.

In what follows, the concepts introduced in the foregoing sections are reformulated to extend the application of the modal superposition and response spectrum methods to three-dimensional structures under the simultaneous effect of three translational ground motion components. Although it is possible to include all six components in this reformulation, for the sake of clarity only the translational components are considered. The extension to all six components is a relatively straightforward task once it is understood how three ground motion components can be incorporated in the analysis.

10.8.2 EQUATION OF MOTION

Consider an irregular, flexible multi-degree-of-freedom structure such as the three-dimensional frame shown in Figure 10.9. Consider, in addition, that the nodes of the structure are free to displace horizontally and vertically and rotate about the three orthogonal axes. That is, consider that the structure has six degrees of freedom per node, three translational and three rotational. Assume that the mass of the structure is lumped at its nodes and that, as a result, the deformed configuration of the structure is fully defined by the displacements and rotations of its nodes. Assume also that the structure is subjected to three components of ground motion and that these ground motion components are directed along the reference axes of the structure. Assume, further, that the structure is completely characterized by its mass matrix $[M]$, damping matrix $[C]$, and stiffness matrix $[K]$. With reference to Figure 10.9, let N denote the number of degrees of freedom of the structure; $u_{gx}(t)$, $u_{gy}(t)$, and $u_{gz}(t)$ be the components along the direction of the x, y, and z axes, respectively, of the displacement of the ground at time t as measured from the structure's original position; $u_j(t)$ represent the jth nodal displacement or rotation at time t when measured relative to the base of the structure; and $y_j(t)$ depict the jth nodal displacement or rotation at time t measured from the structure's original position (i.e., the jth absolute nodal displacement or rotation). Furthermore, let $\{u(t)\}$ be a vector that contains the nodal relative displacements and rotations $u_j(t)$; that is,

$$\{u(t)\} = \begin{Bmatrix} u_1(t) \\ u_2(t) \\ \vdots \\ u_N(t) \end{Bmatrix} \qquad (10.111)$$

Likewise, let $\{y(t)\}$ be a vector that contains the nodal absolute displacements and rotations $y_j(t)$; that is,

$$\{y(t)\} = \begin{Bmatrix} y_1(t) \\ y_2(t) \\ \vdots \\ y_N(t) \end{Bmatrix} \qquad (10.112)$$

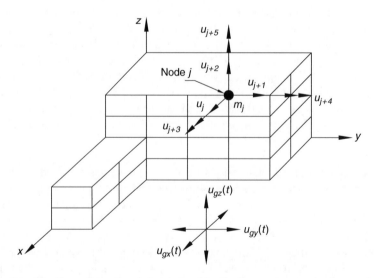

FIGURE 10.9 Irregular three-dimensional structure subject to three components of ground motion.

Structural Response by Response Spectrum Method

As the nodal absolute displacements along any given direction are equal to the relative displacements plus the displacement of the ground along the corresponding direction, consider, in addition, that $\{y(t)\}$ may also be expressed as

$$\{y(t)\} = \begin{Bmatrix} u_1(t) + u_{gx}(t) \\ u_2(t) + u_{gy}(t) \\ u_3(t) + u_{gz}(t) \\ u_4(t) \\ u_5(t) \\ u_6(t) \\ \vdots \\ u_{N-5}(t) + u_{gx}(t) \\ u_{N-4}(t) + u_{gy}(t) \\ u_{N-3}(t) + u_{gz}(t) \\ u_{N-2}(t) \\ u_{N-1}(t) \\ u_N(t) \end{Bmatrix} = \begin{Bmatrix} u_1(t) \\ u_2(t) \\ u_3(t) \\ u_4(t) \\ u_5(t) \\ u_6(t) \\ \vdots \\ u_{N-5}(t) \\ u_{N-4}(t) \\ u_{N-3}(t) \\ u_{N-2}(t) \\ u_{N-1}(t) \\ u_N(t) \end{Bmatrix} + \begin{Bmatrix} 1 \\ 0 \\ 0 \\ 0 \\ 0 \\ 0 \\ \vdots \\ 1 \\ 0 \\ 0 \\ 0 \\ 0 \\ 0 \end{Bmatrix} u_{gx}(t) + \begin{Bmatrix} 0 \\ 1 \\ 0 \\ 0 \\ 0 \\ 0 \\ \vdots \\ 0 \\ 1 \\ 0 \\ 0 \\ 0 \\ 0 \end{Bmatrix} u_{gy}(t) + \begin{Bmatrix} 0 \\ 0 \\ 1 \\ 0 \\ 0 \\ 0 \\ \vdots \\ 0 \\ 0 \\ 1 \\ 0 \\ 0 \\ 0 \end{Bmatrix} u_{gz}(t) \quad (10.113)$$

where it has been assumed that for each node the arrangement of the elements of $\{u(t)\}$ and $\{y(t)\}$ is in the following order: displacement along the x direction first, displacement along the y direction second, displacement along the z direction third, rotation about the x-axis fourth, rotation about the y-axis fifth, and rotation about the z-axis sixth (see Figure 10.9, where rotations are represented by double-head arrows). In compact form, $\{y(t)\}$ may therefore be written as

$$\{y(t)\} = \{u(t)\} + \{J\}_x u_{gx}(t) + \{J\}_y u_{gy}(t) + \{J\}_z u_{gz}(t) \quad (10.114)$$

where $\{J\}_x$, $\{J\}_y$, and $\{J\}_z$ are vectors that contain ones and zeroes, with the ones assigned to the translational degrees of freedom in the direction of the x-axis in the case of the $\{J\}_x$ vector and similarly for the $\{J\}_y$ and $\{J\}_z$ vectors. They are known as *influence vectors* and represent the displacements induced in the structure when a unit ground displacement is applied statically in the x, y, and z directions, respectively.

Applying Newton's second law, the equation of motion of the structure when its base is subjected to the three components of ground motion being considered may be thus expressed as

$$-[C]\{\dot{u}(t)\} - [K]\{u(t)\} = [M]\{\ddot{y}(t)\} \quad (10.115)$$

which in view of Equation 10.114 may also be written as

$$-[C]\{\dot{u}(t)\} - [K]\{u(t)\} = [M][\{\ddot{u}(t)\} + \{J\}_x \ddot{u}_{gx}(t) + \{J\}_y \ddot{u}_{gy}(t) + \{J\}_z \ddot{u}_{gz}(t)] \quad (10.116)$$

or, after rearranging terms, as

$$[M]\{\ddot{u}(t)\} + [C]\{\dot{u}(t)\} + [K]\{u(t)\} = -[M]\{J\}_x \ddot{u}_{gx}(t) - [M]\{J\}_y \ddot{u}_{gy}(t) - [M]\{J\}_z \ddot{u}_{gz}(t) \quad (10.117)$$

10.8.3 Transformation of Equation of Motion

Equation 10.117 represents a coupled system of second-order differential equations in which the independent variable is the time t and the dependent variables are the relative displacements $u_j(t)$. As in the case of a single ground motion component, it may be solved most conveniently by transforming it first into a decoupled system of equations. For this purpose, the transformation of coordinates defined by Equation 10.13 is introduced into Equation 10.117 and the transformed equation is afterward premultiplied by the transpose of the modal matrix $[\Phi]$. The result is

$$[\Phi]^T[M][\Phi]\{\ddot{\eta}(t)\} + [\Phi]^T[C][\Phi]\{\dot{\eta}(t)\} + [\Phi]^T[K][\Phi]\{\eta(t)\}$$
$$= -[\Phi]^T[M]\{J\}_x \ddot{u}_{gx}(t) - [\Phi]^T[M]\{J\}_y \ddot{u}_{gy}(t) - [\Phi]^T[M]\{J\}_z \ddot{u}_{gz}(t) \quad (10.118)$$

which, after considering the orthogonality properties of the mode shapes with respect to the mass and stiffness matrices of the system and assuming that the damping matrix is of the Rayleigh or Caughey type, leads to decoupled equations of the form

$$M_r^* \ddot{\eta}_r(t) + C_r^* \dot{\eta}_r(t) + K_r^* \eta_r(t) = -\{\phi\}_r^T[M]\{J\}_x \ddot{u}_{gx}(t) - \{\phi\}_r^T[M]\{J\}_y \ddot{u}_{gy}(t)$$
$$- \{\phi\}_r^T[M]\{J\}_z \ddot{u}_{gz}(t) \quad r = 1, 2, \ldots, N \quad (10.119)$$

where, as before, M_r^*, K_r^*, and C_r^*, respectively, denote the generalized mass, stiffness, and damping constant in the rth mode of the system. In turn, after dividing through by M_r^* and considering that $K_r^*/M_r^* = \omega_r^2$ and $C_r^*/M_r^* = 2\xi_r\omega_r$, Equation 10.119 yields

$$\ddot{\eta}_r(t) + 2\xi_r\omega_r \dot{\eta}_r(t) + \omega_r^2 \eta_r(t) = -\Gamma_{rx}\ddot{u}_{gx}(t) - \Gamma_{ry}\ddot{u}_{gy}(t) - \Gamma_{rz}\ddot{u}_{gz}(t) \quad r = 1, 2, \ldots, N \quad (10.120)$$

where Γ_{rx}, Γ_{ry}, and Γ_{rz} are modal participation factors defined as

$$\Gamma_{rk} = \frac{\{\phi\}_r^T[M]\{J\}_k}{M_r^*} = \frac{\{\phi\}_r^T[M]\{J\}_k}{\{\phi\}_r^T[M]\{\phi\}} \quad k = x, y, z \quad (10.121)$$

Equation 10.120 represents the desired system of decoupled equations. Note that as in the single ground motion component case, each equation in this decoupled system represents the equation of motion of a single-degree-of-freedom system with natural frequency ω_r and damping ratio ξ_r subjected to a ground motion excitation related to the input ground motion. In this case, however, the excitation has three components as opposed to one in the single ground motion component case. Nonetheless, along each of the three directions being considered, the excitation is also equal to the product of a ground acceleration and a participation factor. Note, also, that the solution of each of the independent equations in the system is no more complicated than the corresponding solution in the single ground motion component case. This is so because each equation can be solved separately for each of the three ground motion components and then the three individual solutions can be added to obtain the solution of the equation.

10.8.4 Solution of rth Independent Equation

As in the single ground motion component case, the solution of Equation 10.120 may be determined using any of the methods available for the solution of the equation of motion of a single-degree-of-freedom system. Accordingly, Duhamel integral may be used for this purpose. In addition, its solution may be obtained by considering, as previously noted, each excitation component separately

Structural Response by Response Spectrum Method

and then adding the three partial solutions. Thus, for zero initial conditions, the solution of Equation 10.120 may be written as

$$\eta_r(t) = \frac{\Gamma_{rx}}{\omega_{dr}} \int_0^t \ddot{u}_{gx}(\tau) e^{-\xi_r \omega_r (t-\tau)} \sin \omega_{dr}(t-\tau) d\tau + \frac{\Gamma_{ry}}{\omega_{dr}} \int_0^t \ddot{u}_{gy}(\tau) e^{-\xi_r \omega_r (t-\tau)} \sin \omega_{dr}(t-\tau) d\tau$$
$$+ \frac{\Gamma_{rz}}{\omega_{dr}} \int_0^t \ddot{u}_{gz}(\tau) e^{-\xi_r \omega_r (t-\tau)} \sin \omega_{dr}(t-\tau) d\tau \qquad (10.122)$$

where all symbols are as previously defined. Alternatively, this solution may be expressed as

$$\eta_r(t) = \Gamma_{rx} z_{rx}(t) + \Gamma_{ry} z_{ry}(t) + \Gamma_{rz} z_{rz}(t) \qquad (10.123)$$

where $z_{rk}(t)$, $k = x, y, z$, denotes the relative displacement response of a single-degree-of-freedom system with natural frequency ω_r and damping ratio ξ_r to a ground acceleration $\ddot{u}_{gk}(t)$.

10.8.5 Structural Response in Terms of Modal Responses

As established for the single ground motion component case, once the generalized coordinates $\eta_r(t)$ are obtained by means of Equation 10.122 or any other similar equation, then the structural displacement vector $\{u(t)\}$ may be determined by substituting the values of the generalized coordinates back into the transformation that relates the displacements $u_j(t)$ to the generalized coordinates $\eta_r(t)$. Accordingly, after substituting Equation 10.123 into Equation 10.13, it is found that

$$\{u(t)\} = \sum_{r=1}^N \{\phi\}_r \eta_r(t) = \sum_{r=1}^N [\Gamma_{rx}\{\phi\}_r z_{rx}(t) + \Gamma_{ry}\{\phi\}_r z_{ry}(t) + \Gamma_{rz}\{\phi\}_r z_{rz}(t)] \qquad (10.124)$$

which represents the displacement response of a structure when its base is subjected to three translational components of earthquake ground motion characterized by the ground accelerations $\ddot{u}_{gx}(t)$, $\ddot{u}_{gy}(t)$, and $\ddot{u}_{gz}(t)$. For convenience, however, this displacement response may be expressed alternatively in terms of modal responses under a single ground motion component as

$$\{u(t)\} = \sum_{r=1}^N [\{u(t)\}_{rx} + \{u(t)\}_{ry} + \{u(t)\}_{rz}] \qquad (10.125)$$

where $\{u(t)\}_{rk}$, $k = x, y, z$, is defined as

$$\{u(t)\}_{rk} = \Gamma_{rk}\{\phi\}_r z_{rk}(t) \qquad (10.126)$$

Note that $\{u(t)\}_{rk}$ represents a vector whose elements are the displacements induced in the structure when only the rth mode is excited and the kth ground motion component is the only input to the structure. This vector may be, thus, considered to be the contribution of the rth mode of vibration and kth component of ground motion to the total displacement response. Conversely, it may be considered that the displacement response of a structure is given by the superposition of the displacement response in each of its modes of vibration under each ground motion component.

In similarity with the single ground motion component case, the solution can be extended to other measures of structural response. Proceeding as in Section 10.2, it may be shown that for the case of three translational components, the relative velocity, relative acceleration, and absolute acceleration responses are given by

$$\{\dot{u}(t)\} = \sum_{r=1}^N [\{\dot{u}(t)\}_{rx} + \{\dot{u}(t)\}_{ry} + \{\dot{u}(t)\}_{rz}] \qquad (10.127)$$

$$\{\ddot{u}(t)\} = \sum_{r=1}^{N}[\{\ddot{u}(t)\}_{rx} + \{\ddot{u}(t)\}_{ry} + \{\ddot{u}(t)\}_{rz}] \qquad (10.128)$$

$$\{\ddot{y}(t)\} = \sum_{r=1}^{N}[\{\ddot{y}(t)\}_{rx} + \{\ddot{y}(t)\}_{ry} + \{\ddot{y}(t)\}_{rz}] \qquad (10.129)$$

where, for $k = x, y, z$

$$\{\dot{u}(t)\}_{rk} = \Gamma_{rk}\{\phi\}_r \dot{z}_{rk}(t) \qquad (10.130)$$

$$\{\ddot{u}(t)\}_{rk} = \Gamma_{rk}\{\phi\}_r \ddot{z}_{rk}(t) \qquad (10.131)$$

$$\{\ddot{y}(t)\}_{rk} = \Gamma_{rk}\{\phi\}_r [\ddot{z}_{rk}(t) + \ddot{u}_{gk}(t)] \qquad (10.132)$$

Similarly, the response in terms of the internal forces in the ith member of the structure results as

$$\{F_i(t)\} = \sum_{r=1}^{N}[\{F_i(t)\}_{rx} + \{F_i(t)\}_{ry} + \{F_i(t)\}_{rz}] \qquad (10.133)$$

in which, for $k = x, y, z$,

$$\{F_i(t)\}_{rk} = \Gamma_{rk}[K_i'][T_i]\{\phi_i\}_r z_{rk}(t) \qquad (10.134)$$

where all variables are as defined before.

10.8.6 Maximum Modal Responses in Terms of Response Spectrum Ordinates

It is also possible to determine the maximum value of each of the modal responses in Equation 10.125 using directly the response spectra of the ground motion components as it was done in Section 10.3 for the single ground motion component case. To this end, one only has to consider that, according to the definition of response spectrum, the maximum value of $z_{rk}(t)$ in Equation 10.126 is equal to the ordinate corresponding to the natural frequency ω_r and damping ratio ξ_r in the displacement response spectrum of the ground acceleration $\ddot{u}_{gk}(t)$. Hence, the maximum values of the displacements in the rth mode of a structure under the kth ground motion component may be determined according to

$$\max\{u(t)\}_{rk} = \Gamma_{rk}\{\phi\}_r SD_k(\omega_r, \xi_r) \qquad (10.135)$$

where $SD_k(\omega_r, \xi_r)$ denotes the aforementioned response spectrum ordinate. In a similar fashion, it is possible to estimate the maximum values of the modal relative velocity and absolute acceleration responses in Equations 10.127 and 10.129 as the maximum values of $\dot{z}_{rk}(t)$ and $\ddot{z}_{rk}(t) + \ddot{u}_{gk}(t)$ in Equations 10.130 and 10.132, respectively, represent ordinates in the velocity and acceleration response spectra of $\ddot{u}_{gk}(t)$. As a result, the maximum values of these responses may be determined according to

$$\max\{\dot{u}(t)\}_{rk} = \Gamma_{rk}\{\phi\}_r SV_k(\omega_r, \xi_r) \qquad (10.136)$$

$$\max\{\ddot{y}(t)\}_{rk} = \Gamma_{rk}\{\phi\}_r SA_k(\omega_r, \xi_r) \qquad (10.137)$$

where $SV_k(\omega_r, \xi_r)$ and $SA_k(\omega_r, \xi_r)$, respectively, denote spectral velocity and spectral acceleration. Finally, if the same argument is applied to Equation 10.134, then it is also possible to obtain the maximum values of the modal internal forces in Equation 10.133 by means of

$$\{F_i(t)\}_{rk} = \Gamma_{rk}[K_i'][T_i]\{\phi_i\}_r SD_k(\omega_r, \xi_r) \qquad (10.138)$$

Structural Response by Response Spectrum Method

Equations 10.135 through 10.137 are useful to determine the maximum modal responses of a structure under one of the ground motion components using the response spectrum of the ground motion component being considered. One cannot, however, obtain the maximum modal responses under the simultaneous effect of the three ground motion components by substituting such maximum values into Equation 10.125, 10.127 through 10.129, or 10.133. The reason is, once again, that it is highly unlikely that the three maxima will occur at the same time. That is, in general one has that

$$\max[\{u(t)\}_{rx} + \{u(t)\}_{ry} + \{u(t)\}_{rz}] \neq \max\{u(t)\}_{rx} + \max\{u(t)\}_{yr} + \max\{u(t)\}_{rz} \quad (10.139)$$

and similarly for the other types of responses. As discussed in Section 10.3.1, one cannot superimpose, either, the maximum modal responses to determine the maximum response. It may be seen, thus, that in the case of multicomponent ground motions one has to resort to approximate relationships to estimate not only the maximum response but also the maxima modal responses. Fortunately, some methods are available for such a purpose. One such method is introduced in the following section.

10.8.7 Approximate Rule to Estimate Maximum System Response

A combination rule to estimate the maximum response of systems subjected to multicomponent ground motions may be obtained using the same technique and assumptions considered in the derivation of the combination rules introduced in Section 10.3.2. That is, it may be obtained by assuming that such a maximum response is proportional to the root mean square of the response and by determining this root mean square on the basis of the equations derived earlier to compute the response of such systems. However, another, simpler, approach may be utilized if advantage is taken of the fact that there exists a set of orthogonal axes, referred to as principal axes, along which the three translational components of ground motions are statistically uncorrelated. The existence of such a set of axes was demonstrated by J. Penzien and M. Watabe in 1975 by considering ground motions as random stationary processes. The advantage of considering the principal axes of a ground motion lies on the fact that, if the three translational components of the ground motion are directed along these axes, then it is possible to combine the maximum responses to each of the ground motion components on the basis of the SRSS rule. Recall that, as shown in Section 10.3.2, the SRSS rule may be applied accurately to combine uncorrelated responses.

Utilizing the latter approach, it may be thus considered that the maximum response of a system subjected to a multicomponent ground motion is given by

$$\max\{R(t)\} = \left\{\sqrt{[\max R_1(t)]^2 + [\max R_2(t)]^2 + [\max R_3(t)]^2}\right\} \quad (10.140)$$

where $R(t)$ is any measure of the system's response and $R_1(t)$, $R_2(t)$, and $R_3(t)$ denote the corresponding responses under each of the ground motion components alone, when these components are directed along the ground motion's principal axes. Henceforth, these principal axes will be identified as axes 1, 2, and 3, where axes 1 and 2 are contained in a horizontal plane and axis 3 is a vertical one. In addition, it may be considered that the maximum values of the responses $R_1(t)$ and $R_2(t)$ may be estimated in terms of the corresponding modal responses using the double-sum combination rule established in Section 10.3.2. That is, it may be considered that

$$\max R_1(t) = \left\{\sum_{m=1}^{N}\sum_{n=1}^{N}\rho_{mn}[\max R_{1m}(t)][\max R_{1n}(t)]\right\}^{1/2} \quad (10.141)$$

$$\max R_2(t) = \left\{\sum_{m=1}^{N}\sum_{n=1}^{N}\rho_{mn}[\max R_{2m}(t)][\max R_{2n}(t)]\right\}^{1/2} \quad (10.142)$$

where $R_{1r}(t)$ and $R_{2r}(t)$, $r = m, n$, represent the modal responses under the ground motion components directed along the principal axes 1 and 2, respectively, and, as before, ρ_{mn} denotes the modal correlation coefficient introduced in Section 10.3.2. Note that max $R_3(t)$ may also be estimated in terms of modal responses, but it will not be considered explicitly so as the vertical response is not affected by the angle of incidence of the horizontal components. Consequently, in terms of modal responses the desired maximum response may be determined according to

$$\max\{R(t)\} = \left\{ \left(\sum_{m=1}^{N} \sum_{n=1}^{N} \rho_{mn} ([\max R_{1m}(t)][\max R_{1n}(t)] + [\max R_{2m}(t)][\max R_{2n}(t)]) + [\max R_3(t)]^2 \right)^{1/2} \right\}$$

(10.143)

Ordinarily, however, the ground motion components are considered in the direction of the structural axes, which in most cases correspond to the structure's axes of symmetry, or are arbitrarily selected. It is convenient, therefore, to express the modal responses in Equation 10.143 in terms of the responses obtained when the ground motion components are applied in the direction of the structural axes. For this purpose, consider the case when an earthquake ground motion with horizontal components along two orthogonal axes and a vertical component excites the base of a structure. With reference to Figure 10.10, let the horizontal components of this ground motion be directed along the ground motion's principal axes. Consider, in addition, that the principal axis 1 makes an angle θ with respect to the structural reference axis x (see Figure 10.10). Furthermore, let $\ddot{u}_{g1}(t)$ and $\ddot{u}_{g2}(t)$, respectively, denote the horizontal ground motions components along the principal directions 1 and 2. For the purpose of simplifying the notation, let R_{1r} and R_{2r}, respectively, represent the maximum values of the modal responses $R_{1r}(t)$ and $R_{2r}(t)$. Similarly, let R_{1xr} and R_{2xr} depict the maximum values of the corresponding modal responses when $\ddot{u}_{g1}(t)$ and $\ddot{u}_{g2}(t)$ are applied in the direction of the x-axis, respectively, and R_{1yr} and R_{2yr} those when $\ddot{u}_{g1}(t)$ and $\ddot{u}_{g2}(t)$ are applied in the direction of the y-axis. Let also $R_3 = \max R_3(t)$.

If $\ddot{u}_{g1}(t)$ is applied along the direction of the x-axis, then the maximum value of the response $R(t)$ in the rth mode of the structure under this component of ground motion alone would be equal to R_{1xr}. Similarly, if $\ddot{u}_{g1}(t)$ is applied along the direction of the y-axis, then such a maximum response would be equal to R_{1yr}. Consequently, if $\ddot{u}_{g1}(t)$ is applied along the direction of the principal axis 1, the corresponding maximum response would be given by

$$R_{1r} = R_{1xr} \cos\theta + R_{1yr} \sin\theta$$

(10.144)

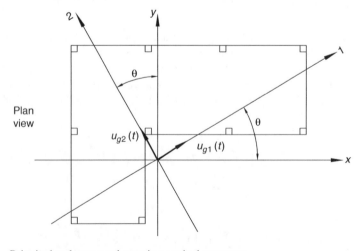

FIGURE 10.10 Principal and structural axes in a typical structure.

such that $R_{1r} = R_{1xr}$ when $\theta = 0°$ and $R_{1r} = R_{1yr}$ when $\theta = 90°$. Similarly, if $\ddot{u}_{g2}(t)$ is applied along the negative direction of the x-axis, then the maximum modal response would be equal to $-R_{2xr}$. If, on the other hand, $\ddot{u}_{g2}(t)$ is applied along the direction of the y-axis, then the maximum modal response would be R_{2yr}. Therefore, if $\ddot{u}_{g2}(t)$ is applied along the principal axis 2, the maximum modal response would be

$$R_{2r} = R_{2yr} \cos\theta - R_{2xr} \sin\theta \tag{10.145}$$

such that $R_{2r} = R_{2yr}$ when $\theta = 0°$ and $R_{2r} = -R_{2xr}$ when $\theta = 90°$.

In terms of modal responses referred to the structural x and y axes, the maximum response of the system may therefore be expressed as

$$\max\{R(t)\} = \left\{ \left(\sum_m \sum_n \rho_{mn} [(R_{1xm} \cos\theta + R_{1ym} \sin\theta)(R_{1xn} \cos\theta + R_{1yn} \sin\theta) \right. \right.$$

$$\left. \left. + (R_{2ym} \cos\theta - R_{2xm} \sin\theta)(R_{2yn} \cos\theta - R_{2xn} \sin\theta)] + R_3^2 \right)^{1/2} \right\}$$

$$= \left\{ \left(\sum_m \sum_n \rho_{mn} [(R_{1xm} R_{1xn} + R_{2ym} R_{2yn}) \cos^2\theta + (R_{1ym} R_{1yn} + R_{2xm} R_{2xn}) \sin^2\theta \right. \right.$$

$$\left. \left. + (R_{1ym} R_{1xn} + R_{1xm} R_{1yn} - R_{2ym} R_{2xn} - R_{2xm} R_{2yn}) \sin\theta \cos\theta] + R_3^2 \right)^{1/2} \right\}$$

$$= \left\{ \left[\left(\sum_m \sum_n \rho_{mn} R_{1xm} R_{1xn} + \sum_m \sum_n \rho_{mn} R_{2ym} R_{2yn} \right) \cos^2\theta \right. \right.$$

$$+ \left(\sum_m \sum_n \rho_{mn} R_{1ym} R_{1yn} + \sum_m \sum_n \rho_{mn} R_{2xm} R_{2xn} \right) \sin^2\theta$$

$$\left. \left. + 2 \left(\sum_m \sum_n \rho_{mn} R_{1xm} R_{1yn} - \sum_m \sum_n \rho_{mn} R_{2xm} R_{2yn} \right) \sin\theta \cos\theta + R_3^2 \right]^{1/2} \right\}$$

$$= \{[(R_{1x}^2 + R_{2y}^2) \cos^2\theta + (R_{1y}^2 + R_{2x}^2) \sin^2\theta + 2(R_{1xy}^2 - R_{2xy}^2) \sin\theta \cos\theta + R_3^2]^{1/2}\}$$

$$= \left\{ \left[(R_{1x}^2 + R_{2y}^2) \frac{1 + \cos 2\theta}{2} + (R_{1y}^2 + R_{2x}^2) \frac{1 - \cos 2\theta}{2} + (R_{1xy}^2 - R_{2xy}^2) \sin 2\theta + R_3^2 \right]^{1/2} \right\}$$

(10.146)

or

$$\max\{R(t)\} = \left\{ \left[\frac{1}{2}(R_{1x}^2 + R_{2y}^2 + R_{1y}^2 + R_{2x}^2) + \frac{1}{2}(R_{1x}^2 + R_{2y}^2 - R_{1y}^2 - R_{2x}^2) \cos 2\theta \right. \right.$$

$$\left. \left. + (R_{1xy}^2 - R_{2xy}^2) \sin 2\theta + R_3^2 \right]^{1/2} \right\} \tag{10.147}$$

where R_{1x}^2, R_{1y}^2, R_{2x}^2, and R_{2y}^2 are given by

$$R_{1x}^2 = \sum_{m=1}^{N} \sum_{n=1}^{N} \rho_{mn} R_{1xm} R_{1xn} \tag{10.148}$$

$$R_{1y}^2 = \sum_{m=1}^{N}\sum_{n=1}^{N} \rho_{mn} R_{1ym} R_{1yn} \tag{10.149}$$

$$R_{2x}^2 = \sum_{m=1}^{N}\sum_{n=1}^{N} \rho_{mn} R_{2xm} R_{2xn} \tag{10.150}$$

$$R_{2y}^2 = \sum_{m=1}^{N}\sum_{n=1}^{N} \rho_{mn} R_{2ym} R_{2yn} \tag{10.151}$$

and represent the square of the estimated maximum values of the response being considered when the ground motion component $\ddot{u}_{g1}(t)$ alone or $\ddot{u}_{g2}(t)$ alone is applied along the direction of the x- or y-axis. Similarly, R_{1xy}^2 and R_{2xy}^2 are defined as

$$R_{1xy}^2 = \sum_{m=1}^{N}\sum_{n=1}^{N} \rho_{mn} R_{1xm} R_{1yn} \tag{10.152}$$

$$R_{2xy}^2 = \sum_{m=1}^{N}\sum_{n=1}^{N} \rho_{mn} R_{2xm} R_{2yn} \tag{10.153}$$

and depict the square of the maximum values of cross-correlated responses, which possess no physical meaning. Note, however, that these cross-correlated responses involve the same terms that are involved in the definitions of R_{1x}^2, R_{1y}^2, R_{2x}^2, and R_{2y}^2.

Equation 10.147 permits the estimation of the maximum response of a system subjected to a multicomponent ground motion when the orientation of the ground motion's principal axes is known. In most situations, however, this orientation is not known. It seems reasonable, therefore, to consider, at least for design purposes, the orientation that yields the largest response. After all, a well-designed structure should be able to resist earthquake ground motions coming from any possible direction. To find, then, the critical angle, denoted here as θ_{cr}, that yields the largest response, one can take the first derivative of max $R(t)$ and set this derivative equal to zero. Proceeding accordingly, one obtains

$$2\max R(t)\frac{d[\max R(t)]}{d\theta} = -(R_{1x}^2 + R_{2y}^2 - R_{2x}^2 - R_{1y}^2)\sin 2\theta + 2(R_{1xy}^2 - R_{2xy}^2)\cos 2\theta = 0 \tag{10.154}$$

from which one gets

$$\tan 2\theta_{cr} = \frac{2(R_{1xy}^2 - R_{2xy}^2)}{R_{1x}^2 + R_{2y}^2 - R_{2x}^2 - R_{1y}^2} \tag{10.155}$$

which leads to

$$\sin 2\theta_{cr} = \frac{R_{1xy}^2 - R_{2xy}^2}{\sqrt{\left[\frac{1}{2}(R_{1x}^2 + R_{2y}^2 - R_{2x}^2 - R_{1y}^2)\right]^2 + (R_{1xy}^2 - R_{2xy}^2)^2}} \tag{10.156}$$

$$\cos 2\theta_{cr} = \frac{\frac{1}{2}(R_{1x}^2 + R_{2y}^2 - R_{2x}^2 - R_{1y}^2)}{\sqrt{\left[\frac{1}{2}(R_{1x}^2 + R_{2y}^2 - R_{2x}^2 - R_{1y}^2)\right]^2 + (R_{1xy}^2 - R_{2xy}^2)^2}} \tag{10.157}$$

Thus, upon substitution of Equations 10.156 and 10.157 into Equation 10.147, one arrives to

$$\max\{R(t)\} = \left\{\left[\frac{1}{2}(R_{1x}^2 + R_{2y}^2 + R_{1y}^2 + R_{2x}^2) + \frac{\frac{1}{2}(R_{1x}^2 + R_{2y}^2 - R_{1y}^2 - R_{2x}^2)\frac{1}{2}(R_{1x}^2 + R_{2y}^2 - R_{1y}^2 - R_{2x}^2)}{\sqrt{\left[\frac{1}{2}(R_{1x}^2 + R_{2y}^2 - R_{2x}^2 - R_{1y}^2)\right]^2 + (R_{1xy}^2 - R_{2xy}^2)^2}}\right.\right.$$

$$\left.\left.+ \frac{(R_{1xy}^2 - R_{2xy}^2)(R_{1xy}^2 - R_{2xy}^2)}{\sqrt{\left[\frac{1}{2}(R_{1x}^2 + R_{2y}^2 - R_{2x}^2 - R_{1y}^2)\right]^2 + (R_{1xy}^2 - R_{2xy}^2)^2}} + R_3^2\right]^{1/2}\right\}$$

(10.158)

which, after simplifying, becomes

$$\max\{R(t)\} = \left\{\left[\frac{1}{2}(R_{1x}^2 + R_{2y}^2 + R_{1y}^2 + R_{2x}^2)\right.\right.$$

$$\left.\left.+ \sqrt{\left[\frac{1}{2}(R_{1x}^2 + R_{2y}^2 - R_{2x}^2 - R_{1y}^2)\right]^2 + (R_{1xy}^2 - R_{2xy}^2)^2} + R_3^2\right]^{1/2}\right\}$$

(10.159)

Equation 10.159 constitutes the desired rule to combine the maximum responses of systems subjected to multicomponent ground motions. Note that its application requires five separate analyses: one under the ground motion component $\ddot{u}_{g1}(t)$ applied along the x-axis, one under this same component applied along the y-axis, one under the ground motion component $\ddot{u}_{g2}(t)$ applied along the x-axis, one under the latter component applied along the y-axis, and one under the vertical component. Its use may thus be computational demanding. The rule, however, may be simplified if it is assumed that the response spectrum of the ground motion component along the direction of the principal axis 2 is a fraction γ of the response spectrum along the direction of the principal axis 1. That is, if it is assumed that

$$R_{2x} = \gamma R_{1x} \qquad R_{2y} = \gamma R_{1y} \qquad R_{2xy} = \gamma R_{1xy} \qquad (10.160)$$

then this assumption, which is common in design practice, implies that the ground motion along the principal axis 2 varies in intensity but not in frequency content from the ground motion along the principal axis 1. Under such an assumption, Equation 10.159 simplifies to

$$\max\{R(t)\} = \left\{\left[\frac{1}{2}(R_{1x}^2 + \gamma^2 R_{1y}^2 + R_{1y}^2 + \gamma^2 R_{1x}^2)\right.\right.$$

$$\left.\left.+ \sqrt{\left[\frac{1}{2}(R_{1x}^2 + \gamma^2 R_{1y}^2 - \gamma^2 R_{1x}^2 - R_{1y}^2)\right]^2 + (R_{1xy}^2 - \gamma^2 R_{1xy}^2)^2} + R_3^2\right]^{1/2}\right\}$$

(10.161)

which leads to

$$\max\{R(t)\} = \left\{\left[(1+\gamma^2)\frac{R_{1x}^2 + R_{1y}^2}{2} + (1-\gamma^2)\sqrt{\left(\frac{R_{1x}^2 - R_{1y}^2}{2}\right)^2 + R_{1xy}^4} + R_3^2\right]^{1/2}\right\}$$

(10.162)

Observe, thus, that by assuming that the response spectrum that defines the ground motion along one of the horizontal axes of the structure is a fraction of the response spectrum specified for the other horizontal direction, the number of required analyses is reduced to three. Observe also that the combination rule defined by Equation 10.162 can be easily implemented in any standard dynamic analysis program because each of the terms in it is routinely calculated in such programs. Observe, further, that if the ground motion horizontal components are assumed to be exactly the same, that is, if $\gamma = 1$, then Equation 10.162 is reduced to

$$\max\{R(t)\} = \left\{\sqrt{R_{1x}^2 + R_{1y}^2 + R_3^2}\right\} \tag{10.163}$$

This means that if the ground motion components along the two principal horizontal directions are the same, then the maximum response under the combined effect of three components of ground motion may be estimated on the basis of the SRSS rule. It also means that a simple but conservative estimate of such a maximum response may be obtained by simply specifying the same response spectrum for the two horizontal directions of the structure, and then combining the maximum responses under each of the three ground motion components using the SRSS rule.

EXAMPLE 10.5 **ACCELERATION RESPONSE OF A ONE-STORY BUILDING UNDER TWO-COMPONENT GROUND MOTION**

Figure E10.5 shows the plan view of a one-story building whose roof height is 12 ft and its structural elements are three shear walls. The roof of the building, which may be idealized as a rigid diaphragm, supports a uniformly distributed vertical load, with its own weight included, of 100 lb/ft². The lateral stiffnesses of the shear walls are 75 kip/ft for shear wall A, 30 kip/ft for shear wall B, and 40 kip/ft for shear wall C. Determine using the response spectrum method the maximum acceleration response of the building when the base of the building is subjected to the ground motion generated during the 1940 El Centro, California, earthquake. Consider that the motion of the roof may be described by the displacements of its center of mass (c.m.) along the x and y axes and its rotation about a vertical axis. For simplicity, neglect the thicknesses of the walls in all calculations as well as the vertical component of the ground motion. Assume a damping ratio of 5% for all modes. The acceleration response spectra of the horizontal components of the 1940 El Centro earthquake are presented in Figures 6.5 and E10.1b.

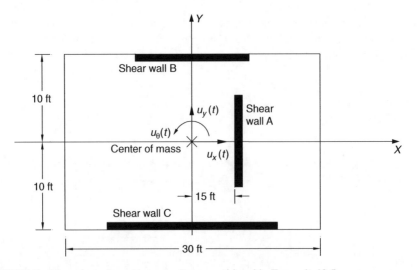

FIGURE E10.5 Plan view of one-story structure considered in Example 10.5.

Structural Response by Response Spectrum Method

Solution

Stiffness Matrix

Using the definition of stiffness influence coefficients and referring to the three degrees of freedom indicated in Figure E10.5, the stiffness matrix of the system is equal to

$$[K] = \begin{bmatrix} 70 & 0 & 100 \\ 0 & 75 & 112.5 \\ 100 & 112.5 & 7168.75 \end{bmatrix} \begin{matrix} u_x \\ u_y \\ u_\theta \end{matrix}$$

Mass Matrix

Considering that the total mass of the system is equal to $0.100(20)(30)/32.2 = 1.863$ kip s²/ft and that the corresponding mass moment of inertia is equal to $1.863\,(20^2 + 30^2)/12 = 201.863$ kip ft s², the mass matrix of the system is given by

$$[M] = \begin{bmatrix} 1.863 & 0 & 0 \\ 0 & 1.863 & 0 \\ 0 & 0 & 201.863 \end{bmatrix} \begin{matrix} u_x \\ u_y \\ u_\theta \end{matrix}$$

Natural Frequencies and Mode Shapes

From the solution of the eigenvalue problem formulated with the determined stiffness and mass matrices, the natural frequencies and mode shapes of the system are

$$\omega_1 = 5.409 \text{ rad/s} \qquad \omega_2 = 6.218 \text{ rad/s} \qquad \omega_3 = 6.739 \text{ rad/s}$$

$$\{\phi\}_1 = \begin{Bmatrix} 0.3523 \\ 0.2996 \\ -0.0546 \end{Bmatrix} \qquad \{\phi\}_2 = \begin{Bmatrix} 0.5715 \\ -0.4421 \\ 0.0116 \end{Bmatrix} \qquad \{\phi\}_3 = \begin{Bmatrix} 0.2933 \\ 0.5015 \\ 0.0429 \end{Bmatrix}$$

Influence Vectors

According to the definition of the influence vectors introduced in Section 10.8.2, for the system under consideration these influence vectors are

$$\{J\}_x = \begin{Bmatrix} 1 \\ 0 \\ 0 \end{Bmatrix} \qquad \{J\}_y = \begin{Bmatrix} 0 \\ 1 \\ 0 \end{Bmatrix}$$

Participation Factors

As for the system under analysis, one has that

$$M_1^* = \{\phi\}_1^T [M] \{\phi\}_1 = \begin{Bmatrix} 0.3523 \\ 0.2996 \\ -0.0546 \end{Bmatrix}^T \begin{bmatrix} 1.863 & 0 & 0 \\ 0 & 1.863 & 0 \\ 0 & 0 & 201.863 \end{bmatrix} \begin{Bmatrix} 0.3523 \\ 0.2996 \\ -0.0546 \end{Bmatrix} = 1.0$$

$$M_2^* = \{\phi\}_2^T [M] \{\phi\}_2 = \begin{Bmatrix} 0.5715 \\ -0.4421 \\ 0.0116 \end{Bmatrix}^T \begin{bmatrix} 1.863 & 0 & 0 \\ 0 & 1.863 & 0 \\ 0 & 0 & 201.863 \end{bmatrix} \begin{Bmatrix} 0.5715 \\ -0.4421 \\ 0.0116 \end{Bmatrix} = 1.0$$

$$M_3^* = \{\phi\}_3^T[M]\{\phi\}_3 = \begin{Bmatrix} 0.2933 \\ 0.5015 \\ 0.0429 \end{Bmatrix}^T \begin{bmatrix} 1.863 & 0 & 0 \\ 0 & 1.863 & 0 \\ 0 & 0 & 201.863 \end{bmatrix} \begin{Bmatrix} 0.2933 \\ 0.5015 \\ 0.0429 \end{Bmatrix} = 1.0$$

then according to Equation 10.121 and the determined influence vectors the modal participation factors of the system are

$$\Gamma_{1x} = \frac{\{\phi\}_1^T[M]\{J\}_x}{M_1^*} = \frac{1}{1.0} \begin{Bmatrix} 0.3523 \\ 0.2996 \\ -0.0546 \end{Bmatrix}^T \begin{bmatrix} 1.863 & 0 & 0 \\ 0 & 1.863 & 0 \\ 0 & 0 & 201.863 \end{bmatrix} \begin{Bmatrix} 1 \\ 0 \\ 0 \end{Bmatrix} = 0.656$$

$$\Gamma_{2x} = \frac{\{\phi\}_2^T[M]\{J\}_x}{M_2^*} = \frac{1}{1.0} \begin{Bmatrix} 0.5715 \\ -0.4421 \\ 0.0116 \end{Bmatrix}^T \begin{bmatrix} 1.863 & 0 & 0 \\ 0 & 1.863 & 0 \\ 0 & 0 & 201.863 \end{bmatrix} \begin{Bmatrix} 1 \\ 0 \\ 0 \end{Bmatrix} = 1.065$$

$$\Gamma_{3x} = \frac{\{\phi\}_3^T[M]\{J\}_x}{M_3^*} = \frac{1}{1.0} \begin{Bmatrix} 0.2933 \\ 0.5015 \\ 0.0429 \end{Bmatrix}^T \begin{bmatrix} 1.863 & 0 & 0 \\ 0 & 1.863 & 0 \\ 0 & 0 & 201.863 \end{bmatrix} \begin{Bmatrix} 1 \\ 0 \\ 0 \end{Bmatrix} = 0.546$$

$$\Gamma_{1y} = \frac{\{\phi\}_1^T[M]\{J\}_y}{M_1^*} = \frac{1}{1.0} \begin{Bmatrix} 0.3523 \\ 0.2996 \\ -0.0546 \end{Bmatrix}^T \begin{bmatrix} 1.863 & 0 & 0 \\ 0 & 1.863 & 0 \\ 0 & 0 & 201.863 \end{bmatrix} \begin{Bmatrix} 0 \\ 1 \\ 0 \end{Bmatrix} = 0.558$$

$$\Gamma_{2y} = \frac{\{\phi\}_2^T[M]\{J\}_y}{M_2^*} = \frac{1}{1.0} \begin{Bmatrix} 0.5715 \\ -0.4421 \\ 0.0116 \end{Bmatrix}^T \begin{bmatrix} 1.863 & 0 & 0 \\ 0 & 1.863 & 0 \\ 0 & 0 & 201.863 \end{bmatrix} \begin{Bmatrix} 0 \\ 1 \\ 0 \end{Bmatrix} = -0.824$$

$$\Gamma_{3y} = \frac{\{\phi\}_3^T[M]\{J\}_y}{M_3^*} = \frac{1}{1.0} \begin{Bmatrix} 0.2933 \\ 0.5015 \\ 0.0429 \end{Bmatrix}^T \begin{bmatrix} 1.863 & 0 & 0 \\ 0 & 1.863 & 0 \\ 0 & 0 & 201.863 \end{bmatrix} \begin{Bmatrix} 0 \\ 1 \\ 0 \end{Bmatrix} = 0.934$$

Note that these participation factors satisfy the relationship (which may be derived in the same way as Equation 10.42)

$$\{J\}_k = \sum_{r=1}^{N} \Gamma_{rk}\{\phi\}_r \quad k = x, y, z$$

as

$$0.656 \begin{Bmatrix} 0.3523 \\ 0.2996 \\ -0.0546 \end{Bmatrix} + 1.065 \begin{Bmatrix} 0.5715 \\ -0.4421 \\ 0.0116 \end{Bmatrix} + 0.546 \begin{Bmatrix} 0.2933 \\ 0.5015 \\ 0.0429 \end{Bmatrix} = \begin{Bmatrix} 1.0 \\ 0 \\ 0 \end{Bmatrix} = \{J\}_x$$

and

$$0.558 \begin{Bmatrix} 0.3523 \\ 0.2996 \\ -0.0546 \end{Bmatrix} - 0.824 \begin{Bmatrix} 0.5715 \\ -0.4421 \\ 0.0116 \end{Bmatrix} + 0.934 \begin{Bmatrix} 0.2933 \\ 0.5015 \\ 0.0429 \end{Bmatrix} = \begin{Bmatrix} 0 \\ 1.0 \\ 0 \end{Bmatrix} = \{J\}_y$$

Structural Response by Response Spectrum Method

Response Spectrum Ordinates

From the response spectra in Figures 6.5 and E10.1b, the spectral ordinates that correspond to the natural frequencies and damping ratios of the system for the N–S component of the specified ground motion are

$$SA_{NS}(5.409, 0.05) = 0.346g \qquad SA_{NS}(6.218, 0.05) = 0.512g \qquad SA_{NS}(6.739, 0.05) = 0.550g$$

while those for the E–W component are

$$SA_{EW}(5.409, 0.05) = 0.288g \qquad SA_{EW}(6.218, 0.05) = 0.285g \qquad SA_{EW}(6.739, 0.05) = 0.313g$$

To follow the notation introduced in Section 10.8.7, it will be considered that the N–S component is applied along the principal direction 1 and the E–W component along the principal direction 2.

Maximum Modal Responses Under Single-Component Ground Motion

The maximum values of the modal accelerations may be obtained from Equation 10.137 and the participation factors, mode shapes, and spectral ordinates determined earlier. Hence, if the N–S component of the ground motion being considered is applied in the direction of the x-axis, these modal accelerations are equal to

$$\{R_{1x1}\} = \max\{\ddot{y}(t)\}_{1x1} = \Gamma_{1x}\{\phi\}_1 SA_{NS}(\omega_1, \xi_1) = 0.656 \begin{Bmatrix} 0.3523 \\ 0.2996 \\ -0.0546 \end{Bmatrix} 0.346g = \begin{Bmatrix} 0.080 \\ 0.068 \\ -0.0124 \end{Bmatrix} g$$

$$\{R_{1x2}\} = \max\{\ddot{y}(t)\}_{1x2} = \Gamma_{2x}\{\phi\}_2 SA_{NS}(\omega_2, \xi_2) = 1.065 \begin{Bmatrix} 0.5715 \\ -0.4421 \\ 0.0116 \end{Bmatrix} 0.512g = \begin{Bmatrix} 0.312 \\ -0.241 \\ 0.0063 \end{Bmatrix} g$$

$$\{R_{1x3}\} = \max\{\ddot{y}(t)\}_{1x3} = \Gamma_{3x}\{\phi\}_3 SA_{NS}(\omega_3, \xi_3) = 0.546 \begin{Bmatrix} 0.2933 \\ 0.5015 \\ 0.0429 \end{Bmatrix} 0.550g = \begin{Bmatrix} 0.088 \\ 0.151 \\ 0.0129 \end{Bmatrix} g$$

and if this same component is applied along the y-axis, the corresponding modal accelerations are

$$\{R_{1y1}\} = \max\{\ddot{y}(t)\}_{1y1} = \Gamma_{1y}\{\phi\}_1 SA_{NS}(\omega_1, \xi_1) = 0.558 \begin{Bmatrix} 0.3523 \\ 0.2996 \\ -0.0546 \end{Bmatrix} 0.346g = \begin{Bmatrix} 0.068 \\ 0.058 \\ -0.0105 \end{Bmatrix} g$$

$$\{R_{1y2}\} = \max\{\ddot{y}(t)\}_{1y2} = \Gamma_{2y}\{\phi\}_2 SA_{NS}(\omega_2, \xi_2) = -0.824 \begin{Bmatrix} 0.5715 \\ -0.4421 \\ 0.0116 \end{Bmatrix} 0.512g = \begin{Bmatrix} -0.241 \\ 0.187 \\ -0.0049 \end{Bmatrix} g$$

$$\{R_{1y3}\} = \max\{\ddot{y}(t)\}_{1y3} = \Gamma_{3y}\{\phi\}_3 SA_{NS}(\omega_3, \xi_3) = 0.934 \begin{Bmatrix} 0.2933 \\ 0.5015 \\ 0.0429 \end{Bmatrix} 0.550g = \begin{Bmatrix} 0.151 \\ 0.258 \\ 0.0220 \end{Bmatrix} g$$

Similarly, if the E–W component is applied along the direction of the x-axis, the modal accelerations result as

$$\{R_{2x1}\} = \max\{\ddot{y}(t)\}_{2x1} = \Gamma_{1x}\{\phi\}_1 SA_{EW}(\omega_1,\xi_1) = 0.656\begin{Bmatrix} 0.3523 \\ 0.2996 \\ -0.0546 \end{Bmatrix} 0.288g = \begin{Bmatrix} 0.067 \\ 0.057 \\ -0.0103 \end{Bmatrix} g$$

$$\{R_{2x2}\} = \max\{\ddot{y}(t)\}_{2x2} = \Gamma_{2x}\{\phi\}_2 SA_{EW}(\omega_2,\xi_2) = 1.065\begin{Bmatrix} 0.5715 \\ -0.4421 \\ 0.0116 \end{Bmatrix} 0.285g = \begin{Bmatrix} 0.173 \\ -0.134 \\ 0.0035 \end{Bmatrix} g$$

$$\{R_{2x3}\} = \max\{\ddot{y}(t)\}_{3x} = \Gamma_{3x}\{\phi\}_3 SA_{EW}(\omega_3,\xi_3) = 0.546\begin{Bmatrix} 0.2933 \\ 0.5015 \\ 0.0429 \end{Bmatrix} 0.313g = \begin{Bmatrix} 0.050 \\ 0.086 \\ 0.0073 \end{Bmatrix} g$$

and if applied along the direction of the y-axis, as

$$\{R_{2y1}\} = \max\{\ddot{y}(t)\}_{2y1} = \Gamma_{1y}\{\phi\}_1 SA_{EW}(\omega_1,\xi_1) = 0.558\begin{Bmatrix} 0.3523 \\ 0.2996 \\ -0.0546 \end{Bmatrix} 0.288g = \begin{Bmatrix} 0.057 \\ 0.048 \\ -0.0088 \end{Bmatrix} g$$

$$\{R_{2y2}\} = \max\{\ddot{y}(t)\}_{2y2} = \Gamma_{2y}\{\phi\}_2 SA_{EW}(\omega_2,\xi_2) = -0.824\begin{Bmatrix} 0.5715 \\ -0.4421 \\ 0.0116 \end{Bmatrix} 0.285g = \begin{Bmatrix} -0.134 \\ 0.104 \\ -0.0027 \end{Bmatrix} g$$

$$\{R_{2y3}\} = \max\{\ddot{y}(t)\}_{2y3} = \Gamma_{3y}\{\phi\}_3 SA_{EW}(\omega_3,\xi_3) = 0.934\begin{Bmatrix} 0.2933 \\ 0.5015 \\ 0.0429 \end{Bmatrix} 0.313g = \begin{Bmatrix} 0.086 \\ 0.147 \\ 0.0125 \end{Bmatrix} g$$

Modal Correlation Coefficients

It may be seen from the values of the determined natural frequencies that the system under analysis possesses closely spaced natural frequencies. Hence, it is necessary to use the double-sum rule to combine the found modal responses and one of the formulas introduced in Section 10.3.2 to determine the modal correlation coefficients in it. For simplicity, the expression developed by Der Kiureghian for equal damping ratios (Equation 10.73) will be employed. Accordingly, as all the damping ratios are equal to 0.05, and as $r_{12} = \omega_1/\omega_2 = 5.409/6.218 = 0.870$, $r_{13} = \omega_1/\omega_3 = 5.409/6.739 = 0.803$, and $r_{23} = \omega_2/\omega_3 = 6.218/6.739 = 0.923$, such modal correlation coefficients are equal to

$$\rho_{12} = \rho_{21} = \frac{8\xi^2(1+r_{12})r_{12}^{3/2}}{(1-r_{12}^2)^2 + 4\xi^2 r_{12}(1+r_{12})^2} = \frac{8(0.05)^2(1+0.870)(0.870)^{3/2}}{[1-(0.870)^2]^2 + 4(0.05)^2(0.870)(1+0.870)^2} = 0.339$$

$$\rho_{13} = \rho_{31} = \frac{8\xi^2(1+r_{13})r_{13}^{3/2}}{(1-r_{13}^2)^2 + 4\xi^2 r_{13}(1+r_{13})^2} = \frac{8(0.05)^2(1+0.803)(0.803)^{3/2}}{[1-(0.803)^2]^2 + 4(0.05)^2(0.803)(1+0.803)^2} = 0.170$$

$$\rho_{23} = \rho_{32} = \frac{8\xi^2(1+r_{23})r_{23}^{3/2}}{(1-r_{23}^2)^2 + 4\xi^2 r_{23}(1+r_{23})^2} = \frac{8(0.05)^2(1+0.923)(0.923)^{3/2}}{[1-(0.923)^2]^2 + 4(0.05)^2(0.923)(1+0.923)^2} = 0.606$$

Structural Response by Response Spectrum Method

Square of Maximum Responses Under Single-Component Ground Motion

According to Equations 10.148 through 10.153 and the aforementioned modal correlation coefficients, the square of the maximum acceleration responses when each of the ground motion components being considered is applied by itself along the direction of the x- or y-axis are given by

$$\{R_{1x}^2\} = \{R_{1x1}^2 + R_{1x2}^2 + R_{1x3}^2 + 2\rho_{12}R_{1x1}R_{1x2} + 2\rho_{13}R_{1x1}R_{1x3} + 2\rho_{23}R_{1x2}R_{1x3}\}$$

$$= \begin{Bmatrix} 0.080^2 \\ 0.068^2 \\ 0.0124^2 \end{Bmatrix} + \begin{Bmatrix} 0.312^2 \\ 0.241^2 \\ 0.0063^2 \end{Bmatrix} + \begin{Bmatrix} 0.088^2 \\ 0.151^2 \\ 0.0129^2 \end{Bmatrix} + 2(0.339)\begin{Bmatrix} (0.080)(0.312) \\ (0.068)(-0.241) \\ (-0.0124)(0.0063) \end{Bmatrix}$$

$$+ 2(0.170)\begin{Bmatrix} (0.080)(0.088) \\ (0.068)(0.151) \\ (-0.0124)(0.0129) \end{Bmatrix} + 2(0.606)\begin{Bmatrix} (0.312)(0.088) \\ (-0.241)(0.151) \\ (0.0063)(0.0129) \end{Bmatrix} = \begin{Bmatrix} 0.1641 \\ 0.0338 \\ 0.351 \times 10^{-3} \end{Bmatrix}$$

$$\{R_{1y}^2\} = \{R_{1y1}^2 + R_{1y2}^2 + R_{1y3}^2 + 2\rho_{12}R_{1y1}R_{1y2} + 2\rho_{13}R_{1y1}R_{1y3} + 2\rho_{23}R_{1y2}R_{1y3}\}$$

$$= \begin{Bmatrix} 0.068^2 \\ 0.058^2 \\ 0.0105^2 \end{Bmatrix} + \begin{Bmatrix} 0.241^2 \\ 0.187^2 \\ 0.0049^2 \end{Bmatrix} + \begin{Bmatrix} 0.151^2 \\ 0.258^2 \\ 0.0220^2 \end{Bmatrix} + 2(0.339)\begin{Bmatrix} (0.068)(-0.241) \\ (0.058)(0.187) \\ (-0.0105)(-0.0049) \end{Bmatrix}$$

$$+ 2(0.170)\begin{Bmatrix} (0.068)(0.151) \\ (0.058)(0.258) \\ (-0.0105)(0.0220) \end{Bmatrix} + 2(0.606)\begin{Bmatrix} (-0.241)(0.151) \\ (-0.187)(0.258) \\ (-0.0049)(0.0220) \end{Bmatrix} = \begin{Bmatrix} 0.0338 \\ 0.1758 \\ 0.444 \times 10^{-3} \end{Bmatrix}$$

$$\{R_{2x}^2\} = \{R_{2x1}^2 + R_{2x2}^2 + R_{2x3}^2 + 2\rho_{12}R_{2x1}R_{2x2} + 2\rho_{13}R_{2x1}R_{2x3} + 2\rho_{23}R_{2x2}R_{2x3}\}$$

$$= \begin{Bmatrix} 0.067^2 \\ 0.057^2 \\ 0.0103^2 \end{Bmatrix} + \begin{Bmatrix} 0.173^2 \\ 0.134^2 \\ 0.0035^2 \end{Bmatrix} + \begin{Bmatrix} 0.050^2 \\ 0.086^2 \\ 0.0073^2 \end{Bmatrix} + 2(0.339)\begin{Bmatrix} (0.067)(0.173) \\ (0.057)(-0.134) \\ (-0.0103)(0.0035) \end{Bmatrix}$$

$$+ 2(0.170)\begin{Bmatrix} (0.067)(0.050) \\ (0.057)(0.086) \\ (-0.0103)(0.0073) \end{Bmatrix} + 2(0.606)\begin{Bmatrix} (0.173)(0.050) \\ (-0.134)(0.086) \\ (0.0035)(0.0073) \end{Bmatrix} = \begin{Bmatrix} 0.0564 \\ 0.0111 \\ 0.153 \times 10^{-3} \end{Bmatrix}$$

$$\{R_{2y}^2\} = \{R_{2y1}^2 + R_{2y2}^2 + R_{2y3}^2 + 2\rho_{12}R_{2y1}R_{2y2} + 2\rho_{13}R_{2y1}R_{2y3} + 2\rho_{23}R_{2y2}R_{2y3}\}$$

$$= \begin{Bmatrix} 0.057^2 \\ 0.048^2 \\ 0.0088^2 \end{Bmatrix} + \begin{Bmatrix} 0.134^2 \\ 0.104^2 \\ 0.0027^2 \end{Bmatrix} + \begin{Bmatrix} 0.086^2 \\ 0.147^2 \\ 0.0125^2 \end{Bmatrix} + 2(0.339)\begin{Bmatrix} (0.057)(-0.134) \\ (0.048)(0.104) \\ (-0.0088)(-0.0027) \end{Bmatrix}$$

$$+ 2(0.170)\begin{Bmatrix} (0.057)(0.086) \\ (0.048)(0.147) \\ (-0.0088)(0.0125) \end{Bmatrix} + 2(0.606)\begin{Bmatrix} (-0.134)(0.086) \\ (0.104)(0.147) \\ (-0.0027)(0.0125) \end{Bmatrix} = \begin{Bmatrix} 0.0111 \\ 0.0590 \\ 0.179 \times 10^{-3} \end{Bmatrix}$$

$$\{R_{1xy}^2\} = \{R_{1x1}R_{1y1} + R_{1x2}R_{1y2} + R_{1x3}R_{1y3} + \rho_{12}R_{1x1}R_{1y2} + \rho_{13}R_{1x1}R_{1y3} + \rho_{23}R_{1x2}R_{1y3}$$
$$+ \rho_{21}R_{1x2}R_{1y1} + \rho_{31}R_{1x3}R_{1y1} + \rho_{32}R_{1x3}R_{1y2}\}$$

$$= \left\{\begin{array}{c}(0.080)(0.068)\\(0.068)(0.058)\\(-0.0124)(-0.0105)\end{array}\right\} + \left\{\begin{array}{c}(0.312)(-0.241)\\(-0.241)(0.187)\\(0.0063)(-0.0049)\end{array}\right\} + \left\{\begin{array}{c}(0.088)(0.151)\\(0.151)(0.258)\\(0.0129)(0.0220)\end{array}\right\}$$

$$+ 0.339\left\{\begin{array}{c}(0.080)(-0.241)\\(0.068)(0.187)\\(-0.0124)(0.0049)\end{array}\right\} + 0.170\left\{\begin{array}{c}(0.080)(0.151)\\(0.068)(0.258)\\(-0.0124)(0.0220)\end{array}\right\}$$

$$+ 0.606\left\{\begin{array}{c}(0.312)(0.151)\\(-0.241)(0.258)\\(0.0063)(0.0220)\end{array}\right\} + 0.339\left\{\begin{array}{c}(0.312)(0.068)\\(-0.241)(0.058)\\(0.0063)(-0.0105)\end{array}\right\}$$

$$+ 0.170\left\{\begin{array}{c}(0.088)(0.068)\\(0.151)(0.058)\\(0.0129)(-0.0105)\end{array}\right\} + 0.606\left\{\begin{array}{c}(0.088)(-0.241)\\(0.151)(0.187)\\(0.0129)(-0.0049)\end{array}\right\} = \left\{\begin{array}{c}-0.0370\\-0.0187\\0.358 \times 10^{-3}\end{array}\right\}$$

$$\{R_{2xy}^2\} = \{R_{2x1}R_{2y1} + R_{2x2}R_{2y2} + R_{2x3}R_{2y3} + \rho_{12}R_{2x1}R_{2y2} + \rho_{13}R_{2x1}R_{2y3} + \rho_{23}R_{2x2}R_{2y3}$$
$$+ \rho_{21}R_{2x2}R_{2y1} + \rho_{31}R_{2x3}R_{2y1} + \rho_{32}R_{2x3}R_{2y2}\}$$

$$= \left\{\begin{array}{c}(0.067)(0.057)\\(0.057)(0.048)\\(-0.0103)(-0.0088)\end{array}\right\} + \left\{\begin{array}{c}(0.173)(-0.134)\\(-0.134)(0.104)\\(0.0035)(-0.0027)\end{array}\right\} + \left\{\begin{array}{c}(0.050)(0.086)\\(0.086)(0.147)\\(0.0073)(0.0125)\end{array}\right\}$$

$$+ 0.339\left\{\begin{array}{c}(0.067)(-0.134)\\(0.057)(0.104)\\(-0.0103)(-0.0027)\end{array}\right\} + 0.170\left\{\begin{array}{c}(0.067)(0.086)\\(0.057)(0.147)\\(-0.0103)(0.0125)\end{array}\right\}$$

$$+ 0.606\left\{\begin{array}{c}(0.173)(0.086)\\(-0.134)(0.147)\\(0.0035)(0.0125)\end{array}\right\} + 0.339\left\{\begin{array}{c}(0.173)(0.057)\\(-0.134)(0.048)\\(0.0035)(-0.0088)\end{array}\right\}$$

$$+ 0.170\left\{\begin{array}{c}(0.050)(0.057)\\(0.086)(0.048)\\(0.0073)(-0.0089)\end{array}\right\} + 0.606\left\{\begin{array}{c}(0.050)(-0.134)\\(0.086)(0.104)\\(0.0073)(-0.0027)\end{array}\right\} = \left\{\begin{array}{c}-0.0083\\-0.0031\\0.153 \times 10^{-3}\end{array}\right\}$$

Maximum Acceleration Response

In terms of the obtained maximum responses and according to Equation 10.159, the maximum acceleration response of the system results thus equal to

$$\max\{\ddot{y}(t)\} = \max\left\{\begin{array}{c}\ddot{u}_x(t) + \ddot{u}_{gx}(t)\\\ddot{u}_y(t) + \ddot{u}_{gy}(t)\\\ddot{u}_\theta(t)\end{array}\right\}$$

Structural Response by Response Spectrum Method

$$= \left\{ \begin{array}{l} \left[\begin{array}{l} \frac{1}{2}(0.1641 + 0.0111 + 0.0338 + 0.0564) \\ + \sqrt{\left[\frac{1}{2}(0.1641 + 0.0111 - 0.0338 - 0.0564)\right]^2 + (-0.0370 + 0.0083)^2} \end{array} \right]^{1/2} \\ \left[\begin{array}{l} \frac{1}{2}(0.0338 + 0.0590 + 0.1758 + 0.0111) \\ + \sqrt{\left[\frac{1}{2}(0.0338 + 0.0590 - 0.1758 - 0.0111)\right]^2 + (-0.0187 + 0.0031)^2} \end{array} \right]^{1/2} \\ \left[\begin{array}{l} \frac{10^{-3}}{2}(0.351 + 0.179 + 0.444 + 0.153) \\ + 10^{-3}\sqrt{\left[\frac{1}{2}(0.351 + 0.179 - 0.444 - 0.153)\right]^2 + (0.358 - 0.153)^2} \end{array} \right]^{1/2} \end{array} \right\}$$

$$= \left\{ \begin{array}{l} 0.429 \\ 0.435 \\ 0.0278 \end{array} \right\} g$$

It is interesting to observe that if the two horizontal ground motion components are assumed to be exactly the same and equal to the N–S component of the specified ground motion, then according to Equation 10.163 the acceleration response would be

$$\max\{\ddot{y}(t)\} = \{\sqrt{R_{1x}^2 + R_{1y}^2}\} = \left\{ \begin{array}{l} \sqrt{0.1641 + 0.0338} \\ \sqrt{0.0338 + 0.1758} \\ \sqrt{10^{-3}(0.351 + 0.444)} \end{array} \right\} = \left\{ \begin{array}{l} 0.445 \\ 0.458 \\ 0.0282 \end{array} \right\} g$$

which, as expected, is slightly larger than the response determined earlier.

10.9 RESPONSE SPECTRUM METHOD FOR SPATIALLY VARYING GROUND MOTIONS

In the derivations presented in Sections 10.2 and 10.8, it was implicitly assumed that ground motions affected equally the entire base of a structure. In other words, it was assumed that the ground motion characteristics at the different support points of a structure were exactly the same. This is an adequate assumption for relatively narrow buildings with a rigid foundation. It is not, however, a satisfactory assumption for long and extended structures such as bridges, dams, and pipelines. The reason is that during an earthquake the motion of the ground changes with distance as the earthquake waves travel away from their source. These changes are caused by the finite values of the velocities with which the waves propagate, the attenuation of earthquake waves with distance, and the different ground conditions commonly observed at two distant points. The finite values of the wave propagation velocities cause that two points on the ground surface receive the earthquake waves at different times. Similarly, the attenuation of earthquake waves with distance makes that such two points receive waves with different amplitudes. In contrast, the different ground conditions produce changes in the propagation medium and thus wave reflections and refractions, changes in the propagation velocity of the waves, and different ground motion characteristics. Ground motions affected by the first factor are sometimes referred to as *asynchronous*, whereas ground motions affected by the other two factors are referred to as *nonuniform*. Accordingly, under an asynchronous motion two support points undergo exactly the same motion but at different times. By contrast, under a nonuniform motion two support points undergo excitations with different amplitude and frequency characteristics. In any actual situation, ground motions are both asynchronous and nonuniform.

The degree of asynchrony and nonuniformity of the ground motion affecting a structure obviously depends on the distance between its supports. Consequently, given the relatively high propagation velocity of earthquake waves and their low attenuation rate, the asynchronous and nonuniform character of ground motions is not an important consideration in the analysis of structures with a small area of contact with the ground. The same cannot be said, however, for large structures with a considerable distance between their supports as in the case of suspension bridges and pipelines. For these structures, it is reasonable to expect substantial differences in the characteristics of the ground motions affecting their various supports and, therefore, a substantial effect in their dynamic response by such ground motion differences. In fact, several studies have shown that the effect of differential support motions may significantly increase the dynamic response of a structure in comparison with the response obtained considering a uniform ground motion. The effect of differential support excitations should therefore be accounted for in the seismic response analysis of long and extended structures.

The formulation and solution of the equations of motion for systems subjected to different support excitations is somewhat different from that for systems under a uniform ground motion. In this section, therefore, the modal superposition and response spectrum methods presented in Section 10.2 are extended to generalize these two methods for the analysis of structures subjected to different support excitations.

When the same excitation is applied simultaneously to the different supports of a structure, the response of the structure is usually considered to consist of two components: a rigid-body component and a flexible component. In the rigid-body component, all the points of the structure undergo exactly the same displacements, which are imposed by the motion of the ground. Hence, no stresses or deformations are induced when the structure is subjected to this rigid-body motion. Consequently, the equations of motion in this case may be formulated on the basis of the flexible component alone. In contrast, when the various supports of a structure are subjected to different ground displacements, it can no longer be considered that there is a rigid-body component of the response as in this case the elements of the structure will move relative to one another. Instead, it needs to be considered that stresses and deformations will be produced by the differential ground motions themselves and that these stresses and deformations will be additive to those produced when the structure deforms as a flexible system. Furthermore, without performing an analysis it is no longer possible to know the magnitude of the displacements imposed on the structure by the ground motion alone. Consequently, the formulation of the equation of motion of a system subjected to differential support excitations requires the consideration of absolute displacements (i.e., displacements referred to a fixed reference frame) and the inclusion of the degrees of freedom of its supports.

As in the case of a uniform ground motion excitation, the equation of motion of a system subjected to differential support excitations may be established by the application of Newton's second law. Accordingly, such an equation may be written as

$$\{F_E(t)\}_t - \{F_s(t)\}_t - \{F_D(t)\}_t = [M]_t\{\ddot{y}(t)\}_t \qquad (10.164)$$

where $\{F_E(t)\}_t$, $\{F_s(t)\}_t$, and $\{F_D(t)\}_t$ represent vectors of external, elastic, and damping forces, respectively, $[M]_t$ is the mass matrix of the system, $\{\ddot{y}(t)\}_t$ is a vector of absolute accelerations, and the subscript t indicates that all (structural and support) degrees of freedom are considered in these vectors and matrices. As before, too, it is possible to express the vector of elastic and damping forces in terms of the stiffness and damping matrices of the system. It is convenient, however, to consider that the displacement, velocity, and acceleration vectors in the equation of motion consist of two subvectors each: (1) a subvector that includes the N degrees of freedom of the superstructure (unconstrained degrees of freedom) and (2) a subvector that includes the N_g degrees of freedom of the supports (Figure 10.11). Thus, if the vectors of elastic and damping forces are written in terms of the stiffness and damping matrices of the system, and if the equation is partitioned to separate

Structural Response by Response Spectrum Method

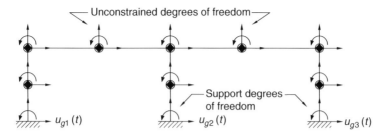

FIGURE 10.11 Identification of unconstrained and support degrees of freedom in a structure subjected to different support excitations.

the structural degrees of freedom from the support degrees of freedom, Equation 10.164 may also be expressed as

$$\begin{bmatrix} [M] & [M_g] \\ [M_g]^T & [M_{gg}] \end{bmatrix} \begin{Bmatrix} \{\ddot{y}(t)\} \\ \{\ddot{u}_g(t)\} \end{Bmatrix} + \begin{bmatrix} [C] & [C_g] \\ [C_g]^T & [C_{gg}] \end{bmatrix} \begin{Bmatrix} \{\dot{y}(t)\} \\ \{\dot{u}_g(t)\} \end{Bmatrix} + \begin{bmatrix} [K] & [K_g] \\ [K_g]^T & [K_{gg}] \end{bmatrix} \begin{Bmatrix} \{y(t)\} \\ \{u_g(t)\} \end{Bmatrix} = \begin{Bmatrix} \{0\} \\ \{P_g(t)\} \end{Bmatrix}$$

(10.165)

where $\{\ddot{y}\}$, $\{\dot{y}\}$, and $\{y\}$ are vectors that contain the structural absolute accelerations, velocities, and displacements, respectively; $\{\ddot{u}_g\}$, $\{\dot{u}_g\}$, and $\{u_g\}$ are vectors whose elements are the known accelerations, velocities, and displacements at the support degrees of freedom; and $\{P_g(t)\}$ depicts a vector that contains the support reactions. It should be noted that the upper subvector of the external force vector is zero because it is assumed that no external forces are applied to the superstructure. The displacements $\{y(t)\}$ and the support reactions $\{P_g(t)\}$ constitute the unknowns in this equation of motion.

To be able to write this equation of motion in a form that is similar to the form of the equation of motion for systems subjected to a uniform support excitation, it is advantageous to consider that the total displacement response of the structure is composed of a pseudostatic component, $\{u_s(t)\}$, and a dynamic component, $\{u(t)\}$, such that

$$\{y(t)\} = \{u_s(t)\} + \{u(t)\}$$

(10.166)

where $\{u_s(t)\}$ is a vector whose elements are the displacements experienced by the structure when the prescribed support displacements at time t are applied statically and is defined by

$$\begin{bmatrix} [K] & [K_g] \\ [K_g]^T & [K_{gg}] \end{bmatrix} \begin{Bmatrix} \{u_s(t)\} \\ \{u_g(t)\} \end{Bmatrix} = \begin{Bmatrix} \{0\} \\ \{P_{gs}(t)\} \end{Bmatrix}$$

(10.167)

where $\{P_{gs}(t)\}$ comprises the reactions that appear at the supports of the structure as result of the imposition of the pseudostatic displacements $\{u_s(t)\}$. Note that these displacements are referred to as pseudostatic because they vary with time and thus are not truly a set of static displacements. Note also that $\{P_{gs}(t)\} = \{0\}$ if the structure is statically determinate, or if the ground displacements induce only a rigid-body motion. The vector $\{u(t)\}$ is referred to as the dynamic displacement vector because these displacements arise as a result of the mass and damping properties of the structure.

With the displacements expressed as indicated by Equation 10.166, the upper part of Equation 10.165 may therefore be expressed as

$$[M](\{\ddot{u}_s(t)\} + \{\ddot{u}(t)\}) + [M_g]\{\ddot{u}_g(t)\} + [C](\{\dot{u}_s(t)\} + \{\dot{u}(t)\}) + [C_g]\{\dot{u}_g(t)\}$$
$$+ [K](\{u_s(t)\} + \{u(t)\}) + [K_g]\{u_g(t)\} = \{0\}$$

(10.168)

which may also be written as

$$[M]\{\ddot{u}(t)\} + [C]\{\dot{u}(t)\} + [K]\{u(t)\} = -[M]\{\ddot{u}_s(t)\} - [M_g]\{\ddot{u}_g(t)\} - [C]\{\dot{u}_s(t)\} - [C_g]\{\dot{u}_g(t)\} \\ - [K]\{u_s(t)\} - [K_g]\{u_g(t)\} \quad (10.169)$$

or as

$$[M]\{\ddot{u}(t)\} + [C]\{\dot{u}(t)\} + [K]\{u(t)\} = \{P_{e\!f\!f}(t)\} \quad (10.170)$$

where $\{P_{e\!f\!f}(t)\}$ is a vector of effective forces defined as

$$\{P_{e\!f\!f}(t)\} = -([M]\{\ddot{u}_s(t)\} + [M_g]\{\ddot{u}_g(t)\} + [C]\{\dot{u}_s(t)\} \\ + [C_g]\{\dot{u}_g(t)\} + [K]\{u_s(t)\} + [K_g]\{u_g(t)\}) \quad (10.171)$$

Note, however, that by virtue of Equation 10.167, one has that

$$[K]\{u_s(t)\} + [K_g]\{u_g(t)\} = \{0\} \quad (10.172)$$

and that

$$\{u_s(t)\} = -[K]^{-1}[K_g]\{u_g(t)\} = [L]\{u_g(t)\} \quad (10.173)$$

where $[L]$ is a matrix that relates the pseudostatic displacements to the ground displacements and describes the influence of the support displacements on the structural displacements. It is called the *influence matrix* and is defined as

$$[L] = -[K]^{-1}[K_g] \quad (10.174)$$

Therefore, the vector of effective forces may also be expressed as

$$\{P_{e\!f\!f}(t)\} = -([M][L] + [M_g])\{\ddot{u}_g(t)\} - ([C][L] + [C_g])\{\dot{u}_g(t)\} \quad (10.175)$$

Equation 10.170, together with Equation 10.175, represents the desired equation of motion for systems subjected to spatially varying ground motions. Note that this equation of motion is of the same form as the conventional equation of motion for systems excited by a uniform ground motion, and thus the same solution techniques may be applied to solve it.

It is of interest to observe that the expression that defines the effective force vector may be simplified further if it is assumed that the damping matrix is proportional to the stiffness matrix. In such a particular case, the term involving the damping matrices in Equation 10.175 vanishes by virtue of Equation 10.172. Also observe that, although in the general case this damping term is not zero, it may be nevertheless neglected in many engineering applications because for small damping values it is usually small relatively to the inertia term. Furthermore, if the structural masses are lumped at the nodes of the structure, then the mass matrix becomes a diagonal matrix and thus $[M_g]$ is a null matrix. Accordingly, if the damping term in Equation 10.175 is neglected and if a lumped-mass model is considered, the effective force vector simplifies to

$$\{P_{e\!f\!f}(t)\} = -[M][L]\{\ddot{u}_g(t)\} \quad (10.176)$$

which, by considering each column of the influence matrix $[L]$ separately, may be expressed alternatively as

Structural Response by Response Spectrum Method

$$\{P_{eff}(t)\} = -[M][\{J\}_1 \{J\}_2 \ldots \{J\}_{N_g}] \begin{Bmatrix} \ddot{u}_{g1}(t) \\ \ddot{u}_{g2}(t) \\ \vdots \\ \ddot{u}_{gN_g}(t) \end{Bmatrix} = -\sum_{l=1}^{N_g} [M]\{J\}_l \ddot{u}_{gl}(t) \quad (10.177)$$

where $\{J\}_l$ denotes the lth column of the influence matrix $[L]$ and is an influence vector associated with the support displacement $u_{gl}(t)$. It represents the displacements induced in the structure resulting from statically applying a unit displacement to the lth support degree of freedom.

Equation 10.176 constitutes a generalization of the effective force vector considered in the case of a uniform ground motion in which the influence matrix $[L]$ substitutes the influence vector $\{J\}$ and the vector $\{\ddot{u}_g(t)\}$ substitutes the scalar $\ddot{u}_g(t)$. The alternative definition given by Equation 10.177 will prove useful in the solution of Equation 10.170.

In terms of Equation 10.177, Equation 10.170 may be put into the form

$$[M]\{\ddot{u}(t)\} + [C]\{\dot{u}(t)\} + [K]\{u(t)\} = -\sum_{l=1}^{N_g} [M]\{J\}_l \ddot{u}_{gl}(t) \quad (10.178)$$

and thus its solution may be obtained by modal superposition. To this end, the equation is first expressed in terms of normal coordinates by applying the transformation

$$\{u(t)\} = [\Phi]\{\eta(t)\} = \sum_{r=1}^{N} \{\phi\}_r \eta_r(t) \quad (10.179)$$

and then premultiplied by $[\Phi]^T$. This leads to

$$[\Phi]^T[M][\Phi]\{\ddot{\eta}(t)\} + [\Phi]^T[C][\Phi]\{\dot{\eta}(t)\} + [\Phi]^T[K][\Phi]\{\eta(t)\} = -\sum_{l=1}^{N_g} [\Phi]^T[M]\{J\}_l \ddot{u}_{gl}(t) \quad (10.180)$$

which, by virtue of the orthogonality properties of the mass and stiffness matrices, and by assuming that the damping matrix of the system is of the Rayleigh or Caughey type, may be reduced to a system of decoupled equations of the form

$$\ddot{\eta}_r(t) + 2\xi_r \omega_r \dot{\eta}_r(t) + \omega_r^2 \eta_r(t) = -\sum_{l=1}^{N_g} \Gamma_{rl} \ddot{u}_{gl}(t) \quad r = 1, 2, \ldots, N \quad (10.181)$$

where Γ_{rl}, $r = 1, \ldots, N$; $l = 1, \ldots, N_g$, are modal participation factors defined as

$$\Gamma_{rl} = \frac{\{\phi\}_r^T [M] \{J\}_l}{M_r^*} = \frac{\{\phi\}_r^T [M] \{J\}_l}{\{\phi\}_r^T [M] \{\phi\}_r} \quad (10.182)$$

Equation 10.181 represents the equation of motion of a single-degree-of-freedom system subjected to the added effect of N_g different excitations. As such, its solution may be obtained by considering each term of the summation separately and then adding the individual solutions. In terms of Duhamel integral, one has thus that

$$\eta_r(t) = \sum_{l=1}^{N_g} \frac{\Gamma_{rl}}{\omega_{dr}} \int_0^t \ddot{u}_{gl}(\tau) e^{-\xi_r \omega_r (t-\tau)} \sin \omega_{dr}(t-\tau) d\tau \quad (10.183)$$

which in compact form may be written as

$$\eta_r(t) = \sum_{l=1}^{N_g} \Gamma_{rl} z_{rl}(t) \tag{10.184}$$

where $z_{rl}(t)$ represents the displacement response of a single degree of freedom with natural frequency ω_r and damping ratio ξ_r to the ground excitation $\ddot{u}_{gl}(t)$.

By combining Equations 10.179 and 10.184, it is possible, thus, to express the dynamic component of the system's displacements as

$$\{u(t)\} = \sum_{l=1}^{N_g} \sum_{r=1}^{N} \Gamma_{rl} \{\phi\}_r z_{rl}(t) \tag{10.185}$$

In turn, by substitution of this equation and Equation 10.173 into Equation 10.166, it is found that the total displacements of the system may be computed according to

$$\{y(t)\} = [L]\{u_g(t)\} + \sum_{l=1}^{N_g} \sum_{r=1}^{N} \Gamma_{rl} \{\phi\}_r z_{rl}(t) \tag{10.186}$$

which, if written explicitly in terms of the influence vectors $\{J\}_l$, may also be put into the form

$$\{y(t)\} = \sum_{l=1}^{N_g} \{J\}_l u_{gl}(t) + \sum_{l=1}^{N_g} \sum_{r=1}^{N} \Gamma_{rl} \{\phi\}_r z_{rl}(t) \tag{10.187}$$

Furthermore, if by analogy with Equations 10.42 and 10.44, it is considered that the influence vectors may also be expressed as

$$\{J\}_l = \sum_{r=1}^{N} \Gamma_{rl} \{\phi\}_r \tag{10.188}$$

then, an alternative expression to determine the system's total displacements is

$$\{y(t)\} = \sum_{l=1}^{N_g} \sum_{r=1}^{N} \Gamma_{rl} \{\phi\}_r [u_{gl}(t) + z_{rl}(t)] \tag{10.189}$$

where the sum $z_{rl}(t) + u_{gl}(t)$ may be interpreted as the absolute displacement response of a single-degree-of-freedom system with natural frequency ω_r and damping ratio ξ_r to a ground acceleration $\ddot{u}_{gl}(t)$.

Equation 10.187 (or Equation 10.189) represents the modal superposition displacement solution of a system subjected to differential support motions. As in the case of systems under a uniform ground excitation and as shown in Section 10.2.5, other measures of the system's response may be obtained directly from this solution. In particular, note that the absolute accelerations are given by

$$\{\ddot{y}(t)\} = \sum_{l=1}^{N_g} \sum_{r=1}^{N} \Gamma_{rl} \{\phi\}_r [\ddot{u}_{gl}(t) + \ddot{z}_{rl}(t)] \tag{10.190}$$

and the internal forces in the ith element of the structure may be computed according to

$$\{F_i(t)\} = [K_i'][T_i]\{y_i(t)\} \tag{10.191}$$

where $\{y_i(t)\}$ is a vector that contains the ith element's nodal absolute displacements and rotations referred to the structure or global system of coordinates, and all other symbols are as defined before. The support reactions may be obtained from the lower part of Equation 10.165.

Structural Response by Response Spectrum Method

As in the case of a uniform ground excitation, it is also possible to determine the maximum value of the single-degree-of-freedom responses $z_{rl}(t)$ from the response spectra of the ground motions $\ddot{u}_{gl}(t)$. As a result, the response of a system subjected to a spatially varying ground motion may also be estimated on the basis of specified response spectra. That is, the peak value of each of the terms in the second summation in Equation 10.187 may be estimated on the basis of the response spectra of the specified support motions. As discussed previously, however, it is highly unlikely that the peak values of such terms will occur at the same time. Therefore, a modal combination rule is needed to combine their peak values. This modal combination rule may be obtained using the same technique and assumptions that were used in the derivation of the combination rules introduced in Section 10.3.2. Accordingly, it may be considered that

$$\max\{y(t)\} = \{c\sqrt{E[y^2(t)]}\} \tag{10.192}$$

where, by virtue of Equation 10.187, $y(t)$ is given by

$$y(t) = \sum_{l=1}^{N_g} a_l u_{gl}(t) + \sum_{l=1}^{N_g}\sum_{r=1}^{N} b_{rl} z_{rl}(t) \tag{10.193}$$

where a_l and b_{rl} are known constants that depend solely on the physical properties of the structure. Thus, the mean square of $y(t)$ may be written as

$$E[y^2(t)] = E[y(t_1)y(t_2)]_{t_1=t_2=t} = E\left[\left\{\sum_{p=1}^{N_g} a_p u_{gp}(t) + \sum_{p=1}^{N_g}\sum_{m=1}^{N} b_{mp} z_{mp}(t)\right\}\left\{\sum_{q=1}^{N_g} a_q u_{gq}(t) + \sum_{q=1}^{N_g}\sum_{n=1}^{N} b_{nq} z_{nq}(t)\right\}\right]$$

$$= E\left[\sum_{p=1}^{N_g}\sum_{q=1}^{N_g} a_p u_{gp}(t) a_q u_{gq}(t) + 2\sum_{p=1}^{N_g}\sum_{q=1}^{N_g}\sum_{r=1}^{N} a_p u_{gp}(t) b_{rq} z_{rq}(t) + \sum_{p=1}^{N_g}\sum_{q=1}^{N_g}\sum_{m=1}^{N}\sum_{n=1}^{N} b_{mp} z_{mp}(t) b_{nq} z_{nq}(t)\right]$$

$$= \sum_{p=1}^{N_g}\sum_{q=1}^{N_g} a_p a_q E[u_{gp}(t) u_{gq}(t)] + 2\sum_{p=1}^{N_g}\sum_{q=1}^{N_g}\sum_{r=1}^{N} a_p b_{rq} E[u_{gp}(t) z_{rq}(t)]$$

$$+ \sum_{p=1}^{N_g}\sum_{q=1}^{N_g}\sum_{m=1}^{N}\sum_{n=1}^{N} b_{mp} b_{nq} E[z_{mp}(t) z_{nq}(t)] \tag{10.194}$$

which, if cross-correlation coefficients are introduced to express the mean square of a product by the product of the root mean squares of the factors, may also be written as

$$E[y^2(t)] = \sum_{p=1}^{N_g}\sum_{q=1}^{N_g} \rho_{pq} a_p a_q \sqrt{E[u_{gp}(t)]}\sqrt{E[u_{gq}(t)]} + 2\sum_{p=1}^{N_g}\sum_{q=1}^{N_g}\sum_{r=1}^{N} \rho_{pqr} a_p b_{rq} \sqrt{E[u_{gp}(t)]}\sqrt{E[z_{rq}(t)]}$$

$$+ \sum_{p=1}^{N_g}\sum_{q=1}^{N_g}\sum_{m=1}^{N}\sum_{n=1}^{N} \rho_{pqmn} b_{mp} b_{nq} \sqrt{E[z_{mp}(t)]}\sqrt{E[z_{nq}(t)]}$$

$$\tag{10.195}$$

where ρ_{pq}, ρ_{pqr}, and ρ_{pqmn} are such cross-correlation coefficients and are defined as

$$\rho_{pq} = \frac{E[u_{gp}(t)u_{gq}(t)]}{\sqrt{E[u_{gp}(t)]}\sqrt{E[u_{gq}(t)]}} \qquad (10.196)$$

$$\rho_{pqr} = \frac{E[u_{gp}(t)z_{rq}(t)]}{\sqrt{E[u_{gp}(t)]}\sqrt{E[z_{rq}(t)]}} \qquad (10.197)$$

$$\rho_{pqmn} = \frac{E[z_{mp}(t)z_{nq}(t)]}{\sqrt{E[z_{mp}(t)]}\sqrt{E[u_{nq}(t)]}} \qquad (10.198)$$

Consequently, the peak displacements may be estimated by means of

$$\max\{y(t)\} = \left\{ \left[\sum_{p=1}^{N_g} \sum_{q=1}^{N_g} \rho_{pq} a_p a_q \max[u_{gp}(t)] \max[u_{gq}(t)] + 2 \sum_{p=1}^{N_g} \sum_{q=1}^{N_g} \sum_{r=1}^{N} \rho_{pqr} a_p b_{rq} \max[u_{gp}(t)] SD_{rq} \right. \right.$$
$$\left. \left. + \sum_{p=1}^{N_g} \sum_{q=1}^{N_g} \sum_{m=1}^{N} \sum_{n=1}^{N} \rho_{pqmn} b_{mp} b_{nq} SD_{mp} SD_{nq} \right]^{1/2} \right\} \qquad (10.199)$$

where $\max[u_{gl}(t)]$, $l = p, q$, represents the peak value of the displacement applied to the lth support degree of freedom, and SD_{rl} denotes the ordinate in the response spectrum of $u_{gl}(t)$ corresponding to the natural frequency and damping ratio of the system in its rth mode of vibration.

In contrast with the case of a uniform support excitation, no explicit expressions have been developed, yet, to evaluate the cross-correlation coefficients in Equation 10.199. A. Der Kiureghian and A. Neuenhofer have suggested a methodology to determine these cross-correlation coefficients on the basis of specified response spectra. However, this methodology requires extensive numerical integrations and an understanding of random vibration theory. As such, it is not discussed here. The interested reader is referred instead to the original article by Der Kiureghian and Neuenhofer, listed at the end of the chapter. Notwithstanding this limitation, note that Equation 10.199 may still be useful to obtain rough estimates of the maximum response of systems subjected to spatially varying ground motions if it is used in combination with arbitrarily selected but conservative values of the cross-correlation coefficients in question. In this regard, it might prove instructive to consult the results of the extensive parametric study presented in the Der Kiureghian and Neuenhofer article to gain an insight into the possible range of the values such cross-correlation coefficients may take.

Although the formulations presented earlier allow one to take into account differential support motions in the analysis of extended structures, it is important to realize that their application requires the adoption of a meaningful ground motion spatial variability model. The definition of such a model, however, is not a straightforward task given the complexity and randomness of the ground motion generation process. As a result, most of the proposed models are based on the consideration of ground motions as random processes whose properties are adjusted to match the spatial variability observed in the ground motions recorded during previous earthquakes. In this regard, note that the simplest representation of ground motion variability between two points is obtained when it is assumed that ground motions are generated by a stationary wave that propagates with a velocity v and attenuates according to an attenuation function $A(r_{pq})$, where r_{pq} denotes the distance

Structural Response by Response Spectrum Method

traveled by the wave. With such a model, it may be considered that the relationship between the ground displacements at two stations p and q is of the form

$$u_{gq}(t) = A(r_{pq})u_{gp}(t - r_{pq}/v) \qquad (10.200)$$

where r_{pq}/v represents a time delay or phase lag and $u_{gp}(t)$ and $u_{gq}(t - r_{pq}/v)$ depict such ground displacements.

EXAMPLE 10.6 BRIDGE RESPONSE UNDER A SPATIALLY VARYING GROUND MOTION

The monolithic, single-span bridge shown in Figure E10.6a is subjected to a spatially varying ground motion. Neglecting axial deformations and lumping its masses as shown in the figure, formulate the equations of motion for this bridge when its left and right supports are, respectively, subjected to horizontal ground displacements $u_{g1}(t)$ and $u_{g2}(t)$, where $u_{g2}(t)$ is equal to $u_{g1}(t - \tau)$, in which τ is a phase lag in seconds. In addition, determine the structural displacements $y_1(t)$ and $y_2(t)$ and the support reactions generated by these two ground motions. Express the answers in terms of the structural mass m, flexural rigidity EI, length L, and $z_{r1}(t)$, the displacement response of a single-degree-of-freedom system to a ground acceleration equal to $\ddot{u}_{g1}(t)$. Assume the structure is undamped.

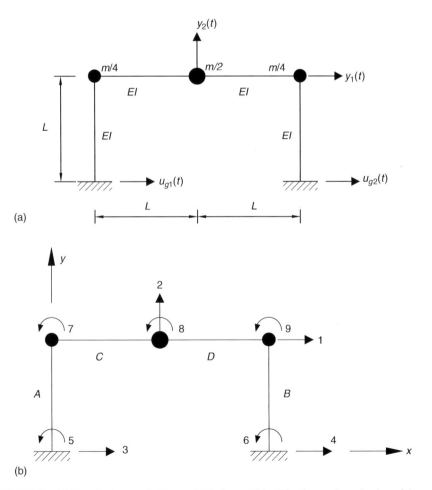

FIGURE E10.6 (a) The single-span bridge and (b) element identification and numbering of degrees of freedom in the bridge considered in Example 10.6.

Solution

Stiffness Matrices of Structural Elements

With reference to the global system of coordinates shown in Figure E10.6b, the stiffness matrices of the two vertical and two horizontal structural elements are given by

$$[K]_A = [K]_B = \frac{EI}{L^3}\begin{bmatrix} 12 & 6L & -12 & -6L \\ 6L & 4L^2 & -6L & 2L^2 \\ -12 & -6L & 12 & 6L \\ -6L & 2L^2 & 6L & 4L^2 \end{bmatrix}$$

$$[K]_C = [K]_D = \frac{EI}{L^3}\begin{bmatrix} 12 & 6L & -12 & 6L \\ 6L & 4L^2 & -6L & 2L^2 \\ -12 & -6L & 12 & -6L \\ 6L & 2L^2 & -6L & 4L^2 \end{bmatrix}$$

Structure Stiffness Matrix

Therefore, with the degrees of freedom numbered as indicated in Figure E10.6b, the stiffness matrix of the entire structure (support degrees of freedom included) results as

$$[\bar{K}] = \frac{EI}{L^3}\begin{bmatrix} 24 & 0 & -12 & -12 & 6L & 6L & 6L & 6L & 0 \\ 0 & 24 & 0 & 0 & 0 & 0 & -6L & 6L & 0 \\ -12 & 0 & 12 & 0 & 0 & -6L & -6L & 0 & 0 \\ -12 & 0 & 0 & 12 & -6L & 0 & 0 & -6L & 0 \\ 6L & 0 & 0 & -6L & 4L^2 & 0 & 0 & 2L^2 & 0 \\ 6L & 0 & -6L & 0 & 0 & 4L^2 & 2L^2 & 0 & 0 \\ 6L & -6L & -6L & 0 & 0 & 2L^2 & 8L^2 & 0 & 2L^2 \\ 6L & 6L & 0 & -6L & 2L^2 & 0 & 0 & 8L^2 & 2L^2 \\ 0 & 0 & 0 & 0 & 0 & 0 & 2L^2 & 2L^2 & 8L^2 \end{bmatrix}\begin{matrix} 1 \\ 2 \\ 3 \\ 4 \\ 5 \\ 6 \\ 7 \\ 8 \\ 9 \end{matrix}$$

which, after eliminating the massless unconstrained rotational degrees of freedom (degrees of freedom 7, 8, and 9) through a static condensation, may be reduced to

$$[\hat{K}] = \frac{EI}{28L^3}\begin{bmatrix} 384 & 0 & -192 & -192 & 120L & 120L \\ 0 & 420 & -126 & 126 & -42L & 42L \\ -192 & -126 & 201 & -9 & 3L & -123L \\ -192 & 126 & -9 & 201 & -123L & 3L \\ 120L & -42L & 3L & -123L & 97L^2 & -L^2 \\ 120L & 42L & -123L & 3L & -L^2 & 97L^2 \end{bmatrix}\begin{matrix} 1 \\ 2 \\ 3 \\ 4 \\ 5 \\ 6 \end{matrix}$$

If it is considered, however, that the supports are fixed against rotations and thus that the rotations at degrees of freedom 5 and 6 are equal to zero, then by partitioning it may be seen that the stiffness matrix that relates the forces and displacements at the first four degrees of freedom is given by

$$[\tilde{K}] = \frac{EI}{28L^3}\begin{bmatrix} 384 & 0 & -192 & -192 \\ 0 & 420 & -126 & 126 \\ -192 & -126 & 201 & -9 \\ -192 & 126 & -9 & 201 \end{bmatrix}\begin{matrix} 1 \\ 2 \\ 3 \\ 4 \end{matrix}$$

Hence, for this case the submatrices $[K]$, $[K_g]$, and $[K_{gg}]$ defined by Equation 10.165 are equal to

$$[K] = \frac{EI}{28L^3}\begin{bmatrix} 384 & 0 \\ 0 & 420 \end{bmatrix} = \frac{3EI}{7L^3}\begin{bmatrix} 32 & 0 \\ 0 & 35 \end{bmatrix}$$

$$[K_g] = \frac{EI}{28L^3}\begin{bmatrix} -192 & -192 \\ -126 & 126 \end{bmatrix} = \frac{3EI}{14L^3}\begin{bmatrix} -32 & -32 \\ -21 & 21 \end{bmatrix}$$

$$[K_{gg}] = \frac{EI}{28L^3}\begin{bmatrix} 201 & -9 \\ -9 & 201 \end{bmatrix} = \frac{3EI}{28L^3}\begin{bmatrix} 67 & -3 \\ -3 & 67 \end{bmatrix}$$

Influence Matrix

As

$$[K]^{-1} = \frac{7L^3}{3EI}\begin{bmatrix} 1/32 & 0 \\ 0 & 1/35 \end{bmatrix}$$

then, according to Equation 10.174, the influence matrix of the system results as

$$[L] = -[K]^{-1}[K_g] = -\frac{7L^3}{3EI}\frac{3EI}{14L^3}\begin{bmatrix} 1/32 & 0 \\ 0 & 1/35 \end{bmatrix}\begin{bmatrix} -32 & -32 \\ -21 & 21 \end{bmatrix} = \begin{bmatrix} 0.5 & 0.5 \\ 0.3 & -0.3 \end{bmatrix}$$

and thus the influence vectors are

$$\{J\}_1 = \begin{Bmatrix} 0.5 \\ 0.3 \end{Bmatrix} \qquad \{J\}_2 = \begin{Bmatrix} 0.5 \\ -0.3 \end{Bmatrix}$$

Mass Matrix

Lumping the mass of the structure as indicated in Figure E10.6a, the mass matrix corresponding to degrees of freedom 1 and 2 is equal to

$$[M] = m\begin{bmatrix} 1 & 0 \\ 0 & 0.5 \end{bmatrix}$$

Equation of Motion

If damping is neglected, then the equation of motion for the structure under consideration is of the form

$$[M]\{\ddot{u}\} + [K]\{u\} = -[M][L]\{\ddot{u}_g(t)\}$$

Hence, after substituting the mass, stiffness, and influence matrices, this equation of motion is given by

$$m\begin{bmatrix} 1 & 0 \\ 0 & 0.5 \end{bmatrix}\begin{Bmatrix} \ddot{u}_1(t) \\ \ddot{u}_2(t) \end{Bmatrix} + \frac{3EI}{7L^3}\begin{bmatrix} 32 & 0 \\ 0 & 35 \end{bmatrix}\begin{Bmatrix} u_1(t) \\ u_2(t) \end{Bmatrix} = -m\begin{bmatrix} 1 & 0 \\ 0 & 0.5 \end{bmatrix}\begin{bmatrix} 0.5 & 0.5 \\ 0.3 & -0.3 \end{bmatrix}\begin{Bmatrix} \ddot{u}_{g1}(t) \\ \ddot{u}_{g2}(t) \end{Bmatrix}$$

$$= -m\begin{bmatrix} 0.5 & 0.5 \\ 0.15 & -0.15 \end{bmatrix}\begin{Bmatrix} \ddot{u}_{g1}(t) \\ \ddot{u}_{g2}(t) \end{Bmatrix}$$

which may also be expressed as

$$\begin{bmatrix} 1 & 0 \\ 0 & 0.5 \end{bmatrix} \begin{Bmatrix} \ddot{u}_1(t) \\ \ddot{u}_2(t) \end{Bmatrix} + \frac{3EI}{7mL^3} \begin{bmatrix} 32 & 0 \\ 0 & 35 \end{bmatrix} \begin{Bmatrix} u_1(t) \\ u_2(t) \end{Bmatrix} = \begin{Bmatrix} 0.5 \\ 0.15 \end{Bmatrix} \ddot{u}_{g1}(t) - \begin{Bmatrix} 0.5 \\ -0.15 \end{Bmatrix} \ddot{u}_{g2}(t)$$

Natural Frequencies and Mode Shapes

As both the mass and stiffness matrices are diagonal matrices, the two unconstrained degrees of freedom being considered are uncoupled. Therefore, the natural frequencies and mode shapes of the system are simply given by

$$\omega_1 = \sqrt{\frac{96EI}{7mL^3}} \text{ rad/s} \qquad \omega_2 = \sqrt{\frac{210EI}{7mL^3}} \text{ rad/s}$$

$$\{\phi\}_1 = \begin{Bmatrix} 1 \\ 0 \end{Bmatrix} \qquad \{\phi\}_2 = \begin{Bmatrix} 0 \\ 1 \end{Bmatrix}$$

Participation Factors

According to Equation 10.182, the participation factors are equal to

$$\Gamma_{11} = \frac{\{\phi\}_1^T[M]\{J\}_1}{\{\phi\}_1^T[M]\{\phi\}_1} = \frac{\begin{Bmatrix} 1 \\ 0 \end{Bmatrix}^T \begin{bmatrix} 1 & 0 \\ 0 & 0.5 \end{bmatrix} \begin{Bmatrix} 0.5 \\ 0.3 \end{Bmatrix}}{\begin{Bmatrix} 1 \\ 0 \end{Bmatrix}^T \begin{bmatrix} 1 & 0 \\ 0 & 0.5 \end{bmatrix} \begin{Bmatrix} 1 \\ 0 \end{Bmatrix}} = \frac{0.5}{1} = 0.5$$

$$\Gamma_{21} = \frac{\{\phi\}_2^T[M]\{J\}_1}{\{\phi\}_2^T[M]\{\phi\}_2} = \frac{\begin{Bmatrix} 0 \\ 1 \end{Bmatrix}^T \begin{bmatrix} 1 & 0 \\ 0 & 0.5 \end{bmatrix} \begin{Bmatrix} 0.5 \\ 0.3 \end{Bmatrix}}{\begin{Bmatrix} 0 \\ 1 \end{Bmatrix}^T \begin{bmatrix} 1 & 0 \\ 0 & 0.5 \end{bmatrix} \begin{Bmatrix} 0 \\ 1 \end{Bmatrix}} = \frac{0.15}{0.5} = 0.3$$

$$\Gamma_{12} = \frac{\{\phi\}_1^T[M]\{J\}_2}{\{\phi\}_1^T[M]\{\phi\}_1} = \frac{\begin{Bmatrix} 1 \\ 0 \end{Bmatrix}^T \begin{bmatrix} 1 & 0 \\ 0 & 0.5 \end{bmatrix} \begin{Bmatrix} 0.5 \\ -0.3 \end{Bmatrix}}{\begin{Bmatrix} 1 \\ 0 \end{Bmatrix}^T \begin{bmatrix} 1 & 0 \\ 0 & 0.5 \end{bmatrix} \begin{Bmatrix} 1 \\ 0 \end{Bmatrix}} = \frac{0.5}{1} = 0.5$$

$$\Gamma_{22} = \frac{\{\phi\}_2^T[M]\{J\}_2}{\{\phi\}_2^T[M]\{\phi\}_2} = \frac{\begin{Bmatrix} 0 \\ 1 \end{Bmatrix}^T \begin{bmatrix} 1 & 0 \\ 0 & 0.5 \end{bmatrix} \begin{Bmatrix} 0.5 \\ -0.3 \end{Bmatrix}}{\begin{Bmatrix} 0 \\ 1 \end{Bmatrix}^T \begin{bmatrix} 1 & 0 \\ 0 & 0.5 \end{bmatrix} \begin{Bmatrix} 0 \\ 1 \end{Bmatrix}} = \frac{-0.15}{0.5} = -0.3$$

Displacement Response

According to Equation 10.187, the displacement response of the system is thus given by

$$\begin{Bmatrix} y_1(t) \\ y_2(t) \end{Bmatrix} = \sum_{l=1}^{2} \{J\}_l u_{gl}(t) + \sum_{l=1}^{2}\sum_{r=1}^{2} \Gamma_{rl}\{\phi\}_r z_{rl}(t)$$

$$= \{J\}_1 u_{g1}(t) + \{J\}_2 u_{g2}(t) + \Gamma_{11}\{\phi\}_1 z_{11}(t) + \Gamma_{21}\{\phi\}_2 z_{21}(t) + \Gamma_{12}\{\phi\}_1 z_{12}(t) + \Gamma_{22}\{\phi\}_2 z_{22}(t)$$

$$= \begin{Bmatrix} 0.5 \\ 0.3 \end{Bmatrix} u_{g1}(t) + \begin{Bmatrix} 0.5 \\ -0.3 \end{Bmatrix} u_{g2}(t) + 0.5 \begin{Bmatrix} 1 \\ 0 \end{Bmatrix} z_{11}(t) + 0.3 \begin{Bmatrix} 0 \\ 1 \end{Bmatrix} z_{21}(t) + 0.5 \begin{Bmatrix} 1 \\ 0 \end{Bmatrix} z_{12}(t) - 0.3 \begin{Bmatrix} 0 \\ 1 \end{Bmatrix} z_{22}(t)$$

However, as $u_{g2}(t) = u_{g1}(t-\tau)$, one has that

$$z_{12}(t) = z_{11}(t-\tau) \qquad z_{22}(t) = z_{21}(t-\tau)$$

and thus,

$$\begin{Bmatrix} y_1(t) \\ y_2(t) \end{Bmatrix} = \begin{Bmatrix} 0.5 \\ 0.3 \end{Bmatrix} u_{g1}(t) + \begin{Bmatrix} 0.5 \\ -0.3 \end{Bmatrix} u_{g1}(t-\tau) + 0.5 \begin{Bmatrix} 1 \\ 0 \end{Bmatrix} z_{11}(t) + 0.3 \begin{Bmatrix} 0 \\ 1 \end{Bmatrix} z_{21}(t)$$

$$+ 0.5 \begin{Bmatrix} 1 \\ 0 \end{Bmatrix} z_{11}(t-\tau) - 0.3 \begin{Bmatrix} 0 \\ 1 \end{Bmatrix} z_{21}(t-\tau)$$

It is interesting to observe that if $\tau = 0$, that is, if the two support motions are exactly the same, then the structural displacements would be equal to

$$\begin{Bmatrix} y_1(t) \\ y_2(t) \end{Bmatrix} = \left(\begin{Bmatrix} 0.5 \\ 0.3 \end{Bmatrix} + \begin{Bmatrix} 0.5 \\ -0.3 \end{Bmatrix}\right) u_{g1}(t) + \left(0.5 \begin{Bmatrix} 1 \\ 0 \end{Bmatrix} + 0.5 \begin{Bmatrix} 1 \\ 0 \end{Bmatrix}\right) z_{11}(t) + \left(0.3 \begin{Bmatrix} 0 \\ 1 \end{Bmatrix} - 0.3 \begin{Bmatrix} 0 \\ 1 \end{Bmatrix}\right) z_{21}(t)$$

$$= \begin{Bmatrix} 1 \\ 0 \end{Bmatrix} u_{g1}(t) + 1.0 \begin{Bmatrix} 1 \\ 0 \end{Bmatrix} z_{11}(t) + 0 \begin{Bmatrix} 0 \\ 1 \end{Bmatrix} z_{21}(t)$$

It may be seen, thus, that no vertical displacement would be induced by a uniform ground motion and that this displacement is different from zero when the ground motion is asynchronous.

Support Reactions

The support reactions may be obtained directly from the lower part of Equation 10.165. However, as for the system under consideration the mass and damping submatrices appearing in the lower part of this equation are all null matrices, the support reactions are simply given by

$$\{P_g(t)\} = [K_g]^T\{y(t)\} + [K_{gg}]\{u_g(t)\}$$

Consequently, upon substitution of the stiffness submatrices and displacement vectors, one obtains

$$\begin{Bmatrix} P_5(t) \\ P_6(t) \end{Bmatrix} = \frac{3EI}{28L^3} \begin{bmatrix} -192 & -126 \\ -192 & 126 \end{bmatrix} \begin{Bmatrix} y_1(t) \\ y_2(t) \end{Bmatrix} + \frac{3EI}{28L^3} \begin{bmatrix} 201 & -9 \\ -9 & 201 \end{bmatrix} \begin{Bmatrix} u_{g1}(t) \\ u_{g1}(t-\tau) \end{Bmatrix}$$

$$= \frac{9EI}{28L^3} \begin{Bmatrix} -64y_1(t) - 42y_2(t) + 67u_{g1}(t) - 3u_{g1}(t-\tau) \\ -64y_1(t) + 42y_2(t) - 3u_{g1}(t) + 67u_{g1}(t-\tau) \end{Bmatrix}$$

which, upon substitution of the developed expressions for $y_1(t)$ and $y_2(t)$, lead to

$$\begin{Bmatrix} P_5(t) \\ P_6(t) \end{Bmatrix} = \frac{9EI}{28L^3} \begin{Bmatrix} 22.4u_{g1}(t) - 22.4u_{g1}(t-\tau) - 32z_{11}(t) - 32z_{11}(t-\tau) - 12.6z_{21}(t) + 12.6z_{21}(t-\tau) \\ -22.4u_{g1}(t) + 22.4u_{g1}(t-\tau) - 32z_{11}(t) - 32z_{11}(t-\tau) + 12.6z_{21}(t) - 12.6z_{21}(t-\tau) \end{Bmatrix}$$

FURTHER READINGS

1. Cheng, F. Y., *Matrix Analysis of Structural Dynamics: Applications and Earthquake Engineering*, Marcel Dekker, Inc., New York, 2001, 997 p.
2. Chopra, A. K., *Dynamics of Structures: Theory and Applications to Earthquake Engineering*, 2nd Edition, Prentice-Hall, Inc., Upper Saddle River, NJ, 2001, 844 p.
3. Der Kiureghian, A., Structural response to stationary excitation, *Journal of Engineering Mechanics, ASCE*, 106, EM6, 1980, 1195–1213.
4. Der Kiureghian, A. and Neuenhofer, A., Response spectrum method for multiple-support seismic excitations, *Earthquake Engineering and Structural Dynamics*, 21, 8, 1992, 713–740.
5. Ghafory-Ashtiany, M. and Singh, M. P., Structural response for six correlated earthquake components, *Earthquake Engineering and Structural Dynamics*, 14, 1, 1986, 103–119.
6. Harichandran, R. S. and Vanmarcke, E. H., Stochastic variation of earthquake ground motion in space and time, *Journal of Engineering Mechanics, ASCE*, 112, 2, 1986, 154–174.
7. Hart, G. C. and Vasudevan, R., Earthquake design of buildings: Damping, *Journal of the Structural Division, ASCE*, 101, ST1, 1975, 11–30.
8. Hart, G. C. and Wong, K., *Structural Dynamics for Structural Engineers*, John Wiley & Sons, Inc., New York, 2000, 591 p.
9. Haviland, R., A Study of the Uncertainties in the Fundamental Translation Periods and Damping Values for Real Buildings, Report No. PB-253, Massachusetts Institute of Technology, 1976.
10. Jeary, A. P., Damping in tall buildings—A mechanism and a predictor, *Earthquake Engineering and Structural Dynamics*, 14, 5, 1986, 733–750.
11. López, O. A. and Torres, R., The critical angle of seismic incidence and the maximum structural response, *Earthquake Engineering and Structural Dynamics*, 26, 9, 1997, 881–894.
12. Price, T. E. and Eberhard, M. O., Effects of spatially varying ground motions on short bridges, *Journal of Structural Engineering, ASCE*, 124, 8, 1998, 948–955.
13. Rosenblueth, E. and Elourdy, J., Response of linear systems in certain disturbances, *Proceedings of Fourth World Conference on Earthquake Engineering*, A-1, Santiago, Chile, 1969, pp. 185–196.
14. Singh, M. P. and Maldonado, G. O., An improved response spectrum method for calculating seismic design response. Part I: Classically damped structures, *Earthquake Engineering and Structural Dynamics*, 20, 7, 1991, 621–635.
15. Smeby, W. and Der Kiureghian, A., Modal combination rules for multicomponent earthquake excitation, *Earthquake Engineering and Structural Dynamics*, 13, 1, 1985, 1–12.

PROBLEMS

10.1 A five-story shear building has the properties shown in Table P10.1. Determine using the response spectrum method the maximum story shears induced in this building when the base of the building is subjected to a ground motion represented by an acceleration response spectrum constructed utilizing the spectral shape corresponding to soil type 3 in Figure 8.17 and a peak ground acceleration of 0.40 g. Consider a damping ratio of 5% in all modes. Recall that the spectral shapes in Figure 8.17 were generated for systems with a damping ratio of 5%.

10.2 Using the response spectrum method and the design spectra shown in Figure P10.2, calculate the maximum lateral forces exerted on a three-story shear building with the properties listed in Table P10.2. Assume a damping matrix proportional to the building's mass and stiffness matrices and a damping ratio of 5% in its first two modes.

10.3 Repeat Problem 10.2 considering a damping matrix proportional to the building's stiffness matrix and a damping ratio of 5% in its fundamental mode.

10.4 Repeat Problem 10.2 considering a damping matrix proportional to the building's mass matrix and a damping ratio of 5% in its fundamental mode.

Structural Response by Response Spectrum Method

TABLE P10.1
Properties of Building in Problem 10.1

Story	Height (m)	Stiffness (MN/m)	Mass (Mg)
1	4	249.88	179
2	3	237.05	170
3	3	224.57	161
4	3	212.09	152
5	3	199.62	143

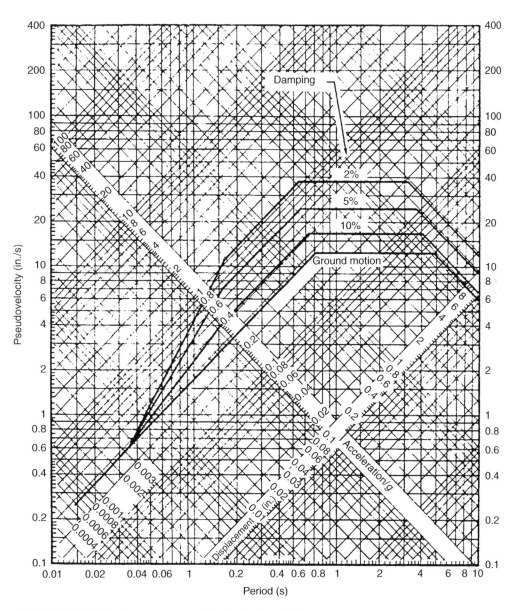

FIGURE P10.2 Design spectrum specified in Problem 10.2.

TABLE P10.2
Mass and Stiffness Properties of Three-Story Shear Building in Problem 10.2

Story	Stiffness (kN/m)	Mass (Mg)
1	355.3	3.0
2	236.9	1.5
3	118.4	1.0

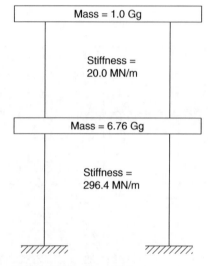

FIGURE P10.5 Two-story shear building in Problem 10.5.

TABLE P10.6
Dynamic Properties of Building in Problem 10.6

Mode	1	2
Frequency (Hz)	0.674	1.112
Participation factor	0.162	0.838
Mode shape	1.000	1.000
	9.747	−0.694

10.5 Estimate the maximum acceleration response of the two-story shear building shown in Figure P10.5 when the base of the building is subjected to the ground motion represented by the elastic response spectrum shown in Figure P8.2a. Assume elastic behavior and a damping ratio of 5% for all modes.

10.6 A system is modeled as an elastic two-degree-of-freedom system with a damping ratio of 10% in each mode and masses for the first and second degrees of freedom are of 6.8 Gg and 1.0 Gg, respectively. The natural frequencies and mode shapes of this two-degree-of-freedom model are given in Table P10.6. Estimate using the response spectrum method the shear force at the base of the system when its base is subjected to a ground motion represented by the response spectrum shown in Figure E10.4b.

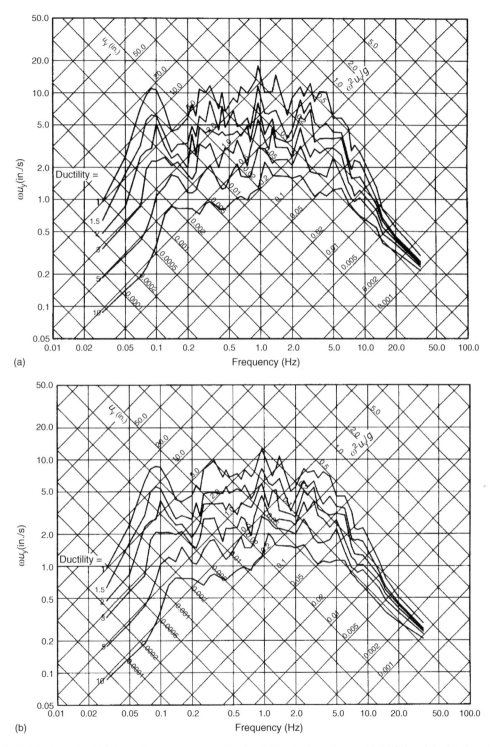

FIGURE P10.7 Yield deformation response spectra for elastoplastic systems with (a) 2% and (b) 5% damping of N10W component of ground accelerations recorded during the 1971 Santiago, Chile, earthquake. (After Riddell, R. and Newmark, N.M., Statistical Analysis of the Response of Nonlinear Systems Subjected to Earthquakes, Structural Research Series No. 468, Department of Civil Engineering, University of Illinois, Urbana, IL, August 1979.)

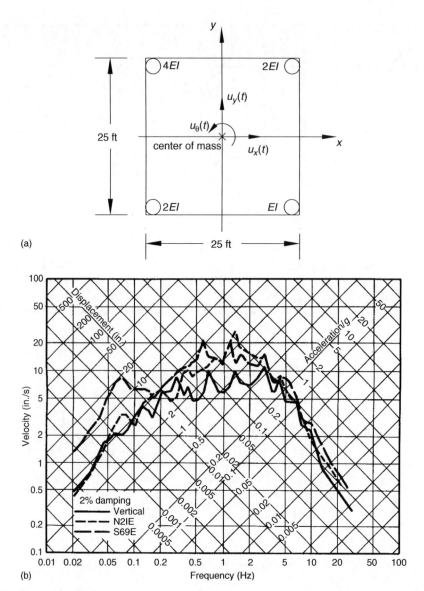

FIGURE P10.10 (a) Plan view of one-story structure in Problem 10.10 and (b) elastic response spectra for 2% damping of the three components of ground acceleration recorded at Taft Lincoln School Tunnel during the 1952 Kern County, California, earthquake. (Reprinted from Newmark, N.M., *Nuclear Engineering and Design*, 20, 303–322, 1972. With permission from Elsevier.)

10.7 The yield deformation response spectra shown in Figures P10.7a and P10.7b are representative of the earthquake ground motions expected at a given area. Using the response spectrum method, determine the maximum story shears in the building described in Problem 10.2 when the building is subjected to a ground motion defined by these response spectra. Consider that the building stories are capable of withstanding deformations that are five times as large as their respective yield deformations. Consider, in addition, that the damping matrix of the building is of the Rayleigh type (i.e., proportional to mass and stiffness matrices) and that the damping ratio in its fundamental mode is 2% of critical. Use linear interpolation to find the spectral ordinates for damping ratios other than 2 and 5%.

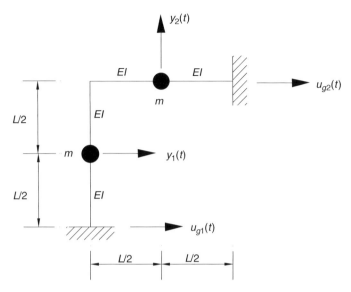

FIGURE P10.11 Two-dimensional pipe in Problem 10.11.

10.8 Determine using the response spectrum method, the maximum acceleration at the top floor of the building-penthouse system analyzed in Example 10.1 when the building is subjected to the same ground motion considered in Example 10.1 but the masses and stiffnesses of the penthouse are reduced by a factor of ten. Use alternatively the following rules to combine the system's modal responses: (a) SRSS, (b) CQC, and (c) double sum in conjunction with the S–M formula to compute the modal correlation coefficients. Assume a damping ratio of 1% for all modes and consider only the first three modes of vibration.

10.9 Determine using the modal acceleration method the displacement response of the five-story building described in Problem 10.1 when the building is subjected to the same ground motion defined therein. Consider alternatively (a) only the first mode of vibration and (b) only the first two modes of vibration.

10.10 Figure P10.10a shows the plan view of a one-story building formed with a perfectly rigid roof slab and four columns. The roof height is 12 ft and the columns are rigidly attached to the roof slab and fixed at their other end. The columns have a circular cross section with the flexural rigidities indicated in Figure P10.10a, where $EI = 2592$ kip ft^2. The roof supports a uniformly distributed vertical load, self weight included, of 144 lb/ft^2. Determine using the response spectrum method and considering the degrees of freedom indicated in Figure P10.10a the maximum acceleration response of the building when the base of the building is subjected to the ground motion recorded during the 1952 Taft earthquake. Figure P10.10b shows the response spectra for the three components of this ground motion. Assume a damping ratio of 2% for all modes.

10.11 The industrial pipe shown in Figure P10.11 is clamped at its two ends and has a 90° bend at midlength. It supports two heavy valves of mass m each as shown in the figure. Neglecting axial deformations and the pipe mass, determine the pipe's acceleration response when its left and right supports are, respectively, subjected to horizontal ground displacements $u_{g1}(t)$ and $u_{g2}(t)$. The ground displacement $u_{g2}(t)$ is equal to the ground displacement $u_{g1}(t)$ except that it begins τ seconds later. Express the answers in terms of the mass m, flexural rigidity EI, length L, phase lag τ, and $\ddot{z}_{r1}(t)$, the acceleration response of a single-degree-of-freedom system to the ground acceleration $\ddot{u}_{g1}(t)$. Assume the pipe has no damping.

11 Structural Response by Step-by-Step Integration Methods

> The search for ways to represent the true nonlinearity of structures goes back to Renaissance times, and present theories are the result of approximately two hundred years of steady development.
>
> W. McGuire, 1994

11.1 INTRODUCTION

The response spectrum method presented in the foregoing chapter is an effective and convenient means of evaluating the seismic response of structures. Still, the method suffers from some shortcomings that make it difficult, if not impossible, to obtain the exact response of a system. Among these shortcomings are (a) the need to make the assumption that the system's mode shapes are also orthogonal with respect to the system's damping matrix and (b) having to use an approximate modal combination rule to estimate the system's maximum response. More importantly, its application is limited to the analysis of linear elastic systems. Fortunately, alternative methods are available that offer the possibility of evaluating a system's seismic response without any of such shortcomings. These alternative methods are numerical procedures that involve (a) dividing the total solution time into a series of small time intervals or *steps* (see Figure 11.1), and (b) integrating the pertinent equations of motion in a sequential manner using the initial and loading conditions at the beginning of each time interval to determine the response of the system at the end of the interval. That is, they involve marching step-by-step along the time axis to integrate the equations of motion and to determine the response of the system at a sequence of specific times. The advantage they offer is that by working with small time intervals it is possible to make suitable assumptions in regard to the variation or characteristics of the response within each of such time intervals without introducing significant errors. Another advantage is that any desired degree of accuracy may be attained by simply reducing the size of the time interval. Furthermore, nonlinear behavior (see Section 6.5.2) may be incorporated into the analysis by merely assuming that the structural properties remain constant during each time interval and reformulating them from one time interval to the next in accordance to an established load-deformation relationship and the solution obtained at the end of the previous time interval.

This chapter will introduce some step-by-step integration procedures and discuss their relative accuracy and the conditions that need to be satisfied to guaranty that the numerical solutions remain bounded despite the inherent approximations and the unavoidable round-off errors. Algorithms for the computer implementation of these integration procedures will also be presented. The discussion will be limited to those methods that are well known or have been used widely in the solution of earthquake engineering problems. Given the vast literature that exists on the subject, no attempt will be made to present even a brief review of all available methods. Instead, the presentation will focus on the fundamental concepts that underlie all such procedures.

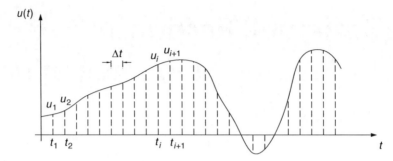

FIGURE 11.1 Displacement solution discretized at a finite number of points separated by time intervals Δt.

11.2 GENERAL CLASSIFICATION

In general, step-by-step integration methods are classified into two broad groups: explicit and implicit. An *explicit method* is one in which the response values calculated at the end of a time interval depend only on the response values obtained in the preceding time step. In explicit methods, therefore, the analysis progresses directly from one step to the next. In contrast, in an *implicit method* the response values at the end of a time interval depend on one or more of the unknown response values at the end of the same time interval. As a result, implicit methods require an iterative procedure in which trial values of the unknown response values are first assumed and then refined by successive iterations. Programming the explicit methods is therefore simpler than programming the implicit ones. However, this advantage is offset by the drawback that the size of the selected time step in the explicit methods normally has to be smaller than in the case of the implicit methods. This is so because in the explicit methods the solution may become numerically unstable (grows without bound as a result of an artificial amplification of the error in one step during the calculations in subsequent steps) if the selected time step is not small enough. By contrast, some of the implicit methods always lead to bounded solutions when the excitation is also bounded, independently of the size of the time step used in the analysis. It is said, thus, that explicit methods are only *conditionally stable*, whereas some implicit methods are *unconditionally stable*. As the primary consideration in selecting a step-by-step integration method is its efficiency measured in terms of the computational effort required to achieve a desired level of accuracy, a conditional stability imposes a serious limitation to the application of the explicit methods. The numerical stability of step-by-step integration methods will be discussed in detail in Section 11.9.

11.3 CENTRAL DIFFERENCE METHOD

As mentioned in Section 11.1, a step-by-step numerical procedure entails discretizing the solution time into small time intervals and determining the response of the system at each time interval using either the initial conditions or the known response parameters at previous time intervals. In the central difference method, the displacements response of the system at the end of a given time interval is determined in terms of the displacements at two previous time steps. It is based on a central difference approximation of the time derivatives that appear in an equation of motion and may be derived as shown later.

As demonstrated in Section 10.2.2, the equation of motion of a linear multi-degree-of-freedom system subjected to a horizontal ground acceleration $\ddot{u}_g(t)$ is given by

$$[M]\{\ddot{u}(t)\} + [C]\{\dot{u}(t)\} + [K]\{u(t)\} = -[M]\{J\}\ddot{u}_g(t) \tag{11.1}$$

where $[M]$, $[C]$, and $[K]$, respectively, denote the mass, damping, and stiffness matrices of the system; $\{u(t)\}$, $\{\dot{u}(t)\}$, and $\{\ddot{u}(t)\}$ are vectors that contain its relative displacements, velocities, and accelerations at time t, and $\{J\}$ is an influence vector that defines the displacements induced in the system when its base is subjected to a static unit displacement in the direction of the excitation. If written for the specific time $t = t_i = i\Delta t$, where i is any integer and Δt a small time interval, Equation 11.1 takes the form

$$[M]\{\ddot{u}\}_i + [C]\{\dot{u}\}_i + [K]\{u\}_i = -[M]\{J\}\ddot{u}_g(t_i) \tag{11.2}$$

where the subscript i indicates that the acceleration, velocity, and displacement vectors are evaluated at the time $t = t_i = i\Delta t$. To solve Equation 11.2 numerically, it may be considered that if expanded into a Taylor series, a displacement $u(t)$ evaluated at times $t_i + \Delta t$ and $t_i - \Delta t$ may be expressed as

$$u(t_i + \Delta t) = u(t_i) + \frac{\dot{u}(t_i)}{1!}\Delta t + \frac{\ddot{u}(t_i)}{2!}\Delta t^2 + \cdots \tag{11.3}$$

$$u(t_i - \Delta t) = u(t_i) - \frac{\dot{u}(t_i)}{1!}\Delta t + \frac{\ddot{u}(t_i)}{2!}\Delta t^2 + \cdots \tag{11.4}$$

where a dot over the function $u(t)$ signifies its derivative with respect to time. Hence, by subtracting Equation 11.4 from Equation 11.3, solving for $\dot{u}(t_i)$ and neglecting high-order terms, one obtains

$$\dot{u}(t_i) \approx \frac{u(t_i + \Delta t) - u(t_i - \Delta t)}{2\Delta t} \tag{11.5}$$

Similarly, adding Equations 11.3 and 11.4, solving for $\ddot{u}(t_i)$ and neglecting high-order terms gives

$$\ddot{u}(t_i) \approx \frac{u(t_i + \Delta t) - 2u(t_i) + u(t_i - \Delta t)}{\Delta t^2} \tag{11.6}$$

Thus, the acceleration and velocity vectors in Equation 11.2 may be considered given approximately by the central differences

$$\{\ddot{u}\}_i = \frac{1}{\Delta t^2}(\{u\}_{i+1} - 2\{u\}_i + \{u\}_{i-1}) \tag{11.7}$$

$$\{\dot{u}\}_i = \frac{1}{2\Delta t}(\{u\}_{i+1} - \{u\}_{i-1}) \tag{11.8}$$

where $\{u\}_{i-1}$, $\{u\}_i$, and $\{u\}_{i+1}$, respectively, denote the displacement vectors evaluated at the times $t = t_i - \Delta t$, $t = t_i$, and $t = t_i + \Delta t$. Likewise, upon substitution of these two relationships into Equation 11.2, the equation of motion at time $t = t_i = i\Delta t$ may be expressed as

$$\frac{1}{\Delta t^2}[M](\{u\}_{i+1} - 2\{u\}_i + \{u\}_{i-1}) + \frac{1}{2\Delta t}[C](\{u\}_{i+1} - \{u\}_{i-1}) + [K]\{u\}_i = -[M]\{J\}\ddot{u}_g(t_i) \tag{11.9}$$

which may also be written as

$$[\hat{M}]\{u\}_{i+1} = \{\hat{F}\}_i \tag{11.10}$$

where $[\hat{M}]$ and $\{\hat{F}\}_i$, respectively, represent an effective mass matrix and an effective load vector defined as

$$[\hat{M}] = \frac{1}{\Delta t^2}[M] + \frac{1}{2\Delta t}[C] \tag{11.11}$$

$$\{\hat{F}\}_i = -[M]\{J\}\ddot{u}_g(t_i) + \left(\frac{2}{\Delta t^2}[M] - [K]\right)\{u\}_i + \left(\frac{1}{2\Delta t}[C] - \frac{1}{\Delta t^2}[M]\right)\{u\}_{i-1} \tag{11.12}$$

It may be noted that all the variables in the right-hand sides of Equations 11.11 and 11.12 are known or have been previously determined. Therefore, one can solve for $\{u\}_{i+1}$ from Equation 11.10 to obtain

$$\{u\}_{i+1} = [\hat{M}]^{-1}\{\hat{F}\}_i \tag{11.13}$$

which represents the solution to the equation of motion given by Equation 11.1 at time $t = t_i + \Delta t$; that is, the solution at the beginning of the next time step. Consequently, Equation 11.13 may be used recursively to obtain the desired solution sequentially from one step to another. Once $\{u\}_{i+1}$ is known, the accelerations and velocities at the beginning of the time step may be obtained, if desired, through the application of Equations 11.7 and 11.8.

In regard to what is needed to get the procedure started, it should be noted that using Equation 11.13 to obtain $\{u\}_1$ requires knowing the solution at $t = 0$ ($i = 0$) and $t = -\Delta t$ ($i = -1$). The solution at $t = 0$ is always known as it represents the specified or assumed vector of initial displacements. However, the solution at $t = -\Delta t$ is not known and needs to be determined as follows:

From Equations 11.7 and 11.8, one has that at $t = 0$ ($i = 0$)

$$\{\ddot{u}\}_0 = \frac{1}{\Delta t^2}(\{u\}_1 - 2\{u\}_0 + \{u\}_{-1}) \tag{11.14}$$

$$\{\dot{u}\}_0 = \frac{1}{2\Delta t}(\{u\}_1 - \{u\}_{-1}) \tag{11.15}$$

Hence, after eliminating $\{u\}_1$ and solving for $\{u\}_{-1}$ from these two equations, it is found that

$$\{u\}_{-1} = \frac{\Delta t^2}{2}\{\ddot{u}\}_0 - \Delta t\{\dot{u}\}_0 + \{u\}_0 \tag{11.16}$$

where $\{u\}_0$, $\{\dot{u}\}_0$, and $\{\ddot{u}\}_0$ are the initial displacement, velocity, and acceleration vectors, respectively. As the initial displacement and velocity vectors are known, and as the initial acceleration vector may be determined directly from the equation of motion in terms of the initial displacement and velocity vectors as

$$\{\ddot{u}\}_0 = -[M]^{-1}([C]\{\dot{u}\}_0 + [K]\{u\}_0) - \{J\}\ddot{u}_g(0) \tag{11.17}$$

then Equation 11.16 may be used to determine the solution at time $t = -\Delta t$ or $i = -1$.

It may be seen from the equations introduced earlier that the central difference method is a simple explicit step-by-step integration method where the most demanding computational task is the inversion of the mass matrix of the system, which is a trivial task when it is a diagonal matrix.

Structural Response by Step-by-Step Integration Methods

The method offers, thus, an excellent computational efficiency. However, the central difference method is only conditionally stable and will lead to exaggeratedly large solutions if the used time interval, Δt, is not short enough. As it will be shown in Section 11.9, the condition for numerical stability for this method is

$$\Delta t \leq \frac{T_N}{\pi} = 0.318 T_N \tag{11.18}$$

where T_N is the natural period of the highest mode of vibration. In those situations where the highest period is not too small, this stability criterion may be satisfied with a reasonable time step. In fact, in such cases it suffices to select the time step that is necessary to define the ground motion function adequately. In the case of multi-degree-of-freedom systems, though, this condition may be overly restrictive and may make the computational cost excessively high if the natural period of the highest mode is too small. Although the higher modes seldom contribute significantly to the response of a system, it is important to realize that the numerical integration should remain stable even for the higher modes because instability of the modal responses in these modes will render the total solution meaningless.

A solution algorithm based on the central difference method is presented in Box 11.1.

Box 11.1   Solution Algorithm Using Central Difference Method

1.0 Initial calculations

  1.1 Formulate the system's stiffness matrix $[K]$, mass matrix $[M]$, and damping matrix $[C]$

  1.2 Establish vectors of initial displacements $\{u\}_0$ and initial velocities $\{\dot{u}\}_0$

  1.3 Compute initial acceleration vector as

$$\{\ddot{u}\}_0 = -[M]^{-1}([C]\{\dot{u}\}_0 + [K]\{u\}_0) - \{J\}\ddot{u}_g(0)$$

  1.4 Select time step Δt

  1.5 Compute displacements at time $t = -\Delta t$ according to

$$\{u\}_{-1} = \{u\}_0 - \Delta t\{\dot{u}\}_0 + \frac{\Delta t^2}{2}\{\ddot{u}\}_0$$

  1.6 Compute effective mass matrix as

$$[\hat{M}] = \frac{1}{\Delta t^2}[M] + \frac{1}{2\Delta t}[C]$$

  1.7 Compute constant matrices $[a]$ and $[b]$ as

$$[a] = \frac{1}{2\Delta t}[C] - \frac{1}{\Delta t^2}[M]$$

$$[b] = \frac{2}{\Delta t^2}[M] - [K]$$

  1.8 Set $i = 0$

2.0 Calculations for ith time step
   2.1 Compute effective load vector at time t_i as

$$\{\hat{F}\}_i = -[M]\{J\}\ddot{u}_g(t_i) + [a]\{u\}_{i-1} + [b]\{u\}_i$$

   2.2 Compute displacement vector at time $t_i + \Delta t$ as

$$\{u\}_{i+1} = [\hat{M}]^{-1}\{\hat{F}\}_i$$

   2.3 If desired, compute velocity and acceleration vectors at time $t_i = i\Delta t$ according to

$$\{\dot{u}\}_i = \frac{1}{2\Delta t}(\{u\}_{i+1} - \{u\}_{i-1})$$

$$\{\ddot{u}\}_i = \frac{1}{\Delta t^2}(\{u\}_{i+1} - 2\{u\}_i + \{u\}_{i-1})$$

3.0 Calculations for next time step
   Replace i by $i + 1$ and repeat steps 2.1–2.3.

EXAMPLE 11.1 DISPLACEMENT RESPONSE OF TWO-DEGREE-OF-FREEDOM SYSTEM BY CENTRAL DIFFERENCE METHOD

An undamped two-degree-of-freedom shear building initially at rest has the following mass and stiffness matrices:

$$[M] = \begin{bmatrix} 2 & 0 \\ 0 & 2 \end{bmatrix} \text{Gg} \quad [K] = \begin{bmatrix} 400 & -200 \\ -200 & 200 \end{bmatrix} \text{MN/m}$$

Determine using the central difference method the displacement response of the building for the first 2 s when the building is subjected to a ground motion characterized by

$$\ddot{u}_g(t) = 3.0\sin(2t) \text{ m/s}^2$$

Use alternatively time steps of 0.05 and 0.15 s.

Solution

Solution at $t = -\Delta t$

As for zero initial conditions, one has that

$$\{u\}_0 = \{0\} \quad \{\dot{u}\}_0 = \{0\}$$

and as for $t = 0$

$$\ddot{u}_g(0) = 0$$

Then, according to Equations 11.16 and 11.17, one obtains

$$\{\ddot{u}\}_0 = \{0\} \qquad \{u\}_{-1} = \{0\}$$

Effective Mass Matrix and Load Vector

According to Equations 11.11 and 11.12 and the mass, damping, and stiffness matrices of the system under consideration, the effective mass matrix, its inverse, and the effective load vector are

$$[\hat{M}] = \frac{1}{\Delta t^2}[M] + \frac{1}{2\Delta t}[C] = \frac{1}{\Delta t^2}\begin{bmatrix} 2 & 0 \\ 0 & 2 \end{bmatrix}$$

$$[\hat{M}]^{-1} = \frac{\Delta t^2}{2}\begin{bmatrix} 1 & 0 \\ 0 & 1 \end{bmatrix}$$

$$\{\hat{F}\}_i = -[M]\{J\}\ddot{u}_g(t_i) + \left(\frac{2}{\Delta t^2}[M] - [K]\right)\{u\}_i + \left(\frac{1}{2\Delta t}[C] - \frac{1}{\Delta t^2}[M]\right)\{u\}_{i-1}$$

$$= -\begin{bmatrix} 2 & 0 \\ 0 & 2 \end{bmatrix}\begin{Bmatrix} 1 \\ 1 \end{Bmatrix}3\sin(2t_i) + \left(\frac{2}{\Delta t^2}\begin{bmatrix} 2 & 0 \\ 0 & 2 \end{bmatrix} - \begin{bmatrix} 400 & -200 \\ -200 & 200 \end{bmatrix}\right)\begin{Bmatrix} u_1 \\ u_2 \end{Bmatrix}_i - \frac{1}{\Delta t^2}\begin{bmatrix} 2 & 0 \\ 0 & 2 \end{bmatrix}\begin{Bmatrix} u_1 \\ u_2 \end{Bmatrix}_{i-1}$$

$$= -\begin{Bmatrix} 2 \\ 2 \end{Bmatrix}3\sin(2t_i) + \frac{4}{\Delta t^2}\begin{Bmatrix} u_1 \\ u_2 \end{Bmatrix}_i - \begin{bmatrix} 400 & -200 \\ -200 & 200 \end{bmatrix}\begin{Bmatrix} u_1 \\ u_2 \end{Bmatrix}_i - \frac{2}{\Delta t^2}\begin{Bmatrix} u_1 \\ u_2 \end{Bmatrix}_{i-1}$$

Displacement Response

In view of Equation 11.13, the displacement response of the system is thus given by

$$\begin{Bmatrix} u_1 \\ u_2 \end{Bmatrix}_{i+1} = [\hat{M}]^{-1}\{\hat{F}\}_i = \frac{\Delta t^2}{2}\begin{bmatrix} 1 & 0 \\ 0 & 1 \end{bmatrix}\{\hat{F}\}_i$$

$$= -\Delta t^2\begin{Bmatrix} 1 \\ 1 \end{Bmatrix}3\sin(2t_i) + 2\begin{Bmatrix} u_1 \\ u_2 \end{Bmatrix}_i - \frac{\Delta t^2}{2}\begin{bmatrix} 400 & -200 \\ -200 & 200 \end{bmatrix}\begin{Bmatrix} u_1 \\ u_2 \end{Bmatrix}_i - \begin{Bmatrix} u_1 \\ u_2 \end{Bmatrix}_{i-1}$$

which leads to the following two algebraic equations:

$$(u_1)_{i+1} = -3\Delta t^2 \sin(2t_i) + 2(u_1)_i - 200\Delta t^2(u_1)_i + 100\Delta t^2(u_2)_i - (u_1)_{i-1}$$

$$(u_2)_{i+1} = -3\Delta t^2 \sin(2t_i) + 2(u_2)_i + 100\Delta t^2(u_1)_i - 100\Delta t^2(u_2)_i - (u_2)_{i-1}$$

The displacements obtained from these two equations for the first 15 steps are listed in Table E11.1 for $\Delta t = 0.05$ s and $\Delta t = 0.15$ s, respectively. The displacement response for the first 2.1 s using a time step of 0.05 s is shown in Figure E11.1. Observe from the results in Table E11.1 how the solution quickly becomes unstable when the time step of 0.15 s is used.

TABLE E11.1
First 15 Steps of Displacement Response of Two-Degree-of-Freedom Building in Example 11.1 Using Central Difference Method

	$\Delta t = 0.05$ s			$\Delta t = 0.15$ s		
Step	Time (s)	u_1 (m)	u_2 (m)	Time (s)	u_1 (m)	u_2 (m)
0	0.00	0.0000	0.0000	0.00	0.0000	0.0000
1	0.05	0.0000	0.000	0.15	0.0000	0.0000
2	0.10	−0.0007	−0.0007	0.30	−0.0199	−0.0199
3	0.15	−0.0028	−0.0030	0.45	−0.0331	−0.0780
4	0.20	−0.0064	−0.0074	0.60	−0.1256	−0.0880
5	0.25	−0.0116	−0.0145	0.75	0.0864	−0.2456
6	0.30	−0.0182	−0.0244	0.90	−0.7102	0.2764
7	0.35	−0.0260	−0.0371	1.05	2.2453	−1.4872
8	0.40	−0.0350	−0.0518	1.20	−8.3075	5.0890
9	0.45	−0.0447	−0.0677	1.35	29.928	−18.522
10	0.50	−0.0549	−0.0837	1.50	−108.21	66.851
11	0.55	−0.0649	−0.0988	1.65	391.02	−241.68
12	0.60	−0.0738	−0.1122	1.80	−1413.1	873.37
13	0.65	−0.0808	−0.1229	1.95	5106.9	−3156.1
14	0.70	−0.0854	−0.1304	2.10	−18455.	11406.
15	0.75	−0.0873	−0.1340	2.25	66696.	−41220.

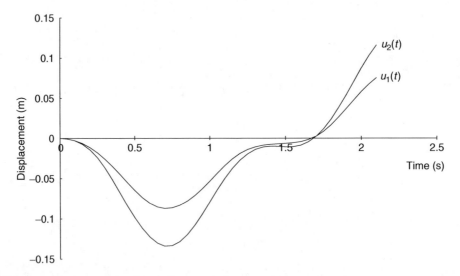

FIGURE E11.1 First 2.1 s of displacement response of two-degree-of-freedom building in Example 11.1 obtained using central difference method with a time step of 0.05 s.

11.4 HOUBOLT METHOD

Houbolt method is another step-by-step integration method that uses finite differences to approximate the time derivatives that appear in an equation of motion. In Houbolt method, however, equilibrium is established at the end of each time interval as opposed to at the beginning of the interval as in the finite difference method. That is, the equation of motion is written for the time $t = t_{i+1} = (i + 1)\Delta t$ as

$$[M]\{\ddot{u}\}_{i+1} + [C]\{\dot{u}\}_{i+1} + [K]\{u\}_{i+1} = -[M]\{J\}\ddot{u}_g(t_{i+1}) \tag{11.19}$$

Structural Response by Step-by-Step Integration Methods

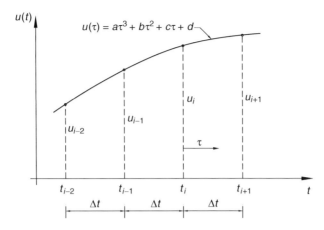

FIGURE 11.2 Assumed variation of displacement response during three consecutive time intervals in Houbolt method.

In addition, it is assumed that within each time interval the displacement response varies with time according to a cubic parabola of the form (see Figure 11.2)

$$u(\tau) = a\tau^3 + b\tau^2 + c\tau + d \tag{11.20}$$

where $\tau = t - t_i$, and a–d are constants. Consequently, the velocities and accelerations within any given time interval vary with time according to

$$\dot{u}(\tau) = 3a\tau^2 + 2b\tau + c \tag{11.21}$$

$$\ddot{u}(\tau) = 6a\tau + 2b \tag{11.22}$$

and thus, by setting $\tau = \Delta t$, the velocity and acceleration at the end of a time interval are given by

$$\dot{u}_{i+1} = 3a\Delta t^2 + 2b\Delta t + c \tag{11.23}$$

$$\ddot{u}_{i+1} = 6a\Delta t + 2b \tag{11.24}$$

The constants a–d in the preceding equations may be determined from the known displacements at the end of three previous time intervals as follows.

With reference to Figure 11.2 and the shown location of the origin of the τ axis, let u_{i+1}, u_i, u_{i-1}, and u_{i-2}, respectively, denote the displacements at the ends of the $(i + 1)$th, ith, $(i - 1)$th, and $(i - 2)$th time intervals. If Equation 11.20 is written specifically for the four points shown in Figure 11.2 (i.e., $\tau = -2\Delta t$, $\tau = -\Delta t$, $\tau = 0$, and $\tau = \Delta t$), one obtains

$$u_{i-2} = -8a\Delta t^3 + 4b\Delta t^2 - 2c\Delta t + d \tag{11.25}$$

$$u_{-1} = -a\Delta t^3 + b\Delta t^2 - c\Delta t + d \tag{11.26}$$

$$u_i = d \tag{11.27}$$

$$u_{i+1} = a\Delta t^3 + b\Delta t^2 + c\Delta t + d \tag{11.28}$$

which represents a system of four simultaneous equations in the constants a–d. Solving this system of simultaneous equations yields

$$a = \frac{1}{6\Delta t^3}(-u_{i-2} + 3u_{i-1} - 3u_i + u_{i+1}) \tag{11.29}$$

$$b = \frac{1}{2\Delta t^2}(u_{i-1} - 2u_i + u_{i+1}) \tag{11.30}$$

$$c = \frac{1}{6\Delta t}(u_{i-2} - 6u_{i-1} + 3u_i + 2u_{i+1}) \tag{11.31}$$

Thus, upon substitution of these expressions into Equations 11.23 and 11.24, the velocity and acceleration at time $t = t_{i+1}$ may be written as

$$\dot{u}_{i+1} = \frac{1}{6\Delta t}(11u_{i+1} - 18u_i + 9u_{i-1} - 2u_{i-2}) \tag{11.32}$$

$$\ddot{u}_{i+1} = \frac{1}{\Delta t^2}(2u_{i+1} - 5u_i + 4u_{i-1} - u_{i-2}) \tag{11.33}$$

and in light of these equations the velocity and acceleration vectors in Equation 11.19 may be considered approximately equal to

$$\{\dot{u}\}_{i+1} = \frac{1}{6\Delta t}(11\{u\}_{i+1} - 18\{u\}_i + 9\{u\}_{i-1} - 2\{u\}_{i-2}) \tag{11.34}$$

$$\{\ddot{u}\}_{i+1} = \frac{1}{\Delta t^2}(2\{u\}_{i+1} - 5\{u\}_i + 4\{u\}_{i-1} - \{u\}_{i-2}) \tag{11.35}$$

Consequently, the equation of motion defined by Equation 11.19 may be expressed approximately as

$$\frac{1}{\Delta t^2}[M](2\{u\}_{i+1} - 5\{u\}_i + 4\{u\}_{i-1} - \{u\}_{i-2}) + \frac{1}{6\Delta t}[C](11\{u\}_{i+1} - 18\{u\}_i + 9\{u\}_{i-1} - 2\{u\}_{i-2}) + [K]\{u\}_{i+1} = -[M]\{J\}\ddot{u}_g(t_{i+1}) \tag{11.36}$$

which, after moving all the known terms to the right-hand side of the equation, may also be written as

$$[\hat{K}_H]\{u\}_{i+1} = \{\hat{F}_H\}_{i+1} \tag{11.37}$$

where $[\hat{K}_H]$ and $\{\hat{F}_H\}_{i+1}$ are an effective stiffness matrix and an effective load vector defined as

$$[\hat{K}_H] = [K] + \frac{2}{\Delta t^2}[M] + \frac{11}{6\Delta t}[C] \tag{11.38}$$

$$\{\hat{F}_H\}_{i+1} = -[M]\{J\}\ddot{u}_g(t_{i+1}) + [M]\left(\frac{5}{\Delta t^2}\{u\}_i - \frac{4}{\Delta t^2}\{u\}_{i-1} + \frac{1}{\Delta t^2}\{u\}_{i-2}\right) + [C]\left(\frac{3}{\Delta t}\{u\}_i - \frac{3}{2\Delta t}\{u\}_{i-1} + \frac{1}{3\Delta t}\{u\}_{i-2}\right) \tag{11.39}$$

and where the subscript H relates these arrays to Houbolt method. The displacement response at the end of the ith time interval may be then obtained in terms of the displacements at the end of the previous three steps according to

$$\{u\}_{i+1} = [\hat{K}_H]^{-1}\{\hat{F}_H\}_{i+1} \tag{11.40}$$

In similarity with the central difference method, Equation 11.40 may be used recursively to obtain the displacement solution of a system at a series of discrete times. If desired, the corresponding velocities and accelerations may also be determined through the application of Equations 11.34 and 11.35. It should be noted, however, that there is no need to compute the velocity and acceleration responses to determine the displacement response as they do not appear in Equations 11.38 and 11.39.

It is important to observe that in Houbolt method it is necessary to know the displacements $\{u\}_2$, $\{u\}_1$, and $\{u\}_0$ to get the step-by-step integration started. The initial displacement vector $\{u\}_0$ is known or specified, but the other two need to be calculated using any other method, such as the central difference method. Thus, Houlbolt method is not a *self-starting* method. Observe also that, differently from the central difference method, Houbolt' method involves the inversion or factorization of the system's stiffness matrix. In contrast from the central difference method, however, Houbolt method is, as it will be shown in Section 11.9, unconditionally stable. This means that there is no limit in the size of the time step that may be used to obtain bounded solutions and that this time step may be selected to be, in general, much larger than the critical one for the central difference method.

The solution algorithm presented in Box 11.2 summarizes Houbolt method.

Box 11.2 Solution Algorithm Using Houbolt Method

1.0 Initial calculations
    1.1 Formulate the system's stiffness matrix $[K]$, mass matrix $[M]$, and damping matrix $[C]$
    1.2 Establish vectors of initial displacements $\{u\}_0$ and initial velocities $\{\dot{u}\}_0$
    1.3 Compute initial acceleration vector as

$$\{\ddot{u}\}_0 = -[M]^{-1}([C]\{\dot{u}\}_0 + [K]\{u\}_0) - \{J\}\ddot{u}_g(0)$$

    1.4 Select time step Δt
    1.5 Compute using an alternative integration method the starting displacement vectors $\{u\}_1$ and $\{u\}_2$
    1.6 Compute effective stiffness matrix as

$$[\hat{K}_H] = [K] + \frac{2}{\Delta t^2}[M] + \frac{11}{6\Delta t}[C]$$

    1.7 Set $i = 2$
2.0 Calculations for ith time step
    2.1 Compute effective load vector at time t_{i+1} as

$$\{\hat{F}_H\}_{i+1} = -[M]\{J\}\ddot{u}_g(t_{i+1}) + [M]\left(\frac{5}{\Delta t^2}\{u\}_i - \frac{4}{\Delta t^2}\{u\}_{i-1} + \frac{1}{\Delta t^2}\{u\}_{i-2}\right)$$

$$+ [C]\left(\frac{3}{\Delta t}\{u\}_i - \frac{3}{2\Delta t}\{u\}_{i-1} + \frac{1}{3\Delta t}\{u\}_{i-2}\right)$$

2.2 Compute displacement vector at time $t = t_i + \Delta t$ as

$$\{u\}_{i+1} = [\hat{K}_H]^{-1}\{\hat{F}_H\}_{i+1}$$

2.3 If desired, compute the velocity and acceleration vectors at time $t = t_i + \Delta t$ according to

$$\{\dot{u}\}_{i+1} = \frac{1}{6\Delta t}\left(11\{u\}_{i+1} - 18\{u\}_i + 9\{u\}_{i-1} - 2\{u\}_{i-2}\right)$$

$$\{\ddot{u}\}_{i+1} = \frac{1}{\Delta t^2}\left(2\{u\}_{i+1} - 5\{u\}_i + 4\{u\}_{i-1} - \{u\}_{i-2}\right)$$

3.0 Calculations for next time step
Replace i by $i + 1$ and repeat steps 2.1–2.3.

EXAMPLE 11.2 DYNAMIC RESPONSE OF TWO-DEGREE-OF-FREEDOM SYSTEM BY HOUBOLT METHOD

Determine for the first 0.75 s the displacement, velocity, and acceleration responses of the two-degree-of-freedom shear building considered in Example 11.1. Use Houbolt method with a time step of 0.05 s.

Solution

Initial Calculations

According to the initial conditions specified for this problem and the calculations made in Example 11.1, one has that

$$\{u\}_0 = \begin{Bmatrix} 0 \\ 0 \end{Bmatrix} \quad \{\dot{u}\}_0 = \begin{Bmatrix} 0 \\ 0 \end{Bmatrix} \quad \{\ddot{u}\}_0 = \begin{Bmatrix} 0 \\ 0 \end{Bmatrix}$$

$$\{u\}_1 = \begin{Bmatrix} 0 \\ 0 \end{Bmatrix} \quad \{u\}_2 = \begin{Bmatrix} -0.00075 \\ -0.00075 \end{Bmatrix}$$

Effective Stiffness Matrix and Load Vector

According to Equations 11.38 and 11.39, the effective stiffness matrix, its inverse, and the effective load vector for the system under consideration are

$$[\hat{K}_H] = [K] + \frac{2}{\Delta t^2}[M] + \frac{11}{6\Delta t}[C]$$

$$= \begin{bmatrix} 400 & -200 \\ -200 & 200 \end{bmatrix} + \frac{2}{0.05^2}\begin{bmatrix} 2 & 0 \\ 0 & 2 \end{bmatrix} = 200\begin{bmatrix} 10 & -1 \\ -1 & 9 \end{bmatrix}$$

$$[\hat{K}_H]^{-1} = \frac{1}{17{,}800}\begin{bmatrix} 9 & 1 \\ 1 & 10 \end{bmatrix}$$

Structural Response by Step-by-Step Integration Methods

$$\{\hat{F}_H\}_{i+1} = -[M]\{J\}u_g(t_{i+1}) + [M]\left(\frac{5}{\Delta t^2}\{u\}_i - \frac{4}{\Delta t^2}\{u\}_{i-1} + \frac{1}{\Delta t^2}\{u\}_{i-2}\right)$$

$$+ [C]\left(\frac{3}{\Delta t}\{u\}_i - \frac{3}{2\Delta t}\{u\}_{i-1} + \frac{1}{3\Delta t}\{u\}_{i-2}\right)$$

$$= -\begin{bmatrix} 2 & 0 \\ 0 & 2 \end{bmatrix}\begin{Bmatrix} 1 \\ 1 \end{Bmatrix}\ddot{u}_g(t_{i+1}) + \begin{bmatrix} 2 & 0 \\ 0 & 2 \end{bmatrix}\left(\frac{5}{0.05^2}\{u\}_i - \frac{4}{0.05^2}\{u\}_{i-1} + \frac{1}{0.05^2}\{u\}_{i-2}\right)$$

$$= -2\begin{Bmatrix} 1 \\ 1 \end{Bmatrix}\ddot{u}_g(t_{i+1}) + 4000\begin{Bmatrix} u_1 \\ u_2 \end{Bmatrix}_i - 3200\begin{Bmatrix} u_1 \\ u_2 \end{Bmatrix}_{i-1} + 800\begin{Bmatrix} u_1 \\ u_2 \end{Bmatrix}_{i-2}$$

Displacement Response

Thus, by virtue of Equation 11.40, the displacement response of the system at time $t = t_{i+1}$ is given by

$$\{u\}_{i+1} = [\hat{K}_H]^{-1}\{\hat{F}_H\}_{i+1} = \frac{1}{17{,}800}\begin{bmatrix} 9 & 1 \\ 1 & 10 \end{bmatrix}\{\hat{F}_H\}_{i+1}$$

$$= -\frac{1}{8900}\begin{Bmatrix} 10 \\ 11 \end{Bmatrix}\ddot{u}_g(t_{i+1}) + \frac{20}{89}\begin{bmatrix} 9 & 1 \\ 1 & 10 \end{bmatrix}\begin{Bmatrix} u_1 \\ u_2 \end{Bmatrix}_i - \frac{16}{89}\begin{bmatrix} 9 & 1 \\ 1 & 10 \end{bmatrix}\begin{Bmatrix} u_1 \\ u_2 \end{Bmatrix}_{i-1} + \frac{4}{89}\begin{bmatrix} 9 & 1 \\ 1 & 10 \end{bmatrix}\begin{Bmatrix} u_1 \\ u_2 \end{Bmatrix}_{i-2}$$

from which one obtains the following two algebraic equations:

$$(u_1)_{i+1} = \frac{1}{89}\left[-0.10\ddot{u}_g(t_{i+1}) + 180(u_1)_i + 20(u_2)_i - 144(u_1)_{i-1} - 16(u_2)_{i-1} + 36(u_1)_{i-2} + 4(u_2)_{i-2}\right]$$

$$(u_2)_{i+1} = \frac{1}{89}\left[-0.11\ddot{u}_g(t_{i+1}) + 20(u_1)_i + 200(u_2)_i - 16(u_1)_{i-1} - 160(u_2)_{i-1} + 4(u_1)_{i-2} + 40(u_2)_{i-2}\right]$$

Velocity and Acceleration Responses

In light of Equations 11.34 and 11.35 and the determined displacement responses, the relative velocity and acceleration responses at time $t = t_{i+1}$ may be then computed according to

$$\begin{Bmatrix} \dot{u}_1 \\ \dot{u}_2 \end{Bmatrix}_{i+1} = \frac{1}{6\Delta t}\left(11\begin{Bmatrix} u_1 \\ u_2 \end{Bmatrix}_{i+1} - 18\begin{Bmatrix} u_1 \\ u_2 \end{Bmatrix}_i + 9\begin{Bmatrix} u_1 \\ u_2 \end{Bmatrix}_{i-1} - 2\begin{Bmatrix} u_1 \\ u_2 \end{Bmatrix}_{i-2}\right)$$

$$= \frac{110}{3}\begin{Bmatrix} u_1 \\ u_2 \end{Bmatrix}_{i+1} - 60\begin{Bmatrix} u_1 \\ u_2 \end{Bmatrix}_i + 30\begin{Bmatrix} u_1 \\ u_2 \end{Bmatrix}_{i-1} - \frac{20}{3}\begin{Bmatrix} u_1 \\ u_2 \end{Bmatrix}_{i-2}$$

$$\begin{Bmatrix} \ddot{u}_1 \\ \ddot{u}_2 \end{Bmatrix}_{i+1} = \frac{1}{\Delta t^2}\left(2\begin{Bmatrix} u_1 \\ u_2 \end{Bmatrix}_{i+1} - 5\begin{Bmatrix} u_1 \\ u_2 \end{Bmatrix}_i + 4\begin{Bmatrix} u_1 \\ u_2 \end{Bmatrix}_{i-1} - \begin{Bmatrix} u_1 \\ u_2 \end{Bmatrix}_{i-2}\right)$$

$$= 800\begin{Bmatrix} u_1 \\ u_2 \end{Bmatrix}_{i+1} - 2000\begin{Bmatrix} u_1 \\ u_2 \end{Bmatrix}_i + 1600\begin{Bmatrix} u_1 \\ u_2 \end{Bmatrix}_{i-1} - 400\begin{Bmatrix} u_1 \\ u_2 \end{Bmatrix}_{i-2}$$

TABLE E11.2
First 15 Steps of Displacement, Velocity, and Acceleration Responses of Two-Degree-of-Freedom Shear Building in Example 11.1 Using Houbolt Method

Step	Time (s)	u_1 (m)	u_2 (m)	$\dot{u}_1$ (m/s)	$\dot{u}_2$ (m/s)	$\ddot{y}_1$ (m/s²)	$\ddot{y}_2$ (m/s²)
0	0.00	0.0000	0.0000	0.0000	0.0000	0.0000	0.0000
1	0.05	0.0000	0.0000	−0.0075	−0.0075	−0.0005	−0.0005
2	0.10	−0.0008	−0.0008	−0.0268	−0.0295	0.1234	0.0161
3	0.15	−0.0027	−0.0029	−0.0533	−0.0632	0.2413	0.0268
4	0.20	−0.0061	−0.0072	−0.0835	−0.1092	0.4909	0.1141
5	0.25	−0.0109	−0.0139	−0.1035	−0.1511	0.7990	0.2947
6	0.30	−0.0173	−0.0232	−0.1052	−0.1737	1.1375	0.5904
7	0.35	−0.0249	−0.0349	−0.0836	−0.1630	1.4866	1.0039
8	0.40	−0.0336	−0.0487	−0.0360	−0.1080	1.8397	1.5167
9	0.45	−0.0429	−0.0639	0.0383	−0.0021	2.2001	2.0926
10	0.50	−0.0526	−0.0794	0.1388	0.1553	2.5723	2.6858
11	0.55	−0.0620	−0.0945	0.2638	0.3590	2.9528	3.2505
12	0.60	−0.0707	−0.1082	0.4099	0.5987	3.3243	3.7480
13	0.65	−0.0781	−0.1196	0.5713	0.8603	3.6572	4.1502
14	0.70	−0.0836	−0.1280	0.7397	1.1277	3.9162	4.4394
15	0.75	−0.0868	−0.1328	0.9045	1.3841	4.0703	4.6053

and the corresponding absolute acceleration response by means of

$$\begin{Bmatrix} \ddot{y}_1 \\ \ddot{y}_2 \end{Bmatrix}_{i+1} = \begin{Bmatrix} \ddot{u}_1 \\ \ddot{u}_2 \end{Bmatrix}_{i+1} + \begin{Bmatrix} 1 \\ 1 \end{Bmatrix} \ddot{u}_g(t_{i+1}) = 800 \begin{Bmatrix} u_1 \\ u_2 \end{Bmatrix}_{i+1} - 2000 \begin{Bmatrix} u_1 \\ u_2 \end{Bmatrix}_i + 1600 \begin{Bmatrix} u_1 \\ u_2 \end{Bmatrix}_{i-1} - 400 \begin{Bmatrix} u_1 \\ u_2 \end{Bmatrix}_{i-2} + \begin{Bmatrix} 1 \\ 1 \end{Bmatrix} \ddot{u}_g(t_{i+1})$$

Results

The application of the foregoing recursive equations leads thus to the results listed in Table E11.2 for the first 15 time steps. It should be noted that the velocity and acceleration responses shown in this table for the first two steps were determined with the central difference method because Houbolt method cannot be used to evaluate these initial responses.

11.5 CONSTANT AVERAGE ACCELERATION METHOD

The constant average acceleration method, also known as the *trapezoidal rule*, is similar to the Houbolt method in that the solution is based on the equation of motion at the end of a time step and that an assumption is made about the variation of a system's response within such a time step. However, in the constant average acceleration method, it is assumed that the acceleration response of the system is constant during any small time interval and equal to the average of the accelerations at the beginning and the end of the interval. That is, it is assumed that the acceleration response at any time τ between the beginning and the end of a time interval is given by

$$\ddot{u}(\tau) = \frac{1}{2}(\ddot{u}_i + \ddot{u}_{i+1}) \tag{11.41}$$

where $\ddot{u}_i$ is the known acceleration at the beginning of the interval, $\ddot{u}_{i+1}$ the unknown acceleration at the end of the interval, and τ a time variable with origin at the beginning of the interval and defined

Structural Response by Step-by-Step Integration Methods

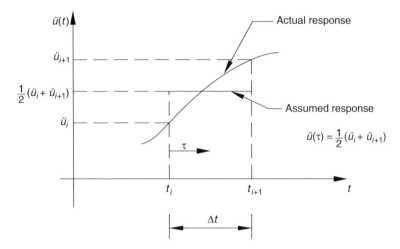

FIGURE 11.3 Assumed variation of acceleration response within a time interval in constant average acceleration method.

as $\tau = t - t_i$ (see Figure 11.3). Also, as the acceleration is equal to the first derivative of the velocity with respect to time, one has that

$$d\dot{u}(\tau) = \frac{1}{2}(\ddot{u}_i + \ddot{u}_{i+1})d\tau \tag{11.42}$$

which, after integrating both sides from $\tau = 0$ to τ, may also be expressed as

$$\int_0^\tau d\dot{u}(\tau) = \frac{1}{2}(\ddot{u}_i + \ddot{u}_{i+1})\int_0^\tau d\tau \tag{11.43}$$

or, after performing the integrations, as

$$\dot{u}(\tau) - \dot{u}(\tau = 0) = \frac{\tau}{2}(\ddot{u}_i + \ddot{u}_{i+1}) \tag{11.44}$$

Thus, under the assumption indicated by Equation 11.41 and considering that $\dot{u}(\tau = 0) = \dot{u}_i$, the velocity response within any time interval is given by

$$\dot{u}(\tau) = \dot{u}_i + \frac{\tau}{2}(\ddot{u}_i + \ddot{u}_{i+1}) \tag{11.45}$$

and the velocity response at the end of the interval, where $\tau = \Delta t$, by

$$\dot{u}_{i+1} = \dot{u}_i + \frac{\Delta t}{2}(\ddot{u}_i + \ddot{u}_{i+1}) \tag{11.46}$$

Similarly, as the velocity is equal to the first derivative of the displacement with respect to time, it may be considered that

$$du(\tau) = \left[\dot{u}_i + \frac{\tau}{2}(\ddot{u}_i + \ddot{u}_{i+1})\right]d\tau \tag{11.47}$$

and that

$$\int_0^\tau du(\tau) = \dot{u}_i \int_0^\tau d\tau + \frac{1}{2}(\ddot{u}_i + \ddot{u}_{i+1})\int_0^\tau \tau d\tau \qquad (11.48)$$

which leads to

$$u(\tau) - u(\tau = 0) = \dot{u}_i \tau + \frac{\tau^2}{4}(\ddot{u}_i + \ddot{u}_{i+1}) \qquad (11.49)$$

from which, after considering that $u(\tau = 0) = u_i$, it is found that the displacement response within a time interval varies according to

$$u(\tau) = u_i + \dot{u}_i \tau + \frac{\tau^2}{4}(\ddot{u}_i + \ddot{u}_{i+1}) \qquad (11.50)$$

Hence, the displacement at the end of the time interval is given by

$$u_{i+1} = u_i + \dot{u}_i \Delta t + \frac{\Delta t^2}{4}(\ddot{u}_i + \ddot{u}_{i+1}) \qquad (11.51)$$

In the constant average acceleration method, therefore, the displacement and velocity vectors that appear in an equation of motion such as that given by Equation 11.19 are approximated by

$$\{\dot{u}\}_{i+1} = \{\dot{u}\}_i + \frac{\Delta t}{2}(\{\ddot{u}\}_i + \{\ddot{u}\}_{i+1}) \qquad (11.52)$$

$$\{u\}_{i+1} = \{u\}_i + \Delta t \{\dot{u}\}_i + \frac{\Delta t^2}{4}(\{\ddot{u}\}_i + \{\ddot{u}\}_{i+1}) \qquad (11.53)$$

However, as the acceleration vector $\{\ddot{u}\}_{i+1}$ is not known in advance (the method is an implicit one), iterations need to be performed at each time step. Accordingly, at each time step the procedure is started by assuming the acceleration values at the end of the time interval and computing a first approximation to the velocity and displacement vectors at the end of the time interval by means of Equations 11.52–11.53. Then, a new acceleration vector is obtained from the equation of motion that establishes the dynamic equilibrium of the system at the end of the time step and the determined velocity and displacement vectors. Afterward, the procedure is repeated with the new acceleration vector until convergence is attained.

Alternatively, the procedure may be converted into an explicit formulation to eliminate the need for iterations. This conversion may be accomplished as follows: First, the acceleration vector $\{\ddot{u}\}_{i+1}$ is solved from Equation 11.53 to obtain the following expression where the only unknown is the displacement vector $\{u\}_{i+1}$:

$$\{\ddot{u}\}_{i+1} = \frac{4}{\Delta t^2}(\{u\}_{i+1} - \{u\}_i) - \frac{4}{\Delta t}\{\dot{u}\}_i - \{\ddot{u}\}_i \qquad (11.54)$$

Then, this expression is substituted into Equation 11.52 to get the following expression for the velocity vector $\{\dot{u}\}_{i+1}$, where, once again, the only unknown is the displacement vector $\{u\}_{i+1}$:

$$\{\dot{u}\}_{i+1} = \{\dot{u}\}_i + \frac{\Delta t}{2}\{\ddot{u}\}_i + \frac{\Delta t}{2}\left[\frac{4}{\Delta t^2}(\{u\}_{i+1} - \{u\}_i) - \frac{4}{\Delta t}\{\dot{u}\}_i - \{\ddot{u}\}_i\right]$$

$$= \frac{2}{\Delta t}(\{u\}_{i+1} - \{u\}_i) - \{\dot{u}\}_i \qquad (11.55)$$

Finally, Equations 11.54 and 11.55 are substituted into the equation of motion that establishes the dynamic equilibrium of the system at the end of the time step (given by Equation 11.19). This yields

$$[M]\left(\frac{4}{\Delta t^2}[\{u\}_{i+1} - \{u\}_i] - \frac{4}{\Delta t}\{\dot{u}\}_i - \{\ddot{u}\}_i\right) + [C]\left(\frac{2}{\Delta t}[\{u\}_{i+1} - \{u\}_i] - \{\dot{u}\}_i\right)$$
$$+ [K]\{u\}_{i+1} = -[M]\{J\}\ddot{u}_g(t_{i+1}) \tag{11.56}$$

which may also be expressed as

$$[\hat{K}_a]\{u\}_{i+1} = \{\hat{F}_a\}_{i+1} \tag{11.57}$$

where $[\hat{K}_a]$ and $\{\hat{F}_a\}_{i+1}$ are an effective stiffness matrix and an effective load vector defined as

$$[\hat{K}_a] = [K] + \frac{2}{\Delta t}[C] + \frac{4}{\Delta t^2}[M] \tag{11.58}$$

$$\{\hat{F}_a\}_{i+1} = -[M]\{J\}\ddot{u}_g(t_{i+1}) + [C]\left(\frac{2}{\Delta t}\{u\}_i + \{\dot{u}\}_i\right)$$
$$+ [M]\left(\frac{4}{\Delta t^2}\{u\}_i + \frac{4}{\Delta t}\{\dot{u}\}_i + \{\ddot{u}\}_i\right) \tag{11.59}$$

Thus, as all the terms on the right-hand sides of Equations 11.58 and 11.59 are known or are functions of the response of the system at the beginning of the time step, which are also known, the following explicit formula may be employed to compute the displacement vector at the end of the time step:

$$\{u\}_{i+1} = [\hat{K}_a]^{-1}\{\hat{F}_a\}_{i+1} \tag{11.60}$$

Similarly, once the displacement vector $\{u\}_{i+1}$ is known, the velocity and acceleration vectors may be determined through the application of Equations 11.54 and 11.55. However, to ensure that dynamic equilibrium is satisfied at the end of the step and to avoid the accumulation of errors from one step to another, it is convenient to compute the acceleration vector directly from the equation of motion as opposed to from Equation 11.54. Hence, the acceleration vector is determined according to

$$\{\ddot{u}\}_{i+1} = -\{J\}\ddot{u}_g(t_{i+1}) - [M]^{-1}([C]\{\dot{u}\}_{i+1} + [K]\{u\}_{i+1}) \tag{11.61}$$

A great advantage of the constant average acceleration method is that it is, as shown in Section 11.9, unconditionally stable. In other words, the error introduced by the method in one step is not amplified in subsequent steps, no matter how large the selected time step is. As a result, the time step may be selected according to the size that is needed to properly represent the input ground motion and the accuracy with which one wants to determine the system's response.

Box 11.3 presents a solution algorithm based on the constant average acceleration method.

Box 11.3 Solution Algorithm Using Constant Average Acceleration Method

1.0 Initial calculations
   1.1 Formulate stiffness matrix $[K]$, mass matrix $[M]$, and damping matrix $[C]$
   1.2 Establish vectors of initial displacements $\{u\}_0$ and initial velocities $\{\dot{u}\}_0$
   1.3 Compute initial acceleration vector as

$$\{\ddot{u}\}_0 = -\{J\}\ddot{u}_g(0) - [M]^{-1}([C]\{\dot{u}\}_0 + [K]\{u\}_0)$$

   1.4 Select time step Δt
   1.5 Compute constant matrices $[a]$ and $[b]$ as

$$[a] = \frac{4}{\Delta t^2}[M] + \frac{2}{\Delta t}[C]$$

$$[b] = \frac{4}{\Delta t}[M] + [C]$$

   1.6 Compute effective stiffness matrix as

$$[\hat{K}_a] = [K] + [a]$$

   1.7 Set $i = 0$

2.0 Calculations for ith time step
   2.1 Compute effective load vector as

$$\{\hat{F}_a\}_{i+1} = -[M]\{J\}\ddot{u}_g(t_{i+1}) + [a]\{u\}_i + [b]\{\dot{u}\}_i + [M]\{\ddot{u}\}_i$$

   2.2 Compute displacement vector at time $t_{i+1} = \Delta t(i+1)$ as

$$\{u\}_{i+1} = [\hat{K}_a]^{-1}\{\hat{F}_a\}_{i+1}$$

   2.3 Compute velocity and acceleration vectors at time $t_{i+1} = \Delta t(i+1)$ according to

$$\{\dot{u}\}_{i+1} = \frac{2}{\Delta t}(\{u\}_{i+1} - \{u\}_i) - \{\dot{u}\}_i$$

$$\{\ddot{u}\}_{i+1} = -\{J\}\ddot{u}_g(t_{i+1}) - [M]^{-1}([C]\{\dot{u}\}_{i+1} + [K]\{u\}_{i+1})$$

3.0 Calculations for next time step
   Replace i by $i+1$ and repeat steps 2.1–2.3.

Structural Response by Step-by-Step Integration Methods

EXAMPLE 11.3 RESPONSE OF TWO-DEGREE-OF-FREEDOM SYSTEM BY CONSTANT AVERAGE ACCELERATION METHOD

The mass, damping, and stiffness matrix of a two-degree-of-freedom shear building are

$$[M] = \begin{bmatrix} 1 & 0 \\ 0 & 1 \end{bmatrix} Gg \quad [C] = \begin{bmatrix} 2 & -1 \\ -1 & 1 \end{bmatrix} \text{MN-s/m} \quad [K] = \begin{bmatrix} 200 & -100 \\ -100 & 100 \end{bmatrix} \text{MN/m}$$

Initially at rest, the building is subjected to a ground motion represented by the bilinear pulse shown in Figure E11.3. Determine the displacement, velocity, and acceleration responses of the system for the first 0.3 s using the constant average acceleration method and a time step of 0.02 s.

Solution

Initial Calculations and Effective Stiffness Matrix

The system is initially at rest and thus it may be established that

$$\{u\}_0 = \begin{Bmatrix} 0 \\ 0 \end{Bmatrix} \quad \{\dot{u}\}_0 = \begin{Bmatrix} 0 \\ 0 \end{Bmatrix} \quad \{\ddot{u}\}_0 = \begin{Bmatrix} 0 \\ 0 \end{Bmatrix}$$

Also, for a time step of 0.02 s, the constants $[a]$ and $[b]$ defined in Box 11.3 result as

$$[a] = \frac{4}{\Delta t^2}[M] + \frac{2}{\Delta t}[C] = \frac{4}{0.02^2}\begin{bmatrix} 1 & 0 \\ 0 & 1 \end{bmatrix} + \frac{2}{0.02}\begin{bmatrix} 2 & -1 \\ -1 & 1 \end{bmatrix}$$

$$= 100\begin{bmatrix} 102 & -1 \\ -1 & 101 \end{bmatrix} \text{MN/m}$$

$$[b] = \frac{4}{\Delta t}[M] + [C] = \frac{4}{0.02}\begin{bmatrix} 1 & 0 \\ 0 & 1 \end{bmatrix} + \begin{bmatrix} 2 & -1 \\ -1 & 1 \end{bmatrix} = \begin{bmatrix} 202 & -1 \\ -1 & 201 \end{bmatrix} \text{MN-s/m}$$

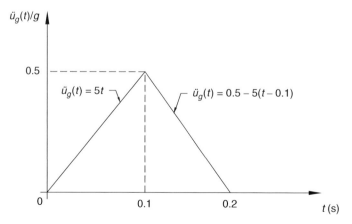

FIGURE E11.3 Ground motion considered in Example 11.3.

Hence, according to Equation 11.58, the effective stiffness matrix of the building is equal to

$$[\hat{K}_a] = [K] + [a] = \begin{bmatrix} 200 & -100 \\ -100 & 100 \end{bmatrix} + 100\begin{bmatrix} 102 & -1 \\ -1 & 101 \end{bmatrix} = 200\begin{bmatrix} 52 & -1 \\ -1 & 51 \end{bmatrix} \text{ MN/m}$$

and its inverse is equal to

$$[\hat{K}_a]^{-1} = \frac{1}{530,200}\begin{bmatrix} 51 & 1 \\ 1 & 52 \end{bmatrix} \text{ m/MN}$$

Effective Load Vector at Time $t = t_{i+1}$

Similarly, according to Equation 11.59, the effective load vector at time $t = t_{i+1}$ results as

$$(\{\hat{F}_a\}_{i+1} = -[M]\{J\}\ddot{u}_g(t_{i+1}) + [a]\{u\}_i + [b]\{\dot{u}\}_i + [M]\{\ddot{u}\}_i)$$

$$= -\begin{bmatrix} 1 & 0 \\ 0 & 1 \end{bmatrix}\begin{Bmatrix} 1 \\ 1 \end{Bmatrix}\ddot{u}_g(t_{i+1}) + 100\begin{bmatrix} 102 & -1 \\ -1 & 101 \end{bmatrix}\begin{Bmatrix} u_1 \\ u_2 \end{Bmatrix}_i + \begin{bmatrix} 202 & -1 \\ -1 & 201 \end{bmatrix}\begin{Bmatrix} \dot{u}_1 \\ \dot{u}_2 \end{Bmatrix}_i + \begin{bmatrix} 1 & 0 \\ 0 & 1 \end{bmatrix}\begin{Bmatrix} \ddot{u}_1 \\ \ddot{u}_2 \end{Bmatrix}_i$$

$$= -\begin{Bmatrix} 1 \\ 1 \end{Bmatrix}\ddot{u}_g(t_{i+1}) + 100\begin{bmatrix} 102 & -1 \\ -1 & 101 \end{bmatrix}\begin{Bmatrix} u_1 \\ u_2 \end{Bmatrix}_i + \begin{bmatrix} 202 & -1 \\ -1 & 201 \end{bmatrix}\begin{Bmatrix} \dot{u}_1 \\ \dot{u}_2 \end{Bmatrix}_i + \begin{Bmatrix} \ddot{u}_1 \\ \ddot{u}_2 \end{Bmatrix}_i$$

Displacement Response at Time $t = t_{i+1}$

In consequence, the displacement response of the system at time $t = t_{i+1}$ is given by

$$\begin{Bmatrix} u_1 \\ u_2 \end{Bmatrix}_{i+1} = [\hat{K}_a]^{-1}\{\hat{F}_a\}_{i+1} = \frac{1}{530,200}\begin{bmatrix} 51 & 1 \\ 1 & 52 \end{bmatrix}\{\hat{F}_a\}_{i+1}$$

$$= \frac{1}{530,200}\left(-\begin{bmatrix} 51 & 1 \\ 1 & 52 \end{bmatrix}\begin{Bmatrix} 1 \\ 1 \end{Bmatrix}\ddot{u}_g(t_{i+1}) + 100\begin{bmatrix} 51 & 1 \\ 1 & 52 \end{bmatrix}\begin{bmatrix} 102 & -1 \\ -1 & 101 \end{bmatrix}\begin{Bmatrix} u_1 \\ u_2 \end{Bmatrix}_i \right.$$

$$\left. + \begin{bmatrix} 51 & 1 \\ 1 & 52 \end{bmatrix}\begin{bmatrix} 202 & -1 \\ -1 & 201 \end{bmatrix}\begin{Bmatrix} \dot{u}_1 \\ \dot{u}_2 \end{Bmatrix}_i + \begin{bmatrix} 51 & 1 \\ 1 & 52 \end{bmatrix}\begin{Bmatrix} \ddot{u}_1 \\ \ddot{u}_2 \end{Bmatrix}_i\right)$$

which, if expanded into algebraic equations, yields

$$(u_1)_{i+1} = \frac{1}{530,200}[-52\ddot{u}_g(t_{i+1}) + 520,100(u_1)_i + 5,000(u_2)_i + 10,301(\dot{u}_1)_i + 150(\dot{u}_2)_i + 51(\ddot{u}_1)_i + (\ddot{u}_2)_i]$$

$$(u_2)_{i+1} = \frac{1}{530,200}[-53\ddot{u}_g(t_{i+1}) + 5,000(u_1)_i + 525,100(u_2)_i + 150(\dot{u}_1)_i + 10,451(\dot{u}_2)_i + (\ddot{u}_1)_i + 52(\ddot{u}_2)_i]$$

Velocity and Acceleration Responses at Time $t = t_{i+1}$

From Equation 11.55 and a time step of 0.02 s, the velocity response of the system at time $t = t_{i+1}$ is given by

$$\begin{Bmatrix} \dot{u}_1 \\ \dot{u}_2 \end{Bmatrix}_{i+1} = \frac{2}{\Delta t}\left(\begin{Bmatrix} u_1 \\ u_2 \end{Bmatrix}_{i+1} - \begin{Bmatrix} u_1 \\ u_2 \end{Bmatrix}_i\right) - \begin{Bmatrix} \dot{u}_1 \\ \dot{u}_2 \end{Bmatrix}_i = 100\begin{Bmatrix} u_1 \\ u_2 \end{Bmatrix}_{i+1} - 100\begin{Bmatrix} u_1 \\ u_2 \end{Bmatrix}_i - \begin{Bmatrix} \dot{u}_1 \\ \dot{u}_2 \end{Bmatrix}_i$$

Structural Response by Step-by-Step Integration Methods

TABLE E11.3
First 15 Steps of Displacement, Velocity, and Acceleration Responses of Two-Degree-of-Freedom Shear Building in Example 11.3 Using Constant Average Acceleration Method

Step	Time (s)	u_1 (m)	u_2 (m)	$\dot{u}_1$ (m/s)	$\dot{u}_2$ (m/s)	$\ddot{u}_1$ (m/s²)	$\ddot{u}_2$ (m/s²)	$\ddot{y}_1$ (m/s²)	$\ddot{y}_2$ (m/s²)
0	0.00	0.0000	0.0000	0.0000	0.0000	0.0000	0.0000	0.0000	0.0000
1	0.02	−0.0001	−0.0001	−0.0096	−0.0098	−0.9621	−0.9806	0.0189	0.0004
2	0.04	−0.0006	−0.0006	−0.0379	−0.0392	−1.8698	−1.9591	0.0922	0.0029
3	0.06	−0.0018	−0.0019	−0.0836	−0.0881	−2.6927	−2.9310	0.2503	0.0120
4	0.08	−0.0041	−0.0043	−0.1446	−0.1563	−3.4080	−3.8886	0.5160	0.0354
5	0.10	−0.0077	−0.0083	−0.2187	−0.2434	−4.0011	−4.8201	0.9039	0.0849
6	0.12	−0.0127	−0.0140	−0.2841	−0.3291	−2.5419	−3.7492	1.3821	0.1748
7	0.14	−0.0188	−0.0212	−0.3202	−0.3928	−1.0658	−2.6229	1.8772	0.3201
8	0.16	−0.0252	−0.0295	−0.3273	−0.4333	0.3575	−1.4298	2.3195	0.5322
9	0.18	−0.0316	−0.0383	−0.3070	−0.4493	1.6680	−0.1642	2.6490	0.8168
10	0.20	−0.0373	−0.0472	−0.2621	−0.4392	2.8191	1.1709	2.8191	1.1709
11	0.22	−0.0420	−0.0557	−0.2058	−0.4117	2.8188	1.5827	2.8188	1.5827
12	0.24	−0.0455	−0.0636	−0.1509	−0.3755	2.6712	2.0321	2.6712	2.0321
13	0.26	−0.0480	−0.0707	−0.1000	−0.3303	2.4108	2.4925	2.4108	2.4925
14	0.28	−0.0496	−0.0767	−0.0551	−0.2760	2.0793	2.9343	2.0793	2.9343
15	0.30	−0.0503	−0.0816	−0.0171	−0.2134	1.7210	3.3268	1.7210	3.3268

Similarly, from Equation 11.61, its relative acceleration response results as

$$\begin{Bmatrix} \ddot{u}_1 \\ \ddot{u}_2 \end{Bmatrix}_{i+1} = -\{J\}\ddot{u}_g(t_{i+1}) - [M]^{-1}([C]\{\dot{u}\}_{i+1} + [K]\{u\}_{i+1})$$

$$= -\begin{Bmatrix} 1 \\ 1 \end{Bmatrix}\ddot{u}_g(t_{i+1}) - \begin{bmatrix} 1 & 0 \\ 0 & 1 \end{bmatrix}^{-1}\left(\begin{bmatrix} 2 & -1 \\ -1 & 1 \end{bmatrix}\begin{Bmatrix} \dot{u}_1 \\ \dot{u}_2 \end{Bmatrix}_{i+1} + \begin{bmatrix} 200 & -100 \\ -100 & 100 \end{bmatrix}\begin{Bmatrix} u_1 \\ u_2 \end{Bmatrix}_{i+1}\right)$$

$$= \begin{Bmatrix} -\ddot{u}_g(t_{i+1}) - 2(\dot{u}_1)_{i+1} + (\dot{u}_2)_{i+1} - 200(u_1)_{i+1} + 100(u_2)_{i+1} \\ -\ddot{u}_g(t_{i+1}) + (\dot{u}_1)_{i+1} - (\dot{u}_2)_{i+1} + 100(u_1)_{i+1} - 100(u_2)_{i+1} \end{Bmatrix}$$

In terms of the relative acceleration response, the absolute acceleration response may then be computed according to

$$\begin{Bmatrix} \ddot{y}_1 \\ \ddot{y}_2 \end{Bmatrix}_{i+1} = \begin{Bmatrix} \ddot{u}_1 \\ \ddot{u}_2 \end{Bmatrix}_{i+1} + \begin{Bmatrix} 1 \\ 1 \end{Bmatrix}\ddot{u}_g(t_{i+1})$$

Results

On the basis of the established recursive equations and starting with the initial conditions, one obtains the responses shown in Table E11.3 for the first 15 steps (0.3 s).

11.6 LINEAR ACCELERATION METHOD

The linear acceleration method is in all respects similar to the constant average acceleration method. The main difference is that in the linear acceleration method it is assumed that the acceleration response of a system varies linearly within each time interval instead of being constant as it is assumed in the

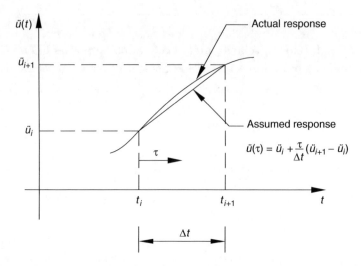

FIGURE 11.4 Assumed variation of acceleration response within a time interval in linear acceleration method.

constant average acceleration method. Consequently, in the linear acceleration method, the acceleration response within a time interval of size Δt varies with time according to (see Figure 11.4)

$$\ddot{u}(\tau) = \ddot{u}_i + \frac{\ddot{u}_{i+1} - \ddot{u}_i}{\Delta t} \tau \tag{11.62}$$

where, as before, $\ddot{u}_i$ is the acceleration at the beginning of the interval, $\ddot{u}_{i+1}$ the acceleration at the end of the interval, and τ a time variable with origin at the beginning of the interval and defined as $\tau = t - t_i$. This assumption, in turn, leads, after integration and proceeding as in the case of the constant average acceleration method, to the following expressions for the corresponding velocity and displacement responses within any time interval:

$$\dot{u}(\tau) = \dot{u}_i + \int_0^\tau \left[\ddot{u}_i + \frac{\ddot{u}_{i+1} - \ddot{u}_i}{\Delta t} \tau \right] d\tau = \dot{u}_i + \ddot{u}_i \tau + \frac{\ddot{u}_{i+1} - \ddot{u}_i}{\Delta t} \frac{\tau^2}{2} \tag{11.63}$$

$$u(\tau) = u_i + \int_0^\tau \left[\dot{u}_i + \ddot{u}_i \tau + \frac{\ddot{u}_{i+1} - \ddot{u}_i}{\Delta t} \frac{\tau^2}{2} \right] d\tau = u_i + \tau \dot{u}_i + \frac{\tau^2}{2} \ddot{u}_i + \frac{\tau^3}{6} \frac{\ddot{u}_{i+1} - \ddot{u}_i}{\Delta t} \tag{11.64}$$

It also leads, after setting $\tau = \Delta t$ into Equations 11.63 and 11.64, to the following expressions to determine the velocity and displacements response at the end of the time interval:

$$\dot{u}_{i+1} = \dot{u}_i + \frac{\Delta t}{2}(\ddot{u}_{i+1} + \ddot{u}_i) \tag{11.65}$$

$$u_{i+1} = u_i + \Delta t \dot{u}_i + \frac{\Delta t^2}{3} \ddot{u}_i + \frac{\Delta t^2}{6} \ddot{u}_{i+1} \tag{11.66}$$

In the linear acceleration method, therefore, the displacement and velocity vectors that appear in the equation of motion defined by Equation 11.19 are approximated by

$$\{\dot{u}\}_{i+1} = \{\dot{u}\}_i + \frac{\Delta t}{2}(\{\ddot{u}\}_{i+1} + \{\ddot{u}\}_i) \tag{11.67}$$

$$\{u\}_{i+1} = \{u\}_i + \Delta t\{\dot{u}\}_i + \frac{\Delta t^2}{3}\{\ddot{u}\}_i + \frac{\Delta t^2}{6}\{\ddot{u}\}_{i+1} \tag{11.68}$$

The linear acceleration method is also an implicit one as the acceleration vector $\{\ddot{u}\}_{i+1}$ is not known beforehand. However, it may be converted into an explicit one following the same approach that led to the explicit formulation for the constant average acceleration method. To this end, $\{\ddot{u}\}_{i+1}$ is solved first from Equation 11.68 to obtain

$$\{\ddot{u}\}_{i+1} = \frac{6}{\Delta t^2}(\{u\}_{i+1} - \{u\}_i) - \frac{6}{\Delta t}\{\dot{u}\}_i - 2\{\ddot{u}\}_i \tag{11.69}$$

Then, this last expression is substituted into Equation 11.67 to arrive to

$$\{\dot{u}\}_{i+1} = \frac{3}{\Delta t}(\{u\}_{i+1} - \{u\}_i) - 2\{\dot{u}\}_i - \frac{\Delta t}{2}\{\ddot{u}\}_i \tag{11.70}$$

Finally, Equations 11.69 and 11.70 are substituted into the equation of motion given by Equation 11.19, leading to

$$[M]\left(\frac{6}{\Delta t^2}(\{u\}_{i+1} - \{u\}_i) - \frac{6}{\Delta t}\{\dot{u}\}_i - 2\{\ddot{u}\}_i\right) + [C]\left(\frac{3}{\Delta t}(\{u\}_{i+1} - \{u\}_i) - 2\{\dot{u}\}_i - \frac{\Delta t}{2}\{\ddot{u}\}_i\right)$$
$$+ [K]\{u\}_{i+1} = -[M]\{J\}\ddot{u}_g(t_{i+1}) \tag{11.71}$$

which may also be written as

$$[\hat{K}_l]\{u\}_{i+1} = \{\hat{F}_l\}_{i+1} \tag{11.72}$$

where

$$[\hat{K}_l] = [K] + \frac{3}{\Delta t}[C] + \frac{6}{\Delta t^2}[M] \tag{11.73}$$

$$\{\hat{F}_l\}_{i+1} = -[M]\{J\}\ddot{u}_g(t_{i+1}) + [M]\left(\frac{6}{\Delta t^2}\{u\}_i + \frac{6}{\Delta t}\{\dot{u}\}_i + 2\{\ddot{u}\}_i\right)$$
$$+ [C]\left(\frac{3}{\Delta t}\{u\}_i + 2\{\dot{u}\}_i + \frac{\Delta t}{2}\{\ddot{u}\}_i\right) \tag{11.74}$$

Thus, in the linear acceleration method, the displacement response at time $t_{i+1} = \Delta t(i+1)$ may be computed according to

$$\{u\}_{i+1} = [\hat{K}_l]^{-1}\{\hat{F}_l\}_{i+1} \tag{11.75}$$

The velocity vector may then be computed using Equation 11.70. As in the constant average acceleration method, it is convenient to determine the acceleration vector directly from the equation of motion. Hence, once the displacement and velocity vectors are evaluated, the acceleration vector may be determined from

$$\{\ddot{u}\}_{i+1} = -\{J\}\ddot{u}_g(t_{i+1}) - [M]^{-1}([C]\{\dot{u}\}_{i+1} + [K]\{u\}_{i+1}) \tag{11.76}$$

As in the methods presented in the foregoing sections, Equations 11.70, 11.75, and 11.76 provide a set of recursive equations by means of which one can compute step-by-step the time history of a system's displacement, velocity, and acceleration responses. The solution algorithm presented in Box 11.4 summarizes the procedure.

Box 11.4   Solution Algorithm Using Linear Acceleration Method

1.0 Initial calculations
   1.1 Formulate stiffness matrix $[K]$, mass matrix $[M]$, and damping matrix $[C]$
   1.2 Establish vectors of initial displacements $\{u\}_0$ and initial velocities $\{\dot{u}\}_0$
   1.3 Compute initial acceleration vector as

$$\{\ddot{u}\}_0 = -\{J\}\ddot{u}_g(0) - [M]^{-1}([C]\{\dot{u}\}_0 + [K]\{u\}_0)$$

   1.4 Select time step Δt
   1.5 Compute constant matrices $[a]$, $[b]$, and $[c]$ as

$$[a] = \frac{6}{\Delta t^2}[M] + \frac{3}{\Delta t}[C]$$

$$[b] = \frac{6}{\Delta t}[M] + 2[C]$$

$$[c] = 2[M] + \frac{\Delta t}{2}[C]$$

   1.6 Compute effective stiffness matrix as

$$[\hat{K}_l] = [K] + [a]$$

   1.7 Set $i = 0$
2.0 Calculations for ith time step
   2.1 Compute effective load vector as

$$\{\hat{F}_l\}_{i+1} = -[M]\{J\}\ddot{u}_g(t_{i+1}) + [a]\{u\}_i + [b]\{\dot{u}\}_i + [c]\{\ddot{u}\}_i$$

   2.2 Compute displacement vector at time $t_{i+1} = \Delta t(i+1)$ as

$$\{u\}_{i+1} = [\hat{K}_l]^{-1}\{\hat{F}_l\}_{i+1}$$

   2.3 Compute velocity and acceleration vectors at time $t_{i+1} = \Delta t(i+1)$ according to

$$\{\dot{u}\}_{i+1} = \frac{3}{\Delta t}(\{u\}_{i+1} - \{u\}_i) - 2\{\dot{u}\}_i - \frac{\Delta t}{2}\{\ddot{u}\}_i$$

$$\{\ddot{u}\}_{i+1} = -\{J\}\ddot{u}_g(t_{i+1}) - [M]^{-1}([C]\{\dot{u}\}_{i+1} + [K]\{u\}_{i+1})$$

3.0 Calculations for next time step
   Replace i by $i+1$ and repeat steps 2.1–2.3.

Structural Response by Step-by-Step Integration Methods

As expected, assuming a linear variation of the acceleration response during each time interval yields a closer approximation to the actual response than assuming that it remains constant as it is considered in the constant average acceleration method. In fact, comparative studies have demonstrated that the linear acceleration method leads to more accurate results than the constant average acceleration method when the same time step is considered. However, the linear acceleration method suffers from the drawback that it is only conditionally stable. As it will be shown in Section 11.9, the condition for numerical stability for the linear acceleration method is

$$\Delta t \leq \frac{\sqrt{3}}{\pi} T_N = 0.551 T_N \tag{11.77}$$

where, as before, T_N denotes the natural period of the system in its highest mode of vibration.

EXAMPLE 11.4 DYNAMIC RESPONSE OF TWO-DEGREE-OF-FREEDOM SYSTEM BY LINEAR ACCELERATION METHOD

Solve the problem considered in Example 11.3 using the linear acceleration method.

Solution

Initial Calculations and Effective Stiffness Matrix

The system is initially at rest and thus, as established in Example 11.3, one has that

$$\{u\}_0 = \begin{Bmatrix} 0 \\ 0 \end{Bmatrix} \quad \{\dot{u}\}_0 = \begin{Bmatrix} 0 \\ 0 \end{Bmatrix} \quad \{\ddot{u}\}_0 = \begin{Bmatrix} 0 \\ 0 \end{Bmatrix}$$

Also, for a time step of 0.02 s and the given system properties, the constant matrices $[a]$, $[b]$, and $[c]$ defined in Box 11.4 are given by

$$[a] = \frac{6}{\Delta t^2}[M] + \frac{3}{\Delta t}[C] = \frac{6}{0.02^2}\begin{bmatrix} 1 & 0 \\ 0 & 1 \end{bmatrix} + \frac{3}{0.02}\begin{bmatrix} 2 & -1 \\ -1 & 2 \end{bmatrix} = 150\begin{bmatrix} 102 & -1 \\ -1 & 101 \end{bmatrix} \text{ MN/m}$$

$$[b] = \frac{6}{\Delta t}[M] + 2[C] = \frac{6}{0.02}\begin{bmatrix} 1 & 0 \\ 0 & 1 \end{bmatrix} + 2\begin{bmatrix} 2 & -1 \\ -1 & 1 \end{bmatrix} = \begin{bmatrix} 304 & -2 \\ -2 & 302 \end{bmatrix} \text{ MN-s/m}$$

$$[c] = 2[M] + \frac{\Delta t}{2}[C] = 2\begin{bmatrix} 1 & 0 \\ 0 & 1 \end{bmatrix} + 0.01\begin{bmatrix} 2 & -1 \\ -1 & 1 \end{bmatrix} = \begin{bmatrix} 2.02 & -0.01 \\ -0.01 & 2.01 \end{bmatrix} \text{ MN-s}^2\text{/m}$$

In view of Equation 11.73, the effective stiffness matrix results then equal to

$$[\hat{K}_t] = [K] + [a] = \begin{bmatrix} 200 & -100 \\ -100 & 100 \end{bmatrix} + 150\begin{bmatrix} 102 & -1 \\ -1 & 101 \end{bmatrix} = 250\begin{bmatrix} 62 & -1 \\ -1 & 61 \end{bmatrix} \text{ MN/m}$$

which has an inverse equal to

$$[\hat{K}_t]^{-1} = \frac{1}{945,250}\begin{bmatrix} 61 & 1 \\ 1 & 62 \end{bmatrix} \text{ m/MN}$$

Effective Load Vector at Time $t = t_{i+1}$

From Equation 11.74 and in terms of the preceding [a], [b], and [c] matrices the effective load vector at time $t = t_{i+1}$ is equal to

$$\{\hat{F}_l\}_{i+1} = -[M]\{J\}\ddot{u}_g(t_{i+1}) + [a]\{u\}_i + [b]\{\dot{u}\}_i + [c]\{\ddot{u}\}_i$$

$$= -\begin{bmatrix} 1 & 0 \\ 0 & 1 \end{bmatrix}\begin{Bmatrix} 1 \\ 1 \end{Bmatrix}\ddot{u}_g(t_{i+1}) + 150\begin{bmatrix} 102 & -1 \\ -1 & 101 \end{bmatrix}\begin{Bmatrix} u_1 \\ u_2 \end{Bmatrix}_i + \begin{bmatrix} 304 & -2 \\ -2 & 302 \end{bmatrix}\begin{Bmatrix} \dot{u}_1 \\ \dot{u}_2 \end{Bmatrix}_i + \begin{bmatrix} 2.02 & -0.01 \\ -0.01 & 2.01 \end{bmatrix}\begin{Bmatrix} \ddot{u}_1 \\ \ddot{u}_2 \end{Bmatrix}_i$$

Displacement Response at Time $t = t_{i+1}$

In light of the determined effective load vector and effective stiffness matrix, the displacement response of the system at time $t = t_{i+1}$ is given by

$$\begin{Bmatrix} u_1 \\ u_2 \end{Bmatrix}_{i+1} = [\hat{K}_l]^{-1}\{\hat{F}_l\}_{i+1} = \frac{1}{945,250}\begin{bmatrix} 61 & 1 \\ 1 & 62 \end{bmatrix}\{\hat{F}_l\}_{i+1}$$

$$= \frac{1}{945,250}\left(-\begin{bmatrix} 61 & 1 \\ 1 & 62 \end{bmatrix}\begin{Bmatrix} 1 \\ 1 \end{Bmatrix}\ddot{u}_g(t_{i+1}) + 150\begin{bmatrix} 61 & 1 \\ 1 & 62 \end{bmatrix}\begin{bmatrix} 102 & -1 \\ -1 & 101 \end{bmatrix}\begin{Bmatrix} u_1 \\ u_2 \end{Bmatrix}_i \right.$$

$$\left. + \begin{bmatrix} 61 & 1 \\ 1 & 62 \end{bmatrix}\begin{bmatrix} 304 & -2 \\ -2 & 302 \end{bmatrix}\begin{Bmatrix} \dot{u}_1 \\ \dot{u}_2 \end{Bmatrix}_i + \begin{bmatrix} 61 & 1 \\ 1 & 62 \end{bmatrix}\begin{bmatrix} 2.02 & -0.01 \\ -0.01 & 2.01 \end{bmatrix}\begin{Bmatrix} \ddot{u}_1 \\ \ddot{u}_2 \end{Bmatrix}_i\right)$$

$$= \frac{1}{945,250}\left(-\begin{Bmatrix} 62 \\ 63 \end{Bmatrix}\ddot{u}_g(t_{i+1}) + 150\begin{bmatrix} 6221 & 40 \\ 40 & 6261 \end{bmatrix}\begin{Bmatrix} u_1 \\ u_2 \end{Bmatrix}_i\right.$$

$$\left. + \begin{bmatrix} 18,542 & 180 \\ 180 & 18,722 \end{bmatrix}\begin{Bmatrix} \dot{u}_1 \\ \dot{u}_2 \end{Bmatrix}_i + \begin{bmatrix} 123.21 & 1.4 \\ 1.4 & 124.61 \end{bmatrix}\begin{Bmatrix} \ddot{u}_1 \\ \ddot{u}_2 \end{Bmatrix}_i\right)$$

which may also be expressed in the form of the following two algebraic equations:

$$(u_1)_{i+1} = \frac{1}{945,250}[-62\ddot{u}_g(t_{i+1}) + 933,150(u_1)_i + 6,000(u_2)_i + 18,542(\dot{u}_1)_i + 180(\dot{u}_2)_i + 123.2(\ddot{u}_1)_i + 1.4(\ddot{u}_2)_i]$$

$$(u_2)_{i+1} = \frac{1}{945,250}[-63\ddot{u}_g(t_{i+1}) + 6,000(u_1)_i + 939,150(u_2)_i + 180(\dot{u}_1)_i + 18,722(\dot{u}_2)_i + 1.4(\ddot{u}_1)_i + 124.61(\ddot{u}_2)_i]$$

Velocity and Acceleration Responses at Time $t = t_{i+1}$

From Equation 11.70 and a time step of 0.02 s, the velocity response of the system at time $t = t_{i+1}$ is given by

$$\begin{Bmatrix} \dot{u}_1 \\ \dot{u}_2 \end{Bmatrix}_{i+1} = \frac{3}{\Delta t}\left(\begin{Bmatrix} u_1 \\ u_2 \end{Bmatrix}_{i+1} - \begin{Bmatrix} u_1 \\ u_2 \end{Bmatrix}_i\right) - 2\begin{Bmatrix} \dot{u}_1 \\ \dot{u}_2 \end{Bmatrix}_i - \frac{\Delta t}{2}\begin{Bmatrix} \ddot{u}_1 \\ \ddot{u}_2 \end{Bmatrix}$$

$$= 150\begin{Bmatrix} u_1 \\ u_2 \end{Bmatrix}_{i+1} - 150\begin{Bmatrix} u_1 \\ u_2 \end{Bmatrix}_i - 2\begin{Bmatrix} \dot{u}_1 \\ \dot{u}_2 \end{Bmatrix}_i - 0.01\begin{Bmatrix} \ddot{u}_1 \\ \ddot{u}_2 \end{Bmatrix}$$

Structural Response by Step-by-Step Integration Methods

TABLE E11.4
First 15 Steps of Displacement, Velocity, and Acceleration Responses of Two-Degree-of-Freedom Shear Building in Example 11.3 Using Linear Acceleration Method

Step	Time (s)	u_1 (m)	u_2 (m)	$\dot{u}_1$ (m/s)	$\dot{u}_2$ (m/s)	$\ddot{u}_1$ (m/s²)	$\ddot{u}_2$ (m/s²)	$\ddot{y}_1$ (m/s²)	$\ddot{y}_2$ (m/s²)
0	0.00	0.0000	0.0000	0.0000	0.0000	0.0000	0.0000	0.0000	0.0000
1	0.02	−0.0001	−0.0001	−0.0097	−0.0098	−0.9652	−0.9807	0.0158	0.0003
2	0.04	−0.0005	−0.0005	−0.0381	−0.0392	−1.8752	−1.9597	0.0868	0.0023
3	0.06	−0.0017	−0.0018	−0.0838	−0.0881	−2.6995	−2.9325	0.2435	0.0105
4	0.08	−0.0040	−0.0042	−0.1449	−0.1564	−3.4151	−3.8914	0.5089	0.0326
5	0.10	−0.0076	−0.0081	−0.2192	−0.2435	−4.0073	−4.8248	0.8977	0.0802
6	0.12	−0.0127	−0.0139	−0.2846	−0.3293	−2.5403	−3.7557	1.3837	0.1683
7	0.14	−0.0188	−0.0212	−0.3206	−0.3932	−1.0572	−2.6307	1.8858	0.3123
8	0.16	−0.0253	−0.0295	−0.3275	−0.4339	0.3714	−1.4376	2.3334	0.5244
9	0.18	−0.0317	−0.0384	−0.3069	−0.4500	1.6846	−0.1706	2.6656	0.8104
10	0.20	−0.0374	−0.0473	−0.2617	−0.4400	2.8356	1.1676	2.8356	1.1676
11	0.22	−0.0421	−0.0558	−0.2051	−0.4125	2.8294	1.5839	2.8294	1.5839
12	0.24	−0.0456	−0.0637	−0.1500	−0.3763	2.6744	2.0384	2.6744	2.0384
13	0.26	−0.0481	−0.0708	−0.0992	−0.3308	2.4062	2.5041	2.4062	2.5041
14	0.28	−0.0496	−0.0769	−0.0545	−0.2763	2.0678	2.9503	2.0678	2.9503
15	0.30	−0.0503	−0.0818	−0.0168	−0.2133	1.7045	3.3460	1.7045	3.3460

Similarly, Equation 11.76 yields the following relative acceleration response:

$$\begin{Bmatrix} \ddot{u}_1 \\ \ddot{u}_2 \end{Bmatrix}_{i+1} = -\{J\}\ddot{u}_g(t_{i+1}) - [M]^{-1}([C]\{\dot{u}\}_{i+1} + [K]\{u\}_{i+1})$$

$$= -\begin{Bmatrix} 1 \\ 1 \end{Bmatrix} \ddot{u}_g(t_{i+1}) - \begin{bmatrix} 1 & 0 \\ 0 & 1 \end{bmatrix}^{-1} \left(\begin{bmatrix} 2 & -1 \\ -1 & 1 \end{bmatrix} \begin{Bmatrix} \dot{u}_1 \\ \dot{u}_2 \end{Bmatrix}_{i+1} + \begin{bmatrix} 200 & -100 \\ -100 & 100 \end{bmatrix} \begin{Bmatrix} u_1 \\ u_2 \end{Bmatrix}_{i+1} \right)$$

$$= \begin{Bmatrix} -\ddot{u}_g(t_{i+1}) - 2(\dot{u}_1)_{i+1} + (\dot{u}_2)_{i+1} - 200(u_1)_{i+1} + 100(u_2)_{i+1} \\ -\ddot{u}_g(t_{i+1}) + (\dot{u}_1)_{i+1} - (\dot{u}_2)_{i+1} + 100(u_1)_{i+1} - 100(u_2)_{i+1} \end{Bmatrix}$$

As before, the absolute acceleration response may be computed according to

$$\begin{Bmatrix} \ddot{y}_1 \\ \ddot{y}_2 \end{Bmatrix}_{i+1} = \begin{Bmatrix} \ddot{u}_1 \\ \ddot{u}_2 \end{Bmatrix}_{i+1} + \begin{Bmatrix} 1 \\ 1 \end{Bmatrix} \ddot{u}_g(t_{i+1})$$

Results

The use of these equations recursively and starting with the initial conditions yields the responses listed in Table E11.4 for the first 15 steps.

11.7 WILSON-θ METHOD

This method, proposed by E. L. Wilson in 1968, is a variation of the linear acceleration method. It is based on the assumption that the acceleration varies linearly over an extended time interval of duration θΔt (see Figure 11.5), where θ is a constant that is always greater than unity. This constant is

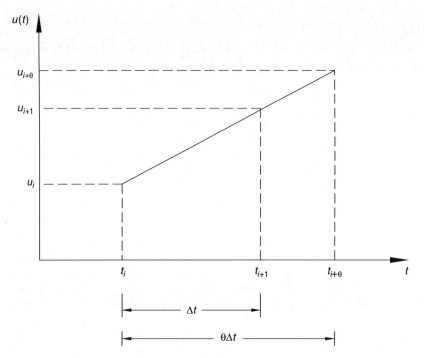

FIGURE 11.5 Assumed variation of acceleration response over extended time interval in Wilson-θ method.

used as a control variable to make the linear acceleration method unconditionally stable. The basic relationships in this method are the same as those derived in the preceding section for the linear acceleration method, except that the subscript $i + 1$ is replaced by $i + \theta$ and Δt is replaced by $\theta \Delta t$. In Wilsons-θ method, therefore, the acceleration, velocity, and displacement at the end of such an extended time interval are given by

$$\{\ddot{u}\}_{i+\theta} = \frac{6}{(\theta \Delta t)^2}(\{u\}_{i+\theta} - \{u\}_i) - \frac{6}{\theta \Delta t}\{\dot{u}\}_i - 2\{\ddot{u}\}_i \tag{11.78}$$

$$\{\dot{u}\}_{i+\theta} = \frac{3}{\theta \Delta t}(\{u\}_{i+\theta} - \{u\}_i) - 2\{\dot{u}\}_i - \frac{\theta \Delta t}{2}\{\ddot{u}\}_i \tag{11.79}$$

$$\{u\}_{i+\theta} = [\hat{K}_W]^{-1}\{\hat{F}_W\}_{i+\theta} \tag{11.80}$$

where

$$[\hat{K}_W] = [K] + \frac{3}{\theta \Delta t}[C] + \frac{6}{(\theta \Delta t)^2}[M] \tag{11.81}$$

$$\{\hat{F}_W\}_{i+\theta} = -[M]\{J\}u_g(t_{i+\theta}) + [M]\left(\frac{6}{(\theta \Delta t)^2}\{u\}_i + \frac{6}{\theta \Delta t}\{\dot{u}\}_i + 2\{\ddot{u}\}_i\right)$$
$$+ [C]\left(\frac{3}{\theta \Delta t}\{u\}_i + 2\{\dot{u}\}_i + \frac{\theta \Delta t}{2}\{\ddot{u}\}_i\right) \tag{11.82}$$

Structural Response by Step-by-Step Integration Methods

In addition, the ground motion is assumed to vary linearly from time $t = t_i = i\Delta t$ to time $t = t_{i+\theta} = (i + \theta)\Delta t$. Therefore, the ground acceleration at the end of the extended time interval is calculated according to

$$\ddot{u}_g(t_{i+\theta}) = \ddot{u}_g(t_i) + \frac{\ddot{u}_g(t_{i+1}) - \ddot{u}_g(t_i)}{\Delta t}\theta\Delta t = (1 - \theta)\ddot{u}_g(t_i) + \theta\ddot{u}_g(t_{i+1}) \tag{11.83}$$

Thus, in Wilson-θ method, Equation 11.80 is used first to determine the displacement vector $\{u\}_{i+\theta}$. Then, this vector is substituted into Equation 11.78 to obtain the corresponding acceleration vector. Afterward, the acceleration vector at the end of the normal time interval Δt is obtained using the interpolation formula

$$\{\ddot{u}\}_{i+1} = \{\ddot{u}\}_i + \frac{1}{\theta}(\{\ddot{u}\}_{i+\theta} - \{\ddot{u}\}_i) \tag{11.84}$$

Finally, this acceleration vector is substituted into Equations 11.67 and 11.68 to determine the velocity and displacement vectors at the end of the normal time interval.

It has been shown through numerical studies that Wilson-θ method is unconditionally stable for $\theta \geq 1.37$ and thus the method is normally used with $\theta = 1.4$. If $\theta = 1.0$, the method reverts to the linear acceleration method, which, as indicated earlier, is only conditionally stable. It has been found, however, that Wilson-θ method introduces excessive artificial damping (a spurious reduction in the amplitude of the response) into the solution (see Section 11.9), and also has a tendency to yield solutions that deviate significantly from the true response (overshoot) in the first few time steps when large time steps are employed.[*]

EXAMPLE 11.5 DYNAMIC RESPONSE OF TWO-DEGREE-OF-FREEDOM SYSTEM BY WILSON-θ METHOD

Solve the problem considered in Example 11.3 using Wilson-θ method.

Solution

Initial Calculations and Effective Stiffness Matrix

As the system is initially at rest, it may be established, as considered in Example 11.3, that

$$\{u\}_0 = \begin{Bmatrix} 0 \\ 0 \end{Bmatrix} \quad \{\dot{u}\}_0 = \begin{Bmatrix} 0 \\ 0 \end{Bmatrix} \quad \{\ddot{u}\}_0 = \begin{Bmatrix} 0 \\ 0 \end{Bmatrix}$$

Constant matrices similar to the $[a]$, $[b]$, and $[c]$ matrices defined in Box 11.4 for the linear acceleration method may also be considered for Wislon's θ method. For a time step of 0.02 s and $\theta = 1.4$, these matrices result as

$$[a] = \frac{6}{(\theta\Delta t)^2}[M] + \frac{3}{\theta\Delta t}[C] = \frac{6}{0.028^2}\begin{bmatrix} 1 & 0 \\ 0 & 1 \end{bmatrix} + \frac{3}{0.028}\begin{bmatrix} 2 & -1 \\ -1 & 1 \end{bmatrix} = \frac{1}{49}\begin{bmatrix} 385,500 & -5250 \\ -5250 & 380,250 \end{bmatrix} \text{MN/m}$$

$$[b] = \frac{6}{\theta\Delta t}[M] + 2[C] = \frac{6}{0.028}\begin{bmatrix} 1 & 0 \\ 0 & 1 \end{bmatrix} + 2\begin{bmatrix} 2 & -1 \\ -1 & 1 \end{bmatrix} = \frac{1}{7}\begin{bmatrix} 1,528 & -14 \\ -14 & 1,514 \end{bmatrix} \text{MN-s/m}$$

$$[c] = 2[M] + \frac{\theta\Delta t}{2}[C] = 2\begin{bmatrix} 1 & 0 \\ 0 & 1 \end{bmatrix} + 0.014\begin{bmatrix} 2 & -1 \\ -1 & 1 \end{bmatrix} = \begin{bmatrix} 2.028 & -0.014 \\ -0.014 & 2.014 \end{bmatrix} \text{MN-s}^2/\text{m}$$

---

[*] See, for instance, Goudreau, G. L. and Taylor, R. L., Evaluation of numerical integration methods in elastodynamics, *Computer Methods in Applied Mechanics and Engineering*, 2, 1972, 69–97.

In view of Equation 11.81, the effective stiffness matrix is then given by

$$[\hat{K}_W] = [K] + [a] = \begin{bmatrix} 200 & -100 \\ -100 & 100 \end{bmatrix} + \frac{1}{49}\begin{bmatrix} 385,500 & -5,250 \\ -5,250 & 380,250 \end{bmatrix} = \frac{1}{49}\begin{bmatrix} 395,300 & -10,150 \\ -10,150 & 385,150 \end{bmatrix} \text{ MN/m}$$

which has an inverse equal to

$$[\hat{K}_W]^{-1} = \begin{bmatrix} 1.24040 \times 10^{-4} & 3.26888 \times 10^{-6} \\ 3.26888 \times 10^{-6} & 1.27309 \times 10^{-4} \end{bmatrix} \text{ m/MN}$$

Effective Load Vector at Time $t = t_{i+\theta}$

Similarly, from Equation 11.82 and the determined [a], [b], and [c] matrices, the effective load vector results as

$$\{\hat{F}_W\}_{i+\theta} = -[M]\{J\}\ddot{u}_g(t_{i+\theta}) + [a]\{u\}_i + [b]\{\dot{u}\}_i + [c]\{\ddot{u}\}_i$$

$$= -\begin{bmatrix} 1 & 0 \\ 0 & 1 \end{bmatrix}\begin{Bmatrix} 1 \\ 1 \end{Bmatrix}\ddot{u}_g(t_{i+\theta}) + \frac{1}{49}\begin{bmatrix} 385,500 & -5,250 \\ -5,250 & 380,250 \end{bmatrix}\begin{Bmatrix} u_1 \\ u_2 \end{Bmatrix}_i$$

$$+ \frac{1}{7}\begin{bmatrix} 1,528 & -14 \\ -14 & 1,514 \end{bmatrix}\begin{Bmatrix} \dot{u}_1 \\ \dot{u}_2 \end{Bmatrix}_i + \begin{bmatrix} 2.028 & -0.014 \\ -0.014 & 2.014 \end{bmatrix}\begin{Bmatrix} \ddot{u}_1 \\ \ddot{u}_2 \end{Bmatrix}_i$$

where, according to Equation 11.83,

$$\ddot{u}_g(t_{i+\theta}) = (1-\theta)\ddot{u}_g(t_i) + \theta\ddot{u}_g(t_{i+1}) = -0.4\ddot{u}_g(t_i) + 1.4\ddot{u}_g(t_{i+1})$$

Displacement and Acceleration Responses at Time $t = t_{i+\theta}$

In light of Equation 11.80 and the established effective load vector and effective stiffness matrix, the displacement response of the system at time $t = t_{i+\theta}$ is therefore given by

$$\begin{Bmatrix} u_1 \\ u_2 \end{Bmatrix}_{i+1} = [\hat{K}_W]^{-1}\{\hat{F}_W\}_{i+1} = \begin{bmatrix} 1.24040 \times 10^{-4} & 3.26888 \times 10^{-6} \\ 3.26888 \times 10^{-6} & 1.27309 \times 10^{-4} \end{bmatrix}\{\hat{F}_W\}_{i+1}$$

$$= -\begin{Bmatrix} 1.27309 \times 10^{-4} \\ 1.30578 \times 10^{-4} \end{Bmatrix}\ddot{u}_g(t_{i+\theta}) + \begin{bmatrix} 9.75519 \times 10^{-1} & 1.20772 \times 10^{-2} \\ 1.20772 \times 10^{-2} & 9.87596 \times 10^{-1} \end{bmatrix}\begin{Bmatrix} u_1 \\ u_2 \end{Bmatrix}_i$$

$$+ \begin{bmatrix} 2.70697 \times 10^{-2} & 4.58935 \times 10^{-4} \\ 4.58935 \times 10^{-4} & 2.75286 \times 10^{-2} \end{bmatrix}\begin{Bmatrix} \dot{u}_1 \\ \dot{u}_2 \end{Bmatrix}_i + \begin{bmatrix} 2.51508 \times 10^{-4} & 4.84696 \times 10^{-6} \\ 4.84696 \times 10^{-6} & 2.56355 \times 10^{-4} \end{bmatrix}\begin{Bmatrix} \ddot{u}_1 \\ \ddot{u}_2 \end{Bmatrix}_i$$

which may also be expressed in the form of the following two algebraic equations:

$$(u_1)_{i+\theta} = [-1.27309\ddot{u}_g(t_{i+\theta}) + 9755.19(u_1)_i + 120.772(u_2)_i + 270.697(\dot{u}_1)_i$$
$$+ 4.58935(\dot{u}_2)_i + 2.51508(\ddot{u}_1)_i + 0.04847(\ddot{u}_2)_i]10^{-4}$$

Structural Response by Step-by-Step Integration Methods

$$(u_2)_{i+\theta} = \left[-1.30578\ddot{u}_g(t_{i+\theta}) + 120.772(u_1)_i + 9875.96(u_2)_i + 4.58935(\dot{u}_1)_i \right.$$
$$\left. + 275.286(\dot{u}_2)_i + 0.04847(\ddot{u}_1)_i + 2.56355(\ddot{u}_2)_i\right]10^{-4}$$

Similarly, from Equation 11.78 and in terms of the determined displacement vector, the acceleration response at time $t = t_{i+\theta}$ may be computed according to

$$\begin{Bmatrix}\ddot{u}_1\\\ddot{u}_2\end{Bmatrix}_{i+\theta} = \frac{6}{0.028^2}\left(\begin{Bmatrix}u_1\\u_2\end{Bmatrix}_{i+\theta} - \begin{Bmatrix}u_1\\u_2\end{Bmatrix}_i\right) - \frac{6}{0.028}\begin{Bmatrix}\dot{u}_1\\\dot{u}_2\end{Bmatrix}_i - 2\begin{Bmatrix}\ddot{u}_1\\\ddot{u}_2\end{Bmatrix}_i$$

$$= 7653.1\left(\begin{Bmatrix}u_1\\u_2\end{Bmatrix}_{i+\theta} - \begin{Bmatrix}u_1\\u_2\end{Bmatrix}_i\right) - 214.3\begin{Bmatrix}\dot{u}_1\\\dot{u}_2\end{Bmatrix}_i - 2\begin{Bmatrix}\ddot{u}_1\\\ddot{u}_2\end{Bmatrix}_i$$

Acceleration, Velocity, and Displacement Responses at Time $t = t_{i+1}$

Once the displacement and acceleration vectors at time $t = t_{i+\theta}$ are known, the acceleration response at time $t = t_{i+1}$ may be determined from Equation 11.84, that is

$$\begin{Bmatrix}\ddot{u}_1\\\ddot{u}_2\end{Bmatrix}_{i+1} = \begin{Bmatrix}\ddot{u}_1\\\ddot{u}_2\end{Bmatrix}_i + \frac{1}{1.4}\left(\begin{Bmatrix}\ddot{u}_1\\\ddot{u}_2\end{Bmatrix}_{i+\theta} - \begin{Bmatrix}\ddot{u}_1\\\ddot{u}_2\end{Bmatrix}_i\right)$$

Similarly, from Equations 11.67 and 11.68 and a time step of 0.02 s, the velocity and displacement responses at time $t = t_{i+1}$ are given by

$$\begin{Bmatrix}\dot{u}_1\\\dot{u}_2\end{Bmatrix}_{i+1} = \begin{Bmatrix}\dot{u}_1\\\dot{u}_2\end{Bmatrix}_i + \frac{1}{100}\begin{Bmatrix}\ddot{u}_1\\\ddot{u}_2\end{Bmatrix}_i + \frac{1}{100}\begin{Bmatrix}\ddot{u}_1\\\ddot{u}_2\end{Bmatrix}_{i+1}$$

$$\begin{Bmatrix}u_1\\u_2\end{Bmatrix}_{i+1} = \begin{Bmatrix}u_1\\u_2\end{Bmatrix}_i + \frac{1}{50}\begin{Bmatrix}\dot{u}_1\\\dot{u}_2\end{Bmatrix}_i + \frac{1}{7{,}500}\begin{Bmatrix}\ddot{u}_1\\\ddot{u}_2\end{Bmatrix}_i + \frac{1}{15{,}000}\begin{Bmatrix}\ddot{u}_1\\\ddot{u}_2\end{Bmatrix}_{i+1}$$

Finally, the absolute acceleration may be computed according to

$$\begin{Bmatrix}\ddot{y}_1\\\ddot{y}_2\end{Bmatrix}_{i+1} = \begin{Bmatrix}\ddot{u}_1\\\ddot{u}_2\end{Bmatrix}_{i+1} + \begin{Bmatrix}1\\1\end{Bmatrix}\ddot{u}_g(t_{i+1})$$

Results

The use of these equations recursively starting with the initial conditions leads to the response values listed in Table E11.5 for the first 15 steps.

TABLE E11.5
First 15 Steps of Displacement, Velocity, and Acceleration Responses of Two-Degree-of-Freedom Shear Building in Example 11.3 Using Wilson-θ Method

Step	Time (s)	u_1 (m)	u_2 (m)	$\dot{u}_1$ (m/s)	$\dot{u}_2$ (m/s)	$\ddot{u}_1$ (m/s²)	$\ddot{u}_2$ (m/s²)	$\ddot{y}_1$ (m/s²)	$\ddot{y}_2$ (m/s²)
0	0.00	0.0000	0.0000	0.0000	0.0000	0.0000	0.0000	0.0000	0.0000
1	0.02	−0.0001	−0.0001	−0.0096	−0.0098	−0.9558	−0.9803	0.0252	0.0007
2	0.04	−0.0005	−0.0005	−0.0377	−0.0392	−1.8579	−1.9577	0.1041	0.0043
3	0.06	−0.0017	−0.0018	−0.0830	−0.0880	−2.6772	−2.9273	0.2658	0.0157
4	0.08	−0.0039	−0.0042	−0.1437	−0.1561	−3.3911	−3.8812	0.5329	0.0428
5	0.10	−0.0075	−0.0081	−0.2175	−0.2430	−3.9851	−4.8079	0.9199	0.0971
6	0.12	−0.0126	−0.0139	−0.2828	−0.3284	−2.5415	−3.7316	1.3825	0.1924
7	0.14	−0.0186	−0.0211	−0.3190	−0.3917	−1.0808	−2.6014	1.8622	0.3416
8	0.16	−0.0251	−0.0294	−0.3265	−0.4318	0.3303	−1.4070	2.2923	0.5550
9	0.18	−0.0315	−0.0382	−0.3068	−0.4473	1.6335	−0.1438	2.6145	0.8372
10	0.20	−0.0373	−0.0471	−0.2627	−0.4369	2.7832	1.1848	2.7832	1.1848
11	0.22	−0.0419	−0.0556	−0.2069	−0.4092	2.7938	1.5866	2.7938	1.5866
12	0.24	−0.0455	−0.0634	−0.1524	−0.3731	2.6613	2.0236	2.6613	2.0236
13	0.26	−0.0481	−0.0705	−0.1016	−0.3282	2.4175	2.4710	2.4175	2.4710
14	0.28	−0.0496	−0.0765	−0.0564	−0.2744	2.1016	2.9008	2.1016	2.9008
15	0.30	−0.0504	−0.0814	−0.0178	−0.2126	1.7554	3.2840	1.7554	3.2840

11.8 NEWMARK-β METHOD

Newmark-β method, introduced in 1959 by N. M. Newmark in what has become a classic paper, constitutes a generalization of the step-by-step integration methods in which an assumption is made in regard to the variation of the acceleration response within a time step. It may be derived as follows:

With reference to Figure 11.6, let the acceleration response within a small time interval Δt be defined by

$$\ddot{u}(\tau) = \ddot{u}_i + f(\tau)(\ddot{u}_{i+1} - \ddot{u}_i) \tag{11.85}$$

where $f(\tau)$ is an arbitrary function of time and, as before, $\ddot{u}_i$ is the acceleration at the beginning of the interval, $\ddot{u}_{i+1}$ is the acceleration at the end of the interval, and $\tau = t - t_i$. Consider, also, that by virtue of the fact that the acceleration is equal to the first derivative of the velocity with respect to time, Equation 11.85 may also be written as

$$\int_0^\tau d\dot{u}(\tau) = \int_0^\tau \left[\ddot{u}_i + f(\tau)(\ddot{u}_{i+1} - \ddot{u}_i)\right] d\tau \tag{11.86}$$

from which one finds that the corresponding velocity response is given by

$$\dot{u}(\tau) = \dot{u}_i + \ddot{u}_i \tau + (\ddot{u}_{i+1} - \ddot{u}_i)\int_0^\tau f(\tau)d\tau \tag{11.87}$$

Thus, after setting $\tau = \Delta t$, it is found that the velocity response at the end of the time interval is given by

$$\dot{u}_{i+1} = \dot{u}_i + \Delta t \ddot{u}_i + \gamma \Delta t (\ddot{u}_{i+1} - \ddot{u}_i) \tag{11.88}$$

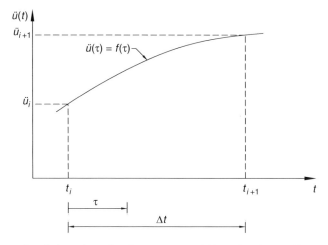

FIGURE 11.6 Assumed variation of acceleration response within a time interval in Newmark-β method.

which may also be written as

$$\dot{u}_{i+1} = \dot{u}_i + \Delta t(1-\gamma)\ddot{u}_i + \gamma \Delta t \ddot{u}_{i+1} \quad (11.89)$$

where γ is defined as

$$\gamma = \frac{1}{\Delta t}\int_0^{\Delta t} f(\tau)d\tau \quad (11.90)$$

Similarly, in light of Equation 11.87 and the fact that the velocity is equal to the first derivative of the displacement with respect to time, one has that

$$\int_0^\tau du(\tau) = \int_0^\tau \left[\dot{u}_i + \ddot{u}_i \tau + (\ddot{u}_{i+1} - \ddot{u}_i)\int_0^\tau f(\tau)d\tau\right]d\tau \quad (11.91)$$

which leads to

$$u(\tau) = u_i + \dot{u}_i \tau + \ddot{u}_i \frac{\tau^2}{2} + (\ddot{u}_{i+1} - \ddot{u}_i)\int_0^\tau\left[\int_0^\tau f(\tau)d\tau\right]d\tau \quad (11.92)$$

and thus, after setting $\tau = \Delta t$, the displacement response at the end of the time interval may be considered equal to

$$u_{i+1} = u_i + \Delta t \dot{u}_i + \frac{\Delta t^2}{2}\ddot{u}_i + \beta \Delta t^2(\ddot{u}_{i+1} - \ddot{u}_i) \quad (11.93)$$

or

$$u_{i+1} = u_i + \Delta t \dot{u}_i + \Delta t^2\left[\left(\frac{1}{2}-\beta\right)\ddot{u}_i + \beta \ddot{u}_{i+1}\right] \quad (11.94)$$

where β is given by

$$\beta = \frac{1}{\Delta t^2}\int_0^{\Delta t}\int_0^\tau f(\tau)d\tau d\tau \quad (11.95)$$

Thus, in Newmark's method the velocity and displacement vectors evaluated at the end of an interval are of the form

$$\{\dot{u}\}_{i+1} = \{\dot{u}\}_i + \Delta t(1-\gamma)\{\ddot{u}\}_i + \gamma\Delta t\{\ddot{u}\}_{i+1} \qquad (11.96)$$

$$\{u\}_{i+1} = \{u\}_i + \Delta t\{\dot{u}\}_i + \Delta t^2\left[\left(\frac{1}{2}-\beta\right)\{\ddot{u}\}_i + \beta\{\ddot{u}\}_{i+1}\right] \qquad (11.97)$$

where γ and β are constants that can be selected arbitrarily to optimize the accuracy and efficiency of the procedure. Once these relationships for the velocity and displacement responses are established, one can then proceed as with the constant average and the linear acceleration methods to obtain an explicit formula to compute the displacement solution. That is, one can solve for the acceleration response from Equation 11.97 to obtain

$$\{\ddot{u}\}_{i+1} = \frac{1}{\beta\Delta t^2}(\{u\}_{i+1} - \{u\}_i - \Delta t\{\dot{u}\}_i) - \left(\frac{1}{2\beta}-1\right)\{\ddot{u}\}_i \qquad (11.98)$$

Then, one can substitute this expression into Equation 11.96 to arrive to the following relationship for the velocity response:

$$\{\dot{u}\}_{i+1} = \frac{\gamma}{\beta\Delta t}(\{u\}_{i+1} - \{u\}_i) + \left(1-\frac{\gamma}{\beta}\right)\{\dot{u}\}_i + \left(1-\frac{\gamma}{2\beta}\right)\Delta t\{\ddot{u}\}_i \qquad (11.99)$$

Lastly, Equations 11.98 and 11.99 can be inserted into the equation of motion written for the time $t_{i+1} = \Delta t(i+1)$ (see Equation 11.19) to get

$$[\hat{K}_N]\{u\}_{i+1} = \{\hat{F}_N\}_{i+1} \qquad (11.100)$$

where

$$[\hat{K}_N] = [K] + \frac{1}{\beta\Delta t^2}[M] + \frac{\gamma}{\beta\Delta t}[C] \qquad (11.101)$$

$$\{\hat{F}_N\}_{i+1} = -[M]\{J\}\ddot{u}_g(t_{i+1}) + [M]\left(\frac{1}{\beta\Delta t^2}\{u\}_i + \frac{1}{\beta\Delta t}\{\dot{u}\}_i + \left(\frac{1}{2\beta}-1\right)\{\ddot{u}\}_i\right)$$
$$+ [C]\left(\frac{\gamma}{\beta\Delta t}\{u\}_i + \left(\frac{\gamma}{\beta}-1\right)\{\dot{u}\}_i + \left(\frac{\gamma}{2\beta}-1\right)\Delta t\{\ddot{u}\}_i\right) \qquad (11.102)$$

Consequently, in Newmark's method the displacement response at time $t_{i+1} = \Delta t(i+1)$ is calculated according to

$$\{u\}_{i+1} = [\hat{K}_N]^{-1}\{\hat{F}_N\}_{i+1} \qquad (11.103)$$

In similarity with the constant average and linear acceleration methods, the velocity and acceleration responses are then, respectively, determined from Equation 11.99 and the equation of motion expressed for the time $t = t_{i+1}$.

It should be realized that the constant average and linear acceleration methods represent particular cases of Newmark's method. To prove this assertion, consider that in light of Equations 11.41 and 11.85 one has that in the constant average acceleration method $f(\tau) = 1/2$. Therefore, according to Equations 11.90 and 11.95, this method corresponds to Newmark's method when the constants γ and β are equal to

$$\gamma = \frac{1}{\Delta t}\int_0^{\Delta t} f(\tau)d\tau = \frac{1}{2\Delta t}\int_0^{\Delta t} d\tau = \frac{1}{2} \tag{11.104}$$

$$\beta = \frac{1}{\Delta t^2}\int_0^{\Delta t}\left[\int_0^{\tau} f(\tau)d\tau\right]d\tau = \frac{1}{2\Delta t^2}\int_0^{\Delta t}\left[\int_0^{\tau} d\tau\right]d\tau = \frac{1}{2\Delta t^2}\int_0^{\Delta t} \tau d\tau = \frac{1}{4} \tag{11.105}$$

Similarly, as in the linear acceleration method (see Equations 11.62 and 11.85) $f(\tau) = \tau/\Delta t$, the linear acceleration method corresponds to Newmark's method when γ and β are equal to

$$\gamma = \frac{1}{\Delta t}\int_0^{\Delta t} f(\tau)d\tau = \frac{1}{\Delta t^2}\int_0^{\Delta t} \tau d\tau = \frac{1}{2} \tag{11.106}$$

$$\beta = \frac{1}{\Delta t^2}\int_0^{\Delta t}\left[\int_0^{\tau} f(\tau)d\tau\right]d\tau = \frac{1}{\Delta t^3}\int_0^{\Delta t}\left[\int_0^{\tau} \tau d\tau\right]d\tau = \frac{1}{2\Delta t^3}\int_0^{\Delta t} \tau^2 d\tau = \frac{1}{6} \tag{11.107}$$

In regard to its stability, it has been shown that Newmark's method is unconditionally stable whenever

$$\gamma \geq \frac{1}{2} \quad \text{and} \quad \beta \geq \frac{1}{4}\left(\frac{1}{2}+\gamma\right)^2 \tag{11.108}$$

However, it has also been found that the method introduces artificial damping when $\gamma > 1/2$ and that this artificial damping is substantial unless very small time steps are used. Traditionally, therefore, the method is used considering $\gamma = 1/2$. The stability of Newmark's method is examined in detail in the following section for the particular case of $\gamma = 1/2$.

The solution algorithm presented in Box 11.4 may also be utilized as the solution algorithm for Newmark's method after the proper substitution is made of the expressions that define the effective stiffness matrix, effective load vector, and velocity response in this method.

11.9 STABILITY AND ACCURACY ISSUES

11.9.1 Introductory Remarks

It should be obvious from the description of the procedures introduced in the foregoing sections that the efficiency and accuracy of a step-by-step integration method critically depend on the size of the time step used. On the one hand, if the time step considered is too large, the solution may become unstable or may be inaccurate. On the other hand, if the time step used is too small, the computational effort and hence the cost of the analysis increase. For efficiency purposes, therefore, it is necessary to use the largest time step that will lead to a stable and accurate solution. But to be able to do this, one needs to be familiar with the accuracy and stability characteristics of the available numerical methods and the influence that the time-step size has in these characteristics. This section will describe thus the accuracy and stability characteristics of the methods presented in the preceding sections and provide useful guidelines for the selection of the appropriate time step in each case.

11.9.2 NUMERICAL STABILITY

As explained in Section 11.2, errors are introduced in a numerical solution as a result of the approximations involved in the development of the numerical procedures used. If the error has the tendency to grow from one step to the next, soon the solution will grow out of bound and become meaningless. In such a case, it is said that the procedure is numerically unstable. A method is unconditionally stable if it leads to solutions that are always bounded, regardless of the size of the time step used. In contrast, a method is conditionally stable if it provides bounded solutions only when the time step used is shorter than a certain size. There are several procedures by means of which one can determine the stability limits of a numerical integration scheme. However, only two of such procedures will be introduced here. One will serve to obtain the stability limit of Newmark-β method, whereas the other will be utilized to find the stability limit of the central difference method. The intention here is to illustrate how such stability limits are determined rather than present the derivation of the stability limits of all the methods described in this chapter.

The stability limit of a step-by-step integration method may be determined from the solution of a difference equation the way Newmark derived the stability limit for his method. It will be thus instructive to review how Newmark established that limit. For this purpose, note, first, that in the derivation of the limit in question, Newmark considered the response of an undamped single-degree-of-freedom system in free vibration. Although damping will undoubtedly affect the stability characteristics of a numerical solution, the consideration of an undamped system is justified on the grounds that the presence of damping will make such stability characteristics less restrictive. Similarly, if a method is unstable for a free vibration solution, then it is likely that it will also be unstable for a forced vibration solution. This is so because instability of the homogeneous solution (free vibration part) of an equation of motion will also make the total solution meaningless. Along the same lines, the consideration of a single-degree-of-freedom system is justified by the argument that the equation that describes the motion of a single-degree-of-freedom system is also the equation that describes the motion of a multi-degree-of-freedom system in one of its natural modes of vibration. Hence, if a method is unstable for a single-degree-of-freedom system, it will also be unstable for one of the modes of vibration of a multi-degree-of-freedom system and, consequently, for the total solution of the multi-degree-of-freedom system. Subsequently, Newmark considered that according to his method the velocity and displacement response of a single-degree-of-freedom system at time $t = t_{i+1}$ are given by (see Equations 11.89 and 11.94)

$$\dot{u}_{i+1} = \dot{u}_i + \Delta t (1 - \gamma)\ddot{u}_i + \gamma \Delta t \ddot{u}_{i+1} \tag{11.109}$$

$$u_{i+1} = u_i + \Delta t \dot{u}_i + \left(\frac{1}{2} - \beta\right) \Delta t^2 \ddot{u}_i + \beta \Delta t^2 \ddot{u}_{i+1} \tag{11.110}$$

and that from the equation of motion of an undamped single-degree-of-freedom system in free vibration the acceleration response is equal to

$$\ddot{u}_{i+1} = -\omega^2 u_{i+1} \tag{11.111}$$

where $\omega = 2\pi/T$ represents the natural frequency of the system and T is the corresponding natural period. Similarly, he considered that the corresponding responses at time $t = t_i$ may be expressed as

$$\dot{u}_i = \dot{u}_{i-1} + \Delta t (1 - \gamma)\ddot{u}_{i-1} + \gamma \Delta t \ddot{u}_i \tag{11.112}$$

$$u_i = u_{i-1} + \Delta t \dot{u}_{i-1} + \left(\frac{1}{2} - \beta\right) \Delta t^2 \ddot{u}_{i-1} + \beta \Delta t^2 \ddot{u}_i \tag{11.113}$$

$$\ddot{u}_i = -\omega^2 u_i \tag{11.114}$$

Hence, by subtracting Equation 11.113 from Equation 11.110 and rearranging terms, he obtained

$$u_{i+1} - 2u_i + u_{i-1} - \Delta t(\dot{u}_i - \dot{u}_{i-1}) - \left(\frac{1}{2} - \beta\right)\Delta t^2(\ddot{u}_i - \ddot{u}_{i-1}) - \beta\Delta t^2(\ddot{u}_{i+1} - \ddot{u}_i) = 0 \quad (11.115)$$

However, from Equation 11.112, one has that

$$\dot{u}_i - \dot{u}_{i-1} = \Delta t(1-\gamma)\ddot{u}_{i-1} + \gamma\Delta t\ddot{u}_i \quad (11.116)$$

Therefore, after substitution of this relationship and the use of Equations 11.111 and 11.114, Equation 11.115 may be rewritten as

$$u_{i+1} - 2u_i + u_{i-1} + \omega^2\Delta t^2(1-\gamma)\ddot{u}_{i-1} + \gamma\omega^2\Delta t^2\ddot{u}_i + \left(\frac{1}{2} - \beta\right)\omega^2\Delta t^2(\ddot{u}_i - \ddot{u}_{i-1})$$
$$+ \beta\omega^2\Delta t^2(u_{i+1} - u_i) = 0 \quad (11.117)$$

which may also be expressed as

$$u_{i+1} - (2 - \alpha^2)u_i + u_{i-1} + \left(\gamma - \frac{1}{2}\right)\alpha^2(u_i - u_{i-1}) = 0 \quad (11.118)$$

where

$$\alpha^2 = \frac{\omega^2 \Delta t^2}{1 + \beta \omega^2 \Delta t^2} \quad (11.119)$$

It was observed by Newmark, however, that from the general relationship between finite differences and derivatives, the *difference* $u_i - u_{i-1}$ in Equation 11.118 was related to the velocity response of the system. Consequently, he concluded that the last term in Equation 11.118 constituted a viscous damping term and that this term was a spurious one as the equation was derived for an undamped system. Hence, he suggested to eliminate this spurious term by selecting $\gamma = 1/2$. With such a choice, Equation 11.118 is thus reduced to

$$u_{i+1} - (2 - \alpha^2)u_i + u_{i-1} = 0 \quad (11.120)$$

Equation 11.120 represents a difference equation whose solution is of the form

$$u = e^{m(i-1)} = M^{i-1} \quad (11.121)$$

where m is a constant and obviously $M = e^m$. When substituted into the difference equation, this solution leads to the characteristic equation

$$M^2 - (2 - \alpha^2)M + 1 = 0 \quad (11.122)$$

from which one finds that

$$M = \frac{1}{2}(2 - \alpha^2) \pm \frac{1}{2}\sqrt{(2-\alpha^2)^2 - 4} \quad (11.123)$$

It may be seen thus that Equation 11.120 will lead to an oscillatory response only when $\alpha^2 < 4.0$, as only in this case the roots of Equation 11.123 are complex valued. Hence, in accordance with the definition of α^2 given by Equation 11.119, Newmark's method provides bounded solutions only when

$$\frac{\omega^2 \Delta t^2}{1 + \beta \omega^2 \Delta t^2} < 4 \tag{11.124}$$

which, after solving for $\omega \Delta t$ and considering that $\omega = 2\pi/T$, leads to

$$\frac{\Delta t}{T} < \frac{1}{\pi\sqrt{1 - 4\beta}} \tag{11.125}$$

Equation 11.125 defines thus the stability limit for Newmark-β method. Observe that for the particular case of the linear acceleration method, which corresponds to $\gamma = 1/2$ and $\beta = 1/6$, the stability limit is

$$\frac{\Delta t}{T} < \frac{\sqrt{3}}{\pi} \tag{11.126}$$

Similarly, for the constant average acceleration method, which corresponds to $\gamma = 1/2$ and $\beta = 1/4$, such a limit is

$$\frac{\Delta t}{T} < \infty \tag{11.127}$$

which shows that, as indicated earlier, the constant average acceleration method is stable independently of the selected time step; that is, the method is unconditionally stable.

The stability of a step-by-step integration method may also be evaluated by examining the amplification matrix of the method (defined below) and the spectral norm of this amplification matrix. This alternative stability evaluation technique will be presented next, but before that it is convenient to review first some concepts from matrix algebra to facilitate the understanding of the technique.

In dealing with the issue of convergence in numerical procedures that involve vectors and matrices, it is sometimes necessary to describe by a single number the *size* of a vector or a matrix. When such a single number depends on the magnitude of all the elements in the array, that number is called a *norm*. Several quantities are used as norms in linear algebra. For example, the *Euclidean norm* of a vector is defined as the length of the vector; that is, if v_i, $i = 1, \ldots, n$, represents the elements of a vector $\{v\}$, the Euclidean norm of the vector $\{v\}$ is defined as

$$\|\{v\}\| = \sqrt{v_1^2 + v_2^2 + \cdots + v_n^2} \tag{11.128}$$

Similarly, if $[A]$ is a square matrix of order n, the spectral norm of $[A]$ is defined as

$$\|[A]\| = \sqrt{\tilde{\lambda}_n} \tag{11.129}$$

where $\tilde{\lambda}_n$ is the largest eigenvalue of $[A]^T[A]$. A useful property of norms is that for any two matrices $[A]$ and $[B]$, or a matrix $[A]$ and a vector $\{v\}$,

$$\|[A][B]\| \leq \|[A]\|\|[B]\| \tag{11.130}$$

$$\|[A]\{v\}\| \leq \|[A]\|\|\{v\}\| \tag{11.131}$$

Another valuable property of norms is related to the eigenvalues of a matrix. If norms are taken on both sides of the equation that defines an eigenvalue problem, that is, $[A]\{v\} = \lambda\{v\}$, where λ represents an eigenvalue of $[A]$, one gets

$$\|[A]\{v\}\| = \|\lambda\{v\}\| \tag{11.132}$$

and thus, by virtue of Equation 11.131, one has that

$$\|[A]\|\|\{v\}\| \geq |\lambda|\|\{v\}\| \tag{11.133}$$

or that

$$|\lambda| \leq \|[A]\| \tag{11.134}$$

Therefore, the absolute value of every eigenvalue of $[A]$ is equal to or smaller than the norm of $[A]$. Hence, if a *spectral radius* $\rho([A])$ is defined as the largest eigenvalue of $[A]$, that is,

$$\rho([A]) = max\{|\lambda_1|, |\lambda_2|, \ldots, |\lambda_n|\} \tag{11.135}$$

it may be concluded that

$$\rho([A]) \leq \|[A]\| \tag{11.136}$$

The alternative stability evaluation technique may now be established as follows. Consider, once again, a free vibration problem with the corresponding initial conditions. If a method is formulated to determine the response of the system at a time step using the value of the response at the previous time step, then it is possible to write for each time step a recurrent expression of the form

$$\{Y\}_{i+1} = [A]\{Y\}_i \tag{11.137}$$

where $\{Y\}_{i+1}$ and $\{Y\}_i$, respectively, represent the response of the system at the times $t = t_{i+1}$ and $t = t_i$, and $[A]$ is a matrix, called the *amplification matrix*, which may be expressed in explicit form for each method. Furthermore, by applying this equation successively for all the previous time steps, one gets

$$\{Y\}_{i+1} = [A]\{Y\}_i = [A]^2\{Y\}_{i-1} = \cdots = [A]^i\{Y\}_0 \tag{11.138}$$

Hence, after taking norms at both sides of the equation and considering Equation 11.131, one obtains

$$\|\{Y\}_{i+1}\| \leq \|[A]^i\|\|\{Y\}_0\| \tag{11.139}$$

However, by virtue of Equation 11.136, Equation 11.139 may also be expressed as

$$\|\{Y\}_{i+1}\| \leq \rho^i([A])\|\{Y\}_0\| \tag{11.140}$$

where $\rho([A])$ is the spectral radius defined by Equation 11.135.

It may be seen, thus, that the solution $\{Y\}_{i+1}$ will be bounded if the spectral radius of the amplification matrix $[A]$, which depends on the particular integration method under consideration, is less than or equal to unity, that is, if

$$\rho([A]) \leq 1.0 \tag{11.141}$$

Consequently, one can establish the stability of a numerical integration method by calculating the spectral radius of its amplification matrix and by comparing this spectral radius against unity. If for a given time step, the spectral radius is <1, then the method is stable when that time step is used. Conversely, if the spectral radius is >1, then the method will yield an unbounded solution. This procedure may also be employed to establish stability limits, as it will be illustrated next for the central difference method.

Inspection of Equations 11.10 through 11.12 reveals that in the central difference method $\{u\}_{i+1}$, the displacement response of a system at time $t = t_{i+1}$ depends on the displacement responses at two previous time steps, that is, $\{u\}_i$ and $\{u\}_{i-1}$. It also reveals that for an undamped single-degree-of-freedom system in free vibration such a response is given by

$$u_{i+1} = \frac{\Delta t^2}{m}\left(\frac{2m}{\Delta t^2} - k\right)u_i - \frac{\Delta t^2}{m}\frac{m}{\Delta t^2}u_{i-1} = (2 - \omega^2\Delta t^2)u_i - u_{i-1} \tag{11.142}$$

Therefore, the recurrence equation for a single-degree-of-freedom system in free vibration using the central difference method is of the form

$$\begin{Bmatrix} u_{i+1} \\ u_i \end{Bmatrix} = [A]\begin{Bmatrix} u_i \\ u_{i-1} \end{Bmatrix} \tag{11.143}$$

where the amplification matrix $[A]$ is given by

$$[A] = \begin{bmatrix} 2 - \omega^2\Delta t^2 & -1 \\ 1 & 0 \end{bmatrix} \tag{11.144}$$

Hence, the eigenvalues λ of the amplification matrix $[A]$ for the central difference method are defined by

$$|[A] - \lambda[I]| = \begin{vmatrix} 2 - \omega^2\Delta t^2 - \lambda & -1 \\ 1 & -\lambda \end{vmatrix} = 0 \tag{11.145}$$

which leads to the characteristic equation

$$\lambda^2 + (\omega^2\Delta t^2 - 2)\lambda + 1 = 0 \tag{11.146}$$

whose two roots are

$$\lambda_{1,2} = \frac{2 - \omega^2\Delta t^2}{2} \pm \sqrt{\left(\frac{2 - \omega^2\Delta t^2}{2}\right)^2 - 1} \tag{11.147}$$

As previously established, a bounded solution is obtained when the spectral radius is equal to or less than unity. Therefore, as by definition the spectral radius corresponds to the absolute value of the largest eigenvalue, for the central difference method stability is attained when

$$\left| \frac{2 - \omega^2 \Delta t^2}{2} \pm \sqrt{\left(\frac{2 - \omega^2 \Delta t^2}{2}\right)^2 - 1} \right| \leq 1.0 \tag{11.148}$$

which, when $\omega^2 \Delta t^2 < 2$, leads to

$$\frac{2 - \omega^2 \Delta t^2}{2} + \sqrt{\left(\frac{2 - \omega^2 \Delta t^2}{2}\right)^2 - 1} \leq 1.0 \tag{11.149}$$

which in turn leads to

$$(2 - \omega^2 \Delta t^2)^2 \leq \omega^4 \Delta t^4 \tag{11.150}$$

or to

$$\Delta t \geq 0 \tag{11.151}$$

In contrast, when $\omega^2 \Delta t^2 > 2$, Equation 11.148 may also be expressed as

$$-\frac{2 - \omega^2 \Delta t^2}{2} - \sqrt{\left(\frac{2 - \omega^2 \Delta t^2}{2}\right)^2 - 1} \leq 1.0 \tag{11.152}$$

which leads to

$$(2 - \omega^2 \Delta t^2)^2 - 4 \leq (4 - \omega^2 \Delta t^2)^2 \tag{11.153}$$

and hence, after considering that $\omega = 2\pi/T$, to

$$\frac{\Delta t}{T} \leq \frac{1}{\pi} = 0.318 \tag{11.154}$$

Thus, as stated in Section 11.3, the central difference method is stable only when the used time step is equal to or less than the highest natural period of the system divided by π.

The stability limits for the other methods may be determined similarly, although explicit expressions are not possible in all cases. It is possible, though, to solve numerically for the eigenvalues of the associated amplification matrices and plot the spectral radii for different time-step sizes. From these plots, it is then possible to determine the limits of the ratio $\Delta t/T$ for which the integration scheme under investigation will lead to bounded solutions. An example of such plots is shown in Figure 11.7, which depicts the variation with time step size of the spectral radius $\rho([A])$ for several of the step-by-step integration methods discussed in this chapter. It may be seen from this figure that Houbolt method, the constant average acceleration method, and Wilson-θ method with $\theta = 1.4$ are all unconditionally stable as their spectral radii are always equal to or less than unity. It may also be seen that, as previously determined, the central difference method is only conditionally stable as for values of $\Delta t/T > 0.318$ the spectral radius for this method is greater than unity.

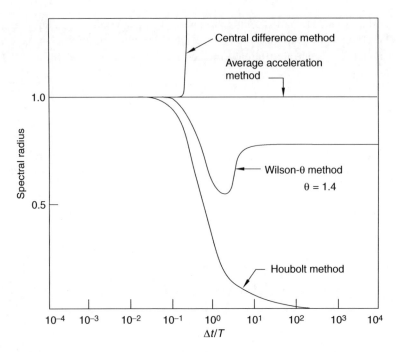

FIGURE 11.7 Variation of spectral radius with time step size over natural period ratio for various numerical integration schemes. (From Bathe, K.-S., *Finite Element Procedures in Engineering Analysis*, Prentice-Hall, Englewood Cliffs, NJ, 1982.)

It is important to keep in mind that the derivations and discussions on the stability limits presented in the preceding paragraphs were limited to linear systems. Thus, the stability properties introduced in this section are strictly valid for linear systems and may not be applicable to nonlinear problems. In fact, it has been found that the unconditional stability of the Houbolt and the constant average acceleration methods, for instance, do not hold for nonlinear systems.[*]

11.9.3 Accuracy

Errors are unavoidable in any numerical procedure, and the methods introduced in this chapter are not an exception. It is important, therefore, to become acquainted with the accuracy capabilities and limitations of these methods. By accuracy it is understood here as the closeness of a numerical solution to the exact solution and should be distinguished from the concept of stability, which solely refers to whether or not bounded solutions will be attained. Clearly, as different methods involve different approximations, it is expected that different methods will provide different accuracy levels when the same time step is used. As discussed before, also, the accuracy of a method will depend on the size of the time step used.

Traditionally, the numerical accuracy of step-by-step integration methods is assessed by measuring the extent to which the methods distort the response of an undamped single-degree-of-freedom system in free vibration. This distortion is measured in terms of amplitude decay (*numerical damping*) and period elongation (*dispersion*) as it has been observed that the response determined

---

[*] See, for example, Stricklin, J. A. et al., Nonlinear dynamic analysis of shells of revolution by matrix displacement method, *Journal of the American Institute of Aeronautics and Astronautics*, 9, 4, 1971, 629–636.

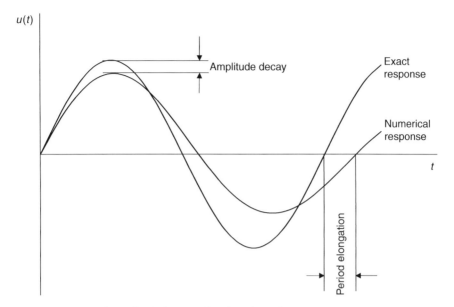

FIGURE 11.8 Definition of amplitude decay and period elongation.

with a numerical procedure deviates from the theoretical solution (a harmonic function) in two ways: (a) a change in the duration of a cycle of response and (b) a reduction in the amplitude of the response. Figure 11.8 clarifies the meaning of these two concepts.

Figure 11.9 shows for several of the numerical integration methods introduced in this chapter the variation of the amplitude decay and period elongation with the size of the time step, expressed as a fraction of the exact amplitude and period. It may be seen from these figures that, in general, the accuracy of the methods increases when the time step size is decreased, and that all the considered methods will yield accurate solutions when $\Delta t/T$ is less than ~0.01. It may also be seen that the central difference, constant average acceleration, and linear acceleration methods introduce no numerical damping. In contrast, Wilson-θ and Houbolt methods introduce a substantial amount of numerical damping if a large time step is used. Additionally, it may be seen that all the methods considered introduce significant period elongation for large time steps. The central difference method introduces the largest period elongation, increasing significantly for time steps larger than the stability limit of $\Delta t/T = 0.318$. In contrast, for $\Delta t/T$ less than its stability limit, the linear acceleration method exhibits the smallest period elongation.

On the basis of the results presented in the preceding paragraph, it has been recommended to use a time step of <1/10 of the shortest natural period of the system under analysis to obtain reasonably accurate results. For example, if the shortest natural period of a system is 0.5 s, then it is recommended to use a time step of 0.05 s. Using a time step equal to 1/10 of the shortest period is not, however, a guaranty of the accuracy of the results. It is thus advisable to solve the problem using the recommended time step, but repeat the analysis with a slightly smaller time step and compare the results. If the results differ considerably, then the analysis is repeated until two successive solutions are close enough. Earthquake ground motions are normally discretized at intervals of 0.02 or 0.01 s. Hence, in the analysis of earthquake engineering problems, it is common to perform numerical integrations using a time step of 0.01 s. The presented results also seem to suggest that the constant average acceleration method offers the best choice as is unconditionally stable and introduces no numerical damping. It should be kept in mind, however, that the most appropriate method always depends on the problem at hand.

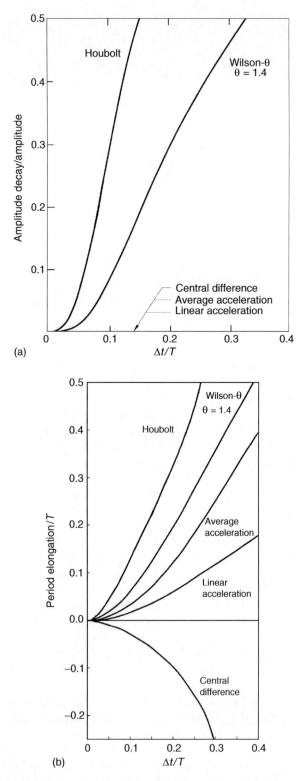

FIGURE 11.9 (a) Amplitude decay and (b) period elongation introduced by different step-by-step integration methods as a function of time step size.

11.10 ANALYSIS OF NONLINEAR SYSTEMS

11.10.1 INTRODUCTION

As described in Section 6.5, earthquake-resistant structures are designed to incur into their nonlinear range of behavior, and thus their analysis requires the consideration of their nonlinear behavior. Fortunately, the step-by-step integration methods described in the previous sections may also be utilized for the seismic analysis of nonlinear structures. It should be noted, however, that these methods cannot be used directly because the properties of nonlinear structures change with the response level and so these properties need to be adjusted during the solution process. Furthermore, the solution in the case of a nonlinear structure depends on the previous loading history. Consequently, in the case of nonlinear structures, it is necessary to (a) write the equations of motion in an incremental form, (b) make the assumption that the properties of the structure remain unchanged during any given time interval, (c) solve the equations of motion for such a time interval considering that the structure behaves linearly, and (d) reformulate its properties on the basis of the obtained solution at the end of the interval to conform to the state of stresses and deformations at that time.

This section will describe how to derive such incremental equations of motion, show how to obtain the response of nonlinear systems using these incremental equations of motion and a step-by-step integration method, illustrate how a system's properties may be adjusted according to the response level, discuss the errors that are always present in the numerical solution of nonlinear systems, and present algorithms tailored for the solution of nonlinear systems using some of the step-by-step integration methods described in the preceding sections.

11.10.2 INCREMENTAL EQUATIONS OF MOTION

Consider the single-degree-of-freedom system shown in Figure 11.10 when the system is subjected to a ground acceleration $\ddot{u}_g(t)$. Let the system be defined by its mass m, stiffness $k(t)$, and damping constant $c(t)$, where $k(t)$ and $c(t)$ will, in general, vary with the displacement and velocity of the system and hence with time. Assume, in addition, that the force in the damper, $F_D(t)$, and the force in the spring, $F_S(t)$, often referred to as the *restoring force* of the system, vary in a nonlinear fashion in the way shown in Figure 11.11. Consideration of the free-body diagram shown in Figure 11.10 and application of Newton's second law at time $t = t_i = i\Delta t$, where, as before, i is any integer and Δt denotes a small time interval, yield

$$m\ddot{u}(t_i) + F_D(t_i) + F_S(t_i) = -m\ddot{u}_g(t_i) \tag{11.155}$$

in which $F_D(t_i)$ and $F_S(t_i)$, respectively, denote the forces acting on the damper and spring of the system at time $t = t_i$, and all other symbols are as defined before. Similarly, consideration of the

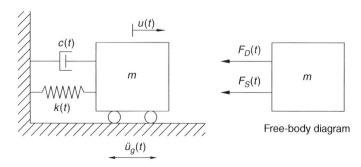

FIGURE 11.10 Single-degree-of-freedom system with variable properties and corresponding free-body diagram.

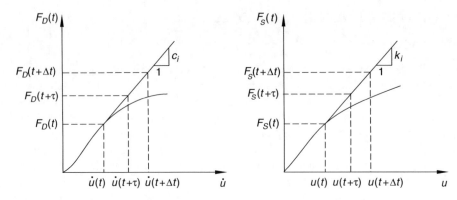

FIGURE 11.11 Nonlinear damping force-velocity and spring-force displacement curves.

dynamic equilibrium of the system at time $t = t_i + \tau$, where, as before, τ is a time variable that varies between 0 and Δt, leads to

$$m\ddot{u}(t_i + \tau) + F_D(t_i + \tau) + F_S(t_i + \tau) = -m\ddot{u}_g(t_i + \tau) \quad (11.156)$$

where $F_D(t_i + \tau)$ and $F_S(t_i + \tau)$, respectively, are the forces acting on the damper and spring of the system at time $t = t_i + \tau$. As in the case of a nonlinear system the force in the spring cannot be considered proportional to the system's displacement, and the force in the damper cannot be considered proportional to the system's velocity; these forces need to be read directly from specified damping force versus velocity and spring force versus displacement curves. If it is assumed, however, that the slopes of these curves remain constant during the time interval Δt and equal to the slopes of the tangents to the curves at the beginning of the interval,* then, as shown in Figure 11.11, these forces may be expressed as

$$F_S(t + \tau) = F_S(t) + k_i \Delta u(\tau) \quad (11.157)$$

$$F_D(t + \tau) = F_D(t) + c_i \Delta \dot{u}(\tau) \quad (11.158)$$

where k_i and c_i are the aforementioned tangent slopes (spring and damping constants) evaluated at time $t = t_i$, and

$$\Delta u(\tau) = u(t_i + \tau) - u(t_i) \quad (11.159)$$

$$\Delta \dot{u}(\tau) = \dot{u}(t_i + \tau) - \dot{u}(t_i) \quad (11.160)$$

As a result, Equation 11.156 may be written as

$$m\ddot{u}(t_i + \tau) + F_D(t_i) + c_i \Delta \dot{u}(\tau) + F_S(t_i) + k_i \Delta u(\tau) = -m\ddot{u}_g(t_i + \tau) \quad (11.161)$$

which may also be expressed as

$$m\ddot{u}(t_i) + m\Delta\ddot{u}(\tau) + F_D(t_i) + c_i \Delta\dot{u}(\tau) + F_S(t_i) + k_i \Delta u(\tau) = -m\ddot{u}_g(t_i) - m\Delta\ddot{u}_g(\tau) \quad (11.162)$$

---

* A more accurate approximation would be attained if the secant slope is used instead of the tangent slope (see Figure 11.11). However, the secant slope cannot be determined because, unless iterations are performed, $u(t_i + \Delta t)$ is not known in advance.

where

$$\Delta \ddot{u}(\tau) = \ddot{u}(t_i + \tau) - \ddot{u}(t_i) \qquad (11.163)$$

$$\Delta \ddot{u}_g(\tau) = \ddot{u}_g(t_i + \tau) - \ddot{u}_g(t_i) \qquad (11.164)$$

Observe, however, that by virtue of Equation 11.155, together the first, third, and fifth terms, that appear on the left-hand side of Equation 11.162 are equal to $-m\ddot{u}_g(t)$. Therefore, Equation 11.162 may be reduced to

$$m\Delta \ddot{u}(\tau) + c_i \Delta \dot{u}(\tau) + k_i \Delta u(\tau) = -m\Delta \ddot{u}_g(\tau) \qquad (11.165)$$

Equation 11.165 represents an incremental equation of motion in which the unknown is the incremental displacement $\Delta u(\tau)$. Therefore, once this equation is solved and the incremental displacement $\Delta u(\tau)$ is determined, the displacement at the end of the time interval may be found from Equation 11.159 by setting $\tau = \Delta t$; namely,

$$u(t_i + \Delta t) = u(t_i) + \Delta u(\Delta t) \qquad (11.166)$$

It may be seen, thus, that within the framework of a step-by-step solution and with the approximations introduced earlier, it is possible to determine the complete time history of the response of any nonlinear system by considering a sequence of time increments starting at $t = 0$ and ending at the end of the desired time length.

It is important to realize that under the assumption of time-invariant damping and spring constants during a time interval Δt, Equation 11.165 represents a linear differential equation with constant coefficients and may therefore be solved using any of the conventional methods employed for the solution of linear systems. It is also important to note that the assumption that the damping and spring constants are equal to the tangents to the force-displacement and force-velocity curves at the beginning of the time interval Δt introduces an error that needs to be kept to a minimum by using sufficiently small time intervals.

An incremental equation of motion may also be formulated for a multi-degree-of-freedom system and derived in much the same way as it was done for the single-degree-of-freedom case. To this end, consider a multi-degree-of-freedom system subjected to a ground acceleration $\ddot{u}_g(t)$ and defined by its mass matrix $[M]$, damping matrix $[C(t)]$, and stiffness matrix $[K(t)]$, where $[C(t)]$ and $[K(t)]$ are matrices that vary with the velocity and displacement response of the system and hence with time. In terms of vectors $\{F_D(t)\}$ and $\{F_S(t)\}$ that respectively contain the damping and restoring forces in the elements of the system, the equation of motion of the system at times $t = t_i = i\Delta t$ and $t = t_i + \tau$, in which, as before, i is any integer, Δt a small time increment, and $0 \leq \tau \leq \Delta t$, may be written as

$$[M]\{\ddot{u}(t_i)\} + \{F_D(t_i)\} + \{F_S(t_i)\} = -[M]\{J\}\ddot{u}_g(t_i) \qquad (11.167)$$

$$[M]\{\ddot{u}(t_i + \tau)\} + \{F_D(t_i + \tau)\} + \{F_S(t_i + \tau)\} = -[M]\{J\}\ddot{u}_g(t_i + \tau) \qquad (11.168)$$

where all symbols are as previously defined. However, by assuming that the properties of the system remain unchanged during the time interval Δt and equal to those at the beginning of the interval, the damping and restoring force vectors may be expressed as

$$\{F_S(t_i + \tau)\} = \{F_S(t_i)\} + [K]_i\{\Delta u(\tau)\} \qquad (11.169)$$

$$\{F_D(t_i + \tau)\} = \{F_D(t_i)\} + [C]_i\{\Delta \dot{u}(\tau)\} \qquad (11.170)$$

where $[K]_i$ and $[C]_i$ represent the stiffness and damping matrices of the system formulated on the basis of the properties of the system at the beginning of the interval, and

$$\{\Delta u(\tau)\} = \{u(t_i + \tau)\} - \{u(t_i)\} \tag{11.171}$$

$$\{\Delta \dot{u}(\tau)\} = \{\dot{u}(t_i + \tau)\} - \{\dot{u}(t_i)\} \tag{11.172}$$

Consequently, Equation 11.168 may also be written as

$$[M]\{\ddot{u}(t_i + \tau)\} + \{F_D(t_i)\} + [C]_i\{\Delta \dot{u}(\tau)\} + \{F_S(t_i)\} + [K]_i\{\Delta u(\tau)\} = -[M]\{J\}\ddot{u}_g(t_i + \tau) \tag{11.173}$$

or as

$$[M]\{\ddot{u}_i(t)\} + [M]\{\Delta \ddot{u}(\tau)\} + \{F_D(t_i)\} + [C]_i\{\Delta \dot{u}(\tau)\} + \{F_S(t_i)\} + [K]_i\{\Delta u(\tau)\}$$
$$= -[M]\{J\}\ddot{u}_g(t_i) - [M]\{J\}\Delta \ddot{u}_g(\tau) \tag{11.174}$$

where

$$\{\Delta \ddot{u}(\tau)\} = \{\ddot{u}(t_i + \tau)\} - \{\ddot{u}(t_i)\} \tag{11.175}$$

and, as before, $\Delta \ddot{u}_g(\tau) = \ddot{u}_g(t_i + \tau) - \ddot{u}_g(t)$. Furthermore, by virtue of Equation 11.167, Equation 11.174 may be reduced to

$$[M]\{\Delta \ddot{u}(\tau)\} + [C]_i\{\Delta \dot{u}(\tau)\} + [K]_i\{\Delta u(\tau)\} = -[M]\{J\}\Delta \ddot{u}_g(\tau) \tag{11.176}$$

In similarity with the single-degree-of-freedom system case, Equation 11.176 represents a system of ordinary differential equations with constant coefficients where the unknown is the incremental displacement vector $\{\Delta u(\tau)\}$. Therefore, it may be solved using the conventional methods for the solution of linear systems. As in the single-degree-of-freedom system case, too, once this incremental displacement vector is determined, the displacement vector at the end of the time interval may be obtained through the application of Equation 11.171 and considering that $\tau = \Delta t$. That is,

$$\{u(t_i + \Delta t)\} = \{u(t_i)\} + \{\Delta u(\Delta t)\} \tag{11.177}$$

11.10.3 Solution of Incremental Equation of Motion

Although, as mentioned in the preceding paragraph, the integration methods used for the solution of linear systems may also be employed for the solution of the incremental equations of motion of nonlinear systems, some modifications are nevertheless necessary to consider some of the features that are unique to the solution of nonlinear systems. As a way to illustrate how any method can be implemented for the analysis of nonlinear systems, this section will present the modifications that are necessary in two of the methods presented in previous sections for such a purpose. Those considered are the central difference and Newmark-β methods.

11.10.3.1 Solution by Central Differences Method

The central difference method can easily be adapted for solving an incremental equation of motion. In this method, however, a solution may be obtained directly from the full equation of motion as at each time step this solution depends on the structural properties evaluated on the basis of the structural response at the previous time step. Thus, for the solution with the central difference method of

Structural Response by Step-by-Step Integration Methods

systems with a nonlinear force-deformation relationship and a constant damping matrix, it is only necessary to replace the restoring force term $[K]\{u\}_i$ in Equation 11.12 by a restoring force vector $\{F_S\}_i$ and express the effective load vector alternatively as

$$\{\hat{F}\}_i = -[M]\{J\}\ddot{u}_g(t_i) + \frac{2}{\Delta t^2}[M]\{u\}_i - \{F_s\}_i + \left(\frac{1}{2\Delta t}[C] - \frac{1}{\Delta t^2}[M]\right)\{u\}_{i-1} \qquad (11.178)$$

where $\{F_S\}_i$ is determined incrementally according to

$$\{F_S\}_i = \{F_S\}_{i-1} + [K]_i(\{u\}_i - \{u\}_{i-1}) \qquad (11.179)$$

in which $[K]_i$ is the stiffness matrix of the system formulated on the basis of the properties of the system at $t = t_i$.

As in the linear case, the central difference method is only conditionally stable and thus the stability of the method requires that the time step used in the solution be less than the T_N/π limit, where T_N is the shortest period of the structure. It should be realized, however, that the natural periods of a nonlinear structure do not remain constant but change with the response level. It is important, therefore, that the selected time step remain smaller than the T_N/π limit throughout the integration process.

The central difference method may be used in the solution of nonlinear systems following the algorithm presented in Box 11.5.

Box 11.5 Solution Algorithm for Nonlinear Systems by Central Difference Method

1.0 Initial calculations

  1.1 Formulate initial stiffness matrix $[K]_0$, mass matrix $[M]$, and damping matrix $[C]$

  1.2 Establish vectors of initial displacements $\{u\}_0$ and initial velocities $\{\dot{u}\}_0$

  1.3 Compute initial acceleration vector as

  $$\{\ddot{u}\}_0 = -[M]^{-1}([C]\{\dot{u}\}_0 + [K]_0\{u\}_0) - \{J\}\ddot{u}_g(0)$$

  1.4 Compute initial restoring force vector as

  $$\{F_S\}_0 = [K]_0\{u\}_0$$

  1.5 Select time step Δt

  1.6 Compute displacement vector at time $t = -\Delta t$ according to

  $$\{u\}_{-1} = \{u\}_0 - \Delta t\{\dot{u}\}_0 + \frac{\Delta t^2}{2}\{\ddot{u}\}_0$$

  1.7 Determine vector of restoring forces at time $t = -\Delta t$ as

  $$\{F_S\}_{-1} = [K]_0\{u\}_{-1}$$

  1.8 Compute effective mass matrix as

  $$[\hat{M}] = \frac{1}{\Delta t^2}[M] + \frac{1}{2\Delta t}[C]$$

1.9 Compute constant matrices $[a]$ and $[b]$ as

$$[a] = \frac{1}{2\Delta t}[C] - \frac{1}{\Delta t^2}[M]$$

$$[b] = \frac{2}{\Delta t^2}[M]$$

1.10 Set $i = 0$

2.0 Calculations for ith time step

2.1 Compute load vector at time $t = t_i = i\Delta t$ as

$$\{F_S\}_i = \{F_S\}_{i-1} + [K]_i\left(\{u\}_i - \{u\}_{i-1}\right)$$

2.2 Compute effective load vector at time $t = t_i$ as

$$\{\hat{F}\}_i = -[M]\{J\}\ddot{u}_g(t_i) + [a]\{u\}_{i-1} + [b]\{u\}_i - \{F_S\}_i$$

2.3 Compute displacement vector at time $t = t_i + \Delta t$ as

$$\{u\}_{i+1} = [\hat{M}]^{-1}\{\hat{F}\}_i$$

2.4 If required, calculate velocity and acceleration vectors at time $t = t_i$ according to

$$\{\dot{u}\}_i = \frac{1}{2\Delta t}(\{u\}_{i+1} - \{u\}_{i-1})$$

$$\{\ddot{u}\}_i = \frac{1}{\Delta t^2}(\{u\}_{i+1} - 2\{u\}_i + \{u\}_{i-1})$$

2.5 Formulate the stiffness matrix $[K]_i$ on the basis of the structural response at time $t = t_i$

3.0 Calculations for next time step

Replace i by $i + 1$ and repeat steps 2.1–2.5.

11.10.3.2 Solution by Newmark-β Method

Newmark-β method may also be used to solve the incremental equation of motion of a nonlinear system. However, it is convenient to reformulate first the method as follows.

From Equations 11.96 and 11.97, the incremental displacement and velocity vectors at time $t = t_i + \Delta t$ may be expressed as

$$\{\Delta \dot{u}\}_{i+1} = \{\dot{u}\}_{i+1} - \{\dot{u}\}_i = \Delta t\{\ddot{u}\}_i + \gamma\Delta t\{\Delta \ddot{u}\}_{i+1} \qquad (11.180)$$

$$\{\Delta u\}_{i+1} = \{u\}_{i+1} - \{u\}_i = \Delta t\{\dot{u}\}_i + \frac{\Delta t^2}{2}\{\ddot{u}\}_i + \beta\Delta t^2\{\Delta \ddot{u}\}_{i+1} \qquad (11.181)$$

Structural Response by Step-by-Step Integration Methods

where $\{\Delta\ddot{u}\}_{i+1} = \{\ddot{u}\}_{i+1} - \{\ddot{u}\}_i$. Then, by solving for the incremental acceleration vector $\{\Delta\ddot{u}\}_{i+1}$ from Equation 11.181, one obtains

$$\{\Delta\ddot{u}\}_{i+1} = \frac{1}{\beta\Delta t^2}\{\Delta u\}_{i+1} - \frac{1}{\beta\Delta t}\{\dot{u}\}_i - \frac{1}{2\beta}\{\ddot{u}\}_i \qquad (11.182)$$

which when substituted into Equation 11.180 leads to

$$\{\Delta\dot{u}\}_{i+1} = \Delta t\left(1 - \frac{\gamma}{2\beta}\right)\{\ddot{u}\}_i + \frac{\gamma}{\beta\Delta t}\{\Delta u\}_{i+1} - \frac{\gamma}{\beta}\{\dot{u}\}_i \qquad (11.183)$$

Hence, upon substitution of Equations 11.182 and 11.183 into the incremental equation of motion given by Equation 11.176 when expressed at time $t = t_i + \Delta t$, one gets

$$[M]\left[\frac{1}{\beta\Delta t^2}\{\Delta u\}_{i+1} - \frac{1}{\beta\Delta t}\{\dot{u}\}_i - \frac{1}{2\beta}\{\ddot{u}\}_i\right]$$

$$+ [C]_i\left[\Delta t\left(1 - \frac{\gamma}{2\beta}\right)\{\ddot{u}\}_i + \frac{\gamma}{\beta\Delta t}\{\Delta u\}_{i+1} - \frac{\gamma}{\beta}\{\dot{u}\}_i\right] \qquad (11.184)$$

$$+ [K]_i\{\Delta u\}_{i+1} = -[M]\{J\}(\Delta\ddot{u}_g)_{i+1}$$

where $(\Delta\ddot{u}_g)_{i+1} = \ddot{u}_g(t_{i+1}) - \ddot{u}_g(t_i)$. In compact form, however, Equation 11.184 may also be written as

$$[\hat{K}_N]_i\{\Delta u\}_{i+1} = \{\Delta\hat{F}_N\}_{i+1} \qquad (11.185)$$

where

$$[\hat{K}_N]_i = [K]_i + \frac{1}{\beta\Delta t^2}[M] + \frac{\gamma}{\beta\Delta t}[C]_i \qquad (11.186)$$

$$\{\Delta\hat{F}_N\}_{i+1} = -[M]\{J\}(\Delta\ddot{u}_g)_{i+1} + [M]\left(\frac{1}{\beta\Delta t}\{\dot{u}\}_i + \frac{1}{2\beta}\{\ddot{u}\}_i\right)$$

$$+ [C]_i\left(\left(\frac{\gamma}{2\beta} - 1\right)\Delta t\{\ddot{u}\}_i + \frac{\gamma}{\beta}\{\dot{u}\}_i\right) \qquad (11.187)$$

Thus, in Newmark-β method the incremental displacement vector at time $t = t_{i+1} = \Delta t(i+1)$ is calculated according to

$$\{\Delta u\}_{i+1} = [\hat{K}_N]^{-1}\{\Delta\hat{F}_N\}_{i+1} \qquad (11.188)$$

whereas the corresponding incremental velocity and acceleration vectors are determined from Equations 11.182 and 11.183. An algorithm for the analysis of nonlinear systems using Newmark-β method is presented in Box 11.6.

Box 11.6 Solution Algorithm for Nonlinear Systems by Newmark-β Method

1.0 Initial calculations
    1.1 Formulate mass matrix $[M]$, initial stiffness matrix $[K]_0$, and initial damping matrix $[C]_0$
    1.2 Establish vectors of initial displacements $\{u\}_0$ and initial velocities $\{\dot{u}\}_0$
    1.3 Compute vector of initial acceleration as

$$\{\ddot{u}\}_0 = -[M]^{-1}([C]_0\{\dot{u}\}_0 + [K]_0\{u\}_0) - \{J\}\ddot{u}_g(0)$$

    1.4 Select time step Δt
    1.5 Select constants γ and β
    1.6 Set $i = 0$

2.0 Calculations for ith time step
    2.1 Compute incremental ground acceleration as

$$(\Delta \ddot{u}_g)_{i+1} = \ddot{u}_g(t_{i+1}) - \ddot{u}_g(t_i)$$

    2.2 Compute effective incremental load vector as

$$\{\Delta \hat{F}_N\}_{i+1} = -[M]\{J\}(\Delta \ddot{u}_g)_{i+1} + \left(\frac{1}{\beta \Delta t}[M] + \frac{\gamma}{\beta}[C]_i\right)\{\dot{u}\}_i + \left(\frac{1}{2\beta}[M] + \left(\frac{\gamma}{2\beta} - 1\right)[C]_i\right)\{\ddot{u}\}_i$$

    2.3 Compute effective stiffness matrix as

$$[\hat{K}_N]_i = [K]_i + \frac{1}{\beta \Delta t^2}[M] + \frac{\gamma}{\beta \Delta t}[C]_i$$

    2.4 Compute incremental displacement, velocity, and acceleration vectors as

$$\{\Delta u\}_{i+1} = [\hat{K}_N]_i^{-1}\{\Delta \hat{F}_N\}_{i+1}$$

$$\{\Delta \dot{u}\}_{i+1} = \Delta t\left(1 - \frac{\gamma}{2\beta}\right)\{\ddot{u}\}_i + \frac{\gamma}{\beta \Delta t}\{\Delta u\}_{i+1} - \frac{\gamma}{\beta}\{\dot{u}\}_i$$

$$\{\Delta \ddot{u}\}_{i+1} = \frac{1}{\beta \Delta t^2}\{\Delta u\}_{i+1} - \frac{1}{\beta \Delta t}\{\dot{u}\}_i - \frac{1}{2\beta}\{\ddot{u}\}_i$$

    2.5 Compute displacement, velocity, and acceleration vectors as

$$\{u\}_{i+1} = \{u\}_i + \{\Delta u\}_{i+1}$$

$$\{\dot{u}\}_{i+1} = \{\dot{u}\}_i + \{\Delta \dot{u}\}_{i+1}$$

$$\{\ddot{u}\}_{i+1} = \{\ddot{u}\}_i + \{\Delta \ddot{u}\}_{i+1}$$

    2.6 Update damping matrix $[C]_i$ and stiffness matrix $[K]_i$ according to current state of stresses and deformations

3.0 Calculations for next time step
    Replace i by $i + 1$ and repeat steps 2.1–2.6.

It should be recognized that with the proper selection of the parameters γ and β, the same algorithm may be used to formulate a solution based on the constant average and linear acceleration methods given that, as demonstrated in Section 11.8, these two methods are special cases of Newmark's method. It should also be noted that, different from the central difference method, Newmark's method requires the inversion or factorization of a different stiffness matrix every time the structural properties change. This may represent a major computational effort if the structure has many degrees of freedom.

11.10.4 Updating of Stiffness Matrix: Elastoplastic Case

As mentioned before, in the analysis of a nonlinear system, it is assumed that the system behaves linearly within each time interval but its properties are changed from one interval to the next according to the response level at the beginning of the interval and the previous loading path. Consequently, in the analysis of a nonlinear system, the stiffness and damping matrices of the system are updated every time there is a change in the properties of one or more of its elements. This updating is regularly performed by tracking the changes in the element properties according to the state of stresses and deformations in the structural elements and whether these stresses and deformations increase or decrease. To illustrate how this tracking and updating may be carried out, this section will describe a technique that may be applied to update the stiffness matrix of a plane-frame structure.

Consider a structure formed with beam elements that exhibit the idealized elastoplastic moment-rotation behavior shown in Figure 11.12, where M_y denotes the bending moment under which the maximum stress in a cross section reaches the yield stress of the material with which the element is made. Assume, in addition, that yielding in an element may take place only at the element ends and thus controlled by the end-bending moments. As is well known, the stiffness matrix of a structure is formed by assembling the stiffness matrices of its elements. Therefore, in the case of a nonlinear structure, the stiffness matrix of the structure is updated by modifying the stiffness matrices of its elements according to the level of flexural deformation in them and by reassembling afterward the elemental stiffness matrices to formulate a new structural or global stiffness matrix. In turn, the stiffness matrix of each element is reformulated by calculating the magnitude of its end-bending moments at the end of each time step and comparing these bending moments to the yield moment for that element. If the two end-bending moments are less than the yield moment, then the element is still in its elastic range. In this case, if reference is made to the local system of

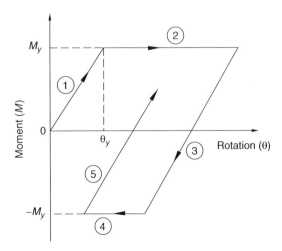

FIGURE 11.12 Typical hysteresis loop in elastoplastic beam element.

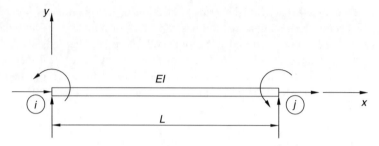

FIGURE 11.13 Local coordinates and degrees of freedom of beam element.

coordinates and the degrees of freedom indicated in Figure 11.13, the stiffness matrix of the element is given by

$$[K]_i = \frac{EI}{L^3}\begin{bmatrix} AL^2/I & 0 & 0 & -AL^2/I & 0 & 0 \\ 0 & 12 & 6L & 0 & -12 & 6L \\ 0 & 6L & 4L^2 & 0 & -6L & 2L^2 \\ -AL^2/I & 0 & 0 & AL^2/I & 0 & 0 \\ 0 & -12 & -6L & 0 & 12 & -6L \\ 0 & 6L & 2L^2 & 0 & -6L & 4L^2 \end{bmatrix} \quad (11.189)$$

where E, I, A, and L, respectively, denote modulus of elasticity, moment of inertia, cross-sectional area, and element length; and the subscript i associates the stiffness matrix to that of the ith element of the structure. However, if the bending moment at end i (see Figure 11.13) reaches the yield moment, then that section starts behaving like a hinge as it cannot support a moment any longer but it continues to rotate. In such a case, the element behaves like a beam with a hinge at one of its ends and restrained at the other. Therefore, the element's stiffness matrix needs to be changed to

$$[K]_i = \frac{EI}{L^3}\begin{bmatrix} AL^2/I & 0 & 0 & -AL^2/I & 0 & 0 \\ 0 & 3 & 0 & 0 & -3 & 3L \\ 0 & 0 & 0 & 0 & 0 & 0 \\ -AL^2/I & 0 & 0 & AL^2/I & 0 & 0 \\ 0 & -3 & 0 & 0 & 3 & -3L \\ 0 & 3L & 0 & 0 & -3L & 3L^2 \end{bmatrix} \quad (11.190)$$

By the same argument, if the bending moment at end j reaches the yield moment but the bending moment at end i is still less than the yield moment, the element's stiffness matrix is given by

$$[K]_i = \frac{EI}{L^3}\begin{bmatrix} AL^2/I & 0 & 0 & -AL^2/I & 0 & 0 \\ 0 & 3 & 3L & 0 & -3 & 0 \\ 0 & 3L & 3L^2 & 0 & -3L & 0 \\ -AL^2/I & 0 & 0 & AL^2/I & 0 & 0 \\ 0 & -3 & -3L & 0 & 3 & 0 \\ 0 & 0 & 0 & 0 & 0 & 0 \end{bmatrix} \quad (11.191)$$

Moreover, if both end moments reach the yield moment, then the two end sections behave like a hinge and the stiffness matrix of the element should be updated to

$$[K]_i = \begin{bmatrix} AE/L & 0 & 0 & -AE/L & 0 & 0 \\ 0 & 0 & 0 & 0 & 0 & 0 \\ 0 & 0 & 0 & 0 & 0 & 0 \\ -AE/L & 0 & 0 & AE/L & 0 & 0 \\ 0 & 0 & 0 & 0 & 0 & 0 \\ 0 & 0 & 0 & 0 & 0 & 0 \end{bmatrix} \qquad (11.192)$$

Finally, it should be noted that the element reverses to elastic behavior whenever the two end moments decrease to values that are less than the yield moment. In such a case, the stiffness matrix of the element is again given by the stiffness matrix defined by Equation 11.189. Whether or not the end of an element has reached the yield moment or has been unloaded back to its elastic range of behavior may be determined as follows.

As shown in Figure 11.12, the elastoplastic moment-rotation behavior of an element's cross section is typically characterized by five branches. Branches 1, 3, and 5 identify regions of elastic behavior and are bi-directional; that is, loading may proceed in either direction along any of these branches. Branches 2 and 4 identify regions of plastic behavior and are only uni-directional. Branch 1 identifies the region of initial loading and is a branch that is no longer used once yielding takes place. If the current state of the section's flexural deformation is defined by branches 1, 3, or 5, then the section's behavior will continue to be defined by branch 1, 3, or 5 until positive or negative yielding occurs. If yielding occurs under positive or negative moment, then the section's behavior will be defined by branch 2 or 4, respectively. In contrast, if the current state of flexural deformation is defined by branch 2 or 4 and the rotational velocity $\dot{\theta}$ remains positive or negative (i.e., the rotation increases or decreases with an increment of time), respectively, then the section's behavior continues to be defined along one of these branches. However, if the rotational velocity changes sign, then the section will be unloaded back to branch 3 or 5, respectively. Thus, one can determine if the end of an element is in its elastic or plastic range of behavior according to the following criteria:

1. When the end rotation θ is in the elastic range at the beginning of the interval, then the element's end remains in the elastic range if at the end of the interval the bending moment M is such that $-M_y < M < M_y$, has incurred into the positive plastic range if at the end of the interval $M > M_y$, or has incurred into the negative plastic range if at the end of the interval $M < -M_y$.
2. If the end rotation θ is in the positive plastic range at the beginning of the interval, then the element's end remains in the positive plastic range if at the end of the interval $\dot{\theta} > 0$, or reverts to the elastic range if at the end of the interval $\dot{\theta} < 0$.
3. If the end rotation θ is in the negative plastic range at the beginning of the interval, then the element's end remains in the negative plastic range if at the end of the interval $\dot{\theta} < 0$, or reverts to the elastic range if at the end of the interval $\dot{\theta} > 0$.

EXAMPLE 11.6   ANALYSIS OF ELASTOPLASTIC TWO-DEGREE-OF-FREEDOM FRAME

The two-story structure shown in Figure E11.6a is formed with braced frames whose columns are hinged at both ends. The braces, which can equally resist tensile and compressive forces, exhibit an elastoplastic behavior with a yield strain of 0.002. All other members remain elastic at all times. Determine the first second of the displacement, velocity, and acceleration responses of the system when its base is excited by the ground motion depicted in Figure E11.6b. Consider that the masses m_1 and m_2 are both equal to 1.0 kip s²/in., that the cross-sectional areas A_1 and A_2 are both equal to 1.25 in.², and that the modulus of elasticity E is equal to 37,500 ksi. Assume that the system is initially at rest. Assume, in addition, that the damping matrix of the system remains unchanged at all times and is proportional to

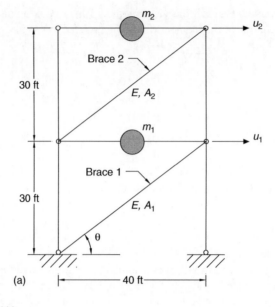

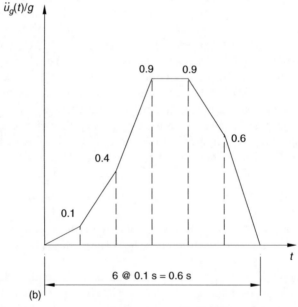

FIGURE E11.6 (a) Structure and (b) ground motion considered in Example E11.6.

the system's initial stiffness matrix with a constant of proportionality equal to 0.005. Use the constant average acceleration method; that is, use Newmark-β method with $\gamma = 1/2$ and $\beta = 1/4$.

Solution

Mass, Stiffness, and Damping Matrices

As the masses of the structure are lumped at its floors, the mass matrix is simply given by the following diagonal matrix:

$$[M] = \begin{bmatrix} m_1 & 0 \\ 0 & m_2 \end{bmatrix} = \begin{bmatrix} 1 & 0 \\ 0 & 1 \end{bmatrix} \text{k s}^2/\text{in.}$$

Also, the stiffness matrix of the structure is initially equal to

$$[K] = \begin{bmatrix} \dfrac{A_1 E}{L_1}\cos^2\theta + \dfrac{A_2 E}{L_2}\cos^2\theta & -\dfrac{A_2 E}{L_2}\cos^2\theta \\ -\dfrac{A_2 E}{L_2}\cos^2\theta & -\dfrac{A_2 E}{L_2}\cos^2\theta \end{bmatrix} = \dfrac{1.25(37{,}500)}{50(12)}\left(\dfrac{4}{5}\right)^2 \begin{bmatrix} 2 & -1 \\ -1 & 1 \end{bmatrix} = \begin{bmatrix} 100 & -50 \\ -50 & 50 \end{bmatrix} \text{ k/in.}$$

but when Brace 1 yields, it becomes

$$[K] = \begin{bmatrix} \dfrac{A_2 E}{L_2}\cos^2\theta & -\dfrac{A_2 E}{L_2}\cos^2\theta \\ -\dfrac{A_2 E}{L_2}\cos^2\theta & -\dfrac{A_2 E}{L_2}\cos^2\theta \end{bmatrix} = \dfrac{1.25(37{,}500)}{50(12)}\left(\dfrac{4}{5}\right)^2 \begin{bmatrix} 1 & -1 \\ -1 & 1 \end{bmatrix} = \begin{bmatrix} 50 & -50 \\ -50 & 50 \end{bmatrix} \text{ k/in.}$$

Similarly, when Brace 2 yields but Brace 1 remains in the elastic range, it changes to

$$[K] = \begin{bmatrix} \dfrac{A_1 E}{L_1}\cos^2\theta & 0 \\ 0 & 0 \end{bmatrix} = \dfrac{1.25(37{,}500)}{50(12)}\left(\dfrac{4}{5}\right)^2 \begin{bmatrix} 1 & 0 \\ 0 & 0 \end{bmatrix} = \begin{bmatrix} 50 & 0 \\ 0 & 0 \end{bmatrix} \text{ k/in.}$$

and when the two braces yield at the same time it is equal to the null matrix; that is,

$$[K] = \begin{bmatrix} 0 & 0 \\ 0 & 0 \end{bmatrix}$$

Lastly, as the damping matrix is supposed to be proportional to the initial stiffness matrix with a proportionality constant of 0.05, this matrix is equal to

$$[C] = 0.005 \begin{bmatrix} 100 & -50 \\ -50 & 50 \end{bmatrix} = \begin{bmatrix} 0.5 & -0.25 \\ -0.25 & 0.25 \end{bmatrix} \text{ k s/in.}$$

Yield Deformations

As the yield strain ε_y of the braces is equal to 0.002, the braces will yield under a lateral relative displacement of

$$u_y = \dfrac{L \varepsilon_y}{\cos\theta} = \dfrac{50(12)(0.002)}{4/5} = 1.5 \text{ in.}$$

As a result, Brace 1 will first yield when

$$u_1(t) = \pm 1.5 \text{ in.}$$

and Brace 2 will first yield when

$$u_2(t) - u_1(t) = \pm 1.5 \text{ in.}$$

Note, however, that after Brace 1 yields for the first time and reverts afterward to elastic behavior (a condition that is indicated by a change in the sign of $\dot{u}_1$), yielding will be attained again when

$$(u_1)_{max} < u_1 < (u_1)_{max} - 2u_y$$

if the brace is in tension, and

$$(u_1)_{min} + 2u_y < u_1 < (u_1)_{min}$$

if the brace is in compression, where $(u_1)_{max}$ and $(u_1)_{min}$ denote the values of the displacement u_1 just before the brace goes back to elastic behavior from the region of plastic behavior in tension and compression, respectively. Similarly, Brace 2 will return to elastic behavior after yielding for the first time whenever there is a change of sign in the velocity of deformation $\dot{u}_2 - \dot{u}_1$, and will yield again whenever

$$(u_2 - u_1)_{max} < u_2 - u_1 < (u_2 - u_1)_{max} - 2u_y$$

if the brace is in tension, and

$$(u_2 - u_1)_{min} + 2u_y < u_2 - u_1 < (u_2 - u_1)_{min}$$

if the brace is in compression, where $(u_2 - u_1)_{max}$ and $(u_2 - u_1)_{min}$ represent the values of $u_2 - u_1$ attained just before the brace reverts to elastic behavior from the region of plastic behavior in tension and compression, respectively.

Time Step

The two natural periods of the structure are

$$T_1 = 1.442\,s \qquad T_2 = 0.5520\,s$$

Hence, according to the recommendations given in Section 11.9.3, an adequate time step for the analysis is $\Delta t = 0.552/10 = 0.055$, which will be rounded to $\Delta t = 0.05$ s.

Effective Incremental Load Vector at Time $t = t_{i+1}$

According to Equation 11.187, for the system under consideration and $\gamma = 1/2$, $\beta = 1/4$, and $\Delta t = 0.05$, the effective incremental load vector is equal to

$$\{\Delta \hat{F}_N\}_{i+1} = -\begin{bmatrix} 1 & 0 \\ 0 & 1 \end{bmatrix} \begin{Bmatrix} 1 \\ 1 \end{Bmatrix} (\Delta \ddot{u}_g)_{i+1} + \left(\frac{4}{0.05} \begin{bmatrix} 1 & 0 \\ 0 & 1 \end{bmatrix} + 2 \begin{bmatrix} 0.50 & -0.25 \\ -0.25 & 0.25 \end{bmatrix} \right) \begin{Bmatrix} \dot{u}_1 \\ \dot{u}_2 \end{Bmatrix}_i + 2 \begin{bmatrix} 1 & 0 \\ 0 & 1 \end{bmatrix} \begin{Bmatrix} \ddot{u}_1 \\ \ddot{u}_2 \end{Bmatrix}_i$$

$$= -\begin{Bmatrix} 1 \\ 1 \end{Bmatrix} (\Delta \ddot{u}_g)_{i+1} + \begin{bmatrix} 81.0 & -0.5 \\ -0.5 & 80.5 \end{bmatrix} \begin{Bmatrix} \dot{u}_1 \\ \dot{u}_2 \end{Bmatrix}_i + \begin{bmatrix} 2 & 0 \\ 0 & 2 \end{bmatrix} \begin{Bmatrix} \ddot{u}_1 \\ \ddot{u}_2 \end{Bmatrix}_i$$

Effective Stiffness Matrix

The effective stiffness matrix is defined by Equation 11.185, which for $\beta = 1/4$ and $\Delta t = 0.05$ becomes

$$[\hat{K}_N]_i = [K]_i + \frac{4}{(0.05)^2}[M] + \frac{2}{0.05}[C] = [K]_i + 1600[M] + 40[C]$$

Therefore, for the system being considered, the effective stiffness matrix is given by

$$[\hat{K}_N^{(1)}]_i = \begin{bmatrix} 100 & -50 \\ -50 & 50 \end{bmatrix} + 1600 \begin{bmatrix} 1 & 0 \\ 0 & 1 \end{bmatrix} + 40 \begin{bmatrix} 0.5 & -0.25 \\ -0.25 & 0.25 \end{bmatrix} = \begin{bmatrix} 1720 & -60 \\ -60 & 1660 \end{bmatrix}$$

when the two braces are in the elastic range; by

$$[\hat{K}_N^{(2)}]_i = \begin{bmatrix} 50 & -50 \\ -50 & 50 \end{bmatrix} + 1600 \begin{bmatrix} 1 & 0 \\ 0 & 1 \end{bmatrix} + 40 \begin{bmatrix} 0.5 & -0.25 \\ -0.25 & 0.25 \end{bmatrix} = \begin{bmatrix} 1670 & -60 \\ -60 & 1660 \end{bmatrix}$$

when Brace 1 has yielded but Brace 2 remains in the elastic range; by

$$[\hat{K}_N^{(3)}]_i = \begin{bmatrix} 50 & 0 \\ 0 & 0 \end{bmatrix} + 1600 \begin{bmatrix} 1 & 0 \\ 0 & 1 \end{bmatrix} + 40 \begin{bmatrix} 0.5 & -0.25 \\ -0.25 & 0.25 \end{bmatrix} = \begin{bmatrix} 1670 & -10 \\ 10 & 1610 \end{bmatrix}$$

when Brace 2 has yielded and Brace 1 is in the elastic range; and by

$$[\hat{K}_N^{(4)}]_i = \begin{bmatrix} 0 & 0 \\ 0 & 0 \end{bmatrix} + 1600 \begin{bmatrix} 1 & 0 \\ 0 & 1 \end{bmatrix} + 40 \begin{bmatrix} 0.5 & -0.25 \\ -0.25 & 0.25 \end{bmatrix} = \begin{bmatrix} 1620 & -10 \\ -10 & 1610 \end{bmatrix} = 10 \begin{bmatrix} 162 & -1 \\ -1 & 161 \end{bmatrix}$$

when both braces have yielded. Their respective inverses are

$$[\hat{K}_N^{(1)}]_i^{-1} = \frac{1}{142{,}580} \begin{bmatrix} 83 & 3 \\ 3 & 86 \end{bmatrix}$$

$$[\hat{K}_N^{(2)}]_i^{-1} = \frac{1}{276{,}860} \begin{bmatrix} 166 & 6 \\ 6 & 167 \end{bmatrix}$$

$$[\hat{K}_N^{(3)}]_i^{-1} = \frac{1}{268{,}860} \begin{bmatrix} 161 & 1 \\ 1 & 167 \end{bmatrix}$$

$$[\hat{K}_N^{(4)}]_i^{-1} = \frac{1}{260{,}810} \begin{bmatrix} 161 & 1 \\ 1 & 162 \end{bmatrix}$$

Incremental Displacement Vector at Time $t = t_{i+1}$

From Equation 11.188 and in terms of the effective incremental load vector and effective stiffness matrices found earlier, the incremental displacement vector at time $t = t_{i+1}$ is thus equal to

$$\begin{Bmatrix} \Delta u_1 \\ \Delta u_2 \end{Bmatrix}_{i+1} = \frac{1}{142{,}580} \left(-\begin{bmatrix} 83 & 3 \\ 3 & 86 \end{bmatrix} \begin{Bmatrix} 1 \\ 1 \end{Bmatrix} (\Delta \ddot{u}_g)_{i+1} + \begin{bmatrix} 83 & 3 \\ 3 & 86 \end{bmatrix} \begin{bmatrix} 81.0 & -0.5 \\ -0.5 & 80.5 \end{bmatrix} \begin{Bmatrix} \dot{u}_1 \\ \dot{u}_2 \end{Bmatrix}_i + \begin{bmatrix} 83 & 3 \\ 3 & 86 \end{bmatrix} \begin{bmatrix} 2 & 0 \\ 0 & 2 \end{bmatrix} \begin{Bmatrix} \ddot{u}_1 \\ \ddot{u}_2 \end{Bmatrix}_i \right)$$

$$= \frac{1}{142{,}580} \left(-\begin{Bmatrix} 86 \\ 89 \end{Bmatrix} (\Delta \ddot{u}_g)_{i+1} + \begin{bmatrix} 6721.5 & 200 \\ 200 & 6921.5 \end{bmatrix} \begin{Bmatrix} \dot{u}_1 \\ \dot{u}_2 \end{Bmatrix}_i + \begin{bmatrix} 166 & 6 \\ 6 & 172 \end{bmatrix} \begin{Bmatrix} \ddot{u}_1 \\ \ddot{u}_2 \end{Bmatrix}_i \right)$$

when the two braces are in the elastic range;

$$\begin{Bmatrix} \Delta u_1 \\ \Delta u_2 \end{Bmatrix}_{i+1} = \frac{1}{276{,}860} \left(-\begin{bmatrix} 166 & 6 \\ 6 & 167 \end{bmatrix} \begin{Bmatrix} 1 \\ 1 \end{Bmatrix} (\Delta \ddot{u}_g)_{i+1} + \begin{bmatrix} 166 & 6 \\ 6 & 167 \end{bmatrix} \begin{bmatrix} 81.0 & -0.5 \\ -0.5 & 80.5 \end{bmatrix} \begin{Bmatrix} \dot{u}_1 \\ \dot{u}_2 \end{Bmatrix}_i + \begin{bmatrix} 166 & 6 \\ 6 & 167 \end{bmatrix} \begin{bmatrix} 2 & 0 \\ 0 & 2 \end{bmatrix} \begin{Bmatrix} \ddot{u}_1 \\ \ddot{u}_2 \end{Bmatrix}_i \right)$$

$$= \frac{1}{276{,}860} \left(-\begin{Bmatrix} 172 \\ 173 \end{Bmatrix} (\Delta \ddot{u}_g)_{i+1} + \begin{bmatrix} 13{,}443 & 400 \\ 402.5 & 13{,}400.5 \end{bmatrix} \begin{Bmatrix} \dot{u}_1 \\ \dot{u}_2 \end{Bmatrix}_i + \begin{bmatrix} 332 & 12 \\ 12 & 334 \end{bmatrix} \begin{Bmatrix} \ddot{u}_1 \\ \ddot{u}_2 \end{Bmatrix}_i \right)$$

when Brace 1 has yielded but Brace 2 remains in the elastic range;

$$\left\{\begin{array}{c}\Delta u_1\\ \Delta u_2\end{array}\right\}_{i+1} = \frac{1}{268{,}860}\left(-\begin{bmatrix}161 & 1\\ 1 & 167\end{bmatrix}\begin{Bmatrix}1\\ 1\end{Bmatrix}(\Delta \ddot{u}_g)_{i+1} + \begin{bmatrix}161 & 1\\ 1 & 167\end{bmatrix}\begin{bmatrix}81.0 & -0.5\\ -0.5 & 80.5\end{bmatrix}\begin{Bmatrix}\dot{u}_1\\ \dot{u}_2\end{Bmatrix}_i\right.$$

$$\left. + \begin{bmatrix}161 & 1\\ 1 & 167\end{bmatrix}\begin{bmatrix}2 & 0\\ 0 & 2\end{bmatrix}\begin{Bmatrix}\ddot{u}_1\\ \ddot{u}_2\end{Bmatrix}_i\right)$$

$$= \frac{1}{268{,}860}\left(-\begin{Bmatrix}162\\ 168\end{Bmatrix}(\Delta \ddot{u}_g)_{i+1} + \begin{bmatrix}13{,}040.5 & 0\\ -2.5 & 13{,}443\end{bmatrix}\begin{Bmatrix}\dot{u}_1\\ \dot{u}_2\end{Bmatrix}_i + \begin{bmatrix}322 & 2\\ 2 & 334\end{bmatrix}\begin{Bmatrix}\ddot{u}_1\\ \ddot{u}_2\end{Bmatrix}_i\right)$$

when Brace 2 has yielded and Brace 1 is in the elastic range; and

$$\left\{\begin{array}{c}\Delta u_1\\ \Delta u_2\end{array}\right\}_{i+1} = \frac{1}{260{,}810}\left(-\begin{bmatrix}161 & 1\\ 1 & 162\end{bmatrix}\begin{Bmatrix}1\\ 1\end{Bmatrix}(\Delta \ddot{u}_g)_{i+1} + \begin{bmatrix}161 & 1\\ 1 & 162\end{bmatrix}\begin{bmatrix}81.0 & -0.5\\ -0.5 & 80.5\end{bmatrix}\begin{Bmatrix}\dot{u}_1\\ \dot{u}_2\end{Bmatrix}_i\right.$$

$$\left. + \begin{bmatrix}161 & 1\\ 1 & 162\end{bmatrix}\begin{bmatrix}2 & 0\\ 0 & 2\end{bmatrix}\begin{Bmatrix}\ddot{u}_1\\ \ddot{u}_2\end{Bmatrix}_i\right)$$

$$= \frac{1}{260{,}810}\left(-\begin{Bmatrix}162\\ 163\end{Bmatrix}(\Delta \ddot{u}_g)_{i+1} + \begin{bmatrix}13{,}040.5 & 0\\ 0 & 13{,}040.5\end{bmatrix}\begin{Bmatrix}\dot{u}_1\\ \dot{u}_2\end{Bmatrix}_i + \begin{bmatrix}322 & 2\\ 2 & 324\end{bmatrix}\begin{Bmatrix}\ddot{u}_1\\ \ddot{u}_2\end{Bmatrix}_i\right)$$

when both braces have yielded.

Incremental Velocity and Acceleration Vectors at Time $t = t_{i+1}$

By virtue of Equations 11.182 and 11.183 and for $\gamma = 1/2$, $\beta = 1/4$, and $\Delta t = 0.05$ s, the incremental velocity and acceleration vectors are similarly given by

$$\left\{\begin{array}{c}\Delta \dot{u}_1\\ \Delta \dot{u}_2\end{array}\right\}_{i+1} = \frac{2}{(0.05)}\left\{\begin{array}{c}\Delta u_1\\ \Delta u_2\end{array}\right\}_{i+1} - 2\left\{\begin{array}{c}\dot{u}_1\\ \dot{u}_2\end{array}\right\}_i = 40\left\{\begin{array}{c}\Delta u_1\\ \Delta u_2\end{array}\right\}_{i+1} - 2\left\{\begin{array}{c}\dot{u}_1\\ \dot{u}_2\end{array}\right\}_i$$

$$\left\{\begin{array}{c}\Delta \ddot{u}_1\\ \Delta \ddot{u}_2\end{array}\right\}_{i+1} = \frac{4}{(0.05)^2}\left\{\begin{array}{c}\Delta u_1\\ \Delta u_2\end{array}\right\}_{i+1} - \frac{4}{0.05}\left\{\begin{array}{c}\dot{u}_1\\ \dot{u}_2\end{array}\right\}_i - 2\left\{\begin{array}{c}\ddot{u}_1\\ \ddot{u}_2\end{array}\right\}_i = 1600\left\{\begin{array}{c}\Delta u_1\\ \Delta u_2\end{array}\right\}_{i+1} - 80\left\{\begin{array}{c}\dot{u}_1\\ \dot{u}_2\end{array}\right\}_i - 2\left\{\begin{array}{c}\ddot{u}_1\\ \ddot{u}_2\end{array}\right\}_i$$

Displacement, Velocity, and Acceleration Vectors at Time $t = t_{i+1}$

Finally, the displacement, velocity, and acceleration vectors may be determined in terms of the incremental vectors formulated earlier according to

$$\left\{\begin{array}{c}u_1\\ u_2\end{array}\right\}_{i+1} = \left\{\begin{array}{c}u_1\\ u_2\end{array}\right\}_i + \left\{\begin{array}{c}\Delta u_1\\ \Delta u_2\end{array}\right\}_{i+1}$$

$$\left\{\begin{array}{c}\dot{u}_1\\ \dot{u}_2\end{array}\right\}_{i+1} = \left\{\begin{array}{c}\dot{u}_1\\ \dot{u}_2\end{array}\right\}_i + \left\{\begin{array}{c}\Delta \dot{u}_1\\ \Delta \dot{u}_2\end{array}\right\}_{i+1}$$

$$\left\{\begin{array}{c}\ddot{u}_1\\ \ddot{u}_2\end{array}\right\}_{i+1} = \left\{\begin{array}{c}\ddot{u}_1\\ \ddot{u}_2\end{array}\right\}_i + \left\{\begin{array}{c}\Delta \ddot{u}_1\\ \Delta \ddot{u}_2\end{array}\right\}_{i+1}$$

TABLE E11.6
First 20 Steps of Displacement, Velocity, and Acceleration Responses of Elastoplastic Two-Degree-of-Freedom Frame in Example 11.6 Using Constant Average Acceleration Method

Step	Time (s)	u_1 (m)	u_2 (m)	$\dot{u}_1$ (m/s)	$\dot{u}_2$ (m/s)	$\ddot{u}_1$ (m/s²)	$\ddot{u}_2$ (m/s²)	$u_2 - u_1$ (m)	$\dot{u}_2 - \dot{u}_1$ (m/s)
0	0.00	0.0000	0.0000	0.0000	0.0000	0.0000	0.0000	0.0000	0.0000
1	0.05	−0.0117	−0.0121	−0.4661	−0.4824	−18.6452	−19.2956	−0.0004	−0.0163
2	0.10	−0.0685	−0.0723	−1.8068	−1.9253	−34.9828	−38.4216	−0.0038	−0.1185
3	0.15	−0.2337	−0.2523	−4.8005	−5.2747	−84.7649	−95.5519	−0.0186	−0.4741
4	0.20	−0.6050	−0.6701	−10.0544	−11.4375	−125.392	−150.963	−0.0650	−1.3831
5	0.25	−1.3057	−1.4871	−17.971	−21.2434	−191.272	−241.272	−0.1814	−3.2724
6	0.30	**−2.4738**	−2.9029	−28.7529	−35.3912	−240.002	−324.642	−0.4292	−6.6384
7	0.35	−4.2218	−5.0598	−41.1698	−50.8825	−256.677	−295.009	−0.8380	−9.7126
8	0.40	−6.6135	−7.9495	−54.495	−64.7044	−276.33	−257.868	−1.3360	−10.2094
9	0.45	−9.6575	−11.4473	−67.2677	−75.2107	−234.575	−162.384	**−1.7898**	−7.9430
10	0.50	−13.2758	−15.3752	−77.4618	−81.9028	−173.191	−105.299	−2.0994	−4.4410
11	0.55	−17.2915	−19.5300	−83.167	−84.2905	−55.0152	9.7912	−2.2385	−1.1235
12	0.60	−21.4457	−23.6603	−83.0008	−80.9229	61.6636	124.9109	−2.2147	**2.0778**
13	0.65	−25.5104	−27.5445	−79.5864	−74.4416	74.9102	134.3438	−2.0341	5.1448
14	0.70	−29.3837	−31.0979	−75.348	−67.6978	94.6259	135.4048	−1.7142	7.6502
15	0.75	−33.0180	−34.3167	−70.0234	−61.0534	118.358	130.3725	−1.2987	8.9700
16	0.80	−36.3561	−37.2112	−63.5006	−54.7258	142.554	122.7311	−0.8551	8.7748
17	0.85	−39.3399	−39.7981	−55.8512	−48.7502	163.4237	116.2924	−0.4582	7.1009
18	0.90	−41.9192	−42.0915	−47.3196	−42.9864	177.8379	114.2624	−0.1724	4.3333
19	0.95	−44.0590	−44.0954	−38.2725	−37.1678	184.048	118.4792	−0.0364	1.1046
20	1.00	−45.7438	−45.7991	−29.1198	−30.9809	182.0592	128.9973	−0.0553	−1.8611

Results

The displacement, velocity, and acceleration responses obtained with the preceding formulas are presented in Table E11.6 for the first 20 steps (1 s). To be able to identify the times at which yielding and unloading of Brace 2 takes place, this table also lists the time history of the difference between the displacements u_2 and u_1, which represents the deformation in this brace, as well as the difference between the velocities $\dot{u}_2$ and $\dot{u}_1$, which represents its deformation velocity. The table also highlights with bold-faced numbers the deformations and velocities that indicate the yielding or unloading of a brace. In particular, note that Brace 1 yields during the sixth step and that Brace 2 yields during the ninth step. It may be noted also that Brace 2 reverts to elastic behavior during the twelfth step.

It is important to bear in mind that the analysis did not include a precise determination of the times at which the braces yielded or were unloaded back to elastic behavior and, thus, did not account for any stiffness matrix changes that might have occurred within a time interval. Note, for example, that, because of the relatively large time step used, the first yielding of Brace 1 is not detected until the displacement decreases to −2.4738 in., which is substantially beyond the yield displacement of −1.5 in. Note, also, that the unloading of Brace 2 is detected only after its deformation velocity reaches the value of 2.0778 in./s despite the fact that the actual unloading took place when this velocity was zero. As a result, the obtained solution may not be a very accurate one. As discussed in the following section, a better accuracy could have been obtained if the time steps during which yielding or unloading occurs would have been subdivided into small time increments to determine with precision the times during which the braces change from elastic to plastic behavior and vice versa, and the modifications to the system's stiffness matrix would have been made at those times.

11.10.5 Sources of Error and Corrective Techniques

Beyond the errors that are inherent in all numerical procedures, the step-by-step integration methods studied in this chapter introduce additional errors when they are employed in the analysis of nonlinear systems. These additional errors may accumulate over a series of time steps and may be thus significantly large unless very small time steps are used or a corrective technique is applied. Such additional errors occur because of (a) the assumption adopted that the properties of the structure remain constant during any given time interval and equal to those at the beginning of the interval and (b) the inability of the methods to detect accurately the time at which there is an abrupt change in the properties of one or more of the elements of a structure, such as when yielding or unloading takes place in an elastoplastic element. Figure 11.14a, which shows a nonlinear load-displacement curve, illustrates the first type of error. It may be seen from this figure that if the displacement at the beginning of a time interval is given by point A, then the consideration of the tangent stiffness at the beginning of the interval would lead to the displacement at the end of the interval identified by point B. However, if one were able to follow the load-displacement curve exactly, one would obtain the displacement indicated by point B′, which may deviate significantly from point B if a large time step is used. Similarly, Figure 11.14b, which also shows a nonlinear load-displacement curve, illustrates the second type of error. With reference to this figure, it may be noted that if the displacement u_i at the beginning of a time interval is given by point A, and if the velocity $\dot{u}_i$ is positive (i.e., the displacement is increasing), then the application of one of the studied numerical integration techniques would predict the displacement at the end of the interval given by point B. However, if the velocity $\dot{u}_{i+1}$ at the end of the interval is negative, the displacement will never reach point B as a negative velocity is an indication that at some point B′ within the time step the velocity became zero, changed sign, and the displacement started to decrease. Hence, if the location of point B′ is not determined, the true displacement will be overshot to point B. In addition, the computation of the displacement at the end of the next time step under the assumption that the displacement started to decrease at point B will erroneously lead to the displacement defined by point C as opposed to the true displacement given by point C′. As a result, the load-displacement path will not be followed closely and may lead to significant errors in the computed solutions.

The error associated with the assumption that the properties of a structure remain constant during a time interval and equal to those at the beginning of the interval may be minimized by the application of Newton–Raphson method, the iterative method used in the solution of algebraic and transcendental equations. To illustrate this approach, consider a single-degree-of-freedom system whose effective load-displacement curve is shown in Figure 11.15a. Suppose that at the beginning of

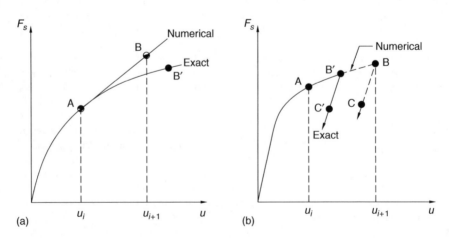

FIGURE 11.14 Illustration of errors introduced in computed solutions of nonlinear systems.

Structural Response by Step-by-Step Integration Methods

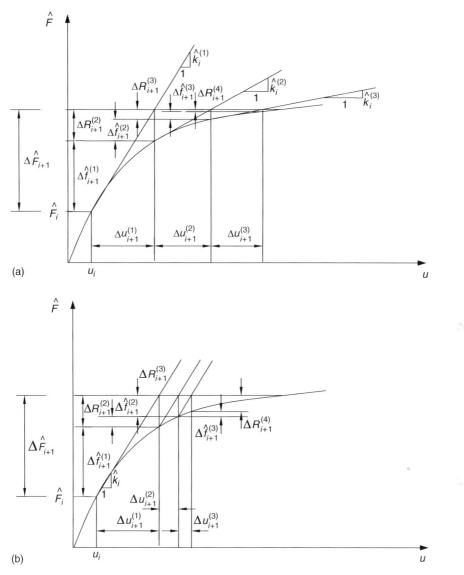

FIGURE 11.15 Iterations performed within a time step using (a) Newton–Raphson and (b) modified Newton–Raphson methods.

a time interval the system is subjected to an effective load increment $\Delta \hat{F}_{i+1}$ and that one is interested in calculating the corresponding incremental displacement Δu_i. Under the assumption of a constant slope equal to the tangent slope at the beginning of the interval, a first estimate of this incremental displacement is

$$\Delta u_{i+1}^{(1)} = \frac{\Delta \hat{F}_{i+1}}{\hat{k}_i^{(1)}} \qquad (11.193)$$

where $\hat{k}_i^{(1)}$ is such a tangent slope and given by Equation 11.186, for example, if Newmark-β method is used. It may be noted, however, that as a result of the use of the tangent stiffness at the beginning of the interval, such a first estimate of the incremental displacement Δu_i will be different from the actual one. Moreover, it may be noted that an incremental effective load $\Delta \hat{f}_{i+1}^{(1)}$ calculated on the basis of Equations 11.185 and 11.186 as

$$\Delta \hat{f}_{i+1}^{(1)} = (\Delta F_s)_{i+1}^{(1)} + (\hat{k}_i - k_i)\Delta u_{i+1}^{(1)} \tag{11.194}$$

where $(\Delta F_s)_{i+1}^{(1)}$ is the incremental restoring force corresponding to the incremental displacement $\Delta u_{i+1}^{(1)}$ obtained directly from the system's force-deformation curve, will be different from $\Delta \hat{F}_{i+1}$ (see Figure 11.15a). A correction may be therefore added to the first estimate of the incremental displacement using the tangent slope $\hat{k}_i^{(2)}$ that corresponds to the displacement $u_i + \Delta u_{i+1}^{(1)}$ and the incremental load imbalance $\Delta R_{i+1}^{(2)} = \Delta \hat{F}_{i+1} - \Delta \hat{f}_{i+1}^{(1)}$. That is, a correction given by

$$\Delta u_{i+1}^{(2)} = \frac{\Delta R_{i+1}^{(2)}}{\hat{k}_i^{(2)}} \tag{11.195}$$

An additional improvement may be similarly obtained by adding a correction based on the tangent slope corresponding to the displacement $u_i + \Delta u_{i+1}^{(1)} + \Delta u_{i+1}^{(2)}$ and the incremental load imbalance $\Delta R_{i+1}^{(3)} = \Delta R_{i+1}^{(2)} - \Delta f_{i+1}^{(2)}$, where $\Delta f_{i+1}^{(2)}$ is an incremental effective load calculated from an equation similar to Equation 11.194 using the aforementioned displacement (see Figure 11.15a). Specifically, a correction equal to

$$\Delta u_{i+1}^{(3)} = \frac{\Delta R_{i+1}^{(3)}}{\hat{k}_i^{(3)}} \tag{11.196}$$

Furthermore, improvements may be continued until the rth correction is negligibly small. The corrected estimate of the desired displacement increment is then calculated as

$$\Delta u_{i+1} = \Delta u_{i+1}^{(1)} + \Delta u_{i+1}^{(2)} + \cdots + \Delta u_{i+1}^{(r)} \tag{11.197}$$

Generalizing the procedure to the case of a multi-degree-of-freedom system and considering its incremental equation of motion in terms of an effective stiffness matrix and effective incremental load vector, one obtains

$$\{\Delta u\}_{i+1}^{(1)} = [\hat{K}^{(1)}]_i^{-1}\{\Delta \hat{F}\}_{i+1} \tag{11.198}$$

$$\{\Delta u\}_{i+1}^{(j)} = [\hat{K}^{(j)}]_i^{-1}\{\Delta R\}_{i+1}^{(j)} \quad j = 2,\ldots,r \tag{11.199}$$

$$\{\Delta u\}_{i+1} = \{\Delta u\}_{i+1}^{(1)} + \{\Delta u\}_{i+1}^{(2)} + \cdots + \{\Delta u\}_{i+1}^{(r)} \tag{11.200}$$

where $[\hat{K}^{(j)}]_i$ is the effective stiffness matrix of the system corresponding to the ith time step and formulated on the basis of the displacements obtained after the jth iteration. Similarly, $\{\Delta R\}_{i+1}^{(j)}$ is an incremental effective load-imbalance vector equal to $\{\Delta \hat{F}\}_{i+1}$ if $j = 1$ and

$$\{\Delta R\}_{i+1}^{(j)} = \{\Delta R\}_{i+1}^{(j-1)} - \{\Delta f\}_{i+1}^{(j-1)} \tag{11.201}$$

if j is different from 1. In this equation, $\{\Delta f\}_{i+1}^{(j-1)}$ is determined according to

$$\{\Delta \hat{f}\}_{i+1}^{(j)} = \{\Delta F_s\}_{i+1}^{(j)} + ([\hat{K}]_i - [K]_i)\{\Delta u\}_{i+1}^{(j)} \tag{11.202}$$

where $\{\Delta F_s\}_{i+1}^{(j)}$ is an incremental restoring force vector whose elements are the incremental restoring forces corresponding to the incremental displacements obtained after the jth iteration.

As it may be observed from the aforementioned equations, Newton–Raphson method requires that the tangent stiffness matrix be formulated and inverted or factorized at each iteration and this may be a time-consuming task if the system has many degrees of freedom. It is thus preferred to use the modified iterative scheme illustrated in Figure 11.15b. In this variation of Newton–Raphson method, the effective stiffness matrix determined on the basis of the displacements at the beginning of the interval is used in all iterations. In comparison with the original Newton–Raphson method, this alternative approach requires many more iterations, but each iteration is faster as it does not require the reinversion or refactorization of the stiffness matrix. This iterative scheme, known as *modified Newton–Raphson method*, is summarized in Box 11.7.

Box 11.7   Algorithm to Improve Solution Estimate at Time Step Using Modified Newton–Raphson Method

1.0 Initial calculations
   1.1 Formulate effective stiffness matrix $[\hat{K}]_i$ and effective load vector $\{\Delta \hat{F}\}_{i+1}$
   1.2 Set $\{\Delta u\}_{i+1}^{(0)} = \{0\}$ and $\{\Delta R\}_{i+1}^{(1)} = \{\Delta \hat{F}\}_{i+1}$
   1.3 Set $j = 1$
2.0 Calculations for jth iteration
   2.1 $\{\Delta u\}_{i+1}^{(j)} = [\hat{K}]_i^{-1} \{\Delta R\}_{i+1}^{(j)}$
   2.2 $\{\Delta \hat{f}\}_{i+1}^{(j)} = \{\Delta F_S\}_{i+1}^{(j)} + ([\hat{K}]_i - [K]_i)\{\Delta u\}_{i+1}^{(j)}$
   2.3 $\{\Delta R\}_{i+1}^{(j+1)} = \{\Delta R\}_{i+1}^{(j)} - \{\Delta f\}_{i+1}^{(j)}$
   2.4 $\{\Delta u\}_{i+1}^{(j)} = \{\Delta u\}_{i+1}^{(j-1)} + \{\Delta u\}_{i+1}^{(j)}$
3.0 Calculations for next iteration
   Replace j by $j + 1$ and repeat steps 2.1–2.4 until convergence is attained.

In some instances, Newton–Raphson method does not converge and does not lead to an improved accuracy with each iteration. Generally, this may happen when the change of curvature in the load-displacement relationship is too pronounced, as is the case in the example shown in Figure 11.16. To deal with such ill-conditioned cases, some other modifications to the original Newton–Raphson methods have been proposed, which for the most part constitute a compromise between the original

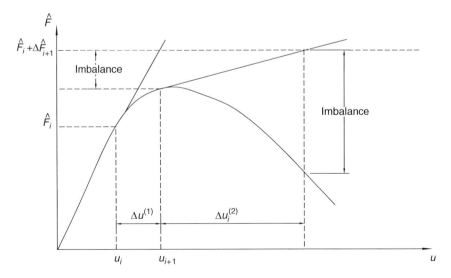

FIGURE 11.16   Ill-conditioned example for which Newton–Raphson method fails to converge.

method and the modified one. In one of such variations, the stiffness matrix is held constant for several iterations or load levels and is updated only after the rate of convergence starts to deteriorate. In another variation, only a single iteration is performed at each load level. These other methods are known as *quasi-Newton methods*.

The use of Netwon–Raphson method requires the establishment of a convergence criterion. One possible convergence criterion may be the requirement that the Euclidean norm of the imbalanced incremental effective load vector be small compared to the norm of the incremental effective load vector. However, this criterion may not be satisfactory for elastoplastic systems for which large displacements may occur for small load increments. An alternative criterion is to use an error measure based on the norm of the incremental displacement vector. A difficulty with using either of these two criteria emerges when different elements of the load and displacement vectors are measured in different units. This may occur, for example, when both rotations and displacements are included in the displacement vector as the units of displacements and rotations are not the same and their numerical values may differ by several orders of magnitude. In such a case, the influence of the rotations would be completely overshadowed if the convergence criterion is based on the norm of the incremental displacements. In such instances, it is convenient to define the convergence criterion on the basis of the work done by the incremental forces during the time step.

The error associated with the occurrence of an abrupt change in the properties of one of the elements of the structure within a time step may be minimized by using smaller time steps in the proximity of such an event. For example, if it is found that the unloading of an element occurs during the time interval from t to $t + \Delta t$, then the integration for that time step may be performed again but this time using a sequence of smaller time steps, as for instance $\Delta t/5$, until the precise moment of unloading is detected. After that, the integration is continued with the adjusted properties and the regular time step. Alternatively, an iterative procedure may be started in which integration is resumed with a smaller step whose size is progressively adjusted until the end of one of such adjusted time steps nearly coincides with the time at which the unloading occurs.

EXAMPLE 11.7 ANALYSIS OF SINGLE-DEGREE-OF-FREEDOM SYSTEM USING MODIFIED NEWTON–RAPHSON METHOD

A single-degree-of-freedom system, initially at rest, has a mass m of 0.3 kip s²/in. and a damping coefficient c of 0.15 kip s/in. The system exhibits the nonlinear force-deformation relationship shown in Figure E11.7a, where the nonlinear part (from $u = -1.0$ in. to $u = 1.0$ in.) is given by $F_S = 12(u - u^3/3)$. Determine the displacement, velocity, and acceleration responses of the system for the first 1.5 s when its base is excited by the ground motion shown in Figure E11.7b. Use the constant average acceleration method in conjunction with the modified Newton–Raphson method. Consider a time step Δt of 0.1 s. Assume convergence has been attained in the Newton–Raphson procedure when the last correction has no significant digit in the fourth decimal place.

Solution

As the constant average acceleration method is given by Newmark-β method when $\gamma = 1/2$ and $\beta = 1/4$, the solution will follow the algorithm established in Box 11.6 considering $\gamma = 1/2$ and $\beta = 1/4$. Note also that the slope of the nonlinear part of the given load-deformation curve is given by

$$k_i = \frac{dF_S}{du} = 12(1 - u^2)$$

which leads to an initial slope (initial stiffness) of 12 kip/in. and a slope equal to zero at $u = +1.0$ in. The slope remains constant and equal to zero afterward until a load reversal occurs. It is constant and equal to 12 kip/in. during unloading.

Structural Response by Step-by-Step Integration Methods

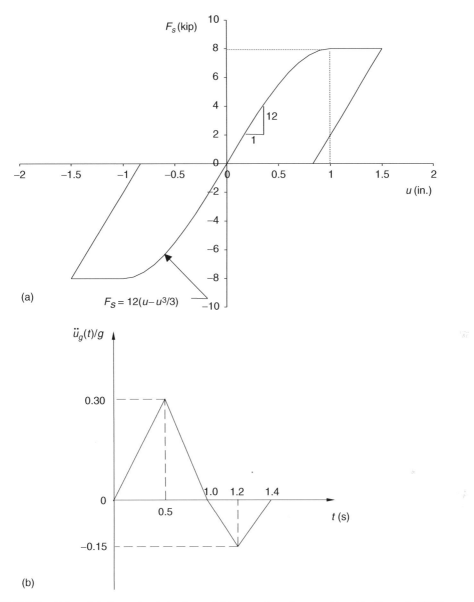

FIGURE E11.7 (a) Load-deformation behavior and (b) ground motion considered in Example E11.7.

Initial Conditions

The system is initially at rest. Therefore, the initial displacement u_0 and initial velocity $\dot{u}_0$ are both equal to zero. Also, as the ground motion at time $t = 0$, $\ddot{u}_g(0)$, is zero, the initial acceleration $\ddot{u}_0$ is also equal to zero.

Effective Stiffness and Effective Load

According to Equation 11.186 and for the given mass and damping coefficients, $\gamma = 1/2$, $\beta = 1/4$, and $\Delta t = 0.1$, the effective stiffness of the system is given by

$$(\hat{k}_N)_i = k_i + \frac{4m}{\Delta t^2} + \frac{2c}{\Delta t} = k_i + \frac{4(0.3)}{0.1^2} + \frac{2(0.15)}{0.1} = k_i + 123$$

where the stiffness k_i varies as indicated earlier. Similarly, according to Equation 11.187, the effective load of the system is equal to

$$(\Delta \hat{F}_N)_{i+1} = -m\Delta \ddot{u}_g(t_{i+1}) + \left(\frac{4m}{\Delta t} + 2c\right)\dot{u}_i + 2m\ddot{u}_i$$

$$= -0.3\Delta \ddot{u}_g(t_{i+1}) + \left[\left(\frac{4(0.3)}{0.1}\right) + 2(0.15)\right]\dot{u}_i + 2(0.3)\ddot{u}_i$$

$$= -0.3\Delta \ddot{u}_g(t_{i+1}) + 12.3\dot{u}_i + 0.6\ddot{u}_i$$

where

$$(\Delta \ddot{u}_g)_{i+1} = \ddot{u}_g(t_{i+1}) - \ddot{u}_g(t_i)$$

Incremental Displacement, Velocity, and Acceleration Responses at Time $t = t_{i+1}$

From Equations 11.182, 11.183, and 11.188 and considering a time step of 0.1 s, the corresponding expressions to calculate the incremental displacement, velocity, and acceleration responses are

$$\Delta u_{i+1} = \frac{(\Delta \hat{F}_N)_{i+1}}{(\hat{k}_N)_i}$$

$$\Delta \dot{u}_{i+1} = \frac{2}{\Delta t}\Delta u_{i+1} - 2\dot{u}_i = \frac{2}{0.1}\Delta u_{i+1} - 2\dot{u}_i = 20\Delta u_{i+1} - 2\dot{u}_i$$

$$\Delta \ddot{u}_{i+1} = \frac{4}{\Delta t^2}\Delta u_{i+1} - \frac{4}{\Delta t}\dot{u}_i - 2\ddot{u}_i = \frac{4}{0.1^2}\Delta u_{i+1} - \frac{4}{0.1}\dot{u}_i - 2\ddot{u}_i = 400\Delta u_{i+1} - 40\dot{u}_i - 2\ddot{u}_i$$

Displacement, Velocity, and Acceleration Response at Time $t = t_{i+1}$

In terms of the foregoing incremental displacement, velocity, and acceleration, the displacement, velocity, and acceleration of the system at time $t = t_{i+1}$ may be thus determined from

$$u_{i+1} = u_i + \Delta u_{i+1}$$

$$\dot{u}_{i+1} = \dot{u}_i + \Delta \dot{u}_{i+1}$$

$$\ddot{u}_{i+1} = \ddot{u}_i + \Delta \ddot{u}_{i+1}$$

Imbalanced Forces and Incremental Displacement Corrections

As discussed in Section 11.10.5, corrections may be added to the incremental displacement determined at any given time interval on the basis of the stiffness of the system at the beginning of the time interval if there is an imbalance between the incremental force determined using such a stiffness and that calculated using the actual force-deformation curve. According to Equation 11.199, these corrections may be calculated from

$$\Delta u_{i+1}^{(j)} = \Delta R_{i+1}^{(j)} / \hat{k}_i$$

where, following the modified Newton–Raphson method, the effective stiffness $\hat{k}_i$ is considered to be the same in all iterations and equal to that determined at the beginning of the time interval. The incremental load imbalance $\Delta R_{i+1}^{(j)}$ is calculated recursively according to

$$\Delta R_{i+1}^{(j+1)} = \Delta R_{i+1}^{(j)} - \Delta f_{i+1}^{(j)}$$

where $\{\Delta R\}_{i+1}^{(j)} = \{\Delta \hat{F}\}_{i+1}$ for $j = 1$, and, by virtue of Equation 11.202 and considering that for the case under consideration $(\hat{k}_N)_i = k_i + 123$, the actual effective incremental force is given by

$$\Delta \hat{f}_{i+1}^{(j)} = (\Delta F_S)_{i+1}^{(j)} + [(\hat{k}_N)_i - k_i]\Delta u_{i+1}^{(j)} = (\Delta F_S)_{i+1}^{(j)} + 123\Delta u_{i+1}^{(j)}$$

in which $(\Delta F_S)_{i+1}^{(j)}$ denotes the incremental restoring force corresponding to the incremental displacement $\Delta u_{i+1}^{(j)}$ obtained directly from the force-deformation relationship of the system.

Results

The displacement, velocity, and acceleration responses obtained using recursively the preceding equations are presented in Table E11.7 for the first 15 steps. The table includes the calculated values at each time step of the stiffness k_i, the restoring force $(F_s)_i$, and the effective incremental force that corresponds to the calculated displacement at the step under consideration and determined from the system's force-deformation relationship. It also includes the values of the incremental load imbalances $\Delta R_{i+1}^{(j)}$ and the calculated corrections to the incremental displacement whenever there is a difference between $\Delta \hat{F}_i$ and $\Delta \hat{f}_i^{(1)}$. The corrected incremental displacements are listed in the sixth column of the table when such corrections are necessary.

It should be noted that a step with a time increment of 0.096725 s was added in Table E11.7 after the 14th step. The addition of this fractional time step was necessary to determine with precision the time at which the system undergoes a load reversal. This load reversal was detected by observing a sign change in the velocity response (from negative to positive) after the 14th step. The exact time increment beyond the 14th step at which the load reversal occurred was obtained by trial and error. That is, different values were considered until the calculated velocity response was exactly zero. It should also be noted that the following step (15th step) was considered with a time increment equal to the difference between 0.1 and 0.096725 to compute the system's response exactly at $t = 1.5$ s.

11.11 FINAL REMARKS

One difficulty that arises in the application of step-by-step integration methods to multi-degree-of-freedom systems is the need to define the damping matrix of the system explicitly rather than in terms of modal damping ratios. This difficulty arises because, as discussed in Section 10.5, it is not possible to determine with any degree of certainty the coefficients that define a damping matrix. Consequently, an approach that is often adopted to circumvent this problem is to generate an orthogonal damping matrix of the Rayleigh or Caughey type on the basis of the assumed damping ratios for some of the modes of the system. Alternatively, one can select appropriate values for the damping ratios in all the modes of the system; substitute these damping ratios into the transformed damping matrix $[C^*] = [\Phi]^T[C][\Phi]$, where, as shown in Section 10.5, the elements along the main diagonal of $[C^*]$ are equal to $2\xi_i\omega_i M_i$ and $[\Phi]$ is the modal matrix; set the off-diagonal elements of $[C^*]$ equal to zero; and then solve for the damping matrix $[C]$. Although both approaches are just an approximation as, in general, damping matrices are not orthogonal, they are more convenient and physically reasonable than trying to evaluate the coefficients of a damping matrix directly. The approximation is equivalent to assume that the off-diagonal coefficients of the transformed damping matrix are negligibly small in comparison with those along the main diagonal, an assumption that may be justified for systems with uniform damping characteristics and damping ratios of <20%. Still another alternative is to formulate different damping matrices for different parts of the system applying any of the procedures just discussed and then assemble these matrices to formulate the system's damping matrix. Although problematic, the need for the consideration of an explicit damping matrix instead of damping ratios in step-by-step integration methods offers, nevertheless, generality and an advantage over the conventional response spectrum method. For with step-by-step integration methods, there is no need to make assumptions about the orthogonality of a damping matrix and thus any damping matrix may be employed in the analysis, including those nonorthogonal damping matrices that result from considering different levels of damping in different parts of a structure.

TABLE E11.7
Displacement, Velocity, and Acceleration Responses of Nonlinear System in Example 11.7 Using Constant Average Acceleration and Modified Newton–Raphson Methods

Step	Time	$\ddot{u}_g(t_i)$ (in./s²)	$\Delta \hat{F}_i$ or $\Delta R_i^{(j)}$ (kip)	Δu_i or $\Delta u_i^{(j)}$ (in.)	$\Sigma \Delta u_i^{(j)}$ (in.)	$\Delta \dot{u}_i$ (in./s)	$\Delta \ddot{u}_i$ (in./s²)	u_i (in.)	$\dot{u}_i$ (in./s)	$\ddot{u}_i$ (in./s²)	k_i (kip/in.)	$(F_S)_i$ (kip)	$\Delta \hat{f}_i^{(j)}$ (kip)
0	0	0	0.0000	0.0000		0.0000	0.0000	0.0000	0.0000	0.0000	12.0000	0.0000	0.0000
1	0.1	23.184	−6.9552	−0.0515		−1.0304	−20.6080	−0.0515	−1.0304	−20.6080	11.9681	−0.6177	−6.9547
			−0.0005	−4.05 × 10⁻⁶	−0.0515			−0.0515					
2	0.2	46.368	−31.9939	−0.2370		−2.6802	−12.3872	−0.2886	−3.7106	−32.9952	11.0007	−3.3667	−31.9059
			−0.0880	−6.52 × 10⁻⁴	−0.2377	−2.6932	−12.6480	−0.2892	−3.7236	−33.2560	10.9962	−3.3739	−0.0874
			−0.0006	−4.68 × 10⁻⁶	−0.2377			−0.2892					
3	0.3	69.552	−72.7091	−0.5426		−3.4052	−1.5921	−0.8318	−7.1288	−34.8482	3.6965	−7.6797	−71.0482
			−1.6610	−0.0124	−0.5550			−0.8442				−7.7240	−1.5689
			−0.0920	−0.0007	−0.5557	−3.6669	−6.8251	−0.8449	−7.3905	−40.0811	3.4333	−7.7263	−0.0868
			−0.0052	−3.87 × 10⁻⁵	−0.5557			−0.8449					
4	0.4	92.736	−121.9065	−0.9642		−4.5030	−9.8980	−1.8091	−11.8935	−49.9791	0.0000	−8.0000	−118.8698
			−3.0367	−0.0240	−0.9882			−1.8331				−8.0000	−2.9542
			−0.0825	−0.0007	−0.9889	−4.9964	−19.7661	−1.8338	−12.3869	−59.8472	0.0000	−8.0000	−0.0802
			−0.0022	0.0000	−0.9889			−1.8338					
5	0.5	115.92	−195.2221	−1.5872		−6.9697	−19.6992	−3.4210	−19.3566	−79.5464	0.0000	−8.0000	−195.2221
6	0.6	92.736	−278.8582	−2.2671		−6.6297	26.4988	−5.6881	−25.9862	−53.0475	0.0000	−8.0000	−278.8582
7	0.7	69.552	−344.5041	−2.8008		−4.0444	25.2062	−8.4889	−30.0307	−27.8413	0.0000	−8.0000	−344.5041
8	0.8	46.368	−379.1270	−3.0823		−1.5853	23.9766	−11.5713	−31.6160	−3.8646	0.0000	−8.0000	−379.1270
9	0.9	23.184	−384.2402	−3.1239		0.7539	22.8071	−14.6952	−30.8621	18.9424	0.0000	−8.0000	−384.2402
10	1.0	0.000	−361.2831	−2.9373		2.9790	21.6945	−17.6324	−27.8831	40.6369	0.0000	−8.0000	−361.2831
11	1.1	−28.98	−309.8863	−2.5194		5.3782	26.2909	−20.1518	−22.5049	66.9278	0.0000	−8.0000	−309.8863
12	1.2	−57.96	−227.9595	−1.8533		7.9432	25.0084	−22.0052	−14.5617	91.9362	0.0000	−8.0000	−227.9595
13	1.3	−28.98	−132.6411	−1.0784		7.5557	−32.7579	−23.0836	−7.0060	59.1783	0.0000	−8.0000	−132.6411
14	1.4	0.000	−59.3604	−0.4826		4.3598	−31.1599	−23.5662	−2.6461	28.0184	0.0000	−8.0000	−59.3604
	1.496725	0.000	−16.8114	−0.1280		2.6461	−1.3230	−23.6941	0.0000	26.6954	0.0000	−8.0000	−16.8114
15	1.5	0.000	16.0058	0.0001		0.0874	−0.0437	−23.6940	0.0873	26.6517	12.0000	−7.9997	16.0061

Another difficulty associated with the application of step-by-step integration methods is the need to consider as excitation a specific ground motion time history. Differently from the response spectrum method, with step-by-step integration methods, it is not possible to specify the input ground motion on the basis of an average or design response spectrum. This is certainly a disadvantage because, given the random nature of earthquake ground motions, an analysis based on a single ground motion is totally meaningless. It is always possible to repeat the analysis using a statistically significant ensemble of earthquake ground motions to obtain meaningful averages, but such an approach is prohibitively costly and time consuming. Another alternative is to use, as described in Chapter 6, an artificial ground motion that matches a specified design spectrum. Although an attractive alternative because it simplifies the analysis, this approach has the drawback that many ground motions can match a given response spectrum but lead to completely different results.

FURTHER READINGS

1. Adeli, H., Gere, J. M., and Weaver, W., Jr., Algorithms for nonlinear structural dynamics, *Journal of the Structural Engineering Division, ASCE*, 104, ST2, February 1978, 263–280.
2. Bathe, K. J. and Wilson, E. L., Stability and accuracy analysis of direct integration methods, *Earthquake Engineering and Structural Dynamics*, 1, 1973, 283–291.
3. Bathe, K. J., *Finite Element Procedures in Engineering Analysis*, Prentice-Hall, Englewood Cliffs, NJ, 1982.
4. Cheng, F. Y., *Matrix Analysis of Structural Dynamics: Applications and Earthquake Engineering*, Marcel Dekker, New York, 2001.
5. Chopra, A. K., *Dynamics of Structures: Theory and Applications to Earthquake Engineering*, 2nd Edition, Prentice-Hall, Upper Saddle River, NJ, 2001.
6. Collatz, L., *The Numerical Treatment of Differential Equations*, Springer-Verlag, New York, 1966.
7. Goudreau, G. L. and Taylor, R. L., Evaluation of numerical integration methods in elastodynamics, *Computer Methods in Applied Mechanics and Engineering*, 2, 1, 1973, 69–97.
8. Hart, G. C. and Wong, K., *Structural Dynamics for Structural Engineers*, Wiley, New York, 2000.
9. Hilber, H. H. and Hughes, T. J. R., Collocation, dissipation, and overshoot for time integration schemes in structural dynamics, *Earthquake Engineering and Structural Dynamics*, 6, 1, 1978, 99–117.
10. Humar, J. L., *Dynamics of Structures*, Prentice-Hall, Englewood Cliffs, NJ, 1990.
11. Newmark, N. M., A method of computation for structural dynamics, *Journal of the Engineering Mechanics Division, ASCE*, 85, EM3, 1959, 67–94.
12. Nickel, R. E., On the stability of approximation operators in problems of structural dynamics, *International Journal of Solids Structures*, 7, 1971, 301–319.
13. Wilson, E. L., Farhoomand, I., and Bathe, K. J., Nonlinear dynamic analysis of complex structures, *Earthquake Engineering and Structural Dynamics*, 1, 1973, 241–252.

PROBLEMS

11.1 Solve the problem in Example 11.3 using the finite difference method with a time step of 0.02 s first and then 0.05 s.

11.2 A damped two-degree-of-freedom shear building initially at rest has the following mass and damping matrices:

$$[M] = \begin{bmatrix} 0.2 & 0 \\ 0 & 0.1 \end{bmatrix} \text{kip s}^2/\text{in.} \qquad [K] = \begin{bmatrix} 45 & -15 \\ -15 & 15 \end{bmatrix} \text{kip/in.}$$

Determine using the central difference method the first half second of the displacement, velocity, and acceleration responses of the building when the building is subjected to a ground motion defined by a piece-wise linear function and the time-acceleration points given in Table P11.2. Use a time step of 0.05 s. Assume the damping matrix of the system is equal to 0.05 times its stiffness matrix.

TABLE P11.2
Ground Motion Specified in Problem P11.2

Time (s)	Ground Acceleration (in./s^2)	Time (s)	Ground Acceleration (in./s^2)
0.00	0.0	0.35	93
0.05	74	0.40	81
0.10	110	0.45	69
0.15	124	0.50	58
0.20	124	0.55	29
0.25	116	0.60	0
0.30	101		

TABLE P11.12
Ground Motion Specified in Problem P11.12

Time (s)	Ground Acceleration (g)	Time (s)	Ground Acceleration (g)
0.00	0.0	0.40	0.0
0.10	−1.0	0.50	0.6
0.20	−0.8	0.60	0.3
0.30	−0.5	0.70	0.0

11.3 Solve the problem in Example 11.3 using Houbolt method and a time step of 0.02 s. Use the solution obtained in Example 11.3 for the first two steps to get the solution started.

11.4 Solve Problem 11.2 using Houbolt method. Use the solution determined in Problem 11.2 for the first two steps to start the solution.

11.5 Solve the problem in Example 11.1 using the constant average acceleration method with a time step of 0.05 s.

11.6 Solve the problem in Example 11.1 using the linear acceleration method with a time step of 0.05 s.

11.7 Solve the problem in Example 11.1 using Wilson-θ method and a time step of 0.05 s.

11.8 Solve Problem 11.2 using the constant average acceleration method.

11.9 Solve the problem in Example 11.6 using the central difference method.

11.10 Solve the problem in Example 11.6 using the linear acceleration method.

11.11 Repeat Example 11.6 incorporating into the solution the exact times at which yielding or unloading of the braces takes place.

11.12 Repeat Example 11.6 considering that the ground motion that excites the structure is a piece-wise linear function defined by the time-acceleration points in Table P11.12.

11.13 Repeat Example 11.7 using a time step of 0.05 s.

11.14 Solve the problem in Example 11.7 using the finite difference method.

11.15 Solve the problem in Example 11.7 using the linear acceleration method.

12 Structural Response by Equivalent Lateral Force Procedure

In anything at all, perfection is finally attained not when there is no longer anything to add, but when there is no longer anything to take away.

Antoine de Saint-Exupery, 1938

12.1 INTRODUCTION

As discussed in Chapters 10 and 11, the seismic response analysis of a structure may be carried out effectively and accurately by either the response spectrum method or a step-by-step integration procedure. Notwithstanding their availability and adequacy, these two methods still have a shortcoming because both require knowing the mass and stiffness matrices of the system being analyzed before they can be applied. This means that their application requires the results from a preliminary design since a system's stiffness and mass matrices depends on the dimensions and characteristics of its structural members, and these dimensions and characteristics depend, in turn, on the results from the seismic response analysis. It is convenient, thus, to count with a simplified approximate method by means of which such preliminary analyses can be performed simply and expeditiously. Such a simplified method may also be convenient for the design of simple structures for which the time and expense of a response spectrum or step-by-step method cannot be justified. The equivalent lateral force procedure described in this chapter is such a method. This procedure is based on the consideration of earthquake effects by means of a series of lateral forces applied statically, and the use of these lateral forces to determine the displacements and internal forces in the system. In turn, the magnitude of such lateral forces is based on an estimate of the fundamental natural period of the structure and some simple formulas. Its simplicity lies on the fact that a dynamic analysis is avoided and thus there is no need to determine the natural frequencies and mode shapes of the system, or to carry out a time-history analysis. This chapter presents in detail the derivation of the procedure and lays out the assumptions on which it is based. The limitations of the method are also discussed in some detail. The procedure is derived specifically for buildings although, in principle, it is possible to extend its application to other structures.

12.2 DERIVATION OF PROCEDURE

12.2.1 Overview

The equivalent lateral force procedure is derived on the basis of the response spectrum method presented in Chapter 10 and the introduction of some simplifying assumptions. In addition, it is assumed that buildings may be idealized as systems with lumped masses and one translational degree of freedom per floor along each of two orthogonal horizontal directions. That is, it is assumed that the building's floors are infinitely rigid and therefore undeformable in their own plane. Furthermore, it is considered that a building may be analyzed independently along each of those two horizontal directions under an in-plane horizontal component of ground motion. The direct outcome of the procedure is lateral forces in the direction of the ground motion being considered.

12.2.2 Maximum Modal Lateral Forces

From the definition of stiffness matrix, the external forces applied to a structure are equal to the product of its stiffness matrix and the vector that contains the displacements induced by the external forces. Hence, if $[K]$ represents the condensed stiffness matrix that relates the lateral forces and displacements in a structure, the maximum values of an equivalent set of lateral external forces that generate the same maximum displacements in the rth mode of the structure as an earthquake ground motion are given by

$$\max\{F_s(t)\}_r = [K]\max\{u(t)\}_r \quad (12.1)$$

where $\{u(t)\}_r$ and $\{F_s(t)\}_r$, respectively, represent vector that contains the displacements induced by the ground motion in the rth mode of the system and vector that contains the equivalent lateral forces.

But, as described in Section 10.3, the maximum displacements induced by an earthquake ground motion in the rth mode of a structure may be expressed in terms of the response spectrum of the earthquake ground motion as

$$\max\{u(t)\}_r = \frac{\Gamma_r}{\omega_r^2}\{\phi\}_r SA(\omega_r,\xi_r) \quad (12.2)$$

where, as defined previously, Γ_r, ξ_r, and $\{\phi\}_r$ denote the participation factor, circular natural frequency, and mode shape of the structure in its rth mode of vibration, respectively, and $SA(\omega_r, \xi_r)$ is the ordinate in the acceleration response spectrum of the specified ground motion corresponding to a natural frequency ω_r and damping ratio ξ_r. Therefore, the maximum values of the set of equivalent lateral forces under consideration may be determined according to

$$\max\{F_s(t)\}_r = \frac{\Gamma_r}{\omega_r^2}[K]\{\phi\}_r SA(\omega_r,\xi_r) \quad (12.3)$$

Note, however, that from the associated eigenvalue problem one has that

$$[K]\{\phi\}_r = \omega_r^2 [M]\{\phi\}_r \quad (12.4)$$

where $[M]$ denotes the system's mass matrix. Consequently, Equation 12.3 may also be written as

$$\max\{F_s(t)\}_r = \Gamma_r [M]\{\phi\}_r SA(\omega_r,\xi_r) \quad (12.5)$$

or, if $[M]$ is considered to be a diagonal matrix, alternatively as

$$\max\{F_s(t)\}_r = \Gamma_r \begin{Bmatrix} m_1\phi_{1r} \\ m_2\phi_{2r} \\ \vdots \\ m_N\phi_{Nr} \end{Bmatrix} SA(\omega_r,\xi_r) \quad (12.6)$$

where m_j, $j = 1, 2, ..., N$, signifies the mass of the jth floor of the structure; ϕ_{jr}, $j = 1, 2, ..., N$, represents the amplitude corresponding to the jth floor of the structure in the structure's rth mode shape; and N denotes the number of stories. As a result, the equivalent lateral force acting on the structure's jth floor may be expressed as

$$F_{sjr} = \max F_{sjr}(t) = \Gamma_r m_j \phi_{jr} SA(\omega_r, \xi_r) \tag{12.7}$$

Note also, in passing, that by combining Equations 12.2 and 12.7 it is possible to express the maximum modal lateral displacements in terms of the maximum lateral forces as

$$u_{jr} = \max u_{jr}(t) = \frac{F_{sjr}}{m_j \omega_r^2} \tag{12.8}$$

where u_{jr} is the displacement corresponding to the jth floor of the structure in its rth mode of vibration.

12.2.3 Modal Base Shear

With reference to the building shown in Figure 12.1 and considering the building as a vertical cantilever beam, the shear force at the base of the building in its rth mode of vibration may be therefore expressed as

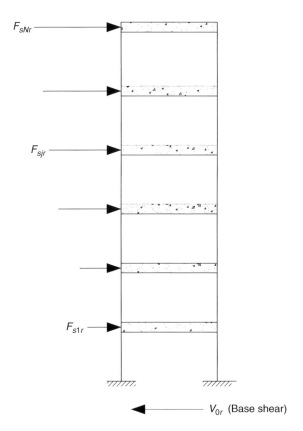

FIGURE 12.1 Lateral forces and base shear in building's rth mode of vibration.

$$V_{0r} = \sum_{j=1}^{N} F_{sjr} = \Gamma_r SA(\omega_r, \xi_r) \sum_{j=1}^{N} m_j \phi_{jr}$$
$$= \Gamma_r \frac{SA(\omega_r, \xi_r)}{g} \sum_{j=1}^{N} w_j \phi_{jr} \qquad (12.9)$$

where w_j and g are weight associated with the building's jth floor and acceleration of gravity
Alternatively, it may be written as

$$V_{0r} = W_r^* \frac{SA(\omega_r, \xi_r)}{g} \qquad (12.10)$$

where W_r^* is defined as

$$W_r^* = \Gamma_r \sum_{j=1}^{N} w_j \phi_{jr} \qquad (12.11)$$

and exhibits the property that

$$\sum_{r=1}^{N} W_r^* = W = \text{total building weight} \qquad (12.12)$$

since from the definition of W_r^* and by virtue of Equation 10.83, one has that

$$\sum_{r=1}^{N} W_r^* = \sum_{r=1}^{N} \Gamma_r \sum_{j=1}^{N} w_j \phi_{jr} = \sum_{j=1}^{N} w_j \sum_{r=1}^{N} \Gamma_r \phi_{jr} = \sum_{j=1}^{N} w_j \qquad (12.13)$$

As such, W_r^* is called the *effective weight* in the rth mode of vibration and is interpreted as the fraction of the total weight of the structure that participates in its rth mode of vibration.

12.2.4 MODAL LATERAL FORCES IN TERMS OF BASE SHEAR

If the participation factor Γ_r is solved from Equation 12.11 and substituted into Equation 12.7, and if, in addition, it is considered that $m_j = w_j/g$, one arrives at

$$F_{sjr} = \frac{w_j \phi_{jr}}{\sum_{j=1}^{N} w_j \phi_{jr}} W_r^* \frac{SA(\omega_r, \xi_r)}{g} \qquad (12.14)$$

By virtue of Equation 12.10, the equivalent lateral forces may be thus expressed in terms of the modal base shear as

$$F_{sjr} = \frac{w_j \phi_{jr}}{\sum_{j=1}^{N} w_j \phi_{jr}} V_{0r} \qquad (12.15)$$

12.2.5 Basic Assumptions

Equation 12.15 in conjunction with Equation 12.10 defines a set of equivalent lateral forces for each of the modes of the structure. Note, however, that to determine these modal lateral forces it is necessary to first determine the mode shapes and natural frequencies of the system. In addition, these modal lateral forces need to be combined according to one of the modal combination rules introduced in Section 10.3, if one wants to obtain an estimate of the maximum values of such lateral forces. Therefore, it is convenient to introduce some approximations to simplify the procedure. In the classical equivalent lateral force procedure, the following simplifying assumptions are made:

1. The structure vibrates in its first mode only; that is, the contribution of the higher modes of vibration to the total response of the structure may be neglected.
2. The fundamental mode shape of the structure varies linearly with height.
3. The effective weight of the structure in its first mode of vibration is equal to the structure's total weight.

The first assumption is based on the fact that the response of a structure is usually dominated by the response in its fundamental mode, although the influence of the higher modes may be significant in long-period buildings. The second approximates in a simple form the fundamental mode shape observed in regular, conventional structures. Finally, the third, besides being a simple and conservative estimate of the effective weight of the system in its fundamental mode, approximately accounts for the effect of the higher modes of vibration. This argument is based on the fact that typical values of W_1^* are between 60 and 80% of the total weight of a building.

12.2.6 Approximate Maximum Lateral Forces

On the basis of the foregoing assumptions and Equation 12.10, the total shear at the base of a structure may be computed approximately according to

$$V_0 = V_{01} = W \frac{SA(\omega_1, \xi_1)}{g} \qquad (12.16)$$

Similarly, with reference to Figure 12.2 and on the basis of the assumption made that the first mode shape of the structure varies linearly with height, one can write

$$\phi_{j1} = \frac{\phi_{N1}}{h_N} h_j \qquad (12.17)$$

where h_j, $j = 1, \ldots, N$, denotes the height of the jth floor of the building measured from the base of the building. Substitution of Equation 12.17 into Equation 12.15 and the consideration that $r = 1$ and $V_0 = V_{01}$ thus lead to

$$F_{sj} = F_{sj1} = \frac{w_j \phi_{j1}}{\sum_{j=1}^{N} w_j \phi_{j1}} V_{01} = \frac{w_j \frac{\phi_{N1}}{h_N} h_j}{\sum_{j=1}^{N} w_j \frac{\phi_{N1}}{h_N} h_j} V_0 \qquad (12.18)$$

from which one obtains

$$F_{sj} = \frac{w_j h_j}{\sum_{j=1}^{N} w_j h_j} V_0 \qquad (12.19)$$

Equation 12.19 represents the desired simplified formula to define a set of equivalent lateral forces for the seismic design of structures. Note that it depends only on the floor weights of the structure,

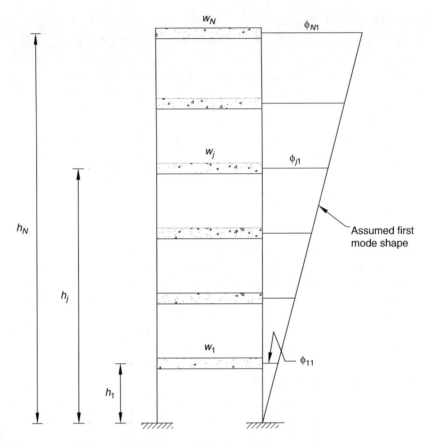

FIGURE 12.2 Assumed first mode shape in equivalent lateral force procedure.

the heights of the floors measured from the base of the structure, and the response spectrum of the ground motion that excites the structure. As such, it constitutes a simple and convenient method to carry out the seismic analysis of those structures for which the assumptions made in its derivation are reasonable. It is also important to note that although the forces determined with Equation 12.19 represent a set of static forces, these forces implicitly account, albeit in an approximate way, for the dynamic nature of the structure. For this reason, the equivalent lateral force procedure is sometimes referred to as the *pseudostatic method*.

12.2.7 Approximate Fundamental Natural Period

It may be seen from Equation 12.16 that the determination of a set of equivalent lateral forces by means of Equation 12.19 requires knowing the fundamental natural period of the structure being considered. However, in a preliminary analysis one cannot solve the corresponding eigenvalue problem to determine this fundamental natural period since in a preliminary analysis the stiffness matrix of the structure is not known. It is convenient, therefore, to use instead an approximate formula to obtain an estimate of it. An approximate formula that is suggested in many building codes for such a purpose is based on the application of Rayleigh's principle and may be derived as follows:

According to Equation 10.30, the square of the fundamental natural frequency of a system is equal to

$$\omega_1^2 = \frac{K_1^*}{M_1^*} = \frac{\{\phi\}_1^T [K] \{\phi\}_1}{\{\phi\}_1^T [M] \{\phi\}_1} \qquad (12.20)$$

where K_1^* and M_1^* denote the generalized mass and stiffness of the system in its fundamental mode of vibration, respectively, and all other symbols are as defined earlier. According to Rayleigh's principle, however, a close approximation to the square of this frequency may be obtained if the mode shape $\{\phi\}_1$ is substituted by an assumed configuration $\{\psi\}$ that roughly resembles the first mode shape of the system. That is,

$$\omega_1^2 \approx \frac{\{\psi\}^T[K]\{\psi\}}{\{\psi\}^T[M]\{\psi\}} \qquad (12.21)$$

an expression known as *Rayleigh quotient*. Thus, if $\{\psi\}$ is selected to be the vector of lateral displacements $\{u\}$ induced by the set of lateral forces $\{F_s\}$, and if by virtue of Equation 12.1 it is considered that $[K]\{u\} = \{F_s\}$, an approximate formula for the square of the fundamental natural frequency of a system is

$$\omega_1^2 \approx \frac{\{u\}^T[K]\{u\}}{\{u\}^T[M]\{u\}} = \frac{\{u\}^T\{F_s\}}{\{u\}^T[M]\{u\}} = g\frac{\sum_{j=1}^{N} F_{sj}u_j}{\sum_{j=1}^{N} w_j u_j^2} \qquad (12.22)$$

from which one obtains the following approximate formula to compute the system's fundamental natural period:

$$T_1 = 2\pi\sqrt{\frac{\sum_{j=1}^{N} w_j u_j^2}{g\sum_{j=1}^{N} F_{sj}u_j}} \qquad (12.23)$$

Customarily, this formula is used in conjunction with the lateral forces that result from the application of Equation 12.19 and the structural displacements induced by these lateral forces. In such a case, it is necessary to perform one or two iterations, beginning with an initial estimate of the natural period being sought or the response spectrum ordinate corresponding to this natural period. For the first trial, one can use one of the available empirical rules to estimate the fundamental natural period of a structure (introduced in Section 17.6.3), or simply use the largest ordinate in the specified response spectrum. Afterward, one can repeat the calculations with the set of forces and displacements determined in the first trial to obtain an improved estimate of the natural period in question.

12.3 APPLICATION OF PROCEDURE

The direct outcome from the application of Equations 12.16 and 12.19 is a set of lateral forces in the direction of the ground motion being considered, applied at the centers of mass of the building's floors. Once these lateral forces are obtained, they can then be used to determine the maximum values of other measures of the building's response such as story drifts and member internal forces. To this end, one can employ any of the methods available for the analysis of structures under static loads. For example, one can use a procedure based on story shears and overturning moments as illustrated in Example 10.3, or a procedure based on the direct stiffness method. Ordinarily, the analysis is performed separately along each of two orthogonal horizontal directions with a two-dimensional model of the building in each case. However, in the case of buildings that possess asymmetries in their geometric configuration or distribution of floor masses, such as those shown in Figure 12.3, it is necessary to consider a three-dimensional model to account for the horizontal floor rotations that are generated in such buildings. As illustrated in Figure 12.4, these floor rotations are generated because in an asymmetric building the line of action of the resultant of the applied lateral forces above one of its stories does not coincide with the line of action of the resultant of the

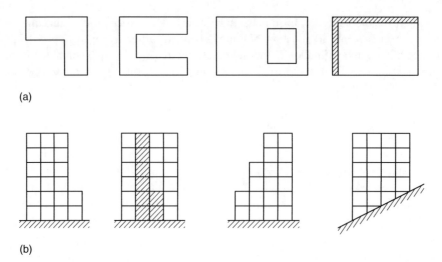

(a)

(b)

FIGURE 12.3 Examples of asymmetric buildings. (a) Asymmetries in plan, (b) asymmetries in elevation.

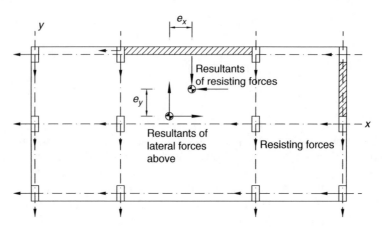

FIGURE 12.4 Plan view of earthquake and resisting forces and eccentricities in $x(e_y)$ and $y(e_x)$ directions in story of asymmetric structure.

resisting forces in the structural elements of that story; that is, it does not pass through the story's *center of twist*.* In such a case, it is said that there is a story *eccentricity*, and that this eccentricity induces a torsional moment whose magnitude is equal to the story shear times the eccentricity. The effect of such a torsional moment will be the rotation of the floor and the part of the structure above, as shown schematically for a simple structure in Figure 12.5. As a result, in an asymmetric building each floor will not only be displaced but also rotated. Also, the story drift will be different at different points within each story, and there will be an uneven distribution of the story shear among the different structural elements in that story. Furthermore, displacements and stresses will be generated in the direction perpendicular to the direction of analysis. That is, a lateral load parallel to the *x*-axis will also induce displacements and internal forces in the resisting elements of the structure parallel to the *y*-axis, and vice versa.

---

* The *center of twist* or *center of rigidity* of a story is defined as the point through which the line of action of the story shear must pass so that the relative motion between the upper and lower ends of the story is in pure translation. Its coordinates may be found by locating the position of the resultant of the forces induced in the structural elements of the story when the story is subjected to a unit relative displacement between its upper and lower ends.

Structural Response by Equivalent Lateral Force Procedure

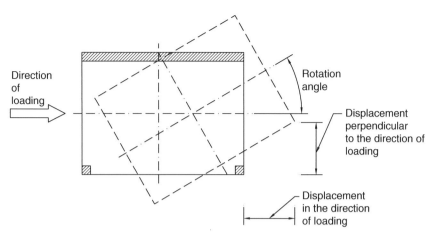

FIGURE 12.5 Plan view of displaced and rotated asymmetric structure.

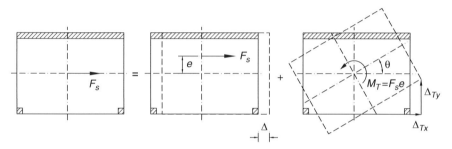

FIGURE 12.6 Superposition of translational and rotational effects in an asymmetric structure.

As illustrated for a single-story structure in Figure 12.6, the floor rotations in asymmetric structures may be considered by superimposing the forces and displacements induced by the torsional moments acting on them to the forces and displacements generated by the equivalent lateral forces when they are applied at the locations where they can only induce translational motions. This entails an additional analysis under the torsional moments alone. In this analysis, it is customary to assume to simplify the analysis that the floors are infinitely rigid in their own planes and that they therefore rotate as rigid bodies. Such an approach is convenient when the calculations are carried out by hand. Alternatively, the floor rotations may be considered directly by working with a three-dimensional model of the structure in which the lateral forces are applied at the floors' centers of mass. An analysis with a three-dimensional model will simultaneously account for the translational and rotational motions of a structure since a three-dimensional model takes into consideration any story eccentricities. This approach offers the advantage that it does not require the calculation of the story eccentricities and torsional moments. It requires, however, the use of a computer program since for structures with more than two stories the calculations can be long and tedious. Nowadays, with the wide availability of simple computer programs to perform a static analysis of three-dimensional structures, this is not an obstacle and thus a direct analysis with a three-dimensional model should be the preferred approach. It is important to keep in mind, however, that any of these procedures only accounts for the static effect of the floor rotations since any possible dynamic magnification of these rotations is neglected. A dynamic magnification of the floor rotations is possible because the floors also possess rotational inertia. It is neglected because such a rotational inertia is not accounted for in the derivation of the procedure. It is also important to keep in mind that in the analysis of asymmetric structures one cannot directly add the effects produced along the direction of analysis by the lateral forces applied along this direction and those applied perpendicular to it. The reason is that, once again, it is unlikely that the maximum values of the two sets of forces, and hence the effects that these forces will produce in the structure, will occur at the same time. Consequently, one should instead use a different method to combine the two effects. I is common to employ the square root of the sum of the squares rule, although, as discussed in Section 10.8.7, this rule is valid only when the two directions correspond to the principal directions of the ground motion, or when the same ground motion is considered along each of two such directions. In all rigor, the combination should be performed according to the rule defined by Equation 10.159.

Another important effect that needs to be considered in the case of flexible buildings is the so-called *P-delta effect*. The P-delta effect refers to the additional bending moments produced by a vertical load in columns or other elements of the lateral-load resisting system as a result of the lateral deflections of such elements. As an illustration of such an effect, consider the column shown in Figure 12.7, which is supporting an axial compressive force P, a shear force V, and bending moments M_A and M_B. Because of the lateral load, the column experiences a relative lateral displacement or drift that leads to a secondary bending moment $M_s = P\Delta$ (hence the name P-delta effect). In turn, this additional bending moment will increase the lateral drift and consequently will produce an increase in the bending moments. Furthermore, this increase in the lateral drifts and bending moments will continue until the column reaches a state of equilibrium.

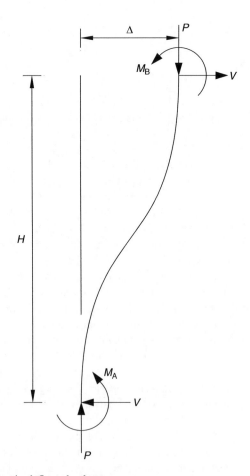

FIGURE 12.7 P-delta effect in deflected column.

Structural Response by Equivalent Lateral Force Procedure

An approximate but acceptable method to estimate the final drift for each story is to add to the drift Δ produced by the primary bending moment M_{xp}, the incremental drift $\Delta\cdot\theta$ due to the secondary moment $M_{xs} = M_{sp}\cdot\theta$, the incremental drift $(\Delta\cdot\theta)\cdot\theta$ due to the secondary moment $M_{xs} = (M_{sp}\cdot\theta)\cdot\theta$, etc., where θ represents the ratio of the first secondary to the primary moment. In this way, the total story drift may be calculated as

$$(\Delta_x)_{\text{total}} = \Delta_x + \Delta_x\theta_x + \Delta_x\theta_x^2 + \Delta_x\theta_x^3 + \cdots$$
$$= \Delta_x(1 + \theta_x + \theta_x^2 + \theta_x^3 + \cdots) \quad (12.24)$$

which is equal to

$$(\Delta_x)_{\text{total}} = \Delta_x\left(\frac{1}{1-\theta_x}\right) \quad (12.25)$$

where

$$\theta_x = \frac{M_{xs}}{M_{xp}} = \frac{P_x\Delta_x}{V_x H_x} \quad (12.26)$$

in which P_x, Δ_x, V_x, and H_x, respectively, denote total weight above story x, drift of story x, shear force at story x, and height of story x.

Consequently, P-delta effects at each story may be considered by multiplying the story drift and story shear determined without the consideration of P-delta effects by the amplification factor $1/(1 - \theta_x)$, and then use the amplified story shear forces to obtain internal forces and other desired seismic effects.

The application of the equivalent lateral force procedure will now be illustrated by means of the following examples.

EXAMPLE 12.1 EQUIVALENT LATERAL FORCES AND LATERAL DISPLACEMENTS IN A THREE-STORY SHEAR BUILDING

Determine using the equivalent lateral force procedure the maximum lateral forces and lateral displacements induced on the three-story shear building considered in Example 10.3 by the N–S component of the ground accelerations recorded during the 1940 El Centro, California, earthquake. The acceleration response spectrum for this acceleration record is shown in Figure 6.5. As in Example 10.3, assume that the damping ratio in the fundamental mode of the building is 2%.

Solution

The total weight of the building in Example 10.3, W, is equal to 4.5(9.81) = 44.1 MN. Also, from Figure 6.5, the response spectrum ordinate corresponding to the fundamental natural period of the structure (1.37 s) and a damping ratio of 2% is 0.24g. Hence, according to Equation 12.16, the structure's base shear is approximately equal to

$$V = W\frac{\text{SA}(\omega_1,\xi_1)}{g} = 44.1(0.24) = 10.6 \text{ MN}$$

Similarly, since

$$\sum_{j=1}^{3} w_j h_j = [2(3) + 1.5(6) + 1.0(9)]g = 24g$$

then according to Equation 12.19, the equivalent lateral forces are

$$F_{s1} = \frac{w_1 h_1}{\sum_{j=1}^{3} w_j h_j} V_0 = \frac{2(3)g}{24g}(10.6) = 0.25(10.6) = 2.65 \text{ MN}$$

$$F_{s2} = \frac{w_2 h_2}{\sum_{j=1}^{3} w_j h_j} V_0 = \frac{1.5(6)g}{24g}(10.6) = 0.375(10.6) = 3.98 \text{ MN}$$

$$F_{s3} = \frac{w_3 h_3}{\sum_{j=1}^{3} w_j h_j} V_0 = \frac{1.0(9)g}{24g}(10.6) = 0.375(10.6) = 3.98 \text{ MN}$$

The points of application of these lateral forces are shown in Figure E12.1.

As described in Section 12.2.2, an estimate of the maximum lateral displacements may be obtained in terms of the equivalent lateral forces acting on the structure using Equation 12.8. Accordingly, after substitution of the determined lateral forces into Equation 12.8, the maximum lateral displacements of the structure are approximately equal to

$$\max\{u(t)\} = \frac{1}{\omega_1^2} \begin{Bmatrix} F_{s1}/m_1 \\ F_{s2}/m_2 \\ F_{s3}/m_3 \end{Bmatrix} = \frac{1}{4.592^2} \begin{Bmatrix} 2.65/2.0 \\ 3.98/1.5 \\ 3.98/1.0 \end{Bmatrix} = \begin{Bmatrix} 0.063 \\ 0.126 \\ 0.189 \end{Bmatrix} \text{m}$$

It is instructive to compare the value of the building's fundamental natural period that one would obtain with Rayleigh's formula and the calculated forces and displacements with the actual value. For this purpose, consider that with the obtained lateral forces and displacements, one has that

$$\sum_{j=1}^{3} w_j u_j^2 = [2.0(0.063)^2 + 1.5(0.126)^2 + 1.0(0.189)^2]g = 0.067g$$

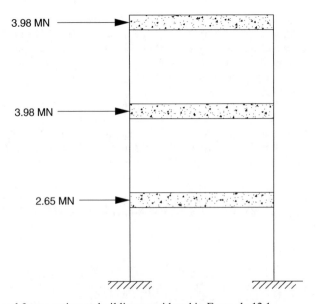

FIGURE E12.1 Lateral forces acting on building considered in Example 12.1.

$$\sum_{j=1}^{3} F_{sj} u_j = 2.65(0.063) + 3.98(0.126) + 3.98(0.189) = 1.42$$

and thus, according to Equation 12.23, an approximate value of the fundamental natural period of the structure is

$$T_1 = 2\pi \sqrt{\frac{\sum_{j=1}^{3} w_j u_j^2}{g \sum_{j=1}^{3} F_{sj} u_j}} = 2\pi \sqrt{\frac{0.067g}{g(1.42)}} = 1.36 \text{ s}$$

which compares well with the actual value given in Table E10.3.

It should be noted that once a set of equivalent lateral forces is determined, the story shears, overturning moments, and internal forces in the structure may be calculated as it was done for each mode in Example 10.3. To illustrate the procedure, suppose one is interested in estimating the maximum values of the story shear at the second story and the shearing forces and bending moments in the columns of this same story. To determine the story shear, one simply calculates the sum of the lateral forces above the second story. That is,

$$V_2 = F_{s3} + F_{s2} = 3.98 + 2.65 = 6.63 \text{ MN}$$

Then, to determine the magnitude of the shearing forces acting at the ends of each column, one divides the story shear by two, the number of columns in the story. The results is

$$V_{c2} = \frac{V_2}{2} = \frac{6.63}{2} = 3.32 \text{ MN}$$

Lastly, the bending moments in the columns may be calculated from the equation of static equilibrium for each column. Accordingly, these bending moments are approximately equal to

$$M_{c2} = \frac{V_{c2} h_2}{2} = \frac{3.32(3.0)}{2} = 4.98 \text{ MN}$$

EXAMPLE 12.2 SEISMIC ANALYSIS OF A TWO-STORY RIGID FRAME BY EQUIVALENT LATERAL FORCE PROCEDURE

Figure E12.2a shows a two-story, one-bay rigid frame whose members all have the same modulus of elasticity, E, and the moments of inertia indicated in the figure. Each floor resists a weight w. Determine using the equivalent lateral force procedure the bending moments and shearing forces in the beams and columns of the first story of this frame when the frame is subjected to a ground motion represented by a base shear V_0. Express the answers in terms of the story height H and base shear V_0. Neglect the axial deformations of all members.

Solution

Equivalent Lateral Forces

According to the given floor weights, story heights, and Equation 12.19, the equivalent lateral forces for this structure are

$$F_{s1} = \frac{wH}{wH + 2wH} V_0 = \frac{V_0}{3}$$

$$F_{s2} = \frac{2wH}{wH + 2wH} V_0 = \frac{2V_0}{3}$$

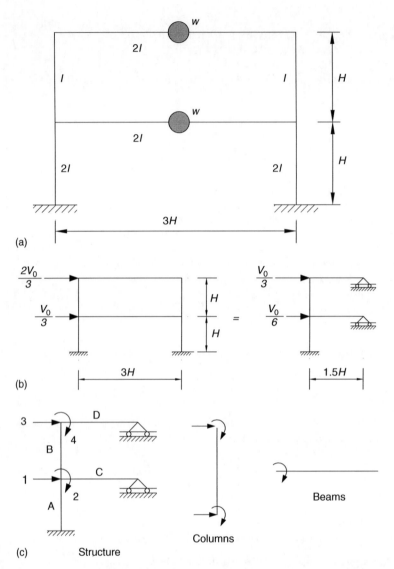

FIGURE E12.2 (a) Two story rigid frame considered in Example 12.2, (b) equivalent lateral forces acting on the rigid frame in Example 12.2 and equivalent structure, and (c) identification of structural and element degrees of freedom.

Therefore, the desired internal forces may be found from the static analysis of the structure under these lateral forces. Note, however, that this analysis may be simplified significantly if one takes advantage of the symmetry of the structure and the antisymmetry of the loading and considers instead the equivalent structure shown on the right-hand side of Figure E12.2b.

Stiffness Matrices

With reference to the degrees of freedom indicated on the right-hand side of Figure E12.2c and the moments of inertia given in Figure E12.2a, the stiffness matrices of the beams and columns of the equivalent structure in Figure E12.2b are given by

$$[K]_\text{C} = [K]_\text{D} = \frac{3E(2I)}{1.5H} = \frac{EI}{H^3}[4H^2]$$

$$[K]_A = \frac{EI}{H^3}\begin{bmatrix} 24 & -12H & -24 & -12H \\ -12H & 8H^2 & 12H & 4H^2 \\ -24 & 12H & 24 & 12H \\ -12H & 4H^2 & 12H & 8H^2 \end{bmatrix}$$

$$[K]_B = \frac{EI}{H^3}\begin{bmatrix} 12 & -6H & -12 & -6H \\ -6H & 4H^2 & 6H & 2H^2 \\ -12 & 6H & 12 & 6H \\ -6H & 2H^2 & 6H & 4H^2 \end{bmatrix}$$

Therefore, with reference to the degrees of freedom identified on the left-hand side of Figure E12.2c, the stiffness matrix of the structure is

$$[K] = \frac{EI}{H^3}\begin{bmatrix} 36 & 6H & -12 & -6H \\ 6H & 16H^2 & 6H & 2H^2 \\ -12 & 6H & 12 & 6H \\ -6H & 2H^2 & 6H & 8H^2 \end{bmatrix}\begin{matrix} 1 \\ 2 \\ 3 \\ 4 \end{matrix}$$

which, after rearranging to first list the translational degrees of freedom and then the rotational degrees of freedom, may also be written as

$$[K] = \frac{EI}{H^3}\begin{bmatrix} 36 & -12 & 6H & -6H \\ -12 & 12 & 6H & 6H \\ 6H & 6H & 16H^2 & 2H^2 \\ -6H & 6H & 2H^2 & 8H^2 \end{bmatrix}\begin{matrix} 1 \\ 3 \\ 2 \\ 4 \end{matrix}$$

Structural Displacements

The nodal displacements and rotations of the structure may be obtained directly from the equilibrium equation

$$\{P\} = [K]\{u\}$$

where $\{u\}$ and $\{P\}$ are displacement and force vectors referred to the structural degrees of freedom, respectively. It is convenient, however, to partition this equation first as

$$\begin{Bmatrix} \{F_s\} \\ \{0\} \end{Bmatrix} = \begin{bmatrix} [K]_{uu} & [K]_{u\theta} \\ [K]_{u\theta}^T & [K]_{\theta\theta} \end{bmatrix} \begin{Bmatrix} \{u\}_L \\ \{\theta\} \end{Bmatrix}$$

where $\{u\}_L$ and $\{\theta\}$ are vectors that contain the nodal lateral displacements and rotations of the structure, respectively,

$$\{F_s\} = \frac{V_0}{3}\begin{Bmatrix} 1 \\ 2 \end{Bmatrix}$$

and

$$[K]_{uu} = \frac{12EI}{H^3}\begin{bmatrix} 3 & -1 \\ -1 & 1 \end{bmatrix} \quad [K]_{u\theta} = \frac{6EI}{H^2}\begin{bmatrix} 1 & -1 \\ 1 & 1 \end{bmatrix} \quad [K]_{\theta\theta} = \frac{2EI}{H}\begin{bmatrix} 8 & 1 \\ 1 & 4 \end{bmatrix}$$

With this partitioning the force–displacement relationship of the system is given by two equations of size two as opposed to a single equation of size four, and the desired displacements and rotations may be determined as follows. First, $\{\theta\}$ is solved from the lower part of the equilibrium equation, to obtain

$$\{\theta\} = -[K]_{\theta\theta}^{-1}[K]_{u\theta}^T \{u\}_L$$

Then, this equation is substituted into the upper part of the equilibrium equation, leading to

$$\{F_s\} = [K]_{uu}\{u\}_L + [K]_{u\theta}\{\theta\} = ([K]_{uu} - [K]_{u\theta}[K]_{\theta\theta}^{-1}[K]_{u\theta}^T)\{u\}_L = [K]_L\{u\}_L$$

where $[K]_L$ is defined as

$$[K]_L = [K]_{uu} - [K]_{u\theta}[K]_{\theta\theta}^{-1}[K]_{u\theta}^T$$

and represents the lateral stiffness of the structure, as it involves only lateral displacements. Thereafter one solves for the nodal lateral displacements from the equation that defines $\{F_S\}$ in terms of $\{u_L\}$ and substitutes the result into the equation that relates $\{\theta\}$ and $\{u_L\}$ to obtain the nodal rotations. Performing all the required matrix inversions and multiplications and proceeding accordingly, one obtains

$$[K]_{\theta\theta}^{-1} = \frac{H}{62EI}\begin{bmatrix} 4 & -1 \\ -1 & 8 \end{bmatrix}$$

$$[K]_{\theta\theta}^{-1}[K]_{u\theta}^T = \frac{H}{62EI}\frac{6EI}{H^2}\begin{bmatrix} 4 & -1 \\ -1 & 8 \end{bmatrix}\begin{bmatrix} 1 & 1 \\ -1 & 1 \end{bmatrix} = \frac{3}{31H}\begin{bmatrix} 5 & 3 \\ -9 & 7 \end{bmatrix}$$

$$[K]_{u\theta}[K]_{\theta\theta}^{-1}[K]_{u\theta}^T = \frac{6EI}{H^2}\frac{3}{31H}\begin{bmatrix} 1 & -1 \\ 1 & 1 \end{bmatrix}\begin{bmatrix} 5 & 3 \\ -9 & 7 \end{bmatrix} = \frac{36EI}{31H^3}\begin{bmatrix} 7 & -2 \\ -2 & 5 \end{bmatrix}$$

$$[K]_L = [K]_{uu} - [K]_{u\theta}[K]_{\theta\theta}^{-1}[K]_{u\theta}^T = \frac{12EI}{H^3}\begin{bmatrix} 3 & -1 \\ -1 & 1 \end{bmatrix} - \frac{36EI}{31H^3}\begin{bmatrix} 7 & -2 \\ -2 & 5 \end{bmatrix} = \frac{12EI}{31H^3}\begin{bmatrix} 72 & -25 \\ -25 & 16 \end{bmatrix}$$

$$[K]_L^{-1} = \frac{H^3}{204EI}\begin{bmatrix} 16 & 25 \\ 25 & 72 \end{bmatrix}$$

$$\{u\}_L = \begin{Bmatrix} u_1 \\ u_2 \end{Bmatrix} = [K]_L^{-1}\{F_s\} = \frac{H^3}{204EI}\begin{bmatrix} 16 & 25 \\ 25 & 72 \end{bmatrix}\begin{Bmatrix} 1 \\ 1/2 \end{Bmatrix}\frac{V_0}{6} = \frac{V_0 H^3}{1224EI}\begin{Bmatrix} 66 \\ 169 \end{Bmatrix} = \begin{Bmatrix} 0.05392 \\ 0.13087 \end{Bmatrix}\frac{V_0 H^3}{EI}$$

$$\{\theta\} = \begin{Bmatrix} \theta_1 \\ \theta_2 \end{Bmatrix} = -[K]_{\theta\theta}^{-1}[K]_{u\theta}^T\{u\}_L = -\frac{3}{31H}\begin{bmatrix} 5 & 3 \\ -9 & 7 \end{bmatrix}\begin{Bmatrix} 66 \\ 169 \end{Bmatrix}\frac{V_0 H^3}{1,224EI}$$

$$= -\frac{V_0 H^2}{12,648EI}\begin{Bmatrix} 837 \\ 589 \end{Bmatrix} = -\begin{Bmatrix} 0.06618 \\ 0.04657 \end{Bmatrix}\frac{V_0 H^2}{EI}$$

Structural Response by Equivalent Lateral Force Procedure

Shearing Forces and Bending Moments

Once the nodal displacements and rotations are obtained, one can determine the internal forces in the structural elements from the force-displacement relationships defined by the elemental stiffness matrices. Accordingly, the bending moments at the ends of the beam of the first story are given by

$$[M_b]_C = [K]_C\{\theta_1\} = \frac{EI}{H^3}[4H^2]\{-0.06618\}\frac{V_0 H^2}{EI} = -0.26470 V_0 H$$

Similarly, the shearing forces and bending moments at the columns of the first story are equal to

$$\begin{Bmatrix} V_{c1} \\ M_{c1} \\ V_{c2} \\ M_{c2} \end{Bmatrix} = [K]_A \begin{Bmatrix} 0 \\ 0 \\ u_1 \\ \theta_1 \end{Bmatrix} = \frac{EI}{H^3}\begin{bmatrix} 24 & -12H & -24 & -12H \\ -12H & 8H^2 & 12H & 4H^2 \\ -24 & 12H & 24 & 12H \\ -12H & 4H^2 & 12H & 8H^2 \end{bmatrix} \begin{Bmatrix} 0 \\ 0 \\ 66 V_0 H^3/1{,}224 EI \\ -837 V_0 H^2/12{,}648 EI \end{Bmatrix} = \begin{Bmatrix} -0.5 \\ 0.38235 H \\ 0.5 \\ 0.11765 H \end{Bmatrix} V_0$$

where V_{c1} and M_{c1} and V_{c2} and M_{c2} denote the shearing forces and bending moments at the lower and upper ends of the columns, respectively.

EXAMPLE 12.3 SHEARING FORCES IN AN ASYMMETRIC TWO-STORY SHEAR-WALL STRUCTURE BY EQUIVALENT LATERAL FORCE PROCEDURE

A two-story shear-wall structure has the dimensions and configuration shown in Figure E12.3a. Each story is 10 ft high and the thickness of all walls is 6 in. The floor weights are 250 and 200 kips for the first and second levels, respectively. Determine using the equivalent lateral force procedure the shearing forces exerted on the longitudinal shear walls of the first and second stories when the structure is subjected to a ground motion along its longitudinal direction represented by a spectral acceleration of 0.7g. Assume the lateral stiffness of the shear walls is given by GA/h, where G is the shear modulus of elasticity, A the cross-sectional area, and h the height.

Solution

Lateral Forces and Story Shears

According to Equation 12.16 and for a spectral acceleration of 0.7g, the base shear for this structure is

$$V_0 = W \frac{SA(\omega_1, \xi_1)}{g} = (250 + 200)0.7 = 315 \text{ kips}$$

Thus, since in this case

$$\sum_{j=1}^{2} w_j h_j = 250(10) + 200(20) = 6500 \text{ kip ft}$$

Equation 12.19 leads to the following equivalent lateral forces:

$$F_{s1} = \frac{w_1 h_1}{\sum_{j=1}^{2} w_j h_j} V_0 = \frac{250(10)}{6500}(315) = 121 \text{ kips}$$

$$F_{s2} = \frac{w_2 h_2}{\sum_{j=1}^{2} w_j h_j} V_0 = \frac{200(20)}{6500}(315) = 194 \text{ kips}$$

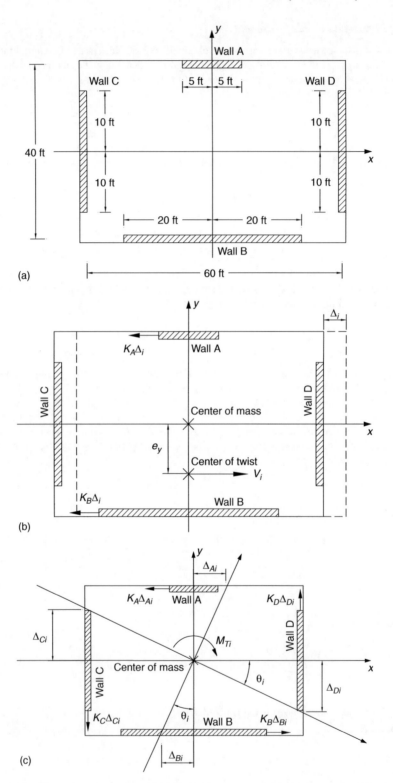

FIGURE E12.3 (a) Floor plan of two-story shear-wall structure considered in Example 12.3, (b) shear forces in shear walls of ith story due to uniform lateral deformation Δ_i, and (c) shear forces and displacements in shear walls of ith story due to torsional moment M_{Ti}.

which in turn lead to the following story shears:

$$V_2 = F_{s2} = 194 \text{ kips}$$

$$V_1 = F_{s2} + F_{s1} = 194 + 121 = 315 \text{ kips}$$

Note that it may be assumed without much error that the structure has a symmetric mass distribution given that the difference between the masses of walls A and B is not large enough compared to the floor masses to deviate appreciably the floors' centers of mass from their geometric centers. In consequence, it may be considered that the two lateral forces and the two-story shears pass through the geometric center of the structure.

Centers of Twist and Torsional Moments

Because of the unequal length of shear walls A and B, the structure is asymmetric in rigidity and thus it will be subjected to translational and torsional motions. To determine the shear force resisted by each shear wall, it is therefore necessary to first determine the location of the centers of twist and the magnitude of the acting torsional moments. To this end, consider that the lateral rigidities of the shear walls are given by

$$K_A = G \frac{A_A}{h_A} = \frac{10(0.5)}{10} G = 0.5G \text{ kip ft}$$

$$K_B = G \frac{A_B}{h_B} = \frac{40(0.5)}{10} G = 2.0G \text{ kip ft}$$

$$K_C = K_D = G \frac{A_C}{h_C} = \frac{20(0.5)}{10} G = 1.0G \text{ kip ft}$$

Consider, in addition, that when the ith story is subjected to a uniform lateral deformation Δ_i in the direction of the x-axis (see Figure E12.3b), from the definition of center of twist, the resultant V_i of the shear forces resisted by the shear walls of the story passes through, precisely, its center of twist. Furthermore, consider that this center of twist is located over the y-axis (by symmetry) at a distance e_y from the x-axis. Thus, from the equilibrium of forces in the x-direction one has that

$$V_i - K_A \Delta_i - K_B \Delta_i = V_i - (K_A + K_B)\Delta_i = 0$$

from which one obtains

$$V_i = (K_A + K_B)\Delta_i$$

Similarly, the equilibrium of moments about the ith floor's center of mass yields

$$V_i e_y + K_A \Delta_i (20) - K_B \Delta_i (20) = V_i e_y + 20(K_A - K_B)\Delta_i = 0$$

from which one finds that the eccentricity e_y is equal to

$$e_y = \frac{20(K_B - K_A)\Delta_i}{V_i} = \frac{20(K_B - K_A)}{K_A + K_B} = \frac{20(2 - 0.5)G}{2.5G} = 12 \text{ ft}$$

As a result, the torsional moments acting on the second and first stories are

$$M_{T2} = F_{s2} e_y = 194(12) = 2328 \text{ kip ft}$$

$$M_{T1} = F_{s2} e_y + F_{s1} e_y = (194 + 121)12 = 3780 \text{ kip ft}$$

Shears Due to Pure Translational Motion

As discussed in Section 12.3, it may be considered that each story is subjected to a story shear applied at its center of twist and a torsional moment. Considering, then, the story shear first, and with reference to Figure E12.3b, note that from the equilibrium of forces in the x-direction established earlier, for any of the two stories

$$\Delta_i = \frac{V_i}{K_A + K_B}$$

Hence, the shearing forces in the shear walls of the second and first stories due to the story shear alone are

$$V'_{A2} = K_A \Delta_2 = \frac{K_A}{K_A + K_B} V_2 = \frac{0.5}{0.5 + 2.0} 194 = 38.8 \text{ kips}$$

$$V'_{B2} = K_B \Delta_2 = \frac{K_B}{K_A + K_B} V_2 = \frac{2.0}{0.5 + 2.0} 194 = 155.2 \text{ kips}$$

$$V'_{A1} = K_A \Delta_1 = \frac{K_A}{K_A + K_B} V_1 = 0.2(315) = 63.0 \text{ kips}$$

$$V'_{B1} = K_B \Delta_1 = \frac{K_B}{K_A + K_B} V_1 = 0.8(315) = 252.0 \text{ kips}$$

Shears Due to Torsional Motion Alone

Consider now only the torsional moments. With reference to Figure E12.3c and from the equilibrium of moments for any of the two stories, one has that

$$M_{Ti} = K_A \Delta_{Ai} y_A + K_B \Delta_{Bi} y_B + K_C \Delta_{Ci} x_C + K_D \Delta_{Di} x_D$$

where y_A, y_B, x_C, and x_D denote the distances from the center of mass to the centers of gravity of walls A, B, C, and D, respectively. However, it may be observed from Figure E12.3c that

$$\Delta_{Ai} = \theta_i y_A \qquad \Delta_{Bi} = \theta_i y_B$$

$$\Delta_{Ci} = \theta_i x_C \qquad \Delta_{Di} = \theta_i x_D$$

where θ_i is the angle of rotation produced by the torsional moment M_{Ti}. Consequently, the preceding equilibrium equation may also be expressed as

$$M_{Ti} = K_A y_A^2 \theta_i + K_B y_B^2 \theta_i + K_C x_C^2 \theta_i + K_D x_D^2 \theta_i$$

from which one finds that

$$\theta_i = \frac{M_{Ti}}{K_A y_A^2 + K_B y_B^2 + K_C x_C^2 + K_D x_D^2}$$

$$= \frac{1}{G} \frac{M_{Ti}}{0.5(20)^2 + 2.0(20)^2 + 1.0(30)^2 + 1.0(30)^2}$$

$$= \frac{1}{G} \frac{M_{Ti}}{2800}$$

Thus, the shearing forces in the shear walls of the second and first stories due to the torsional moments alone are

$$V''_{A2} = K_A \Delta_{A2} = K_A y_A \theta_2 = \frac{K_A y_A}{2800G} M_{T2} = \frac{0.5(20)}{2800} 2328 = 8.3 \text{ kips}$$

$$V''_{B2} = K_{B2} \Delta_{B2} = K_B y_B \theta_2 = \frac{K_B y_B}{2800G} M_{T2} = \frac{2.0(20)}{2800} 2328 = 33.3 \text{ kips}$$

$$V''_{A1} = K_A \Delta_{A1} = K_A y_A \theta_1 = \frac{K_A y_A}{2800G} M_{T1} = \frac{0.5(20)}{2800} 3780 = 13.5 \text{ kips}$$

$$V''_{B1} = K_{B1} \Delta_{B1} = K_B y_B \theta_1 = \frac{K_B y_B}{2800G} M_{T1} = \frac{2.0(20)}{2800} 3780 = 54.0 \text{ kips}$$

Total Shears

Adding the shears determined above for the cases of story shears only and torsional moments only, the shearing forces exerted on the shear walls of the second and first stories are thus equal to (see Figures E12.3b and E12.3c)

$$V_{A2} = V'_{A2} + V''_{A2} = 38.8 + 8.3 = 47.1 \text{ kips}$$

$$V_{B2} = V'_{B2} - V''_{B2} = 155.2 - 33.3 = 121.9 \text{ kips}$$

$$V_{A1} = V'_{A1} + V''_{A1} = 63.0 + 13.5 = 76.5 \text{ kips}$$

$$V_{B1} = V'_{B1} - V''_{B1} = 252.0 - 54.0 = 198.0 \text{ kips}$$

EXAMPLE 12.4 SEISMIC ANALYSIS OF AN ASYMMETRIC BUILDING BY EQUIVALENT LATERAL FORCE PROCEDURE

Figure E12.4a shows a five-story building that is asymmetric in plan and elevation. The lateral force-resisting system of the building is formed with reinforced concrete moment-resisting frames along its x- and y-directions. The floor weights are as indicated in Figure E12.4a. The dimensions of the columns are 0.60 m × 0.60 m for the first and second stories, and 0.50 m × 0.50 m for the other stories. For all the stories, the beams along the x-direction have a width of 0.40 m and a depth of 0.60, whereas the beams along the y-direction have a width of 0.30 m and a depth of 0.50 m. The concrete used has a modulus of elasticity of 18,700 MN/m² and a Poisson ratio of 0.25. Determine using the equivalent lateral force procedure: (a) the equivalent lateral forces acting on the floors of the building parallel to the x-axis; (b) the lateral displacements along the x-direction of columns A, B, C, and D produced by these lateral forces at the level of the fourth floor; (c) the lateral displacements along the y-direction of columns D, E, F, and G produced by the same set of forces at the level of the fourth floor; and (d) the shear forces along the x- and y-directions on columns A, B, C, and D at the level of the first story. Consider a constant acceleration response spectrum with an ordinate of 0.32g.

Solution

Equivalent Lateral Forces

According to the given floor weights, the total weight of the building is 6900 kN. Thus, according to Equation 12.16 and a spectral acceleration of 0.32 g, the building's base shear is equal to

$$V_0 = W \frac{SA(\omega_1, \xi_1)}{g} = (6900)0.32 = 2208 \text{ kN}$$

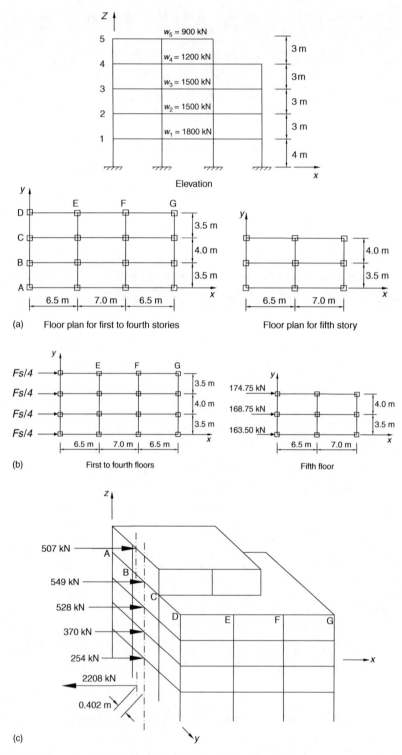

FIGURE E12.4 (a) Five-story building considered in Example 12.4, (b) applied nodal forces, and (c) location of equivalent lateral forces and first-story shear.

Structural Response by Equivalent Lateral Force Procedure

Hence, on the application of Equation 12.19 and organizing the calculations as shown in Table E12.4a, one arrives at the equivalent lateral forces listed in the last column of this table.

Fourth-Floor Lateral Displacements and First-Story Shearing Forces

The lateral displacements and internal forces produced by the lateral forces listed in Table E12.4a may be determined directly and without too many complications using any available computer program for the static analysis of frame structures. It should be noted, however, that the building under consideration is asymmetric and therefore the analysis needs to be performed using a three-dimensional model. It should also be noted that in many computer programs forces may be applied only at the structural nodes. Hence, the analysis is conducted using a computer program for the analysis of space frames considering the forces shown in Figure E12.4b. It may be readily verified that the applied nodal forces lead to resultants that are equal in magnitude to the lateral forces given in Table E12.4a and that these resultants pass through the floors' centers of mass (by symmetry the floors' centers of mass coincide in this case with their geometric centers). The resulting fourth-floor lateral displacements and first-story shear forces are given in Tables E12.4b and E12.4c.

TABLE E12.4a
Calculation of Equivalent Lateral Forces for Building in Example 12.4

Floor	W_j (kN)	h_j (m)	$W_j h_j$ (kN-m)	F_{sj} (kN)
5	900	16	14,400	507
4	1,200	13	15,600	549
3	1,500	10	15,000	528
2	1,500	7	10,500	370
1	1,800	4	7,200	254
Sum	6,900		62,700	2,208

TABLE E12.4b
Lateral Displacements at Level of Fourth Floor Induced by Lateral Forces in Table E12.4a

Column	Displacement x-Direction (mm)	Column	Displacement y-Direction (mm)
A	46.9	D	−3.1
B	48.4	E	−0.9
C	46.1	F	0.9
D	39.6	G	3.2

TABLE E12.4c
Shearing Forces in Columns of First Story Induced by Lateral Forces in Table E12.4a

Column	Shearing Force (kN)	
	x-Direction	y-Direction
A	123	6.5
B	134	7.5
C	136	7.8
D	113	6.9

The displacements listed in Table E12.4b clearly show that the building undergoes horizontal rotations under the effect of the applied lateral forces. Note that the lateral displacement of column A is larger than that of column D, indicating thus that the building's fourth floor experienced a counterclockwise rotation (when seen from above in Figure E12.4b) about a vertical axis. Note, also, that, as expected, lateral displacements are produced along the direction of the y-axis despite the fact that the lateral forces are applied along the direction of the x-axis. The displacements along the y-direction also show that the fourth floor underwent a counterclockwise rotation about a vertical axis. Similarly, the shear forces listed in Table E12.4c manifest the effects of a horizontal rotation of the first story and the fact that shearing forces are generated along the y-direction when the applied lateral forces are applied along the x-direction. Although the shearing forces acting perpendicular to the direction of the applied lateral forces are not significantly large in this example, it should be realized that these shearing forces may be quite large in the case of structures with large story eccentricities.

It is also worthwhile to observe that the detected floor rotations result from the torsional moments that are generated at each of the building's stories by the fifth-floor force, which, as shown in Figure E12.4c, acts at a point that is horizontally at a distance from the centers of mass of the first, second, third, and fourth floors. Thus, for example, there is an internal torsional moment generated at the level of the first story equal to

$$M_{T1} = 507(5.5 - 3.75) = 887.25 \text{ kN m}$$

which corresponds to the sum of the external torsional moments above the story and may be considered produced by the 2208 kN first-story shear, if it is assumed that its line of action passes through a point located at a horizontal distance from the first floor's center of mass (eccentricity) of (see Figure E12.4c)

$$e_1 = \frac{887.25}{2208} = 0.402 \text{ m}$$

12.4 LIMITATIONS OF PROCEDURE

In general, it can be said that the equivalent lateral force procedure is likely to lead to inaccurate results if the dynamic behavior of the analyzed buildings is substantially different from that implied in the derivation of the procedure. According to the assumptions introduced in Section 12.2, this means that, in general, the procedure will be inadequate for the analysis of tall buildings for which their higher modes significantly contribute to their response, buildings that possess sharp discontinuities in elevation, and asymmetric buildings for which their translational and rotational natural frequencies coincide or have similar values. These observations have been verified in comparative studies against solutions obtained by a dynamic analysis. In particular, it has been found that the equivalent lateral force procedure overestimates the base shear for short-period buildings and underestimates it for long-period buildings. This seems to suggest that by using the total weight of a building instead of the fundamental-mode effective weight in the calculation of the base shear is not sufficient to compensate for the higher-mode responses in tall buildings. Similarly, it has been found that the equivalent lateral force procedure yields a distribution of story shears along the height of a building that generally is in agreement with the results from a dynamic analysis. However, it tends to underestimate the story shears in the upper stories of long-period buildings. In contrast, it has been found that the equivalent lateral force procedure overpredicts the overturning moments, particularly at the lower stories of tall buildings. This overprediction occurs because in the equivalent force procedure the overturning moments are determined by statics as the sum of the moments produced by lateral forces that are supposed to represent maximum values, but it is unlikely that these maximum values will all occur at the same time. Other reasons are the fact that the magnitude of the lateral forces in many cases exceed the magnitude of the actual forces and the approximate nature of the distribution of the lateral forces with height. Finally, it has been found that the procedure may underestimate the response of asymmetric buildings whose translational and rotational natural frequencies are close in value. As discussed in Section 12.3, the reason for the underprediction is that in asymmetric buildings the floors are subjected to not only lateral forces but torsional moments as well, and that these torsional moments may be dynamically magnified as a result of the floors' rotational inertia. However, this rotational inertia and the consequent dynamic magnification of the torsional moments are not considered in the derivation of the procedure.

To overcome these limitations and extend the scope of its applicability, the equivalent lateral force procedure is ordinarily used in conjunction with some empirical corrections or adjustments. For example, the base shear is calculated on the basis of a design response spectra whose ordinates are arbitrarily increased in the long-period region to account for the significant contribution of the higher modes in the response of long-period structures. Similarly, the story shears of long-period buildings are calculated considering an additional force with a magnitude equal to a fraction of the base shear and applied at the top of the building to increase the story shears in the upper stories of these buildings. Alternatively, the lateral forces are calculated by assuming that in long-period structures the fundamental mode shape varies with height according to a power law as opposed to linearly as it is assumed in the conventional procedure. Other examples are the use of reduction factors to decrease the values of the overturning moments calculated using the lateral forces determined with the equivalent lateral force procedure, and the use of a dynamic magnification factor to augment the torsional moments determined on the basis of the aforementioned lateral forces and a static analysis. The corrections and adjustments recommended by the International Building Code in connection with the use of the equivalent lateral force procedure will be reviewed in Chapter 17.

12.5 NONLINEAR STATIC PROCEDURE

As most structures are allowed to incur into their nonlinear range of behavior when subjected to the design ground motions, a static procedure has also been developed to estimate the strength and deformation demands in structures when subjected to ground motions that drive them beyond their elastic range of behavior. This nonlinear static procedure is colloquially known as *pushover analysis* and has become a standard tool in engineering practice. It was originally formulated by the Applied Technology Council (ATC) in 1996 as a part of the Federal Emergency Management Agency (FEMA) project on "Guidelines for the Seismic Rehabilitation of Buildings." A current version appears in FEMA Publication No. 356 (listed in the Further Readings section). It is intended to provide information on structural behavior that cannot be obtained from an elastic analysis, such as member ductility demands and formation of failure mechanisms.

In the nonlinear static procedure, a structural model is first formulated considering explicitly the gravity loads and the nonlinear force–deformation behavior of the structural elements. Then, a base shear–lateral displacement relationship is established by subjecting this model to monotonically increasing lateral forces until the displacement of a control node (usually the center of mass of the building's roof) exceeds a target displacement or the structure collapses. These forces are applied with a prescribed distribution along the height of the structure. In this regard, building codes generally recommend the use of a uniform distribution (i.e., forces proportional to the floor masses), the pattern that arises from the application of the equivalent lateral force procedure or the response spectrum method, or both of these distributions hoping to bound the actual response. The target displacement is the maximum displacement likely to be experienced by the structure under the selected seismic hazard level, and may be determined from the linear response spectrum specified for the design of the structure. It may be calculated according to

$$\delta_t = C_0 C_1 C_2 C_3 S_a \frac{T_e^2}{4\pi^2} \tag{12.27}$$

where S_a, T_e, and C_0, C_1, C_2, and C_3, respectively, represent spectral acceleration, fundamental natural period of the structure, and empirical coefficients that relate the displacement response of a multistory nonlinear structure to that of a single-degree-of-freedom linear system.

The demands at this target displacement (element forces, story drifts, or joint rotations) are then compared against a series of prescribed acceptable values. These acceptable values depend on the construction material (steel, reinforced concrete, etc.), member type (beam, column, etc.), member importance (primary or secondary), and a preselected performance level (operational, immediate occupancy, life safety, or collapse prevention) and are normally specified by building codes based on experimental or analytical evidence. A global collapse is assumed to occur whenever the base shear–roof displacement curve attains a negative slope (due to P-delta effects) and reaches afterward a point of zero base shear, as this point implies no lateral resistance and the inability of the structure to resist gravity loads. The performance of the structure is considered satisfactory if the member forces and deformations are all within the available capacities and specified limits and no failure mechanisms are formed.

To illustrate the application of the procedure, consider the six-story frame shown in Figure 12.8a. To evaluate the seismic performance of the frame, an equivalent lateral load pattern and a target

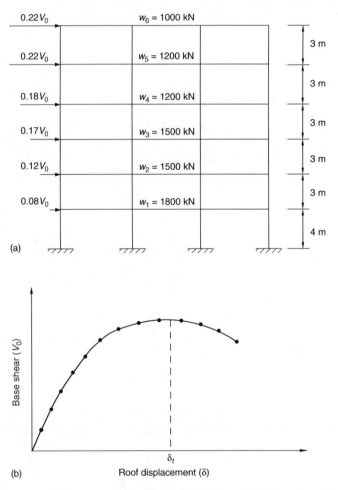

FIGURE 12.8 (a) Six-story frame under lateral forces derived from equivalent lateral force procedure and (b) base shear–roof displacement curve obtained from pushover analysis.

displacement are selected. Then, the structure is analyzed statically under increasing levels of the base shear V_0 until the displacement of the roof is equal to or greater than the selected target displacement. For the case under consideration, the selected load pattern (shown in Figure 12.8a) corresponds to the distribution of lateral forces obtained with Equation 12.19. Similarly, the selected target displacement (in this case equal to 31 cm) is calculated with Equation 12.27 considering a spectral acceleration of $2.5g$, a fundamental natural period of 0.6 s, and coefficients C_0, C_1, C_2, and C_3 equal to, as specified in the aforementioned FEMA publication, 1.4, 1.0, 1.0, and 1.0, respectively. The various values of the base shear are thereafter plotted against the values of the roof displacement induced by the corresponding sets of lateral forces to obtain the curve shown in Figure 12.8b. Once the set of lateral forces that leads to the target roof displacement is identified, the member forces and deformations (e.g., joint rotations) induced by this set of lateral forces are compared against the available member capacities and specified limits. From this comparison, it is found, for example, that plastic hinges form at ends of the beams and columns marked with a solid circle in Figure 12.9 when the frame is subjected to the set of lateral forces that induce the target roof displacement. This information permits an assessment of where damage is likely to occur and whether or not a collapse mechanism is likely.

Although simpler than a true nonlinear dynamic analysis, the procedure lacks, unfortunately, a rigorous theoretical foundation. It is based on the incorrect assumptions that the nonlinear response of a multistory structure can be related to the response of an equivalent single-degree-of-freedom system, and that the distribution over the height of the structure of the equivalent lateral forces remains constant during the entire duration of the structural response. It neglects duration and cyclic effects—the progressive changes in the dynamic properties that take place in a structure as it experiences yielding and unloading during an earthquake, the fact that nonlinear structural behavior is load–path dependent, and the fact that the deformation demands on structural elements depend on ground motion characteristics. In fact, in correlations with observed damage in several of the buildings damaged during the 1994 Northridge earthquake and comparisons with results

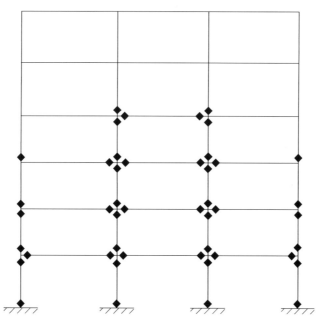

FIGURE 12.9 Distribution of plastic hinges (identified with solid squares) when the frame is subjected to lateral forces that induce target roof displacement.

from nonlinear time-history analyses, several investigators have found that the procedure does not provide an accurate assessment of building behavior. It may lead to gross underestimation of story drifts and may fail to identify correctly the location of formed plastic hinges. This is particularly true for structures that deform far into their inelastic range of behavior and undergo a significant degradation in lateral capacity. In recognition of these deficiencies, improved nonlinear static procedures have been proposed. Among these improved procedures are those that use a time-variant distribution of the equivalent lateral forces and those that consider the contribution of "higher modes." These improved procedures lead to better predictions in some cases, but none has been proven universally applicable. It is doubtful, thus, that nonlinear static methods may be used reliably to evaluate structural performance.

FURTHER READINGS

1. Anderson, A. W., Blume, J. A., Degenkolb, H. B., Hammill, H. B., Knapik, E. M., Marchand, H. L., Powers, H. C., Rinne, J. E., Sedgwick, G. A., and Sjoberg, H. O., Lateral forces of earthquake and wind, *Transactions ASCE*, 117, 1952, 716–754.
2. Chopra, A. K., *Dynamics of Structures: Theory and Applications to Earthquake Engineering*, 2nd Edition, Chapter 21, Prentice-Hall, Upper Saddle River, NJ, 2001, pp. 755–782.
3. Chopra, A. K. and Newmark, N. M., Analysis, Chapter 2 in *Design of Earthquake Resistant Structures*, Rosenblueth, E. ed., Pentech Press, London, 1980, pp. 27–53.
4. Federal Emergency Management Agency, *Prestandard and Commentary for the Seismic Rehabilitation of Buildings*, Publication No. 356, Prepared by the American Society of Civil Engineers for the Federal Emergency Management Agency, Washington, DC, 2000.
5. Newmark, N. M. and Rosenblueth, E., *Fundamentals of Earthquake Engineering*, Chapter 15, Prentice-Hall, Englewoods Cliffs, NJ, 1971, pp. 477–530.

PROBLEMS

12.1 Solve Problem 10.1 using the equivalent lateral force procedure.

12.2 Solve Problem 10.2 using the equivalent lateral force procedure.

12.3 Determine using the equivalent lateral force procedure the axial forces, shearing forces, and bending moments of the columns of the first story of the frame building shown in Figure P12.3. The floor weights of the building are as indicated in Figure P12.3. The building will be designed to resist ground motions represented by the 2%- damping design spectrum shown in Figure P10.2. Assume that the shearing forces at the interior columns are twice those carried by the exterior columns and that the four columns have the same cross-sectional area.

12.4 A two-story shear-wall structure has the dimensions and configuration shown in Figure P12.4. Each story is 12 ft high and the thickness of all walls is 8 in. The floor weights are 350 and 250 kips for the first and second levels, respectively. Determine using the equivalent lateral force procedure the shearing forces exerted on the longitudinal shear walls of the first story when the structure is subjected to a ground motion along its longitudinal direction represented by the 5%-damping design spectrum shown in Figure P10.2. Assume the lateral stiffness of shear walls is equal to GA/h, where G is the shear modulus of elasticity, A the cross-sectional area, and h the height.

12.5 As shown in Figure P12.5, a building has four stories, four bays 4 m wide along its x-direction, and three bays also 4 m wide along its y-direction. The structure of the building is constructed with steel moment-resisting frames along both of these

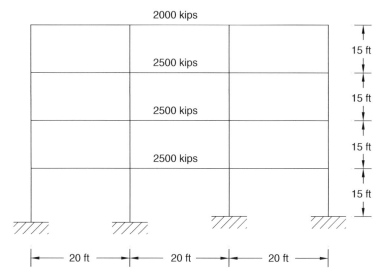

FIGURE P12.3 Frame building considered in Problem 12.3.

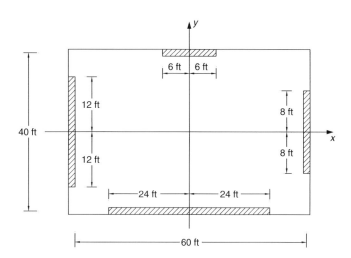

FIGURE P12.4 Floor plan of two-story shear-wall structure considered in Problem 12.4.

directions. The four frames along the x-direction are all identical to the one shown in Figure P12.5ab. These frames are in turn connected at each floor level by beams parallel to the y-axis. The moments of inertia of the columns about their two principal axes are as indicated in Figure P12.5a, where $I = 0.750 \times 10^{-4}$ m^4. The moments of inertia of all beams, longitudinal and transverse, are equal to $2I$. The modulus of elasticity of the steel used in the construction is 200,000 MN/m^2. The floor weights are also as indicated in Figure P12.5a. Determine using the equivalent lateral force procedure: (a) the equivalent lateral forces acting on the floors of the building along the direction of the y-axis; (b) the lateral displacements induced by these lateral forces in columns D and H along the y-direction at the level of the second floor; (c) the lateral displacements

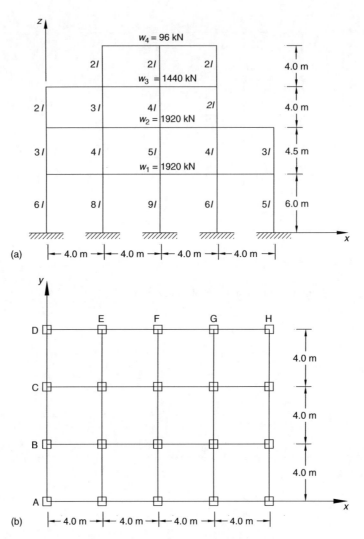

FIGURE P12.5 (a) Elevation and (b) plan views of four-story building considered in Problem 12.5.

generated by the same set of forces in columns A and D along the *x*-direction, also at the level of the second floor; and (d) the shear forces along the *x*- and *y*-directions on columns D and H at the level of the first story. The building will be designed to resist ground motions represented by an acceleration response spectrum constructed utilizing the spectral shape corresponding to soil type 2 in Figure 8.17 and a peak ground acceleration of $0.40g$.

13 Structural Response Considering Soil–Structure Interaction

> With the methods of analysis and the computer programs now available, it is in principle possible to evaluate the dynamic response of any structure–foundation system to any excitation of the base. Such evaluations, however, can be no better than the assumptions underlying the analyses.
>
> **Anestis S. Veletsos, 1992**

13.1 PROBLEM DEFINITION

In the analysis of structures subjected to earthquake ground motions, it is common to assume that the motion that excites the base of a structure is the *free-field* ground motion; that is, the motion that would be observed at the site where the structure stands if the structure were not there. In actuality, a structure always interacts to some extent with its supporting soil during earthquakes. Such an assumption is therefore justified only for structures founded on rock or stiff soil, in which case the high stiffness of the rock or stiff soil would constrain the motion at the base of the structure to be close to the motion observed in the free field. For structures supported on soft soil, however, the foundation motion is in general different from the free-field motion and may include an important rocking component. The reason is that the foundation motion is affected by the change in the geometry and properties of the wave propagation medium introduced by the structure's foundation, the mechanical coupling between the soil and the structure, and the dissipation of energy by radiation and hysteretic damping in the supporting medium—effects that would not exist if the structure were supported by a perfectly rigid medium.

The problem of soil–structure interaction may be thus stated in a simplistic way as follows: If the ground motion induced by an earthquake is known at a site where no structure is present, what would be the characteristics of the ground motion observed at the same site when a structure is present, and what would be the response of the structure to such ground motion? This concept is illustrated graphically in Figure 13.1. Observe from Figure 13.1 that, for the same earthquake, the ground motion at the base of the structure would in general be different from that in the free field, and that the response of the structure will depend on the ground motion that excites the base of the structure, not the ground motion observed in the free field. Soil–structure interaction may be therefore defined as the interdependence between a structure and its foundation soil; that is, the dependence of the dynamic response of a structure on the dynamic response of its supporting soil and the dependence of the dynamic response of a soil deposit on the dynamic response of the structure supported by it.

The effect of soil–structure interaction or soil flexibility on the seismic response of buildings was addressed by R. R. Martel as early as 1940.[*] Ever since, the knowledge and understanding of the phenomenon has grown at a steady pace, with considerable advances made in the 1970s in response to the need to ensure the seismic safety of the large number of nuclear power plants being

---

[*] Martel, R. R., Effect of foundation on earthquake motion, *Civil Engineering*, 10, 1940, 7–10.

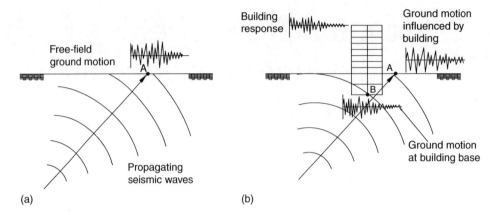

FIGURE 13.1 Schematic representation of ground motion in the (a) absence and (b) presence of a structure.

constructed during that time. At this point in time, it has become clear that soil-interaction effects might not be important for regular, flexible buildings built on rock or stiff soil, but that they could be significant for stiff and massive structures such as nuclear power plants built on soft soils. It has also become clear that the significance of soil–structure interaction effects will depend on the characteristics of the structure and the foundation soil being considered, and that these effects should be evaluated on a case-by-case basis.

A rigorous analysis of soil–structure interaction is a lengthy and complicated process that requires an estimation of key parameters that are difficult to determine, as well as the use of techniques that are by far too intricate to be covered in a nonspecialized textbook like this. This chapter, therefore, only provides a basic description of the phenomenon, some background material that facilitates the comprehension of this phenomenon, an overview of the methods that have been used in the past to account for it, and the introduction of an approximate procedure that can be used to account for soil–structure interaction effects in a simple way. In addition, a simple criterion will be introduced for defining the conditions under which interaction effects are of sufficient importance to warrant consideration in the design of a structure.

13.2 KINEMATIC AND INERTIAL INTERACTION

In accordance with the concepts introduced in Section 13.1, there are two interaction effects that take place as a result of the presence of a structure at a soil site. These two effects are referred to as *kinematic* and *inertial* effects. Kinematic interaction is the effect that occurs as a result of the change in the wave propagation medium in the form of a different density and elasticity, which in turn induces a change in the wave propagation velocity and the reflection and refraction of the incoming seismic waves (scattering effect) as these waves reach the soil–foundation interface. It represents the difference between the structural response that is obtained when this response is computed on the basis of the free-field ground motion and that computed on the basis of the ground motion at the base of the structure when the presence of the structure is considered. It takes place independently of whether or not the structure has any mass and is affected by the geometry and configuration of the structure, the foundation embedment, the composition of the incident free-field waves (e.g., shear or surface waves), and the angle of incidence of these waves. In particular, note that there is no kinematic interaction for foundations built directly on the ground surface and subjected to vertically propagating S waves (see Figure 13.2).

Inertial interaction, however, is the effect that results from the dynamic coupling between a structure and its supporting medium. That is, since a foundation soil or rock has elastic and inertial properties itself, a structure and its supporting medium respond to incoming waves as a single

Structural Response Considering Soil–Structure Interaction

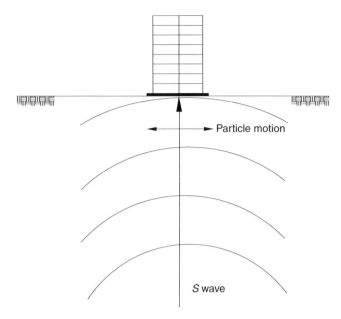

FIGURE 13.2 Building with surface foundation under vertically propagating shear waves.

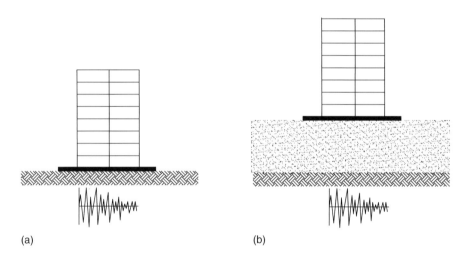

FIGURE 13.3 Schematic representation of structure (a) with a fixed base and (b) on soft soil.

dynamic system. In other words, the flexibility of the supporting medium increases the number of degrees of freedom of the system and makes possible the dissipation of part of the vibrational energy of the structure by the radiation of waves away from the structure and by the hysteretic deformation of the medium itself. Therefore, the ground motion at the base of the structure and the response of the structure to this ground motion are the result of the characteristics of the incoming waves and the dynamic properties of such a soil–structure system.

The inertial interaction effect may be illustrated with the help of Figure 13.3. If the flexibility and energy dissipation capability of a supporting medium is accounted for, then it may be seen from Figure 13.3 that the system that would respond to a given ground motion in the case of the structure supported by the flexible medium would be substantially different from the system that would respond to the same ground motion when the same structure is considered fixed at its base.

It may also be seen that such an interaction effect is affected by the stiffness of the supporting medium relative to the stiffness of the structure. Thus, for example, the inertial interaction effect would be negligible in structures founded on rock or stiff soil since in such a case the response of the structure would be virtually the same as the response that would be obtained considering its base fixed. In contrast, the interaction effect would be highly significant for a structure founded on soft soil. The reason is that in this latter case the response of the structure would approach that of a rigid body supported by a flexible medium. The inertial interaction effect represents, thus, the difference between the structural response when the flexibility of the foundation medium is and is not considered.

It is also important to note that the presence of a deformable soil will make the structure more flexible, decreasing its fundamental natural frequency to a value that will be, in general, below that of the fixed-base structure. The shape of the vibrational modes will also change. Because of the deformability of the soil, the structure will experience rocking motion and this rocking motion will increase the displacements of the structure, particularly at the top.

13.3 VIBRATION OF FOUNDATIONS ON ELASTIC HALFSPACE

13.3.1 Introductory Remarks

A convenient way to represent the global vertical, horizontal, rocking, and torsional stiffness and damping of a foundation soil is by means of an elastic halfspace. Although most soils cannot be modeled accurately with an elastic halfspace, it is nevertheless instructive to learn how to obtain the force–deformation relationship for an elastic halfspace and how to use this force–deformation relationship in a practical soil–structure interaction analysis. From the practical point of view perhaps the greatest value of the elastic halfspace model is the light that it sheds in the identification of the geometric and soil parameters that have a significant influence on the interaction between a structure and its supporting soil. This section is thus devoted to describe the elastic halfspace model. The presentation will follow the chronological order in which the theory was developed, with the purpose of facilitating a thorough understanding of the concepts that are nowadays an essential part of the soil–structure interaction field.

13.3.2 Displacements under Static Point Loads

With reference to Figure 13.4a, consider a semi-infinite elastic and homogeneous halfspace subjected to a vertical point load Q at its surface. The boundary conditions for this problem are (a) normal stresses along the direction of the z-axis are all equal to zero at $z = 0$ and $z = \infty$, except under the load where such a stress is equal to infinity and (b) all shearing stresses at $z = 0$ are equal to zero.

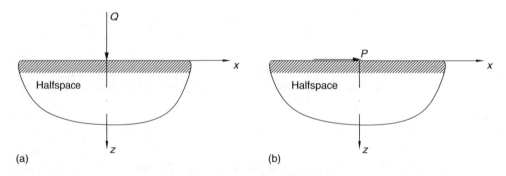

FIGURE 13.4 (a) Vertical and (b) horizontal point loads on the surface of elastic halfspace.

Solving the equations of equilibrium for an elastic solid in conjunction with these boundary conditions, J. Boussinesq found in 1883 that the vertical displacement induced by such a load is given by

$$w = \frac{Q}{4\pi GR}\left[2(1-\mu) + \frac{z^2}{R^2}\right] \quad (13.1)$$

where G denotes the halfspace's shear modulus of elasticity and μ is the corresponding Poisson ratio,

$$R = \sqrt{r^2 + z^2} \quad (13.2)$$

and r and z are the cylindrical coordinates of a point in the three-dimensional medium.

On the basis of Equation 13.1, thus one has that at any point on the surface (i.e., $z = 0$), the vertical displacement induced by the force Q is equal to

$$w\big|_{z=0} = \frac{Q(1-\mu)}{2\pi Gr} \quad (13.3)$$

Similarly, V. Cerruti found in 1927 that the horizontal displacement at any point of the halfspace induced by a horizontal point load P in the direction of the x-axis at the surface of the halfspace (see Figure 13.4b) is given by

$$u = \frac{P}{4\pi GR}\left\{1 + \frac{x^2}{R^2} + (1-2\mu)\left[\frac{R}{R+z} - \frac{x^2}{(R+z)^2}\right]\right\} \quad (13.4)$$

where x is the coordinate in the direction of the x-axis and all other symbols are as previously defined. From Equation 13.4, the horizontal displacement at a point on the surface of the halfspace results thus is equal to

$$u\big|_{z=0} = \frac{P}{4\pi Gr}\left[1 + \frac{x^2}{r^2} + (1-2\mu)\left(1 - \frac{x^2}{r^2}\right)\right] \quad (13.5)$$

which, if it is considered that $x = r \cos \theta$, θ being the angular coordinate in a cylindrical system of coordinates, may also be expressed as

$$u\big|_{z=0} = \frac{1 - \mu \sin^2 \theta}{2\pi Gr} P \quad (13.6)$$

13.3.3 Displacements and Rotations under Static Distributed Loads

Loads are seldom transmitted to the ground as point loads. They are transmitted, rather, through a foundation block or footing that distributes the load over the area of the footing that is in contact with the ground. It should be noted, however, that the shape of such a distributed load depends on the rigidity of the footing and thus it is not always uniform. Explicit expressions to compute the displacements and rotations induced by distributed loads have been obtained for the load distributions shown in Figure 13.5. These load distributions represent the soil pressure under circular footings that are perfectly rigid, perfectly flexible, and those between these two extreme cases. Figure 13.5a depicts distributed loads that may be induced by vertical and horizontal forces and Figure 13.5b those by rocking or torsional moments. The equations that describe the load per unit area in the distributions shown in Figure 13.5a are

$$q = \frac{Q}{2\pi r_0 \sqrt{r_0^2 - r^2}} \quad \text{for rigid footings} \quad (13.7)$$

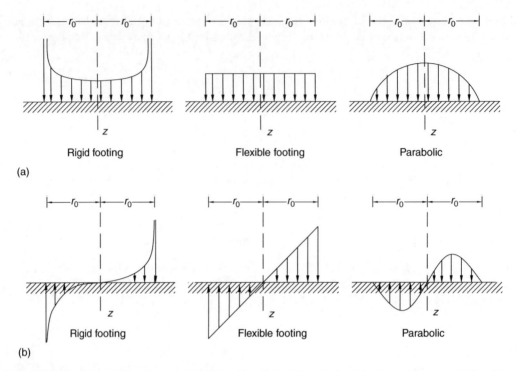

FIGURE 13.5 Three different types of distributed loads for (a) vertical and horizontal forces and (b) rocking and torsional moments.

$$q = \frac{Q}{\pi r_0^2} \text{ for flexible footings} \quad (13.8)$$

$$q = \frac{2(r_0^2 - r^2)Q}{\pi r_0^4} \text{ for parabolic distribution} \quad (13.9)$$

where Q and r_0 are resultant of the distributed load and radius of the loaded area.

Similarly, the equations that define the load per unit area for the distributions in Figure 13.5b, when the distributed load is generated by a moment about the y-axis (perpendicular to the plane of Figure 13.5b), are

$$q = \frac{3r\cos\theta}{2\pi r_0^3 \sqrt{r_0^2 - r^2}} M_y \text{ for rigid footings} \quad (13.10)$$

$$q = \frac{4r\cos\theta}{\pi r_0^4} M_y \text{ for flexible footings} \quad (13.11)$$

$$q = \frac{12r(r_0^2 - r^2)\cos\theta}{\pi r_0^6} M_y \text{ for parabolic distribution} \quad (13.12)$$

whereas the corresponding equations that define the load per unit area generated by a moment about the z-axis have the form

$$q = \frac{3r}{4\pi r_0^3 \sqrt{r_0^2 - r^2}} M_z \text{ for rigid footings} \tag{13.13}$$

$$q = \frac{2r}{\pi r_0^4} M_z \text{ for flexible footings} \tag{13.14}$$

$$q = \frac{6r(r_0^2 - r^2)}{\pi r_0^6} M_z \text{ for parabolic distribution} \tag{13.15}$$

In the preceding equations, r_0 is, as mentioned earlier, the radius of the circular loaded area and Q, M_y, and M_z denote the resultant forces or moments of the distributed loads.

The displacements and rotations induced at the surface of a halfspace by a distributed load may be obtained by considering the load on a differential element of the loaded area as a point load, and by integrating Boussinesq's or Cerruti's solution for point loads over the entire loaded area (see Figure 13.6). Proceeding accordingly, let q represent load per unit area, $r\,dr\,d\theta$ be the area in polar coordinates of such a differential element, and r_0 denote the radius of the loaded area. Consider, then, a point load whose magnitude is given by

$$Q = qr\,dr\,d\theta \tag{13.16}$$

Substitution of Equation 13.16 into Equation 13.3 (Boussinesq's solution) and integration over the entire loaded area thus lead to

$$w_0 = \int_0^{2\pi} \int_0^{r_0} \frac{1-\mu}{2\pi Gr} qr\,dr\,d\theta \tag{13.17}$$

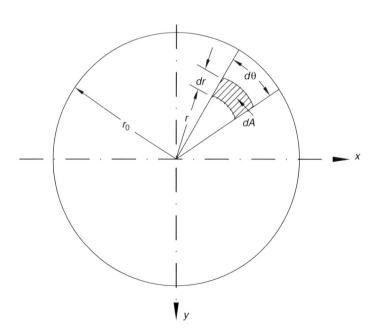

FIGURE 13.6 Circular loaded area and differential element of it.

from which, after substituting the load distributions defined by Equations 13.7 through 13.9 and performing the integration, it is found that the vertical displacement at the center of a circular distributed load acting on the surface of a halfspace is equal to

$$w_0 = \frac{1-\mu}{4r_0 G} Q \text{ for rigid footings} \qquad (13.18)$$

$$w_0 = \frac{1-\mu}{\pi r_0 G} Q \text{ for flexible footings} \qquad (13.19)$$

$$w_0 = \frac{4(1-\mu)}{3\pi r_0 G} Q \text{ for parabolic loads} \qquad (13.20)$$

Using Cerruti's solution (Equation 13.6) and proceeding similarly, it may be shown that the horizontal displacement at the center of a loaded area induced by a horizontal force P is given by

$$u_0 = \frac{2-\mu}{8 G r_0} P \text{ for rigid footings} \qquad (13.21)$$

$$u_0 = \frac{2-\mu}{2\pi G r_0} P \text{ for flexible footings} \qquad (13.22)$$

$$u_0 = \frac{2(2-\mu)}{3\pi G r_0} P \text{ for parabolic loads} \qquad (13.23)$$

When a vertical force Q acts at the origin of the system of coordinates in the halfspace shown in Figure 13.4a, the vertical displacement at a surface point with coordinates (r, θ) is given by Equation 13.3. The rotational angle at that point in the direction of the line that joins the points $(0, 0)$ and (r, θ) is therefore given by

$$\phi_r = \left.\frac{\partial w}{\partial r}\right|_{z=0} = -\frac{1-\mu}{2\pi G r^2} Q \qquad (13.24)$$

and the rotational angle at the same point in the direction of the x-axis is

$$\phi_x = -\frac{(1-\mu)\cos\theta}{2\pi G r^2} Q \qquad (13.25)$$

Consequently, if the vertical force Q acts at point (r, θ), the rotational angle at the origin of the system of coordinates in the direction of the x-axis is of the form

$$\phi = \frac{(1-\mu)\cos\theta}{2\pi G r^2} Q \qquad (13.26)$$

As a result, the rotational angle about the y-axis generated by a circular loaded area with intensity q may be expressed as

$$\phi_0 = \int_0^{2\pi} \int_0^{r_0} \frac{(1-\mu)\cos\theta}{2\pi G r^2} qr \, dr \, d\theta \qquad (13.27)$$

Structural Response Considering Soil–Structure Interaction

which, on substitution of the load-distribution functions defined by Equations 13.10 through 13.12, leads to

$$\phi_0 = \frac{3(1-\mu)}{8Gr_0^3} M_y \text{ for rigid footings} \tag{13.28}$$

$$\phi_0 = \frac{2(1-\mu)}{\pi Gr_0^3} M_y \text{ for flexible footings} \tag{13.29}$$

$$\phi_0 = \frac{4(1-\mu)}{\pi Gr_0^3} M_y \text{ for parabolic loads} \tag{13.30}$$

If it is now considered that a horizontal force P acts at the surface of the halfspace at point $(0, 0)$, then according to Equation 13.6 the horizontal displacement at a surface point with coordinates $(r, \pi/2)$ (a point for which the vertical displacement is equal to zero) is given by

$$u\big|_{z=0} = \frac{1-\mu}{2\pi Gr} P \tag{13.31}$$

As a result, the angle of rotation about the z-axis at this point may be expressed as

$$\Psi = \frac{\partial u}{\partial r} = -\frac{1-\mu}{2\pi Gr^2} P \tag{13.32}$$

from which it may be concluded that the torsional angle at the origin of the system of coordinates in the halfspace shown in Figure 13.4b when a horizontal force P acts on point $(r, \pi/2)$ is equal to

$$\Psi_0 = \frac{1-\mu}{2\pi Gr^2} P \tag{13.33}$$

The torsional angle produced by a circular distributed horizontal load about a vertical axis through the center of the loaded area may be therefore written as

$$\Psi_0 = \int_0^{2\pi} \int_0^{r_0} \frac{1-\mu}{2\pi Gr^2} qr\, dr\, d\theta \tag{13.34}$$

from which, after substitution of the load functions defined by Equations 13.13 through 13.15, one finds that

$$\Psi_0 = \frac{3(1-\mu)}{8Gr_0^3} M_z \text{ for rigid footings} \tag{13.35}$$

$$\Psi_0 = \frac{2(1-\mu)}{\pi Gr_0^3} M_z \text{ for flexible footings} \tag{13.36}$$

$$\Psi_0 = \frac{4(1-\mu)}{\pi Gr_0^3} M_z \text{ for parabolic loads} \tag{13.37}$$

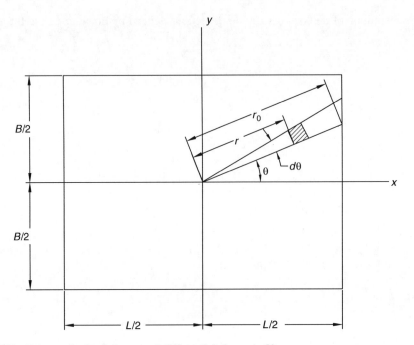

FIGURE 13.7 Rectangular loaded area and differential element of it.

As in the case of a circular footing, the load distribution under a perfectly rigid rectangular footing with length L and width B induced by vertical or horizontal load Q may also be considered to be of the form

$$q = \frac{A_0}{\sqrt{r_0^2 - r^2}} Q \tag{13.38}$$

where, with reference to Figure 13.7, r_0 is now the distance to the edge of the footing corresponding to a given value of the angle θ, and A_0 is a constant that can be determined from the condition that

$$Q = \int_{\text{area}} q \, dA \tag{13.39}$$

or

$$Q = 4A_0 Q \left[\int_0^{\tan^{-1} B/L} \int_0^{L/2 \cos\theta} \frac{r \, dr \, d\theta}{\sqrt{\left(\frac{L/2}{\cos\theta}\right)^2 - r^2}} + \int_0^{\tan^{-1} L/B} \int_0^{B/2 \cos\theta} \frac{r \, dr \, d\theta}{\sqrt{\left(\frac{B/2}{\cos\theta}\right)^2 - r^2}} \right] \tag{13.40}$$

That is, A_0 is a constant equal to

$$A_0 = \frac{1}{4} \left[\frac{L}{2} \ln \tan\left(\frac{\pi}{4} + \frac{1}{2} \tan^{-1} \frac{B}{L}\right) + \frac{B}{2} \ln \tan\left(\frac{\pi}{4} + \frac{1}{2} \tan^{-1} \frac{L}{B}\right) \right]^{-1} \tag{13.41}$$

From Equation 13.17, the vertical displacement under the center of a rigid rectangular footing may be thus expressed as

$$w_0 = 4\int_0^{\tan^{-1}\frac{B/2}{L/2}} \int_0^{\frac{L/2}{\cos\theta}} \frac{1-\mu}{2\pi G} q\, dr\, d\theta + 4\int_0^{\tan^{-1}\frac{L/2}{B/2}} \int_0^{\frac{B/2}{\cos\theta}} \frac{1-\mu}{2\pi G} q\, dr\, d\theta \qquad (13.42)$$

or as

$$w_0 = \frac{2(1-\mu)}{\pi G} A_0 Q \left[\int_0^{\tan^{-1}\frac{B/2}{L/2}} \int_0^{\frac{L/2}{\cos\theta}} \frac{dr\, d\theta}{\sqrt{\left(\frac{L/2}{\cos\theta}\right)^2 - r^2}} + \int_0^{\tan^{-1}\frac{L/2}{B/2}} \int_0^{\frac{B/2}{\cos\theta}} \frac{dr\, d\theta}{\sqrt{\left(\frac{B/2}{\cos\theta}\right)^2 - r^2}} \right] \qquad (13.43)$$

which leads to

$$w_0 = \frac{\pi(1-\mu)}{2G} A_0 Q \qquad (13.44)$$

Similarly, the horizontal displacement under the center of a rigid rectangular footing subjected to a horizontal force Q results as

$$u_0 = \frac{\pi(2-\mu)A_0}{4G} Q \qquad (13.45)$$

Proceeding similarly, it may be shown that the distributed load and rotational angle induced by a rocking moment M_y (parallel to L) in a rigid rectangular footing are given by

$$q = \frac{B_0 r \cos\theta}{\sqrt{r_0^2 - r^2}} M_y \qquad (13.46)$$

$$\phi = \frac{\pi(1-\mu)B_0}{4G} M_y \qquad (13.47)$$

where

$$B_0 = \frac{3}{8}\left\{\left(\frac{L}{2}\right)^3 \left[\ln\tan\left(\frac{\pi}{4} + \frac{1}{2}\tan^{-1}\frac{B}{L}\right)\right] + \left(\frac{B}{2}\right)^3 \left[\frac{1}{2}\frac{L}{B}\sqrt{1+\left(\frac{L}{B}\right)^2} - \frac{1}{2}\ln\tan\left(\frac{\pi}{4} + \tan^{-1}\frac{L}{B}\right)\right]\right\}^{-1} \qquad (13.48)$$

whereas the distributed load and rotational angle for a torsional moment M_z are

$$q = \frac{C_0 r}{\sqrt{r_0^2 - r^2}} M_z \qquad (13.49)$$

$$\Psi = \frac{\pi(1-\mu)C_0}{2G} M_z \qquad (13.50)$$

where

$$C_0 = \frac{3}{4}\left\{\left(\frac{L}{2}\right)^3\left[\frac{B}{L}\sqrt{1+\left(\frac{B}{L}\right)^2}+\ln\tan\left(\frac{\pi}{4}+\frac{1}{2}\tan^{-1}\frac{B}{L}\right)\right]\right.$$
$$\left.+\left(\frac{B}{2}\right)^3\left[\frac{L}{B}\sqrt{1+\left(\frac{L}{B}\right)^2}+\ln\tan\left(\frac{\pi}{4}+\frac{1}{2}\tan^{-1}\frac{L}{B}\right)\right]\right\}^{-1} \quad (13.51)$$

13.3.4 Soil Spring Constants

Since by definition a stiffness coefficient or spring constant is equal to the load divided by the displacement or rotation induced by it, it is possible to use the equations in Section 13.3.3 to define the stiffness or spring constants for an elastic halfspace. On the basis of Equations 13.18 through 13.23, 13.28 through 13.30, and 13.35 through 13.37, the spring constants for the case of a circular footing are thus defined by the expressions listed in Table 13.1. Similarly, based on Equations 13.44, 13.45, 13.47, and 13.50, the spring constants of a halfspace for the case of a rigid rectangular foundation are those listed in Table 13.2, where B and L are the length and width of the foundation, respectively, and

$$F_z = F_x = \frac{L}{2}\ln\tan\left(\frac{\pi}{4}+\frac{1}{2}\tan^{-1}\frac{B}{L}\right)+\frac{B}{2}\ln\tan\left(\frac{\pi}{4}+\frac{1}{2}\tan^{-1}\frac{L}{B}\right) \quad (13.52)$$

TABLE 13.1
Halfspace Spring Constants under Circular Distributed Loads

Direction	Distributed Load Type		
	Rigid Footing	Flexible Footing	Parabolic
Vertical	$\dfrac{4Gr_0}{1-\mu}$	$\dfrac{\pi Gr_0}{1-\mu}$	$\dfrac{3\pi Gr_0}{4(1-\mu)}$
Horizontal	$\dfrac{8Gr_0}{2-\mu}$	$\dfrac{2\pi Gr_0}{2-\mu}$	$\dfrac{3\pi Gr_0}{2(2-\mu)}$
Rocking	$\dfrac{8Gr_0^3}{3(1-\mu)}$	$\dfrac{\pi Gr_0^3}{2(1-\mu)}$	$\dfrac{\pi Gr_0^3}{4(1-\mu)}$
Torsional	$\dfrac{8Gr_0^3}{3(1-\mu)}$	$\dfrac{\pi Gr_0^3}{2(1-\mu)}$	$\dfrac{\pi Gr_0^3}{4(1-\mu)}$

TABLE 13.2
Halfspace Spring Constants under Loads Transmitted through Rigid Rectangular Foundations

Direction			
Vertical	Horizontal	Rocking	Torsional
$\dfrac{8G}{\pi(1-\mu)}F_z$	$\dfrac{16G}{\pi(2-\mu)}F_x$	$\dfrac{32G}{3\pi(1-\mu)}F_\phi$	$\dfrac{8G}{3\pi(1-\mu)}F_\psi$

Note: F_z, F_x, F_ϕ, and F_ψ are defined by Equations 13.52 through 13.54.

$$F_\phi = \left(\frac{L}{2}\right)^3 \ln\tan\left(\frac{\pi}{4} + \frac{1}{2}\tan^{-1}\frac{B}{L}\right) + \left(\frac{B}{2}\right)^3 \left[\frac{1}{2}\frac{L}{B}\sqrt{1+\left(\frac{L}{B}\right)^2} - \frac{1}{2}\ln\tan\left(\frac{\pi}{4} + \tan^{-1}\frac{L}{B}\right)\right] \quad (13.53)$$

$$F_\psi = \left(\frac{L}{2}\right)^3 \left[\frac{B}{L}\sqrt{1+\left(\frac{B}{L}\right)^2} + \ln\tan\left(\frac{\pi}{4} + \frac{1}{2}\tan^{-1}\frac{B}{L}\right)\right]$$
$$+ \left(\frac{B}{2}\right)^3 \left[\frac{L}{B}\sqrt{1+\left(\frac{L}{B}\right)^2} + \ln\tan\left(\frac{\pi}{4} + \frac{1}{2}\tan^{-1}\frac{L}{B}\right)\right] \quad (13.54)$$

It may be noted from the expressions in Table 13.1 that the spring constants for a halfspace change considerably from one pressure distribution to another. The ratio of these spring constants among the three pressure distributions considered is 1:0.785:0.589 for the vertical and horizontal directions and 1:0.589:0.295 for the rocking and torsional directions. The selection of the load distribution form is thus an important consideration in the characterization of soil flexibility. Note also that the expressions for the spring constants in Table 13.1 are determined using the displacement or rotation at the center of the distributed load. Since with the exception of the rigid-base case the displacements and rotations at points other than the center of the loaded area are smaller that those at the center, these soil spring constants tend to exaggerate the flexibility of the halfspace. The consideration of the average displacement over the entire loaded area would thus give a more realistic assessment of this flexibility.

13.3.5 Displacements and Rotations under Dynamic Loads

In 1904, H. Lamb extended Boussinesq's solution to obtain the displacements induced in an elastic halfspace by an oscillating vertical force at the surface of the halfspace—a problem often referred to as the *dynamic Boussinesq problem*. Then in 1936, E. Reissner used Lamb's solution to obtain the displacements generated by an oscillating vertical load uniformly distributed over a circular area on the surface of the halfspace. Reissner considered a harmonic force $P(t) = P_0 e^{i\omega t}$, where P_0 denotes the force amplitude, ω is the frequency of the excitation in radians per second, t depicts time, and i is the unit imaginary number, and obtained the displacement under the center of the load. The expression at which Reissner arrived is of the form

$$w_0(t) = \frac{P_0 e^{i\omega t}}{Gr_0}(f_1 + if_2) \quad (13.55)$$

where $w_0(t)$, G, r_0, and f_1 and f_2, respectively, represent vertical displacement at time t, rigidity of the halfspace, radius of the loaded area, and functions of the excitation frequency and Poisson ratio that involve Fourier-Bessel integrals. The functions f_1 and f_2 are known as "Reissner's displacement functions" and are usually defined in a graphical or tabular form (see Figure 13.10).

Reissner also obtained an expression to determine the displacements generated by a vertical harmonic load $Q(t) = Q_0 e^{i\omega t}$ acting on top of a rigid block with mass m (see Figure 13.8a). For this purpose, he established the dynamic equilibrium of the mass m as (see free-body diagram in Figure 13.8b)

$$Q(t) - P(t) = m\ddot{w}_0(t) \quad (13.56)$$

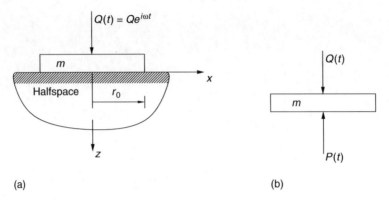

FIGURE 13.8 (a) Circular block of mass m resting on the surface of halfspace and subjected to vertical harmonic force and (b) block's free-body diagram.

where $P(t)$ and $\ddot{w}_0(t)$ are resultant of the reaction of the halfspace against the rigid block and block's acceleration. He considered, in addition, that the block's acceleration was given by

$$\ddot{w}_0(t) = -\omega^2 \frac{P_0 e^{i\omega t}}{Gr_0}(f_1 + if_2) = -\omega^2 w_0(t) \tag{13.57}$$

and that, as determined from Equation 13.55, $P(t)$ in terms of the displacement $w_0(t)$ was equal to

$$P(t) = \frac{Gr_0}{f_1 + if_2} w_0(t) \tag{13.58}$$

After substitution of Equations 13.57 and 13.58 into Equation 13.56, he thus arrived at the equation of motion

$$-\omega^2 m w_0(t) + \frac{Gr_0}{f_1 + if_2} w_0(t) = Q_0 e^{i\omega t} \tag{13.59}$$

whose solution leads to

$$w_0(t) = \frac{(f_1 + if_2) Q_0 e^{i\omega t}}{-\omega^2 m (f_1 + if_2) + Gr_0} \tag{13.60}$$

Equation 13.60, however, may also be written as

$$w_0(t) = \frac{(f_1 + if_2) Q_0 e^{i\omega t}}{Gr_0 \left[\left(1 - \frac{\omega^2 m}{Gr_0} f_1\right) - i\left(\frac{\omega^2 m}{Gr_0} f_2\right)\right]} \tag{13.61}$$

or, by expressing the complex expressions in the numerator and denominator in their polar form, as

$$w_0(t) = \frac{Q_0}{Gr_0} \frac{\sqrt{f_1^2 + f_2^2}\, e^{i(\omega t + \alpha - \beta)}}{\sqrt{\left(1 - \frac{\omega^2 m}{Gr_0} f_1\right)^2 + \left(\frac{\omega^2 m}{Gr_0} f_2\right)^2}} \tag{13.62}$$

where α and β are the phase angles of the aforementioned complex expressions. Reissner thus found that the amplitude of the displacements induced by a dynamic force Q applied through a rigid block of mass m is given by

$$A_z = \frac{Q_0}{Gr_0} \sqrt{\frac{f_1^2 + f_2^2}{\left(1 - \frac{\omega^2 m}{Gr_0} f_1\right)^2 + \left(\frac{\omega^2 m}{Gr_0} f_2\right)^2}} \qquad (13.63)$$

Reissner's theory established the basis for further analytical studies of oscillating forces on the surface of a halfspace. His theory, however, was not accepted immediately because it did not completely agree with the results from field tests. Some of the reasons for the disagreement were (1) permanent settlements developed during the tests, violating the assumption of an elastic medium; (2) the assumption of a uniformly distributed soil pressure was not realistic; and (3) there was a sign error in Reissner's original derivation, which affected the calculation of f_2 and the computed values of the displacement amplitude. Nevertheless, Reissner's theory has now been fully accepted and his contribution has become a classic in the field of soil dynamics.

In 1953, P. M. Quinlan and T. Y. Sung independently extended Reissner's work considering the rigid-footing, uniform, and parabolic distributed loads discussed in Section 13.3.3 to study the effect of load distribution form on the halfspace displacements. Quinlan established the equations that define such load distributions and developed the solution for the rigid-footing case. Sung obtained the solutions for the three load types and determined the corresponding displacements at the center and edges of the circular loaded areas. As expected, Sung found that for the parabolic and uniform distributions the displacements under the center of the load were larger than those at the edges—a displacement pattern that can only be developed by flexible footings. As expected, also, the rigid-footing distribution produced uniform displacements under static loads. However, he assumed that the center of gravity of the footing moved the same distance as the center of the loaded area. This assumption, as pointed out earlier, produced exaggerated displacement responses for the parabolic and uniform loads because in these cases the center of the loaded area undergoes a larger displacement than the average over the loaded area.

Sung also assumed that the pressure distribution at the interface between a footing and a halfspace remained the same for all excitation frequencies. It is now known, however, that the stress distributions in an elastic solid may be different under dynamic loads than under static loads. As a result, a rigid-footing pressure distribution, which correctly predicts a uniform displacement over the loaded area under a static load, may not produce a uniform displacement under a dynamic load.

In 1955, R. N. Arnold, G. N. Bycroft, and G. B. Warburton developed the equations to determine the average displacements over the loaded area of a rigid footing resting on an elastic halfspace. Their investigation included four modes (a) vertical translation, (b) horizontal translation, (c) rocking, and (d) torsion. The considered excitations and the corresponding displacements are shown in Figure 13.9. They obtained these average displacements by integrating the displacements over the circular area, after being weighted in proportion to the corresponding soil pressure intensity. That is, the displacement at a point at any distance r from the center of the footing was weighted by a factor proportional to the soil pressure at that point. They kept, however, the assumption made by Sung that the soil pressure distribution under a dynamic load is the same as that produced by the corresponding static load. The expressions they arrived at are of the form of Equation 13.55, except that the values of the displacement functions differ from those obtained by Reissner. The displacement functions they obtained for the four modes

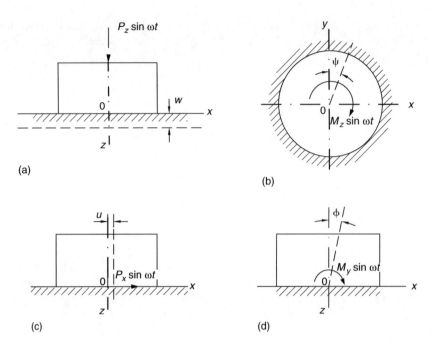

FIGURE 13.9 Exciting forces and corresponding displacements of rigid block on halfspace in four modes of vibration.

considered are shown in Figure 13.10, where the frequency factor a_0 represents a dimensionless frequency defined as

$$a_0 = \frac{\omega r_0}{v_s} = \omega r_0 \sqrt{\frac{\rho}{G}} \qquad (13.64)$$

where, as mentioned earlier, ω denotes the excitation frequency, v_s the shear wave velocity in the halfspace, ρ the mass density, G the shear modulus of the medium, and r_0 the footing's radius. Note that f_1 and f_2 are evaluated only for the range $0 \le a_0 \le 1.5$ for the vertical load and rocking moment cases, and $0 \le a_0 \le 2.0$ for the horizontal load and torsional moment cases. These are practical ranges over which the resonance peaks may occur in response curves and have been proven satisfactory for many practical studies. Furthermore, note that when $a_0 = 0$ (static case), $f_2 = 0$ and thus f_1 alone produces the static displacement. Hence, the function f_2 may be considered as a measure of the damping in the system.

The amplitude of the displacements induced by oscillating vertical and horizontal forces may be thus found using Equation 13.63 in conjunction with the values of f_1 and f_2 given in Figures 13.10a and 13.10c. In the case of rocking and torsional moments, the amplitude of the rotations is given by

$$A_z = \frac{M_0}{Gr_0^3} \sqrt{\frac{f_1^2 + f_2^2}{\left(1 - \frac{\omega^2 I_0}{Gr_0^3} f_1\right)^2 + \left(\frac{\omega^2 I_0}{Gr_0^3} f_2\right)^2}} \qquad (13.65)$$

where M_0 is the magnitude of the exciting moment, I_0 the rigid block's mass moment of inertia about the y- or z-axis, and the displacement functions f_1 and f_2 are defined by Figures 13.10b and 13.10d.

Structural Response Considering Soil–Structure Interaction

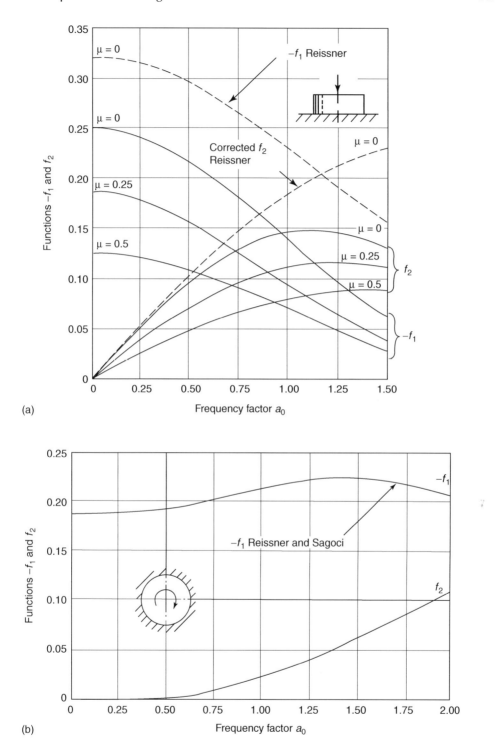

(a)

(b)

FIGURE 13.10 Displacement functions f_1 and f_2 for (a) vertical, (b) torsional, (c) horizontal, and (d) rocking modes of vibration. (Reproduced with permission of ASME from Arnold, R. N., Bycroft, G. N., and Warburton, G. B., *Journal of Applied Mechanics, ASME*, 77, 1955, 391–400.)

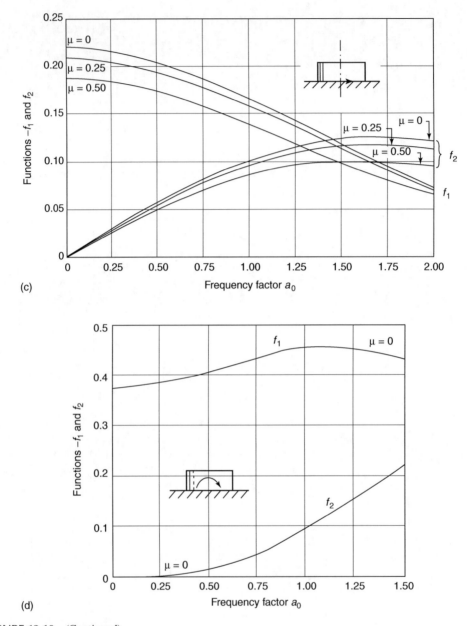

FIGURE 13.10 (Continued)

It should be noted that the solution for horizontal motion is only of academic interest or for approximate solutions, for it presumes horizontal vibration without rocking. This, however, can only happen if the center of gravity of the foundation coincides with the surface of the halfspace, given that only in this particular case the soil reaction against the foundation is colinear with the applied force. In actual foundations, horizontal and rocking vibrations are always coupled. That is, horizontal motion induces rocking and vice versa. The response of foundations subjected to coupled rocking and sliding vibrations is studied in Section 13.5.

Structural Response Considering Soil–Structure Interaction

EXAMPLE 13.1 DISPLACEMENT AMPLITUDE OF A RIGID BLOCK UNDER VERTICAL HARMONIC LOAD

A 16-ft circular foundation block supports a vibrating machine that generates a vertical harmonic force with an amplitude of 1500 lb and a frequency of 180 cycles per minute (cpm). The combined weight of the machine and the foundation is 150,000 lb. The foundation block rests on the surface of a soil deposit that can be idealized as an elastic halfspace. The soil has a unit weight of 115 lb/ft^3, a shear modulus of elasticity of 3000 lb/in.^{2}, and a Poisson ratio of 0.4. Determine the amplitude of the vibrations induced at the soil surface.

Solution

According to Equation 13.64 and the given data, the dimensionless frequency for this soil–foundation system is equal to

$$a_0 = 2\pi(180/60)(8)\sqrt{\frac{115/32.2}{3000(144)}} = 0.43$$

From Figure 13.10a, the displacement functions f_1 and f_2 for this case are thus equal to -0.13 and 0.06, respectively. On substitution of these values and the given data into Equation 13.63, the desired amplitude is therefore equal to

$$A_z = \frac{1{,}500}{(3{,}000 \times 144)(8)} \sqrt{\frac{0.13^2 + 0.06^2}{\left(1 + \frac{2\pi(180/60)(150{,}000/32.2)}{(3{,}000 \times 144)(8)}0.13\right)^2 + \left(\frac{2\pi(180/60)(150{,}000/32.2)}{(3{,}000 \times 144)(8)}0.06\right)^2}}$$

$$= 6.39 \times 10^{-5}\,\text{ft} = 0.00074\,\text{in.}$$

EXAMPLE 13.2 ROTATIONAL AMPLITUDE OF A RIGID BLOCK UNDER ROCKING HARMONIC MOMENT

A circular concrete foundation block, 4 m in diameter and 3 m deep, supports a horizontal compressor that generates a rocking motion. This rocking motion results from a horizontal unbalanced force in the compressor that acts 3.5 m above the base of the foundation. The operating frequency of the compressor is 600 revolutions per minute (rpm) and the amplitude of the unbalanced force is equal to 30 kN. The mass moment of inertia of the compressor about the foundation base is 1.5×10^5 kg m^2. The soil that supports the compressor and the foundation block has a shear modulus of elasticity of 18,000 kN/m^2, a density of 1800 kg/m^3, and a Poisson ratio of 0.35. Determine the amplitude of the rocking vibrations neglecting the coupling with the horizontal motion. Assume the unit weight of the concrete used for the foundation is 23.58 kN/m^3.

Solution

Substituting the given data into Equation 13.64, the dimensionless frequency for this case is equal to

$$a_0 = (2\pi \times 600/60)(2)\sqrt{\frac{1800}{18 \times 10^6}} = 1.26$$

For this value of the dimensionless frequency, Figure 13.10d yields displacement functions f_1 and f_2 equal to (using the curves for a zero Poisson ratio) -0.45 and 0.15, respectively. Similarly, for the given foundation dimensions, the mass moment of inertia is equal to

$$I_f = m\left(\frac{r_0^2}{4} + \frac{h^2}{3}\right) = \frac{25.38 \times 10^3}{9.81}(\pi \times 2^2 \times 3)\left(\frac{2^2}{4} + \frac{3^2}{3}\right) = 390{,}134 \text{ kg m}^2$$

Hence, the mass moment of inertia of the foundation–compressor system results as

$$I_\phi = 390{,}134 + 150{,}000 = 540{,}134 \text{ kg m}^2$$

Additionally, the rocking moment produced by the unbalanced horizontal force is

$$M_0 = 30 \times 10^3 (3.5) = 105{,}000 \text{ N m}$$

Thus, according to Equation 13.65, the desired vibration amplitude is equal to

$$A_z = \frac{105 \times 10^3}{18 \times 10^6 (2)^3} \sqrt{\frac{0.45^2 + 0.15^2}{\left(1 + \frac{(2\pi \times 600/60)^2 (540{,}134)}{18 \times 10^6 (2)^3} 0.45\right)^2 + \left(\frac{(2\pi \times 600/60)^2 (540{,}134)}{18 \times 10^6 (2)^3} 0.15\right)^2}}$$

$$= 4.3 \times 10^{-5} \text{ rad}$$

As pointed out earlier, Quinlan et al. obtained their solutions for rigid footings by assuming a contact stress distribution that produces uniform displacements under a static force or linear displacements under a static moment at the halfspace–footing interface. This assumption, however, is only an approximate one since in reality the stress distribution that is required to maintain uniform or linear displacements is not constant, but varies with the frequency of vibration. The first "rigorous" solution was not obtained until 1964, when N. M. Borodachev analyzed the vertical vibrations of the halfspace as a *mixed boundary value problem*; that is, by prescribing patterns of displacements under the rigid footing and vanishing stresses over the remaining portion of the surface. Borodachev's work was followed by a 1965 comprehensive study by A. D. Awojobi and P. Grootenhuis. Introducing some simplifying assumptions regarding the secondary contact stresses (i.e., a "relaxed boundary" in which it is assumed that no frictional shear tractions are developed during vertical and rocking vibrations, and that the normal tractions are zero during horizontal vibrations), Awojobi and Grootenhuis studied by recourse to integral transform techniques all the possible modes of oscillation of rigid circular and strip footings resting on a halfspace. Armed with the new mathematical techniques available to solve mixed boundary value problems, several investigators have extended afterward the available halfspace solutions for circular foundations to the high-frequency range and to soils with hysteretic damping. For the most part, however, these rigorous solutions have been obtained in the form of impedance or compliance functions; that is, force–displacement relationships for dynamic systems. The discussion of these solutions is therefore deferred to Section 13.8, once the concept of impedance and compliance functions is introduced.

13.3.6 Vibrations of Rectangular Foundations

The methods described in the foregoing sections to estimate the response of foundations to dynamic loads are limited to circular foundations. In principle, therefore, these methods cannot be used to analyze rectangular foundations, the shape of many, if not the majority of, foundations. It is common practice, however, to also estimate the response of rectangular foundations with such methods. For such a purpose, an equivalent radius is defined. This equivalent radius is determined by considering that the stresses exerted on the soil by a foundation block are proportional to the foundation's area in the case of vertical and horizontal forces, proportional to the foundation's moment of inertia in the case of rocking moments, and proportional to the foundation's polar moment of inertia in the case of torsional moments. By thus equating the areas, moments of inertia, and polar moments of inertia of a circular foundation with a radius r_0 and a rectangular foundation with width B and length L, the desired equivalent radius is given by

$$r_0 = \sqrt{\frac{BL}{\pi}} \qquad (13.66)$$

for horizontal and vertical forces,

$$r_0 = \sqrt[4]{\frac{BL^3}{3\pi}} \qquad (13.67)$$

for rocking moments, and

$$r_0 = \sqrt{\frac{BL(B^2 + L^2)}{6\pi}} \qquad (13.68)$$

for torsional moments.

This procedure of transforming a rectangular foundation into an equivalent circular one has been shown to give good results in the response evaluation of foundations with an aspect ratio L/B of less than 2. However, it may lead to significant errors in the case of long and narrow foundations. For foundations with aspect ratios that depart significantly from 2, it is recommended to use the more accurate methods described in Section 13.8.

13.4 LUMPED-PARAMETER MODELS

13.4.1 Hsieh's Equations

In 1962, T. K. Hsieh discovered that a foundation–soil system may be represented in terms of a simple mass–spring–dashpot analog (i.e., a lumped-parameter system), giving rise to an easier method to obtain the response of the foundation–soil system to dynamic loads. In this method, the original foundation–halfspace system is replaced by a single-degree-of-freedom system with an equivalent mass, stiffness, and damping coefficient. To obtain the expressions to determine these equivalent parameters, Hsieh observed that in terms of Reissner's original equations the displacement and velocity induced by a harmonic force $P(t) = P_0 e^{i\omega t}$ acting on a massless rigid disk resting on an elastic halfspace are given by

$$w(t) = \frac{P_0 e^{i\omega t}}{Gr_0}(f_1 + if_2) \qquad (13.69)$$

$$\dot{w}(t) = \frac{P_0 \omega e^{i\omega t}}{Gr_0}(if_1 - f_2) \qquad (13.70)$$

and that, as a result,

$$f_1 \omega w(t) - f_2 \dot{w}(t) = \frac{P(t)\omega}{Gr_0}(f_1^2 + f_2^2) \qquad (13.71)$$

from which he solved for the force $P(t)$ to obtain

$$P(t) = -\frac{Gr_0}{\omega}\frac{f_2}{f_1^2 + f_2^2}\dot{w}(t) + Gr_0\frac{f_1}{f_1^2 + f_2^2}w(t) \qquad (13.72)$$

which can also be expressed as

$$P(t) = c_z \dot{w}(t) + k_z w(t) \qquad (13.73)$$

where

$$c_z = \frac{Gr_0}{\omega}\frac{-f_2}{f_1^2 + f_2^2} \qquad (13.74)$$

and

$$k_z = Gr_0 \frac{f_1}{f_1^2 + f_2^2} \qquad (13.75)$$

Hsieh also considered the case of a rigid block of mass m resting on a halfspace and subjected to a harmonic force $Q(t) = Q_0 e^{i\omega t}$ (see Figure 13.8). Using Newton's second law, he established the equation of dynamic equilibrium for such a mass as

$$Q(t) - P(t) = m\ddot{w}(t) \qquad (13.76)$$

However, $P(t)$ in Equation 13.76 represents the total force with which the soil reacts against the rigid block. Hence, $P(t)$ may be expressed as indicated by Equation 13.73. Consequently, Equation 13.76 may also be written as

$$Q(t) - c_z\dot{w}(t) - k_z w(t) = m\ddot{w}(t) \qquad (13.77)$$

or as

$$m\ddot{w}(t) + c_z\dot{w}(t) + k_z w(t) = Q_0 e^{i\omega t} \qquad (13.78)$$

This is the equation of a single-degree-of-freedom system subjected to a harmonic excitation. Hsieh found thus that the original foundation–halfspace system could be substituted by an equivalent single-degree-of-freedom system with a mass equal to the mass of the foundation and stiffness and damping coefficients given by Equations 13.74 and 13.75 (see Figure 13.11). It should be noted, however, that these constants depend on the excitation frequency as the displacement functions

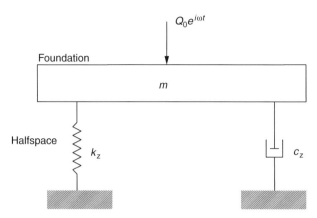

FIGURE 13.11 Single-degree-of-freedom model of foundation–halfspace system.

f_1 and f_2 are both frequency dependent. This poses no problems when the excitation is a harmonic one, but it complicates the analysis when the excitation is of the transient type. That is, an excitation that contains components with different frequencies. It should be noted, also, that the system possesses some damping despite the fact that no damping forces are included in the equation of motion. This damping is due to radiation damping and represents the loss by radiation away from the foundation of the energy generated by the exciting force.

13.4.2 Lysmer and Hall Analogs

In 1965, J. Lysmer went one step ahead toward the simplification of the analysis of a rigid body on a halfspace problem and found frequency-independent values of the spring and damping coefficients in Hsieh's model. He arrived at these values by assuming that the spring constant was equal to the value for the static case and by adjusting the value of the damping constant so as to minimize the error in the entire range of possible Poisson ratios ($0 < \mu < 0.5$) and a range between 0 and 1 for the dimensionless frequency a_0 (i.e., $0 \leq a_0 \leq 1$). The spring and damping constants he determined for a vertical harmonic load applied at the center of a rigid circular block are

$$k_z = \frac{4Gr_0}{1-\mu} \tag{13.79}$$

$$c_z = \frac{3.4r_0^2}{1-\mu}\sqrt{\rho G} \tag{13.80}$$

where all the symbols are as defined earlier. Figure 13.12 shows a comparison between the rigid block displacement response obtained with Lysmer analog for the case under consideration and that obtained with the elastic halfspace theory. It may be seen from Figure 13.12 that the agreement is indeed good for the range $0 \leq a_0 \leq 1$, which is the frequency range of greatest interest.

A single-degree-of-freedom model defined by the foundation mass and the spring and damping constants proposed by Lysmer is known as *Lysmer analog*. With it, it is possible to analyze a soil–structure system without having to use different parameters for the different harmonic components contained in a typical earthquake ground motion. Furthermore, it is possible to obtain the vibration amplitude and resonant frequencies of rigid circular foundations using the simple solutions for a single-degree-of-freedom system.

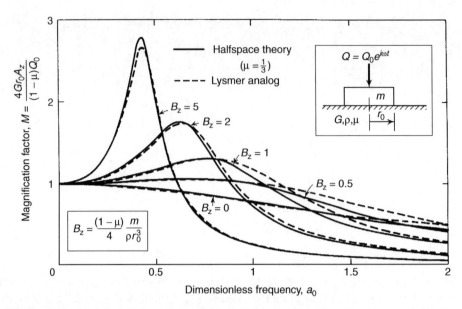

FIGURE 13.12 Normalized response of rigid circular footing to vertical harmonic force as determined by halfspace theory and Lysmer analog. (Reproduced with permission of ASCE from Lysmer, J. and Richarts, F. E., *Journal of the Soil Mechanics and Foundations Division, ASCE*, 92, SM1, 1966, 65–91.)

EXAMPLE 13.3 RESONANT FREQUENCY AND VIBRATION AMPLITUDE OF A RIGID BLOCK UNDER VERTICAL HARMONIC LOAD

Determine using Lysmer analog, the resonant frequency, the amplitude of vibration under operation conditions, and the amplitude of vibration at resonance for the soil–foundation system described in Example 13.1.

Solution

The first step in the solution of this problem using Lysmer analog is the determination of the soil–foundation mass and the equivalent stiffness and damping constants. Thus, according to the given foundation and machine weights and Equations 13.79 and 13.80, the values of these parameters are

$$m = \frac{150{,}000}{32.2} = 4{,}658 \text{ slugs}$$

$$k_z = \frac{4(3{,}000 \times 144)(8)}{1 - 0.4} = 23{,}040{,}000 \text{ lb/ft}$$

$$c_z = \frac{3.4(8)^2}{1 - 0.4}\sqrt{(115/32.2)(3{,}000 \times 144)} = 450{,}475 \text{ lb s/ft}$$

Furthermore, according to these values and the definitions of natural frequency and damping ratio for a single-degree-of-freedom system, one has that

$$f_n = \frac{1}{2\pi}\sqrt{\frac{k_z}{m}} = \frac{1}{2\pi}\sqrt{\frac{23.04 \times 10^6}{4658}} = 11.2 \text{ Hz}$$

$$\xi = \frac{c_z}{(c_z)_{cr}} = \frac{c_z}{2\sqrt{k_z m}} = \frac{450{,}475}{2\sqrt{23.04 \times 10^6 \times 4{,}658}} = 0.69$$

Hence, using the formulas from structural dynamics for a single-degree-of-freedom system under a harmonic excitation, the resonant frequency is equal to

$$f_{res} = f_n\sqrt{1 - 2\xi^2} = 11.2\sqrt{1 - 2(0.69)^2} = 2.4 \text{ Hz}$$

and the amplitude of vibration during the operation of the machine results as

$$A_z = \frac{Q_0}{k_z}\sqrt{\frac{1}{[1-(f/f_n)^2]^2 + [2\xi f/f_n]^2}}$$

$$= \frac{1500}{23.04 \times 10^6}\sqrt{\frac{1}{[1-(3/11.2)^2]^2 + [2(0.69)(3/11.2)]^2}}$$

$$= 6.51 \times 10^{-5} \text{ ft} = 0.00078 \text{ in.}$$

Finally, the amplitude of vibration at resonance is given by

$$(A_z)_{res} = \frac{Q_0}{k_z}\frac{1}{2\xi\sqrt{1-\xi^2}} = \frac{1500}{23.04 \times 10^6}\frac{1}{2(0.69)\sqrt{1-0.69^2}} = 6.51 \times 10^{-5} \text{ ft} = 0.00078 \text{ in.}$$

In 1967, J. R. Hall developed single-degree-of-freedom models for the vibrations of rigid circular footings under horizontal and rocking vibrations, similar to the model developed by Lysmer for vertical excitations. In the case of horizontal vibrations, Hall found that the vibrations could be described by the equation of motion

$$m\ddot{u} + c_x\dot{u} + k_x u = Q_0 e^{i\omega t} \tag{13.81}$$

where u denotes horizontal displacement at the surface of the halfspace, $Q_0 e^{i\omega t} = Q(t)$ represents a harmonic horizontal force applied at the center of the rigid foundation, m is the mass of the foundation, and k_x and c_x are spring and damping constants, respectively, defined as

$$k_x = \frac{32(1-\mu)Gr_0}{7-8\mu} \tag{13.82}$$

$$c_x = \frac{18.4(1-\mu)}{7-8\mu}r_0^2\sqrt{\rho G} \tag{13.83}$$

Similarly, he found that for rocking vibrations the foundation motion could be described by the equation

$$I_\phi\ddot{\phi} + c_\phi\dot{\phi} + k_\phi\phi = M_y e^{i\omega t} \tag{13.84}$$

where ϕ denotes the angle of rotation of the foundation about a horizontal axis through the foundation base, $M_y e^{i\omega t}$ is a harmonic rocking moment about the same axis, I_ϕ is the corresponding mass moment of inertia, and k_ϕ and c_ϕ are spring and damping constants, respectively, given by

$$k_\phi = \frac{8Gr_0^3}{3(1-\mu)} \tag{13.85}$$

$$c_\phi = \frac{0.8r_0^4\sqrt{G\rho}}{(1-\mu)(1+B_\phi)} \tag{13.86}$$

where B_ϕ is a mass ratio defined as

$$B_\phi = \frac{3(1-\mu)}{8} \frac{I_\phi}{\rho r_0^5} \quad (13.87)$$

Figures 13.13 and 13.14 compare the displacement responses obtained with Hall analog for horizontal and rocking excitations against those determined directly with the elastic halfspace theory.

Finally, F. Richart, J. Hall, and R. Woods extended the concept of Lysmer and Hall analogs to the case of circular rigid foundations subjected to a harmonic torsional moment $M_z e^{i\omega t}$. For this case, they established the equation of motion for the equivalent single-degree-of-freedom model as

$$I_\Psi \ddot{\Psi} + c_\Psi \dot{\Psi} + k_\Psi \Psi = M_z e^{i\omega t} \quad (13.88)$$

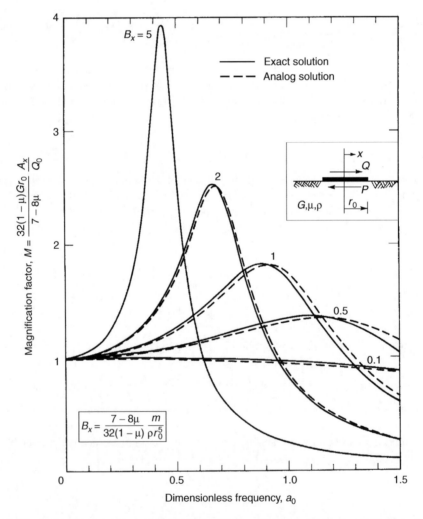

FIGURE 13.13 Normalized response of rigid circular footing to horizontal harmonic force as determined by halfspace theory and Hall's analog. (Reproduced from Hall, J. R., *Proceedings of the International Symposium on Wave Propagation and Dynamic Properties of Earth Materials*, Albuquerque, NM, August 1967.)

Structural Response Considering Soil–Structure Interaction

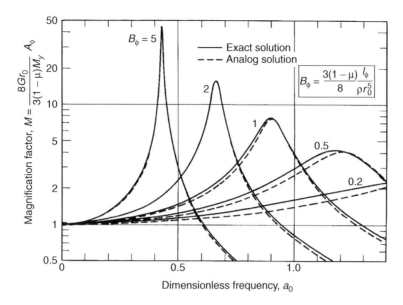

FIGURE 13.14 Response of rigid circular footing to rocking harmonic moment using halfspace theory and Hall's analog. (Reproduced from Hall, J. R., *Proceedings of the International Symposium on Wave Propagation and Dynamic Properties of Earth Materials*, Albuquerque, NM, August 1967.)

where $M_z e^{i\omega t}$ denotes a moment about a vertical axis through the center of the foundation, I_Ψ is the mass moment of inertia about the same axis, Ψ depicts the angle of rotation of the foundation in the direction of M_z, and k_Ψ and c_Ψ are spring and damping constants, respectively, given by

$$k_\Psi = \frac{16}{3} G r_0^3 \tag{13.89}$$

$$c_\Psi = \frac{2.3}{1 + 2B_\varphi} \sqrt{\rho G B_\varphi} \tag{13.90}$$

where B_Ψ is a mass ratio defined as

$$B_\Psi = \frac{I_\Psi}{\rho r_0^5} \tag{13.91}$$

It should be noted at this point that the spring constants used by Lysmer and Hall for the development of their analogs differ from those given in Table 13.1 for the horizontal force and torsional moment cases. The reason for this difference is that the expressions adopted by Lysmer and Hall were derived by making the excitation frequency equal to zero in the solutions obtained for dynamic loads—solutions which, for the most part, were determined introducing some approximations. It is expected, therefore, that the static constants derived with the latter method are also approximate. Note, in particular, that the spring constant for the torsional moment given by Equation 13.89 coincides with that given in Table 13.1 for the specific case of $\mu = 0.5$.

EXAMPLE 13.4 RESONANT FREQUENCY AND VIBRATION AMPLITUDE OF A RIGID BLOCK UNDER ROCKING HARMONIC MOMENT

Determine using Hall analog, the resonant frequency, the amplitude of vibration under the operating frequency, and the amplitude of vibration at resonance for the soil–foundation system described in Example 13.2.

Solution

According to the data in Example 13.2 and Equations 13.85 through 13.87, the parameters of the equivalent single-degree-of-freedom system that represents a rigid block under a rocking moment are

$$I_\phi = 540{,}134 \text{ kg m}^2$$

$$B_\phi = \frac{3(1-0.35)}{8} \frac{540{,}134}{1{,}800(2)^5} = 2.29$$

$$k_\phi = \frac{8(18 \times 10^6)(2)^3}{3(1-0.35)} = 590{,}769{,}231 \text{ N m}$$

$$c_\phi = \frac{0.8(2)^4 \sqrt{(18 \times 10^6)(1{,}800)}}{(1-0.35)(1+2.29)} = 1{,}077{,}391 \text{ N m s}$$

Hence, the natural frequency and damping ratio of the machine–foundation–halfspace system result as

$$f_n = \frac{1}{2\pi} \sqrt{\frac{590{,}769{,}231}{540{,}134}} = 5.3 \text{ Hz}$$

$$\xi = \frac{1{,}077{,}391}{2\sqrt{(590{,}769{,}231)(540{,}134)}} = 0.030$$

which in turn lead to a resonant frequency and amplitude of vibration at the operational frequency equal to

$$A_\phi = \frac{M_y}{k_\phi} \sqrt{\frac{1}{[1-(f/f_n)^2]^2 + [2\xi f/f_n]^2}}$$

$$= \frac{105{,}000}{590{,}769{,}231} \sqrt{\frac{1}{[1-(10/5.3)^2]^2 + [2(0.03)(10/5.3)]^2}}$$

$$= 6.8 \times 10^{-5} \text{ rad}$$

$$f_{res} = f_n \sqrt{1-2\xi^2} = 5.3\sqrt{1-2(0.03)^2} = 5.3 \text{ Hz}$$

Similarly, the amplitude of vibration under resonant conditions results as

$$(A_\phi)_{res} = \frac{M_y}{k_\phi} \frac{1}{2\xi\sqrt{1-\xi^2}} = \frac{105{,}000}{590{,}769{,}231} \frac{1}{2(0.03)\sqrt{1-0.03^2}} = 3 \times 10^{-3} \text{ rad}$$

It should be noted that there is a slight discrepancy between the value of A_ϕ determined here using Hall analog and that obtained in Example 13.2. This discrepancy likely resulted from the use of the curves corresponding to a Poisson ratio of zero to find the values of the displacement functions f_1 and f_2 in Example 13.2 and the use of the actual Poisson ratio of 0.35 in the solution by Hall analog.

13.5 COUPLED ROCKING AND HORIZONTAL VIBRATIONS OF RIGID CIRCULAR FOUNDATIONS

As noted in Section 13.3.5, rocking and horizontal vibrations are always coupled. To find the response of a foundation subjected to a horizontal force or a rocking moment, it is therefore necessary to solve the coupled equations of a two-degree-of-freedom system. The equations of motion of such a coupled system and the solution to these equations may be obtained as follows.

Structural Response Considering Soil–Structure Interaction

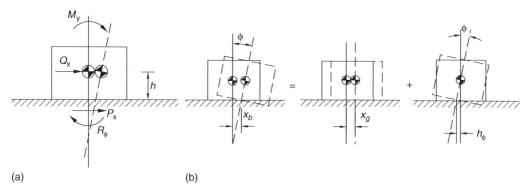

(a)   (b)

FIGURE 13.15 (a) Coupled rocking and (b) horizontal vibration modes.

With reference to Figure 13.15a, consider a foundation block resting on the surface of a halfspace and let it be subjected to a rocking moment $M_y(t) = M_0 e^{i\omega t}$ and a horizontal force $Q_x(t) = Q_0 e^{i\omega t}$, both applied at the foundation's center of gravity. This type of loading arises, for example, when an unbalanced machine produces a horizontal force above the foundation's a center of gravity. Let, then, the block's center of gravity be located on the vertical axis that passes through the center of the foundation base and at a height h above the surface of the halfspace. Note because of the combined effect of the applied horizontal force and rocking moment, the foundation's center of gravity will translate horizontally and rotate as shown in Figure 13.15a.

For convenience, however, it will be considered that the motion of the foundation is given by the superposition of a pure horizontal translation x_g and a pure rotation ϕ about the foundation's center of gravity (see Figure 13.15b). It will also be considered that on the basis of this superposition the horizontal displacement of the foundation base may be expressed as

$$x_b = x_g - \phi h \tag{13.92}$$

By applying Newton's second law for forces, the equation of motion for the horizontal translation of the foundation mass may be thus written as

$$Q_x + P_x = m\ddot{x}_g \tag{13.93}$$

where P_x is the horizontal reaction of the halfspace against the foundation base, which, by analogy with Equation 13.73, may be expressed as

$$P_x = -c_x \dot{x}_b - k_x x_b \tag{13.94}$$

where c_x and k_z represent damping and spring constants that may be determined from Hsieh's equations or Hall analog. On substitution of Equation 13.94 into Equation 13.93, one has thus that

$$Q_x - c_x \dot{x}_b - k_x x_b = m\ddot{x}_g \tag{13.95}$$

which, after substitution of Equation 13.92 and rearranging terms, becomes

$$m\ddot{x}_g + c_x \dot{x}_g + k_x x_g - c_x h\dot{\phi} - k_x h\phi = Q_x \tag{13.96}$$

If Newton's second law for moments is now applied to the rotation of the foundation about its center of gravity, one obtains

$$M_y + R_\phi - P_x h = I_g \ddot{\phi} \tag{13.97}$$

where I_g and R_ϕ denote moment of inertia of the foundation about its center of gravity and reaction of the halfspace against the rotation of the foundation. However, by analogy with Equations 13.73 and 13.94, this reaction may be expressed as

$$R_\phi = -c_\phi \dot{\phi} - k_\phi \phi \tag{13.98}$$

where c_ϕ and k_ϕ denote damping and spring constants for rocking, obtained, as in the case of horizontal forces, from Hsieh's equations or from Hall analog. After the substitution of Equations 13.94 and 13.98, Equation 13.97 may be thus written as

$$M_y - c_\phi \dot{\phi} - k_\phi \phi + (c_x \dot{x}_b + k_x x_b)h = I_g \ddot{\phi} \tag{13.99}$$

which, after incorporation of Equation 13.92 and rearranging terms, leads to

$$I_g \ddot{\phi} + (c_\phi + c_x h^2)\dot{\phi} + (k_\phi + k_x h^2)\phi - c_x h \dot{x}_g - k_x h x_g = M_y \tag{13.100}$$

Equations 13.96 and 13.100 represent the equations of motion that describe the coupled motion of a foundation–halfspace system subjected to a horizontal force and a rocking moment. They demonstrate that coupling between the horizontal and rotational motions occurs because the center of gravity of the foundation system lies above the line of action of the horizontal reaction P_x at the foundation base. Note that if h is zero, there is no coupling effect since in such a case Equation 13.96 would be only a function of the horizontal displacement x_g and Equation 13.100 would be only a function of the rotation ϕ.

The natural frequencies and damping ratios of the foundation–halfspace system may be obtained by setting M_y and Q_x equal to zero in Equations 13.96 and 13.100, and solving the characteristic equation of the resulting system of homogeneous equations. This characteristic equation is determined as follows: First, assume solutions of the form

$$x_g = X e^{\lambda t} \tag{13.101}$$

$$\phi = \Phi e^{\lambda t} \tag{13.102}$$

where, as in the case of a damped single-degree-of-freedom system, λ is a parameter given by

$$\lambda = -\xi \omega + i\omega \sqrt{1 - \xi^2} \tag{13.103}$$

where ω, ξ, and i are natural frequency, damping ratio, and unit imaginary number.

Second, substitute the assumed solutions into the system of homogeneous equations and set the determinant of the resulting algebraic equations equal to zero. Finally, expand the determinant and solve for λ. Since the system has two degrees of freedom, the characteristic equation will be a quadratic one and will thus yield two values of λ. Once these values for λ are obtained, the system's two natural frequencies and damping ratios are obtained from

$$\omega = |\lambda| \tag{13.104}$$

$$\xi = \frac{\operatorname{Re} \lambda}{|\lambda|} \tag{13.105}$$

where "Re" signifies "the real part of."

Structural Response Considering Soil–Structure Interaction

The amplitudes of vibration in the translational and rotational modes are obtained by substituting first into Equations 13.96 and 13.100 solutions of the form

$$x_g = A_x e^{i\omega t} \tag{13.106}$$

$$\phi = A_\phi e^{i\omega t} \tag{13.107}$$

where, once again, ω is the excitation frequency and A_x and A_ϕ are two complex-valued constants. Then, A_x and A_ϕ are solved from the resulting system of two simultaneous equations. The desired vibration amplitudes are given by the absolute values or moduli of these two constants.

EXAMPLE 13.5 NATURAL FREQUENCIES AND VIBRATION AMPLITUDES OF A RIGID BLOCK UNDER HORIZONTAL HARMONIC LOAD

The circular foundation of a machine has a radius of 2.0 m and a depth of 1.5 m. The foundation is subjected to a horizontal harmonic force at a height of 2.0 m from the foundation base. This force has an amplitude of 10 kN and a frequency of 800 cpm. The soil supporting the foundation has a shear modulus of elasticity of 30,000 kN/m², a density of 1700 kg/m³, and a Poisson ratio of 0.2. Determine (a) the natural frequencies in the horizontal and rocking modes of the foundation, and (b) the amplitudes of vibration in these two modes. Use Hall analog to define the stiffness and damping constants of the soil–foundation system and neglect the mass moment of inertia of the machine. The foundation is made of concrete with a unit weight of 23.5 kN/m³.

Solution

a. According to the given data, the mass and mass moment of inertia of the foundation about its center of gravity and about its base are equal to

$$m = \frac{\pi(2.0)^2(1.5)(23.5)}{9.81} = 45.15 \text{ Mg}$$

$$I_g = \frac{45.15}{12}(3 \times 2.0^2 + 1.5^2) = 53.62 \text{ Mg m}^2$$

$$I_0 = 53.62 + 45.15(0.75)^2 = 79.02 \text{ Mg m}^2$$

Similarly, according to Equations 13.82 and 13.83 for the horizontal mode of vibration and Equations 13.85 through 13.87 for the rocking mode of vibration, the system's spring and damping constants are given by

$$k_x = \frac{32(1 - 0.2)(3 \times 10^4)(2)}{7 - 8 \times 0.2} = 284{,}444 \text{ kN/m}$$

$$c_x = \frac{18.41(1 - 0.2)}{7 - 8 \times 0.2}(2)^2 \sqrt{(1.7)(3 \times 10^4)} = 2464 \text{ kN s/m}$$

$$k_\phi = \frac{8(3 \times 10^4)(2)^3}{3(1 - 0.2)} = 800{,}000 \text{ kN m/rad}$$

$$B_\phi = \frac{3(1-0.2)}{8} \frac{79.02}{1.7(2.0)^5} = 0.44$$

$$c_\phi = \frac{0.8(2)^4 \sqrt{1.7 \times 3 \times 10^4}}{(1-0.2)(1+0.44)} = 2509 \text{ kN m s/rad}$$

Thus for the system under consideration, the homogeneous equations of motion are

$$45.15\ddot{x}_g + 2{,}464\dot{x}_g + 284{,}444 x_g - 1{,}848\dot{\phi} - 213{,}333\phi = 0$$

$$53.62\ddot{\phi} + 3{,}895\dot{\phi} + 960{,}000\phi - 1{,}848\dot{x}_g - 213{,}333 x_g = 0$$

which, after substitution of Equations 13.101 and 13.102 and rearranging in matrix form, lead to

$$\begin{bmatrix} 45.15\lambda^2 + 2{,}464\lambda + 284{,}444 & -(1{,}848\lambda + 213{,}333) \\ -(1{,}848\lambda + 213{,}333) & 53.62\lambda^2 + 3{,}895\lambda + 960{,}000 \end{bmatrix} \begin{Bmatrix} X \\ \Phi \end{Bmatrix} = \begin{Bmatrix} 0 \\ 0 \end{Bmatrix}$$

which, in turn, leads to the characteristic equation

$$\begin{vmatrix} 45.15\lambda^2 + 2{,}464\lambda + 284{,}444 & -(1{,}848\lambda + 213{,}333) \\ -(1{,}848\lambda + 213{,}333) & 53.62\lambda^2 + 3{,}895\lambda + 960{,}000 \end{vmatrix} = 0$$

After expanding the determinant, one has thus that

$$(45.15\lambda^2 + 2{,}464\lambda + 284{,}444)(53.62\lambda^2 + 3{,}895\lambda + 960{,}000) - (1{,}848\lambda + 213{,}333)^2 = 0$$

from which one obtains

$$\lambda_1^{(1)} = -46.358 + 127.752i$$

$$\lambda_1^{(2)} = -46.358 - 127.752i$$

$$\lambda_2^{(1)} = -17.2492 + 69.2213i$$

$$\lambda_2^{(2)} = -17.2492 - 69.2213i$$

According to Equations 13.104 and 13.105, the natural frequencies and damping ratios in the system's two modes of vibration are thus equal to

$$\omega_{n1} = 71.338 \text{ rad/s} \qquad \xi_1 = 0.24$$

$$\omega_{n2} = 135.9 \text{ rad/s} \qquad \xi_2 = 0.34$$

For comparison, note that the natural frequencies and damping ratios of the system when the coupling between the horizontal and rocking vibrations is neglected are

$$\omega_{nx} = \sqrt{\frac{284{,}444}{45.15}} = 79.37 \text{ rad/s} \qquad \xi_x = \frac{2{,}464}{2\sqrt{(284{,}444)(45.15)}} = 0.34$$

$$\omega_{n\phi} = \sqrt{\frac{800{,}000}{53.62}} = 122.15 \text{ rad/s} \qquad \xi_\phi = \frac{2{,}509}{2\sqrt{(800{,}000)(53.62)}} = 0.19$$

b. To determine the amplitudes of vibration, it is noted first that the horizontal force and rocking moment applied at the foundation's center of gravity have, respectively, amplitudes equal to

$$Q_0 = 10 \text{ kN}$$

$$M_y = 10(2 - 0.75) = 12.5 \text{ kN m}$$

and that both oscillate with a frequency equal to

$$\omega = 2\pi\left(\frac{800}{60}\right) = 83.78 \text{ rad/s}$$

Then, by substituting into Equations 13.96 and 13.100 these values of Q_0 and M_y, $h = 0.75$ m, and the calculated mass; mass moment of inertia, and spring and damping constants, the system's equations of motion may be written as

$$45.15\ddot{x}_g + 2{,}464\dot{x}_g + 284{,}444 x_g - 1{,}848\ddot{\phi} - 213{,}333\phi = 10e^{i\omega t}$$

$$53.62\ddot{\phi} + 3{,}895\dot{\phi} + 960{,}000\phi - 1{,}848\ddot{x}_g - 213{,}333 x_g = 12.5 e^{i\omega t}$$

which, after substitution of Equations 13.106 and 13.107 with $\omega = 83.78$ rad/s, lead to

$$(-32{,}436 + 206{,}424 i)A_x - (213{,}333 + 154{,}818 i)A_\phi = 10$$

$$-(213{,}333 + 154{,}818 i)A_x + (583{,}674 + 326{,}307 i)A_\phi = 12.5$$

Solving for A_x and A_ϕ from these simultaneous equations, thus one arrives at

$$A_x = (-5.085 - 6.888 i) \times 10^{-5}$$

$$A_\phi = (-3.918 - 384.4 i) \times 10^{-7}$$

from which the desired vibration amplitudes result as

$$|A_x| = 8.56 \times 10^{-5} \text{ m} = 0.0856 \text{ mm}$$

$$|A_\phi| = 3.84 \times 10^{-5} \text{ rad}$$

13.6 VIBRATION OF RIGID CIRCULAR FOUNDATIONS SUPPORTED BY ELASTIC LAYER

In the elastic halfspace theory, it is assumed that a foundation soil is homogeneous throughout. In most cases, however, actual soil deposits are layered and the rigidity of the underlying layers may considerably affect the vibrations of foundations. In an attempt to evaluate the effect of layering in the vibrations of foundations, several investigators have studied the vibrations of a rigid footing

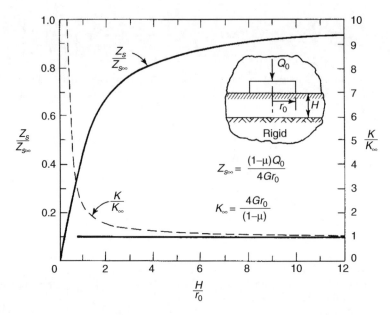

FIGURE 13.16 Static displacement and spring constant versus thickness to radius ratio for rigid circular footing on elastic layer under vertical load. (Reproduced with permission of the Royal Society of London from Bycroft, G. N., *Philosophical Transactions, Royal Society of London, Series A*, 248, 1956, 327–368.)

resting on the surface of an isotropic, homogeneous, elastic layer of thickness H and unbounded in the horizontal direction, when this layer is itself underlain by a halfspace of infinite rigidity.

In 1956, G. N. Bycroft investigated this problem under a vertical oscillating load, assuming a rigid-base-type distribution of vertical pressure under the foundation. He determined the value of the average static displacement for different values of the layer to thickness ratio H/r_0, where, as shown in the inset in Figure 13.16, H is the thickness of the elastic layer and r_0 the radius of the foundation block. He expressed this average displacement in terms of the average displacement for the semi-infinite halfspace. He also determined the ratio of the spring constant for the layered case to the spring constant in the halfspace case. The results are shown in Figure 13.16. It may be observed from Figure 13.16 that the underlying rigid layer introduces a stiffening effect, manifested by a reduced static displacement and an increased spring constant. Note also that for large values of H/r_0, the static stiffness of the elastic layer approximates that of an elastic halfspace. In contrast, for the lower values of this ratio (i.e., $H/r_0 < 1$), the spring constant of the elastic layer increases rapidly as the layer thickness decreases.

Bycroft considered only the case of a massless foundation. In 1957, G. B. Warburton obtained solutions for the case of foundations with a mass different from zero. Both noted that the vibration amplitude becomes infinite at resonance for the case of a foundation mass equal to zero. Bycroft also noted that the vibrations of a massless foundation corresponds to the vibrations of an elastic column fixed at its base, free at the top, and restrained against lateral deformations. For footings that have mass, the amplitude of motion at resonance is large but finite. Warburton determined the maximum dynamic displacements at resonance and expressed them in terms of the magnification over the static displacement. The values listed in Table 13.3 correspond to such magnification factors for different values of the mass ratio $b = m/r_0^3$. These magnification factors, denoted by M_L, are defined as the ratio of the dynamic displacement amplitude to the static displacement in the halfspace case. That is,

$$M_L = \frac{4Gr_0}{(1-\mu)Q_0} A_z \qquad (13.108)$$

where A_z, Q_0, and G and μ, respectively, represent dynamic displacement amplitude, amplitude of the applied force, and shear modulus of elasticity and Poisson ratio for the elastic layer.

TABLE 13.3
Magnification Factors of Vertical Vibrations of Rigid Circular Footings Supported by an Elastic Layer with a Poisson Ratio of 0.25

	M_L				
H/r_0	$b = 0$	$b = 5$	$b = 10$	$b = 20$	$b = 30$
1	∞	5.8	11.4	20.5	28.9
2	∞	8.0	13.1	30.6	40.8
3	∞	4.7	9.5	23.7	36.0
4	∞	3.4	5.9	15.6	27.9
∞	1	1.21	1.60	2.22	2.72

Source: Richards, Jr., F. E., Hall, J. R., and Woods, R. D., *Vibrations of Soils and Foundations*, Prentice-Hall, Englewood Cliffs, NJ, 1970.

It should be noted that the increase in vibration amplitude over the halfspace case is caused by the reflection of waves at the interface between the elastic layer and the rigid halfspace. This reflection of waves brings back to the elastic layer part of the wave energy radiated by the footing and reduces thus its geometric damping. It should also be noted that Warburton's study considered an ideal elastic layer with no internal damping (see Section 13.9). It is likely, therefore, that the consideration of even a small amount of internal damping would significantly reduce the theoretical magnification factors given in Table 13.3.

13.7 VIBRATION OF EMBEDDED FOUNDATIONS

Most of the available solutions for foundation vibrations consider a foundation block as a rigid body attached to the surface of an elastic halfspace. All actual foundations, however, are partially embedded and subjected in consequence to an additional soil resistance on the sides of the foundation. What is more, analytical and experimental studies have shown that this additional soil resistance on the sides of a foundation has a significant influence on the foundation's dynamic response. The consideration of foundation embedment is thus an aspect of foundation vibration that is of significant practical importance. In general, the vibration of embedded foundations is a problem that is more amenable to a solution by the finite element method, but approximate solutions that greatly simplify their treatment have also been proposed. This section presents the approximate analytical solutions developed by V. A. Baranov in 1967 and further studied by M. Novak and his coworkers in the 1970s.

Consider the circular embedded foundation shown in Figure 13.17. The foundation has a radius r_0, rests on the surface of an elastic halfspace with a shear modulus G and density ρ, and is embedded a depth D_f in a soil layer with shear modulus G_s and ρ_s. The foundation, in addition, is subjected to a vertical excitation $Q(t) = Q_0 e^{i\omega t}$ acting along the foundation's vertical axis. It is assumed that there is a perfect bond between the sides of the foundation and the soil, and that the dynamic reaction $R_z(t)$ is independent of the depth of embedment. The equation of dynamic equilibrium for such a system is of the form

$$m\ddot{w}(t) = Q(t) - R_z(t) - N_z(t) \tag{13.109}$$

where $w(t)$, m, and $N_z(t)$ represent vertical displacement of the foundation, foundation mass, and dynamic vertical reaction along the side surface of the foundation block. However, using an elastic

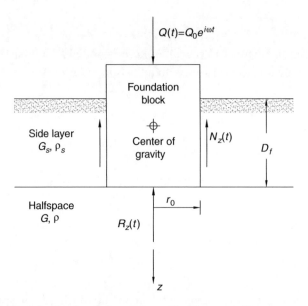

FIGURE 13.17 Embedded circular foundation under vertical harmonic force.

halfspace solution, $R_z(t)$ may be expressed as

$$R_z(t) = Gr_0(C_1 + iC_2)w(t) \tag{13.110}$$

where

$$C_1 = \frac{-f_1}{f_1^2 + f_2^2} \tag{13.111}$$

and

$$C_2 = \frac{f_2}{f_1^2 + f_2^2} \tag{13.112}$$

where f_1 and f_2 are Reissner's displacement functions (see Section 13.3.5). As seen earlier, these displacement functions depend on the dimensionless frequency $a_0 = \omega r_0\sqrt{\rho/G}$, Poisson ratio, and the stress distribution at the base, and may be taken equal to those given in Figure 13.10. Similarly, $N_z(t)$ may be written as

$$N_z(t) = \int_0^{D_f} s(z, t)dz \tag{13.113}$$

where $s(z, t)$ is the dynamic reaction per unit depth of embedment. By assuming, however, that $s(z, t)$ is independent of depth and also using a halfspace solution, Baranov found that

$$s(z, t) = s(t) = G_s(S_1 + iS_2)w(t) \tag{13.114}$$

where

$$S_1 = 2\pi a_0 \frac{J_1(a_0)J_0(a_0) + Y_1(a_0)Y_0(a_0)}{J_0^2(a_0) + Y_0^2(a_0)} \tag{13.115}$$

Structural Response Considering Soil–Structure Interaction

$$S_2 = \frac{4}{J_0^2(a_0) + Y_0^2(a_0)} \tag{13.116}$$

where $J_0(a_0)$ and $J_1(a_0)$ are Bessel functions of the first kind of order 0 and 1, respectively, and $Y_0(a_0)$ and $Y_1(a_0)$ are Bessel functions of the second kind of order 0 and 1, respectively. As a result, Equation 13.113 may be written as

$$N_z(t) = \int_0^{D_f} G_s(S_1 + iS_2)w(t)dz = G_s D_f(S_1 + iS_2)w(t) \tag{13.117}$$

In terms of Equations 13.110 and 13.117, Equation 13.109 may be therefore expressed as

$$m\ddot{w}(t) + Gr_0(C_1 + iC_2)w(t) + G_s D_f(S_1 + iS_2)w(t) = Q(t) \tag{13.118}$$

which, on substitution of a solution of the form

$$w(t) = we^{i\omega t} \tag{13.119}$$

leads to

$$[-m\omega^2 + k_z + i\omega c_z]w = Q_0 \tag{13.120}$$

where k_z and c_z are constants given by

$$k_z = Gr_0\left(C_1 + \frac{G_s D_f}{Gr_0}S_1\right) \tag{13.121}$$

$$c_z = \frac{Gr_0}{\omega}\left(C_2 + \frac{G_s D_f}{Gr_0}S_2\right) \tag{13.122}$$

and may be interpreted as the system's stiffness and damping constants.

Note that as defined by Equations 13.121 and 13.122, the stiffness and damping constants k_z and c_z are frequency dependent since C_1, S_1, C_2, and S_2 depend on the dimensionless frequency a_0. As a result, a trial-and-error procedure needs to be used to determine the natural frequency and damping ratio of a soil–foundation system by means of these equations.

For simplicity, it may be assumed without much loss of accuracy that C_1 and S_1 are equal to frequency-independent constants $\overline{C}_1$ and $\overline{S}_1$, respectively, and that $C_2 = a_0\overline{C}_2$ and $S_2 = a'_0\overline{S}_2$, where a_0 and a'_0 denote the dimensionless frequency for the halfspace and the side layer, respectively, and $\overline{C}_1$, $\overline{S}_1$, $\overline{C}_2$, and $\overline{S}_2$ are the frequency-independent constants given in Table 13.4 for specified ranges of the dimensionless frequency a_0. Thus, when these relationships are substituted into Equations 13.121 and 13.122, one obtains the frequency-independent stiffness and damping constants

$$k_z = Gr_0\left(\overline{C}_1 + \frac{G_s D_f}{Gr_0}\overline{S}_1\right) \tag{13.123}$$

$$c_z = r_0^2\sqrt{\rho G}\left(\overline{C}_2 + \frac{D_f}{r_0}\sqrt{\frac{\rho_s}{\rho}\frac{G_s}{G}}\overline{S}_2\right) \tag{13.124}$$

TABLE 13.4
Values of Constants $\bar{C}_1$ and $\bar{C}_2$ for $0 \leq a_0 \leq 1$ and $\bar{S}_1$ and $\bar{S}_2$ for $0 \leq a_0 \leq 2$

Poisson Ratio	$\bar{C}_1$	$\bar{C}_2$	$\bar{S}_1$	$\bar{S}_2$
0.0	3.9	3.5	2.7	6.7
0.25	5.2	5.0	2.7	6.7
0.5	7.5	6.8	2.7	6.7

TABLE 13.5
Values of Constants $\bar{C}_{x1}$, $\bar{C}_{x2}$, $\bar{S}_{x1}$, and $\bar{S}_{x2}$

	Validity Range					
	$0 \leq a_0 \leq 2.0$		$0 \leq a_0 \leq 1.5$		$0 \leq a_0 \leq 2.0$	
Poisson Ratio	$\bar{C}_{x1}$	$\bar{C}_{x2}$	$\bar{S}_{x1}$	$\bar{S}_{x2}$	$\bar{S}_{x1}$	$\bar{S}_{x2}$
0.0	4.30	2.70	3.6	8.2	—	—
0.25	—	—	—	9.1	4.0	—
0.4	—	—	—	10.6	4.1	—
0.5	5.10	3.15	—	—	—	—

which greatly simplify the analysis of embedded foundations since with these formulas their natural frequencies, damping ratios, and vibration amplitudes may be determined using the conventional solutions from structural dynamics.

Frequency-independent stiffness and damping constants have also been derived for embedded foundations subjected to horizontal forces and rocking and torsional moments. For pure horizontal forces, these constants are

$$k_x = Gr_0 \left(\bar{C}_{x1} + \frac{G_s D_f}{Gr_0} \bar{S}_{x1} \right) \tag{13.125}$$

$$c_x = r_0^2 \sqrt{\rho G} \left(\bar{C}_{x2} + \frac{D_f}{r_0} \sqrt{\frac{\rho_s}{\rho} \frac{G_s}{G}} \bar{S}_{x2} \right) \tag{13.126}$$

where $\bar{C}_{x1}$, $\bar{S}_{x1}$, $\bar{C}_{x2}$, and $\bar{S}_{x2}$ are listed in Table 13.5, whereas for pure rocking moments the corresponding constants are

$$k_\phi = Gr_0^3 \left[\bar{C}_{\phi 1} + \frac{G_s D_f}{Gr_0} \left(\bar{S}_{\phi 1} + \frac{D_f^2}{3r_0^2} \bar{S}_{x1} \right) \right] \tag{13.127}$$

$$c_\phi = r_0^4 \sqrt{\rho G} \left[\bar{C}_{\phi 2} + \frac{G_s}{G} \frac{D_f}{r_0} \left(\bar{S}_{\phi 2} + \frac{D_f^2}{3r_0^2} \bar{S}_{x2} \right) \right] \tag{13.128}$$

where, for a Poisson ratio of zero and $0 \leq a_0 \leq 1.0$, $\bar{C}_{\phi 1} = 2.5$ and $\bar{C}_{\phi 2} = 0.43$, and for any value of Poisson ratio and $0 \leq a_0 \leq 1.5$, $\bar{S}_{\phi 1} = 2.5$ and $\bar{S}_{\phi 2} = 1.8$. As previously mentioned, $\bar{S}_{x1}$ and $\bar{S}_{x2}$

TABLE 13.6
Values of Constants $\overline{C}_{\psi 1}$, $\overline{C}_{\psi 2}$, $\overline{S}_{\psi 1}$, and $\overline{S}_{\psi 2}$

Validity Range					
$0 \leq a_0 \leq 2.0$		$0 \leq a_0 \leq 0.2$		$0.2 \leq a_0 \leq 2.0$	
$\overline{C}_{\psi 1}$	$\overline{C}_{\psi 2}$	$\overline{S}_{\psi 1}$	$\overline{S}_{\psi 2}$	$\overline{S}_{\psi 1}$	$\overline{S}_{\psi 2}$
4.3	0.7	12.4	2.0	10.2	5.4

are listed in Table 13.5. Finally, for torsional moments, the spring and damping constants are given by

$$k_\psi = Gr_0^3 \left(\overline{C}_{\psi 1} + \frac{G_s D_f}{Gr_0} \overline{S}_{\psi 1} \right) \tag{13.129}$$

$$c_\psi = r_0^4 \sqrt{\rho G} \left(\overline{C}_{\psi 2} + \frac{D_f}{r_0} \sqrt{\frac{\rho_s}{\rho} \frac{G_s}{G}} \overline{S}_{\psi 2} \right) \tag{13.130}$$

where $\overline{C}_{\psi 1}$, $\overline{S}_{\psi 1}$, $\overline{C}_{\psi 2}$, and $\overline{S}_{\psi 2}$ are given in Table 13.6.

Foundation responses calculated with the formulas presented earlier are in reasonable agreement with finite element solutions. Comparisons with field experiments have shown significant differences due to the nonlinearities involved in the field tests, as well as the reflection of waves that occur in a soil deposit with layer interfaces and fissures. Another likely factor in the discrepancies may be that the bonding between an actual soil and a foundation may not be as perfect as it is assumed in the theory. There is, however, a good agreement between theory and experiments in regard to the effect of embedment in the response of embedded foundations. In general, it has been observed that neglecting the embedment of a foundation may lead to an unrealistic overestimation of vibration amplitudes. In addition, it has been observed that the effect of embedment depends on embedment depth and the properties of the side soil layer.

EXAMPLE 13.6 NATURAL FREQUENCIES AND VIBRATION AMPLITUDES OF AN EMBEDDED FOUNDATION

A circular foundation is subjected to an oscillating vertical force generated by a machine that operates with a frequency of 600 cpm and weighs 25,000 lb. The amplitude of the force is 2500 lb. The foundation, made of concrete with a unit weight of 150 lb/ft³, has a radius of 4.5 ft, a height of 5 ft, and is embedded into a soil layer of 3.5 ft. The supporting soil has a shear modulus of elasticity of 3000 lb/in.², a Poisson ratio of 0.25, and a unit weight of 122 lb/ft². The side soil layer has a shear modulus of elasticity of 3200 lb/in.², a Poisson ratio of 0.25, and unit weight of 115 lb/ft³. Determine using frequency-independent parameters (a) the damped natural frequency, (b) the amplitude of vertical vibration at resonance, and (c) the amplitude of vibration under operation conditions.

Solution

a. From Table 13.4, the values of the constants $\overline{C}_1$, $\overline{S}_1$, $\overline{C}_2$, and $\overline{S}_2$ for a Poisson ratio of 0.25 are

$$\overline{C}_1 = 5.2 \quad \overline{C}_2 = 5.0 \quad \overline{S}_1 = 2.7 \quad \overline{S}_2 = 6.7$$

According to the given data and Equations 13.123 and 13.124, the mass, stiffness coefficient, and damping constant for this system are thus equal to

$$m = \frac{\pi(4.5^2)(5.0)(150) + 25,000}{32.2} = 2,258 \text{ slugs}$$

$$k_z = (3,000 \times 144)(4.5)(5.2 + \frac{(3,200 \times 144)(3.5)}{(3,000 \times 144)(4.5)} 2.7) = 14,463,360 \text{ lb/ft}$$

$$c_z = (4.5)^2 \sqrt{(122/32.2)(3,000 \times 144)}(5.0 + \frac{3.5}{4.5}\sqrt{\frac{115/32.2}{122/32/2}\frac{3,200 \times 144}{3,000 \times 144}} 6.7) = 264,909 \text{ lb-s/ft}$$

Hence, the undamped natural frequency and damping ratio, respectively, result as

$$\omega_n = \sqrt{\frac{14,463,360}{2,258}} = 80.03 \text{ rad/s}$$

$$\xi = \frac{264,909}{2\sqrt{(14,463,360)(2,258)}} = 0.73$$

from which one finds that the damped natural frequency is equal to

$$\omega_{nd} = 80.03\sqrt{1 - 0.73^2} = 54.70 \text{ rad/s} = 8.71 \text{ Hz}$$

b. Proceeding similarly, it is found that the amplitude of vibration at resonance is given by

$$(A_z)_{res} = \frac{2,500}{14,463,360} \frac{1}{2(0.73)\sqrt{1 - 0.73^2}} = 1.73 \times 10^{-4} \text{ ft} = 0.00208 \text{ in.}$$

c. By the same vein, for an operational frequency of

$$\omega = 600 \text{ cpm} = \frac{2\pi(600)}{60} = 62.83 \text{ rad/s}$$

the amplitude of vibration results as

$$A_z = \frac{2,500}{14,463,360} \frac{1}{\sqrt{\left[1 - \frac{62.83^2}{80.03^2}\right]^2 + 4(0.73)^2\left(\frac{62.83^2}{80.03^2}\right)}} = 1.43 \times 10^{-4} \text{ ft} = 0.00172 \text{ in.}$$

13.8 FOUNDATION VIBRATIONS BY METHOD OF IMPEDANCES

Comparative studies have shown that Lysmer and Hall analogs and Novak's solutions for embedded foundations are reasonably accurate for circular foundations supported by a deep deposit of homogeneous soil and excitations whose frequencies lie in the range $0 \leq a_0 \leq 1$. However, these methods cannot be used for foundations that differ from such cases. The same can be said regarding the use of the equivalent circle to represent large rectangular foundations, which can cause large errors when the aspect ratio is large. For all these cases, solutions have been obtained using instead the concept of the impedance function of a rigid but massless foundation. This section thus introduces the concept of impedance function and describes how the dynamic response of a foundation may be evaluated once the foundation's impedance functions are determined.

Consider a single-degree-of-freedom system with mass m, stiffness k, and damping constant c subjected to a harmonic force $Q(t) = Q_0 e^{i\omega t}$. As is well known, the equation of motion of such a system is given by

$$m\ddot{w}(t) + c\dot{w} + kw(t) = Q_0 e^{i\omega t} \tag{13.131}$$

where $w(t)$ denotes the displacement of the system's mass in the direction of the excitation. Assuming a solution of the form

$$w(t) = w_0 e^{i\omega t} \tag{13.132}$$

and considering that the first and second derivatives of $w(t)$ are given by

$$\dot{w}(t) = i\omega w(t) \tag{13.133}$$

$$\ddot{w}(t) = -\omega^2 w(t) \tag{13.134}$$

then substitution of Equations 13.133 and 13.134 into Equation 13.131 leads to

$$-\omega^2 m w(t) + i\omega c w(t) + k w(t) = Q(t) \tag{13.135}$$

from which one can solve for $w(t)$ to obtain

$$w(t) = \frac{Q(t)}{k - \omega^2 m + ic\omega} = \frac{Q(t)}{\tilde{K}} \tag{13.136}$$

where $\tilde{K}$ is a complex function defined as

$$\tilde{K} = k - \omega^2 m + ic\omega \tag{13.137}$$

and is known as the system's *impedance function* or *dynamic stiffness function*. Alternatively, Equation 13.136 may be expressed as

$$w(t) = \frac{Q(t)}{k - \omega^2 m + ic\omega} = \tilde{F} Q(t) \tag{13.138}$$

where $\tilde{F}$ is given by

$$\tilde{F} = \frac{1}{k - \omega^2 m + ic\omega} \tag{13.139}$$

and is referred to as the system's *compliance function* or *dynamic flexibility function*.

Note, thus, that the displacement response of a system to a harmonic excitation may be obtained by simply dividing the value of the exciting force by a function that is similar to the stiffness constant of a system in the case of a static force. This function depends on the properties of the system and the excitation frequency and is equal to the value of the harmonic force that is needed to induce in the dynamic system a unit displacement in the direction of the force. Its real part characterizes the stiffness and inertia of the system, whereas its imaginary part depicts the damping that is inherent in the system. The response to any arbitrary forcing function may also be determined in terms of such an impedance function by recourse to a Fourier decomposition of the forcing function into harmonic components and the superposition of the system's response to each of such harmonic components.

Observe also that the problem of finding a system's dynamic response is reduced to that of finding its impedance function. Impedance functions may be found by analytical methods (e.g., Reissner method), lumped-parameter methods (e.g., Hsieh method), or the finite element method. For example, the impedance function for an elastic halfspace can be found from Equation 13.55 since this equation may be rewritten as

$$\tilde{K} = \frac{P(t)}{w_0(t)} = \frac{Gr_0}{f_1 + if_2} = \frac{Gr_0}{f_1^2 + f_2^2}(f_1 - if_2) \tag{13.140}$$

or it may be found from Hsieh's equations (Equation 13.73), as these equations similarly lead to

$$\tilde{K} = \frac{P(t)}{w(t)} = k_z + i\omega c_z \tag{13.141}$$

Using the finite element method, the impedance function may be determined as illustrated in Figure 13.18. That is, the finite element method may be used to model the soil–foundation system and determine the displacement at the center of the foundation under a harmonic force $Qe^{i\omega t}$. After considering several values of the excitation frequency, the impedance function is then determined by simply dividing the amplitude of the applied force by the amplitude of the displacement obtained in each case.

Consider now a two-degree-of-freedom system subjected to the harmonic excitations $Q_1(t) = Q_1 e^{i\omega t}$ and $Q_2(t) = Q_2 e^{i\omega t}$ and express the system's equations of motion in the general form

$$m_{11}\ddot{x}_1(t) + m_{12}\ddot{x}_2(t) + c_{11}\dot{x}_1(t) + c_{12}\dot{x}_2(t) + k_{11}x_1(t) + k_{12}x_2(t) = Q_1 e^{i\omega t} \tag{13.142}$$

$$m_{21}\ddot{x}_1(t) + m_{22}\ddot{x}_2(t) + c_{21}\dot{x}_1(t) + c_{22}\dot{x}_2(t) + k_{21}x_1(t) + k_{22}x_2(t) = Q_2 e^{i\omega t} \tag{13.143}$$

where $x_1(t)$ and $x_2(t)$ are the displacements of the system in the direction of the applied forces Q_1 and Q_2, and the coefficients m_{ij}, c_{ij}, and k_{ij} ($i, j = 1, 2$) denote the elements of the system's mass, damping, and stiffness matrices, respectively. These coefficients may be zero and, because of the

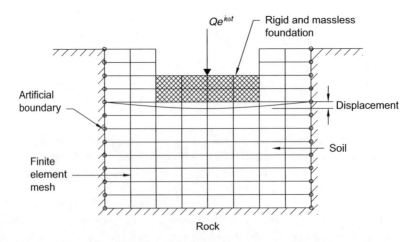

FIGURE 13.18 Schematic illustration of finite element determination of foundation impedance function.

Structural Response Considering Soil–Structure Interaction

symmetry of the matrices, $m_{ij} = m_{ji}$, $c_{ij} = c_{ji}$, and $k_{ij} = k_{ji}$. Express, in addition, the steady-state response of the system in the form

$$x_1(t) = X_1 e^{i\omega t} \tag{13.144}$$

$$x_2(t) = X_2 e^{i\omega t} \tag{13.145}$$

where X_1 and X_2 represent displacement amplitudes and are, in general, complex quantities. On substitution of Equations 13.144 and 13.145 into Equations 13.142 and 13.143, one obtains

$$(-\omega^2 m_{11} + i\omega c_{11} + k_{11})X_1 + (-\omega^2 m_{12} + i\omega c_{12} + k_{12})X_2 = Q_1 \tag{13.146}$$

$$(-\omega^2 m_{21} + i\omega c_{21} + k_{21})X_1 + (-\omega^2 m_{22} + i\omega c_{22} + k_{22})X_2 = Q_2 \tag{13.147}$$

which may also be written as

$$\tilde{K}_{11} X_1 + \tilde{K}_{12} X_2 = Q_1 \tag{13.148}$$

$$\tilde{K}_{12} X_1 + \tilde{K}_{22} X_2 = Q_2 \tag{13.149}$$

where $\tilde{K}_{ij}$, $i, j = 1, 2$, are impedance functions defined as

$$\tilde{K}_{ij} = k_{ij} - \omega^2 m_{ij} + i\omega c_{ij} \qquad i, j = 1, 2 \tag{13.150}$$

In matrix form, Equations 13.148 and 13.149 may be therefore expressed as

$$\begin{bmatrix} \tilde{K}_{11} & \tilde{K}_{12} \\ \tilde{K}_{12} & \tilde{K}_{22} \end{bmatrix} \begin{Bmatrix} X_1 \\ X_2 \end{Bmatrix} = \begin{Bmatrix} Q_1 \\ Q_2 \end{Bmatrix} \tag{13.151}$$

where the matrix that contains the four impedance functions is known as *impedance matrix*. The solution to this matrix equation may then be written as

$$\begin{Bmatrix} X_1 \\ X_2 \end{Bmatrix} = \begin{bmatrix} \tilde{K}_{11} & \tilde{K}_{12} \\ \tilde{K}_{12} & \tilde{K}_{22} \end{bmatrix}^{-1} \begin{Bmatrix} Q_1 \\ Q_2 \end{Bmatrix} \tag{13.152}$$

which, in view of the definition of the inverse of a matrix, may also be expressed as

$$\begin{Bmatrix} X_1 \\ X_2 \end{Bmatrix} = \frac{1}{\tilde{K}_{11}\tilde{K}_{22} - \tilde{K}_{12}^2} \begin{bmatrix} \tilde{K}_{22} & -\tilde{K}_{12} \\ -\tilde{K}_{12} & \tilde{K}_{11} \end{bmatrix} \begin{Bmatrix} Q_1 \\ Q_2 \end{Bmatrix} \tag{13.153}$$

After performing the matrix multiplication, the displacement amplitudes X_1 and X_2 are thus equal to

$$X_1 = \frac{\tilde{K}_{22} Q_1 - \tilde{K}_{12} Q_2}{\tilde{K}_{11}\tilde{K}_{22} - \tilde{K}_{12}^2} \tag{13.154}$$

$$X_2 = \frac{-\tilde{K}_{12} Q_1 + \tilde{K}_{11} Q_2}{\tilde{K}_{11}\tilde{K}_{22} - \tilde{K}_{12}^2} \tag{13.155}$$

It may be observed that as in the single-degree-of-freedom case, the dynamic response of a system with two degrees of freedom is completely defined by its impedance functions. It may also be observed that, as for the one-degree-of-freedom system, these impedance functions may be interpreted as the value of the harmonic force or moment that is needed to produce a unit displacement or rotation. For example, $\tilde{K}_{11}$ may be interpreted as the harmonic force or moment applied in the direction of degree of freedom 1 to produce a unit displacement or rotation in the same direction. Note, however, that in the two-degree-of-freedom case, there is also an impedance function, $\tilde{K}_{12}$, that involves two degrees of freedom. Such a function is known as *cross impedance function* and is interpreted as the value of the harmonic force or moment in the direction of degree of freedom 1 that is needed to produce a unit displacement or rotation in the direction of degree of freedom 2.

As in the single-degree-of-freedom case, the relationship between forces and displacements may also be established instead in terms of compliance functions. For the two-degree-of-freedom case, these compliance functions are defined as the elements of the compliance matrix, defined by the force–displacement relationship

$$\begin{Bmatrix} X_1 \\ X_2 \end{Bmatrix} = \begin{bmatrix} \tilde{F}_{11} & \tilde{F}_{12} \\ \tilde{F}_{12} & \tilde{F}_{22} \end{bmatrix} \begin{Bmatrix} Q_1 \\ Q_2 \end{Bmatrix} \tag{13.156}$$

or, in view of Equation 13.152, by

$$\begin{bmatrix} \tilde{F}_{11} & \tilde{F}_{12} \\ \tilde{F}_{12} & \tilde{F}_{22} \end{bmatrix} = \begin{bmatrix} \tilde{K}_{11} & \tilde{K}_{12} \\ \tilde{K}_{12} & \tilde{K}_{22} \end{bmatrix}^{-1} \tag{13.157}$$

For the particular case of soil–foundation systems, there are six "direct" impedances. These are (1) vertical impedance (force–displacement ratio) for vertical motion, (2) lateral impedance (force–displacement ratio) for horizontal motion in the short direction, (3) longitudinal impedance (force–displacement ratio) for horizontal motion in the long direction, (4) rocking impedance (moment–rotation ratio) for rotational motion about the long centroidal axis of the foundation base, (5) rocking impedance (moment–rotation ratio) for rotation about the short centroidal axis of the foundation base, and (6) torsional impedance (moment–rotation ratio) for rotational motion about a vertical axis. There are also two cross-horizontal-rocking impedances, one for the coupled horizontal and rocking motion along the foundation's longitudinal direction and another for its lateral direction.

Although the impedance functions defined earlier have been derived for systems with a mass, it is advantageous to obtain them for a massless system. This permits their definition on the basis of soil properties and foundation geometry alone and may be used for different structural systems. It also allows their use for the determination of the base input motion from the free-field motion; that is, the quantification of the kinematic effect of soil–structure interaction. The mass of the structure may be incorporated in a later step, at the time the system's equations of dynamic equilibrium are established.

To illustrate how the displacement response of a foundation block may be determined once its impedance functions are known, consider the soil–foundation system shown in Figure 13.19a when the foundation is subjected to a vertical harmonic force acting at its center of gravity. Assume the vertical impedance of the massless system shown in Figure 13.19b is known and denote this impedance function by $\tilde{K}_z(\omega)$. The application of Newton's second law in the vertical direction to the foundation block yields

$$Q(t) - R(t) = m\ddot{w}(t) \tag{13.158}$$

Structural Response Considering Soil–Structure Interaction

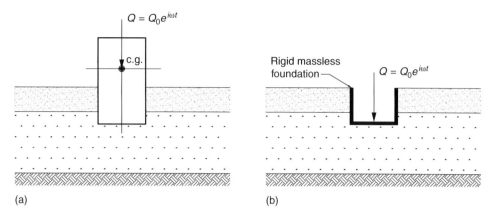

FIGURE 13.19 (a) Soil–foundation system and (b) associated rigid massless foundation.

which may also be expressed as

$$m\ddot{w}(t) + R(t) = Q(t) \tag{13.159}$$

where m, $R(t)$, and $w(t)$ denote mass of the foundation block, vertical reaction of the soil against the foundation, and vertical displacement of the foundation block's center of gravity at time t.

It should be evident that $R(t)$ represents the total soil reaction against the foundation, which is made up of the normal stresses against the foundation base plus the shear stresses along the vertical side walls. However, according to the definition of impedance function, the vertical reaction $R(t)$ may be expressed in terms of the vertical impedance function as

$$R(t) = \tilde{K}_z(\omega)w(t) \tag{13.160}$$

As a result, Equation 13.159 may also be written as

$$m\ddot{w}(t) + \tilde{K}_z(\omega)w(t) = Q(t) \tag{13.161}$$

If the system's steady-state displacement response is expressed as

$$w(t) = we^{i\omega t} \tag{13.162}$$

where w denotes the complex-valued displacement amplitude of the structure's center of gravity, then on substitution of Equation 13.162 into Equation 13.161 one obtains

$$-\omega^2 m w e^{i\omega t} + \tilde{K}_z(\omega)we^{i\omega t} = Q_0 e^{i\omega t} \tag{13.163}$$

from which one can solve for w to obtain

$$w = \frac{Q_0}{\tilde{K}_z(\omega) - \omega^2 m} \tag{13.164}$$

Observe, thus, that the determination of the vibration amplitudes of a soil–structure system is a straightforward task once the system's impedance functions for the frequency of the excitation under consideration are known.

The impedances of a wide variety of foundations have been obtained throughout the years using analytical solutions, the finite element method, boundary-element formulations, or hybrid methods that combine analytical and finite-element solutions. Impedance functions are available for rigid foundations, surface or embedded, on an elastic or viscoelastic halfspace, circular, strip, or rectangular in shape; rigid surface foundations on a homogeneous soil stratum resting over a rigid stratum or an elastic halfspace; rigid foundations of rectangular or arbitrary shape resting on a multilayered viscoelastic halfspace; and even end-bearing and floating single piles. For easy reference, these impedances have been compiled by G. Gazetas in a paper published in 1991 and listed in the Further Readings section. Examples of such available impedance functions are those shown in Figure 13.20, which correspond to a rigid massless circular footing on an elastic halfspace. Note that in Figure 13.20, a_0 is the dimensionless excitation frequency ($a_0 = \omega r_0/v_s$), and k_z and c_z, k_x and c_x, k_ϕ and c_ϕ, and k_ψ and c_ψ denote the dimensionless coefficients in the real and imaginary parts of the impedances for a vertical force, horizontal force, rotational moment, and torsional moment, respectively, when these impedances are expressed in the form

$$\tilde{K}_j = K_j[k_j + ia_0 c_j] \qquad j = z, x, \phi, \Psi \tag{13.165}$$

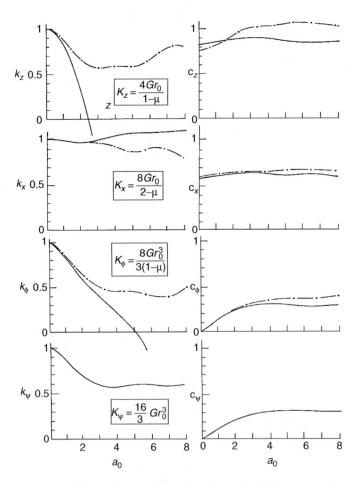

FIGURE 13.20 Values of coefficients in impedance functions of a rigid massless circular footing on an elastic halfspace for two values of Poisson ratio, $\mu = 1/2$ (——) and $\mu = 1/3$ (– · – · –). (Reprinted from Gazetas, G., *International Journal of Soil Dynamics and Earthquake Engineering*, 2, 1, 1983, 2–42. With permission from Elsevier.)

Structural Response Considering Soil–Structure Interaction

In Equation 13.165, K_j is the foundation soil static stiffness defined in the insets of the graphs in Figure 13.20; and, when the soil–foundation system is interpreted as a massless single-degree-of-freedom system, k_j and c_j are given by

$$k_j = 1 - \frac{\omega^2}{\omega_n^2} \qquad c_j = \frac{C_j}{K_j} \frac{v_s}{r_0} \qquad (13.166)$$

where ω_n, v_s, r_0, and C_j, respectively, represent system's natural frequency, velocity of propagation of shear waves in the halfspace, radius of the circular foundation, and system's damping constant.

EXAMPLE 13.7 RESPONSE OF AN INTERACTING SINGLE-DEGREE-OF-FREEDOM STRUCTURE TO EARTHQUAKE GROUND MOTION

The tower shown in Figure E13.7a supports a water tank weighing 2000 kN, is 15 m high, and exhibits, when modeled as a single-degree-of-freedom system with a fixed base, a natural period of 0.25 s and a damping ratio of 2%. The tower's foundation, made of reinforced concrete with a unit weight of 24 kN/m³, is a circular mat with a radius of 6 m and a height of 1 m. The foundation soil has a shear modulus of elasticity of 72 MN/m², a unit weight of 17.7 kN/m³, and a Poisson ratio of 0.5. The tower is excited by vertically propagating seismic shear waves, which in the free field may be characterized by a harmonic displacement function with an amplitude of 0.25 m and a frequency of 26 radians per second. Considering the interaction between foundation soil and structure, determine the amplitude of the induced displacement at the center of the foundation block, the angle of rotation of the foundation block, the structural deformation, and the total horizontal displacement of the water tank. Assume the tower's foundation rests on the surface of the ground, the soil deposit may be represented by a semi-infinite elastic halfspace, there is no slippage between the soil and the base of the foundation, the soil properties do not change during the excitation, and the soil does not experience inelastic strains. Neglect the weight of the tower and the mass moments of inertia of the water tank and the foundation.

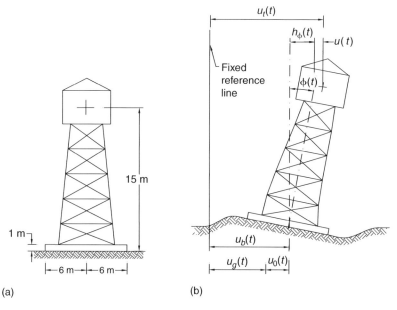

FIGURE E13.7 (a) Original and (b) deformed configuration of water tower in Example 13.7.

Solution

According to the data provided, the basic properties of the structure, when considered with a fixed base, are

Structural mass = m = 2000/9.81 = 204 Mg
Natural frequency = ω_n = 2π/0.25 = 25.13 rad/s
Stiffness constant = k = $(25.13)^2(204)$ = 128,829 kN/m
Damping constant = c = 2(0.02)(25.13)(204) = 205 kN s/m
Foundation mass = m_b = 24(π × 6^2 × 1)/9.81 = 277 Mg

Similarly, the shear wave velocity of the foundation soil is

$$v_s = \sqrt{72{,}000 \times 9.81/17.7} = 200 \text{ m/s}$$

and, according to the formulas in Table 13.1, the static soil stiffnesses in the horizontal and rotational directions are

$$K_x = \frac{8(72{,}000)(6)}{2 - 0.5} = 2.304 \times 10^6 \text{ KN/m}$$

$$K_\phi = \frac{8(72{,}000)(6)^3}{3(1 - 0.5)} = 82.944 \times 10^6 \text{ KN m/rad}$$

Furthermore, from Figure 13.20 for a dimensionless frequency equal to

$$a_0 = 26 \times \frac{6}{200} = 0.78$$

the corresponding dimensionless coefficients k_x, k_ϕ, c_x, and c_ϕ are

$$k_x = 0.98 \quad c_x = 0.6$$

$$k_\phi = 0.85 \quad c_\phi = 0.1$$

As a result, the foundation impedances are equal to

$$\tilde{K}_x = 2.304 \times 10^6[0.98 + i(0.78)(0.6)] = (2.258 + 1.078i) \times 10^6 \text{ kN/m}$$

$$\tilde{K}_\phi = 82.944 \times 10^6[0.85 + i(0.78)(0.1)] = (70.502 + 6.470i) \times 10^6 \text{ kN m/rad}$$

When the soil flexibility is taken into account, it may be considered that the given soil–structure system has three degrees of freedom: the horizontal displacement of the foundation mat, the rotation of the foundation mass, and the horizontal displacement of the structural mass, assumed concentrated at a height h of 15 m aboveground. As shown in Figure E13.7b, let, then, the following variables be defined as follows:

$u_g(t)$ = free-field horizontal ground displacement
$u_b(t)$ = horizontal displacement of the center of the foundation mat
$\phi(t)$ = angle of rotation of the foundation mat about the center of the foundation base
$u(t)$ = deformation of structure
$u_t(t) = u_b(t) + h\phi(t) + u(t)$ = total horizontal displacement of the structural mass
$u_0(t) = u_b(t) - u_g(t)$ = difference between the horizontal displacement of the foundation mat and the free-field horizontal ground displacement

Additionally, let $u_b(t)$, $\phi(t)$, $u(t)$, and $u_0(t)$ be expressed in terms of the ground displacement $u_g(t)$ as

$$u_b(t) = Xu_g(t)$$

$$\phi(t) = Yu_g(t)$$

$$u(t) = Zu_g(t)$$

$$u_0(t) = (X - 1)u_g(t)$$

where X, Y, and Z are the corresponding complex-valued, unknown at this point, transfer functions. Let also the ground motion be represented by the exponential function

$$u_g(t) = 0.25 e^{i26t}$$

such that

$$\dot{u}_g(t) = 26i u_g(t) \quad \text{and} \quad \ddot{u}_g(t) = -676 u_g(t).$$

The equation of motion for the structural mass may be thus written as

$$204\ddot{u}_t(t) + 205\dot{u}(t) + 128{,}829 u(t) = 0$$

which, in terms of the transfer functions and the expressions for the first and second derivatives of $u_g(t)$ introduced earlier, may also be expressed as

$$204(-676)(X + 15Y + Z)u_g(t) + 205(26i)Zu_g(t) + 128{,}829 Z u_g(t) = 0$$

or as

$$X + 15Y + (0.066 - 0.039i)Z = 0$$

Similarly, the dynamic equilibrium of the horizontal forces and moments about the base of the foundation acting on the structure, when the entire structure is considered as a free body, yields

$$277\ddot{u}_b(t) + 204\ddot{u}_t(t) + P(t) = 0$$

$$204(15)\ddot{u}_t(t) + M(t) = 0$$

where $P(t)$ and $M(t)$ are the force and moment exerted by the foundation soil against the base of the foundation mat. Note, however, that in accordance with the definition of impedance function, $P(t)$ and $M(t)$ may be expressed as

$$P(t) = \tilde{K}_x u_0(t) = (2.258 + 1.078i)10^6 u_0(t) \text{ kN}$$

$$M(t) = \tilde{K}_\phi \phi(t) = (70.502 + 6.470i)10^6 \phi(t) \text{ kN m}$$

Hence, the preceding two equations of motion may also be written as

$$277\ddot{u}_b(t) + 204\ddot{u}_t(t) + (2.258 + 1.078i)10^6 u_0(t) = 0$$

$$3060\ddot{u}_t(t) + (70.502 + 6.470i)10^6 \phi(t) = 0$$

or, in terms of the transfer functions X, Y, and Z and the ground displacement $u_g(t)$, as

$$277(-676)Xu_g(t) + 204(-676)(X + 15Y + Z)u_g(t) + (2.258 + 1.078i)10^6(X - 1)u_g(t) = 0$$

$$3060(-676)(X + 15Y + Z)u_g(t) + (70.502 + 6.470i)10^6 Yu_g(t) = 0$$

which, after simplifying, may be reduced to

$$(-14.01 - 7.82i)X + 15Y + Z = -16.37 - 7.82i$$

$$X + (-19.08 - 3.128i)Y + Z = 0$$

Hence, the three equations of motion lead to the matrix equation

$$\begin{bmatrix} 1 & 15 & 0.066 - 0.039i \\ -14.01 - 7.82i & 15 & 1 \\ 1 & -19.08 - 3.128i & 1 \end{bmatrix} \begin{Bmatrix} X \\ Y \\ Z \end{Bmatrix} = \begin{Bmatrix} 0 \\ -16.37 - 7.82i \\ 0 \end{Bmatrix}$$

whose solution yields

$$\begin{Bmatrix} X \\ Y \\ Z \end{Bmatrix} = \begin{Bmatrix} 0.965 - 2.445i \times 10^{-3} \\ -0.055 - 3.962i \times 10^{-3} \\ -2.001 - 0.245i \end{Bmatrix} = \begin{Bmatrix} 0.965 e^{-0.00253i} \\ 0.055 e^{-3.07i} \\ 2.016 e^{-3.02i} \end{Bmatrix}$$

The amplitude of the horizontal displacement of the foundation mat, the rotation of the foundation mat, the structural deformation, and the total displacement of the structural mass induced by the specified ground motion are thus equal to

$$|u_b(t)| = |X||u_g(t)| = 0.965(0.25) = 0.24 \text{ m}$$

$$|\phi(t)| = |Y||u_g(t)| = 0.055(0.25) = 0.014 \text{ rad}$$

$$|u(t)| = |Z||u_g(t)| = 2.016(0.25) = 0.50 \text{ m}$$

$$|u_t(t)| = |u_b(t) + h\phi(t) + u(t)| = |X + hY + Z||u_g(t)|$$
$$= |-1.861 - 0.308i|0.25 = 0.47 \text{ m}$$

It is interesting to note that had the base of the structure been considered fixed, the structural deformation would have been equal to

$$|u(t)| = \left(\frac{26}{25.13}\right)^2 \frac{0.25}{\sqrt{\left[1 - \left(\frac{26}{25.13}\right)^2\right]^2 + \left[2(0.02)\left(\frac{26}{25.13}\right)\right]^2}} = 3.28 \text{ m}$$

Hence, the effect of the soil flexibility was to considerably reduce this deformation.

13.9 MATERIAL OR INTERNAL DAMPING IN SOILS

The damping constants derived from the halfspace theory reflect only the energy dissipation due to the propagation of elastic waves away from the immediate vicinity of the foundation. All real materials, however, also exhibit another type of damping or energy dissipation mechanism. This additional energy dissipation mechanism may be observed when, for example, a soil sample is set into a state of free vibration since in such a case the vibration amplitude will decrease with each cycle and eventually die out. But this reduction in the amplitude of vibration is not the result of viscous damping. It is, rather, caused by friction and the slippage of the soil particles with respect to one another—a phenomenon often referred to as the *hysteretic behavior of soils*. This additional source of damping is known as *material*, *internal*, or *hysteretic damping*. It is precluded from the halfspace theory due to the assumption made that the material that constitutes the halfspace is perfectly elastic. Realistic solutions to the foundation vibration problem should therefore include the material damping of soils, particularly in soils subjected to excitations associated with large strains.

The amount of material or internal damping in a soil deposit depends on the type of soil and varies thus over a wide range. It is primarily a function of strain amplitude and initial confining pressure. Measurements made at the level of the strain levels occurring in machine foundations indicate that internal damping in soils is between 0.01 and 0.10. The average value of 0.05 may be thus added to the damping ratios due to radiation damping to obtain a rough estimate of the combined effects of radiation and internal damping. In general, internal damping may be added to the damping constant found from the halfspace theory in the form

$$c_{tot} = c + \frac{2k}{\omega}\beta \qquad (13.167)$$

where c, k, and β are radiation damping constant, static spring constant, and internal damping ratio.

For horizontal translation, and particularly for vertical translation, internal damping appears to be relatively unimportant compared to radiation damping. For rotational vibrations, however, internal damping becomes a significant part of the total damping, as radiation damping is low in this mode of vibration.

For mathematical convenience, the energy dissipation mechanism of soils is sometimes modeled as in a Kelvin–Voight solid. As shown in Figure 13.21, resistance to shearing deformation in a

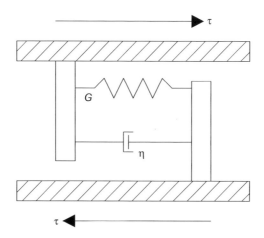

FIGURE 13.21 Elements of Kelvin–Voight solid subjected to horizontal shear stresses.

Kelvin–Voight solid is provided by an elastic part and a viscous part. The stress–strain relationship of a Kelvin–Voight solid in shear is thus expressed as

$$\tau = G\gamma + \eta \frac{\partial \gamma}{\partial t} \tag{13.168}$$

where G and η and τ and γ, respectively, represent solid's shear modulus of elasticity and viscosity coefficient, respectively and shear stress and shear strain, respectively.

To express Equation 13.168 in terms of a damping ratio as opposed to the viscosity coefficient η, it may be considered that when the soil is subjected to a harmonic shear strain of the form

$$\gamma = \gamma_0 e^{i\omega t} \tag{13.169}$$

where γ_0 is the strain amplitude and ω the frequency of the excitation, the shear stress may be written alternatively as

$$\tau = G\gamma_0 e^{i\omega t} + i\omega\eta\gamma_0 e^{i\omega t} = (G + i\omega\eta)\gamma \tag{13.170}$$

Considering then only the real parts of τ and γ, the energy per unit volume dissipated in a single cycle of deformation (given by the area under the stress–strain curve) may be expressed as

$$\Delta W = \int_0^{2\pi/\omega} \tau \dot{\gamma} \, dt = \int_0^{2\pi/\omega} (G\gamma_0 \cos\omega t - \omega\eta\gamma_0 \sin\omega t)(-\omega\gamma_0 \sin\omega t) dt = \pi\eta\omega\gamma_0^2 \tag{13.171}$$

Similarly, the corresponding strain energy stored in a perfectly elastic material, also given by the area under the stress–strain curve, is equal to

$$W = \frac{1}{2} G\gamma_0^2 \tag{13.172}$$

Consider now a single-degree-of-freedom system with mass m, stiffness k, and viscous damping coefficient c, when the system is subjected to a harmonic excitation $F = F_0 e^{i\omega t}$. As is well known from structural dynamics, the response of such a system is of the form

$$u(t) = u_0 e^{i(\omega t - \phi)} = \frac{F_0}{\omega_n^2 m} H(\omega) e^{i(\omega t - \phi)} \tag{13.173}$$

where ω_n, $H(\omega)$, and ϕ are natural frequency of the system, magnification factor, and phase angle.

The energy dissipated by the system in one cycle of oscillation, considering only the real parts of F and u, is thus equal to

$$\Delta W = \int F \, du = \int_0^{2\pi/\omega} F\dot{u} \, dt = \omega u_0 F_0 \int_0^{2\pi/\omega} \cos\omega t \sin(\omega t - \phi) dt = -\pi u_0 F_0 \sin\phi \tag{13.174}$$

However, it is not difficult to show that

$$\sin\phi = \frac{c\omega}{m\omega_n^2} H(\omega) = c\omega \frac{u_0}{F_0} \tag{13.175}$$

Therefore, ΔW may also be expressed as

$$\Delta W = \pi c \omega u_0^2 \tag{13.176}$$

Finally, consider that in the absence of damping, the strain energy stored in the system is given by

$$W = \frac{1}{2} k u_0^2 \tag{13.177}$$

It is also well known that the damping ratio of a single-degree-of-freedom system may be expressed as

$$\xi = \frac{c}{2\sqrt{km}} = \frac{c \omega_n}{2k} \tag{13.178}$$

Therefore, if c and k are solved from Equations 13.176 and 13.177, respectively, and substituted into Equation 13.178, the damping ratio ξ of a linear system with a viscous damper may be expressed in terms of the dissipated and strain energies for the case when $\omega = \omega_n$ as

$$\xi = \frac{1}{4\pi} \frac{\Delta W}{W} \tag{13.179}$$

where the ratio $\Delta W/W$ is known as *specific loss factor*.

Returning to the case of a continuous Kelvin–Voight solid, Equation 13.179 may now be used to find a relationship between viscosity coefficient and damping ratio. Accordingly, if Equations 13.171 and 13.172 are substituted into Equation 13.179, one obtains

$$\xi = \frac{1}{4\pi} \frac{\pi \eta \omega \gamma_0^2}{G \gamma_0^2 / 2} = \frac{\eta \omega}{2G} \tag{13.180}$$

from which it is found that

$$\eta = \frac{2G}{\omega} \xi \tag{13.181}$$

On substitution of Equation 13.181 into Equation 13.170, the shear stress may be thus expressed as

$$\tau = G(1 + 2\xi i)\gamma = G^* \gamma \tag{13.182}$$

where G^* establishes the relationship between stress and strain and is considered to represent the shear modulus of elasticity for viscoelastic solids.

Note that this shear modulus of elasticity is composed of real and imaginary components, where the real component represents the elastic resistance of the solid and the imaginary part its viscous resistance. Note, also, that the phase angle δ between these real and imaginary parts is given by

$$\tan \delta = 2\xi \tag{13.183}$$

from which it may be seen that the damping ratio is related to this phase angle. For this reason, it is customary to express the internal damping in soils by this phase angle, called *loss angle*. By combining Equations 13.183 and 13.179, it is also possible to write this loss angle in terms of the specific loss factor $\Delta W/W$ as

$$\tan \delta = \frac{1}{2\pi} \frac{\Delta W}{W} \tag{13.184}$$

13.10 VIBRATIONS OF FOUNDATIONS ON VISCOELASTIC HALFSPACE

In 1973, A. S. Veletsos and B. Verbic investigated the effect of material damping on the steady-state response of harmonically excited foundations. For this purpose, they modeled the halfspace as a linear viscoelastic solid. They obtained their solutions by application of the *correspondence principle* of the theory of viscoelasticity to the available approximate solutions for the corresponding elastic problem. That is, by considering that the stress–strain relationship for the halfspace is given by

$$\tau = G^*\gamma \tag{13.185}$$

where G^* is the shear modulus of elasticity of the viscoelastic halfspace, which, according to Equation 13.182, is defined as

$$G^* = G(1 + 2\xi i) \tag{13.186}$$

where G is the shear modulus of elasticity of the elastic halfspace and, based on Equations 13.183 and 13.184, ξ is a damping ratio given by

$$\xi = \frac{1}{2}\tan\delta = \frac{1}{4\pi}\frac{\Delta W}{W} \tag{13.187}$$

where ΔW and W are as defined in Section 13.9. In the case of soils, $\Delta W/W$ is normally determined from laboratory tests and is a function of the level of strain. At small strains, $\Delta W/W$ is normally less than 0.1π, whereas at the level of stains associated with high-intensity earthquake ground motions, $\Delta W/W$ may be as high as 0.8π.

By replacing G by G^* in the impedance functions for the purely elastic case and assuming that Poisson ratio is the same for the viscoelastic and elastic cases, Veletsos and Verbic found that the impedance functions of a rigid massless foundation over a viscoelastic halfspace may be expressed as

$$\tilde{K}_j = K_j(k_j^v + ia_0 c_j^v) \tag{13.188}$$

where K_j represents the static stiffness in the j direction, given for the horizontal, rotational, and vertical directions, respectively, by

$$K_x = \frac{8Gr_0}{2-\mu} \tag{13.189}$$

$$K_\phi = \frac{8Gr_0^3}{3(1-\mu)} \tag{13.190}$$

$$K_z = \frac{4Gr_0}{1-\mu} \tag{13.191}$$

and the coefficients k_j^v and c_j^v are real-valued functions of the dimensionless frequency a_0, Poisson ratio μ, and the material damping parameter $\tan\delta$. Figure 13.22 shows for a wide range of the dimensionless frequency a_0 and different values of the damping parameter $\tan\delta$ the values found by Veletsos and Verbic for these coefficients. The coefficients in Figures 13.22a and 13.22b, Figures 13.22c and 13.22d, and Figures 13.22e and 13.22f, respectively, correspond to a horizontal force, a moment about a horizontal axis in the plane of the foundation, and a vertical force. Note that the dashed lines in Figures 13.22a through 13.22e correspond to the case of $\tan\delta = 0$ and thus to the purely elastic halfspace.

Structural Response Considering Soil–Structure Interaction

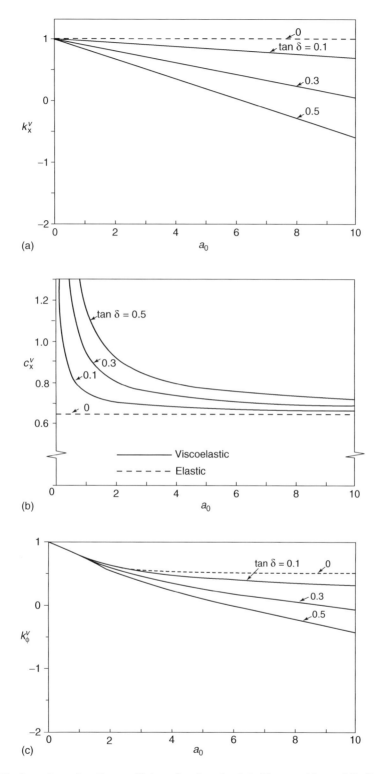

FIGURE 13.22 Impedance function coefficients for viscoelastic halfspace with $\mu = 1/3$. (Reproduced with permission of John Wiley & Sons, Limited, from Veletsos A. S. and Verbic, B., *Earthquake Engineering and Structural Dynamics*, 2, 1973, 87–102.)

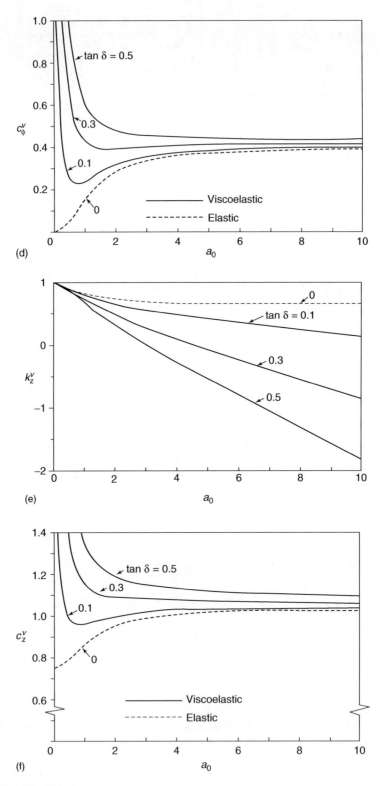

FIGURE 13.22 (Continued)

By interpreting k_j^v as the stiffness constant and c_j^v as the damping coefficient of a massless single-degree-of-freedom system, it may be observed from the graphs in Figure 13.22 that the effect of material damping is to decrease the stiffness of the system and increase its damping coefficient. As previously indicated, this increase in the damping coefficient of a foundation soil may be particularly important for foundations under rocking motions as the effect of radiation damping is generally small for rocking vibrations.

13.11 SIMPLIFIED EQUIVALENT SINGLE-DEGREE-OF-FREEDOM SYSTEM METHOD

Even with the computer equipment and software available nowadays, mathematically precise analyses of complex soil–structure systems still involve a major computational effort, and are generally too costly and too cumbersome for ordinary or preliminary designs. This is particularly true if the responses of the structure and the supporting medium are to be evaluated in the inelastic range. Compounding the difficulty is the fact that many times it is an arduous task to intelligently interpret the results of a number of complex analyses. Moreover, complex analyses are more prone to misinterpretation and misuse than are simpler methods, and may discourage the exercise of engineering judgment, which, as is well known, is an essential component in a successful design. There is thus a need for simple procedures and concepts with which the effects of soil–structure interaction for ground-excited systems may be readily and reliably considered in design.

A method that provides a useful insight into the soil–structure phenomenon and, because of its simplicity, has been adopted by several building codes, consists of substituting a structure and its supporting flexible medium by an equivalent single-degree-of-freedom structure that is fixed at its base. The equivalency of the method is established by modifying the natural frequency and damping ratio of the fixed-base structure so as to obtain the same structural response as the actual soil–structure system. Such an idealized structure may be viewed as either the direct model of a one-story building structure or, more generally, as the model of a multistory structure that responds predominantly in its fundamental mode when considered fixed at its base. This section describes this method in some detail, shows how the equivalency is obtained, identifies some of the method's limitations, and illustrates its application.

For the purpose of obtaining the properties of the equivalent structure, consider the one-story structure shown in Figure 13.23. Assume that this is a linear structure with height h, mass m, and lateral stiffness k when its base is considered fixed. Assume, in addition, that the structure is supported by a circular foundation mat of mass m_0 and radius r_0 resting on the surface of a homogeneous viscoelastic halfspace. Assume, further, that the foundation mat is perfectly rigid, of negligible thickness and bonded to the foundation medium; and that its mass m_0 is negligibly small compared to the mass of the structure. Assume also that the columns of the structure are massless and axially inextensible.

Let the foundation medium be characterized by its shear modulus of elasticity G, mass density ρ, Poisson ratio μ, and specific loss factor $\Delta W/W$. As indicated in Section 13.9, for a material under a harmonic excitation, ΔW is the area under a hysteretic loop in the stress–strain diagram of the material and W is the strain energy stored in a purely elastic material subjected to the same excitation. For a linear material with no internal damping, $\Delta W = 0$. As indicated earlier, too, at small strains, $\Delta W/W$ is normally less than 0.1π, whereas at the level of strains associated with severe earthquake ground motions $\Delta W/W$ may be as high as 0.8π. It will be assumed that $\Delta W/W$ is frequency independent and expressed as (see Equations 13.179 and 13.184)

$$\frac{\Delta W}{W} = 4\pi\xi = 2\pi \tan\delta \quad (13.192)$$

where ξ and δ are damping ratio of an equivalent single-degree-of-freedom system with viscous damping and medium's loss angle (see Section 13.9).

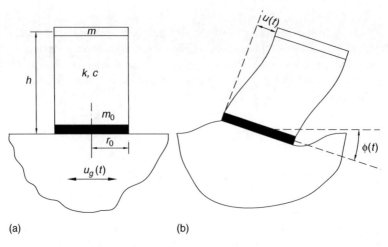

FIGURE 13.23 One-story structure resting on flexible medium.

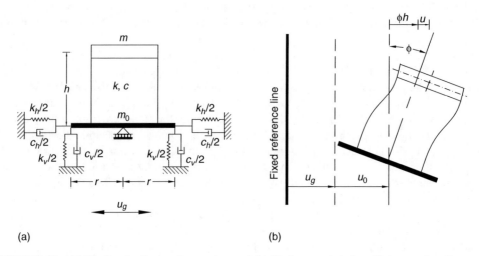

FIGURE 13.24 (a) Idealized soil–structure system and (b) displacements induced by ground motion.

To account for the flexibility of the foundation medium, the base of the structure will be considered free to displace in the horizontal direction and rotate about an axis perpendicular to the plane of the structure passing through the center of the base. In addition, it will be considered restrained by vertical and horizontal springs and viscous dampers as shown in Figure 13.24a. The horizontal springs account for the translational flexibility of the supporting medium, whereas the vertical springs account for the corresponding rotational flexibility. The dashpots account for both the radiation and material damping in the medium.

Adopting the values obtained for a viscoelastic halfspace (see Section 13.10), let the stiffness and damping constants of the horizontal springs and dampers be of the form

$$k_x = \alpha_x K_x \tag{13.193}$$

$$c_x = \beta_x \frac{K_x r_0}{v_s} \tag{13.194}$$

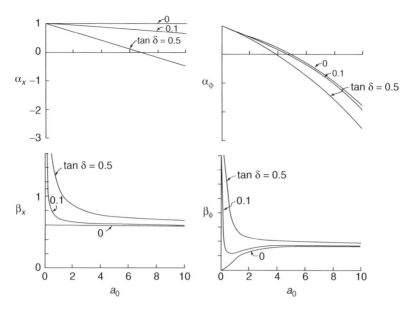

FIGURE 13.25 Coefficients corresponding to a Poisson ratio of 0.45 to determine foundation stiffness and damping constants in Equations 13.193 through 13.196. (Reproduced with permission of ASCE from Veletsos, A. S. and Nair, V. V. D, *Journal of Structural Division, ASCE*, 101, 1975, 109–129.)

and, since no vertical motion is considered, let the stiffness and damping constants for the vertical springs and dampers be obtained in a way that yields a rotational stiffness and damping constant, respectively, equal to

$$k_\phi = \alpha_\phi K_\phi \tag{13.195}$$

$$c_\phi = \beta_\phi \frac{K_\phi r_0}{v_s} \tag{13.196}$$

In the preceding equations, K_x and K_ϕ denote the static stiffnesses of the foundation medium as defined by Equations 13.189 and 13.190, v_s is the shear wave velocity in the elastic halfspace, and r_0 is the radius of the foundation. Similarly, the coefficients α_x, α_ϕ, β_x, and β_ϕ are dimensionless functions that depend on the frequency of the excitation, Poisson ratio, and the material damping tan δ. Figure 13.25 shows the values of these coefficients for a Poisson ratio of 0.45 and several values of tan δ.

Assume now that the motion exciting the base of the structure is the free-field ground motion and that this ground motion has only a horizontal component. Let $u_g(t)$ represent such a horizontal component of the free-field ground motion and consider that, as a result of it, the foundation mat is displaced an amount $u_0(t)$ and rotated an angle $\phi(t)$, and that the structure is deformed an amount $u(t)$ (see Figure 13.24b). Note that if the structure were rigidly supported, $\phi(t)$ and $u_0(t)$ would both be equal to zero. Consider also that the response quantities of interest are the relative displacement of the structure, $u + \phi h$, and the associated base shear, $V = ku$.

Consideration of the free-body diagram shown in Figure 13.26a and the application of Newton's second law in the horizontal direction yield

$$-c\dot{u} - ku = m(\ddot{u}_0 + \ddot{\phi}h + \ddot{u} + \ddot{u}_g) \tag{13.197}$$

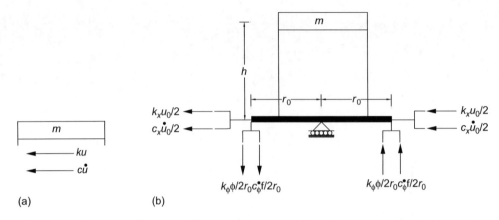

FIGURE 13.26 Free-body diagrams of (a) structural mass and (b) soil–structure system in Figure 13.24a.

Similarly, consideration of the free-body diagram in Figure 13.26b and the dynamic equilibrium of horizontal forces and moments about the center of the foundation mass lead to

$$-c_x \dot{u}_0 - k_x u_0 = m(\ddot{u}_0 + \ddot{\phi}h + \ddot{u} + \ddot{u}_g) \tag{13.198}$$

$$-c_\phi \dot{\phi} - k_\phi \phi = mh(\ddot{u}_0 + \ddot{\phi}h + \ddot{u} + \ddot{u}_g) \tag{13.199}$$

Equations 13.197 through 13.199 represent the coupled equations that describe the motion of the system in Figure 13.24a. In matrix form, Equations 13.197 through 13.199 may be expressed as

$$m\begin{bmatrix}1 & 1 & 1\\ 1 & 1 & 1\\ 1 & 1 & 1\end{bmatrix}\begin{Bmatrix}\ddot{u}\\ \ddot{u}_0\\ \ddot{\phi}h\end{Bmatrix} + \begin{bmatrix}c & 0 & 0\\ 0 & c_x & 0\\ 0 & 0 & c_r\end{bmatrix}\begin{Bmatrix}\dot{u}\\ \dot{u}_0\\ \dot{\phi}h\end{Bmatrix} + \begin{bmatrix}k & 0 & 0\\ 0 & k_x & 0\\ 0 & 0 & k_r\end{bmatrix}\begin{Bmatrix}u\\ u_0\\ \phi h\end{Bmatrix} = -m\begin{Bmatrix}1\\ 1\\ 1\end{Bmatrix}\ddot{u}_g \tag{13.200}$$

where

$$k_r = \frac{k_\phi}{h^2} \quad \text{and} \quad c_r = \frac{c_\phi}{h^2} \tag{13.201}$$

or, alternatively, as

$$[M]\{\ddot{u}\} + [C]\{\dot{u}\} + [K]\{u\} = -m\{J\}\ddot{u}_g \tag{13.202}$$

where $[M]$, $[C]$, and $[K]$ are defined as

$$[M] = m\begin{bmatrix}1 & 1 & 1\\ 1 & 1 & 1\\ 1 & 1 & 1\end{bmatrix} \quad [C] = \begin{bmatrix}c & 0 & 0\\ 0 & c_x & 0\\ 0 & 0 & c_r\end{bmatrix} \quad [K] = \begin{bmatrix}k & 0 & 0\\ 0 & k_x & 0\\ 0 & 0 & k_r\end{bmatrix} \tag{13.203}$$

Structural Response Considering Soil–Structure Interaction

and may be interpreted as the mass, damping, and stiffness matrices of the system. In addition, $\{u\}$ is a displacement vector given by

$$\{u\} = \begin{Bmatrix} u \\ u_0 \\ \phi h \end{Bmatrix} \qquad (13.204)$$

and $\{J\}$ is a unit vector of size three.

Using the mass and stiffness matrices defined by Equation 13.203, the undamped natural frequencies and mode shapes of the interacting system may be thus determined from the eigenvalue problem

$$-m\tilde{\omega}^2 \begin{bmatrix} 1 & 1 & 1 \\ 1 & 1 & 1 \\ 1 & 1 & 1 \end{bmatrix} \begin{Bmatrix} u \\ u_0 \\ \phi h \end{Bmatrix} + \begin{bmatrix} k & 0 & 0 \\ 0 & k_x & 0 \\ 0 & 0 & k_r \end{bmatrix} \begin{Bmatrix} u \\ u_0 \\ \phi h \end{Bmatrix} = \begin{Bmatrix} 0 \\ 0 \\ 0 \end{Bmatrix} \qquad (13.205)$$

which in algebraic form may be written as

$$-m\tilde{\omega}^2(u_0 + \phi h + u) + ku = 0 \qquad (13.206)$$

$$-m\tilde{\omega}^2(u_0 + \phi h + u) + k_x u_0 = 0 \qquad (13.207)$$

$$-m\tilde{\omega}^2(u_0 + \phi h + u) + k_r \phi h = 0 \qquad (13.208)$$

But by subtracting Equation 13.207 from Equation 13.206, one obtains

$$ku - k_x u_0 = 0 \qquad (13.209)$$

from which it may be determined that

$$u_0 = \frac{k}{k_x} u \qquad (13.210)$$

Similarly, by subtracting Equation 13.208 from Equation 13.206, one obtains

$$ku - k_r \phi h = 0 \qquad (13.211)$$

from which it is found that

$$\phi h = \frac{k}{k_r} u \qquad (13.212)$$

Substitution of Equations 13.210 and 13.212 into Equation 13.206 leads therefore to

$$-\tilde{\omega}^2 m \left(1 + \frac{k}{k_h} + \frac{k}{k_r}\right) u + ku = 0 \qquad (13.213)$$

from which one can solve for $\tilde{\omega}$ to obtain

$$\tilde{\omega} = \frac{\tilde{\omega}}{\sqrt{1 + \frac{k}{k_h} + \frac{k}{k_r}}} \tag{13.214}$$

where $\omega = \sqrt{k/m}$ is the undamped natural frequency of the fixed-base structure.

Equation 13.214 gives the value of the undamped natural frequency of the structure supported on flexible soil and is a function of the soil flexibility in the vertical and horizontal directions. Note that since there is only one mass, there is only one natural frequency and only one mode shape. In fact, Equation 13.214 can also be derived by considering the structure–soil system as a single-degree-of-freedom system with a mass m and an equivalent stiffness $\tilde{k}$, by considering that this equivalent stiffness is equal to the stiffness of three springs in series, and by considering that the three springs in question have stiffness constants equal to k, k_x, and k_r. Accordingly, since for springs in series such an equivalent stiffness is given by

$$\frac{1}{\tilde{k}} = \frac{1}{k} + \frac{1}{k_x} + \frac{1}{k_r} \tag{13.215}$$

the natural frequency of the system results as

$$\tilde{\omega} = \sqrt{\frac{\tilde{k}}{m}} = \sqrt{\frac{k/m}{1 + \frac{k}{k_x} + \frac{k}{k_r}}} = \frac{\omega}{\sqrt{1 + \frac{k}{k_x} + \frac{k}{k_r}}} \tag{13.216}$$

which is identical to the expression in Equation 13.214. Note also that according to this formula, the natural frequency of the interacting system is always smaller than the natural frequency of the fixed-based structure. The mode shape corresponding to the natural frequency defined by Equation 13.214 may be obtained by setting $u = 1$ and using Equations 13.210 and 13.212. Accordingly, this mode shape is of the form

$$\{\tilde{X}\} = \begin{Bmatrix} u \\ u_0 \\ \phi h \end{Bmatrix} = \begin{Bmatrix} 1 \\ k/k_h \\ k/k_r \end{Bmatrix} \tag{13.217}$$

Assuming now that the damping matrix of the system is an orthogonal matrix; that is, assuming that the undamped mode shapes of the system uncouple the system's equation of motion, the damping ratio of the system may be obtained in terms of its generalized mass and damping constant as

$$\tilde{\xi} = \frac{1}{2\tilde{\omega}} \frac{C^*}{M^*} \tag{13.218}$$

where C^* and M^* are such generalized mass and damping constants, respectively. Note, however, that in terms of the natural frequency and mode shape given by Equations 13.214 and 13.217, the system's generalized mass and damping constants may be expressed as

$$C^* = \{\tilde{X}\}^T [C] \{\tilde{X}\} = \begin{Bmatrix} 1 \\ k/k_x \\ k/k_r \end{Bmatrix}^T \begin{bmatrix} c & 0 & 0 \\ 0 & c_x & 0 \\ 0 & 0 & c_r \end{bmatrix} \begin{Bmatrix} 1 \\ k/k_x \\ k/k_r \end{Bmatrix} = c + \frac{k^2}{k_x^2} c_x + \frac{k^2}{k_r^2} c_r \tag{13.219}$$

Structural Response Considering Soil–Structure Interaction

$$M^* = \{\tilde{X}\}^T[M]\{\tilde{X}\} = m \begin{Bmatrix} 1 \\ k/k_x \\ k/k_r \end{Bmatrix}^T \begin{bmatrix} 1 & 1 & 1 \\ 1 & 1 & 1 \\ 1 & 1 & 1 \end{bmatrix} \begin{Bmatrix} 1 \\ k/k_x \\ k/k_r \end{Bmatrix} = m\left(1 + \frac{k}{k_x} + \frac{k}{k_r}\right)^2 \quad (13.220)$$

Consider, also, that in view of Equation 13.214, Equation 13.220 may also be written as

$$M^* = m\frac{\omega^4}{\tilde{\omega}^4} \quad (13.221)$$

On substitution of Equations 13.219 and 13.221 into Equation 13.218, the damping ratio of the interacting system may be therefore expressed as

$$\tilde{\xi} = \frac{1}{2m}\frac{\tilde{\omega}^3}{\omega^4}\left(c + \frac{k^2}{k_x^2}c_x + \frac{k^2}{k_r^2}c_r\right) \quad (13.222)$$

or, if it is considered that $k = \omega^2 m$ and $c = 2\xi\omega m$, where ξ represents the damping ratio of the fixed-base structure, as

$$\tilde{\xi} = \left(\frac{\tilde{\omega}}{\omega}\right)^3 \xi + \frac{1}{2}\tilde{\omega}^3 m\left(\frac{c_x}{k_x^2} + \frac{c_r}{k_r^2}\right) \quad (13.223)$$

The damping ratio of the soil–structure system in question may be thus considered to be of the form

$$\tilde{\xi} = \left(\frac{\tilde{\omega}}{\omega}\right)^3 \xi + \tilde{\xi}_0 \quad (13.224)$$

where the first term on the right-hand side of Equation 13.224 represents the contribution of the structural damping, whereas the second term represents the contribution of the foundation damping, including both radiation and material damping. Observe that according to Equation 13.224, $\tilde{\xi}_0$ and ξ cannot be added directly but must be combined in accordance with this equation. Note also that the damping ratio of the equivalent system typically will be larger than the damping ratio of the fixed-base structure. Consequently, Equation 13.224 corroborates that, as indicated earlier, an important effect of soil–structure interaction is to increase the effective damping of the structure.

The shear force at the base of the structure may be obtained either in terms of the displacement $\tilde{u}$ of the equivalent single-degree-of-freedom (SDOF) structure as

$$\tilde{V} = \tilde{k}\tilde{u} = m(\tilde{\omega}^2 \tilde{u}) \quad (13.225)$$

or in terms of the deformation u of the interacting structure as

$$\tilde{V} = ku = m(\omega^2 u) \quad (13.226)$$

It may be seen, thus, that the displacements $\tilde{u}$ and u are interrelated by the equation

$$u = \left(\frac{\tilde{\omega}}{\omega}\right)^2 \tilde{u} \tag{13.227}$$

It is worthwhile to note that Equation 13.224 is merely an approximation as the damping terms that account for the energy dissipation in the foundation medium are of a form that does not permit the uncoupling of the equations of motion using the undamped mode shapes. It is also worthwhile to note that the established method of accounting for soil-structure interaction effects is strictly limited to structures with a single degree of freedom. A reasonable approximation, however, can be obtained for multi-degree-of-freedom structures by assuming that soil–structure interaction influences their response in their fundamental mode only. In such a case, h must be interpreted as the distance from the base of the structure to the centroid of the inertia forces corresponding to the fundamental mode of vibration of the fixed-base structure, and m, k, and c must be interpreted as the system's generalized mass, stiffness, and damping constant, respectively. Along the same lines, it is important to keep in mind that the method is based on the assumption of an elastic halfspace. Actual soils, however, exhibit a nonlinear behavior, particularly at the level of the strains generated by severe earthquake ground motions. It is therefore important that measures be taken in design applications to account for the fact that both v_s and tan δ are a function of strain level. Other things being equal, the higher the severity of ground shaking, the smaller is the effective value of v_s and the greater the value of tan δ. Similarly, it is important to be aware of the fact that the choice of the value for structural damping is normally based on field tests of full-scale structures. Since field test data reflect the overall damping of the soil–structure system, not only the structural damping, the selected value of the damping ratio for the interacting system should never exceed the estimated value of the structural damping ratio. It should also be noted that soil–structure interaction effects tend to reduce the response of a structure, but because the foundation can translate and rotate when these effects are taken into account, they may increase its overall displacements. This increase may be significant for tall, slender structures or for closely spaced structures that may be subjected to pounding when such displacements are large.

It also worthwhile observing that, because of the frequency dependence of the coefficients α_x, α_φ, β_x, and β_φ, the natural frequency and damping ratio of the interacting system must be evaluated by iteration. However, as discussed in Sections 13.4 and 13.7, the use of frequency-independent spring and damping constants may be adequate for many practical applications. This will simplify the calculations, and the solution may not deviate significantly from the "exact" one.

Finally, from studies that have considered the effect of various parameters on the values of the effective natural frequency and damping ratio derived earlier, it has been concluded that soil–structure interaction effects are of insufficient importance to warrant its consideration in design whenever

$$\frac{h}{v_s T}\left(\frac{h}{r_0}\right)^{1/4} \leq 0.125 \tag{13.228}$$

where T denotes the fundamental natural period of the fixed-base structure and all the other variables are as previously defined. When this criterion is satisfied, the structure may be analyzed as if it were fixed at its base. Accordingly, since $T \approx h/30$ for frame buildings (h in meters), the soil–structure interaction may be insignificant whenever $v_s \geq 240$ m/s for short and squatty ($h/r_0 = 1$) buildings, and whenever $v_s \geq 425$ m/s for tall and slender ($h/r_0 = 10$) buildings.

The method is summarized in Box 13.1 and illustrated by the following example.

Structural Response Considering Soil–Structure Interaction

Box 13.1   Equivalent Single-Degree-of-Freedom System Method

1. Estimate the specific loss factor, $\Delta W/W$, for the foundation material and compute the value of $\tan \delta$ from Equation 13.192.
2. With the computed value of $\tan \delta$ and, as a first approximation, the value of the fundamental natural frequency of the fixed-base structure, ω, determine from Figure 13.25 the corresponding values of the dimensionless coefficients α_x, α_ϕ, β_x, and β_ϕ.
3. Using Equations 13.193 through 13.196, compute the values of the translational and rotational spring and damping constants of the equivalent SDOF system.
4. Given the values of the rotational stiffness k_ϕ and damping constant c_ϕ, determine k_r and c_r according to Equation 13.201.
5. Determine the natural frequency $\tilde{\omega}$ and damping ratio $\tilde{\xi}$ of the equivalent SDOF system according to Equations 13.214 and 13.223.
6. If the value of $\tilde{\omega}$ is substantially different from the value of the natural frequency assumed in step 2, repeat steps 2 through 4 using the value of $\tilde{\omega}$ found in step 5. Otherwise, continue with step 7.
7. From the response spectrum of the specified free-field ground motion and corresponding to a natural frequency $\tilde{\omega}$ and damping ratio $\tilde{\xi}$, determine the maximum deformation $\tilde{u}$ of the equivalent SDOF system or the associated pseudoacceleration $\tilde{\omega}^2 \tilde{u}$.
8. Obtain the shear at the base of the interacting structure using the following expression:

$$V = m\tilde{\omega}^2 \tilde{u}$$

9. Compute the deformation of the interacting structure, u, by multiplying $\tilde{u}$ by $(\tilde{\omega}/\omega)^2$.
10. Determine the total displacement of the interacting structure by adding to u the displacement ϕh induced by the rotation of the interacting structure. Or since ϕh is given by Equation 13.212, determine this total displacement as

$$\delta_h = \left(1 + \frac{k}{k_r}\right) u$$

EXAMPLE 13.8   NATURAL FREQUENCY AND DAMPING RATIO OF AN INTERACTING STRUCTURE

In its fundamental mode, a building has an effective height of 80 m, a natural period of 1.8 s, a generalized weight of 35 MN, and a damping ratio of 5%. The building is supported by a foundation whose dimensions in plan are 20 m × 20 m, and the foundation, in turn, rests over a deep deposit of soil with a shear wave velocity of 100 m/s, a shear modulus of 20 MN/m², a Poisson ratio of 0.45, and a specific loss factor equal to 0.6π. Find the fundamental natural frequency and damping ratio of building when the soil–structure interaction is considered. Use the equivalent SDOF system method.

Solution

According to Equation 13.66, the equivalent radius of the building's square foundation is equal to

$$r_0 = \sqrt{\frac{(20)(20)}{\pi}} = 11.28 \text{ m}$$

Also, the natural frequency of the fixed-base building is equal to $2\pi/1.8 = 3.49$ rad/s. Hence, the corresponding dimensionless frequency is equal to

$$a_0 = \frac{3.49(11.28)}{100} = 0.39$$

Similarly, from Equation 13.184 and the given value of the specific loss factor one finds that

$$\tan\delta = \frac{0.6\pi}{2\pi} = 0.3$$

Thus, from Figure 13.25 for $a_0 = 0.39$ and $\tan\delta = 0.3$, one obtains

$$\alpha_x = 0.95 \quad \beta_x = 1.4 \quad \alpha_\phi = 0.95 \quad \beta_\phi = 1.0$$

Likewise, note that according to Equations 13.189 and 13.190 the soil's static stiffness constants for the case under consideration are equal to

$$K_x = \frac{8(20)(11.28)}{1 - 0.45} = 3281 \text{ MN/m}$$

$$K_\phi = \frac{8(20)(11.28)^3}{3(1 - 0.45)} = 139{,}176 \text{ MN m/rad}$$

Hence, in view of Equations 13.193 through 13.196 and Equation 13.201, the equivalent spring and damping constants of the interacting system result as

$$k_x = 0.95(3281) = 3117 \text{ MN/m}$$

$$c_x = \frac{1.4(3281)(11.28)}{100} = 518 \text{ MN s/m}$$

$$k_\phi = 1(139{,}176) = 139{,}176 \text{ MN m/rad}$$

$$c_\phi = \frac{1(139{,}176)(11.28)}{100} = 15{,}699 \text{ MN m s/rad}$$

$$k_r = \frac{139{,}176}{80^2} = 21.75 \text{ MN/m}$$

$$c_r = \frac{15{,}699}{80^2} = 2.45 \text{ MN s/m}$$

Therefore, since the building's generalized stiffness is equal to

$$k = \omega^2 m = (3.49)^2 \left(\frac{35}{9.81}\right) = 43.5 \text{ MN/m}$$

substitution of this value and the determined values of k_x and k_r into Equation 13.214 yields

$$\tilde{\omega} = \frac{3.49}{\sqrt{1 + \dfrac{43.5}{3117} + \dfrac{43.5}{21.75}}} = 2.01 \text{ rad/s}$$

from which it is found that

$$\frac{\tilde{\omega}}{\omega} = 0.576$$

Iterating now with a natural frequency of 2.01 rad/s, one obtains

$$a_0 = \frac{2.01(11.28)}{100} = 0.23$$

$$\alpha_x = 1.0 \quad \beta_x = 1.6 \quad \alpha_\phi = 1.0 \quad \beta_\phi = 1.2$$

$$k_x = 1(3281) = 3281 \text{ MN/m}$$

$$c_x = \frac{1.6(3281)(11.28)}{100} = 592 \text{ MN s/m}$$

$$k_\phi = 1(139,176) = 139,176 \text{ MN m/rad}$$

$$c_\phi = \frac{1.2(139,176)(11.28)}{100} = 18,839 \text{ MN m s/rad}$$

$$k_r = \frac{139,176}{80^2} = 21.75 \text{ MN/m}$$

$$c_r = \frac{18,839}{80^2} = 2.94 \text{ MN s/m}$$

As a result, the new value of the equivalent natural frequency is

$$\tilde{\omega} = \frac{3.49}{\sqrt{1 + \frac{43.5}{3281} + \frac{43.5}{21.75}}} = 2.01 \text{ rad/s}$$

which is the same value found earlier. Hence, no more iterations are needed.
Finally, according to Equation 13.223, the equivalent damping ratio is equal to

$$\tilde{\xi} = (0.576)^3(0.05) + \frac{1}{2}(2.01)^3\left(\frac{35}{9.81}\right)\left(\frac{592}{3,281^2} + \frac{2.94}{21.75^2}\right) = 0.010 + 0.091 = 0.101$$

13.12 ADVANCED METHODS OF ANALYSIS

13.12.1 Introduction

Advanced methods of soil–structure interaction analysis invariably involve the use of the finite element method or any other discretization scheme such as finite differences or the boundary element method. The finite element method is a powerful technique for modeling soil–structure systems since with it one can straightforwardly consider the three-dimensional nature of the problem, irregular geometries, the vertical and horizontal variation of soil stiffness, foundation embedment, foundation flexibility, and the nonlinear behavior of the soil and structural elements. If soil and structure are modeled with finite elements, then a soil–structure interaction analysis simply involves the definition of the ground motion at the base of the soil and the performance of a conventional dynamic analysis of the soil–structure system. A soil–structure interaction analysis on the basis of the finite element method is, however, not problem free. It requires, first, the introduction of an artificial boundary in a medium that is theoretically unbounded. It also requires large amounts of computer time and storage, as the modeling of soil and structure with finite elements involves, more often than not, the consideration of an extraordinarily large number of degrees of freedom. The need to treat an unbounded soil and a desire to reduce the complexity of the problem has given rise to different methods for the solution of soil–structure interaction problems. Broadly, these methods may be classified as *direct* and *substructure methods*. In the direct method, the entire soil–structure system is modeled as one unit and analyzed in a single step. In contrast, in the substructure method this analysis is performed in several steps. Within each of these methods, different versions have proposed, depending on the different modifications or simplifications introduced. In the following sections a brief overview is given of the basic steps in each of two such methods.

13.12.2 Direct Method

As mentioned in the preceding section, in the direct method the soil and the structure system are modeled and analyzed as one unit under a ground motion prescribed at the free field. The method is illustrated schematically in Figure 13.27 for a foundation soil that can be idealized as a horizontally layered system underlain by bedrock and excited by vertically propagating shear waves. To account for the kinematic and inertial interaction between the soil and the structure, the method entails several steps. The first step is concerned with the definition of the ground motion at the bedrock level using the ground motion prescribed at the surface of the free field (often referred to as the control motion). This is accomplished through a process of inverse amplification known as *deconvolution*. Then, by assuming that the ground motion at bedrock directly below the building is equal to the bedrock motion determined in the first step, this motion is used as input to perform a dynamic analysis of the soil–structure system. The characteristics of the ground motion at the base of the structure and the response of the structure at any desired point may then be obtained from this dynamic analysis. If the foundation soil is excited by a wave field other than vertically propagating shear waves or cannot be modeled as a one-dimensional system, then the direct method involves the

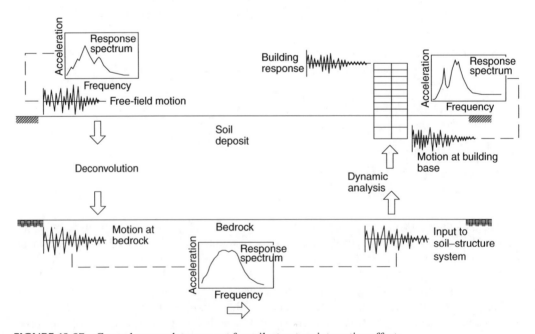

FIGURE 13.27 General approach to account for soil–structure interaction effects.

introduction of artificial boundaries at the sides of the soil and the description of the forces acting against these boundaries. This is in addition to the description of the ground motion at the base of the soil deposit and a finite element analysis of the soil–structure system considering such ground motion and boundary forces (see Figure 13.28).

In the direct method, the equations of motion of the soil–structure system may be, in general, solved by two alternative methods. The first is a solution in the frequency domain and the second a solution in the time domain. In the frequency domain method, the ground motion at bedrock is first transformed to the frequency domain by means of the Fourier transform. Then, the response of the soil–structure system in the frequency domain is obtained by multiplying the transfer function of the structure–soil system by the Fourier transform of the excitation. Thereafter, the response of the structure is obtained in the time domain by taking the inverse Fourier transform of such a product (see Section 8.4.4). The main advantage of a frequency domain solution is that frequency-dependent parameters can be easily incorporated into the analysis. An additional advantage is that once the transfer function of the system is determined, the solution procedure can be easily repeated for different ground excitations. Its main disadvantage is that it is limited to linear systems as it is based on the principle of superposition. For many problems, however, it is adequate to account for the nonlinear behavior of the system in an approximate way by performing an iterative analysis. In this iterative analysis, the soil's rigidity and damping are adjusted according to some characteristic measure of the strain obtained in a previous cycle. The designation "equivalent linear analysis" is used to describe this approach.

A solution in the time domain involves a step-by-step integration of the equations of motion. Its main advantage is that nonlinear stress–strain relationships can easily be incorporated into the solution procedure. With it, the material properties may be adjusted at each time step according to the instantaneous strain level. It has, however, the disadvantage of requiring an iterative procedure since some of the parameters that define a soil–structure system (e.g., complex soil modulus) or its boundaries are frequency dependent. It also presents problems with the generation of the system's damping matrix and problems of stability and accuracy, as there is often a large difference in the value of the soil and structural parameters.

As indicated earlier, the use of the finite element method in a soil–structure interaction analysis requires that the infinite soil medium be truncated along an artificial boundary that separates the

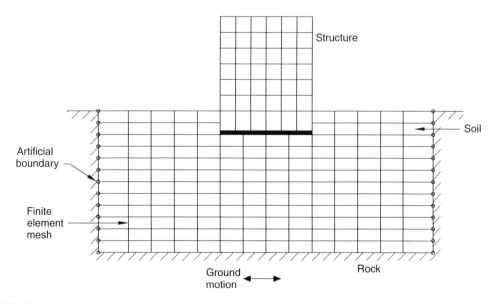

FIGURE 13.28 Direct method of soil–structure interaction analysis.

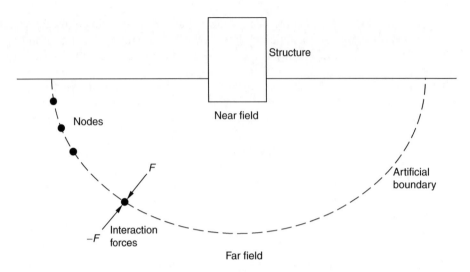

FIGURE 13.29 Artificial boundary and near and far fields in a soil–structure system.

soil–structure system into a bounded near field and an unbounded far field (see Figure 13.29). The need for this artificial boundary is obvious since it is impossible to cover an unbounded solid with a finite number of elements. Such an artificial boundary, however, will introduce an artificial reflection of the outgoing waves, bringing an additional wave field to the base of the structure. For meaningful results, therefore, a finite element analysis of a soil–structure interaction problem requires special considerations to let traveling waves go through the artificial boundary without reflecting back into the near field, or at least to minimize these reflections. These artificial boundaries may also be interpreted as the boundary conditions or interaction forces between the near and far fields (see Figure 13.29).

One simple procedure to minimize the effect of waves reflected at an artificial boundary is to locate these boundaries at a sufficient distance from the structure so that the reflected waves are absorbed by the soil's internal damping before they can reach the structure. Another approach that has been employed in the past is to use boundaries that reproduce the aforementioned interaction forces between the near and far fields. These artificial boundaries are known as *absorbing* or *transmitting boundaries*. With this type of boundary, therefore, the size of the finite element mesh may be reduced significantly. The adequacy of this type of artificial boundary depends, however, on how accurately it reproduces such interaction forces.

Several absorbing boundaries have been proposed. However, two of these boundaries have been the most widely used. These are the *viscous boundary* and the *consistent boundary*. In a viscous boundary, energy-absorbing viscous dampers are added to the boundary of the near field to simulate the action of the far field on such a boundary. That viscous dampers may reproduce the effect of the interacting forces acting at a boundary may be demonstrated for the simple case of a one-dimensional bar as follows.

Consider an infinitely long one-dimensional bar oriented along the direction of the x-axis and assume that a longitudinal wave is traveling along the length of the bar with a velocity v_c. According to the concepts studied in Chapter 4, the displacement and stress induced by the wave at a section located at distance x from the origin of the system of coordinates are given by

$$u = f(x - v_c t) \tag{13.229}$$

$$\sigma_x = E\varepsilon_x = E\frac{\partial u}{\partial x} = Ef' \tag{13.230}$$

where E and f' are bar's modulus of elasticity and first derivative of the function with respect to the argument of the function.

Note, however, that since

$$\dot{u} = \frac{\partial u}{\partial t} = f'(-v_c) \tag{13.231}$$

f' can be written as

$$f' = -\frac{\dot{u}}{v_c} \tag{13.232}$$

Similarly, according to Equation 4.5, one can express E as

$$E = v_c^2 \rho \tag{13.233}$$

where ρ denotes density. On substitution of Equations 13.232 and 13.233 into Equation 13.230, σ_x may be therefore expressed as

$$\sigma_x = -E\frac{\dot{u}}{v_c} = -\rho v_c \dot{u} \tag{13.234}$$

This result shows that if the bar is cut at a section located a distance x from the origin of the system of coordinates and a viscous damper with a constant $C = \rho v_c$ per unit area is placed at the cut end of the bar, then when the wave reaches this end it will generate in the damper a force per unit area equal to $\rho v_c \dot{u}$. This, however, is the value of the force per unit area that would be induced by the wave at the same section if the bar were not cut. The damper, therefore, simulates exactly the interacting force at the boundary produced by the piece cut from the bar and, as a result, the effect of the cut portion of the bar.

In a three-dimensional viscous boundary, three dampers are thus used to simulate the action of the far field: one perpendicular to the boundary and two parallel to it (see Figure 13.30). In the normal and tangential directions, these coefficients are respectively assumed equal to ρv_p and ρv_s, where, as earlier, ρ denotes density and v_p and v_s are the velocities of propagation of P and S waves. This technique, however, is only an approximate one since it is derived for plane waves impinging the artificial boundary at right angles, despite the fact that waves may approach a boundary with any incident angle; that is, waves may reach the boundary from any direction. The approximation improves as the incident angle approaches 90°.

In the consistent boundary, the interaction forces acting on the artificial boundary due to the effect of the far field are imposed as boundary conditions in the analysis of the truncated soil–structure system (see Figure 13.29). For this purpose, these interaction forces are determined as a function of the displacements at the boundary of the near field and in a way that ensures that waves are transmitted from the near to the far field without any reflections. In the context of the finite element method, this has been accomplished by assuming vertical boundaries, a horizontally layered soil system supported on a rigid base, and that the motion in the far field is due only to traveling Love or Rayleigh waves. As the force–displacement relationship for the far field is frequency dependent, the consistent boundary can only be used in a frequency domain formulation.

Among the two boundaries described, the viscous boundary is the most commonly used as it has a simple form and is suitable for a finite element formulation and nonlinear analysis in the time domain. The consistent boundary, however, reproduces better the effect of the far field

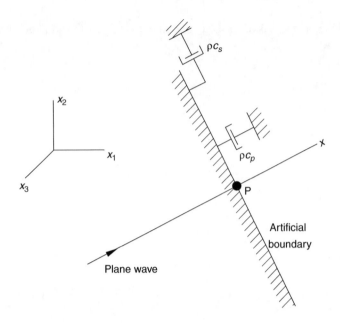

FIGURE 13.30 Viscous dampers at artificial boundary.

and allows the lateral boundary to be placed directly at the side of the foundation, which allows for a considerable reduction in the number of degrees of freedom of the soil–structure model. For this reason, it has gained popularity in recent years and has been implemented in some general-purpose computer programs, notwithstanding the fact that it can only be formulated in the frequency domain.

13.12.3 Substructure Method

In dealing with two different systems with different properties, and furthermore one bounded and the other unbounded, it seems natural to consider the two subsystems separately. For this reason, a variety of substructuring methods have been proposed for the analysis of soil–structure systems in which soil and structure are first analyzed separately. Then the results from these analyses are synthesized to obtain the response of the interacting structure. The technique of substructuring, or separating a dynamic system into several subsystems, has had multiple applications in structural mechanics. It has been used, for example, in the aerospace industry in the dynamic analysis of airplane wings, accounting for the flexibility of the fuselage to which they are connected. The basic idea is to perform the analysis of the interaction problem in several steps and to use the most amenable solution in each step, requiring as a result less computer time and storage than in the case of a complete analysis in a single step. In the case of soil–structure interaction problems, this approach offers the advantage that the soil can be considered as an unbounded continuous solid and the structure as a bounded discrete system. For example, one may analyze first the soil as a halfspace and establish the impedances of this substructure. Then, one may use these impedances as boundary conditions in the dynamic analysis of the structure under an excitation that depends on the ground motion in the free field. In the substructure method, each substructure is thus analyzed by the best-suited technique.

For the case of foundations resting directly on the ground, substructure methods are simple to apply, particularly when the excitation is defined by vertically propagating SH waves. However, most structures have embedded foundations, and simple methods of substructuring cannot treat embedded foundations properly. The main problem is that it is difficult to specify the boundary conditions along the embedded part of the structure. Thus, several substructuring procedures have been proposed specifically for the analysis of embedded structures. According to the number of degrees of

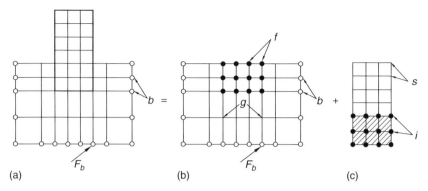

FIGURE 13.31 Soil and structure substructures and corresponding finite element meshes. (a) Soil–structure system, (b) foundation soil, (c) structure minus excavated soil.

freedom considered at the interface between soil and structure, these procedures are classified as (a) rigid boundary method, (b) flexible boundary method, and (c) flexible volume method. As a means to illustrate the concept of substructuring, only the flexible volume method will be described here.

In the flexible volume method, the foundation soil and the structure are divided not at their interface, but in the way shown in Figure 13.31. That is, soil and structure are considered separately, but the soil substructure includes the excavated soil, whereas the structure substructure includes the structure minus the excavated soil; that is, the structure plus an excavated soil with negative properties. In this manner, all the nodes of the structure that are embedded into the soil interact with the soil. This particular substructuring scheme is advantageous because it eliminates the problem of wave scattering due to the presence of the foundation and, having a soil substructure with a regular geometry at the surface, simplifies the calculation of the soil impedances. The response of the structure accounting for soil–structure interaction may be obtained as follows.

Consider the finite element model of the soil–structure system shown in Figure 13.31 and assume that the system is subjected to an arbitrary excitation specified in the form of displacements along the boundary. Let subscripted K's represent dynamic stiffness or impedance submatrices (see Section 13.8) and let subscripted F's and U's denote force and displacement vectors, respectively. Similarly, let subscripts s, f, g, and b identify, as shown in Figure 13.31, parameters associated with the structure, foundation, soil, and boundary nodes, respectively. Furthermore, let subscript i depict the nodes of the region where the soil and the structure interact. Note that the structural nodes s and the soil nodes g exclude the nodes of this interacting region. In terms of such impedance submatrices and partitioning according to the degrees of freedom associated with the interacting nodes i and the structural nodes s, the equation of motion for the structure in Figure 13.31c may be expressed as

$$\begin{bmatrix} K_{ss} & K_{si} \\ K_{is} & K_{ii} - K_{ff} \end{bmatrix} \begin{Bmatrix} U_s \\ U_f \end{Bmatrix} = \begin{Bmatrix} 0 \\ -F_f \end{Bmatrix} \tag{13.235}$$

where it is noted that the submatrix $(K_{ii} - K_{ff})$ represents the impedance of the structure in the region where it interacts with the soil minus the impedance of the excavated soil. It is noted, also, that the equation has incorporated the compatibility condition $U_i = U_f$ and the equilibrium equation $F_i + F_f = 0$. Similarly, the equation of motion for the soil system in Figure 13.31b may be written as

$$\begin{bmatrix} K_{ff} & K_{fg} & K_{fb} \\ K_{gf} & K_{gg} & K_{gb} \\ K_{bf} & K_{bg} & K_{bb} \end{bmatrix} \begin{Bmatrix} U_f \\ U_g \\ U_b \end{Bmatrix} = \begin{Bmatrix} F_f \\ 0 \\ F_b \end{Bmatrix} \tag{13.236}$$

when soil-structure interaction is considered, and as

$$\begin{bmatrix} K_{ff} & K_{fg} & K_{fb} \\ K_{gf} & K_{gg} & K_{gb} \\ K_{bf} & K_{bg} & K_{bb} \end{bmatrix} \begin{Bmatrix} U_f^* \\ U_g^* \\ U_b^* \end{Bmatrix} = \begin{Bmatrix} 0 \\ 0 \\ F_b \end{Bmatrix} \qquad (13.237)$$

when no interaction with the structure is considered, that is, under free-field conditions. Note that in Equation 13.237 an asterisk is used to identify the forces and displacements associated with free-field conditions and that it has been implicitly assumed that the soil boundary (nodes b) is sufficiently away from the structure so that $F_b^* = F_b$.

Subtracting Equation 13.237 from Equation 13.236 leads to

$$\begin{bmatrix} K_{ff} & K_{fg} & K_{fb} \\ K_{gf} & K_{gg} & K_{gb} \\ K_{bf} & K_{bg} & K_{bb} \end{bmatrix} \begin{Bmatrix} U_f - U_f^* \\ U_g - U_g^* \\ U_b - U_b^* \end{Bmatrix} = \begin{Bmatrix} F_f \\ 0 \\ 0 \end{Bmatrix} \qquad (13.238)$$

which, after the elimination of $U_g - U_g^*$ and $U_b - U_b^*$, yields

$$X_f(U_f - U_f^*) = F_f \qquad (13.239)$$

where X_f is the impedance matrix corresponding to the soil region identified by the f nodes, and may be obtained directly, according to the definition of impedance function, from the application of point loads at these nodes and the calculation of the corresponding displacements. On substitution of Equation 13.239, Equation 13.235 may be thus expressed as

$$\begin{bmatrix} K_{ss} & K_{si} \\ K_{is} & K_{ii} - K_{ff} + X_f \end{bmatrix} \begin{Bmatrix} U_s \\ U_f \end{Bmatrix} = \begin{Bmatrix} 0 \\ X_f U_f^* \end{Bmatrix} \qquad (13.240)$$

from which one can obtain the structural displacements U_s when the interaction between soil and structure is considered.

A soil–structure interaction analysis with the substructuring method just described involves thus the following steps:

a. A free-field analysis to determine U_f^*, the displacements under free-field conditions of the soil nodes corresponding to the embedded part of the structure.
b. Determination of the impedance matrix X_f corresponding to the aforementioned soil nodes.
c. Determination of the structure's impedance matrix.
d. Formulation and solution of Equation 13.240 to obtain the structural displacements U_s and $U_i = U_f$.

In essence, the method leads to a set of equations relating the structural response with fictitious forces applied at the soil–foundation interface. These forces in turn are obtained from the free-field displacements and the force–displacements relationships along such an interface.

As noted earlier, a major argument for using substructure methods in design is that they are simpler and less expensive than a direct analysis. This may be obviously the case for a structure at the surface of a uniform halfspace. As seen, however, substructure methods may not be as simple for structures embedded in a layered medium, which constitute the majority of cases in real situations. The problem is that in such cases, the two substructures require a finite element or finite differences analysis, and thus a substructure method may demand nearly the same computational effort as the direct method. It is likely, therefore, that in many cases the substructure method will not be substantially less expensive than the direct method.

13.12.4 Computer Programs

A number of computer programs are available to solve soil–structure interaction problems. For the most part, these programs use either the direct or the substructure method in combination with a finite element formulation. Among the available programs, DYNA4, FLUSH, SASSI, and CLASSI stand out. To learn more about these programs consult the associated reports listed in the Further Readings section.

13.13 EXPERIMENTAL VERIFICATION

To validate the soil–structure interaction methodologies commonly used in the design of nuclear power plants, the Electric Power Research Institute (EPRI) in cooperation with the Taiwan Power Company (TPC) built two reinforced concrete containment models at Lotung, Taiwan—a site of high seismicity and heavily instrumented to record earthquake ground motions (see Section 5.7.5). One of the models was built to a ¼ scale and the other to a ½ scale. Since their completion in 1985, the models have been subjected to forced vibration tests and to a number of earthquakes ranging in magnitude from 4.5 to 7.0. In each of these events, the response of the containments was recorded above, at, and below the surface. In a blind verification program, researchers from industry and academia performed independent calculations to predict the response of one of the containments during some of such events and compared their results with the recorded responses. The calculations were performed without any previous knowledge of the recorded responses and using methods ranging from simple lumped-parameter representations to complex finite-element formulations. Among the findings from this experiment in regard to the validity of the current methodologies used to account for soil–structure interaction during earthquake response are (a) the common assumption of vertical propagating waves is adequate to describe the wave field exciting the structure, (b) equivalent linear analyses of soil response yield acceptable results, (c) a significant but not permanent degradation of the soil modulus occurs during earthquakes, (d) the development of soil stiffness degradation and soil damping as a function of strain requires improvement to reduce variability and uncertainty, (e) backfill stiffness plays an important role in determining impedance functions and input motions, (f) scattering of ground motion due to embedment is an important factor in soil–structure interaction, and (g) differences between observed and predicted responses are owed more to the modeling of the soil–structure system than to the calculation techniques.

Perhaps one of the most significant findings from the verification experiment is that the discrete spring–dashpot models were capable of producing successful predictions with an accuracy that was at least comparable to the accuracy of more complex methods. In these successful predictions the spring and dashpot constants were iteratively adjusted to properly account for the frequency dependence of these constants and the nonlinear relationship between the soil shear modulus and the shear strain level. Still, it is remarkable that such a simple model was capable of providing accurate predictions. Figure 13.32 shows the 5% damping acceleration response spectra of the motions recorded at the Lotung ¼-scale model and those of the motions predicted by the discrete spring–dashpot method. The top spectra correspond to the motions in the north–south direction, whereas those at the bottom correspond to the motions in the east–west direction. The two spectra on the left-hand side of Figure 13.32 are for a location near the roof of the containment structure and the two spectra on the right-hand side are for a basement location. It may be seen from a comparison of the spectra in Figure 13.32 that the prediction by the discrete spring–dashpot method is indeed quite good.

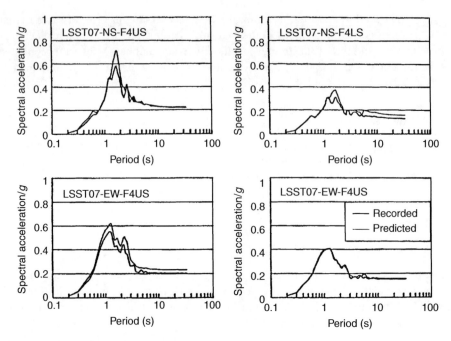

FIGURE 13.32 Five-percent-damping acceleration response spectra of north–south and east–west motions recorded at roof (left-hand side) and basement (right-hand side) of ¼-scale containment model in Lotung, Taiwan, and corresponding predicted motions using discrete spring–dashpot method. (Reproduced with permission of Sociedad Mexicana de Ingeniería Sísmica from Hadjian, A. H., *Memorias del X Congreso Nacional de Ingeniería Sísmica*, Puerto Vallarta, Mexico, 1993.)

FURTHER READINGS

1. Ad-Hoc Group on Soil–Structure Interaction of the Committee on Nuclear Structures and Materials of the Structural Division of ASCE, *Analyses for Soil–Structure Interaction Effects in the Seismic Analysis for Nuclear Power Plants*, American Society of Civil Engineers, New York, 1979, 155 p.
2. Das, B. M., *Principles of Soil Dynamics*, PWS-Kent Publishing Co., Boston, MA, 1993, 570 p.
3. Gazetas, G., Analysis of machine foundation vibrations: state of the art, *Soil Dynamics and Earthquake Engineering*, 2, 1, 1983, 2–42.
4. Gazetas, G., Formulas and charts for impedance of surface and embedded foundations, *Journal of Geotechnical Engineering*, ASCE, 117, 1991, 1363–1381.
5. Hadjian, A. H., The learning from the large scale Lotung soil–structure interaction experiments, *Proceedings 2nd International Conference on Recent Advances in Geotechnical Earthquake Engineering and Soil Dynamics*, Vol. III, St. Louis, MO, March 11–15, 1991, pp. 2047–2070.
6. Lysmer, J., Udaka, T., Tsai, C.F., and Seed, H. B., *FLUSH: A Computer Program for Approximate 3-D Analysis of Soil–Structure Interaction Problems*, Report No. EERC 75-30, University of California, Berkeley, CA, Nov. 1975.
7. Lysmer, J., Ostadan, F., Tabatabaie, M., Vahdani, S., and Tajirian, F., *SASSI—A System for Analysis of Soil–Structure Interaction*, Department of Civil Engineering, University of California, Berkeley, CA, and Bechtel Power Corporation, San Francisco, CA, July 1998.
8. Luco, J. E., *Linear Soil–Structure Interaction* (description of CLASSI computer program), Report UCRL-15272, Lawrence Livermore National Laboratory, Livermore, CA, 1980.
9. Lysmer, J., Analytical procedures in soil dynamics, *Proceedings ASCE Geotechnical Engineering Division Specialty Conference on Earthquake Engineering and Soil Dynamics*, Vol. III, June 19–21, 1978, pp. 1267–1313.
10. Novak, M., Sheta, M., El-Hifnawy, L., El-Marsafawi, H., and Ramadan, O., *DYNA4: A Computer Program for Calculation of Foundation Response to Dynamic Loads*, Report GEOP 93-01, Geotechnical Research Center, The University of Western Ontario, London, Ontario, 1993.

11. Prakash, S. and Puri, V. K., *Foundations for Machines: Analysis and Design*, Wiley, New York, 1988, 656 p.
12. Richards, F. E., Jr., Hall, J. R., Jr., and Woods, R. D., *Vibrations of Soils and Foundations*, Prentice-Hall, Englewood Cliffs, NJ, 1970, 414 p.
13. Veletsos, A. S., Dynamics of structure–foundation systems, in *Structural and Geotechnical Mechanics—A Volume Honoring N. M. Newmark*, Hall, W. J. ed., Prentice-Hall, Englewood Cliffs, NJ, 1977, pp. 333–361.
14. Wolf, J. P., *Dynamic Soil–Structure Interaction*, Prentice-Hall, Englewood Cliffs, NJ, 1985, 466 p.
15. Wolf, J. P., *Soil–Structure Interaction in Time Domain*, Prentice-Hall, Englewood Cliffs, NJ, 1988, 446 p.
16. Wolf, J. P., *Foundation Vibration Analysis Using Simple Physical Models*, Prentice-Hall, Englewood Cliffs, NJ, 1994, 423 p.

PROBLEMS

13.1 A circular reinforced concrete foundation block, 13 ft in diameter and 5 ft deep, is subjected to a vertical harmonic force with an amplitude of 1000 lb and a frequency of 90 cpm. The foundation block rests on the surface of a soil deposit that can be idealized as an elastic halfspace. The soil has a unit weight of 115 lb/ft^3, a shear modulus of elasticity of 3000 lb/in.^{2}, and a Poisson ratio of 0.45. Determine using the elastic halfspace theory the maximum vertical displacement of the foundation block. Assume the unit weight of the reinforced concrete used for the foundation is 150 lb/ft^3.

13.2 A surface, reinforced concrete, foundation block is 6 m × 6 m in plan and 3 m deep. The block supports a machine that generates a horizontal force that acts 4 m above the base of the foundation. The machine operates at a frequency of 500 rpm and the amplitude of the generated force is equal to 50 kN. The machine's mass moment of inertia about the foundation base is equal to 2.5 × 10^5 kg-m^2. The soil that supports the machine and the foundation block has a shear modulus of elasticity of 20,000 kN/m^2, a density of 1800 kg/m^3, and a Poisson ratio of 0.40. Determine using the elastic halfspace theory the amplitude of the rocking vibrations generated by the applied horizontal force. Neglect the coupling with the horizontal motion and assume the reinforced concrete used for the foundation has a unit weight of 23.6 kN/m^3.

13.3 Determine using Lysmer analog the resonant frequency, the amplitude of vibration under operation conditions, and the amplitude of vibration at resonance of the soil–foundation system described in Problem 13.1.

13.4 Determine using Hall analog the resonant frequency, the amplitude of vibration under the operating frequency, and the amplitude of vibration at resonance of the soil–foundation system described in Problem 13.2.

13.5 The circular foundation for a vibrating machine has a radius of 3 m and a depth of 1 m. The foundation is subjected to a horizontal harmonic force at a height of 3 m from the foundation base. The force has an amplitude of 20 kN and oscillates with a frequency of 500 cpm. The soil supporting the foundation block and the machine has a shear modulus of elasticity of 25,000 kN/m^2, a density of 1700 kg/m^3, and a Poisson ratio of 0.25. Find (a) the natural frequencies in the horizontal and rocking modes of the foundation, and (b) the amplitudes of vibration in these two modes. Use Hall analog to define the stiffness and damping constants of the soil–foundation system and neglect the mass moment of inertia of the machine. The foundation is made of reinforced concrete with a unit weight of 23.5 kN/m^3.

13.6 A reinforced concrete rectangular foundation is 14 ft long, 10 ft wide, and 7 ft deep, and is embedded 2 ft into a soil layer with a shear modulus of elasticity of 4000 lb/in.^{2}, a Poisson ratio of 0.30, and unit weight of 120 lb/ft^3. The foundation is subjected to an oscillating vertical force of 7500 lb, which is generated by a machine that operates with a frequency of 600 rpm and weighs 50,000 lb. The soil supporting the foundation and the machine has a shear modulus of elasticity of 8000 lb/in.^{2}, a Poisson ratio of 0.50, and a unit weight of 120 lb/ft². Using a lumped-parameter model and frequency-independent parameters, determine (a) the damped natural frequency of the soil–foundation system, (b) the foundation's maximum vertical displacement at resonance, and (c) the foundation's maximum vertical displacement under operation conditions. Assume the unit weight of the reinforced concrete used for the foundation is 150 lb/ft³.

13.7 A structure that can be modeled as a single-degree-of-freedom system when its base is considered fixed has a natural period of 1 s and a damping ratio of 5%. The total weight of the structure is 5000 kN, which can be assumed concentrated at a height of 18 m aboveground. The structure is supported by a rectangular foundation that is made of reinforced concrete with a unit weight of 24 kN/m³ and is 6 m long, 4 m wide, and 1.5 m deep. The foundation soil has a shear modulus of elasticity of 94 MN/m², a unit weight of 18 kN/m³, and a Poisson ratio of 0.45. The base of the structure is excited by vertically propagating seismic shear waves, which in the free field may be characterized by a harmonic displacement function with an amplitude of 0.45 m and a frequency of 6 rad/s. Using the method of impedances and considering the interaction between soil and structure, determine the amplitude of the induced displacement at the center of the foundation, the angle of rotation of the foundation, the deformation in the structure, and the structure's total horizontal displacement. Assume the foundation rests on the surface of the ground and that the foundation soil may be idealized as a semi-infinite elastic halfspace. Neglect the mass moments of inertia of the structure and the foundation.

13.8 In its fundamental mode, a building has an effective height of 80 m, a natural period of 0.8 s, a generalized weight of 15 MN, and a damping ratio of 2%. The building's foundation is 10 m × 10 m in plan and rests over a deep deposit of soil with a shear wave velocity of 200 m/s, a shear modulus of 18 MN/m², a Poisson ratio of 0.50, and a specific loss factor of 0.8π. Using the equivalent SDOF system method, determine the fundamental natural frequency and damping ratio of the building when the soil flexibility is considered.

14 Seismic Response of Nonstructural Elements

There is no doubt that the cost of damage to contents and to nonstructural features of buildings will far exceed the cost of structural damage.

James L. Stratta, 1987

14.1 INTRODUCTION

Nonstructural elements are those systems and components attached to the floors and walls of a building or industrial facility that are not part of the main or intended load-bearing structural system of the building or industrial facility. Although not part of the main structural system, they may, nevertheless, also be subjected to large seismic forces and depend on their own structural characteristics to resist these seismic forces. In general, nonstructural elements may be classified into three broad categories: (a) architectural components, (b) mechanical and electrical equipment, and (c) building contents. Examples of the first category include elevator penthouses, stairways, partitions, parapets, heliports, cladding systems, signboards, lighting systems, and suspended ceilings. Examples of the second category are storage tanks, pressure vessels, piping systems, ducts, escalators, smokestacks, antennas, cranes, radars and object-tracking devices, computer and data acquisition systems, control panels, transformers, switchgears, emergency power systems, fire protection systems, boilers, heat exchangers, chillers, cooling towers, and machinery such as pumps, turbines, generators, and motors. Among those in the third category are bookshelves, file cabinets, storage racks, decorative items, and any other piece of furniture commonly found in office buildings and warehouses. Nonstructural elements are also known by alternative names such as "appendages;" "nonstructural components;" "building attachments;" "architectural, mechanical, and electrical elements;" "secondary systems;" "secondary structural elements;" and "secondary structures." Perhaps the name that best describes their nature is "secondary structures" since it reflects the fact that they are not part of the main structure but must possess, nonetheless, structural properties to maintain their own integrity.

This chapter explains why the survival of nonstructural elements is an important necessity in the event of a strong earthquake and why they are particularly vulnerable to the effects of earthquakes. It introduces, in addition, two methods by which a rational seismic analysis of nonstructural elements may be carried out.

14.2 IMPORTANCE OF NONSTRUCTURAL ELEMENTS

Despite the fact that they are not part of the main structure, nonstructural elements are far from being secondary in importance. It is nowadays widely recognized that the survival of nonstructural elements is essential to provide emergency services in the aftermath of an earthquake. Past earthquakes have shown that the failure of equipment and the debris caused by falling objects and overturned furniture may critically affect the performance of fire and police stations, emergency command centers, communication facilities, power stations, water supply and treatment plants, food treatment and cold storage plants, hospitals, and collective transportation systems. For example, during the 1994 Northridge earthquake in Los Angeles, California, area, several major hospitals had to be evacuated, not because of structural damage, but because of (a) water damage caused by the failure of water lines and water supply tanks; (b) the failure of emergency power systems and heating, ventilation, and air-conditioning units; and (c) damage to suspended ceilings and light fixtures and some broken windows. Along the same lines, it is now recognized that damage to nonstructural elements represents a threat to life safety, may seriously impair a building's function, and results in major direct and indirect economic losses. Understandably, the collapse of suspended light fixtures, hung ceilings, or partition walls; the fall of cladding components, parapets, signboards, ornaments, or pieces of broken glass; the overturning of heavy equipment, bookshelves, storage racks, and pieces of furniture; and the rupture of pipes and containers with toxic materials are all capable of causing serious injury or death. Figures 14.1 through 14.7 show a few examples

FIGURE 14.1 Collapse of signboard on the street of Kobe, Japan, during the 1995 Great Hanshin earthquake. (Photograph courtesy of Christopher Arnold, Building Systems Development, Inc.)

FIGURE 14.2 Collapse of pipe originally suspended from below floor system of parking structure during the 1987 Whittier Narrows earthquake. (Photograph courtesy of Satwant Rihal, Cal Poly, San Luis Obispo.)

Seismic Response of Nonstructural Elements

FIGURE 14.3 Failure of hospital penthouse in Mexico City during the 1985 Michoacán, Mexico, earthquake.

FIGURE 14.4 Dislocation of window frame in school building during the 1973 Orizaba, Mexico, earthquake.

FIGURE 14.5 Fallen precast element from parking garage, responsible for death of student during the 1987 Whittier Narrows earthquake in Los Angeles, California, area. (Photograph courtesy of Narendra Taly, California State University, Los Angeles; Reproduced with permission of Earthquake Engineering Research Institute from Taly, N., *Earthquake Spectra*, 4, 2, 1988, 277–317.)

FIGURE 14.6 Collapse of building's exterior wall during the 1994 Northridge, California, earthquake. (Photograph courtesy of Yousef Bozorgnia, University of California, Berkeley.)

Seismic Response of Nonstructural Elements

FIGURE 14.7 Dislocation of library bookshelves during the 1987 Whittier Narrows, California, earthquake. (Photograph courtesy of Narendra Taly, California State University, Los Angeles. Reproduced with permission of Earthquake Engineering Research Institute from Taly, N., *Earthquake Spectra*, 4, 2, 1988, 277–317.)

of the failure of nonstructural elements during past earthquakes and vividly illustrate how these failures may cause serious injury and death. A most unfortunate demonstration that this may indeed happen is the death of a student who during the 1987 Whittier Narrows earthquake in California was struck by a falling precast panel while walking out of a parking structure. It is also easy to see that the normal activities that take place in a building may be critically disrupted when some essential equipment fails, or when debris from failed architectural components gets in the way. Typical examples that illustrate the consequences of such an event are the unwanted solidification of a melted metal in an industrial facility, the inaccessibility of financial records in a timely manner in a banking institution, and the failure to fill pending orders in a manufacturing plant. In regard to the economic impact caused by the failure of nonstructural elements, there is plenty of evidence which shows that, because of the loss of the nonstructural elements themselves, loss of inventory, and loss of business income, the cost of such failures may easily exceed the replacement cost of the building. And in today's high technological environment, this cost may be even further as a result of the widespread use of electronic and computer equipment and dependence of industry in this type of equipment. It is clear, thus, that nonstructural elements should be the subject of a rational and careful seismic design in much the same way as their supporting structures are.

14.3 GENERAL PHYSICAL CHARACTERISTICS

Several physical characteristics make nonstructural elements in buildings particularly vulnerable to the effects of earthquakes. Some of these physical characteristics are

1. Nonstructural elements are usually attached to the elevated portions of a building, and thus, they are not subjected to the ground motion generated by an earthquake, but to the amplified motions generated by the dynamic response of the building.
2. They are light compared to the structure to which they are connected, and their stiffness is also much lower than that of the structure as a whole. As a result, their natural frequencies are in many instances close to the natural frequencies of their supporting structures, which means that they may be subjected to severe resonant motions.

3. Their damping ratios may be quite low, much lower than those of the structures that support them, and thus they do not possess the damping characteristics that are necessary for protection against sharp resonant motions.
4. They may be connected to a building at more than one point and may be, therefore, subjected to the distortions induced by the differential motion of their supports.
5. They are designed to perform a function other than to resist forces. As such, they are built with materials that are far from the ideal materials to resist seismic forces and may possess parts that are sensitive to even the smallest level of vibration.

14.4 GENERAL RESPONSE CHARACTERISTICS

The special physical characteristics just described make nonstructural elements not only susceptible to earthquake damage, but also make them react to earthquake ground motions differently from the way building structures do. That is, the response of nonstructural elements to an earthquake ground motion exhibits characteristics that are not common in the observed response of conventional structures. The following are a few of such characteristics:

1. The response of a nonstructural element depends on the response of the structure to which it is connected and depends, thus, not only on the characteristics of the ground motion that excites the base of the structure, but also on the dynamic characteristics of the structure.
2. The response of a nonstructural element depends on its location within its supporting structure. As a result, identical elements respond differently to the effects of an earthquake if they are located at different levels in the same structure.
3. There may be a significant interaction between a nonstructural element and its supporting structure. That is, the motion of the nonstructural element may modify the motion of its supporting structure, and vice versa. In such cases, therefore, one cannot predict the response of the nonstructural element without considering the combined system formed by the nonstructural element and its supporting structure.
4. When a nonstructural element is connected to its supporting structure at more than one point, the element's supports are excited by motions that are different and out of phase.
5. Since the damping forces in a nonstructural element are much lower than those in its supporting structure, the damping properties of the combined system formed by the structure and the nonstructural element, which is the system that characterizes the response of the nonstructural element, are not uniform throughout the system. This means that the response of a nonstructural element is governed by the response of a system without classical modes of vibration; that is, the response of a system whose natural frequencies and mode shapes are complex valued.
6. Since, as mentioned earlier, it is likely that some of the natural frequencies of a nonstructural element may be close in value to those of its supporting structure, the combined structural–nonstructural system may result in a system with closely spaced natural frequencies. As such, the response of a nonstructural element may be controlled by two or more dominant modes of vibration, as opposed to a single mode, which is the case for many building structures.
7. The response of a nonstructural element is affected by its own yielding as well as the yielding of its supporting structure.

14.5 MODELING OF NONSTRUCTURAL ELEMENTS

As it may be inferred from the examples given in Section 14.1, and in contrast to building structures, there exist a vast variety of nonstructural elements. Therefore, there are no general rules or accepted standards for the modeling of nonstructural elements. Furthermore, in many situations the modeling is left to the manufacturer of the particular element in question, who in the majority of cases knows better than anybody else the dynamic properties and significant characteristics that need to be considered in a mathematical model of the element. In general and as far as their modeling is concerned, nonstructural elements may nonetheless be classified into three broad categories: rigid, flexible, and hanging from above. If a nonstructural element is rigid, then its dynamic properties will depend primarily on the flexibility and ductility of its anchors. In this case, it may be modeled as a single-degree-of-freedom system with a mass equal to the mass of the element and stiffness and ductility equal to the stiffness and ductility of the anchors. Examples of this type of nonstructural elements are engines and motors attached to the floors of a structure by means of steel brackets and bolts. If the nonstructural element is flexible, then it is necessary to model it as a multi-degree-of-freedom system with distributed mass, stiffness, and ductility, in much the same way a building structure is normally modeled. In this regard, it should be noted that, different from a building structure, a nonstructural element may be attached to multiple points of its supporting structure and it is important, thus, that the considered model include these multiple attachment points. Typical examples of flexible nonstructural elements are signboards and pipelines, the latter also being an example of a nonstructural element with multiple points of attachment. Finally, if a nonstructural element hangs from above, then it behaves and may be modeled as a single-mass pendulum. Ordinarily, however, hanging nonstructural elements are not analyzed since this type of nonstructural elements is seldom damaged by earthquakes. The exception is when the nonstructural element may impact its supporting structure or any other object nearby as a result of the large oscillations it may be subjected to during an earthquake. In these cases, an analysis may be performed to investigate the amplitude of such oscillations. Examples of hanging nonstructural elements are lighting systems, cable trays, and some decorative items such as chandeliers.

14.6 METHODS OF ANALYSIS

A great research effort has been devoted over the past four decades to develop rational methods for the seismic analysis of nonstructural elements. For the most part, this effort has been fueled by the need to guarantee the survivability of critical equipment such as piping and control systems in nuclear power plants. Therefore, many methods of analysis have been proposed as a result of this research effort, some of them with a strong empirical base and others based on rigorous principles of structural dynamics. Until now, however, none of these methods has become the industry standard or the one that is preferred by most analysts.

In the development of methods of analysis for nonstructural elements, it is generally recognized that they are difficult to analyze accurately and efficiently. It is always possible to consider them in conjunction with the analysis of their supporting structures, but a combined structural–nonstructural system generally results in a system with an excessive number of degrees of freedom and large differences in the values of its various masses, stiffnesses, and damping constants. Such characteristics usually render any conventional method of analysis expensive, inaccurate, and inefficient. For example, a modal analysis exhibits difficulties in the computation of the combined system's natural frequencies and mode shapes, and a step-by-step integration method becomes extraordinarily sensitive to the selected integration time step. Similarly, the consideration of a combined system is impractical since, during the preliminary design of the nonstructural elements, the structure would have to be reanalyzed every time a change is introduced in some of the parameters of the nonstructural elements. Considering that normally structural systems and nonstructural elements are designed by different teams at different times, this approach also brings serious problems of schedule and efficiency. Thus, most of the proposed methods for the analysis of nonstructural elements have been the result of an effort to avoid the analysis of a combined system and overcome the aforementioned difficulties.

The majority of the methods available for the analysis of nonstructural elements are based on concepts not covered in this book or in introductory books on structural dynamics. Therefore, only two of such methods are described in this chapter. One is the floor response spectrum method, and the other is a method based on a modal synthesis. Although with some limitations, these methods are easy to understand and thus useful to introduce the reader to a difficult problem and some of the fundamental concepts involved.

14.7 FLOOR RESPONSE SPECTRUM METHOD

One of the first methods used in the analysis of nonstructural elements is the so-called *systems-in-cascade*, *in-structure response spectrum*, or *floor response spectrum method*. In this method, the excitation at the base of a nonstructural element is first defined in terms of a response spectrum in much the same way as in the case of building structures. Then, the nonstructural element is analyzed, also as in the case of building structures, using this response spectrum. Since, in general, the response spectrum for the excitation at the base of a nonstructural element is different from the response spectrum of the ground motion that excites its supporting structure, the former is called "floor response spectrum" or "in-structure spectrum" to distinguish it from the latter. Note also that a floor response spectrum is needed for each of the points or floors of the structure where there is a nonstructural element attached to it since, as discussed earlier, the motion generated by an earthquake may be markedly different at different points or floors of a structure.

Ordinarily, a floor response spectrum is obtained by means of a time-history analysis (see Chapter 11). That is, given a ground acceleration time history, a step-by-step integration analysis is carried out to determine the acceleration time history of the point or floor to which the nonstructural element under consideration will be attached. Then, this time history is used to generate a response spectrum using any of the conventional methods currently in use for the generation of ground response spectra (see Section 6.4.5). Since the use of a single time history is not acceptable for design purposes, it is necessary to generate floor response spectra for several ground acceleration time histories and use the average or an envelope to all these spectra. As this is a time-consuming procedure that requires lengthy numerical integrations, an alternative commonly employed method is the use of an artificial ground acceleration time history that envelops a given ground design spectrum, such as the design spectrum specified by a building code. However, as discussed in Section 6.9, caution should be exercised with this alternative since such an artificial time history is not uniquely defined. That is, different time histories may envelop the target design spectrum, but give significantly different results. Another alternative is to use one of the methods that are now available to generate floor response spectra directly from a specified ground response spectrum or design spectrum without utilizing a time-history analysis. These methods utilize the response spectrum method introduced in Chapter 10, or any variant of it, to determine in terms of a specified ground response spectrum and the dynamic properties of the structure the maximum acceleration response of a simple oscillator attached to the structural floor for which a floor response spectrum is desired. To learn more about such methods, consult the work by Singh listed in the Further Readings section.

EXAMPLE 14.1   ANALYSIS OF A NONSTRUCTURAL ELEMENT BY THE
                FLOOR RESPONSE SPECTRUM METHOD

A piece of equipment is mounted on the roof of the 10-story shear building shown in Figure E14.1a. This piece of equipment can be modeled as a single-degree-of-freedom system with a natural frequency of 2 Hz and a damping ratio of 0.5%. Determine using the floor response spectrum method the maximum acceleration response of the piece of equipment when the base of the building is subjected to the ground motion shown in Figure E14.1b, which corresponds to the east–west component of the ground accelerations recorded at the Foster City station during the 1989 Loma Prieta, California, earthquake. Assume the building's damping matrix is proportional to its stiffness matrix and that the damping ratio in its fundamental mode is 2%.

Seismic Response of Nonstructural Elements

Solution

The first step in using the floor response spectrum method for the solution of this problem is the generation of the floor response spectrum corresponding to the roof of the building and the given earthquake ground motion. For this purpose, a time-history analysis of the building is first carried out. Then, the desired floor response spectrum is generated using the time history corresponding to the acceleration response of the building at its roof level.

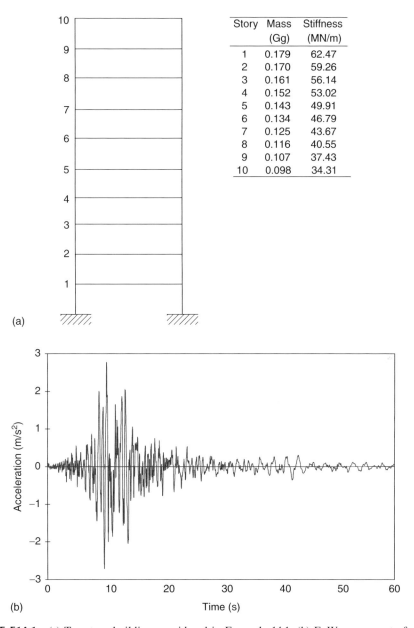

FIGURE E14.1 (a) Ten-story building considered in Example 14.1, (b) E–W component of ground acceleration recorded at Foster City during the 1989 Loma Prieta, California, earthquake, (c) acceleration response at the roof level of the building in Example 14.1 when the building base is subjected to the 1989 E–W Foster City accelerogram, and (d) floor response spectrum for the roof level of the building in Example 14.1 and 0.5% damping when the building base is subjected to the 1989 E–W Foster City accelerogram.

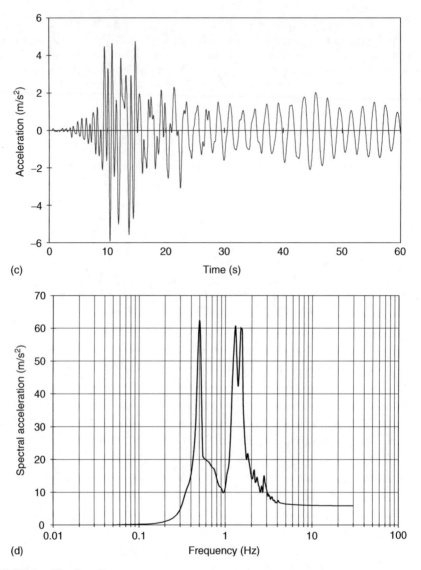

FIGURE E14.1 (Continued)

Proceeding accordingly, the acceleration time history and the floor response spectrum shown in Figures E14.1c and E14.1d are obtained. From this floor response spectrum, one then finds that the ordinate corresponding to the natural frequency of the piece of equipment (2 Hz) is equal to 14.4 m/s². Consequently, the maximum acceleration response of the piece of equipment when the building is subjected to the given excitation is approximately equal to $14.4/9.81 = 1.47g$.

The floor response spectrum method has been proven accurate for nonstructural elements with masses that are much smaller than the masses of their supporting structures and natural frequencies that are not too close to the natural frequencies of their supporting structures. However, it may yield overly conservative results for nonstructural elements that do not have these characteristics. The reason for this overconservatism is that in the floor response spectrum method nonstructural elements are considered separately from their supporting structures, without due consideration to the fact that the response of a nonstructural element may significantly affect the response of its supporting structure and vice versa. That is, the floor response spectrum method neglects the dynamic interaction

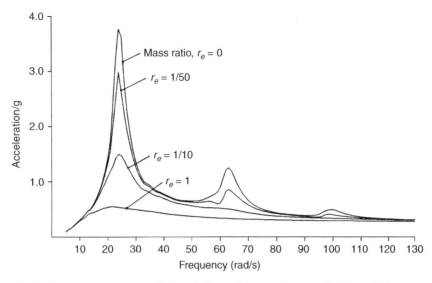

FIGURE 14.8 Floor response spectrum for fourth floor of six-story building in Figure 14.9 corresponding to different nonstructural to structural mass ratios. (Reproduced with permission of John Wiley & Sons, Limited, from Singh, M. P. and Suárez, L. E., *Earthquake Engineering and Structural Dynamics*, 15, 7, 1987, 871–888.)

between a nonstructural element and its supporting structure. An additional reason is that the floor response spectrum method cannot account for the fact that the masses of the structure and the nonstructural element vibrate out of phase, an effect (known as nonclassical damping effect) that arises because, in general, a combined structural–nonstructural system does not possess classical modes of vibration (see Section 15.5.4). There is now ample analytical evidence which demonstrates that ignoring these two effects may lead to gross errors in the response calculation of some nonstructural elements. Another problem with the floor response spectrum method is that it cannot be rationally applied for the analysis of nonstructural elements with multiple points of attachment. This is so because the floor response spectrum method cannot consider that the motions at the different attachment points of a nonstructural element are normally different from one another and out of phase.

The floor response spectra shown in Figure 14.8 illustrate the accuracy involved in the use of the floor response spectrum method. These floor response spectra have been obtained for the fourth floor of the six-story shear building shown in Figure 14.9 for a damping ratio of 6% in the fundamental mode of the building and different values of the mass ratio r_e. The input considered was an ensemble of 75 synthetically generated ground motions. The curve corresponding to a zero mass ratio is obtained using the traditional procedure to generate floor response spectra; that is, ignoring the dynamic interaction between the structure and the nonstructural element. The other curves are obtained considering this interaction fully. The mass ratio r_e was defined as the ratio of the total mass of the nonstructural element to the total mass of the structure.

It may be observed from the curves in Figure 14.8 that neglecting the interaction is always conservative, although in some cases it may be grossly conservative. It may also be seen that the interaction effect becomes important when the mass ratio is not too small and the natural frequency of the nonstructural element is close to one of the dominant natural frequencies of the supporting structure.

Similarly, Tables 14.1 and 14.2 illustrate the importance of considering nonclassical damping effects in the analysis of nonstructural elements. In Tables 14.1 and 14.2, a comparison is made of the absolute acceleration response of the same single-degree-of-freedom nonstructural element described in the preceding paragraphs when the element is analyzed considering the nonclassical damping character of the combined system and when it is assumed that this combined system

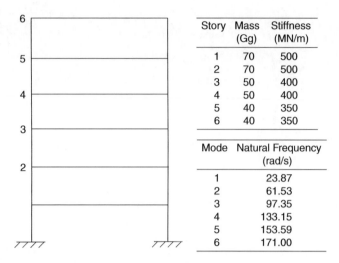

FIGURE 14.9 Properties and natural frequencies of six-story shear building considered in study of interaction and nonclassical damping effects.

TABLE 14.1

Absolute Acceleration Response of Single-Degree-of-Freedom Nonstructural Element Mounted on the Fourth Floor of the Structure Shown in Figure 14.9 Considering and Not Considering the Nonclassical Damping Nature of the Combined System. Damping Ratio in Fundamental Mode of Structure = 6%; Damping Ratio of Nonstructural Element = 1%

	Mass Ratio					
	0.00156			0.000156		
Nonstructural Element Natural Frequency (rad/s)	Classical Damping (g)	Nonclassical Damping (g)	Error (%)	Classical Damping (g)	Nonclassical Damping (g)	Error (%)
24	2.556	3.149	14.8	4.317	3.063	41.0
42	0.446	0.451	1.0	0.449	0.453	1.0
62	0.612	0.312	95.8	0.879	0.313	180.6
97	0.335	0.315	6.4	0.291	0.407	28.6
134	0.281	0.309	8.9	0.386	0.320	20.4
154	0.258	0.257	0.5	0.295	0.258	14.4
162	0.254	0.254	0.0	0.256	0.255	0.3
171	0.252	0.252	0.1	0.252	0.254	0.7

Source: Singh, M. P. and Suárez, L. E., *Earthquake Engineering and Structural Dynamics*, 15, 7, 1987, 871–888.

is classically damped. Table 14.1 summarizes the results for the case when the damping ratios in the fundamental modes of the structure and nonstructural element are 6% and 1%, respectively, whereas Table 14.2 lists those when these damping ratios are 8% and 0.5%.

It may be noted from Tables 14.1 and 14.2 that some large errors may be possible if the nonclassical damping effects in question are not accounted for. It may also be noted that, in general, such an effect becomes important when the nonstructural to structural mass ratio is small, the natural frequency of the nonstructural element is close to one of the dominant frequencies of the supporting structure, and the difference between the damping ratios of the structure and the nonstructural element is large. Furthermore, it may be noted that the discrepancy between the results for the cases

TABLE 14.2
Absolute Acceleration Response of Single-Degree-of-Freedom Nonstructural Element Mounted on the Fourth Floor of the Structure Shown in Figure 14.9 Considering and Not Considering the Nonclassical Damping Nature of the Combined System. Damping Ratio in Fundamental Mode of Structure = 8%; Damping Ratio of Nonstructural Element = 0.5%

	Mass Ratio					
	0.00156			0.000156		
Nonstructural Element Natural Frequency (rad/s)	Classical Damping (g)	Nonclassical Damping (g)	Error (%)	Classical Damping (g)	Nonclassical Damping (g)	Error (%)
24	2.070	2.460	15.9	4.267	2.766	54.3
42	0.451	0.459	1.7	0.455	0.463	1.7
62	0.667	0.299	122.3	1.135	0.301	276.8
97	0.283	0.320	11.7	0.190	0.388	51.2
134	0.252	0.271	6.9	0.359	0.288	24.7
154	0.231	0.231	0.0	0.283	0.234	21.4
162	0.228	0.228	0.1	0.232	0.230	0.9
171	0.225	0.226	0.4	0.225	0.227	0.8

Source: Singh, M. P. and Suárez, L. E., *Earthquake Engineering and Structural Dynamics*, 15, 7, 1987, 871–888.

with classical and nonclassical damping becomes consistently larger as the nonstructural to structural mass ratio becomes smaller.

Finally, it should be noted that traditionally floor response spectra are obtained under the assumption that a structure and its nonstructural elements behave as linear systems. As discussed in Section 6.5, however, conventional structures are designed to resist strong earthquakes by incurring into their nonlinear range of behavior. Similarly, many nonstructural elements have the capability to resist inelastic deformations. Hence, it is likely that a structure and its nonstructural elements will undergo inelastic deformations during a severe earthquake, and that this nonlinear behavior will influence the response of the nonstructural elements significantly.

At present, there is no clear understanding as to how structural nonlinearity may affect a floor response spectrum. In general, response reductions are expected for nonstructural elements with natural frequencies equal to or greater than the fundamental natural frequency of the supporting structure. These reductions result from two main factors: (1) an increase in the damping ratio of the structure caused by its nonlinear behavior (hysteretic damping) and (2) a shift of the fundamental natural frequency of the structure away from the natural frequency of the nonstructural element. A further reduction is attained if the nonstructural element itself is allowed to go nonlinear in much the same way as it is observed in a typical nonlinear response spectrum. Some numerical studies have shown, however, that the floor response spectrum ordinates for a nonlinear structure may actually increase compared to those obtained when the structure is assumed to remain linear at all excitation levels. Moreover, these studies have also shown that this increase is particularly noticeable at high frequencies; that is, at frequencies higher than the fundamental natural frequency of the structure.

Figure 14.10 illustrates the contrasting effect of structural nonlinearity in a floor response spectrum. Figure 14.10 shows floor response spectrum ratios for a nonlinear five-story shear structure, where these ratios are obtained by dividing the ordinates of the floor response spectrum for the nonlinear structure by the corresponding ordinates of the floor response spectrum obtained when the structure is assumed with a linear behavior. Two values of the ductility factor μ are considered for the structure's first story: $\mu = 2$ and $\mu = 4$ (see Section 6.5.4 for the definition of

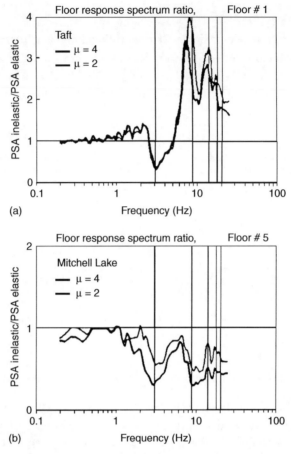

FIGURE 14.10 Floor response spectrum ratios (pseudo spectral acceleration in inelastic structure over pseudo spectral acceleration in elastic structure) corresponding to floor accelerations of a five-story shear building with first-floor story ductilities of 2 and 4. (a) Ratios for first floor when the building is subjected to the S69E Taft acceleration record from the 1952 Kern County, California, earthquake (b) ratios for fifth floor when the building is subjected to the N28E Mitchell Lake acceleration record from the 1982 Brunswick, Canada, earthquake. Vertical lines identify the natural frequencies of the structure. (Reproduced with permission of Robert T. Sewell, from Sewell, R. T., *Damage Effectiveness of Earthquake Ground Motions: Characterizations Based on the Performance of Structures and Equipment*, Ph. D. Thesis, Stanford University, 1988.)

ductility factor). All other stories are assumed to remain in their linear range of behavior. Figure 14.10a shows the aforementioned ratios for the first floor of the structure when the structure is excited by the S69E component of the ground motion recorded at Taft Lincoln School during the 1952 Kern County, California, earthquake. Figure 14.10b shows those for the fifth floor of the structure when the structure is excited by the N28E component of the ground motion recorded at Mitchell Lake Road during the 1982 New Brunswick, Canada, earthquake. Note that in Figures 14.10a and 14.10b a floor response spectrum ratio of less than 1 represents a reduction in the corresponding floor response spectrum ordinate due to the nonlinearity of the structure. Conversely, a floor response spectrum ratio of more than 1 represents an increase.

It may be seen from the curves in Figure 14.10 that in one case (Figure 14.10a) there is a considerable amplification in the high frequency range of the spectrum due to the nonlinearity of the structure. In the other case (Figure 14.10b), a reduction is observed over the entire frequency range of the spectrum. It is also worthwhile to note the fact that in both cases a reduction is attained, as expected, for frequencies that are near the fundamental natural frequency of the structure.

Seismic Response of Nonstructural Elements

14.8 MODAL SYNTHESIS METHOD

14.8.1 Introduction

In view of the limitations of the floor response spectrum method and the impracticality of an analysis based on a combined structural–nonstructural system, several alternative methods that consider the interaction effects between structure and nonstructural element and overcome the problems of practicality associated with a direct analysis of a combined system have been developed. One of such methods is the modal synthesis method. In this method, the response of a nonstructural element is calculated on the basis of a modal analysis of the combined structural–nonstructural system, but the dynamic properties of this combined system are obtained in terms of the dynamic properties of the two components when independently considered. This approach eliminates the main source of error inherent in the floor response spectrum method since, by considering the structure and the nonstructural element together as a single unit, the interaction between the two subsystems and the different and out-of-phase support motions are implicitly considered. It is also a practical approach. By formulating the analysis in terms of the dynamic properties of the independent subsystems, one avoids the numerical difficulties of the conventional methods of analysis associated with the large difference in the values of the parameters of the structure and the nonstructural element. Furthermore, one avoids the need to generate floor spectra since the earthquake input is defined at the ground level, and the need to reanalyze the structure every time the parameters of the nonstructural element are changed.

Methods based on a modal synthesis have been developed for the general case of multi-degree-of-freedom nonstructural elements attached to multiple points of a supporting structure and that, together with their supporting structures, form systems with nonclassical damping. These methods, however, involve concepts and procedures not covered in this book. Therefore, only the case of nonstructural elements with a single point of attachment and classical damping is discussed. The discussion of this case serves to introduce the basic concepts involved, illustrates the application of the modal synthesis approach, and facilitates the understanding of the application of this approach to other types of nonstructural elements. Readers interested in learning more about these methods may find the appropriate references in the state-of-the-art articles listed in the Further Readings section.

The modal synthesis method described in this chapter is based on the application of the conventional response spectrum method to the system that results from considering together as a single unit a nonstructural element and the structure to which it is connected. As such, the method involves four steps: (1) the specification of a ground response spectrum or the consideration of the spectrum specified for the design of the structure; (2) the calculation of the natural frequencies, mode shapes, damping ratios, and participation factors of the combined structural–nonstructural system; (3) the calculation of the nonstructural element's maximum modal responses in terms of the given response spectrum and the natural frequencies, mode shapes, damping ratios, and participation factors of the structural–nonstructural system; and (4) the combination of these maximum modal responses using one of the modal combination rules studied in Chapter 10 to obtain the nonstructural element's maximum response. In the discussion that follows, it will be assumed that both the structure and the nonstructural element are linear elastic systems and that the damping matrix of the combined structural–nonstructural system is of the Rayleigh type (see Section 10.5); that is, proportional to the combined system's mass and stiffness matrices. It will also be assumed that the natural frequencies, mode shapes, damping ratios, and generalized masses of the two independent components are known, and that their mode shapes have been normalized in a way that makes their participation factors equal to unity (see Section 10.4). For convenience, henceforth, the structure will be referred to as the primary system and the nonstructural element as the secondary system.

14.8.2 Mode Shapes and Natural Frequencies of Combined System

Consider the undamped primary and secondary systems shown in Figure 14.11. Assume that the primary system has N degrees of freedom, the secondary system has n degrees of freedom when

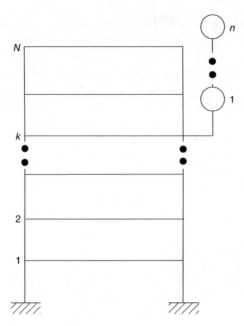

FIGURE 14.11 Combined primary–secondary system.

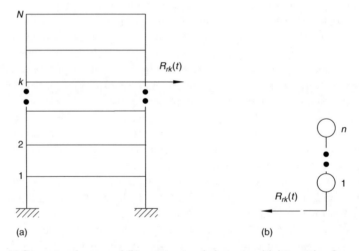

FIGURE 14.12 (a) Separate primary and (b) nonstructural elements with interaction force.

it is considered fixed at the point where it is connected to the primary system, and the secondary system is attached to the kth degree of freedom of the primary system. Assume, in addition, that the primary–secondary system is vibrating freely in its rth mode. If the force that each system exerts on each other is considered, then each system may be considered separately as shown in Figure 14.12. Moreover, the motion of each structure in such an rth mode may be defined by separate equations of motions as shown in the following sections.

14.8.2.1 Primary System

With reference to Figure 14.12a, the equation of motion of the primary structure may be written as

$$[M]\{\ddot{x}_p(t)\}_r + [K]\{x_p(t)\}_r = \{R_p(t)\}_r \tag{14.1}$$

Seismic Response of Nonstructural Elements

where $[M]$ and $[K]$ represent the $N \times N$ mass and stiffness matrices of the primary system, respectively; $\{x_p(t)\}_r$ is a vector that contains the displacements of its masses; and $\{R_p(t)\}_r$ is a vector that contains the reaction of the secondary system acting against the primary one; that is, a vector of the form $\{R_p(t)\}_r^T = \{0\ 0 \ldots R_{rk}(t) \ldots 0\ 0\}^T$, where $R_{rk}(t)$ is the interaction force between the two subsystems (see Figure 14.12). Since the entire primary–secondary system is vibrating freely in its rth mode, it is possible to express its displacement and reaction vectors as

$$\{x_p(t)\}_r = \{\varphi_p\}_r \cos(\omega_r - \theta_r) \tag{14.2}$$

$$\{R_p(t)\}_r = \{R_p\}_r \cos(\omega_r - \theta_r) \tag{14.3}$$

where $\{\varphi_p\}_r$ and $\{R_p\}_r$, ω_r, and θ_r, respectively, represent amplitudes corresponding to $\{x_p(t)\}_r$ and $\{R_p(t)\}_r$, rth natural frequency of the primary–secondary system, and phase angle.

Equation 14.1 may therefore be expressed alternatively as

$$-\omega_r^2[M]\{\varphi_p\}_r + [K]\{\varphi_p\}_r = \{R_p\}_r \tag{14.4}$$

Furthermore, under the modal transformation,

$$\{\varphi_p\}_r = [\Phi]\{Y\}_r \tag{14.5}$$

where $[\Phi]$ denotes the $N \times N$ modal matrix of the independent primary system and $\{Y\}_r$ is a vector of unknown modal coordinates, and if premultiplied throughout by the transpose of $[\Phi]$, Equation 14.4 may be written in terms of modal coordinates as

$$-\omega_r^2[\Phi]^T[M][\Phi]\{Y\}_r + [\Phi]^T[K][\Phi]\{Y\}_r = [\Phi]^T\{R_p\}_r \tag{14.6}$$

However, if the orthogonality properties of the primary system's mode shapes are invoked, Equation 14.6 may be reduced to

$$-\omega_r^2 M_i^* Y_{ri} + K_i^* Y_{ri} = \{\Phi\}_i^T \{R_p\}_r \qquad i = 1, \ldots, N \tag{14.7}$$

where M_i^*, K_i^*, and $\{\Phi\}_i$ and Y_{ri} denote generalized mass, generalized stiffness, and mode shape of the independent primary system in its ith mode and modal coordinate corresponding to such an ith mode. Or if it is considered that $K_i^* = \omega_{pi}^2 M_i^*$ and $\{\Phi\}_i^T \{R\}_r = \Phi_{ki} R_{rk}$, where Φ_{ki} is the amplitude corresponding to the kth degree of freedom in the ith mode shape of the primary system, Equation 14.7 may also be written as

$$M_i^*(\omega_{pi}^2 - \omega_r^2)Y_{ri} = \Phi_{ki} R_{rk} \qquad i = 1, \ldots, N \tag{14.8}$$

where ω_{pi} represents the circular natural frequency of the independent primary system in its ith mode.

14.8.2.2 Secondary System

Consider now the secondary system shown in Figure 14.12b. This is an unrestrained system with $n + 1$ degrees of freedom and subjected to an external force $R_{rk}(t)$. Its equation of motion when the primary–secondary system vibrates in its rth mode takes the form

$$[m]\{\ddot{x}_s(t)\}_r + [k]\{x_s(t)\}_r = -\{R_s(t)\}_r \tag{14.9}$$

where $[m]$ and $[k]$ represent the $(n + 1) \times (n + 1)$ mass and stiffness matrices of the system, respectively; $\{x_s(t)\}$ is its vector of displacements relative to the ground; and $\{R_s(t)\}_r^T = \{R_{rk}(t) \quad 0 \ldots 0\}^T$, where $R_{rk}(t)$ is the previously defined interaction force between the primary and the secondary system. Since the system is unrestrained, note that Equation 14.9 includes the massless degree of freedom corresponding to the point where the secondary system is connected to the primary one.

As in the case of the primary system, the secondary system vibrates with a frequency ω_r when the combined primary–secondary system vibrates in its rth mode. Hence, it is also valid to write $\{x_s(t)\}_r$ and $\{R_s(t)\}_r$ as

$$\{x_s(t)\}_r = \{\bar{\varphi}_s\}_r \cos(\omega_r - \theta_r) \tag{14.10}$$

$$\{R_s(t)\}_r = \{R_s\}_r \cos(\omega_r - \theta_r) \tag{14.11}$$

where $\{\bar{\varphi}_s\}_r$ and $\{R_s\}_r$ are vectors that contain the amplitudes corresponding to $\{x_s(t)\}_r$ and $\{R_s(t)\}_r$, respectively. In view of Equations 14.10 and 14.11, Equation 14.9 may be therefore expressed as

$$-\omega_r^2 [m]\{\bar{\varphi}_s\}_r + [k]\{\bar{\varphi}_s\}_r = -\{R_s\}_r \tag{14.12}$$

Equation 14.12 may also be written in terms of modal coordinates using the modal matrix of the system as the transformation matrix. However, since the system under consideration is unconstrained, such a transformation matrix has to necessarily include a rigid-body mode. Accordingly, in terms of modal coordinates, $\{\bar{\varphi}_s\}_r$ may be expressed as

$$\{\bar{\varphi}_s\}_r = [\bar{\phi}]\{\bar{y}\}_r \tag{14.13}$$

where $\{\bar{y}_s\}_r$ is a vector that includes the system's $n + 1$ modal coordinates (i.e., $y_0, y_1, \ldots, y_n$), and $[\bar{\phi}]$ is the system's $(n + 1) \times (n + 1)$ modal matrix, which in this case is of the form

$$[\bar{\phi}] = [\{\bar{\phi}\}_0 \quad \{\bar{\phi}\}_1 \quad \{\bar{\phi}\}_2 \ldots \{\bar{\phi}\}_n] \tag{14.14}$$

where $\{\bar{\phi}\}_0$ is the aforementioned rigid-body mode; that is, a vector whose entries are all unity for the degrees of freedom in the direction of motion and zero for all the others. Similarly, $\{\bar{\phi}\}_j$, $j = 1, \ldots, n$, are the normal mode shapes of the independent secondary system when its point of connection with the primary system is considered fixed. Since the massless point of connection to the primary system is being considered as an additional degree of freedom, note that the mode shapes $\{\bar{\phi}\}_j$, $j = 1, \ldots, n$, are of the order $n + 1$ and that the first entry in all these vectors is equal to zero.

Substitution of Equation 14.13 into Equation 14.12 and premultiplication by the transpose of the modal matrix $[\bar{\phi}]$ leads to

$$-\omega_r^2 [\bar{\phi}]^T [m][\bar{\phi}]\{\bar{y}\}_r + [\bar{\phi}]^T [k][\bar{\phi}]\{\bar{y}\}_r = -[\bar{\phi}]^T \{R_s\}_r \tag{14.15}$$

Note, however, that if the normal mode shapes are normalized such as to make the corresponding participation factors equal to unity (see Section 10.4), then in view of the definition of participation factor, one can write

$$\{\bar{\phi}\}_0^T [m]\{\bar{\phi}\}_j = \{\bar{\phi}\}_j^T [m]\{\bar{\phi}\}_j = m_j^* \tag{14.16}$$

where m_j^* denotes the jth generalized mass of the independent secondary system. Note also that as a result of the orthogonality properties of the normal mode shapes, one has that

$$\{\bar{\phi}\}_j^T [m] \{\bar{\phi}\}_k = \begin{cases} m_j^* & \text{if } j = k \\ 0 & \text{if } j \neq k \end{cases} \tag{14.17}$$

$$\{\bar{\phi}\}_j^T [k] \{\bar{\phi}\}_k = \begin{cases} k_j^* & \text{if } j = k \\ 0 & \text{if } j \neq k \end{cases} \tag{14.18}$$

where k_j^* represents the jth generalized stiffness constant of the independent secondary system when considered fixed at its point of attachment with the primary system. Furthermore, note that $[k]\{\bar{\phi}\}_0$ is equal to zero since this product represents the internal forces in the structure and these internal forces are zero when the structure is displaced as a rigid body. Consequently, the triple matrix products in Equation 14.15 result as

$$[\bar{\phi}]^T [m][\bar{\phi}] = \begin{bmatrix} m_0^* & m_1^* & m_2^* & \cdots & m_n^* \\ m_1^* & m_1^* & 0 & \cdots & 0 \\ m_2^* & 0 & m_2^* & \cdots & 0 \\ \vdots & \vdots & \vdots & \ddots & \vdots \\ m_n^* & 0 & 0 & \cdots & m_n^* \end{bmatrix} \tag{14.19}$$

where $m_0^* = \{\bar{\phi}\}_0^T [m] \{\bar{\phi}\}_0$ is the total mass of the nonstructural element, and

$$[\bar{\phi}]^T [k][\bar{\phi}] = \begin{bmatrix} 0 & 0 & 0 & \cdots & 0 \\ 0 & k_1^* & 0 & \cdots & 0 \\ 0 & 0 & k_2^* & \cdots & 0 \\ \vdots & \vdots & \vdots & \ddots & \vdots \\ 0 & 0 & 0 & \cdots & k_n^* \end{bmatrix} \tag{14.20}$$

Similarly, it may be observed that the product $[\bar{\phi}]^T \{R_s\}_r$ in Equation 14.15 may be written as

$$[\bar{\phi}]^T \{R_s\}_r = \left[\{\bar{\phi}\}_0^T \{R_s\}_r \quad \{\bar{\phi}\}_1^T \{R_s\}_r \quad \cdots \quad \{\bar{\phi}\}_n^T \{R_s\}_r \right]^T = \left[R_{rk} \quad 0 \cdots 0 \right]^T \tag{14.21}$$

in view of the fact that, except for the rigid-body mode, the first entry in the mode shapes $\{\phi\}_j$ is always zero, and that, except for the first one, all the entries in the vector $\{R_s\}_r$ are also zero.

From Equations 14.19 through 14.21 and the fact that $k_j^* = \omega_{sj}^2 m_j^*$, the transformation of Equation 14.15 thus leads to the $n + 1$ equations

$$\omega_r^2 \left(m_0^* y_{r0} + \sum_{j=1}^n m_j^* y_{rj} \right) = R_{rk} \tag{14.22}$$

$$-\omega_r^2 m_j^* y_{r0} + m_j^* (\omega_{sj}^2 - \omega_r^2) y_{rj} = 0 \quad j = 1, \ldots, n \tag{14.23}$$

where y_{r0} and y_{rj}, $j = 1, 2, \ldots, n$, are the modal coordinates contained in $\{\bar{y}\}_r$.

14.8.2.3 Compatibility Requirements

Since the secondary system is rigidly attached to the primary system, it is necessary to establish to satisfy compatibility requirements that the displacement of the primary system at the degree of freedom to which the secondary system is attached is equal to the displacement of the secondary system at the point where it is attached to the primary structure. Mathematically, this condition may be expressed as

$$\varphi_{pkr} = \varphi_{slr} \tag{14.24}$$

where φ_{pkr} and φ_{slr} are amplitude of the displacement of the kth degree of freedom of the primary structure and amplitude of the displacement of the first degree of freedom of the secondary one, both when the primary–secondary system vibrates in its rth mode.

If (a) Equations 14.2 and 14.10 are considered, (b) Equations 14.5 and 14.13 are expressed as

$$\{\varphi_p\}_r = \{\Phi\}_1 Y_{r1} + \{\Phi\}_2 Y_{r2} + \cdots + \{\Phi\}_N Y_{rN} \tag{14.25}$$

$$\{\overline{\varphi}_s\}_r = \{\overline{\phi}\}_0 y_{r0} + \{\overline{\phi}\}_1 y_{r1} + \cdots + \{\overline{\phi}\}_n y_{rn} \tag{14.26}$$

and (c) it is recalled that the first entry in the mode shape $\{\overline{\phi}\}_0$ is unity and the first entry in the mode shapes $\{\overline{\phi}\}_j$ is zero, then in terms of modal coordinates the relationship given in Equation 14.24 may be expressed as

$$y_{r0} = \sum_{i=1}^{N} \Phi_{ki} Y_{ri} \tag{14.27}$$

14.8.2.4 Eigenvalue Problem

By substitution of Equation 14.22 into Equation 14.8, one obtains

$$M_i^* (\omega_{pi}^2 - \omega_r^2) Y_{ri} = \omega_r^2 \left(\Phi_{ki} m_0^* y_{r0} + \Phi_{ki} \sum_{j=1}^{n} m_j^* y_{rj} \right) \quad i = 1, 2, \ldots, N \tag{14.28}$$

which, after substitution of Equation 14.27, leads to

$$M_i^* (\omega_{pi}^2 - \omega_r^2) Y_{ri} = \omega_r^2 \left(\Phi_{ki} m_0^* \sum_{m=1}^{N} \Phi_{km} Y_{rm} + \Phi_{ki} \sum_{j=1}^{n} m_j^* y_{rj} \right) \quad i = 1, 2, \ldots, N \tag{14.29}$$

Similarly, substitution of Equation 14.27 into Equation 14.23 yields

$$m_j^* (\omega_{sj}^2 - \omega_r^2) y_{rj} = \omega_r^2 m_j^* \sum_{i=1}^{N} \Phi_{ki} Y_{ri} \quad j = 1, \ldots, n \tag{14.30}$$

Seismic Response of Nonstructural Elements

In matrix form, Equations 14.29 and 14.30 may be thus expressed as

$$\begin{bmatrix} \omega_{p1}^2 M_1^* & 0 & \cdots & 0 & 0 & 0 & \cdots & 0 \\ 0 & \omega_{p12}^2 M_2^* & \cdots & 0 & 0 & 0 & \cdots & 0 \\ \vdots & \vdots & \ddots & \vdots & \vdots & \vdots & \cdots & \vdots \\ 0 & 0 & \cdots & \omega_{pN}^2 M_N^* & 0 & 0 & \ddots & 0 \\ 0 & 0 & \cdots & 0 & \omega_{s1}^2 m_1^* & 0 & \cdots & 0 \\ 0 & 0 & \cdots & 0 & 0 & \omega_{s2}^2 m_2^* & 0 & 0 \\ \vdots & \vdots & \ddots & \vdots & \vdots & \vdots & \ddots & \vdots \\ 0 & 0 & \cdots & 0 & 0 & 0 & \cdots & \omega_{sn}^2 m_n^* \end{bmatrix} \begin{Bmatrix} Y_{r1} \\ Y_{r2} \\ \vdots \\ Y_{rN} \\ y_{r1} \\ y_{r2} \\ \vdots \\ y_{rn} \end{Bmatrix}$$

(14.31)

$$= \omega_r^2 \begin{bmatrix} M_1^* + \Phi_{k1}^2 m_0^* & \Phi_{k1}\Phi_{k2} m_0^* & \cdots & \Phi_{k1}\Phi_{kN} m_0^* & \Phi_{k1} m_1^* & \Phi_{k1} m_2^* & \cdots & \Phi_{k1} m_n^* \\ \Phi_{k2}\Phi_{k1} m_0^* & M_2^* + \Phi_{k2}^2 m_0^* & \cdots & \Phi_{k2}\Phi_{kN} m_0^* & \Phi_{k2} m_1^* & \Phi_{k2} m_2^* & \cdots & \Phi_{k2} m_n^* \\ \vdots & \vdots & \ddots & \vdots & \vdots & \vdots & \ddots & \vdots \\ \Phi_{kN}\Phi_{k1} m_0^* & \Phi_{kN}\Phi_{k2} m_0^* & \cdots & M_1^* + \Phi_{kN}^2 m_0^* & \Phi_{kN} m_1^* & \Phi_{kN} m_2^* & \cdots & \Phi_{kN} m_n^* \\ \Phi_{k1} m_1^* & \Phi_{k2} m_1^* & \cdots & \Phi_{kN} m_1^* & m_1^* & 0 & \cdots & 0 \\ \Phi_{k1} m_2^* & \Phi_{k2} m_2^* & \cdots & \Phi_{kN} m_2^* & 0 & m_2^* & \cdots & 0 \\ \vdots & \vdots & \ddots & \vdots & \vdots & \vdots & \ddots & \vdots \\ \Phi_{k1} m_n^* & \Phi_{k2} m_n^* & \cdots & \Phi_{kN} m_n^* & 0 & 0 & \cdots & m_n^* \end{bmatrix} \begin{Bmatrix} Y_{r1} \\ Y_{r2} \\ \vdots \\ Y_{rN} \\ y_{r1} \\ y_{r2} \\ \vdots \\ y_{rj} \end{Bmatrix}$$

or, in short notation, as

$$\begin{bmatrix} [\omega_p^2] & [0] \\ [0] & [\omega_s^2] \end{bmatrix} \begin{bmatrix} [M^*] & [0] \\ [0] & [m^*] \end{bmatrix} \begin{Bmatrix} \{Y\}_r \\ \{y\}_r \end{Bmatrix} = \omega_r^2 \begin{bmatrix} [M^*] + m_0^* \{\Phi_k\}\{\Phi_k\}^T & \{\Phi_k\}\{1\}^T [m^*] \\ [m^*]\{1\}\{\Phi_k\}^T & [m^*] \end{bmatrix} \begin{Bmatrix} \{Y\}_r \\ \{y\}_r \end{Bmatrix} \quad (14.32)$$

where $\{\Phi_k\}$ is a vector that groups together the amplitudes of the kth degree of freedom of the independent primary system in all its mode shapes, that is, $\{\Phi_k\}^T = \{\Phi_{k1} \ \Phi_{k2} \ \ldots \ \Phi_{kN}\}$. In addition, $[M^*]$, $[m^*]$, $[\omega_p^2]$, and $[\omega_s^2]$ are $N \times N$, or $n \times n$, diagonal matrices that contain the natural frequencies squared and generalized masses of the independent primary and secondary systems; $\{1\}$ is a unit vector of order n; as previously established, $\{Y\}_r$ a vector that contains the modal coordinates Y_{ri}, $i = 1, \ldots, N$; and $\{y\}_r$ a vector that contains the modal coordinates y_{rj}, $j = 1, \ldots, n$.

If Equation 14.32 is now put into the form

$$[A]\{Y_c\}_r = \lambda_r \{Y_c\}_r \quad (14.33)$$

where

$$[A] = \begin{bmatrix} [M^*] & [0] \\ [0] & [m^*] \end{bmatrix}^{-1} \begin{bmatrix} [\omega_p^2] & [0] \\ [0] & [\omega_s^2] \end{bmatrix}^{-1} \begin{bmatrix} [M^*] + m_0^* \{\Phi_k\}\{\Phi_k\}^T & \{\Phi_k\}\{1\}^T [m^*] \\ [m^*]\{1\}\{\Phi_k\}^T & [m^*] \end{bmatrix} \quad (14.34)$$

$$\{Y_c\}_r = \begin{Bmatrix} \{Y\}_r \\ \{y\}_r \end{Bmatrix} \quad (14.35)$$

and

$$\lambda_r = \frac{1}{\omega_r^2} \qquad (14.36)$$

then it may be seen that Equation 14.32 represents a standard eigenvalue problem of a nonsymmetric matrix, and that this eigenvalue problem is defined in terms of the combined system natural frequencies ω_r and the modal coordinates Y_r and y_r. Note, therefore, that through the solution to this eigenvalue problem it is possible to find these natural frequencies and, in relative terms, these modal coordinates. Note, also, that in conjunction with Equations 14.5, 14.13, and 14.27, such modal coordinates give the mode shapes of the combined primary–secondary system. Furthermore, note that the coefficient matrix [A] is defined in terms of only the modal properties of the independent primary and secondary systems. Hence, this eigenvalue problem represents a way to obtain the modal properties of a combined primary–secondary system using only the modal properties of its two independent components.

14.8.2.5 Mode Shapes of Combined System in Terms of Modal Coordinates

Once the modal coordinates $\{Y_c\}_r$ are obtained from the solution of the eigenvalue problem established in the preceding section, the mode shapes of the combined primary–secondary system may be determined from a straight application of Equations 14.5 and 14.13. However, to facilitate the calculation of these mode shapes, it is convenient to derive an expression to compute these mode shapes directly in terms of the vectors of modal coordinates $\{Y_c\}_r$ as follows:

Consider, first, that a mode shape of the combined primary–secondary system is a vector that includes the vibrational amplitudes of the masses of the primary and secondary systems and that these amplitudes are contained in the vectors $\{\varphi_p\}_r$ and $\{\bar{\varphi}_s\}_r$. Note, however, that the vector $\{\bar{\varphi}_s\}_r$ includes the amplitude of the massless degree of freedom of the secondary system corresponding to the point where it is attached to the primary system. Thus, to define the mode shapes of the combined system in terms of the modal coordinates included in the vector $\{Y_c\}_r$, it is necessary to first eliminate such a massless degree of freedom from Equation 14.13. For this purpose, consider that according to its definition, $\{\bar{\varphi}_s\}_r$ may be written as

$$\{\bar{\varphi}_s\}_r = \begin{Bmatrix} \varphi_{s1r} \\ \{\varphi_s\}_r \end{Bmatrix} = \begin{bmatrix} 1 & 0 & 0 & \cdots & 0 \\ \{1\} & \{\phi\}_1 & \{\phi\}_2 & \cdots & \{\phi\}_n \end{bmatrix} \begin{Bmatrix} y_{r0} \\ \{y\}_r \end{Bmatrix} \qquad (14.37)$$

where $\{\varphi_s\}_r$ denotes a vector that only includes the amplitudes of the n masses of the nonstructural element when the primary–secondary system vibrates in its rth mode shape; $\{\phi\}_j$, $j = 1, \ldots, n$, represents the jth normal mode shape of the independent secondary system without the massless degree of freedom; that is, one of the ordinary mode shapes of the system when its point where it is attached to the primary system is considered fixed; {1} is a unit vector of size n; and y_{r0} and $\{y\}_r$ are as previously defined. Hence, from the lower part of Equation 14.37 one has that

$$\{\varphi_s\}_r = \{1\}y_{r0} + \{\phi\}_1\{y\}_r + \{\phi\}_2\{y\}_r + \cdots + \{\phi\}_n\{y\}_r = \{1\}y_{r0} + [\phi]\{y\}_r \qquad (14.38)$$

where $[\phi]$ is the $n \times n$ modal matrix of the nonstructural element when the end connected to the primary structure is considered fixed.

On the basis of Equations 14.5 and 14.38, the mode shapes of the combined primary–secondary system may be thus expressed as

$$\{\varphi\}_r = \begin{Bmatrix} \{\varphi_p\}_r \\ \{\varphi_s\}_r \end{Bmatrix} = \begin{Bmatrix} \{0\} \\ \{1\} \end{Bmatrix} y_{r0} + \begin{Bmatrix} [\Phi]\{Y\}_r \\ [\phi]\{y\}_r \end{Bmatrix} \qquad (14.39)$$

Seismic Response of Nonstructural Elements

or simply as

$$\{\varphi\}_r = \begin{Bmatrix} \{0\} \\ \{1\} \end{Bmatrix} y_{r0} + \begin{bmatrix} [\Phi] & [0] \\ [0] & [\phi] \end{bmatrix} \{Y\}_r \qquad (14.40)$$

Furthermore, it is possible to obtain the modal matrix of the combined primary–secondary system, that is, the matrix that contains all its mode shapes, by means of the following simple matrix equation:

$$[\varphi] = \begin{Bmatrix} \{0\} \\ \{1\} \end{Bmatrix} \{y_0\}^T + \begin{bmatrix} [\Phi] & [0] \\ [0] & [\phi] \end{bmatrix} [Y_c] \qquad (14.41)$$

where $[Y_c]$ is a matrix that contains all the eigenvectors obtained from the solution of Equation 14.33 and $\{y_0\}^T = \{y_{10} \quad y_{20} \quad \ldots \quad y_{(N+n)0}\}$ is a vector that groups together the $(N + n)$ y_{r0} coordinates corresponding to the $N + n$ modes of the combined system. In the calculation of these y_{r0} coordinates, it is important to note that from Equations 14.24 and 14.27 they are given by

$$y_{r0} = \varphi_{pkr} \qquad r = 1, \ldots, N + n \qquad (14.42)$$

where, as previously established, φ_{pkr} is the amplitude of the point of the primary structure to which the nonstructural element is connected when the primary–secondary system vibrates in its rth mode.

A procedure that may be used to determine the natural frequencies and mode shapes of a combined primary–secondary system on the basis of the eigenvalue problem discussed in this section is summarized in Box 14.1.

Box 14.1 Procedure to Determine Natural Frequencies and Mode Shapes of a Combined Primary–Secondary System in Terms of Modal Properties of Independent Components

1. Use Equation 14.34 to define matrix $[A]$
2. Solve eigenvalue problem $[A]\{Y_c\}_r = \lambda_r \{Y_c\}_r$
3. Calculate natural frequencies of combined primary–secondary system as

$$\omega_r = \sqrt{\frac{1}{\lambda_r}} \qquad r = 1, \ldots, (N + n)$$

4. Form $[Y]$ matrix as

$$[Y_c] = \begin{bmatrix} \{Y_c\}_1 & \{Y_c\}_2 & \cdots & \{Y_c\}_{N+n} \end{bmatrix}$$

5. Obtain preliminary combined-system modal matrix as

$$[\hat{\varphi}] = \begin{bmatrix} [\Phi] & [0] \\ [0] & [\phi] \end{bmatrix} [Y_c]$$

6. From the row of matrix $[\hat{\varphi}]$ corresponding to the primary degree of freedom to which the secondary system is attached, define a $\{\varphi_{pk}\}$ vector as

$$\{\varphi_{pk}\}^T = \{\varphi_{pk1} \quad \varphi_{pk2} \quad \cdots \quad \varphi_{pk(N+n)}\}$$

where φ_{pkr} signifies the amplitude of vibration of the kth degree of freedom of the primary system when the combined system vibrates in its rth mode.

7. Obtain the combined system modal matrix according to

$$[\varphi] = [\hat{\varphi}] + \begin{Bmatrix} \{0\} \\ \{1\} \end{Bmatrix} \{\varphi_{pk}\}^T$$

EXAMPLE 14.2 NATURAL FREQUENCIES AND MODE SHAPES OF A COMBINED SYSTEM BY MODAL SYNTHESIS

Determine the natural frequencies and mode shapes of the combined primary–secondary system shown in Figure E14.2 using the modal synthesis method described in Section 14.8.2. The primary system is modeled as a three-degree-of-freedom system with the modal properties given in Table E14.2a. Similarly, the secondary system is modeled as a two-degree-of-freedom system with the modal properties listed in Table E14.2b. The secondary system is attached to the first degree of freedom of the primary one. The masses of the two subsystems are lumped at their nodes and equal to the values shown in Figure E14.2.

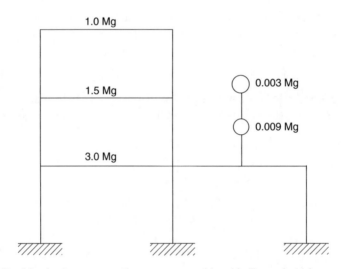

FIGURE E14.2 Combined primary–secondary system considered in Example 14.2.

TABLE E14.2a
Modal Properties of Primary System in Example 14.2

Mode	Natural Frequency (rad/s)	Generalized Mass (Mg)	Damping Ratio	Unit-Participation-Factor Mode Shape		
1	1.0	4.5	0.02	0.5	1.0	1.5
2	2.0	0.9	0.04	0.4	0.2	−0.6
3	3.0	0.1	0.06	0.1	−0.2	0.1

TABLE E14.2b
Modal Properties of Secondary System in Example 14.2 When Lower End Is Considered Fixed

Mode	Natural Frequency (rad/s)	Generalized Mass (Mg)	Damping Ratio	Unit-Participation-Factor Mode Shape	
1	2.0	0.009	0.01	0.5	1.5
2	3.464	0.003	0.017	0.5	−0.5

Solution

Since in this case

$$m_0^* = m_1 + m_2 = 0.012 \text{ Mg},$$

$$[M^*] = \begin{bmatrix} 4.5 & 0 & 0 \\ 0 & 0.9 & 0 \\ 0 & 0 & 0.1 \end{bmatrix} \quad [m^*] = \begin{bmatrix} 0.009 & 0 \\ 0 & 0.003 \end{bmatrix}$$

$$[\omega_p^2] = \begin{bmatrix} 1.0 & 0 & 0 \\ 0 & 4.0 & 0 \\ 0 & 0 & 9.0 \end{bmatrix} \quad [\omega_s^2] = \begin{bmatrix} 4.0 & 0 \\ 0 & 12.0 \end{bmatrix}$$

and

$$\{\Phi_k\}^T = \{0.5 \quad 0.4 \quad 0.1\},$$

then according to Equation 14.34, the [A] matrix for this problem is given by

$$[A] = \begin{bmatrix} 1.00067 & 0.00053 & 0.00013 & 0.001 & 0.00034 \\ 0.00067 & 0.25053 & 0.00013 & 0.001 & 0.00003 \\ 0.00067 & 0.00053 & 0.11124 & 0.001 & 0.00033 \\ 0.125 & 0.10 & 0.025 & 0.25 & 0 \\ 0.04167 & 0.03333 & 0.00833 & 0 & 0.08333 \end{bmatrix}$$

The solution of the eigenvalue problem $[A]\{Y_c\}_r = \lambda_r \{Y_c\}_r$ thus leads to

$$\{\lambda_r\}^T = \{1.00085 \quad 0.26064 \quad 0.24057 \quad 0.11197 \quad 0.08174\}$$

and

$$[Y_c] = \begin{bmatrix} -0.98540 & -0.00143 & 0.00124 & -0.00004 & 0.00036 \\ -0.00130 & 0.10574 & 0.09406 & -0.01422 & 0.01887 \\ -0.00110 & 0.00747 & -0.00679 & 0.95176 & 0.11269 \\ -0.16427 & 0.99417 & -0.99534 & -0.16204 & -0.02823 \\ -0.04481 & 0.01989 & 0.01991 & 0.26019 & -0.99305 \end{bmatrix}$$

In view of Equation 14.36, the following are the natural frequencies of the combined system:

$$\{\omega_r\}^T = \{0.99958 \quad 1.95874 \quad 2.03884 \quad 2.98845 \quad 3.49778\} \text{ rad/s}$$

Similarly, since in this case

$$\begin{bmatrix} [\Phi] & [0] \\ [0] & [\phi] \end{bmatrix} = \begin{bmatrix} 0.5 & 0.4 & 0.1 & 0 & 0 \\ 1.0 & 0.2 & -0.2 & 0 & 0 \\ 1.5 & -0.6 & 0.1 & 0 & 0 \\ 0 & 0 & 0 & 0.5 & 0.5 \\ 0 & 0 & 0 & 1.5 & -0.5 \end{bmatrix}$$

then

$$[\hat{\varphi}] = \begin{bmatrix} [\Phi] & [0] \\ [0] & [\phi] \end{bmatrix} [Y_c] = \begin{bmatrix} -0.49333 & 0.04233 & 0.03756 & 0.08947 & 0.01900 \\ -0.98544 & 0.01822 & 0.02141 & -0.19324 & -0.01840 \\ -1.47743 & -0.06484 & -0.05526 & 0.10364 & 0.00049 \\ -0.10453 & 0.50703 & -0.48772 & 0.04907 & -0.51064 \\ -0.22398 & 1.48130 & -1.50297 & -0.37316 & 0.45419 \end{bmatrix}$$

and from the first (kth) row of this matrix

$$\{\varphi_{pk}\}^T = \{-0.49333 \quad 0.04233 \quad 0.03756 \quad 0.08947 \quad 0.01900\}$$

From Equation 14.41, the mode shapes of the combined system result thus as

$$[\varphi] = \begin{bmatrix} -0.49333 & 0.04233 & 0.03756 & 0.08947 & 0.01900 \\ -0.98544 & 0.01822 & 0.02141 & -0.19324 & -0.01840 \\ -1.47743 & -0.06484 & -0.05526 & 0.10364 & 0.00049 \\ -0.59786 & 0.54936 & -0.45016 & 0.13854 & -0.49164 \\ -0.71731 & 1.52363 & -1.46541 & -0.28369 & 0.47319 \end{bmatrix}$$

14.8.3 Approximate Natural Frequencies and Mode Shapes of Combined System

Although the eigenvalue procedure established in Section 14.8.2 to obtain the natural frequencies and mode shapes of a combined primary–secondary system is simple to formulate and gives accurate answers, it may require, nevertheless, the solution of an excessively large eigenvalue problem if the two subsystems have many degrees of freedom. In such cases, it is convenient to use an approximate method that avoids the solution of an eigenvalue problem. Approximate methods are obviously less accurate than the exact methods, but they are much simpler to use and the accuracy they provide is often sufficient enough for design purposes. Besides, they are useful to visualize the actual behavior of nonstructural elements and identify the main variables that control this behavior. In this section, therefore, one of such approximate methods is derived with the help of the equations developed earlier and some simplifying assumptions.

Consider Equations 14.5 and 14.13, and recall that these equations define the parts corresponding to the primary and secondary systems of the rth mode shape of the combined system. Explicitly as linear combinations of the mode shapes of the two independent components, Equations 14.5 and 14.13 may be written alternatively as

$$\{\varphi_p\}_r = \sum_{i=1}^{N} \{\Phi\}_i Y_{ri} \tag{14.43}$$

Seismic Response of Nonstructural Elements

$$\{\bar{\varphi}_s\}_r = \sum_{j=0}^{n} \{\bar{\phi}\}_j y_{rj} \qquad (14.44)$$

Similarly, consider that an explicit equation to determine the Y_{ri} factors in Equation 14.43 may be obtained from Equation 14.8 by observing that mode shapes can only be defined in relative terms and that it is then possible to normalize the Y_{ri} values in any arbitrary manner. Accordingly, if the modal coordinate corresponding to the Ith mode shape of the primary system is arbitrarily set equal to unity and R_{rk} is solved from Equation 14.8 for the case when $i = I$, one obtains

$$R_{rk} = \frac{\omega_{pI}^2 - \omega_r^2}{\Phi_{kI}} M_I^* \qquad (14.45)$$

Hence, by substitution of Equation 14.45 back into Equation 14.8 and after solving for Y_{ri}, it is found that

$$Y_{ri} = \frac{\omega_r^2 - \omega_{pI}^2}{\omega_r^2 - \omega_{pi}^2} \frac{M_I^*}{M_i^*} \frac{\Phi_{ki}}{\Phi_{kI}} \qquad i = 1, \ldots, N \qquad (14.46)$$

where the subscript I is used to identify the parameters of the independent primary system in the mode for which its modal coordinate Y_i is set equal to unity. Lastly, consider that after solving for y_{rj} from Equation 14.23, the y_{rj} factors in Equation 14.44 may be determined from

$$y_{rj} = \frac{\omega_r^2}{\omega_{sj}^2 - \omega_r^2} y_{r0} \qquad j = 1, 2, \ldots, n \qquad (14.47)$$

where, according to Equation 14.27,

$$y_{r0} = \sum_{i=1}^{N} \Phi_{ki} Y_{ri} \qquad (14.48)$$

It may be seen, thus, that if the natural frequencies ω_r were known, the mode shapes of the combined primary–secondary system could be determined directly from Equations 14.43 and 14.44 in combination with Equations 14.46 through 14.48. Therefore, an approximate expression will be developed next to determine the natural frequencies of the combined system and be able to calculate the desired mode shapes with these equations.

It has been observed in the investigation of the behavior of nonstructural elements that if the masses of a nonstructural element are small compared to the masses of its supporting structure, then the natural frequencies of the combined system only depart slightly from the original natural frequencies of the independent components. This means that when a relatively small nonstructural element with n degrees of freedom is attached to a structure with N degrees of freedom, the combined system results in a system with $N + n$ degrees of freedom, with N of its natural frequencies close in value to the natural frequencies of the structure and n natural frequencies close in value to the natural frequencies of the nonstructural element. From this observation and the inspection of Equations 14.43, 14.44, 14.46, and 14.47, it may be seen that, for every combined-system frequency ω_r, the mode shapes of the structure and nonstructural element that contribute the most to the corresponding mode shape of the combined system are those whose natural frequencies are the closest in value to the frequency ω_r. This fact, in conjunction with Rayleigh's principle, which states that a small deviation from a true mode shape produces only a higher-order variation in the value of the corresponding natural frequency, may be used advantageously to derive the desired approximate formula to compute the natural frequencies of a combined structural–nonstructural system.

Let thus ω_{pI} and ω_{sJ} represent the natural frequencies of the independent primary and secondary systems that are the closest in value to the rth natural frequency of the combined system, respectively. In addition, let Y_{rI} and y_{rJ} denote the corresponding Y_{ri} and y_{rj} factors, which, as deduced from Equations 14.46 and 14.47, are the largest in value among all such factors. If, in accordance with what has been mentioned earlier, one considers only such largest Y_{ri} and y_{rj} factors and neglects all the others, then Equation 14.32 may be approximately reduced to

$$\begin{bmatrix} \omega_{pI}^2 M_I^* & 0 \\ 0 & \omega_{sJ} m_J^* \end{bmatrix} \begin{Bmatrix} Y_{rI} \\ y_{rJ} \end{Bmatrix} = \omega_r^2 \begin{bmatrix} M_I^* + \Phi_{kI}^2 m_0^* & \Phi_{kI} m_J^* \\ \Phi_{kI} m_J^* & m_J^* \end{bmatrix} \begin{Bmatrix} Y_{rI} \\ y_{rJ} \end{Bmatrix} \quad (14.49)$$

where subscripts I and J identify the modal properties of the independent primary and secondary systems in their modes with natural frequencies ω_{pI} and ω_{sJ}, respectively. Hence, if the first equation in this matrix equation is divided by M_I^*, the second one is divided by m_J^*, and it is considered that when the masses of the secondary system are small compared to the masses of the primary one, the term $\Phi_{kI}^2 m_0^*$ is negligibly small compared to M_I^*, the free-vibration equation of motion of a primary–secondary system may be expressed approximately as

$$\begin{bmatrix} \omega_{pI}^2 - \omega_r^2 & -\omega_r^2 \Phi_{kI} \gamma_{IJ} \\ -\omega_r^2 \Phi_{kI} & \omega_{sJ}^2 - \omega_r^2 \end{bmatrix} \begin{Bmatrix} Y_{rI} \\ y_{rJ} \end{Bmatrix} = \begin{Bmatrix} 0 \\ 0 \end{Bmatrix} \quad (14.50)$$

where γ_{JI} is a secondary to primary mass ratio defined as

$$\gamma_{JI} = \frac{m_J^*}{M_I^*} \quad (14.51)$$

Equation 14.50 represents a system of homogeneous equations and, as such, a solution other than the trivial is possible only when the determinant of its coefficient matrix vanishes. That is, when

$$\begin{vmatrix} \omega_{pI}^2 - \omega_r^2 & -\omega_r^2 \Phi_{kI} \gamma_{JI} \\ -\omega_r^2 \Phi_{kI} & \omega_{sJ}^2 - \omega_r^2 \end{vmatrix} = 0 \quad (14.52)$$

After the expansion of this determinant, one arrives thus at

$$(\omega_{pI}^2 - \omega_r^2)(\omega_{sJ}^2 - \omega_r^2) - \omega_r^4 \Phi_{kI}^2 \gamma_{JI} = 0 \quad (14.53)$$

which represents a quadratic equation in ω_r^2. Hence, its solution is of the form

$$\omega_r^2 = \frac{\omega_{pI}^2 + \omega_{sJ}^2 \mp \sqrt{(\omega_{pI}^2 + \omega_{sJ}^2)^2 - 4\omega_{pI}^2 \omega_{sJ}^2 (1 - \Phi_{kI}^2 \gamma_{JI})}}{2(1 - \Phi_{kI}^2 \gamma_{JI})} \quad (14.54)$$

which for small mass ratios may be written simply as

$$\omega_r^2 = \frac{1}{2}(\omega_{pI}^2 + \omega_{sJ}^2) \mp \sqrt{\left[\frac{1}{2}(\omega_{pI}^2 - \omega_{sJ}^2)\right]^2 + \Phi_{kI}^2 \gamma_{JI} \omega_{pI}^2 \omega_{sJ}^2} \quad (14.55)$$

It may be noted from the inspection of Equation 14.55 that when the secondary to primary mass ratio is small (e.g., $\gamma_{JI} \leq 0.01$) and the frequency of the secondary system is well separated from the natural frequency of the primary one (e.g., $\omega_{sJ} \geq 1.5\omega_{pI}$), then the second term within the radical may be neglected. In such a case, one obtains

$$\omega_r^2 = \frac{1}{2}[\omega_{pI}^2 + \omega_{sJ}^2 \mp (\omega_{pI}^2 - \omega_{sJ}^2)] \tag{14.56}$$

Consequently, if the third term in the right-hand side of Equation 14.56 is considered with its positive sign, one obtains $\omega_r = \omega_{pI}$. Conversely, if this term is considered with its negative sign, then $\omega_r = \omega_{sJ}$. This means that, as previously observed, when the masses of a nonstructural element are small compared to the masses of its supporting structure, the natural frequencies of the combined system are almost the same as the natural frequencies of the independent components. It may also be noted that when the second term within the radical in Equation 14.55 is not neglected, the natural frequencies of the combined system will be slightly greater or slightly less than the corresponding natural frequencies of the independent components. If ω_{pI} is greater than ω_{sJ}, then one of the frequencies of the combined system will be slightly greater than ω_{pI} and another will be slightly less than ω_{sJ}. In contrast, if ω_{pI} is smaller than ω_{sJ}, then one of the frequencies of the combined system will be slightly greater than ω_{sJ} and another slightly less than ω_{pI}.

On the basis of the foregoing observations, the following procedure may be established to determine the natural frequencies of a combined primary–secondary system. First, consider that the natural frequencies of the combined system will be close in value to the original natural frequencies of the independent components. Second, use Equation 14.55 to calculate the natural frequencies of the combined system that are close to the natural frequencies of the primary system. In using Equation 14.55, select the positive sign in front of the radical if the value of the closest natural frequency of the secondary system is smaller than the value of the natural frequency of the primary system that is being considered. Otherwise, select the negative sign. Afterward, repeat the procedure to calculate the natural frequencies of the combined system that are close to the natural frequencies of the secondary system. In this case, however, use the radical's positive sign if the value of the closest natural frequency of the primary system is smaller than or equal to the value of the natural frequency of the secondary system that is being considered and the negative one when the opposite is true.

Equation 14.55 and Equations 14.43 and 14.44 in combination with Equations 14.46 through 14.48 represent the desired approximate formulas to determine the natural frequencies and mode shapes of a combined primary–secondary system in terms of the modal properties of its two independent components. When using the equations to determine the combined-system mode shapes, it is important to keep in mind that it is not always necessary to include all the mode shapes of the independent components to obtain accurate estimates of the combined-system mode shapes. In most cases, it will suffice to include only those for which the corresponding natural frequencies are significantly close to the natural frequency of the combined system mode under consideration. This is so because their corresponding Y_{ri} and y_{rj} factors for all the other modes become negligibly small. A procedure to determine the natural frequencies and mode shapes of a combined primary–secondary system on the basis of the aforementioned equations is summarized in Box 14.2.

Box 14.2 Approximate Procedure to Determine Natural Frequencies and Mode Shapes of a Combined Primary–Secondary System

1. For given natural frequencies of the independent primary and secondary systems, assume that the natural frequencies of the combined system are close in value to the given natural frequencies of the independent primary and secondary systems.
2. For each of the natural frequencies of the primary system, use Equation 14.55 to calculate the value of the corresponding natural frequencies of the combined system. In using Equation 14.55, define ω_{pI} as the natural frequency being considered, and ω_{sJ} as the natural

frequency of the secondary system that is the closest in value to ω_{pI}. Use the positive sign in front of the square root sign in Equation 14.55 if $\omega_{pI} > \omega_{sJ}$; otherwise, use the negative sign.
3. For each of the natural frequencies of the secondary system, use Equation 14.55 to calculate the value of the corresponding natural frequencies of the combined system. In using this equation, define ω_{sJ} as the natural frequency being considered, and ω_{pI} as the natural frequency of the primary system that is the closest in value to ω_{sJ}. Use the positive sign in front of the square root sign in Equation 14.55 if $\omega_{sJ} \geq \omega_{pI}$; otherwise, use the negative sign.
4. For each of the calculated values of the combined-system natural frequencies, use Equations 14.46 through 14.48 to determine a few of the Y_{ri} and y_{rj} factors that define the corresponding mode shape of the combined system. Determine only those for which the corresponding natural frequencies of the independent components are significantly close to the natural frequency of the combined-system mode under consideration and neglect the rest of them.
5. Obtain the mode shapes of the combined system using Equations 14.43 and 14.44, the Y_{ri} and y_{rj} factors determined in step 4, and the given mode shapes of the independent primary and secondary systems.

EXAMPLE 14.3 NATURAL FREQUENCIES AND MODE SHAPES OF A COMBINED SYSTEM BY APPROXIMATE METHOD

Determine using the approximate method of Section 14.8.3 the first three natural frequencies and mode shapes of the combined primary–secondary system considered in Example 14.2.

Solution

According to the established procedure, the first three natural frequencies of the combined system under consideration will be close in value to the first two natural frequencies of the primary system and the first one of the secondary system. Therefore, the desired natural frequencies and the corresponding mode shapes may be calculated as follows:

First Mode

The natural frequency in the first mode of the combined system corresponds to the first natural frequency of the primary structure and the closest natural frequency of the nonstructural element to this frequency is the first one. In this case, therefore, one has that

$$\omega_{pI}^2 = 1.0 \quad \omega_{sJ}^2 = 4.0 \quad \Phi_{kI} = 0.5 \quad \gamma_{JI} = \frac{0.009}{4.5} = 0.002$$

and $\omega_{pI} < \omega_{sJ}$. Consequently, for this first mode, Equation 14.55 yields

$$\omega_1^2 = \frac{1}{2}(\omega_{pI}^2 + \omega_{sJ}^2) - \sqrt{\left[\frac{1}{2}(\omega_{pI}^2 - \omega_{sJ}^2)\right]^2 + \Phi_{kI}^2 \gamma_{JI} \omega_{pI}^2 \omega_{sJ}^2}$$

$$= 2.5 - \sqrt{1.5^2 + 0.5^2(0.002)(1.0)(4.0)} = 0.99933$$

from which one obtains

$$\omega_1 = 0.99967 \text{ rad/s}$$

Similarly, if only the component modes corresponding to ω_{pI} and ω_{sJ} are considered, for this first mode Equations 14.46 through 14.48 and Equations 14.43 and 14.44 give

$$Y_{11} = 1.0$$

$$y_{10} = \Phi_{kI} Y_{11} = 0.5(1.0) = 0.5$$

$$y_{11} = \frac{\omega_1^2}{\omega_{sJ}^2 - \omega_1^2} y_{10} = \frac{0.99933}{4.0 - 0.99933}(0.5) = 0.16652$$

$$\{\varphi\}_1 = \begin{Bmatrix} \{\varphi_p\}_1 \\ \{\varphi_s\}_1 \end{Bmatrix} = \begin{Bmatrix} Y_{11}\{\Phi\}_1 \\ y_{10}\{1\} + y_{11}\{\phi\}_1 \end{Bmatrix} = \begin{Bmatrix} 1.0 \begin{Bmatrix} 0.5 \\ 1.0 \\ 1.5 \end{Bmatrix} \\ 0.5\begin{Bmatrix} 1.0 \\ 1.0 \end{Bmatrix} + 0.16652 \begin{Bmatrix} 0.5 \\ 1.5 \end{Bmatrix} \end{Bmatrix} = \begin{Bmatrix} 0.5 \\ 1.0 \\ 1.5 \\ 0.583 \\ 0.750 \end{Bmatrix}$$

Second Mode

The second natural frequency of the combined system corresponds to the second natural frequency of the primary system, and the closest natural frequency of the secondary system is its first one. Thus, for this mode,

$$\omega_{pI}^2 = 4.0 \quad \omega_{sJ}^2 = 4.0 \quad \Phi_{kI} = 0.4 \quad \gamma_{JI} = \frac{0.009}{0.9} = 0.01$$

and $\omega_{pI} \leq \omega_{sJ}$. Proceeding as for the first mode, one has thus that

$$\omega_2^2 = \frac{1}{2}(\omega_{pI}^2 + \omega_{sJ}^2) - \sqrt{\left[\frac{1}{2}(\omega_{pI}^2 - \omega_{sJ}^2)\right]^2 + \Phi_{kI}^2 \gamma_{JI} \omega_{pI}^2 \omega_{sJ}^2}$$

$$= 4.0 - \sqrt{0.4^2(0.01)(4.0)(4.0)} = 3.84$$

$$\omega_2 = 1.95959 \text{ rad/s}$$

$$Y_{22} = 1$$

$$y_{20} = \Phi_{kI} Y_{22} = 0.4(1.0) = 0.4$$

$$y_{21} = \frac{\omega_2^2}{\omega_{sJ}^2 - \omega_2^2} y_{20} = \frac{3.84}{4.0 - 3.84}(0.4) = 9.6$$

$$\{\varphi\}_2 = \begin{Bmatrix} \{\varphi_p\}_2 \\ \{\varphi_s\}_2 \end{Bmatrix} = \begin{Bmatrix} Y_{22}\{\Phi\}_2 \\ y_{20}\{1\} + y_{21}\{\phi\}_1 \end{Bmatrix} = \begin{Bmatrix} 1.0 \begin{Bmatrix} 0.4 \\ 0.2 \\ -0.6 \end{Bmatrix} \\ 0.4\begin{Bmatrix} 1.0 \\ 1.0 \end{Bmatrix} + 9.6\begin{Bmatrix} 0.5 \\ 1.5 \end{Bmatrix} \end{Bmatrix} = \begin{Bmatrix} 0.4 \\ 0.2 \\ -0.6 \\ 5.2 \\ 14.8 \end{Bmatrix}$$

Third Mode

Different from the first and second modes, the third natural frequency of the combined system corresponds to a natural frequency of the secondary system, its first one. Also, the associated closest natural frequency of the primary system is the second one. For this mode, therefore, one has that

$$\omega_{pI}^2 = 4.0 \quad \omega_{sJ}^2 = 4.0 \quad \Phi_{kI} = 0.4 \quad \gamma_{JI} = \frac{0.009}{0.9} = 0.01$$

and $\omega_{sJ} \geq \omega_{pI}$. Making the same assumptions and following the same steps as for of the first and second modes, one obtains

$$\omega_3^2 = \frac{1}{2}(\omega_{pI}^2 + \omega_{sJ}^2) + \sqrt{\left[\frac{1}{2}(\omega_{pI}^2 - \omega_{sJ}^2)\right]^2 + \Phi_{kI}^2 \gamma_{JI} \omega_{pI}^2 \omega_{sJ}^2}$$

$$= 4.0 + \sqrt{0.4^2(0.01)(4.0)(4.0)} = 4.16$$

$$\omega_3 = 2.03961 \text{ rad/s}$$

$$Y_{32} = 1$$

$$y_{30} = \Phi_{kI} Y_{32} = 0.4(1.0) = 0.4$$

$$y_{31} = \frac{\omega_3^2}{\omega_{sJ}^2 - \omega_3^2} y_{20} = \frac{4.16}{4.0 - 4.16}(0.4) = -10.4$$

$$\{\varphi\}_3 = \begin{Bmatrix} \{\varphi_p\}_3 \\ \{\varphi_s\}_3 \end{Bmatrix} = \begin{Bmatrix} Y_{32}\{\Phi\}_2 \\ y_{30}\{1\} + y_{31}\{\phi\}_1 \end{Bmatrix} = \begin{Bmatrix} 1.0 \begin{Bmatrix} 0.4 \\ 0.2 \\ -0.6 \end{Bmatrix} \\ 0.4 \begin{Bmatrix} 1.0 \\ 1.0 \end{Bmatrix} - 10.4 \begin{Bmatrix} 0.5 \\ 1.5 \end{Bmatrix} \end{Bmatrix} = \begin{Bmatrix} 0.4 \\ 0.2 \\ -0.6 \\ -4.8 \\ -15.2 \end{Bmatrix}$$

14.8.4 Damping Ratios of Combined System

In general, damping in a nonstructural element is much lower than the damping traditionally observed in building structures. This is owed to the fact that, by their own nature, nonstructural elements do not possess the same damping mechanisms that generate the damping forces in building structures. Consequently, in the calculation of the damping ratios of a combined primary–secondary system, one cannot assume that these damping ratios are equal to those of the structure alone. Instead, one needs to calculate these damping ratios considering the actual or assumed damping ratios of its two separate components. Procedures have been established to determine the damping ratios of a combined system when the damping ratios of its two components are given (see Section 15.5.4). These procedures involve a free vibration analysis of the damped combined system and, as a result, the consideration of the system's complex-valued natural frequencies. Since a free vibration analysis that involves complex natural frequencies complicates the analysis considerably, these procedures will not be discussed here. However, based on the findings from such a damped free vibration analysis, the following recommendations can be made to estimate the damping ratios of a combined primary–secondary system when the masses of the nonstructural element are small compared to the structural masses:

1. If the natural frequencies of the independent structural and nonstructural systems used in the calculation of the combined-system natural frequency ω_r (i.e., ω_{pI} and ω_{sJ}) are not too close to one another, and if this combined-system natural frequency is close to one of the frequencies of the structure (i.e., $\omega_r \approx \omega_{pI}$), then assume that the damping ratio in the corresponding mode of the combined system is equal to the damping ratio in the Ith mode of the structure.
2. If the natural frequencies of the independent structural and nonstructural systems used in the calculation of the combined system natural frequency ω_r (i.e., ω_{pI} and ω_{sJ}) are not too close to one another, and if this combined system natural frequency is close to one of the frequencies of the nonstructural element (i.e., $\omega_r \approx \omega_{sJ}$), then assume that the damping ratio in the corresponding mode of the combined system is equal to the damping ratio in the Jth mode of the nonstructural element.

3. If the natural frequencies of the independent structural and nonstructural systems used in the calculation of the combined system frequency ω_r (i.e., ω_{pI} and ω_{sJ}) are equal or close to one another, then assume that the damping ratio in the mode that corresponds to this combined system natural frequency is equal to the average of the damping ratios in the Ith mode of the structure and the Jth mode of the nonstructural element.

14.8.5 Maximum Elastic Response of Nonstructural Element

Once the mode shapes, natural frequencies, and damping ratios of a combined structural–nonstructural system are known, the maximum elastic response of the nonstructural element may be calculated straightforwardly using the conventional response spectrum method described in Chapter 10. Specifically, if Γ_r and $SA(\omega_r, \xi_r)$ denote the rth participation factor of the combined system and the ordinate corresponding to the combined system's rth natural frequency and damping ratio in a specified acceleration response spectrum, respectively, then the maximum acceleration response of the nonstructural element in such an rth mode may be calculated as

$$\{\ddot{u}_s\}_r = \Gamma_r \{\varphi_s\}_r SA(\omega_r, \xi_r) \tag{14.57}$$

where, as previously established, $\{\varphi_s\}_r$ is the nonstructural element part of the rth mode shape of the structural–nonstructural system. Similarly, the corresponding vector of maximum accelerations may be estimated by combining these vectors of maximum modal responses according to one of the modal combination rules introduced in Chapter 10. In this regard, it is necessary to keep in mind that structural–nonstructural systems frequently possess closely spaced natural frequencies, and that in such cases it is necessary to use a combination rule of the double sum type.

Example 14.4 Maximum acceleration response of nonstructural element

Determine the maximum accelerations of the masses of the nonstructural element studied in Example 14.2 when the base of its supporting structure is excited by the first 10 s of the north–south component of the ground acceleration recorded during the May 18, 1940, El Centro earthquake. For such a purpose, consider only the first three modes of the combined structural–nonstructural system, utilizing the approximate natural frequencies and mode shapes obtained in Example 14.3. The response spectrum for the specified ground acceleration is given in Figure E14.4.

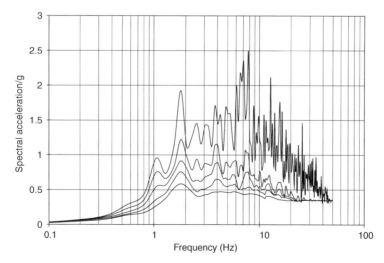

FIGURE E14.4 Acceleration response spectra of first 10 s of N–S 1940 El Centro accelerogram for 0, 2, 5, 10, and 20% damping.

Solution

Since the natural frequencies and mode shapes of the combined system are already known, the solution to this problem requires only four additional steps: (1) calculation of the participation factors and damping ratios of the combined system, (2) definition of the spectral accelerations that correspond to the natural frequencies and damping ratios of the combined system, (3) calculation of maximum modal responses considering only the nonstructural element part of the combined-system mode shapes, and (4) combination of these maximum modal responses to estimate the nonstructural element's maximum acceleration response.

Participation Factors and Damping Ratios

Since according to the results from Example 14.3, the first three mode shapes of the combined system are

$$\{\phi\}_1 = \begin{Bmatrix} 0.5 \\ 1.0 \\ 1.5 \\ 0.583 \\ 0.750 \end{Bmatrix} \quad \{\phi\}_2 = \begin{Bmatrix} 0.4 \\ 0.2 \\ -0.6 \\ 5.2 \\ 14.8 \end{Bmatrix} \quad \{\phi\}_3 = \begin{Bmatrix} 0.4 \\ 0.2 \\ -0.6 \\ -4.8 \\ -15.2 \end{Bmatrix}$$

then according to the definition of participation factor (see Equation 10.32) and the masses given in Figure E14.2, the participation factors corresponding to these mode shapes are

$$\Gamma_1 = 1.00061 \quad \Gamma_2 = 0.55052 \quad \Gamma_3 = 0.45055$$

Similarly, following the guidelines established in Section 14.8.4 and considering the damping ratios given in Tables E14.2a and E14.2b for the independent structural and nonstructural systems, the damping ratios in the first three modes of the combined system may be assumed equal to

$$\xi_1 = \xi_{p1} = 0.02 \quad \xi_2 = \frac{1}{2}(\xi_{p2} + \xi_{s1}) = 0.028 \quad \xi_3 = \frac{1}{2}(\xi_{p2} + \xi_{s1}) = 0.028$$

Spectral Accelerations

With these damping ratios and the natural frequencies obtained in Example 14.3, the spectral accelerations corresponding to the first three modes of the combined system and the specified excitation may now be obtained directly from the response spectrum shown in Figure E14.4. Accordingly, these spectral accelerations are

$$SA(0.99967, 0.02) = 0.017g \quad SA(1.95959, 0.028) = 0.104g \quad SA(2.03961, 0.028) = 0.120g$$

Maximum Modal Responses

Using Equation 14.57 in combination with the determined mode shapes, participation factors, and spectral accelerations, the response of the nonstructural element in the first three modes of the combined system results thus as

$$\{\ddot{u}_s\}_1 = 1.00061 \begin{Bmatrix} 0.583 \\ 0.750 \end{Bmatrix} 0.017g = \begin{Bmatrix} 0.010 \\ 0.013 \end{Bmatrix} g$$

$$\{\ddot{u}_s\}_2 = 0.55052 \begin{Bmatrix} 5.2 \\ 14.8 \end{Bmatrix} 0.104g = \begin{Bmatrix} 0.298 \\ 0.847 \end{Bmatrix} g$$

$$\{\ddot{u}_s\}_3 = 0.45055 \begin{Bmatrix} -4.8 \\ -15.2 \end{Bmatrix} 0.120g = -\begin{Bmatrix} 0.260 \\ 0.822 \end{Bmatrix} g$$

Seismic Response of Nonstructural Elements

Maximum Acceleration Response

These modal responses may now be combined to obtain an estimate of the maximum acceleration response of the nonstructural element. It should be noted, however, that in this case the combined system possesses two natural frequencies that are close to each other. The combination of modal responses should therefore be made using a modal combination rule that accounts for closely spaced natural frequencies. Rosenblueth's rule (see Section 10.3.2) will be thus selected for such a purpose. Accordingly, with a ground motion duration of 10 s and assuming all modal correlation coefficients equal to zero except those that correlate the two closely spaced modes and a single mode (for which the modal correlation coefficient is equal to unity), the values of the parameters needed to use this rule are (see Equations 10.69 and 10.70)

$$\xi_2' = 0.028 + \frac{2}{1.95959(10)} = 0.130 \quad \xi_3' = 0.028 + \frac{2}{2.03961(10)} = 0.126$$

$$\alpha_{23} = \alpha_{32} = \left[1 + \left(\frac{2.03961 - 1.95959}{0.126(2.03961) + 0.130(1.95959)}\right)^2\right]^{-1} = 0.976$$

and thus the maximum acceleration response of the nonstructural element under analysis is given by

$$\{\ddot{u}_s\} = \begin{Bmatrix} \sqrt{0.010^2 + 0.298^2 + 0.260^2 - 2(0.976)(0.298)(0.260)} \\ \sqrt{0.013^2 + 0.847^2 + 0.822^2 - 2(0.976)(0.847)(0.822)} \end{Bmatrix} g = \begin{Bmatrix} 0.073 \\ 0.185 \end{Bmatrix} g$$

14.8.6 Effect of Structural and Nonstructural Nonlinear Behavior

The equations presented in the preceding sections have been derived assuming a linear behavior for both the structure and the nonstructural element, and by considering that $SA(\omega_r, \xi_r)$ represents an ordinate from a linear response spectrum. The established procedure cannot, therefore, be used directly to estimate the maximum response of a nonstructural element under an extreme seismic event since, by design, its supporting structure is supposed to incur into its nonlinear range of behavior in such a case. It cannot consider, either, the often advantageous fact that many nonstructural elements or their anchors are capable of resisting large inelastic deformations. Since the nonlinear behavior of a nonstructural element and its supporting structure may significantly affect the response of the nonstructural element, the use of linear methods in the seismic analysis of nonstructural elements may lead to unrealistic designs.

A simple, albeit approximate, way to account for the nonlinear behavior of a nonstructural element and its supporting structure in the analysis of the nonstructural element is by dividing the ordinates of the linear response spectrum by a reduction factor in much the same way nonlinear behavior is considered in the design of building structures. In other words, the design lateral strength of a nonstructural element may be determined by dividing the lateral strength required to maintain it in its linear range of behavior by a strength reduction factor. In the case of nonstructural elements, however, such a strength reduction factor should be considered equal to the product of two other strength reduction factors. That is, if λ denotes the desired strength reduction factor, then

$$\lambda = RR_p \tag{14.58}$$

where R accounts for the nonlinear behavior of the structure and the fact that the motion at the supports of the nonstructural element is affected by this nonlinear behavior. Similarly, R_p accounts for the nonlinearity of the nonstructural element itself and the fact that it is possible to reduce its design lateral strength when it is capable of resisting inelastic deformations. The first factor is selected based on the capacity of the supporting structure to resist inelastic deformations, and the second based on the capacity of the nonstructural element to withstand inelastic deformations. The validity of this approach has been verified through numerical simulations in which it is observed that the deformation demands on a nonstructural element are reduced when the supporting structure is allowed to incur into its nonlinear range of behavior, and reduced again when the nonstructural component itself is also allowed to go into its nonlinear range of behavior. It should be kept in mind, nonetheless, that such an approach may not be adequate for the few nonstructural components for which the nonlinear behavior of the structure may amplify their response. According to recent studies, these correspond to nonstructural components that have a fundamental natural frequency close to one of the higher natural frequencies of their supporting buildings, are attached to the lower floors of these buildings, and are subjected to a narrowband ground motion with a dominant period close to the fundamental period of their supporting buildings.

As in the case of building structures, it may be assumed that the aforementioned reduction factors are approximately equal to the average values that have been derived for single-degree-of-freedom systems. This assumption is justified on the grounds that, for the purpose of accounting for nonlinear effects, the structure and the nonstructural component may be considered as independent systems, and that the reduction factors for single-degree-of-freedom systems are approximately valid for multi-degree-of-freedom systems that have relatively uniform properties and vibrate predominantly in their fundamental modes. In particular, it may be considered that the strength reduction factors in question are those suggested by Newmark and Hall for the construction of nonlinear response spectra from the linear counterparts, or when more refined values are warranted, those proposed more recently by Miranda (see Section 9.3.3). As indicated in Section 9.3.3, the strength reduction factors proposed by Newmark and Hall are of the form

$$R_\mu = \begin{cases} \mu & \text{if} \leq 2\ \text{Hz} \\ \sqrt{2\mu - 1} & \text{if}\ 2 < f < 8\ \text{Hz} \\ 1 + \dfrac{33 - f}{25}(\sqrt{2\mu - 1} - 1) & \text{if}\ 8 \leq f \leq 33\ \text{Hz} \end{cases} \quad (14.59)$$

where μ and f are predetermined target ductility factor and initial fundamental natural frequency of the system in hertz. The strength reduction factors proposed by Miranda are given by Equations 9.13 through 9.16.

In the application of the suggested strength reduction factors, it should be borne in mind that they represent average values obtained from a statistical analysis with a large number of ground motions and that there is thus an inherent dispersion associated with them. As such, they are useful to estimate, as in the case of building structures and for design purposes, the average forces or accelerations that will make a nonstructural element yield but, at the same time, keep its inelastic deformations within specified limits. They cannot be used to predict the level of the inelastic deformations that would be generated in the element by a single ground motion.

EXAMPLE 14.5 MAXIMUM ACCELERATIONS IN MASSES OF NONSTRUCTURAL ELEMENT CONSIDERING NONLINEAR EFFECTS

Determine using the procedure established in Section 14.8.6 the maximum (yield) accelerations of the masses of the nonstructural element described in Example 14.2 when the base of its supporting structure is excited by the ground motion considered in Example 14.4. Assume elastoplastic behavior for the two systems considering a target ductility factor of six for the structure and two for the nonstructural element. As in Example 14.4, calculate the desired accelerations considering only the first three modes of the combined system.

Solution

From the results in Example 14.2, the first three natural frequencies of the combined structural–nonstructural system are 0.159, 0.312, and 0.324 Hz. Therefore, in accordance with Equation 14.59 and the given ductility factors, the reduction factors that can be used to account for nonlinear effects in the first three modes of the combined structural–nonstructural system are

$$R_1 = R_2 = R_3 = 6$$
$$R_{p1} = R_{p2} = R_{p3} = 2$$

Hence, according to Equation 14.58, the strength reduction factors for the nonstructural element in the three first modes of the system are

$$\lambda_1 = \lambda_2 = \lambda_3 = (6)(2) = 12$$

Consequently, the acceleration responses of the nonstructural element in the first three modes of the combined system may be reduced from those obtained in Example 14.4 to

$$\{\ddot{u}_s\}_1 = 1.00061 \begin{Bmatrix} 0.583 \\ 0.750 \end{Bmatrix} 0.017g/12.0 = \begin{Bmatrix} 0.0008 \\ 0.0011 \end{Bmatrix} g$$

$$\{\ddot{u}_s\}_2 = 0.55052 \begin{Bmatrix} 5.2 \\ 14.8 \end{Bmatrix} 0.104g/12.0 = \begin{Bmatrix} 0.0248 \\ 0.0706 \end{Bmatrix} g$$

$$\{\ddot{u}_s\}_3 = 0.45055 \begin{Bmatrix} -4.8 \\ -15.2 \end{Bmatrix} 0.120g/12.0 = -\begin{Bmatrix} 0.0216 \\ 0.0685 \end{Bmatrix} g$$

which when combined as in the linear case lead to the following maximum accelerations:

$$\{\ddot{u}_s\} = \begin{Bmatrix} \sqrt{0.0008^2 + 0.0248^2 + 0.0216^2 - 2(0.976)(0.0248)(0.0216)} \\ \sqrt{0.0011^2 + 0.0706^2 + 0.0685^2 - 2(0.976)(0.0706)(0.0685)} \end{Bmatrix} g = \begin{Bmatrix} 0.006 \\ 0.015 \end{Bmatrix} g$$

FURTHER READINGS

1. Chen, Y. and Soong, T. T., State-of-the-art-review: Seismic response of secondary systems, *Engineering Structures*, 10, 1988, 218–228.
2. Gupta, A. K., *Response Spectrum Method in Seismic Analysis and Design of Structures*, Blackwell, Boston, MA, 1990.
3. Sankaranarayanan, R. and Medina, R. A., Acceleration response modification factors for nonstructural components attached to inelastic moment-resisting frame structures, *Earthquake Engineering and Structural Dynamics*, 36, 14, 2007, 2189–2210.
4. Singh, M. P., An overview of techniques for analysis of non-structural components, *Proceedings ATC-29 Seminar and Workshop on Seismic Design and Performance of Equipment and Nonstructural Components in Buildings and Industrial Structures*, Irvine, CA, Oct. 3–5, 1990, pp. 215–224.
5. Soong, T. T., Seismic behavior of nonstructural elements—State-of-the-art report, *Proceedings 10th European Conference on Earthquake Engineering*, Vol. 3, Vienna, Austria, Aug. 28–Sept. 2, 1994, pp. 1599–1606.

6. Villaverde, R., Earthquake resistant design of nonstructural elements: State of the art, *Journal of Structural Engineering, ASCE*, 123, 8, 1997, 1011–1019.
7. Chaudhury, S. R. and Villaverde, R., Effect of building nonlinearity on seismic response of nonstructural components: A parametric study, *Journal of Structural Engineering, ASCE*, 134, 4, 2008, 661–670.

PROBLEMS

14.1 Determine using the floor response spectrum method the maximum acceleration response of a piece of equipment rigidly mounted on the fourth floor of a seven-story building. The piece of equipment is modeled as a two-degree-of-freedom system with one of its ends fixed and the other free. Its natural frequencies, mode shapes, and damping ratios are listed in Table P14.1. The 5%-damping floor response spectrum specified for the fourth floor of the building is shown in Figure P14.1.

14.2 A piece of equipment is mounted on the fifth floor of the six-story shear building shown in Figure 14.9. The equipment is modeled as a single-degree-of-freedom system with a natural frequency of 3.8 Hz and a damping ratio of 0.5%. Determine using the floor response spectrum method the maximum acceleration of the piece of equipment when the base of the building is subjected to the accelerogram recorded at the SCT station along the east–west direction during the 1985 Michoacán, Mexico, earthquake. Assume

TABLE P14.1
Modal Properties of Equipment in Problem 14.1

Mode	Natural Frequency (Hz)	Generalized Mass (kip-s²/in.)	Damping Ratio	Unit-Participation-Factor Mode Shape	
1	1.5	0.0045	0.005	0.5	1.5
2	2.6	0.0015	0.005	0.5	−0.5

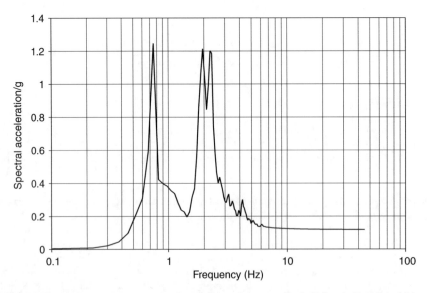

FIGURE P14.1 Floor response spectrum for fourth floor of seven-story building in Problem 14.1 and damping ratio of 0.5%.

the building's damping matrix is proportional to its stiffness matrix and that the damping ratio in its fundamental mode is 8%.

14.3 Determine using the "exact" modal synthesis method of Section 14.8.2 the natural frequencies and mode shapes of the combined primary–secondary system shown in Figure P14.3. The modal properties of the independent systems are listed in Tables P14.3a and P14.3b.

14.4 Determine using the approximate modal synthesis method of Section 14.8.3 the first three natural frequencies and mode shapes of the combined primary–secondary system

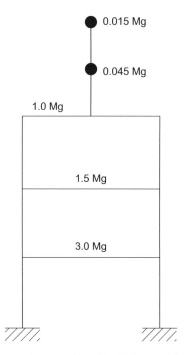

FIGURE P14.3 Primary–secondary system considered in Problem 14.3.

TABLE P14.3a
Modal Properties of Primary System in Problem 14.3

Mode	Natural Frequency (Hz)	Generalized Mass (Mg)	Unit-Participation-Factor Mode Shape		
1	1.0	4.5	0.5	1.0	1.5
2	2.0	0.9	0.4	0.2	−0.6
3	3.0	0.1	0.1	−0.2	0.1

TABLE P14.3b
Modal Properties of Secondary System in Problem 14.3 When Lower End is Considered Fixed

Mode	Natural Frequency (Hz)	Generalized Mass (Mg)	Unit-Participation-Factor Mode Shape	
1	1.0	0.045	0.5	1.5
2	1.732	0.015	0.5	−0.5

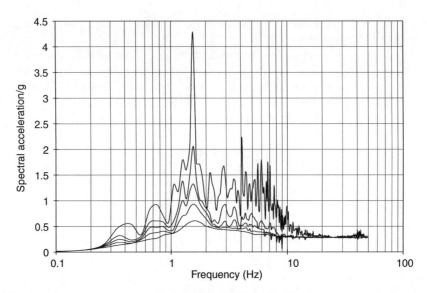

FIGURE P14.8 Acceleration response spectrum of E–W Foster City accelerogram recorded during the 1989 Loma Prieta, California, earthquake for damping ratios of 0, 2, 5, 10, and 20%.

considered in Problem 14.3. Consider all the five component modes in the calculation of the mode shapes.

14.5 Determine using the response spectrum method summarized in Section 14.8.5 the maximum elastic acceleration response of the piece of equipment described in Problem 14.2. For this purpose, calculate the natural frequencies and mode shapes of the combined primary–secondary system using the "exact" modal synthesis method described in Section 14.8.2. Define the earthquake input to the building by the response spectrum given in Figure E10.4b.

14.6 Repeat Problem 14.5 calculating the natural frequencies and mode shapes of the combined primary–secondary system using the approximate modal synthesis method introduced in Section 14.8.3. Consider only the first three modes of the combined primary–secondary system in the calculation of the desired acceleration response. Also, consider only the closest component modes in the calculation of the combined-system mode shapes.

14.7 Estimate using the procedure introduced in Section 14.8.6 the maximum acceleration response of the piece of equipment described in Problem 14.2. Assuming elastoplastic behavior, consider a target ductility factor of six for the building and two for the equipment. The earthquake input to the building is defined by the response spectrum given in Figure E14.4.

14.8 Determine using the procedure introduced in Section 14.8.6 the maximum (yield) accelerations of the masses of the nonstructural element described in Example 14.2 when the base of the building to which it is connected is excited by the east–west component of the ground acceleration recorded at Foster City during the 1989 Loma Prieta, California, earthquake. Assuming elastoplastic behavior, consider a target ductility factor of six for the building and one for the nonstructural element. Obtain the mode shapes of the combined system using only the closest component modes and estimate the desired acceleration response based on only the first three modes of the combined system. The response spectrum for the specified ground acceleration is shown in Figure P14.8.

15 Seismic Protection with Base Isolation

Those who are impatient with progress are reminded that it is better to walk before running. Careful basic research followed by meticulous attention to detail during design, fabrication and construction will assure success in due course. Inspiration must be followed by perspiration if the technology is to be based on a solid foundation. Structural control has a bright future, but one mistake now could set the field back five or more years. Progress may therefore appear to be slow but prudence dictates a cautious approach.

Ian G. Buckle, 1993

15.1 INTRODUCTION

In response to the constant challenge of finding new and better means of protecting new and existing structures against the devastating effects of earthquakes, a number of innovative systems and devices have been suggested over the past four decades. Among these innovative systems, *base isolation* has been implemented in a large number of buildings, bridges, nuclear power plants, and other structures, and it seems, thus, that it has had a wide acceptance (see Section 1.5). In turn, this acceptability has made base isolation a viable structural system in the toolbox of the structural engineer.

This chapter describes this protective system and the methods that may be used to analyze structures implemented with it. The discussion is an introductory one, limited to the basic concepts that are necessary to understand the benefits and limitations of the technique. For a more in-depth coverage, refer to the specialized books listed in the Further Readings section.

15.2 BASIC CONCEPT

Base isolation is a technique whereby a structure is protected from the damaging effects of earthquakes by installing at the base of the structure either flexible elements that elongate the fundamental natural period of the structure to a value that is sufficiently away from the dominant periods of the expected earthquakes, or elements that slide under lateral loads when these lateral loads exceed a predetermined level. This way, the deformations induced by an earthquake will occur at the level of these flexible or sliding elements while the structure essentially moves as a rigid body (see Figure 15.1). In principle, thus, the structure itself is minimally affected by the motions at its base.

In practice, a base-isolation system requires more than flexible or sliding elements. In general, it requires

1. A flexible element that increases the natural period of the structure or a sliding element that prevents the transmission of earthquake forces to the structure above a certain level.
2. A damper or energy dissipation mechanism that reduces the deflection of the flexible or sliding elements to a practical level.
3. A mechanism that provides the building with the necessary rigidity to prevent displacements and vibrations under frequently occurring loads such as wind and minor earthquakes.

The devices commonly used to provide these basic components are described in Section 15.3.

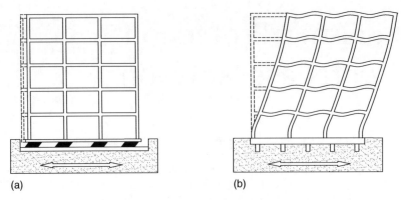

FIGURE 15.1 Schematic behavior of (a) base-isolated structure and (b) conventional structure. (Reproduced with permission of Sociedad Mexicana de Ingeniería Sísmica from Mayes, R. L., *Memorias del VIII Simposio Nacional de Ingeniería Sísmica*, Tlaxcala, Mexico, 2004.)

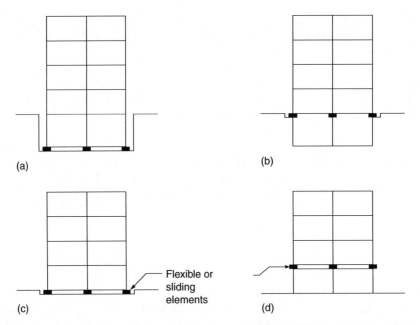

FIGURE 15.2 Possible locations of isolation devices in base-isolated buildings. (a) Bottom of basement columns, (b) top of basement columns, (c) bottom of first-floor columns, (d) top of first-floor columns.

The location of the isolation plane is selected based on site constraints, type of structure, construction costs, and other related factors. Figure 15.2 shows possible locations of this isolation plane for buildings with and without a basement. Some advantages and disadvantages are associated with each of these locations, both related to the added construction costs and the special details required to allow the differential movement of nonstructural components such as stairways and pipelines across the isolation plane. Independently of where the isolation plane is located, the isolation devices are always located where they may be easily inspected and replaced. Also, it is preferable to include a rigid diaphragm above or below these devices to distribute the lateral loads uniformly to each of them.

Seismic isolation may be used not only in building structures but also in bridge structures, where it involves the separation of the superstructure from the substructure at the level of the bent cap.

In the case of bridges, however, the primary intent is the protection of the substructure since this is the part of a bridge that is more vulnerable to the effects of earthquakes. In general, the seismic isolation of a bridge is a straightforward task since for most bridges it only involves the substitution of the isolation devices for the bearings used to accommodate thermal movements. For a new construction, a seismic isolation system substantially reduces the forces in columns, eliminating the need for ductile detailing. For retrofit work, seismic isolation is effective in correcting (a) the inadequate strength of current bearings and connections, (b) the inadequate strength and ductility of columns, and (c) the inadequate girder support length.

15.3 HISTORICAL PERSPECTIVE

The concept of base isolation is far from being a recent development since it has been proposed repeatedly during, at least, the past 100 years. Many variations have been proposed, some of them ingenious and others not so, but all show a persistent and imaginative effort to protect structures from earthquake damage in a distinctive way.

Apparently, the first base-isolation system was proposed by John Milne, an Englishman who was a professor of mining engineering in Tokyo between 1876 and 1895, and whose interest in earthquakes led him to devise a number of seismographs (see Box 2.1). In August 1909, Johannes A. Calantarients, a medical doctor from the northern English City of Scarborough, wrote a letter to the Director of the Seismological Service in Santiago, Chile, describing a method of earthquake protection that he himself had developed. In this letter, Calantarients states that "I made the experiment with balls many years before it was done in Japan, or at any event before any amount of it appeared in the papers about 25 years ago," most likely referring to the experiments performed by Milne. While at the University of Tokyo, Milne built and instrumented an isolated building. It was built on balls in "cast-iron plates with saucer-like edges on the heads of piles. Above the balls and attached to the buildings are cast-iron plates slightly concave but otherwise similar to those below." He described his experiment in a 1885 report to the British Association for the Advancement of Science, with a follow-up report in 1886 in which he suggested a new version. The first balls used were 10 in. in diameter, but he also tried 8 and 1 in. balls. In the final version, the building was rested at each of its piers on a handful of cast-iron shot, each ¼ in. in diameter. The final design was evidently successful under an actual earthquake.

Another early suggestion for the isolation of buildings for earthquake protection came from the Japanese K. Kawai after the 1891 Nobi earthquake that caused 7000 deaths. He introduced his system in his paper "A structural method free from earthquake excitation," published in the *Journal of Architecture and Building Science* in 1891. As shown in Figure 15.3, he proposed the use of log rollers below the foundation mat, laid on several layers along two perpendicular directions.

The method of building construction Calantarients had developed and was referring to in the aforementioned letter was a method by which "substantial buildings can be put up in earthquake countries on this principle with perfect safety since the degree of severity of an earthquake loses its significance through the existence of the lubricated free joint." Calantarients had proposed that buildings be built on a "free joint" lubricated with a layer of fine sand, mica, or talc (see Figure 15.4). This free joint would allow a building to slide during an earthquake and reduce the force transmitted to the building itself. To accommodate the displacements, he also designed a set of ingenious connections for the gas, sewage, and water lines. He submitted a patent application to the British patent office for his construction method in 1909.

In 1906, before Calantarients' patent application, Jacob Bechtold of Munich, Germany, made an application for a U.S. patent for "An earthquake proof building consisting of a rigid base-plate to carry the building and a mass of spherical bodies of hard material to carry the said base-plate freely." A few years later, in 1930, another base-isolation scheme was filed by Robert W. deMontalk of Wellington, New Zealand. In this case, the patent application was for "a means

692 Fundamental Concepts of Earthquake Engineering

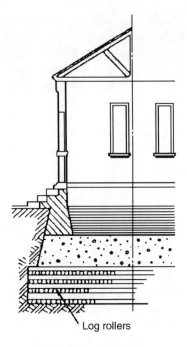

FIGURE 15.3 Isolation system proposed by K. Kawai in 1891.

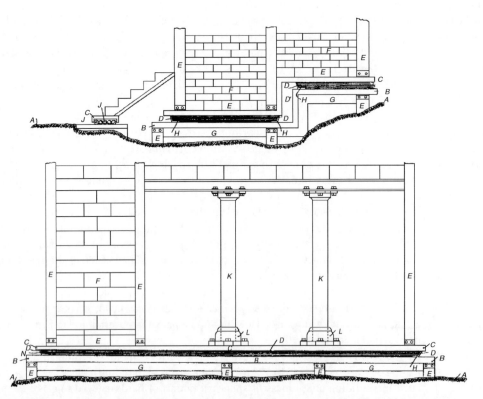

FIGURE 15.4 Base-isolation system patented by J. A. Calantarients in 1909. (Shaded parts depict layers of talc between structure and foundation.)

whereby a bed ... is placed and retained between the base of a building and its solid foundation, the [bed] being composed of material which will absorb or minimize shocks thereby saving the building therefrom."

Base isolation was also considered as a seismic-resistant design strategy by the Italian government after the great Messino–Reggio earthquake of 1908 that killed 58,000 people. After the earthquake, a commission was appointed to make recommendations for rebuilding the area with earthquake-resistant structures that were both economical and safe. Two approaches were considered by the commission to meet this objective. The first was isolating buildings from the ground by either introducing a sand layer in the foundation or using rollers under the columns to let the buildings move freely horizontally. The second approach involved a conventional design with height limitations and a lateral force design requirement (see Section 1.6). Ultimately, the latter approach was recommended over the sliding isolation system.

Although the concept of base isolation was proposed by the end of the nineteenth century, the first two buildings implemented with a base-isolation scheme were designed in 1928 and completed in 1934. These buildings housed the Fudo banks in Himeji and Shimonoseki, Japan. They were designed by R. Oka using rolling bearings. In 1969, another building was constructed using an isolation system developed by K. Stadaucher, a Swiss structural engineer. This building was built for the Pestalozzi Elementary School in Skopje, Macedonia. The three-story, reinforced concrete building was designed using rubber blocks at the base of the building to isolate it from earthquakes. It was the first building ever to use rubber bearings for this purpose. Glass blocks acting as seismic fuses were also introduced to prevent movement of the building under wind, internal foot traffic, and low seismic disturbances, but designed to break when the seismic loading exceeded a certain threshold. A problem encountered with this isolation system was the bulging sideways of the rubber isolators under the weight of the building. Another problem was the likely bouncing and rocking of the building during an earthquake given that the vertical stiffness of the rubber blocks used was about the same as the horizontal one. Soon after the Pestalozzi School, two other isolated buildings were constructed. One was a five-story school completed in 1974 in Mexico City. In this building, each column is supported by two isolation elements, with each element composed of two steel disks separated by over a hundred 1 cm ball bearings. In addition, it has an encircling steel ring that prevents the ball bearings from rolling out of the base plate and limits the displacement between the plates to about 8 cm. The other was a seven-story demonstration building in Sevastopol, Ukraine, built around 1976. This building is supported by ellipsoid-shaped steel bearings. The ellipsoid-shaped bearings make the building rise when displaced, providing thus a restoring force.

In 1974, G. C. Delfosse at the Centre National de la Rechercher Scientifique in Marseille, France, introduced the Gapec seismic isolation system in which laminated rubber bearings were used as isolators. These bearings are made from sheets of rubber strongly bonded during the rubber vulcanization process to thin steel reinforcing plates (often referred to as shims). They are thus very stiff in the vertical direction and can carry the vertical load of the building without bulging and displace laterally without rocking. At the same time, they are horizontally very flexible, thereby enabling the building to move laterally under a strong ground motion. Their development was an extension of the elastomeric bearings used in bridges for thermal expansion and the bearings used to isolate buildings from the vibrations produced by railroads, subway lines, or vehicular traffic. The Gapec system was used around 1977 in the construction of three houses in Saint-Martin de Castillon and a three-story school in the small town of Lambesc, near Marseille. This was the first time steel-laminated rubber bearings were used as seismic isolators. A few years later, in 1984, A. S. Arya, an Indian engineer, proposed a sliding system in which the superstructure in masonry buildings is separated from the foundation by means of a nonbonding membrane such as a black polyethylene sheet or burnt oil. He came up with this idea after observing that the small masonry buildings that slid on their foundations survived the devastating Indian earthquakes of Dhubai in 1930 and

Bihar in 1934, but similar buildings with a fixed base did not. In 1984, also, L. Li reports that Chinese engineers observed the same phenomenon during the catastrophic 1976 Tangshan earthquake and other earthquakes. They noted that buildings that survived the earthquakes had a horizontal crack at the bottom of the walls that allowed them to slip and that this slipping protected the masonry superstructure from damage. They then built a number of small buildings using this isolation technique. The largest was a four-story building for the Earthquake Strong Motion Observatory in Beijing. In this building, the sliding surface is a layer of specially selected sand placed between terrazzo plates above the foundation and under the walls at the ground-floor level. Another sliding isolator but with the capability of producing a restoring force was proposed by Victor A. Zayas in 1985. This isolator, named the friction pendulum system, is built with a spherical sliding surface. Because of the spherical surface, the structure is lifted whenever the sliding surfaces slide relative to each other. In turn, this lifting produces a recentering action in much the same way as the lateral displacement of a pendulum does. The system was used for the first time to retrofit the Marina Apartments, a four-story building in San Francisco, California, damaged during the 1989 Loma Prieta earthquake.

The first base-isolated building in the United States was completed in 1986. This building houses the Foothill Communities Law and Justice Center and is located in Rancho Cucamonga in San Bernardino County, about 97 km east of downtown Los Angeles, California, and 21 km from the San Andreas fault. The building has four stories, a plan of 415 ft × 110 ft, a full basement, and a subbasement where the isolation system is installed. It is structured with steel frames, braced at some bays, and supported by 98 laminated rubber bearings (see Figure 15.5). The isolators measure 30 in. in diameter and 16 in. in height. They were designed to sustain a lateral displacement of 15 in. and individually carry a vertical load of 1200 kips.

Seismic isolation is nowadays a practical reality and a mature technology. It is widely accepted in earthquake-prone regions and has been applied in hundreds of building in Japan, the United States, New Zealand, Italy, Chile, the People's Republic of China, and many other countries. Most of the isolation systems used today incorporate either elastomeric bearings, with the elastomer being natural rubber or neoprene, or sliding bearings, with the sliding surfaces being polytetrafluoroethylene (PTFE or Teflon) and polished stainless steel.

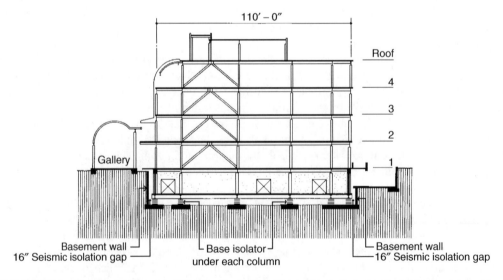

FIGURE 15.5 First base-isolated building in the United States. (Reproduced with permission of Springer Science and Business Media from Kelly, J. M., *Earthquake-Resistant Design with Rubber*, Springer, London, 1997.)

Seismic Protection with Base Isolation

15.4 ISOLATION BEARINGS

15.4.1 Introduction

A variety of systems and devices have been suggested to seismically isolate structures. Rollers, friction slip plates, sleeved piles, and suspension cables are among those that have been proposed. It appears, however, that those that have been actively tested and investigated and used the most in actual implementations are (a) laminated rubber bearings, (b) sliding bearings, and (c) helicoidal steel springs. Thus, the discussion in this section is limited to these systems.

15.4.2 Laminated Rubber Bearings

As described in Section 15.3, a laminated rubber bearing is formed with thin sheets of steel and rubber (or any other elastomer such as neoprene) built up in layers and bonded together by vulcanization. In addition, thick steel plates are bonded to the top and bottom surfaces of the bearing to facilitate the connection of the bearing to the foundation below and the superstructure above. In addition, a rubber cover is used to wrap the bearing as a measure to protect the steel plates from corrosion.

Laminated rubber bearings possess a large vertical load-bearing capacity, but at the same time a large horizontal deformability due to the low shear modulus of rubber (typically between 0.5 and 1.0 MPa [72 and 145 psi] at shear strains of about 50%), which is unaffected by the insertion of the steel plates. In consequence, laminated rubber bearings can easily withstand large lateral deformations (see Figure 15.6). In addition, they have minimum maintenance requirements since they have no moving parts and no components exposed to corrosion. Their failure occurs mainly due to the formation and growth of flaws in the rubber. However, careful manufacturing and quality insurance control can prevent the formation of such flaws. Their cost is relatively high because their manufacturing process is quite elaborate. It requires that the steel plates be cut into exact sizes, sand blasted, and chemically cleaned. Then, they have to be coated with a bounding compound and the rubber sheets laid on them for vulcanization bonding. Thereafter, the interleaved steel plates and rubber sheets are left under pressure for several hours. At the end, fire-resistant coverings are used to protect the isolators from fire hazards.

Several types of laminated rubber bearings have been proposed and implemented in actual buildings. These are (a) low-damping rubber bearings, (b) lead–rubber bearings, and (c) high-damping rubber bearings. A brief description of these bearings follows.

FIGURE 15.6 Laminated rubber bearing sustaining a large shear deformation. (Reproduced with permission of Okumura Corporation from Isolation and response control of nuclear and non-nuclear structures, *Special Issue for the Exhibition of the 11th International Conference on Structural Mechanics in Reactor Technology*, Tokyo, Japan, Aug. 18–23, 1991.)

15.4.2.1 Low-Damping Rubber Bearings

As shown in Figure 15.7, low-damping rubber bearings are constructed in much the same way as the general laminated rubber bearings just described. That is, they are constructed with two thick steel endplates and several thin steel shims interleaved with rubber sheets. The rubber is vulcanized and bonded to the steel in a single operation under heat and pressure in a mold. As mentioned earlier, the steel shims prevent the bulging of the rubber and provide a high vertical stiffness. However, they have no effect on the horizontal stiffness, which is controlled by the low shear modulus of the rubber. The rubber behavior in shear is linear up to shear strains of 100% or more (see Figure 15.8).

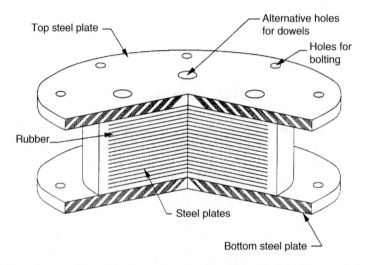

FIGURE 15.7 Components of low-damping rubber bearing. (Reproduced with permission of CISM, the International Center for Mechanical Sciences, Udine, Italy from Constantinou, M. C., Chapter II in *Passive and Active Structural Control in Civil Engineering*, Soong, T. T. and Constantinou, M. C., ed., CIMS Courses and Lectures No. 345, Springer, Vienna, NY, 1994.)

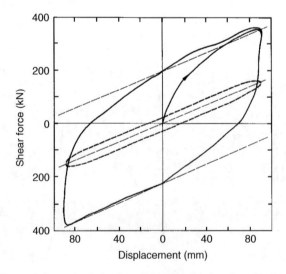

FIGURE 15.8 Typical force–deformation behavior of low-damping bearings (dashed curve) and lead–rubber bearings (solid curve). (Reproduced with permission of John Wiley & Sons, Limited, from Skinner, R. I., Robinson, W. H., and McVerry, G. H., *An Introduction to Seismic Isolation*, John Wiley & Sons, Limited, Chichester, 1993.)

Seismic Protection with Base Isolation

Their damping ratio is between 2 and 3% of critical. Because of their low damping, low-damping rubber bearings are normally used in conjunction with supplementary damping devices such as viscous dampers, steel bars, lead bars, and frictional devices.

The advantages of low-damping rubber bearings are many: they are simple to manufacture, easy to model, and their mechanical properties are unaffected by strain rate, temperature, loading history, or aging. They are not subject to creep, and the long-term stability of their shear modulus is good. The single disadvantage is that a supplementary damping device is generally needed to limit their lateral deformations to reasonable levels. These supplementary devices, however, require elaborate connections and, in the case of metallic dampers, they are prone to low-cycle fatigue.

Assuming a linear elastic behavior and if the effect of the axial force is neglected, the horizontal stiffness of a low-damping rubber bearing is given by (see Section 8.4.3)

$$K_h = \frac{AG}{\Sigma t} \tag{15.1}$$

where G, A, and Σt denote shear modulus of the rubber, full cross-sectional area of the rubber pad, and total rubber thickness.

Similarly, the vertical stiffness is given by

$$K_v = \frac{A_s E_c}{\Sigma t} \tag{15.2}$$

where A_s is the bonded area of the steel shims and E_c is the compression modulus of the rubber–steel compound, which for a circular pad is approximately equal to

$$E_c = \left(\frac{1}{6GS^2} + \frac{4}{3K}\right)^{-1} \tag{15.3}$$

where K is the bulk modulus of rubber (typically about 2000 MPa [290,000 psi]), and S is a shape factor defined as

$$S = \frac{\text{loaded rubber area}}{\text{rubber area free to bulge}} \tag{15.4}$$

The shape factor for a circular pad with diameter R and thickness t is equal to $R/4t$, whereas for a square pad with side dimension a and thickness t it is equal to $a/4t$. An appropriate shape factor for seismic isolation bearings is between 10 and 20.

15.4.2.2 Lead–Rubber Bearings

The lead–rubber bearing was invented in New Zealand in 1975 and has been implemented in isolated buildings in New Zealand, Japan, and the United States. Lead–rubber bearings are similar to low-damping rubber bearings, except that they have a lead plug at the center of the bearing (see Figure 15.9). Lead is a crystalline material that exhibits an elastoplastic force–deformation behavior and yields at a relatively low stress, about 10 MPa (1450 psi). Hence, an inserted lead plug properly confined by the bearing's steel plates allows a lead–rubber bearing to dissipate energy hysteretically after the bearing's shear deformation exceeds the lead plug's shear yield deformation. In consequence, a lead–rubber bearing provides in a single compact unit an energy dissipation capability, an initial high stiffness (about 10 times the postyield stiffness) before the yielding of the

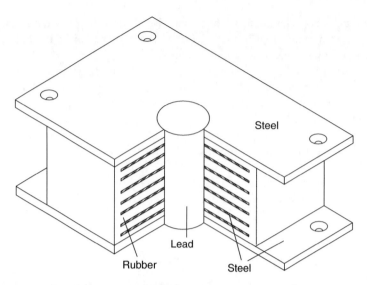

FIGURE 15.9 Components of lead–rubber bearing.

lead plug, and a low postyield stiffness equal to the shear stiffness of the rubber. In addition, the strain energy stored in the lead plug and the rubber during an earthquake provides a restoring force that recenters the structure back to its initial configuration after the earthquake ends. The presence of this restoring force has been validated in shake table tests and in isolated structures mounted on lead–rubber bearings after experiencing an earthquake. In the experimental tests, lead–rubber bearings have returned to positions that are within 2.5 cm (1 in.) from their centered positions. The disadvantage of lead–rubber bearings is that the nonlinearity introduced by the lead plug and the consequent sudden changes in the bearing's stiffness may induce a response in the higher modes of the superstructure, increasing the accelerations in the structural masses over the case where no lead plug is used. These higher-mode responses may affect the performance of the building's contents and nonstructural components.

Lead–rubber bearings may be effectively modeled with a bilinear force–deformation behavior (see Figure 15.8). Therefore, the formulas presented earlier for low-damping bearings are also applicable to lead–rubber bearings, except that the formula for the horizontal stiffness is only valid for the postyield portion of the force–deformation curve. Note, also, that the design of base-isolated structures that use this type of rubber bearings requires a nonlinear analysis. Nonetheless, it is customary to model the behavior of lead–rubber bearings by an equivalent linear, viscously damped system with an effective stiffness and an effective damping ratio (see Section 15.5.5).

15.4.2.3 High-Damping Rubber Bearings

High-damping bearings are laminated bearings constructed with a compounded rubber that intrinsically exhibits a high damping ratio. This high-damping rubber is fabricated by adding extra fine carbons, oils, resins, or other proprietary fillers to natural rubber. The damping in the bearings is neither viscous nor hysteretic. The effective damping is between 10 and 20% at 100% shear strains. The lower level corresponds to rubbers with low hardness (durometer hardness of 50–55) and a low shear modulus (about 0.34 MPa [49 psi]). The higher level corresponds to rubbers with a high hardness (durometer hardness of 70–75) and a high shear modulus (about 1.40 MPa [203 psi]). The methods of vulcanization, bonding, and construction of this type of bearings are the same as those used for any other laminated rubber bearing. The natural rubber compound with enough

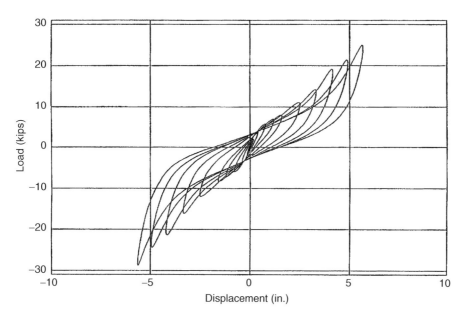

FIGURE 15.10 Typical force–deformation characteristics of high-damping rubber bearings. (Reproduced with permission of John Wiley & Sons, Limited, from Naeim, F. and Kelly, J. M., *Design of Seismic Isolated Structures: From Theory to Practice*, John Wiley & Sons, Limited, New York, 1999.)

inherent damping to eliminate the need for supplementary damping devices was developed in 1982 by the Malaysian Rubber Producers' Research Association of the United Kingdom. Buildings in the United States, Italy, Japan, China, and Indonesia have been isolated using high-damping rubber bearings. The building for the Foothill Communities Law and Justice Center in Rancho Cucamonga, California (described in Section 15.3) was isolated using high-damping rubber bearings.

The force–deformation behavior of high-damping rubber bearings at shear strains of less than 20% is characterized by a high stiffness (see Figure 15.10). Over the range of shear strains between 20 and 120%, the shear modulus is low and fairly constant. At large strains, the modulus increases again due to a strain crystallization process in the rubber. High-damping rubber bearings thus provide an initial high stiffness, which is essential to resist wind loads and minor earthquakes without appreciable motions. As the intensity of the excitation increases, the stiffness is reduced and the isolation system becomes effective. Beyond a shear strain of between 250 and 300%, the stiffness increases again due to hardening effects, providing a safety valve against unexpectedly severe earthquakes. A linear analysis with the initial stiffness may be used to estimate the response of the isolated structure under wind load and other low-level loads. Similarly, a linear analysis may be performed with the stiffness under moderate strains to estimate the response of the system under the design earthquake loads. For earthquake loads beyond the design loads, the system can be analyzed using a bilinear model where the postyield stiffness is much larger than the initial one.

High-damping rubber bearings offer several advantages: (a) they combine in a single element the flexibility and energy dissipation capability that is needed to form an effective isolation system; (b) they are easy to design and manufacture; and (c) they are compact, which simplifies the installation process. However, the material characteristics of these bearings are more sensitive to temperature and frequency than low-damping rubber bearings. They also exhibit a dependence on load history, although some compounds have little or no manifestation of this effect. When subjected to several large-strain cycles, they show a higher effective stiffness and damping in the first cycle than in the following cycles. Generally, the properties of the material stabilize after the third cycle.

15.4.3 SLIDING BEARINGS

In sliding systems, a mechanism is provided to let the structure slide under lateral loads beyond a certain threshold level. Using devices with a low coefficient of friction, these bearings allow the transmission of shear forces from the ground to the superstructure up to a particular level, beyond which sliding occurs and a further transmission is prevented. The magnitude of the forces transmitted to the superstructure during a strong earthquake is thus independent of the severity of the earthquake.

Sliding bearings are very efficient in mitigating the effects of earthquakes. In addition, they are relatively inexpensive and compact in size. However, they do not have the capability to return the structure to its original position after an earthquake since sliding systems, in general, do not generate restoring forces. This means that a permanent offset may result after the occurrence of an earthquake, decreasing the amount of displacement that can be accommodated during the following earthquake. This may produce an unsafe condition whenever two strong earthquakes occur one after the other in a short amount of time, as in the case of a strong aftershock occurring after a severe earthquake. Thus, most sliding systems are designed in combination with a recentering mechanism to avoid this problem. Another disadvantage is that they generate significant higher-mode accelerations and high-frequency vibrations that may affect building contents and equipment due to the sudden sliding that occurs when the frictional forces are overcome. The frictional properties of sliding bearings are affected by composition and condition of the sliding interface, bearing pressure, velocity of sliding, contamination, and temperature.

Sliding bearings that have been implemented in practice are (a) the Electricité-de-France bearing, (b) the Tass bearing, and (c) the friction pendulum bearing. These bearings are briefly described in the following sections.

15.4.3.1 Electricité-de-France Bearing

This sliding bearing was developed by Spie-Batignolles Batiment Travaux Publics and Electricité-de-France. It combines a laminated neoprene bearing with two friction plates that slide with respect to each other. One of the plates is a lead–bronze alloy plate bonded to the bearing; the other is a stainless-steel plate attached to the superstructure (see Figure 15.11). The lead–bronze plate has grooves on the surface to collect debris in the wear process and prevent it from entering the sliding interface. Under these conditions, the sliding interface exhibits essentially a Coulomb-type friction with a friction coefficient between 0.18 and 0.22 at pressures between 2 and 15 MPa (290 and 2200 psi) and all sliding velocities. The bearing is designed so that small to moderate earthquakes are resisted by the elastic deformation of the rubber bearing alone, whereas severe earthquakes are resisted by both the elastic deformation of the rubber bearing and the slip between the lead–bronze and steel plates. The friction between the plates generates an additional damping force over that provided by the elastomeric bearing. The system, however, does not include any recentering devices and may leave permanent displacements after an earthquake.

This system was developed in the early 1970s in an attempt to standardize the design of nuclear power plants. Electricité-de-France developed a standard nuclear power plant qualified to resist a horizontal ground motion with a peak acceleration of 0.2 g. The intention was to develop an isolation system that would give similar design forces regardless of the seismicity of the construction site. This way, the standard plant would simply be isolated with these bearings in regions of high seismicity to keep the plant accelerations below the qualification level. The system has been successful in fulfilling the original objective and has been implemented in the design of four nuclear power plants. These plants are located in Koeberg, South Africa; Karun River, Iran; and Cruas and Le Pellirin, France.

Seismic Protection with Base Isolation

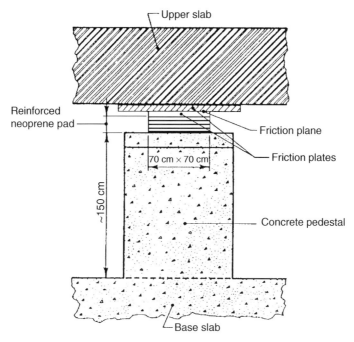

FIGURE 15.11 Electricité-de-France sliding bearing. (Reproduced with permission of Earthquake Engineering Research Institute from Buckle, I. G. and Mayes, R. L., *Earthquake Spectra*, 6, 2, 1990, 161–201.)

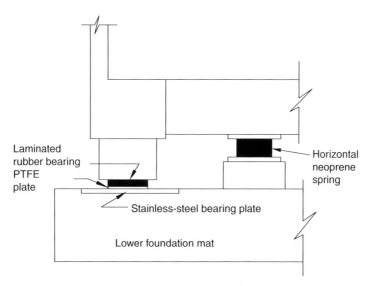

FIGURE 15.12 Configuration of TASS sliding bearing.

15.4.3.2 TASS Bearing

The TASS (TAisei Shake Suppression) sliding bearing was developed by the Taisei Corporation in Japan. It is composed of laminated rubber bearings, stainless-steel bearing plates, and horizontal neoprene springs arranged as shown in Figure 15.12. The laminated rubber bearings have a Teflon plate attached to their lower ends. The horizontal springs are made up of blocks of chloroprene rubber (neoprene) and carry no vertical load. The laminated rubber bearings and bearing plates support the vertical load from the superstructure and reduce the horizontal seismic force by sliding. The horizontal springs restrain the displacement of the bearings with a weak lateral stiffness and provide a restoring force. Under weak or moderate earthquakes, the superstructure displaces by the lateral deformation of the rubber bearings, but sliding does not occur. Under strong and severe earthquakes, sliding occurs in addition to the deformation of the rubber bearings. The coefficient of friction along the sliding surfaces is dependant on pressure and sliding velocity. For very slow sliding rates, the coefficient of friction is typically of the order of 0.05. For operating conditions typical for a seismic isolator, the coefficient of friction ranges from about 0.10 to 0.15. The force–displacement curve of a TASS bearing is close to a bilinear one, with the initial stiffness given by the sum of the stiffnesses of the rubber bearings and the horizontal springs, and the postsliding stiffness given by the stiffness of the horizontal springs. The stiffnesses of the rubber bearings and horizontal springs are selected so that the rigid-body period of the isolated structure before sliding is between 1 and 2 s and the postsliding period is about 5 s.

The strength, initial stiffness, and postsliding stiffness of the system are controlled by different, noninteracting elements. Therefore, a significant abatement of the input ground motion may be achieved by the proper selection of these characteristics. Disadvantages of the system are the possibility of tensile forces in the neoprene block given that they carry no vertical load, and the velocity sensitivity of the friction coefficient on the sliding surface, which complicates the modeling of the system. Three buildings have been implemented with this isolation system. These are a research building in the Technology Research Center of Taisei Corporation in Yokohama, an indoor pool building in Isu peninsula, and an office building in Yokkaichi, all in Japan.

15.4.3.3 Friction Pendulum Bearing

As described in Section 15.3, the friction pendulum bearing is a sliding isolator with a self-centering capability. As shown in Figure 15.13a, it consists of an articulated slider that moves on a concave spherical stainless-steel surface. It also includes an enclosing cylinder that provides a lateral displacement restrain and protects the interior components from environmental contamination. The articulated slider is coated with a low-friction and high-pressure capacity composite material (typically PTFE-based materials with a pressure capacity of up to 275 MPa [39,900 psi]). Because of the friction between the sliding surfaces, a structure supported on this type of bearings responds to low-level forces like a conventional fixed-base structure. That is, it can withstand wind and small earthquake loads without sliding. Once the friction forces are exceeded, however, the structure responds as a free pendulum with the dynamic response controlled by the natural period of this pendulum and the damping generated by the frictional forces. Seismic isolation is achieved by lengthening the natural period of the supported structure. Also, as the slider moves along the spherical surface, it causes the supported structure to rise (see Figure 15.13b), developing a gravity restoring force that

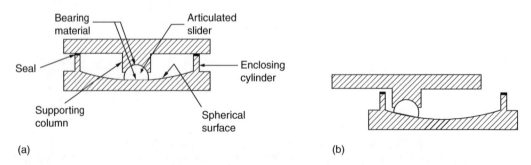

FIGURE 15.13 Sectional views of friction pendulum bearing in (a) centered and (b) displaced positions.

Seismic Protection with Base Isolation

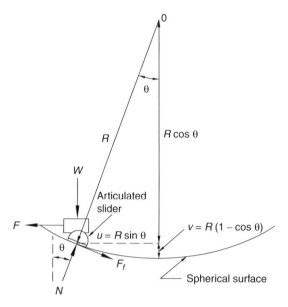

FIGURE 15.14 Free-body diagram of friction pendulum bearing.

helps bring the structure back to its original position. Thus, the friction pendulum bearings combine the basic elements of an isolation system in a compact unit and at the same time have an inherent ability to recenter the supported structure.

The basic principle of operation of a friction pendulum bearing may be illustrated by means of the free-body diagram shown in Figure 15.14. From this free-body diagram, it may be seen that the lateral force needed to induce a lateral displacement in a friction pendulum bearing is equal to the sum of the horizontal component of the normal force and the horizontal component of the friction force at the sliding interface. That is,

$$F = N\sin\theta + \frac{F_f}{\cos\theta} = W\tan\theta + \frac{F_f}{\cos\theta} \tag{15.5}$$

where W and F_f are weight carried by the bearing and aforementioned friction force.

It may also be seen that the horizontal and vertical components of the displacement are, respectively, given by

$$u = R\sin\theta \tag{15.6}$$

$$v = R(1 - \cos\theta) \tag{15.7}$$

where R is the radius of curvature of the spherical surface. In consequence, the force F may also be expressed as

$$F = \frac{W}{R\cos\theta}u + \frac{F_f}{\cos\theta} \tag{15.8}$$

In addition, after noticing that the first term on the right-hand side of Equation 15.8 represents the restoring force of the system, the stiffness of the bearing after the sliding begins is given by

$$K = \frac{W}{R\cos\theta} \quad (15.9)$$

However, for small values of the angle θ, $\cos\theta \approx 1$, and thus F and K are approximately equal to

$$F = \frac{W}{R}u + \mu W \operatorname{sgn}(\dot{u}) \quad (15.10)$$

$$K = \frac{W}{R} \quad (15.11)$$

where the friction force has been replaced by the product of the coefficient of friction μ and the weight W, and $\dot{u}$ is the horizontal component of the sliding velocity. Furthermore, if the superstructure is assumed perfectly rigid, the natural period of the isolated structure is approximately equal to

$$T = 2\pi\sqrt{\frac{W}{Kg}} = 2\pi\sqrt{\frac{R}{g}} \quad (15.12)$$

which shows that the natural period of the isolated structure is independent of its mass and dependent only on the geometry of the bearing, that is, the radius of curvature R. This means that the natural period of the structure will not change if its weight changes or it is different than assumed. A typical force–displacement curve for a friction pendulum bearing is shown in Figure 15.15.

An improved version of the friction pendulum bearing is the so-called triple pendulum bearing. These bearings are composed of independent units with concave spherical surfaces: an inner slider, two sliding concaves, and two main concaves (see Figure 15.16). The inner slider slides along the two sliding concaves, whereas the sliding concaves slide along the two main concaves. The properties of the components are selected so as to activate the inner slider and the sliding concaves sequentially under increasingly stronger ground motions (see Figure 15.17). This way, the natural

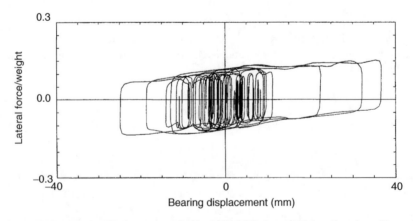

FIGURE 15.15 Typical force–displacement relationship of friction pendulum bearing. (Reproduced with permission of CISM, the International Center for Mechanical Sciences, Udine, Italy from Constantinou, M. C., Chapter VI in *Passive and Active Structural Control in Civil Engineering*, Soong, T. T. and Constantinou, M. C., ed., CIMS Courses and Lectures No. 345, Springer, Vienna, NY, 1994.)

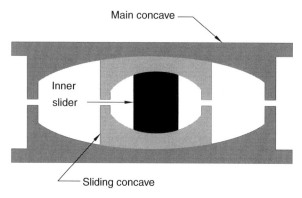

FIGURE 15.16 Components of triple pendulum bearing.

FIGURE 15.17 Sectional views of triple pendulum bearing in centered and displaced positions.

period of the isolated structure is short and the frictional surface (i.e., effective damping) is small under low-intensity, high-frequency ground motions. This results in low in-structure accelerations and less participation of the structural higher modes. However, as the ground motion intensity increases, the natural period of the isolated structure and the frictional surface also increase, resulting in lower base shears and lower bearing displacements. The overall effect is a reduction in the size and cost of the bearings and the size of the gap needed to accommodate the bearings' displacements. When designed for the same ground motion, the plan dimensions of a triple pendulum bearing are approximately 60% of the plan dimensions of a friction pendulum bearing.

Friction pendulum bearings have been used to isolate buildings, bridges, and tanks. Notable structures isolated with these bearings are the new international terminal of San Francisco's International Airport (one of the largest isolated buildings in the United States) and the Benicia–Martinez bridge in Benicia, California, where each bearing has a diameter of 13 ft (4 m) and weighs 40 kips (178 kN) (see Figure 15.18).

15.4.4 Helical Steel Springs

Helical springs have been used in the vibration isolation of equipment for many years, mainly because of their high vertical and horizontal flexibility. They have also been used for seismic isolation, particularly when a three-dimensional isolation is required, as in the case of equipment in nuclear power plants. Since the damping in steel helical springs is negligibly small, they are commonly used in conjunction with a viscous damper to suppress resonance and limit the spring displacements. In addition, spring-isolated structures exhibit a strong coupling between their horizontal and rocking motions, which results from the vertical flexibility of springs. Therefore, this system is practical in situations where the additional horizontal motion produced by rocking is not excessive. Some advantages of using steel springs to build an isolation system are no creep, no deterioration with time, easy inspections, and easy replacements.

A system that employs springs and viscous dampers for seismic isolation is the GERB system developed in Germany originally for the vibration isolation of equipment. It uses large helical steel springs that are flexible both horizontally and vertically and a three-dimensional damper.

FIGURE 15.18 Friction pendulum bearings used in the seismic isolation of the Benicia–Martinez Bridge. (Photograph courtesy of Victor Zayas, Earthquake Protection Systems, Inc.)

The damper is a dashpot with a specially shaped piston moving inside a viscous fluid (a silicon gel). Figure 15.19 shows a typical GERB spring-damper unit. The system was implemented in two steel frame houses in Santa Monica, California, that were strongly affected by the 1994 Northridge earthquake. Their response was monitored by strong motion instruments and, apparently, the isolation system was effective in reducing their floor accelerations compared to the floor accelerations the houses would have experienced without the isolation system.

15.5 METHODS OF ANALYSIS

15.5.1 Linear Two-Degree-of-Freedom Model

To gain insight into the behavior of isolated buildings, it is instructive to use a simple two-degree-of-freedom model with linear springs and linear viscous dampers. The simplicity of the model will allow an intuitive understanding of the system's behavior. It is important to keep in mind, however, that most isolation devices intrinsically exhibit a nonlinear behavior. Therefore, an analysis based on this model will only be approximate for such systems, although good estimates of their response may be obtained through the equivalent linearization technique described in Section 15.5.5.

15.5.1.1 Equations of Motion

Consider the two-degree-of-freedom system shown in Figure 15.20 and let m_s, c_s, and k_s denote the mass, stiffness, and damping constant of the structure, respectively. Similarly, let m_b, c_b, and k_b represent the mass, stiffness, and damping constant of the isolation system, respectively. The two stiffnesses and the two damping constants are assumed to be constant at all times, and the damping forces in the system are supposed to be of the viscous type. That is, it is assumed that the system is a linear one. Furthermore, let u_g, u_b, and u_s represent the absolute displacements of the ground, the mass of the isolation system, and the mass of the structure at any time t, respectively. With reference to the free-body diagram shown in Figure 15.21a, the equation of motion for the structural mass may be written as

$$-c_s(\dot{u}_s - \dot{u}_b) - k_s(u_s - u_b) = m_s \ddot{u}_s \qquad (15.13)$$

Seismic Protection with Base Isolation

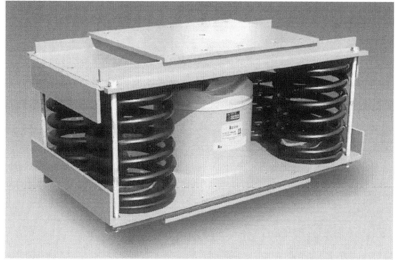

(a)

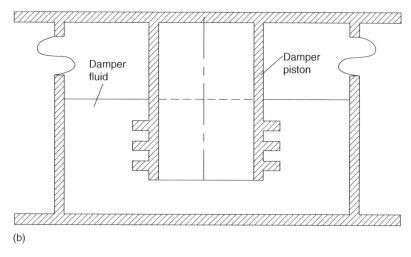

(b)

FIGURE 15.19 (a) GERB four-spring-damper unit, (b) schematic construction of viscous damper. (Photograph courtesy of GERB Vibration Control Systems.)

Similarly, the equation of motion for the isolation mass (see Figure 15.21b) is of the form

$$-c_b(\dot{u}_b - \dot{u}_g) - k_b(u_b - u_g) + c_s(\dot{u}_s - \dot{u}_b) + k_s(u_s - u_b) = m_b\ddot{u}_b \quad (15.14)$$

which in view of Equation 15.13 may also be expressed as

$$-c_b(\dot{u}_b - \dot{u}_g) - k_b(u_b - u_g) = m_b\ddot{u}_b + m_s\ddot{u}_s \quad (15.15)$$

In terms of the relative displacements $v_s = u_s - u_b$ and $v_b = u_b - u_g$, Equations 15.13 and 15.15 may be written alternatively as

$$-c_b\dot{v}_b - k_bv_b = m_b(\ddot{v}_b + \ddot{u}_g) + m_s(\ddot{v}_s + \ddot{v}_b + \ddot{u}_g) \quad (15.16)$$

$$-c_s\dot{v}_s - k_sv_s = m_s(\ddot{v}_s + \ddot{v}_b + \ddot{u}_g) \quad (15.17)$$

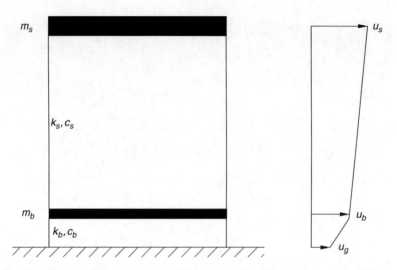

FIGURE 15.20 Two-degree-of-freedom model of base-isolated building and associated displacements.

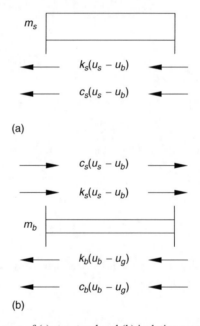

FIGURE 15.21 Free-body diagrams of (a) structural and (b) isolation masses.

or as

$$(m_b + m_s)\ddot{v}_b + m_s\ddot{v}_s + c_b\dot{v}_b + k_b v_b = -(m_s + m_b)\ddot{u}_g \tag{15.18}$$

$$m_s\ddot{v}_b + m_s\ddot{v}_s + c_s\dot{v}_s + k_s v_s = -m_s\ddot{u}_g \tag{15.19}$$

which in matrix form may be expressed as

$$\begin{bmatrix} m & m_s \\ m_s & m_s \end{bmatrix}\begin{Bmatrix} \ddot{v}_b \\ \ddot{v}_s \end{Bmatrix} + \begin{bmatrix} c_b & 0 \\ 0 & c_s \end{bmatrix}\begin{Bmatrix} \dot{v}_b \\ \dot{v}_s \end{Bmatrix} + \begin{bmatrix} k_b & 0 \\ 0 & k_s \end{bmatrix}\begin{Bmatrix} v_b \\ v_s \end{Bmatrix} = -\begin{bmatrix} m & m_s \\ m_s & m_s \end{bmatrix}\begin{Bmatrix} 1 \\ 0 \end{Bmatrix}\ddot{u}_g \tag{15.20}$$

Seismic Protection with Base Isolation

or as

$$[M]\{\ddot{v}\} + [C]\{\dot{v}\} + [K]\{v\} = -[M]\{r\}\ddot{u}_g \qquad (15.21)$$

where

$$[M] = \begin{bmatrix} m & m_s \\ m_s & m_s \end{bmatrix} \quad [C] = \begin{bmatrix} c_b & 0 \\ 0 & c_s \end{bmatrix} \quad [K] = \begin{bmatrix} k_b & 0 \\ 0 & k_s \end{bmatrix} \qquad (15.22)$$

$$\{v\} = \begin{Bmatrix} v_b \\ v_s \end{Bmatrix} \quad \{r\} = \begin{Bmatrix} 1 \\ 0 \end{Bmatrix} \qquad (15.23)$$

and $m = m_b + m_s$ is the total mass of the system.

It is of interest to note that for a perfectly rigid structure, that is, when $v_s = 0$ at all times, the equation of motion of the system is given by

$$m\ddot{v}_b + c_b\dot{v}_b + k_b v_b = -m\ddot{u}_g \qquad (15.24)$$

which corresponds to the equation of motion of the isolation system with the total mass of the structure–isolation system. Similarly, when the isolation system is perfectly rigid, that is, when $v_b = 0$ at all times, the equation of motion of the system is given by

$$m_s\ddot{v}_s + c_s\dot{v}_s + k_s v_s = -m_s\ddot{u}_g \qquad (15.25)$$

which is the equation of motion for the case when the structure is mounted directly on the ground.

15.5.1.2 Natural Frequencies

Let γ denote a mass ratio defined as

$$\gamma = \frac{m_s}{m} \qquad (15.26)$$

and let ω_b and ω_s represent nominal natural frequencies defined by

$$\omega_b^2 = \frac{k_b}{m} \quad \omega_s^2 = \frac{k_s}{m_s} \qquad (15.27)$$

Furthermore, let ξ_b and ξ_s denote nominal damping ratios defined by

$$2\xi_b\omega_b = \frac{c_b}{m} \quad 2\xi_s\omega_s = \frac{c_s}{m_s} \qquad (15.28)$$

By dividing Equation 15.18 by m and Equation 15.19 by m_s, and in terms of the variables defined by Equations 15.26 through 15.28, the equations of motions of the system may also be expressed as

$$\ddot{v}_b + \gamma\ddot{v}_s + 2\xi_b\omega_b\dot{v}_b + \omega_b^2 v_b = -\ddot{u}_g \qquad (15.29)$$

$$\ddot{v}_b + \ddot{v}_s + 2\xi_s\omega_s\dot{v}_s + \omega_s^2 v_s = -\ddot{u}_g \qquad (15.30)$$

which for the undamped, free vibration case reduce to

$$\ddot{v}_b + \gamma \ddot{v}_s + \omega_b^2 v_b = 0 \tag{15.31}$$

$$\ddot{v}_b + \ddot{v}_s + \omega_s^2 v_s = 0 \tag{15.32}$$

which in matrix form may be written as

$$\begin{bmatrix} 1 & \gamma \\ 1 & 1 \end{bmatrix} \begin{Bmatrix} \ddot{v}_b \\ \ddot{v}_s \end{Bmatrix} + \begin{bmatrix} \omega_b^2 & 0 \\ 0 & \omega_s^2 \end{bmatrix} \begin{Bmatrix} v_b \\ v_s \end{Bmatrix} = \begin{Bmatrix} 0 \\ 0 \end{Bmatrix} \tag{15.33}$$

Since in free vibration

$$\begin{Bmatrix} \ddot{v}_b \\ \ddot{v}_s \end{Bmatrix} = -\omega^2 \begin{Bmatrix} v_b \\ v_s \end{Bmatrix} \tag{15.34}$$

then Equation 15.33 may also be written as

$$\left(\begin{bmatrix} \omega_b^2 & 0 \\ 0 & \omega_s^2 \end{bmatrix} - \omega^2 \begin{bmatrix} 1 & \gamma \\ 1 & 1 \end{bmatrix} \right) \begin{Bmatrix} v_b \\ v_s \end{Bmatrix} = \begin{Bmatrix} 0 \\ 0 \end{Bmatrix} \tag{15.35}$$

where ω is one of the natural frequencies of the combined structure–isolation system. In turn, Equation 15.35 leads to the characteristic equation

$$\begin{vmatrix} \omega_b^2 - \omega^2 & -\omega^2 \gamma \\ -\omega^2 & \omega_s^2 - \omega^2 \end{vmatrix} = 0 \tag{15.36}$$

which, after expansion of the determinant, leads to

$$(\omega_b^2 - \omega^2)(\omega_s^2 - \omega^2) - \omega^4 \gamma = 0 \tag{15.37}$$

or to

$$(1 - \gamma)\omega^4 - (\omega_b^2 + \omega_s^2)\omega^2 + \omega_b^2 \omega_s^2 = 0 \tag{15.38}$$

whose solution is

$$\omega^2 = \frac{1}{2(1-\gamma)} \left[\omega_b^2 + \omega_s^2 \mp \sqrt{(\omega_b^2 + \omega_s^2)^2 - 4(1-\gamma)\omega_b^2 \omega_s^2} \right] \tag{15.39}$$

or

$$\omega^2 = \frac{1}{2(1-\gamma)} \left[\omega_b^2 + \omega_s^2 \mp \sqrt{(\omega_b^2 - \omega_s^2)^2 - 4\gamma \omega_b^2 \omega_s^2} \right] \tag{15.40}$$

Seismic Protection with Base Isolation

First-order approximations for the two natural frequencies of the system may be obtained as follows. Let ε represent the ratio between the nominal natural frequencies, squared, of the isolation system and the fixed-based structure. That is, let ε be defined as

$$\varepsilon = \frac{\omega_b^2}{\omega_s^2} \tag{15.41}$$

and consider it to be of the order 10^{-2}. In terms of this new parameter and neglecting higher-order terms, Equation 15.40 may be expressed as

$$\omega^2 = \frac{\omega_s^2}{2(1-\gamma)}\left[1 + \varepsilon \mp \sqrt{1 - 2\varepsilon(1-2\gamma)}\right] \tag{15.42}$$

which, after replacing the radical with its second-order approximation, takes the form

$$\begin{aligned}
\omega^2 &= \frac{\omega_s^2}{2(1-\gamma)}\left\{1 + \varepsilon \mp \left[1 - \varepsilon(1-2\gamma) - \frac{\varepsilon^2}{2}(1-2\gamma)^2\right]\right\} \\
&= \frac{\omega_s^2}{2(1-\gamma)}\left\{1 + \varepsilon \mp \left[1 - \varepsilon(1-2\gamma)\left(1 - \gamma\varepsilon + \frac{\varepsilon}{2}\right)\right]\right\}
\end{aligned} \tag{15.43}$$

Thus,

$$\begin{aligned}
\omega_1^2 &= \frac{\omega_s^2}{2(1-\gamma)}\left\{1 + \varepsilon - \left[1 - \varepsilon(1-2\gamma)\left(1 - \gamma\varepsilon + \frac{\varepsilon}{2}\right)\right]\right\} \\
&= \frac{\omega_s^2 \varepsilon}{(1-\gamma)}(1 - \gamma - \gamma\varepsilon + \gamma^2 \varepsilon) \\
&= \omega_b^2(1 - \gamma\varepsilon)
\end{aligned} \tag{15.44}$$

Similarly,

$$\begin{aligned}
\omega_2^2 &= \frac{\omega_s^2}{2(1-\gamma)}\left\{1 + \varepsilon + \left[1 - \varepsilon(1-2\gamma)\left(1 - \gamma\varepsilon + \frac{\varepsilon}{2}\right)\right]\right\} \\
&= \frac{\omega_s^2}{2(1-\gamma)}(2 + 2\gamma\varepsilon^2 + 2\gamma\varepsilon - 2\gamma^2\varepsilon^2 - \varepsilon^2/2)
\end{aligned} \tag{15.45}$$

which, after neglecting second-order terms, may be reduced to

$$\omega_2^2 = \frac{\omega_s^2}{(1-\gamma)}(1 + \gamma\varepsilon) \tag{15.46}$$

Note, thus, that for low ω_b/ω_s ratios the first natural frequency of the base-isolated structure is slightly less than the nominal natural frequency of the isolation system. In contrast, the second natural frequency is slightly higher than the nominal natural frequency of the fixed-base structure.

In other words, the fundamental natural frequency of the isolated structure is close to the natural frequency of the isolation system and the second natural frequency is close to the natural frequency of the fixed-base structure.

15.5.1.3 Mode Shapes

The undamped mode shapes of the system may be determined from Equation 15.35, after considering that for the ith mode the displacement vector may be expressed as

$$\begin{Bmatrix} v_b \\ v_s \end{Bmatrix} = \begin{Bmatrix} \phi_{bi} \\ \phi_{si} \end{Bmatrix} = \begin{Bmatrix} 1 \\ \phi_{si} \end{Bmatrix} \quad i=1,2 \tag{15.47}$$

That is, they may be determined from

$$\left(\begin{bmatrix} \omega_b^2 & 0 \\ 0 & \omega_s^2 \end{bmatrix} - \omega_i^2 \begin{bmatrix} 1 & \gamma \\ 1 & 1 \end{bmatrix} \right) \begin{Bmatrix} 1 \\ \phi_{si} \end{Bmatrix} = \begin{Bmatrix} 0 \\ 0 \end{Bmatrix} \quad i=1,2 \tag{15.48}$$

from which one finds that

$$\omega_b^2 - \omega_i^2(1+\gamma\phi_{si}) = 0 \quad i=1,2 \tag{15.49}$$

After substituting into Equation 15.49 the simplified expression found earlier for ω_1^2, one has thus that

$$\omega_b^2 - \omega_b^2(1-\gamma\varepsilon)(1+\gamma\phi_{s1}) = 0 \tag{15.50}$$

which leads to

$$\phi_{s1} = \frac{\varepsilon}{1-\gamma\varepsilon} \approx \varepsilon \tag{15.51}$$

and

$$\begin{Bmatrix} \phi_{b1} \\ \phi_{s1} \end{Bmatrix} = \begin{Bmatrix} 1 \\ \varepsilon \end{Bmatrix} \tag{15.52}$$

Similarly, after substituting into Equation 15.49 the simplified expression found earlier for ω_2^2, one obtains

$$\omega_b^2 - \frac{\omega_s^2}{1-\gamma}(1+\gamma\varepsilon)(1+\gamma\phi_{s2}) = 0 \tag{15.53}$$

from which one finds that

$$\phi_{s2} = -\frac{1}{\gamma}\left[1 - \frac{(1-\gamma)\varepsilon}{1+\gamma\varepsilon}\right] \approx -\frac{1}{\gamma}[1-(1-\gamma)\varepsilon] \tag{15.54}$$

Seismic Protection with Base Isolation

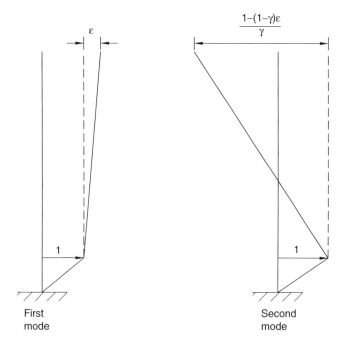

FIGURE 15.22 Mode shapes of two-degree-of-freedom base-isolated model.

and

$$\begin{Bmatrix} \phi_{b2} \\ \phi_{s2} \end{Bmatrix} = \begin{Bmatrix} 1 \\ -\frac{1}{\gamma}[1-(1-\gamma)\varepsilon] \end{Bmatrix} \quad (15.55)$$

The mode shapes depicted by Equations 15.52 and 15.55 are shown in Figure 15.22. It may be noted that in the first mode the deformation of the structure is small, nearly zero. In contrast, the deformation of the structure in the second mode is of the same order of magnitude as the deformation of the isolation system, although in the opposite direction. The relevance of this finding is that in the fundamental mode the deformations are concentrated at the level of the isolation system, but in the second mode both the isolation system and the structure are subjected to significant deformations.

15.5.1.4 Modal Participation Factors

Simplified expressions may also be obtained for the participation factors of the systems as follows:
Recall from Section 10.2.3 that the participation factor in the ith mode of a system with multiple degrees of freedom is given by

$$\Gamma_i = \frac{L_i}{M_i} = \frac{\{\phi\}_i^T[M]\{r\}}{\{\phi\}_i^T[M]\{\phi\}_i} \quad (15.56)$$

where $\{\phi\}_i$ and M_i denote the mode shape and generalized mass in that mode, respectively; $[M]$ represents the mass matrix of the system; and $\{r\}$ is an influence vector that couples the ground displacement to the degrees of freedom of the system. According to Equations 15.22, 15.23, and 15.52, one has thus that for the isolated building

$$M_1 = \begin{Bmatrix} 1 \\ \varepsilon \end{Bmatrix}^T \begin{bmatrix} m & m_s \\ m_s & m_s \end{bmatrix} \begin{Bmatrix} 1 \\ \varepsilon \end{Bmatrix} \approx m + 2m_s\varepsilon = m(1 + 2\gamma\varepsilon) \tag{15.57}$$

$$L_1 = \begin{Bmatrix} 1 \\ \varepsilon \end{Bmatrix}^T \begin{bmatrix} m & m_s \\ m_s & m_s \end{bmatrix} \begin{Bmatrix} 1 \\ 0 \end{Bmatrix} = m + m_s\varepsilon = m(1 + \gamma\varepsilon) \tag{15.58}$$

and

$$\Gamma_1 = \frac{m(1+\gamma\varepsilon)}{m(1+2\gamma\varepsilon)} = \frac{1+2\gamma\varepsilon-\gamma\varepsilon}{1+2\gamma\varepsilon} = 1 - \frac{\gamma\varepsilon}{1+2\gamma\varepsilon} \approx 1 - \gamma\varepsilon \tag{15.59}$$

Similarly,

$$M_2 = \begin{Bmatrix} 1 \\ a \end{Bmatrix}^T \begin{bmatrix} m & m_s \\ m_s & m_s \end{bmatrix} \begin{Bmatrix} 1 \\ a \end{Bmatrix} = m + 2m_s a + m_s a^2 \tag{15.60}$$

$$L_2 = \begin{Bmatrix} 1 \\ a \end{Bmatrix}^T \begin{bmatrix} m & m_s \\ m_s & m_s \end{bmatrix} \begin{Bmatrix} 1 \\ 0 \end{Bmatrix} = m + m_s a \tag{15.61}$$

where, according to Equation 15.55,

$$a = -\frac{1}{\gamma}[1 - (1-\gamma)]\varepsilon \tag{15.62}$$

On substitution of Equation 15.62 into Equations 15.60 and 15.61, one thus obtains

$$M_2 = m - \frac{2m_s}{\gamma}[1-(1-\gamma)\varepsilon] + \frac{m_s}{\gamma^2}[1-(1-\gamma)\varepsilon]^2$$
$$\approx \frac{m(1-\gamma)}{\gamma}[1-2(1-\gamma)\varepsilon] \tag{15.63}$$

$$L_2 = m - \frac{m}{\gamma}[1-(1-\gamma)\varepsilon]$$
$$= m(1-\gamma)\varepsilon \tag{15.64}$$

which leads to

$$\Gamma_2 = \frac{m(1-\gamma)\varepsilon}{\frac{m(1-\gamma)}{\gamma}[1-2(1-\gamma)\varepsilon]} = \frac{\gamma\varepsilon}{1-2(1-\gamma)\varepsilon} \approx \gamma\varepsilon \tag{15.65}$$

Seismic Protection with Base Isolation

Observe, thus, that to a first-order approximation in ε, the participation factor for the first mode is close to unity and for the second one is small. This, together with the shift in the natural frequencies, reveals why a base-isolation system may be effective to reduce the response of a structure. If the frequencies ω_s and ω_b are well separated, then the participation factor for the second mode is negligibly small. This means that the response of the isolated system is controlled by the response in its first mode. But if ω_b is properly selected, the frequency of the isolated system in the first mode will be outside the frequency range where the ground motion has a significant energy content. As a result, the ground motion is not transmitted to the structure.

15.5.1.5 Uncoupled Equations of Motion

With the simplified expressions for the mode shapes and participation factors found in the previous section, it is now possible to derive simplified expressions for the response of the base-isolated building using the modal decomposition method described in Chapter 10. For this purpose, express the vector $\{v\}$ defined by Equation 15.23 in terms of the mode shapes of the system as

$$\{v\} = \{\phi\}_1 \eta_1 + \{\phi\}_2 \eta_2 = [\{\phi\}_1 \ \{\phi\}_2] \begin{Bmatrix} \eta_1 \\ \eta_2 \end{Bmatrix} = [\phi]\{\eta\} \tag{15.66}$$

where $\{\phi\}_1$ and $\{\phi\}_2$ denote the mode shapes of the system, η_1 and η_2 are unknown modal coordinates, and

$$[\phi] = [\{\phi\}_1 \ \{\phi\}_2] \qquad \{\eta\} = \begin{Bmatrix} \eta_1 \\ \eta_2 \end{Bmatrix} \tag{15.67}$$

Substitution of Equation 15.66 into Equation 15.21 and premultiplication by $[\phi]^T$ lead to

$$[\phi]^T[M][\phi]\{\ddot{\eta}\} + [\phi]^T[C]\{\dot{\eta}\} + [\phi]^T[K]\{\eta\} = -[\phi]^T[M]\{r\}\ddot{u}_g \tag{15.68}$$

or to

$$\begin{bmatrix} M_{11} & M_{12} \\ M_{21} & M_{22} \end{bmatrix} \begin{Bmatrix} \ddot{\eta}_1 \\ \ddot{\eta}_2 \end{Bmatrix} + \begin{bmatrix} C_{11} & C_{12} \\ C_{21} & C_{22} \end{bmatrix} \begin{Bmatrix} \dot{\eta}_1 \\ \dot{\eta}_2 \end{Bmatrix} + \begin{bmatrix} K_{11} & K_{12} \\ K_{21} & K_{22} \end{bmatrix} \begin{Bmatrix} \eta_1 \\ \eta_2 \end{Bmatrix} = -\begin{Bmatrix} \{\phi\}_1^T \\ \{\phi\}_2^T \end{Bmatrix}[M]\{r\}\ddot{u}_g \tag{15.69}$$

where, for $i, j = 1, 2$,

$$M_{ij} = \{\phi\}_i^T[M]\{\phi\}_j \quad C_{ij} = \{\phi\}_i^T[C]\{\phi\}_j \quad K_{ij} = \{\phi\}_i^T[K]\{\phi\}_j \tag{15.70}$$

But, in view of the orthogonality properties of mode shapes, one has that for $i \neq j$, $M_{ij} = K_{ij} = 0$. Also, for small damping ratios, it may be assumed that C_{12} and C_{21} are small compared to C_{11} and C_{22} and thus may also be considered equal to zero. As a result, Equation 15.69 may also be written as

$$\begin{bmatrix} M_{11} & 0 \\ 0 & M_{22} \end{bmatrix} \begin{Bmatrix} \ddot{\eta}_1 \\ \ddot{\eta}_2 \end{Bmatrix} + \begin{bmatrix} C_{11} & 0 \\ 0 & C_{22} \end{bmatrix} \begin{Bmatrix} \dot{\eta}_1 \\ \dot{\eta}_2 \end{Bmatrix} + \begin{bmatrix} K_{11} & 0 \\ 0 & K_{22} \end{bmatrix} \begin{Bmatrix} \eta_1 \\ \eta_2 \end{Bmatrix} = -\begin{Bmatrix} \{\phi\}_1^T \\ \{\phi\}_2^T \end{Bmatrix}[M]\{r\}\ddot{u}_g \tag{15.71}$$

or as

$$M_{11}\ddot{\eta}_1 + C_{11}\dot{\eta}_1 + K_{11}\dot{\eta}_1 = -L_1\ddot{u}_g \qquad (15.72)$$

$$M_{22}\ddot{\eta}_2 + C_{22}\dot{\eta}_2 + K_{22}\eta_2 = -L_2\ddot{u}_g \qquad (15.73)$$

which, after dividing through Equation 15.72 by M_{11} and Equation 15.73 by M_{22}, may also be expressed as

$$\ddot{\eta}_1 + 2\xi_1\omega_1\dot{\eta}_1 + \omega_1^2\eta_1 = -\Gamma_1\ddot{u}_g \qquad (15.74)$$

$$\ddot{\eta}_2 + 2\xi_2\omega_2\dot{\eta}_2 + \omega_2^2\eta_2 = -\Gamma_2\ddot{u}_g \qquad (15.75)$$

where ξ_i, ω_i, and Γ_i, $i = 1, 2$, respectively denote the system's damping ratios, natural frequencies, and participation factors.

15.5.1.6 Damping Ratios

Simple approximate expressions may also be derived for the damping ratios of the system in terms of the simplified expressions determined earlier for the natural frequencies and generalized masses. For this purpose, note first that according to the definition for C_{ij} given by Equation 15.70, one has that

$$C_{11} = \begin{Bmatrix} \phi_{b1} \\ \phi_{s1} \end{Bmatrix}^T \begin{bmatrix} c_b & 0 \\ 0 & c_s \end{bmatrix} \begin{Bmatrix} \phi_{b1} \\ \phi_{s1} \end{Bmatrix} \qquad (15.76)$$

$$C_{22} = \begin{Bmatrix} \phi_{b2} \\ \phi_{s2} \end{Bmatrix}^T \begin{bmatrix} c_b & 0 \\ 0 & c_s \end{bmatrix} \begin{Bmatrix} \phi_{b2} \\ \phi_{s2} \end{Bmatrix} \qquad (15.77)$$

which in view of Equations 15.28, 15.52, and 15.55 may be approximated as

$$C_{11} = \begin{Bmatrix} 1 \\ \varepsilon \end{Bmatrix}^T \begin{bmatrix} c_b & 0 \\ 0 & c_s \end{bmatrix} \begin{Bmatrix} 1 \\ \varepsilon \end{Bmatrix} = 2\xi_b\omega_b m + 2\xi_s\omega_s m_s \varepsilon^2 \approx 2\xi_b\omega_b m \qquad (15.78)$$

$$C_{22} = \begin{Bmatrix} 1 \\ a \end{Bmatrix}^T \begin{bmatrix} c_b & 0 \\ 0 & c_s \end{bmatrix} \begin{Bmatrix} 1 \\ a \end{Bmatrix} = 2\xi_b\omega_b m + 2\xi_s\omega_s m_s a^2 \qquad (15.79)$$

where a is given by Equation 15.62. From the definition of damping ratio and in view of Equations 15.44 and 15.57, one has thus that

$$\xi_1 = \frac{C_{11}}{2\omega_1 M_1} = \frac{2\xi_b\omega_b m}{2\omega_b(1-\gamma\varepsilon)^{1/2} m(1+2\gamma\varepsilon)} = \frac{\xi_b(1-2\gamma\varepsilon)}{(1-\gamma\varepsilon)^{1/2}(1-4\gamma^2\varepsilon^2)}$$

$$\approx \frac{\xi_b(1-2\gamma\varepsilon)}{(1-\gamma\varepsilon)^{1/2}} \approx \xi_b(1-2\gamma\varepsilon)^{3/4} \approx \xi_b\left(1-\frac{3}{2}\gamma\varepsilon\right) \qquad (15.80)$$

Seismic Protection with Base Isolation

Similarly, according to Equations 15.46, 15.62, and 15.63, one obtains

$$\xi_2 = \frac{C_{22}}{2\omega_2 M_2} = \gamma \frac{2\xi_b \omega_b m + 2\xi_s \omega_s \gamma m a^2}{2\omega_s(1+\gamma\varepsilon)^{1/2}[1 - 2(1-\gamma)\varepsilon]m(1-\gamma)^{1/2}}$$

$$= \approx \frac{\gamma\varepsilon^{1/2}\xi_b + \xi_s[1 - (1-\gamma)\varepsilon]^2}{(1+\gamma\varepsilon)^{1/2}(1-\gamma)^{1/2}} \approx \frac{\gamma\varepsilon^{1/2}\xi_b + \xi_s}{(1-\gamma)^{1/2}}(1+\gamma\varepsilon)^{-1/2} \quad (15.81)$$

$$\approx \frac{\xi_s + \gamma\varepsilon^{1/2}\xi_b}{(1-\gamma)^{1/2}}\left(1 - \frac{1}{2}\gamma\varepsilon\right)$$

It may be seen, thus, that damping in the first mode (isolation mode) is very close to the damping ratio of the isolation system and is basically unaffected by the damping ratio of the structure. In contrast, the damping ratio in the second mode (structural mode) is affected by both the isolation and the structural damping ratios. Furthermore, it may be seen that damping in the isolation system may increase the damping ratio in the second mode significantly. That is, the term $\gamma\varepsilon^{1/2}\xi_b$ may add significantly to the overall damping ratio, particularly if ξ_s is small.

15.5.1.7 Maximum Response

Equations 15.74 and 15.75 represent the equations of motion of independent single-degree-of-freedom systems. As such, the modal coordinates η_1 and η_2 may be expressed in terms of Duhamel's integral as

$$\eta_1 = -\frac{\Gamma_1}{\omega_{d1}}\int_0^t \ddot{u}_g(t)e^{-\xi_1\omega_1(t-\tau)}\sin\omega_{d1}(t-\tau)d\tau \quad (15.82)$$

$$\eta_2 = -\frac{\Gamma_1}{\omega_{d2}}\int_0^t \ddot{u}_g(t)e^{-\xi_2\omega_2(t-\tau)}\sin\omega_{d2}(t-\tau)d\tau \quad (15.83)$$

where ω_{d1} and ω_{d2} denote the system's damped natural frequencies. Furthermore, the maximum values of these modal coordinates can be obtained from the response spectrum of $\ddot{u}_g$ according to

$$(\eta_1)_{max} = \Gamma_1 SD(\omega_1, \xi_1) \quad (15.84)$$

$$(\eta_2)_{max} = \Gamma_2 SD(\omega_2, \xi_2) \quad (15.85)$$

and an estimate of the maximum response of the system may be obtained through the application of the square-root-of-the-sum-of-the-squares modal combination rule. That is,

$$\{v\}_{max} = \begin{Bmatrix} v_b \\ v_s \end{Bmatrix}_{max} = \sqrt{\{\phi^2\}_1(\eta_1)^2_{max} + \{\phi^2\}_2(\eta_2)^2_{max}} \quad (15.86)$$

or

$$(v_b)_{max} = [\phi_{b1}^2 \Gamma_1^2 SD^2(\omega_1, \xi_1) + \phi_{b2}^2 \Gamma_2^2 SD^2(\omega_2, \xi_2)]^{1/2} \quad (15.87)$$

$$(v_s)_{max} = [\phi_{s1}^2 \Gamma_1^2 SD^2(\omega_1, \xi_1) + \phi_{s2}^2 \Gamma_2^2 SD^2(\omega_2, \xi_2)]^{1/2} \quad (15.88)$$

Simplified expressions for these displacements may be obtained by substituting the simplified expressions found earlier for the mode shapes and modal participation factors. Accordingly, by making use of Equations 15.52, 15.55, 15.59, and 15.65, one obtains

$$(v_b)_{max} = [(1-\gamma\varepsilon)^2 SD^2(\omega_1,\xi_1) + \gamma^2\varepsilon^2 SD^2(\omega_2,\xi_2)]^{1/2} \tag{15.89}$$

$$\begin{aligned}(v_s)_{max} &= \{\varepsilon^2(1-\gamma\varepsilon)^2 SD^2(\omega_1,\xi_1) + [1-(1-\gamma)\varepsilon]^2\varepsilon^2 SD^2(\omega_2,\xi_2)\}^{1/2} \\ &= \varepsilon\{(1-\gamma\varepsilon)^2 SD^2(\omega_1,\xi_1) + [1-(1-\gamma)\varepsilon]^2 SD^2(\omega_2,\xi_2)\}^{1/2}\end{aligned} \tag{15.90}$$

However, for typical ground motions in firm soil, the spectral displacement for high frequencies is always smaller than that for low frequencies. Hence, for those ground motions it is possible to neglect the second term on the right-hand side of Equation 15.89 and approximate $(v_b)_{max}$ as

$$(v_b)_{max} = (1-\gamma\varepsilon)SD(\omega_1,\xi_1) \tag{15.91}$$

Similarly, if higher-order terms are neglected, it is possible to approximate $(v_s)_{max}$ as

$$(v_s)_{max} = \varepsilon[SD^2(\omega_1,\xi_1) + SD^2(\omega_2,\xi_2)]^{1/2} \tag{15.92}$$

In consequence, the base shear coefficient, which is defined as

$$C_s = \frac{k_s(v_s)_{max}}{m_s} = \omega_s^2(v_s)_{max} \tag{15.93}$$

is approximately given by

$$\begin{aligned}C_s &= \omega_s^2\varepsilon[SD^2(\omega_1,\xi_1) + SD^2(\omega_2,\xi_2)]^{1/2} \\ &= \omega_b^2[SD^2(\omega_1,\xi_1) + SD^2(\omega_2,\xi_2)]^{1/2}\end{aligned} \tag{15.94}$$

Furthermore, if only the first terms are retained, one obtains

$$(v_b)_{max} = SD(\omega_1,\xi_1) \tag{15.95}$$

$$(v_s)_{max} = \varepsilon SD(\omega_1,\xi_1) \tag{15.96}$$

$$C_s = \omega_b^2 SD(\omega_1,\xi_1) = \omega_b^2 \frac{SA(\omega_1,\xi_1)}{\omega_1^2} = \frac{1}{1-\gamma\varepsilon}SA(\omega_1,\xi_1) \approx SA(\omega_1,\xi_1) \approx SA(\omega_b,\xi_b) \tag{15.97}$$

which indicates that for small ε and a typical firm-ground response spectrum, in a preliminary design the isolation system may be designed for a relative displacement of $SD(\omega_b,\xi_b)$ and the building for a base shear coefficient of $SA(\omega_b,\xi_b)$. Note also that the reduction in base shear introduced by the isolation system compared to the fixed-base case is approximately equal to $SA(\omega_b,\xi_b)/SA(\omega_s,\xi_s)$.

To summarize, it may be seen that in a base-isolated structure the natural frequency and damping ratio in the first mode depend primarily on the characteristics of the isolation system and are essentially independent of the natural frequency and damping ratio of the structure. In addition,

Seismic Protection with Base Isolation

the form of the first mode shape is close to that of a rigid structure mounted on a flexible element. That is, the displacements in an isolated structure arise largely from the isolator displacement, with little deformations in the structure above the isolation level. In consequence, if an isolation system behaves linearly, the isolated structure may be considered rigid when assessing its seismic response. It may be seen, also, that the isolator displacement may be substantial and consequently it should be an important consideration in the design of base-isolated structures. Finally, it may be seen that the response of a structure may be reduced by a factor of 5–10 if it is base isolated. This clearly shows the benefit of introducing an isolation system into a structure.

EXAMPLE 15.1 MAXIMUM RESPONSE OF BASE-ISOLATED STRUCTURE

Determine using the approximate expressions developed in Section 15.5.1 the dynamic properties and maximum response of an isolated building when the building is subjected to an earthquake ground motion represented by the 5% damping design spectrum in Figure E15.1. When fixed to the ground, the building may be modeled as a single-degree-of-freedom system with a natural period of 0.5 s and a damping ratio of 2%. The isolation system has a foundation slab with a mass equal to 2/3 of the mass of the structure. In addition, it has been designed to exhibit a natural period of 2 s and a damping ratio of 10%. The spectral ordinates for damping values other than 5% may be estimated by multiplying the ordinates in the spectrum for 5% damping by the factor

$$H(\xi) = \frac{1.5}{40\xi + 1} + 0.5$$

where ξ denotes the damping ratio being considered.

Solution

According to the notation introduced in Section 15.5.1, for this example one has that

$$\omega_b = \frac{2\pi}{2.0} = 3.14 \, \text{rad/s} \quad \xi_b = 0.10 \quad \omega_s = \frac{2\pi}{0.5} = 12.57 \, \text{rad/s} \quad \xi_s = 0.02$$

and, according to Equations 15.26 and 15.41, that

$$\gamma = \frac{m_s}{2/3 m_s + m_s} = 0.6 \quad \varepsilon = \frac{(3.14)^2}{(12.57)^2} = 0.062$$

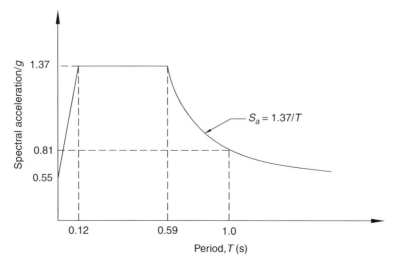

FIGURE E15.1 Design spectrum considered in Example 15.1.

Therefore, if Equations 15.44, 15.46, 15.52, and 15.55 are considered, the natural frequencies, natural periods, and mode shapes of the combined system are

$$\omega_1 = \omega_b\sqrt{1-\gamma\varepsilon} = 3.14\sqrt{1-0.6(0.062)} = 3.08 \text{ rad/s} \quad T_1 = \frac{2\pi}{3.08} = 2.04 \text{ s}$$

$$\omega_2 = \sqrt{\frac{1+\gamma\varepsilon}{1-\gamma}}\omega_s = \sqrt{\frac{1+0.6(0.062)}{1-0.6}}12.57 = 20.24 \text{ rad/s} \quad T_2 = \frac{2\pi}{20.24} = 0.31 \text{ s}$$

$$\begin{Bmatrix}\phi_{b1}\\ \phi_{s1}\end{Bmatrix} = \begin{Bmatrix}1\\ \varepsilon\end{Bmatrix} = \begin{Bmatrix}1\\ 0.062\end{Bmatrix}$$

$$\begin{Bmatrix}\phi_{b2}\\ \phi_{s2}\end{Bmatrix} = \begin{Bmatrix}1\\ -\frac{1}{\gamma}[1-(1-\gamma)\varepsilon]\end{Bmatrix} = \begin{Bmatrix}1\\ -\frac{1}{0.6}[1-(1-0.6)0.062]\end{Bmatrix} = \begin{Bmatrix}1\\ -1.62\end{Bmatrix}$$

Similarly, according to Equations 15.59, 15.65, 15.80, and 15.81, the modal participation factors and damping ratios of the combined system are

$$\Gamma_1 = 1-\gamma\varepsilon = 1-0.6(0.062) = 0.96$$

$$\Gamma_2 = \gamma\varepsilon = 0.6(0.062) = 0.037$$

$$\xi_1 = \xi_b\left(1-\frac{3}{2}\gamma\varepsilon\right) = 0.10\left[1-\frac{3}{2}(0.6)(0.062)\right] = 0.094$$

$$\xi_2 = \frac{\xi_s + \gamma\varepsilon^{1/2}\xi_b}{(1-\gamma)^{1/2}}\left(1-\frac{1}{2}\gamma\varepsilon\right) = \frac{0.02+(0.6)(0.062)^{1/2}(0.10)}{(1-0.6)^{1/2}}\left[1-\frac{1}{2}0.6(0.062)\right] = 0.054$$

Now, for the given design spectrum and the determined natural frequencies and damping ratios, the spectral ordinates corresponding to the two modes of the system are

$$SA(3.08, 0.094) = H(0.094)SA(3.08, 0.05) = \left[\frac{1.5}{40(0.094)+1}+0.5\right]\frac{1.37}{2.04}g = 0.55g$$

$$SA(20.24, 0.054) = H(0.054)SA(3.08, 0.05) = \left[\frac{1.5}{40(0.054)+1}+0.5\right]1.37g = 1.34g$$

In consequence and from Equations 15.91 and 15.92, the relative displacements of the isolation system and structure result as

$$(v_b)_{\max} = (1-\gamma\varepsilon)SD(\omega_1,\xi_1) = (1-0.037)\frac{0.55\,g}{3.08^2} = 0.55 \text{ m}$$

$$(v_s)_{\max} = \varepsilon[SD^2(\omega_1,\xi_1)+SD^2(\omega_2,\xi_2)]^{1/2}$$

$$= 0.062\left[\left(\frac{0.55\,g}{3.08^2}\right)^2+\left(\frac{1.34\,g}{20.24^2}\right)^2\right]^{1/2} = 0.062\,g[0.058^2+0.003^2]^{1/2}$$

$$= 0.035 \text{ m}$$

Furthermore, by virtue of Equation 15.93, the shear force at the base of the isolated structure is

$$V_{bi} = m_s SA(\omega_1,\xi_1) = m_s(0.55\ g) = 0.22\ W_s$$

where W_s denotes the weight of the structure. In comparison, the base shear of the structure when the base is considered fixed is

$$V_{fb} = m_s SA(\omega_s,\xi_s) = m_s SA(12.57, 0.02) = m_s \left[\frac{1.5}{40(0.02)+1} + 0.5 \right] SA(12.57, 0.05)$$

$$= m_s(1.33)1.37g = 1.82\ W_s$$

Thus, the use of the selected isolation system leads to a reduction in the shear force at the base of the structure by a factor of $1.82/0.22 = 8.3$.

15.5.2 Rigid Structure Single-Degree-of-Freedom Model

It may be seen from the discussion in Section 15.5.1 that for isolated structures for which the natural period of the isolation system is much longer than the natural period of the fixed-base structure, the second mode of the combined system contributes little to the response of the system. For those cases, therefore, it is convenient to assume that the structure is perfectly rigid and model the combined system as a single-degree-of-freedom system with the natural frequency and damping ratio of the isolation system. This way, the isolator deformation and the structural base shear coefficient, the parameters of interest in the design of the system, may be estimated in a simple way considering that

$$(v_b)_{max} = SD(\omega_b, \xi_b) \tag{15.98}$$

$$C_s = SA(\omega_b, \xi_b) \tag{15.99}$$

where, as previously established, ω_b and ξ_b denote the natural frequency and damping ratio of the isolation system, respectively. Similarly, the effectiveness of the isolation system may be evaluated from the reduction factor $SA(\omega_s,\xi_s)/SA(\omega_b,\xi_b)$.

Because of its accuracy, this rigid-structure approximation provides a rapid way to estimate the effectiveness of a base-isolation system and estimate the magnitude of the isolator deformation. For this reason, as it will be seen in Chapter 17, code provisions use this approach for the design of base-isolated buildings.

15.5.3 Linear Multi-Degree-of-Freedom Model

The concepts introduced in Section 15.5.1 may be extended to the case when the structure is modeled as a multi-degree-of-freedom system as follows:

15.5.3.1 Equations of Motion

Consider the isolated multi-degree-of-freedom system shown in Figure 15.23 and assume that the structure possesses n degrees of freedom and that the structure and the isolation system are both viscously damped linear systems. Let $[M]$, $[C]$, and $[K]$ represent the mass, damping, and stiffness matrices of the structure when considered fixed at its base, and m_b, c_b, and k_b the mass, stiffness, and damping constant of the isolation system, respectively. Similarly, let u_i, $i = 1, 2, ..., n$, denote the absolute displacements of the structural masses, u_b the absolute displacement of the isolation

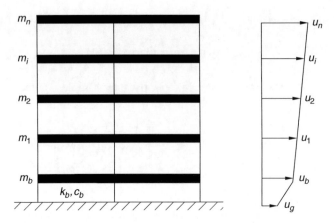

FIGURE 15.23 Multi-degree-of-freedom model of base-isolated building and associated displacements.

mass, and u_g the displacement of the ground, all at any time t. Also, let $v_i = u_i - u_b$ represent the displacement of the ith structural mass with respect to the isolation mass, and $v_b = u_b - u_g$ the displacement of the isolation mass with respect to the ground. By analogy with the equation of motion for a fixed-base multi-degree-of-freedom structure (see Section 10.2.2), the equation of motion of the isolated structure when the structural displacements are referred to the level of the isolation mass may be written as

$$[M]\{\ddot{v}\} + [C]\{\dot{v}\} + [K]\{v\} = -[M]\{r\}\ddot{u}_b \tag{15.100}$$

or as

$$[M]\{r\}\ddot{v}_b + [M]\{\ddot{v}\} + [C]\{\dot{v}\} + [K]\{v\} = -[M]\{r\}\ddot{u}_g \tag{15.101}$$

where $\{v\} = \{v_1\ v_2\ \ldots\ v_n\}^T$ is a vector that contains the displacements of the structure relative to the isolation mass, and $\{r\}$ is an influence vector that couples the degrees of freedom of the fixed base structure to the motion of the isolation mass.

With reference to the free-body diagram shown in Figure 15.24b and by application of Newton's second law, the equation of motion of the isolation mass is given by

$$-k_b v_b - c_b \dot{v}_b - \Sigma m_i \ddot{u}_i = m_b \ddot{u}_b \tag{15.102}$$

which may also be put into the form

$$-k_b v_b - c_b \dot{v}_b - \Sigma m_i (\ddot{v}_i + \ddot{v}_b + \ddot{u}_g) = m_b (\ddot{v}_b + \ddot{u}_g) \tag{15.103}$$

or

$$(m_b + \Sigma m_i)\ddot{v}_b + \Sigma m_i \ddot{v}_i + c_b \dot{v}_b + k_b v_b = -(m_b + \Sigma m_i)\ddot{u}_g \tag{15.104}$$

or

$$m\ddot{v}_b + \Sigma m_i \ddot{v}_i + c_b \dot{v}_b + k_b v_b = -m\ddot{u}_g \tag{15.105}$$

Seismic Protection with Base Isolation

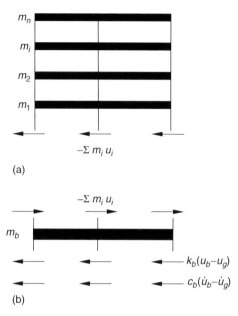

FIGURE 15.24 Free-body diagrams of (a) multi-degree-of-freedom structure and (b) isolation mass.

where $m = m_b + \Sigma m_i$ is the total mass of the isolated building. Thus, if Equations 15.101 and 15.105 are put together into a single matrix equation, one obtains

$$\begin{bmatrix} m & \{r\}^T[M] \\ [M]\{r\} & [M] \end{bmatrix} \begin{Bmatrix} \ddot{v}_b \\ \{\ddot{v}\} \end{Bmatrix} + \begin{bmatrix} c_b & \{0\}^T \\ \{0\} & [C] \end{bmatrix} \begin{Bmatrix} \dot{v}_b \\ \{\dot{v}\} \end{Bmatrix} + \begin{bmatrix} k_b & \{0\}^T \\ \{0\} & [K] \end{bmatrix} \begin{Bmatrix} v_b \\ \{v\} \end{Bmatrix} = -\begin{Bmatrix} m \\ [M]\{r\} \end{Bmatrix} \ddot{u}_g \quad (15.106)$$

$$= -\begin{bmatrix} m & \{r\}^T[M] \\ \{M\}\{r\} & [M] \end{bmatrix} \begin{Bmatrix} 1 \\ \{0\} \end{Bmatrix} \ddot{u}_g$$

where $\{0\}$ is a zero column vector of size n, and it has been considered that $[M]\{r\} = \{m_1\ m_2\ \ldots\ m_n\}^T$.

Thus, the equation of motion of the combined system may be written as

$$[M]_c \{\ddot{v}\}_c + [C]_c \{\dot{v}\}_c + [K]_c \{v\}_c = -[M]_c \{r\}_c \ddot{u}_g \quad (15.107)$$

where

$$[M]_c = \begin{bmatrix} m & \{r\}^T[M] \\ [M]\{r\} & [M] \end{bmatrix} \quad [C]_c = \begin{bmatrix} c_b & \{0\}^T \\ \{0\} & [C] \end{bmatrix} \quad (15.108)$$

$$[K]_c = \begin{bmatrix} k_b & \{0\}^T \\ \{0\} & [K] \end{bmatrix} \quad \{r\}_c = \begin{Bmatrix} 1 \\ \{0\} \end{Bmatrix} \quad (15.109)$$

and

$$\{v\}_c = \{v_b\ v_1\ \cdots\ v_n\}^T \quad (15.110)$$

15.5.3.2 Natural Frequencies and Mode Shapes

For the undamped, free-vibration case, Equation 15.107 is reduced to

$$[M]_c\{\ddot{v}\}_c + [K]_c\{v\}_c = \{0\} \tag{15.111}$$

and thus the natural frequencies and mode shapes of the isolated structure may be determined directly using any standard eigenvalue subroutine and considering that the mass and stiffness matrices of the system are given by

$$[M]_c = \begin{bmatrix} m & \{r\}^T[M] \\ [M]\{r\} & [M] \end{bmatrix} \quad [K]_c = \begin{bmatrix} k_b & \{0\}^T \\ \{0\} & [K] \end{bmatrix} \tag{15.112}$$

This process for finding the eigenproperties of an isolated building will be illustrated by means of the following example.

EXAMPLE 15.2 EIGENPROPERTIES OF BASE-ISOLATED MULTI-DEGREE-OF-FREEDOM STRUCTURE

Determine the natural frequencies and mode shapes of the four-story shear building shown in Figure E15.2a after it is base-isolated as shown in Figure E15.2b. The natural frequencies and unit-participation-factor mode shapes of the fixed-based structure are listed in Table E15.2a. The mass of the selected isolation system is 3 kip s²/in., and its natural period, when the structure is considered perfectly rigid, is 2 s.

Solution

When mounted on the isolation system, the original building may be considered as a shear building with five stories, with the mass and stiffness of the additional story being equal to

$$m_b = 3 \text{ kip s}^2/\text{in.}$$

$$k_b = \left(\frac{2\pi}{2}\right)^2 (11) = 108.57 \text{ kip/in.}$$

As such, its natural frequencies and mode shapes may be obtained by directly using any standard eigenvalue computer program. It is convenient, however, to formulate first the mass and stiffness matrices of the system as established in Section 15.5.3.1 so that the mode shapes of the building are obtained directly in terms of amplitudes relative to the isolation mass.

According to Equation 15.108, the mass matrix of the combined system is given by

$$[M]_c = \begin{bmatrix} m & \{r\}^T[M] \\ [M]\{r\} & [M] \end{bmatrix}$$

But in this case,

$$\{r\}^T = \{1 \quad 1 \quad 1 \quad 1\}$$

$$\{r\}^T[M] = \{3 \quad 2 \quad 2 \quad 1\}$$

Seismic Protection with Base Isolation

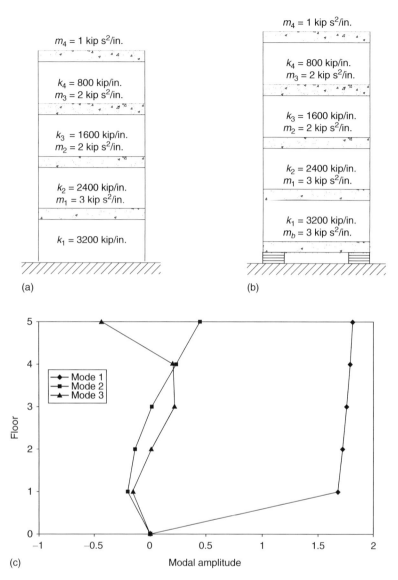

FIGURE E15.2 (a) Fixed-base and (b) base-isolated four-story shear buildings considered in Example 15.2. (c) First three mode shapes of base-isolated building.

TABLE E15.2a
Dynamic Properties of a Fixed-Base Building in Example 15.2

Mode	1	2	3	4
Frequency (Hz)	2.116	4.721	6.538	8.894
Mode shape	0.348	0.320	0.218	0.114
	0.735	0.395	0.049	−0.179
	1.153	0.073	−0.308	0.080
	1.480	−0.731	0.278	−0.028

and

$$[M] = \begin{bmatrix} 3 & 0 & 0 & 0 \\ 0 & 2 & 0 & 0 \\ 0 & 0 & 2 & 0 \\ 0 & 0 & 0 & 1 \end{bmatrix}$$

Therefore, the mass matrix of the combined system results as

$$[M]_c = \begin{bmatrix} 11 & 3 & 2 & 2 & 1 \\ 3 & 3 & 0 & 0 & 0 \\ 2 & 0 & 2 & 0 & 0 \\ 2 & 0 & 0 & 2 & 0 \\ 1 & 0 & 0 & 0 & 1 \end{bmatrix}$$

Similarly and according to Equation 15.109, the stiffness matrix of the combined system is of the form

$$[K]_c = \begin{bmatrix} k_b & \{0\}^T \\ \{0\} & [K] \end{bmatrix}$$

But for the building and isolation system under consideration, k_b is as indicated earlier and $[K]$ is equal to

$$[K] = \begin{bmatrix} 5600 & -2400 & 0 & 0 \\ -2400 & 4000 & -1600 & 0 \\ 0 & -1600 & 2400 & -800 \\ 0 & 0 & -800 & 800 \end{bmatrix}$$

Consequently, for the combined system, one has that

$$[K]_c = \begin{bmatrix} 108.57 & 0 & 0 & 0 & 0 \\ 0 & 5600 & -2400 & 0 & 0 \\ 0 & -2400 & 4000 & -1600 & 0 \\ 0 & 0 & -1600 & 2400 & -800 \\ 0 & 0 & 0 & -800 & 800 \end{bmatrix}$$

The substitution of the matrices $[M]_c$ and $[K]_c$ into Equation 15.111 and the solution of the resulting equation using an eigenvalue computer program leads to the natural frequencies and mode shapes listed in Table E15.2b, where, in accordance with Equation 15.110, the modal amplitudes corresponding to the isolation mass are given relative to the ground, and the modal amplitudes corresponding to the building masses are given relative to the isolation mass. It should also be noted that each of the mode shapes in Table E15.2b has been normalized so as to yield a unit participation factor. Figure E15.2c shows a graphical representation of the first three mode shapes of the isolated structure.

In comparison with the fixed-base case, it may be seen that the isolation system has a large effect on the natural frequencies of the building in its first two modes, but only a small effect on the natural frequencies of the higher modes. It may also be seen that the natural frequency in the first mode of the isolated building is only slightly below the natural frequency of the isolation system when the

TABLE E15.2b
Dynamic Properties of Isolated Building in Example 15.2

Mode	1	2	3	4	5
Frequency (Hz)	0.492	3.109	5.475	7.234	9.183
Mode shape	1.678	−0.200	−0.157	−0.224	−0.096
	0.042	0.065	0.169	0.426	0.298
	0.077	0.216	0.377	0.472	−0.143
	0.109	0.434	0.364	−0.100	0.192
	0.131	0.648	−0.274	0.429	0.066

structure is considered perfectly rigid. As in the case of the single-degree-of-freedom system, it may be seen, further, that in the first mode of the isolated building the deformation of the isolation system is large but the deformations in the structure are small, nearly zero. In the higher modes, the deformations in the structure are of the same order of magnitude as the deformation in the isolation system. However, these deformations are all relatively small. Finally, observe that the beneficial effect of the isolation system is the nearly zero deformations of the structure in its fundamental mode of vibration.

15.5.3.3 Approximate Maximum Response

Let $\{\phi\}_i$ and η_i, $i = 1, 2, \ldots, n$, denote the mode shapes of the fixed-based structure and a set of unknown modal coordinates, respectively. In addition, let $[\phi]$ and $\{\eta\}$ be a modal matrix and a vector of modal coordinates, respectively, defined by

$$[\phi] = [\{\phi\}_1 \ \{\phi\}_2 \cdots \{\phi\}_2] \quad \{\eta\}^T = \{\eta_1 \ \eta_2 \cdots \eta_n\} \tag{15.113}$$

Under the modal transformation

$$\{v\} = [\phi]\{\eta\} \tag{15.114}$$

and after premultiplication by the transpose of $[\phi]$, Equation 15.100 may be expressed as

$$[\phi]^T[M][\phi]\{\ddot{\eta}\} + [\phi]^T[C][\phi]\{\dot{\eta}\} + [\phi]^T[K][\phi]\{\eta\} = -[\phi]^T[M]\{r\}\ddot{u}_b \tag{15.115}$$

which, in view of the orthogonality properties of the mode shapes $\{\phi\}_i$, and if it is assumed that these mode shapes are also orthogonal with respect to the damping matrix $[C]$, may be reduced to the system of decoupled equations

$$M_i\ddot{\eta}_i + C_i\dot{\eta}_i + K_i\eta_i = -L_i\ddot{u}_b \quad i = 1,2,\ldots,n \tag{15.116}$$

where

$$M_i = \{\phi\}_i^T[M]\{\phi\}_i \quad C_i = \{\phi\}_i^T[C]\{\phi\}_i \quad K_i = \{\phi\}_i^T[K]\{\phi\}_i \quad L_i = \{\phi\}_i^T[M]\{r\} \tag{15.117}$$

Furthermore, by dividing through by M_i and considering that $u_b = v_b + u_g$, the system of equations may be put into the form

$$\ddot{\eta}_i + 2\xi_{si}\omega_{si}\dot{\eta}_i + \omega_{si}^2 \eta_i = -\Gamma_i(\ddot{v}_b + \ddot{u}_g) \quad i = 1, 2, \ldots, n \tag{15.118}$$

where ω_{si} and ξ_{si} denote the natural frequency and damping ratio in the ith mode of the fixed-base structure, respectively, and $\Gamma_i = L_i/M_i$ is the corresponding participation factor.

If the modal transformation expressed by Equation 15.114 is applied now to Equation 15.105, and if it is considered once again that $[M]\{r\} = \{m_1\ m_2\ \ldots\ m_n\}^T$, one arrives at

$$m\ddot{v}_b + \{r\}^T[M][\phi]\{\ddot{\eta}\} + c_b\dot{v}_b + k_b v_b = -m\ddot{u}_g \tag{15.119}$$

which may also be expressed as

$$\sum_{i=1}^{n} \{r\}^T[M]\{\phi\}_i \ddot{\eta}_i + m\ddot{v}_b + c_b\dot{v}_b + k_b v_b = -m\ddot{u}_g \tag{15.120}$$

or, after dividing through by m and considering that $\{r\}^T[M]\{\phi\}_i = L_i$ and $\Gamma_i = L_i/M_i$, as

$$\sum_{i=1}^{n} \frac{\Gamma_i M_i}{m} \ddot{\eta}_i + \ddot{v}_b + 2\xi_b\omega_b\dot{v}_b + \omega_b^2 v_b = -\ddot{u}_g \tag{15.121}$$

where, as previously established, ω_b and ξ_b represent the natural frequency and damping ratio of the isolation system when the structure is considered perfectly rigid, respectively.

Equations 15.118 and 15.121 constitute the equations of motion of the isolated structure in modal coordinates. If, as an approximation, only the influence of the fundamental mode of the fixed-base structure is considered, these equations of motion may be simplified into

$$\ddot{\eta}_1 + 2\xi_{s1}\omega_{s1}\dot{\eta}_1 + \omega_{s1}^2 \eta_1 = -\Gamma_1(\ddot{v}_b + \ddot{u}_g) \tag{15.122}$$

$$\frac{\Gamma_1 M_1}{m} \ddot{\eta}_1 + \ddot{v}_b + 2\xi_b\omega_b\dot{v}_b + \omega_b^2 v_b = -\ddot{u}_g \tag{15.123}$$

which, for convenience, may be rewritten as

$$\Gamma_1 \ddot{v}_b + \frac{\Gamma_1^2 M_1}{m} \ddot{\eta}_1 + 2\xi_b\omega_b\Gamma_1\dot{v}_b + \omega_b^2 \Gamma_1 v_b = -\Gamma_1 \ddot{u}_g \tag{15.124}$$

$$\Gamma_1 \ddot{v}_b + \ddot{\eta}_1 + 2\xi_{s1}\omega_{s1}\dot{\eta}_1 + \omega_{s1}^2 \eta_1 = -\Gamma_1 \ddot{u}_g \tag{15.125}$$

By comparison with Equations 15.29 and 15.30, the equations of motion of the isolated one-degree-of-freedom system discussed in Section 15.5.1, it may be seen that the two sets of equations are equivalent if it is considered that

$$\gamma = \frac{\Gamma_1^2 M_1}{m} \tag{15.126}$$

and v_s, v_b, ω_s, ξ_s, and $\ddot{u}_g$ in the equations for the two-degree-of-freedom model are substituted by η_1, $\Gamma_1 v_b$, ω_{s1}, ξ_{s1}, and $\Gamma_1 \ddot{u}_g$, respectively. This equivalency suggests, thus, that the response of an isolated

Seismic Protection with Base Isolation

multi-degree-of-freedom system may be determined approximately using the equations derived for the single-degree-of-freedom system. Specifically, the relative displacement of the isolation mass with respect to the ground may be determined using Equation 15.89, provided $\Gamma_1 v_b$, $\Gamma_1 SD(\omega_1,\xi_1)$, and $\Gamma_1 SD(\omega_2,\xi_2)$ are substituted for v_b, $SD(\omega_1,\xi_1)$, and $SD(\omega_2,\xi_2)$. That is, this relative displacement may be determined from

$$(\Gamma_1 v_b)_{max} = \Gamma_1[(1-\gamma\varepsilon)^2 SD^2(\omega_1,\xi_1) + \gamma^2\varepsilon^2 SD^2(\omega_2,\xi_2)]^{1/2} \tag{15.127}$$

where γ is defined by Equation 15.126; ω_1, ω_2, ξ_1, and ξ_2 are given by Equations 15.44, 15.46, 15.80, and 15.81; and

$$\varepsilon = \frac{\omega_b^2}{\omega_{s1}^2} \tag{15.128}$$

Similarly, the maximum value of the modal coordinate η_1 may be obtained from Equation 15.92 after substituting v_s by η_1, $SD(\omega_1,\xi_1)$ by $\Gamma_1 SD(\omega_1,\xi_1)$, and $SD(\omega_2,\xi_2)$ by $\Gamma_1 SD(\omega_2,\xi_2)$. This leads to

$$(\eta_1)_{max} = \varepsilon\Gamma_1[SD^2(\omega_1,\xi_1) + SD^2(\omega_2,\xi_2)]^{1/2} \tag{15.129}$$

In consequence and according to Equation 15.114 after all higher modes are neglected, the maximum displacements of the structure relative to the isolation mass may be considered approximately equal to

$$\{v\}_{max} = \{\phi\}_1 \eta_1 = \varepsilon[SD^2(\omega_1,\xi_1) + SD^2(\omega_2,\xi_2)]^{1/2} \Gamma_1 \{\phi\}_1 \tag{15.130}$$

Equation 15.130, in turn, may be used to estimate the maximum values of the lateral forces and base shear acting on the structure. That is, the lateral forces may be calculated according to

$$\{F_s\}_{max} = [K]\{v\}_{max} = [K]\{\phi\}_1(\eta_1)_{max} = \omega_{s1}^2[M]\{\phi\}_1(\eta_1)_{max} \tag{15.131}$$

and the base shear coefficient (base shear/weight of structure) according to

$$C_s = \frac{\{r\}^T\{F_s\}_{max}}{m_s g} = \omega_{s1}^2 \frac{\{r\}^T[M]\{\phi\}_1}{m_s g}(\eta_1)_{max} = \omega_{s1}^2 \frac{\Gamma_1 M_1}{m_s g}(\eta_1)_{max}$$

$$= \omega_{s1}^2 \frac{\Gamma_1^2 M_1}{m_s g} \varepsilon[SD^2(\omega_1,\xi_1) + SD^2(\omega_2,\xi_2)]^{1/2} \tag{15.132}$$

where m_s is the total mass of the structure above the isolation mass.

EXAMPLE 15.3 MAXIMUM RESPONSE OF BASE-ISOLATED MULTI-DEGREE-OF-FREEDOM STRUCTURE

Determine using the approximate method described in Section 15.5.3, the maximum displacements, lateral forces, and base shear coefficient of the isolated building considered in Example 15.2 when the building is subjected to an earthquake excitation represented by the 5% damping design spectrum specified in Example 15.1. Assume a damping ratio of 2% for the fundamental mode of the fixed-base structure and 10% for the isolation system. As in Example 15.1, the spectral ordinates for damping ratios other than 5% may be estimated by multiplying the ordinates in the spectrum for 5% damping by the $H(\xi)$ factor specified therein.

Solution

For the fixed-base building considered in Example 15.2, one has that

$$\omega_{s1} = 2\pi(2.116) = 13.295 \text{ rad/s} \quad \xi_{s1} = 0.02 \quad \omega_b = \frac{2\pi}{2.0} = 3.142 \text{ rad/s} \quad \xi_b = 0.10$$

$$\Gamma_1 = 1 \quad m_s = 8 \text{ kip s}^2/\text{in.} \quad m = 11 \text{ kip s}^2/\text{in.}$$

$$M_1 = \{\phi\}_1^T[M]\{r\} = \{0.348 \quad 0.735 \quad 1.153 \quad 1.480\}\begin{Bmatrix} 3 \\ 2 \\ 2 \\ 2 \\ 1 \end{Bmatrix} = 6.300 \text{ kip s}^2/\text{in.}$$

and, according to Equations 15.126 and 15.128,

$$\gamma = \frac{\Gamma_1^2 M_1}{m} = \frac{(1.0)^2(6.300)}{11.0} = 0.573$$

$$\varepsilon = \frac{\omega_b^2}{\omega_{s1}^2} = \left(\frac{3.142}{13.295}\right)^2 = 0.056$$

Therefore, from Equations 15.44 and 15.46, the square of the natural frequencies in the first two modes of the isolated building are approximately equal to

$$\omega_1^2 = \omega_b^2(1 - \gamma\varepsilon) = (3.142)^2[1 - (0.573)(0.056)] = 9.555 \text{ rad}^2/\text{s}^2$$

$$\omega_2^2 = \frac{\omega_{s1}^2}{(1-\gamma)}(1 + \gamma\varepsilon) = \frac{(13.295)^2}{(1-0.573)}[1 + (0.573)(0.056)] = 427.234 \text{ rad}^2/\text{s}^2$$

which, incidentally, compare well with the "exact" values obtained in Example E15.2. Similarly, using Equations 15.80 and 15.81, the damping ratios in such two modes are

$$\xi_1 = \xi_b\left(1 - \frac{3}{2}\gamma\varepsilon\right) = 0.10[1 - 1.5(0.573)(0.056)] = 0.095$$

$$\xi_2 = \frac{\xi_{s1} + \gamma\varepsilon^{1/2}\xi_b}{(1-\gamma)^{1/2}}\left(1 - \frac{1}{2}\gamma\varepsilon\right) = \frac{0.02 + (0.573)\sqrt{0.056}(0.10)}{\sqrt{1-0.573}}[1 - 0.5(0.573)(0.056)] = 0.051$$

Moreover, from the specified design spectrum and considering the $H(\xi)$ factor to convert into spectral ordinates for damping ratios other than 5%, the spectral accelerations corresponding to those two modes are

$$SA(3.09, 0.095) = H(0.095)SA(3.09, 0.05) = \left[\frac{1.5}{40(0.095)+1} + 0.5\right]\frac{1.37}{2.03}g = 0.55g$$

$$SA(20.67, 0.051) = H(0.051)SA(3.08, 0.05) = \left[\frac{1.5}{40(0.051)+1} + 0.5\right]1.37g = 1.36g$$

Seismic Protection with Base Isolation

In consequence and according to Equation 15.127, the maximum displacement of the isolation mass under the specified excitation is

$$(v_b)_{max} = [(1-\gamma\varepsilon)^2 SD^2(\omega_1,\xi_1) + \gamma^2\varepsilon^2 SD^2(\omega_2,\xi_2)]^{1/2}$$

$$= \left\{[1-(0.573)(0.056)]^2\left(\frac{0.55g}{3.09^2}\right)^2 + (0.573)^2(0.056)^2\left(\frac{1.36g}{20.67^2}\right)^2\right\}^{1/2}$$

$$= [21.544^2 + 0.039^2]^{1/2} = 21.5 \text{ in.}$$

Similarly and in view of Equation 15.129, the maximum value of the modal coordinate in the first mode of the isolated building results as

$$(\eta_1)_{max} = \varepsilon\Gamma_1[SD^2(\omega_1,\xi_1) + SD^2(\omega_2,\xi_2)]^{1/2}$$

$$= 0.056(1.0)\left[\left(\frac{0.55g}{3.09^2}\right)^2 + \left(\frac{1.36g}{20.67^2}\right)^2\right]^{1/2} = 0.00323g$$

$$= 1.248 \text{ in.}$$

which, on substitution into Equations 15.130 and 15.131, one finds that the maximum structural displacements and lateral forces are approximately equal to

$$\{v\}_{max} = \{\phi\}_1\eta_1 = 1.248\begin{Bmatrix}0.348\\0.735\\1.153\\1.480\end{Bmatrix} = \begin{Bmatrix}0.434\\0.917\\1.439\\1.847\end{Bmatrix} \text{ in.}$$

$$\{F_s\}_{max} = \omega_{s1}^2[M]\{\phi\}_1(\eta_1)_{max} = (13.295)^2\begin{bmatrix}3.0 & 0 & 0 & 0\\0 & 2.0 & 0 & 0\\0 & 0 & 2.0 & 0\\0 & 0 & 0 & 1.0\end{bmatrix}\begin{Bmatrix}0.348\\0.735\\1.153\\1.480\end{Bmatrix}1.248 = \begin{Bmatrix}230.3\\324.3\\508.7\\326.5\end{Bmatrix} \text{ kips}$$

In turn, after dividing by the respective masses, these lateral forces lead to the following maximum floor accelerations:

$$\{\ddot{u}\}_{max} = \begin{Bmatrix}230.3/3.0\\324.3/2.0\\508.7/2.0\\326.5/1.0\end{Bmatrix} = \begin{Bmatrix}76.8\\162.1\\254.3\\326.5\end{Bmatrix} \text{ in./s}^2 = \begin{Bmatrix}0.20\\0.42\\0.66\\0.84\end{Bmatrix}g$$

Finally and according to Equation 15.132, the base shear coefficient results as

$$C_s = \omega_{s1}^2\frac{\Gamma_1 M_1}{m_s g}(\eta_1)_{max} = (13.295)^2\frac{1.0(6.3)}{8.0g}0.00323g = 0.45$$

For comparison, the base shear coefficient for the fixed-base structure is

$$C_s = \omega_{s1}^2 \frac{\Gamma_1 M_1}{m_s g}(\eta_1)_{max} = \frac{\Gamma_1^2 M_1}{m_s g} SA(13.295, 0.02)$$

$$= \frac{\Gamma_1^2 M_1}{m_s g}\left(\frac{1.5}{40\xi + 1} + 0.5\right) SA(13.295, 0.05)$$

$$= \frac{(1.0)^2(6.3)}{8.0\,g}\left[\frac{1.5}{40(0.02) + 1} + 0.5\right] 1.37g = 1.44$$

It may be seen, thus, that the introduction of an isolation system with a long natural period also reduces significantly the shear force at the base of a multi-degree-of-freedom structure. For the structure considered in this example, the reduction factor is of the order of 1.44/0.45 = 3.2.

15.5.4 Linear Multi-Degree-of-Freedom Model with Nonclassical Damping

In many base-isolated structures, supplemental dampers are added to the isolation system with the purpose of reducing the displacement of the isolation mass. The damping ratios of the isolation system in these cases may be as high as 30%. However, when damping in the isolation system is much higher than the damping in the structure, it cannot be assumed that the combined isolation–structure system is classically damped. That is, it cannot be assumed that after a modal transformation with the undamped mode shapes the damping matrix of the system is transformed into a diagonal matrix. The reason is that in such a case the off-diagonal elements of the transformed damping matrix are of the same order of magnitude as the corresponding diagonal elements and, hence, these off-diagonal elements cannot be considered negligibly small. In such instances, therefore, the analysis of the isolated structure ought to be carried out by a step-by-step integration of the equations of motion or by a complex modal superposition. In this section, the solution based on a complex modal superposition is described.

Consider once again the linear base-isolated structure in Figure 15.23. As shown in Section 15.5.3.1, the equation of motion for this system when the structural displacements are expressed relative to the displacement of the isolation mass is of the form

$$[M]_c\{\ddot{v}(t)\}_c + [C]_c\{\dot{v}(t)\}_c + [K]_c\{v(t)\}_c = -[M]_c\{r\}_c \ddot{u}_g(t) \quad (15.133)$$

where $[M]_c$, $[C]_c$, $[K]_c$, $\{r\}_c$, and $\{v\}_c$ are given by Equations 15.108, 15.109, and 15.110, and denote the mass matrix, damping matrix, stiffness matrix, influence vector, and displacement vector of the system, respectively. As previously established, $\ddot{u}_g(t)$ represents ground acceleration. According to the method suggested by K. A. Foss in 1952 and described in the text by Hurty and Rubinstein listed in the Further Readings section, a modal decomposition is possible if the equation is first reduced to a first-order equation of the form

$$[A]\{\dot{q}(t)\} + [B]\{q(t)\} = \{Q(t)\} \quad (15.134)$$

where

$$[A] = \begin{bmatrix} [0] & [M]_c \\ [M]_c & [C]_c \end{bmatrix} \quad [B] = \begin{bmatrix} -[M]_c & [0] \\ [0] & [K]_c \end{bmatrix} \quad (15.135)$$

$$\{q(t)\} = \begin{Bmatrix} \{\dot{v}(t)\}_c \\ \{v(t)\}_c \end{Bmatrix} \quad \{Q(t)\} = \begin{Bmatrix} \{0\} \\ -[M]_c\{r\}_c \ddot{u}_g(t) \end{Bmatrix} \quad (15.136)$$

Equation 15.134 is of size $2(n + 1) \times 2(n + 1)$, with n being the number of degrees of freedom of the fixed-base structure, and is usually referred to as the reduced equation of motion.

The solution of the homogeneous equation of motion; that is, the solution of

$$[A]\{\dot{q}(t)\} + [B]\{q(t)\} = \{0\} \tag{15.137}$$

is of the form

$$\{q(t)\} = \{s\}e^{\lambda t} \tag{15.138}$$

and, thus, substitution of this solution into Equation 15.137 leads to the eigenvalue problem

$$([B] + \lambda[A])\{s\} = 0 \tag{15.139}$$

whose solution, in turn, leads to a set of $2(n + 1)$ complex-valued eigenvalues λ_i and a set of $2(n + 1)$ complex-valued eigenvectors $\{s\}_i$. It can be shown that when the system is underdamped, these eigenvalues and eigenvectors result in pairs of complex conjugates. It can also be shown that the resulting eigenvectors are orthogonal with respect to the matrices $[A]$ and $[B]$. That is,

$$\{s\}_i^T[A]\{s\}_j = 0 \quad i \neq j \qquad \{s\}_i^T[B]\{s\}_j = 0 \quad i \neq j \tag{15.140}$$

Furthermore, it can be shown that the eigenvalues λ_i and eigenvectors $\{s\}_i$ may be written alternatively as

$$\lambda_i = -\xi_i \omega_i + i\omega_i \sqrt{1 - \xi_i^2} = -\xi_i \omega_i + i\omega_{di} \tag{15.141}$$

$$\{s\}_i = \begin{Bmatrix} \lambda_i \{w\}_i \\ \{w\}_i \end{Bmatrix} \tag{15.142}$$

where ω_i, ω_{di}, and ξ_i represent the natural frequency, damped natural frequency, and damping ratio in the ith mode of the system, respectively; i denotes the unit imaginary number; and $\{w\}_i$ is a complex-valued mode shape of size $(n + 1)$ that defines the relative amplitudes and phase angles of the various masses of the system when the system vibrates freely in its ith mode.

Because of the orthogonality of the eigenvectors $\{s\}_i$ with respect to the matrices $[A]$ and $[B]$, the matrix that contains all such eigenvectors represents a transformation matrix that decouples the reduced equation of motion. Accordingly, if $[s]$ represents the matrix that contains the $2(n + 1)$ eigenvectors of the system, and if $\{z(t)\}$ is a vector of unknown modal coordinates, under the transformation

$$\{q(t)\} = [s]\{z(t)\} = \sum_{i=1}^{2(n+1)} \{s\}_i z_i(t) \tag{15.143}$$

and after premultiplication by the transpose of $[s]$ and the use of the aforementioned orthogonality properties, Equation 15.134 may be transformed into the set of independent equations

$$A_i \dot{z}_i(t) + B_i z_i(t) = Q_i(t) \quad i = 1, 2, \ldots, 2(n+1) \tag{15.144}$$

where z_i is the ith component of $\{z\}$ and A_i, B_i, and $Q_i(t)$ are complex-valued scalars defined by

$$A_i = \{s\}_i^T [A] \{s\}_i \tag{15.145}$$

$$B_i = \{s\}_i^T [B] \{s\}_i \tag{15.146}$$

$$Q_i(t) = \{s\}_i^T \{Q(t)\}_i \tag{15.147}$$

Equation 15.144 is a first-order differential equation with constant coefficients. As such, its solution may be obtained in terms of Duhamel's integral. Accordingly and under the assumption of zero initial conditions, one has that

$$z_i(t) = \frac{1}{A_i} \int_{\tau=0}^{t} Q_i(\tau) e^{\lambda_i (t-\tau)} d\tau \tag{15.148}$$

Equations 15.143 and 15.148 constitute the modal solution of Equation 15.134. Equation 15.143, however, may be written explicitly in terms of complex conjugates as

$$\{q(t)\} = \sum_{i=1}^{n+1} \{s\}_i z_i(t) + \sum_{i=1}^{n+1} \{\bar{s}\}_i \bar{z}_i(t) \tag{15.149}$$

or as

$$\{q(t)\} = 2 \sum_{i=1}^{n+1} \text{Re}[\{s\}_i z_i(t)] \tag{15.150}$$

where a bar above a variable indicates its complex conjugate, and "Re" reads as "the real part of." Furthermore, in view of the first of Equations 15.136 and Equation 15.142, it is possible to express Equation 15.150 alternatively as

$$\begin{Bmatrix} \{\dot{v}(t)\}_c \\ \{v(t)\}_c \end{Bmatrix} = 2 \sum_{i=1}^{n+1} \text{Re} \left[\begin{Bmatrix} \lambda_i \{w\}_i \\ \{w\}_i \end{Bmatrix} z_i(t) \right] \tag{15.151}$$

whose lower part leads to the following explicit solution of Equation 15.133:

$$\{v(t)\}_c = 2 \sum_{i=1}^{n+1} \text{Re}[\{w\}_i z_i(t)] \tag{15.152}$$

Similarly, the substitution of Equation 15.142 and the second of Equations 15.136 into Equation 15.147 yields

$$Q_i(t) = -\{w\}_i^T [M]_c \{r\}_c \ddot{u}_g(t) \tag{15.153}$$

whereas Equation 15.145 in combination with Equations 15.135 and 15.142 leads to

$$A_i = \{w\}_i^T (2\lambda_i [M]_c + [C]_c) \{w\}_i \tag{15.154}$$

Seismic Protection with Base Isolation

which in view of Equation 15.141 may also be written as

$$A_i = 2(-\xi_i\omega_i + i\omega_{di})M_{ci} + C_{ci} \tag{15.155}$$

where M_{ci} and C_{ci} are a generalized mass and a generalized damping constant defined as

$$M_{ci} = \{w\}_i^T[M]_c\{w\}_i \tag{15.156}$$

$$C_{ci} = \{w\}_i^T[C]_c\{w\}_i \tag{15.157}$$

Furthermore, if C_{ci} is expressed in terms of the damping ratio and natural frequency in the ith mode of the system as

$$C_{ci} = 2\xi_i\omega_i M_{ci} \tag{15.158}$$

then A_i may be put into the form

$$A_i = 2(-\xi_i\omega_i + i\omega_{di})M_{ci} + 2\xi_i\omega_i M_{ci} = 2i\omega_{di}M_{ci} \tag{15.159}$$

From Equations 15.153 and 15.159, $z_i(t)$ may be thus expressed as

$$z_i(t) = \frac{1}{A_i}\int_{\tau=0}^{t} Q_i(\tau)e^{\lambda_i(t-\tau)}d\tau = \frac{\Gamma_{ci}}{2i\omega_{di}}\int_{\tau=0}^{t}\ddot{u}_g(\tau)e^{\lambda_i(t-\tau)}d\tau \tag{15.160}$$

where Γ_{ci} is a complex participation factor defined as

$$\Gamma_{ci} = \frac{L_{ci}}{M_{ci}} = \frac{\{w\}_i^T[M]_c\{r\}_c}{\{w\}_i^T[M]_c\{w\}_i} \tag{15.161}$$

Correspondingly, Equation 15.152 may be written as

$$\{v(t)\}_c = -\sum_{i=1}^{n+1}\text{Re}\left[\frac{1}{\omega_{di}}\{w'\}_i\int_{\tau=0}^{t}\ddot{u}_g(\tau)e^{\lambda_i(t-\tau)}d\tau\right] \tag{15.162}$$

where

$$\{w'\}_i = i\Gamma_{ci}\{w\}_i \tag{15.163}$$

However, if $\{w'\}_i$ and λ_i are expressed explicitly in terms of their real and imaginary parts, and if the exponential function is expanded in terms of its sine and cosine components, one has that

$$\{v(t)\}_c = -\sum_{i=1}^{n+1}\text{Re}\left\{\frac{1}{\omega_{di}}(\{u'\}_i + i\{v'\}_i)\int_{\tau=0}^{t}\ddot{u}_g(\tau)e^{-\xi_i\omega_i(t-\tau)}[\cos\omega_{di}(t-\tau) + i\sin\omega_{di}(t-\tau)]d\tau\right\}$$

$$= -\sum_{i=1}^{n+1}\left[\frac{1}{\omega_{di}}(\{u'\}_i\int_{\tau=0}^{t}\ddot{u}_g(\tau)e^{-\xi_i\omega_i(t-\tau)}\cos\omega_{di}(t-\tau)d\tau \right.$$

$$\left. -\frac{1}{\omega_{di}}(\{v'\}_i\int_{\tau=0}^{t}\ddot{u}_g(\tau)e^{-\xi_i\omega_i(t-\tau)}\sin\omega_{di}(t-\tau)d\tau\right] \tag{15.164}$$

where $\{u'\}_i$ and $\{v'\}_i$ represent the real and imaginary parts of $\{w'\}_i$, respectively. Also, according to Equations 6.4 and 6.5, the two integrals in Equation 15.164 may be expressed as

$$\int_{\tau=0}^{t} \ddot{u}_g(\tau) e^{-\xi_i \omega_i (t-\tau)} \cos\omega_{di}(t-\tau) d\tau = -\dot{x}_i(t) - \xi_i \omega_i x_i(t) \quad (15.165)$$

$$\int_{\tau=0}^{t} \ddot{u}_g(\tau) e^{-\xi_i \omega_i (t-\tau)} \sin\omega_{di}(t-\tau) d\tau = -\omega_{di} x_i(t) \quad (15.166)$$

where $x_i(t)$ and $\dot{x}_i(t)$ denote the relative displacement and relative velocity response of a single-degree-of-freedom system with natural frequency ω_i and damping ratio ξ_i at time t, respectively, when the base of the system is subjected to a ground acceleration $\ddot{u}_g(t)$. Therefore, Equation 15.164 may be written in terms of such displacement and velocity responses as

$$\{v(t)\}_c = \sum_{i=1}^{n+1} \left\{ \frac{1}{\omega_{di}} \{u'\}_i [\dot{x}_i(t) + \xi_i \omega_i x_i(t)] - \{v'\}_i x_i(t) \right\} \quad (15.167)$$

or as

$$\{v(t)\}_c = \sum_{i=1}^{n+1} \{v(t)\}_{ci} = \sum_{i=1}^{n+1} [\{a'\}_i \dot{x}_i(t) - \{b'\}_i x_i(t)] \quad (15.168)$$

where $\{v(t)\}_{ci}$ denotes a vector that contains the displacements of the system in its ith mode, and

$$\{a'\}_i = \frac{1}{\omega_{di}} \{u'\}_i \quad \{b'\}_i = \{v'\}_i - \frac{\xi_i}{\sqrt{1-\xi_i^2}} \{u'\}_i \quad (15.169)$$

The maximum values of the displacement $x_i(t)$ and the velocity $\dot{x}_i(t)$ may be determined from the displacement and velocity response spectra of the ground acceleration $\ddot{u}_g(t)$. However, because the maximum values of $x_i(t)$ and $\dot{x}_i(t)$ do not occur at the same time, the maximum values of the modal displacements $\{v(t)\}_{ci}$ cannot be determined directly from those response spectra. It is possible, nonetheless, to obtain an approximation to such maximum values using the response spectra in question if the $x_i(t)$ and $\dot{x}_i(t)$ terms in Equation 15.168 are combined on the basis of the square root of the sum of the squares (see Section 10.3.2). That is, the maximum values of the displacements in each of the modes of the system may be approximated as

$$\max\{v(t)\}_{ci} = [\{a'^2\}_i SV_i^2 + \{b'^2\}_i SD_i^2]^{1/2} \quad (15.170)$$

where $SV_i = SV(\omega_i,\xi_i)$ and $SD_i = SD(\omega_i,\xi_i)$ denote the ordinates corresponding to a natural frequency ω_i and damping ratio ξ_i in the velocity and displacement response spectra of $\ddot{u}_g(t)$, respectively. In turn, such maximum modal responses may be combined again using the square-root-of-the-sum-of-the-squares rule to obtain estimates of the maximum displacements. That is, these maximum displacements may be calculated according to

$$\max\{v(t)\}_c = \left[\sum_{i=1}^{n+1} \max\{v^2(t)\}_{ci} \right]^{1/2} \quad (15.171)$$

It is worthwhile to recall that the applicability of the square-root-of-the-sum-of-the-squares rule is limited to systems with well-separated natural frequencies (see Section 10.3.2). Although most base-isolated structures possess well-separated natural frequencies, it is important to keep in mind that it is necessary to use a combination rule of the double-sum type whenever an isolated structure has, instead, closely spaced natural frequencies. To learn about modal combination rules applicable to systems with nonclassical damping refer to the paper by Maldonado and Singh listed in the Further Readings section.

Equation 15.170 is expressed in terms of a spectral displacement and a spectral velocity, and thus, its application requires knowing the velocity and displacement response spectra of the excitation. In many cases, however, the velocity response spectrum is not directly available. To overcome this problem in these cases, it is possible to use the empirical formula proposed by Sadek et al. and introduced in Section 10.3.2 (see Equation 10.79) to estimate spectral velocities in terms of the corresponding pseudovelocities. That is, if PSV_i denotes the pseudovelocity corresponding to the spectral velocity SV_i, one can consider that

$$SV_i = r_i PSV_i = r_i \omega_{di} SD_i \tag{15.172}$$

where

$$r_i = a_{vi} T_i^{b_{vi}} \tag{15.173}$$

in which

$$a_{vi} = 1.095 + 0.647\xi_i - 0.382\xi_i^2 \tag{15.174}$$

$$b_{vi} = 0.193 + 0.838\xi_i - 0.621\xi_i^2 \tag{15.175}$$

Correspondingly, by substitution of Equation 15.172 into Equation 15.170 and from Equation 15.169, the vector of maximum modal displacements may be calculated approximately from

$$\max\{v(t)\}_{ci} = SD_i [r_i^2 \{u'^2\}_i + \{b'^2\}_i]^{1/2} \tag{15.176}$$

Furthermore, in those cases only the acceleration response spectrum is available, one can determine the required spectral displacements from the equation that relates the spectral acceleration $SA_i = SA(\omega_i, \xi_i)$ to the spectral displacement SD_i and the spectral velocity SV_i. That is, from the equation

$$SA_i = \frac{\omega_{di}}{1-\xi_i^2} \left[\omega_{di}^2 SD_i^2 + 4\xi_i^2(1-\xi_i^2)SV_i^2\right]^{1/2} \tag{15.177}$$

which after incorporating into Equation 15.172 becomes

$$SA_i = \frac{\omega_{di}^2 SD_i}{1-\xi_i^2} \sqrt{1 + 4r_i^2 \xi_i^2 (1-\xi_i^2)} \tag{15.178}$$

Equation 15.177 is derived from Equations 6.4 through 6.6. It is obtained by expressing Equation 6.6 in terms of the displacements and velocities defined by Equations 6.4 and 6.5, and by combining such velocity and displacement terms using the square-root-of-the-sum-of-the-squares rule. Note that for small damping ratios, Equation 15.178 is reduced to the classical relationship $SA_i = \omega_i^2 SD_i$.

To determine the accelerations at the floors of the structure above the isolation system, it should be noted that in the case of high damping ratios the damping forces may not be negligibly small compared to the elastic forces and, thus, the floor accelerations cannot be considered approximately equal to the elastic forces divided by the corresponding masses. Instead, they need to be determined directly from the equation of motion—Equation 15.100. That is, in the case of high damping ratios, the absolute floor accelerations are given by

$$\{\ddot{u}(t)\} = \{\ddot{v}(t)\} + \{r\}_c \ddot{u}_b(t) = -[M]^{-1}([C]\{\dot{v}(t)\} + [K]\{v(t)\}) \tag{15.179}$$

which in terms of modal responses may also be expressed as

$$\{\ddot{u}(t)\} = [M]^{-1} \sum_{i=1}^{n+1} \{F_I(t)\}_i = -[M]^{-1} \sum_{i=1}^{n+1} ([C]\{\dot{v}(t)\}_i + [K]\{v(t)\}_i) \tag{15.180}$$

where $\{F_I\}_i$ is a vector of modal inertia forces equal to

$$\{F_I(t)\}_i = -[C]\{\dot{v}(t)\}_i - [K]\{v(t)\}_i \tag{15.181}$$

But according to Equation 15.168, $\{v(t)\}_i$ and its first derivative are given by

$$\{v(t)\}_i = \{a^*\}_i \dot{x}_i(t) - \{b^*\}_i x_i(t) \tag{15.182}$$

$$\{\dot{v}(t)\}_i = \{a^*\}_i \ddot{x}_i(t) - \{b^*\}_i \dot{x}_i(t) = \{a^*\}_i \ddot{y}_i(t) - \{a^*\}_i \ddot{u}_g(t) - \{b^*\}_i \dot{x}_i(t) \tag{15.183}$$

where $\{a^*\}_i$ and $\{b^*\}_i$ are defined by the n upper elements of $\{a'\}_i$ and $\{b'\}_i$, respectively, and $\ddot{y}_i(t) = \ddot{x}_i(t) + \ddot{u}_g(t)$ represents the absolute acceleration response of a single-degree-of-freedom system with natural frequency ω_i and damping ratio ξ_i to the ground acceleration $\ddot{u}_g(t)$. Thus, $\{F_I(t)\}_i$ may also be expressed as

$$\{F_I(t)\}_i = -[C][\{a^*\}_i \ddot{y}_i(t) - \{a^*\}_i \ddot{u}_g(t) - \{b^*\}_i \dot{x}_i(t)] - [K][\{a^*\}_i \dot{x}_i(t) - \{b^*\}_i x_i(t)] \tag{15.184}$$

or as

$$\{F_I(t)\}_i = -\{p\}_i \ddot{y}_i(t) + \{p\}_i \ddot{u}_g(t) + \{h\}_i \dot{x}_i(t) + \{q\}_i x_i(t) \tag{15.185}$$

where

$$\{p\}_i = [C]\{a^*\}_i \tag{15.186}$$

$$\{q\}_i = [K]\{b^*\}_i \tag{15.187}$$

$$\{h\}_i = [C]\{b^*\}_i - [K]\{a^*\}_i \tag{15.188}$$

An estimate of the maximum values of $\{F_I(t)\}_i$ can now be obtained by combining the maximum values of each of the four terms on the right-hand side of Equation 15.185 using the square-root-of-the-sum-of-the-squares rule. That is, the maximum values of $\{F_I(t)\}_i$ may be considered approximately

Seismic Protection with Base Isolation

given by

$$\max\{F_I(t)\}_i = \{\{p^2\}_i[\max \ddot{y}_i(t)]^2 + \{p^2\}_i[\max \ddot{u}_g(t)]^2 \\ + \{h^2\}_i[\max \dot{x}_i(t)]^2 + \{q^2\}_i[\max x_i(t)]^2\}^{1/2} \tag{15.189}$$

or by

$$\max\{F_I(t)\}_i = [\{p^2\}_i(SA_i^2 + PGA^2) + \{h^2\}_i SV_i^2 + \{q^2\}_i SD_i^2]^{1/2} \tag{15.190}$$

given that the maximum values of $x(t)$, $\dot{x}(t)$, and $\ddot{y}(t)$ are equal to the spectral ordinates SD_i, SV_i, and SA_i, respectively, and that the maximum value of $\ddot{u}_g(t)$ is the peak ground acceleration PGA or zero-period spectral acceleration. Similarly, the maximum modal values determined from Equation 15.190 may be combined using the same rule to obtain an estimate of the maximum values of the inertia forces $\{F_I(t)\}_c$ and, from Equation 15.180, an estimate of the maximum floor accelerations. Proceeding accordingly, one is led to

$$\max\{\ddot{u}(t)\} = [M]^{-1}\left[\sum_{i=1}^{n+1} \max\{F_I^2(t)\}_i\right]^{1/2} \tag{15.191}$$

As with the absolute accelerations, the shear at the base of the structure cannot be determined using only the elastic forces because in the case of high damping ratios the damping forces and the elastic forces may be of the same order of magnitude. In consequence, one needs to consider that in the case of high damping ratios the base shear is given by the sum of the base shear generated by the elastic forces and the base shear generated by the damping forces. Accordingly, the modal base shear at any time t is given by

$$V_i(t) = \{r\}^T[C]\{\dot{v}(t)\}_i + \{r\}^T[K]\{v(t)\}_i = \{r\}^T([C]\{\dot{v}(t)\}_i + [K]\{v(t)\}_i) \tag{15.192}$$

where $\{r\}$ is the influence vector defined in Section 15.5.3. But in view of Equation 15.181, Equation 15.192 may also be expressed as

$$V_i(t) = -\{r\}^T\{F_I(t)\}_i \tag{15.193}$$

and, hence, the maximum value of this modal base shear may be calculated as

$$\max V_i(t) = \{r\}^T \max\{F_I(t)\}_i \tag{15.194}$$

where $\max\{F_I(t)\}_i$ is given by Equation 15.190. Similarly, the maximum base shear may be considered approximately given by

$$\max V(t) = \left[\sum_{i=1}^{n+1} \max V_i(t)\right]^{1/2} \tag{15.195}$$

Consequently, the base shear coefficient, defined once again as base shear over structural weight, may be determined according to

$$C_s = \frac{\max V(t)}{W_s} \tag{15.196}$$

where, as previously established, W_s represents the total weight of the fixed-base structure.

EXAMPLE 15.4 MAXIMUM RESPONSE OF BASE-ISOLATED STRUCTURE WITH HIGH ISOLATION DAMPING

Determine using the method described in Section 15.5.4, the maximum displacements, maximum floor accelerations, and base shear coefficient of the isolated building analyzed in Example 15.3. Assume that the damping ratio of the isolation system is now 50% and that the damping matrix of the fixed-base structure is proportional to its stiffness matrix. Consider only the contribution of the first three modes.

Solution

From the results obtained in Example 15.2, the mass and stiffness matrices of the isolated structure under consideration are

$$[M]_c = \begin{bmatrix} 11 & 3 & 2 & 2 & 1 \\ 3 & 3 & 0 & 0 & 0 \\ 2 & 0 & 2 & 0 & 0 \\ 2 & 0 & 0 & 2 & 0 \\ 1 & 0 & 0 & 0 & 1 \end{bmatrix} \quad [K]_c = \begin{bmatrix} 108.57 & 0 & 0 & 0 & 0 \\ 0 & 5600 & -2400 & 0 & 0 \\ 0 & -2400 & 4000 & -1600 & 0 \\ 0 & 0 & -1600 & 2400 & -800 \\ 0 & 0 & 0 & -800 & 800 \end{bmatrix}$$

To determine the damping matrix of the system, it is noted, first, that since the damping ratio in the fundamental mode of the fixed-base structure is 2%, and since it is assumed that its damping matrix is proportional to its stiffness matrix, the corresponding proportionality constant is equal to (see Equation 10.95)

$$\beta = \frac{2\xi_{s1}}{\omega_{s1}} = \frac{2(0.02)}{2\pi(2.116)} = 0.003$$

Consequently, the damping matrix of the fixed-base structure results as

$$[C] = \beta[K] = 0.003 \begin{bmatrix} 5600 & -2400 & 0 & 0 \\ -2400 & 4000 & -1600 & 0 \\ 0 & -1600 & 2400 & -800 \\ 0 & 0 & -800 & 800 \end{bmatrix} = \begin{bmatrix} 16.8 & -7.2 & 0 & 0 \\ -7.2 & 12.0 & -4.8 & 0 \\ 0 & -4.8 & 7.2 & -2.4 \\ 0 & 0 & -2.4 & 2.4 \end{bmatrix}$$

It is noted, also, that according to Equation 15.28 and the specified damping ratio, the damping constant for the isolation system is given by

$$c_b = 2\xi_b \omega_b m = 2(0.50)\left(\frac{2\pi}{2.0}\right)(11.0) = 34.6 \text{ kip s/in.}$$

As a result and in view of the second of Equations 15.108, the damping matrix of the isolated structure is equal to

$$[C]_c = \begin{bmatrix} c_b & \{0\}^T \\ \{0\} & [C] \end{bmatrix} = \begin{bmatrix} 34.6 & 0 & 0 & 0 & 0 \\ 0 & 16.8 & -7.2 & 0 & 0 \\ 0 & -7.2 & 12.0 & -4.8 & 0 \\ 0 & 0 & -4.8 & 7.2 & -2.4 \\ 0 & 0 & 0 & -2.4 & 2.4 \end{bmatrix}$$

From Equation 15.135, one thus has that

$$[A] = \begin{bmatrix} [0] & [M]_c \\ [M]_c & [C]_c \end{bmatrix} = \begin{bmatrix} 0 & 0 & 0 & 0 & 0 & 11 & 3 & 2 & 2 & 1 \\ 0 & 0 & 0 & 0 & 0 & 3 & 3 & 0 & 0 & 0 \\ 0 & 0 & 0 & 0 & 0 & 2 & 0 & 2 & 0 & 0 \\ 0 & 0 & 0 & 0 & 0 & 2 & 0 & 0 & 2 & 0 \\ 0 & 0 & 0 & 0 & 0 & 1 & 0 & 0 & 0 & 1 \\ 11 & 3 & 2 & 2 & 1 & 34.6 & 0 & 0 & 0 & 0 \\ 3 & 3 & 0 & 0 & 0 & 0 & 16.8 & -7.2 & 0 & 0 \\ 2 & 0 & 2 & 0 & 0 & 0 & -7.2 & 12.0 & -4.8 & 0 \\ 2 & 0 & 0 & 2 & 0 & 0 & 0 & -4.8 & 7.2 & -2.4 \\ 1 & 0 & 0 & 0 & 1 & 0 & 0 & 0 & -2.4 & 2.4 \end{bmatrix}$$

$$[B] = \begin{bmatrix} -[M]_c & [0] \\ [0] & [K]_c \end{bmatrix} = \begin{bmatrix} -11 & -3 & -2 & -2 & -1 & 0 & 0 & 0 & 0 & 0 \\ -3 & -3 & 0 & 0 & 0 & 0 & 0 & 0 & 0 & 0 \\ -2 & 0 & -2 & 0 & 0 & 0 & 0 & 0 & 0 & 0 \\ -2 & 0 & 0 & -2 & 0 & 0 & 0 & 0 & 0 & 0 \\ -1 & 0 & 0 & 0 & -1 & 0 & 0 & 0 & 0 & 0 \\ 0 & 0 & 0 & 0 & 0 & 108.57 & 0 & 0 & 0 & 0 \\ 0 & 0 & 0 & 0 & 0 & 0 & 5600 & -2400 & 0 & 0 \\ 0 & 0 & 0 & 0 & 0 & 0 & -2400 & 4000 & -1600 & 0 \\ 0 & 0 & 0 & 0 & 0 & 0 & 0 & -1600 & 2400 & -800 \\ 0 & 0 & 0 & 0 & 0 & 0 & 0 & 0 & -800 & 800 \end{bmatrix}$$

and from the solution of the eigenvalue problem $[B]\{s\} = -\lambda[A]\{s\}$ and the consideration of Equations 15.141 and 15.142, one obtains the dynamic properties given in Table E15.4a for the five modes of the system whose eigenvalues are not complex conjugates of one another. These dynamic properties include complex natural frequencies, undamped natural frequencies, damping ratios, and complex

TABLE E15.4a
Dynamic Properties of Isolated Building in Example 15.4

Mode	1	2	3	4	5
Eigenvalue, λ_i	-1.51791	-1.99934	-2.74368	-4.39035	-5.51539
	$+2.7493i$	$+19.3311i$	$+34.205i$	$+44.9671i$	$+57.2772i$
Natural frequency (Hz)	0.500	3.093	5.461	7.191	9.158
Damping ratio	0.483	0.103	0.080	0.097	0.096
Mode shape, $\{w\}_i$	0.179	0.00133	0.00142	0.00158	0.000119
	$+0.242i$	$-0.0124i$	$+0.00717i$	$+0.00598i$	$+0.00415i$
	-0.00300	-0.00238	0.000683	-0.000718	-0.00195
	$+0.00703i$	$+0.00385i$	$-0.00805i$	$-0.0117i$	$-0.0126i$
	-0.00559	-0.00446	-0.00143	-0.00301	0.00335
	$+0.0129i$	$+0.0129i$	$-0.0173i$	$-0.0125i$	$+0.00584i$
	-0.00796	-0.00615	-0.00372	-0.00180	-0.00131
	$+0.0182i$	$+0.0261i$	$-0.0162i$	$+0.00282i$	$-0.00811i$
	-0.00958	-0.00693	0.000	-0.000896	0.000606
	$+0.0217i$	$+0.0389i$	$+0.0122i$	$-0.0116i$	$-0.00290i$

mode shapes. Similarly, with the complex mode shapes in Table E15.4a, the consideration of Equation 15.161, and according to the second of Equations 15.109, which for this case leads to

$$\{r\}_c = \begin{Bmatrix} 1 \\ \{0\} \end{Bmatrix} = \{1 \ 0 \ 0 \ 0 \ 0\}^T$$

the complex participation factor for the first mode of the system results as

$$\Gamma_{c1} = \frac{\{w\}_1^T [M]_c \{r\}_c}{\{w\}_1^T [M]_c \{w\}_1} = 1.861 - 2.678i$$

and, similarly, for the second and third modes as

$$\Gamma_{c2} = 1.464 - 2.263i \quad \Gamma_{c3} = -8.071 - 1.551$$

Furthermore, the consideration of Equations 15.163 and 15.169 yields

$$\{w'\}_1 = \{u'\}_1 + i\{v'\}_1 = i\Gamma_{c1}\{w\}_1 = \begin{Bmatrix} 0.0295 \\ -0.0211 \\ -0.0390 \\ -0.0552 \\ -0.0661 \end{Bmatrix} + i \begin{Bmatrix} 0.9825 \\ 0.0133 \\ 0.0242 \\ 0.0339 \\ 0.0404 \end{Bmatrix}$$

$$\{w'\}_2 = \{u'\}_2 + i\{v'\}_2 = i\Gamma_{c2}\{w\}_2 = \begin{Bmatrix} 0.0212 \\ -0.0110 \\ -0.0290 \\ -0.0521 \\ -0.0726 \end{Bmatrix} + i \begin{Bmatrix} -0.0261 \\ 0.0052 \\ 0.0227 \\ 0.0500 \\ 0.0778 \end{Bmatrix}$$

$$\{w'\}_3 = \{u'\}_3 + i\{v'\}_3 = i\Gamma_{c3}\{w\}_3 = \begin{Bmatrix} 0.0601 \\ -0.0639 \\ -0.1419 \\ -0.1369 \\ 0.0989 \end{Bmatrix} + i \begin{Bmatrix} 0.0003 \\ -0.0180 \\ -0.0153 \\ 0.0049 \\ 0.0189 \end{Bmatrix}$$

$$\{a'\}_1 = \frac{1}{\omega_{d1}}\{u'\}_1 = \frac{1}{2.749}\begin{Bmatrix} 0.0295 \\ -0.0211 \\ -0.0390 \\ -0.0552 \\ -0.0661 \end{Bmatrix} = \begin{Bmatrix} 0.0107 \\ -0.0077 \\ -0.0142 \\ -0.0201 \\ -0.0241 \end{Bmatrix}$$

$$\{b'\}_1 = \{v'\}_1 - \frac{\xi_1}{\sqrt{1-\xi_1^2}}\{u'\}_1 = \begin{Bmatrix} 0.9825 \\ 0.0133 \\ 0.0242 \\ 0.0339 \\ 0.0404 \end{Bmatrix} - \frac{0.483}{\sqrt{1-0.483^2}}\begin{Bmatrix} 0.0295 \\ -0.0211 \\ -0.0390 \\ -0.0552 \\ -0.0661 \end{Bmatrix} = \begin{Bmatrix} 0.9662 \\ 0.0249 \\ 0.0457 \\ 0.0644 \\ 0.0769 \end{Bmatrix}$$

Seismic Protection with Base Isolation

and, similarly,

$$\{a'\}_2 = \begin{Bmatrix} 0.0011 \\ -0.0006 \\ -0.0015 \\ -0.0027 \\ -0.0038 \end{Bmatrix} \quad \{b'\}_2 = \begin{Bmatrix} -0.0283 \\ 0.0064 \\ 0.0257 \\ 0.0554 \\ 0.0854 \end{Bmatrix}$$

$$\{a'\}_3 = \begin{Bmatrix} 0.0018 \\ -0.0019 \\ -0.0041 \\ -0.0040 \\ 0.0029 \end{Bmatrix} \quad \{b'\}_3 = \begin{Bmatrix} -0.0051 \\ -0.0129 \\ -0.0040 \\ 0.0158 \\ 0.0110 \end{Bmatrix}$$

Now, from the acceleration design spectrum and damping factor $H(\xi)$ given in Example E15.1, and for the natural frequencies and damping ratios listed in Table E15.4a, the spectral acceleration corresponding to the first mode of the system is

$$SA_1 = SA(T_1, \xi_1) = SA(2.00, 0.483) = H(0.483)SA(2.00, 0.05) = \left[\frac{1.5}{40(0.483)+1} + 0.5\right]\frac{1.37}{2.00}g = 0.39g$$

and, similarly,

$$SA_2 = SA(T_2, \xi_2) = SA(0.26, 0.103) = 1.09g \quad SA_3 = SA(T_3, \xi_3) = SA(0.18, 0.080) = 1.17g$$

To determine the corresponding spectral displacements and velocities, it is necessary to first compute the constants a_{vi}, b_{vi}, and r_i defined by Equations 15.173 through 15.175 and then use Equations 15.172 and 15.178. Proceeding accordingly, one obtains for the first three modes of the system the constants a_{vi}, b_{vi}, and r_i listed in Table E15.4b. Also, the spectral displacement and spectral velocity for the first mode result as

$$SD_1 = \frac{1-\xi_1^2}{\sqrt{1+4r_1^2\xi_1^2(1-\xi_1^2)}}\frac{SA_1}{\omega_{d1}^2} = \frac{1-0.483^2}{\sqrt{1+4(1.805)^2(0.483)^2(1-0.483^2)}}\frac{0.39g}{2.749^2} = 8.38 \text{ in.}$$

$$SV_1 = r_1\omega_{d1}SD_1 = 1.805(2.749)(8.38) = 41.58 \text{ in./s}$$

and, similarly, for the second and third modes as

$$SD_2 = 1.10 \text{ in.} \quad SD_3 = 0.38 \text{ in.}$$

$$SV_2 = 17.05 \text{ in./s} \quad SV_3 = 9.59 \text{ in./s}$$

In consequence and according to Equation 15.176, the displacements of the system in the first mode of the system are approximately equal to

$$\max\{v(t)\}_{c1} = SD_1[r_1^2\{u'^2\}_1 + \{b'^2\}_1]^{1/2} = 8.38\left[1.805^2\begin{Bmatrix} 0.0295^2 \\ -0.0211^2 \\ -0.0390^2 \\ -0.0552^2 \\ -0.0661^2 \end{Bmatrix} + \begin{Bmatrix} 0.9662^2 \\ 0.0249^2 \\ 0.0457^2 \\ 0.0644^2 \\ 0.0769^2 \end{Bmatrix}\right]^{1/2} = \begin{Bmatrix} 8.109 \\ 0.382 \\ 0.703 \\ 0.994 \\ 1.190 \end{Bmatrix} \text{ in.}$$

TABLE E15.4b
Values of Constants a_{vi}, b_{vi}, and r_i

Mode	T_i (s)	ξ_i	a_{vi}	b_{vi}	r_i
1	2.00	0.483	1.318	0.453	1.805
2	0.26	0.103	1.157	0.273	0.802
3	0.18	0.080	1.144	0.256	0.738

and, similarly, the displacements in the second and third modes are approximately equal to

$$\max\{v\}_{c2} = \begin{Bmatrix} 0.036 \\ 0.012 \\ 0.038 \\ 0.076 \\ 0.114 \end{Bmatrix} \text{ in.} \quad \max\{v\}_{c3} = \begin{Bmatrix} 0.017 \\ 0.019 \\ 0.040 \\ 0.039 \\ 0.114 \end{Bmatrix} \text{ in.}$$

From Equation 15.159 and if the contribution of the modes higher than the third is neglected, the maximum displacements of the system are thus equal to

$$\max\{v\}_c = [\max\{v^2\}_{c1} + \max\{v^2\}_{c2} + \max\{v^2\}_{c3}]^{1/2} = \begin{Bmatrix} 8.11 \\ 0.38 \\ 0.71 \\ 1.00 \\ 1.20 \end{Bmatrix} \text{ in.}$$

To compute now the floor accelerations, it may be noted that for the case under consideration and for the first mode of the system the vectors defined by Equations 15.186 through 15.188 result as

$$\{p\}_1 = [C]\{a^*\}_1 = \begin{bmatrix} 16.8 & -7.2 & 0 & 0 \\ -7.2 & 12.0 & -4.8 & 0 \\ 0 & -4.8 & 7.2 & -2.4 \\ 0 & 0 & -2.4 & 2.4 \end{bmatrix} \begin{Bmatrix} -0.0077 \\ -0.0142 \\ -0.0201 \\ -0.0241 \end{Bmatrix} = \begin{Bmatrix} -0.027 \\ -0.018 \\ -0.019 \\ -0.010 \end{Bmatrix}$$

$$\{q\}_1 = [K]\{b^*\}_1 = \begin{bmatrix} 5600 & -2400 & 0 & 0 \\ -2400 & 4000 & -1600 & 0 \\ 0 & -1600 & 2400 & -800 \\ 0 & 0 & -800 & 800 \end{bmatrix} \begin{Bmatrix} 0.0249 \\ 0.0457 \\ 0.0644 \\ 0.0769 \end{Bmatrix} = \begin{Bmatrix} 29.862 \\ 19.934 \\ 19.953 \\ 9.982 \end{Bmatrix}$$

$\{h\}_1 = [C]\{b^*\}_1 - [K]\{a^*\}_1$

$$= \begin{bmatrix} 16.8 & -7.2 & 0 & 0 \\ -7.2 & 12.0 & -4.8 & 0 \\ 0 & -4.8 & 7.2 & -2.4 \\ 0 & 0 & -2.4 & 2.4 \end{bmatrix} \begin{Bmatrix} 0.0249 \\ 0.0457 \\ 0.0644 \\ 0.0769 \end{Bmatrix} - \begin{bmatrix} 5600 & -2400 & 0 & 0 \\ -2400 & 4000 & -1600 & 0 \\ 0 & -1600 & 2400 & -800 \\ 0 & 0 & -800 & 800 \end{bmatrix} \begin{Bmatrix} -0.0077 \\ -0.0142 \\ -0.0201 \\ -0.0241 \end{Bmatrix}$$

Seismic Protection with Base Isolation

$$= \begin{Bmatrix} 0.0896 \\ 0.0598 \\ 0.0599 \\ 0.0299 \end{Bmatrix} - \begin{Bmatrix} -9.0137 \\ -6.1451 \\ -6.2672 \\ -3.1744 \end{Bmatrix} = \begin{Bmatrix} 9.103 \\ 6.205 \\ 6.327 \\ 3.204 \end{Bmatrix}$$

Thus, on substitution of these vectors into Equation 15.190, and if the insignificant contribution of the second and third modes is neglected, the maximum values of the inertia forces are approximately equal to

$$\max\{F_I(t)\} \approx \max\{F_I(t)\}_1 = [\{p^2\}_1(SA_1^2 + PGA^2) + \{h^2\}_1 SV_1^2 + \{q^2\}_1 SD_1^2]^{1/2}$$

$$= \left[\begin{Bmatrix} -0.027^2 \\ -0.018^2 \\ -0.019^2 \\ -0.010^2 \end{Bmatrix} (0.39\,g^2 + 0.55\,g^2) + \begin{Bmatrix} 9.103^2 \\ 6.205^2 \\ 6.327^2 \\ 3.204^2 \end{Bmatrix} 41.58^2 + \begin{Bmatrix} 29.862^2 \\ 19.934^2 \\ 19.953^2 \\ 9.982^2 \end{Bmatrix} 8.38^2 \right]^{1/2}$$

$$= \begin{Bmatrix} 453.9 \\ 307.4 \\ 311.8 \\ 157.4 \end{Bmatrix} \text{kips}$$

where it has been considered that PGA is equal to the zero-period spectral acceleration in the design spectrum given in Example 15.1. Similarly and according to Equation 15.191, an estimate of the maximum floor accelerations is given by

$$\max\{\ddot{u}(t)\} \approx [M]^{-1} \max\{F_I(t)\}_1 = \begin{bmatrix} 3 & 0 & 0 & 0 \\ 0 & 2 & 0 & 0 \\ 0 & 0 & 2 & 0 \\ 0 & 0 & 0 & 1 \end{bmatrix}^{-1} \begin{Bmatrix} 453.9 \\ 307.4 \\ 311.8 \\ 157.4 \end{Bmatrix} = \begin{Bmatrix} 151.3 \\ 153.7 \\ 155.9 \\ 157.4 \end{Bmatrix} \text{in./s}^2 = \begin{Bmatrix} 0.39 \\ 0.40 \\ 0.40 \\ 0.41 \end{Bmatrix} g$$

Finally, according to Equations 15.194 through 15.196, an estimate of the seismic coefficient is

$$C_s = \frac{\max V(t)}{W_s} \approx \frac{\{r\}^T \max\{F_I(t)\}_1}{W_s} = \frac{1}{8g}\{1 \ \ 1 \ \ 1 \ \ 1\} \begin{Bmatrix} 453.9 \\ 307.4 \\ 311.8 \\ 157.4 \end{Bmatrix} = 0.40$$

In comparison with the results obtained in Example 15.3, it is interesting to observe the following:

1. The undamped natural frequencies of the system basically remain the same.
2. The damping ratios are increased significantly in all modes.
3. The displacement of the isolation mass is reduced significantly, from 21.5 in. to 8.11 in.
4. The displacements of the structure are also reduced, but not as significantly as the displacement of the isolation mass.
5. Some of the inertia forces, and correspondingly some of the floor accelerations, are reduced whereas others are increased.
6. The seismic coefficient is slightly reduced, from 0.45 to 0.40.

It is also interesting to observe that structural elastic forces in the first mode of the system are equal to

$$\{F_s\}_{\max} \approx [K]\max\{v\}_1 = \begin{bmatrix} 5600 & -2400 & 0 & 0 \\ -2400 & 4000 & -1600 & 0 \\ 0 & -1600 & 2400 & -800 \\ 0 & 0 & -800 & 800 \end{bmatrix} \begin{Bmatrix} 0.382 \\ 0.703 \\ 0.994 \\ 1.190 \end{Bmatrix} = \begin{Bmatrix} 452.0 \\ 304.8 \\ 308.8 \\ 156.8 \end{Bmatrix} \text{kips}$$

which are very close in value to the computed inertia forces. This shows that the damping forces acting on the structure above the isolation system are small compared to the elastic forces. It also shows that the increase in some of the inertia forces and some of the floor accelerations over the corresponding inertia forces and floor accelerations in the system considered in Example 15.3 is owed to the change in the modal properties of the system and not the contribution of the damping forces.

15.5.5 Equivalent Linear Model

As seen in Section 15.4, some of the bearings used in seismic isolation are characterized by a highly nonlinear behavior and a hysteretic dissipation of energy. Therefore, a proper analysis of structures isolated with this type of bearings requires the use of one of the step-by-step methods of nonlinear analysis described in Chapter 11. However, for the purpose of a preliminary design, it is convenient to linearize such nonlinear behavior through the use of an equivalent linear stiffness and an equivalent damping ratio as this linearization allows the utilization of one of the linear methods of analysis described in the preceding sections. An equivalent stiffness and an equivalent damping ratio for a nonlinear isolation system with a hysteretic dissipation of energy may be determined as follows:

15.5.5.1 Effective Stiffness

Highly nonlinear seismic isolators, such as lead–rubber bearings, exhibit a force–deformation behavior that can be modeled as bilinear one (see Figure 15.8). Consequently, the effective stiffness of such bearings may be defined as the secant stiffness that is obtained by joining the extreme positive and negative displacements in their force–deformation curves (see Figure 15.25). That is, the effective stiffness may be considered given by

$$K_{\text{eff}} = \frac{K_i u_y + K_p(u_{\max} - u_y)}{u_{\max}} = (K_i - K_p)\frac{u_y}{u_{\max}} + K_p \qquad (15.197)$$

where K_i, K_p, u_y, $u_{\max}$, respectively, represent, initial stiffness of the bearing, postyield stiffness, yield deformation, maximum deformation in the bearing during any given deformation cycle.

For design purposes, it is customary to assume that $u_{\max}$ is equal to the expected maximum deformation of the chosen bearings under the design earthquake.

15.5.5.2 Effective Damping Ratio

Similarly, the equivalent damping ratio may be obtained using the concept of loss factor introduced in Section 13.9. That is, the desired equivalent damping ratio may be obtained according to (see Equation 13.179)

$$\xi_{\text{eff}} = \frac{1}{4\pi}h = \frac{1}{4\pi}\frac{\text{area of hysteresis loop}}{\text{strain energy in equivalent linear system}} \qquad (15.198)$$

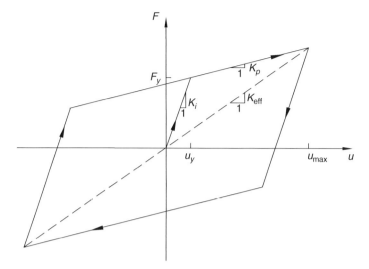

FIGURE 15.25 Bilinear force–deformation curve and effective stiffness.

where h is the aforementioned loss factor and is usually determined experimentally from free vibration tests. The area of a hysteresis loop and, hence, the energy dissipated per cycle in the case of the bilinear force–deformation behavior shown in Figure 15.25 is given by

$$W_d = 4(K_i - K_p)u_y(u_{max} - u_y) \tag{15.199}$$

whereas the strain energy in the equivalent linear system is equal to $K_{eff}u^2_{max}/2$.

Although useful for preliminary designs, it is important to keep in mind that an analysis based on an equivalent linearization is not always a good approximation. The reason is that the equivalent stiffness and the equivalent damping ratio vary considerably with the specific characteristics and intensity of the excitation as well as with the characteristics of the superstructure and the isolation system. Also, the substitution of a hysteretic dissipation of energy with an equivalent viscous damping involves several approximations that may be incorrect in practice. For example, the concept of equivalent damping ratio is based on periodic complete cycles of deformation, whereas an irregular earthquake excitation contains many short deformation cycles with little energy dissipation. This means that an equivalent damping ratio may overestimate the actual energy dissipation during an earthquake. Another problem with an equivalent linearization is that with it one cannot evaluate higher mode effects and the resulting increased floor accelerations due to the bearings' nonlinear behavior. A true nonlinear analysis is needed for this purpose.

15.6 IMPLEMENTATION ISSUES

As it may be appreciated from the foregoing discussion, the main advantage of a base-isolation system is its ability to reduce the earthquake response of a structure and hence its design seismic forces. Equally important is the fact that a base-isolation system may reduce at the same time the floor accelerations, minimizing thus the damage to equipment, contents, and architectural items. Furthermore, a base-isolation system is relatively easy to install and may be implemented in new structures as well in the retrofit of existing ones. It is particularly advantageous in the retrofit of historical buildings and monuments as its impact in their architectural features is minimum. A base-isolation system, however, also has some disadvantages. These are

1. A base-isolation system requires an isolation gap to allow for the free lateral displacement of the isolators. As a result, structures in sites where there is no sufficient clearance for this required gap cannot be base isolated.
2. A base-isolated structure is always in danger of an insufficient isolation gap and thus of unpredictable acceleration pulses that may be generated if the structure pounds against restraining elements.
3. Since in general isolators are incapable of resisting large tensile loads, base isolation is only suitable for structures in which uplifting of the bearings will not occur.
4. Base-isolation systems require devices to resist the maximum lateral loads from wind and other service loads without yielding to avoid unacceptable displacements and vibrations under these loads.
5. Little is known about how the chemical and physical properties of the isolators change over time, and how these properties are affected by environmental factors.
6. The integrity of a base-isolated structure totally depends on the integrity of its isolators.
7. The coefficient of friction in sliding isolators cannot be determined with certainty, particularly after a long period of inactivity.
8. Because of the isolators' nonlinear behavior and the required consideration of the maximum forces before and after the isolators become operational, the analysis of a base-isolated structure is substantially more complicated than the analysis of a conventional one.
9. A base-isolation system may tune the higher modes of a structure to the dominant frequencies of the ground motion that excites it. As a result, it may magnify the response of the structure in its higher modes and, hence, the response of high-frequency nonstructural elements.
10. The nonlinear behavior of the isolators may induce high accelerations in the floors of a building and damage, as a result, its contents.
11. In general, base-isolated structures are more expensive than their fixed-base counterparts because there are not enough savings in the design of the superstructure to offset the cost of the isolators, the double basement slabs and walls, and the special details to provide for the flexibility of the utilities.

It may be said, thus, that, in general, the structures that are suitable for seismic isolation are
- Structures founded on firm soils
- Structures with a short fundamental natural period; that is, squat structures of low to medium height that attract large earthquake forces when they are not isolated
- Structures with significant gravity loads and a height to width ratio that prevents large overturning moments
- Structures on sites that permit the unobstructed horizontal displacement of the isolators
- Structures for which a stringent performance is required, such as in the case of structures which house expensive or vital equipment, or those in need of retrofit and for which a conventional strengthening is more expensive, more disruptive, or architecturally unacceptable

FURTHER READINGS

1. Buckle, I. G. and Mayes, R. L., Seismic isolation: History, application, and performance—A world view, *Earthquake Spectra*, 6, 2, 1990, 161–201.
2. Hurty, W. C. and Rubinstein, M. F., *Dynamics of Structures*, Prentice-Hall, Englewood Cliffs, NJ, 1964, 455 p.
3. Izumi, M., Base isolation and passive seismic response control, *Proceedings 9th World Conference on Earthquake Engineering*, Vol. VIII, Tokyo, Japan, 1988, pp. 385–396.
4. Kelly, J. M., *Earthquake-Resistant Design with Rubber*, 2nd Edition, Springer, London, 1997, 243 p.
5. Komodromos, P., *Seismic Isolation for Earthquake Resistant Structures*, WIT Press, Southampton, 2000, 201 p.

6. Maldonado, G. O. and Singh, M. P., An improved response spectrum method for calculating seismic design response. Part 2: Non-classically damped structures, *Earthquake Engineering and Structural Dynamics*, 20, 7, 1991, 637–649.
7. Naeim, F. and Kelly, J. M., *Design of Seismic Isolated Structures: From Theory to Practice*, John Wiley & Sons, New York, 1999, 289 p.
8. Skinner, R. I., Robinson, W. H., and McVerry, G. H., *An Introduction to Seismic Isolation*, John Wiley & Sons, Chichester, 1993, 354 p.
9. Soong, T. T. and Constantinou, M. C., ed., *Passive and Active Structural Control in Civil Engineering*, CIMS Courses and Lectures No. 345, Springer, Vienna, NY, 1994, 380 p.
10. Taylor, A. W., Lin, A. N., and Martin, J. W., Performance of elastomers in isolation bearings: A literature review, *Earthquake Spectra*, 8, 2, 1992, 279–303.

PROBLEMS

15.1 Determine using the two-degree-of-freedom model introduced in Section 15.5.1 the natural frequencies, mode shapes, and maximum response of an isolated building when the building is subjected to an earthquake ground motion represented by the 5% damping design spectrum shown in Figure E15.1. When considered fixed to the ground, the building may be modeled as a single-degree-of-freedom system with a natural period of 0.4 s and a damping ratio of 5%. The isolation system has a foundation slab with a mass equal to the mass of the structure, a natural period of 2.5 s, and a damping ratio of 15%. As in Example 15.1, estimate the spectral ordinates for damping ratios other than 5% by multiplying the ordinates in the spectrum for 5% damping by the $H(\xi)$ factor specified therein.

15.2 Using the rigid-structure approximation introduced in Section 15.5.2, determine the maximum displacement of the isolation system, the structural base shear coefficient, and the reduction in base shear achieved by the isolation system for the building considered in Problem 15.1.

15.3 Determine the natural frequencies and mode shapes of the isolated shear building shown in Figure P15.3. The isolation system has a mass of 3 Gg and a natural period of 2 s when the structure is considered perfectly rigid. Use the approach introduced in Section 15.5.3.

15.4 Determine the natural frequencies and mode shapes of an isolated five-story shear building with the properties listed in Table P15.4. The isolation system has a mass of 150 Mg and a natural period of 2.5 s when the structure is considered perfectly rigid. Use the approach introduced in Section 15.5.3.

15.5 Determine using the approximate method described in Section 15.5.3, the maximum displacements, lateral forces, and base shear coefficient of the isolated building considered

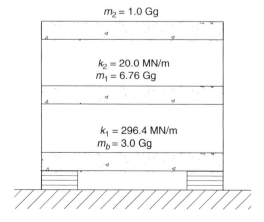

FIGURE P15.3 Base-isolated shear building considered in Problem 15.3.

TABLE P15.4
Properties of Shear Building in Problem 15.4

Story	Height (m)	Stiffness (MN/m)	Mass (Mg)
1	4	249.88	179
2	3	237.05	170
3	3	224.57	161
4	3	212.09	152
5	3	199.62	143

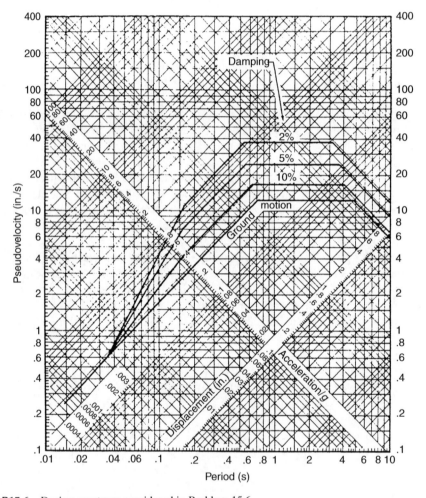

FIGURE P15.6 Design spectrum considered in Problem 15.6.

in Problem 15.3 when the building is subjected to an earthquake excitation represented by the 5% damping design spectrum specified in Example 15.1. Assume a damping ratio of 1% for the fundamental mode of the fixed-base structure and 15% for the isolation system. As in Example 15.1, estimate the spectral ordinates for damping ratios other than 5% by multiplying the ordinates in the spectrum for 5% damping by the $H(\xi)$ factor specified therein.

15.6 Determine using the approximate method described in Section 15.5.3, the maximum displacements, lateral forces, and base shear coefficient of the building considered in

Seismic Protection with Base Isolation

Problem 15.4 when the building is subjected to an earthquake ground motion represented by the design spectrum shown in Figure P15.6. Assume a damping ratio of 2% for the fundamental mode of the fixed-base structure and 10% for the isolation system.

15.7 Using the rigid-structure approximation introduced in Section 15.5.2, determine for the building considered in Problem 15.3 the maximum displacement of the isolation system, the structural base shear coefficient, and the reduction in base shear achieved by the isolation system.

15.8 Using the rigid-structure approximation introduced in Section 15.5.2, determine for the building considered in Problem 15.5 the maximum displacement of the isolation system, the structural base shear coefficient, and the reduction in base shear achieved by the isolation system.

15.9 Determine using the method described in Section 15.5.4 the maximum displacements, maximum floor accelerations, and base shear coefficient of the isolated building considered in Problem 15.5. Assume that the damping ratio of the isolation system is 30% and that the damping matrix of the fixed-base structure is proportional to its stiffness matrix.

15.10 Determine using the method described in Section 15.5.4 the maximum displacements, maximum floor accelerations, and base shear coefficient of the isolated building considered in Problem 15.6. Assume that the damping ratio of the isolation system is 30% and that the damping matrix of the fixed-base structure is proportional to its stiffness matrix. Consider only the contribution of the first two modes.

16 Seismic Protection with Energy Dissipating Devices

> Recent damaging earthquakes and hurricanes provided powerful reminders of how vulnerable we all are to the forces of nature. Even in an advanced industrial nation, our built environment is still quite susceptible to natural disasters. Consequently, one of the principal current challenges in structural engineering concerns the development of innovative design concepts to better protect structures, along with their occupants and contents, from the damaging effects of destructive environmental forces including those due to wind, waves, and earthquakes.
>
> T. T. Soong and G. F. Dargush, 1997

16.1 INTRODUCTION

During a seismic event, energy is input into a structure. This energy deforms and sets the structure into vibration and is thus transformed into kinetic and potential (strain) energy. Eventually, the input energy is absorbed by the structure in the form of permanent deformations or dissipated in the form of heat. If there were no damping in the structure, the structure would vibrate indefinitely. However, there is always an inherent amount of damping in every structure that gradually reduces the amplitude of the vibrations and ultimately suppresses them. The higher this inherent damping is, the smaller the amplitude of vibration would be and the sooner the vibrations would stop. Unfortunately, most structures exhibit only a small amount of damping.

Another design concept that has been suggested and applied in recent years to protect structures against earthquake effects is the incorporation throughout the height of a structure of mechanical devices that augment the amount of damping present in the structure. These mechanical devices resist the motion and absorb or dissipate part of the energy transmitted to a structure during an earthquake by mechanisms that involve the yielding of metallic elements, sliding friction, the motion of a piston within a viscous fluid, or the deformation of viscoelastic materials. This design concept is based on the premise that earthquakes are a dynamic load and that the effect of dynamic loads on a system is reduced by the damping forces present in it. It implies, therefore, an increase in these damping forces through the addition of supplemental dampers or energy-absorbing devices to a level that would keep the structure and its nonstructural components virtually undamaged during a strong earthquake. Different from the base isolation concept, however, energy dissipation devices do not intercept the earthquake energy entering a structure. Instead, they allow the free transfer of this energy into the structure and partially dissipate it in the form of heat. Different from the base-isolation concept, too, these devices can be effective not only against earthquake ground motions but also against wind-induced vibrations. The incorporation of energy dissipation devices has for long been recognized as an effective means of controlling excessive vibration in mechanical systems, with the use of shock absorbers in motor vehicles being perhaps the best-known example.

The purpose of this chapter is to describe the basic concepts behind the supplemental energy dissipation technology. First, a review on the influence damping has in the dynamic behavior of structures and the physical consequences of adding energy dissipation devices to a structure is presented. Then, a description of some of the most commonly used energy dissipation devices is given. Specifically, friction, viscoelastic, fluid, and hysteretic dampers are described. Thereafter, a description of the techniques that may be used to model and analyze structures with added energy dissipation devices is given, illustrating theses techniques with several numerical examples. Finally, issues regarding the effectiveness, practicality, reliability, durability, and cost of adding energy dissipating devices to a structure are discussed. Overall, the objective is to present the essential concepts that will allow the reader to understand the benefits and limitations of such systems in structural applications and introduce some of the tools that may be used for the design of structures that incorporate such systems. As in the case of the base isolation technology presented in Chapter 15, the discussion will be for the most part of an introductory nature, avoiding a detailed treatment of the subject. For an in-depth coverage, refer, once again, to the publications listed in the Further Readings section.

16.2 BASIC CONCEPTS

In general, increasing the amount of damping in a structure reduces the response of the structure to dynamic loads. However, the extent of the reduction varies depending on the characteristics of the excitation, the distribution of the damping increase throughout the structure, the amount of damping in the structure before the damping increase, and the level of the inelastic deformations. Thus, to gain an understanding of how effective the use of energy dissipation devices may be in any given situation, a review of some of the basic concepts that describe the relationship between damping and structural response is presented in this section.

Consider first a linear, viscously damped, single-degree-of-freedom system initially at rest with a natural frequency ω and a damping ratio ξ, and assume that the system is subjected to an impulse that generates an initial velocity v_0. As is well known from structural dynamics, if the system is underdamped, the effect of such an impulse is to set the system into free vibration, with a displacement response given by

$$u(t) = \frac{v_0}{\omega\sqrt{1-\xi^2}} e^{-\xi\omega t} \sin\omega\sqrt{1-\xi^2}\, t \tag{16.1}$$

where $u(t)$ and t denote lateral displacement at time t and time. If Equation 16.1 is plotted for different values of the damping ratio ξ and a specific value of the natural frequency ω, for example, $\omega = 2\pi$, one then obtains the curves shown in Figure 16.1.

The effect of increasing the damping ratio of the system may clearly be seen from Figure 16.1. It may be noted, first, that the peak amplitude of the response decreases with an increase in the damping ratio. For example, an increase from 5 to 30% reduces the normalized displacement response from 0.147 to 0.104, a 29% decrease. Note also that the number of cycles it takes to reduce the displacement amplitude to half its initial value is also reduced with an increase in the damping ratio. For instance, the number of cycles it takes to reduce the initial amplitude by 50% varies from about 11 cycles for a damping ratio of 1% to about half a cycle for a damping ratio of 20%. It may be affirmed, thus, that a damping ratio increase may reduce the response of a system subjected to an impulsive ground motion and may accelerate the decay of this response.

Consider now the single-degree-of-freedom system introduced earlier when subjected to a sinusoidal external load with amplitude P_0 and frequency Ω. As is well known from structural dynamics, the displacement response of the system in such a case is given by

$$u(t) = e^{-\xi\omega t}\left(A\cos\omega\sqrt{1-\xi^2}\,t + B\sin\omega\sqrt{1-\xi^2}\,t\right) + \frac{P_0}{k}\frac{1}{\sqrt{(1-\beta^2)+(2\xi\beta)^2}}\sin(\Omega t - \phi) \tag{16.2}$$

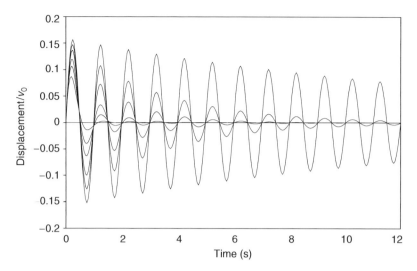

FIGURE 16.1 Displacement response of a single-degree-of-freedom system under initial velocity v_0 for damping ratios of 2, 5, 10, 20, 30, and 50%.

where β is a frequency ratio defined as

$$\beta = \frac{\Omega}{\omega} \quad (16.3)$$

ϕ is a phase angle given by

$$\phi = \tan^{-1} \frac{2\xi\beta}{1-\beta^2} \quad (16.4)$$

and A and B are constants of integration which for zero initial conditions are equal to

$$A = \frac{P_0}{k} \frac{2\xi\beta}{(1-\beta^2)^2 + (2\xi\beta)^2} \quad (16.5)$$

$$B = \frac{P_0}{k} \frac{1}{\sqrt{1-\xi^2}} \left[\frac{2\beta\xi^2 - \beta(1-\beta^2)}{(1-\beta^2)^2 + (2\xi\beta)^2} \right] \quad (16.6)$$

In Equation 16.2, the first term corresponds to the transient component of the response and, because of the exponential term in it, its magnitude diminishes with time as in the free vibration case. The second term corresponds to the steady-state component of the response and it will last as long as the external load keeps acting. Therefore, for moderate amounts of damping only the steady-state component will make a significant contribution to the system's response. Note also that this steady-state response may be expressed as

$$u_{ss}(t) = u_s A_D \sin(\Omega t - \phi) \quad (16.7)$$

where u_s denotes the displacement response of the system when the load is applied statically, and A_D, which represents a magnification factor, is defined as

$$A_D = \frac{1}{\sqrt{(1-\beta^2) + (2\xi\beta)^2}} \qquad (16.8)$$

Furthermore, recall from structural dynamics that the maximum value of this magnification factor is

$$A_D = \frac{1}{2\xi\sqrt{1-\xi^2}} \qquad (16.9)$$

Finally, if the amplitude of the steady-state response, normalized with respect to the static displacement, is plotted as a function of the frequency ratio β and damping ratio ξ, one obtains the curves shown in Figure 16.2.

It may be seen from Figure 16.2 that for all the damping ratios considered, the peak response occurs when the frequency ratio is close to unity; that is, when the frequency of the excitation is approximately equal to the natural frequency of the system. It may also be seen that the value of the damping ratio significantly affects the value of this peak response. For example, if the damping ratio is increased from 2 to 30%, the peak response is reduced from a value of 25 times the static response to a value of 1.75 times the static response. However, the effect of a damping ratio increase becomes less and less as the frequency ratio differs more and more from unity. For example, when the frequency ratio is about 0.8, increasing the damping ratio from 5 to 30% reduces the displacement response by about 38%, compared to about 83% when the frequency ratio is unity. In general, note that for any damping ratio the peak response approaches the static response as the frequency ratio approaches zero. Similarly, the peak response approaches zero for large values of the frequency ratio. This means that if the dominant frequencies of the input motion are not close to the natural frequency of the system, adding damping to the system will not significantly reduce its displacement response. It may be seen, in addition, that the significance of the response reduction owing to a damping increase depends on the initial damping level. For example, an increase in

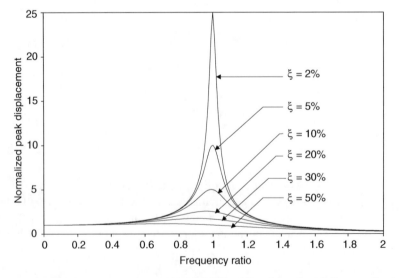

FIGURE 16.2 Variation with frequency ratio and damping ratio of amplitude of normalized steady-state displacement response of a single-degree-of-freedom system under harmonic excitation.

the damping ratio by a factor of 2.5 from 2 to 5% results in a decrease in the peak displacement response from 25 to 10 times the static response; that is, a difference of 15 times the static response. In contrast, an increase in the damping ratio by the same factor of 2.5 from 20 to 50% decreases the peak displacement response from 2.55 to 1.15 times the static response; that is, a difference of only 1.4 times the static response.

Consider next the same single-degree-of-freedom system studied earlier when subjected to an impulsive excitation and let this impulsive excitation be represented by a simple half sinusoidal cycle with frequency Ω. In this case, the response of the system is given by Equation 16.2 when $t \leq t_1$, and by the free vibration solution

$$u(t) = e^{-\xi\omega(t-t_1)} \left[u(t_1)\cos\omega\sqrt{1-\xi^2}(t-t_1) + \frac{\dot{u}(t_1) + \xi\omega u(t_1)}{\omega\sqrt{1-\xi^2}} \sin\omega\sqrt{1-\xi^2}(t-t_1) \right] \quad (16.10)$$

when $t \geq t_1$. Note that in Equation 16.10, $t_1 = \pi/\Omega$ represents the duration of the pulse and $u(t_1)$ and $\dot{u}(t_1)$ denote the displacement and velocity of the system at the end of the pulse, respectively. Note also that $u(t_1)$ and $\dot{u}(t_1)$ may be obtained directly from Equation 16.2.

The variation of the peak displacement response with damping ratio and the duration of the half cycle is as shown in Figure 16.3, where the duration of the half cycle is divided by the natural period of the system and, once again, the displacement response is divided by the static displacement. It may be observed from Figure 16.3 that, as in the case of free vibrations and a sinusoidal excitation, the peak displacement response of the system decreases with an increase in its damping ratio. However, unless the system is highly damped and the pulse duration is about three-quarters of the natural period of the system, the effect of damping in the case of an impulsive load is relatively less significant than in the case of a harmonic excitation. For example, under the sinusoidal excitation, an increase in the damping ratio from 1 to 10% decreases the displacement response of the system by a factor of about 10. In contrast, if the system is excited by a half-cycle sine pulse with duration equal to the natural period of the system, the same increase in the damping ratio decreases the displacement response of the system by a factor of only about 1.45. It is important to keep this observation in mind when considering short or impulsive ground motions, as in the case of near-field ground motions (see Section 5.7.6).

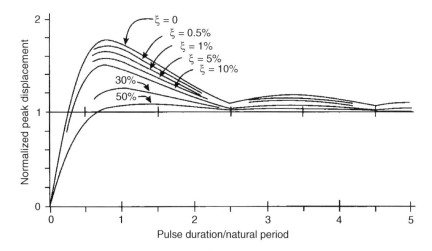

FIGURE 16.3 Normalized peak displacement response of a single-degree-of-freedom system under half-sine pulse. (Reproduced with permission of Earthquake Engineering Research Institute from Hanson, R. D. and Soong, T. T., *Seismic Design with Supplemental Energy Devices*, Earthquake Engineering Research Institute, Oakland, CA, 2001.)

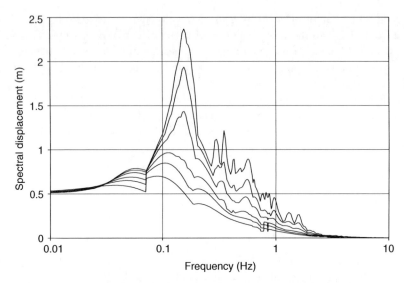

FIGURE 16.4 Displacement response spectrum for damping ratios of 2, 5, 10, 20, 30, and 50% of north–south accelerogram recorded at Wufeng–Taichung Station during the 1999 Chi-Chi, Taiwan, earthquake.

As earthquake ground motions may be considered composed of either a series of short pulses or a series of superimposed harmonic motions with different frequencies, it may be inferred that the effect of damping in single-degree-of-freedom systems under an earthquake ground motion is similar to the effect it has under an impulsive load or a harmonic excitation. This effect may indeed be seen from the displacement response spectrum shown in Figure 16.4, which corresponds to the accelerogram recorded in the north–south direction at the Wufeng–Taichung Station during the 1999 Chi-Chi earthquake in Taiwan. Observe that, as described earlier for the harmonic excitation, the effect of damping is more significant for resonant motions and becomes less significant for the frequencies that are not near to one of the ground motion's dominant frequencies. In addition, as in the case of a harmonic excitation too, an increase in the damping ratio of the system reduces the response of the system more significantly when the initial damping is low than when it is high. In the same way, it may be inferred that the concepts presented earlier for single-degree-of-freedom systems apply equally to linear multi-degree-of-freedom systems. For this, it suffices to recall that the response of a linear multi-degree-of-freedom system may be determined using a modal decomposition and that the response of the system in any given mode of vibration is given by the response of a single-degree-of-freedom system with the natural frequency and damping ratio of the system in that mode. In other words, the effect of damping on a multi-degree-of-freedom system may be visualized by considering the effect of damping in the response in each of its modes of vibration.

Finally, it should be observed that viscous damping is less effective in reducing the response of inelastic systems than it is for linear systems, and that this effectiveness decreases as the inelastic deformations increase. This is so because (a) the input energy imparted to an inelastic system is dissipated through viscous and hysteretic damping (see Section 6.5.2 for a definition of hysteretic damping); (b) as a system incurs into its inelastic range, the system's deformations become relatively more significant than its velocities; and (c) the energy dissipated through viscous damping is consequently less than the energy dissipation through hysteretic damping.

Such an effect may indeed be seen by comparing the way the input energy is dissipated in the two cases presented in Figures 16.5a and 16.5b. Figures 16.5a and 16.5b show the time variation in such two cases of the kinetic and strain energy and the energy dissipated through viscous and hysteretic damping in a 0.3-scale, six-story, concentrically braced steel frame tested in a shaking table. The test is conducted alternatively under different versions of the ground motion recorded

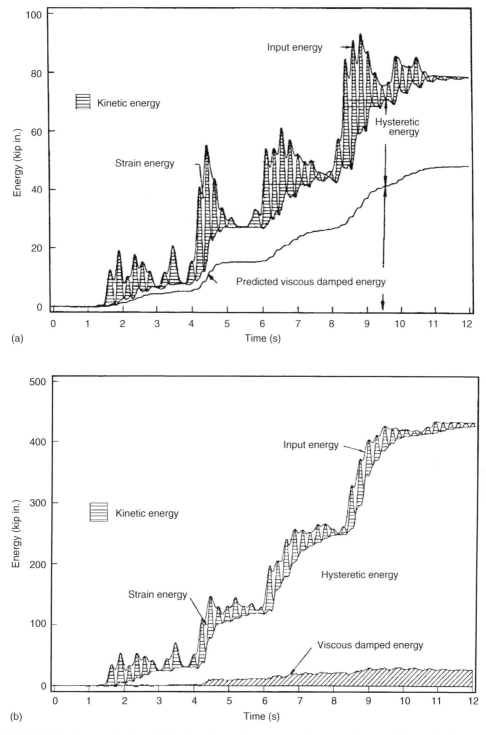

FIGURE 16.5 Time variation of kinetic and strain energy and energy dissipated through viscous and hysteretic damping in specimen tested under ground motion scaled alternatively to a peak acceleration of (a) 0.33*g* and (b) 0.65*g*. (After Uang, C. M. and Bertero, V. V., *Earthquake Simulation Tests and Associated Studies of a 0.3-Scale Model of a Six-Story Concentrically Braced Steel Structure*, Report No. UCB/EERC-86-10, University of California, Berkeley, CA, 1986.)

during the 1978 Miyagi-Ken–Oki earthquake in Japan. In the case shown in Figure 16.5a, the input signal is scaled to a peak acceleration of $0.33g$ and hence induces only moderate inelastic deformations. In contrast, in the case shown in Figure 16.5b the input signal is scaled to a peak acceleration of $0.65g$ and generates in consequence substantial inelastic deformations. It may be noted that in the former case, a significant portion of the input energy is dissipated through both viscous and hysteretic damping. In contrast, most of the input energy in the latter case is dissipated through hysteretic damping, with only an insignificant fraction dissipated through viscous damping. Thus, supplemental dampers may not be an effective way to reduce the response of structures that by design will experience significant inelastic deformations.

16.3 ENERGY DISSIPATING DEVICES

16.3.1 Introduction

Many devices have been suggested over the past few years for the dissipation of energy in structures. However, most of the research and implementation in actual structures have concentrated on four of them: (a) friction dampers, (b) viscoelastic dampers, (c) viscous fluid dampers, and (d) hysteretic dampers. Therefore, this section is devoted to the description of energy dissipation devices, but this description is limited to the four types just mentioned.

16.3.2 Friction Dampers

Friction dampers are those that dissipate kinetic energy through the sliding of surfaces with a high coefficient of friction. They are based on the same principle as the friction brake, which for many years has been successfully used to stop the motion of automobiles, trains, and airplanes. The earliest suggestion for the use of friction dampers to mitigate earthquake effects in buildings appears to have been made by W. O. Keightley in 1977. Keightley's friction dampers are formed with two steel plates slotted and clamped together with bolts and Belleville washers and lubricated to prevent locking. The required normal force is provided by the tension in the bolts. The Belleville washers are used to prevent the loss of this tension. Energy is dissipated after the tension or compression force applied to the plates along their longitudinal direction exceeds the frictional force between the two plates and the plates slide one with respect to the other. Application of a cyclic load with a magnitude greater than the slip force leads to rectangular force–deformation hysteretic loops of the form shown in Figure 16.6.

A similar device is proposed by T. F. Fitzgerald and coworkers in 1989. As shown in Figure 16.7, this device consists of a slotted gusset plate, two back-to-back slotted channel sections, cover plates, and bolts with Belleville washers. The devices are installed at the connection of the braces in concentrically braced frames. In 1993, C. E. Grigorian and coworkers suggested a similar slotted bolted connection, but with a steel-on-brass sliding interface instead of a steel-on-steel interface. In tests conducted with both types of interfaces, they find that the slip force on the steel-on-steel interface is degraded significantly after a few cycles of loading and that this force remains more or less constant on the steel-on-brass interface. That is, the brass-on-steel interface exhibits more stable force–deformation loops than the steel-on-steel one. Another device is the one introduced by Sumitomo Metal Industries of Japan, which originally developed it for railway applications. As shown in Figure 16.8, this device utilizes copper pads impregnated with graphite in contact with the steel casing of the device. The normal force on the friction pads is developed by a series of wedges that act under the compression of Belleville washer springs. The graphite serves the purpose of lubricating the contact surface, ensuring a silent operation, and maintaining a consistent coefficient of friction between the pads and the inner surface of the steel casing.

A friction damper that has been researched extensively and has been implemented in a large number of buildings is that introduced by A. S. Pall in 1979. Pall friction dampers consist of a series of mild steel plates with brake lining pads between them. The plates are clamped together with high-strength

Seismic Protection with Energy Dissipating Devices

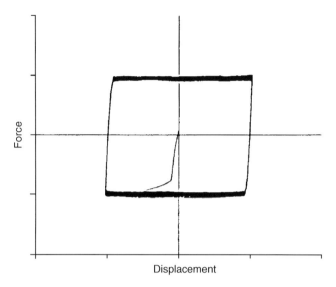

FIGURE 16.6 Typical force–displacement loops generated by friction dampers under cyclic loading. (Reproduced with permission of Earthquake Engineering Research Institute from Constantinou, M. C. et al., *Earthquake Spectra*, 7, 2, 1991, 179–200.)

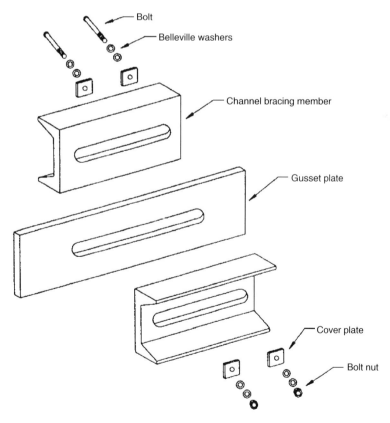

FIGURE 16.7 Construction of bolted slotted connection suggested by T. F. Fitzgerald et al. in 1989. (Reproduced with permission of Earthquake Engineering Research Institute from Fitzgerald, T. F. et al., *Earthquake Spectra*, 5, 2, 1989, 383–391.)

bolts, but allowed to slip under a predetermined load. When subjected to cyclic loading, they exhibit large rectangular hysteretic loops with negligible degradation after several cycles. In addition, their performance is not affected by temperature, velocity, or aging. Furthermore, they neither need maintenance during the lifetime of the building nor replacement after an earthquake. Apparently, their friction properties are predictable and remain stable over the lifetime of the building. They may be installed in single diagonal braces as well as in cross-braces. The installation in cross-braces involves the unique mechanism shown in Figure 16.9. With this mechanism, when an earthquake deforms a structure the compression brace buckles, whereas the tension brace slips at the friction joint. This, in turn, activates the four links, which forces the compression brace to slip too. In this manner, energy is dissipated in both braces despite the fact that the braces are designed to be effective in tension only.

When Pall friction dampers are added, damping in a typical structure may be augmented to exhibit an equivalent damping ratio of up to 30% of critical. Experimental studies have confirmed that these friction dampers indeed enhance the seismic performance of structures as they provide a substantial increase in energy dissipation capacity, and may reduce story drifts by a factor of 2–3. In contrast, the reduction in story shears is only moderate. Pall friction dampers have been

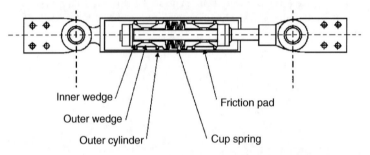

FIGURE 16.8 Sumitomo friction damper. (Reproduced with permission of Earthquake Engineering Research Institute from Aiken, I. D. et al., *Earthquake Spectra*, 9, 3, 1993, 335–370.)

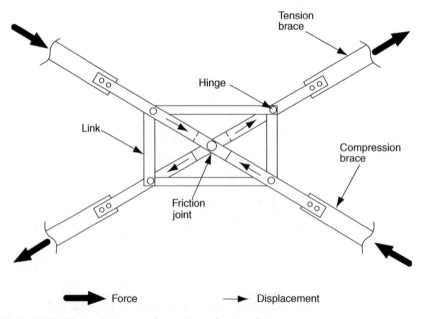

FIGURE 16.9 Pall's friction damper set in tension-only cross-braces.

used worldwide in numerous new and retrofitted buildings as well as in the retrofit of water tanks. An example is the Concordia University library complex in Montreal, Canada, where 143 friction dampers are employed for seismic protection. This complex, designed in 1987 and completed in 1991, consists of two buildings of 6 and 10 stories interconnected by a galleria. The library's exterior and one of the exposed dampers are shown in Figure 16.10.

Friction dampers seem to be an effective way to reduce the seismic response of structures and minimize structural and nonstructural damage. They are easy to construct and install and do not impact the foundation design, features that make them particularly attractive for upgrading existing

(a)

(b)

FIGURE 16.10 (a) Exterior view of Concordia University Library in Montreal, Canada and (b) close-up of one of the added Pall friction dampers. (Photograph courtesy of Avtar S. Pall, Pall Dynamics Limited.)

buildings. In most instances, they are relatively simple to model since the hysteretic loops of a friction damper are similar to the rectangular loops of an ideal elastoplastic material (see Figure 16.6). This means that the slip load of friction dampers may be considered as a fictitious yield force. They also seem to offer savings in the initial cost of new structures and retrofitting of the existing ones compared to conventional solutions. However, it is difficult to maintain their properties over long time intervals. Some of the known problems in this regard are

1. Metal-to-metal interfaces enhance corrosion rates. In particular, common carbon and low-alloy steels suffer enhanced corrosion when in contact with brass, bronze, or copper in all atmospheric environments. Only stainless steel with a high content of chromium does not suffer enhanced corrosion when in contact with brass or bronze. The composition of the interface is thus of paramount importance for ensuring the stability and longevity of the device.
2. The normal load on the sliding interface cannot be reliably maintained and some relaxation of this load should be expected over the years.

Friction dampers also have the following disadvantages:

1. They are effective only for flexible structures; that is, structures for which the dampers will be subjected to relatively large deformations.
2. They encumber the design procedure and make it more expensive. For instance, several alternatives have to be considered to find the optimum slip load, number of dampers, and location of the dampers within a building. Another complication is that their addition to a structure changes the stiffness of the structure and affects its lateral force design. Furthermore, the design of a structure that incorporates friction dampers requires a full nonlinear analysis.
3. The selection of the appropriate slip load is a critical issue in the performance of a structure with friction dampers. Since the devices start working only after their slip load is exceeded, if this load is set too high, then there is the danger of some structural and nonstructural damage before they actually start working. In contrast, if the slip load is set too low, any mild disturbance would unnecessarily make the dampers slip and set the structure into vibrational motion.

Additional issues of concern are (a) the introduction of high-frequency motions owing to the stick–slip behavior of the devices and the consequent potential for damage to equipment and other nonstructural elements, (b) the possibility of a permanent offset of the structure after an earthquake, and (c) the stability of the hysteresis loops under repeated postslip deformations.

16.3.3 Viscoelastic Dampers

Viscoelastic dampers are devices that use the shear deformation of viscoelastic materials (acrylic polymers) as the means to dissipate energy. Viscoelastic materials exhibit the features of an elastic solid as well as those of a viscous liquid. As a result, they return to their original shape after deformation, but with a certain amount of kinetic and strain energy converted into heat in this process. The energy dissipated depends on the induced relative displacement and relative velocity. Viscoelastic dampers have elastic and damping properties and thus not only dissipate energy, but also change the stiffness and natural periods of a structure. As such, they offer a distinct advantage when additional stiffness in addition to damping is beneficial to the structure. A viscoelastic damper developed by the 3M Company is shown in Figure 16.11. This damper is formed with two layers of a copolymer bonded to three steel plates. It dissipates energy through

Seismic Protection with Energy Dissipating Devices

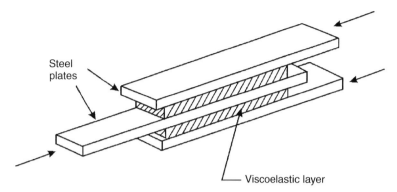

FIGURE 16.11 Viscoelastic damper developed by 3M Company.

the shear deformation of the viscoelastic layers. If properly mounted in a structure (typically in combination with diagonal or chevron braces), energy dissipation takes place when the motion of the structure induces a relative motion between the outer and the center steel plates. Viscoelastic dampers have no force level at which they start to work and thus they dissipate energy at all excitation levels, reducing the structural response under wind loading and small to severe earthquakes. This is in contrast with friction dampers, which for forces less than the slip force do not slide and do not dissipate energy.

The use of viscoelastic materials as a means to dissipate energy dates back to the 1950s when they were first used to control aircraft vibrations through damping surface coatings. The use of viscoelastic dampers in civil engineering structures began in 1969 with the installation of approximately 10,000 of them in each of the 110-story twin towers of the World Trade Center in New York City to mitigate wind-induced vibrations. These dampers were conceived and developed as part of the structural design of these towers. They have also been used for wind control in the 76-story Columbia Seafirst building and the 60-story Two Union Square building in Seattle, Washington; and the 24-story south tower of Shibaura Seavans in Tokyo, Japan. Their use in seismic applications commenced in 1993 with the seismic retrofit of the 13-story Santa Clara County building in San Jose, California. Two viscoelastic dampers per building face per floor level are installed in this building to increase the equivalent damping in its fundamental mode from about 1% to about 17% of critical. In a recent application, they are used for the seismic retrofit of the Treasure Palace in Taipei, Taiwan (see Figure 16.12).

To establish the stress–strain relationship of a viscoelastic material, consider a specimen subjected to a sinusoidal excitation with frequency ω. The shear stress $\tau(t)$ and the shear strain $\gamma(t)$ in the specimen will oscillate with the frequency of the excitation but, because of the viscosity of the material, one will be out-of-phase with respect to the other. Hence, at any time t these two variables may be expressed as

$$\gamma(t) = \gamma_0 \sin \omega t \quad \tau(t) = \tau_0 \sin(\omega t + \delta) \tag{16.11}$$

where γ_0 and τ_0 and δ, respectively, represent peak shear strain and peak shear stress and phase angle.

However, the expression for the shear stress may also be written as

$$\tau(t) = \gamma_0 [G' \sin \omega t + G'' \cos \omega t] \tag{16.12}$$

(a)

(b)

FIGURE 16.12 (a) Exterior view of Treasure Palace in Taipei, Taiwan and (b) view of one of the added viscoelastic dampers. (Courtesy of K. C. Chang, National Taiwan University.)

where

$$G' = \frac{1}{\gamma_0}\tau_0\cos\delta \quad G'' = \frac{1}{\gamma_0}\tau_0\sin\delta \qquad (16.13)$$

In addition, after replacing $\gamma_0 \sin \omega t$ by $\gamma(t)$ and solving for $\cos \omega t$ from Equation 16.12, one obtains

$$\cos\omega t = \frac{1}{G''(\omega)\gamma_0}[\tau(t) - G'(\omega)\gamma(t)] \qquad (16.14)$$

Therefore, with $\sin \omega t$ obtained from Equation 16.11 and using the identity $\sin^2 \omega t + \cos^2 \omega t = 1$, the stress–strain relationship for a viscoelastic material may be alternatively expressed as

$$\left(\frac{\tau(t) - G'(\omega)\gamma(t)}{G''(\omega)\gamma_0}\right)^2 + \left(\frac{\gamma(t)}{\gamma_0}\right)^2 = 1 \tag{16.15}$$

which defines a skewed ellipse as shown in Figure 16.13. It may be seen, thus, that viscoelastic materials exhibit elliptical hysteresis loops with the energy dissipated per unit volume and per cycle of oscillation given by the area of this ellipse. That is, the energy dissipated per unit volume and per cycle of oscillation in a viscoelastic material is given by

$$\begin{aligned} W_D &= \int_0^{2\pi/\omega} \tau(t)\dot{\gamma}(t)dt \\ &= \int_0^{2\pi/\omega} \gamma_0^2 \omega \cos \omega t [G'(\omega)\sin \omega t + G''(\omega)\cos \omega t]dt \\ &= \pi \gamma_0^2 G''(\omega) \end{aligned} \tag{16.16}$$

In view of Equation 16.11 and after considering that $\gamma(t) = \gamma_0 \sin \omega t$ and $\dot{\gamma}(t) = \gamma_0 \omega \cos \omega t$, Equation 16.12 may be rewritten as

$$\tau(t) = G'\gamma(t) + \frac{G''}{\omega}\dot{\gamma}(t) \tag{16.17}$$

which shows that the first term of the shear stress–shear strain relationship represents the in-phase portion of it, with G' characterizing the elastic stiffness (spring constant) of the material. Similarly, the second term represents the out-of-phase portion and energy dissipation component, with G''/ω characterizing its damping constant. Consequently, an equivalent damping ratio for a viscoelastic material is given by

$$\xi = \frac{C}{C_{cr}} = \frac{G''(\omega)}{\omega} \frac{\omega}{2G'} = \frac{G''}{2G'} \tag{16.18}$$

G' is defined as the *shear storage modulus* and is considered a measure of the elastic energy stored and recovered per cycle of deformation. Similarly, G'' is defined as the *shear loss modulus* and is

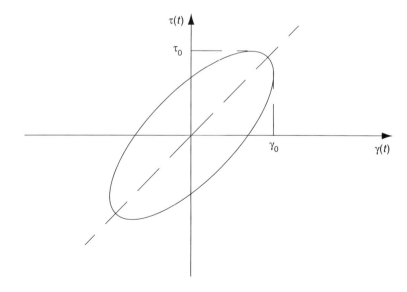

FIGURE 16.13 Shear strain–shear stress relationship of viscoelastic materials.

considered a measure of the energy dissipated per cycle of deformation. A *loss factor* defined as

$$\eta = \frac{G''}{G'} = \tan\delta = 2\xi \tag{16.19}$$

is often used as a measure of the energy dissipation capacity of a viscoelastic material.

The shear storage modulus and shear loss modulus of a viscoelastic material generally depend on the excitation frequency, ambient temperature, and shear strain level. These factors must be therefore accounted for when viscoelastic dampers are incorporated in the design of a building, particularly the temperature dependency. At high frequencies (or low temperatures), the storage modulus G' is large compared to the loss modulus, and the material exhibits a "glassy" behavior, similar to that of a purely elastic material. At low frequencies (or high temperatures), the storage and the loss moduli are both small, and the material exhibits a "rubbery" behavior with little energy dissipation. The loss modulus is the largest at intermediate temperatures and frequencies, and thus the capacity of a viscoelastic material to dissipate energy is the greatest at intermediate temperatures and frequencies. Another factor that may affect the values of the shear moduli is the variation of the temperature within the material itself since energy is dissipated in the form of heat during the deformation of the material and this heat may cause an internal temperature rise.

The variation in the properties of viscoelastic materials with frequency and shear strain may be observed in the force–displacement loops presented in Figure 16.14. These force–displacement loops are obtained from tests where viscoelastic pads are subjected to 20 loading cycles at two different frequencies and two different peak strains. It may be observed from Figure 16.14 that for the test at low frequency (0.10 Hz) the hysteretic loops are regular in shape and show a stable behavior. In contrast, it is seen that during the test at high frequency (3 Hz), the hysteretic loops decrease significantly as the test progresses, although eventually they stabilize to a size that is large enough to dissipate an appreciable amount of energy. These results indicate that under high frequencies and high shear strains, heat is generated rapidly so that the material is unable to radiate this heat fast enough to prevent the material temperature from rising. Consequently, the energy dissipation capacity of the material decreases due to this internal temperature increase. Fortunately, field observations and laboratory experiments have shown that in structural applications this internal temperature increase is typically less than 10° C and has a minor effect on the performance of viscoelastic dampers. It has also been observed that the material properties remain more or less constant with shear strain if this strain is kept below 20%. A further observation is that viscoelastic materials exhibit essentially a linear behavior under cyclic shear strains of less than 125%, provided the temperature rise caused by the generated heat is accounted for. This means that for shear strains of less than 125% a viscoelastic material returns to its initial shape and initial properties after returning to its initial temperature.

In general, the variations of G' and G'' with excitation frequency may be described by straight lines on log–log graphs. Therefore, the effects of temperature can be combined with the effects of frequency in simple log–log graphs such as those shown in Figure 16.15. This depiction provides a simple means for establishing the frequency and temperature dependence of the properties of viscoelastic materials.

Based on Equation 16.17 and the concepts introduced earlier, it may be established that the stress in a viscoelastic material under harmonic motion and for a given ambient temperature and under moderate strain is linearly related to the strain and strain rate. Therefore, the force–displacement relationship for a viscoelastic damper with a total shear area A and thickness h may be expressed as

$$F(t) = k'x(t) + c'\dot{x}(t) \tag{16.20}$$

where $F(t)$ denotes force at time t, $x(t)$ represents the corresponding displacement, $\dot{x}(t)$ is the first time derivative of $x(t)$, and

$$k' = \frac{AG'}{h} \quad c' = \frac{AG''}{\omega h} \tag{16.21}$$

Seismic Protection with Energy Dissipating Devices

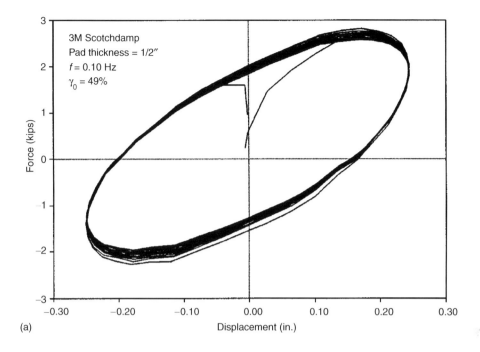

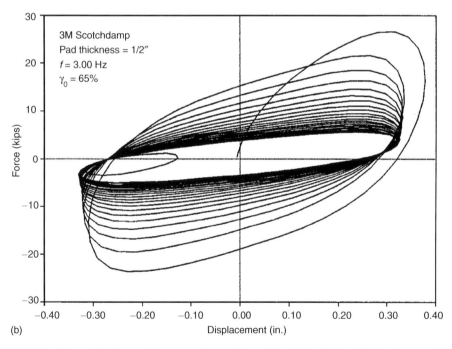

FIGURE 16.14 Force–displacement curves obtained in cyclic tests of viscoelastic dampers: (a) test at frequency of 0.10 Hz and peak shear strain of 49% (b) test at frequency of 3.00 Hz and peak shear strain of 65%. (Reproduced with permission of Earthquake Engineering Research Institute from Bergman, D. M. and Hanson, R. D., *Earthquake Spectra*, 9, 3, 1993, 389–417.)

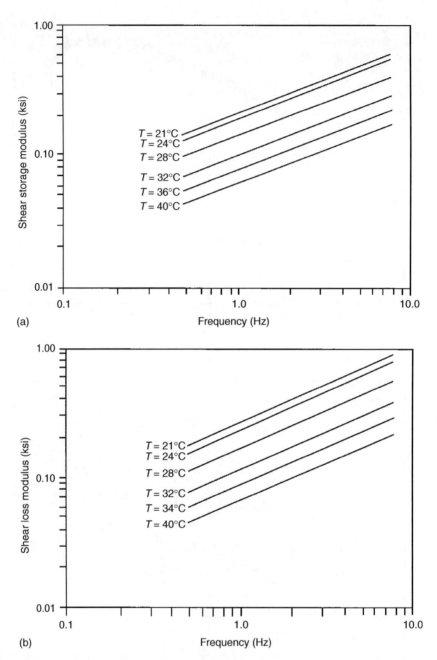

FIGURE 16.15 Approximate variation of (a) shear storage and (b) shear loss moduli with frequency and ambient temperature. (Reproduced with permission of Earthquake Engineering Research Institute from Hanson, R. D. and Soong, T. T., *Seismic Design with Supplemental Energy Devices*, Earthquake Engineering Research Institute, Oakland, CA, 2001.)

This means that a viscoelastic damper may be modeled with a linear spring with stiffness k' and a dashpot with damping constant c'. It should be noted, however, that this is strictly valid for harmonic motion with frequency ω, although it may be a reasonable approximation for general motions if k' and c' are nearly constant within a narrow frequency band.

It may be observed, thus, that the area A of the viscoelastic pads for an effective viscoelastic damper may be calculated straightforwardly using the first expression in Equation 16.21 after (a) determining their shear storage modulus G' and shear loss modulus G'' from the graphs in Figure 16.15 for the design frequency and temperature; (b) a value of the stiffness constant k' is selected (normally by trial and error until a target damping ratio is attained), and (c) a thickness h is chosen based on the largest allowable structural deformation and the desirable maximum strain in the viscoelastic pads. It may also be observed that once the area A is known, the damping constant c' and the associated damping ratio may be determined from the second expression in Equation 16.21.

Extensive analytical and experimental studies have offered proof that viscoelastic dampers may be effectively used to introduce significant extra damping into structures and enhance their seismic performance. They also seem to offer several advantages over other damper types. They provide additional stiffness to the structure, provide significant damping at low intensity vibrations such as those generated by wind motions and moderate seismic events, and are often more cost-effective than other dampers. However, they also have the disadvantage that their properties depend strongly on frequency, ambient temperature, and shear strain level. This disadvantage complicates the analysis of the structure, elevates the cost of this analysis, and encumbers the process for the selection of their optimum number and location within the structure. An additional concern is their long-term reliability and durability. Tests conducted with viscoelastic dampers installed in a number of buildings to reduce wind-induced vibrations revealed that the dampers suffered a stiffness reduction of about 35% in a period of 25 years.

16.3.4 Fluid Dampers

Modern fluid dampers are devices that utilize the flow of a fluid through orifices as a means to dissipate energy. The dissipated energy is transformed into heat, which elevates the temperature of the damper's fluid and mechanical parts, but eventually is harmlessly transferred to the surrounding environment by transport mechanisms such as convection and conduction. They generally consist of a piston housed in a cylinder filled with a viscous fluid. The piston head contains a number of small orifices through which the fluid passes. Energy is dissipated when, on the displacement of the piston, the fluid travels from one side of the piston head to the other. Dampers that operate on the principle of a fluid flowing through orifices are more efficient than the traditional dashpots, which dissipate energy through the motion of a piston in a container filled with a viscous fluid.

Fluid dampers that operated on the principle of fluid flow through orifices (sometimes called inertial fluid dampers) were first used in a French artillery field gun in 1897 with the purpose of arresting the gun recoil. It was a national secret of France until shared with the United States and Great Britain during World War I. By the end of World War I, tens of thousands of fluid dampers were used in field artillery guns, naval guns, coastal guns, and railway guns. With the surging popularity of the automobile in the 1920s and 1930s, manufacturers searched for damping systems that would allow them to produce an automobile that would ride smoothly over all possible road surfaces. In response, fluid inertial dampers were adapted by the automobile industry. In 1925, Ralph Peo of Buffalo, New York, invented a rotatory fluid damper that allowed for about 16,000 km of road travel before needing seal replacement. In 1949, the era of the rotatory dampers ended with the design by the Delco Division of General Motors of a sliding damper that had an adequate life for automotive use. Modern automotive shock absorbers have a design similar to the gun recoil dampers used during World War I, except that the seals in them have a substantially greater life.

Fluid dampers have found numerous applications in the shock isolation of military hardware and shock and wind vibration control of military structures such as missile launching platforms. After the end of the Cold War in the 1990s, security restrictions on the commercial use of the damper technology developed during that era were relaxed, and research related to the use of fluid dampers for seismic and wind protection of civil structures began. Tests were conducted of building and bridge models implemented with the fluid dampers which were developed, for example, for the B-2 Stealth Bomber, the Tomahawk Cruise Missile, and the Glomar Explorer Research Vessel. Eventually, that research led to implementation of fluid dampers in numerous civil structures to enhance their performance under seismic or wind excitations. Use of fluid dampers for seismic energy dissipation commenced in 1993 with the implementation of 186 fluid dampers in the base-isolated buildings of the San Bernandino County Medical Center in California. These dampers were installed in parallel with the isolation bearings to reduce the peak displacement of the bearings. Since then, fluid dampers have been used for seismic protection in a large number of buildings and bridges throughout the world. A most significant application was the installation of 98 fluid dampers in a 57-story office building in Mexico City completed in 2003. The dampers were installed in mega-brace elements that span up to six stories and have a length of up to 20 m each (see Figure 16.16).

A typical fluid damper for seismic applications is shown in Figure 16.17. It contains a stainless-steel piston rod with a pierced bronze head and is filled with silicon oil. Silicon oil is fire-resistant, nontoxic, thermally stable, and does not degrade with age. The piston head orifices are bimetallic, acting thus as a thermostat that compensates for the changes in the fluid properties caused by temperature variations. This allows an effective operation of the device over a temperature range $-40°$ F to $160°$ F ($-40°$ C to $70°$ C). The damper also contains an accumulator and a control valve. The accumulator is provided to compensate for the difference between the volumes of chambers 1 and 2, a difference introduced by the presence of the piston rod in chamber 1. Since the fluid is compressible, this difference in volume would cause the compression of the fluid in chamber 1 when the fluid moves from chamber 2 to chamber 1 and, hence, the development of a springlike restoring force. Alternatively, the damper may be constructed with a run-through rod as shown in Figure 16.18. This configuration prevents the aforementioned fluid compression and the need for an accumulator. The control valve in the damper with the accumulator is used to regulate the amount of fluid entering the accumulator when the damper is compressed. When the damper is extended, the control valve opens to let the fluid freely flow from the accumulator to the chambers. The damper is tightly sealed with seals manufactured with high-strength structural polymers. Tests have demonstrated that these seals do not stick after long periods of inactivity and do not degrade over time. They have a service life of at least 25 years. Observe that if the damper with the accumulator is activated and the piston moves from left to right, for example, then the displacement of the piston compresses the fluid in the central chamber and increases the fluid pressure in this chamber. In turn, this increase in pressure pushes the fluid from the central chamber into the first chamber and into the accumulator. Energy is dissipated as the fluid flows from one chamber to another.

The resistive force provided by a fluid damper may be expressed as

$$F = c_d |\dot{x}|^\alpha \text{ sgn}(\dot{x}) \tag{16.22}$$

where c_d, $\dot{x}$, and α are constant, velocity of the piston rod, and exponent that varies between 0.5 and 2.0. For cylindrical orifices (called Bernoullian orifices because they follow Bernoulli's equation), $\alpha = 2$. A value of α equal to 1 corresponds to the classical linear viscous damper, which facilitates the analysis when dampers are used for seismic energy dissipation. The construction of dampers with an exponent α different from 2 requires specially shaped orifices to attain different flow characteristics. Because of their great efficiency, dampers with an exponent α of 0.5 are useful in applications involving extremely high velocity shocks. Figure 16.19 shows the typical force–displacement loops obtained with a damper for which α is equal to 0.5.

To understand the advantage of using a viscous damper with a small exponent α, consider a fluid damper with the constitutive relation described by Equation 16.22 and subjected to the sinusoidal motion

$$x = x_0 \sin \omega_0 t \tag{16.23}$$

(a)

(b)

FIGURE 16.16 (a) Exterior view of 57-story building in Mexico City implemented with fluid dampers and (b) view of one of the added fluid dampers. (Photograph courtesy of Taylor Devices, Inc.)

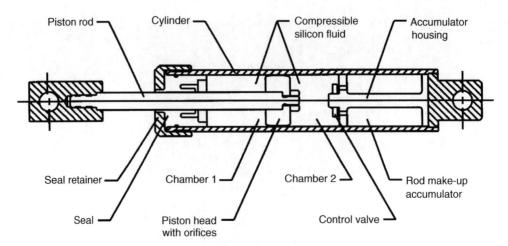

FIGURE 16.17 Construction of modern fluid damper. (Reproduced with permission of Taylor Devices, Inc. from Taylor, D. P. and Constantinou, M. C., *Fluid Dampers for Application of Seismic Energy Dissipation and Seismic Isolation*, Technical Report, Taylor Devices, Inc., 2000.)

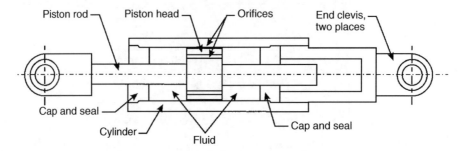

FIGURE 16.18 Fluid damper with run-through rod. (Courtesy of Taylor Devices, Inc.)

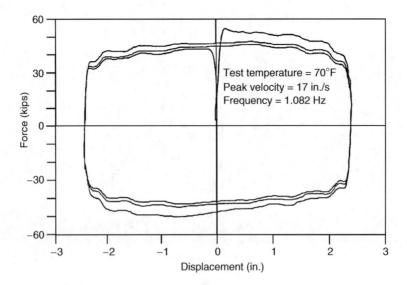

FIGURE 16.19 Force–displacement loops obtained with viscous fluid damper with $\alpha = 0.5$. (Reproduced with permission of Taylor Devices, Inc. from Taylor, D. P. and Constantinou, M. C., *Test Methodology and Procedures for Fluid Viscous Dampers Used in Structures to Dissipate Seismic Energy*, Technical Report, Taylor Devices, Inc., April 1994.)

where x_0 represents displacement amplitude, t denotes time, and $\omega_0 = 2\pi/T$ in which T is the period of the excitation. The energy dissipated by the damper in a cycle of motion is given by

$$W_d = \int_0^T F\dot{x}\,dt \qquad (16.24)$$

which after substitution of Equation 16.22 and integration leads to

$$W_d = 4(2^\alpha)\frac{\Gamma^2\left(1+\dfrac{\alpha}{2}\right)}{\Gamma(2+\alpha)} c_d \omega_0^\alpha x_0^{1+\alpha} \qquad (16.25)$$

where Γ represents the gamma function. Note, however, that in terms of $F_{max} = C_d \omega_0^\alpha x_0^\alpha$ (the peak value of the force F), Equation 16.25 may also be written as

$$W_d = 4(2^\alpha)\frac{\Gamma^2\left(1+\dfrac{\alpha}{2}\right)}{\Gamma(2+\alpha)} F_{max} x_0 \qquad (16.26)$$

It may be seen, thus, that if $\alpha = 2$, $W_d = 2.667 F_{max} x_0$. In contrast, if $\alpha = 0.5$, $W_d = 3.496 F_{max} x_0$. Consequently, a damper with $\alpha = 0.5$ dissipates 31% more energy than a damper with $\alpha = 2.0$. Therefore, dampers with an exponent α of less than 2 are considered more efficient than dampers with an exponent α of 2.

The significance of the influence of the exponent α in the energy dissipation capacity of a fluid damper may be further demonstrated by comparing the displacement time histories shown in Figure 16.20. These time histories correspond to a linear system subjected to an initial displacement and equipped with a fluid damper for which the exponent α is alternatively considered equal to 0.5, 1.0, or 2.0. It may be seen from Figure 16.20 that the reduction in amplitude per cycle of motion and the overall performance are indeed better for the damper with α equal to 0.5.

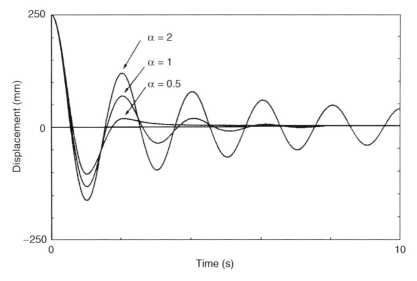

FIGURE 16.20 Displacement responses of linear system equipped with fluid damper for which exponent α in force–velocity relationship is considered alternatively equal to 0.5, 1.0, and 2.0. (Reproduced with permission of CISM, the International Center for Mechanical Sciences, Udine, Italy from Constantinou, M. C., Chapter X in *Passive and Active Structural Control in Civil Engineering*. Soong, T. T. and Constantinou, M. C., editors, CIMS Courses and Lectures No. 345, Springer, Vienna, NY, 1994.)

Fluid dampers have the unique ability to simultaneously reduce both stresses and deflections in a structure subjected to a seismic excitation. This is so because the force in a fluid damper varies only with velocity and velocity is inherently out-of-phase with the deformations that produce the stresses in the structure. Consider, for example, the stresses in the column of a building under an earthquake ground motion. The stresses in the column reach their peak value when the building has flexed a maximum amount from its normal position. However, this is also the point at which the flexed columns reverse direction to move back in the opposite direction and the point at which the velocity response of the column is zero. If a fluid damper is connected to the column, the force in the damper will thus reduce to zero at this point of maximum deflection. Similarly, the maximum damper force will occur at the point of maximum velocity, which occurs when the column flexes back to its undeformed position. But this is also the point when the column stresses are at a minimum. In other words, the peak stress in the column and the peak damper force never occur at the same time. This is an attractive design feature that other types of dampers do not have since the resisting forces generated by these other dampers are a function of displacement or displacement and velocity. Hence, other types of dampers may decrease the deflection of a structure, but at the expense of increasing its stresses. It is important to keep in mind, nonetheless, that the addition of fluid dampers to a structure may change its natural frequencies (see Section 15.5.4) and that this change in natural frequencies may, in turn, increase or decrease the earthquake input to the structure.

Fluid dampers thus seem to be an effective way to enhance the seismic performance of many types of buildings, and some shake table tests have verified this effectiveness. They may be used to predictably increase the damping forces in a building and reduce in consequence the force and deformation demands on the building. In addition, they are easy to install due to their relatively compact size. If used in combination with cross-braces, fluid dampers will usually have a smaller cross-section envelope than the brace. However, as with other types of dampers, they are effective only for flexible structures, given that fluid dampers work only when subjected to relatively large displacements. This means that fluid dampers would be of limited value in short shear-wall buildings, for example. The reason is that under small displacements (2 mm or less) the fluid pressure inside the damper is not high enough to generate a significant fluid flow and hence significant energy dissipation. Because of their nonlinear force–displacement behavior and the need to consider several alternatives to find their optimum number and location, they also encumber the design of the structure and make it more expensive. Further concerns are their high cost, the long-term durability of the damper seals, and the maintenance costs associated with this issue.

16.3.5 Yielding Metallic Dampers

Yielding metallic dampers use the hysteretic properties of metals when deformed into their postelastic range as a means to dissipate energy. The metal employed is usually mild steel, although lead and some other sophisticated materials such as nickel–titanium alloys (nitinol) have been considered. Under moderate earthquakes (small deformations), a yielding metallic damper acts as a stiff member that helps to resist structural deformations, whereas under severe earthquakes (large deformations) it acts as an energy absorber. They require, thus, a relatively large deformation before they start dissipating energy. Nonetheless, when incorporated at strategic locations within a building, yielding metallic dampers are capable of improving the earthquake performance of a structure by increasing its stiffness, its strength, and its ability to dissipate energy. In general, yielding metallic dampers show a stable behavior under a large number of loading cycles, exhibit insignificant age effects, and have an adequate resistance to environmental and temperature factors. However, repeated yielding may lead to a fatigue failure.

The idea of using yielding metallic elements as energy absorbers was first introduced by R. Skinner, J. M. Kelly, and A. J. Heine in 1973, then at the Physics and Engineering Laboratory of New Zealand's Department of Scientific and Industrial Research. These investigators considered the use and tested to examine their performance as dampers torsional beams, flexural beams, and rolling-bending U-strips in the forms shown in Figure 16.21. The basic difference among these elements is that the torsional beam deforms the metal in torsion, the flexural beam in bending, and the U-strips also in bending, but through the rolling motion of the strips. Another one of the earlier metallic dampers is the lead-extrusion damper proposed in 1976 by the New Zealanders W. H. Robinson and L. R. Greenback. This damper is based on the process of extrusion, which consists of forcing a solid element through a hole or orifice with the purpose of changing its shape. As shown in Figure 16.22, the lead-extrusion damper consists of a thick-wall tube coaxial with a shaft that has two pistons. The tube has a constriction between the two pistons, and the space between the pistons is filled with lead. The lead is separated from the tube by a thin layer of lubricant, in turn kept in place by two hydraulic seals. As the shaft is displaced back and forth relative to the tube, the lead is forced to extrude through the orifice formed by the constriction in the tube. In the process, the lead deforms plastically and dissipates energy. The success of this damper is owed to the use of lead, which is a metal that recovers and recrystallizes rapidly at room temperature.

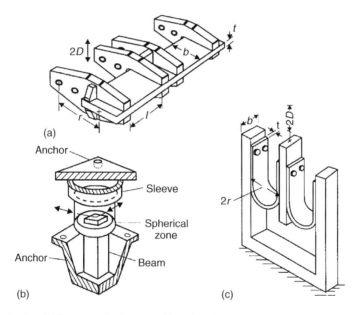

FIGURE 16.21 Earlier yielding metallic dampers: (a) torsional beam, (b) flexural beam, and (c) U-strip. (Reproduced with permission of John Wiley & Sons from Skinner, R. I., Kelly, J. M., and Heine, A. J., *Earthquake Engineering and Structural Dynamics*, 3, 3, 1975, 287–296.)

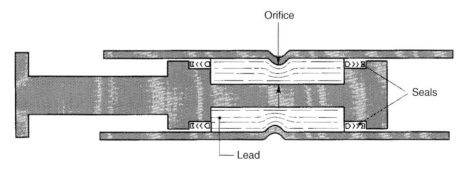

FIGURE 16.22 Longitudinal section of lead-extrusion damper. (Reproduced with permission of John Wiley & Sons from Robinson, W. H. and Greenback, L. R., *Earthquake Engineering and Structural Dynamics*, 4, 3, 1976, 251–259.)

After the introduction of the metallic dampers just described, several other much simpler types have been developed. Among the most prominent are (a) the X-shaped steel plate damper (also known as added damping and stiffness element, or *ADAS element*), (b) the triangular steel plate damper (also known as triangular added damping and stiffness element, or *TADAS element*), and (c) the honeycomb plate damper. All these metallic dampers, illustrated in Figure 16.23, dissipate energy through the bending of simple steel plates. The X-shaped plate damper utilizes multiple mild steel plates with an hourglass form configured in parallel and with their ends fixed. It is an offshoot of the triangular dissipative devices developed in 1978 by R. G. Tyler (another New Zealander) for piping support. The hourglass form is used so that, as each plate deforms in double curvature into its plastic range, yielding of the plate is uniformly distributed along its length. In contrast, a rectangular plate yields only at its ends when deformed plastically in double curvature. This concentration of yielding is undesirable in terms of both the amount of energy that can be dissipated and the stability of the hysteretic force–deformation loops. Figure 16.24 shows the shape of a single plate and a schematic installation of the dampers in a three-story frame, whereas Figure 16.25 depicts the typical force–deformation loops obtained with an X-shaped steel damper. Shake table tests of structural models have demonstrated the effectiveness of these dampers. However, it has been observed that the dampers' stiffness is sensitive to the tightness of the bolts that hold the ends of the plates together, and that, in general, the stiffness obtained in tests is much less than the stiffness calculated assuming that both ends are fixed. X-shaped steel dampers have been implemented in

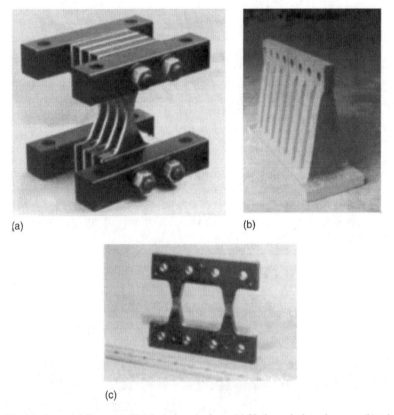

FIGURE 16.23 Modern yielding metallic damping devices: (a) X-shaped plate damper, (b) triangular plate damper (Reproduced from Whittaker, A. and Constantinou, M., *Earthquake Engineering: From Engineering Seismology to Performance-Based Engineering*, CRC Press LLC, Boca Raton, FL, 2004.), and (c) honeycomb plate damper (Reproduced from Kobori, T. et al., *Proceedings 10th World Conference on Earthquake Engineering*, Vol. 4, Madrid, Spain, 1992, pp. 2341–2352.)

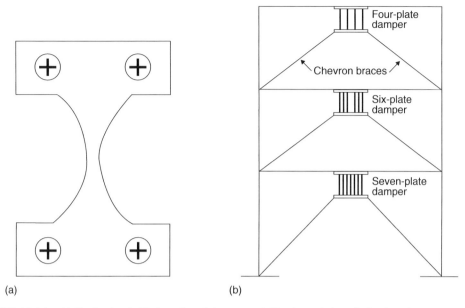

FIGURE 16.24 (a) Single plate in X-shaped steel damper and (b) schematic installation in a three-story frame.

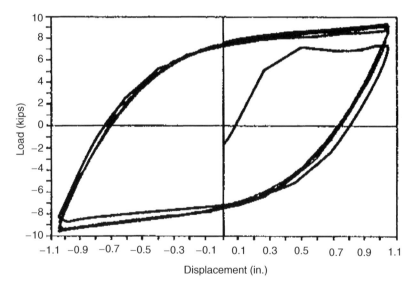

FIGURE 16.25 Typical force–displacement loops obtained with a seven-plate X-shaped steel damper. (After Bergman, D. M. and Goel, S. C., *Evaluation of Cyclic Testing of Steel-Plate Devices for Added Stiffness and Damping*, Technical Report UMCE 87-01, University of Michigan, 1987.)

the retrofitting of a two-story concrete building in the United States and at least four buildings in Mexico City. One of these buildings is shown in Figure 6.26.

Like the X-shaped plate damper, the triangular plate damper is formed with a series of identical triangular steel plates in parallel and typically installed within a frame between chevron braces and the frame's beam. The base of each triangular plate is welded into a rigid plate to attain a fixed condition as close as possible. A slotted pin connection is used at the other end to ensure the relatively free motion of the plates in the vertical direction. Because of this configuration, the damper resists

FIGURE 16.26 (a) Office building in Mexico City repaired in 1997 with X-shaped steel plate dampers mounted on chevron braces and (b) view of one of the added braces. (Reproduced with permission from Sociedad Mexicana de Ingeniería Sísmica from Tena-Colunga, A., *Memorias del VIII Simposio Nacional de Ingeniería Sísmica*, Tlaxcala, Mexico, 2004.)

lateral loads as a cantilever beam with a uniform flexural deformation along the length of the damper (see Figure 16.27). An example of the force–displacement hysteretic loops obtained with this type of damper under cyclic load is shown in Figure 16.28. It may be seen that if properly constructed, triangular steel plate dampers are capable of sustaining a large number of yielding loading cycles without strength and stiffness degradation. Triangular steel plate dampers have been implemented in the construction of the Living Mall in Taipei, Taiwan.

Lastly, the honeycomb damper is formed with a steel plate that has openings in a honeycomb pattern. The force–deformation hysteresis loops obtained with this type of damper are of the form shown in Figure 16.29, which correspond to a test specimen under cyclic load with increasing amplitude. It may be seen that, as the X-shaped and triangular steel dampers, honeycomb steel dampers exhibit a stable energy-absorbing capability with almost square hysteresis loops. Honeycomb dampers were used in the construction of a 29-story hotel and apartment building in Japan (see Figure 16.30).

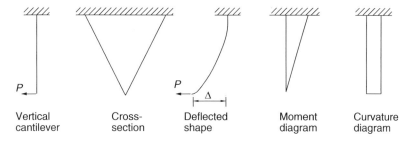

FIGURE 16.27 Deflected shape, bending moment, and curvature of a triangular plate under lateral load.

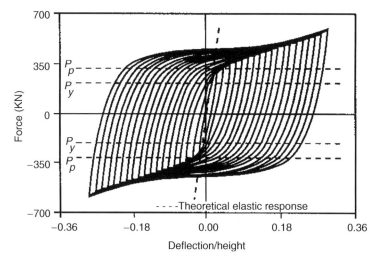

FIGURE 16.28 Typical force–deformation loops obtained with a triangular plate damper. (Reproduced with permission of Earthquake Engineering Research Institute from Tsai, K. C. et al., *Earthquake Spectra*, 9, 3, 1993, 505–528.)

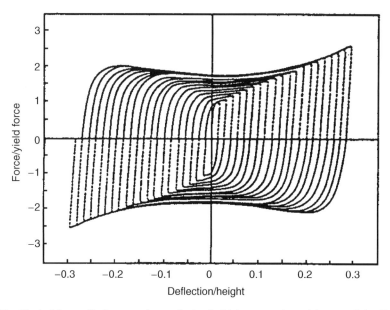

FIGURE 16.29 Typical force–displacement loops obtained with honeycomb steel damper. (After Kobori, T. et al., *Proceedings 10th World Conference on Earthquake Engineering*, Vol. 4, Madrid, Spain, 1992, pp. 2341–2352.)

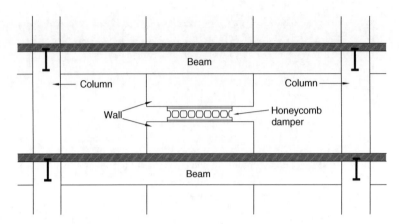

FIGURE 16.30 Installation of honeycomb steel dampers at mid-height of shear walls. (After Kobori, T. et al., *Proceedings 10th World Conference on Earthquake Engineering*, Vol. 4, Madrid, Spain, 1992, pp. 2341–2352.)

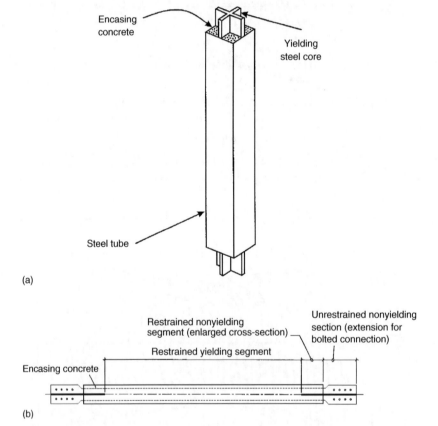

FIGURE 16.31 (a) Schematic and (b) detailed configuration of buckling-restrained steel brace.

Another damper that operates on the same energy dissipation principle as the yielding dampers described earlier is the tension-compression yielding brace, also known as *unbonded brace* or *buckling-restrained brace*. As shown in Figure 16.31, this damper consists of a core steel plate with a rectangular or cruciform cross-section encased in a concrete-filled steel tube. A special coating or unbonding layer of rubber, polyethylene, silicon grease, or mastic tape is used to eliminate or reduce the transfer of the axial force resisted by the core plate to the concrete-filled steel tube and avert the buckling of the tube. The concrete-filled tube prevents the compression buckling of the core steel plate and thus the brace is capable of yielding under tensile and compressive loads. Buckling-restrained braces are typically installed using a chevron configuration.

The buckling-restrained brace was developed by Nippon Steel Corporation in 1977 in an effort to overcome the inadequacies of conventional braced frames. Braces in a conventional braced frame are designed to yield when in tension and buckle when in compression, thus providing a poor energy dissipation capability. In contrast, buckling-restrained braces can yield in tension and compression and provide in consequence large and stable hysteresis loops and an excellent energy dissipation capability. Figure 16.32 schematically illustrates the different force–deformation behaviors of conventional and buckling-restrained braces and why the energy dissipated during a cycle of loading is indeed much greater for a buckling-restrained brace than for a conventional one. Figure 16.33 shows the force–deformation behavior typically observed in cyclic tests of such a brace. In addition to providing a large energy dissipation capability, buckling-restrained braces are capable of furnishing the rigidity needed to satisfy structural drift limits, are easy to install, and seem to be cost-effective. They have been used for seismic applications in a large number of buildings in the United States, Japan, and Taiwan. An example is shown in Figure 16.34.

It appears, thus, that yielding metallic devices are another effective way to augment damping in structures and improve their seismic performance, an effectiveness verified by shake table tests. They also seem to offer cost advantages, particularly when used in the retrofitting of existing structures. Furthermore, their properties are virtually insensitive to environmental factors and age effects. Notwithstanding these advantages, they also have the same disadvantages of friction dampers. Namely, (a) they are only effective for flexible structures, (b) they complicate the design

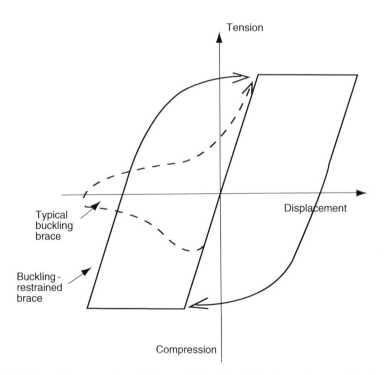

FIGURE 16.32 Schematic comparison of force–deformation behaviors of conventional and buckling-restrained steel braces. (Reproduced with permission of SIE Inc. from Clark, P. et al., *Proceedings 69th Annual Convention of SEAOC*, Sacramento, CA, 1999.)

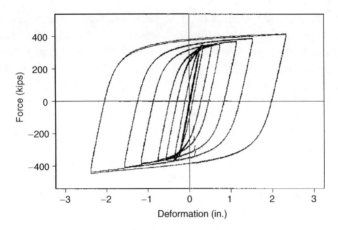

FIGURE 16.33 Typical force–deformation behavior of buckling-restrained steel braces. (Reproduced with permission of SIE Inc. from Clark, P. et al., *Proceedings 69th Annual Convention of SEAOC*, Sacramento, CA, 1999.)

FIGURE 16.34 (a) Application of buckling-restrained braces in Santa Clara Kaiser Permanente Medical Center Replacement Project and (b) example of connection detail. (Photograph courtesy of M. Stoops, Kaiser Permanente; Zigmund Rubel, Anshen & Allen; and E. Ko, Arup.)

procedure as several alternatives need to be considered to find the optimum yield level and number and location of dampers, (c) they make a nonlinear structural analysis necessary, (d) they are capable of generating high-frequency motions due to the sudden change in the stiffness of the structure after the dampers yield, and (e) they may leave a structure with a permanent offset after an earthquake. An additional disadvantage is the possibility of a premature fatigue failure.

16.4 ANALYSIS OF STRUCTURES WITH ADDED DAMPERS

16.4.1 Overview

As far as the analysis is concerned, there are significant differences between a structure with added dampers and a traditional one without dampers, and these differences may complicate the analysis of structures with added dampers. Some of such differences are

1. In a structure that behaves linearly, the addition of dampers causes significant increases in the modal damping ratios of the structure and these increases may make some of these damping ratios reach values that are close to or even greater than the critical ones. In addition, it may cause a redistribution of the damping ratios among its various modes of vibration, such that some modes may contribute little to the total response of the structure before the dampers are added, but may contribute substantially after the dampers are added.
2. Because of the uneven distribution of the damping sources and the high damping ratios involved, structures with added dampers may not be considered as classically damped systems. That is, it may not be assumed without introducing significant errors that the undamped mode shapes of the structure with added dampers would uncouple its equation of motion. This is particularly true when the added dampers are not uniformly distributed throughout the structure. The analysis of structures with added dampers thus requires the consideration of a direct integration method or the complex modal superposition approach described in Section 15.5.4.
3. As seen from the discussion in Section 16.3, most damping devices exhibit nonlinear behavior. Therefore, damper-added structures will behave nonlinearly even when the original structure behaves linearly. This means that the analysis of a structure with added dampers needs to be performed, at least for the final design, considering explicitly such nonlinear behavior.

In any case, the analysis of damper-added structures may be carried out by simply incorporating into the original structure additional structural elements that characterize the mass, stiffness, and damping properties of the added dampers. These elements must reflect the force–deformation behavior observed in experimental tests of the dampers being considered. This behavior, however, may be approximately represented with simple models. For example, a friction damper may be modeled with a rigid-plastic spring or truss element with a yield force equal to the slip load of the damper. That is, it may be modeled as a rigid link before it slips and as an element with zero stiffness after it slips. Similarly, a metallic damper may be modeled with a bilinear (see Section 6.5.2) spring or truss element with a yield force, initial stiffness, and postyield stiffness that reflect the properties of the damper in question. In the case of a fluid damper, the damper may be modeled with a dashpot whose force–velocity relationship is given by Equation 16.22. In this regard, it should be noted that if the exponent α in this relationship is different from 1 for the selected dampers, then a nonlinear analysis is necessary to properly account for the contribution of these dampers. Lastly, a viscoelastic damper may be modeled with a spring and a damper. As mentioned in Section 16.3.3, the spring and damping constants may be determined for the design frequency and temperature using the expressions given by Equation 16.21. Because of the dependence of these constants on the excitation frequency and ambient temperature, the analysis of structures with viscoelastic dampers requires the consideration of the variations in the properties of the dampers due to the possible deviations from the design temperature and frequency. As added dampers are usually installed in combination with one or more braces, the models of damper-added structures should also incorporate additional structural elements to represent the influence of such braces.

The idea of representing added dampers and braces with simple structural elements is illustrated in Figure 16.35 for a simple frame and different brace configurations. In Figure 16.35, k_d and c_d denote the spring and damping constants of the dampers, respectively, whereas k_b represents the stiffness of the braces.

The following sections describe in detail how to perform the analysis of damper-added structures using the ideas just presented. The discussion is first limited to a simple structure to introduce the basic concepts in a clear and simple manner. Then, it is extended to the general case of a multi-degree-of-freedom structure to show how these basic concepts can be applied to realistic structures. Some numerical examples are also presented to further illustrate the introduced concepts.

16.4.2 Seismic Response of Simple Structure with Added Dampers

Consider the simple frame shown in Figure 16.36 and assume the frame can be modeled as a single-degree-of-freedom system with mass m, damping constant c, and stiffness k. Assume, in addition, that the frame is subjected to a ground acceleration $\ddot{u}_g(t)$ and that the initial displacement and initial velocity are both zero. Furthermore, assume that a damper has been added to the frame and that, by design, the structure is to remain linear at all times.

According to the free-body diagram shown in Figure 16.36b, the equation of motion of the frame with the added damper may be written as

$$m\ddot{u}(t) + c\dot{u}(t) + ku(t) + [F_k(t) + F_d(t)]\cos\theta = -m\ddot{u}_g(t) \tag{16.27}$$

where $u(t)$ denotes the horizontal displacement of the frame's mass relative to the ground; $F_k(t)$ and $F_d(t)$ represent the elastic and damping components of the force in the damper, respectively; θ is the inclination angle with respect to the horizontal of the brace where the damper is installed, and a dot above a variable indicates its derivative with respect to time.

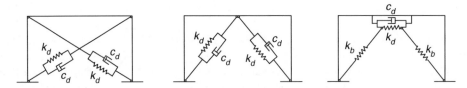

FIGURE 16.35 Structural models of simple frames with added brace-damper assemblies.

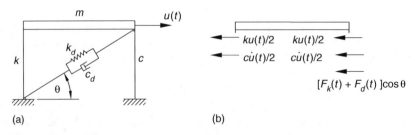

FIGURE 16.36 (a) Simple frame with added damper and (b) free-body diagram of frame's mass.

16.4.2.1 Viscoelastic Damper

If the damper is of the viscoelastic type, then, according to Equation 16.20, $F_k(t)$ and $F_d(t)$ are given by

$$F_k(t) = k'u(t)\cos\theta \tag{16.28}$$

$$F_d(t) = c'\dot{u}(t)\cos\theta \tag{16.29}$$

where, for a selected ambient temperature and dominant excitation frequency, k' and c' are given by Equation 16.21, and where it has been considered that the relative displacement and relative velocity along the direction of the diagonal brace are, respectively, equal to $u(t)\cos\theta$ and $\dot{u}(t)\cos\theta$.

On substitution of Equations 16.28 and 16.29 into Equation 16.27, one has thus that with a viscoelastic damper the equation of motion of the frame is of the form

$$m\ddot{u}(t) + [c + c'\cos^2\theta]\dot{u}(t) + [k + k'\cos^2\theta]u(t) = -m\ddot{u}_g(t) \tag{16.30}$$

This is a linear differential equation and may therefore be solved using any of the known methods for single-degree-of-freedom linear systems. The only difference is that now the damping and stiffness constants of the system are given by

$$\bar{c} = c + c'\cos^2\theta \tag{16.31}$$

$$\bar{k} = k + k'\cos^2\theta \tag{16.32}$$

16.4.2.2 Fluid Damper

If the added damper is of the fluid type, then, according to Equation 16.22, $F_k(t) = 0$ and

$$F_d(t) = c_d |\dot{x}(t)^\alpha| \, \text{sgn}[\dot{x}(t)] \tag{16.33}$$

where $\dot{x}$ and c_d and α denote relative velocity along the longitudinal direction of the damper and constants that depend on the properties of the damper used. Consequently, the equation of motion of the structure in the case of a fluid damper may be expressed as

$$m\ddot{u}(t) + c\dot{u}(t) + \text{sgn}[\dot{u}(t)]c_d|\dot{u}(t)|^\alpha \cos^{\alpha+1}\theta + ku(t) = -m\ddot{u}_g(t) \tag{16.34}$$

This is a nonlinear differential equation and therefore its solution requires a step-by-step numerical integration. To this end, consider the discretized form of the equation of motion, which for the time $t_{i+1} = (i+1)\Delta t$, where Δt is the selected time step of integration and i is any integer, may be written as

$$m\ddot{u}_{i+1} + c\dot{u}_{i+1} + ku_{i+1} = -m(\ddot{u}_g)_{i+1} - \text{sgn}(\dot{u}_{i+1})c_d|\dot{u}_{i+1}|^\alpha \cos^{\alpha+1}\theta \tag{16.35}$$

where a variable with the subscript $i+1$ signifies that the variable is being evaluated at time t_{i+1}.

Equation 16.35 may be solved with any of the direct integration methods introduced in Chapter 11, such as the constant average acceleration method discussed in Section 11.5. It should be noted, however, that in this case, the relative velocity $\dot{u}_{i+1}$ is not known in advance and thus the right-hand side of Equation 16.35 cannot be determined in advance, unless an iterative procedure is adopted. To avoid this problem, one can consider that this velocity is given approximately by the expression that is derived from assuming that the acceleration at any given time interval is constant and equal to the acceleration at the beginning of the interval. That is, it may be considered that the velocity in question is approximately equal to

$$\dot{u}_{i+1} = \dot{u}_i + \ddot{u}_i \Delta t \qquad (16.36)$$

which is a function of only the parameters at the beginning of the interval and thus involves no unknowns. It should be realized that this is just an approximation although it may be an accurate one if small time steps are used. It is in any case a more practical approach than using an iterative procedure. Proceeding accordingly, Equation 16.35 may then be expressed as

$$m\ddot{u}_{i+1} + c\dot{u}_{i+1} + ku_{i+1} = -m(\ddot{u}_g)_{i+1} - \text{sgn}(\dot{u}_i + \ddot{u}_i \Delta t) c_d \left| \dot{u}_i + \ddot{u}_i \Delta t \right|^\alpha \cos^{\alpha+1}\theta \qquad (16.37)$$

In consequence and if the constant average acceleration method is used, the solution to Equation 16.37 may be obtained using the algorithm described in Box 11.3, considering that the effective load vector in this algorithm is instead given by

$$(\hat{F}_a)_{i+1} = -m(\ddot{u}_g)_{i+1} - \text{sgn}(\dot{u}_i + \ddot{u}_i \Delta t) c_d \left| \dot{u}_i + \ddot{u}_i \Delta t \right|^\alpha \cos^{\alpha+1}\theta + a u_i + b \dot{u}_i + m \ddot{u}_i \qquad (16.38)$$

16.4.2.3 Friction or Yielding Steel Damper

If the damper is of the friction or yielding steel type, then $F_d(t) = 0$ and $F_k(t)$ is a force that varies according to the diagram shown in Figure 16.37, if a bilinear force–deformation behavior is assumed. In Figure 16.37, x is the deformation of the damper along its longitudinal axis; k_d the damper's initial stiffness; βk_d the damper's postyield or postslip stiffness; F_{ky} and x_y the damper's yield or slip force and deformation before any yielding or slipping takes place, respectively; x_{yt} and x_{yc} the yield or slip deformation in tension and compression after the damper yields or slips and then unloading occurs, respectively; and x_u the deformation reached at the instant of unloading. In other words, $F_k(t)$ is a force whose magnitude depends on whether the deformation $x(t)$ is increasing (i.e., the relative velocity between the two ends of the damper is greater than zero)

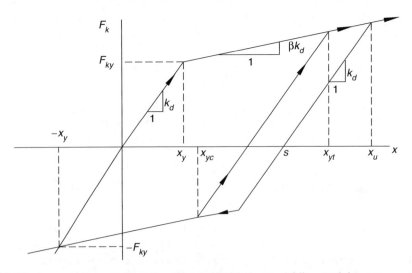

FIGURE 16.37 Assumed force–deformation behavior of friction or yielding steel damper.

Seismic Protection with Energy Dissipating Devices

or decreasing (i.e., the relative velocity between the two ends of the damper is less than zero). Mathematically, it is given by

$$F_k(t) = k_d[x(t) - s] \quad \text{if } x_{yc} < x(t) < x_{yt} \tag{16.39}$$

$$F_k(t) = k_d(x_{yt} - s) + \beta k_d[x(t) - x_{yt}] \quad \text{if } x(t) > x_{yt} \tag{16.40}$$

$$F_k(t) = k_d[x_{yc} - s] + \beta k_d[x(t) - x_{yc}] \quad \text{if } x(t) < x_{yc} \tag{16.41}$$

where s, computed at the instant of unloading, is the permanent set, which may be determined according to

$$s = (1 - \beta)(x_u - x_y) \tag{16.42}$$

following an excursion of yielding or slipping in tension, and

$$s = (1 - \beta)(x_u + x_y) \tag{16.43}$$

following an excursion of yielding or slipping in compression (see Section 6.5.5).

Note that in Equations 16.39 through 16.41, $x_{yt} = x_y$ and $x_{yc} = -x_y$ before any yielding or slipping takes place. However, if unloading occurs after an excursion of yielding or slipping in tension, then

$$x_{yt} = x_u \quad x_{yc} = x_u - 2x_y \tag{16.44}$$

or if unloading occurs after an excursion of yielding or slipping in compression, then

$$x_{yt} = x_u + 2x_y \quad x_{yc} = x_u \tag{16.45}$$

Unloading takes place whenever $\dot{x}(t) < 0$ following an excursion of yielding or slipping in tension, and $\dot{x}(t) > 0$ following an excursion of yielding or slipping in compression.

The equation of motion in the case of a friction or steel yielding damper may be therefore expressed as

$$m\ddot{u}(t) + c\dot{u}(t) + ku(t) + F_k(t)\cos\theta = -m\ddot{u}_g(t) \tag{16.46}$$

where $F_k(t)$ is given by Equations 16.39 through 16.41 after substituting $u(t)\cos\theta$ for $x(t)$. Once again, this is a nonlinear differential equation whose solution requires a step-by-step numerical integration.

To obtain such a solution, consider the discretized form of the equation of motion, which for the time $t_{i+1} = (i+1)\Delta t$ may be expressed as

$$m\ddot{u}_{i+1} + c\dot{u}_{i+1} + ku_{i+1} + F_{k(i+1)}\cos\theta = -m(\ddot{u}_g)_{i+1} \tag{16.47}$$

where

$$F_{k(i+1)} = k_d[u_{i+1}\cos\theta - s] \quad \text{if } x_{yc} < u_{i+1}\cos\theta < x_{yt} \tag{16.48}$$

$$F_{k(i+1)} = k_d(x_{yt} - s) + \beta k_d[u_{i+1}\cos\theta - x_{yt}] \quad \text{if } u_{i+1}\cos\theta > x_{yt} \tag{16.49}$$

$$F_{k(i+1)} = k_d[x_{yc} - s] + \beta k_d[u_{i+1}\cos\theta - x_{yc}] \quad \text{if } u_{i+1}\cos\theta < x_{yc} \tag{16.50}$$

where s, x_{yt}, and x_{yc} are defined as indicated earlier, after considering that, for any i, $x_i = u_i\cos\theta$. If the constant average acceleration method is used, for example, then substitution of Equations 11.54 and 11.55 into Equation 16.47 leads to an equation in which the only unknown is u_{i+1}. Consequently, one can use this equation to solve for u_{i+1}, use Equations 11.55 and 11.61 to determine

$\dot{u}_{i+1}$ and $\ddot{u}_{i+1}$ in terms of the found value of u_{i+1}, and repeat the procedure for successive time steps until the entire solution time history is obtained.

As with any other nonlinear problem, significant errors may be introduced with the indicated procedure if large time steps are used. Significant errors may also arise in this case due to an inaccurate calculation of the damper force $F_{k(i+1)}$ during the abrupt transitions from elastic behavior to yielding and from yielding to unloading. To minimize these errors, it is recommended to reduce the size of the time step in the neighborhood of these transitions.

EXAMPLE 16.1 RESPONSE OF SIMPLE FRAME WITH ADDED FLUID DAMPER

The frame shown in Figure E16.1a will be strengthened with an added damper to improve its seismic performance. In its original configuration, the frame has a mass of 1 k s²/in., an undamped natural period of 0.333 s, and a damping ratio of 2% of critical. The added damper will be of the fluid type with a force–deformation relationship defined by Equation 16.22 with a constant c_d of 6.45 k s$^{1/2}$/in.$^{1/2}$ and an exponent α of 0.5. Determine the response of the frame with and without the added damper when subjected to the first 20 s of the S69E component of the acceleration time history recorded during the 1952 Taft, California, earthquake, shown in Figure E16.1b. Use the constant average acceleration method with a time step Δt of 0.02 s. Assume zero initial conditions.

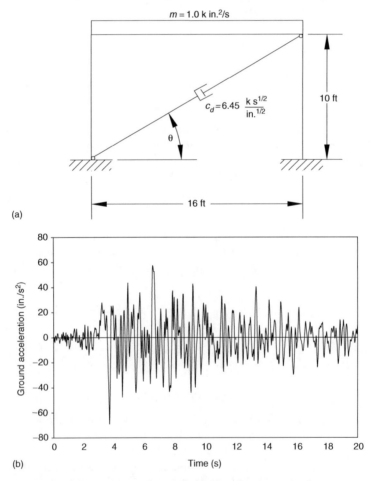

FIGURE E16.1 (a) Frame considered in Example 16.1, (b) first 20 seconds of S69E component of acceleration time history recorded during the 1952 Taft, California, earthquake, (c) displacement, (d) velocity, and (e) absolute acceleration responses of frame with and without fluid damper.

Seismic Protection with Energy Dissipating Devices

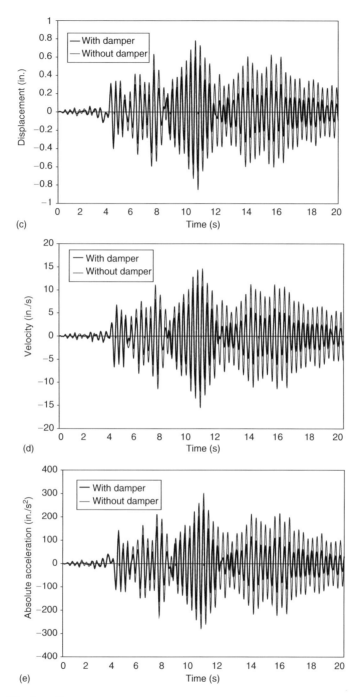

FIGURE E16.1 (Continued)

Solution

If modeled as a single-degree-of-freedom system, then the stiffness and damping constants of the frame are

$$k = \omega^2 m = \frac{4\pi^2}{0.333^2}(1) = 356 \, \text{k/in.} \quad c = 2\xi\omega m = 2(0.02)(18.868)(1.0) = 0.755 \, \text{k s/in.}$$

and, from the geometry of the frame, the angle of inclination with respect to the horizontal of the brace where the damper is mounted equals

$$\theta = \tan^{-1}\frac{10}{16} = 32°$$

Then, according to Equation 16.38 and the algorithm presented in Box 11.3, the constants a and b, the effective stiffness, and the effective load at time $t_{i+1} = (i+1)\Delta t$ for the problem under consideration are

$$a = \frac{4m}{\Delta t^2} + \frac{2c}{\Delta t} = \frac{4(1.0)}{0.02^2} + \frac{2(0.755)}{0.02} = 10{,}075.5$$

$$b = \frac{4m}{\Delta t} + c = \frac{4(1.0)}{0.02} + 0.755 = 200.75$$

$$\hat{k}_a = k + a = 356.0 + 10{,}075.5 = 10{,}431.5 \text{ k/in.}$$

$$(\hat{F}_a)_{i+1} = -m(\ddot{u}_g)_{i+1} - \text{sgn}(\dot{u}_i + \ddot{u}_i\Delta t)c_d|\dot{u}_i + \ddot{u}_i\Delta t|^\alpha \cos^{\alpha+1}\theta + au_i + b\dot{u}_i + m\ddot{u}_i$$

$$= -1.0(\ddot{u}_g)_{i+1} - \text{sgn}(\dot{u}_i + 0.02\ddot{u}_i)6.45|\dot{u}_i + 0.02\ddot{u}_i|^{0.5}\cos^{0.5+1}32° + 10{,}075.5u_i + 200.75\dot{u}_i + 1.0\ddot{u}_i$$

$$= -(\ddot{u}_g)_{i+1} - \text{sgn}(\dot{u}_i + 0.02\ddot{u}_i)5.037|\dot{u}_i + 0.02\ddot{u}_i|^{0.5} + 10{,}075.5u_i + 200.75\dot{u}_i + \ddot{u}_i$$

In consequence, the relative displacement, relative velocity, and relative acceleration of the system at time t_{i+1} is given by

$$u_{i+1} = \frac{(\hat{F}_a)_{i+1}}{\hat{k}_a} = \frac{(\hat{F}_a)_{i+1}}{10{,}431.5}$$

$$\dot{u}_{i+1} = \frac{2}{\Delta t}(u_{i+1} - u_i) - \dot{u}_i = \frac{2}{0.02}(u_{i+1} - u_i) - \dot{u}_i$$
$$= 100(u_{i+1} - u_i) - \dot{u}_i$$

$$\ddot{u}_{i+1} = -\ddot{u}_g(t_{i+1}) - \frac{1}{m}(c\dot{u}_{i+1} + ku_{i+1}) = -\ddot{u}_g(t_{i+1}) - \frac{1}{1.0}(0.755\dot{u}_{i+1} + 356.0u_{i+1})$$
$$= -\ddot{u}_g(t_{i+1}) - 0.755\dot{u}_{i+1} - 356.0u_{i+1}$$

and the absolute acceleration by

$$\ddot{y}_{i+1} = \ddot{u}_{i+1} + (\ddot{u}_g)_{i+1}$$

Similarly, for the case with no damper, the effective force at time t_{i+1} is

$$(\hat{F}_a)_{i+1} = -(\ddot{u}_g)_{i+1} + 10{,}075.5u_i + 200.75\dot{u}_i + \ddot{u}_i$$

whereas the corresponding relative displacement, relative velocity, and relative acceleration are given by the same expressions previously obtained.

Using the foregoing recursive equations, one finds the displacement, velocity, and acceleration responses shown in Figures E16.1c through E16.1e, with the numerical values for the first 15 steps given in Table E16.1a for the frame with the damper. The peak values of these responses with and without the damper are presented in Table E16.1b, together with the reduction ratio (response with damper over

TABLE E16.1a
First 15 Steps of Displacement, Velocity, and Absolute Acceleration Responses of Simple Frame with Added Fluid Damper in Example 16.1

i	t_i (s)	$(\ddot{u}_g)_i$ (in./s²)	u_i (in.)	$\dot{u}_i$ (in./s)	$(\ddot{u}_g)_i$ (in./s²)	$(\hat{F}_a)_{i+1}$ (kips)	$\ddot{y}_i$ (in./s²)
0	0	0.000	0.000	0.000	0.000	3.000	0.000
1	0.02	−3.000	0.000	0.029	2.876	9.797	−0.124
2	0.04	0.749	0.001	0.036	−1.111	16.787	−0.362
3	0.06	−1.560	0.002	0.031	0.964	26.413	−0.596
4	0.08	−3.890	0.003	0.062	2.942	41.641	−0.948
5	0.10	−2.050	0.004	0.084	0.565	56.437	−1.485
6	0.12	0.112	0.005	0.058	−2.081	61.755	−1.969
7	0.14	1.650	0.006	−0.007	−3.753	54.291	−2.103
8	0.16	1.100	0.005	−0.065	−2.904	38.034	−1.804
9	0.18	−0.485	0.004	−0.091	−0.744	20.868	−1.229
10	0.20	−2.090	0.002	−0.074	1.434	9.010	−0.656
11	0.22	−1.540	0.001	−0.040	1.263	1.141	−0.277
12	0.24	1.200	0.000	−0.035	−1.212	−9.758	−0.012
13	0.26	3.350	−0.001	−0.069	−2.965	−28.366	0.385
14	0.28	3.370	−0.003	−0.109	−2.319	−52.636	1.051
15	0.30	2.380	−0.005	−0.123	−0.491	−75.002	1.889

TABLE E16.1b
Peak Values of Frame Responses with and without Fluid Damper and Corresponding Reduction Ratios

Response	With Damper	Without Damper	Reduction Ratio
Displacement (in.)	0.65	0.85	0.76
Velocity (in./s)	11.99	15.54	0.77
Acceleration (in./s²)	231.5	302.4	0.77

response without damper) obtained for each of the three responses. It may be observed that a significant reduction is attained in the displacement, velocity, and acceleration responses of the system with the added damper.

EXAMPLE 16.2 RESPONSE OF A SIMPLE FRAME WITH ADDED FRICTION DAMPER

Solve Example 16.1 considering instead a damper of the friction type with a slip load of 65 kips. Assume the brace where the damper will be installed has an axial stiffness of 278.1 kips/in. and that the brace will be designed to resist the damper's slip force without yielding. Assume also that the damper's force–deformation behavior is rigid-plastic with a yield force equal to its slip load.

Solution

Since a friction damper behaves as a rigid link before it slips and has zero stiffness after it slips, in this case the added brace-damper assembly provides a stiffness equal to the stiffness of the brace before the damper slips and a zero stiffness after the damper slips. In this case, therefore, k_d and βk_d in Equations 16.39 through 16.41 are equal to

$$k_d = 278.1 \text{ kips/in.} \quad \beta k_d = 0$$

and the axial deformation at which the damper slips is

$$x_y = \frac{65.0}{278.1} = 0.234 \text{ in.}$$

In consequence, the discretized equation of motion for the frame under consideration is of the form (see Equation 16.47)

$$\ddot{u}_{i+1} + 0.755\dot{u}_{i+1} + 356.0u_{i+1} + 0.848F_{k(i+1)} = -(\ddot{u}_g)_{i+1}$$

where

$$F_{k(i+1)} = 278.1(0.848u_{i+1} - s) \quad \text{if } 1.179x_{yc} < u_{i+1} < 1.179x_{yt}$$

$$F_{k(i+1)} = 65 \text{ kips} \quad \text{if } u_{i+1} > 1.179x_{yt}$$

$$F_{k(i+1)} = -65 \text{ kips} \quad \text{if } u_{i+1} < 1.179x_{yc}$$

where (see Equations 16.42 through 16.45),

$$s = 0 \quad x_{yt} = 0.234 \text{ in.} \quad x_{yc} = -0.234 \text{ in.}$$

before the damper slips for the first time,

$$s = 0.848u_u - 0.234 \quad x_{yt} = 0.848u_u \quad x_{yc} = 0.848u_u - 0.468$$

if unloading occurs after the damper slips in tension, and

$$s = 0.848u_u + 0.234 \quad x_{yt} = 0.848u_u + 0.468 \quad x_{yc} = 0.848u_u$$

if unloading occurs after the damper slips in compression. In the foregoing equations, u_u is the lateral displacement reached by the frame at the instant of unloading; that is, whenever $\dot{u}(t) < 0$ following slipping of the damper in tension and $\dot{u}(t) > 0$ following slipping of the damper in compression.

Thus, using the constant average acceleration method and considering that according to Equations 11.54 and 11.55 one has that

$$\ddot{u}_{i+1} = \frac{4}{0.02^2}(u_{i+1} - u_i) - \frac{4}{0.02}\dot{u}_i - \ddot{u}_i = 10,000(u_{i+1} - u_i) - 200\dot{u}_i - \ddot{u}_i$$

$$\dot{u}_{i+1} = \frac{2}{0.02}(u_{i+1} - u_i) - \dot{u}_i = 100(u_{i+1} - u_i) - \dot{u}_i$$

the discretized equation of motion of the system for the case when $1.179x_{yc} < u_{i+1} < 1.179x_{yt}$ may be alternatively expressed as

$$10,000(u_{i+1} - u_i) - 200\dot{u}_i - \ddot{u}_i + 0.755[100(u_{i+1} - u_i) - \dot{u}_i]$$
$$+ 356.0u_{i+1} + 0.848[278.1(0.848u_{i+1} - s)] = -(\ddot{u}_g)_{i+1}$$

or as

$$10,631.48u_{i+1} - 10,075.5u_i - 200.75\dot{u}_i - \ddot{u}_i - 235.83s = -(\ddot{u}_g)_{i+1}$$

from which one can solve for u_{i+1} to obtain

$$u_{i+1} = \frac{1}{10,631.48}[-(\ddot{u}_g)_{i+1} + 10,075.5u_i + 200.75\dot{u}_i + \ddot{u}_i + 235.83s] \quad \text{if } 1.179x_{yc} < u_{i+1} < 1.179x_{yt}$$

Similarly, the equation of motion of the system for the case when $u_{i+1} > 1.179x_{yt}$ may be written as

$$10{,}000(u_{i+1} - u_i) - 200\dot{u}_i - \ddot{u}_i + 0.755[100(u_{i+1} - u_i) - \dot{u}_i] + 356.0u_{i+1} + 0.848(65) = -(\ddot{u}_g)_{i+1}$$

and for the case when $u_{i+1} < 1.179x_{yc}$ as

$$10{,}000(u_{i+1} - u_i) - 200\dot{u}_i - \ddot{u}_i + 0.755[100(u_{i+1} - u_i) - \dot{u}_i] + 356.0u_{i+1} + 0.848(-65) = -(\ddot{u}_g)_{i+1}$$

Therefore, in such cases u_{i+1} is given by

$$u_{i+1} = \frac{1}{10{,}431.5}[-(\ddot{u}_g)_{i+1} + 10{,}075.5u_i + 200.75\dot{u}_i + \ddot{u}_i - 55.12] \quad \text{if } u_{i+1} > 1.179x_{yt}$$

and

$$u_{i+1} = \frac{1}{10{,}431.5}[-(\ddot{u}_g)_{i+1} + 10{,}075.5u_i + 200.75\dot{u}_i + \ddot{u}_i + 55.12] \quad \text{if } u_{i+1} < 1.179x_{yc}$$

In addition, the corresponding relative velocity, as established in Example 16.1, may be determined from

$$\dot{u}_{i+1} = 100(u_{i+1} - u_i) - \dot{u}_i$$

and, once the relative displacement and relative velocity are known, the corresponding relative acceleration may be obtained directly from the discretized equation of motion. That is, the relative acceleration may be calculated according to

$$\ddot{u}_{i+1} = -(\ddot{u}_g)_{i+1} - 0.755\dot{u}_{i+1} - 356.0u_{i+1} - 0.848F_{k(i+1)}$$

where $F_{k(i+1)}$ is defined by the expressions derived. The corresponding absolute acceleration is given by

$$\ddot{y}_{i+1} = \ddot{u}_{i+1} + (\ddot{u}_g)_{i+1}$$

Using recursively the preceding equations from $i = 0$ to $i = 1000$, one obtains the time histories presented in Figures E16.2a through E16.2c, which show the relative displacement, relative velocity, and absolute acceleration responses of the system, respectively. The numerical values of these responses for the first 15 time steps are listed in Table E16.2a. For comparison, Figures E16.2a through E16.2c also include the corresponding responses obtained in Example 16.1 for the case when no damper is considered. The peak responses obtained with and without the damper under consideration are given in Table E16.2b, together with the response reduction ratios attained.

It may be observed from the results presented in Table E16.2b that the displacement, velocity, and acceleration responses of the frame are reduced significantly with the added damper. It should be noted, however, that this reduction is partly owed to the stiffening of the frame and the associated change in natural period with the added brace. This may be seen by computing the response of the frame without the damper, but considering a lateral stiffness equal to the stiffness of the original frame plus the lateral stiffness of the brace; that is, a lateral stiffness of $356 + 278.1\cos 32° = 556$ kips/in. In such a case, the peak values of the displacement, velocity, and acceleration responses of the system are equal to 0.44 in., 9.96 in./s, and 244.7 in./s², respectively, which shows that a reduction in response is indeed attained as a result of the addition of the brace alone.

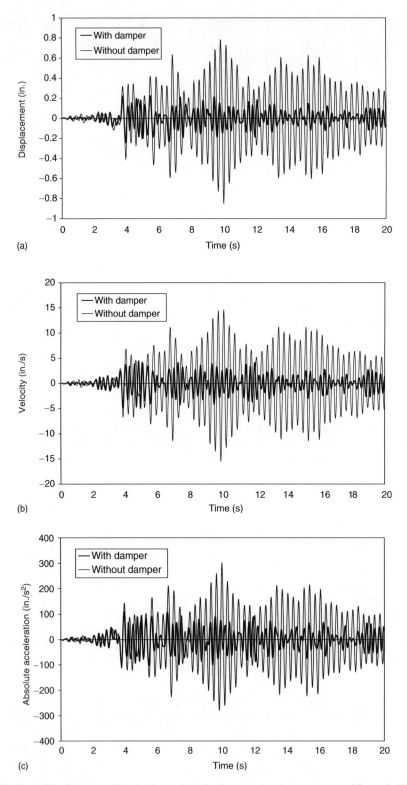

FIGURE E16.2 (a) Displacement, (b) velocity, and (c) absolute acceleration responses of frame in Example 16.2 with and without friction damper.

Seismic Protection with Energy Dissipating Devices

TABLE E16.2a
First 15 Steps of Displacement, Velocity, and Acceleration Responses of Simple Frame with Added Friction Damper in Example 16.2

i	t_i (s)	$(\ddot{u}_g)_i$ (in./s²)	u_i (in.)	$\dot{u}_i$ (in./s)	$\ddot{u}_i$ (in./s²)	F_{ki} (kips)	$\ddot{y}_i$ (in./s²)
0	0	0.000	0.000	0.000	0.000	0.000	0.000
1	0.02	−3.000	0.000	0.028	0.067	2.822	−0.178
2	0.04	0.749	0.001	0.043	0.235	−1.335	−0.586
3	0.06	−1.560	0.002	0.035	0.419	0.545	−1.015
4	0.08	−3.890	0.003	0.064	0.652	2.304	−1.586
5	0.10	−2.050	0.004	0.083	0.998	−0.367	−2.417
6	0.12	0.112	0.006	0.047	1.305	−3.225	−3.113
7	0.14	1.650	0.006	−0.033	1.339	−4.781	−3.131
8	0.16	1.100	0.004	−0.114	0.992	−3.352	−2.252
9	0.18	−0.485	0.002	−0.150	0.367	−0.267	−0.752
10	0.20	−2.090	−0.001	−0.125	−0.281	2.848	0.758
11	0.22	−1.540	−0.003	−0.063	−0.724	3.296	1.756
12	0.24	1.200	−0.004	−0.020	−0.922	0.988	2.188
13	0.26	3.350	−0.004	−0.020	−1.016	−0.939	2.411
14	0.28	3.370	−0.005	−0.036	−1.147	−0.638	2.732
15	0.30	2.380	−0.006	−0.035	−1.313	0.742	3.122

TABLE E16.2b
Peak Values of Frame Responses with and without Friction Damper and Corresponding Reduction Ratios

Response	With Damper	Without Damper	Reduction Ratio
Displacement (in.)	0.25	0.85	0.29
Velocity (in./s)	5.02	15.5	0.32
Acceleration (in./s²)	111.5	302.4	0.37

EXAMPLE 16.3   RESPONSE OF SIMPLE FRAME WITH ADDED VISCOELASTIC DAMPER

Solve Example 16.1 considering that the added damper is of the viscoelastic type. The damper employed will be fabricated with two viscoelastic pads, each with a length of 73 in., a width of 13 in., and a thickness of 0.8 in. For design purposes, assume that the damper will be subjected to an ambient temperature of 21°C and loading with a dominant frequency of 3 Hz. Assume also that the viscoelastic pads will behave linearly at all times.

Solution

From the graphs in Figure 16.15 for an ambient temperature of 21°C and a frequency of 3 Hz, one obtains

$$G' = 0.35 \text{ kip/in.}^2 \quad G'' = 0.50 \text{ kip/in.}^2$$

According to Equation 16.21 and the given dimensions of the viscoelastic pads, the stiffness and damping constants of the damper are thus equal to

$$k' = \frac{AG'}{h} = \frac{(73 \times 13)0.35}{1.6} = 207.6 \text{ kip/in.}$$

$$c' = \frac{AG''}{\omega h} = \frac{(73 \times 13)0.50}{(2\pi/0.333)1.6} = 15.72 \text{ kip s/in.}$$

In consequence, the lateral stiffness and damping constant of the frame with the added damper are (see Equations 16.31 and 16.32)

$$\bar{k} = k + k'\cos^2\theta = 356.0 + 207.6\cos^2 32° = 505.3 \text{ kip/in.}$$

$$\bar{c} = c + c'\cos^2\theta = 0.755 + 15.72\cos^2 32° = 12.06 \text{ kip s/in.}$$

which, incidentally, lead to a natural period and a damping ratio equal to

$$\bar{T} = 2\pi\sqrt{\frac{m}{\bar{k}}} = 2\pi\sqrt{\frac{1.0}{505.3}} = 0.280 \text{ s}$$

$$\bar{\xi} = \frac{\bar{c}}{2\sqrt{\bar{k}m}} = \frac{12.06}{2\sqrt{505.3(1.0)}} = 0.27$$

Using a computer program for the seismic response analysis of linear systems to determine the response of the frame with its original properties first and then with the modified properties, one obtains the relative displacement, relative velocity, and absolute acceleration time histories shown in Figures E16.3a through E16.3c. The peak values in these time histories are summarized in Table E16.3, where the response reduction ratios attained in each case are also given.

It may be seen from Figures E16.3a through E16.3c and the values presented in Table E16.3 that the displacement, velocity, and acceleration responses of the frame are reduced significantly with the

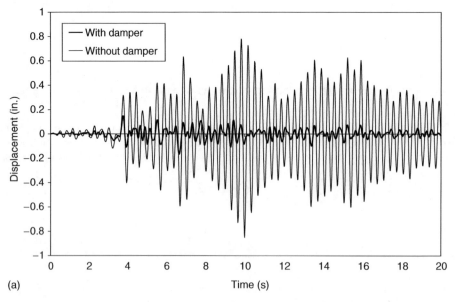

FIGURE E16.3 (a) Displacement, (b) velocity, and (c) absolute acceleration responses of frame in Example 16.3 with and without viscoelastic damper.

Seismic Protection with Energy Dissipating Devices

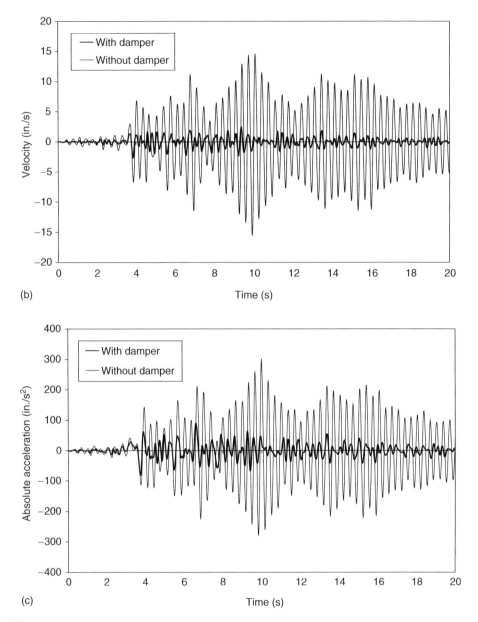

FIGURE E16.3 (Continued)

introduction of the viscoelastic damper. It should be noted, however, that, as in the case of the friction damper, part of the reduction is achieved by stiffening and changing the natural period of the frame, with the additional stiffness provided by the damper. This may be verified by computing the peak responses of the frame for the case when the stiffness constant is equal to the modified stiffness of 505.3 kip/in. and the damping constant is equal to the original damping constant of 0.755 kip s/in. In such a case, the peak values are 0.45 in., 9.31 in./s, and 237.0 in./s^2 for the displacement, velocity, and acceleration responses of the system, respectively, which are all less than the peak values obtained when the frame is considered with its original properties.

TABLE E16.3
Peak Values of Frame Responses with and without Viscoelastic Damper and Corresponding Reduction Ratios

Response	With Damper	Without Damper	Reduction Ratio
Displacement (in.)	0.17	0.85	0.20
Velocity (in./s)	2.73	15.54	0.18
Acceleration (in./s^2)	91.0	302.4	0.30

16.4.3 Seismic Response of Multistory Structures with Added Dampers

Consider a linear n-story structure whose equation of motion, with no added dampers, is given by

$$[M]\{\ddot{u}(t)\} + [C]\{\dot{u}(t)\} + [K]\{u(t)\} = -[M]\{J\}\ddot{u}_g(t) \tag{16.51}$$

where $[M]$, $[C]$, and $[K]$ denote the $n \times n$ mass, damping, and stiffness matrices of the structure, respectively; $\{u(t)\}$, $\{\dot{u}(t)\}$, and $\{\ddot{u}(t)\}$ represent vectors of displacements, velocities, and accelerations relative to the base of the structure; $\{J\}$ is an influence vector that contains the displacements in the system when the base is subjected to a unit static horizontal displacement; and $\ddot{u}_g(t)$ signifies ground acceleration. Consider, in addition, that a damper is added to each of the structure's stories as illustrated in Figure 16.38 for a three-story frame. Consider, further, that the jth damper exerts on the jth and jth – 1 floors of the structure a horizontal force $F_{kj}(t)\cos\theta_j$ that is a function of the relative displacement between the jth and jth –1 floors, and a horizontal force $F_{dj}(t)\cos\theta_j$ that is a function of the corresponding relative velocity. $F_{kj}(t)$ and $F_{dj}(t)$ denote the elastic and damping components of the force exerted by the jth damper along its longitudinal direction, respectively, and θ_j is the angle of inclination of the brace where the jth damper is mounted. Thus, if dampers are added to the structure, the equation of motion is of the form

$$[M]\{\ddot{u}(t)\} + [C]\{\dot{u}(t)\} + [K]\{u(t)\} + [I_d]\{F_a(t)\} = -[M]\{J\}\ddot{u}_g(t) \tag{16.52}$$

where $[I_d]$ and $\{F_a(t)\}$ are defined as

$$[I_d] = \begin{bmatrix} 1 & -1 & 0 & 0 & \cdots & 0 \\ 0 & 1 & -1 & 0 & \cdots & 0 \\ 0 & 0 & 1 & -1 & \cdots & 0 \\ 0 & 0 & 0 & 1 & \cdots & 0 \\ \vdots & \vdots & \vdots & \vdots & \cdots & \vdots \\ 0 & 0 & 0 & 0 & \cdots & 1 \end{bmatrix} \tag{16.53}$$

$$\{F_a(t)\} = \begin{Bmatrix} F_{k1}(t)\cos\theta_1 \\ F_{k2}(t)\cos\theta_2 \\ \vdots \\ F_{kn}(t)\cos\theta_n \end{Bmatrix} + \begin{Bmatrix} F_{d1}(t)\cos\theta_1 \\ F_{d2}(t)\cos\theta_2 \\ \vdots \\ F_{dn}(t)\cos\theta_n \end{Bmatrix} \tag{16.54}$$

Seismic Protection with Energy Dissipating Devices

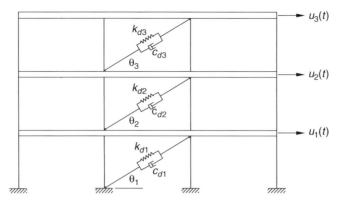

FIGURE 16.38 Multistory frame with added dampers.

16.4.3.1 Viscoelastic Dampers

If the added dampers are of the viscoelastic type, then according to Equation 16.20 $F_{kj}(t)$ and $F_{dj}(t)$ are given by

$$F_{kj}(t) = k'_j[u_j(t) - u_{j-1}(t)]\cos\theta_j \tag{16.55}$$

$$F_{dj}(t) = c'_j[\dot{u}_j(t) - \dot{u}_{j-1}(t)]\cos\theta_j \tag{16.56}$$

where $u_j(t)$ and $\dot{u}_j(t)$ and $u_{j-1}(t)$ and $\dot{u}_{j-1}(t)$ denote the relative lateral displacements and velocities of the jth and jth $-$ 1 floors, respectively. Consequently, for the case of viscoelastic dampers the vector $\{F_a(t)\}$ in Equation 16.52 may be expressed as

$$\{F_a\} = \begin{Bmatrix} k'_1[u_1(t)]\cos^2\theta_1 \\ k'_2[u_2(t) - u_1(t)]\cos^2\theta_2 \\ \vdots \\ k'_n[u_n(t) - u_{n-1}(t)]\cos^2\theta_n \end{Bmatrix} + \begin{Bmatrix} c'_1[\dot{u}_1(t)]\cos^2\theta_1 \\ c'_2[\dot{u}_2(t) - \dot{u}_1(t)]\cos^2\theta_2 \\ \vdots \\ c'_n[\dot{u}_n(t) - \dot{u}_{n-1}(t)]\cos^2\theta_n \end{Bmatrix} \tag{16.57}$$

and the product $[I_d]\{F_a(t)\}$ as

$$[I_d]\{F_a(t)\}$$

$$= \begin{bmatrix} k'_1\cos^2\theta_1 + k'_2\cos^2\theta_2 & -k'_2\cos^2\theta_2 & \cdots & 0 \\ -k'_2\cos^2\theta_2 & k'_2\cos^2\theta_2 + k'_3\cos^2\theta_3 & \cdots & 0 \\ \vdots & \vdots & \ddots & \vdots \\ 0 & 0 & \cdots & k'_n\cos^2\theta_n \end{bmatrix} \begin{Bmatrix} u_1(t) \\ u_2(t) \\ \vdots \\ u_n(t) \end{Bmatrix}$$

$$+ \begin{bmatrix} c'_1\cos^2\theta_1 + c'_2\cos^2\theta_2 & -c'_2\cos^2\theta_2 & \cdots & 0 \\ -c'_2\cos^2\theta_2 & c'_2\cos^2\theta_2 + c'_3\cos^2\theta_3 & \cdots & 0 \\ \vdots & \vdots & \ddots & \vdots \\ 0 & 0 & \cdots & c'_n\cos^2\theta_n \end{bmatrix} \begin{Bmatrix} \dot{u}_1(t) \\ \dot{u}_2(t) \\ \vdots \\ \dot{u}_n(t) \end{Bmatrix} \tag{16.58}$$

which may also be written as

$$[I_d]\{F_a(t)\} = [K']\{u(t)\} + [C']\{\dot{u}(t)\} \tag{16.59}$$

where

$$[K'] = \begin{bmatrix} k'_1 \cos^2\theta_1 + k'_2 \cos^2\theta_2 & -k'_2 \cos^2\theta_2 & \cdots & 0 \\ -k'_2 \cos^2\theta_2 & k'_2 \cos^2\theta_2 + k'_3 \cos^2\theta_3 & \cdots & 0 \\ \vdots & \vdots & \ddots & \vdots \\ 0 & 0 & \cdots & k'_n \cos^2\theta_n \end{bmatrix} \tag{16.60}$$

and

$$[C'] = \begin{bmatrix} c'_1 \cos^2\theta_1 + c'_2 \cos^2\theta_2 & -c'_2 \cos^2\theta_2 & \cdots & 0 \\ -c'_2 \cos^2\theta_2 & c'_2 \cos^2\theta_2 + c'_3 \cos^2\theta_3 & \cdots & 0 \\ \vdots & \vdots & \ddots & \vdots \\ 0 & 0 & \cdots & c'_n \cos^2\theta_n \end{bmatrix} \tag{16.61}$$

The equation of motion of the structure when viscoelastic dampers are added may thus be expressed as

$$[M]\{\ddot{u}(t)\} + [\overline{C}]\{\dot{u}(t)\} + [\overline{K}]\{u(t)\} = -[M]\{J\}\ddot{u}_g(t) \tag{16.62}$$

where $[\overline{C}]$ and $[\overline{K}]$ are defined as

$$[\overline{C}] = [C] + [C'] \tag{16.63}$$

$$[\overline{K}] = [K] + [K'] \tag{16.64}$$

and represent the effective damping and stiffness matrices of the system, respectively.

As in the single-degree-of-freedom case discussed in Section 16.4.2, it may be seen that the equation of motion for a multistory structure with added viscoelastic dampers is linear and may therefore be solved using any of the available solution methods for linear multi-degree-of-freedom systems. In fact, the solution may even be obtained using a modal superposition or the response spectrum method. It should be kept in mind, however, that the undamped mode shapes of the system will not, in general, diagonalize the damping matrix $[\overline{C}]$ and that a solution based on a modal superposition or the response spectrum method will require the use of the complex mode shapes and natural frequencies of the system as described in Section 15.5.4 for base-isolated systems.

16.4.3.2 Fluid Dampers

If the dampers added to the structure are of the fluid type, then according to Equation 16.22, one has that for the jth damper, $j = 1, 2, \ldots, n$, $F_{kj}(t) = 0$ and

$$F_{dj}(t) = c_{dj} \left| [\dot{u}_j(t) - \dot{u}_{j-1}(t)]\cos\theta_j \right|^{\alpha_j} \text{sgn}[\dot{u}_j(t) - \dot{u}_{j-1}(t)] \tag{16.65}$$

where c_{dj} and α_j are the constants that characterize the jth damper. Consequently, in the case of fluid dampers the vector $\{F_a(t)\}$ in Equation 16.52 is given by

$$\{F_a(t)\} = \begin{Bmatrix} c_{d1}|\dot{u}_1(t)|^{\alpha_1} \cos^{\alpha_1+1}\theta_1 \operatorname{sgn}[u_1(t)] \\ c_{d2}|\dot{u}_2(t) - \dot{u}_1(t)|^{\alpha_2} \cos^{\alpha_2+1}\theta_2 \operatorname{sgn}[\dot{u}_2(t) - \dot{u}_1(t)] \\ \vdots \\ c_{dn}|\dot{u}_n(t) - \dot{u}_{n-1}(t)|^{\alpha_n} \cos^{\alpha_n+1}\theta_n \operatorname{sgn}[\dot{u}_n(t) - \dot{u}_{n-1}(t)] \end{Bmatrix} \quad (16.66)$$

from which it may be seen that in such a case Equation 16.52 is, in general, a nonlinear equation and, thus, its solution requires a step-by-step numerical integration.

Consider, then, the discretized form of Equation 16.52 at time $t_{i+1} = (i+1)\Delta t$, where Δt is the selected integration time step and i is any integer. After some rearrangement, this equation may be expressed as

$$[M]\{\ddot{u}\}_{i+1} + [C]\{\dot{u}\}_{i+1} + [K]\{u\}_{i+1} = -[M]\{J\}(\ddot{u}_g)_{i+1} - [I_d]\{F_a\}_{i+1} \quad (16.67)$$

where the subscript $i+1$ indicates that the quantities marked with it are evaluated at time t_{i+1}, and

$$\{F_a\}_{i+1} = \begin{Bmatrix} c_{d1}|\dot{u}_1|_{i+1}^{\alpha_1} \cos^{\alpha_1+1}\theta_1 \operatorname{sgn}(u_1)_{i+1} \\ c_{d2}|\dot{u}_2 - \dot{u}_1|_{i+1}^{\alpha_2} \cos^{\alpha_2+1}\theta_2 \operatorname{sgn}(\dot{u}_2 - \dot{u}_1)_{i+1} \\ \vdots \\ c_{dn}|\dot{u}_n - \dot{u}_{n-1}|_{i+1}^{\alpha_2} \cos^{\alpha_n+1}\theta_n \operatorname{sgn}(\dot{u}_n - \dot{u}_{n-1})_{i+1} \end{Bmatrix} \quad (16.68)$$

Equation 16.68 may be solved, once again, with any of the direct integration methods introduced in Chapter 11, and in particular with the constant average acceleration method discussed there. It is noted, however, that the vector $\{F_a\}_{i+1}$ is a function of the relative velocity vector $\{\dot{u}\}_{i-1}$, which is not known in advance, and thus Equation 16.67 cannot be solved explicitly for $\{u\}_{i+1}$. Nonetheless, one can adopt the same scheme used in the single-degree-of-freedom case to overcome this problem; namely, one can assume that the velocity vector is given approximately by

$$\{\dot{u}\}_{i+1} = \{\dot{u}\}_i + \{\ddot{u}\}_i \Delta t \quad (16.69)$$

which is a function of only the known velocities and accelerations at the beginning of the time step. This allows the straightforward calculation of the vector $\{F_a\}_{i+1}$ and the solution of Equation 16.67. In particular, if the constant average acceleration method is used, this solution may be obtained using the algorithm described in Box 11.3, after considering that the effective load vector and the relative acceleration vector in this algorithm are instead given by

$$\{\hat{F}_a\}_{i+1} = -[M]\{J\}\ddot{u}_g(t_{i+1}) - [I_d]\{F_a\}_{i+1} + [a]\{u\}_i + [b]\{\dot{u}\}_i + [M]\{\ddot{u}\}_i \quad (16.70)$$

$$\{\ddot{u}\}_{i+1} = -\{J\}(\ddot{u}_g)_{i+1} - [M]^{-1}([C]\{\dot{u}\}_{i+1} + [K]\{u\}_{i+1} + [I_d]\{F_a\}_{i+1}) \quad (16.71)$$

16.4.3.3 Friction or Yielding Steel Dampers

If the added dampers are of the friction or yielding steel type, then $F_{dj}(t) = 0$ and $F_{kj}(t)$ is a force that varies according to the diagram shown in Figure 16.39, if it is assumed that the force–deformation behavior of the dampers is approximately bilinear. In Figure 16.39, x_j is the deformation of the jth damper along its longitudinal axis; k_{dj} denotes its initial stiffness; $\beta_j k_{dj}$ represents its postyield or postslip stiffness; F_{kjy} and x_{jy} are the damper's yield or slip force and deformation before any yielding or slipping takes place, respectively; x_{jyt} and x_{jyc} are the yield or slip deformation in tension and compression after the damper yields or slips and then unloading occurs, respectively; and x_{ju} is the deformation attained at the instant of unloading. Specifically, $F_{kj}(t)$ is a force whose magnitude depends on whether the deformation $x_j(t)$ is increasing (i.e., the relative velocity between the two ends of the jth damper is greater than zero) or is decreasing (i.e., the relative velocity between the two ends of the jth damper is less than zero). Mathematically, it is given by

$$F_{kj}(t) = k_{dj}[x_j(t) - s_j] \quad \text{if } x_{jyc} < x_j(t) < x_{jyt} \tag{16.72}$$

$$F_{kj}(t) = k_{dj}(x_{jyt} - s_j) + \beta_j k_{dj}[x_j(t) - x_{jyt}] \quad \text{if } x_j(t) > x_{jyt} \tag{16.73}$$

$$F_{kj}(t) = k_{dj}[x_{jyc} - s_j] + \beta_j k_{dj}[x_j(t) - x_{jyc}] \quad \text{if } x_j(t) < x_{jyc} \tag{16.74}$$

where s_j, computed at the instant of unloading, is the permanent set in the jth damper (see Figure 16.39) and may be calculated according to

$$s_j = (1 - \beta_j)(x_{ju} - x_{jy}) \tag{16.75}$$

following an excursion of yielding or slipping in tension, and according to

$$s_j = (1 - \beta_j)(x_{ju} + x_{jy}) \tag{16.76}$$

following an excursion of yielding or slipping in compression.

Observe that in Equations 16.72 through 16.74, $x_{jyt} = x_{jy}$ and $x_{jyc} = -x_{jy}$ before any yielding or slipping takes place. However, if unloading occurs after an excursion of yielding or slipping in tension, then

$$x_{jyt} = x_{ju} \quad x_{jyc} = x_{ju} - 2x_{jy} \tag{16.77}$$

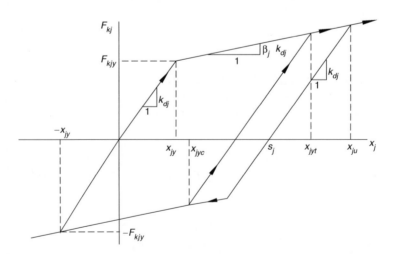

FIGURE 16.39 Assumed force–deformation behavior of jth friction or yielding steel damper.

Similarly, if unloading occurs after an excursion of yielding or slipping in compression, then

$$x_{jyt} = x_{ju} + 2x_{jy} \quad x_{jyc} = x_{ju} \tag{16.78}$$

Observe, also, that unloading takes place whenever $\dot{x}_j(t) < 0$ following an excursion of yielding or slipping in tension, and $\dot{x}_j(t) > 0$ following an excursion of yielding or slipping in compression.

The equation of motion in the case of added friction or steel yielding dampers may therefore be expressed as

$$[M]\{\ddot{u}(t)\} + [C]\{\dot{u}(t)\} + [K]\{u(t)\} + [I_d]\{F_k(t)\} = -[M]\{J\}\ddot{u}_g(t) \tag{16.79}$$

where

$$\{F_k(t)\} = \begin{Bmatrix} F_{k1}(t)\cos\theta_1 \\ F_{k2}(t)\cos\theta_2 \\ \vdots \\ F_{kn}(t)\cos\theta_n \end{Bmatrix} \tag{16.80}$$

where, for $j = 1, 2, \ldots, n$, $F_{kj}(t)$ is given by Equations 16.72 through 16.74, after it is considered that $x_j(t) = [u_j(t) - u_{j-1}(t)]\cos\theta_j$.

As in the case of the fluid dampers, Equation 16.79 is a nonlinear differential equation whose solution requires a step-by-step numerical integration. To obtain this solution, consider then the discretized form of this equation at time $t_{i+1} = (i + 1)\Delta t$, which may be written as

$$[M]\{\ddot{u}\}_{i+1} + [C]\{\dot{u}\}_{i+1} + [K]\{u\}_{i+1} + [I_d]\{F_k\}_{i+1} = -[M]\{J\}(\ddot{u}_g)_{i+1} \tag{16.81}$$

where, as earlier, the subscript $i + 1$ indicates that the quantities marked with it are evaluated at time t_{i+1}. In addition,

$$\{F_k\}_{i+1} = \begin{Bmatrix} F_{k1(i+1)}\cos\theta_1 \\ F_{k2(i+1)}\cos\theta_2 \\ \vdots \\ F_{kn(i+1)}\cos\theta_n \end{Bmatrix} \tag{16.82}$$

where, for $j = 1, 2, \ldots, n$, $F_{kj(i+1)}$ is given by

$$F_{kj(i+1)} = k_{dj}[(u_{j(i+1)} - u_{ji})\cos\theta_j - s_j] \quad \text{if } x_{jyc} < (u_{j(i+1)} - u_{ji})\cos\theta_j < x_{jyt} \tag{16.83}$$

$$F_{kj(i+1)} = k_{dj}(x_{jyt} - s_j) + \beta_j k_{dj}[(u_{j(i+1)} - u_{ji})]\cos\theta_j - x_{jyt}] \quad \text{if} (u_{j(i+1)} - u_{ji})\cos\theta_j > x_{jyt} \tag{16.84}$$

$$F_{kj(i+1)} = k_{dj}(x_{jyc} - s_j) + \beta_j k_{dj}[(u_{j(i+1)} - u_{ji})\cos\theta_j - x_{jyc}] \quad \text{if} [u_{j(i+1)} - u_{ji}]\cos\theta_j < x_{jyc} \tag{16.85}$$

In these equations, s_j, x_{jyt}, and x_{jyc} are determined as indicated earlier, after considering that, for any i, $x_{ji} = (u_{ji} - u_{j(i-1)})\cos\theta_j$.

If, for example, the constant average acceleration method is used to solve Equation 16.81, then substitution of Equations 11.54 and 11.55 into Equation 16.81 yields an equation in which the only unknown is the vector $\{u\}_{i+1}$. Consequently, one can use this equation to solve for $\{u\}_{i+1}$, then use Equations 11.55 and 11.61 to determine $\{\dot{u}\}_{i+1}$ and $\{\ddot{u}\}_{i+1}$, and repeat the procedure for successive time steps until the entire solution is obtained. As in the single-degree-of-freedom case, one should be aware of the errors that may arise due to the inaccurate calculation of the damper forces when the dampers transition from elastic behavior to yielding or slipping and from yielding or slipping to unloading. Once again, it is recommended to reduce the size of the time step in the neighborhood of these transitions to minimize such errors.

EXAMPLE 16.4   RESPONSE OF MULTISTORY FRAME WITH ADDED FLUID DAMPERS

The three-story shear frame shown in Figure E16.4a will be strengthened with fluid dampers to improve its seismic performance. In its original configuration, the frame has mass, stiffness, and damping matrices given, respectively, by

$$[M] = \begin{bmatrix} 15 & 0 & 0 \\ 0 & 10 & 0 \\ 0 & 0 & 5 \end{bmatrix} \text{kip s}^2/\text{in.} \quad [C] = \begin{bmatrix} 30.1 & -12.9 & 0 \\ -12.9 & 21.5 & -8.6 \\ 0 & -8.6 & 8.6 \end{bmatrix} \text{kip s/in.}$$

$$[K] = \begin{bmatrix} 7000 & -3000 & 0 \\ -3000 & 5000 & -2000 \\ 0 & -2000 & 2000 \end{bmatrix} \text{kip/in.}$$

which lead to natural periods of 0.675 s, 0.300 s, and 0.217 s and damping ratios of 0.020, 0.045, and 0.062 in the first, second, and third modes of the frame, respectively. The added dampers will all have a force–velocity relationship defined by Equation 16.22 with a constant c_d of 320 kip s$^{1/2}$/in.$^{1/2}$ and an exponent α of 0.5. Determine the response of the frame with and without the added dampers when subjected to the first 20 s of the north–south component of the acceleration time history recorded at the Takatori Station during the 1995 Kobe earthquake, shown in Figure E16.4b. Use the constant average acceleration method with a time step Δt of 0.01 s and assume zero initial conditions.

Solution

For the given geometry of the frame, the angles of inclination of the braces where the dampers will be installed are equal to

$$\theta_1 = \theta_2 = \theta_3 = \tan^{-1} \frac{12}{20} = 31°$$

and thus for the case under consideration the vector $\{F_a\}_{i+1}$ defined by Equation 16.68 becomes

$$\{F_a\}_{i+1} = \begin{Bmatrix} 320 \mid \dot{u}_1 \mid^{0.5}_{i+1} \cos^{0.5+1} 31° \text{sgn}(u_1)_{i+1} \\ 320 \mid \dot{u}_2 - \dot{u}_1 \mid^{0.5}_{i+1} \cos^{0.5+1} 31° \text{sgn}(\dot{u}_2 - \dot{u}_1)_{i+1} \\ 320 \mid \dot{u}_3 - \dot{u}_2 \mid^{0.5}_{i+1} \cos^{0.5+1} 31° \text{sgn}(\dot{u}_3 - \dot{u}_2)_{i+1} \end{Bmatrix}$$

$$= \begin{Bmatrix} 254 \mid \dot{u}_1 \mid^{0.5}_{i+1} \text{sgn}(u_1)_{i+1} \\ 254 \mid \dot{u}_2 - \dot{u}_1 \mid^{0.5}_{i+1} \text{sgn}(\dot{u}_2 - \dot{u}_1)_{i+1} \\ 254 \mid \dot{u}_3 - \dot{u}_2 \mid^{0.5}_{i+1} \text{sgn}(\dot{u}_3 - \dot{u}_2)_{i+1} \end{Bmatrix}$$

where, utilizing Equation 16.69 and in accordance with the foregoing discussion, the velocity vector at time t_{i+1} may be considered given approximately by

$$\{\dot{u}\}_{i+1} = \{\dot{u}\}_i + 0.01\{\ddot{u}\}_i$$

Seismic Protection with Energy Dissipating Devices

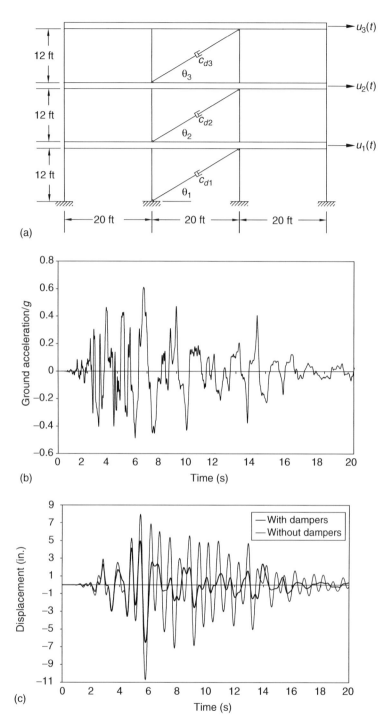

FIGURE E16.4 (a) Multistory shear frame with fluid dampers in Example 16.4, (b) ground acceleration recorded in north–south direction at the Takatori Station during the 1995 Kobe earthquake, (c) displacement, (d) velocity, and (e) absolute acceleration responses of frame's third floor with and without fluid dampers.

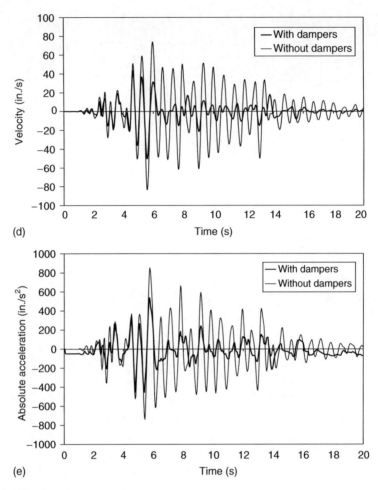

(d)

(e)

FIGURE E16.4 (Continued)

Also, according to the formulas in Box 11.3 and for a time step of 0.01 s, one has that

$$[a] = \frac{4}{0.01^2}[M] + \frac{2}{0.01}[C]$$

$$= 40{,}000 \begin{bmatrix} 15 & 0 & 0 \\ 0 & 10 & 0 \\ 0 & 0 & 5 \end{bmatrix} + 200 \begin{bmatrix} 30.1 & -12.9 & 0 \\ -12.9 & 21.5 & -8.6 \\ 0 & -8.6 & 8.6 \end{bmatrix} = \begin{bmatrix} 606{,}020 & -2{,}580 & 0 \\ -2{,}580 & 404{,}300 & -1{,}720 \\ 0 & -1{,}720 & 201{,}720 \end{bmatrix}$$

$$[b] = \frac{4}{0.01}[M] + [C]$$

$$= 400 \begin{bmatrix} 15 & 0 & 0 \\ 0 & 10 & 0 \\ 0 & 0 & 5 \end{bmatrix} + \begin{bmatrix} 30.1 & -12.9 & 0 \\ -12.9 & 21.5 & -8.6 \\ 0 & -8.6 & 8.6 \end{bmatrix} = \begin{bmatrix} 6030.1 & -12.9 & 0 \\ -12.9 & 4021.5 & -8.6 \\ 0 & -8.6 & 2008.6 \end{bmatrix}$$

$[\hat{K}_a] = [K] + [a]$

$$= \begin{bmatrix} 7{,}000 & -3{,}000 & 0 \\ -3{,}000 & 5{,}000 & -2{,}000 \\ 0 & -2{,}000 & 2{,}000 \end{bmatrix} + \begin{bmatrix} 606{,}020 & -2{,}580 & 0 \\ -2{,}580 & 404{,}300 & -1{,}720 \\ 0 & -1{,}720 & 201{,}720 \end{bmatrix}$$

$$= \begin{bmatrix} 613{,}020 & -5{,}580 & 0 \\ -5{,}580 & 409{,}300 & -3{,}720 \\ 0 & -3{,}720 & 203{,}720 \end{bmatrix}$$

Therefore, in this case the effective load vector (see Equation 16.70) results as

$$\{\hat{F}_a\}_{i+1} = -[M]\{J\}(\ddot{u}_g)_{i+1} - [I_d]\{F_a\}_{i+1} + [a]\{u\}_i + [b]\{\dot{u}\}_i + [M]\{\ddot{u}\}_i$$

$$= -\begin{Bmatrix} 15 \\ 10 \\ 5 \end{Bmatrix}(\ddot{u}_g)_{i+1} - \begin{bmatrix} 1 & -1 & 0 \\ 0 & 1 & -1 \\ 0 & 0 & 1 \end{bmatrix}\{F_a\}_{i+1} + \begin{bmatrix} 606{,}020 & -2{,}580 & 0 \\ -2{,}580 & 404{,}300 & -1{,}720 \\ 0 & -1{,}720 & 201{,}720 \end{bmatrix}\{u\}_i$$

$$+ \begin{bmatrix} 6{,}030.1 & -12.9 & 0 \\ -12.9 & 4{,}021.5 & -8.6 \\ 0 & -8.6 & 2{,}008.6 \end{bmatrix}\{\dot{u}\}_i + \begin{bmatrix} 15 & 0 & 0 \\ 0 & 10 & 0 \\ 0 & 0 & 5 \end{bmatrix}\{\ddot{u}\}_i$$

Using once again the formulas given in Box 11.3 together with Equation 16.71, the vectors of relative displacements, relative velocities, and relative accelerations at time t_{i+1} may be thus calculated as

$$\{u\}_{i+1} = [\hat{K}_a]^{-1}\{\hat{F}_a\}_{i+1} = \begin{bmatrix} 1.63147\text{E-}06 & 2.224561\text{E-}08 & 4.06212\text{E-}10 \\ 2.22456\text{E-}08 & 2.4439\text{E-}06 & 4.46266\text{E-}08 \\ 4.06212\text{E-}10 & 4.46266\text{E-}08 & 4.90951\text{E-}06 \end{bmatrix}\{\hat{F}_a\}_{i+1}$$

$$\{\dot{u}\}_{i+1} = \frac{2}{0.01}(\{u\}_{i+1} - \{u\}_i) - \{\dot{u}\}_i$$

$$= 200(\{u\}_{i+1} - \{u\}_i) - \{\dot{u}\}_i$$

$$\{\ddot{u}\}_{i+1} = -\{J\}(\ddot{u}_g)_{i+1} - [M]^{-1}[C]\{\dot{u}\}_{i+1} - [M]^{-1}[K]\{u\}_{i+1} - [M]^{-1}[I_d]\{F_a\}_{i+1}$$

$$= -\begin{Bmatrix} 1 \\ 1 \\ 1 \end{Bmatrix}(\ddot{u}_g)_{i+1} - \begin{bmatrix} 2.01 & -0.86 & 0 \\ -1.29 & 2.15 & -0.86 \\ 0 & -1.72 & 1.72 \end{bmatrix}\{\dot{u}\}_{i+1}$$

$$- \begin{bmatrix} 466.67 & -200 & 0 \\ -300 & 500 & -200 \\ 0 & -400 & 400 \end{bmatrix}\{u\}_{i+1} - \begin{bmatrix} 0.067 & -0.067 & 0 \\ 0 & 0.100 & -0.100 \\ 0 & 0 & 0.200 \end{bmatrix}\{F_a\}_{i+1}$$

Similarly, one can compute the vector of absolute accelerations as

$$\{\ddot{y}\}_{i+1} = \{\ddot{u}\}_{i+1} + \begin{Bmatrix} 1 \\ 1 \\ 1 \end{Bmatrix}(\ddot{u}_g)_{i+1}$$

TABLE E16.4
Peak Third-Floor Responses and Peak Base Shear of Frame in Example 16.4 with and without Dampers and Corresponding Reduction Ratios

Response	With Dampers	Without Dampers	Reduction Ratio
Displacement (in.)	6.50	10.73	0.61
Velocity (in./s)	50.31	83.39	0.60
Acceleration (in./s^2)	545.3	854.3	0.64
Base shear (kips)	12,624	19,863	0.64

and, if the contribution of the damping forces is neglected, the corresponding base shear as

$$V_{i+1} = \{15 \quad 10 \quad 5\}\{\ddot{y}\}_{i+1}$$

The displacement, velocity, and acceleration responses of the frame when the frame is considered with the added fluid dampers can then be determined using the developed recursive relationships successively from $i = 0$ to $i = 2000$, after it is considered that for zero initial conditions

$$\{u\}_0 = \{\dot{u}\}_0 = \{\ddot{u}\}_0 = \{0\}$$

Using these same relationships but considering that $\{F_a\}_{i+1} = \{0\}$, one can also determine the same responses for the case when the frame is considered in its original configuration with no dampers. Proceeding accordingly, one obtains the relative displacement, relative velocity, and absolute acceleration time histories shown in Figures E16.4c through E16.4e for the frame's third floor. The peak values of these responses as well as the peak values of the base shear obtained in each case are listed in Table E16.4. The reduction ratios for each of the measured responses, calculated each as the peak response with the added dampers over the peak response when no dampers are considered, are also listed in Table E16.4. It may be seen that in this example too a significant reduction in response is achieved with the addition of the dampers in question.

16.5 IMPLEMENTATION ISSUES

It is clear from the discussion in the preceding sections that the reviewed dampers may be used effectively as energy dissipation devices and, thus, they represent an attractive alternative to improve the seismic performance of structures. It should be realized, however, that not every damper may be applied to every building and that their applicability might be limited by issues such as structural flexibility, performance objective, secondary effects, environmental effects, architectural considerations, and cost. The following observations can be made in this regard:

Structural Flexibility

As mentioned earlier, energy dissipation devices require significant relative displacements or velocities to dissipate energy. Therefore, added dampers will perform well in buildings with a flexible structural system such as steel moment-resisting frames. By the same vein, dampers may not perform effectively if installed in stiff reinforced concrete or masonry shear-wall systems. A solution that seems to alleviate this problem is the use of a toggle or scissor-jack bracing of the type shown in Figure 16.40, which may magnify the lateral deformations of a structure. These alternative bracing systems, however, complicate the installation of the dampers.

Performance Objective

Energy dissipation devices primarily reduce the displacements of a structure and may also reduce the forces induced in the structure if the structure responds elastically to the exciting motion. Thus, these devices may be an attractive alternative for projects where the performance objective is the

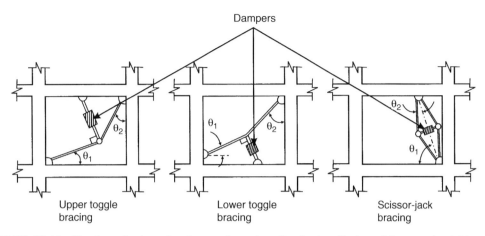

FIGURE 16.40 Toggle and scissor bracing configurations for the installation of dampers in rigid structures. (Reproduced with permission of Earthquake Engineering Research Institute from Hanson, R. D. and Soong, T. T., *Seismic Design with Supplemental Energy Devices*, Earthquake Engineering Research Institute, Oakland, CA, 2001.)

safety of nonstructural components, for example. However, they have limited applicability in projects where the primary performance objective is collapse prevention since, as pointed out in Section 16.2, structural damping plays a minor role at high levels of inelastic response.

Secondary Effects

The addition of energy dissipating devices may generate secondary effects that need to be considered in the selection of the damper type and deciding whether or not added dampers may be an effective solution. Consider, for example, the addition of dampers and the necessary bracing system to an existing moment-resisting frame. The result is reduced story drifts and reduced inelastic action. However, the structure is changed from a moment-resisting frame to a braced frame and thus the forces that develop in the dampers will induce an additional axial force in the frame's columns. Depending on the damper type, this additional axial force may be in phase with the bending moments induced by the story drifts and may, therefore, affect the safety of the loaded column. If friction or steel-yielding devices are used, then the peak brace forces will occur at the same time as the peak story drifts. Accordingly, the additional column forces will be in phase with the column bending moments induced by the story drifts. Similarly, if viscoelastic devices are used, a major portion of the additional column forces will be in phase with the drift-induced bending moments. In contrast, if viscous devices are used, the additional column forces will be out-of-phase with these moments. For viscous dampers, therefore, this secondary effect is not significant. The consideration of such a secondary effect is particularly important in the seismic retrofitting of structures that have suffered damage in previous earthquakes since it may not always be possible to upgrade the seismic resistance of such structures with the addition of dampers alone. It may also be necessary to strengthen the columns.

Environmental Effects

As discussed in Section 16.3, dampers may be sensitive to the effects of ambient temperature, humidity, corrosion, dust, ultraviolet light exposure, and fatigue-causing vibrations. Therefore, these effects should be an important consideration in the selection of the damper type or deciding if the addition of energy dissipating devices to a structure is a viable solution in locations where harsh conditions may significantly affect their performance.

Architectural Considerations

These may limit the number of dampers that may be added to a building and their location within the building. Because of their relatively small size and since they are usually integrated into the framing system, they can be hidden behind walls and partitions. They, nevertheless, obstruct openings and may interfere with the building's architectural functions. If exposed, they may have an adverse visual effect, although some architects consider them an esthetic enhancement.

Cost

This is certainly an important consideration in deciding whether or not to use dampers in a given project and what type of dampers to use. The cost of analysis, design, fabrication, and installation of dampers in a structure depends on factors such as structure type; damper type; damper location; desired structural damping; performance objective; maintenance, inspection, and replacement cost; and whether the application is for a new or an existing structure. Therefore, minimizing the cost and making a supplemental damping solution an attractive option requires a complete understanding of damper properties and a comprehensive design where all such factors are examined.

FURTHER READINGS

1. Hanson, R. D. and Soong, T. T., *Seismic Design with Supplemental Energy Devices*, Monograph MNO-8, Earthquake Engineering Research Institute, Oakland, CA, 2001, 135 p.
2. Hanson, R. D., ed., Passive energy dissipation, theme issue, *Earthquake Spectra*, 9, 3, 1993, 319–641.
3. Hart, G. C. and Wong, K., *Structural Dynamics for Structural Engineers*, John Wiley & Sons, New York, 2000, 591 p.
4. Pall, A. S. and Pall, R. T., Performance-based design using pall friction dampers—An economical design solution, Paper No. 1955, *Proceedings of 13th World Conference on Earthquake Engineering*, Vancouver, Canada, 2004.
5. Soong, T. T. and Constantinou, M. C., ed., *Passive and Active Structural Control in Civil Engineering*, CIMS Courses and Lectures No. 345, Springer, Vienna, NY, 1994, 380p.
6. Soong, T. T. and Dargush, G. F., *Passive Energy Dissipation Systems in Structural Engineering*, John Wiley & Sons, Chichester, England, 1997, 356 p.
7. Soong, T. T. and Spencer, B. F., Jr., Supplemental energy dissipation: State of the art and state of the practice, *Engineering Structures*, 24, 3, 2002, 243–259.
8. Special Scientific Events Subcommittee of the Executive Committee for the 11th International Conference for Structural Mechanics in Reactor Technology, *Seismic Isolation and Response Control for Nuclear and Non-Nuclear Structures*, Tokyo, Japan, 1991.
9. Taylor, D. P., History, design, and applications of fluid dampers in structural engineering, http://www.taylordevices.com/papers/history.design.htm.
10. Uang, C. M. and Nakashima, M., Steel buckling-restrained braced frames, Chapter 16 in *Earthquake Engineering: From Engineering Seismology to Performance-Based Design*, Bozorgnia, Y. and Bertero, V. V., ed., CRC Press, Boca Raton, FL, 2004.

PROBLEMS

16.1 Solve Example 16.1 considering instead a fluid damper with a constant c_d of 6.45 k s$^{1/2}$/in.$^{1/2}$ and an exponent α of 1.

16.2 Solve Example 16.1 considering that the frame's natural period is instead equal to 2 s.

16.3 Solve Example 16.1 considering a different ground motion.

16.4 Solve Example 16.2 considering that the slip load of the friction damper is instead equal to 90 kips.

16.5 Solve Example 16.2 considering that the frame's natural period is instead equal to 2 s.

16.6 Solve Example 16.2 considering a different ground motion.

16.7 Solve Example 16.3 assuming that the damper will be subjected instead to an ambient temperature of 32°C and a motion with a dominant frequency of 2 Hz.

16.8 Solve Example 16.3 considering that the frame's natural period is instead equal to 2 s.

16.9 Solve Example 16.3 considering a different ground motion.

16.10 Solve Example 16.4 considering that only one of the specified dampers is added to the frame and that this damper is installed at the third story.

16.11 Solve Example 16.4 considering that only one of the specified dampers is added to the frame and that this damper is installed at the first story.

16.12 Solve Example 16.4 considering a different ground motion.

16.13 Solve Example 16.4 considering instead friction dampers at each of the frame's three stories, each with a slip load of 1500 kips. The braces where the dampers will be installed will have an axial stiffness of 4500 kips/in. and will be designed to resist the dampers' slip forces without yielding. Assume the dampers' force–deformation behavior is rigid-plastic with a yield force equal to their slip load.

16.14 Solve Example 16.4 considering instead one viscoelastic damper at each of the frame's three stories. The dampers will be manufactured with two viscoelastic pads, each with a length of 30 in., a width of 15 in., and a thickness of 1 in. For design purposes, assume the dampers will be subjected to an ambient temperature of 21°C and a motion with a dominant frequency of 1.5 Hz. Assume also that the viscoelastic pads will exhibit linear behavior.

17 Seismic Code Provisions

The art of designing for earthquakes does not consist in producing structures capable of withstanding given sets of forces, although that capability is part of a sound design. It involves producing systems characterized by an optimum combination of properties such as strength, stiffness, energy-absorption, and ductile-deformation capacities that will enable them to respond to frequent, moderate earthquakes without suffering significant damage, and to exceptional, severe earthquakes without endangering their own stability, their contents, or human life and limb. Achievement of this purpose means much more than application of codified rules; it demands understanding of the basic factors that determine the seismic response of structures, as well as ingenuity to produce systems with the required properties.

Luis Esteva, 1980

17.1 INTRODUCTION

Building codes constitute an essential tool for the design of structures. They comprise a set of legal requirements for a particular geographical area or political jurisdiction to regulate and control the design, construction, quality of materials, use and occupancy, location and maintenance of buildings and other structures, and provide minimum standards to safeguard life or limb, health, property, and public welfare. Building codes have existed for thousands of years. The earliest known legal requirement for the construction of buildings is the code decreed by Hammurabi, the king of Babylon, in the eighteenth century BC. Although not a building code in the modern sense, it served to stipulate the penalties to builders of unsafe constructions:

> If a builder build a house for a man and do not make its construction firm and the house which he has built collapse and cause the death of the owner of the house, that builder shall be put to death.
> If it cause the death of the son of the owner of the house, they shall put to death a son of that builder.
> If it cause the death of a slave of the owner of the house, he shall give to the owner of the house a slave of equal value.
> If it destroy property, he shall restore whatever it destroyed, and because he did not make the house he built firm and it collapsed, he shall rebuild the house which collapsed at his own expense.
> If a builder build a house for a man and does not make its construction meet the requirements and a wall falls in, that builder shall strengthen the wall at his own expense.

Seismic codes, however, have been in existence for only a relatively short time (see a historical account in Section 1.6), mainly because an understanding of earthquakes and their effect on structures has come about only during the past few decades. Currently, seismic codes are the main mechanism through which past experience and current knowledge about earthquake characteristics and structural performance is synthesized and disseminated. They provide designers with rules and guidelines for the construction of structures that will have the strength and resilience to resist adequately any earthquake that might affect them during their intended useful life.

Until recently, different building codes were in use in different regions of the United States. These were (a) the BOCA Basic Building Code, issued by the Building Officials and Code Administrators, Homewood, Illinois; (b) the National Building Code, issued by the American Insurance Association, New York; (c) the Standard Building Code, issued by the Southern Building Code Congress, Birmingham, Alabama; and (d) the Uniform Building Code, issued by the International Conference of Building Officials, Whittier, California. The Uniform Building Code was used mostly in the West, the BOCA Basic Building Code in the Midwest, the Standard Building Code in the South, and the National Building Code in the Northeast. The Uniform Building Code was well known around the world and was used for many years as a model for the development of seismic codes in many other countries. It was first enacted in 1927 and revised every 3 years (except during World War II). The last edition was issued in 1997. Traditionally, however, the Uniform Building Code used to adopt the standards issued by the Structural Engineers Association of California (SEAOC) in their publication *Recommended Lateral Force Requirements and Commentary*. Presently, the aforementioned codes have merged into a single one with the purpose of achieving national uniformity. This code is the *International Building Code*, issued by the International Code Council and first published in the year 2000. It is based on the provisions contained in the *NEHRP Recommended Provisions for Seismic Regulations for New Buildings*, issued by the Building Seismic Safety Council under the sponsorship of the Federal Emergency Management Agency (FEMA), which used to lead the National Earthquake Hazard Reduction Program (NEHRP). This publication is the outcome of a program initiated by the federal government of the United States to develop a modern set of seismic provisions that could be adopted by building authorities all across the nation. Seismic provisions are also included in the ASCE Standard 7, *Minimum Design Loads for Buildings and Other Structures*, published by the American Society of Civil Engineers, and in the *Tri-Services Manual*, issued jointly by the departments of the Army, Navy, and Air Force for the design of military installations. The 2006 version of the International Building Code makes extensive references to the provisions contained in the ASCE Standard ASCE 7-05, which are in turn based on the 2000 edition of the NEHRP provisions.

As mentioned earlier, the requirements and recommendations in modern seismic codes are based on the current knowledge about earthquakes and structural performance. It is thus instructive to correlate such requirements and recommendations with the concepts and procedures presented in the preceding chapters. With this purpose in mind, this chapter reviews the seismic provisions contained in the 2006 edition of the International Building Code and provides numerical examples to clarify its use. Although many other seismic provisions are available and worth of discussion, such as the Eurocode 8 used by the European Community, no other provisions are described here mainly because it is believed that, given the similarities among the different codes, one set of provisions is enough is to bring out the aforementioned correlation with the material presented in previous chapters. The selection of the International Building Code is based on the belief that this code will be the prevalent "model code" in the near future in the United States.

The International Building Code, henceforth referred to simply as "the code," requires, with a few exceptions, that all structures be designed and constructed to resist the effects of earthquake ground motions. It requires, therefore, that all structures be provided with lateral- and vertical-force-resisting systems capable of providing adequate strength, stiffness, and energy dissipation capacity to withstand the design earthquake ground motions within prescribed deformation limits. To that end, the code specifies the type of resisting system that may be used in a seismic design, the ground motions for which structures should be designed, the maximum story drifts they can experience under the effect of the specified seismic loads, the methods of analysis that can be used in their design, and the load combinations they need to resist. In the following sections, therefore, these specifications are described. However, only those closely related to the material covered in this book are considered.

17.2 GENERAL REQUIREMENTS

17.2.1 ALLOWABLE SEISMIC-FORCE-RESISTING SYSTEMS

The code requires that the structural system used to resist horizontal and vertical seismic forces conform to one of the types listed in Table 17.1. It should be noted, however, that some of the listed systems have height limitations and not all systems are allowed for all seismic design categories

TABLE 17.1
Allowable Seismic-Force-Resisting Systems and Corresponding Design Coefficients and Factors

Seismic-Force-Resisting System	Response Modification Coefficient, R	System Overstrength Factor, Ω_0	Deflection Amplification Factor, C_d
A. Bearing wall systems			
1. Special reinforced concrete shear walls	5	2½	5
2. Ordinary reinforced concrete shear walls	4	2½	4
3. Detailed plain concrete shear walls	2½	2½	2
4. Ordinary plain concrete shear walls	1½	2½	1½
5. Intermediate precast shear walls	4	2½	4
6. Ordinary precast shear walls	3	2½	3
7. Special reinforced masonry shear walls	5	2½	3½
8. Intermediate reinforced masonry shear walls	3½	2½	2¼
9. Ordinary reinforced masonry shear walls	2	2½	1¾
10. Detailed plain masonry shear walls	2	2½	1¾
11. Ordinary plain masonry shear walls	1½	2½	1¼
12. Prestressed masonry shear walls	1½	2½	1¾
13. Light-framed walls sheathed with wood structural panels rated for shear resistance or steel sheets	6½	3	4
14. Light-framed walls with shear panels of all other materials	2	2½	2
15. Light-framed walls systems using flat strap bracing	4	2	3½
B. Building frame systems			
1. Steel eccentrically braced frames, moment-resisting, connections at columns away from links	8	2	4
2. Steel eccentrically braced frames, non-moment resisting, connections at columns away from links	7	2	4
3. Special steel concentrically braced frames	6	2	5
4. Ordinary steel concentrically braced frames	3¼	2	3¼
5. Special reinforced concrete shear walls	6	2½	5
6. Ordinary reinforced concrete shear walls	5	2½	4½
7. Detailed plain concrete shear walls	3	2½	2
8. Ordinary plain concrete shear walls	1½	2½	1½
9. Intermediate precast shear walls	5	2½	4½
10. Ordinary precast shear walls	4	2½	4
11. Composite steel and concrete eccentrically braced frames	8	2	4
12. Composite steel and concrete concentrically braced frames	5	2	4½
13. Ordinary composite braced frames	3	2	3
14. Composite steel plate shear walls	6½	2½	5½
15. Special composite reinforced concrete shear walls with steel elements	6	2½	5
16. Ordinary composite reinforced concrete shear walls with steel elements	5	2½	4½
17. Special reinforced masonry shear walls	5½	2½	4
18. Intermediate reinforced masonry shear walls	4	2½	4
19. Ordinary reinforced masonry shear walls	2	2½	2
20. Detailed plain masonry shear walls	2½	2½	2
21. Ordinary plain masonry shear walls	1	2½	1¼
22. Prestressed masonry shear walls	1½	2½	1¾

(Continued)

TABLE 17.1 (Continued)

Seismic-Force-Resisting System	Response Modification Coefficient, R	System Overstrength Factor, Ω_0	Deflection Amplification Factor, C_d
23. Light-framed sheathed with wood structural panels rated for shear resistance or steel sheets	7	2½	4½
24. Light-framed walls with shear panels of all other materials	2½	2½	2½
25. Buckling-restrained braced frames, non-moment-resisting beam-column connections	7	2	5½
26. Buckling-restrained braced frames, moment-resisting beam-column connections	8	2½	5
27. Special steel plate shear wall	7	2	6
C. Moment-resisting frame systems			
1. Special steel moment frames	8	3	5½
2. Special steel truss moment frames	7	3	5½
3. Intermediate steel moment frames	4½	3	4
4. Ordinary steel moment frames	3½	3	3
5. Special reinforced concrete moment frames	8	3	5½
6. Intermediate reinforced concrete moment frames	5	3	4½
7. Ordinary reinforced concrete moment frames	3	3	2½
8. Special composite steel and concrete moment frames	8	3	5½
9. Intermediate composite moment frames	5	3	4½
10. Composite partially restrained moment frames	6	3	5½
11. Ordinary composite moment frames	3	3	2½
D. Dual systems with special moment frames capable of resisting at least 25% of prescribed seismic forces			
1. Steel eccentrically braced frames	8	2½	4
2. Special steel concentrically braced frames	7	2½	5½
3. Special reinforced concrete shear walls	7	2½	5½
4. Ordinary reinforced concrete shear walls	6	2½	5
5. Composite steel and concrete eccentrically braced frames	8	2½	4
6. Composite steel and concrete concentrically braced frames	6	2½	5
7. Composite steel plate shear walls	7½	2½	6
8. Special composite reinforced concrete shear walls with steel elements	7	2½	6
9. Ordinary composite reinforced concrete shear walls with steel elements	6	2½	5
10. Special reinforced masonry shear walls	5½	3	5
11. Intermediate reinforced masonry shear walls	4	3	3½
12. Buckling-restrained braced frame	8	2½	5
13. Special steel plate shear walls	8	2½	6½
E. Dual systems with intermediate moment frames capable of resisting at least 25% of prescribed seismic forces			
1. Special steel concentrically braced frames	4½	2½	5
2. Special reinforced concrete shear walls	6	2½	5
3. Ordinary reinforced masonry shear walls	3	3	2½
4. Intermediate reinforced masonry shear walls	5	3	4½
5. Composite steel and concrete concentrically braced frames	5	2½	4½
6. Ordinary composite braced frames	3½	2½	3

(Continued)

TABLE 17.1 (Continued)

Seismic-Force-Resisting System	Response Modification Coefficient, R	System Overstrength Factor, Ω_0	Deflection Amplification Factor, C_d
7. Ordinary composite reinforced concrete shear walls with steel elements	5	3	4½
8. Ordinary reinforced concrete shear walls	5½	2½	4½
F. Shear wall-frame interactive systems with ordinary reinforced concrete moment frames and ordinary reinforced concrete shear walls	4½	2½	4
G. Inverted pendulum systems and cantilevered column systems			
1. Special steel moment frames	2½	1¼	2½
2. Intermediate steel moment frames	1½	1¼	2½
3. Ordinary steel moment frames	1¼	1¼	1¼
4. Special reinforced concrete moment frames	2½	1¼	2½
5. Intermediate concrete moment frames	1½	1¼	1½
6. Ordinary concrete moment frames	1	1¼	1
7. Timber frames	1½	1½	1½
H. Structural steel systems not specifically detailed for seismic resistance, excluding cantilever column systems	3	3	3

Source: Reprinted with permission of American Society of Civil Engineers from *Minimum Design Loads for Buildings and Other Structures*, ASCE/SEI 7-05, Reston, VA, 2006.

(seismic design categories are defined in Section 17.4.3). There are also additional requirements that apply to dual systems and systems formed with a combination of different structural types. The code should be consulted directly to find out what these height limitations and additional requirements are. Table 17.1 also includes the response modification coefficient R, system overstrength factor Ω_0, and deflection amplification factor C_d for each of the systems considered, which, as shown in Sections 17.2 and 17.5 through 17.7, the code uses in its specifications for the calculation of base shear, load combinations, and design story drifts.

17.2.2 Load Combinations

Both allowable stress and load and resistance factor designs are allowed by the code for the seismic design of the elements and components of a structure. When load and resistant factor design is used, the load combinations specified by the code are

$$1.2D + 1.0E + f_1L + f_2S \tag{17.1}$$

$$0.9D + 1.0E + 1.6H \tag{17.2}$$

However, for conditions requiring a direct recognition of structural overstrength and when specifically required by the code, the load combinations are

$$1.2D + f_1L + E_m \tag{17.3}$$

$$0.9D + E_m \tag{17.4}$$

In the preceding equations,

D = Dead load
L = Live load, which excludes roof live load
S = Snow load
H = Load due to lateral earth pressure, groundwater pressure, or pressure of bulk materials
f_1 = 1.0 for floors in places of public assembly, for live loads in excess of 100 lb/ft² (4.79 kN/m²), and for parking garage live load
   = 0.5 for other live loads
f_2 = 0.7 for roof configurations that do not shed snow off the structure
   = 0.2 for other roof configurations

In addition, E and E_m denote earthquake loads, defined by

$$E = \rho Q_E + 0.2 S_{DS} D \tag{17.5}$$

$$E_m = \Omega_0 Q_E + 0.2 S_{DS} D \tag{17.6}$$

when the effects of seismic and gravity loads are additive, and by

$$E = \rho Q_E - 0.2 S_{DS} D \tag{17.7}$$

$$E_m = \Omega_0 Q_E - 0.2 S_{DS} D \tag{17.8}$$

when gravity loads counteract the seismic effects. In these expressions, Q_E denotes the effect of horizontal earthquake forces, D represents the effect of dead load, and S_{SD} is the design spectral acceleration for short periods defined in Section 17.3.5. The $0.2 S_{DS} D$ term accounts for the effects of earthquake-induced vertical accelerations. The factor $0.2 S_{DS}$ represents an estimate of such vertical acceleration expressed as a fraction of the corresponding horizontal acceleration. It is not intended to represent a peak value in recognition of the fact that the peak values of the vertical and horizontal accelerations are unlikely to occur at the same time. The $0.2 S_{DS} D$ term need not be included in Equations 17.5 through 17.8 when S_{DS} is equal to or less than 0.125 or in Equation 17.7 when considering the demands on the soil–structure interface of foundations.

In Equations 17.6 and 17.8, Ω_0 is a system overstrength factor determined from Table 17.1 in accordance to the structural type. Similarly, in Equations 17.5 and 17.7, ρ is a redundancy coefficient that reflects the degree of redundancy in the system and is introduced as a penalty to encourage system redundancy. It is equal to 1.3 for structures in seismic design categories D, E, and F (seismic design categories are defined in Section 17.4.3), unless one of two conditions are met (consult the code directly for a detailed description of these conditions), in which case it may be considered equal to 1.0. It may also be considered equal to 1.0:

1. For structures belonging to seismic design category B or C and structures with damping systems
2. In the calculation of drifts and P-delta effects
3. In the design of nonstructural components and nonbuilding structures
4. In the design of members, connections, collector elements, and splices for which the load combinations that account for overstrength are required
5. In the determination of diaphragm loads

TABLE 17.2
Allowable Story Drifts, Δ_a

Structure	Occupancy Category		
	I or II	III	IV
Structures, other than masonry shear wall structures, four stories or less with interior walls, partitions, ceilings, and exterior wall systems that have been designed to accommodate the story drifts	$0.025h_{sx}$	$0.020h_{sx}$	$0.015h_{sx}$
Masonry cantilever shear wall structures	$0.010h_{sx}$	$0.010h_{sx}$	$0.010h_{sx}$
Other masonry shear wall structures	$0.007h_{sx}$	$0.007h_{sx}$	$0.007h_{sx}$
All other structures	$0.020h_{sx}$	$0.015h_{sx}$	$0.010h_{sx}$

Note: h_{sx} = story height.

Source: Reprinted with permission of American Society of Civil Engineers from *Minimum Design Loads for Buildings and Other Structures*, ASCE/SEI 7-05, Reston, VA, 2006.

17.2.3 STORY DRIFT LIMITS

With the purpose of avoiding large P-delta effects and minimizing damage to nonstructural elements, the code limits the drift building stories can endure under the design seismic forces. These limits are specified in Table 17.2 in accordance to structure type and occupancy category (occupancy categories are defined in Section 17.4.2).

17.2.4 DIRECTION OF EARTHQUAKE LOADS

In regard to the direction of the design seismic forces, the code stipulates that these forces be applied in the direction that will produce the most critical effect in the structural elements. However, for structures assigned to seismic design category B (seismic design categories are defined in Section 17.4.3) and regular structures in all the other categories (see Section 17.4.4 for the definition of a regular structure), the code considers that this requirement is satisfied if the design seismic forces are applied separately and independently along each of two orthogonal directions. Similarly, for structures assigned to seismic design categories C, D, E, and F that have a horizontal structural irregularity type 5 as defined in Table 17.9, the requirement will be satisfied if the elements of the structure are designed for one of the following orthogonal load combinations:

1. 100% of the forces in one direction plus 30% of the forces in the perpendicular direction, considering the combination that requires the maximum element strength.
2. The simultaneous application of orthogonal pairs of ground motion acceleration time histories.

17.3 DESIGN GROUND MOTION

17.3.1 OVERVIEW

The code requires that structures be designed to resist code-specified earthquake ground motions portrayed in the form of a design response spectrum (see Chapter 9 for a definition of a design response spectrum). The specified design response spectrum, however, varies according to the geographical location and characteristics of the site of interest. Consequently, the code provides a series of ground motion intensity maps and amplification factors to account for such variations in regional seismicity and site characteristics. It also prescribes a response spectrum shape and rules to construct the required design spectra in terms of factors or parameters that account for regional seismicity and site characteristics.

17.3.2 Ground Motion Intensity Maps

As discussed in Chapter 2, earthquakes occur with different frequency and intensity in different geographical regions, and thus the earthquake ground motion considered for design purposes depend on the geographical location and seismicity of the site of interest. To account for this dependence of earthquake ground motions on geographic location and site seismicity, the code provides contour maps that define, for different regions of the country and its possessions, the earthquake ground motion intensity that has a 2% probability of being exceeded in 50 years. The maps were developed by the U.S. Geological Survey and may be downloaded directly from the U.S. Geological Survey World Wide Web site: http://earthquake.usgs.gov/research/hazmaps/. The earthquake ground motion intensity in these maps is characterized by the spectral accelerations corresponding to the natural periods of 0.2 and 1.0 s in response spectra for 5% damping and rock sites. The spectral acceleration for a natural period of 0.2 s serves to define the short-period region of the specified design response spectrum, whereas that for a natural period of 1.0 s serves to define its long-period region. The given intensities are supposed to represent the most severe ground motion considered by the code, to which the code refers to as the "*maximum considered earthquake*," or the 2500-year return period event (see Section 7.6.2 for a definition of return period). Contour maps are used as opposed to seismic zone maps to permit a smooth transition between intensity values and a direct consideration of near-source effects (see Section 5.7.6 for a description of near-source effects). Figures 17.1 through 17.4 are examples of the code-provided contour maps. Figures 17.1 and 17.2 show the maps for the states of California and Nevada and Figures 17.3 and 17.4 the maps for the state of Alaska.

17.3.3 Site-Specific Response Spectra

The code also permits, and requires in some cases, the definition of the design earthquake ground motion at a site on the basis of site-specific studies following procedures similar to those described in Chapter 7. In these site-specific studies, the code requires that an account be made of (1) regional seismicity, tectonic setting, and geology; (2) expected recurrence rates and maximum magnitude of events on known seismic sources; (3) site location with respect to known seismic sources and ground motion attenuation with distance; (4) near-source effects; and (5) subsurface characteristics. It also requires that the maximum considered earthquake ground motion be represented by, whichever is less, a 5%-damping acceleration response spectrum determined probabilistically for a 2% probability of exceedance in 50 years, or that obtained deterministically by augmenting by 150% the largest of the response spectra for the characteristic earthquakes (see Section 7.5.4) on all known active faults within the region of interest. However, the ordinates in the deterministic response spectrum should not be less than the corresponding ordinates in the limit response spectrum shown in Figure 17.5, where the site coefficients F_a and F_v are determined as indicated in Section 17.3.4, but considering S_s equal to 1.5 and S_1 equal to 0.6. Site-specific studies are required for sites with weak or very soft soils, that is, sites classified as class F according to the site classification introduced in the following section.

17.3.4 Site Classes and Site Coefficients

As described in Chapter 8, soft soil deposits have the capability of amplifying or attenuating earthquake ground motions. For the purpose of accounting for such possible amplification and attenuation effects, the code classifies sites into six classes in accordance to the rigidity of the rock

Seismic Code Provisions

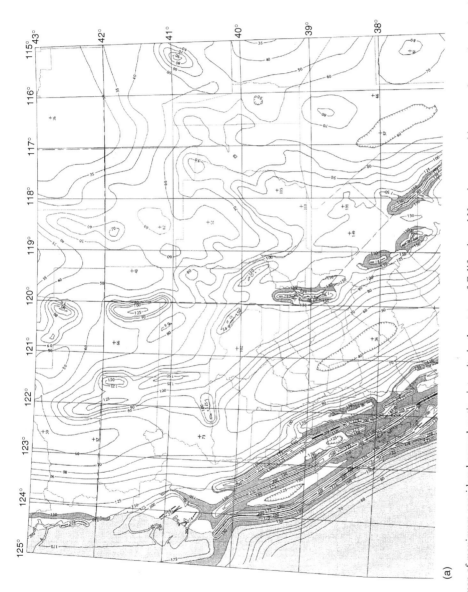

FIGURE 17.1 Contour map of maximum considered ground motion intensity in the states of California and Nevada measured in terms of spectral accelerations for a natural period of 0.2 s and a damping ratio of 5% and expressed as a percentage of the acceleration of gravity. (Reproduced with permission of American Society of Civil Engineers from *Minimum Design Loads for Buildings and Other Structures*, ASCE/SEI 7-05, Reston, VA, 2006.)

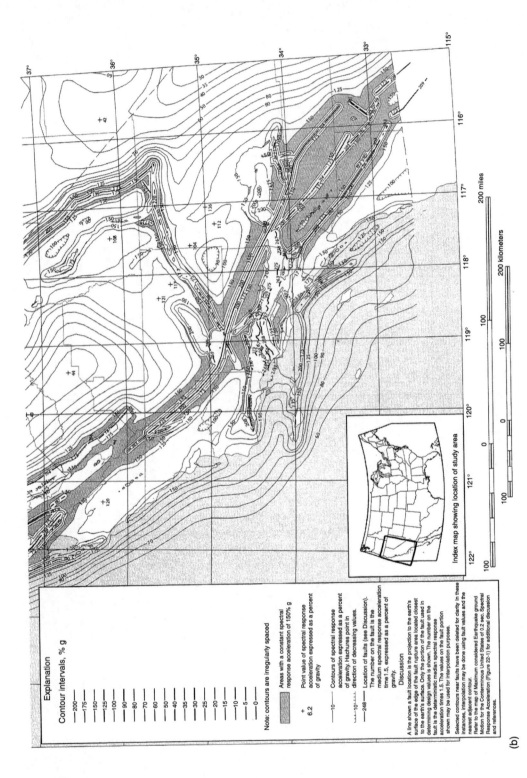

FIGURE 17.1 (Continued)

Seismic Code Provisions

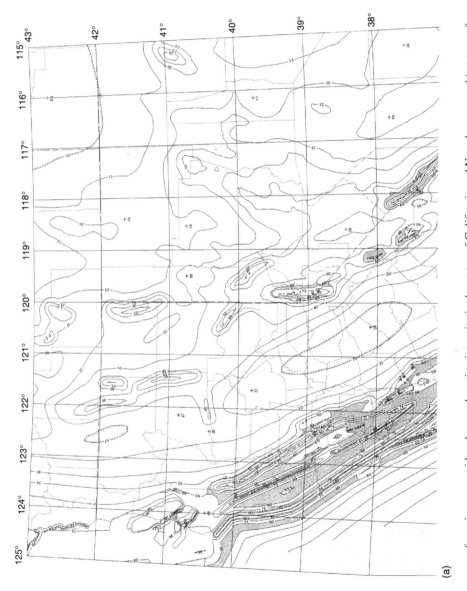

FIGURE 17.2 Contour map of maximum considered ground motion intensity in the states of California and Nevada measured in terms of spectral accelerations for a natural period of 1 s and a damping ratio of 5% and expressed as a percentage of the acceleration of gravity. (Reproduced with permission of American Society of Civil Engineers from *Minimum Design Loads for Buildings and Other Structures*, ASCE/SEI 7-05, Reston, VA, 2006.)

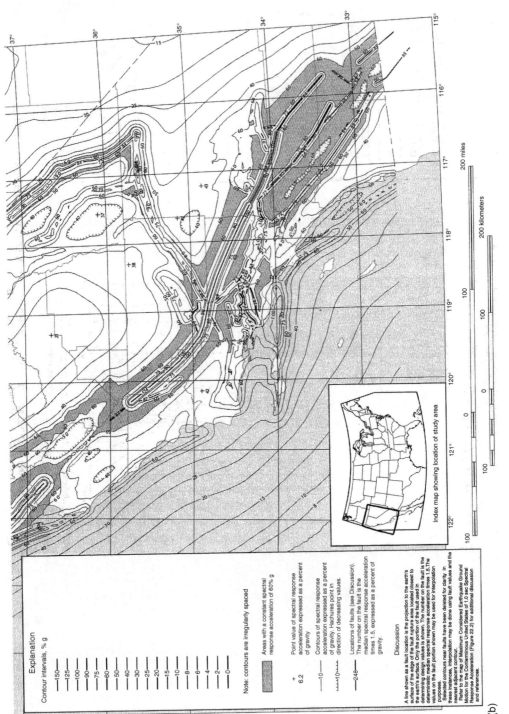

FIGURE 17.2 (Continued)

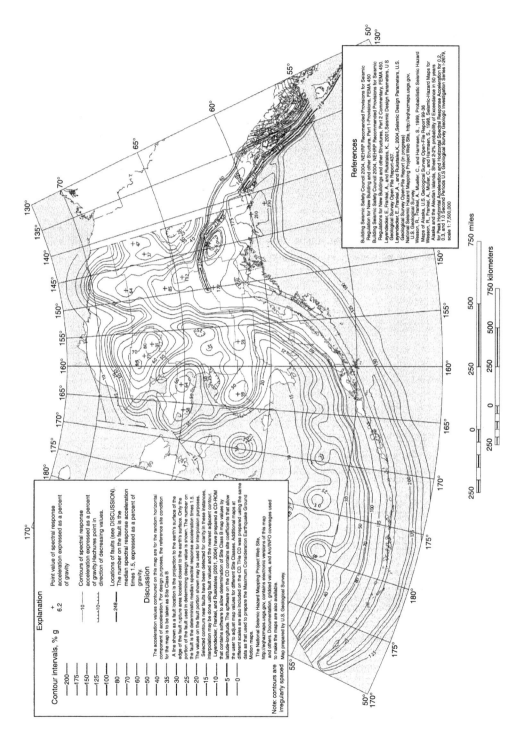

FIGURE 17.3 Contour map of maximum considered ground motion intensity in the state of Alaska measured in terms of spectral accelerations for a natural period of 0.2 s and a damping ratio of 5% and expressed as a percentage of the acceleration of gravity. (Reproduced with permission of American Society of Civil Engineers from *Minimum Design Loads for Buildings and Other Structures*, ASCE/SEI 7-05, Reston, VA, 2006.)

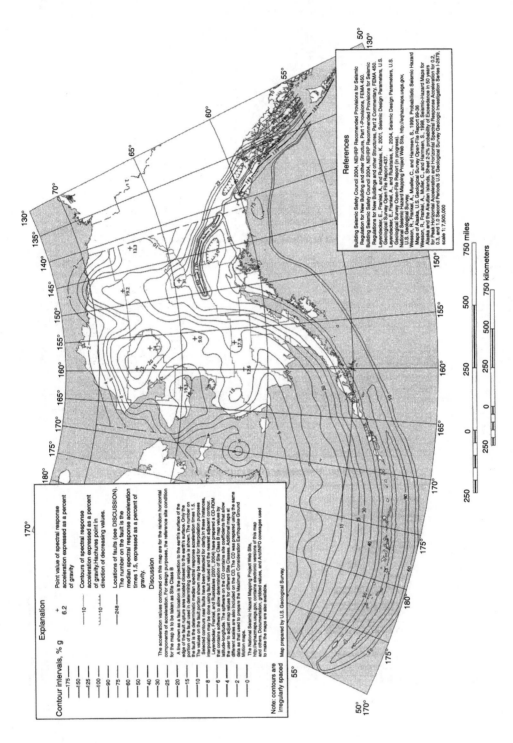

FIGURE 17.4 Contour map of maximum considered ground motion intensity in the state of Alaska measured in terms of spectral accelerations for a natural period of 1 s and a damping ratio of 5% and expressed as a percentage of the acceleration of gravity. (Reproduced with permission of American Society of Civil Engineers from *Minimum Design Loads for Buildings and Other Structures*, ASCE/SEI 7-05, Reston, VA, 2006.)

Seismic Code Provisions

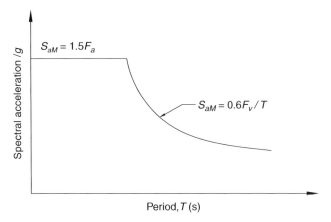

FIGURE 17.5 Limit response spectrum for maximum considered earthquake ground motion.

TABLE 17.3
Site Classes

		Average Properties over Top 100 ft		
Site Class	Soil Profile	Shear Wave Velocity, $\bar{v}_s$ (ft/s)	Standard Penetration Resistance, $\bar{N}$	Undrained Shear Strength, $\bar{s}_u$ (psf)
A	Hard rock	$\bar{v}_s > 5000$	Not applicable	Not applicable
B	Rock	$2500 < \bar{v}_s \leq 5000$	Not applicable	Not applicable
C	Very dense soil and soft rock	$1200 < \bar{v}_s \leq 2500$	$\bar{N} > 50$	$\bar{s}_u > 2{,}000$
D	Stiff soil	$600 \leq \bar{v}_s \leq 1200$	$15 \leq \bar{N} \leq 50$	$1{,}000 \, \bar{s}_u \, 2{,}000$
E	Soft soil	$v < 600$	$\bar{N} < 15$	$\bar{s}_u < 1{,}000$
		Any profile with more than 10 ft of soil having the following characteristics: 1. Plasticity index $PI > 20$ 2. Water content $w \geq 40\%$ 3. Undrained shear strength $\bar{s}_u < 500$ psf		
F		Any profile containing soils having one or more of the following characteristics: 1. Soils vulnerable to potential failure or collapse under seismic loading such as liquefiable soils, quick and highly sensitive clays, or collapsible weakly cemented soils 2. Peats or highly organic clays with a thickness of more than 10 ft 3. Very high plasticity clays with a thickness of more than 25 ft and a plasticity index greater than 75 4. Soft or medium stiff clays with a thickness of more than 120 ft		

Note: 1 ft = 0.3048 m, 1 ft² = 0.0929 m², 1 lb/ft² = 0.0479 kPa.

Source: Reprinted with permission of American Society of Civil Engineers from *Minimum Design Loads for Buildings and Other Structures*, ASCE/SEI 7-05, Reston, VA, 2006.

formation or soil deposit underlying a site. These six site classes are defined in Table 17.3, where the rock or soil rigidity is described in terms of the average shear wave velocity, standard penetration resistance, or undrained shear strength of the top 100 ft (30 m) of the rock formation or soil deposit at the site. In the case of class E and F soils, such rigidity is described in terms of soil thickness and

TABLE 17.4
Site Coefficient F_a as a Function of Site Class and Mapped Spectral Acceleration at Short Periods

Site Class	Mapped Spectral Acceleration at Short Periods				
	$S_S \leq 0.25$	$S_S = 0.50$	$S_S = 0.75$	$S_S = 1.00$	$S_S \geq 1.25$
A	0.8	0.8	0.8	0.8	0.8
B	1.0	1.0	1.0	1.0	1.0
C	1.2	1.2	1.1	1.0	1.0
D	1.6	1.4	1.2	1.1	1.0
E	2.5	1.7	1.2	0.9	0.9
F	Value needs to be determined from site response analysis				

Source: Reprinted with permission of American Society of Civil Engineers from *Minimum Design Loads for Buildings and Other Structures*, ASCE/SEI 7-05, Reston, VA, 2006.

TABLE 17.5
Site Coefficient F_v as a Function of Site Class and Mapped Spectral Acceleration at 1 s Period

Site Class	Mapped Spectral Acceleration at 1 s Period				
	$S_1 \leq 0.1$	$S_1 = 0.2$	$S_1 = 0.3$	$S_1 = 0.4$	$S_1 \geq 0.5$
A	0.8	0.8	0.8	0.8	0.8
B	1.0	1.0	1.0	1.0	1.0
C	1.7	1.6	1.5	1.4	1.3
D	2.4	2.0	1.8	1.6	1.5
E	3.5	3.2	2.8	2.4	2.4
F	Value needs to be determined from site response analysis				

Source: Reprinted with permission of American Society of Civil Engineers from *Minimum Design Loads for Buildings and Other Structures*, ASCE/SEI 7-05, Reston, VA, 2006.

some common soil properties. When the soil properties at a site are not known in sufficient detail to determine the site class, the code stipulates that the site be considered as class D, unless geotechnical data indicates that class E or F soils are likely to be present. In addition, the code requires that the spectral accelerations determined from the provided ground motion intensity maps that define the maximum considered earthquake be affected by the site coefficients given in Tables 17.4 and 17.5 according to

$$S_{MS} = F_a S_s \qquad (17.9)$$

$$S_{M1} = F_v S_1 \qquad (17.10)$$

where
 S_{MS} = Maximum considered spectral acceleration for short periods adjusted for site effects
 S_{M1} = Maximum considered spectral acceleration for 1 s period adjusted for site effects

Seismic Code Provisions

F_a = Site coefficient obtained from Table 17.4
F_v = Site coefficient obtained from Table 17.5
S_s = Spectral acceleration for short periods determined from provided ground motion intensity maps
S_1 = Spectral acceleration for 1 s period determined from provided ground motion intensity maps

It is worthwhile to note that the factors F_a and F_b in Tables 17.4 and 17.5 are all equal to 1.0 for site class B. This is so because the spectral accelerations given in the intensity maps were all determined for sites classified as class B. Note also that the amplification factors for soft soils may be as large as 2.5 for short periods and as much as 3.5 for long periods. Furthermore, note that the amplification factors considered are less for large spectral accelerations than for small ones. The code recognizes thus that under large spectral accelerations soils may incur into their nonlinear range of behavior and that this nonlinear behavior inhibits resonant effects and ground motion amplifications.

17.3.5 Design Spectral Accelerations

The specified design spectral accelerations at short periods and at a 1 s period are

$$S_{DS} = \frac{2}{3} S_{MS} \tag{17.11}$$

$$S_{D1} = \frac{2}{3} S_{M1} \tag{17.12}$$

where S_{MS} and S_{M1} are the site-adjusted spectral accelerations defined earlier. The factor of 2/3 in Equations 17.11 and 17.12 is introduced to account for an assumed inherent level of building overstrength.

17.3.6 Design Response Spectrum

For those cases in which the use of a design spectrum is required, the code stipules that this spectrum be constructed in the form of the acceleration response spectrum shown in Figure 17.6, defining the spectral accelerations S_a as follows:

1. For periods less than or equal to T_0,

$$S_a = S_{DS}\left(0.4 + 0.6\frac{T}{T_0}\right) \tag{17.13}$$

2. For periods greater than or equal to T_0 and less than or equal to T_S,

$$S_a = S_{DS} \tag{17.14}$$

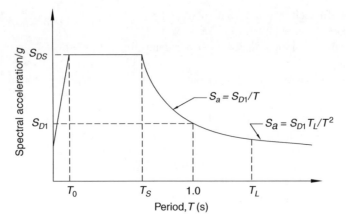

FIGURE 17.6 Design response spectrum.

3. For periods greater than T_S and less than or equal to T_L

$$S_a = \frac{S_{D1}}{T} \tag{17.15}$$

4. For periods greater than T_L,

$$S_a = \frac{S_{D1}T_L}{T^2} \tag{17.16}$$

In the foregoing expressions and Figure 17.6, T is the fundamental natural period of the structure in seconds,

$$T_0 = 0.2\frac{S_{D1}}{S_{DS}} \tag{17.17}$$

$$T_S = \frac{S_{D1}}{S_{DS}} \tag{17.18}$$

and T_L is a long-period transition period given for different regions of the country in maps provided by the code (as an example, see the maps shown in Figure 17.7); S_{DS} and S_{D1} are the design spectral accelerations defined by Equations 17.11 and 17.12, respectively.

When studies are conducted to define the design ground motion for a site, the code stipulates the use of a design spectrum equal to 2/3 of the site-specific response spectrum determined on the basis of the maximum considered ground motion. It also stipulates that the design spectrum so determined be always greater than or equal to 80% of the design spectrum shown in Figure 17.6. To this end, it requires that S_{DS} be taken as the spectral acceleration S_a at a period of 0.2 s, except that it should not be less than 90% of the peak spectral acceleration at any period longer than 0.2 s. Similarly, it requires that S_{D1} be taken as the greater of the spectral acceleration at a period of 1 s or two times the spectral acceleration at a period of 2 s. S_{MS} and S_{M1} are to be taken as 1.5 times S_{DS} and S_{D1}, respectively. However, the values so obtained are not to be less than 80% of the values obtained as specified in Section 17.3.4 for S_{MS} and S_{M1} and Section 17.3.5 for S_{DS} and S_{D1}.

Seismic Code Provisions

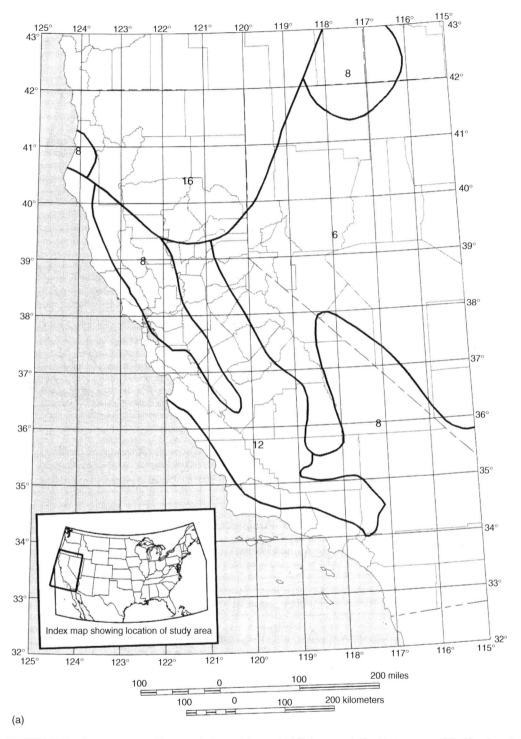

(a)

FIGURE 17.7 Contour maps of long-period transition period T_L in seconds for (a) the states of California and Nevada, and (b) the state of Alaska. (Reprinted with permission of American Society of Civil Engineers from *Minimum Design Loads for Buildings and Other Structures*, ASCE/SEI 7-05, Reston, VA, 2006.)

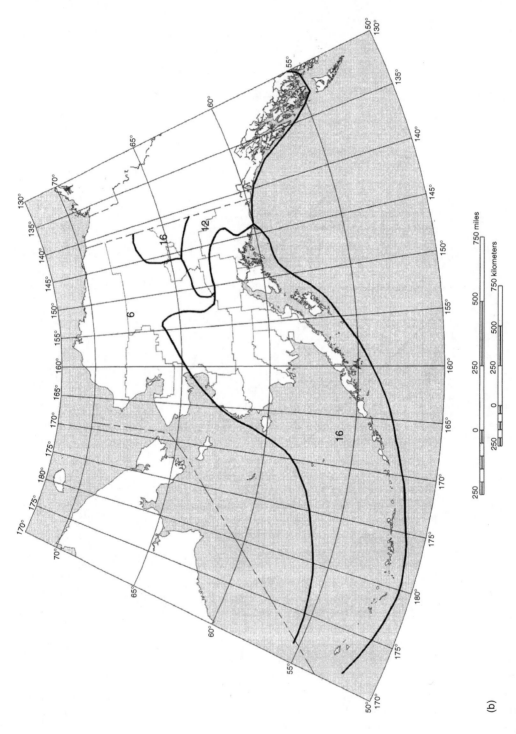

FIGURE 17.7 (Continued)

17.4 ANALYSIS PROCEDURES

17.4.1 OVERVIEW

For the purpose of determining the seismic effects E and E_m that need to be considered in the load combinations described in Section 17.2.2, the code allows the use of several methods of analysis. However, not every method can be used in every case. Instead, the code prescribes the minimum method that can be used in accordance to the use or type of occupancy of the building, the expected level of seismic intensity, and the height and configuration of the structure. To this end, the code defines several occupancy categories and classifies structures into different seismic design categories based on the expected level of seismic intensity and occupancy category. It also delineates what makes a structure irregular either in plan or in elevation.

17.4.2 OCCUPANCY CATEGORIES AND IMPORTANCE FACTORS

The code classifies buildings and other structures based on the nature of occupancy. It also assigns *importance factors* to each *occupancy category* with the intention of increasing the safety levels in essential facilities and those that represent a substantial hazard to human life in the event of failure. These occupancy categories and the importance factors assigned to each category are identified in Table 17.6. Examples of buildings in occupancy category I are agricultural buildings and temporary facilities. Examples of buildings in occupancy category III are those where more than 300 people congregate and power-generating stations. Examples of buildings in occupancy IV are hospitals, fire stations, and emergency shelters.

17.4.3 SEISMIC DESIGN CATEGORY

The code also classifies buildings into the different *seismic design categories* shown in Tables 17.7 and 17.8 to specify permissible structural systems, limitations on height and irregularity, the structural components that must be designed for earthquake loads, and, as mentioned earlier, the allowable analysis procedures. As it may be seen from Tables 17.7 and 17.8, a seismic design category is selected on the basis of occupancy category and the level of the design spectral accelerations S_{DS} and S_{D1}. Because the seismic design category determined from Table 17.7 may differ from that determined from Table 17.8, the code stipulates that the most severe seismic design category be selected irrespective of the fundamental period of the structure whenever S_{DS} and S_{D1} lead to two different design categories. It is permitted, however, to determine the seismic design category from Table 17.7 alone when (a) S_1 is less than 0.75, (b) the approximate period of the structure along each of two orthogonal directions is less than 0.8 times the T_S period defined by Equation 17.18, (c) Equation 17.23 is used to determine the seismic response coefficient C_s, and (d) the diaphragms of the structure are rigid or flexible with a separation between vertical lateral-load-resisting elements

TABLE 17.6
Occupancy Categories and Seismic Importance Factors

Occupancy Category	Nature of Occupancy	Importance Factor, I
I	Buildings and other structures that represent a low hazard to human life in the event of failure	1.00
II	Buildings and other structures not included in occupancy categories I, III, and IV	1.00
III	Buildings and other structures that represent a substantial hazard to human life in the event of failure	1.25
IV	Buildings and other structures designated as essential facilities (needed for emergency response and postearthquake recovery, or have critical national defense functions)	1.50

TABLE 17.7
Seismic Design Category Based on Short-Period Spectral Acceleration and Occupancy Category

	Occupancy Category		
S_{DS}	I or II	III	IV
$S_{DS} < 0.167$	A	A	A
$0.167 \leq S_{DS} < 0.33$	B	B	C
$0.33 \leq S_{DS} < 0.50$	C	C	D
$S_{DS} \geq 0.50$	D	D	D

Source: Reprinted with permission of American Society of Civil Engineers from *Minimum Design Loads for Buildings and Other Structures*, ASCE/SEI 7-05, Reston, VA, 2006.

TABLE 17.8
Seismic Design Category Based on 1 s Period Spectral Acceleration and Occupancy Category

	Occupancy Category		
S_{D1} or S_1	I or II	III	IV
$S_{D1} < 0.067$	A	A	A
$0.067 \leq S_{D1} < 0.133$	B	B	C
$0.133 \leq S_{D1} < 0.20$	C	C	D
$S_{D1} \geq 0.20$	D	D	D
$S_1 \geq 0.75$	E	E	F

Source: Reprinted with permission of American Society of Civil Engineers from *Minimum Design Loads for Buildings and Other Structures*, ASCE/SEI 7-05, Reston, VA, 2006.

of no more than 40 ft (12.2 m). When S_1 is less than or equal to 0.04 and S_s is less than or equal to 0.15, the structure may be assigned to design category A.

17.4.4 STRUCTURAL IRREGULARITIES

For the purpose of specifying different requirements for buildings that exhibit horizontal and vertical irregularities, the code classifies buildings as irregular if they possess one of the structural irregularities defined in Tables 17.9 and 17.10. Buildings having one or more of the features listed in Table 17.9 are designated as having a horizontal structural irregularity. Similarly, buildings having one or more of the features listed in Table 17.10 are designated as having a vertical structural irregularity. However, vertical irregularities 1a, 1b, or 2 need not be considered when, without considering the top two stories, no story drift ratio under the design lateral seismic forces is greater than 130% of the story drift ratio of the next story above. Similarly, vertical irregularities 1a, 1b, and 2 need not be considered for one-story buildings in any seismic design category or for two-story buildings in seismic design categories B, C, or D. However, the code does not allow the use of structures in design category E or F having horizontal irregularity 1b or vertical irregularities 1b, 5a, or 5b.

TABLE 17.9
Definition of Horizontal Structural Irregularities

Identifying Number	Type	Definition
1a	Torsional irregularity (considered when diaphragms are rigid or semirigid)	Torsional irregularity exists when the maximum story drift at one end of the structure transverse to an axis, computed with accidental torsion included, is more than 1.2 times the average of the story drifts at the two ends of the structure.
1b	Extreme torsional irregularity (considered when diaphragms are rigid or semirigid)	Extreme torsional irregularity exists when the maximum story drift at one end of the structure transverse to an axis, computed with accidental torsion included, is more than 1.4 times the average of the story drifts at the two ends of the structure.
2	Re-entrant corners	Reentrant corners exist in a structure when both projections of the structure beyond a reentrant corner are greater than 15% of the plan dimension of the structure in the given direction.
3	Diaphragm discontinuity	Diaphragm discontinuity exists when a diaphragm possesses an abrupt discontinuity or exhibits a stiffness variation, including a cutout or open area greater than 50% of the gross enclosed diaphragm area, or changes in effective stiffness of more than 50% from one story to the next.
4	Out-of-plane offsets	Out-of-plane offsets exist when there is a discontinuity in a lateral-force-resisting path, such as out-of-plane offsets of the vertical elements.
5	Nonparallel resisting systems	A nonparallel resisting system is one in which the vertical lateral-force-resisting elements are not parallel to or symmetric about the major orthogonal axes of the lateral-force-resisting system.

Source: Reprinted with permission of American Society of Civil Engineers from *Minimum Design Loads for Buildings and Other Structures*, ASCE/SEI 7-05, Reston, VA, 2006.

Similarly, it does not allow the use of structures in seismic design category D having vertical irregularity 5b. Figure 17.8 provides some examples of vertical and horizontal irregularities.

17.4.5 ALLOWABLE ANALYSIS PROCEDURES

The code requires that all structures be the subject of a seismic analysis to determine the seismic forces Q_E that appear in the load combinations described in Section 17.2.2 and the corresponding story drifts. To this end, the code allows the use of three different analysis procedures: (a) equivalent lateral-force procedure, (b) modal response spectrum analysis, and (c) response time-history analysis. However, not every procedure can be used in every case. The procedures that can be used depend, instead, on the seismic design category, structural system, structural configuration, and natural period, as indicated in Table 17.11. The code also allows the use of the simplified procedure described in Section 17.5, provided the following conditions are met:

1. The structure qualifies for occupancy category I or II.
2. The site class is not E or F.
3. The structure does not exceed three stories in height above grade.
4. The lateral-force-resisting system is a bearing-wall or building frame system.
5. The structure has at least two lines of lateral resistance along each of two major-axis directions.

TABLE 17.10
Definition of Vertical Structural Irregularities

Identifying Number	Type	Definition
1a	Stiffness irregularity: soft story	A soft story is one in which the lateral stiffness is less than 70% of the lateral stiffness in the story above or less than 80% of the average stiffness of the three stories above.
1b	Stiffness irregularity: extreme soft story	An extreme soft story is one in which the lateral stiffness is less than 60% of the lateral stiffness of the story above or less than 70% of the average stiffness of the three stories above.
2	Weight (mass) irregularity	Mass irregularity is considered to exist when the effective mass of any story is more than 150% of the effective mass of an adjacent story. A roof that is lighter than the floor below need not be considered.
3	Vertical geometric irregularity	Vertical geometric irregularity is considered to exist when the horizontal dimension of the lateral-force-resisting system in any story is more than 130% of that in the adjacent story.
4	In-plane discontinuity in vertical lateral-force-resisting elements	It is considered to exist when there is an in-plane offset of the lateral-force-resisting elements greater than the length of the elements or a reduction in stiffness of the resisting element in the story below.
5a	Discontinuity in lateral strength: weak story	A weak story is one in which the story lateral strength is less than 80% of the lateral strength of the story above. (Story strength is the total strength of seismic-resisting elements sharing the story shear for the direction under consideration.)
5b	Discontinuity in lateral strength: extreme weak story	A weak story is one in which the story lateral strength is less than 65% of the lateral strength of the story above. (Story strength is the total strength of seismic-resisting elements sharing the story shear for the direction under consideration.)

Source: Reprinted with permission of American Society of Civil Engineers from *Minimum Design Loads for Buildings and Other Structures*, ASCE/SEI 7-05, Reston, VA, 2006.

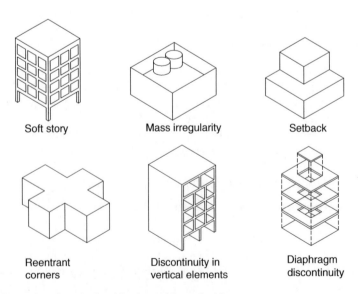

FIGURE 17.8 Examples of vertical and horizontal irregularities.

TABLE 17.11
Permitted Analytical Procedures

Seismic Design Category	Structural Characteristics	Equivalent Lateral Force Procedure	Modal Response Spectrum Analysis	Seismic Response History Procedures
B, C	Occupancy category I or II buildings of light-framed construction not exceeding three stories in height	P	P	P
	Other occupancy categories I or II buildings not exceeding two stories in height	P	P	P
	All other structures	P	P	P
D, E, F	Occupancy category I or II buildings of light-framed construction not exceeding three stories in height	P	P	P
	Other occupancy category I or II buildings not exceeding two stories in height	P	P	P
	Regular structures with $T < 3.5T_S$, and all structures of light-frame construction	P	P	P
	Irregular structures with $T < 3.5T_S$, and having only horizontal irregularities Type 2, 3, 4, or 5, or vertical irregularities Type 4, 5a, or 5b	P	P	P
	All other structures	NP	P	P

Note: P = permitted, NP = not permitted.

Source: Reprinted with permission of American Society of Civil Engineers from *Minimum Design Loads for Buildings and Other Structures*, ASCE/SEI 7-05, Reston, VA, 2006.

6. At least one line of resistance is provided on each side of the center of mass in each direction.
7. For buildings with flexible diagrams, the overhangs beyond the outside line of shear walls or braced frames do not exceed one-fifth of the depth of the diaphragm parallel to the forces being considered at the line of vertical resistance closest to the diaphragm edge.
8. For buildings with nonflexible diaphragms, the distance between the center of rigidity and center of mass parallel to each major axis does not exceed 15% of the greatest width of the diaphragm parallel to that axis. In addition, for each major-axis direction,

$$\sum_{i=1}^{m} k_{1i}d_i^2 + \sum_{j=1}^{n} k_{2j}d_j^2 \geq 2.5\left(0.05 + \frac{e_1}{b_1}\right)b_1^2 \sum_{i=1}^{m} k_{1i} \qquad (17.19)$$

where

k_{1i}, k_{2j} = Lateral stiffness of wall or braced frame i, j parallel to major axis 1, 2
d_{1i}, d_{2j} = Distance from wall or braced frame i, j to center of rigidity, perpendicular to major axis 1, 2
e_1 = Distance perpendicular to major axis 1 between center of rigidity and center of mass
b_1 = Width of diaphragm perpendicular to major axis 1
m, n = Number of resisting walls and braced frames in direction 1, 2

9. Lines of resistance of lateral-force-resisting system are oriented at angles of no more than 15° from alignment with the building's major orthogonal horizontal axes.
10. The simplified procedure is used for each of the two major horizontal directions of the building.

11. The elements of the lateral-force-resisting system have no in-plane or out-of-plane offsets.
12. The lateral-force resistance of any story is not less than 80% of the resistance of the story above.

17.4.6 MATHEMATICAL MODEL

For the purpose of analysis, the codes requires the use of a mathematical model that represents the spatial distribution of mass and stiffness throughout the structure, and includes the stiffness and strength of the elements that are significant to the distribution of forces and deformations in the structure. For regular structures with independent orthogonal seismic-force-resisting systems, it is possible to construct two independent two-dimensional models to represent each system. However, for irregular systems without independent orthogonal systems, it is necessary to construct a three-dimensional model that incorporates a minimum of three degrees of freedom per floor: two translational degrees of freedom along two orthogonal horizontal directions and a torsional one about the structure's vertical axis. When the floor diaphragms are not rigid in comparison with the vertical elements of the lateral-force-resisting system, the model needs to include additional degrees of freedom to account for the diaphragms' flexibility. The code also stipulates that the used mathematical model consider the (1) the effect of cracked sections in the determination of the stiffness properties of concrete and masonry elements, and (2) the contribution of panel zone deformations in steel moment frame systems. It allows, nonetheless, that the structure be considered fixed at its base.

17.5 SIMPLIFIED METHOD

In the simplified method, equivalent lateral forces are applied at each building level. In any given direction, these equivalent lateral forces are calculated according to

$$F_x = \frac{w_x}{W} V \qquad (17.20)$$

where W and w_x are total gravity load computed as indicated in Section 17.6.2 and fraction of total gravity load assigned to level x, and V is the base shear, considered equal to

$$V = \frac{F S_{DS}}{R} W \qquad (17.21)$$

where

$$F = \begin{cases} 1.0 & \text{for one-story buildings} \\ 1.1 & \text{for two-story buildings} \\ 1.2 & \text{for three-story buildings} \end{cases}$$

S_{DS} = Design spectral acceleration for short periods determined as indicated in Section 17.3.5
R = Response modification factor selected from Table 17.1 in accordance to structure type

In the calculation of S_{DS} according to Equations 17.9 and 17.11, it is permitted to take F_a as 1.0 for rock sites and 1.4 for soil sites. For this purpose, it may be considered that a rock site is one in which

Seismic Code Provisions

there is no more than 10 ft (3 m) of soil between the rock surface and the bottom of the spread footing or mat foundation. Similarly, it is permitted to limit the value of S_s to 1.5 and select the seismic design category based on the value of S_{DS} alone.

For structures designed using this procedure, it is not necessary to compute and evaluate story drift. If needed for the calculation of building separation, the design of cladding, or other design requirements, the code allows the consideration of a design story drift Δ equal to 1% of the story height, unless it is determined by calculation to be less.

17.6 EQUIVALENT LATERAL FORCE PROCEDURE

17.6.1 Overview

The equivalent lateral force procedure recommended by the code is closely based on the equivalent lateral force procedure described in Chapter 12. It involves the computation of a natural period, a base shear, equivalent static lateral forces, and the story drifts, overturning moments, and torsional moments induced by these lateral forces. When this procedure is used, the code requires that the building be designed to resist these lateral forces and overturning and torsional moments. It is also required to comply with story drift limits. The procedure is applied under the assumption that the building is fixed at its base.

17.6.2 Base Shear

The formula specified by the code to calculate the shear force at the base of a structure in any given direction is

$$V = C_s W \tag{17.22}$$

where W is the effective seismic weight, which includes the dead load and

1. A minimum of 25% of the reduced floor live load in areas used for storage (floor live load in public garages and open parking structures need not be included)
2. Actual partition weight or a minimum weight of 10 lb/ft² (0.48 kN/m²) of floor area, whichever is greater
3. Weight of permanent equipment
4. 20% of flat-roof snow load when the flat-roof snow load exceeds 30 lb/ft² (1.44 kN/m²), regardless of actual roof slope

In addition, C_s is a seismic response coefficient defined as

$$C_s = \frac{S_{DS}}{R/I} \tag{17.23}$$

which is not to be less than

$$(C_s)_{min} = 0.01 \tag{17.24}$$

or less than

$$(C_s)_{min} = \frac{0.5 S_1}{R/I} \tag{17.25}$$

if the structure is located in a region where S_1 is equal or greater than 0.6. Furthermore, it need not exceed

$$(C_s)_{max} = \frac{S_{D1}}{T(R/I)} \quad \text{if } T \leq T_L \tag{17.26}$$

$$(C_s)_{max} = \frac{S_{D1}T_L}{T^2(R/I)} \quad \text{if } T > T_L \tag{17.27}$$

For regular structures five stories or less in height and having a natural period T of 0.5 s or less, it is permitted to calculate the seismic coefficient C_s considering S_s equal to 1.5.

In the preceding equations,

- I = Occupancy importance factor selected from Table 17.6
- R = Response modification factor selected from Table 17.1
- S_{DS} = Design spectral acceleration for short periods determined as indicated in Section 17.3.5
- S_{D1} = Design spectral acceleration at 1 s period determined as indicated in Section 17.3.5
- S_1 = Maximum considered spectral acceleration at 1 s period determined as indicated in Section 17.3.2
- T = Fundamental natural period of building in seconds
- T_L = Long-period transition period determined as indicated in Section 17.3.6.

17.6.3 Natural Period

The code offers two options to estimate the fundamental natural period T of a structure. The first is the use of a properly substantiated analysis such as an eigenvalue analysis. The second is the use of given approximate formulas. For structures with moment-resisting frames, the given approximate formula is

$$T_a = C_t h_n^x \tag{17.28}$$

where h_n, C_t, and x represent height of the building in feet or meters about its base, building period coefficient selected from Table 17.12, and exponent whose value is also selected from Table 17.12.

TABLE 17.12
Building Period Coefficient C_t and Exponent x for Moment-Resisting Frame Systems in Which the Frames Resist 100% of the Required Seismic Forces and Are Not Enclosed or Adjoined by Rigid Components That Will Prevent the Frames from Deflecting When Subjected to Seismic Forces

Structure Type	C_t (h_n in feet)	C_t (h_n in meters)	x
Steel moment-resisting frames	0.028	0.072	0.8
Concrete moment-resisting frames	0.016	0.047	0.9
Eccentrically braced steel frames	0.030	0.073	0.75
All other structural systems	0.020	0.049	0.75

Equation 17.28 has been empirically derived from correlations between numerous structures and their natural periods calculated by rigorous means. For concrete and steel moment-resisting frame buildings not exceeding 12 stories in height and having a minimum story height of 10 ft (3.05 m), the code also permits the use of the approximate formula

$$T_a = \frac{N}{10} \tag{17.29}$$

where N denotes the number of stories. For structures with masonry or concrete shear walls, the given approximate formula is

$$T_a = \frac{0.0019}{\sqrt{C_W}} h_n \tag{17.30}$$

where C_W is given by

$$C_W = \frac{100}{A_B} \sum_{i=1}^{x} \left(\frac{h_n}{h_i}\right)^2 \frac{A_i}{1 + 0.83\left(\frac{h_i}{D_i}\right)^2} \tag{17.31}$$

where
A_B = Area of the base of the structure in square feet
A_i = Web area of shear wall i in square feet
h_i = Height of shear wall i in feet
D_i = Length of shear wall i in feet
x = Number of shear walls in the building effective in resisting lateral forces in the direction of analysis

There is, however, a restriction imposed by the code on the calculation of the natural period T using rigorous means. It specifies that the calculated value of T is not to exceed an upper limit equal to the product of the approximate value T_a and a coefficient c_u, where c_u is selected from Table 17.13. This provision is established to avoid the possibility of an unreasonably low base shear because of an excessively long natural period.

TABLE 17.13
Coefficient for Calculation of Upper Limit on Natural Period Determined by Rigorous Means

Design Spectral Acceleration at 1 s Period, S_{D1}	c_u
≥ 0.4	1.4
0.3	1.4
0.2	1.5
0.15	1.6
≤ 0.1	1.7

17.6.4 Equivalent Lateral Forces

The equivalent lateral forces acting on the floor levels of a building are calculated as the product of a vertical distribution factor C_{vx} and the base shear determined from Equation 17.22. That is,

$$F_x = C_{vx} V \qquad (17.32)$$

where V is the total design lateral shear at the building base, and

$$C_{vx} = \frac{w_x h_x^k}{\sum_{i=1}^{N} w_i h_i^k} \qquad (17.33)$$

where k, h_i, h_x, and w_i, w_x are distribution exponent determined from Table 17.14, height from building base to level i, x, and fraction of total gravity load W assigned to level i, x.

The formula specified by the code to determine equivalent lateral forces is essentially the same formula introduced in Chapter 12 for the same purpose. The main difference is that for flexible structures the code formula considers that the fundamental mode shape of the structure varies parabolically with height as opposed to linearly, as it was assumed in Chapter 12. The parabolic assumption is made in an effort to account for the fact that in flexible structures the fundamental mode shape of flexible structures deviates significantly from a straight line. Figure 17.9 shows the distribution of lateral forces for the cases when $k = 1$ and $k = 2$.

17.6.5 Horizontal Shear Distribution

For structures with rigid diaphragms, the code specifies that the horizontal shear force at a given story be distributed to the vertical elements of the seismic-resisting system at that story based on the relative lateral stiffnesses of these vertical elements. For structures with flexible diaphragms, it requires that the aforementioned shear force be distributed to the resisting vertical elements at the story under consideration in proportion to the tributary area of the diaphragm to each of these vertical elements. In this latter case, it is permitted to consider that vertical elements are in the same line of resistance if the maximum out-of-plane offset between such elements is less than 5% of the building dimension perpendicular to the direction of the applied lateral loads. When the diaphragms are rigid, it is required to include the *torsional moments* that result from the difference between the locations of the centers of mass and the centers of stiffness. In addition, it is required to include the *accidental torsional moment* that results, along each direction, from a dislocation of the center of mass from its actual location equal to 5% of the dimension of the building perpendicular to the direction of the applied lateral loads. In other words, it is required

TABLE 17.14
Exponent for Calculation of Vertical Distribution Factor C_{vx}

Natural Period, T (s)	k
$T \leq 0.5$	1.0
$T \geq 2.5$	2.0
$0.5 < T < 2.5$	2.0 or $1.0 + \frac{1}{2}(T - 0.5)$

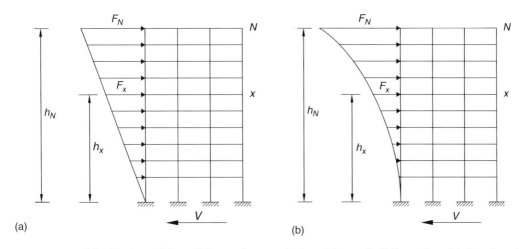

FIGURE 17.9 Distribution of lateral forces in a typical multistory building when (a) $k = 1$ and (b) $k = 2$.

to consider along each direction an accidental eccentricity equal to 5% of the dimension of the building in the direction perpendicular to the direction of analysis. If the lateral forces are applied simultaneously along two orthogonal directions, the required 5% dislocation of the center of mass need not be applied in the two directions at the same time, but needs to be applied in the direction that produces the largest effect.

For structures in the seismic design categories C, D, E, and F for which there is a type 1a or 1b horizontal torsional irregularity as defined in Table 17.9, it is required to multiply the accidental torsional moments by the torsional amplification factor

$$A_x = \left(\frac{\delta_{max}}{1.2\delta_{avg}}\right)^2 \leq 3 \qquad (17.34)$$

where δ_{max} and δ_{avg} are maximum lateral displacement at level x and average of the displacements at the extreme points of the structure at level x both computed assuming $A_x = 1$. However, the accidental torsional moments need not be amplified for structures of light-frame construction. For the purpose of horizontal shear distribution, a diaphragm is considered flexible if the in-plane deflection of the diaphragm is more than two times the average story drift of the associated story.

17.6.6 Overturning Moment

The code requires that structures be designed to resist the overturning moments generated by the equivalent lateral forces specified in Section 17.6.4. These overturning moments may be calculated according to

$$M_x = \sum_{i=x}^{N} F_i(h_i - h_x) \qquad (17.35)$$

where M_x, F_i, and h_i, h_x denote overturning moment at level x, lateral force at level i, and height from base to level i, x.

The code also specifies that the overturning moments so calculated be distributed to the resisting vertical elements in the same proportion with which the horizontal shears are distributed to these same elements. No reduction is allowed to compensate for the fact that, as discussed in Chapter 12, the overturning moments are likely to be overpredicted because the considered lateral forces are supposed to represent maximum values and these maximum values are unlikely to occur all at the same time. However, with the exception of those for structures of the inverted-pendulum type, foundations may be designed for 75% of the foundation overturning moment computed at the foundation–soil interface.

17.6.7 Story Drift Determination

For the purpose of complying with story drift limitations and minimum building separations, the code defines story drift as the difference between the displacements of the centers of mass of the floors immediately above and below the story under consideration. However, for structures in the seismic design categories C–F for which there exists a Type 1a or Type 1b horizontal irregularity, story drift is defined as the largest difference between the displacements along the edges of the structure at the floors immediately above and below the story being considered. In addition, the code specifies that the floor displacements be calculated according to

$$\delta_x = \frac{C_d \delta_{xe}}{I} \qquad (17.36)$$

where C_d, δ_{xe}, and I are displacement amplification factor specified in Table 17.1, displacement induced by the equivalent lateral forces at level x of the structure determined by an elastic analysis, and occupancy importance factor determined from Table 17.6.

It should be noted that the structural displacements computed on the basis of the specified equivalent lateral forces are multiplied by the amplification factor C_d because the specified equivalent lateral forces represent the forces that just make the structure incur into its inelastic range of behavior and thus these displacements correspond to just the elastic components of the total displacements. Consequently, by multiplying by the amplification factor C_d, which is related to the ductility factor introduced in Section 6.5.4, estimates of the corresponding total displacements are obtained.

For the calculation of story drifts, the code further stipulates that

1. There is no need to consider the upper limit to the fundamental natural period of the structure specified in Section 17.6.3.
2. P-delta effects be accounted for by rational analysis when the stability coefficient θ defined in the following section is greater than 0.10. Alternatively, it is allowed to consider P-delta effects by multiplying the design story drifts by the incremental factor $1/(1 - \theta)$.
3. The redundancy coefficient ρ may be considered equal to 1.0.

17.6.8 P-Delta Effects

In regard to P-delta effects, the code specifies that the P-delta effects on story shears and moments, member forces, and story drifts need not be considered when the structure's stability coefficient θ is equal to or less than 0.10, where for each story this stability coefficient is defined as the ratio of the secondary moment to the primary moment; that is,

$$\theta = \frac{P_x \Delta}{V_x h_{sx} C_d} \qquad (17.37)$$

which cannot exceed the limit

$$\theta_{max} = \frac{0.5}{\beta C_d} \leq 0.25 \qquad (17.38)$$

In the preceding equations,

- P_x = Total unfactored vertical design load at and above level x
- V_x = Seismic shear force acting between levels x and $x-1$
- Δ = Design story drift occurring simultaneously with V_x
- h_{sx} = Story height below level x
- C_d = Deflection amplification factor determined from Table 17.1

and β is the ratio of shear demand to shear capacity of the story, which may be conservatively considered equal to 1 if not calculated. The product βC_d is supposed to represent an adjusted ductility demand, whereas β represents a measure of the structure's overstrength due to drift control, wind design, or other design aspects. When P-delta effects are included in an automated analysis, the limit established by Equation 17.38 still needs to be satisfied. However, the value of θ computed using Equation 17.37 and the results from the automated P-delta analysis may be divided by $(1+\theta)$ before checking whether or not θ complies with such a limit. Structures with a stability coefficient greater than θ_{max} are considered potentially unstable and should be, therefore, redesigned.

It should be recognized that the stability coefficient and amplification factor specified by the code to account for P-delta effects are based on the secondary to primary moment ratio and amplification factors introduced in Chapter 12.

EXAMPLE 17.1 ANALYSIS OF SIX-STORY STRUCTURE BY EQUIVALENT LATERAL FORCE PROCEDURE

The building shown in Figure E17.1a is square in plan, will be located in the city of Anchorage, Alaska, will be used as an office building, and will be built with steel ductile moment-resisting frames. The proposed construction site overlies a deposit of very dense soil. The weight at each floor, which includes the floor dead load; the weight of the beams, curtain walls, and partitions; and 25% of the floor live load, is 2.2 MN. The weight at the roof is also 2.2 MN. Determine using the equivalent lateral force procedure prescribed by the International Building Code (a) the design seismic forces for the building, and (b) the story drifts induced by these forces when the sections for the beams and columns of the building are as indicated in Figure E17.1a.

Solution

Total Building Weight

As the gravity load per floor and the roof gravity load are both 2.2 MN, the total weight of the building is

$$W = 2.2(6) = 13.2 \text{ MN}$$

Natural Period

Using Equation 17.28 and considering a building height h_n of 19.8 m and the coefficient C_t and exponent x given in Table 17.12 for buildings with steel frames, an approximate value of the fundamental natural period of the building is

$$T_a = C_t(h_n)^{0.8} = 0.072(19.8)^{0.8} = 0.78 \text{ s}$$

It may be noted that in this case T_a is less than the long-period transition period T_L since from Figure 17.7 it is found that for Anchorage, Alaska, $T_L = 16$ s.

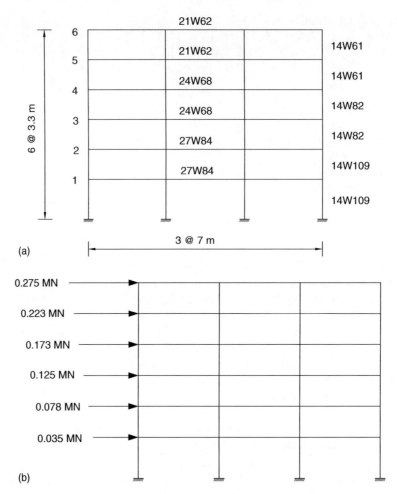

FIGURE E17.1 (a) Six-story building considered in Example 17.1 and (b) magnitude and location of equivalent lateral forces.

Base Shear

The base shear is calculated using Equations 17.22 through 17.27. The use of these equations, however, requires knowing the maximum considered spectral accelerations S_S and S_1, the site coefficients F_a and F_v, the response modification coefficient R, and the importance factor I. Accordingly, these parameters will be determined first.

From the seismic intensity maps shown in Figures 17.3 and 17.4, the maximum considered spectral accelerations for short periods and a period of 1 s for Anchorage, Alaska (expressed as a fraction of the acceleration of gravity) are

$$S_S = 1.50$$

$$S_1 = 0.50$$

Similarly, from Tables 17.4 and 17.5, the site coefficients for a site with very dense soils (class C), $S_S \geq 1.25$, and $S_1 \geq 0.5$ are

$$F_a = 1.0$$

$$F_v = 1.3$$

Therefore, according to Equations 17.9 and 17.10, the maximum considered spectral accelerations adjusted for site effects are

$$S_{MS} = F_a S_s = 1.0(1.50) = 1.50$$

$$S_{M1} = F_v S_1 = 1.3(0.50) = 0.65$$

and, according to Equations 17.11 and 17.12, the corresponding design spectral accelerations are

$$S_{DS} = \frac{2}{3} S_{MS} = \frac{2}{3}(1.50) = 1.00$$

$$S_{D1} = \frac{2}{3} S_{M1} = \frac{2}{3}(0.65) = 0.43$$

Similarly, it may be considered that the building belongs to occupancy category II since it will be used for office space (see Table 17.6). As such, the importance factor in this case is

$$I = 1$$

Also, according to the classifications in Tables 17.7 and 17.8, the building belongs to seismic design category D. Note, in addition, that it may be assumed that the steel frames will be detailed as special moment frames. Hence, from Table 17.1 the response modification coefficient for the structure under consideration may be taken as

$$R = 8$$

According to Equation 17.23, the seismic response coefficient for the building is thus equal to

$$C_s = \frac{S_{DS}}{R/I} = \frac{1.00}{8/1} = 0.125$$

which is greater than the minimum value (see Equation 17.24)

$$(C_s)_{min} = 0.01$$

but greater than the maximum value required by the code when $T \leq T_L$ (see Equation 17.26), which for the building in this example is

$$(C_s)_{max} = \frac{S_{D1}}{T(R/I)} = \frac{0.43}{0.78(8/1.0)} = 0.069$$

Consequently, C_s will be considered equal to

$$C_s = 0.069$$

and, correspondingly, the base shear equal to

$$V = C_s W = 0.069(13.2) = 0.910 \text{ MN}$$

Lateral Forces and Story Shears

Once the base shear is determined, the magnitude of the equivalent lateral forces acting on the floors of the building may be determined using Equations 17.32 and 17.33. For this purpose, consider that for a natural period of 0.78 s the distribution exponent k is equal to (see Table 17.14)

$$k = 1.0 + \frac{1}{2}(T - 0.5) = 1.0 + \frac{1}{2}(0.78 - 0.5) = 1.14$$

TABLE E17.1a
Lateral Forces and Story Shears in Building Considered in Example 17.1

Level	w_x (MN)	h_{sx} (m)	h_x (m)	$w_x h_x^k$ (MN-m)	$C_{vx} = \dfrac{w_x h_x^k}{\sum_{i=1}^{N} w_i h_i^k}$	$F_x = C_{vx} V$ (MN)	V_x (MN)
6	2.2	3.3	19.8	68.16	0.302	0.275	0.275
5	2.2	3.3	16.5	53.75	0.245	0.223	0.498
4	2.2	3.3	13.2	41.68	0.190	0.173	0.671
3	2.2	3.3	9.9	30.02	0.137	0.125	0.796
2	2.2	3.3	6.6	18.91	0.086	0.078	0.874
1	2.2	3.3	3.3	8.58	0.039	0.035	0.909
Sum				219.10	1.000	0.909	

TABLE E17.1b
Lateral Displacements and Story Drifts in Building Considered in Example 17.1

Level	h_{sx} (m)	δ_{xe} (m)	$\delta_x = \dfrac{C_d \delta_{xe}}{I}$ (m)	Δ_x (m)	$\Delta_a = 0.020\, h_{sx}$ (m)	Is $\Delta_x \leq \Delta_a$?
6	3.3	0.079	0.435	0.050	0.066	Yes
5	3.3	0.070	0.385	0.077	0.066	No
4	3.3	0.056	0.308	0.088	0.066	No
3	3.3	0.040	0.220	0.094	0.066	No
2	3.3	0.023	0.127	0.072	0.066	No
1	3.3	0.010	0.055	0.055	0.066	Yes

Consider also that the calculation of the desired forces and the corresponding story shears may be organized in a tabular form as shown in Table E17.1a. By substitution of the base shear and distribution exponent determined above into Equations 17.32 and 17.33, one arrives thus to the lateral forces and story shears listed in Table E17.1a. Figure E17.1b shows the distribution along the height of the building of the obtained lateral forces.

Lateral Displacements and Story Drifts

With the lateral forces listed in Table E17.1a, the given beam and column sections, and using a computer program for the static analysis of frames, the elastic floor displacements shown in the third column of Table E17.1b are obtained. Table E17.1b also shows the corresponding inelastic displacements obtained using Equation 17.36, an importance factor I of 1.0, and a C_d value of 5.5, the displacement amplification factor listed in Table 17.1 for special moment-resisting frames. It shows, in addition, the story drifts and the maximum story drifts allowed by the code as specified in Table 17.2 for structures belonging to occupancy category II. It may be noted from the values listed in Table 17.2 that some of the calculated story drifts exceed the allowable ones and thus the building needs to be redesigned (i.e., increase beam and column sizes) to reduce the story drifts below the maximum values permitted by the code.

17.7 MODAL RESPONSE SPECTRUM ANALYSIS

17.7.1 Modal Properties

For a modal response spectrum analysis, the code requires that the mode shapes, natural periods, and participations factors of the structure be determined using established methods of structural analysis, elastic properties, and fixed-base conditions. It also requires including as many modes

Seismic Code Provisions

as necessary to obtain a combined participating mass of at least 90% of the actual building mass in each of the two orthogonal horizontal directions of the structure. For the purpose of complying with this latter requirement, the building mass that participates in each mode of vibration may be determined from Equation 17.41 below, which in turn represents the definition of modal effective weight introduced in Chapter 12 (see Equation 12.11).

17.7.2 Modal Base Shears

For the calculation of the base shear when a modal response spectrum analysis is performed, the code stipulates that the fraction of the total base shear that is contributed by the mth mode of vibration be calculated according to

$$V_m = C_{sm} W_m \tag{17.39}$$

where C_m is a modal seismic response coefficient defined as

$$C_{sm} = \frac{S_{am}}{R/I} \tag{17.40}$$

and W_m is a modal effective weight given by

$$W_m = \frac{\left(\sum_{i=1}^{N} w_i \phi_{im}\right)^2}{\sum_{i=1}^{N} w_i \phi_{im}^2} \tag{17.41}$$

where w_i, ϕ_{im}, S_{am}, R, and I, respectively, represent fraction of building's total gravity load W at level i, displacement amplitude at the ith building level when the building vibrates in its mth mode, design spectral acceleration determined from either a design spectrum constructed as indicated in Section 17.3.6 or a site-specific response spectrum that conforms to the requirements described in Sections 17.3.3 and 17.3.6, response modification factor determined from Table 17.1, and occupancy importance factor determined from Table 17.6.

17.7.3 Modal Lateral Forces and Drifts

In a modal response spectrum analysis, the lateral forces in each mode of vibration are determined according to

$$F_{xm} = C_{vxm} V_m \tag{17.42}$$

where V_m is the modal base shear given by Equation 17.39 and C_{vxm} is a vertical distribution factor for the mth mode of vibration defined as

$$C_{vxm} = \frac{w_x \phi_{xm}}{\sum_{i=1}^{N} w_i \phi_{im}} \tag{17.43}$$

where w_x, w_i, and ϕ_{xm}, ϕ_{im} are fraction of total gravity load W at level i, x and displacement amplitude at the x, i level of the building when vibrating in its mth mode.

Similarly, the story drifts in each mode of vibration may be computed as the difference between the displacements δ_{xm} at the floors immediately above and below the story in question, considering that these modal displacements are equal to

$$\delta_{xm} = \frac{C_d \delta_{xem}}{I} \tag{17.44}$$

where δ_{xem}, the elastic displacement at level x in the mth mode of vibration, may be determined according to

$$\delta_{xem} = \left(\frac{g}{4\pi^2}\right)\left(\frac{T_m^2 F_{xm}}{w_x}\right) \qquad (17.45)$$

In the preceding expressions,

C_d = Deflection amplification factor determined from Table 17.1
F_{xm} = Modal lateral force at level x defined by Equation 17.42
g = Acceleration of gravity
w_x = Fraction of total gravity load W at level x
T_m = Natural period of the building in its mth mode of vibration

Note that Equation 17.42 is equivalent to Equation 12.15. Note also that Equation 17.45 is another way to express Equation 12.8, which is the equation derived in Section 12.2.2 to determine the maximum modal displacements in a building in terms of equivalent lateral forces.

17.7.4 Design Values

For the determination of the design values of the base shear, story shears, story moments, story drifts, and floor displacements, the code stipulates that the modal values be combined using the square root of the sum of the squares (SRSS) rule or the complete quadratic combination (CQC) rule (see Section 10.3.2 for a description of these rules). The CQC rule is to be used when the natural periods in the translational and torsional modes of the structure are closely spaced. The code, however, imposes a limit to the calculated design values using this procedure. It specifies that if the base shear calculated on the basis of a modal response spectrum analysis, V_t, is less than 85% of the base shear V obtained using the equivalent lateral force procedure described in Section 17.6 (with the fundamental period considered as $T = C_u T_a$ when the calculated fundamental period exceeds the product $C_u T_a$), then the lateral forces, story shears, and story moments (but not the displacements and story drifts) obtained from the modal response spectrum analysis need to be multiplied by the modification factor

$$C_m = 0.85\frac{V}{V_t} \qquad (17.46)$$

Recall that C_u is the coefficient given in Table 17.13 to establish an upper limit to the calculated natural period and T_a is the approximate natural period determined as described in Section 17.6.3. Finally, the code requires that the horizontal shear distribution be carried out and P-delta effects and story drifts be calculated in the same way it is required for the equivalent lateral force procedure. However, it does not require the magnification of the torsional moments by the amplification factor defined by Equation 17.34 when the accidental torsional effects are included in the mathematical model used in the analysis.

17.8 LINEAR RESPONSE TIME-HISTORY ANALYSIS

According to the code, a linear response time-history analysis consists of an analysis in which methods of numerical integration are used to determine the response of a linear mathematical model of the structure to a suite of ground motion acceleration time histories compatible with the specified

Seismic Code Provisions

design response spectrum for the site under consideration. The code provisions pertaining to the use of this procedure are

1. The mathematical model used needs to conform to the requirements stipulated in Section 17.4.6.
2. The analysis needs to be performed with not less than three ground motions.
3. When a two-dimensional analysis is performed, each ground motion will consist of a horizontal acceleration time history selected from an actual recorded event.
4. When a three-dimensional analysis is performed, each ground motion will consist of pairs of horizontal acceleration time histories selected from individual recorded events.
5. The time histories selected for the analysis should be from earthquakes that have magnitudes, source mechanisms, and fault distances that are consistent with those that control the maximum considered earthquake.
6. Simulated ground motion time histories may be used when recorded ground motions with the desired characteristics are not available.
7. When a two-dimensional analysis is performed, the selected ground motions will be scaled such that the average of the 5%-damping response spectra for the suite of motions will not be less than the design response spectrum for the site determined according to the guidelines introduced in Section 17.3.6 for each period between $0.2T$ to $1.5T$, where T is the fundamental natural period of the structure in seconds in the direction of analysis.
8. When a three-dimensional analysis is performed, each pair of ground motions will be scaled such that the average of the square root of the sum of the squares (SRSS) response spectra for the selected pairs will not be less (with a tolerance of 10%) than 1.3 times the 5%-damping design response spectrum determined according to the guidelines introduced in Section 17.3.6 for each period between $0.2T$ and $1.5T$, where T is the fundamental natural period of the structure in seconds. For each pair of orthogonal ground motions, the SRRS response spectrum is constructed by combining on the basis of the square root of the sum of the squares the 5%-damping response spectra of the scaled individual components.
9. If the analysis is performed with seven or more ground motions, then the design member forces and the design story drifts may be considered to be the average of the values obtained under each ground motion. If the analysis is performed with only three ground motions, then the design member forces and design story drifts will be considered to be the largest of the obtained values.
10. The response parameters obtained under each individual ground motion will be divided by R/I, where I and R denote the importance factor and response modification coefficient of the structure under consideration, respectively.
11. When the maximum scaled base shear, V_i, is less than the minimum specified in Section 17.6.2 for structures analyzed with the equivalent lateral force procedure, the member forces will be additionally multiplied by the factor

$$C_m = \frac{V}{V_i} \qquad (17.47)$$

where V is the minimum base shear determined as indicated in Section 17.6.2.

12. In the consideration of the special load combinations introduced in Section 17.2.2, the value of $\Omega_0 Q_E$ need not be larger than the maximum of the unscaled values of Q_E (i.e., before dividing them by R/I) obtained under each of the considered ground motions.

17.9 NONLINEAR RESPONSE TIME-HISTORY ANALYSIS

In a similar way, the code defines a nonlinear response time-history analysis as an analysis in which numerical integration methods are used to determine the response to a suite of ground motion acceleration time histories compatible with the design response spectrum for the site under consideration of a mathematical model of the structure that directly accounts for the nonlinear hysteretic behavior of the structure's components. The requirements related to the use of this procedure may be summarized as follows:

1. The models used to represent the hysteretic behavior of the structural elements must be consistent with suitable test data and should account for all the significant yielding, stiffness and strength degradation, and hysteretic pinching indicated by such data.
2. The strength of the elements is to be evaluated based on expected values and considering material overstrength, strain hardening, and hysteretic strength degradation.
3. The ground motions used in the analysis must conform to the requirements established for a linear response time-history analysis.
4. The structure need be analyzed under the combined effects of the selected ground motions, dead load, and not less than 25% of the required live loads.
5. If the analysis is performed with seven or more ground motions, then the design member forces, member inelastic deformations, and story drifts are to be taken as the average of the values obtained under each ground motion. If the analysis is performed with less than seven ground motions, then the design member forces, member inelastic deformations, and story drifts are to be taken as the largest of the values obtained under each of the considered ground motions.
6. It is not necessary to evaluate the adequacy of the members' strengths against the seismic load combinations effects described in Section 17.2.2.
7. The adequacy of individual members and their connections to resist the design deformation values obtained from the analysis need be evaluated based on laboratory test data for similar elements.
8. The deformation in each member must not exceed 2/3 of the value that results in the inability of the member to carry gravity loads, or the value that results in the deterioration of the member's strength to less than 67% of the peak value.
9. The design story drifts obtained from the analysis should not exceed 125% of the drift limits specified in Table 17.2.

EXAMPLE 17.2 ANALYSIS OF FOUR-STORY STRUCTURE BY MODAL ANALYSIS PROCEDURE

A four-story reinforced concrete frame building has the dimensions shown in Figure E17.2. The exterior and interior columns of the lower two stories have cross-sections of 12 in. $\times$ 20 in. and 12 in. $\times$ 24 in., respectively, whereas the corresponding columns of the top two stories have cross-sections of 12 in. $\times$16 in. and 12 in. $\times$ 20 in. All the beams have a cross-section of 12 in. $\times$24 in. The stories are all 12 ft high. For all members, the modulus of elasticity is considered equal to 3000 kips/in.^{2}. The floor dead load per unit area, which includes the weight of the floor slab, half the weight of the columns above and below the floor, and partition walls, is estimated to be 140 lb/ft^2. The live load is assumed to be 125 lb/ft^2. The foundation overlies a deposit of stiff soil. The building will be built on a site located in Palm Springs, California, and will be used as a warehouse. Perform the seismic analysis of the building along its transverse direction using the modal response spectrum analysis procedure prescribed by the International Building Code.

Solution

Floor Weights

For a warehouse, the code requires including 25% of the live load in the calculation of the gravity load, but no live load need be considered at the roof. Therefore, the floor weights of the building are

$$w_1 = w_2 = w_3 = (0.140 + 0.25 \times 0.125)(48 \times 96) = 789.1 \text{ kips}$$

and the roof weight is

$$w_4 = 0.140(48 \times 96) = 645.1 \text{ kips}$$

Correspondingly, the total weight of the buildings is

$$W = 789.1 \times 3 + 645.1 = 3012.4 \text{ kips}$$

Mass and Stiffness Matrices

Concentrating the mass of the building at its four floors, the mass matrix is of the form

$$[M] = \begin{bmatrix} 2.042 & 0 & 0 & 0 \\ 0 & 2.042 & 0 & 0 \\ 0 & 0 & 2.042 & 0 \\ 0 & 0 & 0 & 1.670 \end{bmatrix} \text{kip s}^2/\text{in.}$$

Similarly, by modeling the building as a two-bay, four-story two-dimensional frame in which the flexural rigidity of each member is equal to the sum of the flexural rigidities of the corresponding members in the nine parallel frames of the building, the condensed stiffness matrix of the building referenced to the building's four lateral degrees of freedom results as

$$[K] = \begin{bmatrix} 5611 & -2916 & 458 & -45 \\ -2916 & 3863 & -1885 & 269 \\ 458 & -1885 & 2893 & -1398 \\ -45 & 269 & -1398 & 1168 \end{bmatrix} \text{kip/in.}$$

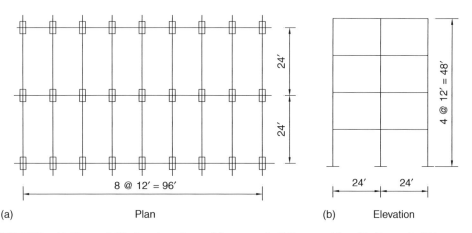

FIGURE E17.2 (a) Plan and (b) elevation views of four-story building considered in Example 17.2.

It should be noted that for the sake of simplicity, this stiffness matrix has been derived considering the moments of inertia of the gross cross-sections of the frame's beams and columns instead of, as required by the code, the cracked sections.

Natural Periods and Mode Shapes

From the solution of the eigenvalue problem

$$([K] - \omega^2[M])\{\phi\} = \{0\}$$

it is found that the eigenvalues, natural periods, and modal matrix of the system are

$$\omega_1^2 = 77.584 \text{ rad}^2/\text{s}^2 \quad \omega_2^2 = 679.26 \text{ rad}^2/\text{s}^2$$

$$\omega_3^2 = 1941.6 \text{ rad}^2/\text{s}^2 \quad \omega_4^2 = 4057.3 \text{ rad}^2/\text{s}^2$$

$$T_1 = 0.713 \text{ s} \quad T_2 = 0.241 \text{ s}$$

$$T_3 = 0.143 \text{ s} \quad T_4 = 0.099 \text{ s}$$

$$[\Phi] = \begin{bmatrix} 0.11295 & -0.31081 & -0.35336 & 0.50546 \\ 0.27101 & -0.46810 & -0.11508 & -0.42885 \\ 0.43145 & -0.06912 & 0.50340 & 0.21301 \\ 0.51554 & 0.45488 & -0.34651 & -0.07773 \end{bmatrix}$$

Effective Modal Weights

According to Equation 17.41, the modal effective weights of the building are given by

$$W_m = \frac{\left(\sum_{i=1}^{4} w_i \phi_{im}\right)^2}{\sum_{i=1}^{4} w_i \phi_{im}^2}$$

Therefore, after substitution of the determined mode shape amplitudes and floor weights, the modal effective weights are equal to the values listed in Table E17.2a. It may be noted from the percentages shown in Table E17.2a that the effective weights in the first two modes alone add up to more than 90% of the total weight of the building. Consequently, in accordance with the code recommendations (see Section 17.7.1), only the first two modes are considered in the subsequent analysis.

Design Response Spectrum

For design purposes, the code stipulates the use of a design spectrum of the form shown in Figure 17.6, with its four different regions defined by Equations 17.13 through 17.16. The use of these equations,

TABLE E17.2a
Modal Effective Weights of Building Considered in Example 17.2

Mode	Effective Weight, W_m (kips)	Percentage of Total Weight
1	2465.5	81.8
2	365.4	12.1
3	99.4	3.3
4	82.3	2.7
Total	3012.6	99.9

Seismic Code Provisions

however, requires knowing the maximum considered spectral accelerations S_1 and S_D and the site coefficients F_a and F_v. Accordingly, these spectral accelerations and site coefficients are determined first.

From the seismic hazard maps provided by the code (see Figure 17.1), the spectral accelerations, expressed as a fraction of the acceleration of gravity, for short periods and at a period of 1 s for Palm Springs, California, are

$$S_S = 2.00$$

$$S_1 = 0.75$$

Similarly, from Tables 17.4 and 17.5, the site coefficients for a site with stiff soils (class D) are

$$F_a = 1.0$$

$$F_v = 1.5$$

After adjusting for site effects, the maximum considered spectral accelerations for the system under consideration are thus equal to (see Equations 17.9 and 17.10)

$$S_{MS} = F_a S_s = 1.0(2.00) = 2.00$$

$$S_{M1} = F_v S_1 = 1.5(0.75) = 1.12$$

and the corresponding design spectral accelerations are (see Equations 17.11 and 17.12)

$$S_{DS} = \frac{2}{3} S_{MS} = \frac{2}{3}(2.00) = 1.33$$

$$S_{D1} = \frac{2}{3} S_{M1} = \frac{2}{3}(1.12) = 0.75$$

Furthermore, by virtue of Equations 17.17 and 17.18 and the map in Figure 17.7a, the periods that delimit the four regions of the design spectrum are

$$T_0 = 0.2 \frac{S_{D1}}{S_{DS}} = 0.2 \frac{0.75}{1.33} = 0.11 \text{ s}$$

$$T_S = \frac{S_{D1}}{S_{DS}} = \frac{0.75}{1.33} = 0.56 \text{ s}$$

$$T_L = 8.00 \text{ s}$$

Using Equations 17.13 through 17.16, the spectra accelerations in the design spectrum for the system under analysis may be defined thus as follows:

a. For periods less than or equal to 0.11 s,

$$S_a = S_{DS}\left(0.4 + 0.6\frac{T}{T_0}\right) = 1.33\left(0.4 + 0.6\frac{T}{0.11}\right) = 0.53 + 7.25\,T$$

b. For periods greater than or equal to 0.11 s, but less than or equal to 0.56 s,

$$S_a = S_{DS} = 1.33$$

c. For periods greater than or equal to 0.56 s, but less than or equal to 8.0 s,

$$S_a = \frac{S_{D1}}{T} = \frac{0.75}{T}$$

d. For periods greater than or equal to 8.0 s,

$$S_a = \frac{S_{D1}T_L}{T^2} = \frac{0.75(8.00)}{T^2} = \frac{6.00}{T^2}$$

Spectral Accelerations for First Two Modes

In accordance with the foregoing formulas, the spectral accelerations corresponding to the natural periods of the system in its first two modes of vibration are

$$S_{a1} = \frac{0.75}{T} = \frac{0.75}{0.713} = 1.05$$

$$S_{a2} = 1.33$$

Modal Base Shears

As described in Section 17.7.2, the code requires that the modal base shears be calculated using Equations 17.39 and 17.40, which are expressed in terms of the modal effective weights of the building, its importance factor, and its response modification coefficient. For the purpose of computing the modal base shears, consider thus that according to the values given in Table 17.6 for an ordinary warehouse the importance factor is equal to

$$I = 1.0$$

Similarly, consider that if it is assumed that the reinforced concrete frames of the building will be detailed as special moment-resisting frames, then from Table 17.1 the response modification coefficient for the structure is equal to

$$R = 8.0$$

From Equation 17.40 and the spectral accelerations determined earlier, the modal seismic response coefficients for the first two modes of vibration are therefore equal to

$$C_{s1} = \frac{S_{a1}}{R/I} = \frac{1.05}{8.0/1.0} = 0.13$$

$$C_{s2} = \frac{S_{a2}}{R/I} = \frac{1.33}{8.0/1.0} = 0.17$$

and, correspondingly, the modal base shears for the first two modes of the system are

$$V_1 = C_{s1}W_1 = 0.13(2465.5) = 320.5 \text{ kips}$$

$$V_2 = C_{s2}W_2 = 0.17(365.4) = 62.1 \text{ kips}$$

Total Base Shear

As the natural periods of the building are well separated from one another, it is valid to combine the modal base shears using the SRSS rule to determine the total base shear. Proceeding accordingly, one obtains

$$V_t = \sqrt{V_1^2 + V_2^2} = \sqrt{320.5^2 + 62.1^2} = 326.5 \text{ kips}$$

It should be recalled, however, that the code imposes a limit to the value of the base shear that may be considered when the modal analysis procedure is used. As described in Section 17.7.4, this limit is given by 85% of the value of the base shear obtained with the equivalent lateral force procedure considering a natural period equal to $T = C_u T_a$, where C_u is determined from Table 17.13 and T_a is the approximate value of the fundamental natural period of the system computed with Equation 17.28, whenever the calculated period exceeds the product of C_u and T_a. To obtain such a limit, consider then that for the system under analysis

$$T_a = C_t(h_n)^x = 0.016(48)^{0.9} = 0.52\,\text{s}$$

$$T = C_u T_a = 1.4(0.52) = 0.73\,\text{s}$$

and that, in consequence, the calculated period of 0.713 s does not exceed the upper limit imposed by the code for the fundamental natural period. From Equation 17.23, the seismic response coefficient calculated using the equivalent lateral force procedure is thus equal to

$$C_s = \frac{S_{DS}}{R/I} = \frac{1.33}{8.0/1.0} = 0.17$$

which is greater than (see Equation 17.25)

$$(C_s)_{min} = \frac{0.5 S_1}{R/I} = \frac{0.5(0.75)}{8.0/1.0} = 0.05$$

but it need not exceed (see Equation 17.26)

$$(C_s)_{max} = \frac{S_{D1}}{T(R/I)} = \frac{0.75}{0.713(8.0/1.0)} = 0.13$$

This upper limit applies and hence, from Equation 17.22, one has that

$$V = C_s W = 0.13(3012.4) = 391.6\,\text{kips}$$

which leads to a lower limit for the base shear of $0.85 \times 391.6 = 332.9$ kips. This value, in turn, exceeds the value of V_t determined earlier. The lateral forces, story shears, and story moments obtained by the modal analysis procedure need thus be scaled up by the factor (see Equation 17.46)

$$C_m = 0.85 \frac{V}{V_t} = 0.85 \frac{391.6}{326.5} = 1.02$$

Lateral Forces and Story Shears

The modal lateral forces are given by Equations 17.42 and 17.43, and their calculation may be organized in tabular form as shown in Tables E17.2b and E17.2c. The values of the vertical distribution factor C_{vxm}, the lateral forces F_{xm}, and the story shears V_{xm} for the two modes being considered are thus as indicated in these tables. It should be noted that in these calculations the modal base shears V_1 and V_2 are augmented by the factor C_m of 1.02 determined earlier. The design values of the lateral forces and story shears, obtained by combining the modal values with the SRSS rule, are shown in Table E17.2d.

Lateral Displacements and Story Drifts

The modal lateral displacements are given by Equations 17.44 and 17.45. Since from Tables 17.1 and 17.6 the deflection amplification factor C_d and the importance factor I for the system being considered are equal to 5.5 and 1.0, respectively, the modal lateral displacements may be calculated according to

$$\delta_{xm} = \frac{C_d \delta_{xem}}{I} = 5.5 \delta_{xem}$$

where

$$\delta_{xem} = \left(\frac{g}{4\pi^2}\right)\left(\frac{T_m^2 F_{xm}}{w_x}\right)$$

TABLE E17.2b
Lateral Forces and Story Shears in First Mode

Level, x	w_x (kips)	ϕ_{x1}	$w_x\phi_{x1}$ (kips)	$C_{vx1} = \dfrac{w_x\phi_{x1}}{\sum_{i=1}^{4} w_i\phi_{i1}}$	$F_{x1} = C_{vx1}V_1$ (kips)	V_{x1} (kips)
4	645.1	0.51554	332.6	0.341	111.5	111.5
3	789.1	0.43145	340.4	0.349	114.1	225.6
2	789.1	0.27101	213.9	0.219	71.6	297.2
1	789.1	0.11295	89.1	0.091	29.7	326.9
Sum	3012.4		976.0	1.000	326.9	

TABLE E17.2c
Lateral Forces and Story Shears in Second Mode

Level, x	w_x (kips)	ϕ_{x2}	$w_x\phi_{x2}$ (kips)	$C_{vx2} = \dfrac{w_x\phi_{x2}}{\sum_{i=1}^{4} w_i\phi_{i2}}$	$F_{x2} = C_{vx2}V_2$ (kips)	V_{x2} (kips)
4	645.1	0.45488	293.4	−0.781	−49.5	−49.5
3	789.1	−0.06912	−54.5	0.145	9.2	−40.3
2	789.1	−0.46810	−369.4	0.983	62.3	22.0
1	789.1	−0.31081	−245.3	0.653	41.4	63.4
Sum	3012.4		−375.8	1.000	63.4	

TABLE E17.2d
Design Values of Lateral Forces and Story Shears

	Lateral Forces (kips)			Story Shears (kips)		
Level	Mode 1	Mode 2	Design Value	Mode 1	Mode 2	Design Value
4	111.5	−49.5	122.0	111.5	−49.5	122.0
3	114.1	9.2	114.5	225.6	−40.3	229.2
2	71.6	62.3	94.9	297.2	22.0	298.0
1	29.7	41.4	51.0	326.9	63.4	333.0

After substituting into this expression the floor weights and the values of the natural periods and lateral forces determined earlier (divided by the factor $C_m = 1.02$ since the lower limit for the base shear does not apply for displacements and drifts), one obtains the values listed in Table E17.2e for the first two modes of the system. The modal story drifts, also listed in Table E17.2e, are determined from the computed modal lateral displacements considering that at each story the drift is given by the difference between the lateral displacements of the floors immediately above and below. Table E17.2f shows the design values of the lateral displacements and story drifts, which, once again, are obtained by combining the modal values with the SRSS rule.

It may be observed that all the calculated story drifts are below the limit imposed by the code. This is so because, as indicated in Table 17.2, the maximum allowed story drift for buildings in the occupancy category II is 2% of the story height h_{sx}, which for the building in this example turns out to be equal to $0.02 \times 144 = 2.88$ in.

Seismic Code Provisions

TABLE E17.2e
Modal Lateral Displacements and Story Drifts in First and Second Modes

		First Mode				Second Mode			
Level	w_x (kips)	F_{x1}/C_m (kips)	δ_{xe1} (in.)	δ_{x1} (in.)	Δ_{x1} (in.)	F_{x2}/C_m (kips)	δ_{xe2} (in.)	δ_{x2} (in.)	Δ_{x2} (in.)
4	645.1	109.3	0.84	4.62	0.71	−48.5	−0.04	−0.24	−0.28
3	789.1	111.9	0.71	3.91	1.49	9.0	0.01	0.04	−0.20
2	789.1	70.2	0.44	2.42	1.43	61.1	0.04	0.24	0.08
1	789.1	29.1	0.18	0.99	0.99	40.6	0.03	0.16	0.16

TABLE E17.2f
Design Values of Lateral Displacements and Story Drifts

	Lateral Displacements (in.)			Story Drifts (in.)		
Level	Mode 1	Mode 2	Design Value	Mode 1	Mode 2	Design Value
4	4.62	−0.24	4.63	0.71	−0.28	0.76
3	3.91	0.04	3.91	1.49	−0.20	1.50
2	2.42	0.24	2.43	1.43	0.08	1.43
1	0.99	0.16	1.00	0.99	0.16	1.00

Overturning Moments

The modal overturning moment M_{xm} at level x of the building may be calculated as the sum of the moments produced by the modal lateral forces above that level; that is,

$$M_{xm} = \sum_{i=x}^{4} F_{im}(h_i - h_x)$$

where h_i and h_x are, the heights above ground of levels i and x, respectively. The design values are then obtained by combining the modal overturning moments using the SRSS rule. Proceeding accordingly with the lateral forces shown in Tables E17.2b and E17.2c, the modal and design values shown in Table E17.2g are obtained.

Torsional Moments

As mentioned in Section 17.7.4, the code requires that torsional effects be accounted for in the same way as in the equivalent lateral force procedure. In this case, however, the building is symmetric. Therefore, torsional effects may be accounted for by simply considering the torsional moments produced at each story when the computed lateral forces are considered offset by an accidental eccentricity equal to 5% of the floor dimension perpendicular to the direction of analysis. Accordingly, if the modal lateral forces in Tables E17.2b and E17.2c are considered, and since for the building under analysis the accidental eccentricity is equal to $0.05 \times 96 = 4.8$ ft, the modal torsional moments shown in Table E17.2h are obtained. Combining, then, these modal torsional moments using the SRSS rule leads to the design values listed in the last column of Table E17.2h.

P-delta Effects

Similarly, the code specifies that P-delta effects be quantified in the same way as in the equivalent lateral force procedure. Consequently, since in the equivalent lateral force procedure the code stipulates that P-delta effects need not be considered when the stability coefficient defined by Equation 17.37 is

TABLE E17.2g
Modal and Design Values of Overturning Moments

Story	h_{sx} (ft)	Mode 1		Mode 2		Design Value M_x (kip ft)
		F_{x1} (kips)	M_{x1} (kip ft)	F_{x2} (kip)	M_{x2} (kip ft)	
4	12	111.5	1,338	−49.5	−594	1,464
3	12	114.1	4,045	9.2	−1,078	4,186
2	12	71.6	7,612	62.3	−814	7,655
1	12	29.7	11,534	41.4	−53	11,534

TABLE E17.2h
Modal and Design Values of Torsional Moments

Story	e_y (ft)	Mode 1		Mode 2		Design Value M_{tx} (kip ft)
		F_{x1} (kips)	M_{tx1} (kip ft)	F_{x2} (kips)	M_{tx2} (kip ft)	
4	4.8	111.5	535	−49.5	−238	586
3	4.8	114.1	1083	9.2	−193	1100
2	4.8	71.6	1427	62.3	106	1431
1	4.8	29.7	1569	41.4	304	1598

TABLE E17.2I
Stability Coefficients

Level	Floor Weight, w_x (kips)	Weight Above, P_x (kips)	Story Drift, Δ_x (in.)	Story Shear, V_x (kip)	Story Height, h_{sx} (in.)	$\theta = \dfrac{P_x \Delta_x}{V_x h_{sx} C_d}$
4	645.1	645	0.76	122.0	144	0.005
3	789.1	1,434	1.50	229.2	144	0.012
2	789.1	2,223	1.43	298.0	144	0.013
1	789.1	3,012	1.00	333.0	144	0.011

less than 0.10, it is necessary to determine first the value of this coefficient for each story. Substituting, then, the gravity loads, story shears, and story drifts determined above and a deflection amplification factor C_d equal to 5.5 into Equation 17.37, one obtains the stability coefficients shown in Table E17.2i. It may be seen from the results listed in this table that all the stability coefficients are well below the code limit of 0.10. In this case, therefore, there is no need to account for P-delta effects.

17.10 SOIL–STRUCTURE INTERACTION EFFECTS

17.10.1 Introduction

The code specifies a simplified procedure that may optionally be used to incorporate the effects of soil–structure interaction in the determination of the design seismic forces and displacements of a structure if the mathematical model employed in the analysis does not directly account for these effects. This procedure is based on the simplified method introduced in Section 13.11. As discussed

Seismic Code Provisions

in Section 13.11, this simplified method is based on the fact that the fundamental effect of foundation flexibility on the response of a structure is (a) the introduction of a rotational component of motion that increases the flexibility of the structure, and (b) an increase in the energy dissipation capability of the system as a result of the radiation of waves away from the structure and the hysteretic deformation of the soil. It is therefore based on the premise that the response of a system on flexible soil may be characterized by the response of an equivalent system on rigid soil, but with its properties modified to account for the changes in its natural periods and damping ratios. In accordance with this approach, the code provides provisions to modify the base shear and the lateral displacements in a structure determined under the assumption of a fixed base. In this regard, it is important to note that the use of the code requirements to account for soil–structure interaction effects will decrease the design values of the base shear, lateral forces, and overturning moments, but may increase the lateral displacements and the forces associated with P-delta effects. Separate requirements are specified by the code to use with the equivalent lateral force procedure and a modal response spectrum analysis.

17.10.2 Modifications to Equivalent Lateral Force Procedure

17.10.2.1 Base Shear

To account for soil–structure interaction effects, the code allows the reduction of the base shear to a value $\tilde{V}$ given by

$$\tilde{V} = V - \Delta V \tag{17.48}$$

where

$$\Delta V = \left[C_s - \tilde{C}_s \left(\frac{0.05}{\tilde{\beta}} \right)^{0.4} \right] \bar{W} \tag{17.49}$$

which cannot exceed $0.3V$. In other words, the code does not allow the base shear in the structure on flexible soil to be less than 70% of the base shear in the fixed-base structure.

In the preceding equations,

- C_s = Seismic response coefficient computed as indicated in Section 17.6.2 considering the fundamental natural period of the fixed-base structure
- $\tilde{C}_s$ = Seismic response coefficient computed as indicated in Section 17.6.2, but considering the fundamental natural period of the structure supported on flexible soil
- $\tilde{\beta}$ = Fraction of critical damping for the structure–foundation system determined as indicated below
- $\bar{W}$ = Effective gravity load of the structure, taken as $0.7W$ for multilevel buildings and equal to W for single-level buildings

and, as previously established, V and W denote base shear and total gravity load, respectively.

Equations 17.48 and 17.49 are derived as follows: According to Equation 17.22, the base shear in a fixed-base structure is given by

$$V = C_s W \tag{17.50}$$

where W is the total gravity load, and C_s a seismic response coefficient or spectral acceleration determined as indicated in Section 17.6.2 and is, in general, a function of the natural period and damping ratio of the structure. If the structure is supported on flexible soil, and if it is assumed that only the first mode of the structure is affected by the interaction between the soil and the structure, then the base shear of the structure on flexible soil may be expressed as

$$\tilde{V} = C_s(\tilde{T},\tilde{\beta})\overline{W} + C_s(T,\beta)(W - \overline{W}) \tag{17.51}$$

where $\tilde{T}$ and $\tilde{\beta}$, T and β, $\overline{W}$ denote fundamental natural period and damping ratio of the structure on flexible soil, fundamental natural period and damping ratio of the fixed-base structure, and part of the gravity load of the structure (effective weight) that participates when the structure vibrates in its fundamental mode.

It may be noted, however, that by virtue of Equation 17.50, Equation 17.51 may also be written as

$$\tilde{V} = V - [C_s(T,\beta) - C_s(\tilde{T},\tilde{\beta})]\overline{W} \tag{17.52}$$

or as

$$\tilde{V} = V - \Delta V \tag{17.53}$$

where

$$\Delta V = [C_s(T,\beta) - C_s(\tilde{T},\tilde{\beta})]\overline{W} \tag{17.54}$$

Furthermore, if it is assumed that the ratio between the spectral ordinates for two different damping ratios is approximately equal to the ratio between the two damping ratios in question elevated to the power of 0.4, then $C_s(\tilde{T}, \tilde{\beta})$ may be considered to be approximately equal to

$$C_s(\tilde{T},\tilde{\beta}) = C_s(\tilde{T},\beta)\left(\frac{\beta}{\tilde{\beta}}\right)^{0.4} \tag{17.55}$$

Thus, after considering that for fixed-base structures the code assumes $\beta = 0.05$, Equation 17.54 becomes

$$\Delta V = \left[C_s(T,0.05) - C_s(\tilde{T},0.05)\left(\frac{0.05}{\tilde{\beta}}\right)^{0.4}\right]\overline{W} \tag{17.56}$$

which in turn leads to Equation 17.49 since $\tilde{C}_s = C_s(\tilde{T},0.05)$.

17.10.2.2 Effective Building Period

To compute the effective building period, the code specifies the formula

$$\tilde{T} = T\sqrt{1 + \frac{\overline{k}}{K_y}\left(1 + \frac{K_y \overline{h}^2}{K_\theta}\right)} \tag{17.57}$$

where
T = Fundamental natural period of the fixed-base structure
$\overline{k}$ = Effective stiffness of the fixed-base structure, defined as

$$\overline{k} = 4\pi^2\left(\frac{\overline{W}}{gT^2}\right) \tag{17.58}$$

$\bar{h}$ = Effective structure height, taken as $0.7h_n$ for multilevel buildings and as h_n for single-level buildings, h_n being the building height

K_y = Lateral stiffness of foundation soil defined as the horizontal force at the level of the foundation necessary to produce a unit displacement at that level, applying the force and measuring the displacement in the direction of analysis (see Equation 13.193)

K_θ = Rocking stiffness of foundation soil defined as the moment necessary to produce a unit average rotation of the foundation, applying the moment and measuring the rotation in the direction of analysis (see Equation 13.195)

g = Acceleration of gravity

The code requires that K_y and K_θ be computed using soil properties with the soil strain associated with the design ground motion. To this end, it specifies that the average shear modulus at large strains, G, and the associated shear wave velocity, v_s, for the soils beneath the foundation be determined using the G/G_0 and v_s/v_{s0} ratios given in Table 17.15 (see Section 8.5 for a review of the dependence of shear modulus on shear strain), where $G_0 = \gamma v_{s0}^2/g$ is the average shear modulus of soils beneath the foundation at small strain levels ($10^{-3}\%$ or less), v_{s0} is the average shear wave velocity for soils beneath the foundation at small strain levels, and γ is the average unit weight of the soils beneath the foundation.

For structures supported on mat foundations that rest at or near the ground surface or are embedded in such a way that the side wall contact with the soil cannot be considered to remain effective during the design ground motion, the code provides an alternative formula to compute the effective period of a structure. This alternative formula is

$$\tilde{T} = T\sqrt{1 + \frac{25\alpha r_a \bar{h}}{v_s^2 T^2}\left(1 + \frac{1.12 r_a \bar{h}^2}{\alpha_\theta r_m^3}\right)} \tag{17.59}$$

where

α = Relative effective weight of the structure, defined as

$$\alpha = \frac{\bar{W}}{\gamma A_0 \bar{h}} \tag{17.60}$$

α_θ = Dynamic modifier for rocking stiffness determined from Table 17.16

r_a, r_m = Foundation characteristic lengths (radii of equivalent circular foundation, see Equations 13.66 and 13.67), defined as

$$r_a = \sqrt{\frac{A_0}{\pi}} \tag{17.61}$$

$$r_m = \sqrt[4]{\frac{4 I_0}{\pi}} \tag{17.62}$$

where A_0 and I_0 are foundation area in contact with the soil and moment of inertia of foundation area in contact with the soil about a horizontal centroidal axis normal to the direction of analysis.

Note that Equation 17.57 may be obtained directly from Equation 13.214 after the relationship between natural period and circular natural frequency is used and the actual building height h is substituted by the effective height $\bar{h}$. Note also that the foundation characteristic lengths r_a and r_m defined by Equations 17.61 and 17.62 represent, as described in Section 13.3.6, the radii of equivalent circular foundations for horizontal forces and rocking moments, respectively. Finally, note that Equation 17.59 may be obtained by substituting Equations 13.193 and 13.195 into

TABLE 17.15
Values of G/G_0 and v_s/v_{s0} As a Function of Spectral Acceleration Level

Spectral Acceleration, S_{D1}	G/G_0	v_s/v_{s0}
≤0.10	0.81	0.90
0.15	0.64	0.80
0.20	0.49	0.70
≥0.30	0.42	0.65

Source: Reprinted with permission of American Society of Civil Engineers from *Minimum Design Loads for Buildings and Other Structures,* ASCE/SEI 7-05, Reston, VA, 2006.

TABLE 17.16
Values of Dynamic Modifier for Rocking Stiffness

$r_m/v_s T$	α_θ
<0.05	1.00
0.15	0.85
0.35	0.70
0.50	0.60

Source: Reprinted with permission of American Society of Civil Engineers from *Minimum Design Loads* for Buildings and Other Structures, ASCE/SEI 7-05, Reston, VA, 2006.

Equation 17.57 and considering that (a) the lateral and rotational static stiffnesses for surficial rigid foundations are given by

$$K_y = \frac{8Gr_a}{2-\mu} \quad (17.63)$$

$$K_\theta = \frac{8Gr_m^3}{3(1-\mu)} \quad (17.64)$$

(b) $\mu = 0.4$, (c) the dynamic modifier α_y for the lateral stiffness is equal to unity, (d) $G = \rho v_s^2$ (see Equation 4.48), (e) $\bar{k}$ is given by Equation 17.58, and (f) A_0 and r_a are related as indicated by Equation 17.61.

17.10.2.3 Effective Damping Ratio

For the computation of the effective damping ratio, the code provides the formula

$$\tilde{\beta} = \beta_0 + \frac{0.05}{(\tilde{T}/T)^3} \quad (17.65)$$

where the first term on the right-hand side of Equation 17.65 represents the foundation damping ratio (radiation and hysteretic damping due to soil flexibility) and the second term the structural damping. This formula is obtained in the same way as Equation 13.224, except that ξ, the damping ratio of the fixed-base structure, is considered equal to 5%. For the determination of β_0, the code provides the graph shown in Figure 17.10. The curves in this graph are based on the results of extensive parametric studies and represent average values. However, for structures supported on point bearing piles and in all other cases where the foundation soil consists of a soft stratum of reasonably uniform properties underlain by a much stiffer, rocklike deposit, and if $4D_s/v_s\tilde{T} < 1$, where D_s is the depth of the soft stratum, β_0 in Equation 17.65 needs to be replaced by the factor

$$\beta_0' = \left(\frac{4D_s}{v_s\tilde{T}}\right)^2 \beta_0 \quad (17.66)$$

The intent of this formula is to reduce the amount of additional damping that can be considered in structures for which the existence of a hard stratum below causes the reflection of outgoing waves

Seismic Code Provisions

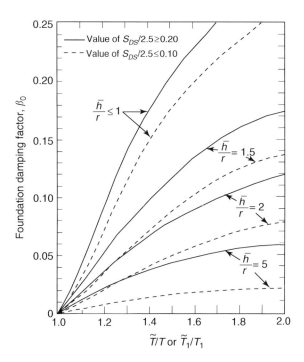

FIGURE 17.10 Foundation damping factor. (Reproduced with permission of American Society of Civil Engineers from *Minimum Design Loads for Buildings and Other Structures*, ASCE/SEI 7-05, Reston, VA, 2006.)

back into the structure and a consequent reduction in the radiation damping. Finally, it should be noted that the code stipulates that in no case the damping ratio of an interacting structure should be taken as less than 5% or greater than 20%.

In Figure 17.10, r is considered equal to either r_a or r_m and is determined according to the following criterion:

$$r = r_a \quad \text{if} \quad \frac{\bar{h}}{L_0} \leq 0.5 \tag{17.67}$$

$$r = r_m \quad \text{if} \quad \frac{\bar{h}}{L_0} \geq 1.0 \tag{17.68}$$

where L_0 denotes the overall length of the foundation in the direction of analysis. For intermediate values of $\bar{h}/L_0$, r is obtained by linear interpolation. This criterion is established in the belief that r_a should be used when the lateral motion is dominant, as in the case of short structures, and that r_m should be used when the rocking motion is dominant, as in the case of tall and slender structures.

17.10.2.4 Vertical Distribution of Base Shear

For the distribution of the base shear $\tilde{V}$ over the height of the building, the code requires that the same procedure specified for fixed-base structures be used.

17.10.2.5 Story Shears, Overturning Moments, and Torsional Moments

The code stipulates that story shears, overturning moments, and torsional moments be determined as for fixed-base structures but using the reduced lateral forces.

17.10.2.6 Lateral Displacements

The lateral displacements in structures founded on flexible soil are to be determined according to

$$\tilde{\delta}_x = \frac{\tilde{V}}{V}\left(\frac{M_0 h_x}{K_\theta} + \delta_x\right) \tag{17.69}$$

where M_0, h_x, and δ_x denote, overturning moment at the base of the building determined using the unmodified lateral forces, but without including the reduction permitted for the design of the foundation, height above the base of the building of level x, and lateral displacement at level x of the fixed-base structure using the unmodified lateral forces.

It may be noted that M_0/K_θ represents the rotation of the building when founded on flexible soil and thus the first term within the parentheses in Equation 17.69 accounts for the increase in the displacements of the building that results from the foundation rotation. Similarly, the factor $\tilde{V}/V$ accounts for the reduction in the displacements that results from the increase in the damping of the system that is caused by the radiation of waves away from the structure and the hysteretic deformation of the soil.

EXAMPLE 17.3 ANALYSIS OF 20-STORY BUILDING FOUNDED ON FLEXIBLE SOIL

The 20-story building shown in Figure E17.3 is located on a site for which the design spectral accelerations S_{DS} and S_{D1} are equal to 1.37 and 0.70, respectively. The building is supported by a 30 m × 30 m mat foundation that rests near the surface of a homogeneous soil deposit with the properties indicated in Figure E17.3. It resists a total gravity load W of 88,000 kN and an overturning moment M_0 at its base of 115,000 kN m. From a fixed-base analysis using the equivalent lateral force procedure stipulated by the code, it is found that the fundamental natural period of the building is 2 s, its base shear, V, is 8800 kN, and the lateral displacement at its roof level, δ_n, is 0.20 m. Determine the reduction in the base shear of the building and the increase or decrease in its roof displacement when the interaction between the foundation soil and the building is considered. Follow the provisions of the International Building Code.

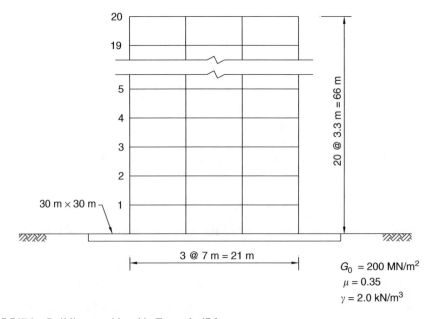

FIGURE E17.3 Building considered in Example 17.3.

Solution

Effective Period

The effective natural period of the building may be computed using Equation 17.57. This equation, however, depends on the effective stiffness and the effective height of the building. It also depends on the lateral and rotational stiffness of the foundation soil, which in turn depend on the soil's shear modulus of elasticity and the characteristic foundation lengths. Consequently, all these parameters will be determined first.

For a multilevel building, the code specifies that the effective gravity load and effective height be considered equal to $0.7W$ and $0.7h_n$, respectively, where W and h_n denote total gravity load and total building height. Therefore, for the building under consideration these parameters are equal to

$$\bar{W} = 0.7W = 0.7(88,000) = 61,600 \text{ kN}$$

$$\bar{h} = 0.7h_n = 0.7(66) = 46.2 \text{ m}$$

Similarly, the code specifies that the effective stiffness and characteristic foundation lengths be calculated using Equations 17.58, 17.61, and 17.62. As a result, for the building in this example one has that

$$\bar{k} = 4\pi^2 \left(\frac{\bar{W}}{gT^2} \right) = 4\pi^2 \left(\frac{61,600}{9.81 \times 2.0^2} \right) = 61,974 \text{ kN/m}$$

$$r_a = \sqrt{\frac{A_0}{\pi}} = \sqrt{\frac{30 \times 30}{\pi}} = 16.93 \text{ m}$$

$$r_m = \sqrt[4]{\frac{4I_0}{\pi}} = \sqrt[4]{\frac{4 \times 30^4/12}{\pi}} = 17.12 \text{ m}$$

Furthermore, the code requires that the shear modulus G employed in the computation of the soil stiffnesses be adjusted for strain level using the G/G_0 ratios given in Table 17.15. Hence, since $S_{D1} = 0.70$, and since from Table 17.15 it is found that $G/G_0 = 0.42$ for $S_{D1} \geq 0.30$, the adjusted shear modulus for the foundation soil is

$$G = 0.42G_0 = 0.42(200) = 84 \text{ GN/m}^2$$

Consequently, Equations 17.63 and 17.64 lead to the following lateral and rotational static soil stiffnesses:

$$K_y = \frac{8Gr_a}{2-\mu} = \frac{8(84)(16.93)}{2-0.35} = 6895 \text{ GN/m}$$

$$K_\theta = \frac{8Gr_m^3}{3(1-\mu)} = \frac{8(84)(17.12)^3}{3(1-0.35)} = 1,729,203 \text{ GN-m/rad}$$

The corresponding dynamic stiffnesses (see Equations 13.193 and 13.195) may be determined by multiplying these static stiffnesses by the coefficients a_x and a_ϕ given in Figure 13.25 as a function of the dimensionless frequency a_0 defined by Equation 13.64 and the material damping $\tan \delta$, which according to Equation 13.192 is equal to twice the value of the damping ratio ξ. However, for $\tan \delta = 2(0.05) = 0.10$ and the dimensional frequencies corresponding to the lateral and rotational characteristic lengths, which in this case are, respectively, equal to

$$a_0 = \frac{2\pi r_a}{T} \sqrt{\frac{\gamma}{Gg}} = \frac{2\pi(16.93)}{2.0} \sqrt{\frac{20}{(84,000)(9.81)}} = 0.26$$

$$a_0 = \frac{2\pi r_m}{T} \sqrt{\frac{\gamma}{Gg}} = \frac{2\pi(17.12)}{2.0} \sqrt{\frac{20}{(84,000)(9.81)}} = 0.26$$

the coefficients a_x and a_ϕ are both very close to unity and, thus, it will be assumed that the dynamic soil stiffnesses are equal to the static stiffnesses.

On substitution of the determined values of the effective stiffness, effective height, and soil stiffnesses into Equation 17.57, the effective natural period of the building results equal to

$$\tilde{T} = T\sqrt{1 + \frac{\bar{k}}{K_y}\left(1 + \frac{K_y \bar{h}^2}{K_\theta}\right)} = T\sqrt{1 + \frac{61.974}{6,895}\left(1 + \frac{6,895 \times 46.2^2}{1,729,203}\right)}$$

$$= 1.042T = 1.042(2.0) = 2.08 \text{ s}$$

As the building foundation is a mat foundation and rests near the surface, the effective natural period of the building may be computed alternatively using Equation 17.59. To this end, consider that for the building and foundation soil under consideration,

$$\alpha = \frac{\bar{W}}{\gamma A_o \bar{h}} = \frac{61,600}{20(900)(46.2)} = 0.074$$

$$v_s = \sqrt{\frac{G}{\rho}} = \sqrt{\frac{84,000}{20/9.81}} = 203 \text{ m/s}$$

$$\frac{r_m}{v_s T} = \frac{17.12}{203(2.0)} = 0.042$$

and that, according to Table 17.16, for this value of $r_m/v_s T$, $\alpha_\theta = 1.0$. As a result, Equation 17.59 yields

$$\tilde{T} = T\sqrt{1 + \frac{25\alpha r_a \bar{h}}{v_s^2 T^2}\left(1 + \frac{1.12 r_a \bar{h}^2}{\alpha_\theta r_m^3}\right)} = T\sqrt{1 + \frac{25(0.074)(16.93)(46.2)}{(203)^2(2.0)^2}\left(1 + \frac{1.12(16.93)(46.2)^2}{1.0(17.12)^3}\right)}$$

$$= 1.039T = 1.039(2.0) = 2.08 \text{ s}$$

which coincides with the value found earlier.

Effective Damping Ratio

The effective damping ratio may be determined from Equation 17.65, where β_0 is determined from Figure 17.10. The use of Figure 17.10, however, requires knowing the spectral acceleration S_{DS}, the ratio $\tilde{T}/T$, and the ratio $\bar{h}/r$, h where r depends on the ratio $\bar{h}/L_0$, L_0 being the foundation length in the direction of analysis. Hence, since in this case

$$\frac{S_{DS}}{2.5} = \frac{1.37}{2.5} = 0.55 \geq 0.20$$

$$\frac{\bar{h}}{L_0} = \frac{46.2}{30} = 1.54 > 1.0$$

$$r = r_m = 17.12 \text{ m}$$

$$\frac{\bar{h}}{r} = \frac{46.2}{17.12} = 2.7$$

$$\frac{\tilde{T}}{T} = \frac{2.08}{2.0} = 1.04$$

Seismic Code Provisions

from Figure 17.10, one obtains

$$\beta_0 \approx 0.01$$

Consequently, the effective damping ratio of the building results as

$$\tilde{\beta} = \beta_0 + \frac{0.05}{(\tilde{T}/T)^3} = 0.01 + \frac{0.05}{1.04^3} = 0.054$$

which is greater than the specified minimum of 0.05 and less than the specified maximum of 0.20.

Base Shear Reduction

In accordance with Section 17.6.2, the seismic response coefficient of the fixed-base structure is given by

$$C_s = \frac{S_{D1}}{T(R/I)}$$

and thus the seismic response coefficient for the structure on flexible soil may be expressed as

$$\tilde{C}_s = \frac{S_{D1}}{\tilde{T}(R/I)} = \frac{S_{D1}}{1.04T(R/I)} = 0.96 C_s$$

From Equation 17.49, the reduction in base shear for the building being considered due to soil–structure interaction effects is therefore equal to

$$\Delta V = \left[C_s - \tilde{C}_s \left(\frac{0.05}{\beta} \right)^{0.4} \right] \overline{W} = C_s \left[1 - 0.96 \left(\frac{0.05}{0.054} \right)^{0.4} \right] 0.7W = 0.048 C_s W$$

which represents a reduction of 4.8% over the base shear on the fixed-base building. The corresponding reduced base shear is

$$\tilde{V} = V - \Delta V = C_s W - 0.048 C_s W = (1 - 0.048) C_s W = 0.952 V$$

$$= 0.952(8800) = 8378 \text{ kN}$$

Roof Displacement Increase or Decrease

The lateral displacement at level x of a building when soil–structure interaction effects are accounted for is given by Equation 17.69. The roof displacement of the building on flexible soil is thus equal to

$$\tilde{\delta}_n = \frac{\tilde{V}}{V} \left(\frac{M_0 h_n}{K_\theta} + \delta_x \right) = \frac{8,378}{8,800} \left(\frac{115 \times 66}{1,729,203} + 0.20 \right) = 0.195 \text{ m}$$

which represents a reduction of 2.5% over the value obtained when the building is considered with a fixed base.

17.10.3 Modifications to Modal Analysis Procedure

17.10.3.1 Modal Base Shears

To account for soil–structure interaction effects when a modal response spectrum analysis is carried out, the code permits the reduction of the base shear in the fundamental mode of the fixed-base structure according to

$$\tilde{V}_1 = V_1 - \Delta V_1 \tag{17.70}$$

where ΔV_1 is given by Equation 17.49, except that $\overline{W}$ needs to be taken as the effective gravity load defined by Equation 17.41, and C_s and $\tilde{C}_s$ are to be computed from Equation 17.40 using, respectively, the fundamental period T_1 of the fixed-base structure, the fundamental period $\tilde{T}_1$ of the structure on flexible soil, and the corresponding spectral accelerations from the specified design spectrum. The period $\tilde{T}_1$ is to be determined from Equation 17.57, or Equation 17.59 when applicable, evaluating $\bar{k}$ using Equation 17.58 with $\overline{W} = \overline{W}_1$ and computing $\bar{h}$ using the following formula, which represents the distance from the base to the centroid of the equivalent lateral forces in the fundamental mode of the structure:

$$\bar{h} = \frac{\sum_{i=1}^{N} w_1 \phi_{i1} h_i}{\sum_{i=1}^{N} w_i \phi_{i1}} \qquad (17.71)$$

The values of $\overline{W}$, $\bar{h}$, $\tilde{T}$, and T so determined are also to be used to evaluate the factor α defined by Equation 17.60 and the factor β_0 given in Figure 17.10. No reduction is allowed for the base shears in the higher modes of vibration. In addition, the code specifies that in no case the reduced base shear $\tilde{V}_1$ be less than $0.7V_1$.

17.10.3.2 Modal Lateral Forces, Story Shears, and Overturning Moments

To account for soil–structure interaction effects in the calculation of the lateral forces, story shears, and overturning moments that correspond to the fundamental natural period of the structure, the code stipulates that these be determined as for fixed-base structures, except that the modified base shear $\tilde{V}_1$ should be used instead of V_1.

17.10.3.3 Modal Lateral Displacements

To determine the modal lateral displacements that include soil–structure interaction effects, the code specifies the formulas

$$\tilde{\delta}_{x1} = \frac{\tilde{V}_1}{V_1}\left(\frac{M_{01} h_x}{K_\theta} + \delta_{x1}\right) \qquad (17.72)$$

$$\tilde{\delta}_{xm} = \delta_{xm} \qquad m = 2, 3, \ldots, N \qquad (17.73)$$

where M_{01} and δ_{xm} are overturning moment at the base of the fixed-base structure in its fundamental mode and lateral displacement at level x in the mth mode of the fixed-base structure and all other symbols are as defined earlier.

17.10.3.4 Design Values

The code specifies that the design values of the modified lateral forces, story shear, overturning moments, and lateral displacements be obtained as for fixed-base structures by taking the square root of the sum of the squares of the respective modal values. Also as in the case of fixed-base structures, it is permitted to reduce the overturning moment at the level of the foundation–soil interface by 10%.

17.11 NONSTRUCTURAL COMPONENTS

The International Building Code also includes provisions for the seismic design of equipment and other nonstructural elements in buildings. In general, these provisions are based on concepts and techniques that are similar to those used for the design of building structures. The intention is to ensure that nonstructural elements designed according to the code provisions will be able to

Seismic Code Provisions

withstand the accelerations and deformations generated by the design earthquake without fracturing, shifting, or toppling. They are based on ultimate strength design principles and establish minimum design criteria for architectural, mechanical, electrical, and nonstructural systems; components; and elements permanently attached to structures, including supporting structures and components (henceforth generically referred to as "components"). They recognize ground motion and structural amplifications, component toughness and weight, and performance expectations.

The provisions require that nonstructural components be able to resist a minimum equivalent static force and, in the case of components connected to their supporting structures at multiple points, a minimum relative displacement demand. The required minimum static force is given by

$$F_p = \frac{0.4 a_p S_{DS} W_p}{R_p/I_p}\left(1 + 2\frac{z}{h}\right) \tag{17.74}$$

except that F_p need not be greater than

$$(F_p)_{max} = 1.6 S_{DS} I_p W_p \tag{17.75}$$

and need not be less than

$$(F_p)_{min} = 0.3 S_{DS} I_p W_p \tag{17.76}$$

In the foregoing equations,

F_p = Seismic design force considered acting at the component's center of mass and distributed according to the component's mass distribution
S_{DS} = Design spectral acceleration for short periods defined in Section 17.3.5
I_p = Component importance factor determined from Table 17.17 according to component importance (equal to either 1.00 or 1.50)
a_p = Component amplification factor ranging from 1.0 to 2.5 and determined from Tables 17.18 or 17.19 according to type and flexibility of component
R_p = Component response modification factor ranging from 1.0 to 5.0 and determined from Tables 17.18 or 17.19 according to type and deformability of component
z = Height above grade of highest point of component attachment (for items at or below structure base, z is considered equal to 0)
h = Average height of structure's roof relative to grade elevation
W_p = Component weight

The force F_p is to be applied independently in at least two orthogonal horizontal directions in combination with the service loads acting on the component. For vertically cantilevered systems,

TABLE 17.17
Component Importance Factor, I_p

Component Importance	I_p
Component required to function after an earthquake (e.g., fire protection sprinkler system)	1.5
Component containing hazardous materials	1.5
Component in or attached to occupancy category IV structure, and needed for continued operation of facility; failure could impair continued operation of facility	1.5
All other components	1.0

TABLE 17.18
Coefficients a_p and R_p for Architectural Components

Architectural Component	$a_p{}^a$	$R_p{}^b$
1. Interior nonstructural walls and partitions		
a. Plain (unreinforced) masonry walls	1.0	1.5
b. All other walls and partitions	1.0	2.5
2. Cantilever elements (unbraced or braced to a structural frame below its center of mass)		
a. Parapets and cantilever interior nonstructural walls	2.5	2.5
b. Chimneys and stacks when laterally braced or supported by a structural frame	2.5	2.5
3. Cantilever elements (braced to a structural frame above its center of mass)		
a. Parapets	1.0	2.5
b. Chimneys and stacks	1.0	2.5
c. Exterior nonstructural walls	1.0	2.5
4. Exterior nonstructural wall elements and connections		
a. Wall element	1.0	2.5
b. Body of wall panel connections	1.0	2.5
c. Fasteners of connecting system	1.25	1.0
5. Veneer		
a. Limited deformability elements and attachments	1.0	2.5
b. Low deformability and attachments	1.0	1.5
6. Penthouses (except when framed by an extension of the building frame)	2.5	3.5
7. Ceilings		
a. All	1.0	2.5
8. Cabinets		
a. Storage cabinets and laboratory equipment	1.0	2.5
9. Access floors		
a. Special access floors	1.0	2.5
b. All other	1.0	1.5
10. Appendages and ornamentations	2.5	2.5
11. Signs and billboards	2.5	2.5
12. Other rigid components		
a. High-deformability elements and attachments	1.0	3.5
b. Limited deformability elements and attachments	1.0	2.5
c. Low-deformability elements and attachments	1.0	1.5
13. Other flexible components		
a. High-deformability elements and attachments	2.5	3.5
b. Limited deformability elements and attachments	2.5	2.5
c. Low-deformability elements and attachments	2.5	1.5

[a] A lower value for a_p cannot be used unless justified by detailed dynamic analysis. The value for a_p shall not be less than 1.0. The value of $a_p = 1$ is for equipment generally regarded as rigid and rigidly attached. The value of $a_p = 2.5$ is for flexible components or flexibly attached components.

[b] Where flexible diaphragms provide lateral support for walls and partitions, the design forces for anchorage to the diaphragm shall be as specified in Section 12.11.2 of ASCE Standard ASCE-7.

Source: Reprinted with permission of American Society of Civil Engineers from *Minimum Design Loads for Buildings and Other Structures*, ASCE/SEI 7-05, Reston, VA, 2006.

TABLE 17.19
Coefficients a_p and R_p for Mechanical and Electrical Components

Mechanical or Electrical Component	a_p[a]	R_p
1. General mechanical or electrical equipment		
a. Air-side HVAC, fans, air handlers, air-conditioning units, cabinet heaters, air distribution boxes, and other mechanical components constructed of sheet metal framing	2.5	6.0
b. Wet-side HVAC, boilers, furnaces, atmospheric tanks and bins, chillers, water heaters, heat exchangers, evaporators, air separators, manufacturing or process equipment, and other mechanical components constructed of high-deformability materials	1.0	2.5
c. Engines, turbines, pumps, compressors, and pressure vessels not supported on skirts and not considered as nonbuilding structures	1.0	2.5
d. Skirt-supported pressure vessels not considered as nonbuilding structures	2.5	2.5
e. Elevator and escalator components	1.0	2.5
f. Generators, batteries, inverters, motors, transformers, and other electrical components constructed of high-deformability materials	1.0	2.5
g. Motor control centers, panel boards, switch gears, instrumentation cabinets, and other components constructed of sheet metal framing	2.5	6.0
h. Communication equipment, computers, instrumentation, and controls	1.0	2.5
g. Roof-mounted chimneys, stacks, cooling and electrical towers laterally braced below their center of mass	2.5	3.0
i. Roof-mounted chimneys, stacks, cooling and electrical towers laterally braced above their center of mass	1.0	2.5
j. ALighting fixtures	1.0	1.5
k. Other mechanical or electrical equipment	1.0	1.5
2. Vibration-isolated components and systems[b]		
a. Components and systems using neoprene elements and neoprene-isolated floors with built-in or separate elastomeric snubbing devices or resilient perimeter stops	2.5	2.5
b. Spring-isolated components and systems and vibration-isolated floors closely restrained using built-in or separate elastomeric snubbing devices or resilient perimeter stops	2.5	2.0
c. Internally isolated components and systems	2.5	2.0
d. Suspended vibration-isolated equipment including in-line duct devices and suspended internally isolated components	2.5	2.5
3. Distribution systems		
a. Piping in accordance with ASME B31, including in-line components, with joints made by welding or brazing	2.5	12.0
b. Piping in accordance with ASME B31, including in-line components constructed of high- or limited deformability materials, with joints made by threading, bonding, compression couplings, or grooved couplings	2.5	6.0
c. Piping and tubing not in accordance with ASME B31, including in-line components, constructed of high-deformability materials, with joints made by welding or brazing	2.5	9.0
d. Piping and tubing not in accordance with ASME B31, including in-line components, constructed of high- or limited deformability materials, with joints made by threading, bonding, compression couplings, or grooved couplings	2.5	4.5
e. Piping and tubing constructed of low-deformability materials such as cast iron, glass, and nonductile plastics	2.5	3.0
f. Ductwork, including in-line components, constructed of high-deformability materials, with joints made by welding or brazing	2.5	9.0
g. Ductwork, including in-line components, constructed of high-deformability materials, with joints made by means other than welding or brazing	2.5	6.0

(Continued)

TABLE 17.19 (Continued)

Mechanical or Electrical Component or Element	$a_p{}^a$	$R_p{}^b$
h. Ductwork, including in-line components, constructed of low-deformability materials, such as cast iron, glass, and nonductile plastics	2.5	3.0
i. Electric conduit, bus ducts, rigidly mounted cable trays, and plumbing	1.0	2.5
j. Manufacturing or process conveyors (nonpersonnel)	2.5	3.0
k. Suspended cable trays	2.5	6.0

[a] A lower value for a_p is permitted where justified by a detailed dynamic analysis. The value for a_p shall not be less than 1.0. The value of $a_p = 1.0$ is for rigid components and rigidly attached components. The value of $a_p = 2.5$ is for flexible components and flexibly attached components.

[b] Components mounted on vibration-isolation systems shall have a bumper restraint or snubber in each horizontal direction. The design force shall be taken as $2F_p$ if the nominal clearance (air gap) between the equipment support frame and restraint is greater than 0.25 in. If the nominal clearance specified on the construction documents is not greater than 0.25 in., the design force may be taken as F_p.

Source: Reprinted with permission of American Society of Civil Engineers from *Minimum Design Loads for Buildings and Other Structures*, ASCE/SEI 7-05, Reston, VA, 2006.

however, F_p needs to be assumed acting in any horizontal direction. Horizontal and vertical earthquake effects are to be combined according to

$$E = \rho F_p + 0.2 S_{DS} D \tag{17.77}$$

when such vertical and horizontal effects are additive, and according to

$$E = \rho F_p - 0.2 S_{DS} D \tag{17.78}$$

when the vertical effect counteracts the horizontal one. In Equations 17.77 and 17.78,

$E = $ Effect of horizontal and vertical earthquake-induced forces
$D = $ Effect of component dead load
$\rho = $ Reliability factor, allowed to be taken equal to 1.0

When nonseismic loads exceed the value of F_p, such loads will govern the design, but the detailing requirements and limitations for the seismic loads still apply.

The code also allows the use of an alternative formula to determine the lateral force F_p. This alternative formula is of the form

$$F_p = \frac{a_i a_p W_p}{R_p / I_p} A_x \tag{17.79}$$

where a_i and A_x are acceleration at floor i determined from a modal analysis with R considered equal to 1.0 and torsional amplification factor defined by Equation 17.34 and all other symbols are as defined earlier. This formula is also limited by the upper and lower limits indicated by Equations 17.75 and 17.76.

The required minimum relative displacement demand between two of the connection points of a nonstructural component with multiple connection points is determined as follows.

For two connection points on the same structure or same structural system (Structure A), one point at level x and the other at level y, the relative displacement is determined according to

$$D_p = \delta_{xA} - \delta_{yA} \tag{17.80}$$

except that it need not be greater than

$$(D_p)_{max} = (h_x - h_y)\frac{\Delta_{aA}}{h_{sx}} \qquad (17.81)$$

Alternatively, D_p may be calculated from a modal analysis considering the difference in story deflections in each mode and then combining these modal differences using an appropriate modal combination rule.

For two connection points on separate structures or structural systems, Structure A and Structure B, one at level x and the other at level y, D_p is determined according to

$$D_p = |\delta_{xA}| + |\delta_{yB}| \qquad (17.82)$$

except that it need not be greater than

$$(D_p)_{max} = \frac{h_x \Delta_{aA}}{h_{sx}} + \frac{h_y \Delta_{aB}}{h_{sx}} \qquad (17.83)$$

In the foregoing equations,

D_p = Relative seismic displacement the component needs to resist
$\delta_{xA}, \delta_{yA}, \delta_{yB}$ = Deflection at level x of Structure A, deflection at level y of Structure A, and deflection at level y of Structure B, respectively, all determined from an elastic analysis as indicated in Section 17.6.7
h_x = Height measured from the base of the structure of level x to which the upper connection point is attached
h_y = Height measured from the base of the structure of level y to which the lower connection point is attached
Δ_{aA}, Δ_{aB} = Allowable story drift for Structure A and Structure B as specified in Table 17.2
h_{sx} = Story height used in the definition of the allowable drift Δ_a

When the seismic relative displacements are calculated using the preceding equations, the code requires that the effect of these displacements be considered in combination with the displacements induced by other loads, such as those generated by thermal and static loads.

The specified formulas to determine equivalent lateral forces for nonstructural components are developed empirically on the basis of floor acceleration data recorded in buildings during strong California earthquakes. It represents a trapezoidal distribution of the floor accelerations within the structure, varying linearly from the ground to the roof, with the acceleration at the ground level intended to be the ground acceleration used as the input for the structure itself. From the examination of the recorded data, it is found reasonable to assume a maximum value for the roof acceleration equal to three to four times the ground acceleration. The response modification factor R_p is supposed to account for the overstrength and the inelastic deformation capability of nonstructural components or their anchors. The inelastic behavior of the supporting structure is not included because it is believed that (a) the extent of inelastic behavior is usually minor for structures designed by modern building codes, as their design is in many cases governed by drift limits or other loads; (b) nonstructural components are often designed without the knowledge of the structure's composition; and (c) it is a conservative consideration. Altogether, the formulas are intended to account for some of the most important factors that influence the response of nonstructural components, but without unduly burdening designers with complicated formulations. The equations for the computation of the relative displacement demands are introduced in recognition that components with multiple points of connection such as cladding, stairwells, windows, and piping systems need to be designed to resist the relative displacement between their attachment points.

For the purpose of complying with the code requirements, components are considered to have the same seismic design category as that of the structure they occupy or to which they are attached. Similarly, flexible components are those that together with their attachments have a fundamental natural period of 0.06 s or greater. Exempted from the code requirements are

1. All components belonging to seismic design category A
2. Architectural components belonging to seismic design category B other than parapets supported by bearing walls or shear walls, provided their importance factor I_p is equal to 1.0
3. Mechanical and electrical components belonging to seismic design category B
4. Mechanical and electrical components belonging to seismic design category C when their importance factor I_p is equal to 1.0
5. Mechanical and electrical components that belong to seismic design categories D–F when they have an importance factor I_p of 1.0 and are mounted at 4 ft (1.22 m) or less above floor level and weigh 400 lb (1780 N) or less, or when flexible connections are provided between them and the associated ductwork, piping, and conduit
6. Mechanical and electrical components belonging to seismic design categories D–F that have a weight of 20 lb (95 N) or less (5 lb/ft [7 N/m] or less for distributed systems), have an importance factor I_p of 1.0, and are provided with flexible connections between them and the associated ductwork, piping, and conduit

EXAMPLE 17.4 DESIGN LATERAL FORCE FOR EQUIPMENT UNIT

The reciprocating chiller shown in Figure E17.4 is mounted on the roof of a 10-story hospital building. The building is located in the city of San Francisco, California, over a deposit of stiff soils and has a height of 180 ft. The chiller weighs 15 kips and is mounted on four flexible isolators to damp the vibrations generated during the operation of the unit. Determine the shear and tension demands on the vibration isolators using the provisions of the International Building Code.

Solution

From the seismic intensity map in Figure 17.1, the maximum considered earthquake spectral acceleration for short periods corresponding to the city of San Francisco, California, and site class B is equal to

$$S_s = 1.5$$

Similarly, from Table 17.4 the value of the coefficient F_a to adjust for site class D, the site class corresponding to stiff soils, is equal to

$$F_a = 1.0$$

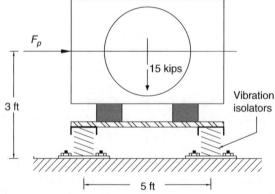

FIGURE E17.4 Equipment considered in Example 17.4.

Seismic Code Provisions

As a result, for the location under consideration, the maximum considered earthquake spectral acceleration for short periods, adjusted for class site, is (see Equation 17.9)

$$S_{MS} = F_a S_s = 1.0(1.5) = 1.5$$

and the corresponding design spectral acceleration for short periods is (see Equation 17.11)

$$S_{DS} = \frac{2}{3} S_{MS} = \frac{2}{3}(1.5) = 1.0$$

Similarly, from Table 17.19, the component amplification a_p and the component response modification factor R_p for spring-isolated equipment are equal to

$$a_p = 2.5$$
$$R_p = 2.0$$

and from Table 17.17, for equipment that is required to function after an earthquake (as is the case of hospitals),

$$I_p = 1.5$$

Accordingly, since for the building and equipment location under consideration,

$$z = h = 180 \text{ ft}$$

from Equation 17.74, one obtains

$$F_p = \frac{0.4 a_p S_{DS} W_p}{R_p / I_p}\left(1 + 2\frac{z}{h}\right)$$

$$= \frac{0.4(2.5)(1.00)}{2.0/1.5}\left(1 + 2\frac{180}{180}\right) W_p = 2.25 W_p$$

$$= 2.25(15) = 33.7 \text{ kips}$$

which is greater than

$$(F_p)_{min} = 0.3 S_{DS} I_p W_p = 0.3(1.00)(1.5) W_p = 0.45 W_p = 0.45(15) = 6.7 \text{ kips}$$

and less than

$$(F_p)_{max} = 1.6 S_{DS} I_p W_p = 1.6(1.00)(1.5) W_p = 2.40 W_p = 2.40(15) = 36.0 \text{ kips}$$

It should be noted, however, that the code requires that the force F_p be doubled for components mounted on vibration isolators if the nominal clearance between the equipment support frame and restraint is greater than 0.25 in. (see footnote in Table 17.19). Therefore, since no restrain is specified, the chiller's support system should be designed for a lateral force equal to

$$F_p = 2(33.7) = 67.4 \text{ kips}$$

and, correspondingly, each isolator should be designed for a shear force equal to

$$V = \frac{67.4}{4} = 16.9 \text{ kips}$$

Similarly, the overturning moment due to the lateral force is equal to

$$M_0 = 67.4(3) = 202.2 \text{ kip ft}$$

From Equation 17.77, earthquake effects alone produce thus an uplift force on each isolator equal to

$$(F_t)_E = \frac{1}{2}\frac{202.2}{5.0} + \frac{0.2(1.00)(15)}{4} = 21.0 \text{ kips}$$

Consequently, when combined with the gravity load in accordance with Equation 17.2, each isolator is subjected to and should be designed for an uplift force of

$$(F_t)_{D+E} = 21.0 - 0.9\frac{15}{4} = 17.6 \text{ kips}$$

17.12 BASE-ISOLATED STRUCTURES

17.12.1 Overview

The International Building Code also provides provisions for the design of seismically isolated structures. In general, these provisions require that the components of the isolation system and the structure above the isolation system be both designed to resist the deformations and stresses produced by the effects of specified earthquake ground motions. In addition, they specify the methods of analysis that can be used to determine these stresses and deformations, story drift limits, requirements to verify the stability of the isolation system, and the minimum separation between the isolated structure and surrounding obstructions. Furthermore, they require experimental tests to determine the characteristics of the isolation units and peer reviews of the isolation system design and related test programs.

17.12.2 General Requirements

In general, the code specifies that

1. A base-isolated structure be classified as regular or irregular on the basis of the structural configuration above the isolation system.
2. All portions of the structure, including those above the isolation system, be assigned an occupancy category as indicated in Section 17.4.2, with the occupancy importance factor taken as 1.0 regardless of the occupancy category classification.
3. The isolation system be configured to produce a restoring force such that the lateral force at the total design displacement is at least $0.025W$ greater than the lateral force at 50% of the total design displacement, where W is the effective seismic weight of the structure above the isolation system (calculated as indicated in Section 17.6.2) and the total maximum displacement is computed as indicated in the following section.
4. A restrain system be provided to limit the lateral displacement of the isolation system under the design wind loads to a value equal to the story drift limit specified for the structure above the isolation interface in accordance with Section 17.12.4.
5. Each element of the isolation system be stable under the design vertical load and a lateral seismic displacement equal to the total maximum displacement defined in the following section.
6. No uplifting be allowed in the individual isolation elements unless the resulting deflections do not cause overstress or instability of the isolation elements or other structural elements.

7. A dynamic analysis be performed for the determination of the design displacements and forces. This dynamic analysis may be in the form of a time-history analysis or, if certain conditions are met, a response spectrum analysis. In some instances, the equivalent lateral force procedure may also be used. The conditions for the use of a response spectrum analysis or the equivalent lateral force procedure are described in the following section.

17.12.3 EFFECTIVE NATURAL PERIODS AND MINIMUM DISPLACEMENTS

The code uses two different effective natural periods and two different minimum displacements to specify some requirements. These effective natural periods and minimum displacements correspond to the alternative consideration of the design earthquake and maximum considered earthquake defined in Section 17.3. The design of the isolation system and the structure above the isolation system, for example, is based on displacements calculated under the design earthquake. In contrast, the stability check of the isolation elements, the calculation of the axial forces on these units, and the minimum separation of the isolated structure from any obstructing element are all based on displacements calculated under the maximum considered earthquake.

The alternative effective natural periods are defined as

$$T_D = 2\pi \sqrt{\frac{W}{k_{D\min} g}} \tag{17.84}$$

$$T_M = 2\pi \sqrt{\frac{W}{k_{M\min} g}} \tag{17.85}$$

where
T_D = Effective period of the isolation system under the design earthquake in the direction under consideration (in short, the *effective period at the design displacement*)
T_M = Effective period of the isolation system under the maximum considered earthquake in the direction under consideration (in short, the *effective period at the maximum displacement*)
W = Effective seismic weight of the structure above the isolation system, calculated as indicated in Section 17.6.2
$k_{D\min}$ = Minimum effective stiffness of the isolation system at the design displacement in the horizontal direction under consideration, determined as specified in Section 17.12.7
$k_{M\min}$ = Minimum effective stiffness of the isolation system at the maximum displacement in the horizontal direction under consideration, determined as specified in Section 17.12.7

Similarly, the alternative minimum displacements, identified respectively as the *design displacement* and the *maximum displacement*, are calculated according to

$$D_D = \frac{g S_{D1} T_D}{4\pi^2 B_D} \tag{17.86}$$

$$D_M = \frac{g S_{M1} T_M}{4\pi^2 B_M} \tag{17.87}$$

where
D_D = Displacement at the center of rigidity of the isolation system under the design earthquake
D_M = Displacement at the center of rigidity of the isolation system under the maximum considered earthquake

g = Acceleration due to gravity

S_{D1} = 5%-Damping spectral acceleration at 1 s period corresponding to the design earthquake (see Equation 17.12)

S_{M1} = 5%-Damping spectral acceleration at 1 s period corresponding to the maximum considered earthquake (see Equation 17.10)

$B_D = B_M$ = Numerical coefficient determined from Table 17.20 that adjusts the spectral acceleration for a damping ratio of 5% to another damping ratio value

Note that D_D and D_M are simply the spectral displacements corresponding to the curved portion (for long periods) of the design spectrum in Figure 17.6. Similarly, note that the values of $k_{D\min}$ and $k_{M\min}$ are not known during the preliminary design phase of the isolation system. Therefore, the design procedure usually starts with assumed values obtained from the results of tests on similar components or the typical characteristics of the selected components. After the preliminary design is satisfactorily completed, prototype isolators are tested and the values of $k_{D\min}$ and $k_{M\min}$ are determined from the tests results.

To account for the additional displacements produced by actual and accidental eccentricities in the design of the elements of the isolation system, the code also requires the consideration of a *total design displacement*, D_{TD} and a *total maximum displacement*, D_{TM}. These displacements are calculated according to the spatial distribution of the lateral stiffness of the isolation system and the most unfavorable location of an eccentric center of mass. For isolation systems with a uniform spatial distribution of lateral stiffness, these displacements need to be at least equal to

$$D_{TD} = D_D \left[1 + y \frac{12e}{b^2 + d^2} \right] \quad (17.88)$$

$$D_{TM} = D_M \left[1 + y \frac{12e}{b^2 + d^2} \right] \quad (17.89)$$

where
y = Distance between the center of rigidity of the isolation system and the element of interest, measured perpendicularly to the direction of the seismic loading being considered
e = Actual eccentricity in plan between the center of mass of the structure above the isolation system and the center of rigidity of the isolation system, plus an accidental eccentricity

TABLE 17.20
Damping Coefficients B_D and B_M

Damping Percentage	B_D or B_M
≤ 2	0.8
5	1.0
10	1.2
20	1.5
30	1.7
40	1.9
≥ 50	2.0

Source: Reprinted with permission of American Society of Civil Engineers from *Minimum Design Loads for Buildings and Other Structures*, ASCE/SEI 7-05, Reston, VA, 2006.

taken as 5% of the longest dimension of the structure perpendicular to the direction of the lateral force under consideration

d = Longest plan dimension of the structure

b = Shortest plan dimension of the structure, measured perpendicularly to d

It should be noted that Equations 17.88 and 17.89 are obtained by considering that the additional displacement caused by a torsional moment $V_b e$ is equal to ϕy, where V_b is the shear force at the base of the structure, y is as previously defined, and ϕ is the angle of twist produced by such torsional moment. In turn, this angle of twist is given by the elementary formula from strength of materials for a rectangular section with sides b and d.

17.12.4 Equivalent Lateral Force Procedure

17.12.4.1 Conditions for Use

The equivalent lateral force procedure can be used for the analysis of a base-isolated structure if

1. The structure is located at a site with an S_1 of less than or equal to 0.60, where S_1 is the spectral acceleration for a 1 s period determined for the site from the code-provided ground motion intensity maps.
2. The structure is located on a site with a soil classified as class A, B, C, or D (see Section 17.3.4).
3. The structure above the isolation interface is not more than four stories or 65 ft (19.8 m) in height.
4. The effective natural period at maximum displacement of the isolated structure, T_M, does not exceed 3 s.
5. The effective natural period at the design displacement, T_D, of the isolated structure is greater than three times the natural period of the elastic, fixed-base structure above the isolation system, as determined by Equation 17.28 or Equation 17.29.
6. The structure above the isolation system is a regular one.
7. The isolation system meets all of the following criteria:
   a. The effective stiffness of the isolation system at the design displacement is greater than one-third of the effective stiffness at 20% of the design displacement.
   b. The restoring force is such that the lateral force at the total design displacement is at least $0.025W$ greater than the lateral force at 50% of the total design displacement, where W is the total effective seismic dead load of the structure above the isolation system.
   c. The isolation system does not limit the maximum considered earthquake displacement to a value that is less than the total maximum displacement, D_{TM}.

Note that, in practical terms, the code requires that the equivalent lateral force procedure be used for the analysis of all base-isolated structures. This is so because the code establishes the minimum levels of the design displacements and forces for all base-isolated structures based on the results from an analysis with that procedure. Note, also, that the equivalent lateral force procedure is also useful for preliminary designs when a dynamic analysis is required and to conduct rapid design reviews.

17.12.4.2 Minimum Displacements

The code requires that the isolation system be designed and constructed to withstand a lateral displacement at least equal to the design displacement D_D defined by Equation 17.86 and assumed to occur at the center of rigidity in the direction of each of the main horizontal axes of the structure. As noted in Section 17.12.3, the individual components of the isolation system are to be designed to resist the total design displacements that account for stiffness and mass eccentricities in the isolation system.

17.12.4.3 Minimum Lateral Forces

The isolation system, the foundation, and all structural elements below the isolation system are to be designed and constructed to withstand a minimum lateral force V_b considering all the appropriate provisions for nonisolated structures. The value of V_b is given by

$$V_b = k_{D\max} D_D \tag{17.90}$$

where $k_{D\max}$ and D_D are maximum effective stiffness of the isolation system at the design displacement in the horizontal direction under consideration, determined as specified in Section 17.12.7 and design displacement at the center of rigidity of the isolation system in the direction under consideration, calculated in accordance with Equation 17.86.

V_b is not to be taken as less than the maximum force in the isolation system at any displacement up to and including the design displacement.

For the structural elements above the isolation system, the code requires that they be designed to withstand a minimum base shear V_s considering all the appropriate provisions for nonisolated structures. This minimum base shear is to be determined from

$$V_s = \frac{k_{D\max} D_D}{R_I} \tag{17.91}$$

where $k_{D\max}$ and D_D are as previously defined and R_I is a factor that depends on the type of lateral-force-resisting system considered for the structure above the isolation system, analogous to the reduction factor R used in the design of nonisolated structures. It is equal to 3/8 of the R-value specified for nonisolated structures in Table 17.1, and need not be greater than 2 nor less than 1. In addition, the code requires that V_s be not less than

1. The base shear in the fixed-base structure considering the same seismic weight W and a natural period equal to the isolated period T_D
2. The base shear corresponding to the factored design wind load
3. The lateral seismic force required to fully activate the isolation system (e.g., the yield level of a softening system, the ultimate capacity of a sacrificial wind-restrain system, or the breakaway friction force of a sliding system) multiplied by 1.5

The base shear V_s is to be distributed over the height of the structure above the isolation interface according to

$$F_x = \frac{w_x h_x}{\sum_{i=1}^{n} w_i h_i} V_s \tag{17.92}$$

where F_x, w_x, w_i, and h_x, h_i denote lateral force at level x, portion of W located at or assigned to level x, i, and height above isolation interface of level x, i.

Note that to account for the effect of higher modes, the code conservatively prescribes the distribution of the base shear over the height of the structure using the same formula that prescribes for fixed-base structures, despite the fact that the theory of seismically isolated structures suggests a more or less uniform distribution.

17.12.4.4 Drift Limits

The maximum story drift (difference between the displacements of two adjacent floors) permitted by the code for base-isolated structures is a function of the method of analysis used. When the equivalent lateral force is employed, the maximum story drift allowed is 0.015 h_{sx}, where h_{sx} is the height of the story below level x. For the purpose of complying with this limitation, the code specifies that the floor displacements be calculated as in the case of nonisolated structures (i.e., Equation 17.36), except that the C_d factor need be considered equal to the R_I factor considered in Equation 17.91.

17.12.5 Dynamic Analysis

17.12.5.1 Choice of Procedure

The dynamic analysis of a base-isolated structure may be performed using a response spectrum analysis or a time-history analysis. A response spectrum analysis may be used if the requirements 2 and 7 listed in Section 17.12.4 for the equivalent lateral force procedure are satisfied. In addition, the use of a damping ratio of more than 30% for the isolation system is not permitted in a response spectrum analysis even if the system is provided with a higher damping ratio. A time-history analysis may be used in all cases.

17.12.5.2 Input Earthquake

If a response spectrum analysis is performed, then the analysis is carried out with the same design spectrum specified for fixed-base structures; that is, the design spectrum shown in Figure 17.6. However, the analysis is to be performed in conjunction with a site-specific design spectra if

1. The structure is located on a class F site
2. The structure is located at a site with an S_1 greater than 0.60

If a site-specific design spectrum is used, this spectrum should not be taken as less than 80% of the design spectrum shown in Figure 17.6. For the purpose of computing the total maximum displacement of the isolation system, a design spectrum is to be constructed, too, for the maximum considered earthquake. This design spectrum need be at least equal to 1.5 times the design spectrum for the design earthquake.

17.12.5.3 Response Spectrum Analysis

If a response spectrum analysis is performed, then

1. The damping ratio considered for the fundamental mode in the direction of interest is not to be greater than the effective damping of the isolation system or 30%, whichever is less. Similarly, the damping ratios considered for the higher modes need to be selected as would be appropriate for a response spectrum analysis of the structure above the isolation system assuming a fixed base.
2. The total design displacement and total maximum displacement are to be calculated considering the simultaneous application of 100% of the ground motion in the most critical direction and 30% of the ground motion in the perpendicular, horizontal direction. The maximum displacement of the isolation system is to be calculated as the vectorial sum of the two orthogonal displacements.
3. The design shear at any story is not to be less than the story shear determined with the lateral forces given by Equation 17.92 in which the value of V_s is equal to the base shear obtained from the response spectrum analysis in the direction of interest.

17.12.5.4 Time-History Analysis

If a time-history analysis is performed, then

1. Pairs of horizontal ground motion time histories need to be selected and scaled from at least three recorded events. The selected time histories are to be consistent with the magnitude,

fault distance, and source mechanisms that control the maximum considered event for the site under consideration. If appropriate recorded time histories are not available, simulated time histories may be employed.
2. For each pair of ground motion time histories, a 5%-damping response spectrum is to be constructed for the selected components, combining the individual spectra of the components with the square root of the sum of the squares rule (SRSS spectrum).
3. The time histories are to be scaled such that the average value of the SRSS spectra does not fall by more than 10% below 1.3 times the design response spectrum over the range of periods between $0.5T_D$ and $1.25T_M$, where T_D and T_M are the effective periods of the isolation system at the design displacement and maximum displacement, respectively, defined in Section 17.12.3.
4. Each pair of time histories is to be applied simultaneously to the structure considering the most unfavorable location of an eccentric center of mass. The maximum displacement of the isolation system is to be calculated from the vectorial sum of the two orthogonal components at each time step.
5. For design, the parameters of interest are to be determined as follows: If fewer than seven time-history analyses are performed, then the maximum values among those obtained under each time-history analyses are to be considered. If seven or more time histories are performed, then the average values may be considered.

17.12.5.5 Minimum Displacements

If a dynamic analysis is performed, then the isolation system and the structural elements below the isolation system are to be designed considering the total design displacement and total maximum displacement obtained from the dynamic analysis. However, the total design displacement cannot be less than 90% of D_{TD} and the total maximum displacement cannot be less than 80% of D_{TM}, where D_{TD} and D_{TM} are the total design displacement and total maximum displacement determined, respectively, as specified in Section 17.12.3. In the calculation of these limits, the code permits the replacement of the values of D_D and D_M determined from Equations 17.86 and 17.87 by the reduced values, D'_D and D'_M, obtained from

$$D'_D = \frac{D_D}{\sqrt{1 + (T/T_D)^2}} \tag{17.93}$$

$$D'_M = \frac{D_D}{\sqrt{1 + (T/T_M)^2}} \tag{17.94}$$

where T and D_D and D_M are elastic, fixed-base natural period of the structure above the isolation system and design displacement and maximum displacement defined by Equations 17.86 and 17.87, respectively.

Equations 17.93 and 17.94 are simple approximations that result from considering that (a) the base-isolated structure may be modeled as a two-degree-of-freedom system, (b) its response is given by the square root of the sum of the squares of the responses in its two modes, and (c) the natural period in the first mode is equal to the natural period of the isolation system and the natural period of the second mode is equal to the natural period of the fixed-base structure (see Section 15.5.1).

17.12.5.6 Minimum Lateral Forces

If a dynamic analysis is performed, it is possible to design the isolation system and the structural elements below and above the isolation system considering the forces obtained from the dynamic analysis. However, the code requires that the isolation system and the structural elements below the isolation system be designed for a base shear of not less than 90% of the base shear calculated with Equation 17.90. Similarly, it requires that the structure above the isolation system, if regular in configuration, be designed for a base shear of not less than 80% of the base shear calculated with Equation 17.91 and not less than the limits established for the equivalent lateral force procedure. As an exception, the code allows a base shear of less than the aforementioned limit when a time-history analysis is performed, but it cannot be less than 60% of the base shear calculated with Equation 17.91. In the case of structures with an irregular configuration, the base shear should not be less than the base shear calculated with Equation 17.91, unless a time-history analysis is conducted, in which case the limit is 80% of the base shear calculated with Equation 17.91.

17.12.5.7 Drift Limits

As in the case of the equivalent lateral force procedure, the code requires that the story drift at any level x determined from a response spectrum analysis should not be greater than $0.015h_{sx}$. However, if a time-history analysis is performed in which the nonlinear behavior of the isolating elements is considered, the code limits the maximum story drift to $0.020h_{sx}$, where, as previously established, h_{sx} is the height of the story below level x. As in the equivalent lateral force procedure, the floor displacements are calculated with Equation 17.36, but considering a C_d factor equal to the R_I factor introduced in Section 17.12.4. If the story drift ratio exceeds $0.010/R_I$, the code requires that the secondary (P-delta) effects of the displacements calculated with the maximum considered earthquake in combination with gravity loads be investigated.

17.12.6 Required Testing of Isolation System Components

The code requires that the deformation characteristics and damping values of an isolation system be based on tests of a selected sample of the components. To this end, it specifies prototype tests of two full-size specimens of each predominant type and size of the elements used in the isolation system. It also specifies the sequence of the tests that need to be performed and the horizontal displacements and vertical loads that need to be applied in each test. If the mechanical characteristics of the isolation system are dependent on the rate of loading, additional tests must be performed to characterize this dependence. Rate-dependence behavior is exhibited by most sliding isolation systems (velocity rate dependence) and some elastomeric bearings (strain rate dependence). Reduced-scale models may be used to capture rate effects provided the reduced-scale models are fabricated using the same process and same quality control program as the full-size elements. The tests may be skipped whenever prototypes of similar size and the same type and material of the isolation elements used in the design have been previously tested using the specified sequence of tests. The details of the required tests may be learned directly from the code.

17.12.7 Effective Stiffness and Effective Damping of Isolating System

As discussed in Chapter 15, most seismic isolators exhibit a nonlinear force–deformation behavior. Therefore, the code requires that if nonlinear isolating elements are used in a design, the effective stiffness and the effective damping ratio of the isolating elements be calculated according to

$$k_{\text{eff}} = \frac{|F^+| + |F^-|}{|\Delta^+| + |\Delta^-|} \tag{17.95}$$

$$\beta_{\text{eff}} = \frac{2}{\pi} \frac{E_{\text{loop}}}{k_{\text{eff}}(|\Delta^+| + |\Delta^-|)^2} \tag{17.96}$$

where Δ^+ and Δ^- are the peak positive and negative displacements attained during a full cycle of loading, respectively; F^+ and F^- are the positive and negative forces corresponding to Δ^+ and Δ^-;

E_{loop} is the energy dissipated per cycle of loading, ordinarily calculated as the area enclosed by a single force–deformation loop in a complete cycle of loading where the peak positive and negative displacements are Δ^+ and Δ^-. The peak displacements and forces in these equations are to be determined from the loading tests required by the code. It should be evident that Equations 17.95 and 17.96 correspond to the secant stiffness and the equivalent viscous damping ratio introduced in Chapter 15 (see Equations 15.197 and 15.198).

Recognizing that the force–displacement relationship of an isolation system may vary during an earthquake, the code requires the consideration of a maximum and a minimum effective stiffness for the isolation system. For instance, it requires that the maximum effective stiffness be used in the calculation of the maximum force transmitted to the isolating elements, and the minimum effective stiffness be used to calculate the effective natural period of the isolated building. At the design displacement, D_D, these maximum and minimum effective stiffnesses of the isolation system are to be calculated based on the results from the required prototype tests and according to

$$k_{D\max} = \frac{\sum |F_D^+|_{\max} + \sum |F_D^-|_{\max}}{2D_D} \tag{17.97}$$

$$k_{D\min} = \frac{\sum |F_D^+|_{\min} + \sum |F_D^-|_{\min}}{2D_D} \tag{17.98}$$

with similar expressions given to determine the maximum and minimum stiffnesses at the maximum displacement, D_M. In the preceding expressions, $\Sigma|F_D^+|_{\max}$ and $\Sigma|F_D^+|_{\min}$ denote the sum for all the isolating elements of the absolute values of the maximum and minimum forces on the elements when subjected to a positive displacement D_D, respectively and, similarly, $\Sigma|F_D^-|_{\max}$ and $\Sigma|F_D^-|_{\min}$ denote the sum for all the isolating elements of the absolute values of the maximum and minimum values of the forces on the elements when subjected to a negative displacement D_D. Similarly, the effective damping of the isolation system at the design displacement is to be calculated according to

$$\beta_D = \frac{\sum E_D}{2\pi k_{D\max} D_D^2} \tag{17.99}$$

where ΣE_D denotes the total energy dissipated per cycle of design displacement and is equal to the sum for all the elements in the isolation system of the energy dissipated per cycle by each of the isolating elements. As previously established, D_D is the design displacement. A similar expression is specified to determine the effective damping of the isolation system at the maximum displacement.

17.12.8 Peer Review

The code mandates that the design of an isolation system and the related tests be reviewed by an independent engineering team composed of licensed design professionals with experience in seismic analysis methods and theory and application of seismic isolation. The rationale for this review is an awareness that (a) the consequences of a failure in the isolation system could be catastrophic; (b) the design, fabrication, and testing of isolation elements is evolving rapidly and may be thus unfamiliar to many design professionals; and (c) the analysis and design of an isolation system involves complex procedures that may be significantly sensitive to the assumptions and idealizations made. The scope of the required peer review includes but is not limited to

1. Review of seismic design criteria such as the development of site-specific response spectra and ground motion time histories, as well as other project-specific design criteria

2. Review of the preliminary design, including the determination of the total design displacement, total maximum displacement, and lateral forces
3. Overview and observation of the prototype testing program
4. Review of the final design of the entire structural system and all supporting analyses
5. Review of the isolation system quality-control testing program

EXAMPLE 17.5.   DESIGN DISPLACEMENTS AND BASE SHEAR FOR AN ISOLATION SYSTEM AND A BASE-ISOLATED STRUCTURE

A new four-story hospital building is being implemented with a base-isolation system. The building will be located on a site for which the ground motion intensity maps specify spectral accelerations S_s and S_1 of 1.5 and 0.60, respectively. The site soil is classified as class C. The building has a regular configuration with no vertical or horizontal irregularities. Its plan dimensions are 150 × 150 ft and its weight for seismic design is 9000 kips. The lateral-load-resisting system consists of ordinary steel concentrically braced frames. The distance between the center of mass and the center of rigidity at each floor is 5 ft. The fundamental natural period of the building when its base is considered fixed is 0.50 s. It is estimated that for an effective performance the isolation system should have effective isolated periods at the design and maximum displacements of 2.0 s and 2.5 s, respectively, and an effective damping ratio of 20%. A variation of ±10% from the mean in the stiffness values of the isolation elements is considered acceptable. For the purpose of a preliminary design, determine using the provisions of the International Building Code: (a) the minimum design displacement for the isolation system; (b) the base shear for designing the isolation system and the structural elements below the isolation system; and (c) the base shear for the design of the superstructure.

Solution

From Tables 17.4 and 17.5, the value of the site coefficients F_a and F_v for site class C and $S_s \geq 1.25$ and $S_1 \geq 0.5$ are

$$F_a = 1.0$$

$$F_v = 1.3$$

Hence, for the location under consideration and according to Equations 17.11 and 17.12, the spectral accelerations for short periods and for a 1 s period corresponding to the maximum considered earthquake and adjusted for class site are

$$S_{MS} = F_a S_s = (1.0)(1.5) = 1.5$$

$$S_{M1} = F_v S_1 = (1.3)(1.5) = 1.95$$

and, from Equations 17.13 and 17.14, the associated design spectral accelerations are

$$S_{DS} = \frac{2}{3}S_{MS} = \frac{2}{3}(1.5) = 1.0$$

$$S_{D1} = \frac{2}{3}S_{M1} = \frac{2}{3}(1.95) = 1.3$$

In addition, from Table 17.20 for a damping ratio of 20%, one obtains

$$B_D = B_M = 1.5$$

Thus, considering Equations 17.86 and 17.87, the calculated values of S_{D1} and S_{M1}, and the assumed values for the design and maximum natural periods of the isolation system (T_D and T_M), the design displacement and maximum displacement at the center of rigidity of the isolation system are

$$D_D = \frac{gS_{D1}T_D}{4\pi^2 B_D} = \frac{386.4(1.3)(2.0)}{4\pi^2(1.5)} = 17.0 \text{ in.}$$

$$D_M = \frac{gS_{M1}T_M}{4\pi^2 B_M} = \frac{386.4(1.95)(2.5)}{4\pi^2(1.5)} = 31.8 \text{ in.}$$

To compute now the maximum displacements of the individual isolation elements, one needs to account for the actual floor eccentricities of 5 ft (given) and the code-mandated mass eccentricity of 5% of the building dimension perpendicular to the direction of analysis. That is, the total displacements of the isolation elements need to be calculated for a total eccentricity of

$$e = 5 + 0.05(150) = 12.5 \text{ ft}$$

Also, if it is assumed that the isolation system will have a uniform spatial distribution of lateral stiffness, one can use Equations 17.88 and 17.89 to determine the total displacements for the isolation elements at the building edges. In consequence, estimates of these total displacements are

$$D_{TD} = D_D\left[1 + y\frac{12e}{b^2 + d^2}\right] = 17.0\left[1 + \frac{150}{2}\frac{12(12.5)}{150^2 + 150^2}\right] = 17.0(1.25) = 21.2 \text{ in.}$$

$$D_{TM} = D_M\left[1 + y\frac{12e}{b^2 + d^2}\right] = 31.8(1.25) = 39.7 \text{ in.}$$

Now, from Equations 17.84 and 17.85, the assumed values of T_D and T_M, and the seismic weight W, one finds that

$$k_{D\min} = \frac{4\pi^2}{T_D^2}\frac{W}{g} = \frac{4\pi^2}{2.0^2}\frac{9000}{386.4} = 229.9 \text{ kips/in.}$$

$$k_{M\min} = \frac{4\pi^2}{T_M^2}\frac{W}{g} = \frac{4\pi^2}{2.5^2}\frac{9000}{386.4} = 147.1 \text{ kips/in.}$$

which, after consideration of the assumed ±10% variation about the mean stiffness values, lead to

$$k_{D\max} = 1.10 k_{D\text{mean}} = 1.10\frac{k_{D\min}}{0.90} = \frac{1.10}{0.90}(229.9) = 281.0 \text{ kips/in.}$$

$$k_{M\max} = \frac{1.10}{0.90}(147.1) = 179.8 \text{ kips/in.}$$

Thus, from Equation 17.90, the design shear force for the isolation system and the structural elements below is

$$V_b = k_{D\max}D_D = 281.0(17.0) = 4777 \text{ kips}$$

which corresponds to a base shear coefficient of 4777/9000 = 0.53. Similarly, since from Table 17.1 it is found that the strength reduction factor for ordinary steel concentrically braced frames is 3.25, and that, in consequence, the corresponding reduction factor for the base-isolated structure is

$$R_I = \frac{3}{8}(3.25) = 1.22 \leq 2.0$$

then the base shear for the design of the superstructure results as (see Equation 17.91)

$$V_s = \frac{k_{D\max}D_D}{R_I} = \frac{281.0(17.0)}{1.22} = 3915 \text{ kips}$$

which is tantamount to a seismic coefficient of 3915/9000 = 0.43. The design base shear for a fixed-base structure with the same weight, a natural period equal to $T_D = 2.0$ s, and an importance factor of 1.5 (for a hospital) is equal to (see Equations 17.22 and 17.26)

$$V = C_s W = \frac{S_{D1}}{T(R/I)} W = \frac{1.3}{2.0(3.25/1.5)} 9000 = 0.30(9000) = 2700 \text{ kips}$$

which indicates that the base shear of 3915 kips complies with the code requirement that the base shear for the isolated structure should not be less than the base shear for a fixed-base structure with the same weight and same natural period.

To summarize, the elements of the isolation system should be designed to resist a maximum lateral deformation of 21.2 in. and a shear force of 4777 kips. Similarly, the vertical stability of the isolating elements should be verified for the design vertical load and a horizontal displacement of 39.7 in. The superstructure should be designed to resist a base shear of 3915 kips and story drift ratios of less than 0.015. The minimum clearance between the isolated structure and surrounding obstructions should be 39.7 in.

17.13 STRUCTURES WITH ADDED DAMPING DEVICES

17.13.1 Overview

The International Building Code also contains provisions for the design of structures with added damping devices. In general, these provisions require that the seismic-force-resisting system comply with the strength and drift requirements for conventional structures, except that the damping system may be used to meet the drift requirements. The provisions also specify the methods of analysis that may be used, the way the reduction in response may be calculated in accordance to the damping devices used, and the way to determine the design forces in the damping devices. As in the case of base-isolated structures, the provisions also require experimental tests to determine the characteristics of the damping devices and peer reviews of the overall design and test programs. The provisions are applicable to all types of linear or nonlinear damping devices including displacement-dependent devices (friction and hysteretic dampers), velocity-dependent devices (viscous dampers), and displacement-and-velocity-dependent devices (viscoelastic dampers). For the application of these provisions, the code defines a *damping system* as the collection of structural elements that includes the damping devices and the structural elements or bracing required to transfer the forces from the damping devices to the base of the structure or the seismic-force-resisting system. Similarly, it defines a *damping device* as a flexible element that dissipates energy as one end of the device moves relative to the other. Pins, bolts, gusset plates, brace extensions, and all other components needed for connection to other elements of the structure are considered to be part of a damping device.

17.13.2 General Requirements

In general, the code specifies that

1. The seismic-force-resisting system of structures with added damping systems conform to one of the types listed in Table 17.1.
2. Structures with added damping systems be designed using a linear procedure, a nonlinear procedure, or a combination of a linear and a nonlinear procedure. But regardless of the procedure used, it is necessary to verify the obtained results through a nonlinear time-history

procedure whenever the structure is located at a site where the specified spectral acceleration S_1 is greater than or equal to 0.6.

3. The seismic base shear of the seismic-force-resisting system should not be less than the greater of

$$V_{min} = \frac{V}{B_{V+1}} \qquad (17.100)$$

or

$$V_{min} = 0.75V \qquad (17.101)$$

where V is the base shear in the direction of interest, determined as indicated in Section 17.6.2; and B_{V+1} is a numerical coefficient determined as indicated in Section 17.13.7 for an effective damping ratio equal to the sum of the viscous damping in the fundamental mode of the structure in the direction of interest and the inherent damping of the structure (see Section 17.13.8). However, the seismic base shear is not to be less than V if the damping system has less than two damping devices at each floor level, or if the seismic-force-resisting system has a horizontal or vertical irregularity Type 1b as defined in Tables 17.9 and 17.10.

4. The elements of the damping system be designed to remain elastic under the design loads, unless it is shown by analysis or tests that the inelastic behavior of the elements would not adversely affect the damping system's function and the forces in the elements, calculated with an overstrength factor Ω_0 of 1, do not exceed the strength required to satisfy the load combinations specified in Section 17.2.2.

17.13.3 Methods of Analysis

The methods of analysis allowed by the code include static and response spectrum methods specifically developed for structures with added damping devices. The code also allows, and mandates in some cases, the use of nonlinear response time-history methods in which the properties of the structure are based on the explicit modeling of the postyield behavior of its structural and damping elements. As with conventional structures, the code specifies the way a nonlinear response time-history procedure need be performed. The static and response spectrum methods are based on the following premises:

1. A nonlinear multi-degree-of-freedom structure with supplemental dampers may be subjected to a conventional modal decomposition and the response in the fundamental mode of the structure may be considered given by the response of an equivalent single-degree-of-freedom system with an effective (secant) stiffness and an effective damping ratio.
2. The force–deflection properties of the structure may be characterized by the force–deflection curve obtained from a pushover analysis (see Section 12.5), and the effective stiffness and effective damping of the aforementioned equivalent single-degree-of-freedom system may be determined based on this pushover curve. Furthermore, the force–deflection curve from the pushover analysis may be idealized as an elastoplastic curve with the two curves sharing a common point at the peak displacement under the design earthquake (see Figure 17.11). This idealized curve permits the consideration of the peak displacement response divided by the yield displacement as a measure of the global ductility demand and uses this global ductility demand to evaluate whether or not the global response of the building is within the limits established for conventional buildings.
3. As in the case of conventional structures, the overall behavior of a structure with added damping devices may be characterized by a response modification factor R, a deflection

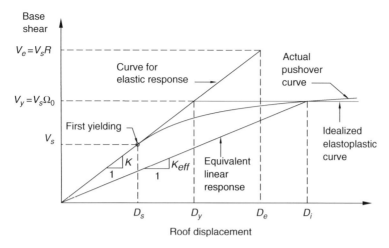

FIGURE 17.11 Actual and idealized base shear versus roof displacement curves.

amplification factor C_d, and an overstrength factor Ω_0. It may be considered, in addition, that in terms of the base shears and deflections defined in Figure 17.11 these factors and the displacement ductility ratio μ and the ductility factor R_μ are given by

$$\mu = \frac{D_i}{D_y} \quad R = \frac{V_e}{V_s} = \Omega_0 R_\mu \quad \Omega_0 = \frac{V_y}{V_s} \quad R_\mu = \frac{V_e}{V_y} \quad C_d = \frac{D_i}{D_s} = \mu\Omega_0 \qquad (17.102)$$

where V_s and D_s, V_y and D_y, and V_e and D_e represent the base shear and roof deflection at first yield, yield strength and yield displacement, and the elastic base shear and displacement under the design earthquake if the structure had unlimited strength, respectively; and D_i is the maximum inelastic displacement under the design earthquake. Furthermore, it may be assumed that $D_i = D_e$ for the constant velocity segment of a response spectrum (see Section 9.3.3) and thus

$$\mu = R_\mu = \frac{R}{\Omega_0} \qquad (17.103)$$

Similarly, it may be assumed that the energy is preserved for the constant acceleration segment of a response spectrum (see Section 9.3.3) and thus

$$\mu = \frac{1}{2}\left[\left(\frac{R}{\Omega_0}\right)^2 + 1\right] \qquad (17.104)$$

4. Response spectra for damping ratios different from 5% may be obtained by dividing the response spectra for 5% damping by a damping coefficient B that is a function of the damping ratio β and is independent of the natural period T for periods longer than $0.2T_s$; that is,

$$S_a(T,\beta) = \frac{S_a(T,0.05)}{B(\beta)} \qquad (17.105)$$

where S_a signifies spectral acceleration.

In the static and response spectrum methods the response of a structure with added dampers is determined in terms of a spectral acceleration and modal properties. In the static method, two modes are considered: the fundamental mode and a so-called residual mode. The residual mode lumps in a single mode the effect of all the higher modes and is used to account in an approximate way for this effect. The effect of the higher modes is considered in both methods because these modes may contribute significantly to story velocity and in consequence to the forces in velocity-dependent damping devices. In the response spectrum method, the properties of the structure in the higher modes are evaluated explicitly. In both the static and response spectrum methods, the nonlinearity of the structure and the damping elements is accounted for in terms of equivalent linear properties. This characterization is necessary since, in contrast to isolated structures, structures with added dampers are expected to yield under strong ground shaking.

17.13.4 EQUIVALENT LATERAL FORCE PROCEDURE

17.13.4.1 Conditions for Use

The code permits the use of the equivalent lateral force procedure herein described in the analysis of structures with added damping devices if

1. The damping system in the direction of interest has at least two damping devices and these devices are configured to resist torsional motions
2. The effective damping ratio in the fundamental mode of the structure in the direction of interest is not greater than 35%
3. The seismic-force-resisting system possesses neither horizontal irregularity Type 1a or 1b, as defined in Table 17.9, nor vertical irregularity Type 1a, 1b, 2, or, 3, as defined in Table 17.10
4. Floor diagrams are rigid as defined in Section 17.6.5
5. The height of the structure above its base does not exceed 100 ft (30 m)

17.13.4.2 Modeling

The mathematical model employed to represent the elements of the seismic-force-resisting system is to be constructed as described in Section 17.4.6. The elements of the damping system are to be modeled as required to be able to determine the forces transmitted to the ground and the seismic-force-resisting system. However, the damping devices need not be explicitly modeled if an effective damping ratio is used to modify the response of the structure and this effective damping ratio is calculated as indicated in Section 17.13.8.

17.13.4.3 Base Shear

The base shear, V, on the structural system with added damping devices in any given direction is determined according to

$$V = \sqrt{V_1^2 + V_R^2} \geq V_{min} \quad (17.106)$$

where V_1 is the design value of the base shear in the fundamental mode of the structure, V_R the design value of the base shear in the residual mode of the structure, and V_{min} is as defined in Section 17.13.2. V_1 is given by

$$V_1 = C_{S1}\overline{W}_1 \quad (17.107)$$

where $\overline{W}_1$ is the seismic weight of the structure in the fundamental mode and given by (see Equation 12.11)

$$\overline{W}_1 = \frac{\left(\sum_{i=1}^{n} w_i \phi_{i1}\right)^2}{\sum_{i=1}^{n} w_i \phi_{i1}^2} \quad (17.108)$$

where w_i and ϕ_{i1} are portion of the total seismic weight located at level i, and displacement amplitude at level i in the fundamental mode of the structure when the mode is normalized to unity at roof level, which may be determined from a dynamic analysis with the elastic properties of the structure, or according to

$$\phi_{i1} = \frac{h_i}{h_r} \tag{17.109}$$

where h_i and h_r are the heights above the base of the structure of level i and roof, respectively. C_{s1} is a seismic coefficient of the form

$$C_{s1} = \left(\frac{R}{C_d}\right)\frac{S_{DS}}{\Omega_0 B_{1D}} \quad \text{if } T_{1D} < T_S \tag{17.110}$$

or

$$C_{s1} = \left(\frac{R}{C_d}\right)\frac{S_{D1}}{T_{1D}(\Omega_0 B_{1D})} \quad \text{if } T_{1D} \geq T_S \tag{17.111}$$

where

T_S = Spectral corner period defined by Equation 17.18
S_{DS} and S_{D1} = Design spectral accelerations for short periods and 1 s period defined in Section 17.3.5
R, C_d, Ω_0 = Response modification coefficient, deflection amplification factor, and overstrength factor, respectively, given in Table 17.1
B_{1D} = Numerical coefficient determined as indicated in Section 17.13.7 for an effective damping ratio equal to β_{1D} and natural period equal to T_{1D}
T_{1D} = Effective fundamental natural period of the structure at the design earthquake
β_{1D} = Effective damping ratio in the fundamental mode of the structure, obtained as indicated in Section 17.13.8

The effective fundamental natural period T_{1D} may be calculated based on an explicit consideration of the postyield characteristics of the structure or determined according to

$$T_{1D} = T_1\sqrt{\mu_D} \tag{17.112}$$

where T_1 is the fundamental natural period of the structure calculated on the basis of elastic properties and μ_D the effective ductility demand on the seismic-force-resisting system due to the design earthquake. This effective ductility demand is determined as indicated in Section 17.13.9.

It should be realized that the seismic coefficients given by Equations 17.110 and 17.111 are derived directly from the expression that defines the seismic coefficient for $m = 1$ in the response spectrum procedure for conventional structures (see Equation 17.40). However, in this case the spectral acceleration is divided by B_{1D} to allow the use of damping ratios other than 5% (see Equation 17.105). In addition, it is divided by the overstrength factor Ω_0 (conservatively reduced by the ratio C_d/R) instead of the response modification factor R to account for the fact that the system being considered is explicitly elastoplastic, as reflected by the use of the effective period T_{1D} instead of the elastic period T_1 (see Figure 17.11). It should also be realized that Equation 17.112 is obtained from the consideration that for an elastoplastic system with an initial natural period T_1, a peak displacement demand D_i, yield base shear V_y, and yield displacement D_y (see Figure 17.11),

$$\frac{T_{\text{eff}}}{T_1} = \sqrt{\frac{K}{K_{\text{eff}}}} = \sqrt{\frac{V_y/D_y}{V_y/D_i}} = \sqrt{\frac{D_i}{D_y}} = \sqrt{\mu} \tag{17.113}$$

V_R in Equation 17.106 is given by

$$V_R = C_{sR}\overline{W}_R \qquad (17.114)$$

where $\overline{W}_R$ is the effective weight; of the structure in the residual mode and is equal to

$$\overline{W}_R = W - \overline{W}_1 \qquad (17.115)$$

in which W denotes total seismic weight; C_{sR} is the seismic coefficient in this residual mode, given by

$$C_{sR} = \left(\frac{R}{C_d}\right)\frac{S_{DS}}{\Omega_0 B_R} \qquad (17.116)$$

where S_{DS} is the design spectral acceleration for short periods defined in Section 17.3.5, and B_R is a numerical coefficient determined as indicated in Section 17.13.7 for an effective damping ratio β_R and a natural period T_R, with T_R being equal to

$$T_R = 0.4T_1 \qquad (17.117)$$

and β_R is determined as indicated in Section 17.13.8.

It may be noted that the effective weight in the residual mode is simply the difference between the total weight and the effective weight in the fundamental mode of the structure. This definition is based on the fact that, as shown in Section 12.2.3, the sum of the effective weights in all the modes of a structure is equal to its total weight, and hence the total weight in the higher modes is simply the difference between the total weight and the effective weight in the fundamental mode. Note, also, that the natural period in the residual mode is intended to be the natural period in the second mode, but arbitrarily selected equal to 0.4 times the effective fundamental mode of the structure. The selection might have been based on the fact that the natural period in the second mode of a continuous shear beam equals 0.33 times the natural period in its fundamental mode (see Equation 8.25).

17.13.4.4 Design Lateral Forces

The design lateral force acting at the ith level of the seismic-force-resisting system in the fundamental mode and the residual mode are, respectively, given by

$$F_{i1} = w_i \phi_{i1} \frac{\Gamma_1}{\overline{W}_1} V_1 \qquad (17.118)$$

$$F_{iR} = w_i \phi_{iR} \frac{\Gamma_R}{\overline{W}_R} V_R \qquad (17.119)$$

where Γ_1 and Γ_R denote the participation factors in the fundamental and residual modes, respectively, and are defined as

$$\Gamma_1 = \frac{\overline{W}_1}{\sum_{i=1}^{n} w_i \phi_{i1}} \qquad (17.120)$$

$$\Gamma_R = 1 - \Gamma_1 \qquad (17.121)$$

ϕ_{iR} denotes the displacement amplitude in the residual mode, defined as

$$\phi_{iR} = \frac{1 - \Gamma_1 \phi_{i1}}{1 - \Gamma_1} \tag{17.122}$$

and all other symbols are as defined earlier. The design forces are given by the square root of the sum of the squares of the forces in the fundamental mode and residual modes.

Once again, it may be noted that the expressions for the lateral forces given by Equations 17.118 and 17.119 are derived directly from the expression for the modal lateral forces in the response spectrum procedure for conventional structures (see Equations 17.42 and 17.43). It may be noted, also, that the definition of Γ_R is based on the fact that, as shown in Section 10.4, the sum of all the participation factors is equal to unity when the modal amplitudes at one of the levels of the structure are all set equal to 1. Furthermore, note that the derivation of Equation 17.122 is based on Equation 10.83.

17.13.4.5 Floor Deflections, Story Drifts, and Story Velocities

The floor deflections, story drifts, and story velocities are obtained for both the design earthquake and the maximum considered earthquake. The parameters under the design earthquake are used for the design of the seismic-force-resisting system. The parameters under the maximum considered earthquake are used for the design and testing of the damping devices.

The design floor deflections in the direction of interest are calculated according to

$$\delta_{iD} = \sqrt{\delta_{i1D}^2 + \delta_{iRD}^2} \tag{17.123}$$

where δ_{i1D} and δ_{iRD} represent the floor deflections in the fundamental and residual modes of the structure, respectively, and are given by

$$\delta_{i1D} = D_{1D} \phi_{i1} \tag{17.124}$$

$$\delta_{iRD} = D_{RD} \phi_{iR} \tag{17.125}$$

where D_{1D} and D_{RD} denote the fundamental- and residual-mode design displacements of the roof at the center of rigidity in the direction of interest, respectively. In turn, these displacements are given by

$$D_{1D} = \left(\frac{g}{4\pi^2}\right) \Gamma_1 \frac{S_{DS} T_{1D}^2}{B_{1D}} \geq \left(\frac{g}{4\pi^2}\right) \Gamma_1 \frac{S_{DS} T_1^2}{B_{1E}} \quad \text{if } T_{1D} < T_s \tag{17.126}$$

$$D_{1D} = \left(\frac{g}{4\pi^2}\right) \Gamma_1 \frac{S_{D1} T_{1D}}{B_{1D}} \geq \left(\frac{g}{4\pi^2}\right) \Gamma_1 \frac{S_{D1} T_1}{B_{1E}} \quad \text{if } T_{1D} \geq T_s \tag{17.127}$$

$$D_{RD} = \left(\frac{g}{4\pi^2}\right) \Gamma_R \frac{S_{D1} T_R}{B_R} \leq \left(\frac{g}{4\pi^2}\right) \Gamma_R \frac{S_{DS} T_R^2}{B_R} \tag{17.128}$$

Similarly, the design story drifts, Δ_D, in the direction of interest are determined according to

$$\Delta_D = \sqrt{\Delta_{1D}^2 + \Delta_{RD}^2} \tag{17.129}$$

where Δ_{1D} and Δ_{RD} are the design story drifts in the fundamental and residual modes of the structure, respectively, determined as the difference between the deflections at the top and lower floors of the story under consideration using the formulas introduced above. Observe that the expressions established above for the modal displacements are based on the approximate relationship between spectral displacement and spectral acceleration (see Equation 6.11).

Similarly, the design story velocities are calculated according to

$$V_D = \sqrt{V_{1D}^2 + V_{RD}^2} \tag{17.130}$$

where V_{1D} and V_{RD} represent the design story velocities in the fundamental and residual modes of the structure, respectively, and are obtained according to

$$V_{1D} = 2\pi \frac{\Delta_{1D}}{T_{1D}} \tag{17.131}$$

and

$$V_{RD} = 2\pi \frac{\Delta_{RD}}{T_R} \tag{17.132}$$

These relationships are based on the relationship between pseudovelocity and spectral displacement (see Equation 16.11) and the assumption that the spectral velocity is approximately equal to the pseudovelocity. It should be noted, however, that this assumption is inadequate and may lead to significant errors.

The total and modal floor deflections, story drifts, and story velocities corresponding to the maximum considered earthquake are given by the same equations presented earlier, except that the maximum roof displacements in the fundamental and residual modes are calculated according to

$$D_{1M} = \left(\frac{g}{4\pi^2}\right) \Gamma_1 \frac{S_{MS} T_{1M}^2}{B_{1M}} \geq \left(\frac{g}{4\pi^2}\right) \Gamma_1 \quad \text{if } T_{1M} < T_s \tag{17.133}$$

$$D_{1M} = \left(\frac{g}{4\pi^2}\right) \Gamma_1 \frac{S_{M1} T_{1M}}{B_{1M}} \geq \left(\frac{g}{4\pi^2}\right) \Gamma_1 \frac{S_{M1} T_1}{B_{1E}} \quad \text{if } T_{1M} \geq T_s \tag{17.134}$$

$$D_{RM} = \left(\frac{g}{4\pi^2}\right) \Gamma_R \frac{S_{M1} T_R}{B_R} \leq \left(\frac{g}{4\pi^2}\right) \Gamma_R \frac{S_{MS} T_R^2}{B_R} \tag{17.135}$$

where S_{MS} and S_{M1} are maximum considered spectral accelerations for short periods and 1 s period, respectively, defined in Section 17.3.4, and B_{1M} is a numerical coefficient determined as indicated in Section 17.13.7 for an effective damping ratio equal to β_{1M} and a natural period equal to T_{1M}.

The effective damping ratio β_{1M} is determined for the fundamental natural period T_{1M} as indicated in Section 17.13.8 and T_{1M} is given by

$$T_{M1} = T_1 \sqrt{\mu_M} \tag{17.136}$$

where μ_M is the effective ductility demand on the seismic-force-resisting system due to the maximum considered earthquake. As in the design earthquake case, this effective ductility demand is obtained as indicated in Section 17.13.9.

17.13.4.6 Story Drift Limits

The code requires that the design story drifts, Δ_D, calculated as indicated earlier and considering torsional effects as specified in Section 17.6.7, should not be greater than (R/C_d) times the allowable story drifts listed in Table 17.2.

17.13.4.7 Design Forces in Damping System

The design forces in the damping devices and other elements of the damping system are to be determined on the basis of the specified floor deflections, story drifts, and story velocities. In the calculation of these forces, the angle of inclination of the damping devices with respect to the horizontal and the possible response increase due to torsional effects need to be considered. The effects due to gravity loads and seismic forces are to be combined using the criteria specified in Section 17.2.2, but considering the effect of horizontal seismic forces, Q_E, as established in Section 17.13.10. The redundancy factor ρ is to be taken as 1 in all cases and the load combination with the overstrength factor Ω_0 need not be considered. The damping devices and their connections need to be sized to resist the forces, displacements, and velocities under the maximum considered earthquake.

17.13.5 Response Spectrum Procedure

17.13.5.1 Conditions for Use

The code allows the use of the response spectrum procedure described below in the analysis of structures with added damping devices whenever

1. The damping system has in the direction of interest at least two damping devices in each story
2. The damping system is configured to resist torsional motions
3. The total effective damping in the fundamental mode of the structure, β_{1D}, in the direction of interest is not greater than 35% of critical

17.13.5.2 Modeling

The mathematical model to represent the seismic-force-resisting system is to be constructed as described in Section 17.4.6. However, the stiffness and damping properties of the damping devices used are to be based on or verified by tests as specified in Section 17.13.11. In addition, the elastic stiffness of the elements of the damping system other than the damping devices needs to be modeled explicitly. Depending on the type, the stiffness of the damping devices needs to be modeled as follows:

1. If the device is displacement dependent, the device is to be modeled with an effective stiffness that represents the damping device force at the response displacement of interest (e.g., design story drift). It is permitted to exclude the stiffness of hysteretic and friction devices from the mathematical model of the structure if the design forces in the devices are applied as external loads.
2. If the device is velocity dependent and has a stiffness component (e.g., viscoelastic dampers), the device needs to be modeled with an effective stiffness corresponding to the amplitude and frequency of interest.

17.13.5.3 Base Shear

The base shear, V, on the structural system with added damping devices in any given direction is to be determined by combining the modal base shears, V_m, with the SRSS or the CQC rule, provided it satisfies the limit established in Section 17.13.2. The modal base shears are calculated in accordance with

$$V_m = C_{sm}\overline{W} \qquad (17.137)$$

where

$$\bar{W}_m = \frac{\left(\sum_{i=1}^{n} w_i \phi_{im}\right)^2}{\sum_{i=1}^{n} w_i \phi_{im}^2} \tag{17.138}$$

where ϕ_{im} is the displacement amplitude at level i in the mth mode of the structure when this mode is normalized to unity at roof level, and C_{sm} is a seismic response coefficient, which for the fundamental mode is given by Equations 17.110 and 17.111, whereas for the higher modes is given by

$$C_{sm} = \left(\frac{R}{C_d}\right) \frac{S_{DS}}{\Omega_0 B_{mD}} \quad \text{if } T_m < T_s \tag{17.139}$$

and

$$C_{sm} = \left(\frac{R}{C_d}\right) \frac{S_{D1}}{T_m (\Omega_0 B_{mD})} \quad \text{if } T_m \geq T_s \tag{17.140}$$

where T_m and B_{mD} are natural period of the structure in its mth mode of vibration in the direction under consideration and numerical coefficient determined as indicated in Section 17.13.7 for an effective damping ratio equal to β_{mD} and a natural period equal to T_m. The effective damping ratio β_{mD} is obtained as indicated in Section 17.13.8.

17.13.5.4 Design Lateral Forces

The design lateral force acting at the ith level of the seismic-force-resisting system due to the mth mode of vibration is given by

$$F_{im} = w_i \phi_{im} \frac{\Gamma_m}{\bar{W}_m} V_m \tag{17.141}$$

where Γ_m is the participation factor in the mth mode of the structure, calculated according to

$$\Gamma_m = \frac{\bar{W}_m}{\sum_{i=1}^{n} w_i \phi_{im}} \tag{17.142}$$

and all other symbols are as previously defined. The design lateral forces are obtained by combining the modal lateral forces by the SRSS or the CQC rule.

17.13.5.5 Floor Deflections, Story Drifts, and Story Velocities

The floor deflections, story drifts, and story velocities are calculated for the design earthquake and the maximum considered earthquake as follows:

The design floor deflection at level i in the mth mode of the structure and the direction of interest is calculated in accordance with

$$\delta_{imD} = D_{mD} \phi_{i1} \tag{17.143}$$

where D_{mD} is the design displacement of the roof in the mth mode of vibration at the center of rigidity and in the direction of interest; it is given by Equations 17.126 and 17.127 for the fundamental mode and by

$$D_{mD} = \left(\frac{g}{4\pi^2}\right)\Gamma_m \frac{S_{D1}T_m}{B_{mD}} \leq \left(\frac{g}{4\pi^2}\right)\Gamma_m \frac{S_{DS}T_m^2}{B_{mD}} \quad (17.144)$$

for the higher modes. The design floor displacement, δ_{iD}, is obtained by combining the modal floor displacements using either the SRSS or the CQC rule. As in the case of Equations 17.126 and 17.127, observe that Equation 17.144 is also based on the approximate relationship between spectral displacement and spectral acceleration established by Equation 6.11.

The story drift, Δ_{mD}, in the mth mode of the structure and the story and direction of interest is obtained as the difference between the deflections at the top and lower floors of the story calculated with the foregoing formulas. The design story drift, Δ_D, is determined by combining the modal story drifts with the SRSS or the CQC rule. The design story velocities are similarly determined, except that the modal story velocities are calculated according to Equation 17.131 for the fundamental mode, and according to

$$\nabla_{mD} = 2\pi \frac{\Delta_{mD}}{T_m} \quad (17.145)$$

for the higher modes. As in the case of Equation 17.131, Equation 17.145 is obtained by considering the relationship between pseudovelocity and displacement and assuming that the spectral velocity is approximately equal to the pseudovelocity.

The total and modal floor deflections, story drifts, and story velocities corresponding to the maximum considered earthquake are given by the same equations presented earlier, except that the maximum roof displacement in the fundamental mode is calculated with Equation 17.133 or Equation 17.134, whereas the corresponding displacements in the higher modes are calculated in accordance with

$$D_{mM} = \left(\frac{g}{4\pi^2}\right)\Gamma_m \frac{S_{M1}T_m}{B_{mM}} \leq \left(\frac{g}{4\pi^2}\right)\Gamma_m \frac{S_{MS}T_m^2}{B_{mM}} \quad (17.146)$$

where B_{mM} is a numerical coefficient determined as indicated in Section 17.13.7 for an effective damping ratio equal to β_{1M} and a natural period equal to T_m.

17.13.5.6 Story Drift Limits

As in the equivalent lateral force procedure, the code requires that the design story drift, Δ_D, calculated as indicated in Section 17.13.5.5 and considering torsional effects as indicated in Section 17.6.7, should not be greater than (R/C_d) times the allowable story drifts listed in Table 17.2.

17.13.5.7 Design Forces in Damping System

The design forces in the damping devices and other elements of the damping system are to be determined as established for the equivalent lateral force procedure, except that the pertinent floor deflections, story drifts, and story velocities need to be calculated as indicated in this section.

17.13.6 NONLINEAR PROCEDURES

The code permits the use of two nonlinear procedures for the analysis of any structure with added damping devices: (a) a nonlinear response time-history procedure, and (b) a static nonlinear procedure. In both procedures, it is necessary to model the nonlinear force–deformation characteristics of the damping devices as required to explicitly account for the devices' dependency on frequency, amplitude, and duration of the seismic loading.

The following are the code provisions pertaining to the response time-history procedure:

1. The mathematical model used needs to conform to the requirements established in Section 17.9. In addition, the model needs to directly account for the nonlinear hysteretic behavior of the elements of the structure and the damping devices. Furthermore, it needs to account for the elements that connect the damping devices to the structure.
2. Any element of the seismic-force-resisting system may be modeled with a linear force–deformation behavior if the calculated force in the element does not exceed 1.5 times its nominal strength.
3. The mathematical models of displacement-dependent damping devices need to account for the hysteretic behavior of the devices and the characteristics of this hysteretic behavior need to be consistent with the test data. Moreover, they need to account for all significant changes in strength, stiffness, and hysteretic loop shape. Similarly, the mathematical models of velocity-dependent damping devices need to consider velocity coefficients that are consistent with the test data. If these coefficients change with time and temperature, these changes then need to be modeled explicitly.
4. The analysis needs to be conducted in accordance to the requirements established in Section 17.9. However, the inherent damping of the structure is not to be taken greater than 5% of critical, unless the test data consistent with levels of deformation at or just below the effective yield displacement of the seismic-force-resisting system support higher values. The response parameters of interest will include, as required, the maximum values of the forces, displacements, and velocities in the damping devices.
5. If the properties of the damping devices are expected to change during the duration of the analysis, it is permitted to envelop the dynamic response of the system by the upper and lower limits of such properties.
6. As in the case of conventional structures, the design values of the forces, displacements, and velocities in the damping devices are permitted to be taken as the average of the values obtained in the response time-history analysis if at least seven ground motions are used in this analysis. If less than seven ground motions are used, then the aforementioned maximum values need to correspond to the maximum values obtained in the response time-history analysis. Also as in the case of conventional structures, at least three ground motions need to be considered.

In the static nonlinear procedure allowed by the code, a mathematical model of the seismic-force-resisting system that conforms to the requirements established in Section 17.9 is subjected to lateral forces with a vertical distribution equal to the distribution obtained using the equivalent lateral force procedure or the response spectrum procedure to produce a base shear-roof displacement curve. This curve is then used instead of Equation 17.160 to define a yield displacement and calculate the effective ductility demands μ_D and μ_M in accordance with Equations 17.158 and 17.159, respectively. Equations 17.158 through 17.160 are introduced in Section 17.13.9. The value of R/C_d is to be taken equal to 1 when the static nonlinear procedure is used in combination with the equivalent lateral force and response spectrum procedures described in Sections 17.13.4 and 17.13.5.

When a nonlinear procedure is used, the seismic-force-resisting system needs to satisfy the strength requirements using the minimum base shear specified in Section 17.13.2. The story drifts need to be determined using the design earthquake. However, the damping devices and their connecting elements need to be sized to resist the forces, displacements, and velocities corresponding to the maximum considered earthquake. The effects on the damping system due to gravity loads and seismic forces need to be combined as indicated in Section 17.2.2, considering the effect of the horizontal seismic forces, Q_E, as the effect obtained with the nonlinear procedure. The redundancy factor ρ is to be taken equal to 1 in all cases, and the seismic load effect with overstrength need not be considered.

17.13.7 Response Modification Due to Damping Increase

As indicated in Section 17.13.3, the reduction in the response of a structure introduced by an added damping system is determined by dividing the response when the structure is considered without the damping system by a damping coefficient. For structures with a natural period greater than or equal to the natural period T_0 defined by Equation 17.17, this damping coefficient is selected from Table 17.21 according to the effective damping in the structure. For structures with a natural period of less than T_0, the coefficient is calculated by linear interpolation between the value of 1 for a zero-second natural period and the value from Table 17.21 corresponding to a period T_0. The effective damping of a structure is determined as indicated in Section 17.13.8.

The values in Table 17.21 are obtained empirically from the average response spectra for a suite of 20 horizontal accelerograms from 6 earthquakes with magnitudes larger than 6.5 in the western United States. The accelerograms were recorded at epicentral distances between 10 and 20 km at sites classified as class C or D. For this study, the records were scaled to match a design spectrum with $S_{DS} = 1.0$, $S_{D1} = 0.6$, and $T_S = 0.6$ s. Records from near-field or soft-soil sites were not included in the study and thus the values presented in Table 17.21 might not be valid for such sites.

17.13.8 Effective Damping

In general, the effective damping in a structure with supplemental dampers is a combination of three components: (a) inherent damping, (b) hysteretic damping, and (c) added viscous damping.

TABLE 17.21
Damping Coefficients B_{V+1}, B_{1D}, B_R, B_{mD}, or B_{mM} for Natural Periods Greater Than $0.2T_S$

Damping Ratio (%)	B_{V+1}, B_{1D}, B_R, B_{mD}, or B_{mM}
≤ 2	0.8
5	1.0
10	1.2
20	1.5
30	1.8
40	2.1
50	2.4
60	2.7
70	3.0
80	3.3
90	3.6
≥100	4.0

Source: Reprinted with permission of American Society of Civil Engineers from *Minimum Design Loads for Buildings and Other Structures*, ASCE/SEI 7-05, Reston, VA, 2006.

The inherent damping is the existing damping in the structure before the supplemental dampers are added. Hysteretic damping corresponds to the dissipation of energy that results from the postyield hysteretic force–deformation behavior of the structure and the elements of the damping system; it is taken as zero at or below yield. Added viscous damping corresponds to the dissipation of energy by added dampers; it is taken as zero for hysteretic or friction dampers. Both hysteretic damping and added viscous damping are amplitude dependent and therefore their relative contributions to the total effective damping changes with the extent of postyield displacements. This means, for example, that adding viscous dampers to a structure may decrease the postyield displacements of the structure and in consequence may also decrease the amount of hysteretic damping provided by the seismic-force-resisting system. For the linear methods of analysis, the determination of the effective damping is based on the nonlinear force–deflection properties of the structure through the use of a secant stiffness or effective natural period. For the nonlinear methods of analysis, it is based on the explicit modeling of the postyield behavior of the structural and damping elements.

In accordance with the foregoing concepts, the code requires that the effective damping in the mth mode of a structure be determined as

$$\beta_{mD} = \beta_I + \beta_{Vm}\sqrt{\mu_D} + \beta_{HD} \tag{17.147}$$

at the peak displacement under the design earthquake, and as

$$\beta_{mM} = \beta_I + \beta_{Vm}\sqrt{\mu_M} + \beta_{HM} \tag{17.148}$$

at the peak displacement under the maximum considered earthquake. In Equations 17.147 and 17.148,

β_I = Component of the effective damping due to the inherent dissipation of energy by the elements of the seismic-force-resisting system at, or just below, the effective yield displacement

β_{Vm} = Component of the effective damping in the mth mode of vibration due to the dissipation of energy by added viscous dampers at, or just below, the effective yield displacement of the seismic-force-resisting system

β_{HD}, β_{HM} = Component of the effective damping due to the postyield hysteretic behavior of the seismic-force-resisting systems and elements of the damping system at an effective ductility demand equal to μ_D, μ_M

μ_D, μ_M = Effective ductility demand on the seismic-force-resisting system in the direction of interest under the design, maximum considered earthquake and determined as indicated in Section 17.13.9

It may be noted that β_{Vm} in the expressions for effective damping is multiplied by the square root of an effective ductility demand because the viscous damping component is determined for an elastoplastic structure with an equivalent (secant) stiffness or effective natural period. That is, since the natural period of the equivalent structure is equal to the product of the natural period of the elastic structure and the square root of the ductility demand (see Equation 17.113), from the definition of damping ratio one can express the effective damping ratio for a system with damping constant C, mass m, and natural frequency ω_{eff} as

$$(\beta_{Vm})_{\text{eff}} = \frac{C}{2\omega_{\text{eff}}m} = \frac{C\sqrt{\mu}}{2\omega m} = \beta_{Vm}\sqrt{\mu} \tag{17.149}$$

The damping components in Equations 17.147 and 17.148 are determined as indicated in the following sections.

17.13.8.1 Inherent Damping

Unless analysis or test data supports a different value, for all modes the inherent damping is not to be taken greater than 5% of critical.

17.13.8.2 Hysteretic Damping

The hysteretic damping may be determined from tests or analysis, or calculated according to

$$\beta_{HD} = q_H(0.64 - \beta_I)\left(1 - \frac{1}{\mu_D}\right) \tag{17.150}$$

$$\beta_{HM} = q_H(0.64 - \beta_I)\left(1 - \frac{1}{\mu_M}\right) \tag{17.151}$$

where q_H is a hysteresis-loop adjustment factor that accounts for pinching and other effects that reduce the area of the hysteresis loops during repeated cycles of loading. Unless analysis or test data support other values, it is to be taken equal to the empirical factor

$$q_H = 0.67 \frac{T_S}{T_1} \tag{17.152}$$

where T_S is the period defined by Equation 17.18 and T_1 is the fundamental natural period of the structure in the direction of interest. The value of q_H is not to be taken greater than 1.0 and need not be less than 0.5. Unless test data or analysis supports a different value, the hysteretic damping is to be taken as zero for the higher modes of vibration.

Equations 17.150 and 17.151 are based on the concept of effective damping ratio introduced in Section 15.5.5. They may be derived using Equation 15.198, after considering that the area of a hysteresis loop and the strain energy in an elastoplastic system subjected to a peak displacement D are, respectively, given by

$$W_D = 4F_y(D - D_y) \tag{17.153}$$

$$W_S = \frac{1}{2}F_y D \tag{17.154}$$

and thus

$$\beta_H = \frac{1}{4\pi}\frac{W_D}{W_S} = \frac{1}{4\pi}\frac{4F_y(D - D_y)}{F_y D/2} = \frac{2}{\pi}\left(1 - \frac{1}{\mu}\right) \tag{17.155}$$

Further considerations are the multiplication by the factor q_H to approximately account for the degradation of the hysteresis loops and the substitution of the factor $2/\pi$ (≈ 0.64) by $(0.64 - \beta_I)$. This substitution is introduced to avoid overestimating the inherent damping in the structure, as the commonly assumed values for the inherent damping may account for the nonlinear behavior of the structure that may not be activated when damping devices are added.

17.13.8.3 Viscous Damping

The viscous damping ratio in the mth mode of the structure due to the added damping system is to be calculated as (see Equation 15.198)

$$\beta_{Vm} = \frac{\sum_j W_{mj}}{4\pi W_m} \tag{17.156}$$

where W_{mj} is the work done by the jth damping device in the mth mode of the structure and the direction of interest in one complete response cycle of vibration with a displacement amplitude of δ_{im}. Similarly, W_m is the maximum strain energy in the mth mode of vibration of the structure in the direction of interest corresponding to a modal displacement δ_{im} and determined according to

$$W_m = \frac{1}{2}\sum_i F_{im}\delta_{im} \qquad (17.157)$$

where F_{im} and δ_{im} are inertial force at level i of the structure and deflection of level i at the center of rigidity of the structure in the mth mode of vibration and the direction of interest.

For displacement-dependent devices (e.g., viscoelastic dampers), the viscous damping is to be based on a response amplitude equal to the effective yield displacement of the structure (determined as indicated in Section 17.13.9).

In the calculation of the work done by individual damping devices, it is required to reduce such work as necessary to account for the flexibility of pins, bolts, gusset plates, brace extensions, and any other elements used to connect the damping devices to the structure.

17.13.9 Effective Ductility Demand

Some of the formulas introduced in the preceding sections are expressed in terms of an effective ductility demand. This effective ductility demand is a measure of the extent of the inelastic deformations in the structural system. The code defines this effective ductility demand as

$$\mu_D = \frac{D_{1D}}{D_Y} \geq 1 \qquad (17.158)$$

for the design earthquake, and as

$$\mu_M = \frac{D_{1M}}{D_Y} \geq 1 \qquad (17.159)$$

for the maximum considered earthquake. In Equations 17.158 and 17.159, D_{1D} and D_{1M} represent the design and maximum considered displacements at the center of rigidity of the roof level in the fundamental mode of the structure and in the direction under analysis, respectively. They are determined as indicated in Section 17.13.4 or Section 17.13.5. Similarly, D_Y is the displacement at the center of rigidity of the roof level of the structure at the effective yield point of the seismic-force-resisting system and is calculated according to

$$D_Y = \left(\frac{g}{4\pi^2}\right)\left(\frac{\Omega_0 C_d}{R}\right)\Gamma_1 C_{s1} T_1^2 \qquad (17.160)$$

where C_{s1} is the seismic response coefficient given by Equation 17.110 or Equation 17.111 and all other variables are as previously defined.

The code, however, requires that the design effective ductility demand be no greater than the maximum value given by (see Equations 17.103 and 17.104)

$$\mu_{max} = \frac{1}{2}\left[\left(\frac{R}{\Omega_0 I}\right)^2 + 1\right] \quad \text{if } T_{1D} \leq T_S \qquad (17.161)$$

$$\mu_{max} = \frac{R}{\Omega_0 I} \quad \text{if } T_1 \geq T_S \qquad (17.162)$$

Seismic Code Provisions

and by linear interpolation between the values obtained with Equations 17.161 and 17.162 if $T_1 < T_S < T_{1D}$. The symbol I in Equations 17.161 and 17.162 denotes the importance factor introduced in Section 17.4.2.

17.13.10 Combination of Load Effects

The equivalent lateral force and response spectrum procedures prescribed by the code are based on the calculation of the peak displacement response of the system using a specified acceleration response spectrum, the modification of this spectrum to account for the damping increase due to the added damping system, and the use of the conventional approximate relationship between spectral displacements and spectral accelerations. The information obtained using the equivalent lateral force and response spectrum procedures is thus the value of the maximum displacement, the value of the acceleration at the time the system attains this maximum displacement, and the value of the maximum velocity if it is assumed (such as the code does) that the spectral velocity is equal to the pseudovelocity. This information may be used directly to compute the peak force in displacement-dependent damping devices. However, it is not sufficient to determine the peak force in velocity-dependent devices since this peak force does not correspond to the point of maximum displacement (see Figure 17.12). A procedure is therefore necessary to calculate such peak force in terms of the peak displacement response and peak velocity response. This procedure may be established as follows.

Consider a single-degree-of-freedom system with mass m, stiffness k, and a nonlinear viscous damper whose force–velocity relationship is defined by a constant C_N and an exponent α (see Equation 16.22). Consider, also, that according to Equations 15.198 and 16.25, the effective damping ratio of the system corresponding to the aforementioned damper and a peak displacement D is given by

$$\beta_V = \frac{W_D}{4\pi W_s} = \frac{C_N \lambda \omega_n^\alpha D^{1+\alpha}}{4\pi(\omega_n^2 m D^2 / 2)} = \frac{C_N \lambda \omega_n^{\alpha-2}}{2\pi m} D^{\alpha-1} \tag{17.163}$$

and, thus, in terms of such an effective damping ratio, the constant C_N may be expressed as

$$C_N = \frac{2\pi m \beta_V}{\lambda \omega_n^{\alpha-2} D^{\alpha-1}} \tag{17.164}$$

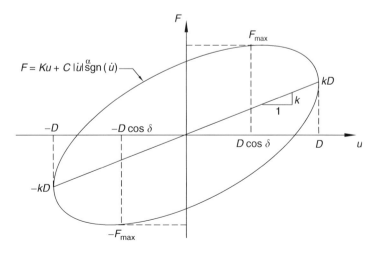

FIGURE 17.12 Force–deformation relationship of linear structure with nonlinear viscous dampers.

where ω_n is the natural frequency of the system and

$$\lambda = 4(2^\alpha)\frac{\Gamma^2(1+\alpha/2)}{\Gamma(2+\alpha)} \qquad (17.165)$$

where Γ represents the gamma function.

The equation of motion of such a system may be written as

$$mA(t) + C_N \mid \dot{u}(t) \mid^\alpha \operatorname{sgn}(\dot{u}) + ku(t) = 0 \qquad (17.166)$$

where $u(t)$ denotes the displacement of the system at time t, $A(t)$ is the corresponding absolute acceleration, and all other symbols are as previously defined. Consequently, the force on the mass of the system is given by

$$F(t) = mA(t) = -[C_N \mid \dot{u}(t) \mid^\alpha \operatorname{sgn}(\dot{u}) + ku(t)] \qquad (17.167)$$

which, on substitution of Equation 17.164 and considering that $k = \omega_n^2 m$, may also be expressed as

$$F(t) = -m\left[\frac{2\pi\beta_V}{\lambda\omega_n^{\alpha-2}D^{\alpha-1}} \mid \dot{u}(t) \mid^\alpha \operatorname{sgn}(\dot{u}) + \omega_n^2 u(t)\right] \qquad (17.168)$$

If it is assumed, however, that the system is undergoing harmonic motion with frequency ω_n and amplitude D, one has that

$$u(t) = D\cos\omega_n t \qquad (17.169)$$

$$\dot{u}(t) = -D\omega_n \sin\omega_n t \qquad (17.170)$$

and thus

$$\begin{aligned} F(t) &= -m\left[-\frac{2\pi\beta_V}{\lambda\omega_n^{\alpha-2}D^{\alpha-1}} \mid D\omega_n \mid^\alpha \sin^\alpha\omega_n t + \omega_n^2 D\cos\omega_n t\right] \\ &= -m\omega_n^2 D\left[-\frac{2\pi\beta_V}{\lambda}\sin^\alpha\omega_n t + \cos\omega_n t\right] \end{aligned} \qquad (17.171)$$

The maximum value of $F(t)$ occurs at time t_m, which may be found by making the derivative of the right-hand side of Equation 17.171 with respect to time equal to zero. That is, the time t_m at which the force $F(t)$ reaches its peak may be found from

$$\frac{\sin^{2-\alpha}\omega_n t_m}{\cos\omega_n t_m} = -\frac{2\pi\alpha\beta_V}{\lambda} \qquad (17.172)$$

except that Equation 17.172 cannot be solved explicitly for t_m. However, an approximate solution may be obtained if it is assumed that

$$\omega_n t_m = \pi - \delta \qquad (17.173)$$

Seismic Code Provisions

where δ is a small phase lag (see Figure 17.12). With this assumption, one has that $\sin \omega_n t_m = \sin \delta \approx \delta$ and $\cos \omega_n t_m = -\cos \delta \approx -1$, and thus from Equation 17.172 one obtains

$$\delta = \left(\frac{2\pi\alpha\beta_V}{\lambda}\right)^{1/(2-\alpha)} \tag{17.174}$$

In consequence, the maximum value of $F(t)$ is approximately equal to

$$F_{max} = -m\omega_n^2 D \left[\frac{2\pi\beta_V}{\lambda}\sin^\alpha \delta + \cos\delta\right] \tag{17.175}$$

where δ is given by Equation 17.174. Furthermore, if Equation 17.163 is substituted back into Equation 17.175, F_{max} may be alternatively expressed as

$$F_{max} = -\left[\frac{2\pi}{\lambda}\frac{C_N \lambda \omega_n^\alpha}{2\pi} D^\alpha \sin^\alpha \delta + D\omega_n^2 m \cos\delta\right] \tag{17.176}$$

$$= -[C_N(\omega_n D)^\alpha \sin^\alpha \delta + kD\cos\delta]$$

or as

$$F_{max} = -[C_{FV}F_D + C_{FD}F_s] \tag{17.177}$$

where

$$F_D = C_N(\omega_n D)^\alpha \tag{17.178}$$

$$F_s = kD \tag{17.179}$$

which may be interpreted, respectively, as the force in the damper and the restoring force at the time the system attains its maximum displacement, and

$$C_{FV} = \sin^\alpha \delta \tag{17.180}$$

$$C_{FD} = \cos\delta \tag{17.181}$$

which may be interpreted as load combination factors that may be used to calculate the peak response of the system in terms of response parameters evaluated at the time the maximum displacement occurs.

Equation 17.177 has been derived assuming elastic behavior; that is, it is valid for $D < D_y$. Nonetheless, it may still be used for elastoplastic systems if it is considered that, as inelastic action takes place, $D = D_y\mu$ and the value of $C_{FD}F_s$ increases until it reaches the limit of kD_y, at which point $C_{FD} = 1$. Moreover, the natural frequency of the system decreases and in consequence the effective damping ratio increases. For elastoplastic systems, therefore, F_{max} is also given by Equation 17.177, except that in this case

$$\delta = \left(\frac{2\pi\alpha\beta_{Veff}}{\lambda}\right)^{1/(2-\alpha)} \tag{17.182}$$

where

$$\beta_{Veff} = \beta_V \mu^{1-(\alpha/2)} \qquad (17.183)$$

and

$$C_{FD} = \begin{cases} \cos\delta & \text{if } D \le D_y \\ \mu\cos\delta & \text{if } D > D_y \quad \text{and} \quad \mu\cos\delta < 1 \\ 1 & \text{if } D > D_y \quad \text{and} \quad \mu\cos\delta \ge 1 \end{cases} \qquad (17.184)$$

where $\mu = D/D_y$. C_{FV} remains defined by Equation 17.180.

Based on the concepts discussed above, the code requires that the seismic design force, Q_E, in each element of the damping system due to horizontal earthquake load be taken as the largest among the values obtained under the following three loading conditions:

1. Force at the stage of maximum displacement, calculated according to

$$Q_E = \Omega_0 \sqrt{\sum_m (Q_{mSFRS})^2} \pm Q_{DSD} \qquad (17.185)$$

where Q_{mSFRS} is the force in the element equal to the design seismic force in the mth mode of vibration of the seismic-force-resisting system in the direction of interest, and Q_{DSD} is the force in the element required to resist the design seismic forces of displacement-dependent damping devices. The forces Q_{DSD} are to be calculated by imposing the design forces of displacement-dependent devices on the damping system as pseudostatic forces.

2. Force at the stage of maximum velocity, calculated according to

$$Q_E = \sqrt{\sum_m (Q_{mDSV})^2} \qquad (17.186)$$

where Q_{mDSV} is the force in the element required to resist the design damping forces of velocity-dependent damping devices in the mth mode of vibration and the direction of interest. The forces Q_{mDSV} are to be obtained by imposing the modal design forces of velocity-dependent devices on the undeformed damping system as pseudostatic forces. Horizontal restraint forces are to be applied at each floor level concurrent with the design forces in velocity-dependent damping devices such as to make the horizontal floor displacements zero. These forces are to be proportional and applied to each floor mass.

3. Force at the stage of maximum acceleration, calculated according to

$$Q_E = \sqrt{\sum_m (C_{mFD}\Omega_0 Q_{mSFRS} + C_{mFV} Q_{mDSV})^2} \pm Q_{DSD} \qquad (17.187)$$

where the coefficients C_{mFD} and C_{mFV} (defined by Equations 17.180 and 17.184) are obtained from Tables 17.22 and 17.23 for given values of effective damping, velocity exponent α, and displacement ductility demand. For the fundamental mode (i.e., for $m = 1$), the coefficients C_{mFD} and C_{mFV} are to be determined based on the actual value of the velocity exponent α. Also, the effective damping is to be taken equal to the total effective damping in the fundamental mode minus the hysteretic component at the response level of interest (i.e., $\beta_{1D} - \beta_{HD}$, or $\beta_{1M} - \beta_{HM}$). For the residual mode or the higher modes (i.e., $m > 1$), they are to be

TABLE 17.22
Force Coefficient C_{mFD}

Effective Damping	$\mu \leq 1.0$				$C_{mFD} = 1.0^a$
	$\alpha \leq 0.25$	$\alpha = 0.5$	$\alpha = 0.75$	$\alpha \geq 1.0$	
≤0.05	1.0	1.0	1.0	1.0	$\mu \geq 1.0$
0.1	1.0	1.0	1.0	1.0	$\mu \geq 1.0$
0.2	1.0	0.95	0.94	0.93	$\mu \geq 1.1$
0.3	1.0	0.92	0.88	0.86	$\mu \geq 1.2$
0.4	1.0	0.88	0.81	0.78	$\mu \geq 1.3$
0.5	1.0	0.84	0.73	0.71	$\mu \geq 1.4$
0.6	1.0	0.79	0.64	0.64	$\mu \geq 1.6$
0.7	1.0	0.75	0.55	0.58	$\mu \geq 1.7$
0.8	1.0	0.70	0.50	0.53	$\mu \geq 1.9$
0.9	1.0	0.66	0.50	0.50	$\mu \geq 2.1$
≥1.0	1.0	0.62	0.50	0.50	$\mu \geq 2.2$

Note: Unless analysis or test data support other values, $C_{mFD} = 1.0$ for viscoelastic systems. Use interpolation for intermediate values of α and μ.

[a] $C_{mFD} = 1.0$ for μ greater than or equal to the shown values.

Source: Reprinted with permission of American Society of Civil Engineers from *Minimum Design Loads for Buildings and Other Structures*, ASCE/SEI 7-05, Reston, VA, 2006.

TABLE 17.23
Force Coefficient C_{mFV}

Effective Damping	$\mu \leq 1.0$			
	$\alpha \leq 0.25$	$\alpha = 0.5$	$\alpha = 0.75$	$\alpha \geq 1.0$
≤0.05	1.0	0.35	0.20	0.10
0.1	1.0	0.44	0.31	0.20
0.2	1.0	0.56	0.46	0.37
0.3	1.0	0.64	0.58	0.51
0.4	1.0	0.70	0.69	0.62
0.5	1.0	0.75	0.77	0.71
0.6	1.0	0.80	0.84	0.77
0.7	1.0	0.83	0.90	0.81
0.8	1.0	0.90	0.94	0.90
0.9	1.0	1.00	1.00	1.00
≥1.0	1.0	1.00	1.00	1.00

Note: Unless analysis or test data support other values, $C_{mFV} = 1.0$ for viscoelastic systems. Use interpolation for intermediate values of α.

Source: Reprinted with permission of American Society of Civil Engineers from *Minimum Design Loads for Buildings and Other Structures*, ASCE/SEI 7-05, Reston, VA, 2006.

determined based on a value of α equal to 1 and an effective modal damping equal to the total effective damping in the mode of interest (i.e., β_{mD} or β_{mM}). The ductility demand is to be taken equal to the ductility demand in the fundamental mode.

17.13.11 REQUIRED TESTING OF DAMPING DEVICES

As in the case of base-isolation systems, the code requires that the force–velocity–displacement characteristics and damping properties of the damping devices used in a design be based on prototype tests. To this end, it specifies that two full-size specimens of each type and size considered in the design be tested. It also stipulates that at least five fully reversed, sinusoidal cycles at the maximum displacement and response frequency be conducted. If the devices have characteristics that vary with temperature, then tests at a minimum of three temperatures covering the operating range of the expected temperatures need to be conducted. The code specifies, further, the criteria by which it may be considered that a tested damping device exhibits a satisfactory performance. Reduced-scale models may be used provided they are of the same type and material as the full-size devices, are fabricated using the same process and quality control program, and the loading cycles are applied with a similitude-scaled frequency that is representative of the full-scale loading rates. The tests may be skipped whenever similar damping devices have been previously tested and all pertinent test data are available. The code may be consulted directly to learn about the details of the required tests and the aforementioned acceptability criteria.

17.13.12 PEER REVIEW

Also as in the case of base-isolation systems, the code mandates that the design of a damping system and related tests be reviewed by an independent engineering team composed of registered design professionals with experience in seismic analysis methods and theory and application of energy dissipation systems. The scope of the review includes but is not limited to

1. Review of seismic design criteria including the development of site-specific response spectra and ground motion time histories, as well as other project-specific design criteria
2. Review of the preliminary design of the seismic-force-resisting system and damping system, including the design parameters of the damping devices
3. Review of the final design of the seismic-force-resisting system and damping system and all supporting analyses
4. Review of damping device test requirements, device manufacturing quality control and assurance, and scheduled maintenance and inspection requirements.

EXAMPLE 17.6. ANALYSIS OF STRUCTURE WITH ADDED DAMPERS

The three-story frame shown in Figure E17.6 is to be retrofitted by adding to the frame three nonlinear viscous dampers with a velocity exponent of 0.5. The frame is one of two special steel moment-resisting frames of an office building located at a site characterized by a design spectrum with $S_{DS} = 1.0$, $S_{D1} = 0.6$, $T_s = 0.6$ s, and soil properties classified as class B. The building has a regular configuration, plan dimensions of 41.15 m × 41.15 m, and the structural sections and floor weights indicated in Figure E17.6. The structural steel has a nominal yield strength of 345 MPa. The dynamic properties in the first three modes of the building without the dampers are listed in Table E17.6. Determine the size of the dampers needed to augment the damping ratio in the fundamental mode of the frame by 20%. Determine also the base shear, lateral forces, and story drifts of the frame with the added dampers and whether or not it complies with the drift limits imposed by the code. Use the equivalent lateral force procedure prescribed in the International Building Code assuming an inherent damping ratio of 5%.

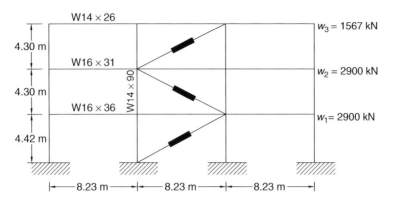

FIGURE E17.6 Frame with viscous dampers considered in Example 17.6.

TABLE E17.6
Dynamic Properties of Frame without Added Dampers in Example 17.6

Mode	Natural Frequency (rad/s)	Natural Period (s)	Participation Factor	Effective Weight (%)	Mode Shape		
					Floor 1	Floor 2	Roof
1	4.4487	1.412	1.388	80.9	0.265	0.675	1.000
2	14.439	0.435	−0.526	14.7	−0.739	−0.510	1.000
3	27.088	0.232	0.136	4.30	1.757	−1.497	1.000

Solution

Dynamic Properties in Residual Mode

According to Equations 17.115, 17.117, 17.121, and 17.122, the effective weight, natural period, participation factor, and mode shape in the residual mode of the structure are, respectively, equal to

$$\bar{W}_R = W - \bar{W}_1 = 3683.5(1 - 0.809) = 703.5 \text{ kN}$$

$$T_R = 0.4T_1 = 0.4(1.412) = 0.565 \text{ s}$$

$$\Gamma_R = 1 - \Gamma_1 = 1 - 1.388 = -0.388$$

$$\{\phi_R\} = \frac{1}{1-\Gamma_1}\begin{Bmatrix} 1-\Gamma_1\phi_{11} \\ 1-\Gamma_1\phi_{21} \\ 1-\Gamma_1\phi_{31} \end{Bmatrix} = \frac{1}{-0.388}\begin{Bmatrix} 1-1.388(0.265) \\ 1-1.388(0.675) \\ 1-1.388(1.000) \end{Bmatrix} = \begin{Bmatrix} -1.629 \\ -0.163 \\ 1.000 \end{Bmatrix}$$

Response Modification Coefficient and Deflection Amplification, Overstrength, and Importance Factors

From Table 17.1, the response modification coefficient R, deflection amplification factor C_d, and overstrength factor Ω_0 for a special moment-resisting steel frame are

$$R = 8.0 \quad C_d = 5.5 \quad \Omega_0 = 3.0$$

Similarly, from Table 17.6 for a structure (office building) that belongs to occupancy category II, the importance factor I is

$$I = 1.0$$

Minimum Base Shear

The minimum base shear stipulated by the code for structures with added dampers is a function of the base shear in the structure without the added dampers. Hence, it is necessary to determine this base shear first. To this end, consider that the minimum natural period that can be used in the calculation of such base shear is

$$T_{min} = 1.4T_a = 1.4c_t h_n^x = 1.4(0.072)(13.02)^{0.8} = 0.786 \text{ s}$$

and, thus, according to Equation 17.23, one has that

$$C_s = \frac{S_{DS}}{R/I} = \frac{1.0}{8/1.0} = 0.125$$

which is greater than (see Equation 17.25)

$$(C_s)_{min} = \frac{0.5 S_1}{R/I} = \frac{0.5(0.6 \times 1.5)}{8/1} = 0.056$$

but exceeds (see Equation 17.26)

$$(C_s)_{max} = \frac{S_{D1}}{T(R/I)} = \frac{0.6}{0.786(8/1)} = 0.095$$

Hence, for the structure without the added dampers (see Equation 17.22)

$$V = C_s W = 0.095(3683.5) = 350 \text{ kN}$$

In this case, the sum of the inherent damping ratio (assumed equal to 5% in all modes) and the viscous damping ratio due to the added dampers (considered equal to 20%) in the fundamental mode of the frame equals 0.25. Also, from Table 17.21 one has that $B_{V+I} = 1.65$ for a damping ratio of 0.25. As a result, the minimum base shear required by the code for the structure with the added dampers is the largest of

$$V_{min} = 0.75V = 0.75(350) = 262.5 \text{ kN}$$

$$V_{min} = \frac{V}{B_{V+I}} = \frac{350}{1.65} = 212.1 \text{ kN}$$

that is, $V_{min} = 262.5$ kN.

Effective Fundamental Period and Damping Ratio

The effective fundamental period and the effective damping ratios in the fundamental and residual modes of the structure with the added dampers depend on the effective ductility demand. However, this ductility demand is not known at this point. In consequence, it will be assumed first that $\mu_D = 2.0$ and then iterations will be performed if necessary until the assumed and calculated values approximately coincide.

Proceeding accordingly, Equation 17.112 yields the following value for the effective fundamental period:

$$T_{1D} = T_1 \sqrt{\mu_D} = 1.412\sqrt{2.0} = 2.0 \text{ s}$$

Similarly, since for the case under consideration one has that (see Equation 17.152)

$$q_H = 0.67 \frac{T_S}{T_1} = 0.67 \frac{0.6}{1.412} = 0.285$$

which is less than the minimum of 0.5 and thus $q_H = 0.5$, Equation 17.150 gives

$$\beta_{HD} = q_H(0.64 - \beta_I)\left(1 - \frac{1}{\mu_D}\right) = 0.5(0.64 - 0.05)\left(1 - \frac{1}{2.0}\right) = 0.148$$

In consequence and in accordance with Equation 17.147, the effective damping ratio in the first mode of the system is

$$\beta_{1D} = \beta_I + \beta_{V1}\sqrt{\mu_D} + \beta_{HD} = 0.05 + 0.20\sqrt{2.0} + 0.148 = 0.481$$

Base Shear in Fundamental Mode

From Table 17.21, one has that $B_{1D} = 2.34$ for a damping ratio of 48.1%. Therefore, with the assumed value of μ_D and Equations 17.111 and 17.107, one obtains

$$C_{s1} = \left(\frac{R}{C_d}\right)\frac{S_{D1}}{T_{1D}(\Omega_0 B_{1D})} = \left(\frac{8.0}{5.5}\right)\frac{0.6}{2.0(3.0)(2.34)} = 0.062$$

$$V_1 = C_{S1}\overline{W}_1 = 0.062(0.809)(3683.5) = 184.7 \text{ kN}$$

Roof Design Displacement, Yield Roof Displacement, and Effective Ductility Demand in Fundamental Mode

For the structure under consideration, the roof design displacement is given by Equation 17.127. For the structure under consideration, this displacement is therefore equal to

$$D_{1D} = \left(\frac{g}{4\pi^2}\right)\Gamma_1 \frac{S_{D1}T_{1D}}{B_{1D}} = \left(\frac{9.81}{4\pi^2}\right)1.388\frac{0.6(2.0)}{2.34} = 0.177 \text{ m}$$

which is equal to or greater than

$$(D_{1D})_{min} = \left(\frac{g}{4\pi^2}\right)\Gamma_1 \frac{S_{D1}T_1}{B_{1E}} = \left(\frac{9.81}{4\pi^2}\right)1.388\frac{0.6(1.412)}{1.65} = 0.177$$

Similarly, the yield displacement is given by Equation 17.160, and thus for the structure under consideration this yield displacement is

$$D_Y = \left(\frac{g}{4\pi^2}\right)\left(\frac{\Omega_0 C_d}{R}\right)\Gamma_1 C_{s1}T_1^2 = \left(\frac{9.81}{4\pi^2}\right)\left[\frac{3.0(5.5)}{8.0}\right](1.388)(0.062)(1.412^2) = 0.088 \text{ m}$$

In consequence and in view of Equation 17.158, the effective ductility demand in the fundamental mode of the structure is

$$\mu_D = \frac{D_{1D}}{D_Y} = \frac{0.177}{0.088} = 2.01$$

which is less than the allowed maximum of (see Equation 17.162)

$$\mu_{max} = \frac{R}{\Omega_0 I} = \frac{8.0}{3.0(1.0)} = 2.67$$

It may be noted that the assumed and calculated values of the design ductility demand are almost the same and, thus, there is no need for iterations.

Base Shear in Residual Mode

According to the code provisions, the inherent damping is the same in all the modes and the effective ductility demands in the higher modes are all equal to 1. Therefore, if it is considered that the viscous damping for the residual mode is the same as for the fundamental mode and that the effective damping in the residual mode is given by Equation 17.147, this effective damping is equal to

$$\beta_{RD} = \beta_I + \beta_{VR} + \beta_{HD} = 0.05 + 0.20 + 0 = 0.25$$

which, from Table 17.21, corresponds to a B_R value of 1.65. In consequence and according to Equations 17.114 and 17.116, the seismic coefficient and base shear in the residual mode of the structure are equal to

$$C_{sR} = \left(\frac{R}{C_d}\right)\frac{S_{DS}}{\Omega_0 B_R} = \left(\frac{8.0}{5.5}\right)\frac{1.0}{3.0(1.65)} = 0.294$$

$$V_R = C_{sR}\bar{W}_R = 0.294(703.5) = 206.8\,\text{kN}$$

Total Base Shear

From Equation 17.106 and the calculated modal base shears, the total base shear is

$$V = \sqrt{V_1^2 + V_R^2} = \sqrt{(184.7)^2 + (206.8)^2} = 277.3\,\text{kN}$$

which is greater than the minimum of 262.5 kN obtained earlier and meets thus the limit imposed by the code.

Design Lateral Forces

Using Equations 17.118 and 17.119 and the determined modal base shears, the design lateral forces in the fundamental and residual modes are

$$\{F_1\} = \{w_i\phi_{i1}\}\frac{\Gamma_1}{\bar{W}_1}V_1 = 1.388 \begin{Bmatrix} 1450.0(0.265) \\ 1450.0(0.675) \\ 783.5(1.000) \end{Bmatrix} 0.062 = \begin{Bmatrix} 33.1 \\ 84.2 \\ 67.4 \end{Bmatrix}\text{kN}$$

$$\{F_R\} = \{w_i\phi_{iR}\}\frac{\Gamma_R}{\bar{W}_R}V_R = -0.388 \begin{Bmatrix} 1450.0(-1.629) \\ 1450.0(-0.163) \\ 783.5(1.000) \end{Bmatrix} 0.294 = \begin{Bmatrix} 269.4 \\ 27.0 \\ -89.4 \end{Bmatrix}\text{kN}$$

and thus the total lateral forces are

$$\{F_x\} = \begin{Bmatrix} (33.1^2 + 269.4^2)^{1/2} \\ (84.2^2 + 27.0^2)^{1/2} \\ (67.4^2 + 89.4^2)^{1/2} \end{Bmatrix} = \begin{Bmatrix} 271.4 \\ 88.42 \\ 112.0 \end{Bmatrix}\text{kN}$$

Floor Deflections, Story Drifts, and Story Velocities

According to Equation 17.128 and the value of B_R found earlier, the roof deflection in the residual mode is equal to

$$D_{RD} = \left(\frac{g}{4\pi^2}\right)\Gamma_R\frac{S_{D1}T_R}{B_R} = \left(\frac{9.81}{4\pi^2}\right)(-0.388)\frac{0.6(0.565)}{1.65} = -0.020\,\text{m}$$

which is greater than

$$(D_{RD})_{\max} = \left(\frac{g}{4\pi^2}\right)\Gamma_R\frac{S_{DS}T_R^2}{B_R} = \left(\frac{9.81}{4\pi^2}\right)(-0.388)\frac{1.0(0.565)^2}{1.65} = -0.019\,\text{m}$$

and thus $D_{RD} = -0.019$ m. In consequence, the modal and total floor deflections are (see Equations 17.123 through 17.125)

$$\{\delta_{1D}\} = D_{1D}\{\phi_{i1}\} = 0.177 \begin{Bmatrix} 0.265 \\ 0.675 \\ 1.000 \end{Bmatrix} = \begin{Bmatrix} 0.047 \\ 0.119 \\ 0.177 \end{Bmatrix} \text{m}$$

$$\{\delta_{RD}\} = D_{RD}\{\phi_{iR}\} = -0.019 \begin{Bmatrix} -1.629 \\ -0.163 \\ 1.000 \end{Bmatrix} = \begin{Bmatrix} 0.031 \\ 0.003 \\ -0.019 \end{Bmatrix} \text{m}$$

$$\{\delta_D\} = \begin{Bmatrix} (0.047^2 + 0.031^2)^{1/2} \\ (0.119^2 + 0.003^2)^{1/2} \\ (0.177^2 + 0.019^2)^{1/2} \end{Bmatrix} = \begin{Bmatrix} 0.056 \\ 0.119 \\ 0.178 \end{Bmatrix} \text{m}$$

Similarly, the modal and total story drifts are

$$\{\Delta_{1D}\} = \begin{Bmatrix} 0.047 \\ 0.119 - 0.047 \\ 0.177 - 0.119 \end{Bmatrix} = \begin{Bmatrix} 0.047 \\ 0.072 \\ 0.058 \end{Bmatrix} \text{m}$$

$$\{\Delta_{RD}\} = \begin{Bmatrix} 0.031 \\ 0.003 - 0.031 \\ -0.019 - 0.003 \end{Bmatrix} = \begin{Bmatrix} 0.031 \\ -0.028 \\ -0.022 \end{Bmatrix} \text{m}$$

$$\{\Delta_D\} = \begin{Bmatrix} (0.047^2 + 0.031^2)^{1/2} \\ (0.072^2 + 0.028^2)^{1/2} \\ (0.058^2 + 0.022^2)^{1/2} \end{Bmatrix} = \begin{Bmatrix} 0.056 \\ 0.077 \\ 0.062 \end{Bmatrix} \text{m}$$

Drift Limits

The story drift limit stipulated by the code for a structure with added dampers is R/C_d times the limit indicated in Table 17.2 for conventional structures, which for a steel building with occupancy category II is 0.020 times the story height. Therefore, for the building under consideration, the drift limit for the first story is $(8.0/5.5)(0.020)(4.42) = 0.129$ m and the drift limit for the second and third stories is $(8.0/5.5)(0.020)(4.30) = 0.125$ m. Thus, the building meets the story drift limits when it is considered with the added dampers.

Dampers' Damping Constant

The dampers' damping constant, C_N, may be determined by equating the damping ratio β_{V1} calculated using Equation 17.156 to the specified viscous damping ratio of 20% for the fundamental mode of the system; that is, by considering that

$$\frac{\sum_j W_{1j}}{4\pi W_1} = 0.20$$

where, as defined in Section 17.13.8, β_{V1} represents the damping ratio in the first mode of the system due to the added damping system when the structure vibrates at or just below the effective yield displacement; and ΣW_{1j} and W_1 are, respectively, the dissipated energy in one cycle of response and the maximum strain energy in the first mode of the structure under the displacements

$$\{\delta_{i1}\} = \{\phi_{i1}\}D_Y = \begin{Bmatrix} 0.265 \\ 0.675 \\ 1.000 \end{Bmatrix} 0.088 = \begin{Bmatrix} 0.023 \\ 0.059 \\ 0.088 \end{Bmatrix} \text{m}$$

However, according to Equation 16.25, ΣW_{1j} is given by

$$\sum_j W_{1j} = \lambda C_N \omega_1^\alpha [(\Delta_{11} \cos\theta_1)^{\alpha+1} + \Delta_{12} \cos\theta_2)^{\alpha+1} + \Delta_{13} \cos\theta_3)^{\alpha+1}]$$

where λ is defined by Equation 17.165, θ_1, θ_2, and θ_3 are the angles the dampers in the first, second, and third stories make with respect to the horizontal, and Δ_{11}, Δ_{12}, and Δ_{13} are the story drifts of the first, second, and third stories when the structure is subjected to the displacement $\{\delta_{i1}\}$. For the case under consideration, one has that $\alpha = 0.5$, $\theta_1 = 28.2°$, $\theta_2 = \theta_3 = 27.6°$,

$$\lambda = 4(2^\alpha) \frac{\Gamma^2(1+\alpha/2)}{\Gamma(2+\alpha)} = 4(2^{0.5}) \frac{\Gamma^2(1+0.5/2)}{\Gamma(2+0.5)} = 3.496$$

and thus

$$\sum_j W_{1j} = C_N(3.496)(4.4487)^{0.5}\{(0.023\cos 28.2°)^{1.5} + [(0.059 - 0.023)\cos 27.6°]^{1.5}$$
$$+ [(0.088 - 0.059)\cos 27.6°]^{1.5}\}$$
$$= 0.094 C_N$$

Similarly, according to Equation 17.157, the maximum strain energy is given by

$$W_1 = \frac{1}{2}\sum_i F_{i1}\delta_{i1}$$

where F_{i1} denotes the previously calculated design lateral forces in the fundamental mode of the structure and δ_{i1} represents the elements of the displacement vector $\{\delta_{i1}\}$. Consequently, for the system being considered, the maximum strain energy is equal to

$$W_1 = \frac{1}{2}[33.1(0.023) + 84.2(0.059) + 67.4(0.088)] = 5.83 \text{ kN m}$$

The dampers' constant is thus equal to

$$C_N = \frac{4\pi(0.20)(5.830)}{0.094} = 155.9 \text{ kN s}^{0.5}/\text{m}^{0.5}$$

Maximum Damper Forces

As the code specifies that the damping devices be designed based on the maximum considered earthquake, the calculation of the maximum forces in the dampers requires that the story velocities under the maximum considered earthquake be determined first. For this purpose, assume that $\mu_M = 3.0$. By virtue of Equation 17.136, one obtains

$$T_{1M} = T_1\sqrt{\mu_M} = 1.412\sqrt{3.0} = 2.45 \text{ s}$$

and by virtue of Equations 17.151 and 17.148, one gets

$$\beta_{HM} = q_H(0.64 - \beta_I)\left(1 - \frac{1}{\mu_M}\right) = 0.5(0.64 - 0.05)\left(1 - \frac{1}{3.0}\right) = 0.197$$

$$\beta_{1M} = \beta_I + \beta_{V1}\sqrt{\mu_M} + \beta_{HM} = 0.05 + 0.20\sqrt{3.0} + 0.197 = 0.593$$

Seismic Code Provisions

which according to Table 17.21 corresponds to a B_{1M} value of 2.679. Recalling, then, that $S_{M1} = 1.5S_{D1} = 1.5(0.6) = 0.9$ and $S_{MS} = 1.5S_{DS} = 1.5(1.0) = 1.5$, from Equation 17.134 one finds that

$$D_{1M} = \left(\frac{g}{4\pi^2}\right)\Gamma_1\frac{S_{M1}T_{1M}}{B_{1M}} = \left(\frac{9.81}{4\pi^2}\right)1.388\frac{0.9(2.45)}{2.679} = 0.284\,\text{m}$$

which satisfies the lower limit imposed by the code since it is greater than

$$(D_{1M})_{min} = \left(\frac{g}{4\pi^2}\right)\Gamma_1\frac{S_{M1}T_1}{B_{1E}} = \left(\frac{9.81}{4\pi^2}\right)1.388\frac{0.9(1.412)}{1.65} = 0.266\,\text{m}$$

Thus, since $D_Y = 0.095$ m, the effective ductility demand due to the maximum considered earthquake results as

$$\mu_M = \frac{D_{1M}}{D_Y} = \frac{0.284}{0.095} = 2.99$$

which is sufficiently close to the assumed value of 3.0 and thus the calculated value of D_{1M} may be considered the correct one.

Turning now to the residual mode, Equation 17.135 leads to

$$D_{RM} = \left(\frac{g}{4\pi^2}\right)\Gamma_R\frac{S_{M1}T_R}{B_R} = \left(\frac{9.81}{4\pi^2}\right)0.388\frac{0.9(0.565)}{1.65} = 0.030\,\text{m}$$

which is greater than

$$(D_{RM})_{max} = \left(\frac{g}{4\pi^2}\right)\Gamma_R\frac{S_{MS}T_R^2}{B_R} = \left(\frac{9.81}{4\pi^2}\right)0.388\frac{1.5(0.565)^2}{1.65} = 0.028\,\text{m}$$

and thus $D_{RM} = 0.028$ m. Consequently, the modal floor deflections under the maximum considered earthquake are equal to

$$\{\delta_{1M}\} = D_{1M}\{\phi_{i1}\} = 0.284\begin{Bmatrix}0.265\\0.675\\1.000\end{Bmatrix} = \begin{Bmatrix}0.075\\0.192\\0.284\end{Bmatrix}\text{m}$$

$$\{\delta_{RM}\} = D_{RM}\{\phi_{iR}\} = -0.028\begin{Bmatrix}-1.629\\-0.163\\1.000\end{Bmatrix} = \begin{Bmatrix}0.046\\0.005\\-0.028\end{Bmatrix}\text{m}$$

and the corresponding modal story drifts equal to

$$\{\Delta_{1M}\} = \begin{Bmatrix}0.075\\0.192-0.075\\0.284-0.192\end{Bmatrix} = \begin{Bmatrix}0.075\\0.117\\0.092\end{Bmatrix}\text{m}$$

$$\{\Delta_{RM}\} = \begin{Bmatrix}0.046\\0.005-0.046\\-0.028-0.005\end{Bmatrix} = \begin{Bmatrix}0.046\\-0.041\\-0.033\end{Bmatrix}\text{m}$$

Hence, the modal story velocities under the maximum considered earthquake result as

$$\{\nabla_{1M}\} = \frac{2\pi}{T_{1M}}\{\Delta_{1M}\} = \frac{2\pi}{2.45}\begin{Bmatrix} 0.075 \\ 0.117 \\ 0.092 \end{Bmatrix} = \begin{Bmatrix} 0.192 \\ 0.300 \\ 0.236 \end{Bmatrix} \text{m/s}$$

$$\{\nabla_{RM}\} = \frac{2\pi}{T_R}\{\Delta_{RM}\} = \frac{2\pi}{0.565}\begin{Bmatrix} 0.046 \\ -0.041 \\ -0.033 \end{Bmatrix} = \begin{Bmatrix} 0.512 \\ -0.456 \\ -0.367 \end{Bmatrix} \text{m/s}$$

According to Equation 16.22 and these horizontal story velocities, the maximum forces in the dampers in the fundamental and residual modes are then given by

$$\{F_{1D}\} = C_N \begin{Bmatrix} (\nabla_{11M}\cos\theta_1)^{0.5} \\ (\nabla_{21M}\cos\theta_2)^{0.5} \\ (\nabla_{31M}\cos\theta_3)^{0.5} \end{Bmatrix} = 155.9 \begin{Bmatrix} (0.192\cos 28.2°)^{0.5} \\ (0.300\cos 27.6°)^{0.5} \\ (0.236\cos 27.6°)^{0.5} \end{Bmatrix} = \begin{Bmatrix} 64.13 \\ 80.38 \\ 71.30 \end{Bmatrix} \text{kN}$$

$$\{F_{RD}\} = C_N \begin{Bmatrix} (\nabla_{1RM}\cos\theta_1)^{0.5} \\ (\nabla_{2RM}\cos\theta_2)^{0.5} \\ (\nabla_{3RM}\cos\theta_3)^{0.5} \end{Bmatrix} = 155.9 \begin{Bmatrix} (0.512\cos 28.2°)^{0.5} \\ (0.456\cos 27.6°)^{0.5} \\ (0.367\cos 27.6°)^{0.5} \end{Bmatrix} = \begin{Bmatrix} 104.7 \\ 99.10 \\ 88.91 \end{Bmatrix} \text{kN}$$

which lead to the following maximum damper forces:

$$\{F_D\} = \begin{Bmatrix} (64.13^2 + 104.7^2)^{1/2} \\ (80.38^2 + 99.10^2)^{1/2} \\ (71.30^2 + 88.91^2)^{1/2} \end{Bmatrix} = \begin{Bmatrix} 122.8 \\ 127.6 \\ 114.0 \end{Bmatrix} \text{kN}$$

It may be seen, thus, that three dampers with an output force of at least 125 kN on each of the two seismic-force-resisting frames will provide the desired additional damping of 20% and reduce the story drifts to levels below the limits specified by the code. It should be noted, however, that the code requires that these results be confirmed with a response time-history analysis since the structure under consideration is located at a site with a spectral acceleration S_1 equal to or greater than 0.6 (see Section 17.13.2).

FURTHER READINGS

1. Building Seismic Safety Council, *NEHRP Recommended Provisions for Seismic Regulations for New Buildings and Other Structures* (FEMA 450), 2003 Edition, Part 1: Provisions, Washington, DC, 2004, available at www.bssconline.org.
2. Building Seismic Safety Council, *NEHRP Recommended Provisions for Seismic Regulations for New Buildings and Other Structures* (FEMA 450), 2003 Edition, Part 2: Commentary, Washington, DC, 2004, available at www.bssconline.org.
3. *Tri-Services Manual: Seismic Design of Buildings*, Departments of the Army (TM 5-809-10), Navy (NAVFAC 355), and Air Force (AFM 88-3), Washington, DC, 1992.
4. American Society of Civil Engineers, *Minimum Design Loads for Buildings and Other Structures*, ASCE/SEI 7-05, Reston, VA, 2006.
5. International Code Council, Inc., *International Building Code*, 2006.
6. Berg, G. V., *Seismic Design Codes and Procedures*, Monograph MNO-6, Earthquake Engineering Research Institute, Oakland, CA, 1983.
7. Applied Technology Council, *Tentative Provisions for the Development of Seismic Regulations for Buildings*, ATC 3-06, National Bureau of Standards, Special Publication 510, U.S. Government Printing Office, Washington, DC, 1978.
8. Seismological Committee, *Recommended Lateral Force Requirements and Commentary*, Structural Engineers Association of California, Sacramento, CA, 1999.

PROBLEMS

17.1 Calculate the base shear, story shears, overturning moments, and story drifts for the shear building shown in Figure P17.1 using the equivalent lateral force procedure prescribed by the International Building Code. The building will be used for offices; will be located in Los Angeles, California, over a deposit of stiff soil; and will be constructed with moment-resistant steel frames. Consider $m_1 = 2$ kip s²/in. and $k_1 = 4000$ kip/in.

17.2 Compute the lateral displacements and the overturning moment at the base of the two-story shear building shown in Figure P17.2 using the equivalent lateral force procedure prescribed by the International Building Code. The building will be located in San Francisco,

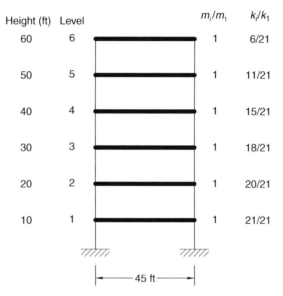

FIGURE P17.1 Six-story building considered in Problem 17.1.

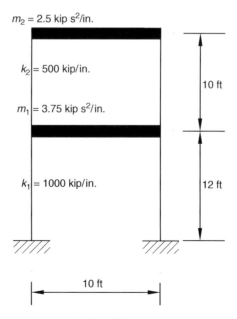

FIGURE P17.2 Two-story building considered in Problem 17.2.

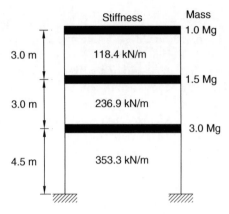

FIGURE P17.3 Three-story building considered in Problem 17.3.

California, over a deep deposit of soft clay; will be used as a hospital; and will be structured with moment-resisting reinforced concrete frames. Determine the fundamental natural period of the building using an eigenvalue analysis.

17.3 Calculate the design shearing forces for the columns of the first story of the shear building shown in Figure P17.3 using the equivalent lateral force procedure prescribed by the International Building Code. The building will be located in Reno, Nevada, over stiff soil and will be used as a hotel. Its lateral-load-resisting system will be constructed with special reinforced concrete moment-resisting frames. The given floor masses include dead and live load.

17.4 Determine the design axial forces in the columns of the first story of the building shown in Figure P17.4 using the equivalent lateral force procedure prescribed by the International Building Code. The building will be located in Las Vegas, Nevada, over a deep deposit of soft clay; will be used as a hospital; and will be structured with special moment-resistant reinforced concrete frames.

17.5 Determine the design story drifts of the building described in Problem P17.4 using the equivalent lateral force procedure prescribed by the International Building Code. Indicate, in addition, whether the calculated story drifts are within the limits established by the code. The inverse of the structure's stiffness matrix is equal to

$$[K]^{-1} = \frac{1}{250} \begin{bmatrix} 0.250 & 0.250 & 0.250 & 0.250 \\ 0.250 & 0.583 & 0.583 & 0.583 \\ 0.250 & 0.583 & 0.917 & 0.917 \\ 0.250 & 0.583 & 0.917 & 1.417 \end{bmatrix} \text{in./kip}$$

17.6 The building shown in Figure P17.6 will be built in Anchorage, Alaska, on a parcel of land classified as class B using special moment-resisting steel frames. The building height is 135 ft and the plan dimensions are 100 ft × 170 ft. The dead load at each floor is 120 lb/ft². Determine using the equivalent lateral force procedure prescribed by the International Building Code the design base shear, lateral forces, and story shears along the transverse direction of the building.

17.7 The 12-story moment-resisting steel frame shown in Figure P17.7 is used to model a building for seismic analysis. The effective weight on all levels of the building is 90 lb/ft², except on the roof where it is 80 lb/ft². The tributary area for the frame has a width of 30 ft. The overall dimension of the building in the direction perpendicular to the direction of analysis is 180 ft. The building is located in Valdez, Alaska, on a site classified as

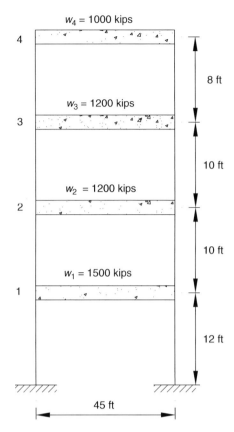

FIGURE P17.4 Four-story building considered in Problem 17.4.

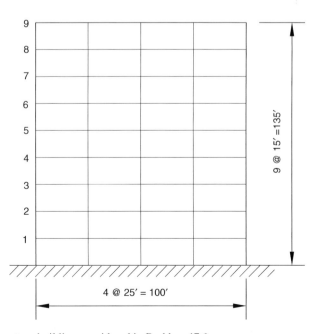

FIGURE P17.6 Nine-story building considered in Problem 17.6.

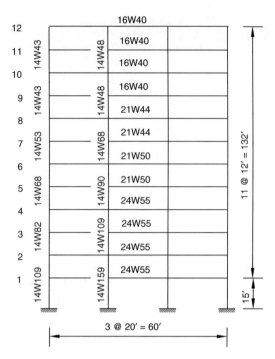

FIGURE P17.7 Twelve-story building considered in Problem 17.7.

class C. The building will be used for office space. Determine using the equivalent lateral force procedure prescribed by the International Building Code the (a) equivalent lateral forces, (b) story shears, (c) torsional moments, (d) overturning moments, (e) lateral displacements, and (f) story drifts. Consider that the modulus of elasticity for the steel elements is 30,000 kips/in.²

17.8 Figure P17.8 shows the elevation and plan views of a three-story reinforced concrete warehouse that will be built in Carson City, Nevada. As shown in Figure P17.8, the building is constructed with shear walls with a height of 20 ft for the first story and 15 ft for the second and third stories. The total gravity load for the first, second, and third floors is 300 kips, 250 kips, and 220 kips, respectively. Determine using the equivalent lateral force procedure prescribed by the International Building Code the design lateral forces, story shears, lateral displacements, and story drifts along the transverse direction of the building. Assume a modulus of elasticity of 3600 kips/in.² for the concrete used in the construction of the shear walls. Neglect the bending deformation of the shear walls.

17.9 The building shown in Figure P17.9 will be located in Fresno, California; will be used for hazardous and water treatment facilities; and will be constructed with ordinary moment-resisting frames. The building site has been classified as class B. The dead loads for the first, second, and third floors are 609.4 kips, 590.9 kips, and 217.2 kips, respectively. Determine using the modal response spectrum analysis procedure prescribed by the International Building Code the base shear, lateral forces, story shears, lateral displacements, and story drifts of the building along its transverse direction. Assume that the moments of inertia of the columns are $I_x = 1530$ in.⁴ and $I_y = 548$ in.⁴, and those of the beams are $I_x = 1170$ in.⁴ and $I_y = 60.3$ in.⁴ Similarly, assume that the modulus of elasticity of the steel used in the construction of the building structure is 29,000 kips/in.²

17.10 The three-story building shown in Figure P17.10 will be located in San Diego, California, and constructed with intermediate moment-resisting steel frames. An evaluation of the

Seismic Code Provisions

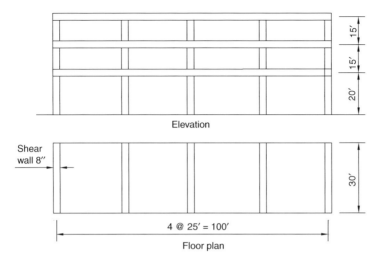

FIGURE P17.8 Three-story building considered in Problem 17.8.

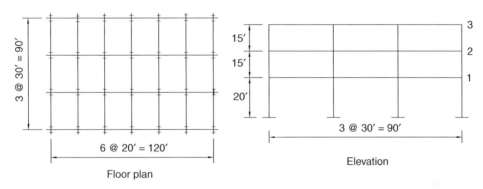

FIGURE P17.9 Three-story building considered in Problem 17.9.

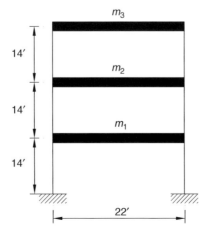

FIGURE P17.10 Three-story building considered in Problem 17.10.

building site shows that the underlying soil deposit has a depth of about 250 ft and an average shear wave velocity of 1500 ft/s. The masses, natural frequencies, and modal matrix of the structure are as shown below. Determine using the modal response spectrum analysis procedure prescribed by the International Building Code the design base shear, story shear, overturning moments, and story drifts.

$$m_1 = m_2 = m_3 = 1132.2141 \text{ lb s}^2/\text{ft}$$

$$\omega_1 = 7.30916 \text{ rad/s} \quad \omega_2 = 23.58844 \text{ rad/s} \quad \omega_3 = 40.04150 \text{ rad/s}$$

$$[\Phi] = \begin{bmatrix} 0.2555 & 0.9413 & -0.4655 \\ 0.6752 & 1.0 & 1.0 \\ 1.0 & -0.9421 & -0.6895 \end{bmatrix}$$

17.11 Repeat Problem 17.1 using the modal response spectrum analysis procedure prescribed by the International Building Code.

17.12 Repeat Problem 17.5 using the modal response spectrum analysis procedure prescribed by the International Building Code.

17.13 Repeat Problem 17.7 using the modal response spectrum analysis procedure prescribed by the International Building Code.

17.14 A 25-story building is located in the city of Santa Monica, California; has a height of 300 ft; is supported by a 100 ft × 100 ft mat foundation that rests near the surface of a homogeneous soil deposit; and supports a total gravity load of 15,000 kips. The soil deposit exhibits a shear modulus of elasticity of 2.4×10^6 lb/ft², a unit weight of 120 lb/ft³, and a Poisson ratio of 0.45. From a fixed-base analysis using the equivalent lateral force procedure in the International Building Code, it is found that the building has a fundamental natural period of 3 s, and is subjected to a base shear of 1200 kips, an overturning moment at its base of 240,000 kip ft, and a lateral displacement at its roof level of 3 in. Determine following the provisions of the International Building Code the reduction in the base shear of the building and the increase or decrease in its roof displacement when the flexibility of the foundation soil is considered.

17.15 The tower shown in Figure P17.15 is located in the city of Irvine, California, over a site classified as class E. It supports a water tank that weighs 2000 kN. The tower foundation

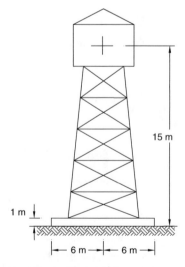

FIGURE P17.15 Water tower considered in Problem 17.15.

is a reinforced concrete circular mat with a diameter of 12 m, a depth of 1 m, and a unit weight of 24 kN/m³. The foundation soil has a shear modulus of elasticity of 72 MN/m³, a unit weight of 17.7 kN/m³, and a Poisson ratio of 0.45. When modeled as a single-degree-of-freedom system with a fixed base, the tower exhibits a natural period of 0.25 s. Modeling it as a single-degree-of-freedom system and using the equivalent lateral force procedure prescribed by the International Building Code, determine the base shear and maximum lateral displacement in the tower considering the flexibility of the foundation soil. Assume a response modification coefficient of 2.5, a deflection amplification factor of 2.5, and an importance factor of 1.0. Neglect the weight of the tower.

17.16 The steam boiler shown in Figure P17.16 will be installed on the roof of a four-story building that is 40 ft high. The boiler weighs 20 kips and will be anchored to the roof slab using four bolts, 1 in. in diameter each. The bolts have an embedment length of 6 in. The fundamental period of the boiler and its anchors is 0.04 s. The building is located in Fairbanks, Alaska, over a site classified as class C and will be used as a hotel. Determine using the provisions of the International Building Code the shear and tension demands on the anchor bolts.

17.17 The electric generator shown in Figure P17.17 will be installed in the third floor of a five-story emergency command center. The command center is located in the downtown area of Crescent City, California, over a site classified as class E. The fundamental natural period of the building is 0.4 s and the height of each of its stories is 12 ft. The generator weighs 15 kips and will be mounted on four vibration isolators, each with a lateral stiffness of

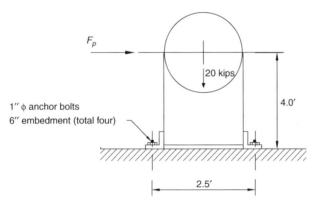

FIGURE P17.16 Steam boiler considered in Problem 17.16.

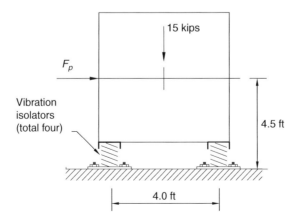

FIGURE P17.17 Electric generator considered in Problem 17.17.

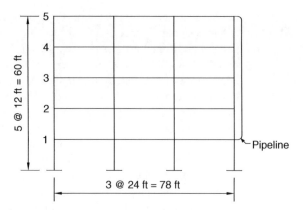

FIGURE P17.18 Building and pipeline considered in Problem 17.18.

3 kips/in. Determine using the provisions of the International Building Code the shear and tension demands on the vibration isolators.

17.18 A vertical steel pipeline is connected to a five-story building as shown in Figure P17.18. The building is located in Long Beach, California, over a deposit of stiff sandy soil and is used as a hospital. It is constructed with ordinary moment-resisting steel frames and carries a total gravity load of 1000 kips per floor. The total weight of the pipeline is 6 kips. Determine using the provisions of the International Building Code the design lateral force and the relative displacement demand for the pipeline. Assume that the moments of inertia of the beams and columns of the building are equal to 1170 in.^{4} and 1530 in.^{4}, respectively. Similarly, consider that the modulus of elasticity of the structural steel is 29,000 kips/in.^{2}

17.19 A three-story office building will be implemented with a base-isolation system. The building will be located on Anchorage, Alaska, on a site soil classified as class D. The building has a regular configuration with no vertical or horizontal irregularities. Its plan dimensions are 150 ft × 100 ft and its weight for seismic design is 6200 kips. The lateral-load-resisting system consists of ordinary reinforced concrete moment frames. The center of mass and the center of rigidity coincide at each floor. The fundamental natural period of the building when its base is considered fixed is 0.35 s. It is estimated that for an effective performance the isolation system should have effective isolated periods at the design and maximum displacements equal to 2.5 and 3.0 times the fixed-base natural period, respectively, and an effective damping ratio of 15%. A variation of ±15% from the mean stiffness values of the isolation elements is considered acceptable. For the purpose of a preliminary design, determine using the provisions of the International Building Code: (a) the minimum design displacement for the isolation system, (b) the base shear for designing the isolation system and the structural elements below the isolation system, and (c) the base shear for the design of the superstructure.

17.20 Repeat Example 17.6 considering instead linear viscous dampers.

17.21 Repeat Example 17.6 using instead the response spectrum procedure prescribed in the provisions of the International Building Code.

Appendix

Catastrophic and Notable Earthquakes

Year	Date	Region	Deaths	Magnitude
342		Antioch, Turkey	40,000	
365	July 21	Crete, Greece	50,000	
551	July 9	Beirut, Lebanon		7.2
684	November 29	Nankaido, Japan		8.4
745	June 5	Mino, Japan		7.9
818	August 10	Tokyo, Japan		7.9
856	December	Corinth, Greece	45,000	
869	July 13	Osju, Japan		8.6
887	August 26	Mino, Japan		8.6
893		India	180,000	
1038	January 9	Shanxi Province, China	23,000	7.3
1057		Chihli, China	25,000	
1068	March 18	Palestine	25,000	
1138	September 8	Aleppo, Syria	100,000	
1202	May 20	Middle East	30,000	
1268		Seyhan Province, Turkey	60,000	
1290	September 27	Chihli, China	100,000	
1293	May 20	Kamakura, Japan	30,000	7.1
1303	September 17	Shanxi Province, China	15,000	8.0
1361	August 3	Kinai, Japan		8.4
1455	December 5	Naples, Italy	40,000	
1498	September 20	Totomi, Japan		8.6
1531	January 26	Lisbon, Portugal	30,000	
1556	January 23	Shanxi Province, China	830,000	8.0
1570	February 8	Concepción, Chile		8.0
1575	December 16	Valdivia, Chile		8.5
1604	November 24	Arica, Chile		8.5
1605	January 31	Shikoku Island, Japan		7.9
1611	December 2	Sendai, Japan		8.1
1626	July 30	Naples, Italy	70,000	
1638	June 11	Plymouth, Massachusetts		
1647	May 13	Santiago, Chile	1,000	8.5
1653	February 23	Izmir, Turkey	15,000	
1657	March 15	Concepción, Chile		8.0
1663	February 5	St. Lawrence River, Canada		7.0
1667	November	Shemaka, Caucasia	80,000	
1668	July 5	Turkey	15,000	
1668	July 25	Shantung Province, China	50,000	8.5
1677	April 13	Tsugaru, Japan		8.1
1692	June 7	Port Royal, Jamaica	2,000	
1693	January 11	Catania, Italy	60,000	
1695	May 18	Shanxi Province, China	30,000	8.0
1703	December 30	Tokyo, Japan	5,200	8.2
1707	October 28	Shikoku Island, Japan	4,900	8.4

(*Continued*)

(Continued)

Year	Date	Region	Deaths	Magnitude
1715	May	Algeirs, Algeria	20,000	
1727	November 18	Tabriz, Iran	77,000	
1730	December 30	Hokkaido Island, Japan	137,000	
1730	July 8	Valparaiso, Chile		8.5
1737	October 11	Calcutta, India	300,000	
1739	January 3	Ningxia Province, China	50,000	8.0
1746	October 28	Lima, Peru	5,000	8.4
1751	May 20	Echigo, Japan	2,000	6.6
1751	May 25	Concepción, Chile		8.5
1755	June 7	Northern Persia	40,000	
1755	November 1	Lisbon, Portugal	70,000	8.6
1755	November 18	Cape Ann, Massachusetts		6.0
1757	August 6	Sicily, Italy	10,000	
1759	October 30	Baalbek, Lebanon	30,000	
1773	July 29	Guatemala	100	
1783	February 4	Calabria, Italy	50,000	
1792	May 21	Kyushu Island, Japan	15,000	
1797	February 4	Quito, Ecuador	41,000	
1811	December 16	New Madrid, Missouri	1	7.5
1812	January 23	New Madrid, Missouri		7.3
1812	February 7	New Madrid, Missouri		7.8
1812	March 25	Caracas, Venezuela	20,000	
1815	November 27	Bali, Indonesia	10,200	
1819	April 3	Copiapó, Chile		8.0
1819	June 16	Kutch, India	1,543	
1822	September 5	Aleppo, Syria	22,000	
1822	November 19	Valparaiso, Chile	72	8.0
1828	December 18	Echigo (Niigata), Japan	30,000	6.9
1835	February 20	Concepción, Chile		8.5
1836	June 10	Hayward, California		8.0
1837		Syria and Palestine	5,000	
1838	June	San Francisco, California		8.2
1843	April 25	Yedo, Japan		8.4
1847	May 8	Zenkoji, Japan	12,000	7.4
1850	September 12	Sichuan Province, China	21,000	7.5
1854	December 23	Simoda, Japan	3,000	8.4
1855	January 23	Wairarapa, New Zealand	5	8.1
1855	November 11	Sagami, Japan	7,000	6.9
1857	January 9	Fort Tejon, California	1	8.3
1857	December 16	Naples, Italy	12,000	6.5
1859	June 2	Erzurum, Turkey	15,000	6.1
1861	March 10	Mendoza, Argentina	8,000	7.5
1863	July 2	Manila, Philippines	10,000	
1868	April 2	Hawaii	46	7.7
1868	August 13	Arica, Chile	25,000	8.5
1868	August 16	Ibarra, Ecuador	70,000	
1868	October 21	Hayward, California	30	7.5
1870	October 20	Montreal, Quebec, Canada		6.5
1872	March 26	Owens Valley, California	27	8.5
1875	February 11	Jalisco, Mexico		7.5
1875	May 16	Cúcuta, Colombia	10,000	

(Continued)

Year	Date	Region	Deaths	Magnitude
1877	May 9	Iquique, Chile		8.0
1883	August 26	Java, Indonesia	36,000	
1883	October 15	Cesme–Urla, Turkey	15,000	
1886	August 31	Charleston, South Carolina	110	7.0
1891	October 28	Mino–Owari Provinces (Nobi), Japan	7,000	7.9
1896	June 15	Riku-Ugo, Japan	22,000	7.5
1897	June 12	Assam, India	1,500	8.7
1897	September 21	Basilan, Philippines		8.7
1897	October 18	Samar, Philippines		8.1
1898	April 29	Nicaragua		7.9
1899	January 29	Oaxaca, Mexico		8.9
1899	September 3	Yakutat Bay, Alaska		7.8
1899	September 10	Yakatut Bay, Alaska		8.6
1900	June 21	Cayman Islands		7.9
1900	October 9	Chugach Mountains, Alaska		8.6
1901	December 13	Andreanof Islands, Alaska		7.8
1901	December 13	Unimak Island, Alaska		7.8
1902	April 19	Quetzaltenango, Guatemala		8.3
1902	September 23	Chiapas, Mexico		7.8
1903	January 4	Tonga		8.0
1903	January 14	Oaxaca, Mexico		8.3
1903	February 27	Java		8.1
1903	June 2	Aleutian Islands, Alaska		8.3
1903	August 11	Thessaly, Greece		8.3
1904	June 7	Japan		7.9
1904	June 25	Kamchatka Peninsula		8.3
1904	August 27	Kolyma, Siberia		8.3
1904	December 20	Costa Rica		8.3
1905	February 14	Andreanof Islands, Alaska		7.9
1905	April 4	Kangra, India	19,000	8.6
1905	July 23	Lake Baikal, Russia		8.7
1905	September 8	Calabria, Italy	557	7.9
1906	January 21	Honshu Island, Japan		8.4
1906	January 31	Colombia–Ecuador		8.9
1906	March 16	Kagi, Taiwan	1,258	7.1
1906	April 18	San Francisco, California	700	8.3
1906	August 16	Valparaiso, Chile	1,500	8.6
1906	August 17	Rat Islands, Alaska		8.3
1906	September 14	New Guinea		8.4
1906	December 22	Sikiang Province, China		8.3
1907	January 16	Kingston, Jamaica	1,000	6.5
1907	April 15	Acapulco, Mexico		8.3
1907	September 2	Near Islands, Alaska		7.8
1907	October 21	Karatag, Tadzhikistan	12,000	8.1
1908	March 26	Guerrero, Mexico		8.1
1908	December 28	Messina, Italy	58,000	7.5
1909	January 23	Iran (west-central)	5,500	7.3
1909	February 22	Fiji		7.9
1909	March 13	Honshu Island, Japan		8.3
1909	July 7	Hindu Kush		8.1

(*Continued*)

(Continued)

Year	Date	Region	Deaths	Magnitude
1909	July 31	Guerrero, Mexico		7.8
1910	April 12	Ryukyu Islands, Japan		8.3
1910	June 16	Loyalty Island, New Caledonia		8.6
1911	January 3	Tyan Shan, Turkistan		8.7
1911	February 18	Ferghana (Pamir), Russia		7.8
1911	June 7	Jalisco, Mexico	45	8.0
1911	June 15	Ryukyu Islands, Japan		8.7
1912	May 23	Burma		8.0
1913	March 14	Molucca Islands		8.3
1914	January 30	San Luis, Argentina		8.2
1914	May 26	New Guinea		8.0
1914	November 24	Mariana Islands		8.7
1915	January 13	Avezzano, Italy	32,600	7.5
1915	May 1	Kamchatka Peninsula, Russia		8.1
1915	October 2	Pleasant Valley, Nevada		7.6
1916	January 13	New Guinea		8.1
1917	January 21	Bali, Indonesia	15,000	
1917	May 1	Kamchatka Peninsula, Russia		8.1
1917	June 26	Samoa Islands		8.7
1918	February 13	Guangdong Province, China	10,000	7.3
1918	August 15	Carolina Islands		8.3
1918	September 7	Kuril Islands		8.3
1918	October 11	Puerto Rico	116	7.5
1919	January 1	Tonga		8.3
1919	April 17	Guatemala		7.0
1919	April 30	Tonga		8.4
1920	January 3	Veracruz, Mexico	648	7.8
1920	June 5	Taiwan		8.3
1920	June 5	Fiji		8.3
1920	December 16	Ningxia Province, China	200,000	8.6
1921	November 15	Hindu Kush		8.1
1921	December 18	Peru		7.9
1922	March 10	Cholame Valley, California		6.5
1922	November 11	Atacama, Chile		8.4
1923	February 3	Kamchatka Peninsula, Russia		8.4
1923	March 24	Sichuan Province, China	5,000	7.3
1923	May 26	Iran (northeastern)	2,200	5.5
1923	September 1	Kanto (Tokyo–Yokohama), Japan	143,000	8.3
1924	April 14	Mindanao, Philippines		8.3
1925	February 28	St. Lawrence River, Canada		7.0
1925	March 16	Yunnan Province, China		7.1
1925	June 27	Manhattan, Montana		6.7
1925	June 29	Santa Barbara, California	13	6.3
1926	June 26	Rhodes, Dodecanese Islands		8.3
1927	March 7	Tango Peninsula, Japan	3,020	7.8
1927	May 23	Kansu Province, China	41,000	8.3
1927	November 4	Santa Barbara, California		7.5
1928	June 17	Oaxaca, Mexico		7.9
1928	December 1	Talca, Chile		8.3
1929	March 7	Fox Islands, Alaska		8.6
1929	May 1	Shirwan, Iran	5,800	7.2

Appendix

(Continued)

Year	Date	Region	Deaths	Magnitude
1929	June 16	Buller, New Zealand	17	7.8
1929	June 27	South Sandwich Islands		8.3
1929	October 5	Hawaii		6.5
1930	May 7	Iran (northwestern)	2,510	7.2
1930	July 23	Campania, Italy	1,425	6.5
1931	January 15	Oaxaca, Mexico		7.9
1931	February 2	Hawkes Bay, New Zealand	256	7.9
1931	October 3	Solomon Islands		8.1
1932	May 14	Celebes, Indonesia		8.3
1932	May 26	Tonga		7.9
1932	June 18	Jalisco, Mexico		7.9
1932	June 22	Colima, Mexico		7.9
1932	December 20	Cedar Mountain, Nevada		7.3
1932	December 26	Kansu, China	70,000	7.6
1933	March 3	Sanriku Coast, Japan	3,000	8.9
1933	March 10	Long Beach, California	120	6.3
1934	January 15	Bihar, India	10,700	8.4
1934	March 12	Kosmo, Utah	2	6.6
1934	July 18	Santa Cruz Islands		8.1
1934	December 31	Baja California, Mexico		7.1
1935	April 21	Shinchiku, Taiwan	3,276	7.1
1935	June 1	Quetta, Pakistan	25,000	7.6
1935	August 7	Pasto, Colombia		7.5
1935	October 18	Helena, Montana	4	6.2
1935	December 28	Sumatra		8.1
1937	April 16	Tonga		8.1
1937	August 20	Luzon, Philippines		7.5
1938	February 1	Java		8.6
1938	November 10	Alaska		8.3
1939	January 24	Chillán, Chile	28,000	8.3
1939	April 30	Solomon Islands		8.1
1939	April 30	Celebes, Indonesia		8.6
1939	December 27	Erzincan, Turkey	32,700	8.0
1940	May 18	El Centro, California	9	7.1
1940	May 24	Peru (central)		8.4
1940	July 14	Rat Islands, Alaska		7.9
1940	November 10	Bucharest, Romania	1,000	7.4
1941	June 26	Burma		8.7
1941	June 27	Australia (central)		6.8
1941	November 25	Portugal		8.4
1942	May 14	Ecuador		8.3
1942	August 6	Guatemala		8.3
1942	August 24	Nazca, Peru		8.6
1943	February 22	Guerrero, Mexico		7.5
1943	April 6	Illapel, Chile	11	8.3
1943	May 25	Philippines		8.1
1943	July 23	Java		8.1
1943	September 6	Macquarie Islands		7.8
1943	September 10	Tottori, Japan	1,190	7.4
1943	November 26	Turkey (central)	4,000	7.6

(*Continued*)

(Continued)

Year	Date	Region	Deaths	Magnitude
1944	January 15	San Juan, Argentina	5,000	7.8
1944	December 7	Tonankai, Japan	1,000	8.3
1945	January 13	Aichi Prefecture, Japan	1,960	7.1
1945	November 28	Pakistan	4,100	8.3
1946	April 1	Aleutian Islands, Alaska	173	7.5
1946	August 4	Santo Domingo, Dominican Republic		8.1
1946	November 10	Ancash, Peru	1,400	7.4
1946	December 21	Nankaido, Japan	1,360	8.4
1948	January 24	Panay, Philippines	20	8.3
1948	June 28	Fukui Prefecture, Japan	5,400	7.3
1948	October 6	Ashkhabad, Turkmenistan	19,800	7.5
1948	December 4	Islas Marias, Mexico		7.2
1949	April 13	Olympia, Washington	8	7.3
1949	July 10	Tadzhikstan		8.0
1949	August 5	Pelileo, Ecuador	6,000	6.8
1949	August 22	Queen Charlotte Island, Alaska		8.1
1950	February 28	Hokkaido Island, Japan		7.9
1950	August 15	Assam, India	1,530	8.7
1950	December 9	Andes, Argentina		8.3
1950	December 12	Vanuatu		8.1
1951	August 21	Hawaii		6.8
1952	March 4	Tokachi-Oki (Hokkaido), Japan	31	8.3
1952	July 21	Kern County, California	12	7.7
1952	November 4	Kamchatka Peninsula, Russia		8.4
1953	March 18	Turkey (northwestern)	1,100	7.5
1953	November 25	Honshu Island, Japan		8.0
1954	September 9	El-Asnam, Algeria	1,240	6.8
1954	December 16	Fairview Peak, Nevada		7.1
1954	December 21	Eureka, California	1	6.6
1955	February 18	Quetta, Pakistan		6.8
1955	April 1	Lanao, Philippines	400	7.6
1956	February 9	Baja California (San Miguel), Mexico		6.8
1956	June 10	Kabul, Afghanistan	2,000	7.7
1956	July 9	Santorin, Greece		7.7
1957	March 9	Aleutian Islands, Alaska		8.3
1957	July 2	Abegharm, Iran	2,500	7.4
1957	July 28	Guerrero, Mexico	54	7.5
1957	December 4	East Altai, Mongolia	20	8.3
1957	December 13	Kangavar, Iran	2,000	7.1
1958	January 19	Colombia–Ecuador		7.8
1958	April 7	Alaska (central)		7.3
1958	July 9	Lituya Bay, Alaska	5	7.9
1958	November 6	Kuril Islands		8.7
1959	May 4	Kamchatka Peninsula, Russia		8.2
1959	August 17	Hebgen Lake, Montana	28	7.7
1960	February 29	Agadir, Morocco	12,000	5.7
1960	May 22	Valdivia, Chile	2,230	8.3
1961	August 31	Peru–Brazil		7.8
1962	September 1	Quazin, Iran	12,230	7.3
1963	July 26	Skopje, Yugoslavia	1,070	6.0
1963	October 13	Kuril Islands		8.2

(Continued)

Year	Date	Region	Deaths	Magnitude
1964	March 28	Prince William Sound, Alaska	131	8.4
1964	June 16	Niigata, Japan	26	7.5
1965	February 4	Rat Island, Alaska		7.9
1965	March 28	La Ligua, Chile		7.5
1965	March 30	Amchitka Island, Alaska		7.5
1965	April 29	Puget Sound, Washington	6	6.6
1965	May 3	San Salvador, El Salvador	120	6.2
1966	March 12	Taiwan (northeastern)		7.5
1967	July 29	Caracas, Venezuela	266	6.5
1967	December 11	Koyna Dam, India	177	6.4
1968	January 14	Sicily, Italy	740	6.1
1968	May 16	Tokachi–Oki, Japan	48	8.6
1968	August 1	Casiguran, Philippines		7.7
1968	August 31	Khorasan, Iran	12,100	7.3
1968	October 14	Meckering, Australia		7.2
1969	July 26	Guangdong Province, China	3,000	5.9
1970	March 28	Gediz, Turkey	1,100	7.3
1970	April 7	Luzon, Philippines		7.7
1970	May 31	Chimbote, Peru	67,000	7.8
1971	January 10	Iran (western)		8.0
1971	February 9	San Fernando, California	65	6.6
1971	July 9	Illapel, Chile		7.7
1971	July 14	Solomon Islands		8.1
1972	January 25	Taiwan		7.5
1972	April 10	Ghir, Iran	5,400	7.1
1972	July 30	Sitka, Alaska		7.6
1972	December 23	Managua, Nicaragua	5,000	6.2
1973	January 30	Michoacán, Mexico	56	7.5
1973	April 26	Hawaii		6.2
1973	June 17	Hokkaido, Japan		7.7
1973	August 28	Oaxaca, Mexico	530	7.2
1974	May 11	Yunnan Province, China	20,000	7.1
1974	October 3	Lima, Peru	78	7.6
1974	December 28	Kashmir, Pakistan	5,300	6.2
1975	February 11	Guatemala		7.5
1975	February 14	Haicheng, Liaoning Province, China	1,300	7.4
1975	June 30	Yellowstone, Wyoming		6.4
1975	July 11	Solomon Islands		7.9
1975	September 6	Lice, Turkey	2,370	6.7
1975	November 29	Hawaii	2	7.2
1976	January 14	Kermadec Islands		8.0
1976	February 4	Guatemala City, Guatemala	23,000	7.5
1976	May 6	Friuli (Gemona), Italy	965	6.5
1976	June 26	New Guinea (west)	6,000	7.1
1976	July 14	Bali, Indonesia	560	6.5
1976	July 27	Tangshan, Hebei Province, China	243,000	7.8
1976	August 17	Moro Gulf, Philippines	6,500	8.0
1976	November 24	Turkey (eastern)	5,000	7.3
1977	March 4	Bucharest, Romania	1,570	7.2
1977	August 19	Sumbawa, Indonesia	100	8.0

(Continued)

(Continued)

Year	Date	Region	Deaths	Magnitude
1977	November 23	San Juan, Argentina	70	7.4
1978	June 12	Sendai, Japan	27	7.5
1978	September 16	Tabas, Iran	15,000	7.7
1978	November 29	Oaxaca, Mexico	8	7.8
1979	April 15	Montenegro, Yugoslavia	56	7.0
1979	September 12	New Guinea (west)	100	8.1
1979	October 15	Imperial Valley, California		6.7
1979	December 12	Colombia (southwestern)	600	7.7
1980	October 10	El-Ansam, Algeria	3,500	7.7
1980	November 8	Eureka, California		7.4
1980	November 23	Italy (southern)	3,100	7.2
1981	June 11	Iran (southeastern)	3,000	6.9
1981	July 28	Iran (southeastern)	2,500	7.3
1982	June 7	Guerrero, Mexico	9	7.2
1982	June 19	El Salvador	16	7.0
1982	December 13	Yemen (north)	2,800	6.0
1982	December 16	Afghanistan	510	6.8
1983	March 31	Popayán, Colombia	350	5.5
1983	May 2	Coalinga, California	1	6.5
1983	May 25	Honshu Island, Japan	106	7.7
1983	October 28	Idaho	2	6.9
1983	October 30	Turkey (eastern)	2,000	6.9
1985	March 3	Valparaíso, Chile	147	7.8
1985	September 19	Michoacán, Mexico	9,500	8.1
1986	January 31	Painesville, Ohio	0	5.0
1986	July 8	North Palm Springs, California	0	5.9
1986	October 10	San Salvador, El Salvador	1,000	5.4
1987	March 6	Napo Province, Ecuador	1,000	6.9
1987	October 1	Whittier Narrows, California	3	5.3
1988	August 20	Nepal–India	1,450	6.6
1988	November 6	Yunnan Province, China	730	7.6
1988	November 25	Quebec, Canada	0	6.0
1988	December 7	Spitak–Leninakan, Armenia	25,000	7.0
1989	May 23	Macquaire Islands	0	8.2
1989	October 17	Loma Prieta, California	67	7.1
1989	December 28	Newcastle, Australia	13	5.6
1990	June 20	Manjil, Iran	14,000	7.7
1990	July 16	Luzon, Philippines	1,700	7.8
1990	June 20	Iran (northern)	40,000	7.7
1991	April 22	Costa Rica, Limon	53	7.5
1991	October 20	Uttarkashi, India	768	6.6
1992	March 13	Erzincan, Turkey	800	6.8
1992	April 25	Petrolia, California	0	7.1
1992	June 28	Landers, California	1	7.5
1993	August 8	Mariana Islands, Guam	0	8.1
1993	July 12	Hokkaido–Nansei–Oki, Japan	231	7.8
1993	September 20	Klamath Falls, Oregon	2	5.7
1994	January 17	Northridge, California	57	6.6
1995	January 17	Hyogo-Ken Nambu (Kobe), Japan	5,090	6.8

(Continued)

Year	Date	Region	Deaths	Magnitude
1995	October 9	Colima (Manzanillo), Mexico	40	7.6
1997	May 10	Qaen, Iran	2,400	7.1
1999	August 17	Kocaeli, Turkey	17,000	7.8
1999	September 21	Chi-Chi, Taiwan	2400	7.6
1999	October 16	Hector Mine, California	0	7.1
1999	November 12	Duzce, Turkey	800	7.2
2000	June 4	Sumatra, Indonesia	120	7.9
2001	January 13	El Salvador	700	7.6
2001	January 26	Ahmedabad, India	18,600	7.9
2001	June 23	Arequipa, Peru	102	8.4
2002	January 21	Colima (Tecomán), Mexico	21	7.6
2002	November 3	Denali, Alaska	0	7.9
2003	December 26	Bam, Iran	26,500	6.6
2004	December 26	Indonesia, Sumatra	187,000	9.1
2005	October 8	Kashmir, Indo-Pakistan	86,000	7.6
2006	November 15	Kuril Islands	0	8.3
2007	April 2	Solomon Islands	40	8.1
2007	August 15	Pisco, Peru	540	8.0
2007	September 12	Sumatra, Indonesia	25	8.4
2008	May 12	Wenchuan, China	69,170	8.0

Source: Gere, J. M. and Shah, H. C., *Terra Non Firma*, W. H. Freeman and Co., New York, 1984; Bolt, B. A., *Earthquakes*, 5th Edition, W. H. Freeman and Co., New York, 2004; Eiby, G. A., *Earthquakes*, Van Nostrand Reinhold Co., New York, 1980; Lomnitz, C., *Global Tectonics and Earthquake Risk*, Elsevier, Amsterdam, 1974; Lay, T. and Wallace, T. C., *Modern Global Seismology*, Academic Press, San Diego, CA, 1995; Earthquake Engineering Research Institute.

Index

A

Absolute sum of the maxima rule, 391
Absorbing or transmitting boundaries, 640
Acceleration, zero-period, 354
Acceleration method, linear, 489–91, 493, 495–7, 503, 511
Acceleration response spectrum, 185
Accelerograms, 159
Accelerographs, 154–6, 159
    strong motion, 154
Acceptable probability of exceedance, 268
Accidental torsional moment, 844
Accuracy, 510
Acrylic polymers, 764
Active control systems, 26
ADAS element, 778
Added viscous damping, 903–4
Aftershocks, 53–4
Aggarwal, Y., 57
Alaska 1964 earthquake, 3
Aleutian Islands, Alaska, 1946 earthquake, 11
Aliasing, 217
Allowable analysis procedures, 837
Allowable seismic-force resisting systems, 816
Alpine Fault, 44
Alpine-Himalayan Belt, 31
Amplification and attenuation effects, 822
Amplification factors, 551, 847
Amplification function, 320–2, 327
Amplitude decay, 511
Amplitude Fourier spectrum, 153, 211–2, 214, 217, 223
Analysis of asymmetric structures, 549
Analysis of damper-added structures, 785
Analysis of linear systems, 513
Analysis of structures with added dampers, 785
Analysis procedures, 835
Aoyama, H., 413
Apparent wave, 84, 86
Appendages, 649
Applied Technology Council (ATC), 565
Approximate fundamental natural period, 546
Approximate natural frequencies and mode shapes of combined systems, 674
Approximate rules to estimate maximum system response, 391
Araya, R., 228
Araya-Saragoni intensity, 228
Architectural, mechanical, and electrical elements, 649
Architectural components, 649
Architectural considerations, 812
Arias, A., 226
Arias intensity, 226, 228
Armenian 1988 earthquake, 49
Arnold, R. N., 585
Artificial damping, 497, 503
Arya, A. S., 693
ASCE Standard, 7, 816

Asthenosphere, 40, 42–3
Asynchronous motion, 449
Attenuation laws, 246
Attenuation relationships, 246–7, 251, 259
Average P-wave velocity, 141–2
Average response spectra, 353–4
Average slip rate in some known faults, 54
Awojobi, A. D., 590

B

b value, 261
Baranov, V. A., 605
Base-isolated buildings, 694, 721, 772
Base-isolated structures, 737, 748, 880
Base Isolation, 689, 691, 693
Base-isolation system, 691, 721, 747–8
Base shear, 729, 841
    modal, 543–4, 851, 858
Base shear coefficient, 718
Base shear-lateral displacement relationship, 565
Base shear-roof displacement curves, 566
Baseline correction, 162
Bay mud, 297
Bechtold, J., 691
Bedrock flexibility, 325
Benicia-Martinez bridge, 706
Benioff, H., 170, 183
Benioff zone, 43
Bernoullian orifices, 772
Bessel functions, 607
Biot, M. A., 183
Black, W. L., 336
BOCA Basic Building Code, 815
Body waves, 98, 100, 115
Bolt, B. A., 182
Boore, D. M., 303
Boore, Joyner, and Fumal equations, 247
Borah Peak, Idaho, 1983 earthquake, 241
Borodachev, N. M., 590
Borrego Mountain, California, 1968 earthquake, 240
Boundaries
    artificial, 638–41
    consistent, 640–1
    relaxed, 590
Bounded Gutenberg-Richter recurrence law, 262
Bounded near field, 640
Boussinesq, J., 575
Boussinesq's solution, 577, 583
Box 1.1 Soil Liquefaction, 8
Box 1.2 Tsunamis, 12
Box 1.3 Indian Ocean Tsunami of December, 26, 2004, 13
Box 2.1 The Global Seismographic Network, 29
Box 3.1 Earthquakes Triggered by Filling of Reservoirs and Wells, 38
Box 3.2 Successful Earthquake Predictions, 57

Box 5.1 Modified Mercalli Intensity (MMI) Scale, 130
Box 5.2 Japanese Seismic Intensity Scale, 130
Box 5.3 Medvedev-Sponheuer-Karnik Intensity (MKS) Scale, 131
Box 6.1 Algorithm to Determine Response Spectrum by Nigam-Jennings Approach, 196
Box 6.2 Algorithm to Determine Nonlinear Response Spectrum, 205
Box 7.1 Procedure to Fit a Type II Extreme-Value Probability Distribution to Given Peak Ground Acceleration Data, 272
Box 7.2 Procedure to Generate Frequency-Intensity Curves for a Given Site Using Semiprobabilistic Approach, 272
Box 7.3 Procedure to Generate Frequency-Intensity Curves for a Given Site Using Probabilistic Approach, 280
Box 9.1 Procedure to Construct Design Spectrum Using Peak Ground Acceleration and Response Spectrum Shape, 358
Box 9.2 Procedure to Construct Elastic Design Spectrum Using Newmark-Hall Approach, 362
Box, 9.3 Procedure to Construct Inelastic Design Spectrum Using Newmark-Hall Approach, 367
Box 10.1 Response Spectrum Method—Summary, 416
Box 11.1 Solution Algorithm Using Central Difference Method, 473
Box 11.2 Solution Algorithm Using Houbolt Method, 479
Box 11.3 Solution Algorithm Using Constant Average Acceleration Method, 486
Box 11.4 Solution Algorithm Using Linear Acceleration Method, 492
Box 11.5 Solution Algorithm for Nonlinear Systems by Central Difference Method, 517
Box 11.6 Solution Algorithm for Nonlinear Systems by Newmark-Beta Method, 520
Box 11.7 Algorithm to Improve Solution Estimate at Time Step Using Modified Newton-Raphson Method, 533
Box 13.1 Equivalent Single-Degree-of-Freedom System Method, 635
Box 14.1 Procedure to Determine Natural Frequencies and Mode Shapes of Combined Primary-Secondary System, 671
Box 14.2 Approximate Procedure to Determine Natural Frequencies and Mode Shapes of Combined Primary-Secondary System, 677
Brace, W., 56
Brady. A. G., 182
Buckling-restrained braces, 782–3
Building attachments, 649
Building contents, 649, 700
Building period, effective, 864
Buildings, asymmetric, 547–8, 564–5
Bycroft, G. N., 585, 604

C

Calantarients, J. A., 691
Campbell equations, 253
Caughey damping, 410–1
Cause of plate movement, 45
Cecchi, F., 134
Center of twist, 548
Centers of mass, 547
Central Chile 1971 earthquake, 201
Central difference method, 470, 472–3, 479, 509, 511, 516–7
Central frequency, 223
Cerruti, V., 575
Cerruti's solution, 578
Characteristic earthquake recurrence law, 263
Chi-Chi, Taiwan, 1999 earthquake, 17–8, 167, 176, 758
Chiba Experimental Station, 166
Chile 1985 earthquake, 299
Chilean 1960 earthquake, 53, 150, 154
Chimbote 1970 earthquake, 5
Circum-Pacific seismic belt, 31
CLASSI, 645
Classically damped systems, 785
Closely spaced natural frequencies, 393–5, 399, 654, 681, 737
Closest distance to surface projection of rupture zone, 247
Clough-Penzien spectral density function, 225
Coefficients
    redundancy, 820, 846
    response modification, 819
    viscosity, 622
Collapse mechanism, 567
Combination of load effects, 907
Combined isolation-structure system, 732
Combined primary-secondary system, 666, 670–1, 674–5, 677, 680
Complex impedance ratio, 326
Complex modal superposition, 785
Complex participation factor, 735
Compliance function, 611
Component amplification factor, 873
Component importance factor, 873
Component response modification factor, 873
Components
    dynamic, 451
    pseudostatic, 451
    steady-state, 755
    transient, 755
Computation of nonlinear response spectrum, 202
Computer programs, 645
Concordia University Library, Montreal, Canada, 763
Condition for numerical stability, 473, 493
Conditionally stable, 470, 493, 504
Constant average acceleration method, 482, 484–5, 509–11
Constrained modulus of elasticity, 64
Continental drift theory, 40
Contour map, 822
Convection current theory, 45
Convergence criterion, 534
Cooley, J. W., 218
Coordinates, normal, 383
Coppersmith, K. J., 263
Core, inner, 39
Correspondence principle, 624
Cost, 812
Couple rocking and horizontal vibrations, 598
Cross-correlated responses, 440
Cross impedance function, 614
Crust, 39

D

Damaging earthquakes in United States, 32
Damped natural frequency, 386
Dampers
  fluid, 771, 776, 787, 802
  fluid inertial, 771
  hysteretic, 26, 760
  lead-extrusion, 777
  metallic, 777–8, 785
  supplemental, 753
  triangular steel plate, 778–80
  tuned mass, 26
Damping
  geometric, 122
  internal, 123, 321, 621, 623, 640
  material, 123, 624
  numerical, 510–11
  radiation, 122, 621, 627
  viscous, 905
Damping constants, generalized, 385
Damping devices, 891, 912
Damping Evaluation Committee of Architectural Institute of Japan, 413
Damping forces, 379
Damping matrix of the Rayleigh type, 409
Damping ratios, 334
  equivalent, 333, 394, 637, 746–7, 762, 767
Damping ratios of combined system, 680
Damping system, 891
Davis, L. L., 303
de Rossi, M. S., 129
Dead Sea Fault, 44
Deconvolution, 638
Deflection amplification factor, 819, 847, 852
Deformation spectrum, 200–1
Deformations, panel zone, 840
Degrees of freedom of superstructure, 450
Degrees of freedom of supports, 450
Delfosse, G. C., 693
deMontalk, W. R., 691
Der Kiureghian, A., 395, 456
Design displacement, 881
  total, 882
Design forces in damping system, 899, 901
Design response spectrum, 353, 821–2, 831, 853–4
Design spectral acceleration at 1 s period, 842
Design spectral acceleration for short periods, 842
Design spectral accelerations, 831, 840, 842
Design spectrum, 353
  elastic, 354
  nonlinear, 364
  target, 228–30
Design story drifts, 846–7, 901
Design story velocities, 898
Design values, 852, 872
Determination of modal damping ratios, 407
Determination of shear modulus and damping ratio, 334
Devices, energy-absorbing, 753
Difference equation, 505
Dilatation, 89
Dimensionless frequency, 586
Direction of earthquake loads, 821
Dispersion, 84, 510
Displacement amplification factor, 846
Displacement response spectrum, 185
Displacement vectors, incremental, 516, 519
Displacements
  absolute, 381
  design floor, 901
Displacements and rotations under dynamic loads, 583
Displacements and rotations under static distributed loads, 575
Displacements under static point loads, 574
Distance
  closest, 247–8, 259
  shortest, 253
Distribution exponent, 844
Distribution of horizontal shear, 844–5, 852
Distribution of lateral forces with height, 565
Distribution of story shears, 564
Distribution of the equivalent lateral forces, 568
Dobry, R., 342
Double-couple source model, 152
Double-sum combination rule, 391, 393–4
Ductility, 200, 232, 234, 465, 655
Ductility demand
  effective, 906
  global, 892
Ductility factor, 199–201, 205, 356, 371
Duhamel's integral, 386, 734
Duration, 395
  bracketed, 182
Durometer hardness, 698
DYNA4, 645
Dynamic analysis of a base-isolated structure, 885
Dynamic Boussinesq problem, 583
Dynamic flexibility function, 611
Dynamic magnification factor, 565
Dynamic magnification of floor rotations, 549
Dynamic stiffness function, 611

E

Earth structure, 39
Earthquake epicenter, 133, 142, 144, 146, 148, 154
Earthquake faults, 2, 46–7, 54, 60, 151, 239
Earthquake focus, 141
Earthquake ground motion intensity, 822
Earthquake hazard, 25
Earthquake hypocenter, 140–1, 144
Earthquake magnitude, 146–8, 170, 246, 277
Earthquake precursors, 59–60
Earthquake predictions, 55–7, 59, 61
Earthquake recurrence, 133
Earthquake size, 23, 53, 129, 148, 151, 153
Earthquake sources, 108
Earthquakes
  characteristic, 263, 822
  intraplate, 46, 50
  intraslab, 258–9
  maximum considered, 822
  subduction-zone, 257, 259
  zero-magnitude, 146–7
Earthquakes triggered by filling of reservoirs and wells, 38
East Pacific Rise, 42
Eccentricity, accidental, 845

Effect of cracked sections, 840
Effect of differential support excitations, 450
Effect of structural and nonstructural nonlinear
    behavior, 683
Effect of topographic conditions, 303
Effective damping, 892, 903–4
Effective damping ratio, 698, 746, 866, 871, 900–1, 905
Effective force vector, 453
Effective load vector, 472, 478, 485, 498, 503, 517, 788
Effective natural periods, 881
Effective period at design displacement, 881
Effective period at maximum displacement, 881
Effective period of isolation system, 881
Effective stiffness, 698, 746, 837, 881, 887, 892
Effective stiffness and effective damping of isolation
    system, 887
Effective stiffness matrix, 478, 485, 498, 503, 532–3
Effective weight, 544–5
    modal, 851
El Centro, California
    1940 earthquake, 215, 224, 360–1, 365–6, 400
    1979 earthquake, 21, 48
El Centro array, 166
Elastic design spectra, 360, 362
Elastic halfspace, 574, 604, 612, 624
Elastic halfspace model, 574
Elastic rebound theory, 49
Elasticity, shear modulus of, 334
Elastoplastic moment-rotation behavior, 521
Electricité-de-France Bearing, 700
Empirical rules to estimate fundamental natural period, 547
Energy dissipating devices, 760
Energy dissipation, 772
Energy dissipation devices, 753–4, 760, 810
Environmental effects, 811
Epicenter, 5, 14, 53, 62, 133–4, 141–2, 144–5, 147, 159,
    167, 171, 239, 245
Epicentral distances, 142, 144, 147–9, 247
Equivalent ground motion duration, 395
Equivalent lateral force procedure, 541, 546, 564, 837,
    841, 863, 883, 894
Equivalent lateral forces, 542, 544–6, 549, 567–8, 840,
    844, 846, 852, 877
Equivalent lateral forces for nonstructural components, 877
Equivalent lateral load pattern, 566
Equivalent linear model, 746
Equivalent linear stiffness, 746
Equivalent linearization technique, 333
Equivalent radius, 591
Equivalent single-degree-of-freedom system method, 627
Estimation of fundamental natural period, 842
Euclidean norm, 506, 534
Eureka, California, 1954 earthquake, 191
Eurocode, 8, 816
Evaluation of site effects using analytical techniques, 306
Evaluation of site effects using statistical correlations, 304
Ewing, J., 135
Exceedance probability curves, 267
Excitation frequency, dimensionless, 616
Experimental verification, 645
Explicit method, 470
Exponential Fourier series, 208
Exposure time, 267
External forces, lateral, 542

F

Fatigue failure, 776
    premature, 785
Fault, 53
    inactive, 47
    transform, 43–5
Fault displacement, 53–4, 175, 243
Fault plane, 48, 53, 151, 170–1
Fault rupture, 53–4, 149, 153, 170–1, 251
Fault-rupture zones, 253
Fault slip, 53, 153
Fault zone, 60, 152
Fault's scarp, 49
FEMA (Federal Emergency Management Agency), 565, 816
Fermat's principle, 116–7
Field Act, 27
Fires, 10, 16
Fitzgerald, T. F., 760
Flexible boundary method, 643
Flexible components, 878
Flexible volume method, 643
Flexural beam, 776
Fling, 166
Floor diaphragms, 840
Floor displacements, 846
Floor response spectra, 656, 659, 661
Floor response spectrum method, 656, 658–9, 663
Floor rotations, horizontal, 547
Floor rotations in asymmetric structures, 549
Floors' centers of mass, 549
Floors' rotational inertia, 565
Flow failures, 10
Fluid damper for seismic applications, 772
Fluid dampers, 771–2, 775–6, 785, 787, 803
    inertial, 771
    viscous, 760
Fluid flow through orifices, 771
FLUSH, 645
Focal depth, 53, 144, 149, 259
Focus, 53
Folding frequency, 217
Foothill zone, 288
Force-deformation behavior, 783
Force-deformation loops, stable, 760
Forel, F., 129
Foreshocks, 54
Foss, K. A., 732
Foundation-halfspace system, 591–2, 600
Foundation vibration by method of impedances, 610
Foundations, massless, 611
Foundations on viscoelastic halfspace, 624
Fourier, 79, 207–8, 215, 319
Fourier-Bessel integrals, 583
Fourier series, 207–8, 319
Fourier series decomposition, 79
Fourier spectrum, 181, 207, 210–1, 215–7
Fourier transform, 210
    discrete, 216–20, 222
    fast, 217–8, 220, 222, 319
Frequency
    dimensionless, 593, 606–7, 624
    dominant, 321
Frequency analysis, 211

Frequency content, 211, 215, 223, 226
Frequency domain, 210–11, 319
Frequency-domain approach, 319
Frequency factor, 586
Frequency-intensity curves, 268–72
Frequency of earthquake occurrence, 32, 34, 37
Frequency of occurrence of earthquakes in United States, 34
Frequency of occurrence of worldwide earthquakes, 33
Frequency ratio, 756
Frequency response functions, 295
Friction dampers, 26, 760, 762–5, 785
Friction or yielding steel damper, 788, 804
Friction pendulum, 700, 702–5
Friction pendulum bearing, 700, 702–5
Friction pendulum system, 694
Function
    compliance, 614
    potential, 92

G

Galitzin, B., 135
Gapec seismic isolation system, 693
Gastaldi scale, 129
Gazetas, G., 616
Generalized coordinates, 383
Geological evidence, 239
GERB system, 705
Global collapse, 566
Global seismographic network, 29
Goodman, L. E., 393
Great Hanshin, Japan, 1995 earthquake, 650
Greenback, L. R., 777
Grigorian, C. E., 760
Ground failures, 1, 2, 10, 287
Ground modification techniques, 10
Ground motion duration, 214–5, 226, 395
Ground motion intensity maps, 821–2, 830–1, 889
Ground motion spatial variability model, 456
Ground motions
    free-field, 571
        multicomponent, 431, 437, 441
        near-field, 757
Ground response spectrum, 663
Ground shaking, 16
Ground subsidence, 2, 10, 24
Group velocity, 86
Guatemala 1976 earthquake, 18
Gutenberg, B., 148, 261, 295
Gutenberg-Richter law, 261, 263
Gutenberg-Richter recurrence Law, 261

H

Haicheng 1975 earthquake, 56
Hall, J. R., 595–6
Hall, W. J., 360, 415
Hall analog, 596, 600
Hammurabi, king of Babylon, 815
Hanging nonstructural elements, 655
Hanks, T. C., 153
Hardin, B. O., 336

Harmonic components, 84, 86, 109, 208, 211, 214, 217
Harmonic waves, 78–83, 102, 120
Hart, G. C., 412
Haviland, R., 414
Hebgen Lake, Montana, 1959 earthquake, 2
Heine, A. J., 776
Helical steel springs, 705
High-damping bearings, 698
High-damping rubber bearings, 695, 698–9
Hill site, 297
Historical accounts, 239
Historical records, 239
Hokkaido-Nansei-Oki, Japan, 1993 earthquake, 4, 7, 11
Homogeneous damped soil on flexible rock, 325
Homogeneous damped soil on rigid rock, 321
Homogeneous undamped soil on rigid rock, 319
Honeycomb dampers, 780
Honeycomb steel dampers, 780
Horizontal displacement under center of rigid rectangular
    footing, 581
Houbolt method, 476, 479
Housner, G. W., 225, 353
Housner spectral intensity, 225
Hsieh, T. K., 591
Hsieh's equations, 591
Hypocentral distances, 141, 144, 246–7, 259
Hysteresis-loop adjustment factor, 905
Hysteresis loops, stable, 783
Hysteretic damping, 198, 621, 661, 758, 760, 903–5
Hysteretic loops, 197

I

Idriss, I. M., 336
Impedance matrix, 613
Impedances
    lateral, 614
    longitudinal, 614
    vertical, 614
Imperial Valley, California, 1979 earthquake, 164, 172, 190, 302
Implementation issues, 747, 810
Implicit methods, 470
Importance factors, 835
Incident wave, 68, 70, 73, 115–6, 118, 120, 330
Incremental equation of motion, 513, 515–6, 518–9
Inelastic design spectrum, 355, 358
Inertial interaction, 573–4
Influence matrix, 452
Influence vector associated with support displacement, 453
Influence vectors, 381, 433
Inherent damping, 753, 903–5
Integration procedures, step-by-step, 469
Intensity scales, 129, 132–4, 146
Interaction
    dynamic, 658–9
    inertial, 572
    kinematic, 572
Interface, soil-rock, 325–6
Internal damping ratio, 621
Internal forces, 388, 547
International Building Code, 305, 565, 816, 891
Inverse Fourier transform, 210

Irregularities
    horizontal structural, 836
    vertical structural, 836
Ishibashi, I., 343
Isolation bearings, 695
Isolation gap, 748
Isoseismal map, 133
Isoseisms, 133

J

Jeary, A. P., 413
Jennings, P. C., 193

K

Kanai-Tajimi spectral density function, 224–5, 396
Kanamori, H., 153
Kasai, K., 399
Katayama, T., 415
Kawai, K., 691
Keightley, W. O., 760
Kelly, J. M., 776
Kelvin-Voight solid, 621–3
Kelvin-Voight model, 123
Kern County, California, 1952 earthquake, 662
Kobe, Japan, 1995 earthquake, 8, 16, 19–20, 159
Kondner, R. L., 335
Koto, B., 49

L

Lake zone, 288
Lamb, H., 583
Lame's constants, 89
Laminated rubber bearing, 695
Landers, California, 1992 earthquake, 171, 175
Landslides, 2, 4–5, 38, 131
Laplacian operator, 91
Largest earthquakes in the world, 33
Largest earthquakes in United States, 34
Lateral displacements, modal, 872
Lateral displacements in structures founded on flexible soil, 868
Lateral forces, modal, 544, 851
Lateral spreading, 10
Lateral strength of nonstructural element, 683
Layered damped soil on flexible rock, 327
Lead-rubber bearings, 695, 697–8, 746
Left lateral strike-slip fault, 48
Linear multi-degree-of-freedom model, 721
Linear multi-degree-of-freedom model with nonclassical damping, 732
Linear two-degree-of-freedom model, 706
Liquefaction, 2, 24
Lithosphere, 40, 43
Load combinations, 819
Load-deformation curve, 198
Load-deformation curves under severe cyclic loading, 197
Load distribution under perfectly rigid rectangular footing, 580
Load-path dependent, 567

Load pattern, selected, 567
Load vector, effective incremental, 532
Local site conditions, 299–300, 304–5
Local system of coordinates, 388
Location of isolation plane, 690
Loma Prieta, California, 1989 earthquake, 6, 55, 133, 186, 213, 297–9, 363, 371
Long-period transition period, 832, 842
Longitudinal waves, 63, 76–7, 95
Loops, force-displacement, 768
Loss factor, 746, 768
Lotung, Taiwan, 645
Love waves, 109–10, 112–4
Lumped-mass model, 315, 317–8, 334
Lysmer, J., 593
Lysmer analog, 593

M

MacMurdo, J., 295
Magnification, dynamic, 549
Magnitude
    body-wave, 149–50
    local, 146
Magnitude-recurrence relationships, 261
Magnitude scales, 138, 146, 149–51, 153
Main faults in State of California, 47
Maison, B. F., 399
Major tectonic plates, 41
Malaysian Rubber Producers' Research Association, 699
Maldonado, G. O., 395
Mallet, R., 27, 295
Mantle, 39, 46
Maps, microzonation, 24–5, 287
Marianas Trench, 43
Martel, R. R., 571
Mass, generalized, 384
Mass matrix, effective, 472
Mass method, consistent, 380
Material, internal, or hysteretic damping, 621
Material damping, 624
Material or internal damping in soils, 621
Mathematical model, 840
Matrix, amplification, 506–8
Maximum considered spectral acceleration for 1 s period, 830
Maximum displacements, 881
    total, 882
Maximum elastic response of nonstructural element, 681
Maximum lateral forces, 545
Maximum modal lateral displacements, 543
Maximum modal lateral forces, 542
Maximum modal responses in terms of response spectrum ordinates, 389
McCann, M. W., 182
McNally, K., 58
Mean occurrence rate, 266
Mean return period, 266
Mean square of modal response, 392
Mean square of total response, 392
Mechanical and electrical equipment, 649
Member internal forces, 547
Memory, pre-event, 155

Mercalli, G., 129
Messino-Reggio 1908 earthquake, 27, 693
Methods
    direct, 638
    static, 893
Mexico City 1985 earthquake, 17, 186, 212
Michoacán, Mexico, 1985 earthquake, 3, 185, 259, 295, 297–8, 651
Microseisms, 140
Mid-Atlantic ridge, 31, 42
Milne, J., 30, 134, 691
Milne seismograph, 135
Minimum displacements, 881
Minimum effective stiffness of isolation system, 881
Minimum relative displacement demand, 876
Mino-Owari, 1891 earthquake, 27
Miranda, E., 370
Miranda's reduction factors, 371
Missing mass effect, 423
Mixed boundary value problems, 590
Miyagi-Ken-Oki 1978 earthquake, 760
Modal acceleration method, 423–6
Modal analysis, 387
Modal coordinates, 670, 731
Modal correlation coefficients, 392, 394–6, 400, 402
Modal matrix, 383
Modal participation factors, 434, 453
Modal response spectrum analysis, 837, 850–1
Modal seismic response coefficient, 851
Modal superposition, complex, 732
Modal synthesis, 656
Modal synthesis method, 663
Mode shape with unit participation factor, 407
Mode shapes
    combined-system, 677
    complex-valued, 733
Mode shapes and natural frequencies of combined system, 663
Mode shapes of combined primary-secondary system, 670
Modeling of nonstructural elements, 654
Models based on reliability theory and decision analysis, 25
Models of damper-added structures, 786
Modes
    high-frequency, 402
    rigid-body, 666–7
Modifications to equivalent lateral force procedure, 863
Modifications to modal analysis procedure, 871
Modified Newton-Raphson method, 533
Mohraz, B., 362
Moment magnitude, 153
Motagua Fault, 44, 48
MSK (Medvedev-Sponheuer-Karnik) scale, 129
Multiple points of attachment, 659

N

Naples 1857 earthquake, 27
National Building Code, 815
Natural frequencies, combined-system, 680
Near-source effects, 822
NEHRP (National Earthquake Hazard Reduction Program), 285, 816

NEHRP Recommended Provisions for Seismic Regulations for New Buildings, 816
Neuenhofer, A., 456
Neuss, C. F., 399
New Brunswick, Canada, 1982 earthquake, 662
Newman, F., 129
Newmark, N. M., 360, 363, 369, 393, 415, 500
Newmark-beta method, 500, 506, 518–9
Newmark-Hall amplification factors, 363
Newmark-Hall approach, 360, 369
Newmark-Hall reduction factors, 370
Newton-Raphson method, 530, 533
Nickel-titanium alloys, 776
Nigam, N. C., 193
Nigam-Jennings approach, 193
Niigata, Japan, 1964 earthquake, 6–7
Nitinol, 776
Nobi, Japan, 1891 earthquake, 691
Nonclassical damping effects, 659–60
Nonlinear procedures, 902
    static, 902
Nonlinear response spectrum, 196
Nonlinear site response, 332, 347
Nonlinear static methods, 568
Nonlinear stress-strain behavior, 332
Nonstructural components, 649, 872
Nonstructural elements, 649, 654–5, 681, 821
Nonuniform motion, 449–50
Normal dip-slip fault, 48
Normal faults, 49
Norms, 506
    spectral, 506
North Palm Springs, California, 1996 earthquake, 21
Northridge, California, 1994 earthquake, 4, 22, 134, 159, 161, 171, 174, 567, 649, 652, 706
Novak, M., 605
Nuclear Regulatory Commission, 401
Numerical accuracy of step-by-step integration methods, 510
Numerically unstable, 504
Nur, A., 56
Nyquist frequency, 217

O

Occupancy category, 821, 835
Occupancy importance factor, 842
Occurrence rate, 263, 266
Offset, permanent, 764
Oils, silicon, 772
Oka, R., 693
Olympia, Washington, 1949 earthquake, 200
On-line seismicity information, 34
One-dimensional continuous model, 306, 319
Orientation of ground motion's principal axes, 440
Orizaba, Mexico, 1973 earthquake, 651
Orthogonality properties of mode shapes, 382
Ouality factor, 123
Outer core, 39
Overconsolidation ratio, 340
Overshoot, 497
Overturning moments, 564–5, 845–6

P

P-delta effects, 550–1, 821, 846–7
P wave, 95
Pacific tsunami warning system, 13
Pacoima dam, 299
Pacoima dam abutment, 303
Paleoseismology, 239
Pall, A. S., 760
Pall friction dampers, 760, 762
Pangaea, 40
Parkfield, California, 1966 earthquake, 168
Parseval's theorem, 227
Participation factors, 385
Particle motion, 95, 98, 100, 105–6, 109, 113
Peak ground acceleration and response spectrum shape method, 354
Peer review, 888, 912
Penzien, J., 437
Performance objective, 810
Period elongation, 510–11
Peru-Chile Trench, 43
Phase velocity, 86
Plane waves, 98, 100, 109, 116
Plastic hinges, 567
Plasticity index, 340–3, 829
Plate boundary, 58–9
Plate tectonics theory, 40
Poisson probabilistic model, 265
Poisson process, 265–7
Polytetrafluoroethylene (PTFE or Teflon), 694
Postearthquake damage field investigations, 24
Power spectral density function, 223
Power spectrum, 223
Primary-secondary system, 665, 668
Primary system, 663–8, 670, 672, 675, 677
Principal axes, 437–8, 440, 570
Principal axes of ground motion, 437
Principal directions of ground motion, 550
Probability distribution, extreme-value, 271
Probability distribution of earthquake magnitude, 277
Probability distribution of source-to-site distance, 275
Probability of exceedance, 268, 822
Properties of modal participation factors, 406
Protective system, 689
Pseudoacceleration, 189
Pseudoacceleration response spectrum ordinates, 248
Pseudodisplacement, 189
Pseudostatic method, 546
Pseudovelocity, 189, 191, 463
    spectral, 189
Pushover analysis, 892

Q

Quality factor, 123
Quality factors for various rocks, 124
Quasi-Newton methods, 534
Quinlan, P. M., 585

R

Ralph, P., 771
Ray, 116
Ray path, 116
Ray theory, 116
Rayleigh quotient, 547
Rayleigh waves, 100, 105–6, 108–9, 112, 114, 142
Rayleigh's principle, 546–7, 675
Rectangular foundations, 590
Reduction factors, 371
Reflected wave, 68–9, 71–2, 118, 319, 325, 328
Reflection and refraction of body waves, 115
Region of constant elastic spectral accelerations, 366
Region of constant elastic spectral displacements, 366
Region of constant elastic spectral velocities, 366
Region of initial loading, 523
Regions of elastic behavior, 523
Regions of plastic behavior, 523
Reid, H. F., 49, 295
Reissner, E., 583
Reissner's displacement functions, 583, 606
Reissner's theory, 585
Relative displacement, 381
Reliability and engineering decision models, 24
Requirements, compatibility, 668
Residual modes, 894, 896–8, 916
Resistance, standard penetration, 829
Response modification due to damping increase, 903
Response modification factor, 840, 842
Response spectrum, 182, 185
    definition of, 196
    floor, 656, 661
    in-structure, 656
Response spectrum analysis, 885
Response spectrum method, 379, 423, 426, 431, 449, 656, 663
Response spectrum method based on modal acceleration method, 423
Response spectrum method for multicomponent ground motions, 431
Response spectrum method for spatially varying ground motions, 449
Response spectrum procedures, 899
Response time-history analysis, 837
Restoring force, 513
Restoring force vector, 517
Return period, 266, 822
Reverse or thrust dip-slip fault, 48
Reverse or thrust fault, 49
Richart, F., 596
Richter, C. F., 146, 261
Richter magnitude, 146, 150
Riddell, R., 363, 369
Rift, 42
Right lateral strike-slip fault, 48
Rigid boundary method, 643
Rigid circular foundations supported by elastic layer, 603
Rigid diaphragms, 844
Rigid structure single-degree-of-freedom model, 721
Riley Act, 27
Ring of Fire, 31
Robinson, W. H., 777
Rocking impedance, 614
Rogers, A. M., 303
Roof displacement, target, 567
Root mean square acceleration, 223, 225
Rosenblueth, E., 393–4

Index

Rosenblueth's Rule, 394
Rotational inertia, 549
Round-off errors, 469
Rubber, chloroprene, 702
Rubber bearings
  laminated, 693–6, 701–2
  low-damping, 695–7, 699
Rules to combine modal responses, 391
Run-through rod, 772
Rupture directivity, 170
Rupture surface, 53, 259
Rupture zone, 244, 247, 258

S

S wave, 98
Sag pond, 241
San Andreas Fault, 43, 47–8, 50, 54, 57, 59, 243, 261
San Fernando, California, 1971 earthquake, 20, 168, 299, 303, 363, 400, 412
San Francisco, 1906 earthquake, 295
San Francisco Bay mud, 288
Sand boils, 9
Santa Clara Kaiser Permanente medical center, 784
Saragoni, G. R., 228
SASSI, 645
Scale saturation, 149
Scarps, 241
Scattering effect, 572
Scholz, C., 56
Sea-floor spreading, 42
Secondary bending moment, 550
Secondary effects, 811
Secondary structural elements, 649
Secondary structures, 649
Secondary systems, 649, 663–6, 668, 670, 674, 676–7, 680
Seed, H. B., 336
Seiches, 10, 15
Seismic analysis of nonlinear structures, 513
Seismic analysis of nonstructural elements, 655
Seismic coefficient, 358
Seismic design categories, 816, 819–21, 835–7, 841, 845–6
Seismic energy dissipation, 772
Seismic-force-resisting system, 816
Seismic gaps, 58–9
Seismic hazard assessment, 238
Seismic hazard curves, 268
Seismic isolation, 690
Seismic moment, 151
Seismic moment and magnitude of great earthquakes, 151
Seismic-protective systems, 24, 26
Seismic quiescence, 59
Seismic response coefficient, 841
Seismic response of multistory structures with added dampers, 800
Seismic response of simple structures with added dampers, 786
Seismic weight, effective, 841
Seismic zonation maps, 25, 282, 285–7
Seismicity, 29
Seismicity maps, 29
Seismicity of United States, 32
Seismograms, 134, 139–42, 146–9

Seismographs, 13, 30, 131, 134–6, 138–40, 146–9, 154–6
Self-starting method, 479
SH waves, 98, 109, 113, 118, 121, 303
Shah, H. C., 182
SHAKE, 334
Shaking tables, 24–6
Shallow earthquakes, 146, 248, 257
Shape factor, 223
Shear loss modulus, 767–8, 771
Shear modulus, complex, 321
Shear modulus and damping ratio
  gravels, 338
  sands, 336
  saturated clays, 340
Shear modulus of elasticity of viscoelastic halfspace, 624
Shear storage modulus, 767–8, 771
Shear strength, undrained, 829
Shear waves, 74, 97
Shims, 693
Significance of response reduction, 756
Silicon gel, 706
Simplified method, 840
Simplified procedure, 837
Singh, M. P., 395
Site class, 822, 829
Site coefficient
  long periods, 305
  short periods, 305
Site coefficients, 305, 822, 830
Site-specific response spectra, 822
Skinner, R., 776
Slider, articulated, 702
Sliding bearings, 694–5, 700
Sliding isolators, 702, 748
Slip force, 760, 765
Slip load, 764, 785
Slip rate, 54
Slopes, tangent, 514, 531
Small time intervals, 469
SMART-1, 165
Snell's law, 117
Soil liquefaction, 2
Soil spring constants, 582
Soil-structure interaction, 571–2, 645
Soil-structure interaction effects, 862
Solution in frequency domain, 319
Source-to-site distance, 253, 255, 259, 270, 275
Sources of error, 530
South Indian Rise, 42
Spaced natural frequencies, 393–5, 399, 654, 681, 737
Specific loss factor, 623
Spectral acceleration for 1 s period, 831
Spectral acceleration for short periods, 831
Spectral ordinates, 185
Spectral radius, 507–9
Spectrum, in-structure, 656
Spring constants
  circular footing, 582
  rigid rectangular foundation, 582
Spring-isolated structures, 705
Square root of the sum of the squares (SRSS) rule, 393–4, 852
SRRS response spectrum, 853

Stability
    numerical, 504
    unconditional, 510
Stability and accuracy issues, 503
Stability coefficients, 846–7
Stability limit, 504, 506, 511
Stadaucher, K., 693
Standard Building Code, 815
Static displacement, 170, 451
Static effect of floor rotations, 549
Static procedure, nonlinear, 565
Statistical correlations, direct, 375
Steel springs, helicoidal, 695
Step-by-step integration methods, 469–70, 476, 500, 503–4, 506, 509–10, 513, 530, 537, 539
Stick-slip behavior, 764
Stiffness
    equivalent, 747
    generalized, 384
    tangent, 530–1
Stiffness degradation, 198
Stiffness matrices, elemental, 521
Stiffness method, direct, 547
Story drift determination, 846
Story drift limits, 821
Story drifts, 547, 846
Story eccentricities, 548–9
Story shears, 564
Strains, volumetric, 89
Strength degradation, 198
Strike-slip faults, 171
Strong-motion duration, 182
Strong motion instrument arrays, 165
Structural Engineers Association of California (SEAOC), 816
Structural flexibility, 810
Structural irregularities, 836
Structural-nonstructural systems, 681
    combined, 659, 663, 675
Structural or global stiffness matrix, 521
Structure or global system of coordinates, 388
Structures suitable for seismic isolation, 748
Subduction zones, 42–3, 58, 247, 257–9
Substructure methods, 638, 642, 644–5
Successful earthquake predictions, 57
Sung, T. Y., 585
Superposition method, modal, 379, 382, 387
Support reactions, 451
Surface faulting, 2
Surface-wave magnitude, 148–50, 154
Surface waves, 100
Suyehiro, K., 182
SV wave, 98
System identification, 26
System identification techniques, 24
System of uncoupled equations, 383, 386
System overstrength factor, 819–20
System without classical modes of vibration, 654
Systems-in-cascade, 656
Systems with nonclassical damping, 411, 663, 737
Systems without classical modes of vibration, 411

T

TADAS element, 778
Tangshan, China, 1976 earthquake, 694
Target displacement, 565–7
    selected, 567
TASS bearing, 700, 702
Technology, supplemental energy dissipation, 753
Teflon, 701
Tension-compression yielding brace, 782
Testing of damping devices, 912
Testing of isolation system components, 887
Thomas, G., 135
Time histories
    design-spectrum-compatible, 228, 230
    ground motion acceleration, 852, 854
Time-history analysis, 885
    linear response, 852
    nonlinear response, 854
Time lag, 141–4
Time step size, 511
Tonga Trench, 44
Topographic irregularities, 299
Torsional amplification factor, 845
Torsional beams, 776
Torsional impedance, 614
Torsional irregularity, horizontal, 845
Torsional moments, 548–9, 565, 844
Transcurrent or strike-slip fault, 48
Transfer function, 319, 619, 639
Transformation of coordinates, 383
Transition zone, 288
Transmitted wave, 68
Trapezoidal rule, 482
Travel-time curves, 142
Treasure Palace, Taipei, Taiwan, 766
Trenching, 243
Tri-Services Manual, 816
Triangular dissipative devices, 778
Trifunac, M. D., 182
Tripartite representation, 189
Triple pendulum bearing, 704–5
Tsunami warning, 13
Tsunami watch, 13
Tsunamis, 10–4
    distant, 12
    local, 12
Tukey, J. W., 218
Tyler, R. G., 778
Type II extreme-value probability distribution, 271
Types of earthquakes, 38
Typical force-displacement loops, 772, 779, 781

U

U-strips, rolling-bending, 776
Unbonded braces, 782
Unbounded far field, 640
Unconditionally stable, 470, 479, 485, 496–7, 504, 506, 509
Undamped soil on rigid rock, 319
Uneven distribution of story shear, 548
Uniform Building Code, 27, 282, 355, 815

Index

Unit-participation-factor mode shape, 407
Updating of stiffness matrix, 521
U.S. Geological Survey, 822

V

Variation of modal damping ratios with natural frequency, 409
Vasudevan, R., 412
Vector of effective forces, 452
Veletsos, A. S., 624
Velocities of P waves and S waves, rock, 99
Velocities of P waves and S waves, soil, 99
Velocity, particle, 67–8
Velocity of wave propagation, 72, 81, 86, 97, 100, 105, 109, 112–3, 141
Velocity response spectrum, 185
Verbic, B., 624
Vertical displacement under center of rigid rectangular footing, 581
Vertical distribution factor, 844
Vertical distribution of base shear, 867
Vibration, ambient, 26
Vibration of embedded foundations, 605
Vibration of rectangular foundations, 590
Viscoelastic dampers, 26, 760, 764–5, 768, 770–1, 785, 787, 801
Viscoelastic materials, 764–5, 767–8
Viscoelastic pads, 768, 770–1
Viscous boundary, 640
Viscous dampers, 26
Vucetic, M., 342

W

Wadati, K., 146
Wadati diagram, 142
Warburton, G. B., 585, 604
Watabe, M., 437
Wave equation, one-dimensional, 65
Wave front, 116
Wave number, 80
Wave propagation velocity, 67, 117
Wave velocity, 67–9, 77, 98, 110
Wavelength, 80–1, 84–6, 102, 105, 112–4, 150
Waves
    harbor, 12
    primary, 95, 97–8
    secondary, 98
    seismic sea, 12–3
Wegener, A., 40
West, L. R., 303
Whitcomb, J. H., 57
Whittier-Narrows, California, 1987 earthquake, 371, 650, 652–3
Wiechert, E., 135
Williams, D., 424
Wilson, E. L., 495
Wilson-θ method, 495, 497, 509
Wood, H. O., 129
Wood-Anderson seismograph, 138, 148
Woods, R., 596
World seismicity, 30
Worldwide earthquakes, 33
Worldwide Standardized Seismographic Station Network, 30

X

X-shaped steel dampers, 778

Y

Yield acceleration inelastic design spectrum, 355
Yield acceleration spectrum, 357
Yield deformation spectrum, 200–1
Yielding metallic dampers, 776
Youngs, Chiou, Silva, and Humphrey equations, 257
Youngs, R. R., 263

Z

Zayas, V. A., 694
Zhang, X., 343
Zonation maps, 25, 285–6